PRINCIPLES OF COMMUNICATIONS NETWORKS AND SYSTEMS

PRINCIPLES OF COMMUNICATIONS NETWORKS AND SYSTEMS

N. Benvenuto, M. Zorzi
University of Padova, Italy

JOHN WILEY & SONS, LTD

This edition first published 2010

Registered office
John Wiley & Sons Ltd, The Atrium, Southern Gate, Chichester, West Sussex, PO19 8SQ, United Kingdom

For details of our global editorial offices, for customer services and for information about how to apply for permission to reuse the copyright material in this book please see our website at www.wiley.com.

ISBN 9780470971918 (P/B)

Printed and Bound in Great Britain

CONTENTS

Preface xv

Acronyms xix

Symbols xxv

1 Introduction to Telecommunication Services, Networks and Signaling 1

1.1 Telecommunication Services 1

1.1.1 Definition 1

1.1.2 Taxonomies according to different criteria 2

1.1.3 Taxonomies of information sources 4

1.2 Telecommunication Networks 4

1.2.1 Introduction 4

1.2.2 Access network and core network 9

1.3 Circuit switched and packet switched communication modes 11

1.4 Introduction to the ISO/OSI Model 13

1.4.1 The layered model 13

1.4.2 The ISO/OSI model 16

1.5 Signaling 19

1.5.1 Introduction 19

1.5.2 Channel Associated and Common Channel 19
1.5.3 SS7 20
1.5.4 PDH networks 21
1.5.5 SDH networks 23
References 25

2 Deterministic and random signals 27

2.1 Time and frequency domain representation 27
2.1.1 Continuous time signals 27
2.1.2 Frequency domain representation for periodic signals 32
2.1.3 Discrete time signals 36
2.2 Energy and power 39
2.2.1 Energy and energy spectral density 39
2.2.2 Instantaneous and average power 41
2.3 Systems and transformations 45
2.3.1 Properties of a system 46
2.3.2 Filters 46
2.3.3 Sampling 49
2.3.4 Interpolation 50
2.4 Bandwidth 53
2.4.1 Classification of signals and systems 55
2.4.2 Uncertainty principle 56
2.4.3 Practical definitions of band 57
2.4.4 Heaviside conditions 58
2.4.5 Sampling theorem 60
2.4.6 Nyquist criterion 63
2.5 The space of signals 66
2.5.1 Linear space 66
2.5.2 Signals as elements in a linear space 69
2.5.3 Gram-Schmidt orthonormalization in signal spaces 70
2.5.4 Vector representation of signals 75
2.5.5 Orthogonal projections onto a signal space 77
2.6 Random variables and vectors 79
2.6.1 Statistical description of random variables 80
2.6.2 Expectation and statistical power 82
2.6.3 Random vectors 86
2.6.4 Second order description of random vectors and Gaussian vectors 91
2.6.5 Complex-valued random variables 94
2.7 Random processes 96

2.7.1 Definition and properties 96
2.7.2 Point and Poisson processes 98
2.7.3 Stationary and ergodic random processes 105
2.7.4 Second order description of a WSS process 107
2.7.5 Joint second order description of two random processes 111
2.7.6 Second order description of a cyclostationary process 113
2.8 Systems with random inputs and outputs 114
2.8.1 Filtering of a WSS random process 115
2.8.2 Filtering of a cyclostationary random process 118
2.8.3 Sampling and interpolation of stationary random processes 119
Appendix: The complementary normalized Gaussian distribution function 121
References 124
Problems 125

3 Sources of Digital Information 131

3.1 Digital representation of waveforms 131
3.1.1 Analog to digital converter (ADC) 132
3.1.2 Digital to analog converter (DAC) 134
3.1.3 Quantizer 136
3.1.4 Uniform quantizers 137
3.1.5 Quantization error 138
3.1.6 Quantizer SNR 142
3.1.7 Non-uniform quantizers 144
3.1.8 Companding techniques and SNR 148
3.2 Examples of application 153
3.3 Information and entropy 157
3.3.1 A measure for information 157
3.3.2 Entropy 159
3.3.3 Efficiency and redundancy 165
3.3.4 Information rate of a message 166
3.4 Source coding 167
3.4.1 The purpose of source coding 167
3.4.2 Entropy coding 168
3.4.3 Shannon theorem on source coding 171
3.4.4 Optimal source coding 174
3.4.5 Arithmetic coding 178
References 181
Problems 184

4 Characterization of transmission media and devices **191**

4.1 Two-terminal devices 192
4.1.1 Electrical representation of a signal source 192
4.1.2 Electrical power 193
4.1.3 Measurement of electrical power 194
4.1.4 Load matching and available power 195
4.1.5 Thermal noise 197
4.1.6 Other sources of noise 199
4.1.7 Noise temperature 199
4.2 Two-port networks 200
4.2.1 Reference model 200
4.2.2 Network power gain and matched network 201
4.2.3 Power gain in terms of electrical parameters 202
4.2.4 Noise temperature 203
4.2.5 Noise figure 204
4.2.6 Cascade of two-port networks 206
4.3 Transmission system model 209
4.3.1 Electrical model 209
4.3.2 AWGN model 211
4.3.3 Signal-to-noise ratio 211
4.3.4 Narrowband channel model and link budget 214
4.4 Transmission media 217
4.4.1 Transmission lines and cables 217
4.4.2 Power-line communications 222
4.4.3 Optical fibers 227
4.4.4 Radio links 231
4.4.5 Underwater acoustic propagation 236
References 243
Problems 245

5 Digital modulation systems **251**

5.1 Introduction 251
5.2 Digital modulation theory for an AWGN channel 252
5.2.1 Transmission of a single pulse 252
5.2.2 Optimum detection 254
5.2.3 Statistical characterization of random vectors 255
5.2.4 Optimum decision regions 257
5.2.5 Maximum a posteriori criterion 261
5.2.6 Maximum likelihood criterion 262

5.2.7 Minimum distance criterion 262
5.2.8 Implementation of minimum distance receivers 264
5.2.9 The theorem of irrelevance 267
5.3 Binary modulation 268
5.3.1 Error probability 268
5.3.2 Antipodal and orthogonal signals 273
5.3.3 Single filter receivers 276
5.4 M-ary modulation 279
5.4.1 Bounds on the error probability 279
5.4.2 Orthogonal and bi-orthogonal modulations 283
5.5 The digital modulation system 287
5.5.1 System overview 287
5.5.2 Front-end receiver implementation 292
5.5.3 The binary channel 293
5.5.4 The inner numerical channel 294
5.5.5 Realistic receiver structure 296
5.6 Examples of digital modulations 298
5.6.1 Pulse amplitude modulation (PAM) 298
5.6.2 Quadrature amplitude modulation (QAM) 304
5.6.3 Phase shift keying (PSK) 314
5.6.4 Frequency shift keying (FSK) 319
5.6.5 Code division modulation 323
5.7 Comparison of digital modulation systems 326
5.7.1 Reference bandwidth and link budget 326
5.7.2 Comparison in terms of performance, bandwidth and spectral efficiency 328
5.8 Advanced Digital Modulation Techniques 329
5.8.1 Orthogonal frequency division multiplexing 329
5.8.2 Spread spectrum techniques 332
5.9 Digital transmission of analog signals 334
5.9.1 Transmission through a binary channel 335
5.9.2 Evaluation of the overall SNR 336
5.9.3 Digital versus analog transmission 338
5.9.4 Digital transmission over long distances: analog versus regenerative repeaters 343
References 344
Problems 347

6 Channel Coding and Capacity 361

6.1 Principles of channel coding 361
6.1.1 The purpose of channel coding 361
6.1.2 Binary block codes 363
6.1.3 Decoding criteria. Minimum distance decoding 364
6.2 Linear block codes 370
6.2.1 Construction of linear codes 370
6.2.2 Decoding of linear codes 373
6.2.3 Cyclic codes 377
6.2.4 Specific classes of linear block codes 379
6.2.5 Performance of linear codes 382
6.3 Convolutional codes 384
6.3.1 Construction and properties 384
6.3.2 Decoding of convolutional codes and the Viterbi algorithm 388
6.4 Channel capacity 392
6.4.1 Capacity of a numerical channel 392
6.4.2 Capacity of the additive white Gaussian noise (AWGN) channel 398
6.5 Codes that approach capacity 406
6.5.1 Soft decoding 406
6.5.2 Concatenated codes 408
6.5.3 Low density parity check codes 409
References 410
Problems 412

7 Markov chains theory 417

7.1 Introduction 418
7.2 Discrete-time Markov chains 418
7.2.1 Definition of discrete-time MC 418
7.2.2 Transition probabilities of discrete-time Markov chain (MC) 421
7.2.3 Sojourn times of discrete-time MC 423
7.2.4 Chapman-Kolmogorov equations for discrete-time MC 425
7.2.5 Transition diagram of discrete-time MC 426
7.2.6 State probability of discrete-time MC 426
7.2.7 Classification of discrete-time Markov chains 431
7.2.8 Asymptotic behavior of discrete-time MC 441
7.3 Continuous-time Markov Chains 451
7.3.1 Definition of continuous-time MC 451
7.3.2 Transition probabilities of continuous-time MC 452
7.3.3 Sojourn times of continuous-time MC 453

7.3.4 Chapman-Kolmogorov equations for continuous-time MC 457
7.3.5 The infinitesimal generator matrix Q 458
7.3.6 Forward and Backward equations for continuous-time MC 459
7.3.7 Embedded Markov chain 462
7.3.8 Flow diagram of continuous-time MC 465
7.3.9 State probability of continuous-time MC 465
7.3.10 Classification of continuous-time MC 471
7.3.11 Asymptotic behavior of continuous-time MC 472
7.4 Birth-death processes 475
7.4.1 Definition of birth-death process (BDP) 475
7.4.2 Time-dependent behavior of BDP 477
7.4.3 Asymptotic behavior of BDP 483
References 491
Problems 492

8 Queueing Theory 499

8.1 Objective of queueing theory 500
8.2 Specifications of a queueing system 500
8.2.1 The arrival process 503
8.2.2 The service process 505
8.2.3 The queueing structure 506
8.2.4 The service discipline 507
8.2.5 Kendall notation 508
8.3 Performance characterization of a queueing system (QS) 510
8.3.1 Occupancy measures 510
8.3.2 Time measures 512
8.3.3 Traffic measures 513
8.4 Little's law 516
8.5 Markovian queueing models 519
8.5.1 The $\mathrm{M/M/1}$ queueing system 520
8.5.2 The $\mathrm{M/M/m}$ queuing system 532
8.5.3 The $\mathrm{M/M/1}/K$ queueing system 543
8.5.4 The $\mathrm{M/M/m/m}$ queuing system 551
8.5.5 The $\mathrm{M/M/m}/K$ queuing system 555
8.6 The $\mathrm{M/G/1}$ queuing system 559
8.7 The $\mathrm{M/D/1}$ queuing system 566
References 568
Problems 571

9 Data Link Layer **577**

9.1 Introduction 578
9.2 Medium Access Control 579
9.2.1 Deterministic Access: TDMA and FDMA 584
9.2.2 Time Division Multiple Access 584
9.2.3 Frequency Division Multiple Access 586
9.2.4 Comparison between TDMA and FDMA 587
9.2.5 Demand-based Access: Polling and Token Ring 588
9.2.6 Random Access Protocols: ALOHA and Slotted ALOHA 603
9.2.7 Carrier Sense Multiple Access 612
9.2.8 Performance Comparison of Channel Access Schemes 637
9.3 Automatic Retransmission reQuest 640
9.3.1 Stop and Wait ARQ 644
9.3.2 Go Back N ARQ 646
9.3.3 Selective Repeat ARQ 649
9.3.4 Performance Comparison of ARQ Schemes 653
9.3.5 Optimal PDU Size for ARQ 654
9.4 Examples of LAN Standards 656
9.4.1 Ethernet 657
9.4.2 Wireless Local Area Networks 660
9.4.3 IEEE 802.11 661
9.4.4 Bluetooth 668
References 674
Problems 677

10 Network layers **683**

10.1 Introduction 684
10.1.1 Switching and connecting 684
10.1.2 Networks and network topology 686
10.2 Routing 690
10.2.1 Routing objectives 691
10.2.2 Routing algorithms 698
10.2.3 Technical aspects of routing implementation 708
10.2.4 Routing strategies 714
10.2.5 Routing protocols 720
10.3 The Internet and IP 723
10.3.1 The Internet Protocol 724
10.3.2 IP addressing system 726
10.3.3 Control protocols 733

10.4 The Transport Layer 736
10.4.1 User Datagram Protocol 738
10.4.2 Transmission Control Protocol 739
10.5 The Application Layer 746
10.5.1 Domain Name Server 746
10.5.2 Email exchange and the World Wide Web 748
References 751

Topic Index 753

PREFACE

This book addresses the fundamentals of communications systems and networks, providing models and analytical methods for evaluating their performance. It is divided in ten chapters, which are the result of a joint effort of the authors and some contributors. The authors and the contributors have a long history of collaboration, both in research and in teaching, which makes this book very consistent in both approach and notation.

The uniqueness of this textbook lies in the fact that topics from the physical layer (digital transmission and modulation) to the networking layers (MAC, routing, and transport) are addressed. Moreover, quality of service concepts belonging to the different layers of the protocol stack are covered, and the relationships between the metrics at different layers are discussed. This should help the student in the analysis and design of modern communications systems, which are better understood at a system level, rather than adopting a narrow viewpoint.

A suitable selection of chapters from this book provides the necessary teaching material for a one-semester undergraduate course on the basics of telecommunications, but some advanced material can be also used in graduate classes as complementary reading. The book is also a useful comprehensive reference for telecommunications professionals. Each chapter provides a number of problems to test the reader's understanding of the material. The numerical solutions for these exercises are provided in a companion website.

The content of each chapter is briefly summarized below

Chapter 1 Overview of modern communications services, along with the presentation of the OSI/ISO model.

Chapter 2 Analyzes both deterministic signals and random processes. In particular, the basics of signal theory are recalled by introducing both continuous and discrete time signals, and their Fourier transform with its properties. Concepts of energy, power and bandwidth are also reviewed, together with the vector representation of signals by use of linear space methodologies. Lastly, this chapter introduces the basics of random variables and random processes, their common statistical description, and how statistical parameters are modified by linear systems.

Chapter 3 Describes how information produced by a source, either analog or digital, can be effectively encoded into a digital message for efficient transmission. We will first introduce the reference scheme for analog to digital conversion, and focus on *quantization*, the operation of approximating an analog value with a finite digit representation. We will also present the fundamentals of Information Theory with the notion of information carried by a digital message. Lastly we introduce the principles of source coding, state the fundamental performance bounds and discuss some coding techniques.

Chapter 4 Models a transmission medium. Firstly, a description of the two-port network is given. A model of noise sources is then provided and its description in terms of parameters such as the noise temperature and the noise figure is also presented. A characterization, especially in terms of power attenuation, of transmission lines, power lines, optical fibers, radio propagation and underwater propagation concludes this chapter.

Chapter 5 Deals with digital modulations. The general theory is first presented, relying on concepts of linear spaces and hypothesis testing. Performance measure is given by the bit error probability, which again can be expressed in terms of system parameters. Then, the most important modulations are presented as exemplifications of the general concept. These include pulse amplitude modulation, phase shift keying, quadrature amplitude modulation, frequency shift keying and others. A comparison between the different modulation schemes is carefully drawn. Then, more advanced modulation techniques are briefly presented, namely orthogonal frequency division multiplexing and spread spectrum. Finally, performance of the digital approach is also compared against the analog transmission, explaining why digital transmission is so widely used.

Chapter 6 Investigates how an information message can be robustly encoded into the signal for reliable transmission over a noisy channel. We describe the principles of channel coding techniques, where robustness is obtained at the price of a reduced information rate and complexity of the decoding process. Then, based upon the fundamentals of Information Theory introduced in Chapter 3, we aim to establish upper bounds on the amount of information that can be effectively carried through a noisy channel, by introducing the concept of channel capacity. Lastly, we briefly decribe how recently devised coding schemes allow to closely approach such upper bounds while maintaining a moderate complexity.

Chapter 7 Introduces some basic statistical methods that are widely used in the performance analysis of telecommunication networks. The topics covered are: the elementary theory of discrete-time and continuous-time Markov chains and birth-death processes. Next, in Chapter 9 we will apply the Markov chains theory to the performance analysis of the channel access, and retransmission protocols. In Chapter 8, instead, the theory of birth-death processes will be applied to the analysis of simple queueing systems.

Chapter 8 Presents some basic principles of queueing theory. It starts with the definition of a queueing system in terms of arrival, departure, service processes, queueing process and service discipline. After that, we define some basic performance metrics for queueing systems, which are classified in occupancy measures, i.e., number of customers in the different parts of the system; time measures, i.e., system and queueing times; and traffic measures, i.e., average rate at which customer arrive and leave the system. We discuss the concept of system stability, both for blocking and non-blocking queueing systems. Thus, we illustrate the Little's law, a simple but fundamental result of queueing theory. The chapter is concluded with the performance analysis of fundamental queueing models, featuring Markovian as well as non-Markovian statistics for the service process. Examples of application of the developed theory to practical problems, not only related to telecommunications, are provided throughout the chapter.

Chapter 9 Presents and analyzes link layer algorithms, especially focusing on aspects related to channel access and retransmission of lost or corrupted data packets over point-to-point links, a technique often referred to as Automatic Retransmission reQuest (ARQ). Channel access protocols are subdivided into the following classes: deterministic access, demand-based access, random access and carrier sense multiple access. Mathematical models are given for each class of protocols, thus obtaining the related performance in terms of throughput and delay. The different channel access schemes are then compared as a function of the traffic load of the system. Next, the chapter presents three standard ARQ schemes, namely, stop and wait, go back N and selective repeat ARQ, which are characterized analytically in terms of their throughput performance. The chapter ends with the presentation of some relevant LAN standards: Ethernet, IEEE 802.11 and Bluetooth, with emphasis on their link layer techniques.

Chapter 10 Describes all the layers above the data link layer dealing with the interconnection of distinct devices so as to form a communication network. In detail, the chapter starts reviewing a useful mathematical tool for network analysis, namely, graph theory. Routing methodologies, i.e., how to find efficient paths in the network, are identified and discussed within this framework. Subsequently, we review how the network layer is implemented in the Internet detailing the Internet Protocol (IP), as well as its addressing issues, including Address Resolution Protocol (ARP) and Network Address Translation (NAT). Hence, the chapter describes the implementations of transport Transport Control Protocol (TCP) and the User Datagram Protocol (UDP)) and application (Domain Name Server (DNS)) layers for the Internet. Last, some examples of application protocols, i.e., HyperText Transport Protocol (HTTP) and Simple Mail Transfer Protocol (SMTP) are given.

ACRONYMS

1P-CSMA	one persistent CSMA
1P-CSMA/CD	one persistent CSMA/CD
*p*P-CSMA	*p* persistent CSMA
ACL	asynchronous connection-less
ACK	acknowledgment
ADPCM	adaptive differential pulse coded modulation
ADSL	asymmetric digital subscriber loop
AM	amplitude modulation
AMR	adaptive multi rate
ARQ	automatic repeat request
ARCNET	attached resource computer network
ARR	adaptive round robin
a.s.	almost surely
ATM	asynchronous transfer mode
AWGN	additive white Gaussian noise
BDP	birth-death process

BER	bit error rate
BMAP	bit map
BPF	bandpass filter
BPSK	binary phase shift keying
BSC	binary symmetric channel
BSS	basic service set
CA	collision avoidance
CBR	constant bit rate
CD	collision detection
CDF	cumulative distribution function
CDM	code division modulation
CDMA	code division multiple access
CTS	clear to send
DSLAM	digital subscriber line access multiplexer
ETSI	european telecommunications standards institute
CRC	cyclic redundancy check
CSMA	carrier sense multiple access
CSMA/CA	CSMA with collision avoidance
CSMA/CD	CSMA with collision detection
DBPSK	differential binary phase shift keying
DC	direct current
DCF	distributed coordination function
DIFS	distributed inter frame space
DNS	domain name server
DQPSK	differential quadrature phase shift keying
DSSS	direct sequence spread spectrum
DPSK	differential phase shift keying
DSB	double side band
DVB	digital video broadcasting
ERR	exhaustive round robin
ESS	extended service set
FCC	federal communication commission
FDDI	fiber distributed data interface
FDM	frequency division multiplexing
FDMA	frequency division multiple access

FEC	forward error correction
FH	frequency hopping
FHS	frequency hopping synchronization packet
FHSS	frequency hopping spread spectrum
FM	frequency modulation
FSK	frequency shift keying
Ftf	Fourier transform
GBN-ARQ	go back N ARQ
gcd	greatest common divider
GD	gap detection
GPS	global positioning system
GSM	global system for mobile communications
HPF	highpass filter
HTML	hyper-text markup language
HTTP	hyper-text transfer protocol
IANA	Internet assigned numbers authority
IBMAP	inverse bit map
IBSS	independent basic service set
IEEE	institute for electrical and electronic engineers
iid	independent and identically distributed
IP	Internet protocol
ISI	intersymbol interference
ISM	industrial scientific and medical
ISDN	integrated services digital network
JPEG	joint photographic experts group
LAN	local area network
lcm	least common multiplier
LDPC	low density parity check
LLC	logical link control
LPF	lowpass filter
LRR	limited round robin
LTI	linear time invariant
m.s.	mean square
MAC	medium access control
MAP	maximum a posteriori

MC	Markov chain
MIMO	multiple-input multiple-output
ML	maximum likelihood
MSDU	medium access control service data unit
NAK	negative acknowledgment
NAT	network address translation
NAV	network allocation vector
NBF	narrowband filter
NP-CSMA	nonpersistent CSMA
NP-CSMA/CD	nonpersistent CSMA/CD
NTF	notch filter
OFDM	orthogonal frequency division multiplexing
OFDMA	orthogonal frequency division multiple access
PAM	pulse amplitude modulation
PBX	private branch exchange
PCF	point coordination function
PCI	protocol control information
PCM	pulse code modulation
PDA	personal digital assistant
PDH	plesiochronous digital hierarchy
POTS	plain old telecommunications network
PDF	probability density function
PDU	packet data unit
PM	phase modulation
PMD	probability mass distribution
PRR	pure round robin
P/S	parallel-to-serial
PSD	power spectral density
PSK	phase shift keying
PSTN	public switched telephone network
QAM	quadrature amplitude modulation
QM	quadrature modulation
QoS	quality of service
QPSK	quaternary phase shift keying
QS	queueing system

rp	random process
RTS	request to send
RTP	real time protocol
RTT	round trip time
rv	random variable
rve	random vector
SAP	service access point
SCO	synchronous connection oriented
SDH	synchronous digital hierarchy
SDM	spatial division multiplexing
SDU	service data unit
SIFS	short inter frame space
SMTP	simple mail transfer protocol
SNR	signal-to-noise ratio
SONET	synchronous optical networking
S/P	serial-to-parallel
SSB	single side band
SR-ARQ	selective repeat ARQ
SW-ARQ	stop and wait ARQ
TDD	time division duplex
TDMA	time division multiple access
TCP	transmission control protocol
UMTS	universal mobile telecommunications system
UWB	ultra wide band
UDP	user datagram protocol
UMTS	universal mobile telecommunications system
UTP	unshielded twisted pair
VBR	variable bit rate
VCS	virtual carrier sensing
VSB	vestigial side band
Wi-Fi	wireless fidelity
WLAN	wireless local area network
WSS	wide sense stationary
WGN	white Gaussian noise
WWW	world wide web

SYMBOLS

$\angle z$ and $\arg(z)$	phase of the complex number z
$\langle x, y \rangle$	inner product between x and y
$(x * y)(t)$	convolution between x and y
$\mathcal{A}$	alphabet
$\mathsf{a}(f)$	power attenuation
$\text{argmax}_x f(x)$	value of x which maximizes $f(x)$
$\text{argmin}_x f(x)$	value of x which minimizes $f(x)$
B_{min}	minimum bandwidth for a given modulation
B	bandwidth of a signal (measure of $\mathcal{B}$)
$\mathcal{B}$	band of a signal
$\chi\{A\}$	indicator function of the event A
d_{min}	minimum distance
$d_{\text{H}}(\boldsymbol{x}, \boldsymbol{y})$	Hamming distance between $\boldsymbol{x}$ and $\boldsymbol{y}$
$D(\rho; n)$	probability function for determining optimum decision regions
$\delta(t)$	Dirac delta function (impulse)
δ_m	Kronecker delta function

$\mathrm{E}[x]$	expectation of the random variable (rv) x
E_b	average energy per bit
E_s	average signal energy
E_x	energy of the signal $x(t)$
E_{xy}	cross energy of the signals $x(t)$ and $y(t)$
$\mathcal{E}_x(f)$	energy spectral density of the signal $x(t)$
η	throughput
F_{s}	sampling frequency
$\mathcal{F}$	Fourier transform operator
$\mathtt{F}$	noise figure of a two-port network
Γ	reference signal-to-noise ratio
$\phi_i(t)$	ith signal of an orthonormal basis
f_0	carrier frequency
$\mathtt{G}$	offered traffic
$\mathcal{G}_{\mathrm{Ch}}(f)$	transmission channel frequency response
$g_{\mathrm{Ch}}(t)$	transmission channel impulse response
$\mathtt{g}(f)$	power gain
$h_{\mathrm{Tx}}(t)$	filter/waveform at transmission
H_n	hypothesys in signal detection
$H(x)$	entropy of the rv x
$H(x_1, \ldots)$	joint entropy of the rvs $x_1, \ldots$
$H(x\|y)$	conditional entropy of the rv x given the rv y
$\Im$	imaginary part of a complex quantity
I	signal space dimension
$I(x, y)$	mutual information between the rvs x and y
$\mathtt{k}_{xy}$	covariance of the rvs x and y
$\boldsymbol{k}_{\boldsymbol{x}}$	covariance matrix of the random vector (rve) $\boldsymbol{x}$
λ	average customers arrival rate
Λ	signal-to-noise ratio (SNR)
$\Lambda_{\mathtt{M}}$	SNR between statistical powers
$\Lambda_{\mathtt{P}}$	SNR between electrical powers
Λ_q	signal-to-quantization noise ratio
M	alphabet cardinality
$\mathtt{M}_x$	average power of the signal $x(t)$
$\mathtt{M}_x$	statistical power of the rv x or the random process (rp) $x(t)$

$\mathtt{M}_{xy}$	cross power of the signals $x(t)$ and $y(t)$
$\mathtt{M}_{xy}$	cross power of the rps $x(t)$ and $y(t)$
m	number of servers in a queueing system
$\mathtt{m}_x$	mean (expectation) of the rv x or the rp $x(t)$
$\boldsymbol{m}_{\boldsymbol{x}}$	mean vector, expectation of the rve $\boldsymbol{x}$
$\frac{N_0}{2}$	power spectral density of white noise at the receiver input
ν	spectral efficiency
Ω	sample space of a probability space
P_{bit}	bit error probability
$\mathrm{P}[A]$	probability of the event A
$\mathrm{P}[A\|C]$	conditional probability of the event A, given the event C
$\mathrm{P}[\mathrm{C}]$	probability of correct decision
$\mathrm{P}[\mathrm{E}]$	probability of error
P_e	probability of error
$\mathcal{P}_x(f)$	power spectral density (PSD) of the rp $x(t)$
$\mathcal{P}_{xy}(f)$	cross PSD of the rps $x(t)$ and $y(t)$
P	average power (electrical)
$\mathsf{p}(f)$	average power density (electrical)
$P_x(a)$	cumulative distribution function (CDF) of the rv x
$p_x(a)$	probability density function (PDF), or probability mass distribution (PMD), of the rv x
$p_{x\|y}(a\|b)$	conditional PDF, or PMD, of the rv x given y
$p_{\boldsymbol{x}}(\boldsymbol{a})$	PDF, or PMD, of the rve $\boldsymbol{x}$
$p_x(a;t)$	first order PDF, or PMD, of the rp $x(t)$
$\boldsymbol{P}(n)$	transition matrix of a discrete-time Markov chain in n steps
$\boldsymbol{P}$	transition matrix of a discrete-time Markov chain in one step
$P_{i,j}(n)$	transition probability of a discrete-time Markov chain from state i to state j in n steps
$P_{i,j}$	transition probability of a discrete-time Markov chain from state i to state j in one step
$\boldsymbol{P}(t)$	transition matrix of a continuous-time Markov chain in a time interval of length t
$P_{i,j}(t)$	transition probability of a continuous-time Markov chain from state i to state j in the time interval of length t
P_{blk}	call-blocking probability
$\boldsymbol{p}_x$	(row) vector of the asymptotic PMD of $x(t)$ as $t \to \infty$
$\boldsymbol{\pi}$	asymptotic state probability vector of a MC

π_j	asymptotic probability of state $j \in \mathcal{S}$ of a MC
$\mathrm{Q}(a)$	normalized Gaussian complementary distribution function
$\boldsymbol{Q}$	infinitesimal generator matrix of a continuous-time Markov chain
$\Re$	real part of a complex quantity
R_n	decision region associated to the transmitted waveform $s_n(t)$
R_b	bit rate
$\mathtt{r}_{xy}$	cross correlation of the rvs x and y
$\mathtt{r}_x(\tau)$	autocorrelation of the rp $x(t)$
$\mathtt{r}_{xy}(\tau)$	cross correlation of the rps $x(t)$ and $y(t)$
$\boldsymbol{r}_{\boldsymbol{x}}$	correlation matrix of the rve $\boldsymbol{x}$
ρ	correlation coefficient
S	useful traffic
$s_n(t)$	transmitted waveform associated to the symbol $a_0 = n$
$\boldsymbol{s}_n$	vector representation of waveform $s_n(t)$
$s_{\mathrm{Tx}}(t)$	transmitted signal
$s_{\mathrm{Rc}}(t)$	received signal (useful component)
σ_x^2	variance of the rv x or the rp $x(t)$
T_s	sampling period
T	temperature, in K
T_0	standard temperature of 290 K
$\mathrm{T}_{\mathrm{eff}}$	effective noise temperature of a system
$w(t)$	noise, typically Gaussian
$\boldsymbol{w}$	vector representation of noise $w(t)$
$w_{\mathrm{Rc}}(t)$	noise at the receiver input
$\mathcal{X}(f)$	Fourier transform of the signal $x(t)$
x^*	complex conjugate of x
$\boldsymbol{x}^{\mathrm{T}}$	transpose of $\boldsymbol{x}$
$\boldsymbol{x}^{\mathrm{H}}$	conjugate transpose (Hermitian) of $\boldsymbol{x}$

CHAPTER 1

INTRODUCTION TO TELECOMMUNICATION SERVICES, NETWORKS AND SIGNALING

Lorenzo Vangelista

The goal of telecommunication architectures is to provide people (and machines as well) with *telecommunication services*. In this chapter we give the basic definition of telecommunication services and we provide an introduction to telecommunication networks and signaling as well as to the well known ISO/OSI model.

1.1 TELECOMMUNICATION SERVICES

1.1.1 Definition

The definition of telecommunication service is quite broad; basically, it can be considered as transfer of information. In a telecommunication service at least three actors are usually involved:

1. one or more sources
2. a carrier
3. one or more receivers.

Principles of Communications Networks and Systems N. Benvenuto, M. Zorzi

The source(s) and the receiver(s) are not necessarily human beings; they can also be, for example, computers. Usually, they are also called "customers" or "users", when we do not need to distinguish between the two roles. The carrier is normally a company owning equipment and resources, e.g., frequencies, in case of radio services. The equipment (hardware and software) that allows the information exchange is called *telecommunication network*, and according to its extension can be classified as "local area network" when serving a certain specific small geographic area or "wide area network" when serving a large geographic area. As an example the network serving an university campus is a local area network, while a large Internet provider network is a wide area network.

Telecommunication networks can be categorized in many different ways. The most common one refers to the services the networks carry. We have then

- telephony networks carrying voice related services;
- data networks.

Using another criterion, still based on services, we have

- dedicated networks to one specific service, e.g., the plain old telephone network (POTS);
- integrated services networks designed to carry different services at the same time, e.g., the integrated services digital network (ISDN).

1.1.2 Taxonomies according to different criteria

In this section we classify telecommunication services according to different criteria, including symmetry, configuration, and initialization. We would like to remark that the role of source and receiver in a communication are not exclusive, i.e., each party involved in the communication clearly can act as

1. a source only
2. receiver only
3. both source and receiver, at different time intervals
4. both source and receiver, at the same time (think of the ordinary phone calls).

Regarding the *symmetry* characteristic of telecommunication services we can distinguish between

- *unidirectional* services, e.g., television broadcasting;
- *bidirectional asymmetric* services, e.g., Web browsing, [1] where there is a bidirectional exchange of information between the user and the Web server but the amount of information flowing from the server to the user is by far larger than that flowing from the user to the Web server;
- *Bidirectional symmetric*, e.g., the usual telephone conversation.

[1] Actually here we refer to the so-called Web 1.0; with Web 2.0, user-generated content plays a major role.

If we turn our attention to the *configuration* of services we can make the following distinction:

- *point to point* services, where only two users are involved in the telecommunication service, which in turn can be either unidirectional or bidirectional.
- *multipoint* services, involving multiple users, some acting as sources and some as receivers, possibly changing their roles dynamically.
- *broadcast* services, where a single source of information transmits to many receivers.

Telecommunications services can also be categorized based on their *initialization*; there are basically three categories

- *call based* services; any user needs to have an *identifier* to activate its service and this can be established at any time. A classical example is a telephone call or conference call.
- *reservation based* services; similar to the call based services, except that there must be a kind of booking of the service made to the carrier. This is often the case for conference call services nowadays or for bulk data transfers, usually scheduled overnight.
- *permanent mode* services; service is negotiated by a contract between the carrier and the parties involved (two or more) and it is always running from the contract signature on. Examples include data connection between two data centers of a large company or, for a wireless operator without fixed infrastructure, a connection between two of its switches.

From the above examples, we would like to remark that in modern telecommunication networks there must be a mechanism to establish a service as easily as possible; this fundamental mechanism is actually hidden from users. Also, it is a network on its own, with users (usually the real end users) and specific protocols[2] called *signaling protocols*, which will be discussed in detail in Section 1.5.

Finally, we consider the classification of telecommunication services according to their *communication mode* and we have *interactive* services and *broadcast* services. As far as the interactive services are concerned we can distinguish between

- *conversational* services, usually bidirectional, where there must be a very low delay in delivering the information; consider, for example, a telephone call and how annoying a delay could be – which is sometimes happening for internet based calls or for calls using satellite communication links.
- *messaging* services, where there are no strict real time constraints but still the delivery time matters. An example is instant messaging, which is very popular nowadays.
- *information retrieval* services, where the interactivity and real time constraints can be mild, as is the case for a database search.

As far as broadcast services are concerned, considering them from the communication mode point of view, we may distinguish between services

[2]We have not introduced the notion of protocol yet; we can think of a protocol as a set of procedures and rules to exchange data or to let two entities agree on certain actions to be performed

- *without session control* where a user has no control on what is delivered.
- *with session control* where a user can have some control on how the information is presented to her. An example could be the delivery of the same movie on a broadcast channel using different subchannels starting at different times, so that the user can somehow control the start of the movie at her convenience.

1.1.3 Taxonomies of information sources

In this section we classify information sources according to their time-varying characteristics and there are two types of sources:

- constant bit rate (CBR): the bit rate[3] emitted by the source is constant over time;
- variable bit rate (VBR): the bit rate may vary over time.

Example 1.1 A The digital encoding of speech according to the ITU G.711 standard (also known as pulse code modulation or PCM encoding) produces 64000 bit/s; this values is constant over time regardless of whether the source user actually speaks or not; we can conclude that a voice signal encoded according to the ITU G.711 standard is a constant bit rate source.

Example 1.1 B adaptive multi rate (AMR) is a voice encoding standard employed mainly in cellular networks like universal mobile telecommunications system (UMTS) and it is designed to exploit both the silence periods of the speaker and to adapt to the different conditions of the transmission channels over which the bits are transmitted. If the speaker is silent a special sequence of bits is sent and then nothing is emitted until the speaker starts speaking again; at the receiving side when this sequence is received noise is randomly reproduced, as a comfort feature, just to keep the receiver aware that the link is not broken. The AMR is basically made of eight source encoders with bit rates of 12.2, 10.2, 7.95, 7.40, 6.70, 5.90, 5.15, and 4.75 kbit/s. Indeed, even if the link between the transmitter and the receiver is fed at a constant bit rate, when the transmission channel at some point in time deteriorates, AMR is informed and selects a lower bit rate. Hence, more redundancy bits can be used (resorting to *channel coding* techniques) for protecting the voice encoded bits from channel errors. We can easily conclude that a voice signal encoded according to the AMR standard is a variable rate source.

1.2 TELECOMMUNICATION NETWORKS

1.2.1 Introduction

In this section we give a general description of a communication network as an infrastructure for exchanging information.

As illustrated in Fig. 1.1 a *network* is a collection of *nodes* connected by *links*. Nodes can

[3]The term "bit rate" has not been defined yet; the reader should rely on an intuitive meaning as the number of binary digits or bits per second (bit/s) sent over the channel.

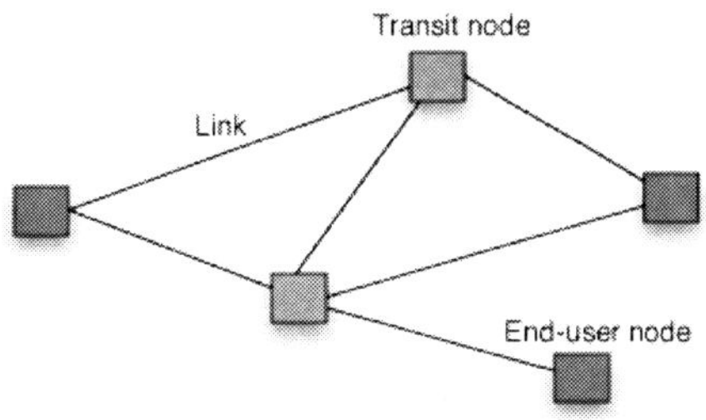

Figure 1.1 A generic network.

be either *end-user* or *transit* nodes. Links connect two nodes and represent the potential of exchanging telecommunication services. When discussing the nature of the links, we may distinguish between *physical* and *logical* links; the former represent the presence of an actual resource which connects two nodes and may be used for the communication, the latter are instead a more general abstraction of the capability of two nodes to communicate. This is to say, there could be many logical links conveyed by a physical link, [4] whereas many physical links could be needed to establish a logical link.

Example 1.2 A In a telephone network the *end-user* nodes are the the usual telephones while the *transit* nodes include local exchanges and switches.

Example 1.2 B Let's consider a basic ISDN connection; it usually consists of *physical* link made of a copper twisted pair of cables and three *logical* links: two links (called *B links*) carrying user information and one link (called *D link*) carrying ancillary information used by the network to set–up and control the calls. The bit rate of each B link is 64 kbit/s while the D link bit rate is 16 kbit/s.

It is common practice, when dealing with communication networks, to abstract from many details of the implementation, including both physical aspects of the network elements and the policies used to make them communicate with each other. On the one hand, this simply means that we adopt a layered analysis of the network, as will be discussed later in this chapter. In particular, one layer considers the physical devices and another separated layer deals with the network operations from an abstract point of view. At the same time, this enables us to look at the problem from a different perspective, more theoretical and founded on that branch of mathematics called *topology*.

A historical background In year 1735 the mathematician Leonhard Euler solved a popular puzzle of the time, the problem of the seven bridges of Königsberg [1].

This was a sort of networking problem, even though of course it did not involve any transmission or data exchange, but rather it was about finding a specific walk through the

[4] An example of this case, which will be discussed in the following, is the *bus*, i.e., a single communication line which connects several nodes and enable communication among them, but only for one pair at a time.

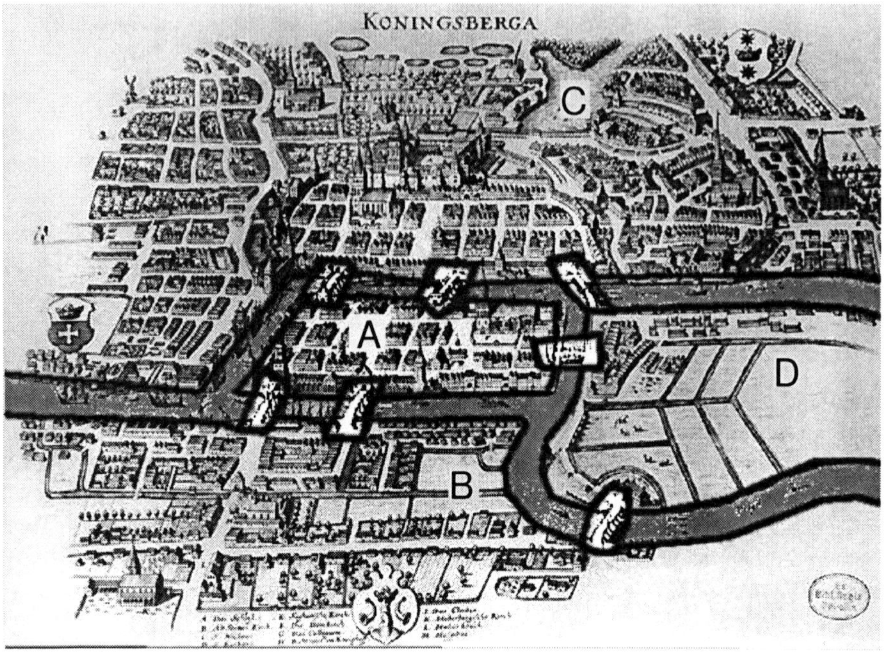

Figure 1.2 A map of Königsberg and its seven bridges. The bridges are highlighted in white with black border. The city districts are labeled A, B, C, D, as done in the original formulation by Euler.

bridges of a city built over a river. The city was Königsberg, the then-capital of the Province of Prussia, which consisted of four districts (two sides of the river and two small islands in between), connected through seven bridges, as shown in Fig. 1.2. More precisely, the problem consisted of finding a walk visiting every bridge, and doing so exactly once. The details of the actual problem can be found in [2]. The full tale of Euler's solution of the problem is very educational, and is recommended as a further reading.

Aside from the specific problem, we are more interested in which instruments to use in order to approach and solve it, so that we can use the same instruments for information networks. Euler found a clever way to translate the bridge problem into a much simpler symbolic representation which could be analyzed from a theoretical perspective.

First of all, it is clear that an immediate and cheap way to approach the problem is, rather than visiting the city, to use a map; instead of walking through the bridges, one can simply look at the graphical representation. However, Euler realized that a good mathematical model did not need unnecessary details. Thus, instead of a precise map, in which the geometrical proportion of the bridges and the city districts were respected, it was possible to use a simpler scheme. What does matter is just which parts of the city are connected, and how.

Thus, Euler used a simple tabular representation, where he took note of which parts of the city were interconnected, and by how many bridges. This is a simple topological representation, where only the strictly necessary details are considered. It is commonly said and believed that in this circumstance Euler invented the branch of mathematics known as *topology*, and more specifically its part called *graph theory*. However, the original representation he gave of the problem involved a table, not a graph [1]. Yet, later re-arrangements made by other scholars proposed a graphical representation, which is equivalent to Euler's table. Such a representation is reported in Fig. 1.3.

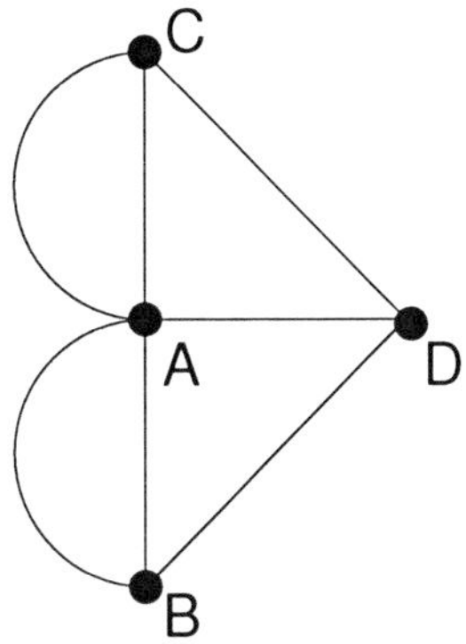

Figure 1.3 The graph resulting from Königsberg's bridge problem.

Through this formulation of the problem, the whole city is modeled as a *graph*, where the districts and the bridges are the *vertices* and the *edges* between vertices, respectively. The problem of traversing the bridges could be analyzed by means of this representation, e.g., by marking, or specifically *coloring*, the edges corresponding to traversed bridges. It is worthwhile to note that this representation preserves very few characteristics of the original problem. Indeed, the only elements given by the graph are those which are topological invariants of the network, i.e., which cannot be altered without substantially changing the problem. In this problem, the topological invariants are the connections between districts along the bridges, whereas it does not matter how large the bridges are or what is the physical distance between the districts.

More in general, formulating a problem in terms of network topology means to abstract from many aspects and mostly consider just connectivity properties, e.g., whether two vertices are connected through an edge or not. Once we solve a problem formalized in this manner (or, as is the case for Euler's original problem, we prove it does not admit any solution), we do the same for any problem which is *topologically equivalent*, i.e., that involves the same questions on a network with identical connectivity properties.

Notation and terminology The topological representations we will use for communication networks do not significantly differ from that used for Königsberg's bridge problem. Network devices which are able to interconnect with each other are generically called *nodes*; they are represented as the vertices of a graph, which are in turn linked through edges representing connection means, i.e., cables, radio links, optical fibers or something similar.

The whole network is thus represented as a graph $\mathcal{G}$, which is often written as $\mathcal{G} = (\mathcal{N}, \mathcal{E})$, where $\mathcal{N}$ is the set of the nodes and $\mathcal{E}$ is the set of the edges. The number of elements in set $\mathcal{N}$ is $|\mathcal{N}|$. According to the context, the elements of $\mathcal{N}$ can be denoted in many ways; it is usual to use integer numbers (in which case the nodes are labeled , e.g., as $1, 2, \ldots, |\mathcal{N}|$), or capital Latin letters, which is the notation we will employ in this chapter; we will use instead lower-case Latin letters to denote a generic node. In some cases, the edges are also called *arcs*, *links*, *hops*. We will use all these terms as synonyms.

The set of edges $\mathcal{E}$ is a subset of $\mathcal{N}^2$. Thus, a typical element of $\mathcal{E}$ can be written as (A, B), where $\mathrm{A}, \mathrm{B} \in \mathcal{N}$. In this case, node A is said to be the source (or the transmitter, or the tail) and node B to be the destination (or the receiver, or the head) of the edge. However, it is commonly avoided to include in $\mathcal{E}$ the pairs of the form (i, i), i.e., with both elements equal to the same node. This would correspond to a single hop looping to and from the same node, which is usually unrealistic.

Notable network topologies. Several network topologies are actually possible. Some of them are rather complicate, since, as will be discussed in Chapter 10, they can be nested. In other words, a larger network can be formed by connecting two or more networks, so as to create a hierarchical structure. However, there are certain network topologies which are, in a sense, considered as the most basic and frequently studied. In other words, they are standard cases, and there are proper names for them.

These notable topologies include the following cases.

- *star* topology: a center node is connected to every other node; there are no links between the other nodes. This case physically represents an interconnection where several end-users can communicate only through a transit node (which is the only one to be connected to all of them);

- *mesh* topology: in this kind of network, every node is connected to every other node; sometimes, this full interconnection is better emphasized by calling it "full mesh;"

- *tree* topology: one way to obtain this topology is to have one node, called *root*, connected with multiple nodes (not connected to each other), called *leaves*; this is actually the same as the star topology defined above, so the star topology is also a tree topology; however, we can also have a tree if we replace any of the leaves with the root of another tree topology, so that the tree is obtained as a hierarchy of nodes; more in general, as will be discussed in Chapter 10, the term "tree" actually applies to all the topologies where there are no loops (as they will be defined in the Chapter);

- *ring* topology: each node is connected to two other nodes only; moreover, assume we arbitrarily select one node A of the network; then, consider either of the nodes A is connected to (call it B) and then take the node B is connected to besides A (call it C). Now, repeat the procedure indefinitely by replacing (A,B) with (B,C), respectively. Sooner or later, A will be "touched" again by this procedure. This means that the network contains a loop; actually, it is only a single loop.

- *bus* topology: from the strict topological point of view, this is the same as a mesh topology; that is, every node is connected to each other; however, in a bus topology the connection is achieved with a special communication line which is shared by all nodes. This line, called a *bus* admits all multiple ways to interconnect nodes, but only one at the same time, since the bus is only one and can be used only by a pair of nodes at a time. Actually, the definition of a bus topology introduces a further element than the mere placement of nodes and links, since it requires a characterization also in terms of

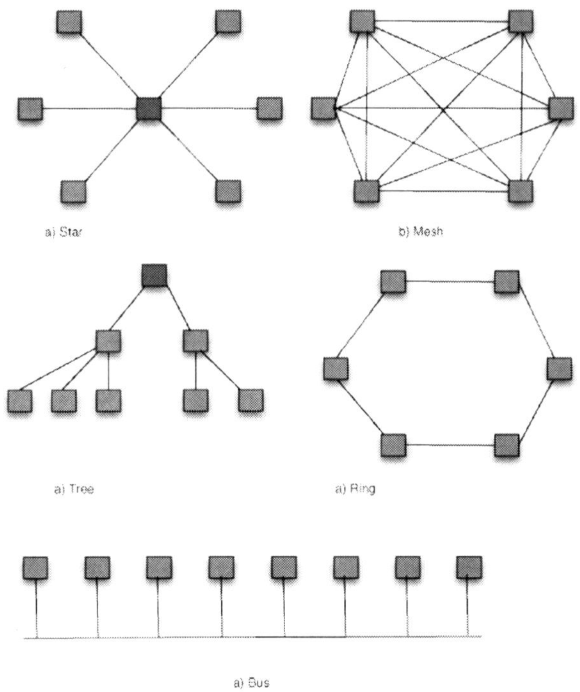

Figure 1.4 Five network topologies.

medium access. These differences will become clearer when the medium access control will be discussed in Chapter 9.

1.2.2 Access network and core network

The topological representation of the network gives an abstract perspective describing a mathematical structure. From an implementation point of view, most telecommunication networks are further subdivided into two parts:

- the *access network*, which is the part interacting with the end-user, collecting and transferring the information to the end-user; it is the edge of the network itself and enables users to access the network.
- the *core network*, which is responsible for the transmission of the information over long distances, routing the information from one user to the other, and so on; it is the core of the network itself.

The general architecture is shown in Fig. 1.5.

There are networks for which the distinction between access and core networks is not valid: these are the *peer to peer* networks where all nodes have the same characteristics and there is no hierarchy. Two remarkable examples of peer to peer networks are:

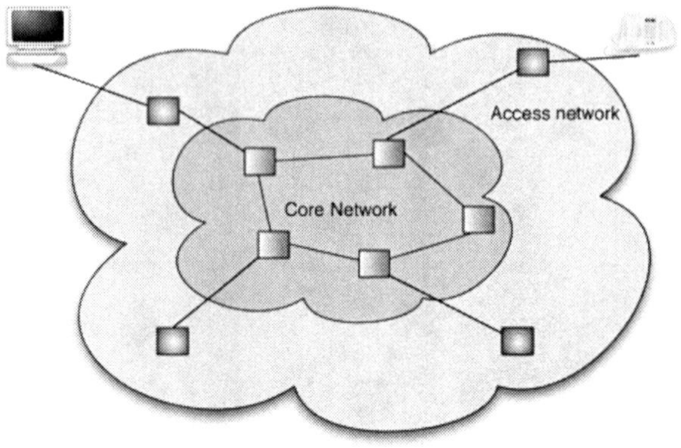

Figure 1.5 Core and access networks

- the *Skype*™[3] network, where the voice traffic is routed through a series of nodes and there is no "central server" to/from which everything is sent from/to.
- some *sensor networks*, especially *multi-hop sensor* networks where sensors (for example, measuring the temperature) send their measurements – according to specific protocols – to the neighbors and the information is propagating hop by hop to the *information sink*

Example 1.2 C Consider a very important service nowadays, the asymmetric digital subscriber loop (ADSL), whose architecture is based on three components (Fig. 1.6 provides a generic scheme for this architecture):

- a user equipment called ADSL modem to which the user attaches a PC or other networking equipment like a router to form a local area network. Incidentally, equipment combining modem and router in one piece of hardware modem is widespread.
- digital subscriber line access multiplexer (DSLAM) at the local exchange on the customer side. It intermixes voice traffic and ADSL, i.e., it collects data from many modem ports and aggregates their voice and data traffic into one complex composite signal via multiplexing. On the core network side it connects the data signals to the appropriate carrier's core network, and the phone signals to the voice switch. The DSLAM equipment aggregates the ADSL lines over its asynchronous transfer mode (ATM), frame relay, and/or Internet Protocol network port. The aggregated traffic at up to 10 Gbit/s is then directed to a carrier's backbone switch (in the core network) to the Internet backbone.
- the Internet backbone.

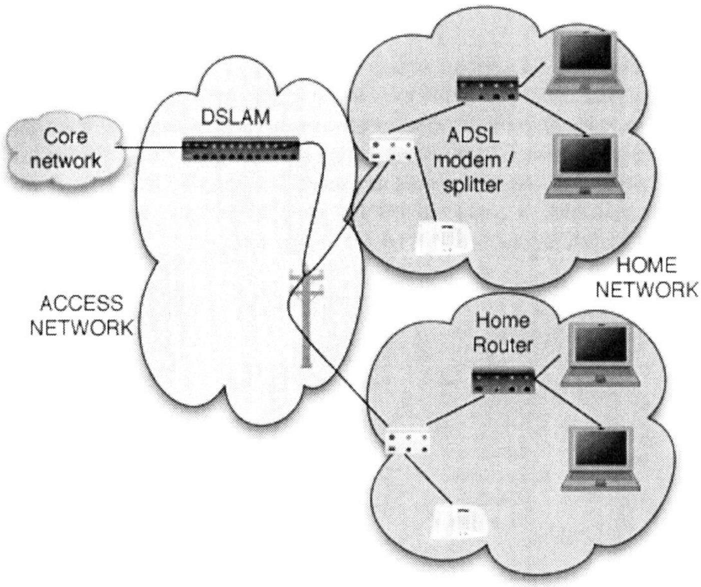

Figure 1.6 Core and access networks for the ADSL service

1.3 CIRCUIT SWITCHED AND PACKET SWITCHED COMMUNICATION MODES

In this section we briefly discuss the two basic modes in which communication can occur in a network, i.e., the *circuit switched* and the *packet switched* paradigms. Further insight can be found in Chapter 10.

We say that a *circuit switched* communication mode is used when, in a communication between the user A and the user B there is a uninterrupted physical route over the information is continuously flowing; the route, traversing the home network, the access network and the core network and again the access network and the home network is established at the beginning of the communication and not changed along the communication session. This communication mode is also often referred to as "connection orientated".

Conversely, *packet switched* communication mode is adopted when, in a communication between the user A and the user B, the information (usually digital or digitized) is grouped into packets, each one having embedded the *address* of the user B, and these packets are sent across the networks without the constraint that each and everyone follow the same route. The route that each packet is following may be determined by the nodes it is traversing and can of course vary during the session. This communication mode is often referred to as "connectionless".

The two communication modes have advantages and disadvantages, which will be highlighted later, although the general trend is towards packet switched communication networks. The following example tries to clarify the differences between the two modes.

Example 1.3 A The plain old telecommunications network (POTS) network (that is, the common and basic form of voice-base telephony still in use for residential services nowadays) of Fig. 1.7 is probably the most classical example of a circuit switched network; at the establishment of the phone call a physical deterministic route is established from user A, along the local switches and the transit switches (of the core network) of possibly different operators, to user B. The phone of user B rings only when this route is found and it is available. It could happen that some of the links are completely busy and cannot accommodate another call; in this case, user A gets a busy tone.

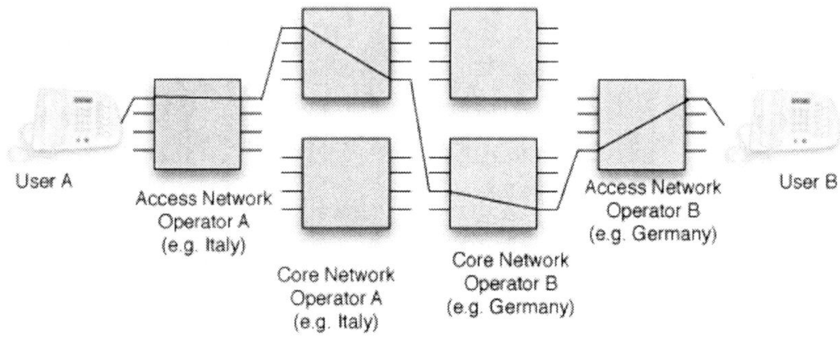

Figure 1.7 Circuit switched network

Example 1.3 B An example of packet switched network is the Internet (see Fig. 1.8), which is based on the Internet protocol (IP) where each "user" has an identifier, called *IP address*. When user A wants to send some digital information to user B, it puts the information in *packets* containing the IP address of user B and presents the packet to the first edge *router* (almost equivalent to the switch of a POTS network) of the network. The rule is that any router, based on already known information, forwards the packet to the router attached to it, which is in the best position to get the packet delivered to user B; the choice of the router can change over time, so that two consecutive packets can take different routes and it might as well happen that the latter packet arrives before the former. It is up to the receiver to re-order the packets, check that everything is delivered as expected, and so on.

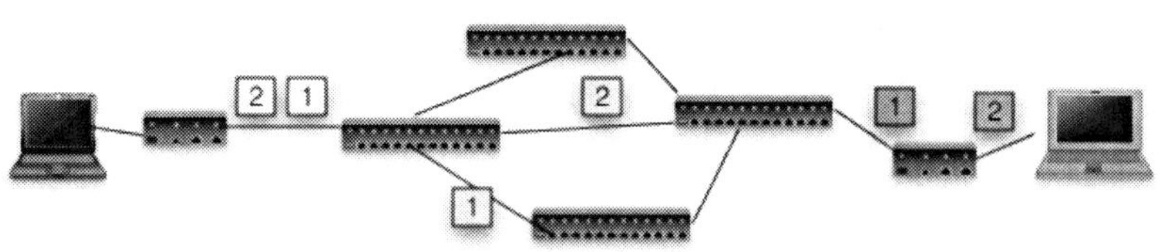

Figure 1.8 Packet switched network - different gray tones of the same box number represent the same packet at different time instants

1.4 INTRODUCTION TO THE ISO/OSI MODEL

In this section we introduce the ISO/OSI model and provide some examples of application

1.4.1 The layered model

First, we introduce the notion of *protocol* as a set of rules defining telecommunication systems interact. To reduce the unavoidable confusion that could be generated by describing protocols without any guideline, a kind of template has been defined to enable

i a consistent definition of different protocols,

ii an easier comparison between them and – last but not least –

iii a *translation* of one protocol into another, through a suitable conversion, to.

This template is the *layered model* and the *ISO/OSI model* is a particular instance of this model.

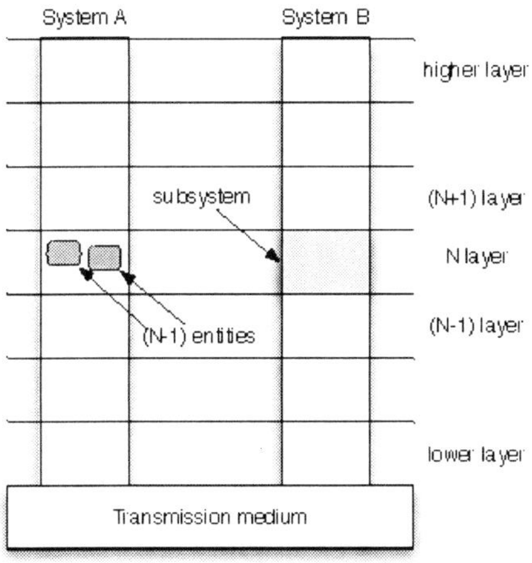

Figure 1.9 Layered architecture

According to the layered model a communication, unidirectional or bidirectional, between two systems A and B is made of *layers* and *entities* (see Fig. 1.9). For each system, each *layer-(N-1) entity* provide a *service* to a *layer-N entity*. The dialog between the two entities carried out via an exchange of service messages called *primitives* and data messages called packet data unit (PDU) through the service access point (SAP). In this manner, a layer-N entity of system A gets to talk to a peer layer-N entity of system B, as messages get down on system A and then up on system B. The way the two peer entities talk is called *layer-N protocol* and is shown in Fig. 1.10.

An insight on how the data message is actually processed is given in Fig. 1.11, where we see that through the layer-N SAP a layer-(N+1) PDU is passed down; this PDU contains

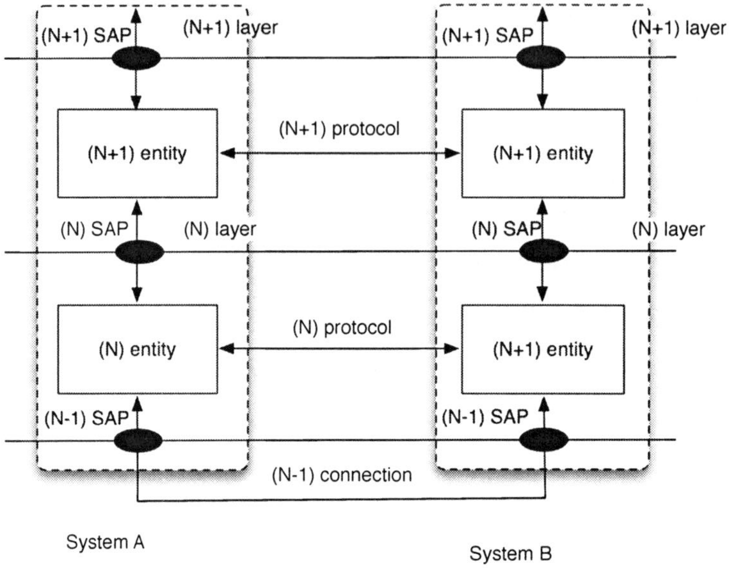

Figure 1.10 SAPs, services and protocols in a layered architecture.

the actual pieces of information transmitted from one system to the other at a certain layer. The underlying layer, however, does not simply carry the layer-(N+1) PDU to the other side but clearly needs to add some information needed for the layer-N protocol to work. As a consequence, a protocol control information (PCI) piece of information is prepended to the layer-(N+1) PDU (which in the context of layer-N is called *layer-N service data unit (SDU)* to highlight that it is coming from the SAP) to make the layer-N PDU, as shown in Fig. 1.11. We would like to remark that exactly one PDU fitting into one lower layer PDU is a particular case: actually, it is very common that the layer-(N+1) PDU is split by the layer-N protocol in smaller pieces so that the transmission of a layer-(N+1) PDU is accomplished via the transmission of several layer-N PDUs.

As said before, the service interaction through SAPs is carried out through the exchange of primitives, which usually are of four types:

1. *request*
2. *confirm*
3. *indication*
4. *response*

This is shown pictorially in Fig. 1.12, where 2 primitives are shown on one side and two on the other side, grouped as they usually are.

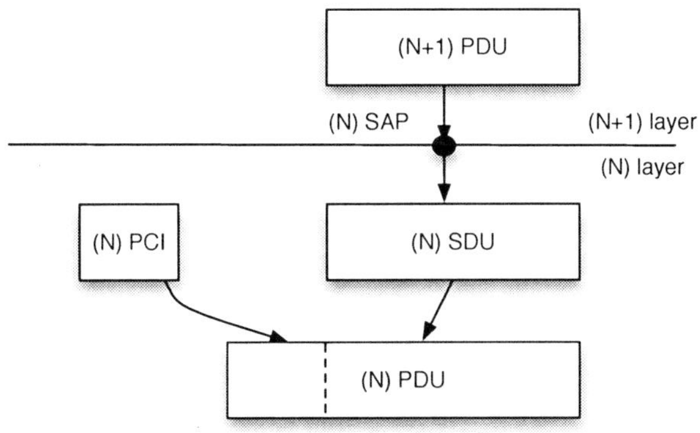

Figure 1.11 PDU, PCI and SDU in a layered architecture

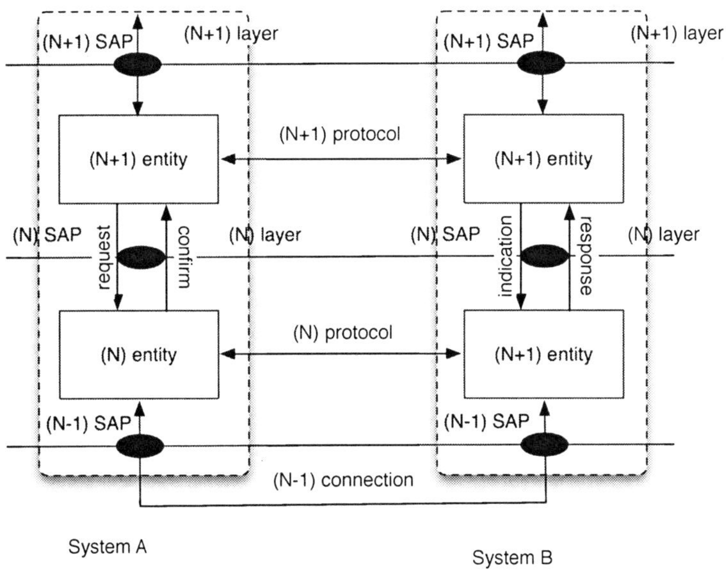

Figure 1.12 Primitives in a layered architecture

We would like to point out that many protocols do not follow the layered architecture and instead violate it by having, for example, the $(N+1)$th layer sending primitives to the $(N-1)$th layer. Furthermore, recently, there is research area called *cross layer* design, which is deliberately designing protocols to avoid each layer working in an isolated way, as implied by the layered approach.

The interaction between layers described above reflects in what is called *encapsulation* of information. In practice, this corresponds to the inclusion of a layer-N PDU as a part of a layer-$(N$-1) PDU. That is, a piece of information of a higher layer is complemented with some control information, then the resulting data are treated by the lower layer as a whole piece of information. To that, another control component is appended (often as a header, but sometimes as a trailer). This information hiding process has pros and cons: surely it enables modularity of the data, so that each layer only has to manage the control information related to it. However, it may cause a considerable growth of the bits sent over the network: in fact, since lower layers are blind to the content of the data they are exchanged, they are often forced to duplicate control information parts. Finally, there are cases, such as the interaction between TCP/UDP and IP for what concerns the pairing of TCP/UDP port and IP address, which do not respect this concept faithfully, as will be discussed in Chapter 10.

1.4.2 The ISO/OSI model

The ISO/OSI model (Fig. 1.13) is a particular instance of the layered protocol architecture specifying

- the number of layers, set to 7;
- the function and naming of each layer.

In the following we give a short description of each layer.

1. The *physical layer* is responsible for the actual delivery of the information over the physical medium; for example, at the transmitter's side, by employing a modulator, a given waveform is sent over the medium when the bit is 0 and a different one is sent when the bit is 1; therefore, at the receiver's side, a demodulator can reconstruct the 0/1 sequence based on the received waveform.

2. The *data link layer* is mainly responsible for enabling the local transfer of information. Its intervention is necessary in order to enable the actual exchange of bits through the physical layer. Indeed, the functionalities of the data link layer are several and also related to different needs. For this reason, it is common, especially within the IEEE 802 standards, to distinguish between two *sublayers* which both belong to the data link layer: (i) the medium access control (MAC), which manages the presence of multiple users on the same medium, and (ii) the logical link control (LLC), which provides the interaction between the upper layers and the MAC as well as performing flow and error control on a local basis. Indeed, the classification of these two entities as two subsets of the second layer, instead of being two separate layers themselves, is more of a conventional matter. Moreover, it is worthwhile noting that their features may not always be required. For example, there might be one single user in the network, in which case multiplexing through MAC could be avoided (or better, simplified, since the

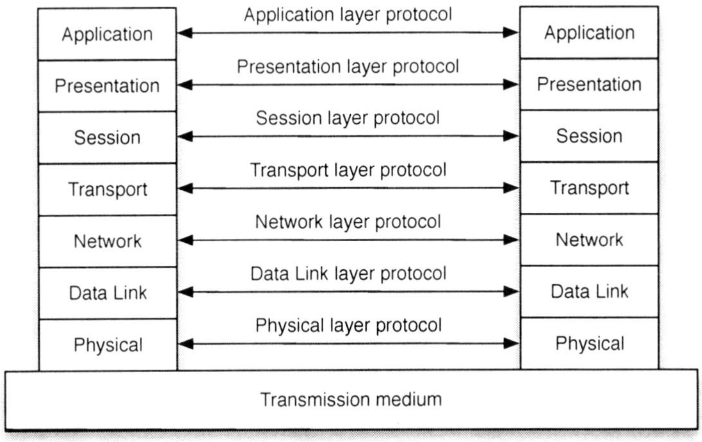

Figure 1.13 The ISO/OSI model.

availability of the medium still has to be checked). Also, if the network is very simple or the communication medium is guaranteed to be extremely reliable, one may avoid performing error and flow control on a local basis, thus simplifying the LLC. However, in most cases the whole operations of the data link layer are quite complex and will be discussed in detail within Chapter 9.

3. The *network layer* is responsible for setting up, maintaining and closing the connection between the network layer entities of different subsystems. It makes it invisible to the upper layers how the network resources are utilized: referring to Fig. 1.14, the network layer entity on system A does not know whether the flow of information is going through the systems B or C while exchanging information with system D. To achieve this goal, the network layer has the responsibility of performing *host addressing*, i.e., being able to identify all the nodes in the network in a unique manner, and *routing*, that is, to identify the desirable path the data has to follow throughout the network. It is worth mentioning that, while the data link layer is concerned with local communication exchanges, the network layer has a global view. This is not surprising and somehow implicit in their names: layer 2 is obviously concerned with single links, while layer 3 considers the whole network.

4. The *transport layer* is responsible for creating a one to one *transparent* virtual pipe between the two ends of the communication flow. It does not deal with the network topology, instead it hides the presence of multiple terminals interconnected in a network fashion which was the key aspect of the third layer. At layer four, communication proceeds as though only two entities existed in the network; when the communication session is established the transport layer sets up this virtual pipe according to the requests

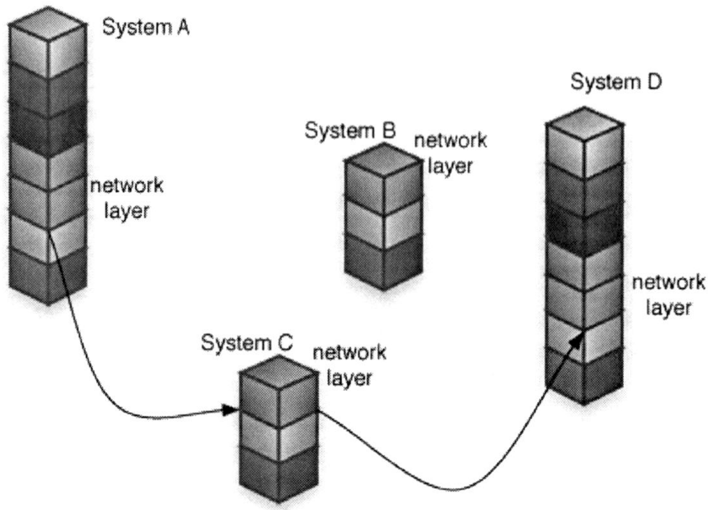

Figure 1.14 The Network Layer.

of the session layer. Also, it may include features to deliver an error-free communication, e.g., requesting the network layer to retransmit the pieces of information which are in error. Finally, the transport layer can perform *flow control*, i.e., it prevents the transport layer on one side from sending information faster than the rate that the receiving transport layer on the other side can handle.

5. The *session Layer* The session layer is responsible for establishing, suspending and tearing down in an orderly fashion any communication session the presentation layer might need. In a POTS network, people adopt, as a matter of fact, a *call control* protocol which is controlling the session. In fact, the dialer takes the phone off-hook, waits for the dial tone, dials the number, waits for the ringing tone and subsequently for the receiver to answer, talks, then closes the communication by putting the phone on hook. If during any of these operations a busy tone is heard, the communication is dropped and the dialer is invited to try again later. When using a cellular phone, a slightly different protocol is adopted since the network availability may be known in advance. Furthermore, even a simple POTS call may have multiple concurrent sessions, i.e., a call can be put on waiting, meanwhile another call goes on and finally the previous call is resumed.

6. The *presentation layer* is responsible for providing directly to the application layer the session layer services as requested by the applications and for processing the data to make them usable to the applications: e.g., the encryption and decryption of the data are usually a responsibility of the presentation layer.

7. The *application layer* is self explaining: it is the realm of the applications; normally users interact with networks at this level.

1.5 SIGNALING

In this section we provide an introduction to the concept of signaling, giving some background and indicating why it is essential in every network. A little historic and geographic survey is given, listing different types of signalling systems in time and in different regions.

1.5.1 Introduction

In almost every system, with some remarkable exceptions e.g., somehow, the Internet, the communication between two end-points involves two *planes*: a *user* plane and a *control* plane. Both of them are structured according to the layered architecture previously described, but usually, we, as users, are exposed only to the *user* plane. Users just send and receive useful data, voice, video; to them, it is totally transparent that there is a kind of cooperating network behind the fourth wall which is taking care of setting up the communication links, monitoring them, reconfiguring them if we are moving (as in a cellular network) etc. This is what is called the *control* plane, which is normally implemented according to the layered architecture i.e.,there is a physical layer (which can be shared somehow with the *user* plane), a data link layer, a network layer and so on. Of course there are SAPs and primitives as well.

To have an idea of how complicated the control plane can be we refer to Fig. 1.15 which shows the *control* plane of a GSM (Global System for Mobile communications: originally from Groupe Spècial Mobile) network (for the sake of simplicity the figure only includes the elements related to the voice service).

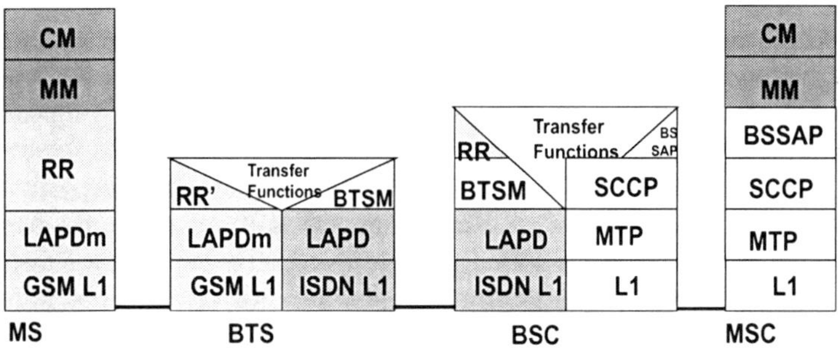

Figure 1.15 The GSM protocol architecture

The *control* flow of information is called *signaling* and the *control* plane protocols are called *signaling protocols*. A coordinated set of signaling protocols and entities is called *signaling system*.

1.5.2 Channel Associated and Common Channel

There are two types of signaling systems:

- *Channel Associated* signaling

- *Common Channel* signaling

All types of signaling can be divided in two parts: the access network signaling, or *user signaling*, and the core network signaling signaling.

Channel Associated signaling has been the first to be utilized and is divided in *in-band* signaling and *out of band* signaling, where it is intended that the *band* is the voice band. There are 3 *Channel Associated* systems still in use:

1. the Bell System multi frequency signaling very similar to the CCITT R1 system: used in the USA since after the second World War;
2. the CCITT R2 system used in Europe since 1960;
3. the CCITT signaling system 5, defined in 1964 and used in intercontinental analog links.

For reason of space we do get into details about these systems; they are very likely the ones we interact with on a daily basis in the user signaling (e.g., DTMF); for more in depth information the reader is referred to [4]. Next subsection is instead devoted to the most important *Common Channel* signaling system: the CCITT signaling system 7.

1.5.3 SS7

In this section we provide a brief overview of the CCITT signaling system 7. This system is replacing step by step the *Channel associated* signaling systems.

The main innovation and difference of this system with respect to the previous ones is that while in the other systems the nodes and the links for the signaling are in common with the ones carrying the user information in CCITT signaling system 7 the signaling network (somehow ironically) is not in common with the user information network. In other words, the *user plane* is completely separated from the control plane. The reason for using the word "common" when talking about the CCITT signaling system 7 is that e.g., a certain signaling link is not dedicated to a specific user link: it is used by the signaling network for the whole signaling so is common to all user links. The kind of revolutionary concept brought in by the CCITT signaling system 7 is that there is an overlay network with respect to the one carrying the user information, not synchronous with it, perhaps using different physical layer transmission technologies, that controls the user information network. Fig. 1.16 gives an illustration of this concept.

The overlay network is called *Common Channel Signaling* (CCS) network. It is made of two different types of nodes:

- *Signaling Point* (SP) nodes, which are at the edge of the network and are either the sources or the destinations of the signaling information; normally they are one-to-one mapped to the switches of the user plane network and co-located with them; there is an actual connection between them via suitable interfaces; in the diagrams they are usually represented with circles.
- *Signaling Transfer Point* (STP) nodes, which are the core of the signaling network and their purpose is to switch the signaling messages in the network; in the diagrams they are usually represented with squares.

We would like to remark that some equipment could act as both SP and STP at the same time.

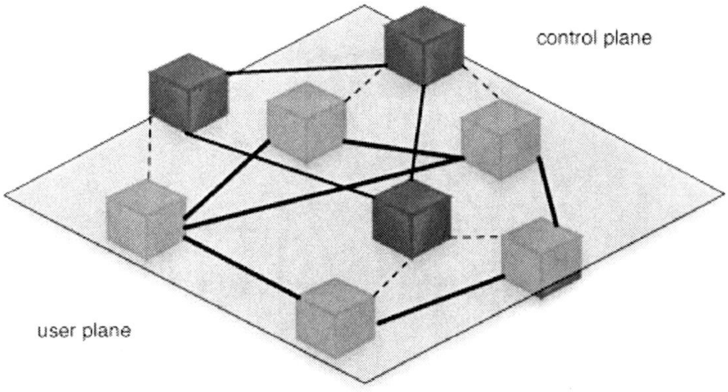

Figure 1.16 The CCITT signaling system 7 concept

The links connecting the nodes of the CCS network are called *signaling links* (SL) and are commonly made of ISDN (e.g., 64 kbit/s) links which are always on. The SL are usually represented with dashed lines.

In Fig. 1.17 an example of a network with the CCITT signaling system 7 is shown.

In Fig. 1.18 the CCITT signaling system 7 protocol stack is shown. It is made of different components and in its specification divided in 4 layers, shown in Fig. 1.18 and compared again in Fig. 1.18 with the ISO/OSI layers definition. According to this latest definition the first three are handled by the STPs, whereas level 4 (i.e.,all the rest) is end-to-end and performed by the SPs.

Without going into full details, we now go through the different protocols bottom-up. The *Message Transfer Part* is the part of the protocol in charge of delivering and routing the messages from and to the different entities. The ISO/OSI level 4 is basically divided in two parts: one related to a kind of "signaling-in-the signaling" control plane (*control part*) and one related the "user plane" (*user part*) in the signaling part. The first one is needed to establish, control and tear down the logical connection needed for the signaling and comprises the *Signaling Connection Control Part* (SCCP), the *Intermediate Service Part* (ISP) and the *Transaction Capability Application Part* (TCAP). The user plane instead depends on the specific network and can be the plain *Telephone User Part* (TUP), the *Data User Part* (DUP) or finally the *ISDN User Part* (ISUP).

1.5.4 PDH networks

The plesiochronous digital hierarchy (PDH) is a technology used in telecommunications networks to transport large quantities of data over digital transport equipment such as optic fibers and microwave radio systems. The word "plesiochronous" comes from the Greek, more precisely from $\pi\lambda\eta\sigma\iota o\varsigma$ (near) and $\chi\rho\acute{o}\nu o\varsigma$ (time), so it means "more or less at the same time:" this is to say that operations are not perfectly synchronous, but are not asynchronous either.

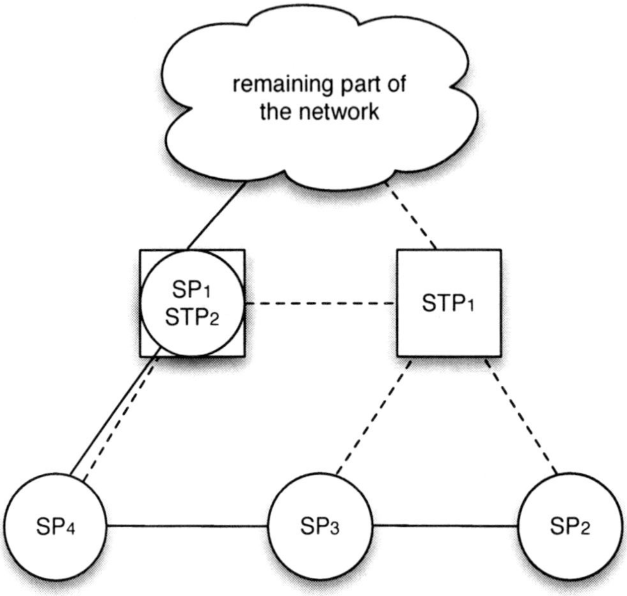

Figure 1.17 Example of a network using the CCITT signaling system 7

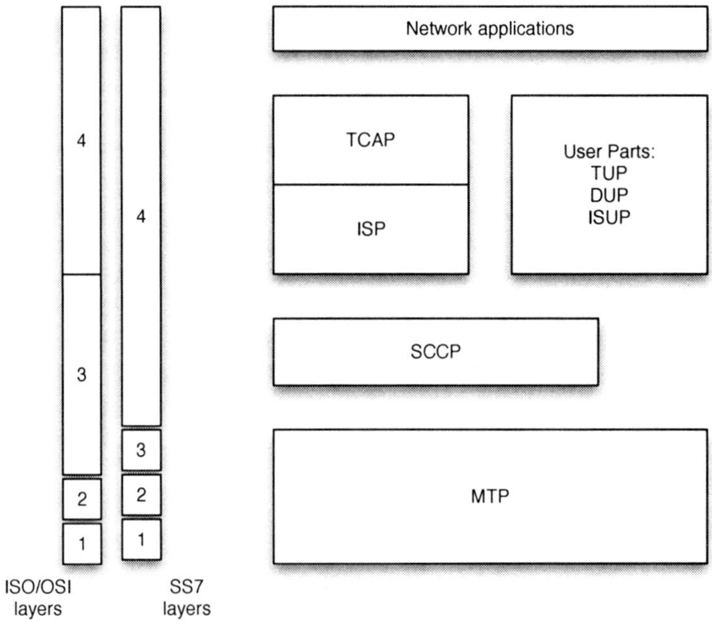

Figure 1.18 CCITT signaling system 7 Protocol Stack

The European *E-carrier system* is described below. The basic data transfer rate is a data stream of 2048 kbit/s (which approximately corresponds to 2 Mbit/s if a megabit is considered as 1024 kilobits). For speech transmission, this is broken down into thirty 64 kbit/s channels plus two 64 kbit/s channels used for signalling and synchronization. Alternatively, the whole 2 Mbit/s may be used for non speech purposes, for example, data transmission. The exact data rate of the 2 Mbit/s data stream is controlled by a clock in the equipment generating the data. The exact rate is allowed to vary by some small amount ($\pm 5 \cdot 10^{-5}$) on either side of an exact 2.048 Mbit/s. This means that multiple 2 Mbit/s data streams can be running at slightly different rates. In order to move several 2 Mbit/s data streams from one place to another, they are combined together, i.e., multiplexed in groups of four. This is done by taking 1 bit from stream number 1, followed by 1 bit from stream number 2, then number 3, then number 4. The transmitting multiplexer also adds additional bits in order to allow the far end receiving demultiplexer to distinguish which bits belong to which 2 Mbit/s data stream, and so correctly reconstruct the original data streams. These additional bits are called *justification* or *stuffing* bits. Because each of the four 2 Mbit/s data streams is not necessarily running at the same rate, some compensation has to be made. The transmitting multiplexer combines the four data streams assuming that they are running at their maximum allowed rate. This means that occasionally (unless the 2 Mbit/s is really running at the maximum rate) the multiplexer will look for the next bit but this will not yet have arrived. In this case, the multiplexer signals to the receiving multiplexer that a bit is missing. This allows the receiving multiplexer to correctly reconstruct the original data for each of the four 2 Mbit/s data streams, and at the correct, different, plesiochronous rates. The resulting data stream from the above process runs at 8448 kbit/s (about 8 Mbit/s). Similar techniques are used to combine four 8 Mbit/s together, plus bit stuffing, giving 34 Mbit/s.[5] By taking four 34 Mbit/s flows and adding bit stuffing, a 140 Mbit/s flow is achieved. Four 140 Mbit/s flows are further combined (again, plus bit stuffing) into a 565 Mbit/s flow. 565 Mbit/s is the rate typically used to transmit data over an optic fiber system for long distance transport. Recently, many telecommunication companies have been replacing their PDH equipment with synchronous digital hierarchy (SDH) equipment capable of much higher transmission rates.

The Japanese and American versions of the PDH system slightly differ in minor details, but the principles are the same. In Tab. 1.1 the details of these hierarchies are shown.

1.5.5 SDH networks

Synchronous networking differs from PDH in that the exact rates that are used to transport the data are tightly synchronized across the entire network. This synchronization system allows entire inter-country networks to operate synchronously, greatly reducing the amount of buffering required between elements in the network. Both SDH and synchronous optical networking (SONET) (the North American counterpart of SDH, which is mainly a European technology) can be used to encapsulate earlier digital transmission standards, such as the PDH standard, or directly employed to support either ATM or so-called Packet over SONET/SDH (POS) networking. As such, it is inaccurate to think of SDH or SONET as communications protocols in and of themselves, but rather they should be thought of as generic and all-purpose

[5]In this and in the following computations, it is implicitly visible that bit stuffing increases the amount of bits exchanged. For example, 34 Mbit/s is strictly larger than what is coming from the four 8 Mbit/s flows.

Table 1.1 PDH hierarchies: in brackets the rate in number of basic user channels at 64 kbit/s, DS stands for Digital Signal and E for Europe.

	North America	*Japan*	*Europe*
Level zero	64 kbit/s (DS0)	64 kbit/s	64 kbit/s
First Level	1.544 Mbit/s (DS1) (24) (T1)	1.544 Mbit/s (24)	2.048 Mbit/s (32)
Intermediate Level	3.152 Mbit/s (DS1C) (48)	–	–
Second Level	6.312 Mbit/s (DS2) (96)	6.312 Mbit/s (96), or 7.786 Mbit/s (120)	8.448 Mbit/s (128) (E2)
Third Level	44.736 Mbit/s (DS3) (672) (T3)	32.064 Mbit/s (480)	34.368 Mbit/s (512) (E3)
Fourth Level	274.176 Mbit/s (DS4) (4032)	97.728 Mbit/s (1440)	139.264 Mbit/s (2048) (E4)
Fifth Level	400.352 Mbit/s (DS5) (5760)	565.148 Mbit/s (8192)	565.148 Mbit/s (8192) (E5)

transport containers for moving both voice and data. The basic format of an SDH signal allows it to carry many different services in its Virtual Container (VC) because it is bandwidth-flexible.

The basic unit of framing in SDH is a STM-1 (Synchronous Transport Module level - 1), which operates at 155.52 Mbit/s. SONET refers to this basic unit as an STS-3c (Synchronous Transport Signal - 3, concatenated), but its high-level functionality, frame size, and bit-rate are the same as STM-1. SONET offers an additional basic unit of transmission, the STS-1 (Synchronous Transport Signal - 1), operating at 51.84 Mbit/s - exactly one third of an STM-1/STS-3c. Some manufacturers also support the SDH equivalent STM-0, but this is not part of the standard.

SONET and SDH are both using a frame based transmission. The frame is organized by grouping 8 bits at a time (octet-based) and is represented as a bi-dimensional matrix, organized in rows and columns, where some parts are devoted to overhead and the rest is left for actual payload. This bi-dimensional payload is then scanned by a pointer which selects the octet to be transmitted.

In Tab. 1.2 the details of the SONET and SDH hierarchies are given. In Fig. 1.19 a typical architecture for an SDH network is shown: the main equipment is the *Add-Drop Multiplexer* (ADM) which is capable to extract from an higher rate stream a slower rate one and, at the same time, insert a slower rate stream into a higher rate one.

Table 1.2 SDH and SONET hierarchies.

SONET Optical carrier level	SONET Frame Format	SDH level and Frame Format	Payload kbit/s	Line Rate kbit/s
OC-1	STS-1	*STM-0*	50,112	51,840
OC-3	STS-3	STM-1	150,336	155,520
OC-12	STS-12	STM-4	601,344	622,080
OC-24	STS-24	–	1,202,688	1,244,160
OC-48	STS-48	STM-16	2,405,376	2,488,320
OC-192	STS-192	STM-64	9,621,504	9,953,280
OC-768	STS-768	STM-256	38,486,016	39,813,120
OC-3072	STS-3072	STM-1024	153,944,064	159,252,240

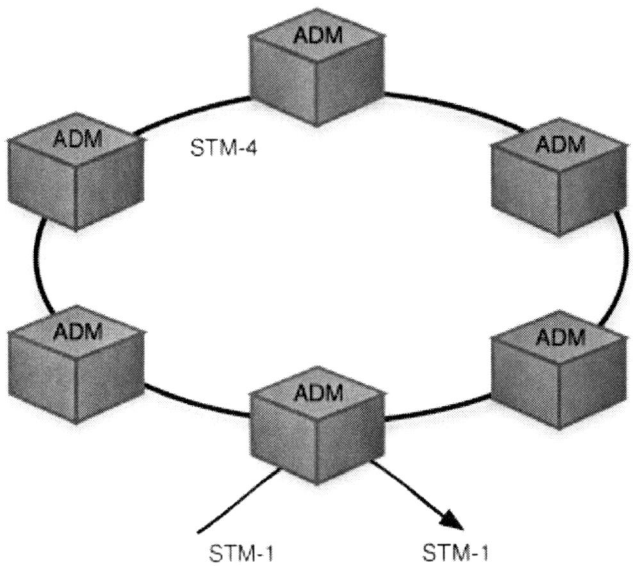

Figure 1.19 SDH ring with ADMs.

REFERENCES

1. B. Hopkins and R. J. Wilson, *The truth about Königsberg*, College Mathematics Journal, Vol. 35, pp. 198–207, 2004.

2. G. L. Alexanderson, *Euler and Königsberg's bridges: A historical view*, Bulletin (New Series) of the American Mathematical Society, Vol. 43, pp. 567–573, 2006.

3. Skype, `www.skype.com`

4. van Bosse, John G., *Signaling in Telecommunication Networks*, New York: John Wiley & Sons Inc., 1998.

CHAPTER 2

DETERMINISTIC AND RANDOM SIGNALS

Nicola Laurenti and Tomaso Erseghe

Signals are the object of communication systems and as such the vehicle for communications. In particular, it is only through random processes, which are unpredictable to some extent, that we can carry information from source to destination. In this chapter we review some basic results about deterministic and random processes that the reader should be familiar with. The abridgements of such results in this chapter is also the occasion to establish the notation we will use throughout the book.

2.1 TIME AND FREQUENCY DOMAIN REPRESENTATION

2.1.1 Continuous time signals

Definition 2.1 *In our context a* continuous time signal *is a complex-valued function of a real variable*

$$x : \mathbb{R} \mapsto \mathbb{C} . \tag{2.1}$$

The independent real variable often has the meaning of *time* and is thus denoted by the letter t and measured in seconds (s) or their multiples or sub-multiples. On the contrary the function $x(t)$ can represent any quantity that possibly evolves with time, such as the voltage across the terminals of an electrical device, or the current in a mesh, the temperature at some point in

Principles of Communications Networks and Systems N. Benvenuto, M. Zorzi

space, or the optical power emitted by a *laser* source, and so on; its value will then accordingly be expressed in the proper measurement units: volt (V), ampere (A), kelvin (K), watt (W), etc. Most of these quantities are intrinsically real, and the signals that represent them will only take real values. Such signals are called *real-valued* or simply *real* signals. The graphical representation of a real signal x is the plot of $x(t)$ versus t, as illustrated in Fig. 2.1.

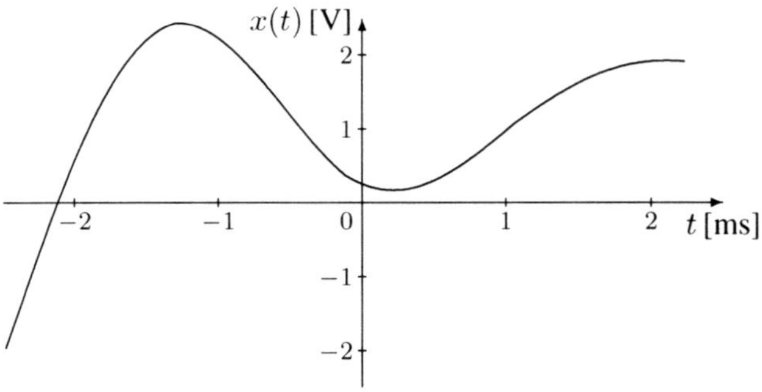

Figure 2.1 Graphical representation of a real-valued signal representing a voltage.

Example 2.1 A A few real-valued functions that are very useful in the definition of signals are:

- The *Heaviside step function*

$$1(x) = \begin{cases} 1 & , \; x > 0 \\ 1/2 & , \; x = 0 \\ 0 & , \; x < 0 \end{cases}$$

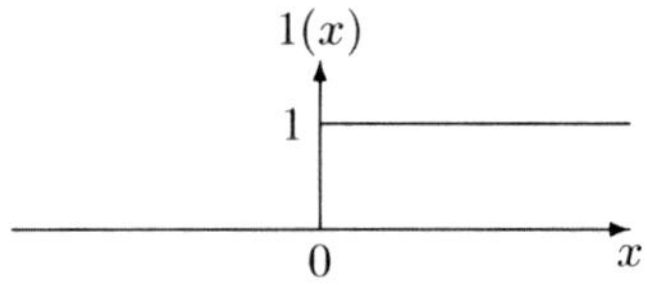

- We also make use of

$$1_0(x) = \begin{cases} 1 & , \; x \geq 0 \\ 0 & , \; x < 0 \end{cases}$$

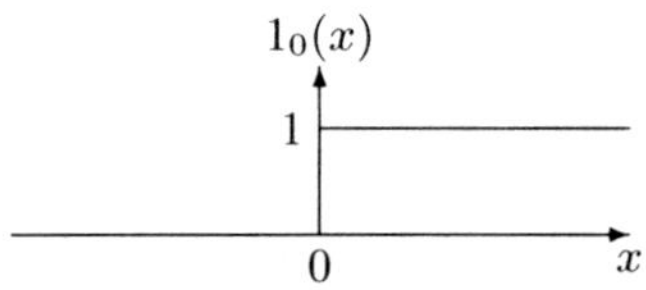

- The *sign function*

$$\operatorname{sgn}(x) = \begin{cases} -1 & , \; x < 0 \\ 0 & , \; x = 0 \\ 1 & , \; x > 0 \end{cases}$$

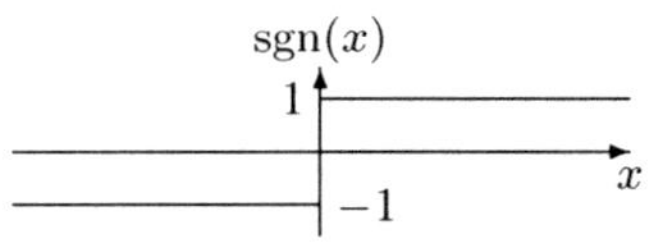

- The *rectangle function*

$$\operatorname{rect}(x) = \begin{cases} 1 & , \; |x| < 1/2 \\ 1/2 & , \; |x| = 1/2 \\ 0 & , \; |x| > 1/2 \end{cases}$$

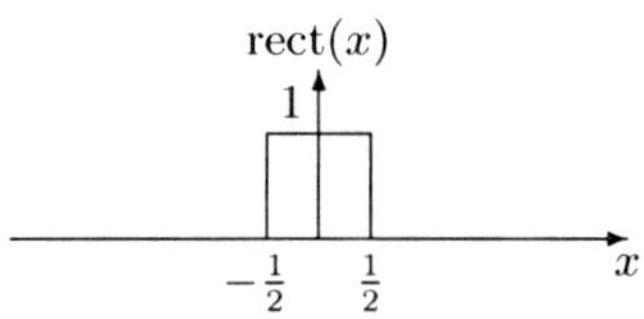

- The *triangle function*

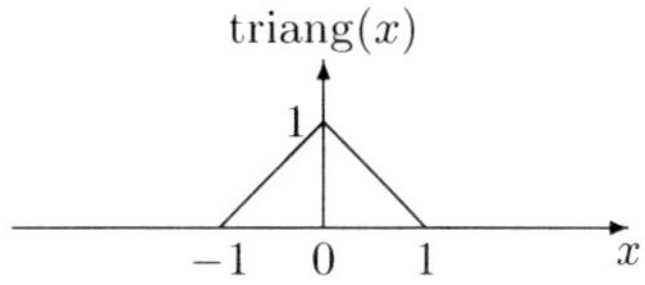

$$\text{triang}(x) = \begin{cases} 1 - |x| & , \ |x| < 1 \\ 0 & , \ |x| \geq 1 \end{cases}$$

- The *Woodward sine cardinal function*

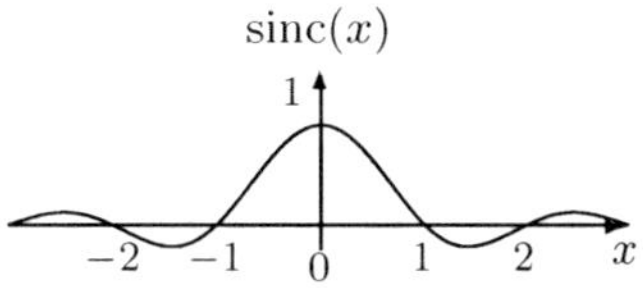

$$\text{sinc}(x) = \begin{cases} \dfrac{\sin(\pi x)}{\pi x} & , \ x \neq 0 \\ 1 & , \ x = 0 \end{cases}$$

- The *raised cosine function*

$$\text{rcos}(x, \rho) = \begin{cases} 1 & , \ 0 \leq |x| \leq \frac{1-\rho}{2} \\ \cos^2\left(\dfrac{\pi}{2}\dfrac{|x| - \frac{1-\rho}{2}}{\rho}\right) & , \ \frac{1-\rho}{2} < |x| \leq \frac{1+\rho}{2} \\ 0 & , \ |x| > \frac{1+\rho}{2} \end{cases} \tag{2.2}$$

- The *inverse raised cosine function*

$$\text{ircos}(x, \rho) = \text{sinc}\,(x) \cos\,(\pi \rho x) \frac{1}{1 - (2\rho x)^2} \tag{2.3}$$

- The *square root raised cosine function*

$$\text{rrcos}(x, \rho) = \sqrt{\text{rcos}\,(x, \rho)} \tag{2.4}$$

- The *inverse square root raised cosine function*

$$\text{irrcos}(x, \rho) = \frac{\sin\left[\pi\,(1 - \rho)\,x\right] + 4\rho x \cos\left[\pi(1 + \rho)x\right]}{\pi\left[1 - (4\rho x)^2\right] x} \tag{2.5}$$

Many signals can be obtained from the above basic functions with operations of time shifting, time scaling, and amplitude scaling. For example we can define the step signal starting at t_1 with amplitude A_1 shown in Fig. 2.2a as $x_1(t) = A_1\, 1(t - t_1)$ and the rectangular pulse with amplitude A_2 and width T_2 shown in Fig. 2.2b as $x_2(t) = A_2 \,\text{rect}(t/T_2)$.

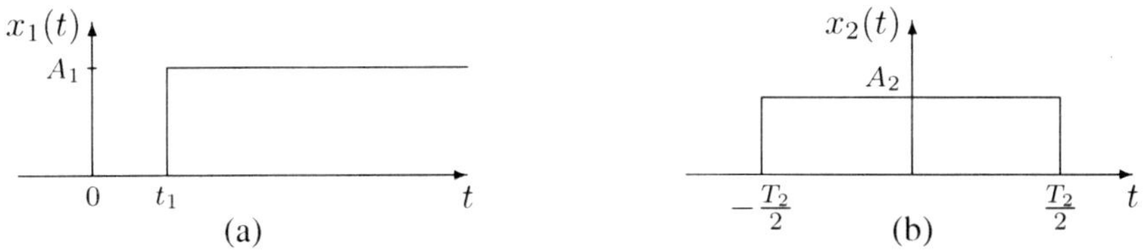

Figure 2.2 Plots of: (a) the signal $x_1(t) = A_1\, 1(t - t_1)$; (b) the signal $x_2(t) = A_2 \,\text{rect}(t/T_2)$.

Example 2.1 B The *Dirac delta function* or *impulse* $\delta(t)$ is also very useful in defining signals, although it is not a "function" strictly speaking. It is characterized by the following properties

- it vanishes outside the origin

$$\delta(t) = 0 \quad , \quad t \neq 0$$

- its integral over any interval including the origin is unity

$$\int_a^b \delta(t)\,\mathrm{d}t = \begin{cases} 1 & , \ 0 \in (a,b) \\ 0 & , \ 0 \notin (a,b) \end{cases}$$

- the *sifting property*

$$\int_{-\infty}^{+\infty} x(t)\delta(t-t_0)\,\mathrm{d}t = x(t_0)$$

It is easy to see that $\delta(0)$ can not have a finite value, since the above integrals would vanish in all cases: thus, it is commonly said that the Dirac impulse is infinite at the origin.

It is also easy to verify that the Dirac impulse and the step function are related by

$$\int_{-\infty}^{t} \delta(\tau)\,\mathrm{d}\tau = 1(t) \quad , \quad \delta(t) = \frac{d}{dt}1(t)$$

A very common graphical representation of the Dirac impulse is given by an arrow pointing upward, as in Fig. 2.3.

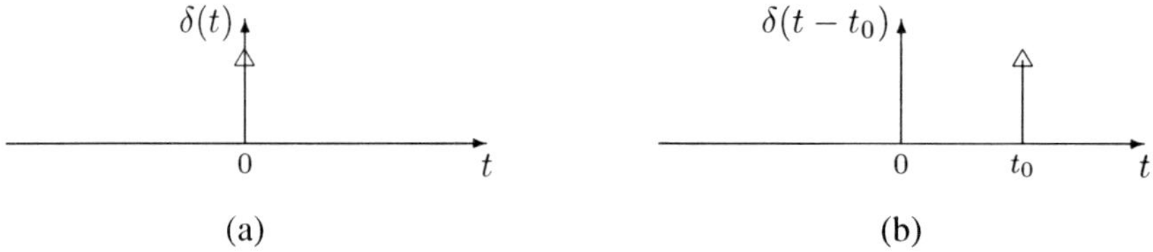

Figure 2.3 Plots of: (a) the Dirac impulse; (b) the same impulse shifted in time.

As we show in Fig. 2.4, for a complex signal x two different representations are commonly used, each combining two plots: the real and imaginary parts representation, combining plots of the real part $\Re[x(t)]$ and imaginary part $\Im[x(t)]$ versus t as in Fig. 2.4b, and the amplitude and phase representation, combining plots of the amplitude

$$|x(t)| = \sqrt{(\Re[x(t)])^2 + (\Im[x(t)])^2} \tag{2.6}$$

and phase[1]

$$\angle x(t) = \arg x(t) = \arctan\left(\frac{\Im[x(t)]}{\Re[x(t)]}\right) \tag{2.7}$$

versus t as in Fig. 2.4c. Hence

$$x(t) = \Re[x(t)] + j\,\Im[x(t)] \tag{2.8}$$

or

$$x(t) = |x(t)|\,e^{j\angle x(t)} \tag{2.9}$$

[1] We assume that the range of the arctan is $(-\pi, \pi)$ by inspection of the sign of $\Re[x(t)]$.

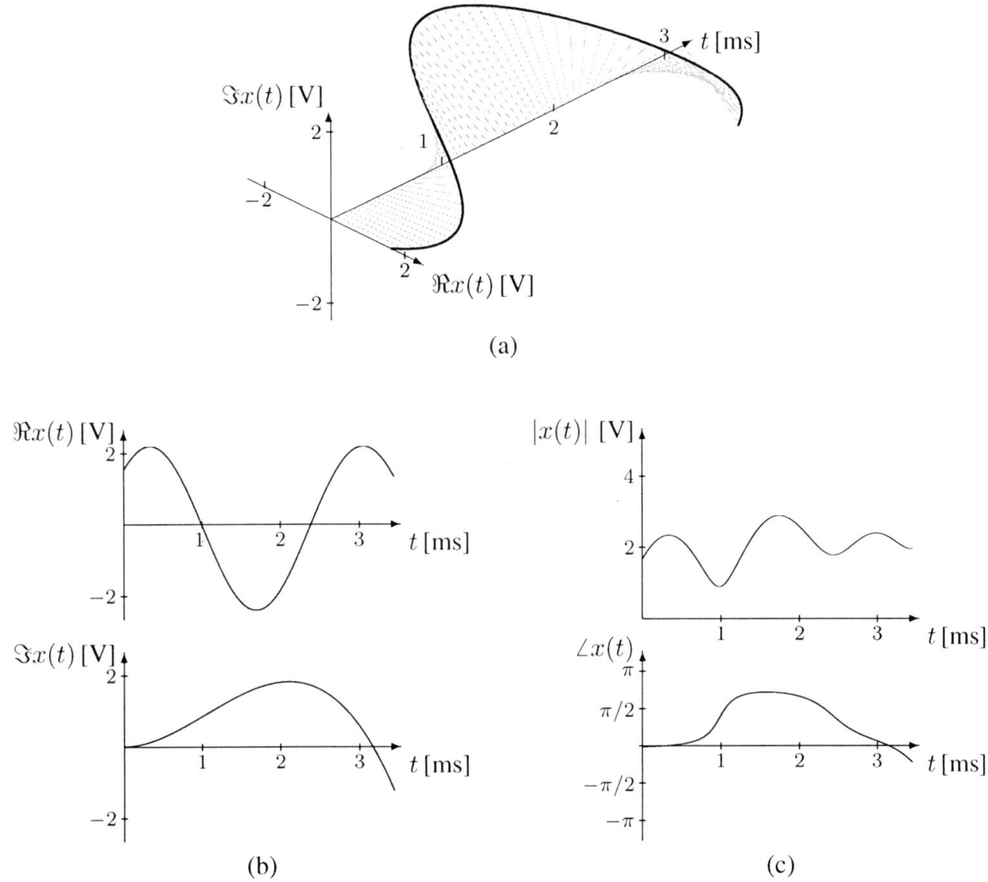

Figure 2.4 Graphical representations of a complex-valued signal representing a voltage: (a) 3-D representation; (b) real/imaginary part representation; (c) amplitude/phase representation.

It is far less common to use the three-dimensional representation of Fig. 2.4a.

A continuous time signal can also be identified through its Fourier transform (Ftf) that is itself a complex function of a real variable, defined as

$$\mathcal{X} : \mathbb{R} \mapsto \mathbb{C} \quad , \quad \mathcal{X}(f) = \mathcal{F}[x|f] = \int_{-\infty}^{+\infty} x(t) e^{-j2\pi f t} \, \mathrm{d}t \tag{2.10}$$

The variable denoted with f has the meaning of frequency, the inverse of time, and is thus measured in hertz (Hz), whereas the dimension of $\mathcal{X}$ is the product of the dimensions of x and t. For example, if the signal x represents a voltage evolving with time, its Fourier transform will be measured in volts times seconds (V·s) or volts per Hz (V/Hz). Moreover the transform is usually complex, even for a real signal, and it is quite common to choose the amplitude/phase plots for its graphical representation.

The Fourier transformation (2.10), yielding $\mathcal{X}$ from x can be inverted allowing to unambiguously recover the signal x from its transform $\mathcal{X}$ by the following relation

$$x(t) = \int_{-\infty}^{+\infty} \mathcal{X}(f) e^{j2\pi ft} \, \mathrm{d}f \tag{2.11}$$

Since knowledge of $\mathcal{X}$ is equivalent to the knowledge of x, $\mathcal{X}$ can be seen as an alternative way to represent the signal x and thus it is usually referred to as the *frequency domain* representation of x. In this context, the function $x(t)$ is called the *time domain* representation of the signal.

The reader should be already familiar with the remarkable concept of Fourier transform. Here we limit ourselves to recall some of its properties in Tab. 2.1, and collect a few signal/transform pairs in Tab. 2.2.

Example 2.1 C The signal with constant amplitude A and support from t_1 to t_2

$$x(t) = \begin{cases} A & , \ t_1 \le t < t_2 \\ 0 & , \ \text{elsewhere} \end{cases}$$

has the following Ftf

$$\mathcal{X}(f) = \int_{-\infty}^{+\infty} x(t) e^{-j2\pi ft} \, \mathrm{d}t = \int_{t_1}^{t_2} A e^{-j2\pi ft} \, \mathrm{d}t = jA \frac{e^{-j2\pi ft_2} - e^{-j2\pi ft_1}}{2\pi f}$$

The same result can be obtained by writing $x(t)$ as a rectangular pulse, centered at $t_0 = (t_1 + t_2)/2$ and having width $T = t_2 - t_1$, that is

$$x(t) = A \,\mathrm{rect}\left(\frac{t - t_0}{T}\right)$$

with the rect function introduced in Example 2.1 A. Starting from the rect $\leftrightarrow$ sinc signal/transform pair given in Tab. 2.2, we write

$$x_1(t) = \mathrm{rect}\left(\frac{t}{T}\right) \xrightarrow{\mathcal{F}} \mathcal{X}_1(f) = T \,\mathrm{sinc}(Tf)$$

then by applying the time shift rule in Tab. 2.1

$$x_2(t) = x_1(t - t_0) = \mathrm{rect}\left(\frac{t - t_0}{T}\right) \xrightarrow{\mathcal{F}} \mathcal{X}_2(f) = e^{-j2\pi ft_0} \mathcal{X}_1(f) = T \,\mathrm{sinc}(Tf) e^{-j2\pi ft_0}$$

Eventually, linearity yields

$$x(t) = A x_2(t) = A \,\mathrm{rect}\left(\frac{t - t_0}{T}\right) \xrightarrow{\mathcal{F}} \mathcal{X}(f) = A \mathcal{X}_2(f) = AT \,\mathrm{sinc}(Tf) e^{-j2\pi ft_0}$$

The signal and its transform are shown in Fig. 2.5.

2.1.2 Frequency domain representation for periodic signals

For periodic signals, that is signals having the property

$$x(t) = x(t + T_\mathrm{p}) \tag{2.12}$$

Table 2.1 Some general properties of the Fourier transform.

property	*signal*	*Fourier transform*
	$x(t)$	$\mathcal{X}(f)$
linearity	$a\,x(t) + b\,y(t)$	$a\,\mathcal{X}(f) + b\,\mathcal{Y}(f)$
duality	$\mathcal{X}(t)$	$x(-f)$
time inverse	$x(-t)$	$\mathcal{X}(-f)$
complex conjugate	$x^*(t)$	$\mathcal{X}^*(-f)$
real part	$\Re[x(t)] = \frac{1}{2}[x(t) + x^*(t)]$	$\frac{1}{2}[\mathcal{X}(f) + \mathcal{X}^*(-f)]$
imaginary part	$\Im[x(t)] = \frac{1}{2j}[x(t) - x^*(t)]$	$\frac{1}{2j}[\mathcal{X}(f) - \mathcal{X}^*(-f)]$
scaling	$x(at),\, a \neq 0$	$\frac{1}{\lvert a\rvert}\mathcal{X}\left(\frac{f}{a}\right)$
time shift	$x(t - t_0)$	$e^{-j2\pi f t_0}\,\mathcal{X}(f)$
frequency shift	$x(t)\,e^{j2\pi f_0 t}$	$\mathcal{X}(f - f_0)$
modulation	$x(t)\cos(2\pi f_0 t + \varphi)$	$\frac{1}{2}\left[e^{j\varphi}\mathcal{X}(f - f_0) + e^{-j\varphi}\mathcal{X}(f + f_0)\right]$
	$x(t)\sin(2\pi f_0 t + \varphi)$	$\frac{1}{2j}\left[e^{j\varphi}\mathcal{X}(f - f_0) - e^{-j\varphi}\mathcal{X}(f + f_0)\right]$
	$\Re[x(t)\,e^{j(2\pi f_0 t + \varphi)}]$	$\frac{1}{2}\left[e^{j\varphi}\mathcal{X}(f - f_0) + e^{-j\varphi}\mathcal{X}^*(-f - f_0)\right]$
differentiation	$\frac{d}{dt}\,x(t)$	$j2\pi f\,\mathcal{X}(f)$
	$-j2\pi t\,x(t)$	$\frac{d}{df}\,\mathcal{X}(f)$
integration	$\int_{-\infty}^{t} x(\tau)\,d\tau = (1 * x)(t)$	$\frac{1}{j2\pi f}\,\mathcal{X}(f) + \frac{1}{2}\mathcal{X}(0)\,\delta(f)$
convolution	$(x * y)(t)$	$\mathcal{X}(f)\,\mathcal{Y}(f)$
product	$x(t)\,y(t)$	$(\mathcal{X} * \mathcal{Y})(f)$
real signal	$x(t) = x^*(t)$	$\mathcal{X}(f) = \mathcal{X}^*(-f)$, $\mathcal{X}$ Hermitian, $\Re[\mathcal{X}(f)]$ even, $\Im[\mathcal{X}(f)]$ odd, $\lvert\mathcal{X}(f)\rvert$ even, $\angle\mathcal{X}(f)$ odd
imaginary signal	$x(t) = -x^*(t)$	$\mathcal{X}(f) = -\mathcal{X}^*(-f)$
even signal	$x(t) = x(-t)$	$\mathcal{X}(f) = \mathcal{X}(-f)$, $\mathcal{X}$ even
odd signal	$x(t) = -x(-t)$	$\mathcal{X}(f) = -\mathcal{X}(-f)$, $\mathcal{X}$ odd
Parseval theorem	$E_x = \int_{-\infty}^{+\infty} \lvert x(t)\rvert^2\,dt = \int_{-\infty}^{+\infty} \lvert\mathcal{X}(f)\rvert^2\,df = E_{\mathcal{X}}$	
Poisson sum formula	$\sum_{k=-\infty}^{+\infty} x(kT_c) = \frac{1}{T_c}\sum_{\ell=-\infty}^{+\infty} \mathcal{X}(\ell/T_c)$	

for some period T_p, the frequency domain representation (2.10) fails, since the integral over the whole time axis would not converge, at least in the ordinary sense. Indeed for periodic signals it is common to consider the expansion of $x(t)$ into its *Fourier series* of complex exponentials with frequencies that are multiples of $F_p = 1/T_p$ (called the *fundamental frequency* of the

Table 2.2 Examples of signal/Fourier transform pairs.

signal	*Fourier transform*
$x(t)$	$\mathcal{X}(f)$
$\delta(t)$	1
1	$\delta(f)$
$e^{j2\pi f_0 t}$	$\delta(f - f_0)$
$\cos(2\pi f_0 t)$	$\frac{1}{2}\left[\delta(f - f_0) + \delta(f + f_0)\right]$
$\sin(2\pi f_0 t)$	$\frac{1}{2j}\left[\delta(f - f_0) - \delta(f + f_0)\right]$
$1(t)$	$\frac{1}{2}\,\delta(f) + \frac{1}{j2\pi f}$
$\text{sgn}(t)$	$\frac{1}{j\pi f}$
$\text{rect}\left(\frac{t}{T}\right)$	$T\,\text{sinc}(fT)$
$\text{sinc}\left(\frac{t}{T}\right)$	$T\,\text{rect}(fT)$
$\text{triang}\left(\frac{t}{T}\right)$	$T\,\text{sinc}^2(fT)$
$e^{-at}\,1(t)$, with $a > 0$	$\frac{1}{a+j2\pi f}$
$t\,e^{-at}\,1(t)$, with $a > 0$	$\frac{1}{(a+j2\pi f)^2}$
$e^{-a\lvert t\rvert}$, with $a > 0$	$\frac{2a}{a^2+(2\pi f)^2}$
e^{-at^2} , with $a > 0$	$\sqrt{\frac{\pi}{a}}\,e^{-\pi^2 f^2/a}$

signal). Thus we can write

$$x(t) = \sum_{\ell=-\infty}^{+\infty} \mathcal{X}_\ell \, e^{j2\pi \ell F_p t} \tag{2.13}$$

where the *Fourier coefficients* $\{\mathcal{X}_\ell\}$ are determined as

$$\mathcal{X}_\ell = \frac{1}{T_p}\int_0^{T_p} x(t) e^{-j2\pi \ell F_p t}\, dt \tag{2.14}$$

Observe that the Fourier coefficients of a signal have the same physical dimension as the signal.

Example 2.1 D A *square pulse repetition* signal with amplitude A, period T_p and duty cycle d, with $0 < d < 1$, has the expression

$$x(t) = \sum_{n=-\infty}^{+\infty} A\,\text{rect}\left(\frac{t - nT_p}{dT_p}\right)$$

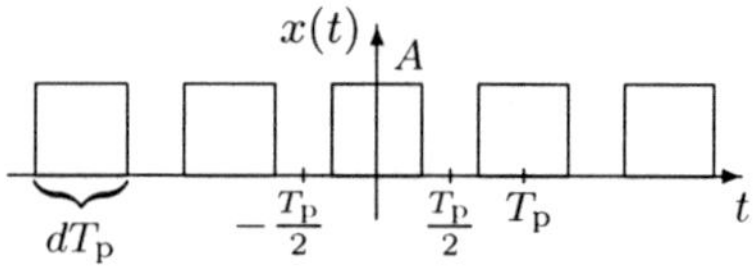

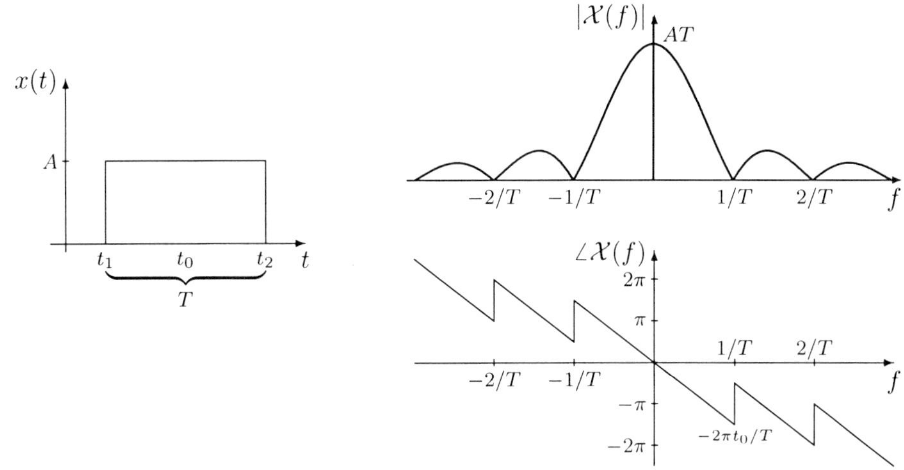

Figure 2.5 Graphical representations of the signal and its transform in Example 2.1 C, in the case $t_2 = 5t_1$.

Its Fourier coefficients are calculated as the integral over any period, that is any interval of length T_p,

$$\begin{aligned}\mathcal{X}_\ell &= \frac{1}{T_p}\int_{-T_p/2}^{T_p/2} x(t)e^{-j2\pi\ell F_p t}\,dt = \frac{1}{T_p}\int_{-dT_p/2}^{dT_p/2} Ae^{-j2\pi\ell F_p t}\,dt \\ &= j\frac{A}{T_p}\frac{e^{-j\pi\ell F_p dT_p} - e^{j\pi\ell F_p dT_p}}{2\pi\ell F_p} = A\frac{\sin(\ell\pi d)}{\ell\pi} \\ &= Ad\,\mathrm{sinc}(\ell d)\end{aligned}$$

For the particular case $d = 1/2$ we get

$$\mathcal{X}_\ell = \begin{cases} A/2 & , \ \ell = 0 \\ (-1)^{\frac{\ell-1}{2}}\dfrac{A}{\ell\pi} & , \ \ell \text{ odd} \\ 0 & , \ \ell \text{ even}\end{cases}$$

Based on the Fourier series expansion (2.13) and the Fourier transform of the complex exponential in Tab. 2.2, it is possible to circumvent the convergence problem in (2.10) and define a Fourier transform for periodic signals as a sum of Dirac impulses, regularly spaced by F_p, each weighted by the corresponding Fourier coefficient

$$\mathcal{X}(f) = \sum_{\ell=-\infty}^{+\infty} \mathcal{X}_\ell\,\delta(f - \ell F_p)\,. \tag{2.15}$$

Conversely, if the Ftf of a signal only contains Dirac impulses at multiples of a frequency F_p as in (2.15), we can write the signal as a linear combination of complex exponentials as in (2.13). This implies that the signal is periodic with period $T_p = 1/F_p$.

2.1.3 Discrete time signals

Analogously to (2.1) we define

Definition 2.2 *A* discrete time signal *or* sequence *is a complex function of an integer variable*

$$x : \mathbb{Z} \mapsto \mathbb{C} . \tag{2.16}$$

The integer variable has no physical dimensions attached to it, but it is usually associated with the flowing of time through some constant T called *quantum*. Thus, we write $x(nT)$ or equivalently x_n to represent the value of the quantity x observed at the instant nT. As in continuous time signals, x may represent a physical quantity (voltage, current, temperature, pressure), or unlike continuous time signals it may represent just a number. The discrete nature of the time domain may either be inherent to the signal (think for example of a sequence of bits representing a text file), or come from the observation of an analog quantity at regularly spaced instants (see Section 2.3.3 on sampling) in which case T is called the *sampling* period.

A common graphical representation for a real-valued discrete time signal $\{x(nT)\}$ is via a sequence of points at coordinates (nT, x_n) stemming out from the horizontal (time) axis, as in Fig. 2.6. We stress the fact that x is not defined at instants that are not multiples of T.

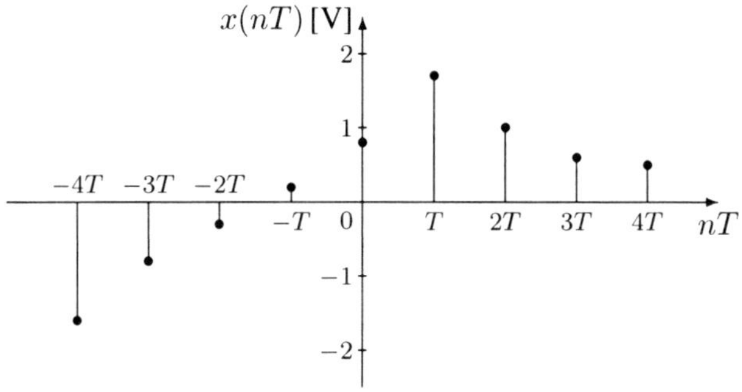

Figure 2.6 Graphical representation of a real-valued discrete time signal.

For complex-valued signals two-plot representations are used, either real/imaginary part or amplitude/phase plots as in Fig. 2.7.

The frequency domain representation of a discrete time signal $\{x(nT)\}$ is given by its Fourier transform

$$\mathcal{X} : \mathbb{R} \mapsto \mathbb{C} \quad , \quad \mathcal{X}(f) = \sum_{n=-\infty}^{+\infty} x(nT) e^{-j2\pi f nT} \tag{2.17}$$

which turns out to be a periodic function of frequency with period $F = 1/T$, as it is obtained by a linear combination of complex exponentials with periods that are sub-multiples of F. The inverse transform in this case is given by the formula

$$x(nT) = T \int_0^F \mathcal{X}(f) e^{j2\pi nTf} \, df \tag{2.18}$$

The following example is the discrete time analogous of Example 2.1 C.

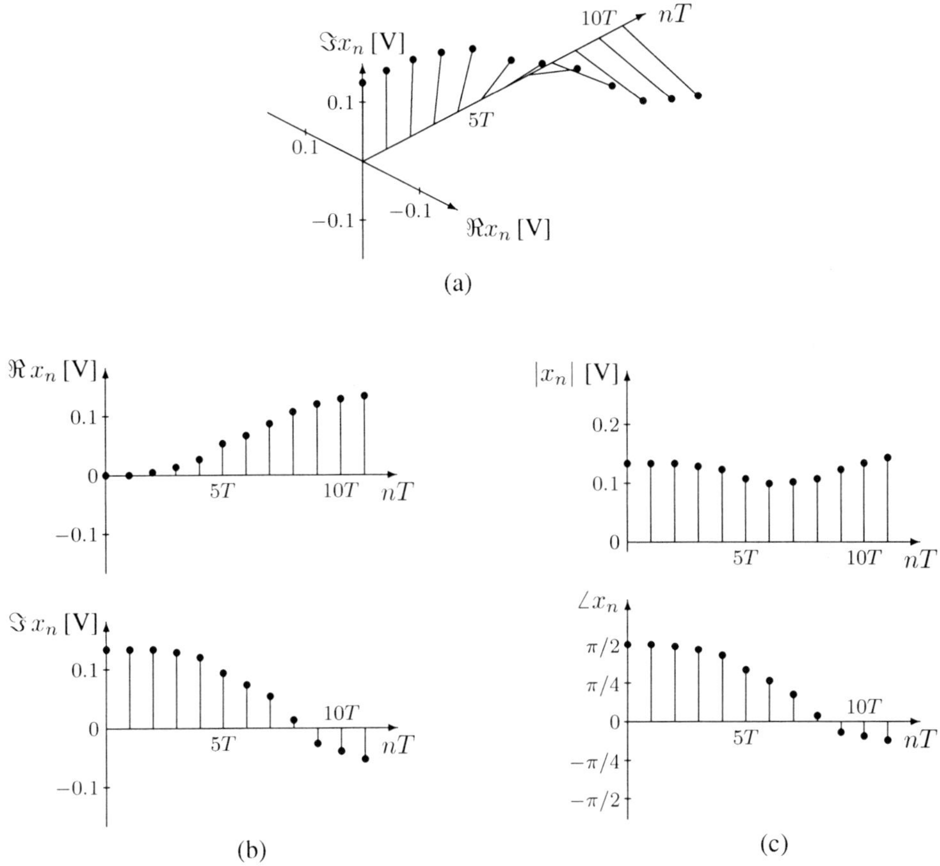

Figure 2.7 Graphical representations of a complex-valued discrete time signal representing a voltage: (a) 3-D representation; (b) real/imaginary part representation; (c) amplitude/phase representation.

Example 2.1 E The signal with constant amplitude A and support from n_1T to n_2T

$$x(nT) = \begin{cases} A & , \ n_1 \leq n \leq n_2 \\ 0 & , \ \text{elsewhere} \end{cases}$$

has the following Fourier transform

$$\mathcal{X}(f) = \sum_{n=-\infty}^{+\infty} x(nT)e^{-j2\pi fnT} = \sum_{n=n_1}^{n_2} A\, e^{-j2\pi fnT} = A \sum_{n=n_1}^{n_2} \left(e^{-j2\pi fT}\right)^n$$

By the algebraic identity, holding for any $z \in \mathbb{C}$

$$\sum_{n=n_1}^{n_2} z^n = \frac{z^{n_2+1} - z^{n_1}}{z-1}$$

we obtain the result

$$\begin{aligned}\mathcal{X}(f) &= A\frac{e^{-j2\pi f(n_2+1)T} - e^{-j2\pi f n_1 T}}{e^{-j2\pi fT} - 1} \\ &= Ae^{-j2\pi n_0 fT}\frac{\sin(\pi NfT)}{\sin(\pi fT)}\end{aligned}$$

where $n_0 = (n_1 + n_2)/2$ is the midpoint between n_1 and n_2, and $N = n_2 - n_1 + 1$ is the number of nonzero samples of the signal. The signal and its transform are shown in Fig. 2.8.

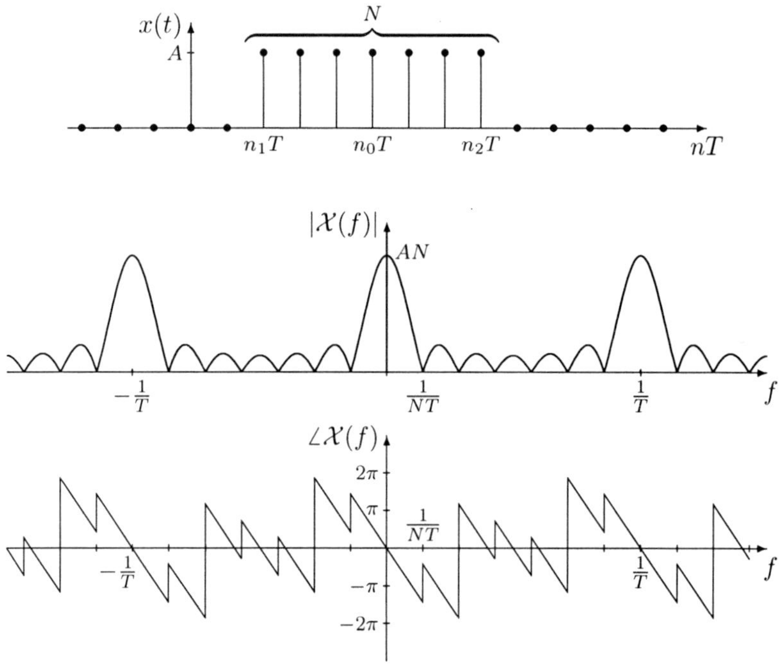

Figure 2.8 Graphical representations of the signal and its Fourier transform in Example 2.1 E, in the case $n_1 = 2, n_2 = 8, n_0 = 5, N = 7$.

Example 2.1 F In discrete time signals the equivalent role of the Dirac delta function for continuous time signals is played by the Kronecker delta,

$$\delta_m = \begin{cases} 1, & m = 0 \\ 0, & m \neq 0 \end{cases} \tag{2.19}$$

having as Ftf the constant value 1.

2.2 ENERGY AND POWER

2.2.1 Energy and energy spectral density

The *energy* of a continuous time signal $x(t)$ is defined as

$$E_x = \int_{-\infty}^{+\infty} |x(t)|^2 \, dt \quad , \tag{2.20}$$

providing the integral exists and is finite.

From the definition (2.20) we see that the energy of a signal is always *real and nonnegative*, even for complex signals. Moreover, if $x(t)$ represents a voltage, E_x will be measured in $V^2 \cdot s$.

For a pair of signals x, y their *cross energy* is defined as

$$E_{xy} = \int_{-\infty}^{+\infty} x(t)y^*(t) \, dt \tag{2.21}$$

providing the integral exists and is finite. Unlike the energy of a signal, the cross energy *may* be negative, or even complex for complex signals. Moreover, since

$$E_{yx} = \int_{-\infty}^{+\infty} y(t)x^*(t) \, dt = E_{xy}^* \tag{2.22}$$

the cross energy is not commutative in general. However, if both x and y are real-valued, then $E_{xy} = E_{yx}$. Observe that the definition of energy (2.20) can be obtained as a particular case of cross energy by choosing $y(t) = x(t)$ in the definition (2.21).

In Section 2.5 we will see that the cross energy between two signals satisfies all the axioms of the *inner product* in a signal space over the complex field $\mathbb{C}$. In this contest, we now state a very useful property: if x and y have finite energy, their cross energy is guaranteed to be finite and is bounded by the *Schwarz inequality*

$$|E_{xy}| \leq \sqrt{E_x E_y} \; . \tag{2.23}$$

Moreover, two distinct signals whose cross energy is zero are said to be *orthogonal*.

A fundamental result relating energies and Fourier transforms is given next, whose proof can be found in [6].

Theorem 2.1 (Parseval theorem) *Let x, y be two continuous time signals with finite energy and $\mathcal{X}, \mathcal{Y}$ their Fourier transforms. Then the cross energy of the signals equals the cross energy of their transforms, that is*

$$E_{xy} = \int_{-\infty}^{+\infty} x(t)y^*(t) \, dt = \int_{-\infty}^{+\infty} \mathcal{X}(f)\mathcal{Y}^*(f) \, df = E_{\mathcal{X}\mathcal{Y}} \tag{2.24}$$

In particular for the energy of a signal we have

$$E_x = \int_{-\infty}^{+\infty} |x(t)|^2 \, dt = \int_{-\infty}^{+\infty} |\mathcal{X}(f)|^2 \, df = E_{\mathcal{X}} \; . \tag{2.25}$$

The above results (2.24) and (2.25) allow us to define the *energy spectral density* of a signal and the *cross energy spectral density* of two signals as

$$\mathcal{E}_x(f) = |\mathcal{X}(f)|^2 \quad , \quad \mathcal{E}_{xy}(f) = \mathcal{X}(f)\mathcal{Y}^*(f) \tag{2.26}$$

respectively, so that the energies are obtained by integrating such densities

$$E_x = \int_{-\infty}^{+\infty} \mathcal{E}_x(f)\,\mathrm{d}f \quad , \quad E_{xy} = \int_{-\infty}^{+\infty} \mathcal{E}_{xy}(f)\,\mathrm{d}f \; . \tag{2.27}$$

Remark We observe that the function $\mathcal{E}_x(f)$ ($\mathcal{E}_{xy}(f)$) describes how the signal energy (cross energy) is distributed in the frequency domain. It thus provides a more complete information than just the total of energy given by E_x (E_{xy}).

Example 2.2 A The energy of a causal decreasing exponential signal

$$x(t) = A\,e^{-t/\tau_1}1(t) \quad , \quad t \in \mathbb{R}$$

with $A = 20\,\mathrm{mV}$ and $\tau_1 = 100\,\mu\mathrm{s}$ can be calculated as

$$E_x = \int_0^\infty A^2\,e^{-2t/\tau_1}\,\mathrm{d}t = \frac{A^2\tau_1}{2} \; .$$

By substituting the parameter values, the above expression yields $E_x = 2 \cdot 10^{-8}\,\mathrm{V}^2\mathrm{s}$. We can also check Parseval theorem in this particular case, as

$$\mathcal{X}(f) = \frac{A\tau_1}{1 + j2\pi f\tau_1} \quad \Rightarrow \quad \mathcal{E}_x(f) = \frac{A^2\tau_1^2}{1 + (2\pi f\tau_1)^2}$$

and indeed we get

$$\begin{aligned}\int_{-\infty}^{+\infty} \frac{A^2\tau_1^2}{1+(2\pi f\tau_1)^2}\,\mathrm{d}f &= A^2\tau_1^2 \left[\frac{1}{2\pi\tau_1}\arctan(2\pi f\tau_1)\right]_{-\infty}^{+\infty} \\ &= A^2\tau_1^2\,\frac{1}{2\pi\tau_1}\,\pi = \frac{A^2\tau_1}{2}\end{aligned}$$

This example is illustrated in Fig. 2.9.

We observe that in the above example, both the integrals in the time and frequency domain were equally straightforward to carry out. This is not the case in general, and the Parseval theorem often is a very useful tool for the solution of the integral, as can be seen in the following example.

Example 2.2 B Suppose we wish to calculate the energy of the signal

$$x(t) = A\,\mathrm{sinc}(t/T) \quad , \quad t \in \mathbb{R}$$

with $A = 3\,\mathrm{V}$ and $T = 1\,\mathrm{ms}$. By the change of variable $u = \pi t/T$ we can write

$$E_x = \int_{-\infty}^{+\infty} A^2\,\mathrm{sinc}^2(t/T)\,\mathrm{d}t = A^2\frac{T}{\pi}\int_{-\infty}^{+\infty}\frac{\sin^2 u}{u^2}\,\mathrm{d}u \tag{2.28}$$

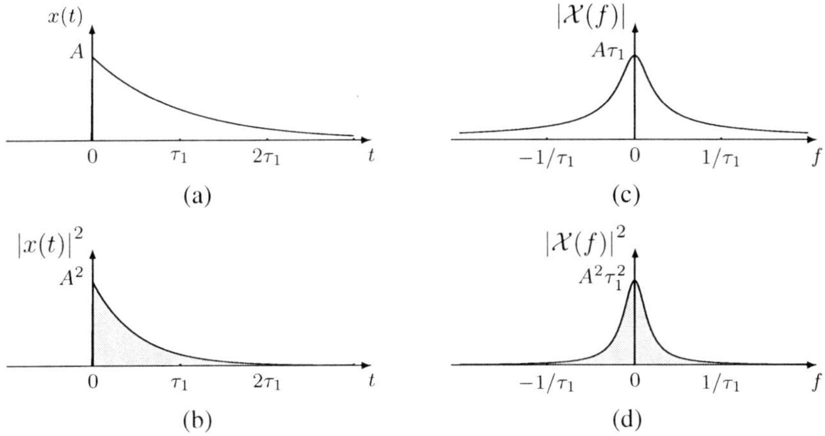

Figure 2.9 Illustration of Example 2.2 A: (a)-(b) in the time domain; (c)-(d) in the frequency domain. The Parseval theorem assures that the shaded areas on the (b) and (d) plots are equal.

However the integral on the right-hand side is not straightforward to evaluate, since an integral function of $\sin^2 u/u^2$ is not easily expressed in terms of elementary functions.

Observing from Tab. 2.2 that

$$\mathcal{X}(f) = AT \operatorname{rect}(Tf)$$

we get

$$\mathcal{E}_x(f) = A^2T^2 \operatorname{rect}(Tf) .$$

Hence, by resorting to the Parseval theorem, we get

$$E_x = \int_{-\infty}^{+\infty} A^2T^2 \operatorname{rect}(Tf)\, \mathrm{d}f = A^2T^2 \int_{-\frac{1}{2T}}^{\frac{1}{2T}} \mathrm{d}f = A^2T . \tag{2.29}$$

By substituting the values for A and T, the result is $E_x = 9 \cdot 10^{-3}\ \mathrm{V}^2\mathrm{s}$.

2.2.2 Instantaneous and average power

For a signal $x(t)$, the quantity $|x(t)|^2$ represents the *instantaneous power* of the signal. The *average power* is defined as the average over time of the instantaneous power, that is

$$\mathtt{M}_x = \lim_{u\to\infty} \frac{1}{2u} \int_{-u}^{u} |x(t)|^2\, \mathrm{d}t \tag{2.30}$$

as illustrated in Fig. 2.10. The reader should not confuse this concept with the notion of power from Physics. Here, the power of a signal has the the same physical dimension as the square signal (e.g., V^2 for a voltage signal), and not W.

The following proposition gives a very important result.

Proposition 2.2 *The average power of a signal with finite energy is zero.*

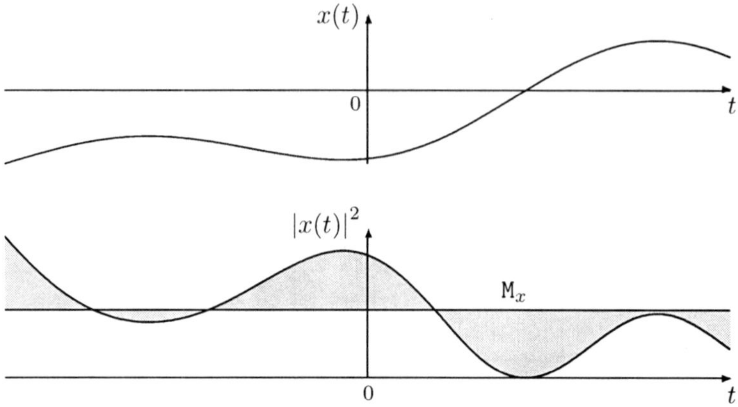

Figure 2.10 Illustration of the average power of a signal. The shaded areas above and below the $\mathtt{M}_x$ line are equal.

Proof Let E_x be the finite energy of the signal x. For any u, we have

$$0 \leq \int_{-u}^{u} |x(t)|^2 \, \mathrm{d}t \leq E_x$$

and if we divide all sides by $2u$ and take the limit as in (2.30) we get

$$0 \leq \mathtt{M}_x \leq \lim_{u \to \infty} \frac{E_x}{2u}$$

Since E_x is finite and fixed, the limit on the right-hand side is zero, and so is $\mathtt{M}_x$ □

This proposition shows that, for signals with finite energy it is meaningless to consider the average power, as it is null. On the other hand, for signals with nonzero average power it is meaningless to consider the energy, as it will turn out to be infinite.

Extending (2.30) we can define the *average cross power* for a pair of signals x and y as

$$\mathtt{M}_{xy} = \lim_{u \to \infty} \frac{1}{2u} \int_{-u}^{u} x(t) y^*(t) \, \mathrm{d}t \tag{2.31}$$

The average cross power has the same physical dimension as the product of the two signals.

For *periodic signals* with period T_p, the above definitions simplify and lead to consider the average power within a single period

$$\mathtt{M}_x = \frac{1}{T_\mathrm{p}} \int_{0}^{T_\mathrm{p}} |x(t)|^2 \, \mathrm{d}t \tag{2.32}$$

and similarly for the cross power between periodic signals x and y *with the same period* T_p,

$$\mathtt{M}_{xy} = \frac{1}{T_\mathrm{p}} \int_{0}^{T_\mathrm{p}} x(t) y^*(t) \, \mathrm{d}t \tag{2.33}$$

We apply the above result in the following example.

Example 2.2 C We want to find the average power of the sinusoidal signal

$$x(t) = A\cos(2\pi f_0 t + \varphi_0)$$

with $A = 2\,\text{V}$, $f_0 = 60\,\text{Hz}$ and $\varphi_0 = \pi/3$.

The signal x is periodic with $T_p = 1/f_0$, thus by (2.32) we get

$$\mathtt{M}_x = f_0 \int_0^{T_p} A^2 \cos^2(2\pi f_0 t + \varphi_0)\,\mathrm{d}t$$

By applying the trigonometric formula $\cos^2 x = (1 + \cos 2x)/2$ we get

$$\mathtt{M}_x = f_0 \int_0^{T_p} \frac{A^2}{2}\,\mathrm{d}t + f_0 \int_0^{T_p} \frac{A^2}{2} \cos(4\pi f_0 t + 2\varphi_0)\,\mathrm{d}t$$

As the second integral vanishes we get

$$\boxed{\mathtt{M}_x = \frac{A^2}{2}\,,} \tag{2.34}$$

and substituting the value of A, $\mathtt{M}_x = 2\,\text{V}^2$.

Observe that in the case of a sinusoidal signal the average power has a particularly simple expression that depends only on the amplitude A, and does not depend either on its frequency f_0 or phase φ_0.

The result in the previous example can be extended to an arbitrary sum of sinusoids, where the overall power is the sum of the power of each sinusoid. For example, if

$$x(t) = \sum_{i=1}^{N} A_i \cos(2\pi f_i t + \varphi_i)$$

with distinct frequencies, $f_i \neq f_j$, for $i \neq j$, it is

$$\mathtt{M}_x = \sum_{i=1}^{N} \frac{A_i^2}{2} \tag{2.35}$$

In general, for a periodic signal we have the useful theorem

Theorem 2.3 (Parseval theorem for periodic signals) *Let x, y be two periodic signals with the same period T_p and let, $\{\mathcal{X}_\ell\}, \{\mathcal{Y}_\ell\}$ be the coefficients of their Fourier series. Then the average cross power between x and y is*

$$\mathtt{M}_{xy} = \sum_{\ell=-\infty}^{+\infty} \mathcal{X}_\ell \mathcal{Y}_\ell^* \tag{2.36}$$

Correspondingly for a periodic signal

$$\mathtt{M}_x = \sum_{\ell=-\infty}^{+\infty} |\mathcal{X}_\ell|^2 \tag{2.37}$$

Proof We start from expression (2.33) of average cross power in a period. By replacing $x(t)$ and $y(t)$ with their Fourier series we get

$$\begin{aligned}\mathtt{M}_{xy} &= \frac{1}{T_{\mathrm{p}}}\int_0^{T_{\mathrm{p}}}\left(\sum_{\ell=-\infty}^{+\infty}\mathcal{X}_\ell e^{j2\pi\ell F_{\mathrm{p}}t}\right)\left(\sum_{m=-\infty}^{+\infty}\mathcal{Y}_m e^{j2\pi mF_{\mathrm{p}}t}\right)^* \mathrm{d}t \\ &= \frac{1}{T_{\mathrm{p}}}\int_0^{T_{\mathrm{p}}}\sum_{\ell=-\infty}^{+\infty}\sum_{m=-\infty}^{+\infty}\mathcal{X}_\ell\mathcal{Y}_m^* e^{j2\pi(\ell-m)F_{\mathrm{p}}t}\,\mathrm{d}t\end{aligned}$$

Then by exchanging the order of sums and integral,

$$\mathtt{M}_{xy} = \frac{1}{T_{\mathrm{p}}}\sum_{\ell=-\infty}^{+\infty}\sum_{m=-\infty}^{+\infty}\mathcal{X}_\ell\mathcal{Y}_m^*\int_0^{T_{\mathrm{p}}} e^{j2\pi(\ell-m)F_{\mathrm{p}}t}\,\mathrm{d}t$$

and making use of the fact that

$$\int_0^{T_{\mathrm{p}}} e^{j2\pi(\ell-m)F_{\mathrm{p}}t}\,\mathrm{d}t = \begin{cases} T_{\mathrm{p}} & , \ \ell = m \\ 0 & , \ \ell \neq m\end{cases}$$

we prove the statement. □

Example 2.2 D Consider the two complex exponential signals

$$x(t) = A_1 e^{j(2\pi f_0 t+\varphi_1)} \quad , \quad y(t) = A_2 e^{j(2\pi f_0 t+\varphi_2)}$$

with the parameters $A_1, A_2, f_0, \varphi_1, \varphi_2$ all real-valued. Both signals are periodic with fundamental frequency $F_{\mathrm{p}} = f_0$ so that their Fourier coefficients are easily seen by inspection to be

$$\mathcal{X}_\ell = \begin{cases} A_1 e^{j\varphi_1} & , \ \ell = 1 \\ 0 & , \ \ell \neq 1\end{cases} \quad , \quad \mathcal{Y}_\ell = \begin{cases} A_2 e^{j\varphi_2} & , \ \ell = 1 \\ 0 & , \ \ell \neq 1\end{cases}$$

Then by applying (2.37) we get

$$\mathtt{M}_x = |\mathcal{X}_1|^2 = A_1^2 \quad , \quad \mathtt{M}_y = |\mathcal{Y}_1|^2 = A_2^2$$

and by (2.36)

$$\mathtt{M}_{xy} = \mathcal{X}_1\mathcal{Y}_1^* = A_1 A_2 e^{j(\varphi_1-\varphi_2)}$$

Example 2.2 E Consider the following sum of sinusoids

$$x(t) = A_1\cos(2\pi f_1 t+\varphi_1) + A_2\cos(2\pi f_2 t+\varphi_2) + A_3\sin(2\pi f_1 t+\varphi_3)$$

with $f_2 = 2f_1$. $x(t)$ is a periodic signal with period $T_{\mathrm{p}} = \mathrm{lcm}(1/f_1, 1/f_2) = 1/f_1$ and hence $F_{\mathrm{p}} = 1/T_{\mathrm{p}} = f_1$, or equivalently its fundamental frequency is $F_{\mathrm{p}} = \gcd(f_1, f_2) = f_1$. By using the Euler identities we can rewrite $x(t)$ as

$$x(t) = \frac{A_1}{2}\left(e^{j2\pi f_1 t+j\varphi_1} + e^{-j2\pi f_1 t-j\varphi_1}\right) + \frac{A_2}{2}\left(e^{j2\pi f_2 t+j\varphi_2} + e^{-j2\pi f_2 t-j\varphi_2}\right)$$

$$
\begin{aligned}
&-j\frac{A_3}{2}\left(e^{j2\pi f_1 t+j\varphi_3}-e^{-j2\pi f_1 t-j\varphi_3}\right)\\
&=\frac{1}{2}(A_1e^{j\varphi_1}-jA_3e^{j\varphi_3})e^{j2\pi F_p t}+\frac{1}{2}(A_1e^{-j\varphi_1}+jA_3e^{-j\varphi_3})e^{-j2\pi F_p t}\\
&+\frac{1}{2}A_2e^{j\varphi_2}e^{j2\pi 2F_p t}+\frac{1}{2}A_2e^{-j\varphi_2}e^{-j2\pi 2F_p t}
\end{aligned}
$$

The only nonzero Fourier coefficients are then

$$
\mathcal{X}_1=\frac{1}{2}(A_1e^{j\varphi_1}-jA_3e^{j\varphi_3})\ ,\quad \mathcal{X}_{-1}=\mathcal{X}_1^*\ ,\quad \mathcal{X}_2=\frac{1}{2}A_2e^{j\varphi_2}\ ,\quad \mathcal{X}_{-2}=\mathcal{X}_2^*
$$

Thus from (2.37) we get

$$
\begin{aligned}
\mathtt{M}_x &= |\mathcal{X}_1|^2+|\mathcal{X}_{-1}|^2+|\mathcal{X}_2|^2+|\mathcal{X}_{-2}|^2\\
&=2|\mathcal{X}_1|^2+2|\mathcal{X}_2|^2\\
&=2\frac{1}{4}\left[A_1^2+A_3^2+2\Re(jA_1A_3e^{j\varphi_1-j\varphi_3})\right]+2\frac{1}{4}A_2^2\\
&=\frac{1}{2}\left[A_1^2+A_2^2+A_3^2-2A_1A_3\sin(\varphi_1-\varphi_3)\right]
\end{aligned}
$$

We note that the above result is different from (2.35) since in $x(t)$ there are two sinusoids with the same frequency.

2.3 SYSTEMS AND TRANSFORMATIONS

The *transformation*, or *system*, is the mathematical model of many devices where we can apply a signal at a certain point in the device, named input, and correspondingly observe a signal at some other point, named output. The output signal is also called the *response* of the system to the input signal. An example would be an electrical network where we can apply a time-varying voltage at two terminals (the input) and measure the current flowing through a conductor cross section at some point (the output). In this sense the transformation can be thought of as acting on its input signal to produce the output signal. It is then identified by the class of possible input signals $\mathcal{I}$, the class of possible output signals $\mathcal{O}$, and by the map

$$
\mathcal{M}\ :\ \mathcal{I}\mapsto\mathcal{O} \tag{2.38}
$$

so that if $x\in\mathcal{I}$ is the input signal, the output is

$$
y=\mathcal{M}[x] \tag{2.39}
$$

Equation (2.39) which describes the mapping $\mathcal{M}$ is named the *input-output relationship* of the system.

In our context we will use the terms system and transformation as synonyms, although in other contexts a distinction is made between the two. Graphically we will represent a system as in Fig. 2.11 with a box, with input and output signals indicated on top of an incoming and an outgoing arrow, respectively.

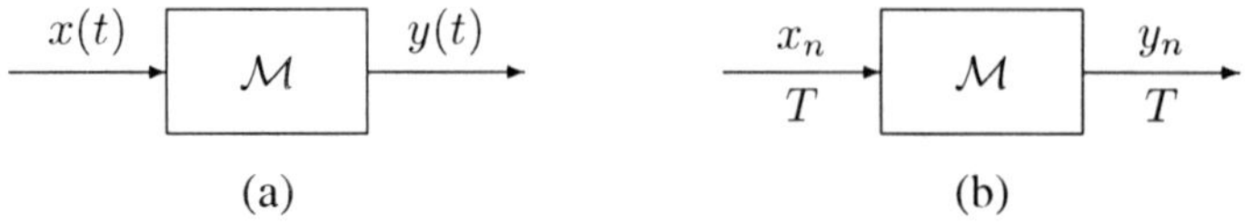

Figure 2.11 Graphical representation of systems: (a) continuous time; (b) discrete time.

2.3.1 Properties of a system

According to the map $\mathcal{M}$, a system may have certain properties, such as

Linearity A system is said to be linear if it satisfies both the following properties

i) *additivity:* for all pairs of input signals x_1, x_2

$$\text{if} \begin{cases} \mathcal{M}[x_1] = y_1 \\ \mathcal{M}[x_2] = y_2 \end{cases} \quad \text{then } \mathcal{M}[x_1 + x_2] = y_1 + y_2 \tag{2.40}$$

ii) *homogeneity:* for any input signal x and any complex constant α

$$\text{if } \mathcal{M}[x] = y \quad \text{then } \mathcal{M}[\alpha x] = \alpha y \tag{2.41}$$

In other words we can say that in a linear system superposition and multiplication by a complex scalar of inputs and outputs hold true. An example of a linear system is the Fourier transformation, with x regarded as input, and $\mathcal{X}$ as output.

Time-invariance A system is said to be time-invariant if any time shift in the input signal is reflected to an identical shift in the output. That is, given any input signal $x(t)$ and any shift $t_0 \in \mathbb{R}$, if $y = \mathcal{M}[x]$ is the system response to $x(t)$, then the system response to $x_1(t) = x(t - t_0)$ is $\mathcal{M}[x_1] = y_1$ with $y_1(t) = y(t - t_0)$. In other words a time-invariant system will always respond in the same manner to a given input signal, independently of the instant in which the input signal is applied, that is the system behaviour remains identical over time.

Causality A system is said to be causal if the output value at an arbitrary instant t_0 is determined by the input values $\{x(t)\}$ for $t \leq t_0$ only. In other words, in a causal system the present value of the response only depends on past and present values of the input signal and not on future input values.

Memorylessness A system is said to be memoryless if the output value at an arbitrary instant t_0 is determined by the input value at the same instant $x(t_0)$ only. In other words, in a memoryless system the present value of the response only depends on the present value of the input signal and not on past or future input values.

2.3.2 Filters

A system that is both linear and time-invariant (LTI) is called a *filter*. Its input-output relationship can always be written as

$$y(t) = \int_{-\infty}^{+\infty} g(t - u)x(u)\, du \tag{2.42}$$

The operation performed by the integral in (2.42) is named the *convolution* between the signals x and g and is more compactly indicated as

$$y = g * x \ . \tag{2.43}$$

Hence, the value of y at instant t is also written as

$$y(t) = (g * x)\ (t)\ . \tag{2.44}$$

The function g in (2.42) uniquely identifies the filter and fully describes it from an input-output point of view: g is called the *impulse response* of the system. The graphical representation of a filter by its input-output map (Fig. 2.11) is often replaced by the impulse response (Fig. 2.12).

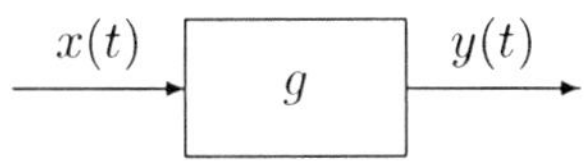

Figure 2.12 Block-diagram representation of a continuous time filter with impulse response g.

Some important properties of convolution are summarized in Tab. 2.3. In particular associativity ensures that the cascade of two filters is equivalent to a single filter having as impulse response the convolution of the two impulse responses of the original filters. Also, the Dirac impulse δ is the unit element of convolution. So, a filter with impulse response δ represents the identity transformation as the output coincides with the input. Moreover, the application of the impulse δ as input to any filter yields a response y coinciding with g, thus justifying the name "impulse response" for g.

Table 2.3 Some properties of convolution.

associativity	$(x * g) * h = x * (g * h)$
commutativity	$x * g = g * x$
unit element	$x * \delta = x \quad , \quad \delta * g = g$
distributivity	$(x_1 + x_2) * g = (x_1 * g) + (x_2 * g)$
homogeneity	$(\alpha x) * g = \alpha(x * g)\ ,\ \alpha \in \mathbb{C}$
shift-invariance	$\left.\begin{array}{r} y = g * x \\ x_1(t) = x(t - t_0) \end{array}\right\} \Rightarrow \left\{\begin{array}{l} y_1 = g * x_1 \\ y_1(t) = y(t - t_0) \end{array}\right.$

In the frequency domain the input-output relationship (2.42) becomes (see the convolution property in Tab. 2.1)

$$\mathcal{Y}(f) = \mathcal{G}(f)\mathcal{X}(f) \tag{2.45}$$

so that the Ftf of the output at any frequency f can be obtained by multiplying the Ftf of the input and of the filter impulse response. The Fourier transform $\mathcal{G}$ of the impulse response is called the *frequency response* of the filter. The reason for this name is well understood by considering the following examples.

Example 2.3 A Consider a filter with frequency response $\mathcal{G}(f)$ and suppose the input signal is a complex exponential with complex amplitude $A \in \mathbb{C}$ and frequency $f_0 \in \mathbb{R}$

$$x(t) = Ae^{j2\pi f_0 t}$$

The input Fourier transform is thus

$$\mathcal{X}(f) = A\delta(f - f_0)$$

and by applying (2.45) and the properties of the Dirac impulse we get

$$\mathcal{Y}(f) = \mathcal{G}(f)A\delta(f - f_0) = A\mathcal{G}(f_0)\delta(f - f_0) .$$

Taking the inverse Fourier transform of $\mathcal{Y}(f)$ yields the output signal

$$y(t) = Be^{j2\pi f_0 t} \quad , \quad \text{with } B = A\mathcal{G}(f_0) .$$

Thus, we see that the filter responds to a complex exponential at frequency f_0 with another complex exponential at the same frequency and with complex amplitude given by the product of the input amplitude and the value of the frequency response at f_0.

Mathematically we can say that a complex exponential is an eigenfunction of the filter with eigenvalue $\mathcal{G}(f_0)$.

A filter can be seen as a device that in the frequency domain analyses each frequency component of the input signal and weights it by the corresponding value of the frequency response $G(f)$. Hence, by properly designing $\mathcal{G}(f)$ we can amplify or attenuate the different frequency components of the input signal.

A filter whose impulse response g is real-valued will always respond to a real input with a real output. Such a filter is called *real*, and its frequency response will always exhibit the Hermitian symmetry

$$\mathcal{G}(-f) = \mathcal{G}^*(f) . \tag{2.46}$$

Equivalently, we see that its frequency response exhibits even symmetry in its amplitude

$$|\mathcal{G}(-f)| = |\mathcal{G}(f)| \tag{2.47}$$

and odd symmetry in its phase

$$\angle\mathcal{G}(-f) = -\angle\mathcal{G}(f) . \tag{2.48}$$

Example 2.3 B Suppose the input to a real filter with frequency response $\mathcal{G}(f)$ is a sinusoid with amplitude $A > 0$, frequency $f_0 \geq 0$ and phase φ_0,

$$x(t) = A\cos(2\pi f_0 t + \varphi_0) \tag{2.49}$$

Its Ftf is

$$\mathcal{X}(f) = \frac{A}{2}\left[e^{j\varphi_0}\delta(f - f_0) + e^{-j\varphi_0}\delta(f + f_0)\right]$$

and, similarly to what we did in Example 2.3 A, we get

$$\mathcal{Y}(f) = \frac{A}{2}\left[e^{j\varphi_0}\mathcal{G}(f_0)\delta(f - f_0) + e^{-j\varphi_0}\mathcal{G}(-f_0)\delta(f + f_0)\right]$$

If we express $\mathcal{G}(f_0)$ in amplitude and phase by writing $\mathcal{G}(f_0) = |\mathcal{G}(f_0)|\, e^{j\angle\mathcal{G}(f_0)}$ and take into account that since the filter is real, $\mathcal{G}$ has the Hermitian symmetry, we get

$$\mathcal{G}(-f_0) = \mathcal{G}^*(f_0) = |\mathcal{G}(f_0)|\, e^{-j\angle\mathcal{G}(f_0)}$$

Thus we can write

$$\mathcal{Y}(f) = \frac{A\,|\mathcal{G}(f_0)|}{2}\left[e^{j[\varphi_0+\angle\mathcal{G}(f_0)]}\delta(f-f_0) + e^{-j[\varphi_0+\angle\mathcal{G}(f_0)]}\delta(f+f_0)\right]$$

and obtain the output signal by inverse Fourier transform as

$$y(t) = B\cos(2\pi f_0 t + \varphi_1) \quad , \quad \text{with } B = A\,|\mathcal{G}(f_0)| \ , \ \varphi_1 = \varphi_0 + \angle\mathcal{G}(f_0) \tag{2.50}$$

A real filter responds to a sinusoid having frequency f_0 with a sinusoid having the same frequency but with amplitude multiplied by the amplitude of the frequency response at f_0, and phase increased by the phase of the frequency response at f_0.

Energy spectral densities From the frequency domain relation (2.45) we can relate input and output energy spectral densities (2.26) as follows

- cross energy spectral density between input and output signals

$$\begin{aligned} \mathcal{E}_{xy}(f) &= \mathcal{X}(f)\mathcal{Y}^*(f) = \mathcal{X}(f)\mathcal{X}^*(f)\mathcal{G}^*(f) = \mathcal{G}^*(f)\mathcal{E}_x(f) \\ \mathcal{E}_{yx}(f) &= \mathcal{Y}(f)\mathcal{X}^*(f) = \mathcal{G}(f)\mathcal{X}(f)\mathcal{X}^*(f) = \mathcal{G}(f)\mathcal{E}_x(f) \end{aligned} \tag{2.51}$$

- energy spectral density of the output signal

$$\mathcal{E}_y(f) = |\mathcal{Y}(f)|^2 = |\mathcal{G}(f)|^2\,|\mathcal{X}(f)|^2 = \mathcal{E}_g(f)\mathcal{E}_x(f) \tag{2.52}$$

  that is the energy spectral density of the output signal is the product between the energy spectral density of the input signal and energy spectral density of the filter impulse response.

Remark In most applications (2.52) is a very useful tool for computing the energy of a filter output given the input energy spectral density. In fact, given $\mathcal{E}_x$ and $\mathcal{E}_g$ we need to evaluate first $\mathcal{E}_y$ by (2.52) and then determine E_y by integration of $\mathcal{E}_y$. The alternative would be to evaluate first y as the convolution $(x * g)$ and then determine E_y by integration of $|y|^2$. However, it turns out that most of the times the convolution $(x * g)$ is very difficult to evaluate, especially for continuous time systems.

2.3.3 Sampling

Sampling is the operation of obtaining a discrete time signal $\{y_n\}$ from a continuous time signal $x(t)$ by simply looking up the values of x at instants that are multiples of a quantum T_s, called the *sampling period*

$$y_n = x(nT_s) \quad , \quad n \in \mathbb{Z}\,. \tag{2.53}$$

It can be viewed as a transformation in which the input signal is continuous time and the output is discrete time, as illustrated in Fig. 2.13, and it is the mathematical model of the acquisition or measurement of a quantity at regularly spaced time instants. The inverse of the sampling period, $F_s = 1/T_s$, is called the *sampling frequency* or *sampling rate* and represents the number of samples per unit time, and is thus measured in Hz.

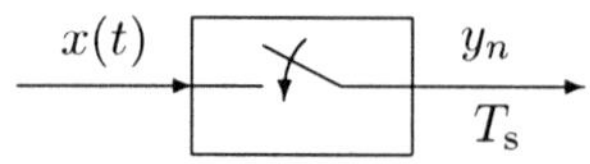

Figure 2.13 Block-diagram representation of a sampler as a signal transformation from continuous time to discrete time.

In the frequency domain the input-output relation is given by the following duality theorem

Theorem 2.4 *If the discrete time signal $\{y_n\}$ is obtained by sampling the continuous time signal $x(t)$, $t \in \mathbb{R}$, with period T_s, its Fourier transform can be obtained from $\mathcal{X}$ as*

$$\mathcal{Y}(f) = F_s \sum_{\ell=-\infty}^{+\infty} \mathcal{X}(f + \ell F_s) \tag{2.54}$$

Proof We prove the theorem by taking the inverse Ftf of (2.54). From (2.18), and by exploiting the relation $F_s = 1/T_s$, we have

$$\begin{aligned} y_n &= T_s \int_0^{F_s} F_s \sum_{\ell=-\infty}^{+\infty} \mathcal{X}(f + \ell F_s)\, e^{j2\pi f n T_s}\, df \\ &= \sum_{\ell=-\infty}^{+\infty} \int_0^{F_s} \mathcal{X}(f + \ell F_s)\, e^{j2\pi f n T_s}\, df \\ &= \sum_{\ell=-\infty}^{+\infty} \int_{\ell F_s}^{(\ell+1)\, F_s} \mathcal{X}(f)\, e^{j2\pi f n T_s}\, e^{-j2\pi \ell n}\, df \\ &= \int_{-\infty}^{+\infty} \mathcal{X}(f)\, e^{j2\pi f n T_s}\, df = x(nT_s) \end{aligned}$$

which proves the result. □

Observe that the sum on the right hand side of (2.54) is made up of shifted versions, also called *images*, of the input Ftf for all shifts multiple of the sampling rate. The relation between x and y is illustrated in Fig. 2.14, both in the time and frequency domains.

2.3.4 Interpolation

Interpolation is the operation of constructing a continuous time signal $y(t)$ from a discrete time signal $\{x_n\}$ with quantum T. If we want the transformation to satisfy both linearity and

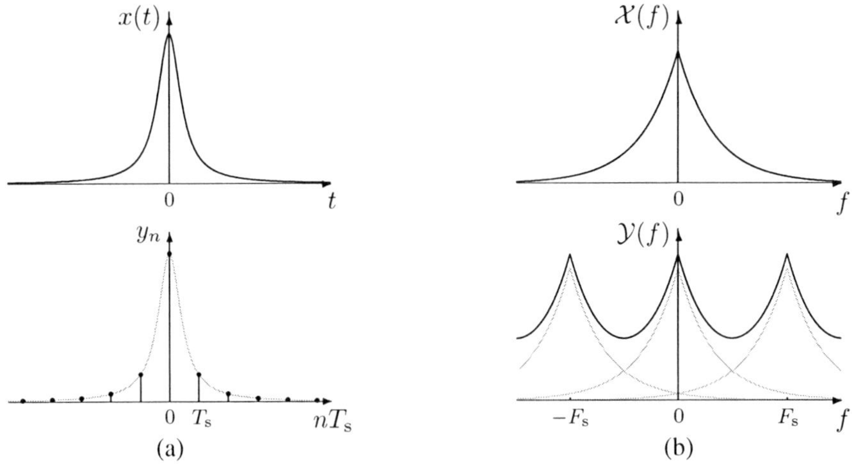

Figure 2.14 The relation between continuous time input and discrete time output in sampling: (a) in the time domain; (b) in the frequency domain.

time-invariance, the input-output relationship must be of the type

$$y(t) = \sum_{n=-\infty}^{+\infty} g(t - nT)x_n \tag{2.55}$$

The transformation (2.55) is called *interpolate filter* and the continuous time function g is its impulse response. In the frequency domain (2.55) corresponds to the following result.

Theorem 2.5 *If $y(t)$ is obtained from $\{x_n\}$ through the interpolate filter g, its Fourier transform is*

$$\mathcal{Y}(f) = \mathcal{G}(f)\mathcal{X}(f) \tag{2.56}$$

Proof We proceed analogously to the proof of Theorem 2.4 and obtain

$$\begin{aligned}
\mathcal{Y}(f) &= \int_{-\infty}^{+\infty} y(t)e^{-j2\pi ft}\,\mathrm{d}t \\
&= \int_{-\infty}^{+\infty} \sum_{n=-\infty}^{+\infty} g(t-nT)x_n e^{-j2\pi ft}\,\mathrm{d}t \\
&= \sum_{n=-\infty}^{+\infty} x_n \int_{-\infty}^{+\infty} g(t-nT)\,e^{-j2\pi ft}\,\mathrm{d}t
\end{aligned}$$

By use of the time shift property of Tab. 2.1, it is

$$\mathcal{Y}(f) = \sum_{n=-\infty}^{+\infty} x_n\,\mathcal{G}(f)\,e^{-j2\pi fnT} = \mathcal{G}(f)\,\mathcal{X}(f)$$

□

The relation (2.56), analogous to (2.45), justifies both denominations of interpolate filter for the transformation, and of frequency response for $\mathcal{G}(f)$. Observe that in (2.56) $\mathcal{X}(f)$ is a periodic function with period $1/T$, whereas $\mathcal{G}(f)$ and $\mathcal{Y}(f)$ are not periodic, in general. Thus, the multiplication by $\mathcal{G}(f)$ destroys the periodicity of $\mathcal{X}(f)$. Fig. 2.15 shows the input and output signals in time and frequency domains.

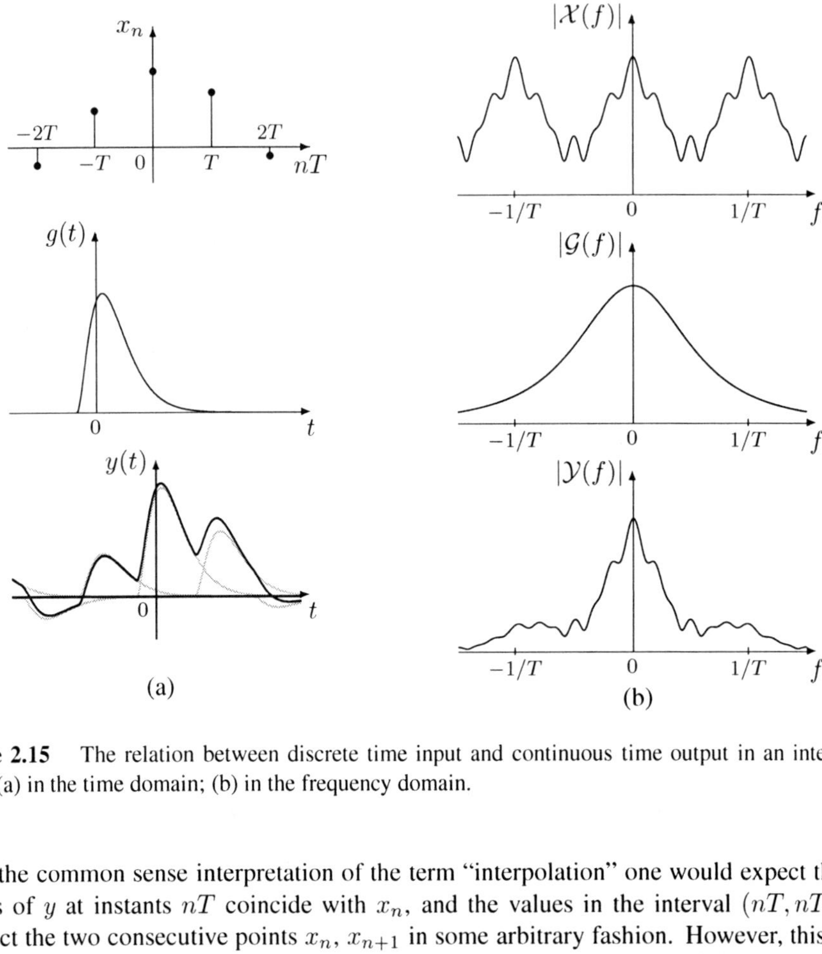

Figure 2.15 The relation between discrete time input and continuous time output in an interpolate filter: (a) in the time domain; (b) in the frequency domain.

In the common sense interpretation of the term "interpolation" one would expect that the values of y at instants nT coincide with x_n, and the values in the interval $(nT, nT + T)$ connect the two consecutive points x_n, x_{n+1} in some arbitrary fashion. However, this is not guaranteed when using an arbitrary interpolate filter. A necessary and sufficient condition on g for the output to coincide with the input at instants nT for any input signal is given by the following proposition.

Proposition 2.6 *Consider an interpolate filter with input-output relation (2.55). Then, for any input signal $\{x_n\}$, it is*

$$y(nT) = x_n \quad , \quad n \in \mathbb{Z} \tag{2.57}$$

if and only if $g(t)$ is such that

$$g(nT) = \begin{cases} 1 & , \quad n = 0 \\ 0 & , \quad n \neq 0 \end{cases} \tag{2.58}$$

Proof We first prove that condition (2.58) is sufficient. Assume (2.58) holds, then for any x we have from (2.55)

$$y(nT) = \sum_{k=-\infty}^{+\infty} g(nT - kT)x_k = g(0)x_n + \sum_{i \neq 0} g(iT)x_{n-i} = x_n$$

To prove that (2.58) is also necessary assume that (2.57) holds for any input signal x, then it must hold in particular for the signal

$$x_n = \begin{cases} 1 & , \; n = 0 \\ 0 & , \; n \neq 0 \end{cases}$$

and since

$$y(nT) = \sum_{k=-\infty}^{+\infty} g(nT - kT)x_k = g(nT)x_0 = g(nT)$$

it must be

$$g(nT) = x_n = \begin{cases} 1 & , \; n = 0 \\ 0 & , \; n \neq 0 \end{cases}$$

□

Observe that condition (2.58) does not uniquely determine the function g since its values at instants that are not multiples of T are not specified. A few examples of interpolate filters that satisfy (2.58) are shown in Fig. 2.16.

2.4 BANDWIDTH

Definition 2.3 *The* full band *of a continuous time signal x is the support of its Fft, i.e. the set of frequencies where its Ftf is nonzero*

$$\overline{\mathcal{B}}_x = \{f \in \mathbb{R} : \mathcal{X}(f) \neq 0\} \tag{2.59}$$

For real-valued signals, due to the Hermitian symmetry of the Fourier transform, $\mathcal{X}(-f) \neq 0$ if and only if $X(f) \neq 0$ so that $\overline{\mathcal{B}}_x$ is a symmetric set with respect to the origin. It is therefore common to use the following definition

Definition 2.4 *The* band *of a continuous time real-valued signal x is the subset of nonnegative frequencies where its Ftf is nonzero*

$$\mathcal{B}_x = \{f \geq 0 : \mathcal{X}(f) \neq 0\} \tag{2.60}$$

Clearly, for real-valued signals we have $\overline{\mathcal{B}}_x = \mathcal{B}_x \cup (-\mathcal{B}_x)$.

When dealing with discrete time signals, due to the periodicity of their Ftf it is convenient to define the band within one period, as

Definition 2.5 *The full band of a discrete time signal x with quantum T is the subset of $\left[-\frac{1}{2T}, \frac{1}{2T}\right]$ where its Ftf is nonzero*

$$\overline{\mathcal{B}}_x = \left\{f \in \left[-\frac{1}{2T}, \frac{1}{2T}\right] : \mathcal{X}(f) \neq 0\right\} \tag{2.61}$$

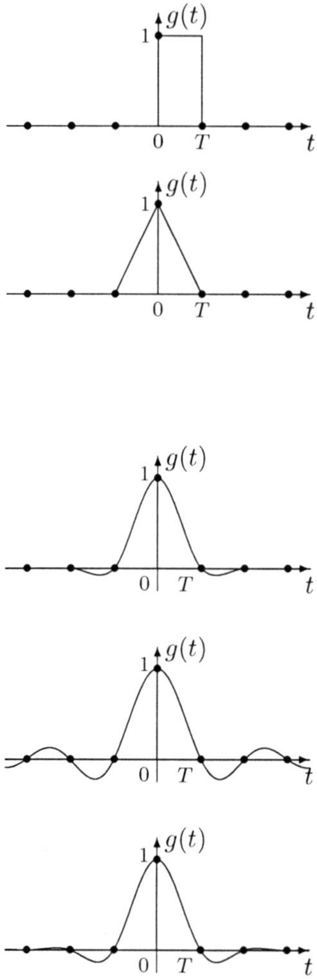

holding interpolation

$$g(t) = \text{rect}\left(\frac{t-T/2}{T}\right)$$

linear interpolation

$$g(t) = \text{triang}\left(\frac{t}{T}\right)$$

cubic interpolation

$$g(t) = \begin{cases} p_1\left(\frac{|t|}{T}\right) & , \ 0 \le |t| < T \\ p_2\left(\frac{|t|}{T}\right) & , \ T \le |t| < 2T \\ 0 & , \ |t| \ge 2T \end{cases}$$

$$p_1(x) = \tfrac{3}{2}x^3 - \tfrac{5}{2}x^2 + 1$$
$$p_2(x) = -\tfrac{1}{2}x^3 + \tfrac{5}{2}x^2 - 4x + 2$$

ideal lowpass interpolation

$$g(t) = \text{sinc}\left(\frac{t}{T}\right)$$

raised cosine interpolation

$$g(t) = \text{ircos}\left(\frac{t}{T}, \rho\right)$$

Figure 2.16 Examples of interpolate filters that satisfy condition (2.58).

Again, for discrete time real-valued signals, thanks to the Hermitian symmetry, we define their band as a subset of $\left[0, \frac{1}{2T}\right]$.

Once established the band $\mathcal{B}_x$ of a real-valued signal x, we can define its *bandwidth* B_x as the measure of $\mathcal{B}_x$, i.e.

$$B_x = \int_{\mathcal{B}_x} df$$

We notice that for a discrete time signal the bandwidth can not be greater than half the rate, $B_x \le \frac{1}{2T}$, whereas for continuous time signals the bandwidth may be arbitrarily large and even infinite.

The concepts of band and bandwidth can also be introduced with regard to systems. In particular the band of a filter is defined as the band of its impulse response g and it is easy to see that it includes the bands of all possible output signals y. Therefore the filter bandwidth is also the maximum bandwidth of all filter output signals.

2.4.1 Classification of signals and systems

Signals and systems are usually grouped into classes according to their band as follows

Definition 2.6 *A signal x is said to be*

bandlimited, *if its band $\mathcal{B}_x$ is a limited set;*

baseband, *if $\mathcal{B}_x$ is a limited set containing a neighbourhood of the origin;*

passband, *if $\mathcal{B}_x$ is a limited set excluding a neighbourhood of the origin;*

narrowband, *if it is passband and its bandwidth is much smaller than its maximum frequency, i.e. $B_x \ll \max\{f; f \in \mathcal{B}_x\}$.*

Fig. 2.17 shows examples of the above definitions.

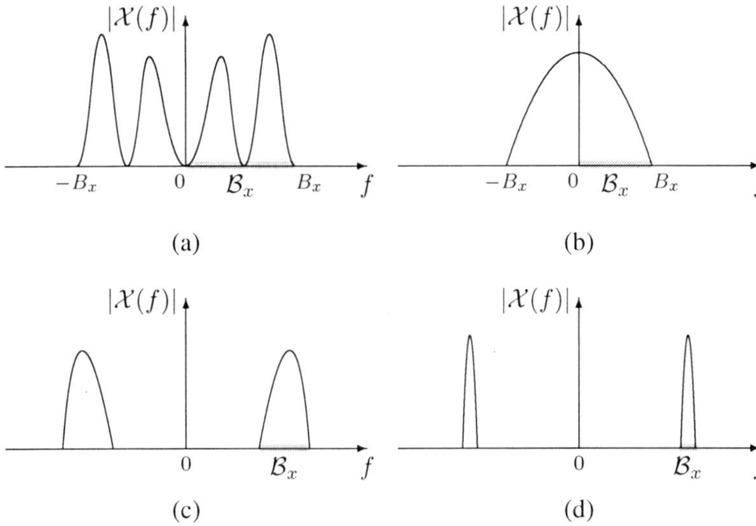

Figure 2.17 Examples of frequency-domain representation of real signals: (a) bandlimited; (b) baseband; (c) passband; (d) narrowband.

Definition 2.7 *A filter with impulse response g is said to be*

lowpass (LPF), *if its band $\mathcal{B}_g$ is a limited set containing a neighbourhood of the origin;*

highpass (HPF), *if $\mathcal{B}_g$ includes a neighbourhood of infinity, excluding a neighbourhood of the origin;*

bandpass (BPF), *if $\mathcal{B}_g$ is a limited set excluding a neighbourhood of the origin;*

narrowband (NBF), *if it is bandpass and its bandwidth is much smaller than its maximum frequency, i.e.* $B_g \ll \max_{f \in \mathcal{B}_g} |f|$;

notch (NTF), *if its band is the complement to that of a narrowband filter;*

allpass, *if its gain* $|\mathcal{G}(f)|$ *is nearly constant over the whole frequency axis.*

The above definitions are illustrated in Fig. 2.18.

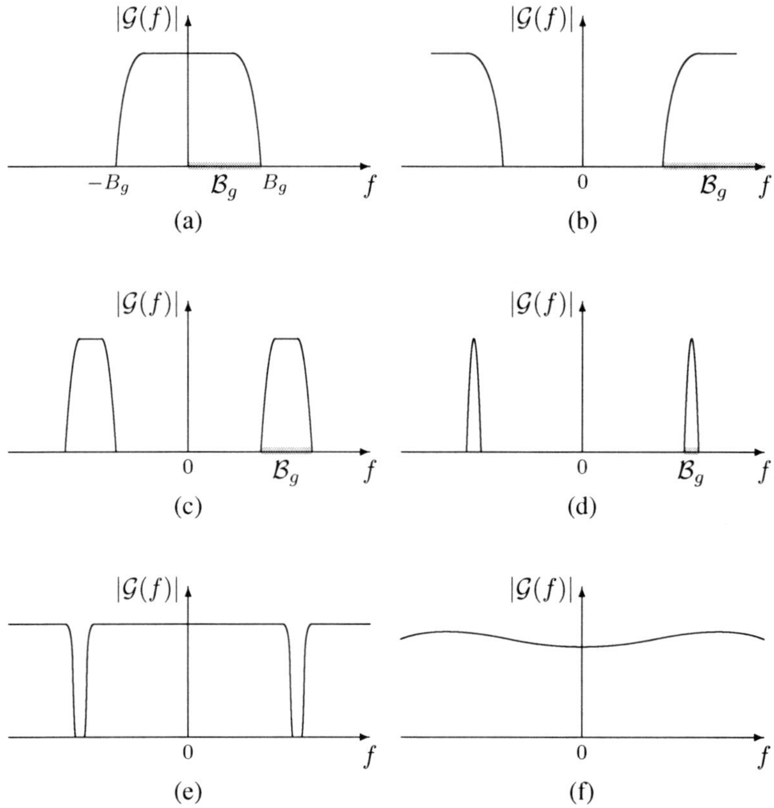

Figure 2.18 Examples of frequency-domain representation of real filters: (a) lowpass; (b) highpass; (c) bandpass; (d) narrowband; (e) notch; (f) allpass.

2.4.2 Uncertainty principle

An important result in Signal Theory is the so called *uncertainty principle* for continuous time signals, which states

Theorem 2.7 *A continuous time signal* x *can not have both a limited support and a limited band.*

The proof of this theorem is far beyond the scope of this book. It relies on the properties of analytical functions. The interested reader can see [6] or books on complex analysis.

The above result states that a continuous time signal can be either time-limited, band-limited, or none of the two, but not both.

2.4.3 Practical definitions of band

Most real life signals (and system impulse responses) have a finite support due to the fact that physical phenomena start and expire within a finite time interval. Strictly speaking, such signals cannot have a limited band, as stated by Theorem 2.7. However, it is quite convenient to assume that their band is "practically" limited in the sense that their Ftf is negligible (i.e. sufficiently small) outside a properly chosen interval (f_1, f_2) that is called *practical band* of the signal. The corresponding *practical bandwidth* of the signal is given by $B = f_2 - f_1$. Such a definition is not unique, depending on what we mean by "negligible". Some common criteria for real-valued signals are given in the following.

Amplitude Given a parameter ε, with $0 < \varepsilon < 1$, we define the band of the passband signal x as the smallest interval (f_1, f_2) such that the amplitude of $\mathcal{X}(f)$ outside (f_1, f_2) does not exceed a fraction ε of its maximum, that is

$$|\mathcal{X}(f)| \leq \varepsilon \mathcal{X}_{\max} \quad , \quad f \geq 0, f \notin (f_1, f_2) \tag{2.62}$$

where

$$\mathcal{X}_{\max} = \max_f |\mathcal{X}(f)| \tag{2.63}$$

Typical values of ε in different applications may be $1/\sqrt{2}$ (3 dB band), 10^{-2} (40 dB band), or 10^{-3} (60 dB band).

Energy Given a parameter ε, with $0 < \varepsilon < 1$, we define the band of the passband signal x as the smallest interval (f_1, f_2) such that the signal energy within (f_1, f_2), obtained by integrating the energy spectral density, differs by no more than a fraction ε from the total signal energy, that is

$$2 \int_{f_1}^{f_2} \mathcal{E}_x(f)\, \mathrm{d}f \geq (1 - \varepsilon) \int_{-\infty}^{+\infty} \mathcal{E}_x(f)\, \mathrm{d}f = (1 - \varepsilon) E_x \tag{2.64}$$

Typical values of ε in applications are 10%, or 1%.

First zero Let f_0 be the frequency corresponding to the maximum value of $|\mathcal{X}(f)|$,

$$f_0 = \operatorname*{argmax}_{f \geq 0} |\mathcal{X}(f)| \ . \tag{2.65}$$

We define the band of the passband signal x as the interval (f_1, f_2) containing f_0 where the transform is nonzero

$$\begin{aligned} f_1 &= \max\{0 \leq f < f_0, |\mathcal{X}(f)| = 0\} \\ f_2 &= \min\{f > f_0, |\mathcal{X}(f)| = 0\} \end{aligned} \tag{2.66}$$

These three criteria can be particularized to baseband signals by considering $f_1 = 0$. In fact, they are illustrated in Fig. 2.19 for the signal $x(t) = \mathrm{rect}(t/T)$ that would not be bandlimited

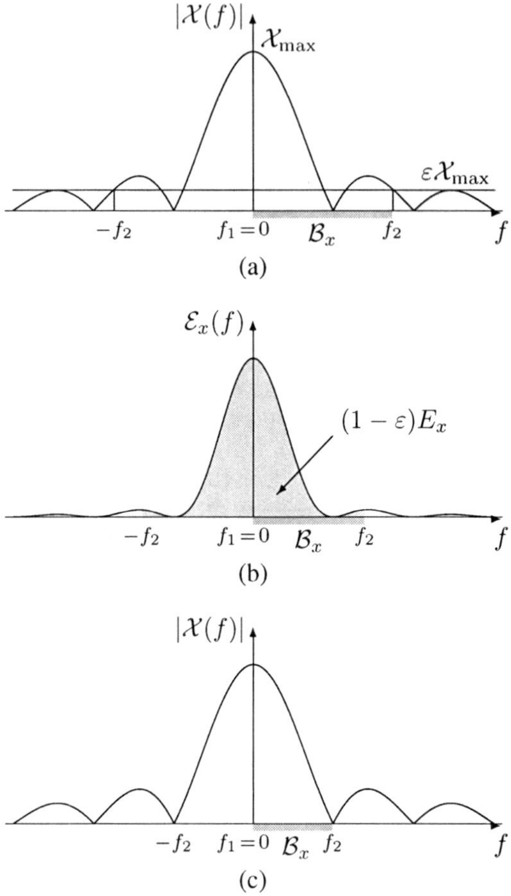

Figure 2.19 Illustration of the practical bandwidth of the signal $x(t) = \text{rect}(t/T)$ according to: (a) the amplitude criterion; (b) the energy criterion; (c) the first zero criterion.

in a strict sense. Observe that it can be called a baseband signal in the practical sense, as its lower frequency is always $f_1 = 0$ and its practical bandwidth is $B = f_2$.

The same definitions can clearly be applied to the band of filters, by means of their impulse response. We underline that the value of the practical bandwidth B will depend on the criterion we choose and the value of the parameter ε. In general, a smaller ε corresponds to a larger bandwidth.

2.4.4 Heaviside conditions

Consider a continuous time or discrete time system with input signal x and output signal y, which may represent the cascade of many transformations that the input signal undergoes along a communication system. Suppose we want the output y to resemble the input x as close as possible, that is without *distortion*. In particular, if we require y to be a possibly attenuated or amplified and/or shifted version of x, i.e.

$$y(t) = A_0 x(t - t_0) \tag{2.67}$$

with $A_0 > 0$, $t_0 \in \mathbb{R}$, we say that we are seeking for the Heaviside conditions for the absence of distortion in the system.

It is evident that if we want (2.67) to hold for *any possible input signal*, the system must be linear and time-invariant, i.e. a filter, and must have the following impulse response

$$h(t) = A_0 \delta(t - t_0) \tag{2.68}$$

or equivalently, by taking the Ftf of (2.68), with frequency response

$$\mathcal{H}(f) = A_0 e^{-j2\pi f t_0} \tag{2.69}$$

Such a filter has

i) *constant amplitude*

$$|\mathcal{H}(f)| = A_0 \quad , \quad f \in \mathbb{R} \tag{2.70}$$

ii) *linear phase* (better, proportional to the frequency)

$$\angle\mathcal{H}(f) = -2\pi f t_0 \quad , \quad f \in \mathbb{R} \tag{2.71}$$

iii) *constant group delay*, also called *envelope delay*

$$\tau(f) = -\frac{1}{2\pi}\frac{d}{df}\angle\mathcal{H}(f) = t_0 \quad , \quad f \in \mathbb{R} \tag{2.72}$$

Conditions i)–ii) are known as the *Heaviside conditions* for the absence of distortion, whereas iii) is simply a consequence of ii). These are quite strong requirements to be met by real-life systems, and are too restrictive indeed.

However, for a system with *bandlimited input signals* having a full band $\overline{\mathcal{B}}$, we may require absence of distortion only within $\overline{\mathcal{B}}$. In fact, by writing the requirement (2.67) in the frequency domain, it is

$$\mathcal{Y}(f) = A_0 e^{-j2\pi f t_0} \mathcal{X}(f) \,, \tag{2.73}$$

and equating it with the general filter relation $\mathcal{Y}(f) = \mathcal{H}(f)\mathcal{X}(f)$, we get that the system frequency response $\mathcal{H}(f)$ must be such that

$$\mathcal{H}(f)\mathcal{X}(f) = A_0 e^{-j2\pi f t_0} \mathcal{X}(f) \,. \tag{2.74}$$

We observe that, for values $f \notin \overline{\mathcal{B}}$, it is $\mathcal{X}(f) = 0$ and (2.74) is an identity, satisfied by any $\mathcal{H}(f)$. Hence the requirement (2.69) becomes

$$\mathcal{H}(f) = \begin{cases} A_0\, e^{-j2\pi f t_0} & , \ f \in \overline{\mathcal{B}} \\ \text{arbitrary} & , \ f \notin \overline{\mathcal{B}} \end{cases} \tag{2.75}$$

Then also the Heaviside conditions i)–ii) can be relaxed to hold only for $f \in \overline{\mathcal{B}}$. Clearly, in this case the impulse response h does not necessarily have the expression (2.68).

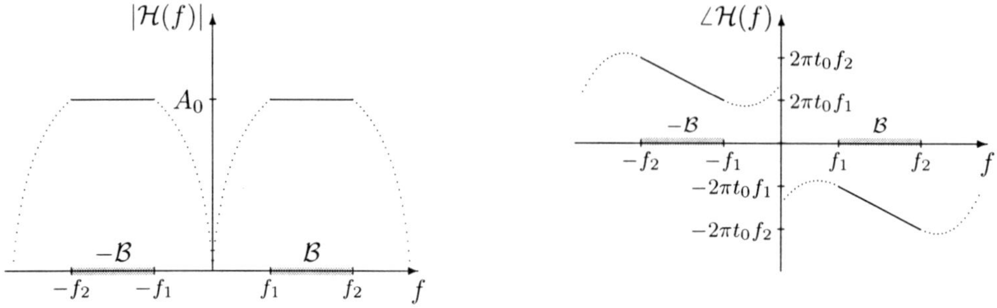

Figure 2.20 Frequency response of a real filter satisfying the conditions for the absence of signal distortion in the full band $\overline{\mathcal{B}} = \mathcal{B} \cup (-\mathcal{B}) = (f_1, f_2) \cup (-f_2, -f_1)$. The behaviour shown with dotted lines is arbitrary.

In other words, if we are only interested in transmitting undistorted signals with full band $\overline{\mathcal{B}}$, it is sufficient that the Heaviside conditions i)–ii) are verified within $\overline{\mathcal{B}}$; outside $\overline{\mathcal{B}}$, the system frequency response may be arbitrary. We show in Fig. 2.20 the frequency response of a real bandpass filter that satisfies the Heaviside conditions within the band $\mathcal{B} = (f_1, f_2)$.

Further insight Indeed, to satisfy Heaviside conditions for the absence of distortion for signals that are band-limited within $\overline{\mathcal{B}}$, it is not necessary that the whole system be a filter. It is sufficient that it *acts* as a filter on these signals. An example which clarifies this observation can be found in Problem 2.8.

2.4.5 Sampling theorem

Consider the problem of sampling a continuous time signal, and subsequently reconstruct all its original values through interpolation, as illustrated in Fig. 2.21.

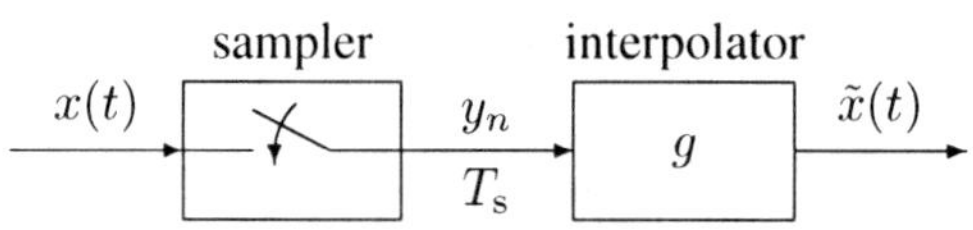

Figure 2.21 Sampling of a continuous time signal and subsequent reconstruction of its values through interpolation.

Sufficient conditions for this problem to be effectively solved are given by the following

Theorem 2.8 (Shannon-Whittaker sampling theorem) *Consider the system of Fig. 2.21 where the input $x(t)$ is a real-valued continuous time signal. If*

i) $x(t)$ is bandlimited with $\mathcal{B}_x \subset [0, B)$

ii) the sampling rate is $F_s \geq 2B$

iii) the interpolate filter has frequency response

$$\mathcal{G}(f) = \begin{cases} T_s & , \quad |f| < B \\ \text{arbitrary} & , \quad B \le |f| \le F_s - B \\ 0 & , \quad |f| > F_s - B \end{cases} \tag{2.76}$$

then the input signal is perfectly reconstructed at the output, that is

$$\tilde{x}(t) = x(t) \quad , \qquad t \in \mathbb{R} \tag{2.77}$$

Proof We proceed to work out the proof in the frequency domain. By combining (2.56) with (2.54) the Ftf of the output signal $\tilde{x}$ is written as

$$\tilde{\mathcal{X}}(f) = \mathcal{G}(f)\mathcal{Y}(f) = F_s \sum_{\ell=-\infty}^{+\infty} \mathcal{G}(f)\mathcal{X}(f - \ell F_s) \tag{2.78}$$

Now if conditions i)–iii) are satisfied, all the terms with $\ell \neq 0$ in the sum vanish, since $\mathcal{G}(f)$ and $\mathcal{X}(f - \ell F_s)$ have disjoint supports (see the illustration in Fig. 2.22) and we are left with the only term for $\ell = 0$, i.e.

$$\tilde{\mathcal{X}}(f) = \mathcal{G}(f)\mathcal{X}(f) \ . \tag{2.79}$$

By observing that $\mathcal{G}$ has constant gain T_s in the band of x, we get

$$\tilde{\mathcal{X}}(f) = \mathcal{X}(f) \ , \tag{2.80}$$

which corresponds to (2.77) in the frequency domain. □

Some comments on the statement and proof of the theorem are appropriate. The interpolate filter gain is set to T_s to obtain equality (2.77); a different gain would yield as $\tilde{x}$ a scaled version of x, but in practice the difference is immaterial. Hypothesis ii) is called the *non-aliasing condition* and assures that in the summation (2.54) the different shifted versions of $\mathcal{X}$ are kept spaced apart and do not overlap with each other. Lastly, condition iii) on the frequency response allows the interpolate filter to gather the desired unshifted component undistorted while rejecting all other images. Observe that if $F_s > 2B$, condition iii) allows a certain degree of freedom in selecting the filter specifications. For example the choice that yields the minimum bandwidth filter is given by the ideal filter

$$\mathcal{G}(f) = T_s \operatorname{rect}\left(\frac{f}{2B}\right) \quad , \qquad g(t) = 2BT_s \operatorname{sinc}(2Bt) \ . \tag{2.81}$$

Specifications (2.81) are also the necessary choice when $F_s = 2B$. In general, other choices for the interpolate filter are possible, as listed in Fig. 2.16. In general, when choosing the sampling frequency for a bandlimited signal one must consider the trade-off between choosing F_s as small as possible to minimize the number of samples per unit time, and allowing some freedom in the design of the interpolate filter.

Further insight Indeed, the practical implementation of the interpolate filter requires a *transition band* between the passband, where $\mathcal{G}(f) \simeq 1$, and the stopband, where $\mathcal{G}(f) \simeq 0$. Moreover, the narrower the transition band, the more challenging the implementation of the

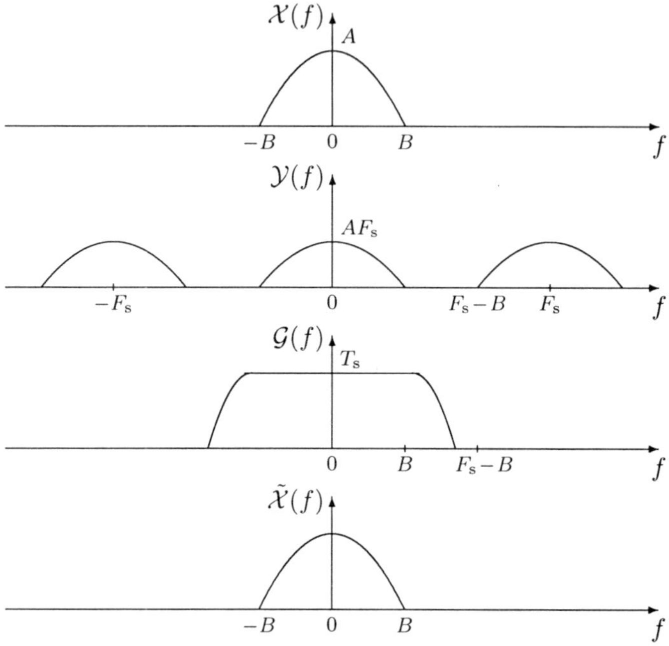

Figure 2.22 Frequency-domain illustration of the sampling and perfect reconstruction of the original signal according to the Shannon-Whittaker sampling theorem.

filter. Given the constraint (2.76), the transition band must be included within the interval $(B, F_s - B)$ and its width is bounded by $F_s - 2B$.

It is understood that if either i) or ii) are not satisfied, i.e. the input signal is not even bandlimited or the sampling frequency is too small, perfect reconstruction (2.77) is not possible and we must take into account a reconstruction error. For this reason, as illustrated in Fig. 2.23, the sampler is preceded by an *anti-aliasing filter* d with band $[0, B]$ where $2B \leq F_s$. Ideally, the frequency response is

$$\mathcal{D}(f) = \text{rect}\left(\frac{f}{2B}\right) . \tag{2.82}$$

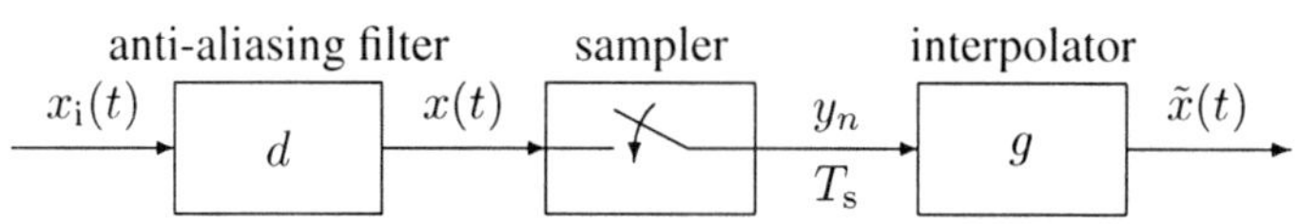

Figure 2.23 Insertion of an anti-aliasing filter in sampling and interpolation transformations.

Thus, the output x of the anti-aliasing filter is bandlimited in $(-B, B)$ and can be perfectly reconstructed by the interpolate filter to yield $\tilde{x}(t) = x(t)$. However, with respect to the initial input $x_{\rm i}$ we have introduced an error

$$e(t) = \tilde{x}(t) - x_{\rm i}(t) = x(t) - x_{\rm i}(t) \tag{2.83}$$

given by the difference between x and $x_{\rm i}$, and caused by the anti-aliasing filter removing the out of band components of $x_{\rm i}$. In the case of an ideal anti-aliasing filter (2.82) and assuming $\tilde{x} = x$, the energy spectral density of e is given by

$$\begin{aligned}\mathcal{E}_e(f) &= |\mathcal{X}_{\rm i}(f) - \mathcal{X}(f)|^2 \\ &= |1 - \mathcal{D}(f)|^2 \, |\mathcal{X}_{\rm i}(f)|^2 \\ &= \begin{cases} 0 & , \ |f| \leq B \\ \mathcal{E}_{x_{\rm i}}(f) & , \ |f| > B \end{cases} ,\end{aligned}$$

and we can calculate the error energy

$$E_e = \int_{-\infty}^{+\infty} \mathcal{E}_e(f) \, \mathrm{d}f = 2 \int_{B}^{+\infty} \mathcal{E}_{x_{\rm i}}(f) \, \mathrm{d}f \ . \tag{2.84}$$

2.4.6 Nyquist criterion

Given the discrete time signal $\{x_n\}$ with quantum T, in this paragraph we consider the problem of interpolating it and subsequently recover its original discrete time sample values through filtering and sampling, as illustrated in Fig. 2.24. It is evident that if the interpolate filter g satisfies condition (2.58), then $y(nT) = x_n$ and we only need to sample y with period T to get back the original values. In fact $\tilde{x}_n = z(nT) = y(nT)$. In this case we can remove the filter h.

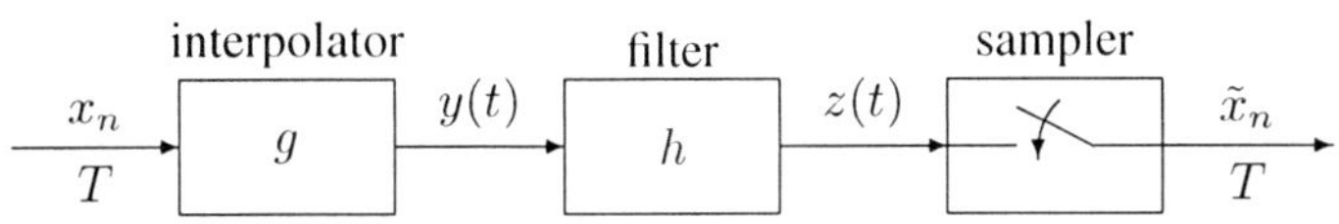

Figure 2.24 Interpolation of a discrete time signal and subsequent recovery of its values through filtering and sampling.

In general, we must consider that the cascade of the interpolate filter g and the filter h is equivalent to a single interpolate filter with impulse response $g_1(t) = (g * h)\ (t)$ and frequency response

$$\mathcal{G}_1(f) = \mathcal{G}(f)\mathcal{H}(f) \ . \tag{2.85}$$

This can be seen by examining the frequency domain input-output relationship of the cascade, which by (2.45) and (2.56) yields

$$\mathcal{Z}(f) = \mathcal{H}(f)\mathcal{Y}(f) = \mathcal{H}(f)\mathcal{G}(f)\mathcal{X}(f) = \mathcal{G}_1(f)\mathcal{X}(f) \ . \tag{2.86}$$

We can therefore state the following theorem which gives a necessary and sufficient condition for the perfect recovery of any discrete time input signal.

Theorem 2.9 (Nyquist criterion, time domain formulation) *Consider the system in Fig. 2.24 with $g_1(t) = (g * h)(t)$. Perfect reconstruction is assured for all discrete time input signals $\{x_n\}$ with quantum T if and only if*

$$g_1(nT) = \begin{cases} 1 & , \ n = 0 \\ 0 & , \ n \neq 0 \end{cases} \tag{2.87}$$

Proof It is sufficient to apply Proposition 2.6 to the equivalent interpolate filter having impulse response g_1. □

In the frequency domain the above theorem can be formulated as follows

Theorem 2.10 (Nyquist criterion, frequency domain formulation) *In the system of Fig. 2.24, perfect reconstruction is assured for all discrete time input signals $\{x_n\}$ with quantum T if and only if*

$$\sum_{k=-\infty}^{+\infty} \mathcal{G}(f + k/T)\mathcal{H}(f + k/T) = T \ , \quad f \in \mathbb{R} \tag{2.88}$$

Proof We prove that condition (2.88) is equivalent to (2.87). Let $q_n = g_1(nT)$ as in (2.87). Its Ftf is $\mathcal{Q}(f) = 1$. Moreover, from (2.54) we have

$$\mathcal{Q}(f) = \frac{1}{T} \sum_{\ell=-\infty}^{+\infty} \mathcal{G}_1(f + \ell/T) = \frac{1}{T} \sum_{\ell=-\infty}^{+\infty} \mathcal{G}(f + \ell/T)\mathcal{H}(f + \ell/T) \tag{2.89}$$

and the theorem is proved. □

Although apparently more cumbersome, the frequency domain formulation (2.88) lends itself to an effective characterization of the filter h. A possible two-step procedure is as follows

1. Choose the frequency response of the equivalent filter $\mathcal{G}_1(f)$ so that

$$\mathcal{B}_{g_1} \subset \mathcal{B}_g \tag{2.90}$$

   and

$$\sum_{\ell=-\infty}^{+\infty} \mathcal{G}_1(f + \ell/T) = 1 \ . \tag{2.91}$$

   For example it could be $\mathcal{G}_1(f) = \mathrm{rect}(Tf)$ or $\mathcal{G}_1(f) = \mathrm{rcos}(Tf, \rho)$

2. Let (see (2.85))

$$\mathcal{H}(f) = \begin{cases} \dfrac{\mathcal{G}_1(f)}{\mathcal{G}(f)} & , \ f \in \mathcal{B}_g \\ \text{arbitrary} & , \ f \notin \mathcal{B}_g \end{cases} \tag{2.92}$$

Observe that from (2.90), it must be $B_{g_1} \leq B_g$. Moreover, for $\mathcal{Q}(f)$ in (2.89) to be nonzero over the whole frequency axis, the bandwidth of g_1 must be at least $\frac{1}{2T}$, as shown in Fig. 2.25.

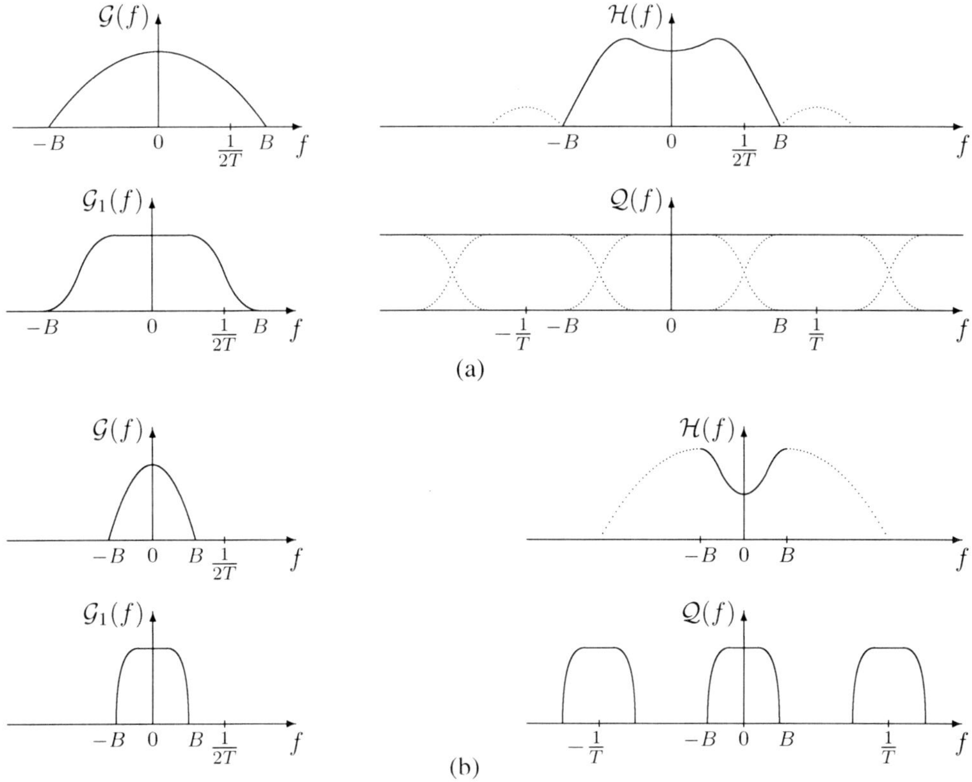

Figure 2.25 Illustration of the Nyquist criterion in the frequency domain: in case (a) it is possible to find a filter h which statisfies (2.88) for the given g; in case (b) the interpolate filter g does not satisfy the necessary condition (2.93), so g_1 does not satisfy the Nyquist criterion.

Therefore, a necessary condition for the Nyquist criterion to hold in the system of Fig. 2.24 for some choice of h is that the interpolate filter g has bandwidth

$$B_g \geq \frac{1}{2T} \cdot \tag{2.93}$$

For the same reasons, also h must have a bandwidth greater than $\frac{1}{2T}$. The Nyquist criterion has important applications in the field of digital transmissions, where the input $\{x_n\}$ represents the transmitted data sequence, the interpolator g represents the cascade of the transmit filter and channel, whereas h represents the receive filter.

2.5 THE SPACE OF SIGNALS

2.5.1 Linear space

We assume that the reader is familiar with the theory of linear spaces and summarize in the following the basic notation and results that will be used in this text (see also [3] for further details).

We recall that a *linear space* is a set of elements called *vectors*, together with two operators defined over the elements of the set, the sum between vectors ($+$) and the multiplication ($\cdot$) of a vector by a *scalar* (Fig. 2.26).

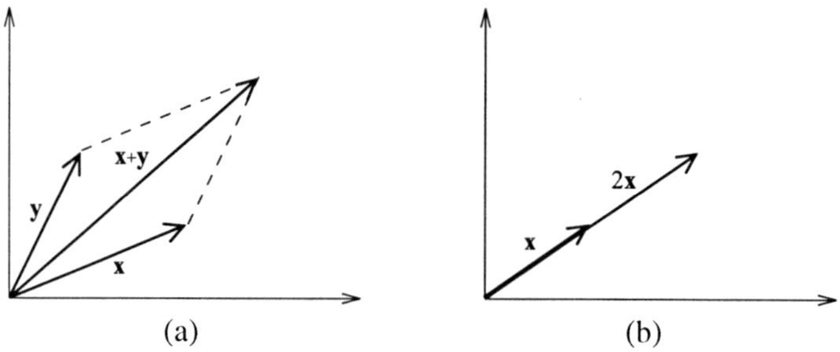

Figure 2.26 Geometrical interpretation in the two-dimensional space of: (a) the sum of two vectors and (b) the multiplication of a vector by a scalar.

More formally, we have the following definition.

Definition 2.8 *A* linear space *is defined by the set* $(V, \mathbb{K}, +, \cdot)$ *where* V *is the non-empty set collecting the elements of the linear space,* $\mathbb{K}$ *is the field of scalars,* $+$ *is an operation* $V \times V \to V$ *called sum, and* $\cdot$ *is an operation* $\mathbb{K} \times V \to V$ *called product by a scalar. In order to identify a linear space, the set* $(V, \mathbb{K}, +, \cdot)$ *must satisfy the following properties:*

1. *V is a commutative group under addition of vectors, that is:*
   (a) *there exists an additive identity element* $\mathbf{0}$ *in* V*, such that for all elements* $\boldsymbol{x} \in V$ *we have* $\boldsymbol{x} + \mathbf{0} = \boldsymbol{x}$;
   (b) *for all* $\boldsymbol{x} \in V$*, there exists an element* $\boldsymbol{y} \in V$*, such that* $\boldsymbol{x} + \boldsymbol{y} = \mathbf{0}$;
   (c) *vector addition is associative, that is* $\boldsymbol{x} + (\boldsymbol{y} + \boldsymbol{z}) = (\boldsymbol{x} + \boldsymbol{y}) + \boldsymbol{z}$;
   (d) *vector addition is commutative, that is* $\boldsymbol{x} + \boldsymbol{y} = \boldsymbol{y} + \boldsymbol{x}$.
2. *Scalar multiplication is associative, that is* $\alpha \cdot (\beta\, \boldsymbol{x}) = (\alpha\, \beta) \cdot \boldsymbol{x}$ *for* $\alpha, \beta \in \mathbb{K}$.
3. *The following distributive laws hold:*

$$\alpha \cdot (\boldsymbol{x} + \boldsymbol{y}) = \alpha \cdot \boldsymbol{x} + \alpha \cdot \boldsymbol{y}$$
$$(\alpha + \beta) \cdot \boldsymbol{x} = \alpha \cdot \boldsymbol{x} + \beta \cdot \boldsymbol{x}\,.$$

One example of signal space is given by the so called *Euclidean space* of dimension N, that is the set of real-valued vectors of length N, denoted by $V = \mathbb{R}^N$, where the generic element

$\boldsymbol{x} \in V$ is the row-vector $\boldsymbol{x} = [x_1, x_2, \ldots, x_N]$ with $x_i \in \mathbb{R}$, $i = 1, \ldots, N$. In an Euclidean space the reference field of scalars is again the group of real-valued numbers, $\mathbb{K} = \mathbb{R}$, while vector addition and multiplication by a scalar are widely known vector operations. We instead talk of an *Hermitian space* when real-valued numbers are replaced by complex numbers that is we have $V = \mathbb{C}^N$ and $\mathbb{K} = \mathbb{C}$. Along with the above complete Euclidean and Hermitian spaces, subspaces generated by a linear combination of vectors are further examples of linear spaces. For instance, the Hermitian subspace *spanned* by the M vectors $\boldsymbol{x}_1, \boldsymbol{x}_2, \ldots, \boldsymbol{x}_M$, where $\boldsymbol{x}_i \in \mathbb{C}^N$, is itself a linear space. Incidentally, we recall the following definition.

Definition 2.9 *Let $(V, \mathbb{K}, +, \cdot)$ be a linear space. The* span *of a set of M vectors $\boldsymbol{x}_1$, $\boldsymbol{x}_2$, ..., $\boldsymbol{x}_M$, belonging to V is the set*

$$V' = \left\{ \boldsymbol{x} \,\middle|\, \boldsymbol{x} = \sum_{i=1}^{M} c_i \, \boldsymbol{x}_i \;,\; c_i \in \mathbb{K} \right\} \subset V \;.$$

Another interesting aspect of linear spaces is given by *inner product spaces*, that is linear spaces with an inner product.

Definition 2.10 *The* inner product *is any function $V \times V \to \mathbb{K}$ that maps two vectors $\boldsymbol{x}$ and $\boldsymbol{y}$ into one scalar k (we write $k = \langle \boldsymbol{x}, \boldsymbol{y} \rangle$), and that satisfies:*

1. *the property $\langle \boldsymbol{x}, \boldsymbol{y} \rangle = \langle \boldsymbol{y}, \boldsymbol{x} \rangle^*$ (that is we need $\mathbb{K} = \mathbb{C}$);*
2. *the distributive law $\langle \alpha \cdot \boldsymbol{x} + \beta \cdot \boldsymbol{y}, \boldsymbol{z} \rangle = \alpha \langle \boldsymbol{x}, \boldsymbol{z} \rangle + \beta \langle \boldsymbol{y}, \boldsymbol{z} \rangle$;*
3. *the property $\langle \boldsymbol{x}, \boldsymbol{x} \rangle \geq 0$, with $\langle \boldsymbol{x}, \boldsymbol{x} \rangle = 0$ if and only if $\boldsymbol{x} = \mathbf{0}$.*

We recall that the square-root of inner product of a vector with itself is referred to as the *norm* or *length* of the vector $\|\boldsymbol{x}\| = \sqrt{\langle \boldsymbol{x}, \boldsymbol{x} \rangle}$, and that the following inequality (Cauchy-Schwartz inequality) holds in general

$$|\langle \boldsymbol{x}, \boldsymbol{y} \rangle| \leq \|\boldsymbol{x}\| \, \|\boldsymbol{y}\| \;. \tag{2.94}$$

In a N-dimensional Hermitian space, $V = \mathbb{C}^N$, the inner product is defined as

$$\langle \boldsymbol{x}, \boldsymbol{y} \rangle = \sum_{i=1}^{N} x_i \, y_i^* \;, \tag{2.95}$$

and so the norm is

$$\|\boldsymbol{x}\| = \sqrt{\sum_{i=1}^{N} |x_i|^2} \;. \tag{2.96}$$

The same definitions hold for all the subspaces of $\mathbb{C}^N$.

If $\langle \boldsymbol{x}, \boldsymbol{y} \rangle$ is real, there is an important geometrical interpretation of the inner product in the Euclidean space, represented in Fig. 2.27, that is obtained from the relation

$$\langle \boldsymbol{x}, \boldsymbol{y} \rangle = \|\boldsymbol{x}\| \, \|\boldsymbol{y}\| \, \cos\varphi \tag{2.97}$$

where φ denotes the angle between the two vectors. Note that, in figure, the vector $\boldsymbol{y}$ is represented as the sum

$$\boldsymbol{y} = \boldsymbol{y}_p + \boldsymbol{y}_\perp \tag{2.98}$$

where $\boldsymbol{y}_p$ is the *projection* of $\boldsymbol{y}$ onto $\boldsymbol{x}$, that is the component of $\boldsymbol{y}$ which is parallel to $\boldsymbol{x}$, having length $\|\boldsymbol{y}\| \cos\varphi$, that is

$$\boldsymbol{y}_p = \frac{\|\boldsymbol{y}\| \cos\varphi}{\|\boldsymbol{x}\|} \boldsymbol{x} = \frac{\langle \boldsymbol{y}, \boldsymbol{x} \rangle}{\|\boldsymbol{x}\|^2} \boldsymbol{x} \tag{2.99}$$

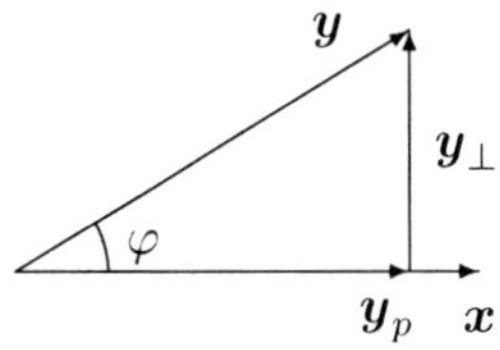

Figure 2.27 Geometrical representation of the projection of $\boldsymbol{y}$ onto $\boldsymbol{x}$.

Instead, $\boldsymbol{y}_\perp$ has length $\|\boldsymbol{y}\| \sin\varphi$, and is the component of $\boldsymbol{y}$ which is *orthogonal* to $\boldsymbol{x}$ where orthogonality is defined as

Definition 2.11 *Two vectors $\boldsymbol{x}$ and $\boldsymbol{y}$ of an inner product space are said* orthogonal *if their inner product is zero, and we write*

$$\boldsymbol{x} \perp \boldsymbol{y} \qquad \Longleftrightarrow \qquad \langle \boldsymbol{x}, \boldsymbol{y} \rangle = 0\,. \tag{2.100}$$

The fact that $\boldsymbol{y}_\perp$ is orthogonal to $\boldsymbol{x}$ is easily proved by substitution. Moreover, the above derivation, (2.98) and (2.99), is valid also when $\langle \boldsymbol{x}, \boldsymbol{y} \rangle$ is complex, with the unique difference of not being able to identify an angle between the two vectors.

Inner product spaces also allow to identify *orthonormal bases.*

Definition 2.12 *An* orthonormal basis *of an inner product space V is any collection of I vectors, $\{\boldsymbol{\varphi}_1, \boldsymbol{\varphi}_2, \ldots, \boldsymbol{\varphi}_I\}$, such that*

1. *are pairwise orthogonal, $\boldsymbol{\varphi}_i \perp \boldsymbol{\varphi}_j$, if $i \neq j$, $i, j = 1, \ldots, I$;*
2. *have unit norm, $\|\boldsymbol{\varphi}_i\| = 1$, $i = 1, \ldots, I$;*
3. *the space V is the span of the I vectors of the basis.*

I is called the dimension *of V, and it is independent of the chosen basis.*

With orthonormal bases, each vector $\boldsymbol{x} \in V$ can be expressed as a linear combination of the basis vectors,

$$\boldsymbol{x} = \sum_{i=1}^{I} c_i \boldsymbol{\varphi}_i\,, \qquad c_i = \langle \boldsymbol{x}, \boldsymbol{\varphi}_i \rangle \tag{2.101}$$

where the coefficients c_i are simply derived through an inner product with the orthonormal basis. Incidentally, $c_i \boldsymbol{\varphi}_i$ represents the projection of $\boldsymbol{x}$ onto $\boldsymbol{\varphi}_i$, in the sense of Fig. 2.27.

2.5.2 Signals as elements in a linear space

Although the concept of linear space is commonly associated to vectors (Euclidean and Hermitian spaces), the same concepts can be as well applied to signals.

So, a noticeable example of linear space is given by the (finite energy) *continuous time signal space*, that is the collection of continuous time signals with finite energy $x(t)$, $t \in \mathbb{R}$. The inner product is defined through the integral

$$\langle x(t), y(t) \rangle = \int_{-\infty}^{+\infty} x(t)\, y^*(t)\, dt \,, \tag{2.102}$$

and for the energy we have

$$\langle x(t), x(t) \rangle = \int_{-\infty}^{+\infty} |x(t)|^2\, dt = E_x \leq +\infty \,. \tag{2.103}$$

We let the reader verify that all the properties of a linear space and of the inner product are verified.

Example 2.5 A One orthonormal basis for the complete (finite energy) continuous time signal space is given by the so called *Hermite-Gauss functions*

$$\varphi_i(t) = \sqrt{\frac{\sqrt{2}}{2^i\, i!}}\, H_i\left(\sqrt{2\pi}\, t\right) e^{-\pi t^2} \,, \qquad i = 0, 1, 2, \ldots \tag{2.104}$$

where $H_i(x)$ identifies the *Hermite polynomial* of order i which can be derived from the recursion

$$H_{i+1}(x) = 2x\, H_i(x) - 2i\, H_{i-1}(x) \tag{2.105}$$

starting from $H_0(x) = 1$ and $H_1(x) = 2x$. Hermite-Gauss functions are shown in Fig. 2.28 for the first orders i.

Example 2.5 B A further example of interest to us is given by the class of continuous time band-limited signals with finite bandwidth B. In this case, the *Sampling Theorem* assures that an orthonormal basis is given by the pulses

$$\varphi_i(t) = \frac{1}{\sqrt{T}}\, \text{sinc}\left(\frac{t - iT}{T}\right) \tag{2.106}$$

with $T = 1/(2B)$.

Example 2.5 C An orthonormal basis for the class of continuous time signals periodic of period T_p is given by the functions

$$\varphi_i(t) = \frac{1}{\sqrt{T_p}}\, e^{j2\pi i\, t/T_p} \,, \tag{2.107}$$

provided that the inner product between signals is defined as

$$\langle x(t), y(t) \rangle = \int_0^{T_p} x(t)\, y^*(t)\, dt \,. \tag{2.108}$$

This result is the consequence of properties of the Fourier series expansion.

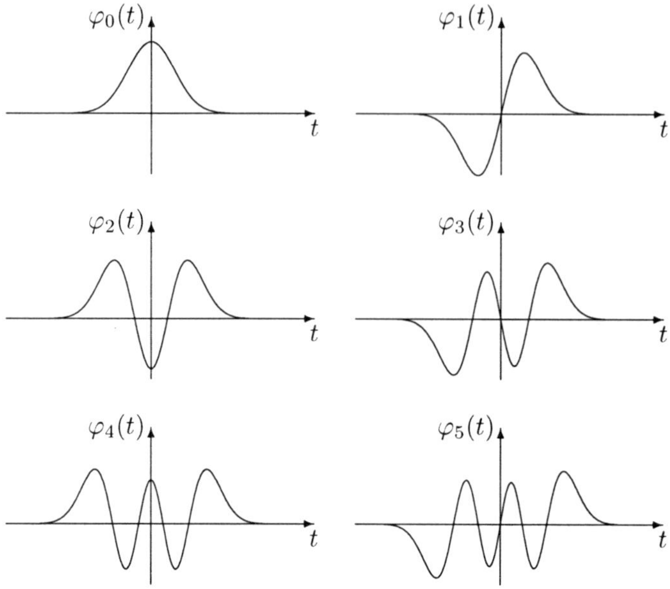

Figure 2.28 First order Hermite-Gauss functions.

Signal spaces are also identified by linear combinations of (finite energy) signals. For example, in the continuous time signal case the M signals $s_1(t), s_2(t), \ldots, s_M(t)$ identify the linear space

$$V = \left\{ s(t) \;\middle|\; s(t) = \sum_{i=1}^{M} c_i\, s_i(t)\ ,\ c_i \in \mathbb{C} \right\}$$

which is a space of dimension $I \leq M$, with $I = M$ if the generating signals are linearly independent, and $I < M$ if some linear dependencies exist between the generating signals, i.e. when one signal can be written as a linear combination of the remaining ones.

2.5.3 Gram-Schmidt orthonormalization in signal spaces

When the signal space of interest is the span of given signals, $\{s_n(t)\}$, $n = 1, \ldots, M$, an orthonormal basis $\{\varphi_i(t)\}$, $i = 1, \ldots, I$, can be identified through a *Gram-Schmidt orthonormalization* procedure. The procedure steps are now carefully outlined.

Step 1 The first element of the basis is chosen proportional to the first signal, that is we set $\varphi'_1(t) = s_1(t)$ and derive $\varphi_1(t)$ as the normalized version of $\varphi'_1(t)$, that is

$$\varphi_1(t) = \frac{\varphi'_1(t)}{\sqrt{E'_1}}$$

with $E'_1 = \|\varphi'_1(t)\|^2 = \langle \varphi'_1(t), \varphi'_1(t) \rangle$ the energy of $\varphi'_1(t)$.

Step 2 The second element of the basis is derived from the second signal, $s_2(t)$, once the component proportional to $\varphi_1(t)$ has been removed. As illustrated in Fig. 2.29a, we choose

$$\varphi_2'(t) = s_2(t) - c_{2,1}\,\varphi_1(t)\;, \qquad c_{2,1} = \langle s_2(t), \varphi_1(t)\rangle$$

to guarantee that $\varphi_2'(t)$ is orthogonal to $\varphi_1(t)$, that is $\langle \varphi_2'(t), \varphi_1(t)\rangle = 0$. This can be easily proved by substitution, to obtain

$$\langle \varphi_2'(t), \varphi_1(t)\rangle = \langle s_2(t), \varphi_1(t)\rangle - c_{2,1}\,\langle \varphi_1(t), \varphi_1(t)\rangle = c_{2,1} - c_{2,1} = 0\;.$$

From Fig. 2.29a, note that $c_{2,1}\,\varphi_1(t)$ is the projection of $s_2(t)$ onto the space spanned by $\varphi_1(t)$. The second element of the basis is finally obtained through normalization, that is

$$\varphi_2(t) = \frac{\varphi_2'(t)}{\sqrt{E_2'}}\;.$$

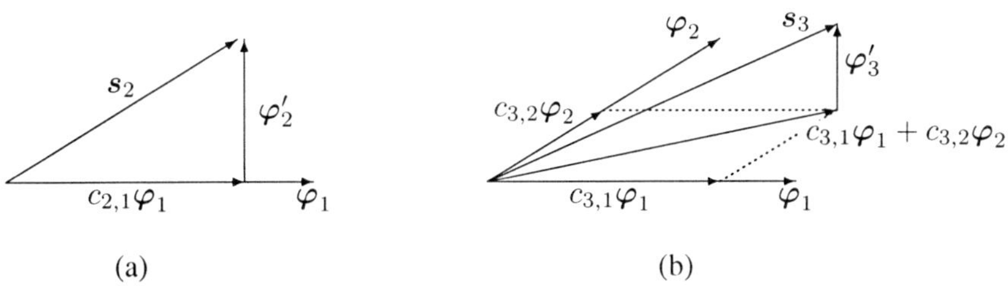

Figure 2.29 Geometrical interpretation of the Gram-Schmidt orthonormalization procedure: (a) derivation of φ_2; (b) derivation of φ_3.

Step 3 The third element of the basis is derived from the third signal, $s_3(t)$, once the components proportional to $\varphi_1(t)$ and $\varphi_2(t)$ have been removed. More specifically, we choose (see Fig. 2.29b)

$$\varphi_3'(t) = s_3(t) - c_{3,1}\,\varphi_1(t) - c_{3,2}\,\varphi_2(t)\;, \qquad c_{3,i} = \langle s_3(t), \varphi_i(t)\rangle$$

and then normalize the result to

$$\varphi_3(t) = \frac{\varphi_3'(t)}{\sqrt{E_3'}}\;.$$

Step n In general, for the nth step, $n \le M$, we have

$$\varphi_n'(t) = s_n(t) - \sum_{i=1}^{n-1} c_{n,i}\,\varphi_i(t)\;, \qquad c_{n,i} = \langle s_n(t), \varphi_i(t)\rangle \tag{2.109}$$

and the result is normalized to obtain

$$\varphi_n(t) = \frac{\varphi_n'(t)}{\sqrt{E_n'}}\;. \tag{2.110}$$

Evidently, in the case that $s_n(t)$ is a linear combination of preceding signals, we would then have $\varphi'_n(t) = 0$, in which case the basis is not updated with a new vector $\varphi_n(t)$. Obviously, a null signal cannot be an element of the basis. In this case, the final dimension I of the basis $\{\varphi_i(t)\}$ will be lower than M. Hence, in general we have

$$\boxed{I \leq M} \tag{2.111}$$

Remark We underline that the basis $\{\varphi_i(t)\}$ obtained through a Gram-Schmidt procedure is in general not unique, but strictly depends on the order given to the signals $\{s_n(t)\}$, $n = 1, \ldots, M$.

Remark The gram-Schmidt procedure given by (2.109) and (2.110) allows us to express $s_n(t)$ in terms of the basis functions $\{\varphi_i(t)\}$ as

$$s_n(t) = \sum_{i=1}^{n-1} c_{n,i}\,\varphi_i(t) + \sqrt{E'_n}\varphi_n(t)\,. \tag{2.112}$$

Example 2.5 D Consider the three continuous signals defined as

$$s_1(t) = \begin{cases} A\sin(2\pi t/T) & ,\ 0 < t < \frac{1}{2}T \\ 0 & ,\ \text{elsewhere} \end{cases}$$

$$s_2(t) = \begin{cases} A\sin(2\pi t/T) & ,\ 0 < t < T \\ 0 & ,\ \text{elsewhere} \end{cases}$$

$$s_3(t) = \begin{cases} A\sin(2\pi t/T) & ,\ \frac{1}{2}T < t < T \\ 0 & ,\ \text{elsewhere} \end{cases}$$

which are depicted in Fig. 2.30a.

We want to identify an orthonormal basis through the Gram-Schmidt procedure. We then immediately set $\varphi'_1(t) = s_1(t)$ from which we obtain

$$\varphi_1(t) = \frac{s_1(t)}{\sqrt{A^2T/4}} = \begin{cases} \frac{2}{\sqrt{T}}\sin(2\pi t/T) & ,\ 0 < t < \frac{1}{2}T \\ 0 & ,\ \text{elsewhere.} \end{cases}$$

The projection of $s_2(t)$ onto $\varphi_1(t)$ is instead given by

$$c_{2,1} = \int s_2(t)\varphi_1(t)^*\,dt = \int_0^{\frac{1}{2}T} \frac{2A}{\sqrt{T}}\sin^2(2\pi t/T)\,dt = \tfrac{1}{2}A\sqrt{T}$$

so that

$$\varphi'_2(t) = s_2(t) - c_{2,1}\,\varphi_1(t)$$

and

$$\varphi_2(t) = \frac{\varphi'_2(t)}{\sqrt{A^2T/4}} = \begin{cases} \frac{2}{\sqrt{T}}\sin(2\pi t/T) & ,\ \frac{1}{2}T < t < T \\ 0 & ,\ \text{elsewhere} \end{cases}$$

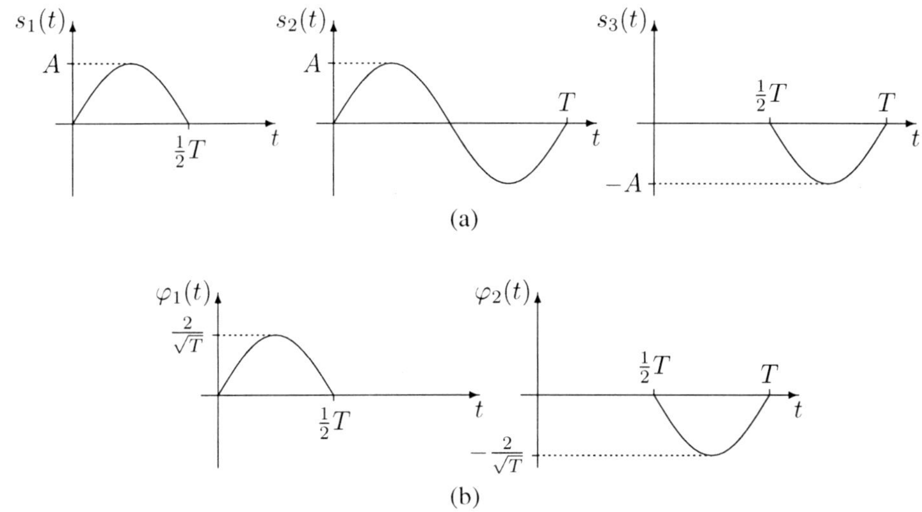

Figure 2.30 (a) The $M = 3$ signals of Example 2.5 D and (b) their complete orthonormal basis $I = 2$.

Since $s_3(t) = \varphi_2'(t)$ the orthonormal basis is $\{\varphi_1(t), \varphi_2(t)\}$, and it is shown in Fig. 2.30b.

Example 2.5 E (4-PSK) Consider the *quaternary phase shift keying* (QPSK or 4-PSK) signal set

$$s_n(t) = \begin{cases} A\cos\left(2\pi f_0 t + (n - \frac{1}{2})\frac{\pi}{2}\right) & , \ 0 < t < T \\ 0 & , \ \text{elsewhere} \end{cases}$$

with $n = 1, 2, 3, 4$, illustrated in Fig. 2.31, of which we want to identify an orthonormal basis.

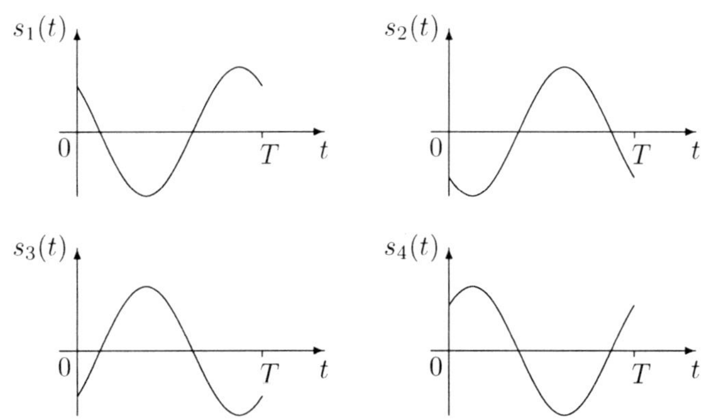

Figure 2.31 4-PSK signal set for $f_0 = 1/T$.

Instead of directly applying the Gram-Schmidt procedure, we preliminarily rewrite the signals by expanding the cosine, to obtain (in the interval $0 < t < T$)

$$s_n(t) = A \cos(2\pi f_0 t) \cos\left((n - \tfrac{1}{2})\tfrac{\pi}{2}\right) - A \sin(2\pi f_0 t) \sin\left((n - \tfrac{1}{2})\tfrac{\pi}{2}\right)$$

which emphasizes that the signals are linear combinations of

$$v_1(t) = \begin{cases} \cos(2\pi f_0 t) & , \ 0 < t < T \\ 0 & , \ \text{elsewhere} \end{cases}$$

$$v_2(t) = \begin{cases} -\sin(2\pi f_0 t) & , \ 0 < t < T \\ 0 & , \ \text{elsewhere} \end{cases}$$

Unfortunately, $v_1(t)$ and $v_2(t)$ are not orthogonal in general. In fact

$$\begin{aligned} \langle v_2(t), v_1(t) \rangle &= -\int_0^T \cos(2\pi f_0 t) \sin(2\pi f_0 t)\, dt \\ &= -\tfrac{1}{2} \int_0^T \sin(2\pi 2 f_0 t)\, dt = \frac{T}{2} \frac{\cos(2\pi 2 f_0 T) - 1}{2\pi 2 f_0 T} . \end{aligned}$$

However, in practical situations where $f_0 \gg \frac{1}{T}$, the inner product $\langle v_2(t), v_1(t) \rangle$ becomes very small. The same result is obtained by assuming $f_0 = K/(2T)$ (with K integer). Hence, by considering that one of the following conditions holds

$$f_0 \gg \frac{1}{T} \quad \text{or} \quad f_0 = \frac{K}{2T} \ (K \text{ integer}) \tag{2.113}$$

the signals $v_1(t)$ and $v_2(t)$ are orthogonal, and we can thus set $\varphi_1'(t) = v_1(t)$ and $\varphi_2'(t) = v_2(t)$. The final result is obtained once the energy of both $v_1(t)$ and $v_2(t)$ are known. We have

$$E_1 = \int_0^T \cos^2(2\pi f_0 t)\, dt = \tfrac{1}{2} \int_0^T \Big[1 + \cos(2\pi 2 f_0 t)\Big]\, dt = \tfrac{1}{2} T \Big[1 + \text{sinc}(4 f_0 T)\Big]$$

$$E_2 = \int_0^T \sin^2(2\pi f_0 t)\, dt = \tfrac{1}{2} \int_0^T \Big[1 - \cos(2\pi 2 f_0 t)\Big]\, dt = \tfrac{1}{2} T \Big[1 - \text{sinc}(4 f_0 T)\Big]$$

which simply become $E_1 = E_2 = \frac{1}{2} T$ once we have taken into account (2.113).

So, under the condition (2.113), the orthonormal basis is given by the signals

$$\varphi_1(t) = \sqrt{\frac{2}{T}}\, v_1(t) = \begin{cases} \sqrt{\frac{2}{T}} \cos(2\pi f_0 t) & , \ 0 < t < T \\ 0 & , \ \text{elsewhere} \end{cases}$$

$$\varphi_2(t) = \sqrt{\frac{2}{T}}\, v_2(t) = \begin{cases} -\sqrt{\frac{2}{T}} \sin(2\pi f_0 t) & , \ 0 < t < T \\ 0 & , \ \text{elsewhere} \end{cases}$$

We let the reader derive the expression for the orthonormal basis when (2.113) does not hold, in which case the Gram-Schmidt procedure must be applied.

2.5.4 Vector representation of signals

We have seen that in a signal space generated by the signals $\{s_n(t)\}$, $n = 1, \ldots, M$, where an orthonormal basis $\{\varphi_i(t)\}$, $i = 1, \ldots, I$, is known, each signal of the set can be expressed as a linear combination of basis vectors, that is

$$s_n(t) = \sum_{i=1}^{I} s_{n,i}\,\varphi_i(t) \;, \qquad s_{n,i} = \langle s_n(t), \varphi_i(t)\rangle \,. \tag{2.114}$$

In a sense, there is a one-to-one relation between the signal $s_n(t)$ and the collection of its coefficients, that is

$$s_n(t) \qquad \Longleftrightarrow \qquad \boldsymbol{s}_n = \left[s_{n,1}, s_{n,2}, \ldots, s_{n,I}\right] . \tag{2.115}$$

Expression (2.115) is thus stating the possibility to represent a continuous time signal $s_n(t)$ through a vector $\boldsymbol{s}_n$ (vector representation of $s_n(t)$).

Definition 2.13 *The vector representation of a set of M signals is often called the* signal constellation.

Remark Using the Gram-Schmidt procedure, from (2.112) it turns out

$$s_n(t) \qquad \Longleftrightarrow \qquad \boldsymbol{s}_n = \left[c_{n,1},\, c_{n,2},\, \ldots,\, c_{n,n-1},\, c_{n,n}, 0, \ldots 0\right] \tag{2.116}$$

with $c_{n,i} = \langle s_n(t), \varphi_i(t)\rangle$ and $\varphi_n(t)$ as defined in (2.110). Note that $c_{n,n}$ is also equal to $\sqrt{E'_n} = \|\varphi'_n(t)\|$, where $\varphi'_n(t)$ is defined in (2.109).

Example 2.5 F The vector representation for the signals of Example 2.5 D, with respect to the basis derived in the example, are

$$\boldsymbol{s}_1 = \left[\tfrac{A}{2}\sqrt{T}, 0\right] \qquad \boldsymbol{s}_2 = \left[\tfrac{A}{2}\sqrt{T}, \tfrac{A}{2}\sqrt{T}\right] \qquad \boldsymbol{s}_3 = \left[0, \tfrac{A}{2}\sqrt{T}\right] .$$

The signal constellation is given in Fig. 2.32a.

Example 2.5 G The vector representation for the signals of Example 2.5 E, with respect to the basis derived in the example, are

$$\begin{aligned} \boldsymbol{s}_1 &= \left[+\tfrac{A}{2}\sqrt{T}, +\tfrac{A}{2}\sqrt{T}\right] & \boldsymbol{s}_2 &= \left[-\tfrac{A}{2}\sqrt{T}, +\tfrac{A}{2}\sqrt{T}\right] \\ \boldsymbol{s}_3 &= \left[-\tfrac{A}{2}\sqrt{T}, -\tfrac{A}{2}\sqrt{T}\right] & \boldsymbol{s}_4 &= \left[+\tfrac{A}{2}\sqrt{T}, -\tfrac{A}{2}\sqrt{T}\right] . \end{aligned}$$

The signal constellation is given in Fig. 2.32b.

The convenience of managing vectors instead of signals is many-fold, and is particularly evident when the signal space dimension I is small. One of the properties of main interest to us is given by the relation between the inner product in the signal domain and in the vector

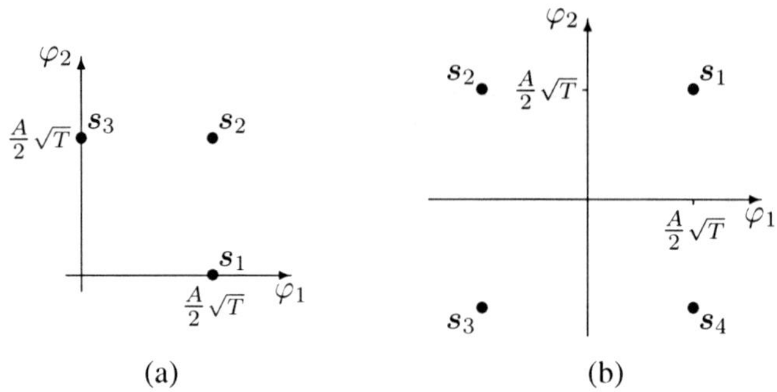

Figure 2.32 Two signal constellations: (a) constellation of Example 2.5 D; (b) constellation of Example 2.5 E.

domain. Given two signals $x(t)$ and $y(t)$ belonging to V, and given their vector representations $\boldsymbol{x}$ and $\boldsymbol{y}$, we have

$$\langle x(t), y(t)\rangle = \int_{-\infty}^{+\infty} x(t)\, y^*(t)\, dt = \sum_{i=1}^{I} x_i\, y_i^* = \langle \boldsymbol{x}, \boldsymbol{y}\rangle \ . \tag{2.117}$$

This can be easily proved by substituting signals with their expansions (2.114). In fact

$$\begin{aligned}\langle x(t), y(t)\rangle &= \left\langle \sum_{i=1}^{I} x_i\, \varphi_i(t), \sum_{j=1}^{I} y_i\, \varphi_j(t) \right\rangle \\ &= \sum_{i,j=1}^{I} x_i\, y_j^*\, \langle \varphi_i(t), \varphi_j(t)\rangle = \sum_{i,j=1}^{I} x_i\, y_j^*\, \delta_{i-j}\end{aligned} \tag{2.118}$$

where δ_k is a Kronecker delta (2.19), and (2.117) holds.

The inner product property (2.117) immediately implies a further equivalence between energies in signal and vector domains. We have

$$E_x = \int_{-\infty}^{+\infty} |x(t)|^2\, dt = \sum_{i=1}^{I} |x_i|^2 = \|\boldsymbol{x}\|^2 = E_{\boldsymbol{x}} \ . \tag{2.119}$$

Another worth mentioning equivalence is related to the concept of Euclidean distance. In vector terms, the *distance between two vectors* is defined as the norm of their difference, that is

$$d(\boldsymbol{x}, \boldsymbol{y}) = \|\boldsymbol{x} - \boldsymbol{y}\| = \sqrt{E_{\boldsymbol{x}-\boldsymbol{y}}} = \sqrt{\sum_{i=1}^{I} |x_i - y_i|^2} \tag{2.120}$$

where we underline the relation between the norm and the square-root of energy. A similar definition is usually applied to continuous time signals for which the *distance* is defined as

$$d(x(t), y(t)) = \sqrt{\int_{-\infty}^{+\infty} |x(t) - y(t)|^2\, dt} \tag{2.121}$$

and we have

$$d(x(t), y(t)) = d(\boldsymbol{x}, \boldsymbol{y}) . \tag{2.122}$$

Note also that the square distance can be expanded to give

$$\begin{aligned} d^2(x(t), y(t)) &= \langle x(t) - y(t), x(t) - y(t) \rangle \\ &= \langle x(t), x(t) \rangle - \langle x(t), y(t) \rangle - \langle y(t), x(t) \rangle + \langle y(t), y(t) \rangle \\ &= E_x + E_y - 2\Re[\langle x(t), y(t) \rangle] \end{aligned} \tag{2.123}$$

when using a signal notation, or as

$$d^2(\boldsymbol{x}, \boldsymbol{y}) = E_{\boldsymbol{x}} + E_{\boldsymbol{y}} - 2\Re[\langle \boldsymbol{x}, \boldsymbol{y} \rangle] \tag{2.124}$$

when using a vector notation.

Remark Quantities such as energy, internal product and distance between signals of a space do not depend on the chosen basis $\{\varphi_i(t)\}$, $i = 1, \ldots, I$. In other words, let $\boldsymbol{x}$ be the vector representation of $x(t)$ with basis $\{\varphi_i(t)\}$. If an alternative basis $\{\tilde{\varphi}_i(t)\}$ is chosen, leading to the vector representation $\tilde{\boldsymbol{x}}$, we have

$$\|\tilde{\boldsymbol{x}}\| = \|\boldsymbol{x}\| , \qquad \langle \tilde{\boldsymbol{x}}, \tilde{\boldsymbol{y}} \rangle = \langle \boldsymbol{x}, \boldsymbol{y} \rangle , \qquad d(\tilde{\boldsymbol{x}}, \tilde{\boldsymbol{y}}) = d(\boldsymbol{x}, \boldsymbol{y}) . \tag{2.125}$$

Example 2.5 H By using their vector representation, for the signals of Example 2.5 D the energies are

$$E_1 = E_3 = \frac{A^2 T}{4} , \qquad E_2 = \frac{A^2 T}{2}$$

while their relative distances $d_{i,j} = d(s_i(t), s_j(t))$ are

$$d_{1,2} = d_{2,3} = \tfrac{A}{2}\sqrt{T} , \qquad d_{1,3} = A\sqrt{\tfrac{T}{2}} .$$

Example 2.5 I By using their vector representation, for the signals of Example 2.5 E the energies are

$$E_1 = E_2 = E_3 = E_4 = \tfrac{1}{2} A^2 T$$

while their relative distances $d_{i,j} = d(s_i(t), s_j(t))$ are

$$d_{1,2} = d_{2,3} = d_{3,4} = d_{4,1} = A\sqrt{T} , \qquad d_{1,3} - d_{2,4} = A\sqrt{2T} .$$

2.5.5 Orthogonal projections onto a signal space

The results illustrated so far imply that all the continuous time signals we are dealing with belong to the signal space V generated by an orthonormal basis $\{\varphi_i(t)\}$, $i = 1, \ldots, I$. A signal $s(t)$ does not necessarily belong to the reference signal space V. In this case it is useful to identify the signal

$$\hat{s}(t) = \sum_{i=1}^{I} s_i \, \varphi_i(t) \tag{2.126}$$

that belongs to V and is closer to $s(t)$ as possible. In mathematical terms, we define the error signal

$$e(t) = s(t) - \hat{s}(t) \tag{2.127}$$

and translate the request of *being close* into the request of minimizing the energy of $e(t)$, that is

$$\{s_{i,\text{opt}}\} = \arg\min_{\{s_i\}} E_e \; . \tag{2.128}$$

The solution to (2.128) is found by expanding the error energy by use of (2.126) and (2.127). By recalling the property (2.123) of signal distances, we have

$$\begin{aligned} E_e &= d^2(s(t), \hat{s}(t)) = E_s + E_{\hat{s}} - 2\,\text{Re}[\langle s(t), \hat{s}(t)\rangle] \\ &= E_s + \sum_{i=1}^{I} \left(|s_i|^2 - 2\,\text{Re}[s_i \, c_i^*] \right) \end{aligned} \tag{2.129}$$

where $c_i = \langle s(t), \varphi_i(t)\rangle$. By then adding and subtracting $|c_i|^2$ it is

$$E_e = E_s + \sum_{i=1}^{I} \left(|s_i - c_i|^2 - |c_i|^2 \right) \tag{2.130}$$

as $|s_i|^2 - 2\,\text{Re}[s_i \, c_i^*] + |c_i|^2 = |s_i|^2 - s_i c_i^* - s_i^* c_i + |c_i|^2 = (s_i - c_i)(s_i - c_i)^*$. Thus (2.128) becomes

$$\{s_{i,\text{opt}}\} = \arg\min_{\{s_i\}} E_s + \sum_{i=1}^{I} |s_i - c_i|^2 - \sum_{i=1}^{I} |c_i|^2 \tag{2.131}$$

whose minimum is obtained when the second term is zero, that is

$$s_{i,\text{opt}} = c_i = \langle s(t), \varphi_i(t)\rangle = \int_{-\infty}^{+\infty} s(t)\varphi_i^*(t)\, dt \; . \tag{2.132}$$

Note that (2.132) is the projection (in the sense of Fig. 2.27) of $s(t)$ onto the space spanned by the orthonormal basis $\{\varphi_i(t)\}$, $i = 1, \ldots, I$.

Besides guaranteeing the minimization of the error energy, the condition (2.132) also satisfies the property

$$\langle e(t), \varphi_i(t)\rangle = 0 \, , \qquad i = 1, \ldots, I \tag{2.133}$$

that is the error is *orthogonal to each of the basis signals*, and so it is also *orthogonal to* $\hat{s}(t)$,

$$\langle e(t), \hat{s}(t)\rangle = 0 \, , \tag{2.134}$$

since $\hat{s}(t)$ is a linear combination of the basis signals. Moreover, the resulting (minimum) error energy is

$$E_e = E_s - \sum_{i=1}^{I} |c_i|^2 = E_s - E_{\hat{s}} \tag{2.135}$$

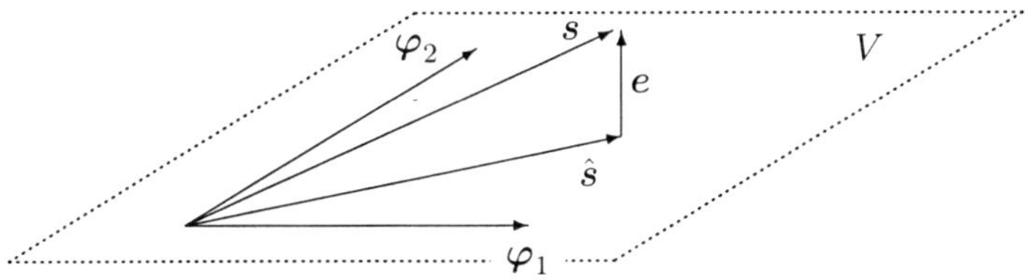

Figure 2.33 Geometrical representation of the projection of $s(t)$ onto the signal space V spanned by the orthonormal basis $\{\varphi_1(t), \varphi_2(t)\}$.

which is in accordance with the orthogonality condition (2.134). This is illustrated in Fig. 2.33.

Definition 2.14 *If* $E_e = 0$, *then* $\{\varphi_i(t)\}$, $i = 1, \ldots, I$, *is called a* complete *basis for* $s(t)$, *and* $s(t)$ *can be expressed as*

$$s(t) = \sum_{i=1}^{I} c_i \, \varphi_i(t) \tag{2.136}$$

where equality must be intended in terms of quadratic norm.

2.6 RANDOM VARIABLES AND VECTORS

The present section is an abridgement of useful results, by no means complete, and is ancillary to the introduction of random processes (which model probabilistic phenomena) in Section 2.7. We start with the following definition.

Definition 2.15 *A* random variable (rv) x *is a real-valued function defined on a probability space*

$$x \, : \, \Omega \mapsto \mathbb{R} \tag{2.137}$$

where Ω *is called the sample space.*

We should therefore write $x(\omega)$ with ω an arbitrary point in Ω, but it is customary to drop the dependence on ω from the notation. Each of the possible values of x corresponding to a given ω is called a *realization* of the variable x.

The fact that x be defined on a probability space means that there is a probability function $\mathrm{P}[\cdot]$ defined on the subsets of Ω (called the *events* of the probability space) so that for any reasonable subset A of $\mathbb{R}$ (made as union, possibly countable, of intervals and single points) we can determine the probability that the quantity x takes values in A as

$$\mathrm{P}[x(\omega) \in A] = \mathrm{P}\left[x^{-1}(A)\right]$$

where, for simplicity, $x^{-1}(A)$ denotes the inverse range of A.

A rv is usually classified, according to its possible values, as

discrete-valued or discrete if x can only take a countable number of values. The set $\mathcal{A}_x$ of possible values is called the *alphabet* of the rv x;

continuous-valued or continuous otherwise.

Examples of discrete rvs are: the number of correctly received bits in the transmission of a digital message over a noisy channel, the intensity level of a pixel in a digital picture, etc. For examples of continuous rvs one can take a noisy voltage at some instant, the pitch frequency of a voiced segment in an audio signal, etc.

2.6.1 Statistical description of random variables

A complete statistical description for a *discrete rv* x is given by its *probability mass distribution* (PMD) which is a real function defined on the alphabet of x

$$p_x \; : \; \mathcal{A}_x \mapsto \mathbb{R} \tag{2.138}$$

so that $p_x(a)$ gives the probability that x takes exactly the value a

$$p_x(a) = \mathrm{P}\left[x = a\right] = \mathrm{P}\left[x \in \{a\}\right] \tag{2.139}$$

It is evident from (2.139) that p_x can only take values in the interval $[0, 1]$, since it represents a probability. Then from the knowledge of the PMD we can calculate the probability that x takes value in any set $C \subset \mathbb{R}$ as

$$\mathrm{P}\left[x \in C\right] = \sum_{a \in \mathcal{A}_x \cap C} p_x(a) \tag{2.140}$$

thus justifying the expression "complete statistical description". Besides being limited between 0 and 1, p_x must also satisfy the following *normalization condition*

$$\sum_{a \in \mathcal{A}_x} p_x(a) = \mathrm{P}\left[x \in \mathcal{A}_x\right] = 1 \tag{2.141}$$

A peculiar case is that of a rv taking a single value a_0 with probability 1: such a variable is called *almost surely (a.s.) constant*. In terms of the above definitions it is

$$\mathcal{A}_x = \{a_0\} \quad , \quad p_x(a_0) = 1 \tag{2.142}$$

For *continuous rvs* a complete description is given by the *probability density function* (PDF), a real function of real variable

$$p_x \; : \; \mathbb{R} \mapsto \mathbb{R} \, , \tag{2.143}$$

defined as

$$p_x(a) = \frac{d}{da} \mathrm{P}\left[x \le a\right] \, . \tag{2.144}$$

Then, the probability of x taking values in some set $C \subset \mathbb{R}$ can be calculated as

$$\mathrm{P}\left[x \in C\right] = \int_C p_x(a) \, \mathrm{d}a \tag{2.145}$$

The peculiarities of this function are less evident than for its discrete counterpart. However, as the probability $\mathrm{P}\left[x \le a\right]$ can be shown to be a monotonically nondecreasing function of a, we must have

$$p_x(a) \ge 0 \quad , \quad a \in \mathbb{R} \tag{2.146}$$

and the normalization condition now reads

$$\int_{-\infty}^{+\infty} p_x(a)\,\mathrm{d}a = \mathrm{P}\left[x \in \mathbb{R}\right] = 1 \tag{2.147}$$

Example 2.6 A A rv x is called *Gaussian* or *normal* if its PDF has the form

$$p_x(a) = \frac{1}{\sqrt{2\pi\sigma^2}} e^{-\frac{(a-\mathtt{m})^2}{2\sigma^2}}$$

where $\mathtt{m}$ and σ are real parameters (with $\sigma > 0$). It is said to be *normalized* if $\mathtt{m} = 0$ and $\sigma = 1$. We want to find the probability that x takes values in the interval (b, c). By (2.145) we get

$$\mathrm{P}\left[b < x < c\right] = \int_b^c \frac{1}{\sqrt{2\pi\sigma^2}} e^{-\frac{(a-\mathtt{m})^2}{2\sigma^2}}\,\mathrm{d}a\;.$$

We face the non trivial problem of evaluating the above integral. By the change of variable $u = (a - \mathtt{m})/\sigma$, and defining the *normalized complementary Gaussian distribution* function as the integral of the normalized Gaussian PDF

$$\mathrm{Q}\left(y\right) = \int_y^{+\infty} \frac{1}{\sqrt{2\pi}} e^{-\frac{u^2}{2}}\,\mathrm{d}u\;, \tag{2.148}$$

we can express the above probability as

$$\mathrm{P}\left[b < x < c\right] = \int_{(b-\mathtt{m})/\sigma}^{(c-\mathtt{m})/\sigma} \frac{1}{\sqrt{2\pi}} e^{-\frac{u^2}{2}}\,\mathrm{d}u = \mathrm{Q}\left(\frac{b-\mathtt{m}}{\sigma}\right) - \mathrm{Q}\left(\frac{c-\mathtt{m}}{\sigma}\right)\;, \tag{2.149}$$

as illustrated in Fig. 2.34. The same expression can be used for evaluating the probabilities on unlimited intervals, by using the limit values

$$\mathrm{Q}\left(-\infty\right) = 1\;,\quad \mathrm{Q}\left(+\infty\right) = 0\;. \tag{2.150}$$

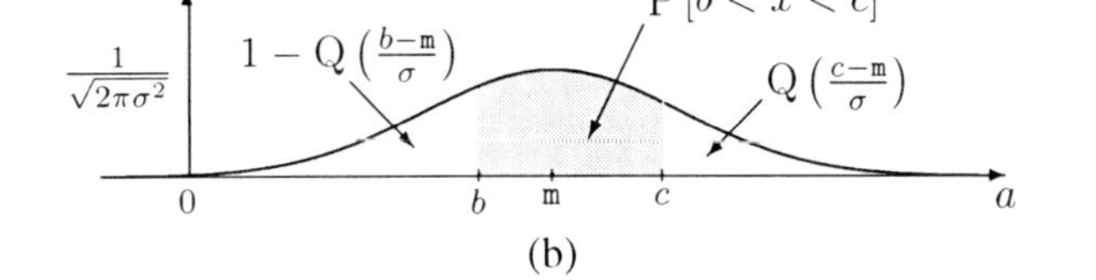

Figure 2.34 Plots of: (a) the normalized Gaussian PDF; (b) a Gaussian PDF with $\mathtt{m} = 2$, $\sigma^2 = 1/2$.

Gaussian rvs, introduced in the above example, play a very important role in the theory of transmission over noisy channels. To indicate that a rv x is Gaussian with parameters $\mathtt{m}$ and σ we will write $x \sim \mathcal{N}(\mathtt{m}, \sigma^2)$. Values of $\mathrm{Q}\left(y\right)$ are given in Tab. 2.A.1 of the Appendix.

2.6.2 Expectation and statistical power

An important statistical parameter for a rv x is its *mean* (also called *mean value* or *expectation*) indicated with $\mathtt{m}_x$ or $\mathrm{E}[x]$. It is a real value, dimensionally homogeneous with x, representing the mass center of the distribution p_x and it is calculated as

$$\mathrm{E}[x] = \begin{cases} \displaystyle\int_{-\infty}^{+\infty} a\, p_x(a)\, \mathrm{d}a & , \text{ for continuous rvs} \\ \displaystyle\sum_{a \in \mathcal{A}_x} a\, p_x(a) & , \text{ for discrete rvs} \end{cases} \tag{2.151}$$

In other words, expectation represents an integration of the variable (in ω over Ω) weighted by the probability measure. The reader should not be misled by the term "expectation": the mean is not the most likely value of the random variable x, it is just a weighted average of its possible values. Indeed, for a discrete x, $\mathtt{m}_x$ may not even belong to the alphabet $\mathcal{A}_x$.

Given an arbitrary real function of real variable $g : \mathbb{R} \mapsto \mathbb{R}$ we can compose it with a rv x obtaining a new rv y

$$y(\omega) = g(x(\omega)) \tag{2.152}$$

whose statistical description can be obtained from that of x. In particular, for discrete rvs we get

$$\mathcal{A}_y = g(\mathcal{A}_x) \quad , \quad p_y(b) = \sum_{a \in g^{-1}(b)} p_x(a) \quad , \quad b \in \mathcal{A}_y \tag{2.153}$$

whereas for continuous rvs, by exploiting the identity

$$\mathrm{P}[y \leq b] = \mathrm{P}\left[x \in g^{-1}(C_b)\right] \quad , \quad \text{where } C_b = (-\infty, b] \tag{2.154}$$

we get

$$p_y(b) = \frac{d}{db} \int_{g^{-1}(C_b)} p_x(a)\, \mathrm{d}a \,. \tag{2.155}$$

However, sometimes we might be interested only in determining the mean of the new rv $y = g(x)$, and not in finding its complete statistical description. The following theorem, that we just state, gives a method to calculate the mean of y, directly from the statistical description of x.

Theorem 2.11 (Fundamental theorem for expectation) *Given the statistical description of a rv x, we can calculate the expectation of any other rv $y = g(x)$ that is a function of x as*

$$\mathrm{E}[y] = \mathrm{E}[g(x)] = \begin{cases} \displaystyle\int_{-\infty}^{+\infty} g(a)\, p_x(a)\, \mathrm{d}a & , \textit{ for continuous rvs} \\ \displaystyle\sum_{a \in \mathcal{A}_x} g(a)\, p_x(a) & , \textit{ for discrete rvs} \end{cases} \tag{2.156}$$

We illustrate the above theorem with a couple of examples.

Example 2.6 B Let x be a discrete rv with alphabet

$$\mathcal{A}_x = \{-2, -1, 0, 1, 2\}$$

and PMD

$$p_x(-2) = p_x(2) = 1/8\ ,\ p_x(-1) = p_x(1) = p_x(0) = 1/4$$

and $y = g(x) = x^2$. Then by applying Theorem 2.11 we can calculate

$$\mathrm{E}[y] = \mathrm{E}[x]^2 = \sum_{a \in \mathcal{A}_x} a^2 p_x(a) = 4 \cdot \frac{1}{8} + 1 \cdot \frac{1}{4} + 0 + 1 \cdot \frac{1}{4} + 4 \cdot \frac{1}{8} = \frac{3}{2}$$

Example 2.6 C Let φ be a continuous rv with PDF

$$p_\varphi(a) = \frac{1}{2\pi} \operatorname{rect}\left(\frac{a}{2\pi}\right)$$

and $y = g(\varphi) = \cos^2 \varphi$. By Theorem 2.11 we get

$$\mathrm{E}[y] = \mathrm{E}\left[\cos^2 \varphi\right] = \int_{-\infty}^{+\infty} \cos^2(a)\, p_x(a)\, \mathrm{d}a = \int_{-\pi}^{\pi} \frac{1}{2\pi} \cos^2(a)\, \mathrm{d}a = \frac{1}{2}$$

Two important statistical parameters for x are obtained from Theorem 2.11 by respectively choosing $g(x) = x^2$ and $g(x) = (x - \mathtt{m}_x)^2$, that is

- *statistical power* of x

$$\mathtt{M}_x = \mathrm{E}\left[x^2\right] = \begin{cases} \displaystyle\int_{-\infty}^{+\infty} a^2 p_x(a)\, \mathrm{d}a & \text{, for continuous rvs} \\ \displaystyle\sum_{a \in \mathcal{A}_x} a^2\, p_x(a) & \text{, for discrete rvs} \end{cases} \tag{2.157}$$

- *variance* of x

$$\sigma_x^2 = \mathrm{E}\left[(x - \mathtt{m}_x)^2\right] = \begin{cases} \displaystyle\int_{-\infty}^{+\infty} (a - \mathtt{m}_x)^2 p_x(a)\, \mathrm{d}a & \text{, for continuous rvs} \\ \displaystyle\sum_{a \in \mathcal{A}_x} (a - \mathtt{m}_x)^2\, p_x(a) & \text{, for discrete rvs} \end{cases} \tag{2.158}$$

The square root of the variance, σ_x, is also an important parameter and is called the *standard deviation* of x. It can be seen by exploiting the linearity of the expectation that statistical mean, power and variance of a rv are related by

$$\sigma_x^2 + \mathtt{m}_x^2 = \mathtt{M}_x\ . \tag{2.159}$$

Example 2.6 D An *exponential rv* takes nonnegative values with probability 1 and has

$$\mathrm{P}\left[x > a\right] = \begin{cases} e^{-\lambda a} & , \; a > 0 \\ 1 & , \; a \leq 0 \end{cases} \tag{2.160}$$

Its PDF can then be derived from (2.144) as

$$p_x(a) = \frac{d}{da}\left(1 - \mathrm{P}\left[x > a\right]\right) = \lambda e^{-\lambda a} 1(a) \tag{2.161}$$

that is characterized by the parameter λ. Its mean can be derived as

$$\mathtt{m}_x = \int_0^{\infty} \lambda a e^{-\lambda a} \, da = \frac{1}{\lambda} \tag{2.162}$$

Example 2.6 E The *indicator function* $\chi\{A\}$ of an event $A \subset \Omega$ is a binary rv defined by

$$\chi\{A\}(\omega) = \begin{cases} 1 & , \; \omega \in A \\ 0 & , \; \omega \notin A \end{cases}$$

or, by dropping the ω notation,

$$\chi\{A\} = \begin{cases} 1 & , \; \text{if the event } A \text{ occurs} \\ 0 & , \; \text{otherwise} \end{cases} \tag{2.163}$$

From (2.163) it immediately follows that the PDF of the rv is given by

$$p_{\chi\{A\}}(0) = 1 - \mathrm{P}\left[A\right] , \quad p_{\chi\{A\}}(1) = \mathrm{P}\left[A\right] , \tag{2.164}$$

whereas its mean, statistical power, as well as the expectation of the nth power of x (called the *statistical moment* of order n) are all given by

$$\mathrm{E}\left[\chi\{A\}^n\right] = \mathrm{P}\left[A\right] , n = 1, 2, \ldots \; . \tag{2.165}$$

Example 2.6 F A discrete rv with alphabet $\mathcal{A} = \{0, 1, 2, \ldots\}$ and PMD

$$p_x(a) = e^{-\beta} \frac{\beta^a}{a!} \quad , \quad a \in \mathcal{A} \tag{2.166}$$

is said to be a Poisson rv with parameter β, which is both the mean and the variance of x. In fact,

$$\mathtt{m}_x = \sum_{a=0}^{\infty} a e^{-\beta} \frac{\beta^a}{a!} = e^{-\beta} \sum_{a=1}^{\infty} a \frac{\beta^a}{a!} = e^{-\beta} \sum_{a=1}^{\infty} \frac{\beta^a}{(a-1)!}$$

and with the change of variable $b = a - 1$ we get

$$\mathtt{m}_x = e^{-\beta} \sum_{b=0}^{\infty} \frac{\beta^{b+1}}{b!} = \beta e^{-\beta} \sum_{b=0}^{\infty} \frac{\beta^b}{b!} = \beta \tag{2.167}$$

Similarly,

$$\mathtt{M}_x = \sum_{a=0}^{\infty} a^2 e^{-\beta}\frac{\beta^a}{a!} = e^{-\beta}\sum_{a=1}^{\infty}\frac{\beta^a}{(a-1)!} = e^{-\beta}\sum_{b=0}^{\infty}(b+1)\frac{\beta^{b+1}}{b!}$$

$$= e^{-\beta}\beta\left(\sum_{b=0}^{\infty} b\frac{\beta^b}{b!} + \sum_{b=0}^{\infty}\frac{\beta^b}{b!}\right) = \beta(\beta+1) = \beta^2+\beta \tag{2.168}$$

and hence $\sigma_x^2 = \mathtt{M}_x - \mathtt{m}_x^2 = \beta$

Example 2.6 G A *geometric* rv is a discrete rv x with alphabet $\mathcal{A} = \{0, 1, 2, \ldots\}$, and PMD

$$p_x(a) = (1-q)q^a \quad , \quad a \in \mathcal{A}$$

where q is a parameter characterizing the distribution.

The probability of x lying in an arbitrary interval can be easily expressed as

$$\mathrm{P}\left[a_1 \leq x \leq a_2\right] = \sum_{a=a_1}^{a_2}(1-q)q^a = q^{a_1} - q^{a_2+1} \quad , \quad a_1, a_2 \in \mathcal{A}$$

and in particular we have

$$\mathrm{P}\left[x \leq a\right] = \mathrm{P}\left[0 \leq x \leq a\right] = 1 - q^{a+1} \quad , \quad a \in \mathcal{A} \tag{2.169}$$

and

$$\mathrm{P}\left[x \geq a\right] = 1 - \mathrm{P}\left[x \leq a-1\right] = q^a \quad , \quad a \in \mathcal{A} \tag{2.170}$$

Its mean is given by

$$\mathtt{m}_x = \sum_{a=0}^{\infty} a(1-q)q^a = (1-q)q\sum_{a=0}^{\infty} aq^{a-1}$$

where we recognize that the general term in the series is the derivative of the power function, and by exchanging the two operations

$$\mathtt{m}_x = (1-q)q\sum_{a=0}^{\infty}\frac{d}{dq}q^a = (1-q)q\frac{d}{dq}\sum_{a=0}^{\infty}q^a = (1-q)q\,\frac{d}{dq}\frac{1}{1-q}$$

$$= (1-q)q\,\frac{1}{(1-q)^2} = \frac{q}{1-q}$$

Similarly we can derive its power as

$$\mathtt{M}_x = \sum_{a=0}^{\infty} a^2(1-q)q^a = (1-q)\sum_{a=0}^{\infty}[a(a-1)+a]q^a$$

$$= (1-q)q^2\sum_{a=0}^{\infty} a(a-1)q^{a-2} + (1-q)q\sum_{a=0}^{\infty} aq^{a-1}$$

$$= (1-q)q^2\sum_{a=0}^{\infty}\frac{d^2}{dq^2}q^a + (1-q)q\sum_{a=0}^{\infty}\frac{d}{dq}q^a$$

$$= (1-q)q^2 \frac{d^2}{dq^2} \frac{1}{1-q} + (1-q)q \frac{d}{dq} \frac{1}{1-q}$$

$$= \frac{2q^2}{(1-q)^2} + \frac{q}{1-q} = \frac{q+q^2}{(1-q)^2}$$

and it variance as $\sigma_x^2 = \mathtt{M}_x - \mathtt{m}_x^2 = q/(1-q)^2$. Note that for a geometric rv holds the relation $\sigma_x^2 = \mathtt{m}_x + \mathtt{m}_x^2$.

2.6.3 Random vectors

We extend the notion of a random variable and introduce a *random vector (rve)* $\boldsymbol{x}$ as an N-ple of random variables, i.e. $\boldsymbol{x} = [x_1, x_2, \ldots, x_N]^{\mathrm{T}}$ (the superscript ^{T} denotes transposition). The relevant case is when the rvs that make the vector are related to each other, in some sense; therefore, we consider the variables in the vector to be defined on a common probability space. In this sense, the rve is a function

$$\boldsymbol{x} : \Omega \mapsto \mathbb{R}^N \tag{2.171}$$

and we are interested in a *joint* statistical description. As was done for rvs, we can apply the distinction into discrete and continuous rves, as well as the notion of alphabet, which is now a countable subset of $\mathbb{R}^N$, namely $\mathcal{A}_{\boldsymbol{x}}$. Thus we can define the PMD as

$$p_{\boldsymbol{x}}(\boldsymbol{a}) = \mathrm{P}[\boldsymbol{x} = \boldsymbol{a}] = \mathrm{P}[x_1 = a_1, \ldots, x_N = a_N] \ , \tag{2.172}$$

and the PDF as

$$p_{\boldsymbol{x}}(\boldsymbol{a}) = \frac{d}{d\boldsymbol{a}} \mathrm{P}[\boldsymbol{x} \le \boldsymbol{a}] = \frac{\partial^N}{\partial a_1 \cdots \partial a_N} \mathrm{P}[x_1 \le a_1, \ldots, x_N \le a_N] \ . \tag{2.173}$$

For example, the PDF allows to calculate the probability that $\boldsymbol{x}$ takes values in a subset $C \in \mathbb{R}^N$ in the same manner as described by (2.145) where the integral is now N-dimensional.

Example 2.6 H Given the PDF of the rve $\boldsymbol{x} = [x_1, x_2]^T$

$$p_{\boldsymbol{x}}(\boldsymbol{a}) = \beta\gamma\, e^{-\beta a_1 - \gamma a_2} 1(a_1)\, 1(a_2)$$

we calculate the probability $\mathrm{P}[x_1 > x_2]$. First we rewrite the event as

$$\{x_1 > x_2\} = \{x \in C\} \quad , \quad \text{with } C = \left\{(a_1, a_2) \in \mathbb{R}^2 : a_1 > a_2\right\}$$

Then we write the two-dimensional integral of $p_{\boldsymbol{x}}$ over C as

$$\mathrm{P}[x_1 > x_2] = \int_C p_{\boldsymbol{x}}(\boldsymbol{a})\, \mathrm{d}\boldsymbol{a} = \int_{-\infty}^{+\infty} \int_{-\infty}^{a_1} p_{\boldsymbol{x}}(a_1, a_2)\, \mathrm{d}a_2\, \mathrm{d}a_1$$

and we observe that $p_{\boldsymbol{x}}(\boldsymbol{a})$ is nonzero only for $a_1 \ge 0, a_2 \ge 0$, so we can calculate

$$\mathrm{P}[x_1 > x_2] = \int_0^{+\infty} \int_0^{a_1} \beta\gamma\, e^{-\beta a_1 - \gamma a_2}\, \mathrm{d}a_2\, \mathrm{d}a_1$$

$$
\begin{aligned}
&= \int_0^{+\infty} \beta\, e^{-\beta a_1} \int_0^{a_1} \gamma\, e^{-\gamma a_2}\, \mathrm{d}a_2\, \mathrm{d}a_1 \\
&= \int_0^{+\infty} \beta\, e^{-\beta a_1} \left(1 - e^{-\gamma a_1}\right) \mathrm{d}a_1 \\
&= 1 - \frac{\beta}{\beta+\gamma} = \frac{\gamma}{\beta+\gamma} \,.
\end{aligned}
$$

An important special case is that of rves made of independent rvs.

Definition 2.16 *N rvs $x_1, \ldots, x_N$ are said to be statistically independent if their joint statistical description can be factored into the product of the N single rv descriptions. For example using the joint PDF it results*

$$
\begin{aligned}
p_{\boldsymbol{x}}(\boldsymbol{a}) &= p_{x_1 \cdots x_N}(a_1, \ldots, a_N) = \prod_{i=1}^{N} p_{x_i}(a_i) \\
&= p_{x_1}(a_1) p_{x_2}(a_2) \cdots p_{x_N}(a_N)
\end{aligned}
\tag{2.174}
$$

In this case, also the probability that $\boldsymbol{x} \in C$, with C the Cartesian product $C = C_1 \times \cdots \times C_N$ and each $C_i \subset \mathbb{R}$, can be factored as

$$
\mathrm{P}\left[\boldsymbol{x} \in C\right] = \mathrm{P}\left[x_1 \in C_1, \ldots, x_N \in C_N\right] = \prod_{i=1}^{N} \mathrm{P}\left[x_i \in C_i\right] \,. \tag{2.175}
$$

An example follows.

Example 2.6 I Suppose that two continuous rvs have joint PDF

$$
p_{x_1 x_2}(a_1, a_2) = \frac{1}{2\pi\sigma_1\sigma_2} e^{-\frac{(a_1-\mathtt{m}_1)^2}{2\sigma_1^2} - \frac{(a_2-\mathtt{m}_2)^2}{2\sigma_2^2}}
$$

Since we can factor the PDF as

$$
p_{x_1 x_2}(a_1, a_2) = p_{x_1}(a_1) p_{x_2}(a_2) \quad , \quad \text{with } p_{x_i}(a_i) = \frac{1}{\sqrt{2\pi}\sigma_i} e^{-\frac{(a_i-\mathtt{m}_i)^2}{2\sigma_i^2}}
$$

we realize that x_1 and x_2 are two statistically independent Gaussian rvs, with mean $\mathtt{m}_1$, $\mathtt{m}_2$ and variance σ_1^2, σ_2^2, respectively. To calculate the probability $\mathrm{P}\left[x_1 > a, b < x_2 < c\right]$ we can either integrate their joint PDF over the set

$$
C = \{(a_1, a_2) \,:\, a_1 > a, b < a_2 < c\} = (a, +\infty) \times (b, c)
$$

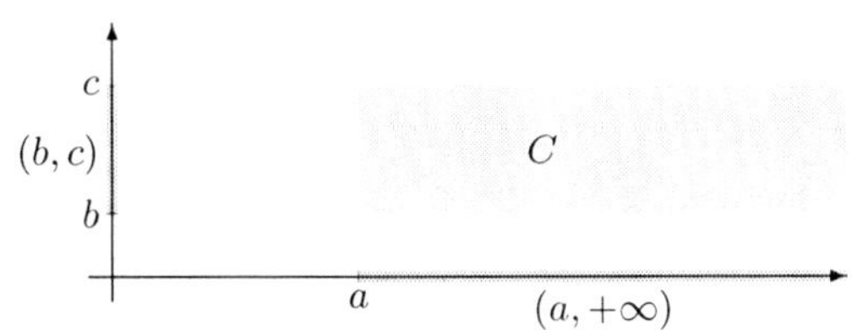

$$\mathrm{P}\left[x_1 > a, b < x_2 < c\right] = \int_a^{+\infty} \int_b^c p_{x_1 x_2 (a_1, a_2)}\, \mathrm{d}a_2\, \mathrm{d}a_1$$

or simply by making use of independence

$$\mathrm{P}\left[x_1 > a, b < x_2 < c\right] = \mathrm{P}\left[x_1 > a\right] \mathrm{P}\left[b < x_2 < c\right]$$

and using the results in Example 2.6 A provide

$$\mathrm{P}\left[x_1 > a, b < x_2 < c\right] = \mathrm{Q}\left(\frac{a - \mathtt{m}_1}{\sigma_1}\right) \left[\mathrm{Q}\left(\frac{b - \mathtt{m}_2}{\sigma_2}\right) - \mathrm{Q}\left(\frac{c - \mathtt{m}_2}{\sigma_2}\right)\right]$$

When two rvs are not statistically independent we are interested in expressing how the knowledge of the realization of one rv may influence our statistical information on the other rv. This can be formally expressed by using the *conditional PDF* of a rv x, given that the rv y is known to take a given value b, defined as

$$p_{x|y}(a|b) = \frac{p_{xy}(a, b)}{p_y(b)} \tag{2.176}$$

The conditional PDF enjoys all the properties seen in Section 2.6.1 for a PDF, such as the normalization condition

$$\int_{-\infty}^{+\infty} p_{x|y}(a|b)\, \mathrm{d}a = \mathrm{P}\left[x \in \mathbb{R} | y = b\right] = 1 \tag{2.177}$$

and allows calculation of probabilities of the rv x conditioned by the value of y as

$$\mathrm{P}\left[x \in C | y = b\right] = \int_C p_{x|y}(a|b)\, \mathrm{d}a\ . \tag{2.178}$$

We can observe that if x and y are independent rvs, then by factoring their joint PDF, the conditional PDF becomes

$$p_{x|y}(a|b) = \frac{p_x(a) p_y(b)}{p_y(b)} = p_x(a)$$

so that knowledge of the value taken by y does not change the probability of the values that x can take. This intuitively justifies the adjective "independent" for the two rvs.

We recall an important result relating unconditioned and conditional probabilities.

Theorem 2.12 (Total probability theorem) *Given two rvs x and y, the PDF of x can be derived by averaging the joint PDF of x and y with respect to the values taken by y*

$$p_x(a) = \begin{cases} \displaystyle\int_{-\infty}^{+\infty} p_{xy}(a, b)\, \mathrm{d}b & , \text{ for } y \text{ a continuous rv} \\ \displaystyle\sum_{b \in \mathcal{A}_y} p_{xy}(a, b) & , \text{ for } y \text{ a discrete rv} \end{cases} \tag{2.179}$$

Equivalently, by (2.176), we can also derive p_x from the conditional PDF $p_{x|y}$ and p_y as

$$p_x(a) = \begin{cases} \int_{-\infty}^{+\infty} p_{x|y}(a|b)p_y(b)\,\mathrm{d}b & , \text{ for } y \text{ a continuous rv} \\ \sum_{b\in\mathcal{A}_y} p_{x|y}(a|b)p_y(b) & , \text{ for } y \text{ a discrete rv} \end{cases} \tag{2.180}$$

The relationships (2.179) are also known as *marginal rules.*

The following important result allows to exchange the roles between the conditioning and conditioned rvs.

Proposition 2.13 (Bayes rule) *The conditional PDF of y given $x = a$ and the conditional PDF of x, given $y = b$, are related by*

$$p_{y|x}(b|a) = p_{x|y}(a|b)\frac{p_y(b)}{p_x(a)} \,. \tag{2.181}$$

Proof Multiply both sides of (2.181) by $p_x(a)$ to obtain

$$p_{y|x}(b|a)p_x(a) = p_{x|y}(a|b)p_y(b)$$

where the two sides represent two equivalent expressions for the joint PDF $p_{xy}(a, b)$. □

In a communication context, often the problem arises of finding the statistical description of a rv that is obtained as the sum of two or more statistically independent rvs, of known description. Such is the case for example of the input signal at the receiver, modeled as the superposition of a delayed and attenuated version of the transmitted useful signal, other interfering signal entering the channel, and the noise generated in the receiver circuitry. Or the total number of connection requests at a node coming from a certain number of distinct and independent terminals. To this aim it is useful to introduce the following result.

Proposition 2.14 *Let x, y be two independent rvs and $z = x + y$ be their sum. The PDF of z can be written as the convolution integral*

$$p_z(b) = \int_{-\infty}^{+\infty} p_y(b-a)p_x(a)\,\mathrm{d}a \tag{2.182}$$

and the conditional PDF of z given $x = a$ as

$$p_{z|x}(b|a) = p_y(b-a) \tag{2.183}$$

Proof We start from the conditional probability

$$\mathrm{P}\left[z \le b|x = a\right] = \mathrm{P}\left[x + y \le b|x = a\right] = \mathrm{P}\left[y \le b - a|x = a\right] = \mathrm{P}\left[y \le b - a\right] \tag{2.184}$$

where the last step is due to the independence of y from x. Then by differentiating the leftmost and rightmost sides of (2.184) with respect to a we get (2.183). From it, (2.182) can be derived through (2.180) in the total probability Theorem 2.12. □

The above results holds also for discrete rvs with the obvious substitutions of PMDs for PDFs and of the sum for the integral in (2.182).

Example 2.6 J Let x_1, x_2 be two independent, Gaussian rvs, as in Example 2.6 I. Their sum z has PDF given by

$$\begin{aligned}
p_z(b) &= \int_{-\infty}^{+\infty} p_{x_1}(b-a)p_{x_2}(a)\,\mathrm{d}a = \int_{-\infty}^{+\infty} \frac{1}{2\pi\sigma_1\sigma_2} e^{-\frac{(b-a-\mathtt{m}_1)^2}{2\sigma_1^2}-\frac{(a-\mathtt{m}_2)^2}{2\sigma_2^2}}\,\mathrm{d}a \\
&= \int_{-\infty}^{+\infty} \frac{1}{2\pi\sigma_1\sigma_2} e^{-\left(\frac{1}{2\sigma_1^2}+\frac{1}{2\sigma_2^2}\right)a^2+2a\left(\frac{b-\mathtt{m}_1}{2\sigma_1^2}+\frac{m_2}{2\sigma_2^2}\right)-\left(\frac{(b-\mathtt{m}_1)^2}{2\sigma_1^2}+\frac{m_2^2}{2\sigma_2^2}\right)}\,\mathrm{d}a \\
&= \frac{1}{\sqrt{2\pi(\sigma_1^2+\sigma_2^2)}} e^{-\frac{(b-\mathtt{m}_1-\mathtt{m}_2)^2}{2(\sigma_1^2+\sigma_2^2)}}
\end{aligned}$$

By letting $\mathtt{m}_z = \mathtt{m}_1 + \mathtt{m}_2$ and $\sigma_z^2 = \sigma_1^2 + \sigma_2^2$ we get

$$p_z(b) = \frac{1}{\sqrt{2\pi}\sigma_z} e^{-\frac{(b-\mathtt{m}_z)^2}{2\sigma_z^2}}$$

so that z is itself a Gaussian rv with mean $\mathtt{m}_z$ and variance σ_z^2.

Example 2.6 K Let x, y be independent Poisson rvs (see Example 2.6 F), with parameters β_x, β_y respectively. Their sum $z = x + y$ has the same alphabet $\mathcal{A} = \{0, 1, 2, \ldots\}$ and we can derive its PMD as

$$\begin{aligned}
p_z(b) &= \sum_{a=0}^{b} p_y(b-a)p_x(a) \\
&= \sum_{a=0}^{b} e^{-\beta_x-\beta_y} \frac{\beta_y^{b-a}}{(b-a)!}\frac{\beta_x^a}{a!} \\
&= e^{-(\beta_x+\beta_y)} \frac{1}{b!} \sum_{a=0}^{b} \binom{b}{a} \beta_y^{b-a}\beta_x^a \\
&= e^{-(\beta_x+\beta_y)} \frac{(\beta_x+\beta_y)}{b!}
\end{aligned}$$

so that z is itself a Poisson rv with parameter $\beta_z = \beta_x + \beta_y$

Example 2.6 L Consider the sum $y = x_1 + x_2 + \ldots + x_n$ where $\{x_i\}$ are n independent and identically distributed (iid) binary rvs, each with alphabet $\mathcal{A}_x = \{0, 1\}$ and PMD $p_x(1) = p$, $p_x(0) = (1-p)$. The PMD of y can be obtained through repeated application of (2.182) (with a sum replacing the integral) as

$$p_y(b) = \binom{n}{b} p^b(1-p)^{n-b} \quad , \quad b = 0, 1, \ldots, n \tag{2.185}$$

and is known as a *binomial* distribution with parameters n and p. Its mean and variance can be calculated by summing the mean and variances of the x_n as

$$\mathtt{m}_y = n\mathtt{m}_x = np \quad , \quad \sigma_y^2 = n\sigma_x^2 = n(p-p^2)$$

Example 2.6 M Consider an infinite sequence of rvs $\{y_n\}$ defined as

$$y_n = \sum_{m=1}^{n} x_{m,n}$$

where for each n the rvs $x_{1,n}, \ldots, x_{m,n}$ are n iid binary rvs, each with alphabet $\mathcal{A}_x = \{0, 1\}$ and $p_x(1) = \beta/n$. Then by the results in Example 2.6 L y_n has a binomial distribution with parameters n and β/n, while its mean is $\mathtt{m}_{y_n} = \beta$ irrespective of n. Here we are interested in the limit distribution of y_n as $n \to \infty$, which is given by

$$\begin{aligned}
\lim_{n\to\infty} p_{y_n}(a) &= \lim_{n\to\infty} \binom{n}{a} \left(\frac{\beta}{n}\right)^a \left(1 - \frac{\beta}{n}\right)^{n-a} \\
&= \lim_{n\to\infty} \frac{n!}{a!(n-a)!} \frac{\beta^a}{n^a} \left(1 - \frac{\beta}{n}\right)^{n-a} = \frac{\beta^a}{a!} \lim_{n\to\infty} \frac{\prod_{i=0}^{a-1}(n-i)}{n^a} \left(1 - \frac{\beta}{n}\right)^{n-a} \\
&= \frac{\beta^a}{a!} \left[\lim_{n\to\infty} \prod_{i=0}^{a-1} \left(1 - \frac{i}{n}\right)\right] \left[\lim_{n\to\infty} \left(1 - \frac{\beta}{n}\right)^n\right] \left[\lim_{n\to\infty} \left(1 - \frac{\beta}{n}\right)^{-a}\right] = \frac{\beta^a}{a!} e^{-\beta}
\end{aligned}$$

where we have used the well known fact $\lim_{n\to\infty}(1 + x/n)^n = e^x$. Thus, we see that the limit distribution of y_n is that of a Poisson rv with parameter β.

A significant interpretation of this result is obtained by considering n independent events $\{A_{mn}\}, m = 1, \ldots, n$, each with probability $p = \beta/n$. The number of such events that actually occur is a rv given by the sum of their indicator functions $\sum_m \chi\{A_m n\}$ (see Example 2.6 E). Based on the limit PMD above, one can say that for large n (and hence for small p) the number of occurrences is well approximated by a Poisson rv with parameter $\beta = np$. This result is thus often called the *law of rare events*.

Example 2.6 N Consider the sum $y = x_1 + x_2 + \ldots + x_n$ where $\{x_i\}$ are n iid exponential rvs with parameter β. Then the PDF of y has the expression

$$p_y(a) = \begin{cases} \dfrac{\beta^n a^{n-1}}{(n-1)!} e^{-\beta a} & , \ a \geq 0 \\ 0 & , \ a < 0 \end{cases} \quad . \tag{2.186}$$

which is called the Erlang distribution with index n and paramenter β. Such a rv has mean given by the sum of the means for x_i, i.e. $\mathtt{m}_y = n/\beta$.

2.6.4 Second order description of random vectors and Gaussian vectors

For a pair of rvs, with alphabet $\mathcal{A}_{xy}$, we can consider their second-order description that, besides their means $\mathtt{m}_x, \mathtt{m}_y$, statistical powers $\mathtt{M}_x, \mathtt{M}_y$ and variances σ_x^2, σ_y^2, includes also two joint statistical parameters,

- *correlation*

$$\mathtt{r}_{xy} = \mathrm{E}\left[xy\right] = \begin{cases} \displaystyle\int_{-\infty}^{+\infty}\int_{-\infty}^{+\infty} ab\, p_{xy}(a,b)\,\mathrm{d}b\,\mathrm{d}a & \text{, for continuous rvs} \\ \displaystyle\sum_{(a,b)\in\mathcal{A}_{xy}} ab\, p_{xy}(a,b) & \text{, for discrete rvs} \end{cases} \tag{2.187}$$

- *covariance*

$$\mathtt{k}_{xy} = \mathrm{E}\left[(x-\mathtt{m}_x)(y-\mathtt{m}_y)\right] = \begin{cases} \displaystyle\int_{-\infty}^{+\infty}\int_{-\infty}^{+\infty} (a-\mathtt{m}_x)(b-\mathtt{m}_y)\, p_{xy}(a,b)\,\mathrm{d}b\,\mathrm{d}a & \text{, for continuous rvs} \\ \displaystyle\sum_{(a,b)\in\mathcal{A}_{xy}} (a-\mathtt{m}_x)(b-\mathtt{m}_y)\, p_{xy}(a,b) & \text{, for discrete rvs} \end{cases} \tag{2.188}$$

From (2.188) it is easy to see that correlation and covariance of two rvs are related to their means as

$$\mathtt{r}_{xy} = \mathtt{k}_{xy} + \mathtt{m}_x\mathtt{m}_y \ . \tag{2.189}$$

Definition 2.17 *If* $\mathtt{r}_{xy} = \mathtt{m}_x\mathtt{m}_y$ *or equivalently* $\mathtt{k}_{xy} = 0$, *the two rvs are said* uncorrelated. *If* $\mathtt{r}_{xy} = 0$ *they are said* orthogonal. *Obviously, the two definitions coincide when at least one of the two means is zero.*

An important property regarding the second-order description of statistically independent rvs is the following

Proposition 2.15 *If two rvs x and y are statistically independent they are also uncorrelated.*

Proof By factoring the joint PDF $p_{xy}(a,b) = p_x(a)p_x(b)$ we can also factor the two-dimensional integral (2.187) into

$$\mathtt{r}_{xy} = \int_{-\infty}^{+\infty}\int_{-\infty}^{+\infty} ab\, p_x(a)p_y(b)\,\mathrm{d}b\,\mathrm{d}a = \int_{-\infty}^{+\infty} a\, p_x(a)\,\mathrm{d}a \int_{-\infty}^{+\infty} b\, p_y(b)\,\mathrm{d}b = \mathtt{m}_x\mathtt{m}_y \ .$$

□

Further insight Also related to the 2nd order description of a pair of variables is the concept of *equality in the mean square sense*

Definition 2.18 *Two rvs x, y are said to be* equal in the mean square sense *(we will write $x \overset{\text{m.s.}}{=} y$) if*

$$\mathrm{E}\left[|x-y|^2\right] = 0 \tag{2.190}$$

The above definition states that $e = x - y$, representing the difference between x and y, has null statistical power. This does not mean that e is identically zero, but rather that it is almost surely null as $\mathrm{P}[e=0]=1$. Hence in this case x and y take the same value with probability 1.

By extending the above description to rves of arbitrary length N we can obtain their 2nd order description, made of

- a *mean vector*, $\boldsymbol{m_x} = \mathrm{E}[\boldsymbol{x}] = [\mathtt{m}_1, \ldots, \mathtt{m}_N]^{\mathrm{T}}$
- a *correlation matrix* holding power and correlation values

$$\boldsymbol{r_x} = \mathrm{E}\left[\boldsymbol{x}\boldsymbol{x}^{\mathrm{T}}\right] = \begin{bmatrix} \mathtt{M}_{x1} & \mathtt{r}_{x1x2} & \cdots & \mathtt{r}_{x1xN} \\ \mathtt{r}_{x2x1} & \mathtt{M}_{x2} & \cdots & \mathtt{r}_{x2xN} \\ \vdots & \vdots & \ddots & \vdots \\ \mathtt{r}_{xNx1} & \mathtt{r}_{xNx2} & \cdots & \mathtt{M}_{xN} \end{bmatrix}$$

- a *covariance matrix* holding variance and covariance values

$$\boldsymbol{k_x} = \mathrm{E}\left[(\boldsymbol{x}-\boldsymbol{m_x})(\boldsymbol{x}-\boldsymbol{m_x})^{\mathrm{T}}\right] = \begin{bmatrix} \sigma^2_{x1} & \mathtt{k}_{x1x2} & \cdots & \mathtt{k}_{x1xN} \\ \mathtt{k}_{x2x1} & \sigma^2_{x2} & \cdots & \mathtt{k}_{x2xN} \\ \vdots & \vdots & \ddots & \vdots \\ \mathtt{k}_{xNx1} & \mathtt{k}_{xNx2} & \cdots & \sigma^2_{xN} \end{bmatrix}$$

Although quite informative, the 2nd order description is not a complete statistical description for a vector, which would be given by the PDF. An exceptional case to this respect is given by *Gaussian vectors*, where the PDF can always be written in terms of the mean vector and covariance matrix as

$$p_{\boldsymbol{x}}(\boldsymbol{a}) = \frac{1}{\sqrt{2^N \pi^N \det \boldsymbol{k_x}}} e^{-\frac{1}{2}(\boldsymbol{a}-\boldsymbol{m_x})^{\mathrm{T}} \boldsymbol{k_x}^{-1} (\boldsymbol{a}-\boldsymbol{m_x})} \tag{2.191}$$

As an example, for $N=2$ we get

$$p_{x_1x_2}(a_1, a_2) = \frac{1}{2\pi\sqrt{\sigma_1^2\sigma_2^2 - \mathtt{k}^2}} e^{-\frac{(a_1-\mathtt{m}_1)^2\sigma_2^2 + (a_2-\mathtt{m}_2)^2\sigma_1^2 - 2(a_1-\mathtt{m}_1)(a_2-\mathtt{m}_2)\mathtt{k}}{2\sigma_1^2\sigma_2^2 - 2\mathtt{k}^2}} \tag{2.192}$$

where for compactness we have written $\mathtt{m}_1 = \mathtt{m}_{x_1}$, $\mathtt{m}_2 = \mathtt{m}_{x_2}$, $\sigma_1^2 = \sigma_{x_1}^2$, $\sigma_2^2 = \sigma_{x_2}^2$, $\mathtt{k} = \mathtt{k}_{x_1x_2}$.

Beside being an excellent model for noise phenomena, as will be seen in the next chapters, Gaussian vectors are particularly welcome for both their mathematical tractability and the following properties.

Proposition 2.16 *Any linear transformation $\boldsymbol{y} = \boldsymbol{A}\boldsymbol{x} + \boldsymbol{b}$ of a Gaussian vector $\boldsymbol{x}$ is itself a Gaussian vector with mean $\boldsymbol{m_y} = \boldsymbol{A}\boldsymbol{m_x} + \boldsymbol{b}$ and covariance matrix $\boldsymbol{k_y} = \boldsymbol{A}\boldsymbol{k_x}\boldsymbol{A}^{\mathrm{T}}$.*

Observe that in the above proposition, if $\boldsymbol{x}$ and $\boldsymbol{y}$ are $N-$ and $M-$dimensional vectors, respectively, then $\boldsymbol{A}$ is a $M \times N$ real matrix and $\boldsymbol{b}$ a column of $\mathbb{R}^N$.

Proposition 2.17 *If a Gaussian vector has uncorrelated rvs (that is $\mathtt{k}_{x_ix_j} = 0$, for all $i \neq j$) then its rvs are also statistically independent.*

Proof In the case of a rve with uncorrelated rvs the covariance matrix is diagonal and

$$\det \boldsymbol{k_x} = \prod_{i=1}^{N} \sigma_i^2 \quad , \quad \boldsymbol{k_x}^{-1} = \begin{bmatrix} 1/\sigma_{x_1}^2 & 0 & \cdots & 0 \\ 0 & 1/\sigma_{x_2}^2 & \ddots & 0 \\ \vdots & \ddots & \ddots & \vdots \\ 0 & 0 & \cdots & 1/\sigma_N^2 \end{bmatrix}$$

so that we can write (2.191) as

$$\begin{aligned} p_{\boldsymbol{x}}(\boldsymbol{a}) &= \frac{1}{\sqrt{2^N \pi^N \sigma_{x_1}^2 \cdots \sigma_{x_N}^2}} e^{-\frac{(a_1 - \mathtt{m}_{x_1})^2}{2\sigma_{x_1}^2} - \cdots - \frac{(a_N - \mathtt{m}_{x_N})^2}{2\sigma_{x_N}^2}} \\ &= \prod_{i=1}^{N} \frac{1}{\sqrt{2\pi}\sigma_{x_i}} e^{-\frac{(a_i - \mathtt{m}_{x_i})^2}{2\sigma_{x_i}^2}} \end{aligned}$$

and thus obtain the factorization (2.174) □

The above result represents the converse of Proposition 2.15, and it only holds for Gaussian vectors.

2.6.5 Complex-valued random variables

A *complex-valued rv* is a function

$$x : \Omega \mapsto \mathbb{C} \, . \tag{2.193}$$

Actually, it is modeled as a pair of real-valued rvs, $x_{\mathrm{I}} = \Re[x]$ and $x_{\mathrm{Q}} = \Im[x]$, and thus a real-valued 2-dimensional rve. The PDF of x is defined as the joint PDF of x_{I} and x_{Q}, which for convenience we write as

$$p_x(a) = p_{x_{\mathrm{I}} x_{\mathrm{Q}}}(a_{\mathrm{I}}, a_{\mathrm{Q}}) \, . \tag{2.194}$$

Example 2.6 O Suppose we want to find the probability $\mathrm{P}\left[|x| < 1\right]$ with x a continuous complex-valued rv. Then we must solve the integral

$$\mathrm{P}\left[|x| < 1\right] = \int_{|a|<1} p_x(a) \, da \, . \tag{2.195}$$

By writing the subset as

$$\{|a| < 1\} = \left\{a_{\mathrm{I}}^2 + a_{\mathrm{Q}}^2 < 1\right\} \, , \tag{2.196}$$

we can calculate the probability as the double integral

$$\mathrm{P}\left[|x| < 1\right] = \int_{-1}^{1} \int_{-\sqrt{1-a_{\mathrm{Q}}^2}}^{\sqrt{1-a_{\mathrm{Q}}^2}} p_{x_{\mathrm{I}} x_{\mathrm{Q}}}(a_{\mathrm{I}}, a_{\mathrm{Q}}) \, da_{\mathrm{I}} \, da_{\mathrm{Q}} \, . \tag{2.197}$$

Alternatively we can resort to the PDF of $|x|$

$$\mathrm{P}\left[|x| < 1\right] = \int_{0}^{1} p_{|x|}(\rho) \, d\rho \, . \tag{2.198}$$

Example 2.6 P To calculate the expectation of a continuous complex-valued rv we have

$$\mathrm{E}[x] = \int_{\mathbb{C}} a p_x(a)\, da = \int_{-\infty}^{+\infty} \int_{-\infty}^{+\infty} (a_\mathrm{I} + j a_\mathrm{Q}) p_{x_\mathrm{I} x_\mathrm{Q}}(a_\mathrm{I}, a_\mathrm{Q})\, da_\mathrm{I}\, da_\mathrm{Q} \tag{2.199}$$

and by applying the marginal rule (2.179)

$$\mathrm{E}[x] = \int_{-\infty}^{+\infty} a_\mathrm{I} p_{x_I}(a_\mathrm{I})\, da_\mathrm{I} + j \int_{-\infty}^{+\infty} a_\mathrm{Q} p_{x_Q}(a_\mathrm{Q})\, da_\mathrm{Q} = \mathrm{E}[x_\mathrm{I}] + j\,\mathrm{E}[x_\mathrm{Q}] \tag{2.200}$$

Observe that the result in Example 2.6 P could also be obtained by applying the linearity of expectation

$$\mathtt{m}_x = \mathrm{E}[x_\mathrm{I} + j x_\mathrm{Q}] = \mathtt{m}_{x_\mathrm{I}} + j \mathtt{m}_{x_\mathrm{Q}} \,. \tag{2.201}$$

The second-order description of a complex-valued rv includes means, statistical powers, variances, correlation and covariance of the pair $x_\mathrm{I}, x_\mathrm{Q}$. However, usually we are interested in the statistical power and variance of the complex rv x given by

$$\mathtt{M}_x = \mathrm{E}\left[|x|^2\right] \quad , \quad \sigma_x^2 = \mathrm{E}\left[|x - \mathtt{m}_x|^2\right] \tag{2.202}$$

Observe that both parameters are real and nonnegative. It is easy to derive the relations

$$\mathtt{M}_x = \mathrm{E}\left[x_\mathrm{I}^2 + x_\mathrm{Q}^2\right] = \mathtt{M}_{x_\mathrm{I}} + \mathtt{M}_{x_\mathrm{Q}} \tag{2.203}$$

$$\sigma_x^2 = \mathrm{E}\left[(x_\mathrm{I} - \mathtt{m}_{x_\mathrm{I}})^2 + (x_\mathrm{Q} - \mathtt{m}_{x_\mathrm{Q}})^2\right] = \sigma_{x_\mathrm{I}}^2 + \sigma_{x_\mathrm{Q}}^2 \tag{2.204}$$

$$\mathtt{M}_x = \sigma_x^2 + |\mathtt{m}_x|^2 \,. \tag{2.205}$$

Example 2.6 Q A complex-valued rv x is called *complex Gaussian* if its components x_I and x_Q form a Gaussian vector. Hence, its PDF is of the type (2.192). In particular we are interested in complex Gaussian rvs that have uncorrelated components ($\mathtt{k}_{x_\mathrm{I} x_\mathrm{Q}} = 0$) with equal variance $\sigma_{x_\mathrm{I}}^2 = \sigma_{x_\mathrm{Q}}^2 = \frac{1}{2}\sigma_x^2$. Hence, the PDF can be written as

$$\begin{aligned} p_x(a) &= \frac{1}{\sqrt{2\pi\sigma_{x_\mathrm{I}}^2}} e^{-\frac{(a_\mathrm{I} - \mathtt{m}_{x_\mathrm{I}})^2}{2\sigma_{x_\mathrm{I}}^2}} \frac{1}{\sqrt{2\pi\sigma_{x_\mathrm{Q}}^2}} e^{-\frac{(a_\mathrm{Q} - \mathtt{m}_{x_\mathrm{Q}})^2}{2\sigma_{x_\mathrm{Q}}^2}} \\ &= \frac{1}{\pi\sigma_x^2} \exp\left(-\frac{|a - \mathtt{m}_x|^2}{\sigma_x^2}\right) \quad , \quad a \in \mathbb{C} \,. \end{aligned} \tag{2.206}$$

A rv with above PDF is called *circularly symmetric* since p_x exhibits a circular symmetry in the complex plane around the point $\mathtt{m}_x$. We indicate that a rv x is circularly symmetric complex Gaussian by writing $x \sim \mathcal{CN}(\mathtt{m}_x, \sigma_x^2)$.

In this case, if $\mathtt{m}_x = 0$, then $|x|$ is Rayleigh distributed with PDF

$$p_{|x|}(\rho) = \frac{\rho}{\sigma_x^2} e^{-\left(\frac{\rho}{\sigma_x}\right)^2} 1(\rho) \,. \tag{2.207}$$

As an example we calculate the probability $\mathrm{P}[|x| > R]$ with $R > 0$, for $x \sim \mathcal{CN}(0, \sigma_x^2)$. We get

$$\mathrm{P}[|x| > R] = \int_{\rho > R} p_{|x|}(\rho)\, \mathrm{d}\rho = e^{-\frac{R^2}{\sigma_x^2}} . \tag{2.208}$$

Analogously, we can define the second-order description of *a pair of complex-valued rvs* x and y by giving their correlation

$$\mathtt{r}_{xy} = \mathrm{E}[xy^*] = (\mathtt{r}_{x_\mathrm{I}y_\mathrm{I}} + \mathtt{r}_{x_\mathrm{Q}y_\mathrm{Q}}) + j(\mathtt{r}_{x_\mathrm{Q}y_\mathrm{I}} - \mathtt{r}_{x_\mathrm{I}y_\mathrm{Q}}) , \tag{2.209}$$

and covariance

$$\mathtt{k}_{xy} = \mathrm{E}[(x - \mathtt{m}_x)(y - \mathtt{m}_y)^*] = (\mathtt{k}_{x_\mathrm{I}y_\mathrm{I}} + \mathtt{k}_{x_\mathrm{Q}y_\mathrm{Q}}) + j(\mathtt{k}_{x_\mathrm{Q}y_\mathrm{I}} - \mathtt{k}_{x_\mathrm{I}y_\mathrm{Q}})$$

which are both complex-valued, in general. Moreover,

$$\mathtt{r}_{xy} = \mathtt{k}_{xy} + \mathtt{m}_x \mathtt{m}_y^* . \tag{2.210}$$

In a similar way the description of a N-dimensional complex-valued rve $\boldsymbol{x} = [x_1, \ldots, x_N]^{\mathrm{T}}$ can be derived from that of the corresponding $2N$-dimensional real-valued rve

$$\boldsymbol{x}' = [x_{1,\mathrm{I}}, x_{1,\mathrm{Q}}, \ldots, x_{N,\mathrm{I}}, x_{N,\mathrm{Q}}]^{\mathrm{T}}$$

2.7 RANDOM PROCESSES

Random signals, more properly called random processes, are a key topic in the study of communication systems, where the amount of information is related to the degree of unpredictability of the transmitted signals. We therefore devote this section to the definition and analysis of random processes. In this context the signals defined according to (2.1) or (2.11) will be called *deterministic* as the counterpart of random processes.

2.7.1 Definition and properties

Definition 2.19 *A continuous time real-valued* rp *is a function of two variables*

$$x : \mathbb{R} \times \Omega \mapsto \mathbb{R} . \tag{2.211}$$

where Ω is the sample space of some probability space.

It should thus be denoted as $x(t, \omega)$, $t \in \mathbb{R}$, $\omega \in \Omega$ but it is customary to drop the dependence on ω in the notation and simply write $x(t)$, $t \in \mathbb{R}$. Analogously we introduce discrete time random processes.

Definition 2.20 *A discrete time real-valued rp is a function of two variables*

$$x : \mathbb{Z} \times \Omega \mapsto \mathbb{R} . \tag{2.212}$$

Again, we point out that it is customary to omit the ω dependence and write $x(nT)$ or x_n, $n \in \mathbb{Z}$.

A rp can be interpreted as a collection of deterministic signals, each one identified by a different point ω in the probability space Ω, as illustrated in Fig. 2.35. Each of the deterministic signals composing x is called a *realization* of x. From another point of view a rp can be interpreted as a collection of rvs, each one identified by a different instant $t \in \mathbb{R}$, or index $n \in \mathbb{Z}$.

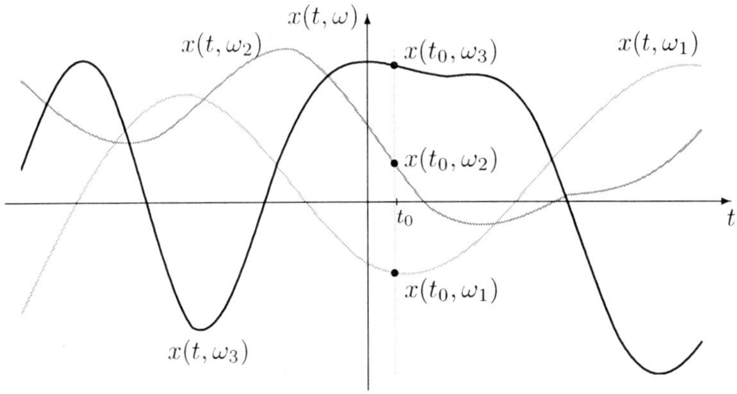

Figure 2.35 Three realizations of a continuous time real-valued random signal.

The *first-order* statistical description of a rp is given by the first order PDF

$$p_x(a;t) = p_{x(t)}(a) \quad , \quad t \in \mathbb{R} \quad , \quad a \in \mathbb{R}$$

Definition 2.21 *A continuous time* complex-valued rp *is a function*

$$x \,:\, \mathbb{R} \times \Omega \mapsto \mathbb{C}\,. \tag{2.213}$$

As we did for complex rv, we can view it as a pair of real-valued rps, its real and imaginary part

$$x_{\mathrm{I}}(t) = \Re x(t) \quad , \quad x_{\mathrm{Q}}(t) = \Im x(t) \quad , \quad x(t) = x_{\mathrm{I}}(t) + jx_{\mathrm{Q}}(t)$$

and its first order statistical description is given by the function

$$p_x(a;t) = p_{x(t)}(a) = p_{x_{\mathrm{I}}(t),x_{\mathrm{Q}}(t)}(a_I, a_Q) \quad , \quad t \in \mathbb{R} \quad , \quad a \in \mathbb{C} \quad , \quad a_I, a_Q \in \mathbb{R}\,. \tag{2.214}$$

i.e. by the joint description of $x_{\mathrm{I}}(t)$ and $x_{\mathrm{Q}}(t)$. From now on, for the sake of generality, all rps we deal with are assumed to be complex-valued unless otherwise stated.

Definition 2.22 *The* mean of a rp x *is a deterministic signal, that at each time instant t takes the value of the statistical mean of the rv* $x(t)$

$$\mathtt{m}_x \,:\, \mathbb{R} \mapsto \mathbb{C} \quad , \quad \mathtt{m}_x(t) = \mathrm{E}\left[x(t)\right]\,. \tag{2.215}$$

Analogously, we have the following definitions.

Definition 2.23 *The* statistical power of a rp x *is a deterministic signal, that at each time instant t takes the value of the statistical power of the rv* $x(t)$

$$\mathtt{M}_x : \mathbb{R} \mapsto \mathbb{R} \ , \quad \mathtt{M}_x(t) = \mathrm{E}\left[|x(t)|^2\right] \ , \tag{2.216}$$

and it turns out that $\mathtt{M}_x(t)$ is a real-valued nonnegative signal.

It is also interesting to consider some joint statistical descriptions of pairs of rvs taken from the same rp at different instants. Thus we define

Definition 2.24 *The* autocorrelation function *of a rp x is a function of two real variables, the reference time t and the delay time (lag) τ. It takes the value of the correlation between two rvs in the process, one taken at the instant t, and the other at the instant t $- \tau$*

$$\mathtt{r}_x(t,\tau) = \mathrm{E}\left[x(t)x^*(t-\tau)\right] \ . \tag{2.217}$$

It is easy to notice that the autocorrelation function, evaluated at $\tau = 0$ yields the statistical power as

$$\mathtt{r}_x(t,0) = \mathrm{E}\left[x(t)x^*(t)\right] = \mathrm{E}\left[|x(t)|^2\right] = \mathtt{M}_x(t) \tag{2.218}$$

An important class is that of *Gaussian rps.*

Definition 2.25 *A real-valued rp is Gaussian if all the rves that can be built by taking an arbitrary number of variables in the rp are Gaussian vectors in the sense of (2.191).*

In particular all the rvs of a Gaussian process are Gaussian, $x(t) \sim \mathcal{N}(\mathtt{m}_x(t), \sigma_x^2(t))$, so that, for example, we can calculate, starting from the mean $\mathtt{m}_x(t)$ and power $\mathtt{M}_x(t)$, the probability

$$\mathrm{P}\left[b < x(t) < c\right] = Q\left(\frac{b - \mathtt{m}_x(t)}{\sigma_x(t)}\right) - Q\left(\frac{c - \mathtt{m}_x(t)}{\sigma_x(t)}\right) \ , \tag{2.219}$$

where $\sigma_x(t) = \sqrt{\mathtt{M}_x(t) - \mathtt{m}_x^2(t)}$ and Q is defined in (2.148).

Definition 2.26 *A complex-valued rp $x(t)$ is said to be circularly symmetric Gaussian if $x(t)$ is a circularly symmetric Gaussian rv for any t, $x(t) \sim \mathcal{CN}(\mathtt{m}_x(t), \sigma_x^2(t))$.*

2.7.2 Point and Poisson processes

A special class of continuous time rps is given by *point processes.* Such processes can be written as sequences of Dirac impulses, centered at random times, e. g.,

$$x'(t) = \sum_{n=-\infty}^{+\infty} \delta(t - t_n) \tag{2.220}$$

and can effectively model the distribution of discrete "events" in time. The elements in the random sequence $\{t_n\}$ are called *arrival times*, and are said to be *ordered* if $t_n \leq t_{n+1}$ for all $n \in \mathbb{Z}$. The time elapsing between two consecutive arrivals is called *interarrival time*, and is itself a random variable. Once the arrival times are ordered, we can then define the nth interarrival time as $\tau_n = t_n - t_{n-1}$.

Closely related to a point process is its *counting process*[2]

$$x(t) = \int_{u_0^+}^{t^+} x'(u)\, du \qquad (2.221)$$

which is a continuous-time rp with the following properties

1. it only takes integer values, i.e. $x(t) \in \mathbb{Z}$
2. any realization of $x(t)$ is a non decreasing function of t

Conversely, the point process $x'(t)$ can be thought as the time-derivative of $x(t)$, as any of its realizations is the derivative of the corresponding realization of $x(t)$. It is customary to set $u_0 = 0$ in (2.221), so that the counting process has $x(0) = 0$. A possible realization of a point process and of the associated counting process is shown in Fig. 2.36.

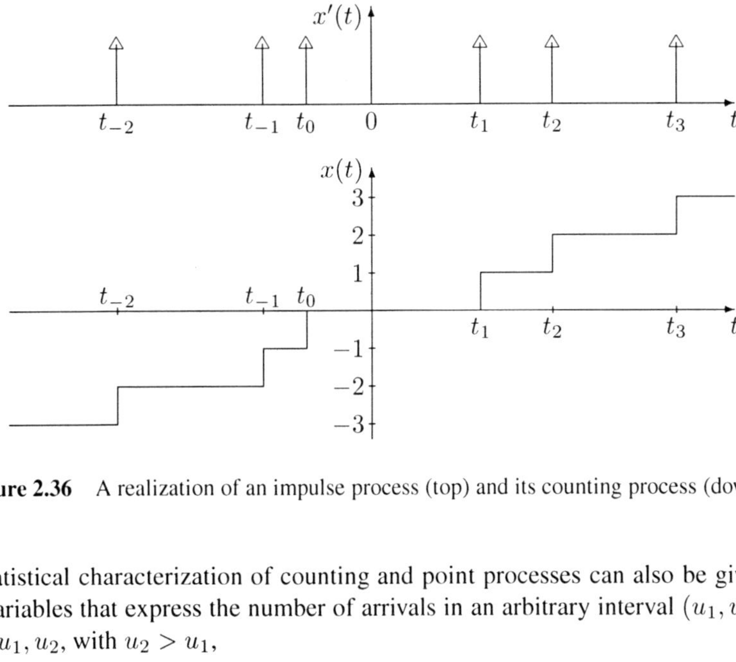

Figure 2.36 A realization of an impulse process (top) and its counting process (down).

The statistical characterization of counting and point processes can also be given via the random variables that express the number of arrivals in an arbitrary interval $(u_1, u_2]$, for any choice of u_1, u_2, with $u_2 > u_1$,

$$N_x(u_1, u_2) = x(u_2) - x(u_1) = \int_{u_1}^{u_2} x'(u)\, du\,. \qquad (2.222)$$

[2]The compact notation

$$\int_{a^+}^{b^+} y(u)\, du$$

indicates the limit

$$\lim_{\varepsilon \to 0^+} \int_{a+\varepsilon}^{b+\varepsilon} x(u)\, du$$

and is used for clarity when the signal $x(u)$ contains impulses centered in b (whose area must be counted in the integral) and/or in a (whose area should not).

By differentiating the mean $\mathtt{m}_x(t)$ of the counting process we obtain the *arrival rate* which represents the expected number of arrivals per unit time at time t

$$\lambda(t) = \frac{d}{dt}\mathtt{m}_x(t) = \lim_{\Delta t \to 0} \frac{\mathrm{E}\left[N_x(t, t+\Delta t)\right]}{\Delta t} . \tag{2.223}$$

The arrival rate can also be seen as the mean of the point process, as can be shown by formally exchanging the order of differentiation and expectation operators[3]

$$\mathrm{E}\left[x'(t)\right] = \mathrm{E}\left[\frac{d}{dt}x(t)\right] = \frac{d}{dt}\,\mathrm{E}\left[x(t)\right] = \frac{d}{dt}\mathtt{m}_x(t) = \lambda(t) . \tag{2.224}$$

If the arrival rate is constant, with $\lambda(t) = \lambda$, the point and counting processes are said to be *homogeneous*.

A *Poisson process* can be defined as a counting process in which the following conditions are met

1. The number of arrivals in n *disjoint intervals* $N_x(u_1, u_2), \ldots, N_x(u_{2n-1}, u_{2n})$, with $u_1 \leq u_2 \leq \ldots \leq u_{2n}$, are *independent* rvs
2. The number of arrivals in the interval $(u_1, u_2]$ has a Poisson distributon (see Example 2.6 K)

$$\mathrm{P}\left[N_x(u_1, u_2) = k\right] = e^{-\Lambda}\frac{\Lambda^k}{k!} \;, \quad \text{with } \Lambda = \int_{u_1}^{u_2} \lambda(u)\, du \;, \quad k = 0, 1, 2, \ldots \tag{2.225}$$

The associated point process is called a *Poisson point process*. The statistical description of a Poisson process is therefore completely determined by its arrival rate $\lambda(t)$, $t \in \mathbb{R}$. Similarly, a homogenous Poisson process is fully specified by its constant rate λ.

Example 2.7 A Consider a homogeneous Poisson process $x(t)$, with arrival rate λ. Suppose we want to find the probability that the second ordered arrival after time $u_0 = 0$ is later than time $T = 1/\lambda$. This is equivalent to require that $N_x(0, T] < 2$. Since $N_x(0, T]$ is Poisson distributed with parameter $\lambda T = 1$ we get

$$\begin{aligned}\mathrm{P}\left[t_2 > T\right] = \mathrm{P}\left[N_x(0, T] < 2\right] &= \sum_{k=0}^{1} e^{-\lambda T}\frac{(\lambda T)^k}{k!} \\ &= e^{-\lambda T}(1 + \lambda T) = \frac{2}{e} \simeq 0.74\end{aligned}$$

Now suppose we want to find out the probability that there are *exactly* 2 arrivals in $(0, T]$ *and exactly* 3 arrivals in $(0, 2T]$, or equivalently $N_x(0, T) = 2$ and $N_x(0, 2T) = 3$. If we equivalently express the event in terms of arrivals in disjoint intervals:

$$\mathrm{P}\left[N_x(0, T) = 2, N_x(0, 2T) = 3\right] = \mathrm{P}\left[N_x(0, T) = 2, N_x(T, 2T) = 1\right]$$

[3]The exchanging of differentiation (with respect to time t) and expectation is not allowed for random processes in general. A thorough investigation on necessary and/or sufficient conditions for it to hold is beyond the scope of this book. Suffices to say here that the result in (2.224) holds for all point processes of interest.

we can make use of the independence between the two rvs and obtain

$$\begin{aligned} \mathrm{P}\left[N_x(0,T)=2, N_x(0,2T)=3\right] &= \mathrm{P}\left[N_x(0,T)=2\right]\mathrm{P}\left[N_x(T,2T)=1\right] \\ &= \frac{1}{2}(\lambda T)^2\, e^{-\lambda T}\, \lambda T\, e^{-\lambda T} \\ &= \frac{1}{2}e^{-2} \simeq 6.77\cdot 10^{-2} \end{aligned}$$

Poisson processes enjoy a number of interesting properties that besides allowing for mathematical tractability, make them a very suitable model for the distribution of a large number of independent 'events'.

Proposition 2.18 (Superposition property) *If $x(t)$ and $y(t)$ are* independent Poisson processes *with arrival rates $\lambda_x(t), \lambda_y(t)$, respectively, their sum $z(t) = x(t) + y(t)$ is itself a Poisson process with arrival rate $\lambda_z(t) = \lambda_x(t) + \lambda_y(t)$*

Proof First observe that the arrival times for $z(t)$ are the union of the arrival times of $x(t)$, and $y(t)$, and hence we can write for any interval (u_1, u_2)

$$N_z(u_1,u_2) = N_x(u_1,u_2) + N_y(u_1,u_2)$$

We check that condition 1 is met, by considering that if the $2n$ rvs $N_x(u_1,u_2), N_y(u_1,u_2), \ldots, N_x(u_{2n-1},u_{2n}), N_y(u_{2n-1},u_{2n})$ are independent and so are the n rvs $N_z(u_1,u_2), \ldots, N_z(u_{2n-1},u_{2n})$, obtained by summing the above rvs in pairs.

As regards condition 2, we know from Example 2.6 K that the sum of the rvs $N_x(u_1,u_2)$ and $N_y(u_1,u_2)$, having a Poisson distribution with mean

$$\Lambda_x = \int_{u_1}^{u_2} \lambda_x(u)\,\mathrm{d}u \quad , \quad \Lambda_y = \int_{u_1}^{u_2} \lambda_y(u)\,\mathrm{d}u$$

respectively, has itself a Poisson distribution with mean

$$\int_{u_1}^{u_2} \lambda_z(u)\,\mathrm{d}u = \Lambda_z = \Lambda_x + \Lambda_y = \int_{u_1}^{u_2} (\lambda_x(u) + \lambda_y(u))\,\mathrm{d}t = \int_{u_1}^{u_2} \lambda_z(u)\,\mathrm{d}u$$

thus proving that the condition is met. □

The following result gives a further strong motivation for the study of Poisson processes. It represents the rp counterpart of Example 2.6 M, and can be proved similarly, albeit with more formal technicalities.

Proposition 2.19 (Law of rare events) *Consider a sequence of point processes $\{x_n(t)\}$, $n = 1, 2, \ldots$, where each $x_n(t) = \sum_{m=1}^{n} y_{mn}(t)$ is the sum of n independent point processes, each with arrival rate $\lambda(t)/n$, so that the arrival rate for $x_n(t)$ is $\lambda(t)$, independent of n.*

Then, as $n \to \infty$, the statistical description of $x_n(t)$ converges to that of a Poisson process with rate $\lambda(t)$.

We remark the generality of the above result, which does not set forth any hypothesis[4] on the arrival distribution in the processes $y_{mn}(t)$. Further generalization can be made by allowing

[4]Actually, further hypotheses are needed to rule out pathological cases, such as processes with likely multiple concurrent arrivals, etc. We deliberately such hypotheses, for the sake of compactness and clarity in the statement.

different arrival rates for $y_{mn}(t), m = 1, \ldots, n$, under hypothesis on their vanishing as $n \to \infty$. Thus, the law of rare events justifies the use of Poisson rps in modeling the aggregate statistics of a large number of independent phenomena, such as the arrival of packets from multiple sources at a common gateway.

Proposition 2.20 (Thinning property) *If $x(t)$ is a Poisson process with arrival rate $\lambda_x(t)$, the point process*

$$y'(t) = \sum_n \chi_n \delta(t - t_n) \tag{2.226}$$

where $\{t_n\}$ are the arrival times of $x(t)$, and $\{\chi_n\}$ is a sequence of iid binary rvs with alphabet $\mathcal{A} = \{0, 1\}$, independent of the rp $x(t)$ with PMD

$$p_\chi(1) = q \quad , \quad p_\chi(0) = 1 - q \tag{2.227}$$

is a Poisson point process with arrival rate $\lambda_y(t) = q\lambda_x(t)$

Proof Observe that the arrival times for $y(t)$ are obtained from those of $x(t)$ by either keeping (with probabiliy q) or dropping (with probability $1 - q$) each arrival time independently.

Hence, condition 1 is easily met, whereas to derive the distribution of $N_y(u_1, u_2)$ we apply the total probability theorem 2.12 by writing

$$\mathrm{P}\left[N_y(u_1, u_2) = k\right] = \sum_{n=0}^{+\infty} \mathrm{P}\left[N_y(u_1, u_2) = k | N_x(u_1, u_2) = n\right] \mathrm{P}\left[N_x(u_1, u_2) = n\right]$$

Since the probability that exactly k out of the n arrival times for $x(t)$ are kept in $y(t)$ can be written as

$$\mathrm{P}\left[N_y(u_1, u_2) = k | N_x(u_1, u_2) = n\right] = \begin{cases} \binom{n}{k} q^k (1 - q)^{n-k} & , \quad n \geq k \\ 0 & , \quad n = 0, \ldots, k - 1 \end{cases}$$

we get

$$\begin{aligned} \mathrm{P}\left[N_y(u_1, u_2) = k\right] &= \sum_{n=k}^{+\infty} \binom{n}{k} q^k (1 - q)^{n-k} e^{-\Lambda_x} \frac{\Lambda_x^n}{n!} \\ &= e^{-\Lambda_x} q^k \frac{1}{k!} \sum_{n=k}^{+\infty} \frac{\Lambda_x^n (1 - q)^{n-k}}{(n - k)!} \\ &= e^{-\Lambda_x} q^k \frac{\Lambda_x^k}{k!} e^{\Lambda_x(1-q)} = e^{-q\Lambda_x} \frac{(q\Lambda_x)^k}{k!} \end{aligned}$$

where in the last line we have used the exponential series $e^z = \sum_{h=0}^{+\infty} z^h/h!$. Thus, $N_y(u_1, u_2)$ is a Poisson rv with mean

$$\int_{u_1}^{u_2} \lambda_y(u)\, du = \Lambda_y = q\Lambda_x = \int_{u_1}^{u_2} q\lambda_x(u)\, du$$

□

The above property can be generalized into the following result, which we state without proof.

Proposition 2.21 (Splitting property) *If $x(t)$ is a Poisson process with arrival rate $\lambda_x(t)$, and $\{\chi_n\}$ is a sequence of iid binary rvs with alphabet $\mathcal{A} = \{1, 2, \ldots, M\}$ and PMD $p_\chi(m)$, independent of the rp $x(t)$ then the M point processes*

$$y'_m(t) = \sum_{n\,:\,\chi_n = m} \delta(t - t_n) \quad , \quad m = 1, 2, \ldots, M \tag{2.228}$$

are independent Poisson point processes with arrival rates $\lambda_{y_m}(t) = p_\chi(m)\lambda_x(t)$.

An interesting property holds for the interarrival times of *homogeneous* Poisson processes.

Proposition 2.22 (Interarrival times) *The interarrival times $\{\tau_n\}$ of a homogeneous Poisson process $x(t)$ with arrival rate λ are iid rvs with exponential PDF (see Example 2.6 D)*

$$p_\tau(a) = \lambda e^{-\lambda a} 1(a) \tag{2.229}$$

Proof We start by expressing the conditional probability

$$\begin{aligned} &\mathrm{P}\left[\tau_n \le a | t_{n-1} = b_{n-1}, t_{n-2} = b_{n-2}, \ldots\right] \\ &= \begin{cases} \mathrm{P}\left[N_x\left(b_{n-1}, b_{n-1} + a\right) > 0 | t_{n-1} = b_{n-1}, t_{n-2} = b_{n-2}, \ldots\right] & , \ a \ge 0 \\ 0 & , \ a < 0 \end{cases} \end{aligned}$$

As $N_x\left(b_{n-1}, b_{n-1} + a\right)$ is independent of what happened in the interval $(-\infty, b_{n-1}]$, we can remove the conditioning on the right-hand side, obtaining

$$\begin{aligned} \mathrm{P}\left[\tau_n \le a | t_{n-1} = b_{n-1}, t_{n-2} = b_{n-2}, \ldots\right] &= \begin{cases} 1 - \mathrm{P}\left[N_x\left(b_{n-1}, b_{n-1} + a\right) = 0\right] & , \ a \ge 0 \\ 0 & , \ a < 0 \end{cases} \\ &= \begin{cases} 1 - e^{-\lambda a} & , \ a \ge 0 \\ 0 & , \ a < 0 \end{cases} \end{aligned}$$

Note that the last expression does not depend on the $\{b_i\}$, so that this is also the value for the unconditional probability $\mathrm{P}\left[\tau_n \le a\right]$, nor does it depend on n, so that all interarrival times are identically distributed. Then we derive the common PDF of τ_n by differentiating with respect to a

$$p_\tau(a) = \frac{d}{da} \mathrm{P}\left[\tau_n \le a\right] = \lambda e^{-\lambda a} 1(a)$$

□

Observe that in the proof for (2.229) we have not used the fact that t_{n-1} is an arrival time, so the same distribution holds for the random time that elapses before the next arrival, regardless of the initial instant t_{n-1}[5]. The importance of this result is best understood by stating that, in a homogeneous Poisson process, interarrival times are *memoryless rvs*, in the sense that the residual time before the next arrival always has the same statistics, no matter how long has

[5]The same is true if the initial instant t_{n-1} is random, providing it only depends on previous events.

passed since the last arrival. In fact

$$\begin{aligned} \mathrm{P}\left[\tau_n \le a | \tau_n > a_0\right] &= \frac{\mathrm{P}\left[a_0 < \tau_n \le a\right]}{\mathrm{P}\left[\tau_n > a_0\right]} \\ &= \begin{cases} \dfrac{e^{-\lambda a_0} - e^{-\lambda a}}{e^{-\lambda a_0}} & , \ a > a_0 \\ 0 & , \ a \le a_0 \end{cases} \\ &= \begin{cases} 1 - e^{-\lambda(a-a_0)} & , \ a > a_0 \\ 0 & , \ a \le a_0 \end{cases} \\ &= \mathrm{P}\left[\tau_n \le a - a_0\right] \end{aligned} \tag{2.230}$$

so that the conditional probability is a shifted version of the unconditional one. We illustrate this property of the exponential PDF in Fig. 2.37.

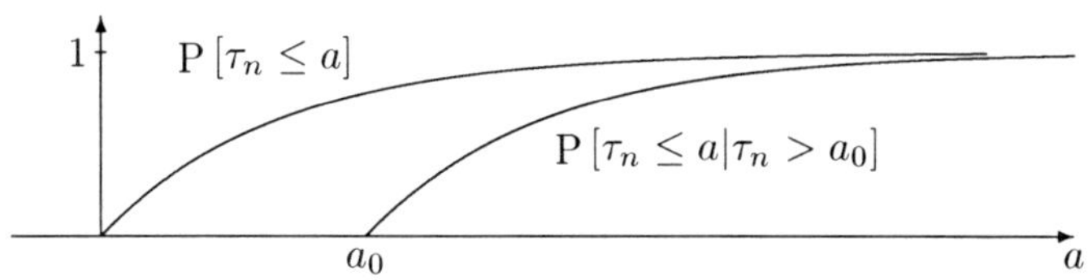

Figure 2.37 Illustration of the memoryless property for interarrival times.

From the expression of the mean in Example 2.6 D we get that for a homogeneous Poisson process[6]

$$\mathrm{E}\left[\tau_n\right] = 1/\lambda \tag{2.231}$$

which formally states the intuitive result that the mean interarrival time is the inverse of the arrival rate.

The following proposition (which we give without proof) gives further insight on the notion of homogeneity for a Poisson process.

Proposition 2.23 (Uniform arrivals) *Conditioned on the fact that there are exactly k arrivals in the same interval* $(u_1, u_2]$ *and considering them regardless of their order, the* k *arrival times are distributed as* k *independent rvs, uniform in* $(u_1, u_2]$.

Example 2.7 B Consider again the process $x(t)$ of Example 2.7 A. Suppose we know that 2 arrivals of $y(t)$ take place in the interval $(0, T]$, and want to find the probability that they both take place in the first half of the interval $(0, T/2]$. The unordered arrivals t_1 e t_2 are independent and uniform in $(t, t + T_0]$, so we can factor the probability as

$$\begin{aligned} \mathrm{P}\left[t < t_1 \le t + T_0/2, t < t_2 \le t + T_0/2\right] &= \mathrm{P}\left[t < t_1 \le t + T_0/2\right] \mathrm{P}\left[t < t_2 \le t + T_0/2\right] \\ &= \tfrac{1}{2} \cdot \tfrac{1}{2} = \tfrac{1}{4} \end{aligned}$$

---

[6]It should be noted that (2.231) holds in general for a wider class of homogeneous point processes. However, the proof for the general case is beyond the scope of this book.

2.7.3 Stationary and ergodic random processes

Definition 2.27 *A rp $x(t)$ is said to be* stationary *with respect to a statistical description (e.g. mean, autocorrelation or PDF) if such description is invariant to any time shift of the signal, that is all rps $x_1(t) = x(t-t_0)$ obtained by an arbitrary time shift t_0 have the same description.*

In particular, it can be easily seen that for a rp to be stationary in its mean, $\mathtt{m}_x(t)$ is a constant since it must be

$$\mathtt{m}_x(t) = \mathtt{m}_{x_1}(t) = \mathrm{E}\left[x_1(t)\right] = \mathrm{E}\left[x(t-t_0)\right] = \mathtt{m}_x(t-t_0) \quad , \quad t, t_0 \in \mathbb{R} \ . \tag{2.232}$$

Analogously, for a rp stationary in statistical power or variance, these quantities are constant, so that we can drop the time dependence in the notation and write simply $\mathtt{m}_x$, $\mathtt{M}_x$ and σ_x^2 for the mean, power and variance, respectively.

In a similar manner, for a rp stationary in its autocorrelation, $\mathtt{r}_x(t,\tau)$ is only a function of the lag τ and independent of the reference time instant t, so that we can simply write $\mathtt{r}_x(\tau)$.

Definition 2.28 *A rp that is stationary both in its mean and autocorrelation (and consequently in its power and variance also) is called* wide sense stationary (WSS). *Hence, for a WSS rp x it is*

$$\begin{aligned} \mathrm{E}\left[x(t)\right] &= \mathtt{m}_x \\ \mathrm{E}\left[x(t)x^*(t-\tau)\right] &= \mathtt{r}_x(\tau) \ . \end{aligned} \tag{2.233}$$

Remark The reader should have clear in mind that stationarity does not imply that the signal realizations are constant over time, only that the signal statistical description does not depend on time. A constant rp is of course also stationary, but not the converse. A rough illustration of the realizations of a stationary process (in mean) is given in Fig. 2.38.

A generalization of the stationarity property is *cyclostationarity*, defined as follows

Definition 2.29 *A rp $x(t)$ is said to be cyclostationary with respect to a statistical description if such description is invariant to a time shift of the signal by some fixed quantity T_c, which is called the* cyclostationarity period *of x.*

In other words the definition states that the rp $x_1(t) = x(t-T_c)$ has the same description as $x(t)$. By mimicking (2.232) we can see that for a rp that is cyclostationary in its mean (or similarly in its power, or variance) with period T_c, the mean must satisfy

$$\mathtt{m}_x(t) = \mathtt{m}_{x_1}(t) = \mathrm{E}\left[x_1(t)\right] = \mathrm{E}\left[x(t-T_c)\right] = \mathtt{m}_x(t-T_c) \quad , \quad t \in \mathbb{R} \ , \tag{2.234}$$

and thus it is a periodic signal with period T_c. Similarly for a rp that is cyclostationary in its correlation, the function $\mathtt{r}_x(t,\tau)$ is periodic in t with period T_c. Again, cyclostationarity of a rp does not imply its realizations are periodic of T_c, only that its statistical description is periodic. A rp with periodic realizations is of course also cyclostationary, but not the converse. Fig. 2.39 gives an intuitive illustration of this property.

Example 2.7 C Let $a(t)$ be a WSS rp, and $c(t)$ be a deterministic periodic signal with period T_0. Then their product $x(t) = a(t)c(t)$, $t \in \mathbb{R}$ is a rp with mean

$$\mathtt{m}_x(t) = \mathrm{E}\left[a(t)c(t)\right] = c(t)\,\mathrm{E}\left[a(t)\right] = \mathtt{m}_a c(t)$$

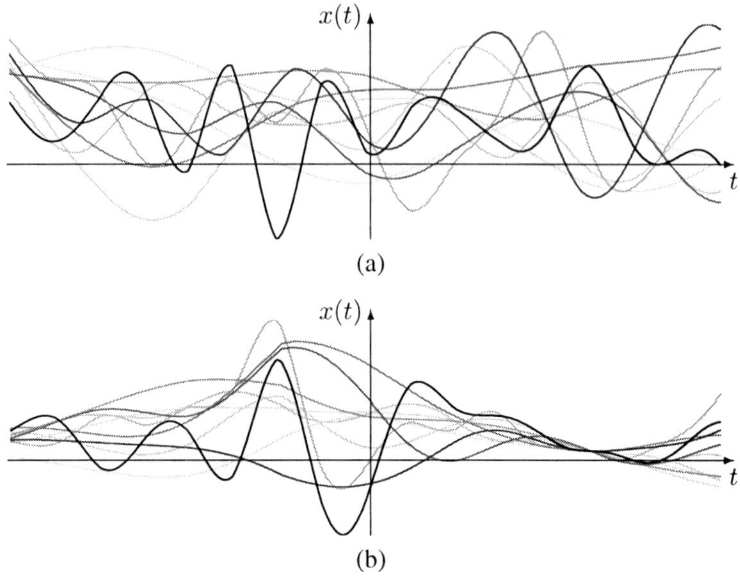

Figure 2.38 Realizations of: (a) a stationary process; (b) a non-stationary process.

and statistical power

$$\mathtt{M}_x(t) = \mathrm{E}\left[|a(t)c(t)|^2\right] = |c(t)|^2\,\mathrm{E}\left[|a(t)|^2\right] = \mathtt{M}_a\,|c(t)|^2$$

Since $c(t)$ and $|c(t)|^2$ are periodic with period T_0, so are both $\mathtt{m}_x(t)$ and $\mathtt{M}_x(t)$. As regards the autocorrelation

$$\begin{aligned}\mathtt{r}_x(t,\tau) &= \mathrm{E}\left[a(t)c(t)a^*(t-\tau)c^*(t-\tau)\right]\\ &= c(t)c^*(t-\tau)\,\mathrm{E}\left[a(t)a^*(t-\tau)\right]\\ &= \mathtt{r}_a(\tau)c(t)c^*(t-\tau)\end{aligned}$$

it is periodic in t with period T_0. Thus $x(t)$ is cyclostationary in mean, power and autocorrelation with period T_0.

Another property that is strongly linked to stationarity is *ergodicity*.

Definition 2.30 *A stationary rp (in mean) is said to be* ergodic *in its mean if the time average converges to its statistical mean, in some sense to be specified*

$$\frac{1}{2u}\int_{-u}^{u} x(\omega,t)\,\mathrm{d}t \underset{u\to\infty}{\longrightarrow} \mathtt{m}_x \quad , \qquad \omega\in\Omega\,. \tag{2.235}$$

Observe that the quantity on the left-hand side of (2.235), call it $\bar{x}_u$, is itself a rv, its value depending on the particular realization of $x(t)$, that is on the sample point $\omega \in \Omega$. On the contrary, thanks to the stationarity of $x(t)$, $\mathtt{m}_x$ is a constant value. As defined in (2.235), ergodicity states that in some sense, as u grows, $\bar{x}_u$ will converge to $\mathtt{m}_x$.

Similarly we can state the property of ergodicity in correlation or statistical power for rps that are stationary in these descriptions.

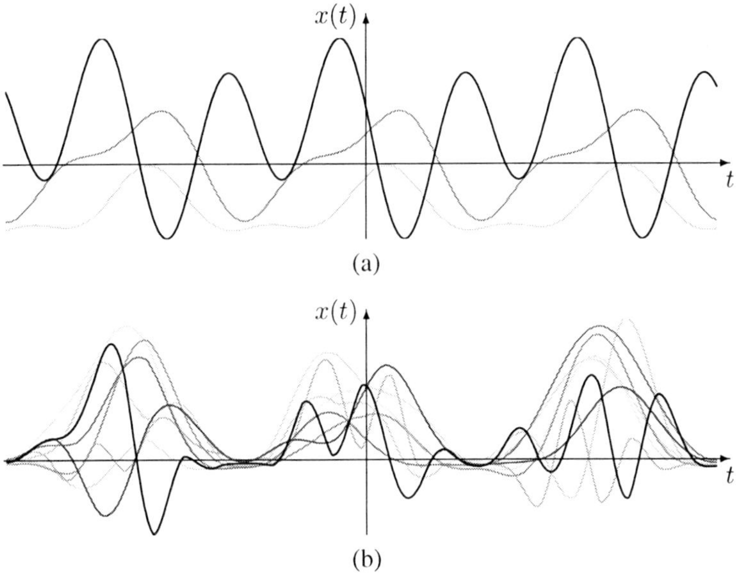

Figure 2.39 Realizations of: (a) a periodic process; (b) a cyclostationary process.

We observe that ergodicity of a rp is a very welcome property since it relates a *statistical description* of the rp to a measure on a *single realization*. All rps we will deal with in this book can be assumed ergodic unless otherwise stated. That is why we use the same notation for the "average power" of a deterministic signal (see (2.30)) and the "statistical power" of a rp. Indeed ergodicity is the property that justifies the procedure of gathering values from a realization as time elapses, to empirically infer some statistical properties of the associated rp.

2.7.4 Second order description of a WSS process

We recall that for a WSS random process the autocorrelation function $\mathtt{r}_x(\tau)$ is only a function of the lag τ. Thanks to this property we can make the following definition.

Definition 2.31 *The* power spectral density *(*PSD*) of a WSS rp is the Fourier transform of its autocorrelation function*

$$\mathcal{P}_x(f) = \begin{cases} \displaystyle\int_{-\infty}^{+\infty} \mathtt{r}_x(\tau)\, e^{-j2\pi f\tau}\, \mathrm{d}\tau & , \textit{for continuous time signals} \\ \displaystyle T\sum_{k=-\infty}^{+\infty} \mathtt{r}_x(kT)\, e^{-j2\pi fkT} & , \textit{for discrete time signals} \end{cases} \tag{2.236}$$

Observe that for discrete time rps, the Fourier transform defined in (2.17) is multiplied by the quantum T.

The PSD enjoys the following properties:

1. It is real-valued, $\mathcal{P}_x(f) \in \mathbb{R}$, even for complex-valued rps.

2. It is non-negative, $\mathcal{P}_x(f) \geq 0$.

3. Its integral yields the rp statistical power

$$\begin{aligned}
&\int_{-\infty}^{+\infty} \mathcal{P}_x(f)\, \mathrm{d}f = \mathtt{M}_x \quad , \qquad \text{for continuous time rps} \\
&\int_{0}^{1/T} \mathcal{P}_x(f)\, \mathrm{d}f = \mathtt{M}_x \quad , \qquad \text{for discrete time rps with quantum } T \; .
\end{aligned} \tag{2.237}$$

4. For real rps it is even, $\mathcal{P}_x(f) = \mathcal{P}_x(-f)$.

The first and fourth properties are straightforward consequences of the symmetries in the correlation function and properties of the Fourier transform. The second property is not easy to prove and we provide some justifications for it later on (see (2.285)). The third property comes from relating $\mathtt{r}_x(0)$ with $\mathcal{P}_x(f)$ and recalling (2.218). This property intuitively justifies the name "statistical power spectral density". A deeper motivation for this name will be seen in Section 2.8.1.

Mean and correlation (or PSD) give the second-order description of a WSS rp. It is of course an incomplete statistical description, nevertheless it is very useful. For Gaussian rps for example, it allows to derive any statistical description, as illustrated in the following example.

Example 2.7 D Let $x(t)$ be a continuous time real-valued stationary Gaussian rp with mean $\mathtt{m}_x = 0$ and power spectral density $\mathcal{P}_x(f) = P_0 e^{-T_0|f|}$, with $P_0 = 10^{-4}\,\mathrm{V}^2/\mathrm{Hz}$, and $T_0 = 100\,\mu\mathrm{s}$. We want to find the probabilities $\mathrm{P}[x(t) > A]$ and $\mathrm{P}[x(t) > x(t+T_0) + A]$ with $A = 1$ V. From the PSD and (2.237) we calculate the statistical power

$$\mathtt{M}_x = 2\, P_0/T_0 \tag{2.238}$$

which coincides with the variance since $x(t)$ has zero mean. Therefore, in terms of the complementary Gaussian distribution function (2.148) we have

$$\mathrm{P}[x(t) > A] = \mathrm{Q}\left(\frac{A - \mathtt{m}_x}{\sigma_x}\right) = \mathrm{Q}\left(A\sqrt{\frac{T_0}{P_0}}\right) = \mathrm{Q}(1) \simeq 0.159 \; . \tag{2.239}$$

To calculate the second probability we consider the rp $z(t) = x(t) - x(t+T_0)$: being a linear combination of jointly Gaussian variables, $z(t)$ is itself a Gaussian rp with mean

$$\mathrm{E}[x(t)] - \mathrm{E}[x(t+T_0)] = \mathtt{m}_x - \mathtt{m}_x = 0 \; , \tag{2.240}$$

i.e. $\mathtt{m}_z = 0$, and variance

$$\mathrm{E}\left[x^2(t)\right] + \mathrm{E}\left[x^2(t+T_0)\right] - 2\,\mathrm{E}[x(t)x(t+T_0)] = 2[\mathtt{M}_x - \mathtt{r}_x(-T_0)] \; , \tag{2.241}$$

i.e. $\sigma_z^2 = \mathtt{M}_z = 2[\mathtt{M}_x - \mathtt{r}_x(-T_0)]$. By taking the inverse Fourier transform of $\mathcal{P}_x(f)$ we obtain the autocorrelation function

$$\mathtt{r}_x(\tau) = \frac{2P_0T_0}{T_0^2 + (2\pi\tau)^2} \quad \Rightarrow \quad \mathtt{r}_x(-T_0) = \frac{2P_0}{T_0(1+4\pi^2)} \tag{2.242}$$

Hence we get

$$\sigma_z^2 = \frac{4P_0}{T_0}\left(1 - \frac{1}{1+4\pi^2}\right) \tag{2.243}$$

and the probability we look for is

$$\mathrm{P}\left[x(t) > x(t+T_0) + A\right] = \mathrm{P}\left[z(t) > A\right] = \mathrm{Q}\left(\frac{A - \mathtt{m}_z}{\sigma_z}\right) = \mathrm{Q}\left(A\sqrt{\frac{T_0}{4P_0}\,\frac{4\pi^2+1}{4\pi^2}}\right)$$

giving

$$\mathrm{P}\left[x(t) > x(t+T_0) + A\right] = \mathrm{Q}\left(\sqrt{\frac{1}{4}\,\frac{4\pi^2+1}{4\pi^2}}\right) \simeq 0.308 \;.$$

Example 2.7 E Consider a WSS discrete time rp $x(nT)$, whose rvs are statistically independent and have the same statistical distribution. Such variables are called iid, and the PSD of x takes a particular form. Indeed, as distinct rvs are uncorrelated, the autocorrelation function is

$$\mathtt{r}_x(mT) = \begin{cases} \mathrm{E}\left[|x(nT)|^2\right] = \mathtt{M}_x & , \quad m = 0 \\ \mathrm{E}\left[x(nT)\right]\mathrm{E}\left[x^*(nT - mT)\right] = |\mathtt{m}_x|^2 & , \quad m \neq 0 \end{cases} \tag{2.244}$$

which can be written in a more compact form as

$$\mathtt{r}_x(mT) = |\mathtt{m}_x|^2 + \sigma_x^2 \delta_m \;. \tag{2.245}$$

By taking its Fourier transform times T we obtain

$$\mathcal{P}_x(f) = T\sigma_x^2 + |\mathtt{m}_x|^2 \sum_{\ell=-\infty}^{+\infty} \delta(f - \ell/T)\;, \tag{2.246}$$

which is composed of a constant part $\sigma_x^2 T$ and a periodic sequence of Dirac impulses (also called *spectral lines*) spaced by $1/T$. Correlation and PSD for such a discrete time rp with iid variables are both illustrated in Fig. 2.40.

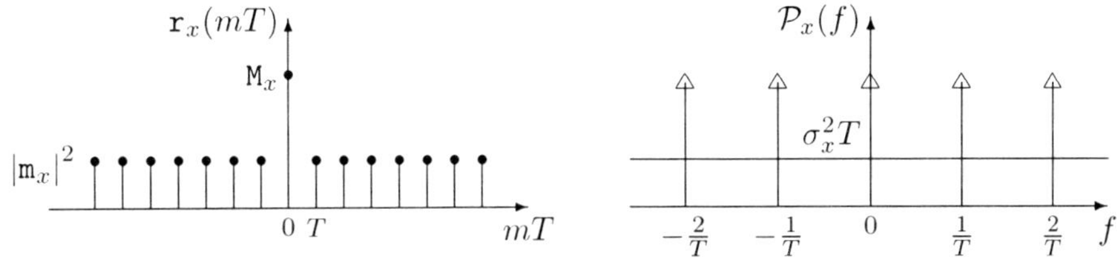

Figure 2.40 Graphical representation of (a) the autocorrelation function and (b) the PSD for a discrete time rp $x(nT)$ with iid rvs.

Spectral lines in the PSD In many applications it is important to detect the presence of sinusoidal components in a rp. With this intent we give the following theorem.

Theorem 2.24 (Spectral lines in the PSD) *The PSD of a WSS rp x, $\mathcal{P}_x$, can be uniquely decomposed into a component $\mathcal{P}_x^{(c)}$ with no impulses and a discrete component consisting of Dirac impulses (spectral lines) $\mathcal{P}_x^{(d)}$, so that*

$$\mathcal{P}_x(f) = \mathcal{P}_x^{(c)}(f) + \mathcal{P}_x^{(d)}(f)\;, \tag{2.247}$$

where $\mathcal{P}_x^{(c)}$ is an ordinary (piecewise continuous) function and

$$\mathcal{P}_x^{(d)}(f) = \sum_{i \in \mathcal{I}} \mathtt{M}_i \, \delta(f - f_i) \, , \tag{2.248}$$

where $\mathcal{I}$ identifies a discrete set of frequencies $\{f_i\}$, $i \in \mathcal{I}$.

The inverse Fourier transform of (2.247) yields the relation

$$\mathtt{r}_x(\tau) = \mathtt{r}_x^{(c)}(\tau) + \mathtt{r}_x^{(d)}(\tau) \, , \tag{2.249}$$

with

$$\mathtt{r}_x^{(d)}(\tau) = \sum_{i \in \mathcal{I}} \mathtt{M}_i \, e^{j2\pi f_i \tau} \, . \tag{2.250}$$

The most interesting consideration is that the following rp decomposition corresponds to the decomposition (2.247) of the PSD:

$$x(t) = x^{(c)}(t) + x^{(d)}(t) \, , \tag{2.251}$$

where $x^{(c)}$ and $x^{(d)}$ are *orthogonal* rps having PSD functions

$$\mathcal{P}_{x^{(c)}}(f) = \mathcal{P}_x^{(c)}(f) \qquad \text{and} \qquad \mathcal{P}_{x^{(d)}}(f) = \mathcal{P}_x^{(d)}(f) \, . \tag{2.252}$$

Moreover, $x^{(d)}$ is given by

$$x^{(d)}(t) = \sum_{i \in \mathcal{I}} x_i e^{j2\pi f_i t} \, , \tag{2.253}$$

where $\{x_i\}$ are orthogonal rvs with statistical powers $\{\mathtt{M}_i\}$.

Definition 2.32 *A WSS rp x is said to be asymptotically uncorrelated if the following two properties hold:*

$$\begin{aligned} &1) \quad \lim_{\tau \to \infty} \mathtt{r}_x(\tau) = |\mathtt{m}_x|^2 \, , \\ &2) \quad \mathtt{k}_x(\tau) = \mathtt{r}_x(\tau) - |\mathtt{m}_x|^2 \qquad \textit{is absolutely integrable} \end{aligned} \tag{2.254}$$

The property 1) denotes that $x(t)$ and $x(t-\tau)$ become uncorrelated for $\tau \to \infty$.

For such processes, it is

$$\mathtt{r}_x^{(c)}(\tau) = \mathtt{k}_x(\tau) \qquad \text{and} \qquad \mathtt{r}_x^{(d)}(\tau) = |\mathtt{m}_x|^2 \, . \tag{2.255}$$

Remark From (2.255), it is $\mathcal{P}_x^{(d)}(f) = |\mathtt{m}_x|^2 \delta(f)$, and an asymptotically uncorrelated process exhibits at most one spectral line at the origin. In other words, an asymptotically uncorrelated process can have a spectral line only if $\mathtt{m}_x \neq 0$.

By making use of the PSD, we extend the notion of band to rps as follows.

Definition 2.33 *The* full band *of a WSS rp x is the support of its PSD*

$$\overline{\mathcal{B}}_x = \{f \in \mathbb{R} : \mathcal{P}_x(f) > 0\} \tag{2.256}$$

For real-valued rps the autocorrelation function $\mathtt{r}_x$ is itself real-valued, and its PSD has even symmetry. It is therefore customary, as we did in Section 2.4, to define its band as follows.

Definition 2.34 *The* band *of a real-valued WSS rp x is the subset of nonnegative frequencies where its PSD is nonzero*

$$\mathcal{B}_x = \{f \geq 0 : \mathcal{P}_x(f) > 0\} \tag{2.257}$$

An important class of rps has constant PSD.

Definition 2.35 *A WSS rp x, whose power spectral density is constant over the whole frequency axis, $\mathcal{P}_x(f) = \mathcal{P}_0$, is called* white noise, *or simply* white.

Observe that in order to have constant PSD, a white rp must have an impulsive autocorrelation

$$\begin{aligned} \mathtt{r}_x(\tau) &= \mathcal{P}_0\delta(\tau) \ , \quad \text{for continuous time rps} \\ \mathtt{r}_x(mT) &= P_0/T\ \delta_m \ , \quad \text{for discrete time rps} \end{aligned}$$

and in any case $\mathtt{r}_x$ must vanish outside the origin, that is any two distinct rvs belonging to the same white rp are orthogonal (see Definition 2.17). From (2.255), if x is asymptotically incorrelated then a white process has also zero mean and all its rvs are uncorrelated. In particular, a zero-mean white Gaussian rp is called *white Gaussian noise (WGN)* and is made of independent rvs, whereas any zero-mean rp with iid rvs (not necessarily Gaussian) is white.

A peculiarity of continuous time white rps is that their statistical power turns out to be infinite, since $\mathtt{M}_x = \int_{-\infty}^{+\infty} \mathcal{P}_0 \, \mathrm{d}f = \infty$, and one may wonder whether it makes sense to consider rps with infinite power in applications. However, we will see in the following chapter that continuous time white rps are a very convenient model for many random phenomena. On the contrary, this problem is not present for discrete time rps, where $\mathtt{M}_x = \int_0^{1/T} \mathcal{P}_0 \, \mathrm{d}f = \mathcal{P}_0/T$ is finite.

2.7.5 Joint second order description of two random processes

For a pair of rps $x(t)$ and $y(t)$, we consider their cross correlation function

$$\mathtt{r}_{xy}(t, \tau) = \mathrm{E}\left[x(t)y^*(t-\tau)\right] \ . \tag{2.258}$$

By extension of Definition 2.17 two rps are said to be *orthogonal* if $\mathtt{r}_{xy}(t, \tau) = 0$, for all t, τ and *uncorrelated* if $\mathtt{r}_{xy}(t, \tau) = \mathtt{m}_x \mathtt{m}_y^*$, for all t, τ.

Definition 2.36 *Two WSS rps x and y are said to be* jointly WSS *if their cross correlation function $\mathtt{r}_{xy}(t, \tau)$ does not depend on t.*

In the case of jointly stationary rps we can thus drop the dependence on t and simply write $\mathtt{r}_{xy}(\tau)$. Correspondingly, the Fourier transform of the cross correlation, $\mathcal{P}_{xy}(f)$, is called the *cross power spectral density* of the two rps. Unlike the PSD of a single rp, the cross PSD of two rps may also be negative or complex-valued, similarly to the cross energy spectral density of two deterministic signals.

Further insight The PSD $\mathcal{P}_x(f)$ and the cross PSD $\mathcal{P}_{xy}(f)$ satisfy the following properties that we give without proof

$$\mathcal{P}_{xy}(f) = \mathcal{P}_{yx}^*(f) \tag{2.259}$$

$$\mathcal{P}_{x^*}(f) = \mathcal{P}_x(-f) \tag{2.260}$$

$$0 \leq |\mathcal{P}_{xy}(f)|^2 \leq \mathcal{P}_x(f)\mathcal{P}_y(f) \, . \tag{2.261}$$

From (2.261) we deduce that if two rps x and y have PSDs with disjoint bands, then they are orthogonal.

A case that we frequently encounter in practice is that of a zero mean WSS (complex-valued) rp x that is *orthogonal with its complex conjugate* x^*, that is

$$\mathtt{r}_{xx^*}(t,\tau) = \mathrm{E}\left[x(t)x(t-\tau)\right] = 0 \quad , \quad \text{for all } t, \tau \, . \tag{2.262}$$

Proposition 2.25 *Let* $x(t) = x_{\mathrm{I}}(t) + jx_{\mathrm{Q}}(t)$ *be a zero mean WSS rp,* orthogonal with its conjugate x^*. *Then the auto and cross correlations of its components* $x_{\mathrm{I}}, x_{\mathrm{Q}}$ *are*

$$\mathtt{r}_{x_{\mathrm{I}}}(\tau) = \mathtt{r}_{x_{\mathrm{Q}}}(\tau) = \frac{1}{2}\Re\left[\mathtt{r}_x(\tau)\right] \tag{2.263}$$

$$\mathtt{r}_{x_{\mathrm{Q}}x_{\mathrm{I}}}(\tau) = -\mathtt{r}_{x_{\mathrm{I}}x_{\mathrm{Q}}}(\tau) = \frac{1}{2}\Im\left[\mathtt{r}_x(\tau)\right] \tag{2.264}$$

In particular their cross power is zero,

$$\mathtt{r}_{x_{\mathrm{I}}x_{\mathrm{Q}}}(0) = 0 \tag{2.265}$$

Proof To prove (2.263) and (2.264), we simply write

$$x_{\mathrm{I}}(t) = \frac{x(t) + x^*(t)}{2} \quad , \quad x_{\mathrm{Q}}(t) = \frac{x(t) - x^*(t)}{2j}$$

and apply the linearity of expectation and the hypothesis $\mathtt{r}_{xx^*}(t,\tau) = 0 = \mathtt{r}_{x^*x}(t,\tau)$. Finally, from (2.264) it follows

$$\mathtt{r}_{x_{\mathrm{Q}}x_{\mathrm{I}}}(0) = \frac{1}{2}\Im\left[\mathtt{r}_x(0)\right] = \frac{1}{2}\Im[\mathtt{M}_x] = 0$$

□

Proposition 2.26 *A zero-mean WSS complex-valued Gaussian rp* x *that is orthogonal to* x^* *is also circularly symmetric, with* $x(t) \sim \mathcal{CN}(0, \sigma_x^2)$.

Proof Consider the two rvs $x_{\mathrm{I}}(t)$ and $x_{\mathrm{Q}}(t)$. They both have zero mean and from (2.265) their covariance is null, while from (2.263)

$$\mathtt{M}^2_{x_{\mathrm{I}}} = \mathtt{r}_{x_{\mathrm{I}}}(0) = \frac{1}{2}\Re\left[\mathtt{r}_x(0)\right] = \frac{1}{2}\Re\mathtt{M}_x = \frac{1}{2}\mathtt{M}_x \tag{2.266}$$

$$\mathtt{M}^2_{x_{\mathrm{Q}}} = \mathtt{r}_{x_{\mathrm{Q}}}(0) = \frac{1}{2}\Re\left[\mathtt{r}_x(0)\right] = \frac{1}{2}\Re\mathtt{M}_x = \frac{1}{2}\mathtt{M}_x \, . \tag{2.267}$$

Hence, $x_{\mathrm{I}}(t)$ and $x_{\mathrm{Q}}(t)$ are Gaussian (with zero mean), uncorrelated, and hence statistically independent, rvs. Moreover, they have the same variance, so that $x(t)$ is a circularly symmetric rv for all t. □

2.7.6 Second order description of a cyclostationary process

For cyclostationary rps, whose statistical description is periodic in t, with period T_c, we consider average values over a period for mean, statistical power, autocorrelation and PSD

$$\mathtt{m}_x = \frac{1}{T_c}\int_0^{T_c} \mathtt{m}_x(t)\,dt \ , \quad \mathtt{M}_x = \frac{1}{T_c}\int_0^{T_c} \mathtt{M}_x(t)\,dt \ , \quad \mathtt{r}_x(\tau) = \frac{1}{T_c}\int_0^{T_c} \mathtt{r}_x(t,\tau)\,dt \quad (2.268)$$

$$\mathcal{P}_x(f) = \frac{1}{T_c}\int_0^{T_c} \mathcal{P}_x(t,f)\,dt \ , \tag{2.269}$$

where $\mathcal{P}_x(t,f)$ is the Ftf of $\mathtt{r}_x(t,\tau)$ with respect to τ. It is seen that the Ftf of the average autocorrelation yields the *average power spectral density*, $\mathcal{P}_x(f)$, whose integral yields the *average statistical power* $\mathtt{M}_x$.

Example 2.7 F Let $a(t)$ be a continuous time WSS process and let $x(t) = a(t)\cos(2\pi f_0 t + \varphi_0)$. By the results in Example 2.7 C, with $c(t) = \cos(2\pi f_0 t + \varphi_0)$ we know that $x(t)$ is cyclostationary with period $T_c = 1/f_0$. It has time-varying mean

$$\mathtt{m}_x(t) = \mathtt{m}_a \cos(2\pi f_0 t + \varphi_0) \ ,$$

and autocorrelation

$$\mathtt{r}_x(t,\tau) = \mathtt{r}_a(\tau)\cos(2\pi f_0 t + \varphi_0)\cos(2\pi f_0(t-\tau) + \varphi_0) \ .$$

Then, by taking the average of the mean we get

$$\mathtt{m}_x = \frac{1}{T_c}\int_0^{T_c} \mathtt{m}_x(t)\,dt = \frac{\mathtt{m}_a}{T_c}\int_0^{T_c} \cos(2\pi f_0 t + \varphi_0)\,dt = 0 \ ,$$

since the integral of the cosine function over a period is null. As regards the autocorrelation, by means of the trigonometric formula

$$\cos\alpha\cos\beta = \frac{1}{2}[\cos(\alpha+\beta) + \cos(\alpha-\beta)] \ ,$$

we can calculate

$$\begin{aligned}
\mathtt{r}_x(\tau) &= \frac{1}{T_0}\int_0^{T_0} \mathtt{r}_x(t,\tau)\,dt \\
&= \mathtt{r}_a(\tau)\frac{1}{2T_0}\int_0^{T_0} \cos(2\pi f_0(2t-\tau) + 2\varphi_0) + \cos(2\pi f_0\tau)\,dt \\
&= \frac{1}{2}\mathtt{r}_a(\tau)\cos(2\pi f_0\tau)
\end{aligned}$$

where we made use of the fact that $\cos(4\pi f_0 t - 2\pi f_0 \tau - 2\varphi_0)$ is periodic in t with period $T_c/2$, and its integral vanishes. By taking the Ftf of the last result we get the average PSD

$$\mathcal{P}_x(f) = \frac{1}{4}\left[\mathcal{P}_a(f-f_0) + \mathcal{P}_a(f+f_0)\right] \tag{2.270}$$

The average power can either be calculated as $\mathtt{M}_x = \mathbf{r}_x(0)$ or by integrating $\mathcal{P}_x(f)$: in any case we get

$$\boxed{\mathtt{M}_x = \frac{1}{2}\mathtt{M}_a} \tag{2.271}$$

This result extends (2.34) and is illustrated in Fig. 2.41.

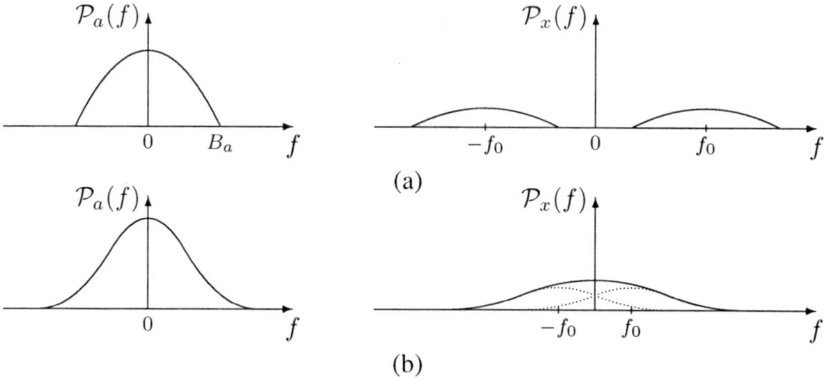

Figure 2.41 Illustration of the PSDs in Example 2.7 F: (a) in the case that a is bandlimited with bandwidth $B_a < f_0$; (b) in the general case.

For a cyclostationary rp the definition of band is still given by (2.256), with reference to its *average* PSD. An apparently surprising result related to the band of a cyclostationary rp is the following, which we give without proof.

Proposition 2.27 *If a cyclostationary rp with period T_c is* bandlimited *with bandwidth $B < \frac{1}{2T_c}$, then it is also stationary.*

2.8 SYSTEMS WITH RANDOM INPUTS AND OUTPUTS

Consider a system with map $\mathcal{M}$ and input the rp $x(t)$. Pick any ω_0 in the sample space Ω on which $x(t, \omega)$ is defined, let the realization $x(t, \omega_0)$ be the input to $\mathcal{M}$, and call the corresponding observed output $y(t, \omega_0)$. By repeating this procedure for all $\omega \in \Omega$ we obtain a collection of deterministic outputs that are the realizations of the output random process $y(t, \omega)$.

The rp $y(t)$ is then defined on the same probability space as $x(t)$, since each realization of the input corresponds to a realization of the output. In this context we face the problem of deriving the statistical description of the output y from that of the input x. We will shortly see that this problem has an easy solution in the case of linear systems. A noteworthy result, dealing with Gaussian rps is the following, which somehow resembles Proposition 2.16

Proposition 2.28 *The output of a linear system with input a Gaussian rp is itself a Gaussian rp.*

2.8.1 Filtering of a WSS random process

In the case of a filter we have two important results

Theorem 2.29 *In a filter with impulse response $g(t)$ and input a WSS rp $x(t)$, the output rp $y(t)$ is itself WSS and jointly stationary with the input, having*
mean

$$\mathtt{m}_y = \mathtt{m}_x \int_{-\infty}^{+\infty} g(t)\,\mathrm{d}t \,, \tag{2.272}$$

cross correlation

$$\mathtt{r}_{yx}(\tau) = (g * \mathtt{r}_x)\,(\tau) \,, \tag{2.273}$$

and autocorrelation

$$\mathtt{r}_y(\tau) = \left(g_-^* * \mathtt{r}_{yx}\right)\,(\tau) = \left[\left(g_-^* * g\right) * \mathtt{r}_x\right]\,(\tau) \,, \tag{2.274}$$

where $g_-^*(t) = g^*(-t)$.

Proof We first prove (2.272). From the input-output relationship of the filter we have

$$\mathrm{E}\left[y(t)\right] = \mathrm{E}\left[\int_{-\infty}^{+\infty} x(t-u)g(u)\,\mathrm{d}u\right]$$

Now the key step is to exchange the order of integration in t and expectation (which is a linear operator, see Section 2.6.2) to obtain

$$\mathrm{E}\left[y(t)\right] = \int_{-\infty}^{+\infty} \mathrm{E}\left[x(t-u)g(u)\right]\,\mathrm{d}u$$

and observe that the signal $g(t)$ is *not* random, so we can factor it out of the expectation

$$\mathrm{E}\left[y(t)\right] = \int_{-\infty}^{+\infty} \mathrm{E}\left[x(t-u)\right] g(u)\,\mathrm{d}u$$

We now observe that $\mathrm{E}\left[x(t-u)\right] = \mathtt{m}_x$ by the hypothesis on stationarity of the input rp and obtain the result

$$\mathrm{E}\left[y(t)\right] = \int_{-\infty}^{+\infty} \mathtt{m}_x g(u)\,\mathrm{d}u$$

which is independent of t. To prove (2.273) we proceed analogously by writing

$$\begin{aligned}
\mathrm{E}\left[y(t)x^*(t-\tau)\right] &= \mathrm{E}\left[\int_{-\infty}^{+\infty} x(t-u)g(u)\,\mathrm{d}u\, x^*(t-\tau)\right] \\
&= \mathrm{E}\left[\int_{-\infty}^{+\infty} x(t-u)g(u)x^*(t-\tau)\,\mathrm{d}u\right] \\
&= \int_{-\infty}^{+\infty} \mathrm{E}\left[x(t-u)g(u)x^*(t-\tau)\right]\,\mathrm{d}u \\
&= \int_{-\infty}^{+\infty} g(u)\,\mathrm{E}\left[x(t-u)x^*(t-\tau)\right]\,\mathrm{d}u \\
&= \int_{-\infty}^{+\infty} g(u)\mathtt{r}_x(\tau-u)\,\mathrm{d}u
\end{aligned}$$

The proof of (2.274) is similar, and we omit it here. □

In applications, it is important to consider the frequency domain equivalents of the above results.

Proposition 2.30 *Given a filter with frequency response $\mathcal{G}(f)$, in terms of the input statistical description, the second order output statistical description is given by*
mean

$$\mathtt{m}_y = \mathtt{m}_x \mathcal{G}(0) , \tag{2.275}$$

cross PSDs

$$\mathcal{P}_{yx}(f) = \mathcal{G}(f)\mathcal{P}_x(f) , \quad \mathcal{P}_{xy}(f) = \mathcal{G}^*(f)\mathcal{P}_x(f) , \tag{2.276}$$

output PSD

$$\mathcal{P}_y(f) = \mathcal{G}^*(f)\mathcal{P}_{yx}(f) \tag{2.277}$$

or, equivalently,

$$\mathcal{P}_y(f) = |\mathcal{G}(f)|^2 \mathcal{P}_x(f) . \tag{2.278}$$

Proof Equation (2.275) is just a rewriting of (2.272), with $\mathcal{G}(0) = \int_{-\infty}^{+\infty} g(t)\, dt$. To obtain (2.276) and (2.278) we just take Fourier transforms of both sides in equations (2.273) and (2.274), respectively. The second of (2.276) follows from (2.259). □

The result given by (2.278) is by far the most relevant: it states that the PSD of the output rp is the product between the PSD of the input and the energy spectral density $\mathcal{E}_g(f) = |\mathcal{G}(f)|^2$ of the filter impulse response.

Remark Equation (2.278), relating the PSD of WSS rps in a filter, is similar to (2.52), relating the energy spectral density of deterministic signals.

Remark Observe that to evaluate the statistical power at the output of a cascade of filters, we can either calculate the autocorrelation functions or the PSD of the processes along the cascade. Most commonly, the frequency domain approach with calculation of the PSD is much easier, since it requires multiplications (see (2.278)) instead of convolution integrals (see (2.274)).

Further insight An application of the above result (2.278) is given by the

Theorem 2.31 (Irrelevance principle) *If a WSS rp x with full band $\overline{\mathcal{B}}_x$ is the input to a filter whose frequency response has unit gain over $\overline{\mathcal{B}}_x$*

$$\mathcal{G}(f) = \begin{cases} 1 & , \ f \in \overline{\mathcal{B}}_x \\ \textit{arbitrary} & , \ f \notin \overline{\mathcal{B}}_x \end{cases} \tag{2.279}$$

then the output y coincides with the input in the mean-square sense, that is

$$\mathrm{E}\left[|y(t) - x(t)|^2\right] = 0 , \tag{2.280}$$

and by extending Definition 2.18 we can write $y(t) \overset{\text{m.s.}}{=} x(t)$. *A filter satisfying (2.279) is thus called* irrelevant *for the rp* $x(t)$.

Proof Let $e(t) = y(t) - x(t)$. Since for any realization we can write $\mathcal{E}(f) = \mathcal{Y}(f) - \mathcal{X}(f) = [\mathcal{G}(f) - 1]\mathcal{X}(f)$, we can think of $e(t)$ as output of a filter with input $x(t)$ and frequency response

$$\mathcal{G}_e(f) = \mathcal{G}(f) - 1 = \begin{cases} 0 & , \ f \in \overline{\mathcal{B}}_x \\ \text{unspecified} & , \ f \notin \overline{\mathcal{B}}_x \end{cases}$$

e is therefore stationary and its power spectral density is

$$\mathcal{P}_e(f) = |\mathcal{G}_e(f)|^2 \, \mathcal{P}_x(f) = 0 \ , \quad \text{for all } f \in \mathbb{R}$$

vanishing at all frequencies, since $\mathcal{G}_e(f)$ is zero on the support of $\mathcal{P}_x$. Its power

$$\mathtt{M}_e = \mathrm{E}\left[|y(t) - x(t)|^2\right] = \int_{-\infty}^{+\infty} \mathcal{P}_e(f) \, df \tag{2.281}$$

is thus null. □

In the above theorem, $e(t)$ represents the difference (or the error signal) between $y(t)$ and $x(t)$. The theorem does not state that all the realizations of $e(t)$ are identically zero so that all the realizations of x and y coincide. It rather states that at any instant t, $e(t)$ is an almost surely null rv, so that the rvs $x(t)$ and $y(t)$ coincide with probability 1.

It is also noteworthy to consider the result of applying (2.278) when $\mathcal{G}$ is an ideal filter with full band $\overline{\mathcal{B}}$,

$$\mathcal{G}(f) = \begin{cases} 1 & , \ f \in \overline{\mathcal{B}} \\ 0 & , \ f \notin \overline{\mathcal{B}} \end{cases} \tag{2.282}$$

with an arbitrary input rp $x(t)$. Then each realization in $y(t)$ will contain only those frequency components of the corresponding realization of $x(t)$ which fall in $\overline{\mathcal{B}}$. From (2.278) we also obtain the PSD of $y(t)$ as

$$\mathcal{P}_y(f) = \begin{cases} \mathcal{P}_x(f) & , \ f \in \overline{\mathcal{B}} \\ 0 & , \ f \notin \overline{\mathcal{B}} \end{cases} \tag{2.283}$$

and its power as

$$\mathtt{M}_y = \int_{-\infty}^{+\infty} \mathcal{P}_y(f) \, df = \int_{\overline{\mathcal{B}}} \mathcal{P}_x(f) \, df \tag{2.284}$$

so that (2.284) represents the statistical power associated with the components of x whose frequency lies in the filter full band $\overline{\mathcal{B}}$. In particular when $\overline{\mathcal{B}} = [f_0 - \Delta F/2, f_0 + \Delta F/2]$ is a sufficiently narrow interval around its central frequency f_0, so that $\mathcal{P}_x(f) \simeq \mathcal{P}_x(f_0)$, for all $f \in \overline{\mathcal{B}}$, equation (2.284) becomes

$$\mathtt{M}_y = \Delta F \, \mathcal{P}_x(f_0) \tag{2.285}$$

so that $\mathcal{P}_x(f_0)$ represents the statistical power per unit frequency of the components of the rp x at the frequency f_0. This justifies furthermore the name "power spectral density" for $\mathcal{P}_x$. Moreover, since the statistical power is a real nonnegative quantity, so must be the PSD $\mathcal{P}_x(f_0)$ at any frequency f_0, as stated by property 2 at page 107.

2.8.2 Filtering of a cyclostationary random process

On the other hand, it is also reasonable to ask what happens when the filter input is a cyclostationary rp. In this case we have the following.

Theorem 2.32 *In a filter with impulse response $g(t)$ having as input a cyclostationary rp $x(t)$ with period T_c, the output rp $y(t)$ is cyclostationary (also jointly with the input) with the same period, having*
average mean

$$\mathtt{m}_y = \mathtt{m}_x \int_{-\infty}^{+\infty} g(t)\,\mathrm{d}t \tag{2.286}$$

average cross correlation

$$\mathtt{r}_{yx}(\tau) = (g * \mathtt{r}_x)(\tau) \tag{2.287}$$

average autocorrelation

$$\mathtt{r}_y(\tau) = \left(g_-^* * \mathtt{r}_{yx}\right)(\tau) = \left[\left(g_-^* * g\right) * \mathtt{r}_x\right](\tau) \tag{2.288}$$

Proof We only give proof of (2.286). The other relationships can be derived similarly (see also Theorem 2.29).

The time varying mean of $y(t)$ can be calculated as

$$\mathtt{m}_y(t) = \mathrm{E}\left[y(t)\right] = \int_{-\infty}^{+\infty} g(u)\,\mathrm{E}\left[x(t-u)\right]\,\mathrm{d}u = (g * \mathtt{m}_x)(t)$$

and it is therefore periodic with period T_c. Its average within a period is

$$\mathtt{m}_y = \frac{1}{T_c}\int_0^{T_c} \mathtt{m}_y(t)\,\mathrm{d}t = \int_{-\infty}^{+\infty} g(u)\frac{1}{T_c}\int_0^{T_c} \mathrm{E}\left[x(t-u)\right]\,\mathrm{d}t\,\mathrm{d}u = \int_{-\infty}^{+\infty} g(u)\mathtt{m}_x\,\mathrm{d}u$$

where we have used the fact that the average of $\mathtt{m}_x(t)$ over any period is $\mathtt{m}_x$. □

Again by writing these results into the frequency domain we obtain the
average mean

$$\mathtt{m}_y = \mathtt{m}_x \mathcal{G}(0) \tag{2.289}$$

average cross PSD

$$\mathcal{P}_{yx}(f) = \mathcal{G}(f)\mathcal{P}_x(f)\,, \qquad \mathcal{P}_{xy}(f) = \mathcal{G}^*(f)\mathcal{P}_x(f) \tag{2.290}$$

average output PSD

$$\mathcal{P}_y(f) = \mathcal{G}^*(f)\mathcal{P}_{yx}(f) = \left|\mathcal{G}(f)\right|^2 \mathcal{P}_x(f)\,. \tag{2.291}$$

These relationships coincide with those for a stationary input given by (2.275)–(2.278).

2.8.3 Sampling and interpolation of stationary random processes

After examining the effect of filters on stationary rps, we now move to the other two linear transformations that were introduced in Section 2.3, sampling and interpolation. Firstly, we consider a continuous time stationary rp $x(t)$, with mean $\mathtt{m}_x$ and autocorrelation $\mathtt{r}_x(\tau)$ is sampled with period T_s yielding a discrete time signal $\{y_n\}$. Then, by the identity $y_n = y(nT_s) = x(nT_s)$ we get straightforwardly the following result.

Proposition 2.33 *The discrete time rp $\{y_n\}$ obtained by sampling a WSS continuous time signal $x(t)$ is itself WSS. Its (constant) mean is the same as that of the input*

$$\mathtt{m}_y = \mathtt{m}_x \,, \tag{2.292}$$

and its autocorrelation function is obtained by sampling the autocorrelation function of x,

$$\mathtt{r}_y(mT_s) = \mathtt{r}_x(mT_s) \,. \tag{2.293}$$

Proof By exploiting the input-output relationship for sampling (2.53) and the stationarity of the input we can write

$$\mathrm{E}\left[y_n\right] = \mathrm{E}\left[x(nT_s)\right] = \mathtt{m}_x \,, \tag{2.294}$$

and

$$\mathrm{E}\left[y_n y_{n-m}^*\right] = \mathrm{E}\left[x(nT_s)x^*(nT_s - mT_s)\right] = \mathtt{r}_x(mT_s) \,. \tag{2.295}$$

□

By taking Fourier transforms of both sides of (2.293), times T_s, we get

Proposition 2.34 *The (periodic) PSD of the discrete time output rp y can be obtained from the PSD of the continuous time input rp x by*

$$\mathcal{P}_y(f) = \sum_{\ell=-\infty}^{+\infty} \mathcal{P}_x(f + \ell F_s) \tag{2.296}$$

For an interpolator, the framework is complicated by the fact that even with a stationary discrete time input rp, the continuous time output rp is cyclostationary in general.

Theorem 2.35 *The continuous time rp $y(t)$ obtained by interpolating a discrete time WSS rp $\{x_n\}$ with quantum T through a filter g (see (2.55)) is cyclostationary with period T, and it has average mean*

$$\mathtt{m}_y = \frac{\mathtt{m}_x}{T} \int_{-\infty}^{+\infty} g(t)\,\mathrm{d}t \,, \tag{2.297}$$

and average autocorrelation

$$\mathtt{r}_y(\tau) = \frac{1}{T} \sum_{m=-\infty}^{+\infty} \left(g_-^* * g\right)(\tau - mT)\, \mathtt{r}_x(mT) \,. \tag{2.298}$$

Proof We only show the proof of (2.297). The proof of (2.298) is similar. From the input-output relation (2.55), the time varying mean of $y(t)$ can be calculated as

$$\begin{aligned}
\mathtt{m}_y(t) &= \mathrm{E}\left[y(t)\right] \\
&= \mathrm{E}\left[\sum_{n=-\infty}^{+\infty} g(t-nT)x_n\right] \\
&= \sum_{n=-\infty}^{+\infty} g(t-nT)\,\mathrm{E}\left[x_n\right] \\
&= \mathtt{m}_x \sum_{n=-\infty}^{+\infty} g(t-nT)
\end{aligned}$$

and the series in the last step yields a function that is periodic with period T. Averaging the above result within a period gives

$$\begin{aligned}
\mathtt{m}_y &= \frac{1}{T}\int_0^T \mathtt{m}_y(t)\,\mathrm{d}t \\
&= \frac{1}{T}\mathtt{m}_x \sum_{n=-\infty}^{+\infty} \int_0^T g(t-nT)\,\mathrm{d}t \\
&= \frac{1}{T}\mathtt{m}_x \sum_{n=-\infty}^{+\infty} \int_{nT}^{(n+1)T} g(u)\,\mathrm{d}u \\
&= \frac{1}{T}\mathtt{m}_x \int_{-\infty}^{+\infty} g(u)\,\mathrm{d}u
\end{aligned}$$

where in the third step we have used the change of variable $u = t - nT$ in the integral. □

The corresponding frequency domain result reads

Proposition 2.36 *The cyclostationary output of an interpolate filter with frequency response $\mathcal{G}(f)$ driven by a WSS input has average mean*

$$\mathtt{m}_y = \mathtt{m}_x \frac{1}{T}\mathcal{G}(0) \tag{2.299}$$

and average power spectral density

$$\mathcal{P}_y(f) = \left|\frac{1}{T}\mathcal{G}(f)\right|^2 \mathcal{P}_x(f) = \frac{\mathcal{E}_g(f)}{T^2}\mathcal{P}_x(f) \tag{2.300}$$

Proof The derivation of (2.299) from (2.297) is straightforward. To derive (2.300) we observe that (2.298) is the time domain input-output relation of an interpolate filter having $\mathtt{r}_x$ as input, $\mathtt{r}_y$ as output and $\frac{1}{T}g * g_-^*$ as impulse response. The corresponding frequency domain relationship is obtained by Theorem 2.5

$$\mathcal{F}[\mathtt{r}_y] = \frac{1}{T}\mathcal{F}[g * g_-^*]\,\mathcal{F}[\mathtt{r}_x]\,. \tag{2.301}$$

Then, since $\mathcal{P}_y$ is the Fourier transform of $\mathtt{r}_y$, whereas $\mathcal{P}_x$ is the Fourier transform of $\mathtt{r}_x$ multiplied by T we get

$$\mathcal{P}_y(f) = \frac{1}{T} |\mathcal{G}(f)|^2 \frac{1}{T} \mathcal{P}_x(f) , \tag{2.302}$$

and hence the result. □

Observe that equations (2.297)–(2.300) are analogous to the corresponding relations (2.275)–(2.278) for ordinary filters, although in the interpolator case we deal with a cyclostationary output rp and therefore average its statistics within a period. Interesting applications of the above results will be seen in the context of digital pulse amplitude modulation and bandlimited interpolate filters in Chapter 5.

A peculiar case is when the interpolate filter g is lowpass with bandwidth $B_g \leq \frac{1}{2T}$, driven by a WSS input. Then the average output PSD will turn out to be baseband with bandwidth $B_y \leq \frac{1}{2T}$, and by Proposition 2.27 $y(t)$ will be *stationary* rather than cyclostationary. In particular, for an *ideal lowpass* interpolate filter

$$\mathcal{G}(f) = T \operatorname{rect}(Tf) = \begin{cases} T & , \ |f| \leq \frac{1}{2T} \\ 0 & , \ |f| > \frac{1}{2T} \end{cases} \tag{2.303}$$

the output is WSS with statistical power

$$\mathtt{M}_y = \int_{-\infty}^{+\infty} \mathcal{P}_y(f) \, \mathrm{d}f = \int_{-\infty}^{+\infty} \frac{|\mathcal{G}(f)|^2}{T^2} \mathcal{P}_x(f) \, \mathrm{d}f = \int_{-\frac{1}{2T}}^{\frac{1}{2T}} \mathcal{P}_x(f) \, \mathrm{d}f = \mathtt{M}_x \, . \tag{2.304}$$

coinciding with that of the input.

Appendix: The complementary normalized Gaussian distribution function

We give possible solutions to the problem of evaluating the complementary Gaussian distribution function $\mathrm{Q}(y)$, which was defined in Example 2.6 A.

In the following Tab. 2.A.1 we give values of $\mathrm{Q}(y)$ with 4 significant digits, for many values of y that lie between 0 and 30.

For other values of $\mathrm{Q}(y)$ one can resort to good approximations as follows:

- For negative values of y, use the complementary symmetry illustrated in Fig. 2.A.1a, $\mathrm{Q}(y) = 1 - \mathrm{Q}(-y)$.
- If y is small, use linear interpolation, that is pick the values y_1, y_2 that are the closest to y among those given in Tab. 2.A.1, so that $y_1 < y < y_2$ and obtain the approximation

$$\mathrm{Q}(y) \simeq \frac{y - y_1}{y_2 - y_1} \mathrm{Q}(y_2) + \frac{y_2 - y}{y_2 - y_1} \mathrm{Q}(y_1) \, . \tag{2.A.1}$$

Since $\mathrm{Q}(y)$ is a convex function, the linear interpolation (2.A.1) is always an upper bound. Moreover, let y_i be the closest value to y between y_1 and y_2, a lower bound is given by

$$\mathrm{Q}(y) > \mathrm{Q}(y_i) - |y - y_i| \frac{1}{\sqrt{2\pi}} e^{-\frac{1}{2}y_i^2} \tag{2.A.2}$$

Table 2.A.1 Table of values for the complementary normalized Gaussian distribution function $\mathrm{Q}(y)$.

y	$\mathrm{Q}(y)$	y	$\mathrm{Q}(y)$	y	$\mathrm{Q}(y)$
0.00	$5.000 \cdot 10^{-1}$	2.0	$2.275 \cdot 10^{-2}$	6.0	$9.866 \cdot 10^{-10}$
0.05	$4.801 \cdot 10^{-1}$	2.1	$1.786 \cdot 10^{-2}$	6.2	$2.823 \cdot 10^{-10}$
0.10	$4.602 \cdot 10^{-1}$	2.2	$1.390 \cdot 10^{-2}$	6.4	$7.769 \cdot 10^{-11}$
0.15	$4.404 \cdot 10^{-1}$	2.3	$1.072 \cdot 10^{-2}$	6.6	$2.056 \cdot 10^{-11}$
0.20	$4.207 \cdot 10^{-1}$	2.4	$8.198 \cdot 10^{-3}$	6.8	$5.231 \cdot 10^{-12}$
0.25	$4.013 \cdot 10^{-1}$	2.5	$6.210 \cdot 10^{-3}$	7.0	$1.280 \cdot 10^{-12}$
0.30	$3.821 \cdot 10^{-1}$	2.6	$4.661 \cdot 10^{-3}$	7.2	$3.011 \cdot 10^{-13}$
0.35	$3.632 \cdot 10^{-1}$	2.7	$3.467 \cdot 10^{-3}$	7.4	$6.809 \cdot 10^{-14}$
0.40	$3.446 \cdot 10^{-1}$	2.8	$2.555 \cdot 10^{-3}$	7.6	$1.481 \cdot 10^{-14}$
0.45	$3.264 \cdot 10^{-1}$	2.9	$1.866 \cdot 10^{-3}$	7.8	$3.095 \cdot 10^{-15}$
0.50	$3.085 \cdot 10^{-1}$	3.0	$1.350 \cdot 10^{-3}$	8.0	$6.221 \cdot 10^{-16}$
0.55	$2.912 \cdot 10^{-1}$	3.1	$9.676 \cdot 10^{-4}$	8.2	$1.202 \cdot 10^{-16}$
0.60	$2.743 \cdot 10^{-1}$	3.2	$6.871 \cdot 10^{-4}$	8.4	$2.232 \cdot 10^{-17}$
0.65	$2.578 \cdot 10^{-1}$	3.3	$4.834 \cdot 10^{-4}$	8.6	$3.986 \cdot 10^{-18}$
0.70	$2.420 \cdot 10^{-1}$	3.4	$3.369 \cdot 10^{-4}$	8.8	$6.841 \cdot 10^{-19}$
0.75	$2.266 \cdot 10^{-1}$	3.5	$2.326 \cdot 10^{-4}$	9.0	$1.129 \cdot 10^{-19}$
0.80	$2.119 \cdot 10^{-1}$	3.6	$1.591 \cdot 10^{-4}$	9.2	$1.790 \cdot 10^{-20}$
0.85	$1.977 \cdot 10^{-1}$	3.7	$1.078 \cdot 10^{-4}$	9.4	$2.728 \cdot 10^{-21}$
0.90	$1.841 \cdot 10^{-1}$	3.8	$7.235 \cdot 10^{-5}$	9.6	$3.997 \cdot 10^{-22}$
0.95	$1.711 \cdot 10^{-1}$	3.9	$4.810 \cdot 10^{-5}$	9.8	$5.629 \cdot 10^{-23}$
1.00	$1.587 \cdot 10^{-1}$	4.0	$3.167 \cdot 10^{-5}$	10	$7.620 \cdot 10^{-24}$
1.05	$1.469 \cdot 10^{-1}$	4.1	$2.066 \cdot 10^{-5}$	11	$1.911 \cdot 10^{-28}$
1.10	$1.357 \cdot 10^{-1}$	4.2	$1.335 \cdot 10^{-5}$	12	$1.776 \cdot 10^{-33}$
1.15	$1.251 \cdot 10^{-1}$	4.3	$8.540 \cdot 10^{-6}$	13	$6.117 \cdot 10^{-39}$
1.20	$1.151 \cdot 10^{-1}$	4.4	$5.413 \cdot 10^{-6}$	14	$7.794 \cdot 10^{-45}$
1.25	$1.056 \cdot 10^{-1}$	4.5	$3.398 \cdot 10^{-6}$	15	$3.671 \cdot 10^{-51}$
1.30	$9.680 \cdot 10^{-2}$	4.6	$2.112 \cdot 10^{-6}$	16	$6.389 \cdot 10^{-58}$
1.35	$8.851 \cdot 10^{-2}$	4.7	$1.301 \cdot 10^{-6}$	17	$4.106 \cdot 10^{-65}$
1.40	$8.076 \cdot 10^{-2}$	4.8	$7.933 \cdot 10^{-7}$	18	$9.741 \cdot 10^{-73}$
1.45	$7.353 \cdot 10^{-2}$	4.9	$4.792 \cdot 10^{-7}$	19	$8.527 \cdot 10^{-81}$
1.50	$6.681 \cdot 10^{-2}$	5.0	$2.867 \cdot 10^{-7}$	20	$2.754 \cdot 10^{-89}$
1.55	$6.057 \cdot 10^{-2}$	5.1	$1.698 \cdot 10^{-7}$	21	$3.279 \cdot 10^{-98}$
1.60	$5.480 \cdot 10^{-2}$	5.2	$9.964 \cdot 10^{-8}$	22	$1.440 \cdot 10^{-107}$
1.65	$4.947 \cdot 10^{-2}$	5.3	$5.790 \cdot 10^{-8}$	23	$2.331 \cdot 10^{-117}$
1.70	$4.457 \cdot 10^{-2}$	5.4	$3.332 \cdot 10^{-8}$	24	$1.390 \cdot 10^{-127}$
1.75	$4.006 \cdot 10^{-2}$	5.5	$1.899 \cdot 10^{-8}$	25	$3.057 \cdot 10^{-138}$
1.80	$3.593 \cdot 10^{-2}$	5.6	$1.072 \cdot 10^{-8}$	26	$2.476 \cdot 10^{-149}$
1.85	$3.216 \cdot 10^{-2}$	5.7	$5.990 \cdot 10^{-9}$	27	$7.389 \cdot 10^{-161}$
1.90	$2.872 \cdot 10^{-2}$	5.8	$3.316 \cdot 10^{-9}$	28	$8.124 \cdot 10^{-173}$
1.95	$2.559 \cdot 10^{-2}$	5.9	$1.818 \cdot 10^{-9}$	29	$3.290 \cdot 10^{-185}$
2.00	$2.275 \cdot 10^{-2}$	6.0	$9.866 \cdot 10^{-10}$	30	$4.907 \cdot 10^{-198}$

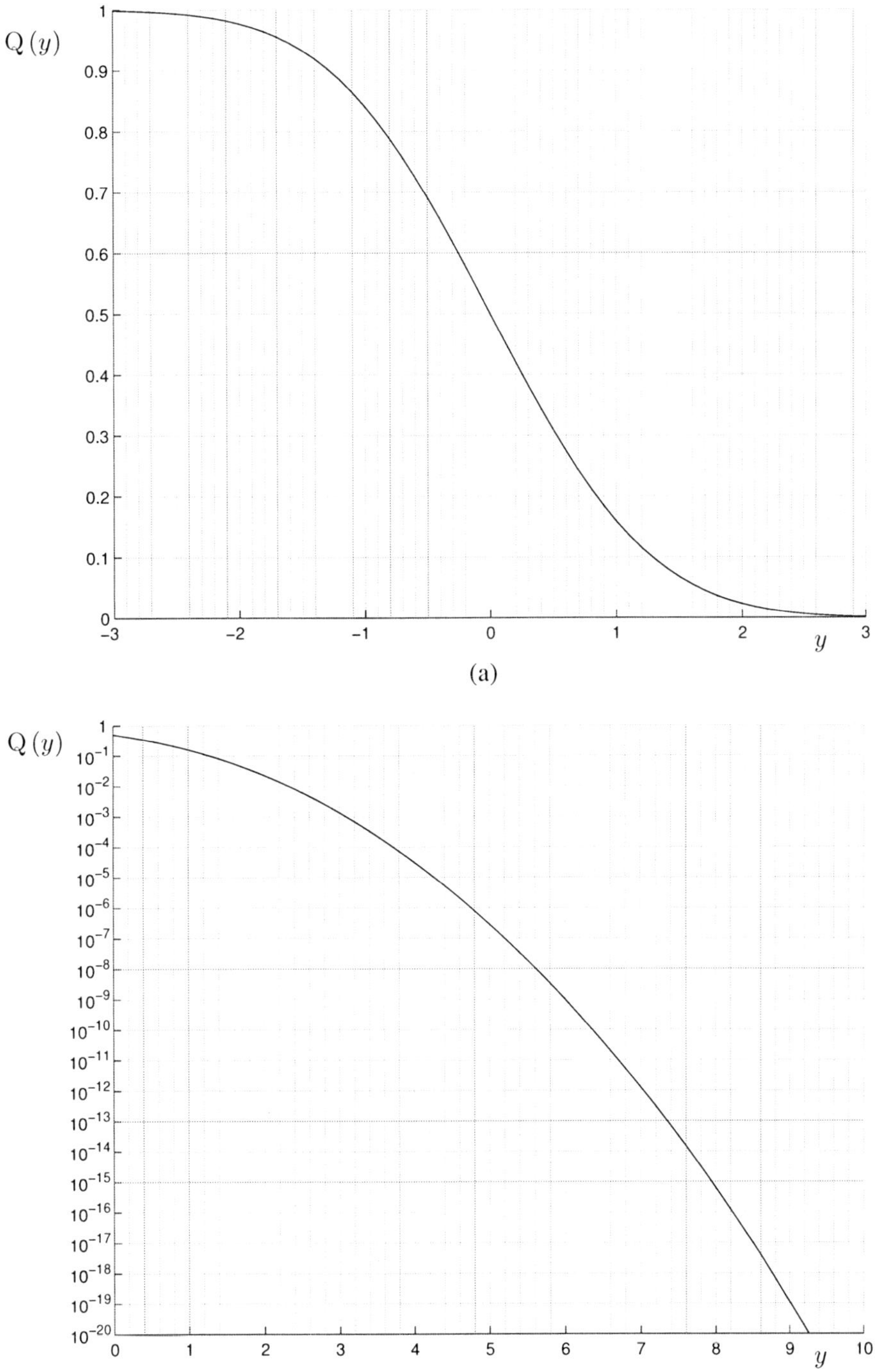

Figure 2.A.1 Plots of the complementary normalized Gaussian distribution function $Q(y)$: (a) on a linear scale; (b) on a logarithmic scale.

By letting $\Delta y = y_2 - y_1$ and $\Delta Q = \mathrm{Q}(y_1) - \mathrm{Q}(y_2)$, the absolute error in the approximation is thus bounded by

$$|y - y_i| \left(\frac{\Delta Q}{\Delta y} - \frac{e^{-\frac{1}{2}y_i^2}}{\sqrt{2\pi}} \right) \tag{2.A.3}$$

For example, to evaluate $\mathrm{Q}(3.77)$ we find $y_1 = 3.7$ and $y_2 = 3.8$, and the approximation (2.A.1) yields $\mathrm{Q}(3.77) \simeq 0.7 \cdot \mathrm{Q}(3.8) + 0.3 \cdot \mathrm{Q}(3.7) \simeq 8.298 \cdot 10^{-5}$. The error is less than $1.88 \cdot 10^{-6}$ (about 2% relative error) so that the exact value lies between $8.11 \cdot 10^{-5}$ and $8.3 \cdot 10^{-5}$.

- If y is large, use the approximation

$$\mathrm{Q}(y) \simeq \frac{1}{y\sqrt{2\pi}} e^{-\frac{1}{2}y^2} \tag{2.A.4}$$

The relative accuracy of the approximation is assured by the fact that $\mathrm{Q}(y)$ obeys the following bounds

$$\frac{1}{y\sqrt{2\pi}} e^{-\frac{1}{2}y^2} \left(1 - \frac{1}{y^2}\right) < \mathrm{Q}(y) < \frac{1}{y\sqrt{2\pi}} e^{-\frac{1}{2}y^2} \quad , \quad y > 0 \tag{2.A.5}$$

so that the approximation (2.A.4) is always an upper bound and the relative error is lower than $1/y^2$.

For example, to evaluate $\mathrm{Q}\left(\sqrt{200}\right)$ we calculate $e^{-100}/\sqrt{400\pi} \simeq 1.05 \cdot 10^{-45}$, with a relative error lower than 0.5%. The absolute error is smaller than $5 \cdot 10^{-48}$, that is the exact value is between $1.045 \cdot 10^{-45}$ and $1.055 \cdot 10^{-45}$.

In Fig. 2.A.1 we give two plots of the function $\mathrm{Q}(y)$ on a linear and a logarithmic scale respectively.

REFERENCES

1. G.R. Grimmett, D.R. Stirzaker, *Probability and random processes.* Oxford, United Kingdom: Clarendon Press, 2nd ed., 1992.
2. A.V. Oppenheim, R.W. Schafer, *Discrete time signal processing.* Englewood Cliffs, NJ: Prentice-Hall, 2nd ed., 1998.
3. Strang, Gilbert, *Introduction to linear algebra*, 3rd ed. Wellesley, MA: Wellesley-Cambridge Press, March 2003.
4. A. Papoulis, *Probability, random variables and stochastic processes.* New York, NY: McGraw-Hill, 3rd ed., 1991.
5. A. Papoulis, *Signal analysis.* New York: McGraw-Hill, 1984.
6. A. Papoulis, *The Fourier integral and its applications.* New York, NY: McGraw-Hill, 1962.

Problems

Energy and power

2.1 Plot the signal

$$x(t) = Ae^{-t/T}1(t+T) + Ae^{-t/T}1(t-T) \quad , \quad A = 1\,\text{V}\,,\; T = 1\,\text{ms}$$

and find its energy E_x and energy spectral density $\mathcal{E}_x$.

2.2 Given the continuous time signals

$$x(t) = A\,\text{sinc}^2(Ft)e^{j2\pi Ft} \quad , \quad y(t) = A\,\text{sinc}(2Ft)$$

let $z = x * y$,

a) Find the expression of $\mathcal{Z}(f)$.

b) Find the energy of z.

c) Find the cross energies E_{xy} and E_{yz}.

2.3

a) Find the average power of the following signals

$$x(t) = \cos 2\pi t/T_p + |\cos 2\pi t/T_p|\,, \qquad y(t) = \sum_{\ell=0}^{+\infty} \frac{1}{(3+j)^{\ell}}\, e^{j2\pi F\ell t}\,.$$

b) Find the average power of the signal with Fourier transform

$$\mathcal{X}(f) = f^2\,\text{rect}\left(\frac{f}{2B}\right)\,.$$

Bandwidth

2.4 The real valued signal $x(t)$, $t \in \mathbb{R}$, has bandwidth B. Determine the bandwidth of the following signals.

a) $y_1(t) = x(t)\,\cos(2\pi f_0 t)$,

b) $y_2(t) = (x(t) + A)\,\cos(2\pi f_0 t)$,

c) $y_3(t) = [x(t)\,\cos(2\pi f_1 t)] * [\text{sinc}(t/T)\,e^{j2\pi t/T}\,\cos(2\pi f_1 t)]$.

Consider $B = 1/T = 100\,\text{Hz}$, $f_0 = 20\,\text{Hz}$, $f_1 = 30\,\text{Hz}$.

2.5 Find the band and bandwidth of the following signals

a) $x(t) = A\,\text{sinc}(Ft)e^{j\pi Ft/2}$

b) $y(t) = A\,\text{sinc}^k(t/T_1)$

c) $s(nT) = Ae^{-|n|}$

d) $z(t) = A\sin^3(2\pi t/T_p)$

2.6 Consider the continuous time real-valued signal $x(t) = e^{(t_0-t)/T}\,1(t-t_0)$, $t \in \mathbb{R}$.

a) Find the expression of its practical bandwidth B based on the amplitude criterion. Calculate the value of B with $\varepsilon = -40\,\text{dB}$, $t_0 = 30\,\text{ms}$ and $T = 1\,\text{ms}$.

b) Find the expression of its practical bandwidth B based on the energy criterion. Calculate the value of B with $\varepsilon = 1\%$, $t_0 = 30\,\text{ms}$ and $T = 1\,\text{ms}$.

Discuss the dependence of the results on the signal parameters t_0 and T.

2.7 Consider the continuous time real signal

$$x(t) = \frac{A}{(T^2 + t^2)} \quad , \quad t \in \mathbb{R}$$

a) Find the expression of its practical bandwidth B based on the amplitude criterion. Calculate the value of B with $\varepsilon = -20\,\text{dB}$, $A = 1$ V, $T = 1$ ms.

b) Find the expression of its practical bandwidth B based on the energy criterion. Calculate the value of B with $\varepsilon = 1\%$, $A = 1$ V, $T = 1$ ms.

c) Repeat 2.7.a for the signal $y(t) = x(t)\cos 2\pi f_0 t$, with $f_0 T > 1$, and calculate B for the same parameters as 2.7.a and $f_0 = 10\,\text{kHz}$. Discuss the dependence of the results on the signal parameters T and f_0.

2.8 Consider the transformation of Fig. 2.23 where $T_s = 0.1$ ms and both the filter and the interpolate filter have frequency response

$$\mathcal{H}(f) = \mathcal{G}(f) = \frac{1}{2}\,\text{rcos}(f/F_1, \rho) \quad , \quad \text{with } \rho = \frac{1}{4}\,, \; F_1 = 9\,\text{kHz}$$

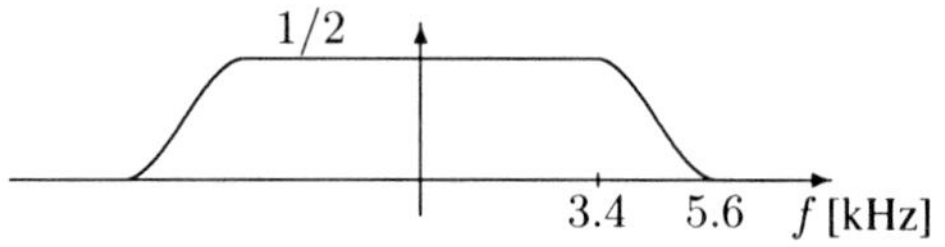

a) Prove that the overall transformation satisfies the Heaviside conditions $\tilde{x}(t) = A x_i(t - t_0)$ for all bandlimited signals x_i with bandwidth $B_x \leq (1-\rho)F_1/2 = 3375\,\text{Hz}$ and give the appropriate values for A and t_0.

b) Prove that the overall transformation is not in general equivalent to a filter for signals with bandwidth greater than $F_s - B_g = 4375$ Hz.

c) Suppose the filter h and the sampler are assigned as above, the task is to design the interpolate filter so that Heaviside conditions are met for all signals $x_i(t)$ with bandwidth up to $F_1/2$. Determine the frequency response of both the interpolate filter with minimum bandwidth and with maximum bandwidth.

Vector representation of signals

2.9 Consider the signal set

$$s_1(t) = \begin{cases} 2A & , 0 \leq t < 2 \\ A & , 2 \leq t < 3 \\ 0 & , \text{otherwise} \end{cases} \qquad s_2(t) = \begin{cases} 2A & , 0 \leq t < 1 \\ A & , 1 \leq t < 3 \\ 0 & , \text{otherwise} \end{cases}$$

Identify an orthonormal basis and the corresponding vector representation using the Gram-Schmidt procedure.

2.10 Evaluate an orthonormal basis for the signal set

$$s_1(t) = \text{rect}(t)\,, \quad s_2(t) = \text{sgn}(t)\,\text{rect}(t)\,, \quad s_3(t) = \text{triang}(2t)\,.$$

What is the corresponding signal vector representation? What are the signal energies?

2.11 Evaluate, for the signals displayed in figure, an orthonormal basis and the corresponding signal vector representation. Also evaluate energies and distances between signals.

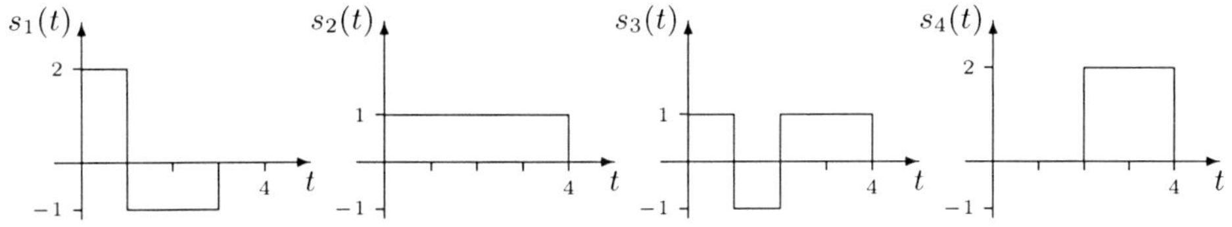

2.12 Express the projection of the signal $s(t) = e^{-t}\,\mathrm{rect}(t)$ onto the signal space spanned by the basis with $\varphi_1(t) = \mathrm{rect}(t)$ and

$$\varphi_i(t) = A_i\,\sin(2\pi f_i t)\,\mathrm{rect}(t)\,, \qquad f_i = i - 1\,, \quad i = 2, 3, 4, 5\,.$$

What is the relevance of the projection error in terms of percentage of the signaling energy?

2.13 Let the signal

$$s(t) = \sin(\pi t/4)\,, \qquad 0 \le t < 8$$

be projected onto the space spanned by the signals illustrated in figure. Evaluate the projection $\hat{s}(t)$ and the energy of the projection error.

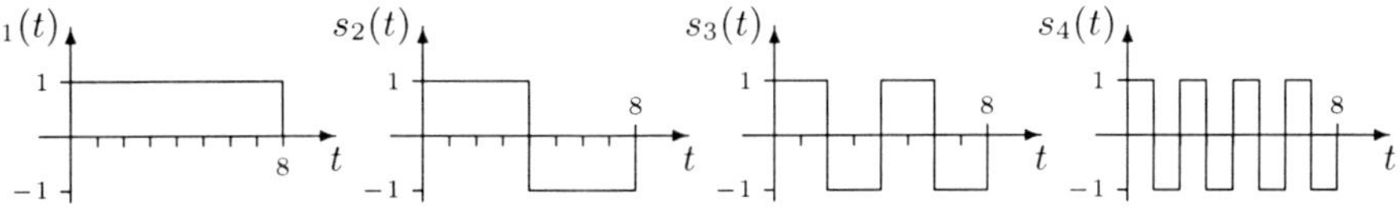

Random variables and vectors

2.14 Let $y = x + w$ with w a Gaussian rv with zero mean and variance σ_w^2. Determine the probability $\mathrm{P}\left[y > 0 | x = -1\right]$.

2.15 Let x_1 and x_2 be two independent Gaussian rvs with zero mean and variance σ_1^2 and σ_2^2, respectively. Determine the probability P_1 that the random vector $[x_1, x_2]$ is within the upper right sub-plane having lower left corner $(1, 1)$. Compute the probability P_2 that $[x_1, x_2]$ is outside of the above sub-plane.

2.16 Let x a Poisson rv, with alphabet $\mathcal{A}_x = \{0, 1, 2, \ldots\}$ and PMD $p_x(a) = e^{-\Lambda}\Lambda^a/(a!)$.

a) Find the value of Λ for which $\mathrm{P}\left[x = 2\right]$ is maximized.

b) Show that the probability $\mathrm{P}\left[x > 2\right]$ as a function of Λ is continuous and monotonically increasing.

c) Calculate $\mathrm{E}\left[e^x\right]$.

2.17 Let x a Gaussian rv $x \sim \mathcal{N}(0, \sigma^2)$.

a) Find the PDF of $y = x^2 + 1$.

b) Calculate $\mathrm{E}\left[x^2(x + 1 + \sin x)\right]$.

2.18 In GSM cellular phone newtworks, the cells covering wide open environments are approximately circular with a maximum radius of 35 km. Assume the propagation speed of the waves in air be $c = 3 \cdot 10^8$ m/s. Consider a terminal randomly placed within such a cell, so that with x and y the spatial coordinates of the terminal with respect to the base station, the PDF of the rve $\boldsymbol{x} = [x, y]$ is constant within a circle centered at the origin and having a radius of 35 km, and zero outside.

a) Write the expression of $p_{\boldsymbol{x}}(a_1, a_2)$.

b) Find the PDF of the distance r between terminal and base station.

c) With τ the propagation delay over the distance r, find $\mathtt{m}_\tau$, σ_τ^2 and $\mathrm{P}[\tau > \mathtt{m}_\tau]$.

2.19 Let $\boldsymbol{x}$ a Gaussian rve with mean vector and covariance matrix

$$\boldsymbol{m_x} = \begin{bmatrix} 2 \\ -1 \end{bmatrix} \;, \quad \boldsymbol{k_x} = \begin{bmatrix} 1 & 0 \\ 0 & 3 \end{bmatrix} \;,$$

and let $z = x_1 + x_2$. Are x_1 and x_2 statistically independent? Write the expression of $p_z(a)$.

2.20 Consider the rve $\boldsymbol{x} = [x_1, x_2]$ having PDF

$$p_{\boldsymbol{x}}(a_1, a_2) = \begin{cases} e^{\lambda(a_1+a_2)+c} & , \; |a_1| + |a_2| \le 1 \\ 0 & , \; |a_1| + |a_2| > 1 \end{cases}$$

a) State whether x_1, x_2 are statistically independent and give the expression of the conditional PDF $p_{x_1|x_2}(a_1|a_2)$.

b) Find the value of c for $\lambda = 1$.

c) Does it exist a linear invertible transformation $\boldsymbol{y} = \boldsymbol{A}\boldsymbol{x}$ such that $\boldsymbol{y}$ is a Gaussian rve?

2.21 Consider K independent rvs y_k, $k = 1, 2, \ldots, K$, with exponential distribution of parameter λ. Let $y_{\max} = \max_k\{y_k\}$ and $y_{\min} = \min_k\{y_k\}$. Find the PDF of $y_{\max}$ and $y_{\min}$ and determine whether they are exponentially distributed themselves.

2.22 Let y be an exponential rv of parameter λ. Consider another rv v that, conditioned on $y = a$, is Poisson distributed with parameter a. Find the PMD of v.

2.23 Let y_1, y_2, ... iid rvs with exponential distribution of parameter λ. Let x be a discrete rv defined as follows

$$x = \max\left\{k : \sum_{i=1}^{k} y_i \le 1\right\}$$

where it is understood that $x = 0$ if $y_1 > 1$. Prove that x has a Poisson distribution of parameter λ. Note that this method can be used for the computer-aided generation of instances of a Poisson-distributed rv. **Hint.** The rv g_k resulting from the sum of k iid rvs with exponential distribution has gamma distribution with PDF

$$p_{g_k}(a) = \frac{\lambda^k a^{k-1}}{(k-1)!} e^{-\lambda a}; \; a \ge 0 \, .$$

Random processes

2.24 Let $x(t)$ be a Poisson process with rate $\lambda = 10^3\,\mathrm{s}^{-1}$, and fix instants $t_1 = 0\,\mathrm{ms}$, $t_2 = 3\,\mathrm{ms}$, $t_3 = 4\,\mathrm{ms}$, $t_4 = 5\,\mathrm{ms}$

a) Find the probability that exactly 2 arrivals lie in the interval (t_2, t_4), given that there are none in the interval (t_1, t_3).

b) Let τ_1', τ_2' be two unordered arrivals in the interval (t_3, t_4), find the mean and variance of their difference $z = \tau_1' - \tau_2'$.

2.25 Let $x(t)$ be a Poisson process with arrival rate $\lambda = 0.5\,\mathrm{s}^{-1}$.

a) Find the probability that $x(t)$ has at least 2 arrivals in the interval (t_1, t_3) and exactly 1 of them lies in (t_2, t_3), where $t_1 = 3\,\text{s}$, $t_2 = 4\,\text{s}$, $t_3 = 6\,\text{s}$.

b) With τ_1, τ_2, τ_3 three consecutive and ordered arrival times for $x(t)$, find the probability $\mathrm{P}\left[\tau_3 - \tau_1 > 3\,\text{s} | \tau_2 - \tau_1 = 1\,\text{s}\right]$

2.26 The random process $x(t)$ is defined as

$$x(t) = \sum_{n=-\infty}^{+\infty} \text{rect}\left(\frac{t - \tau_n}{T}\right) \quad , \quad t \in (R)$$

where $\{\tau_n\}$ are the ordered arrival times of a homogeneous Poisson process with rate $\lambda = 0.8/T$. Plot a realization of $x(t)$ and find the probability $\mathrm{P}\left[x(t) \leq 1\right]$.

2.27 Let $x(t)$ and $y(t)$, be two independent and homogeneous Poisson processes with rates $\lambda_x = 5 \cdot 10^3\,\text{s}^{-1}$ and $\lambda_y = 2 \cdot 10^3\,\text{s}^{-1}$, respectively

a) Find mean and statistical power of the total number of arrivals for the two processes in $(0, T]$ with $T = 3\,\text{ms}$.

b) Find the probability that the first arrival of $y(t)$ preceeds the first arrival of $x(t)$.

2.28 Let $x(t)$ denote a Poisson counting process, homogenous with arrival rate λ. Find its autocorrelation $\mathbf{r}_x(t, \tau)$.

2.29 Let $\{x_i(t)\}$, $i = 1, 2, \ldots, n$, be a family of statistically independent Poisson processes with equal parameter λ. Determine the PDF of the random variable T define as the first instant ≥ 0 at which all the processes have seen at least one arrival , i.e.,

$$T = \arg\min_{t \geq 0} \{x_i(t) > 0, \ \forall i\}.$$

2.30 For what values of the parameters $A, B \in \mathbb{R}$ the functions below can be

a) PSDs of stationary rps with finite power

$$\mathcal{P}_1(f) = A\,\text{rect}(Tf) + B\,\text{triang}(Tf/2)$$

$$\mathcal{P}_2(f) = e^{A|f|} - e^{B|f|}$$

b) Autocorrelations of stationary rps

$$\mathbf{r}_3(mT) = \begin{cases} A & , \quad m = 0 \\ B & , \quad m = \pm 1 \\ 0 & , \quad |m| \geq 2 \end{cases}$$

$$\mathbf{r}_4(\tau) = A\,\text{sinc}(\tau/T - B)$$

Hint: Check whether their Fourier transform can be PSDs.

2.31 Starting from the binary random process x_n, $n \in \mathbb{Z}$, with iid random variables and probability mass distribution $p_x(1) = p_x(-1) = \frac{1}{2}$.

a) Calculate the mean, autocorrelation, PSD and the first order PMD of the process y_n

$$y_n = \frac{x_n + x_{n-1}}{2}$$

and determine whether it has iid random variables itself.

b) Determine whether the signal

$$z_n = x_n \, x_{n-1}$$

has iid random variables.

2.32 Consider the Gaussian random process $s(t)$ having zero mean and PSD

$$\mathcal{P}_s(f) = \mathrm{triang}(fT) \; ,$$

where $T = 2$. Evaluate the probability $\mathrm{P}\left[s(t) > 1\right]$.

2.33 Consider a continuous time WSS Gaussian rp $x(t)$ with mean $\mathtt{m_x} = 1$ and autocorrelation $\mathtt{r}_x(\tau) = 1 + \mathrm{triang}(\tau/T)$.

a) Write the expression of its PSD.
b) Are the rvs $x(0)$ and $x(T)$ statistically independent?
c) Find the probability $\mathrm{P}\left[x(0) + x(T) < 1\right]$.

2.34 The continuous time Gaussian process x has zero mean and autocorrelation

$$\mathtt{r}_x(t,\tau) = e^{-f_0|\tau|} \left|\cos 2\pi f_0 t\right|$$

a) Find $\mathrm{P}\left[x(T_1) > 1\right]$ and $\mathrm{P}\left[x(T_1) > x(2T_1)\right]$, with $T_1 = 1/(8f_0)$.
b) Write the expression of the average PSD of x.

Systems with random inputs and outputs

2.35 Let $s(t)$ be a Gaussian random process with zero mean and PSD

$$\mathcal{P}_s(f) = \frac{10^{-2}}{2B} \, \mathrm{rect}\left(\frac{f}{2B}\right) \qquad B = 10^6 \; \mathrm{Hz} \; .$$

The process is filtered by a filter with frequency response

$$\mathcal{G}_{\mathrm{Ch}}(f) = \mathcal{G}_0 \, \mathrm{rect}\left(\frac{f}{2B_{\mathrm{Ch}}}\right) \; ,$$

with $\mathcal{G}_0 = 0.3$, $B_{\mathrm{Ch}} = 10^9$ Hz. Compute the probability that the amplitude of the signal at the filter output, $y(t)$, lies in the range $[-0.6, 0.6]$.

2.36 The continuous time random process x, WSS with mean $\mathtt{m}_x = 2$ V and PSD

$$\mathcal{P}_x(f) = \frac{P_0}{B} \, \mathrm{triang}(f/B) + P_0 \delta(f - B) + \mathtt{m}_x^2 \delta(f) \quad , \quad P_0 = 8\,\mathrm{V}^2$$

is input to a filter with frequency response $\mathcal{G}(f) = \mathcal{G}_0 \, \mathrm{rect}(f/B)$.

a) Find the mean and statistical power of the filter ouput y for $\mathcal{G}_0 = -3$ dB.
b) For $\mathcal{G}_0 = 1$, define the rp z as $z(t) = x(t) - y(t)$. Find mean, PSD and statistical power of z. Is z a real valued rp?

CHAPTER 3

SOURCES OF DIGITAL INFORMATION

Nicola Laurenti

In this chapter we describe how information produced by a source, either analog or digital, can be effectively encoded into a digital message for efficient transmission. We will introduce the reference scheme for analog to digital conversion, as well as the fundamentals of Information Theory with a quantitative notion of the information carried by a message, and source coding techniques.

3.1 DIGITAL REPRESENTATION OF WAVEFORMS

As outlined in the Introduction, the objective of a communication system is to transmit an information signal, which is typically analog. One possibility is to transmit the information signal by means of an *analog modulation* scheme, as shown in Fig. 3.1. Typically the information signal is passband, but its low frequency is much smaller than its bandwidth, making it more similar to a baseband signal. On the other hand, in most practical cases, the channel is characterized by a bandpass medium whose band does not in general include that of the information signal. Here the information signal is simply matched to the channel characteristics, by replacing it with an associated passband signal through an analog modulator. At the receiver side the original information signal is reproduced by a suitable demodulator.

Alternatively, the transmission may take place by converting the analog information signal into a digital (usually binary) message by means of an *analog-to-digital converter* (ADC). Next,

Principles of Communications Networks and Systems N. Benvenuto, M. Zorzi

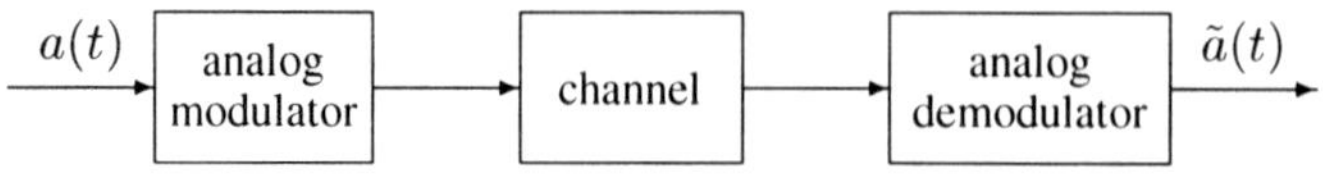

Figure 3.1 Analog modulation system.

the binary message is converted by a digital modulator into an analog signal that is suitable for transmission over the analog channel. At the receiver, the reverse process occurs: firstly, a digital demodulator reproduces the transmitted binary message, whereas the conversion of the sequence of bits into a replica of the analog information signal is performed by a *digital-to-analog converter* (DAC), as illustrated in Fig. 3.2. We observe that the system from the input of the digital modulator to the output of the digital demodulator consitutes a *digital channel* (see Section 5.5.3).

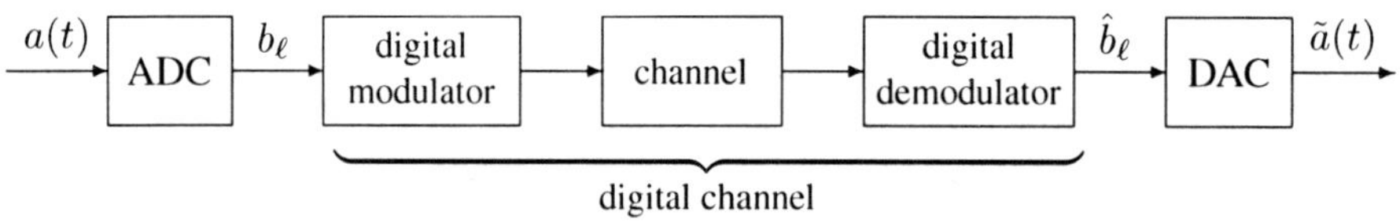

Figure 3.2 Digital transmission of an analog information signal.

To summarize the overall system is composed of:

- an ADC
- a digital channel
- a DAC

In the following, the principle of the conversion blocks (ADC and DAC) will be presented, whereas the performance of the overall system will be derived in Chapter 5.

3.1.1 Analog to digital converter (ADC)

The basic blocks to convert an analog signal into a digital (binary) sequence are illustrated in Fig. 3.3. These are as follows:

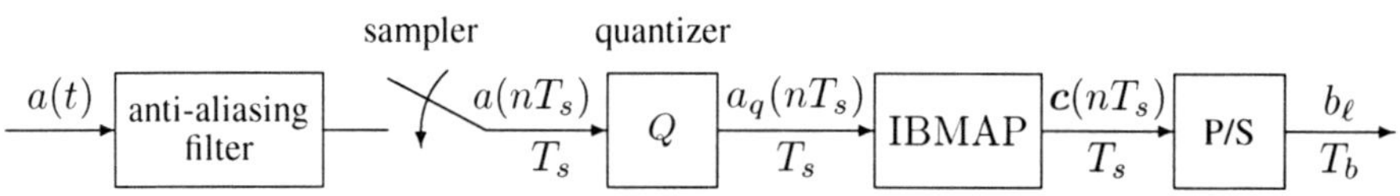

Figure 3.3 Basic scheme of an ADC for the digital coversion of an analog signal.

- A *sampler* transforms a continuous time signal $a(t)$ into a discrete time signal $\{a(nT_s)\}$.

- A *quantizer* approximates each continuous value sample $a(nT_s)$, with a quantized value sample $a_q(nT_s)$ taking values in a *finite alphabet* $\mathcal{A}_q$ with L elements

$$\mathcal{A}_q = \{\mathcal{Q}_0, \mathcal{Q}_1, \ldots, \mathcal{Q}_{L-1}\}, \tag{3.1}$$

where $\{\mathcal{Q}_i\}$ are the quantizer output values, also denoted as *quantizer levels*.

- An *inverse bit map* (IBMAP) represents each element of $\mathcal{A}_q$ with a binary word of $\mathtt{b}$ bits, with $2^{\mathtt{b}} \geq L$. Hence to each quantized value sample $a_q(nT_s)$, taking values in $\mathcal{A}_q$, we obtain a corresponding binary representation

$$\boldsymbol{c}(nT_s) = [c_{\mathtt{b}-1}, \ldots, c_0] \ , \ c_j \in \{0, 1\}, \tag{3.2}$$

of $\mathtt{b}$ bits. The sequence of words is then parallel-to-serial (P/S) converted into a binary stream $\{b_\ell\}$ with period

$$T_b = \frac{T_s}{\mathtt{b}} . \tag{3.3}$$

The signal transformations of Fig. 3.3 are illustrated in Fig. 3.4 for $L = 8$, where the samples obtained every T_s seconds are represented by crosses, while the quantized values are represented by bold dots.

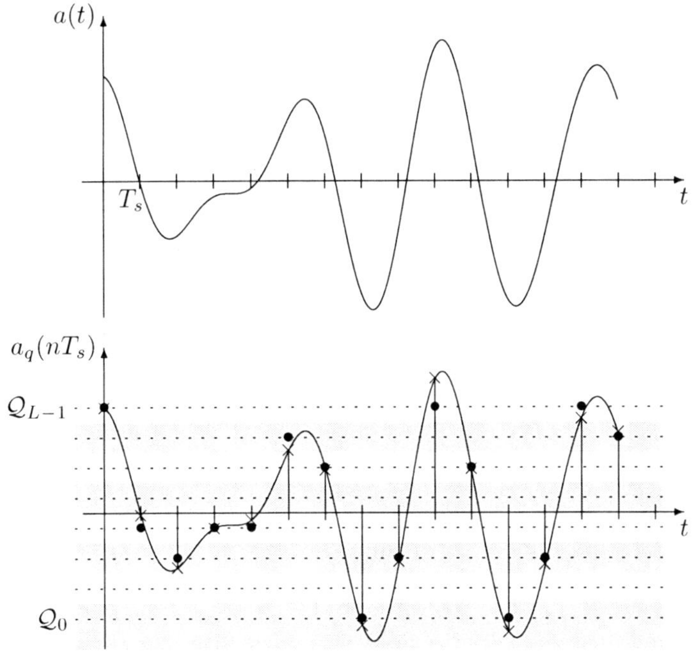

Figure 3.4 Signal transformations in an ADC. The dotted lines denote quantizer levels, while stripes denote quantization regions.

From the Sampling Theorem 2.8, the choice of the sampling frequency $F_s = 1/T_s$ is related to the bandwidth B of the information signal $a(t)$, i.e. we must have

$$F_s = \frac{1}{T_s} \geq 2B \,. \tag{3.4}$$

In practice, we observe that F_s is chosen greater than $2B$ to simplify the realization of the interpolate filter at the receiver. As will be further discussed in Section 3.3.4, one parameter characterizing the binary message $\{b_\ell\}$ in Fig. 3.2 is the nominal information rate (also called the bit rate) $R_b = 1/T_b$, where T_b is the bit period. From (3.3), it is

$$R_b = \mathtt{b}\, F_s \text{ [bit/s]} \,. \tag{3.5}$$

In general, the transformation from an analog signal into a binary stream is called *encoding*. In fact, the ADC can be considered as the simplest encoder. Indeed, regarding the system of Fig. 3.3, further processing can be applied to the output bits, such as *source coding* (described in Section 3.4) to reduce the bit-rate and hence obtain a more efficient system in terms of the required resources (e.g. the channel bandwidth).

3.1.2 Digital to analog converter (DAC)

As shown in Fig. 3.5, restoring the analog signal from a sequence of bits requires mapping each binary word of $\mathtt{b}$ bits into the corresponding element of the alphabet $\mathcal{A}_q$. Indeed, the binary mapping (BMAP) realizes the inverse function of IBMAP. Hence, starting from the binary stream $\{\hat{b}_\ell\}$, firstly a serial-to-parallel conversion (S/P) yields a sequence of words $\{\tilde{\mathtt{c}}(nT_s)\}$ of $\mathtt{b}$ bits. Then each word is mapped into an element of $\mathcal{A}_q$, $\tilde{a}_q(nT_s)$. All these sample values are finally time interpolated by an interpolate filter whose ideal frequency response is (see (2.76))

$$\mathcal{G}(f) = T_s \operatorname{rect}\left(\frac{f}{F_s}\right) \,. \tag{3.6}$$

In general, the transformation from a binary stream into an analog signal is called *decoding*.

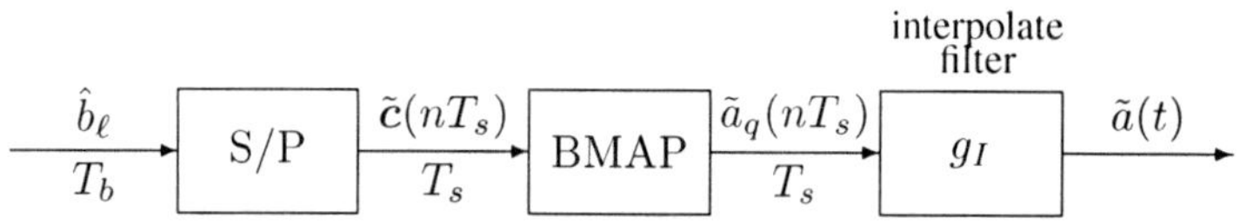

Figure 3.5 Basic scheme of a DAC for the analog conversion of a digital signal.

The signal transformations of Fig. 3.5 are illustrated in Fig. 3.6. Since the input samples to the DAC are quantized, we note that the reconstructed analog signal $\tilde{a}(t)$ is different from the original signal $a(t)$.

Typically, the DAC employs a much simpler interpolate filter than (3.6), that is an *holder* that holds each input value for T_s seconds as illustrated in Fig. 3.7. In this case the interpolate filter has impulse response

$$g(t) = \operatorname{rect}\left(\frac{t - T_s/2}{T_s}\right) , \tag{3.7}$$

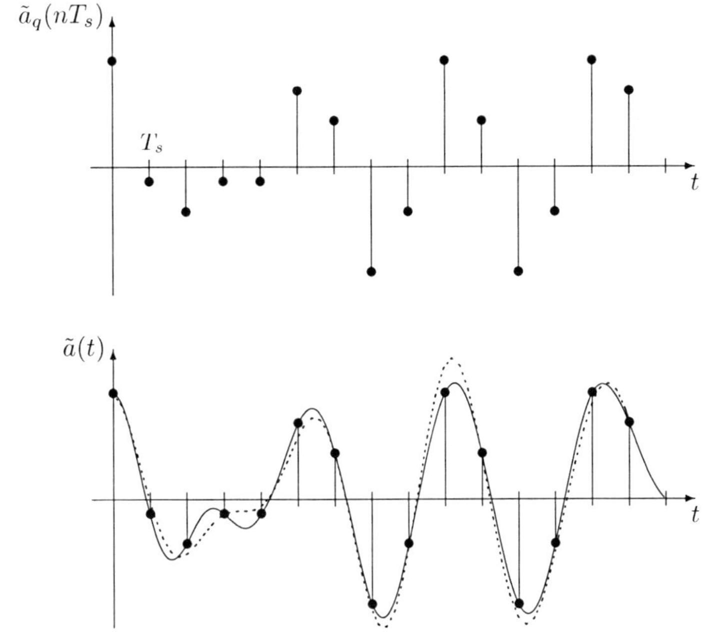

Figure 3.6 Signal transformations in a DAC. The dotted line represents the original signal $a(t)$, while the solid line represents the reconstructed signal $\tilde{a}(t)$.

and frequency response

$$\mathcal{G}(f) = T_s \operatorname{sinc}\left(\frac{f}{F_s}\right) e^{-j2\pi f T_s/2} . \tag{3.8}$$

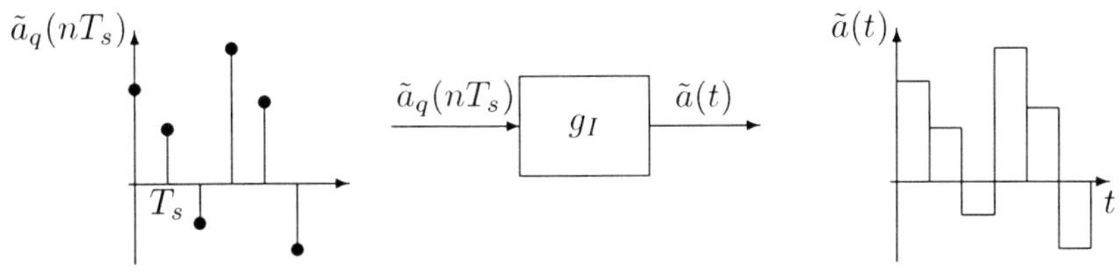

Figure 3.7 Interpolate filter as a holder.

Unless the sampling frequency F_s has been chosen sufficiently higher than twice the bandwidth of $a(t)$, we see that the interpolate filter (3.8), besides not attenuating adequately the images of $\tilde{a}_q(nT_s)$ (i.e. the frequency components of $\tilde{a}_q(nT_s)$ around the frequencies ℓF_s, $\ell \neq 0$) introduces also distortion in the band of the desired signal. A solution to this problem consists in introducing, before interpolation, a digital equalizer filter g_{comp} with a frequency response equal to $1/\operatorname{sinc}(f/F_s)$ in the band of $a(t)$. Fig. 3.8 illustrates this solution.

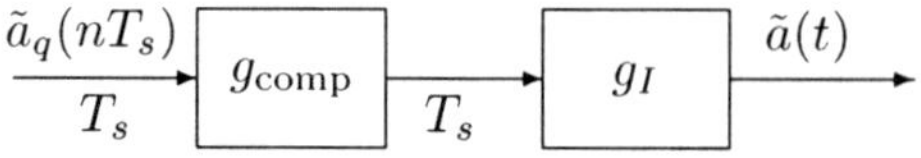

Figure 3.8 Interpolate filter as a holder preceded by a digital equalizer.

Remark In practice, while processing is done at the slowest rate with a sampling frequency F_s just slightly higher than $2B$, before the 'analog interpolator' a 'digital interpolator' can be used to increase the sampling rate by a factor at least four, to alleviate the specifications of the analog interpolate filter.

Let us consider the quantizer in detail.

3.1.3 Quantizer

With reference to the scheme of Fig. 3.3, for a quantizer with L output values we have:

- input sample $a(nT_s) \in \mathbb{R}$;
- quantized sample $a_q(nT_s) \in \mathcal{A}_q = \{\mathcal{Q}_0, \ldots, \mathcal{Q}_{L-1}\}$;
- codeword $\mathbf{c}(nT_s)$ of b bits representing the value of $a_q(nT_s)$.

The system with input $a(nT_s)$ and output $\mathbf{c}(nT_s)$ is also called a *pulse code modulation (PCM) encoder*, where each codeword $\mathbf{c}(nT_s)$ represents a sample of the input signal.

The quantizer can be described by the function

$$\mathcal{Q} : \mathbb{R} \longrightarrow \mathcal{A}_q \,. \tag{3.9}$$

In other words, specifying the quantizer characteristic is equivalent to determining a partition of the real axis in the regions $\{R_i\}$, $i = 0, \ldots, L-1$, such that $\mathbb{R} = \bigcup_{i=0}^{L-1} R_i$ and $R_i \bigcap R_j = \emptyset$, $i \neq j$, and the quantization rule is

$$\text{if} \quad a(nT_s) \in R_i \qquad \text{then} \qquad \mathcal{Q}[a(nT_s)] = a_q(nT_s) = \mathcal{Q}_i \,. \tag{3.10}$$

A common choice for the quantization regions is given by adjacent intervals:

$$R_i = (v_i, v_{i+1}] \;, \;\text{for}\;\; i = 0, \ldots, L-1 \;, \tag{3.11}$$

where $v_0 = -\infty$ and $v_L = \infty$. We note that $L-1$ *quantization thresholds* $\{v_i\}$ need to be determined, as v_0 and v_L are fixed. The mapping function (3.9) is called the *quantizer characteristic* and is illustrated in Fig. 3.9 for $L = 8$.

It is often useful to compactly represent the quantizer characteristic on a single axis, as illustrated in Fig. 3.10, where the values of the quantization thresholds are indicated by lines, and the quantizer output values by dots.

Now each value of the alphabet $\mathcal{A}_q$ and hence each value of $a_q(nT_s)$ can be represented by a binary representation with $\lceil \log_2 L \rceil$ bits[1]. Hence usually it is $L = 2^{\mathsf{b}}$. For the quantizer characteristic of Fig. 3.9, a binary representation is adopted that goes from 000 in correspondence

[1] $\lceil x \rceil$ is the ceiling function and denotes the smallest integer higher than x

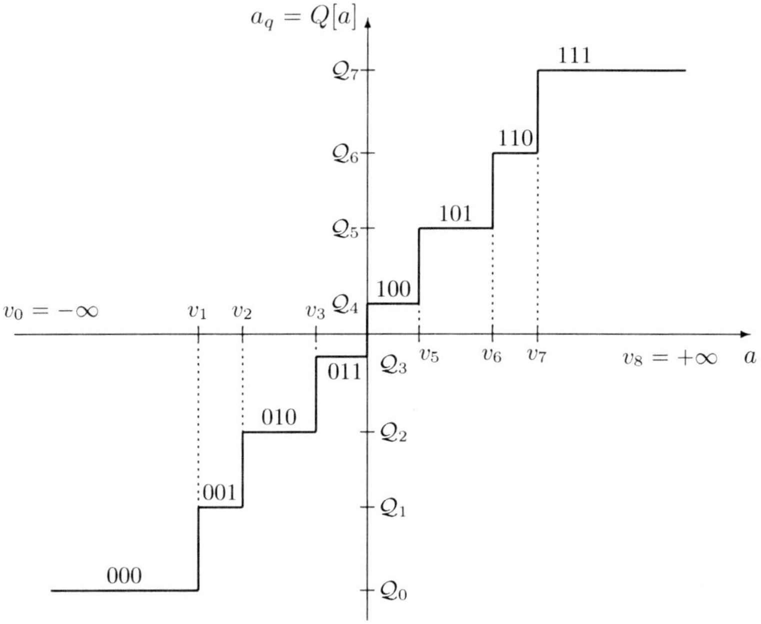

Figure 3.9 Quantizer characteristic for $L = 8$.

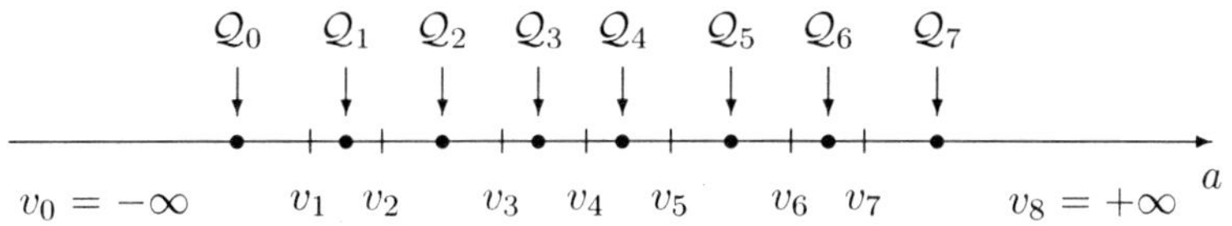

Figure 3.10 Compact quantizer characteristic for $L = 8$.

of the minimum quantizer level, to 111 in correspondence of the maximum level. Another representation will be used in Section 3.1.4. In this example, from (3.5), the rate of the produced binary stream is equal to $R_b = 3F_s$ bit/s.

3.1.4 Uniform quantizers

A quantizer with L equally spaced output levels and thresholds is called *uniform*. We have:

$$\begin{aligned} v_i - v_{i-1} &= \Delta\,, \qquad i = 2, \ldots, L-1\,, \\ Q_i - Q_{i-1} &= \Delta\,, \qquad i = 1, \ldots, L-1\,, \end{aligned} \tag{3.12}$$

where Δ is the *quantization step size*. Since it is usually $L = 2^{\mathrm{b}}$, typically the characteristic is symmetric with $L/2$ positive and $L/2$ negative output values. This characteristic is called *mid-riser* and the zero output level does not belong to $\mathcal{A}_{\mathrm{q}}$.

Mid-riser characteristic The quantizer characteristic is given in Fig. 3.11 for $L = 8$: in this case the smallest value, in magnitude, assumed by $a_q(nT_s)$ is $\Delta/2$, even for a very small input value a.

For this quantizer we define the *saturation value* as

$$v_{sat} = v_{L-1} + \Delta \ , \tag{3.13}$$

or equivalently

$$v_{sat} = \mathcal{Q}_{L-1} + \Delta/2 \ , \tag{3.14}$$

that is shifted by Δ with respect to the maximum finite threshold value v_{L-1} or by $\Delta/2$ with respect to the maximum level $\mathcal{Q}_{L-1}$. Having defined v_{sat}, we can consider the uniform quantizer as described by the rule of dividing the input value range $(-v_{sat}, v_{sat})$ in L equal intervals and associating to each input value the mid-value of the interval it belongs to. This is clearly shown in the compact representation of the uniform quantizer of Fig. 3.12. The interval $(-v_{sat}, v_{sat})$ is called *quantizer dynamic range.*

A possible binary representation of $a_q(nT_s)$ into $\mathbf{c}(nT_s)$, can be defined according to the following rule: the most significant bit of the binary representation denotes the sign (± 1) of $a_q(nT_s)$, whereas the remaining bits denote its amplitude. From (3.2) the relation between the quantized value $a_q(nT_s)$ and its codeword representation $\mathbf{c}(nT_s)$ is given by

$$a_q(nT_s) = (1 - 2c_{\mathsf{b}-1})\,\Delta \left(\frac{1}{2} + \sum_{j=0}^{\mathsf{b}-2} c_j\, 2^j \right) , \qquad c_j \in \{0, 1\} \ . \tag{3.15}$$

where $(1 - 2c_{\mathsf{b}-1})$ determines the sign of $a_q(nT_s)$, according to the most significant bit, while $\Delta \left(\sum_{j=0}^{\mathsf{b}-2} c_j\, 2^j + \frac{1}{2} \right)$ determines the amplitude of $a_q(nT_s)$.

To summarize, the uniform quantizer is specified by: (i) the dynamic range $(-v_{sat}, v_{sat})$; (ii) the step size Δ; and (iii) the number of levels L, or equivalently, by the number of bits b. The relation among these three parameters is the following:

$$L\Delta = 2v_{sat} \ . \tag{3.16}$$

If $L = 2^{\mathsf{b}}$, it is

$$2^{\mathsf{b}-1}\Delta = v_{sat} \ . \tag{3.17}$$

Remark In the quantizer characteristic of Fig. 3.11, zero is not an output level, so that a sample close to zero is represented either by $\Delta/2$ or $-\Delta/2$. An alternative, in which zero is an output level, is a quantizer with a *mid-tread characteristic*, as shown in Fig. 3.13.

3.1.5 Quantization error

By representing a continuous-amplitude signal with a discrete set of values, an error is introduced; indeed, the input-output characteristic (see for example Fig. 3.9) differs from the identity. Let

$$e_q(nT_s) = a_q(nT_s) - a(nT_s) = \mathcal{Q}[a(nT_s)] - a(nT_s) \tag{3.18}$$

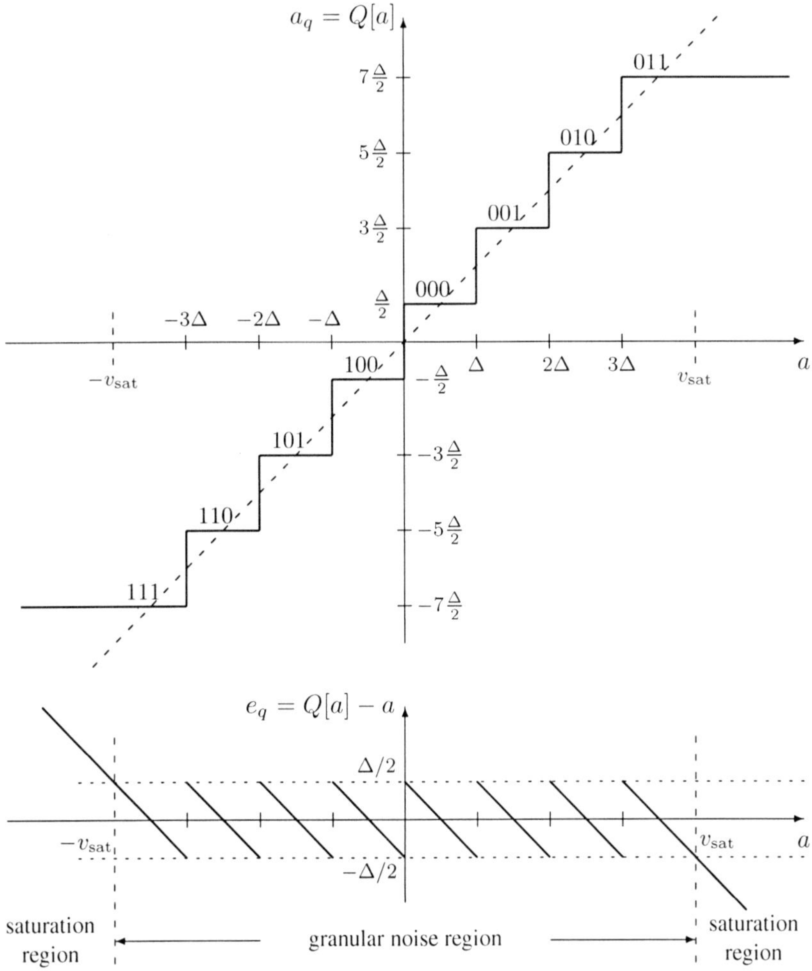

Figure 3.11 Mid-riser characteristic of an uniform quantizer with $L = 8$ levels ($\mathtt{b} = 3$): input a, quantized value a_q and corresponding quantization error e_q.

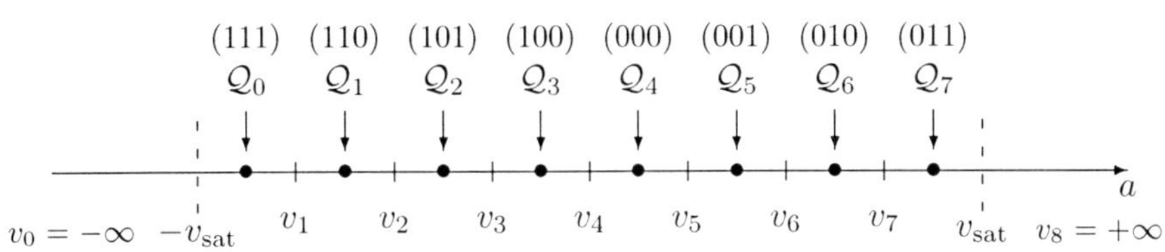

Figure 3.12 Uniform quantizer with $L = 8$ ($\mathtt{b} = 3$).

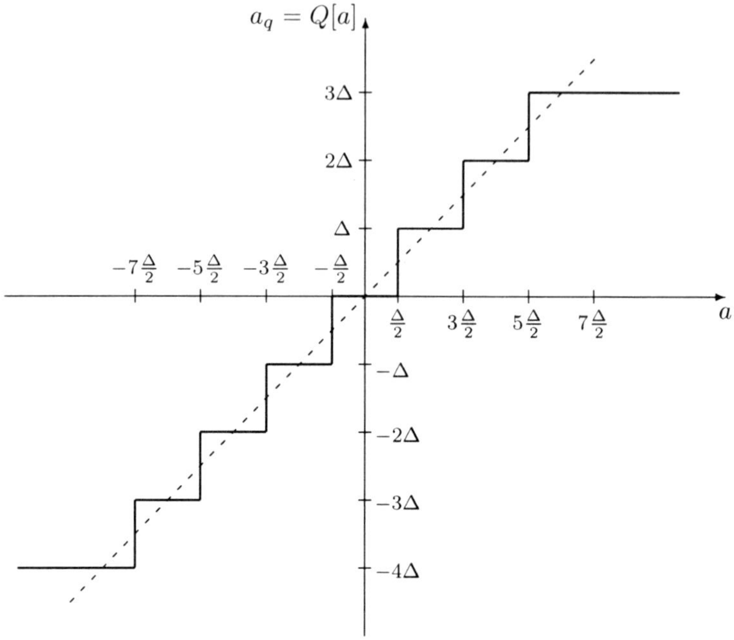

Figure 3.13 Mid-tread characteristic of an uniform quantizer with $L = 8$ levels ($\mathtt{b} = 3$).

be the quantization error, that is the difference between the quantizer output and input values. From (3.18) we write $a_q(nT_s) = a(nT_s) + e_q(nT_s)$ and we can model the effect of quantization as introducing an error $e_q(nT_s)$ on the input signal, as illustrated in Fig. 3.14.

In general, the problem of designing a quantizer is equivalent to determining the minimum number of bits $\mathtt{b}$ (to minimize the bit rate) that keeps the quantization error below a certain level (to guarantee a certain quality of the quantized signal).

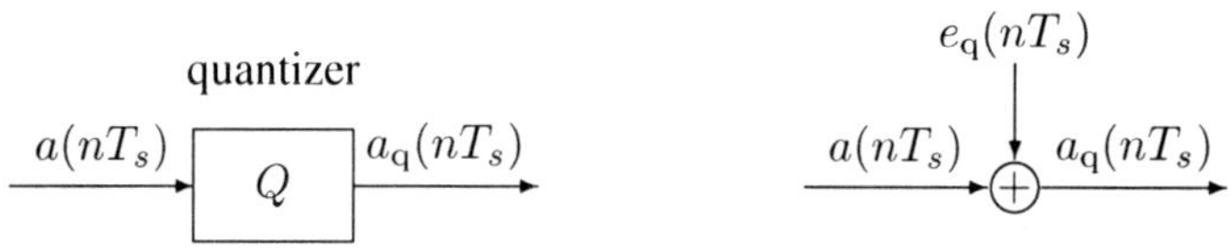

Figure 3.14 Equivalent model of a quantizer.

We will refer to uniform symmetrical quantizers, with a mid-riser characteristic as given in Fig. 3.11. If for each value of a we compute the corresponding error $e_q = \mathcal{Q}(a) - a$, we obtain the quantization error characteristic of Fig. 3.11. It turns out that

$$|e_q| \leq \frac{\Delta}{2} \qquad \text{for } |a| < v_{\text{sat}} \;, \tag{3.19}$$

while

$$e_q = \begin{cases} \mathcal{Q}_{L-1} - a & , \text{ for } a > v_{sat} \\ \mathcal{Q}_0 - a & , \text{ for } a < -v_{sat} \end{cases}, \tag{3.20}$$

with $|e_q| > \Delta/2$ for $|a| > v_{sat}$. These results suggests to split the real axis into two regions:

1. the region $a \in (-\infty, -v_{sat}) \cup (v_{sat}, +\infty)$, where e_q is called *saturation* or *overload error* e_{sat};
2. the region $a \in [-v_{sat}, v_{sat}]$, where e_q is called *granular error* e_{gr}.

If $|a(nT_s)| < v_{sat}$, from (3.19), we have that e_q is granular with values in the range

$$-\frac{\Delta}{2} \leq e_q(nT_s) \leq \frac{\Delta}{2} . \tag{3.21}$$

On the other hand, if $|a(nT_s)| > v_{sat}$ the saturation error can assume large values: therefore v_{sat} must be chosen so that the probability of the event $|a(nT_s)| > v_{sat}$ is small. For a fixed number of bits b, and consequently for a fixed number of levels L, from (3.16) it is immediate to verify that increasing v_{sat} also Δ increases and hence, from (3.21), also the *granular error* increases; on the other hand, choosing a small Δ leads to a considerable *saturation error*. As a result for each value of b there will be an optimum choice of v_{sat} and Δ. In any case, to decrease both errors we must increase b with consequent increase of the bit rate R_b.

Statistical description of the quantization noise In Fig. 3.14 we presented an equivalent model of a quantizer where the quantization error is additive. Now, if the input $a_q(nT_s)$ is a rp, the *quantization error* obtained through the memoryless transformation $\mu_e(\cdot)$ is a rp itself.

In particular, if $a(nT_s)$ is strict-sense stationary, so is $e_q(nT_s)$. Moreover, if (3.21) holds, that is for the *granular noise* with $e_q = e_{gr}$, and the quantization step Δ is sufficiently small (that is b is large), the following simplifications can be assumed by a fairly good approximation.

1. If Δ is small in comparison to the typical variations between samples of $a(nT_s)$, the quantization noise is white

$$\mathrm{E}\left[e_q(nT_s)e_q(nT_s - mT_s)\right] = \begin{cases} \mathtt{M}_{e_q} & , \ m = 0 \\ 0 & , \ m \neq 0 \end{cases} \tag{3.22}$$

and is uncorrelated[2] with the input signal

$$\mathrm{E}\left[a(nT_s)e_q(nT_s)\right] = 0 , \text{ for all } n . \tag{3.23}$$

2. With good approximation in all cases of practical interest[3], e_q has a uniform distribution as shown in Fig. 3.15:

$$p_{e_q}(a) = \frac{1}{\Delta} \operatorname{rect}\left(\frac{a}{\Delta}\right) . \tag{3.24}$$

[2]We remark that uncorrelation (3.23) is a second-order property and does not imply statistical independence (which must be excluded here since $e_q(nT_s)$ is a deterministic function of $a(nT_s)$). Hence, although the additive model of Fig. 3.14 is very useful for the purpose of second-order analysis, the quantization error should not be treated as an external noise source.

[3]A sufficient condition (by no means necessary) for (3.24) to hold with excellent approximation is if Δ is small with respect to the input PDF (that is $p_a(x)$ is approximately constant in each interval $[v_i, v_{i+1}]$). Even if the above condition is not met, however, in the presence of certain symmetries (e.g., a triangular PDF) in $p_a(x)$, (3.24) may still hold, so that its applicability is quite general.

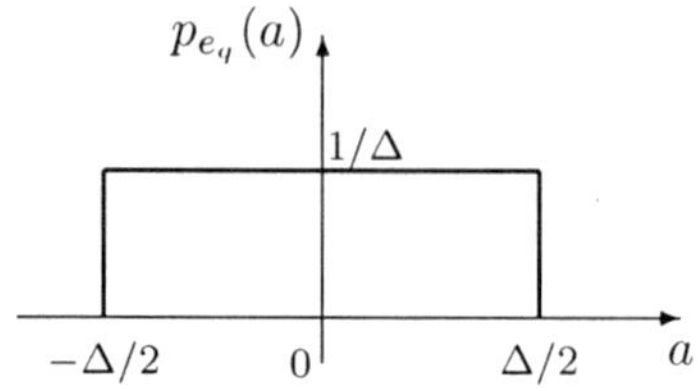

Figure 3.15 Probability density function of e_q.

Fig. 3.16 illustrates the quantization error for a 8-level quantized signal. No relation between the error $e_q(t)$ and the input signal $a(t)$ is evident, and the above assumptions are plausible.

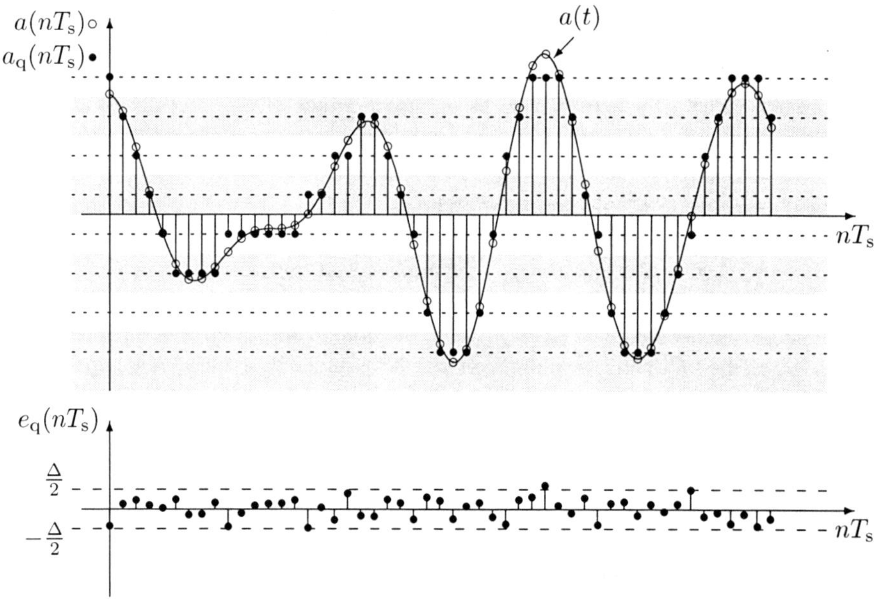

Figure 3.16 Signal and quantization error. Uniform quantizer with $L = 8$ levels.

3.1.6 Quantizer SNR

With reference to the model of Fig. 3.14, a simple measure of the quality of a quantizer is the signal-to-quantization noise ratio

$$\Lambda_q = \frac{\mathrm{E}\left[a^2(nT_s)\right]}{\mathrm{E}\left[e_q^2(nT_s)\right]} . \tag{3.25}$$

Choosing $v_{\rm sat}$ so that $e_{\rm q}$ is mainly granular, from (3.24) it results

$$\mathtt{M}_{e_{\rm q}} \simeq \mathtt{M}_{e_{\rm gr}} \simeq \frac{\Delta^2}{12} \,. \tag{3.26}$$

Further insight For an exact computation that includes also the saturation noise we need to know the probability density function of a. From (3.18) the statistical power of $e_{\rm q}$ is given by

$$\begin{aligned} \mathtt{M}_{e_{\rm q}} &= \mathrm{E}\left[e_{\rm q}^2(nT_{\rm s})\right] = \int_{-\infty}^{+\infty} \left[\mathcal{Q}(u) - u\right]^2 p_a(u)\, du \\ &= \int_{-v_{\rm sat}}^{v_{\rm sat}} \left[\mathcal{Q}(u) - u\right]^2 p_a(u)\, du \\ &\quad + \int_{-\infty}^{-v_{\rm sat}} \left[\mathcal{Q}_0 - u\right]^2 p_a(u)\, du + \int_{v_{\rm sat}}^{\infty} \left[\mathcal{Q}_{L-1} - u\right]^2 p_a(u)\, du \,, \end{aligned} \tag{3.27}$$

where

$$\int_{-v_{\rm sat}}^{v_{\rm sat}} \left[\mathcal{Q}(u) - u\right]^2 p_a(u)\, du = \sum_{i=0}^{L-1} \int_{\mathcal{Q}_i-\Delta/2}^{\mathcal{Q}_i+\Delta/2} \left[\mathcal{Q}_i - u\right]^2 p_a(u)\, du \tag{3.28}$$

(3.29)

accounts for the statistical power of the granular noise $e_{\rm gr}$, while the other two terms in (3.27) represent the statistical power of the saturation noise $e_{\rm sat}$.

Assuming the input $a(nT_{\rm s})$ has zero mean and variance σ_a^2, and defining the parameter

$$k_f = \frac{\sigma_a}{v_{\rm sat}} \,, \tag{3.30}$$

we have that the signal-to-quantization noise ratio can be written as

$$\Lambda_{\rm q} = \frac{\mathtt{M}_a}{\mathtt{M}_{e_{\rm q}}} \simeq \frac{\sigma_a^2}{\Delta^2/12} = 3k_f^2 L^2 \,. \tag{3.31}$$

The inverse of (3.30), $1/k_f = v_{\rm sat}/\sigma_a$ is called *loading factor*.

For $L = 2^{\mathsf{b}}$, (3.31) in dB yields,

$$\boxed{(\Lambda_{\rm q})_{\rm dB} \simeq 6.02\,\mathsf{b} + 4.77 + 20\log\left(\frac{\sigma_a}{v_{\rm sat}}\right)} \tag{3.32}$$

Recalling that this law considers *only granular noise*, for a given loading factor $v_{\rm sat}/\sigma_a$, if we double the number of quantizer levels, i.e. increase by one the number of bits b, the signal-to-quantization noise ratio increases by 6 dB.

Design of a uniform quantizer Note that the design of a uniform quantizer requires to specify the number of levels $L = 2^{\mathsf{b}}$ and the dynamic range $(-v_{\rm sat}, v_{\rm sat})$, that is the range of input values where the granular error occurs. The procedure of designing a uniform quantizer consists of two steps.

1. Determine $v_{\rm sat}$ so that the saturation probability is sufficiently small:

$$P_{\rm sat} = {\rm P}\left[|a(nT_{\rm s})| > v_{\rm sat}\right] \ll 1 \ . \tag{3.33}$$

This guarantees that $e_{\rm sat} \simeq 0$ with very high probability.

2. Based on (3.31) or (3.32), choose L or $\mathtt{b}$ so that the signal-to-quantization noise ratio $\Lambda_{\rm q}$ assumes the desired value.

Given $v_{\rm sat}$ and L, we can obtain the quantization step size Δ from (3.16),

$$\Delta = \frac{2v_{\rm sat}}{L} = \frac{2\sigma_a}{k_f L} \ . \tag{3.34}$$

Example 3.1 A For a Gaussian signal, $a(nT_{\rm s}) \sim \mathcal{N}(0, \sigma_{\rm a}^2)$, it is

$$P_{\rm sat} = 2Q\left(\frac{v_{\rm sat}}{\sigma_a}\right) = \begin{cases} 0.046 & , v_{\rm sat} = 2\sigma_a \\ 0.0027 & , v_{\rm sat} = 3\sigma_a \\ 0.000063 & , v_{\rm sat} = 4\sigma_a \end{cases} \tag{3.35}$$

Example 3.1 B For a uniform signal, $a(nT_{\rm s}) \sim \mathcal{U}\left([-a_{\max}, a_{\max}]\right)$, setting $v_{\rm sat} = a_{\max}$, we get

$$\frac{v_{\rm sat}}{\sigma_a} = \frac{a_{\max}}{\sigma_a} = \sqrt{3} \Longrightarrow (\Lambda_{\rm q})_{\rm dB} = 6.02\,\mathtt{b} \ . \tag{3.36}$$

Example 3.1 C For a sinusoidal signal, $a(nT_{\rm s}) = a_{\max}\cos(2\pi f_0 T_{\rm s} k + \varphi_0)$, setting $v_{\rm sat} = a_{\max}$, we get

$$\frac{v_{\rm sat}}{\sigma_a} = \frac{a_{\max}}{\sigma_a} = \sqrt{2} \Longrightarrow (\Lambda_{\rm q})_{\rm dB} = 6.02\,\mathtt{b} + 1.76 \ . \tag{3.37}$$

Example 3.1 D For $a(nT_{\rm s})$ not limited in amplitude, and assuming $P_{\rm sat}$ negligible for $v_{\rm sat} = 4\sigma_a$, we get

$$(\Lambda_{\rm q})_{\rm dB} = 6.02\,\mathtt{b} - 7.2 \ . \tag{3.38}$$

The plot of $\Lambda_{\rm q}$ in dB, as given by (3.32), versus $\sigma_a/v_{\rm sat}$ in dB, is illustrated in Fig. 3.17 in dash-dot lines for various values of $\mathtt{b}$. Indeed they are straight lines with unit slope.

However, we recall that (3.32) measures only the granular noise, hence for a generic signal when σ_a is near $v_{\rm sat}$, i.e. $(\sigma_a/v_{\rm sat})_{\rm dB}$ close to zero, the approximation $\mathtt{M}_{e_{\rm q}} \simeq \mathtt{M}_{e_{\rm gr}}$ is no longer valid because the statistical power of the saturation noise $\mathtt{M}_{e_{\rm sat}}$ becomes non negligible. For the computation of $\mathtt{M}_{e_{\rm sat}}$ we need to know the probability density function of a and apply (3.27). Assuming a Laplacian input signal we obtain the curves also shown in Fig. 3.17, that coincide with the curves given by (3.32) for $\sigma_a \ll v_{\rm sat}$. Let us note the effect of the saturation noise, which deteriorates the value of $\Lambda_{\rm q}$, as soon as σ_a becomes close to $v_{\rm sat}$.

3.1.7 Non-uniform quantizers

Two considerations suggest the choice of non-uniform quantizers. The first refers to signals with a non-uniform probability density function: for such signals uniform quantizers are

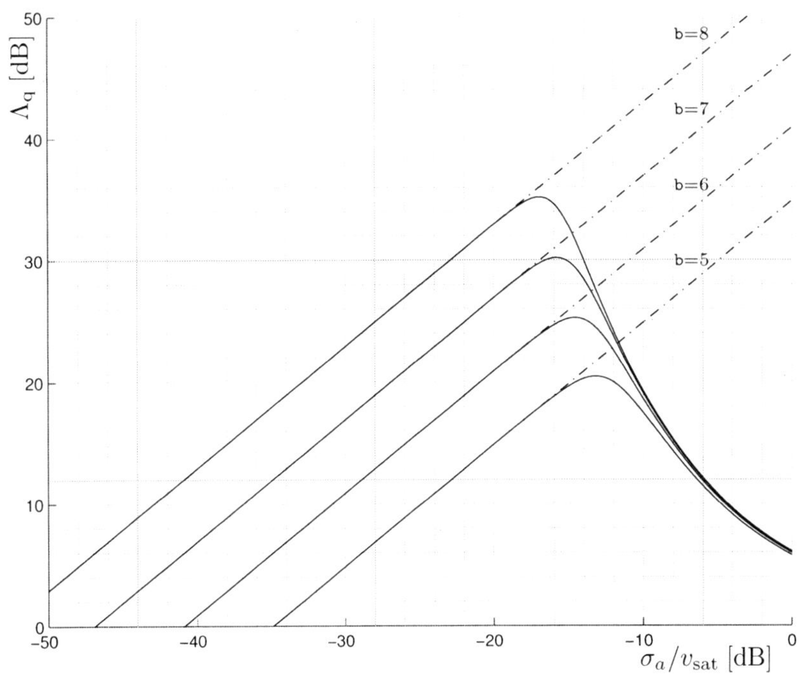

Figure 3.17 Signal-to-quantization noise ratio versus $\sigma_a/v_{\rm sat}$ of an uniform quantizer for granular noise only (dashed-dot lines) and for a Laplacian signal (solid lines). $L = 2^{\rm b}$ is the number of quantizer levels. The expression of $\Lambda_{\rm q}$ for granular noise only is given by (3.32).

suboptimal, i.e. they yield a lower signal-to-quantization noise ratio for a given number of bits. In fact, the quantization error should be kept small for the most probable input amplitudes and may be allowed to increase for the less probable amplitudes, thus gaining in the overall signal-to-quantization noise ratio.

The second aspect refers to non stationary signals, where the short term and the long term powers differ significantly; an example is speech, where the power estimated over windows of tenths of milliseconds can exhibit variations of several dB. In such conditions the idea is to obtain a quantizer SNR that is nearly independent of the short term input power. A quantizer with a non-uniform characteristic is depicted in Fig. 3.18: in this case, the quantization step, and correspondingly the quantization error, is small when the input signal is small and large when the signal is large: therefore the ratio $\Lambda_{\rm q}$ tends to remain constant for a wide dynamic range of the input signal, and is almost independent of the short term input power.

Design methods to determine the quantizer thresholds and levels, according to the probability density function of the input signal, can be found in [1] and lead to sophisticated coding schemes. Here we consider three possible implementations, assuming the non-uniform characteristic is given.

1. The characteristic of Fig. 3.18 can be implemented directly: the input sample value is compared with the non equally spaced thresholds to determine the output value and

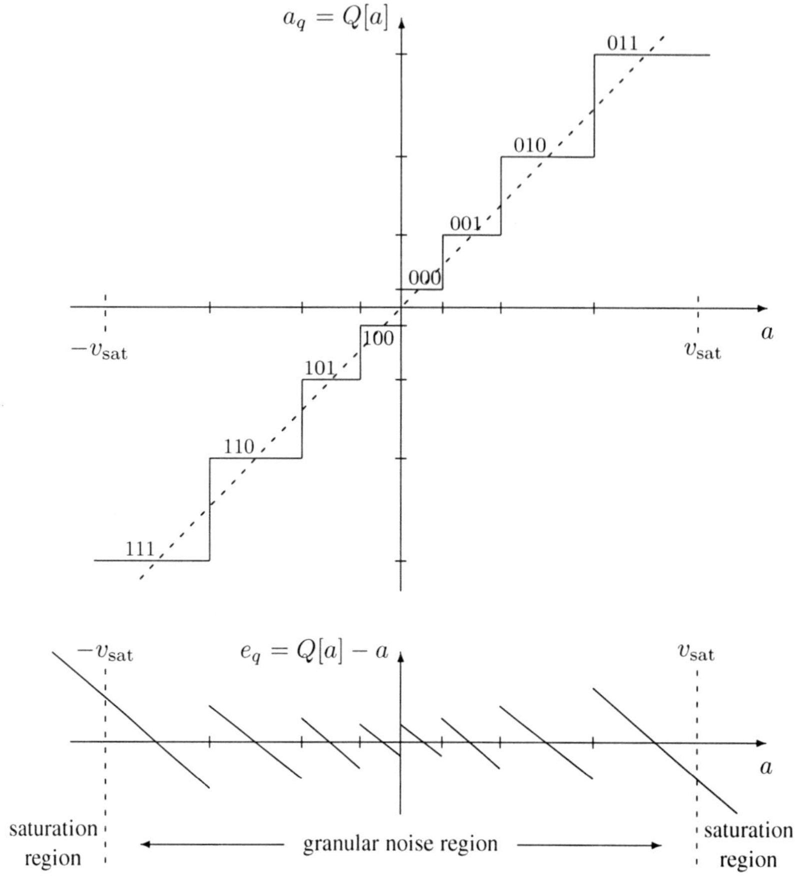

Figure 3.18 Non-uniform quantizer characteristic with $L = 8$ levels and corresponding quantization error.

correspondingly its binary representation. We note that now the correspondence between values and binary representation is non linear.

2. As shown in Fig. 3.19, a *compression* function precedes a uniform quantizer. In turn, to reproduce the input signal an *expansion* of the quantized sample is needed (see Section 3.1.8).

3. The most popular method, depicted in Fig. 3.20, employs (i) a uniform quantizer having a larger number of levels, with a step size equal to the minimum step size of the desired non-uniform quantizer, followed by (ii) a look-up table to associate the uniformly quantized samples to the actual non-uniform quantizer levels. The look-up table corresponding to the quantizer of Fig. 3.20 is presented in Tab. 3.1.

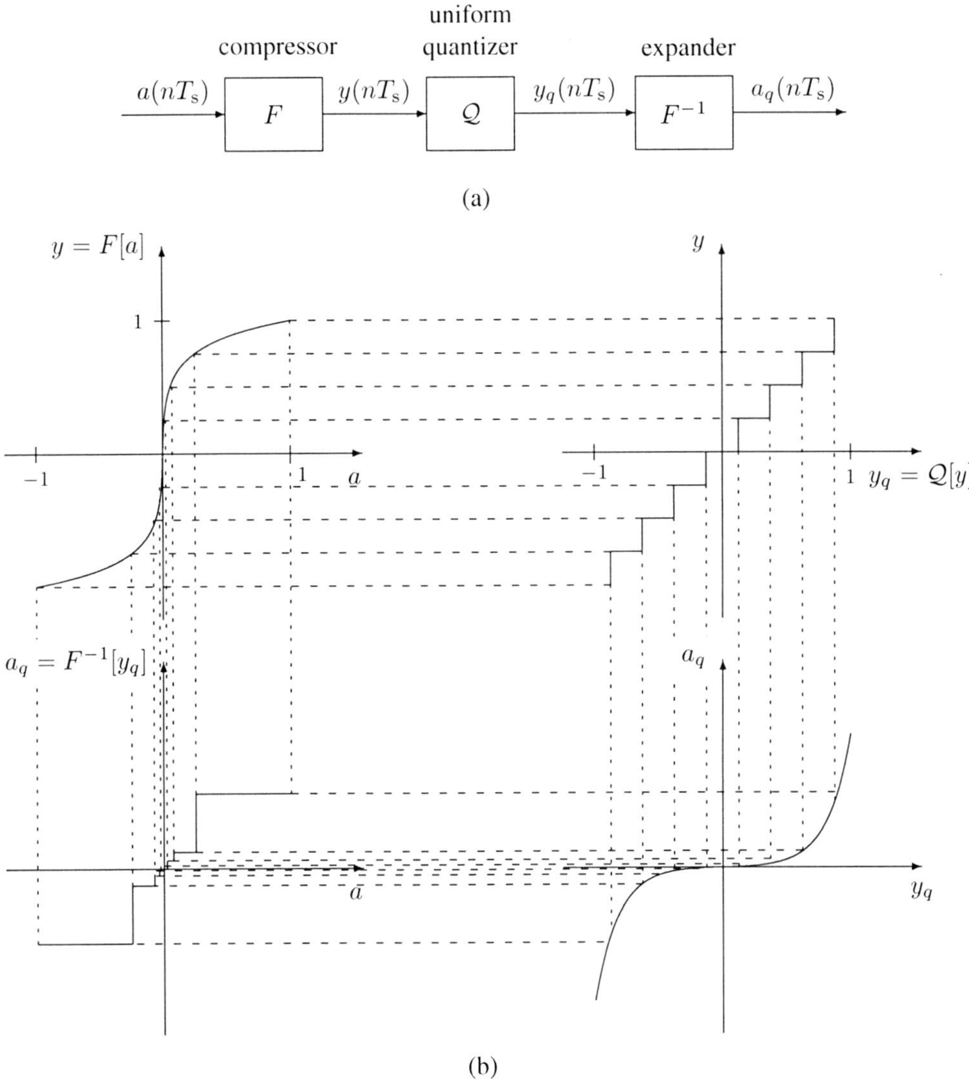

Figure 3.19 (a) Use of a compression function F to implement a non uniform quantizer; (b) non uniform quantizer characteristic implemented by companding and uniform quantization. Here $v_{\text{sat}} = 1$ is assumed.

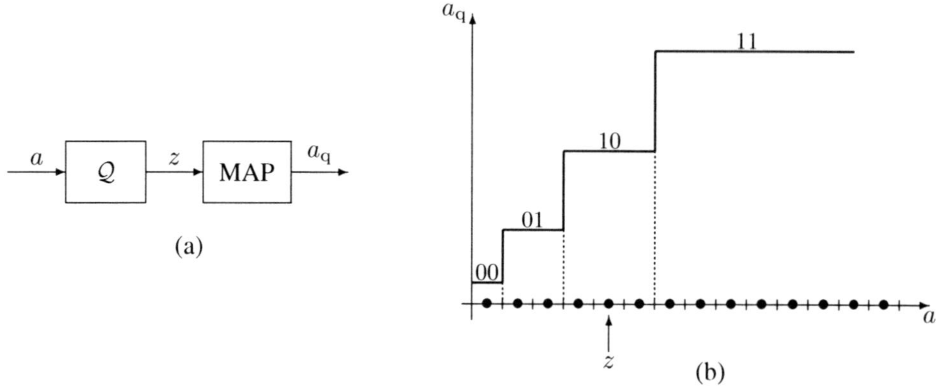

Figure 3.20 Non uniform quantizer digitally implemented using an uniform quantizer with a small step size followed by a look-up table (MAP).

Table 3.1 Example of non linear PCM from 4 to 2 bits (sign omitted).

coding of z	*coding of* a_q
0000	00
0001 0010	01
0011 0100 0101	10
1000 1001 1010 1011 1100 1101 1110 1111	11

3.1.8 Companding techniques and SNR

Fig. 3.19b illustrates in detail the principle of Fig. 3.19a. The sampled signal is first compressed through a non linear function F, that yields the signal

$$y = F(a) \; . \tag{3.39}$$

In Fig. 3.19 we assume $v_{sat} = 1$: in other words, the input signal a is normalized to v_{sat}. The signal y is uniformly quantized to give y_q, that must be expanded by the transformation F^{-1}

to yield the non-uniformly quantized version of a

$$a_q = F^{-1}[Q[y]] \, . \tag{3.40}$$

This quantization technique takes the name of *companding* from a contraction of *compressing* and *expanding*. It can be seen that the cascade of transformations shown in Fig. 3.19b yields an overall non-uniform quantizer characteristic, as shown in the bottom left graph. Indeed, the thresholds $\{v_i^{(U)}\}$ and levels $\{\mathcal{Q}_i^{(U)}\}$ of the uniform quantizer become correspondingly the non-uniform thresholds $\{v_i^{(NU)}\}$ and levels $\{\mathcal{Q}_i^{(NU)}\}$ of, respectively the input and output signal. In particular, by companding, it is

$$v_i^{(NU)} = F^{-1}\left[v_i^{(U)}\right] \tag{3.41}$$

and

$$\mathcal{Q}_i^{(NU)} = F^{-1}\left[\mathcal{Q}_i^{(U)}\right] \, . \tag{3.42}$$

Remark It is interesting to show that in order to obtain a signal-to-quantization noise ratio Λ_q that is independent of the input signal power, the compressing function F must be logarithmic,

$$F[a] = \ln a \qquad a > 0 \, . \tag{3.43}$$

We consider the two schemes shown in Fig. 3.21: the encoder applies the logarithmic function to the absolute value of the input sample, prior to the uniform quantization, while the sign is considered separately. At the decoder side the exponential function is applied to the quantized sample; again the sign is considered separately.

Let

$$a(nT_s) = e^{y(nT_s)} \text{sgn}[a(nT_s)] \, , \tag{3.44}$$

that is

$$y(nT_s) = \ln |a(nT_s)| \, , \tag{3.45}$$

while the $\text{sgn}[a(nT_s)]$ is considered separately. Uniform quantization of $y(nT_s)$ gives

$$y_q(nT_s) = Q[y(nT_s)] = \ln |a(nT_s)| + e_q(nT_s) \, . \tag{3.46}$$

At the decoder, the inverse of (3.44), with y replaced by y_q, yields the quantized version of $a(nT_s)$

$$\begin{aligned} a_q(nT_s) &= e^{y_q(nT_s)} \, \text{sgn}[a(nT_s)] \\ &= |a(nT_s)| \, \text{sgn}[a(nT_s)] e^{e_q(nT_s)} \\ &= a(nT_s) e^{e_q(nT_s)} \, . \end{aligned} \tag{3.47}$$

If the uniform quantizer has a sufficiently high number of levels, so that the quantization error can be considered small, $e_q \ll 1$, then the approximation of the exponential to the first term of its power series expansion holds

$$e^{e_q(nT_s)} \simeq 1 + e_q(nT_s) \, , \tag{3.48}$$

and

$$a_q(nT_s) = a(nT_s) + e_q(nT_s) a(nT_s) \, , \tag{3.49}$$

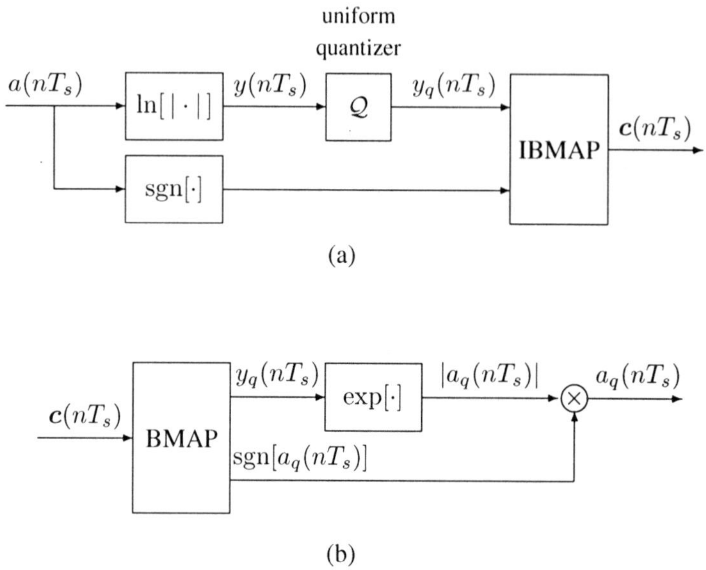

Figure 3.21 Non uniform quantization by companding and uniform quantization: (a) encoder; (b) decoder.

where $e_{\mathrm{q}}(nT_{\mathrm{s}})a(nT_{\mathrm{s}})$ represents the output error of the system. In terms of the signal-to-quantization noise ratio we get

$$\Lambda_{\mathrm{q}} = \frac{\mathtt{M}_a}{\mathrm{E}\left[e_{\mathrm{q}}^2(nT_{\mathrm{s}})a^2(nT_{\mathrm{s}})\right]} \,. \tag{3.50}$$

As $e_{\mathrm{q}}(nT_{\mathrm{s}})$ is uncorrelated with $a(nT_{\mathrm{s}})$ (see (3.23)), we get

$$\mathrm{E}\left[e_{\mathrm{q}}^2(nT_{\mathrm{s}})a^2(nT_{\mathrm{s}})\right] = \mathrm{E}\left[e_{\mathrm{q}}^2(nT_{\mathrm{s}})\right] \mathtt{M}_a$$

and

$$\Lambda_{\mathrm{q}} = \frac{1}{\mathrm{E}\left[e_{\mathrm{q}}^2(nT_{\mathrm{s}})\right]} = \frac{1}{\mathtt{M}_{e_{\mathrm{q}}}} \,. \tag{3.51}$$

Consequently Λ_{q} does *not* depend on $\mathtt{M}_a$.

We note that a logarithmic compression function generates a signal y with unbounded amplitude, thus an approximation of the logarithmic law is usually adopted for small values of a.

Regulatory bodies have defined two compression functions:

1. *A-Law* ($A = 87.56$). For $v_{\mathrm{sat}} = 1$,

$$y = F[a] = \begin{cases} \frac{A|a|}{1+\ln(A)} & , \; 0 \leq |a| \leq \frac{1}{A} \\ \frac{1+\ln(A|a|)}{1+\ln(A)} & , \; \frac{1}{A} \leq |a| \leq 1 \end{cases} \tag{3.52}$$

This law, illustrated in Fig. 3.22a, is adopted in Europe. The sign is considered separately:

$$\mathrm{sgn}[y] = \mathrm{sgn}[a] \ . \tag{3.53}$$

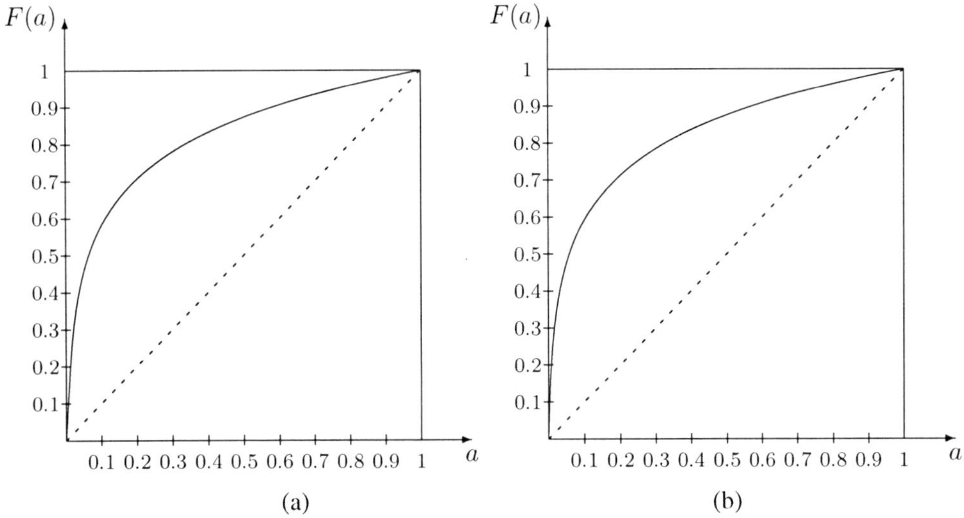

Figure 3.22 (a) A Law; (b) μ Law.

2. *μ-Law* ($\mu = 255$). For $v_{\rm sat} = 1$,

$$y = F[a] = \frac{\ln(1 + \mu|a|)}{\ln(1 + \mu)} \ . \tag{3.54}$$

This law, illustrated in Fig. 3.22b, is adopted in the United States and Canada. The compression increases for higher values of μ and the standard value is $\mu = 255$.

Remark For the μ-law, when $\mu a \gg 1$, we have

$$F[a] = \frac{\ln[\mu a]}{\ln(1 + \mu)} \ , \tag{3.55}$$

with a behaviour similar to the ideal function (3.45). A similar behavior is exhibited by (3.52).

Signal-to-quantization noise ratio Derivation of the signal-to-quantization noise ratio in this case is less direct. If we assume that within each quantization interval the error can be represented by a uniform rv, then, for the μ-law, we have

$$\boxed{\begin{aligned}(\Lambda_{\rm q})_{\rm dB} &= 6.02\mathsf{b} + 4.77 - 20\log_{10}\Big[\ln(1+\mu)\Big] \\ &\quad - 10\log_{10}\left\{1 + \left(\frac{v_{\rm sat}}{\mu\sigma_a}\right)^2 + \sqrt{3}\left(\frac{v_{\rm sat}}{\mu\sigma_a}\right)\right\}\end{aligned}} \tag{3.56}$$

We note that this expression considers only the granular noise and holds if the number of levels is sufficiently high. Curves of Λ_q versus σ_a/v_{sat}, the normalized standard deviation of the input signal, are plotted for $\mu = 255$ in Fig. 3.23.

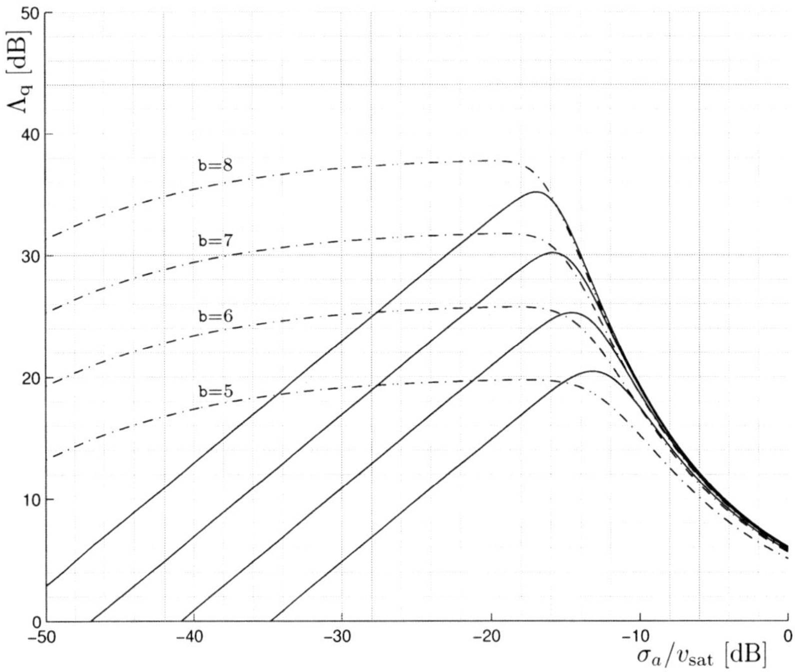

Figure 3.23 Signal-to-quantization noise ratio versus σ_a/v_{sat} of an uniform quantizer for a Laplacian signal (solid lines), and of a μ-law ($\mu = 255$) quantizer (dash-dot lines). $L = 2^b$ is the number of quantizer levels.

Again, for values of σ_a/v_{sat} close to one, evaluation of Λ_q must also take into account the saturation noise and formula (3.27) should be used, which needs the input probability density function. Here a Laplacian input was assumed. Note that, with respect to the uniform quantizer, Λ_q is almost independent of the input power (for a wide range of values). Moreover, in the saturation region the curves coincide with the curves obtained for the uniform quantizer. We emphasize that also in this case Λ_q increases by 6 dB with an increase of b by one. Moreover, if $b = 8$, $\Lambda_q \simeq 38$ dB for a wide range of values of σ_a/v_{sat}. An effect not shown in Fig. 3.23 is that, by increasing μ, the plot of Λ_q becomes "flatter", but the maximum value decreases.

Remark The binary representation associated to a non-uniform quantizer is also called *non linear PCM*. In the standard *non linear PCM* for telephone voice applications, a quantizer with 128 levels (7 bit/sample) is employed after the compression; including also the sign we have 8 bit/sample. For a sampling frequency of $F_s = 8$ kHz, this leads to a bit rate $R_b = 64$ kbit/s.

3.2 EXAMPLES OF APPLICATION

Speech coding Speech coding addresses person-to-person communications and is strictly related to the transmission, for example over the public network, and storage of speech signals. The aim is to encode speech as a digital signal that requires the lowest possible bit rate to recreate the speech signal at the decoder [10].

For audio signals, the sampling frequency F_s depends on the signal quality that we wish to maintain and therefore it depends on the application (see Tab. 3.2).

Table 3.2 Sampling frequency of the audio signal in four applications.

Application	*Band* [Hz]	F_s [Hz]
telephone	$300 - 3400$ (narrow band speech)	8000
broadcasting	$50 - 7000$ (wide band speech)	16000
audio, compact disc	$10 \div 20000$	44100
digital audio tape	$10 \div 20000$	48000

Some examples of speech waveforms for an interval of 0.25 s are given in Fig. 3.24. From these plots, we can obtain a speech model as a succession of *voiced speech* spurts (see Fig. 3.25a), or *unvoiced speech* spurts (see Fig. 3.25b). In the first case, the signal is strongly correlated and almost periodic, with a period that is called *pitch*, and exhibits large amplitudes; conversely in an *unvoiced speech* spurt the signal is weakly correlated and has small amplitudes. We note moreover that the average level of speech changes in time: indeed speech is a non stationary signal. In Fig. 3.26 it is interesting to observe the instantaneous frequency behaviour of some voiced and unvoiced sounds; we also note that the latter may have a bandwidth larger than 10 kHz.

Concerning the *amplitude distribution* of speech signals, we observe that over short time intervals, of the order of a few tenths of milliseconds (or of a few hundreds of samples at a sampling frequency of 8 kHz), the amplitude statistic is Gaussian with good approximation; over long time intervals, because of the numerous pauses in speech, it tends to exhibit a gamma or Laplacian distribution. We give here the probability density functions of the amplitude that are usually adopted. Let σ_a be the standard deviation of the signal $a(t)$; then we have

$$\begin{aligned}
\text{gamma}: \quad & p_a(u) = \left(\frac{\sqrt{3}}{8\pi\sigma_a|u|}\right)^{\frac{1}{2}} e^{-\frac{\sqrt{3}|u|}{2\sigma_a}}, \\
\text{Laplacian}: \quad & p_a(u) = \frac{1}{\sqrt{2}\sigma_a} e^{-\frac{\sqrt{2}|u|}{\sigma_a}}, \\
\text{Gaussian}: \quad & p_a(u) = \frac{1}{\sqrt{2\pi}\sigma_a} e^{-\frac{1}{2}\left(\frac{u}{\sigma_a}\right)^2}.
\end{aligned} \tag{3.57}$$

Coding techniques and applications At the output of an ADC, the PCM encoded samples, after suitable transformations, can be further quantized in order to reduce the bit rate.

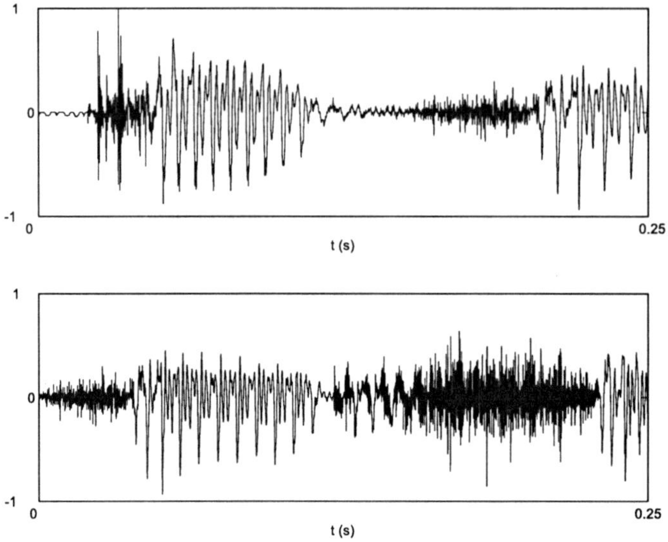

Figure 3.24 Speech waveforms.

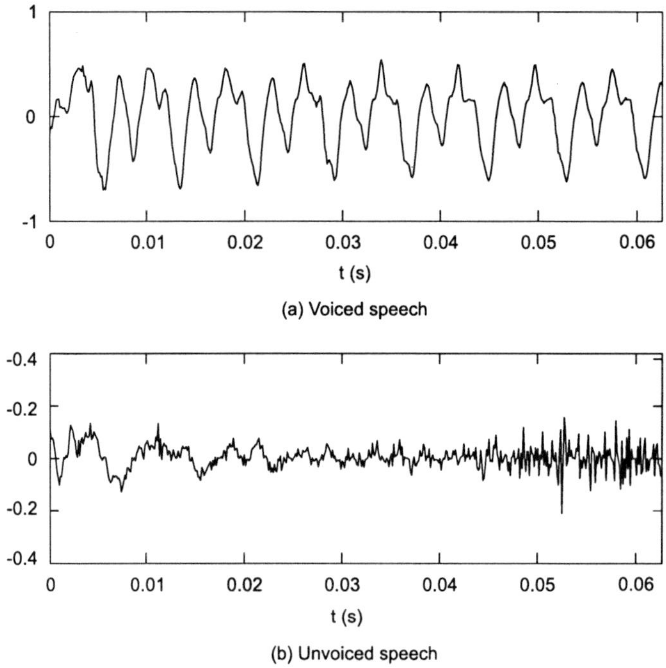

Figure 3.25 Voiced and unvoiced speech spurts.

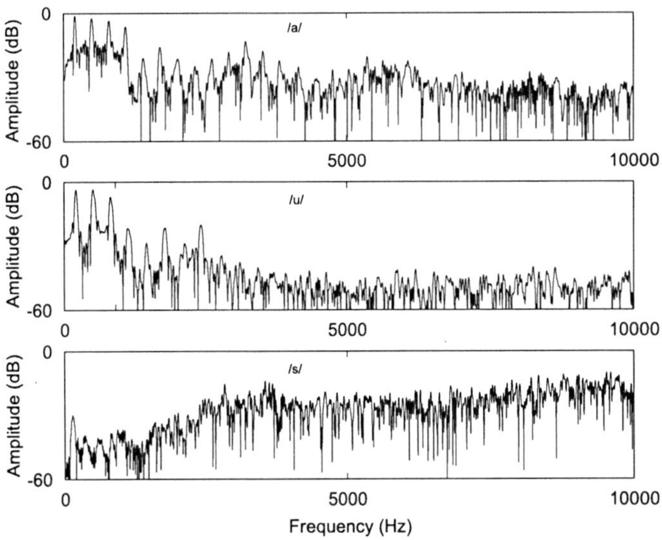

Figure 3.26 Fourier transform (amplitude only) of voiced and unvoiced sounds, with a sampling frequency of 20 kHz.

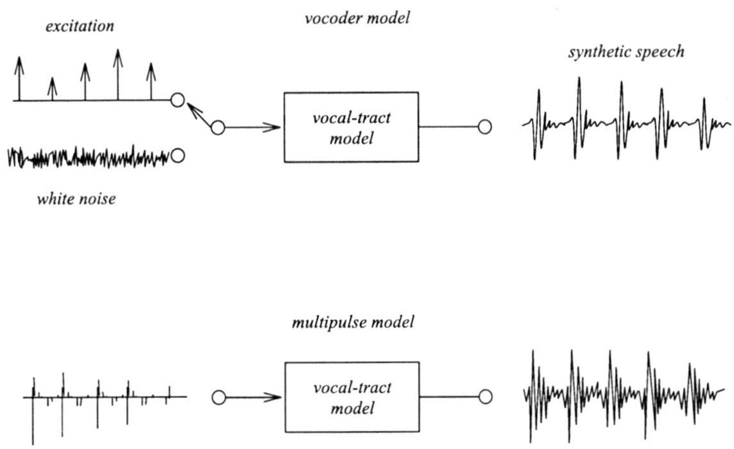

Figure 3.27 Vocoder and multipulse models for speech synthesis.

Essentially, the various coding techniques are divided in three groups, that essentially exploit two elements:

- redundancy of speech,
- sensitivity of the ear as a function of the frequency.

Waveform coding *Waveform encoders* attempt to reproduce the waveform as closely as possible. This type of coding is applicable to any type of signal; two examples are the PCM and adaptive differential PCM (ADPCM) schemes.

Coding by modelling In this case coding is not related to signal samples, but to the parameters of the source that generates them. Assuming the *voiced/unvoiced* speech model, an example of a classical encoder (*vocoder*) is given in Fig. 3.27, where a periodic excitation, or white noise segment, filtered by a suitable filter, yields a synthesized speech segment. A more sophisticated model uses a more articulated *multipulse* excitation.

Frequency domain coding In this case coding occurs after signal transformation into a domain different from time, usually frequency: examples are subband coding and transform coding.

As reported in Tab. 3.3 for most common speech coders we remark that for more sophisticated encoders the implementation complexity expressed in millions of instructions per second (MIPS), as well as the delay introduced by the encoder (*latency*), can be considerable.

Table 3.3 Parameters of a few speech coders.

Coder	*Bit rate* (kbit/s)	*Computational complexity* (MIPS)	*Latency* (ms)
PCM	64	0.0	0
ADPCM	32	0.1	0
ASBC	16	1	25
MELP	8	10	35
CELP	4	100	35
LPC	2	1	35

The various coding techniques are different in quality and cost of implementation: obviously a higher implementation complexity is expected for encoders with low bit rate and good quality. We go from a bit rate in the range from 4.4 to 9.6 kbit/s for cellular radio systems, to a bit rate in the range from 16 to 64 kbit/s for transmission over the public network.

Video coding It is also interesting to compare the various standards to code video signals given in Tab. 3.4 with those of Tab. 3.3 for speech and audio.

Table 3.4 Bit rates for video standards.

Application	*Target bit rate*
ISDN Video Telephone	64÷128 kbit/s
ISDN Video Conferencing	128 kbit/s
MPEG1 CD-Rom Video	1.5 Mbit/s
MPEG2 TV (Broadcast Quality)	6 Mbit/s
HDTV (Broadcast Quality)	24 Mbit/s
TV (Studio Quality, Compressed)	34 Mbit/s
HDTV (Studio Quality, Compressed)	140 Mbit/s
TV (Studio Quality, Uncompressed)	216 Mbit/s
HDTV (Studio Quality, Uncompressed)	1 Gbit/s

3.3 INFORMATION AND ENTROPY

In the previous sections we have derived methods for the representation of an analog signal via a digital stream, and characterized its bit rate. Now we wonder whether the bit rate of a digital signal is representative of its information content and how much it can be lowered (trough a suitable transformation) for a more efficient exploitation of transmission resources, without losing information (i.e. the transformation must be reversible). The answer to this question is provided by source coding, which will be seen in Section 3.4, but first let us address the relationship between the bit rate and the information in a message. In this section we introduce a measure for the average information per time unit carried by a digital message, modeled as a rp with a known statistical description.

3.3.1 A measure for information

Any event can be thought of as bearing some information, that is related to its being partly unexpected, on the basis of a statistical description.

Formally, given a probability space $(\Omega, \mathcal{F}, \mathrm{P}[\cdot])$, we wish to express how informative an event $A \in \mathcal{F}$ is, through a quantity $i(A)$, named *information* of A, that depends only on the event probability $\mathrm{P}[A]$. This means defining a function

$$i : \mathcal{F} \mapsto \mathbb{R} \quad , \quad i(A) = g(\mathrm{P}[A]) \tag{3.58}$$

for a suitable map $g : [0, 1] \mapsto \mathbb{R}$.

Based on our intuitive notion of information we require $i(A)$ to satisfy the following axioms

A1. The information carried by any event is non negative: $i(A) \geq 0$, for all A.

A2. The information carried by the sure event (that is the whole sample space Ω) is null: $i(\Omega) = 0$.

A3. The less likely an event is, the more informative it is: if $\mathrm{P}[A] \leq \mathrm{P}[B]$, then $i(A) \geq i(B)$.

A4. The information jointly carried by two *independent* events (that is the information of their intersection) is the sum of their informations: $i(A \cap B) = i(A) + i(B)$.

Each of the axioms above can be translated into a corresponding property for the function g as follows

P1. $g(\alpha) \geq 0$, for all $\alpha \in [0, 1]$;

P2. $g(1) = 0$;

P3. g is monotonically nonincreasing;

P4. $g(\alpha\beta) = g(\alpha) + g(\beta)$.

Property P4 derives from Axiom A4 by considering that if two independent events have probabilities $\mathrm{P}[A] = \alpha$ and $\mathrm{P}[B] = \beta$, respectively, the probability of their intersection is $\mathrm{P}[A \cap B] = \alpha\beta$.

It can be shown that the only functions that enjoy the above properties are of the type

$$g(\alpha) = \log_b \frac{1}{\alpha} \quad , \quad 0 < \alpha \leq 1 \tag{3.59}$$

with an arbitrary base $b > 1$.

As $\log_c \alpha = \log_c b \cdot \log_b \alpha$, the definitions of information with a base or another would only differ by a multiplicative constant. It is customary to choose $b = 2$ so that an event with probability $\alpha = 1/2$ carries unit information, called *bit*.[4] Thus we obtain the following

Definition 3.1 *The information of an event A, having* $\mathrm{P}[A] > 0$, *is given by*

$$i(A) = \log_2 \frac{1}{\mathrm{P}[A]} \quad \text{[bit]} \tag{3.60}$$

Example 3.3 A The probability of extracting hearts out of a regular 52-card pack is $1/4$. Hence the information of such event is 2 bit.

Example 3.3 B Given a message with iid binary symbols having equally likely values 0 and 1, the probability of the string '010' (as any given triplet) is $1/8$, and the information is 3 bit.

Observe that according to the definition above it is not possible to define the information of an event with null probability. It is also said that an event with null probability carries infinite information.

[4]The name "bit" as contraction of "binary digit" was suggested in the 1940s by J. W. Tukey (who would later be the inventor of the Fast Fourier Transform computational algorithm, together with J. W. Cooley), but it was C. E. Shannon in its milestone paper [12] who gave it the meaning of information unit.

So far in this book we have used the word "bit" to indicate a binary symbol in a message with no reference to its information content, as it is common in the computer and communication literature. Through the rest of this chapter, for better clarity, we will keep the distinction and reserve the word "bit" only to indicate the information unit. We will use use the expression "binary symbol" or simply "symbol" to identify an element of a binary message.

A far less common choice, made by some authors, is the base $b = e$ so that $g(\alpha) = \ln 1/\alpha$. The resulting measurement unit for information is called *nat*, contraction for "natural unit".

3.3.2 Entropy

In the following analysis we will consider *discrete* rvs and rves. Extension to continuous rvs can be found in more advanced texts [5].

Entropy of a rv Given a discrete rv x with alphabet $\mathcal{A}_x = \{a_1, \ldots, a_M\}$, we define its *information function* as the function that maps any point $a \in \mathcal{A}_x$ into the information of the event $\{x = a\}$

$$i_x : \mathcal{A}_x \mapsto \mathbb{R} \quad , \quad i_x(a) = i(\{x = a\}) = \log_2 \frac{1}{p_x(a)} \tag{3.61}$$

We can apply i_x to x itself and calculate its expectation to obtain a quantity that represents the *average information* carried by x, and is called the *entropy* of x.

Definition 3.2 *The* entropy $H(x)$ *of a discrete rv* x *is the expectation of its information function*

$$H(x) = \mathrm{E}\left[i_x(x)\right] \tag{3.62}$$

By the fundamental Theorem 2.11 for expectation, the entropy can be calculated as

$$H(x) = \sum_{a \in \mathcal{A}_x} p_x(a) i_x(a) = \sum_{a \in \mathcal{A}_x} p_x(a) \log_2 \frac{1}{p_x(a)} \tag{3.63}$$

From (3.63) we see that $H(x)$ does not depend on the possible values (the alphabet) of x, but only on its probability mass distribution.[5]

Example 3.3 C A binary rv with alphabet $\mathcal{A}_x = \{a_1, a_2\}$ and PMD

$$p_x(a_1) = p \quad , \quad p_x(a_2) = 1 - p$$

has entropy given by

$$\begin{aligned} H(x) &= p_x(a_1) i_x(a_1) + p_x(a_2) i_x(a_2) \\ &= p \log_2 \frac{1}{p} + (1 - p) \log_2 \frac{1}{(1 - p)} \ . \end{aligned}$$

In particular, for $p = 1/2$ we get $H(x) = 1$ bit, since both values a_1 and a_2 carry an information of 1 bit. On the other hand, as it can be seen by the plot of $H(x)$ as a function of p in Fig. 3.28, any unbalancing of the probabilities (either $p > 1/2$ or $p < 1/2$) leads to a decrease in the entropy, with the limit value of null entropy for an a.s. constant rv (either $p = 0$ or $p = 1$).

[5]The notation $H(x)$ is somehow improper, since entropy does not represent a transformation of the variable x, but rather a parameter obtained from its statistical description, just like its mean, power or variance. In this sense it would be more proper to write H_x or $H(\boldsymbol{p})$ with $\boldsymbol{p} = [p_x(a_1), \cdots, p_x(a_M)]$, as Shannon pointed out in his paper [12]. However, the notation $H(x)$ is universally adopted.

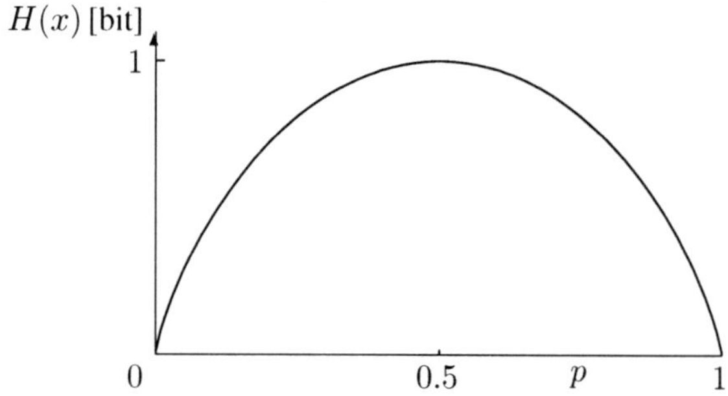

Figure 3.28 Entropy of a binary rv x versus the probability $p = p_x(a_1)$.

In the following we will determine bounds on the entropy $H(x)$. We start by recalling the definition of a strictly concave/convex function.

Definition 3.3 *A function* $h : \mathbb{R} \mapsto \mathbb{R}$ *is said to be* strictly concave *in the interval* $\mathcal{I}$ *if for any* m*-ple of distinct points* $a_1, \ldots, a_m \in \mathcal{I}$ *and real coefficients* $k_1, \ldots, k_m \in [0, 1)$ *such that* $\sum_{i=1}^{m} k_i = 1$*, it is*

$$\sum_{i=1}^{m} k_i h(a_i) < h\left(\sum_{i=1}^{m} k_i a_i\right) \tag{3.64}$$

On the contrary, h *is said to be* strictly convex *if it is*

$$\sum_{i=1}^{m} k_i h(a_i) > h\left(\sum_{i=1}^{m} k_i a_i\right) \tag{3.65}$$

Definition 3.3 is illustrated in Fig. 3.29. Observe that (3.64) states that any linear combination of points $[a_i, h(a_i)]$ in the graph of a strictly concave function h lies below the graph itself. Similarly, (3.65) states that any convex linear combination of points in the graph of a strictly convex h lies above the graph. For example, the function $h(a) = \log_b a$, with $b > 1$, is strictly concave in the interval $(0, +\infty)$, while $h(a) = \log_{1/b} a$ is strictly convex.

Figure 3.29 Illustration of (a) a strictly concave and (b) a strictly convex function in the interval $\mathcal{I}$, according to Definition 3.3. The point coordinates in both plots are $\boldsymbol{c}_i = [a_i, h(a_i)]$, for $i = 1, 2, 3$, $\boldsymbol{d} = [\sum_i k_i a_i, h(\sum_i k_i a_i)]$, and $\boldsymbol{q} = \sum_i k_i \boldsymbol{c}_i = [\sum_i k_i a_i, \sum_i k_i h(a_i)]$ is a linear combination of the $\{\boldsymbol{c}_i\}$.

We now give a fundamental result on the expectation of strictly concave and convex functions of rvs.

Proposition 3.1 (Jensen's inequality) *Let z be a rv that is not a.s. constant, taking values in the interval $\mathcal{I}$. Then, for any function h strictly concave in $\mathcal{I}$*

$$\mathrm{E}\left[h(z)\right] < h\left(\mathrm{E}\left[z\right]\right) \tag{3.66}$$

whereas, for any function h strictly convex in $\mathcal{I}$

$$\mathrm{E}\left[h(z)\right] > h\left(\mathrm{E}\left[z\right]\right) \tag{3.67}$$

provided all the expectations do exist.

Proof We give the proof in the case z is a discrete rv with a finite alphabet $\mathcal{A}_z = \{a_1, \ldots, a_M\}$ and $\mathcal{A}_z \subset \mathcal{I}$. Then, since z is not a.s. constant, we have $0 < p_z(a_i) < 1$. For this reason and the normalization condition (2.141), if h is concave we can apply (3.64) with $k_i = p_z(a_i)$ and $m = M$, obtaining

$$\sum_{i=1}^{M} p_z(a_i)h(a_i) < h\left(\sum_{i=1}^{M} p_z(a_i)a_i\right) \tag{3.68}$$

which by Theorem 2.11, yields (3.66). Analogously, (3.67) is proved by using (3.65). □

Observe that if the rv z is a.s. constant taking the value a_1, the inequalities (3.66)-(3.67) must be replaced by the equality

$$\mathrm{E}\left[h(z)\right] = h(a_1) = h\left(\mathrm{E}\left[z\right]\right) \tag{3.69}$$

holding by Theorem 2.11 for *any* function h, be it concave, convex or neither of the two.

In proving many of the following results we will make extensive use of Jensen's inequality, together with the concavity of the function $h(a) = \log_2 a$.

Proposition 3.2 *Let x be a discrete rv.*

i) If x is a.s. constant, then $H(x) = 0$.

ii) Otherwise, $H(x) > 0$.

Proof To prove i) consider that the alphabet of x is composed only of a_1 and $i_x(a_1) = 0$. The only term in the sum (3.63) is therefore null.

To prove ii), we observe that in this case for all $a \in \mathcal{A}_x$ we have $0 < p_x(a) < 1$ and $i_x(a) > 0$. Thus the terms in the sum (3.63) are strictly positive. □

Proposition 3.3 *Let x be a discrete rv with a finite alphabet $\mathcal{A}_x$ having cardinality M.*

i) If all $a \in \mathcal{A}_x$ are equally likely with probability $p_x(a) = 1/M$, then $H(x) = \log_2 M$.

ii) Otherwise, $H(x) < \log_2 M$.

Proof The proof of i) is straightforward. Since all values in the alphabet of x bear the same information

$$i_x(a) = \log_2 M \quad , \quad \text{for all } a \in \mathcal{A}_x$$

by taking the expectation we get the result. Note that the variable $i_x(x)$ is a.s. constant.

The proof of ii) is based on Jensen's inequality, with the function $h(z) = \log_2 z$ being strictly concave. Then, since the rv $z = 1/p_x(x)$ is not a.s. constant, we get from (3.62) and (3.66)

$$\begin{aligned} H(x) &= \mathrm{E}\left[\log_2 \frac{1}{p_x(x)}\right] \\ &< \log_2 \mathrm{E}\left[\frac{1}{p_x(x)}\right] \\ &= \log_2 \left(\sum_{a \in \mathcal{A}_x} p_x(a) \frac{1}{p_x(a)}\right) \\ &= \log_2 M \; . \end{aligned}$$

□

By combining the results of Propositions 3.2 and 3.3 we see that the entropy of a discrete rv x is bounded by

$$0 \leq H(x) \leq \log_2 M \; , \tag{3.70}$$

with the minimum value corresponding to a rv whose outcome is almost certain (with probability 1), and the maximum value corresponding to a rv that is most uncertain, with each of its M possible outcomes being equally likely. This intuitively justifies the notion of entropy of a rv as a *measure of the uncertainty* of its outcome.

Joint entropy of a rve Analogously to (3.61) and (3.62) we can define the information function $i_{\boldsymbol{x}}$ and entropy $H(\boldsymbol{x})$ of a discrete rve $\boldsymbol{x} = [x_1, \ldots, x_n]$ as

$$i_{\boldsymbol{x}} : \mathcal{A}_{\boldsymbol{x}} \mapsto \mathbb{R} \quad , \quad i_{\boldsymbol{x}}(\boldsymbol{a}) = i\left(\{\boldsymbol{x} = \boldsymbol{a}\}\right) = \log_2 \frac{1}{p_{\boldsymbol{x}}(\boldsymbol{a})} \tag{3.71}$$

$$H(\boldsymbol{x}) = \mathrm{E}\left[i_{\boldsymbol{x}}(\boldsymbol{x})\right] \; . \tag{3.72}$$

Observe that the entropy calculation can again be performed according to (3.63), with obvious modifications.

The vector entropy $H(\boldsymbol{x}) = H(x_1, \ldots, x_n)$ is also called the *joint entropy* of the rvs $x_1, \ldots, x_n$ making up the rve $\boldsymbol{x}$. We now relate $H(\boldsymbol{x})$ to the entropies of the single rvs, $\{H(x_i)\}$, starting with the case of two variables.

Proposition 3.4 *Let x, y be two discrete rvs.*

i) If y is a.s. a function of x (that is for all $a \in \mathcal{A}_x$, $\exists b : p_{xy}(a, b) = p_x(a)$), then $H(x, y) = H(x)$.

ii) Otherwise, $H(x, y) > H(x)$.

Proof Let $\boldsymbol{x} = [x, y]$, with alphabet $\mathcal{A}_{\boldsymbol{x}}$. To prove i) observe that in this case $i_{\boldsymbol{x}}(a, b) = i_x(a)$, for all $[a, b] \in \mathcal{A}_{\boldsymbol{x}}$ and we get

$$H(x, y) = \mathrm{E}\left[i_{\boldsymbol{x}}(x, y)\right] = \mathrm{E}\left[i_x(x)\right] = H(x)$$

For the proof of ii) consider that when y is not a function of x, there exist points $[a, b] \in \mathcal{A}_{\boldsymbol{x}}$ such that

$$p_{\boldsymbol{x}}(a, b) < p_x(a)$$

for which $i_{\boldsymbol{x}}(a, b) > i_x(a)$. For all other points in $\mathcal{A}_{\boldsymbol{x}}$, $i_{\boldsymbol{x}}(a, b) \geq i_x(a)$. Then

$$\begin{aligned} H(x, y) = \mathrm{E}\left[i_{\boldsymbol{x}}(x, y)\right] &= \sum_{[a,b]\in\mathcal{A}_{\boldsymbol{x}}} p_{\boldsymbol{x}}(a, b) i_{\boldsymbol{x}}(a, b) \\ &> \sum_{[a,b]\in\mathcal{A}_{\boldsymbol{x}}} p_{\boldsymbol{x}}(a, b) i_x(a) = \mathrm{E}\left[i_x(x)\right] = H(x) \ . \end{aligned}$$

Observe in the last step above that the expectation of $i_x(x)$ (as that of any function of the rv x) can be calculated by interpreting $i_x(x)$ as a function of the rve $[x, y]$. □

Clearly, the above proposition can also be stated with the roles of x and y swapped.

The following result sets an upper bound to the joint entropy of two rvs.

Proposition 3.5 *Let x and y be two discrete rvs.*

i) If x and y are statistically independent, then $H(x, y) = H(x) + H(y)$.

ii) Otherwise, $H(x, y) < H(x) + H(y)$.

Proof The proof of i) is straightforward. The statistical independence between the events $\{x = a\}$ and $\{y = b\}$ assures by Axiom A4 on page 157 that

$$\begin{aligned} i_{\boldsymbol{x}}(a, b) &= i(\{x = a, y = b\}) \\ &= i(\{x = a\}) + i(\{y = b\}) = i_x(a) + i_y(b) \quad , \qquad \text{for all } [a, b] \in \mathcal{A}_{\boldsymbol{x}} \end{aligned}$$

and by the linearity of expectation

$$\begin{aligned} H(x, y) &= \mathrm{E}\left[i_{\boldsymbol{x}}(x, y)\right] \\ &= \mathrm{E}\left[i_x(x)\right] + \mathrm{E}\left[i_y(y)\right] = H(x) + H(y) \ . \end{aligned}$$

For the proof of ii) we consider the quantity $H(x, y) - H(x) - H(y)$ and prove that it is strictly negative. Indeed

$$\begin{aligned} H(x, y) - H(x) - H(y) &= \mathrm{E}\left[i_{\boldsymbol{x}}(x, y) - i_x(x) - i_y(y)\right] \\ &= \mathrm{E}\left[\log_2 \frac{p_x(x) p_y(y)}{p_{\boldsymbol{x}}(x, y)}\right] \ . \end{aligned}$$

Since x and y are not statistically independent, $p_{\boldsymbol{x}}(a, b)$ will be greater than the product $p_x(a) p_y(b)$ for some pairs (a, b) and smaller for others. Thus the ratio $p_x(x) p_y(y) / p_{\boldsymbol{x}}(x, y)$ is not a.s. constant and we can apply Jensen's inequality (3.66) to the concave function $\log_2$ and get

$$\mathrm{E}\left[\log_2 \frac{p_x(x) p_y(y)}{p_{\boldsymbol{x}}(x, y)}\right] < \log_2 \mathrm{E}\left[\frac{p_x(x) p_y(y)}{p_{\boldsymbol{x}}(x, y)}\right]$$

$$= \log_2 \left[\sum_{[a,b] \in \mathcal{A}_{\boldsymbol{x}}} p_{\boldsymbol{x}}(a,b) \frac{p_x(a) p_y(b)}{p_{\boldsymbol{x}}(a,b)} \right]$$

$$= \log_2 \left[\sum_{[a,b] \in \mathcal{A}_{\boldsymbol{x}}} p_x(a) p_y(b) \right] .$$

Now, considering that $\mathcal{A}_{\boldsymbol{x}}$ is a subset of the Cartesian product $\mathcal{A}_x \times \mathcal{A}_y$, we get

$$\log_2 \left[\sum_{[a,b] \in \mathcal{A}_{\boldsymbol{x}}} p_x(a) p_y(b) \right] \leq \log_2 \left[\sum_{[a,b] \in \mathcal{A}_x \times \mathcal{A}_y} p_x(a) p_y(b) \right]$$

$$= \log_2 \left[\left(\sum_{a \in \mathcal{A}_x} p_x(a) \right) \left(\sum_{b \in \mathcal{A}_y} p_y(b) \right) \right]$$

$$= \log_2 1 = 0 .$$

By combining the above steps we obtain

$$H(x,y) - H(x) - H(y) < 0 .$$

thus completing the proof. □

From Propositions 3.4 and 3.5 we obtain the following bounds for the joint entropy

$$\max \{H(x), H(y)\} \leq H(x,y) \leq H(x) + H(y) . \tag{3.73}$$

The lower bound is attained when one of the two rvs is a.s. a function of the other (i.e. the value of one is a.s. determined by the value of the other). The upper bound is reached when the two rvs are statistically independent.

A generalization of the above propositions and hence (3.73) to the case of n rvs is straightforward and yields

$$\max_i \{H(x_i)\} \leq H(x_1, \ldots, x_n) \leq \sum_{i=1}^{n} H(x_i) . \tag{3.74}$$

Conditional entropy Starting from a discrete rve $\boldsymbol{x} = [x, y]$ and the conditional statistical description of x given y, we can define the conditional information and conditional entropy as

$$i_{x|y} : \mathcal{A}_{\boldsymbol{x}} \mapsto \mathbb{R} \quad , \quad i_{x|y}(a|b) = \log_2 \frac{1}{p_{x|y}(a|b)} \tag{3.75}$$

$$H(x|y) = \mathrm{E}\left[i_{x|y}(x|y) \right] . \tag{3.76}$$

Since $i_{x|y}(x|y)$ is a function of both rvs x and y, the entropy calculation via Theorem 2.11 must be performed taking the expectation with respect to their joint statistical description, that is

$$H(x|y) = \sum_{[a,b] \in \mathcal{A}_{\boldsymbol{x}}} p_{\boldsymbol{x}}(a,b) i_{x|y}(a|b) = \sum_{[a,b] \in \mathcal{A}_{\boldsymbol{x}}} p_{\boldsymbol{x}}(a,b) \log_2 \frac{1}{p_{x|y}(a|b)} . \tag{3.77}$$

Observe that the conditional PMD is used in the logarithm and the joint PMD in the expectation.

It is easy to see how conditional information and entropies are related to joint and individual statistics, as given in the following proposition.

Proposition 3.6 *Given a discrete rve* $\boldsymbol{x} = [x, y]$, *we have*

$$i_{x|y}(a|b) = i_{\boldsymbol{x}}(a, b) - i_y(b) \tag{3.78}$$

$$H(x|y) = H(x, y) - H(x) \, . \tag{3.79}$$

Proof From (2.176) we get

$$\begin{aligned} i_{x|y}(a|b) &= \log_2 \frac{1}{p_{x|y}(a|b)} = \log_2 \frac{p_y(b)}{p_{xy}(a, b)} \\ &= \log_2 \frac{1}{p_{xy}(a, b)} - \log_2 \frac{1}{p_y(b)} = i_{\boldsymbol{x}}(a, b) - i_y(b) \, , \end{aligned}$$

proving (3.78). Then by writing

$$i_{x|y}(x|y) = i_{\boldsymbol{x}}(x, y) - i_y(y) \, ,$$

and taking the expectation of both sides, (3.79) is proved. □

If we combine (3.79) with the upper bound (3.73) for the joint entropy we get that the conditional entropy is bounded by

$$0 \leq H(x|y) \leq H(x) \, , \tag{3.80}$$

where the lower bound is due to the fact that $H(x|y)$ is an entropy. Thus, the conditional entropy of x given y is null when the conditioned variable x is a.s. a function of y, and it is equal to the entropy of x when the two variables are statistically independent. We can therefore think of $H(x|y)$ as a measure of the *average information* (uncertainty) carried by x once we know y.

3.3.3 Efficiency and redundancy

For a rv x with a finite alphabet of size M, the value $\log_2 M$ is called the *nominal information* and represents the maximum value for entropy. The ratio between the actual entropy $H(x)$ and the nominal information is called *efficiency* of the rv x

$$\text{efficiency:} \quad \eta_x = \frac{H(x)}{\log_2 M} \, . \tag{3.81}$$

Thus, a rv with M equally likely values has unit efficiency. From (3.81) we also define the *redundancy* of x as

$$\text{redundancy:} \quad 1 - \eta_x \, . \tag{3.82}$$

For a rve $\boldsymbol{x} = [x_1, \ldots, x_n]$, it is common to consider as alphabet the Cartesian product of the alphabets of each rv, $\mathcal{A}_{\boldsymbol{x}} = \mathcal{A}_{x_1} \times \cdots \times \mathcal{A}_{x_n}$. Moreover, when all the rvs of the rve have the same alphabet $\mathcal{A}_x$ of size M, it is common to refer to the rve $\boldsymbol{x}$ as a *word* and the vector alphabet $\mathcal{A}_{\boldsymbol{x}} = \mathcal{A}_x^n$ is called *dictionary* to avoid confusion with $\mathcal{A}_x$. Also, the single rvs are usually called *symbols*.

Thus, for a word with n symbols, all with the same alphabet, the cardinality of the dictionary is M^n and the efficiency is given by

$$\eta_{\boldsymbol{x}} = \frac{H(\boldsymbol{x})}{\log_2 M^n} = \frac{H(x_1, \ldots, x_n)}{n \log_2 M} . \tag{3.83}$$

Moreover we define the *average entropy per symbol* of the word $\boldsymbol{x}$ as

$$H_s(\boldsymbol{x}) = \frac{1}{n} H(\boldsymbol{x}) \tag{3.84}$$

If all the rvs $\{x_i\}$ have the same distribution, and hence the same entropy $H(x_i)$, we can apply (3.73) and (3.70) to get the inequalities

$$\frac{1}{n} H(x) \leq H_s(\boldsymbol{x}) \leq H(x) \leq \log_2 M \tag{3.85}$$

If $x_1, \ldots, x_n$ are statistically independent

$$H_s(\boldsymbol{x}) = H(x) \tag{3.86}$$

3.3.4 Information rate of a message

We now move on to consider *infinite* messages, modeled as stationary discrete time rps with finite alphabet $\mathcal{A}_x$. As it is made of infinite rvs, the entropy of a message may in general be infinite. However, we can define its entropy per symbol as follows.

Let $\{x_n\}$ be a stationary rp with quantum T. For any word $\boldsymbol{x}$ that can be built with a finite choice of symbols $\{x_n\}$, we can find its average entropy per symbol according to (3.84). We define the *average entropy per symbol* of the message $\{x_n\}$ as the limit

$$H_s(x) = \lim_{N \to \infty} \frac{1}{2N+1} H(x_{-N}, \cdots, x_N) . \tag{3.87}$$

From the bounds (3.85) we get the inequalities

$$0 \leq H_s(x) \leq H(x) \leq \log_2 M \tag{3.88}$$

For a message with iid symbols it follows from (3.86) that $H_s(x) = H(x)$.

Similarly, for a pair of messages $\{x_n\}, \{y_n\}$ we can define their *joint* average entropy per symbol as

$$H_s(x, y) = \lim_{N \to \infty} \frac{1}{2N+1} H(x_{-N}, y_{-N}, \ldots, x_N, y_N) . \tag{3.89}$$

If we multiply $H_s(x)$, the average information carried by each symbol, by the symbol rate, we can think of the result as the average information *per unit time* carried by $\{x_n\}$.

Definition 3.4 *The* information rate *of an infinite message $\{x_n\}$ with symbol rate $F = 1/T$ is defined as*

$$R(x) = F H_s(x) \quad [bit/s] \tag{3.90}$$

Observe that the information rate is measured (as the ratio between information and time units) in bit/s. If the message $\{x_n\}$ is produced by a source x emitting symbols at rate F, we call $R(x)$ the information rate of the source. If $\{x_n\}$ has iid rvs we say that the source is *memoryless*.

Correspondingly, we define the *nominal* information rate, or simply *nominal rate*, of the message $\{x_n\}$ as

$$F \log_2 M \tag{3.91}$$

which coincides with the information rate of a memoryless source with equally likely symbol values.

The efficiency of an information source x emitting symbols with rate F and M-ary alphabet is then measured as

$$\eta_x = \frac{R(x)}{F \log_2 M} = \frac{H_s(x)}{\log_2 M} \tag{3.92}$$

Example 3.3 D As seen in Section 3.2, the speech signal in digital telephony is sampled at rate $F_s = 8\,\text{kHz}$, and quantized with $2^8 = 256$ levels. Hence the nominal information rate of the PCM encoder (here seen as the source) for speech signal is 64 kbit/s. We will show in Example 3.4 A that its efficiency is quite low.

3.4 SOURCE CODING

3.4.1 The purpose of source coding

Source coding is a transformation that replaces a (possibly infinite) sequence $\{x_n\}$, with symbol rate F_x and alphabet size M_x, with another (encoded) sequence $\{y_\ell\}$, with symbol rate F_y and alphabet size M_y, bearing the same information rate but with a lower nominal rate, and therefore higher efficiency. In turn, a lower nominal rate allows a more efficient exploitation of the channel.

In source coding *without distortion* the aim is to maintain in y all the information that was originally in x, therefore it must be $R(y) = R(x)$. Since the information rate can not exceed the nominal rate of a message, $R(x)$ represents also a lower bound to nominal rate of y. In practice the coder is designed so that the nominal rate of y approaches the information rate of x as closely as possible, that is

$$R(x) \leq F_y \log_2 M_y < F_x \log_2 M_x \tag{3.93}$$

In source coding *with distortion*, the nominal rate of the encoded message $\{y_\ell\}$ is further lowered, possibly below the source information rate $R(x)$, by allowing its information rate to be $R(y) < R(x)$. This implies some loss of information, which can not be recovered at decoding, however it is admitted as long as the quality that is perceived by the receiver does not degrade. The choice on what level of distortion can be allowed, and what nominal rate reduction is sought is a trade-off problem in the system design.

In this chapter we only consider source coding without distortion.

Example 3.4 A In the global system for mobile communications (GSM) standard for mobile digital telephony, the speech signal is source coded and transmitted (with slight distortion) at a nominal rate of 13 kbit/s. Thus we can say that the information rate of speech is lower than 13 kbit/s, showing the redundancy of the standard PCM encoder for speech (as in Example 3.3 D) is at least 80%.

Besides transmission, source coding has important applications also in the storing of information, with the aim of reducing the amount of used memory space as much as possible. Source coding is commonly known also as *data compression*.

3.4.2 Entropy coding

A possible scheme for source coding is illustrated in Fig. 3.30. The input information symbol stream $\{x_n\}$ at rate $F_x = 1/T_x$ is segmented into words $\boldsymbol{x}_m$, through a serial-to-parallel (S/P) transformation. Each information word $\boldsymbol{x}_m$ is mapped into a codeword $\boldsymbol{y}_m$ with symbols taken from an alphabet $\mathcal{A}_y$. The coded symbols are then transmitted sequentially.

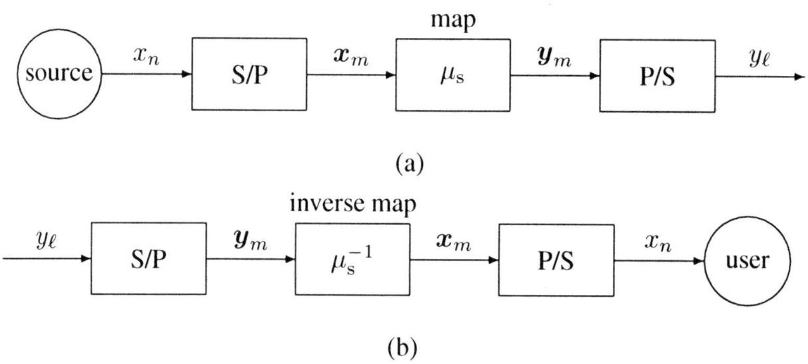

Figure 3.30 Block diagram representation of source coding: (a) encoder; (b) decoder.

Within the scheme shown in Fig. 3.30, entropy coding allows the nominal rate of y to asymptotically achieve the lower bound in (3.93) as the length of information words $\boldsymbol{x}_m$ becomes sufficiently large. The strategy to accomplish this task is by mapping the most likely information sequences into short codewords, while the least likely information sequences are mapped into long codewords. In this way we will transmit shorter codewords more often than longer codewords, with the effect of reducing the average word length, and the nominal rate of y. However, the formal description of this procedure bears a more cumbersome notation, since the length of the codeword $\boldsymbol{y}_m$ depends on the particular realization of the information word $\boldsymbol{x}_m$. This requires a formal definition of variable length rves.

Variable length words A variable length rve (or word) $\boldsymbol{x}$ is defined by a triplet $\{\mathcal{L}_{\boldsymbol{x}}, \mathcal{A}_x, \mathcal{D}_{\boldsymbol{x}}\}$ where

- $\mathcal{L}_{\boldsymbol{x}} \subset \mathbb{N}$ is the set of possible word lengths.
- $\mathcal{A}_x$ is the alphabet (having cardinality M_x) of each symbol in the word.
- $\mathcal{D}_{\boldsymbol{x}} \subset \bigcup_{L \in \mathcal{L}_{\boldsymbol{x}}} \mathcal{A}_x^L$ is the dictionary of $\boldsymbol{x}$ and is a subset of the set of all words with lengths in $\mathcal{L}_{\boldsymbol{x}}$ that can be built from the elements in $\mathcal{A}_x$.

It is evident that a fixed length word of L symbols can always be seen as a variable length word with $\mathcal{L}_{\boldsymbol{x}} = \{L\}$,.

The statistical description of $\boldsymbol{x}$ is given by its PMD

$$p_{\boldsymbol{x}} : \mathcal{D}_{\boldsymbol{x}} \mapsto [0, 1] \tag{3.94}$$

from which we can derive the information function and entropy, in a manner analogous to (3.71)-(3.72). If we also define the *average length* of $\boldsymbol{x}$ as the sum of the lengths $L(\boldsymbol{a})$ of the words in $\mathcal{D}_{\boldsymbol{x}}$, weighted by their probabilities

$$L_{\boldsymbol{x}} = \sum_{\boldsymbol{a} \in \mathcal{D}_{\boldsymbol{x}}} p_{\boldsymbol{x}}(\boldsymbol{a}) L(\boldsymbol{a}) = \mathrm{E}\left[L(\boldsymbol{x})\right] \tag{3.95}$$

we can introduce the average entropy per symbol and the efficiency, respectively, as

$$H_{\mathrm{s}}(\boldsymbol{x}) = \frac{1}{L_{\boldsymbol{x}}} H(\boldsymbol{x}) \quad , \quad \eta_{\boldsymbol{x}} = \frac{H_{\mathrm{s}}(\boldsymbol{x})}{\log_2 M_x} = \frac{H(\boldsymbol{x})}{L_{\boldsymbol{x}} \log_2 M_x} \tag{3.96}$$

With reference to (3.96) observe that $L_{\boldsymbol{x}} \log_2 M$ represents the *nominal information* of $\boldsymbol{x}$.

Variable length coding Given two variable length rves $\boldsymbol{x}$, with $\{\mathcal{L}_{\boldsymbol{x}}, \mathcal{A}_x, \mathcal{D}_{\boldsymbol{x}}\}$ and $\boldsymbol{y}$, with $\{\mathcal{L}_{\boldsymbol{y}}, \mathcal{A}_y, \mathcal{D}_{\boldsymbol{y}}\}$, a variable length coder is a map

$$\mu_{\mathrm{s}} : \mathcal{D}_{\boldsymbol{x}} \mapsto \mathcal{D}_{\boldsymbol{y}} \tag{3.97}$$

and the *code* is the range of μ_{s}, that is $\mathcal{C} = \mu_{\mathrm{s}}(\mathcal{D}_{\boldsymbol{x}})$, whith some abuse of notation since $\mathcal{D}_{\boldsymbol{x}}$ is a set.

The statistical description of the codeword $\boldsymbol{y}$ can be obtained from that of $\boldsymbol{x}$ in the same way as for a map between ordinary fixed length rves, that is

$$p_{\boldsymbol{y}}(\boldsymbol{b}) = \sum_{\boldsymbol{a} \in \mu_{\mathrm{s}}^{-1}(\boldsymbol{b})} p_{\boldsymbol{x}}(\boldsymbol{a}) \tag{3.98}$$

A code is said to be *decodable* if we can uniquely recover the stream of original symbols $\{x_n\}$ from the stream of coded symbols $\{y_\ell\}$.

A *necessary* condition for decodability is that the map μ_{s} be one-to-one (i.e. injective), however this is not sufficient in general. In fact, the parallel-to-serial (P/S) converter at the decoder front end in Fig. 3.30b must be able to parse the incoming coded message $\{y_\ell\}$ into the words $\boldsymbol{y}_m$. For a fixed length code this would only require synchronization between transmitter and receiver, but in the case of a variable length code the decoder does not know the length of the current codeword $\boldsymbol{y}_m$.

A solution to this problem is to make $\mathcal{C}$ a *prefix code*, where no codeword $\boldsymbol{b}$ is the initial part (prefix) of another codeword $\boldsymbol{b}'$. In other words, by indicating with $(\boldsymbol{b})_1^k = [b_1, \ldots, b_k]$ the k-symbol prefix of the word $\boldsymbol{b} = [b_1, \ldots, b_L]$, a prefix code $\mathcal{C}$ verifies

$$\text{for all } \boldsymbol{b}, \boldsymbol{b}' \in \mathcal{C} \text{ such that } L(\boldsymbol{b}') \geq L(\boldsymbol{b}) \quad , \quad \boldsymbol{b} \neq (\boldsymbol{b}')_1^{L(\boldsymbol{b})} . \tag{3.99}$$

In this case the decoder can start from the first received symbol, say y_0, and keep reading incoming symbols one at a time until they make up a codeword $\boldsymbol{b}$. Since this codeword can not be the prefix of another codeword, it must be the first transmitted codeword, $\boldsymbol{y}_0 = \boldsymbol{b}$. Then the decoder starts building another codeword $\boldsymbol{y}_1$ from the next symbol, and so on. Therefore, a *sufficient* condition for decodability is that the coding map be one-to-one *and* yield a prefix code.

Example 3.4 B Consider a variable length coder with:

- a quaternary input message that is coded symbol by symbol

$$\mathcal{A}_x = \{0, 1, 2, 3\} \quad , \quad \mathcal{L}_x = \{1\} \quad , \quad \mathcal{D}_x = \mathcal{A}_x \ ;$$

- a binary variable length output rve

$$\mathcal{A}_y = \{0, 1\} \quad , \quad \mathcal{L}_y = \{1, 2\} \quad , \quad \mathcal{D}_y = \{0, 1, 10, 01\} \ ;$$

- map

$$\mu_s(0) = 0 \quad , \quad \mu_s(1) = 1 \quad , \quad \mu_s(2) = 10 \quad , \quad \mu_s(3) = 01 \ .$$

The map is one-to-one, but the code $\mathcal{C} = \mathcal{D}_y$ is not a prefix code, and indeed the code is not decodable. As a matter of fact, the received symbol sequence 011011 ... can be parsed into words as

$$01, 10, 1, 1, \ldots \quad \text{and decoded as } 3, 2, 1, 1, \ldots$$

or

$$0, 1, 1, 01, 1, \ldots \quad \text{and decoded as } 0, 1, 1, 3, 1, \ldots$$

or

$$01, 1, 01, 1, \ldots \quad \text{and decoded as } 3, 1, 3, 1, \ldots$$

and so on.

Example 3.4 C Consider the same input words as in the previous example, but the codewords are now

$$\mathcal{A}_y = \{0, 1\} \quad , \quad \mathcal{L}_y = \{2, 3\} \quad , \quad \mathcal{D}_y = \{00, 11, 010, 101\}$$

with the map

$$\mu_s(0) = 00 \quad , \quad \mu_s(1) = 11 \quad , \quad \mu_s(2) = 010 \quad , \quad \mu_s(3) = 101 \ .$$

Now $\mathcal{D}_y$ is a prefix code. The received sequence 10100010 ... has the unique parsing

$$101, 00, 010, \ldots \quad \text{and is decoded as } 3, 0, 2, \ldots$$

Example 3.4 D The reader may have been led by the discussion so far to think erroneously that a prefix code is also a necessary condition for decodability. In this example we show that this is not the case. Let

$$\mathcal{A}_x = \{0, 1, 2\} \quad , \quad \mathcal{L}_x = \{1\} \quad , \quad \mathcal{D}_x = \mathcal{A}_x$$

$$\mathcal{A}_y = \{0,1\} \quad , \quad \mathcal{L}_y = \{2,3\} \quad , \quad \mathcal{D}_y = \{01, 011, 110\}$$
$$\mu_s(0) = 01 \quad , \quad \mu_s(1) = 011 \quad , \quad \mu_s(2) = 110$$

Albeit not a prefix code, this code is decodable: for example, the sequence 01101110 ... can be uniquely parsed into

$$011, 01, 110, \ldots \quad \text{and is decoded as } 1, 0, 2, \ldots$$

3.4.3 Shannon theorem on source coding

We precede the statement of the Shannon Theorem by the following lemma, that we give without proof [4, 5].

Theorem 3.7 (Kraft-McMillan Theorem) *For* any decodable code $\mathcal{C}$ *it must be*

$$\sum_{\boldsymbol{b}\in\mathcal{C}} \frac{1}{M_y^{L(\boldsymbol{b})}} \leq 1 \tag{3.100}$$

Conversely, consider a source with dictionary $\mathcal{D}_{\boldsymbol{x}} = \{\boldsymbol{a}_1, \ldots, \boldsymbol{a}_N\}$ *of size* N. *If* $\ell_1, \ldots, \ell_N$ *and* M *are integers such that*

$$\sum_{i=1}^{N} \frac{1}{M^{\ell_i}} \leq 1 \tag{3.101}$$

then it is possible to find a prefix code *with alphabet cardinality* $M_y = M$ *so that each codeword* $\boldsymbol{b}_i = \mu_s(\boldsymbol{a}_i)$ *has length* $L(\boldsymbol{b}_i) = \ell_i$.

The inequality (3.100) is known as *Kraft inequality* and it can also be written as

$$\mathrm{E}\left[\frac{1}{M_y^{L(\boldsymbol{y})} p_{\boldsymbol{y}}(\boldsymbol{y})}\right] \leq 1 \tag{3.102}$$

Theorem 3.8 (Shannon theorem on source coding) *For* any decodable code $\mathcal{C}$ *with alphabet cardinality* M_y, *the average length* $L_{\boldsymbol{y}}$ *satisfies*

$$L_{\boldsymbol{y}} \geq \frac{H(\boldsymbol{x})}{\log_2 M_y} \tag{3.103}$$

Conversely, it is possible to find prefix codes with average length

$$L_{\boldsymbol{y}} < \frac{H(\boldsymbol{x})}{\log_2 M_y} + 1 \tag{3.104}$$

Proof Since the code is decodable it must satisfy Kraft inequality and by taking logarithms on both sides of (3.102) we get

$$\log_{1/M_y} \mathrm{E}\left[\frac{1}{M_y^{L(\boldsymbol{y})} p_{\boldsymbol{y}}(\boldsymbol{y})}\right] \geq 0 \tag{3.105}$$

Moreover, since $\log_{1/M_y}$ is a convex function we can apply Jensen's inequality (3.67) and obtain

$$\begin{aligned}\log_{1/M_y} \mathrm{E}\left[\frac{1}{M_y^{L(\boldsymbol{y})} p_{\boldsymbol{y}}(\boldsymbol{y})}\right] &\leq \mathrm{E}\left[\log_{1/M_y} \frac{1}{M_y^{L(\boldsymbol{y})} p_{\boldsymbol{y}}(\boldsymbol{y})}\right] \\ &= \mathrm{E}\left[\log_{1/M_y} \frac{1}{M_y^{L(\boldsymbol{y})}} - \log_{1/M_y} p_{\boldsymbol{y}}(\boldsymbol{y})\right] \\ &= \mathrm{E}\left[L(\boldsymbol{y})\right] - \frac{1}{\log_2 M_y} \mathrm{E}\left[\log_{1/2} p_{\boldsymbol{y}}(\boldsymbol{y})\right] \\ &= L_{\boldsymbol{y}} - \frac{H(\boldsymbol{x})}{\log_2 M_y}\end{aligned}$$

By combining the above with (3.105) we obtain the statement (3.103)

To prove that it is possible to find a prefix code that satisfies (3.104), we let $\mathcal{D}_{\boldsymbol{x}} = \{\boldsymbol{a}_1, \ldots, \boldsymbol{a}_N\}$ be the information words and select the corresponding codeword lengths as

$$\ell_i = \lceil \log_{1/M_y} p_{\boldsymbol{x}}(\boldsymbol{a}_i) \rceil \; . \tag{3.106}$$

First we prove that it is indeed possible to build a prefix code with lengths $\{\ell_1, \ldots, \ell_N\}$, as they satisfy Kraft inequality (3.101). As a matter of fact, with the choice (3.106) we have

$$\ell_i \geq \log_{1/M_y} p_{\boldsymbol{x}}(\boldsymbol{a}_i)$$

or

$$\frac{1}{M_y^{\ell_i}} \leq p_{\boldsymbol{x}}(\boldsymbol{a}_i) \quad , \quad i = 1, \ldots, N$$

and by summing over the N information words, due to the probability normalization condition we obtain

$$\sum_{i=1}^{N} \frac{1}{M_y^{\ell_i}} \leq \sum_{i=1}^{N} p_{\boldsymbol{x}}(\boldsymbol{a}_i) = 1 \; .$$

To conclude the proof we show that from (3.106)

$$L(\mu_{\mathrm{s}}(\boldsymbol{a}_i)) = \ell_i < \log_{1/M_y} p_{\boldsymbol{x}}(\boldsymbol{a}_i) + 1 \quad , \quad \boldsymbol{a}_i \in \mathcal{D}_{\boldsymbol{x}} \tag{3.107}$$

thus we have

$$L(\boldsymbol{y}) < \frac{1}{\log_2 M_y} \log_{1/2} p_{\boldsymbol{x}}(\boldsymbol{x}) + 1 \tag{3.108}$$

By taking expectation of both sides we get

$$\mathrm{E}\left[L(\boldsymbol{y})\right] < \frac{1}{\log_2 M_y} \mathrm{E}\left[i_{\boldsymbol{x}}(\boldsymbol{x})\right] + 1 \tag{3.109}$$

and hence (3.104) □

By the result (3.104) of Theorem 3.8, we see that for an information message with a given dictionary $\mathcal{D}_{\boldsymbol{x}}$ it is always possible to construct a variable length prefix code so that the efficiency of the coded message is lower bounded by

$$\eta_y > 1 - \frac{1}{L_{\boldsymbol{y}}} \geq 1 - \frac{\log_2 M_y}{H(\boldsymbol{x})} \tag{3.110}$$

where the second inequality is obtained by use of (3.103).

In order to achieve an efficiency of the coded message that is as close as possible to one, (3.110) suggests two strategies, that can also be combined:

- choose M_y as small as possible, and indeed most source codes use binary symbols $(M_y = 2)$;
- choose a large value of $H(\boldsymbol{x})$ by increasing the length of the input words and hence the size of the input dictionary; this allows to get arbitrarily close to unit efficiency, but also increases the complexity of the encoder and decoder.

Further insight An alternative form of the Shannon source coding theorem [12], formulated for memoryless sources, states that we can bound the codeword length close to $H(\boldsymbol{x})/\log_2 M_y$ not only on average, but, more strongly, also with arbitrarily high probability, that is

Theorem 3.9 *Let x be a memoryless source with alphabet cardinality M_x and entropy per symbol $H(x)$. Then, for any $\varepsilon, \delta > 0$ and a sufficiently large L, it is possible to find a prefix code that maps L-symbol source words $\boldsymbol{x}$ into codewords $\boldsymbol{y}$ whose lenght $L(\boldsymbol{y})$ statisfies*

$$\mathrm{P}\left[L(\boldsymbol{y}) \geq \frac{L(H(x)+\varepsilon)}{\log_2 M_y} + 2\right] < \delta \tag{3.111}$$

while their average length $L_{\boldsymbol{y}}$ satisifies

$$L_{\boldsymbol{y}} < \frac{L(H(x)+\varepsilon)}{\log_2 M_y} + 2 + \delta L \log_{M_y} M_x \ . \tag{3.112}$$

By allowing ε and δ to be negligible, we obtain from (3.111)

$$\frac{L(\boldsymbol{y})}{L} \underset{L\to\infty}{\longrightarrow} \frac{H(x)}{\log_2 M_y} \ , \quad \text{in probability} \tag{3.113}$$

and from (3.112)

$$\lim_{L\to\infty} \frac{L_{\boldsymbol{y}}}{L} = \frac{H(x)}{\log_2 M_y} \ . \tag{3.114}$$

Moreover (3.114) allows to derive the asymptotic efficiency of the code as

$$\lim_{L\to\infty} \eta_{\boldsymbol{y}} = \lim_{L\to\infty} \frac{H(\boldsymbol{x})}{L_{\boldsymbol{y}} \log_2 M_y} = \lim_{L\to\infty} \frac{L}{L_{\boldsymbol{y}}} \frac{H(x)}{\log_2 M_y} = 1 \ .$$

3.4.4 Optimal source coding

The code constructing procedure described in the proof of Theorem 3.8 yields what is called the *Shannon-Fano code*. Although it satisfies property (3.110) and hence asymptotically it approaches unit efficiency as $L_{\boldsymbol{x}} \to \infty$, it does not achieve unit efficiency in general for finite values of $L_{\boldsymbol{x}}$. A particular case in which the Shannon-Fano code achieves unit efficiency is if $p_{\boldsymbol{x}}(\boldsymbol{a}_i)$ is an integer power of $1/M_y$ for all $\boldsymbol{a}_i \in \mathcal{D}_{\boldsymbol{x}}$. Then, we can choose $\ell_i = \log_{1/M_y} p_{\boldsymbol{x}}(\boldsymbol{a}_i)$ in (3.106) so that (3.108) is replaced by

$$L(\boldsymbol{y}) = \frac{1}{\log_2 M_y} \log_{1/2} p_{\boldsymbol{x}}(\boldsymbol{x}) \tag{3.115}$$

and (3.104) becomes

$$L_{\boldsymbol{y}} = \frac{H(\boldsymbol{x})}{\log_2 M_y} \,. \tag{3.116}$$

Apart from this case, the Shannon-Fano code does not in general yield maximum efficiency, among prefix codes.

Huffman coding The Huffman procedure [7] allows to construct an *optimal* prefix code in the sense that it has *minimum average length* $L_{\boldsymbol{y}}$ among all the decodable codes with variable length that use the same input dictionary $\mathcal{D}_{\boldsymbol{x}}$ and the corresponding PMD, and the same code alphabet $\mathcal{A}_y$. The code alphabet is always binary, while typically the input words have fixed length $L_{\boldsymbol{x}}$, and are often single symbols (that is $L_{\boldsymbol{x}} = 1$) so that $\mathcal{D}_{\boldsymbol{x}} = \mathcal{A}_x$. The optimum code $\mathcal{C}_{\text{opt}}$ and the corresponding map μ_s^{opt} are built from the PMD of $\boldsymbol{x}$, and the results of Propositions 3.10 and 3.11 below.

Proposition 3.10 *In an* optimal *code longer codewords have lower probabilities*

$$\text{if } p_{\boldsymbol{y}}(\boldsymbol{b}') < p_{\boldsymbol{y}}(\boldsymbol{b}'') \quad \text{then } L(\boldsymbol{b}') \geq L(\boldsymbol{b}'') \tag{3.117}$$

Proof The statement is proved by contradiction.

Suppose there exist in $\mathcal{C}_{\text{opt}}$ two words $\boldsymbol{b}'$ and $\boldsymbol{b}''$ for which (3.117) does not hold, that is

$$p_{\boldsymbol{y}}(\boldsymbol{b}') < p_{\boldsymbol{y}}(\boldsymbol{b}'') \quad \text{and} \quad L(\boldsymbol{b}') < L(\boldsymbol{b}'')$$

and let $\boldsymbol{a}'$ and $\boldsymbol{a}''$ be the corresponding input words so that $p_{\boldsymbol{x}}(\boldsymbol{a}') < p_{\boldsymbol{x}}(\boldsymbol{a}'')$. Consider the coding map μ_s' obtained from μ_s^{opt} by exchanging $\boldsymbol{b}'$ and $\boldsymbol{b}''$ while leaving all the other codewords unchanged, that is

$$\mu_s'(\boldsymbol{a}') = \boldsymbol{b}'' \quad , \quad \mu_s'(\boldsymbol{a}'') = \boldsymbol{b}' \quad , \quad \mu_s'(\boldsymbol{a}) = \mu_s^{\text{opt}}(\boldsymbol{a}) \ \text{ for all } \boldsymbol{a} \neq \boldsymbol{a}', \boldsymbol{a}'' \,.$$

Then the average length of $\boldsymbol{y}$ with the coding μ_s' is given by

$$\begin{aligned} L'_{\boldsymbol{y}} &= \sum_{\boldsymbol{b} \neq \boldsymbol{b}', \boldsymbol{b}''} p_{\boldsymbol{y}}(\boldsymbol{b}) L(\boldsymbol{b}) + p_{\boldsymbol{x}}(\boldsymbol{a}') L(\boldsymbol{b}'') + p_{\boldsymbol{x}}(\boldsymbol{a}'') L(\boldsymbol{b}') \\ &= L_{\boldsymbol{y}} - p_{\boldsymbol{x}}(\boldsymbol{a}') L(\boldsymbol{b}') + p_{\boldsymbol{x}}(\boldsymbol{a}'') L(\boldsymbol{b}'') + p_{\boldsymbol{x}}(\boldsymbol{a}') L(\boldsymbol{b}'') + p_{\boldsymbol{x}}(\boldsymbol{a}'') L(\boldsymbol{b}') \\ &= L_{\boldsymbol{y}} - [p_{\boldsymbol{x}}(\boldsymbol{a}') - p_{\boldsymbol{x}}(\boldsymbol{a}'')][L(\boldsymbol{b}') - L(\boldsymbol{b}'')] \\ &< L_{\boldsymbol{y}} \end{aligned}$$

so that $\mathcal{C}_{\rm opt}$ would no longer be optimal. □

Proposition 3.11 *In $\mathcal{C}_{\rm opt}$ there are two words $\boldsymbol{b}'$ and $\boldsymbol{b}''$ that have maximum length and only differ in their last symbol,*

$$L(\boldsymbol{b}') = L(\boldsymbol{b}'') = \max_{\boldsymbol{b}\in\mathcal{C}_{\rm opt}} L(\boldsymbol{b}) = L_{\max} \quad , \quad (\boldsymbol{b}')_1^{L_{\max}-1} = (\boldsymbol{b}'')_1^{L_{\max}-1} \tag{3.118}$$

Proof This result will be proved by contradiction.

Let $\boldsymbol{b}'$ be a word in $\mathcal{C}_{\rm opt}$ with length $L_{\max}$ and assume no other maximal length word has the same prefix $(\boldsymbol{b}')_1^{L_{\max}-1}$. Moreover, $(\boldsymbol{b}')_1^{L_{\max}-1}$ is not a codeword itself, since $\mathcal{C}_{\rm opt}$ is a prefix code. Hence, we can replace $\boldsymbol{b}'$ with its own prefix $(\boldsymbol{b}')_1^{L_{\max}-1}$ and still obtain a prefix code with average length

$$\begin{aligned} L'_{\boldsymbol{y}} &= \sum_{\boldsymbol{b}\neq\boldsymbol{b}'} p_{\boldsymbol{y}}(\boldsymbol{b})L(\boldsymbol{b}) + p_{\boldsymbol{y}}(\boldsymbol{b}')(L_{\max}-1) \\ &= \sum_{\boldsymbol{b}\in\mathcal{C}_{\rm opt}} p_{\boldsymbol{y}}(\boldsymbol{b})L(\boldsymbol{b}) - p_{\boldsymbol{y}}(\boldsymbol{b}') = L_{\boldsymbol{y}} - p_{\boldsymbol{y}}(\boldsymbol{b}') \\ &< L_{\boldsymbol{y}} \end{aligned}$$

so that $\mathcal{C}_{\rm opt}$ would no longer be optimal. □

Remark By combining Propositions 3.10 and 3.11 we get that two codewords of maximal length $L_{\max}$, $\boldsymbol{b}_{N-1}$ and $\boldsymbol{b}_N$ will have the two lowest probabilities, and the same prefix

$$\boldsymbol{b}'_{N-1} = (\boldsymbol{b}_{N-1})_1^{L_{\max}-1} = (\boldsymbol{b}_N)_1^{L_{\max}-1} \tag{3.119}$$

The probability of the prefix will be

$$\mathrm{P}\left[(\boldsymbol{y})_1^{L_{\max}-1} = \boldsymbol{b}'_{N-1}\right] = p_{\boldsymbol{y}}(\boldsymbol{b}_{N-1}) + p_{\boldsymbol{y}}(\boldsymbol{b}_N) \tag{3.120}$$

The procedure devised by Huffman to construct an optimal binary prefix code for an input dictionary $\mathcal{D}_{\boldsymbol{x}}$ of N words with probabilities $p_i = p_{\boldsymbol{x}}(\boldsymbol{a}_i)$, $i = 1, \ldots, N$, is therefore expressed in recursive form as follows.

1. If $N = 2$, an optimal binary code is trivially found as

$$\mathcal{C} = \{0, 1\} \quad , \quad \mu(\boldsymbol{a}_1) = 0 \quad , \quad \mu(\boldsymbol{a}_2) = 1 \quad , \quad L_{\boldsymbol{y}} = 1 \tag{3.121}$$

2. If $N > 2$, proceed as follows
   (a) Sort the N probabilities p_i in decreasing order, $p_1 \geq p_2 \geq \ldots \geq p_N$.
   (b) Since we will assign probabilities p_{N-1} and p_N to the two longest codewords $\boldsymbol{b}_{N-1}$ and $\boldsymbol{b}_N$ with the same prefix $\boldsymbol{b}'_{N-1}$, the probability of the prefix from (3.120) is $p'_{N-1} = p_{N-1} + p_N$.

(c) Find an optimal binary prefix code $\mathcal{C}'$ of $N-1$ words for the probabilities $p_1, p_2, \ldots, p'_{N-1}$.

(d) Obtain the code $\mathcal{C}$ from $\mathcal{C}'$, by replacing $\boldsymbol{b}'_{N-1}$ with the pair of words $\boldsymbol{b}_{N-1} = [\boldsymbol{b}'_{N-1}, 0]$ and $\boldsymbol{b}_N = [\boldsymbol{b}'_{N-1}, 1]$.

The key point in the procedure is the recursion in step 2c. In turn, at each iteration the optimal code $\mathcal{C}$ is obtained by expanding the previous optimal code $\mathcal{C}'$ with the addition of one word.

We clarify the Huffman procedure with an example.

Example 3.4 E Consider a source of information words $\boldsymbol{x}$ with the following dictionary

$$\mathcal{A}_x = \{1, 2, 3\} \quad , \quad \mathcal{D}_{\boldsymbol{x}} = \{11, 12, 13, 21, 23\}$$

and PMD given by

$$p_{\boldsymbol{x}}(11) = 0.25 \ , \quad p_{\boldsymbol{x}}(12) = 0.45 \ , \quad p_{\boldsymbol{x}}(13) = 0.05 \ , \quad p_{\boldsymbol{x}}(21) = 0.09 \ , \quad p_{\boldsymbol{x}}(23) = 0.16$$

We wish to find an optimal prefix code for $\boldsymbol{x}$ via the Huffman procedure.

At the first step, after ordering

$$p_1 = 0.45 \quad , \quad p_2 = 0.25 \quad , \quad p_3 = 0.16 \quad , \quad p_4 = 0.09 \quad , \quad p_5 = 0.05$$

we replace p_5 and p_4 with $p'_4 = p_4 + p_5 = 0.14$. At the next step we reorder

$$p'_1 = 0.45 \quad , \quad p'_2 = 0.25 \quad , \quad p'_3 = 0.16 \quad , \quad p'_4 = 0.14$$

and replace p'_4 and p'_3 with $p''_3 = p_4 + p_3 = 0.3$. In the last step we reorder

$$p''_1 = 0.45 \quad , \quad p''_2 = 0.3 \quad , \quad p''_3 = 0.25$$

and obtain $p'''_2 = p''_2 + p''_3 = 0.55$.

Now we are left with only two probabilities, and associate the symbol 0 to $p'''_2 = 0.55$ and the symbol 1 to $p'''_1 = 0.45$. By proceeding backwards, we observe that p'''_2 was obtained as the sum of $p''_2 = 0.3$ and $p''_3 = 0.25$, so we associate the pair [10] to p''_2 and [11] to p''_3. Then, since $p'''_2 = 0.3$ was obtained as the sum of $p'_3 = 0.16$ and $p'_4 = 0.14$, we construct the two corresponding words by adding a trailing 0 and a trailing 1 to the word [10], and associate [100] to p'_3 and [101] to p'_4. Finally, as $p'_4 = 0.14$ was obtained as the sum of $p_4 = 0.09$ and $p_5 = 0.05$, we construct the two corresponding words by adding a trailing 0 and a trailing 1 to the word [101], and associate [1010] to p_4 and [1011] to p_5.

The steps described above are illustrated in Fig. 3.31, and are summarized in the tree representation of Fig. 3.32.

Huffman coding is used in the data compression utility `pack` on UNIX systems with binary input and output alphabets, and fixed length input dictionary with 8-symbol words (i.e. bytes). Modified versions of Huffman coding are also used in fax transmission, and in the *joint photographic expert group* (JPEG) and MP3 standards for lossy compression of digital images and audio files, respectively.

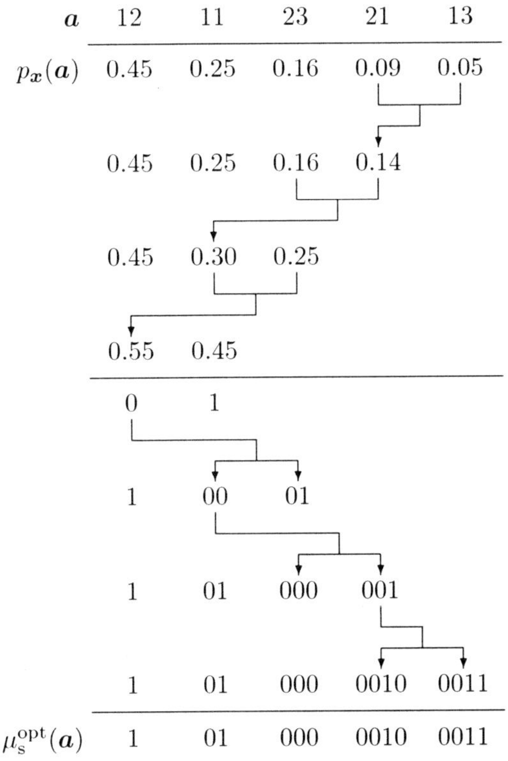

Figure 3.31 Construction of the Huffman code for Example 3.4 E.

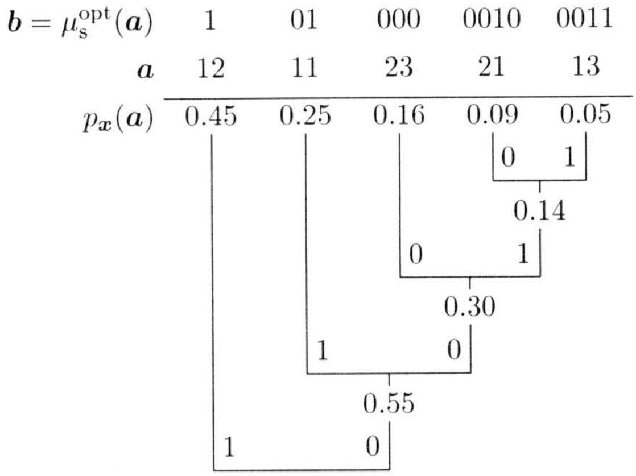

Figure 3.32 Tree representation of the Huffman code in Example 3.4 E.

3.4.5 Arithmetic coding

Arithmetic coding aims at constructs a code $\mathcal{C}$ (that asymptotically achieves unit efficency) and the corresponding map μ_s by processing the incoming information symbols $\{x_n\}$ one at a time.

The coding procedure for a memoryless M_x-ary source and a binary code alphabet ($M_y = 2$) requires to choose an ordering for the values in the input alphabet $\mathcal{A}_x$, so that the *lexicographical* ordering can be established between any two input sequences $\boldsymbol{a}'$ and $\boldsymbol{a}$ of length L, that is

$$\boldsymbol{a}' < \boldsymbol{a} \quad \text{if and only if} \quad \begin{cases} a'_i = a_i \, , & \text{for all } i < j \\ a'_j < a_j \end{cases} \tag{3.122}$$

Then the coding of a sequence $\boldsymbol{a}$ is obtained as follows:

1. Associate the interval $[q_{\boldsymbol{a}}, q_{\boldsymbol{a}} + p_{\boldsymbol{x}}(\boldsymbol{a}))$ to $\boldsymbol{a}$, where

$$q_{\boldsymbol{a}} = \sum_{\substack{\boldsymbol{a}' \in \mathcal{A}_x^L \\ \boldsymbol{a}' < \boldsymbol{a}}} p_{\boldsymbol{x}}(\boldsymbol{a}') \tag{3.123}$$

2. Choose a number $N_{\boldsymbol{a}}$ in $[q_{\boldsymbol{a}}, q_{\boldsymbol{a}} + p_{\boldsymbol{x}}(\boldsymbol{a}))$ whose binary representation $0.b_1 b_2 \ldots b_{L(\boldsymbol{b})}$ has $L(\boldsymbol{b}) \simeq i_{\boldsymbol{x}}(\boldsymbol{a})$ digits after the point, and let $\mu_s(\boldsymbol{a}) = \boldsymbol{b} = \left[b_1, b_2, \ldots, b_{L(\boldsymbol{b})}\right]$.

Observe that the intervals associated to all the possible information words of length L in step 1 make a partition of $[0, 1)$. Their being disjoint assures that the coding map is one-to-one.

A proper choice of $\boldsymbol{b}$ in step 2 assures $\mathcal{C}$ be a prefix code. For example in the *Elias code* [4, 8], $\boldsymbol{b}$ is the binary representation of the interval midpoint $m_{\boldsymbol{a}} = q_{\boldsymbol{a}} + \frac{1}{2} p_{\boldsymbol{x}}(\boldsymbol{a})$ rounded down to $L(\boldsymbol{b})$ binary digits, with

$$L(\boldsymbol{b}) = \lceil i_{\boldsymbol{x}}(\boldsymbol{a}) \rceil + 1 \, . \tag{3.124}$$

Thus $N_{\boldsymbol{a}}$ is the highest multiple of $1/2^{L(\boldsymbol{b})}$ not exceeding $m_{\boldsymbol{a}}$, and indeed it lies inside the interval since it is trivially $N_{\boldsymbol{a}} \leq m_{\boldsymbol{a}}$, and from (3.124)

$$N_{\boldsymbol{a}} > m_{\boldsymbol{a}} - \frac{1}{2^{L(\boldsymbol{b})}} \geq m_{\boldsymbol{a}} - \frac{1}{2^{i_{\boldsymbol{x}}(\boldsymbol{a})+1}} = m_{\boldsymbol{a}} - \frac{1}{2} p_{\boldsymbol{x}}(\boldsymbol{a}) = q_{\boldsymbol{a}} \, . \tag{3.125}$$

Proposition 3.12 *The Elias code is a prefix code.*

Proof The proof is by contradiction. Assume there exist two input words $\boldsymbol{a}$ and $\boldsymbol{a}'$ of the same length such that

$$\mu_s(\boldsymbol{a}) = \boldsymbol{b} \quad , \quad \mu_s(\boldsymbol{a}') = \boldsymbol{b}' \quad \text{and} \quad \boldsymbol{b} = (\boldsymbol{b}')_1^{L(\boldsymbol{b})} \tag{3.126}$$

Then it must be $N_{\boldsymbol{a}'} \geq N_{\boldsymbol{a}} > q_{\boldsymbol{a}}$ and

$$N_{\boldsymbol{a}'} = N_{\boldsymbol{a}} + \sum_{i=L(\boldsymbol{b})+1}^{L(\boldsymbol{b}')} \frac{b'_i}{2^i} < N_{\boldsymbol{a}} + \frac{1}{2^{L(\boldsymbol{b})}} \leq N_{\boldsymbol{a}} + \frac{1}{2} p_{\boldsymbol{x}}(\boldsymbol{a})$$

Moreover, since $N_{\boldsymbol{a}} \leq m_{\boldsymbol{a}}$ we also get

$$N_{\boldsymbol{a}'} < m_{\boldsymbol{a}} + \frac{1}{2} p_{\boldsymbol{x}}(\boldsymbol{a}) = q_{\boldsymbol{a}} + p_{\boldsymbol{x}}(\boldsymbol{a})$$

so that $N_{\boldsymbol{a}'}$ would lie in the interval associated to $\boldsymbol{a}$ instead of that for $\boldsymbol{a}'$. □

Example 3.4 F Consider a memoryless source with alphabet and PMD given by

$$\mathcal{A}_x = \{1, 2, 3\} \quad , \quad p_x(1) = \frac{3}{10} \quad , \quad p_x(2) = 1/2 \quad , \quad p_x(3) = 1/5$$

We want to find the Elias code of the input sequence $\boldsymbol{a} = [122]$. Its probability is

$$p_{\boldsymbol{x}}(\boldsymbol{a}) = p_x(1)p_x^2(2) = 3/40$$

and its information is $i_{\boldsymbol{x}}(\boldsymbol{a}) \simeq 3.74$ bit, so by (3.124) it will be coded into a word of length $L(\boldsymbol{b}) = 5$.

The set of 3-symbol words that precede $\boldsymbol{a}$ in the lexicographical order is

$$\{111, 112, 113, 121\}$$

and the sum of their probabilities is $q_{\boldsymbol{a}} = 27/200$. The interval midpoint is $m_{\boldsymbol{a}} = 69/400$, and the largest multiple of $1/2^5$ not exceeding $m_{\boldsymbol{a}}$ is $N_{\boldsymbol{a}} = 5/32$ which has the binary representation 0.00101. The codeword is therefore $\boldsymbol{b} = \mu_s(\boldsymbol{a}) = 00101$.

The Elias code for the source in this example in shown in Fig. 3.33

Arithmetic codes are asymptotically optimal in the sense that the average code length $L_{\boldsymbol{y}}$ is asymptotic to the lower bound (3.103), as the length L of the information word $\boldsymbol{x}$ grows to infinity. As a matter of fact, for the Elias code by (3.124) it is

$$i_{\boldsymbol{x}}(\boldsymbol{x}) + 1 \le L(\boldsymbol{y}) < i_{\boldsymbol{x}}(\boldsymbol{x}) + 2 \tag{3.127}$$

so that by taking expectation we obtain

$$H(\boldsymbol{x}) + 1 \le L_{\boldsymbol{y}} < H(\boldsymbol{x}) + 2 \,. \tag{3.128}$$

Since the source is memoryless, we divide all sides by $H(\boldsymbol{x}) = LH(x)$

$$1 + \frac{1}{LH(x)} \le \frac{L_{\boldsymbol{y}}}{H(\boldsymbol{x})} < 1 + \frac{2}{LH(x)} \tag{3.129}$$

and hence

$$\lim_{L\to\infty} \frac{L_{\boldsymbol{y}}}{H(\boldsymbol{x})} = 1 \,. \tag{3.130}$$

As its optimality is only asymptotic, arithmetic coding must use long information words to prove its effectiveness.

3.4.5.1 *Implementation aspects* The knowledge of the coding map for arithmetic coding is not needed before transmission as it is built along with the encoding process, starting from the source PMD. Therefore, when long information words are used, arithmetic coding is preferred over other methods that require precomputing and storing a large coding table at the transmitter and receiver. Moreover, in its *adaptive* version, arithmetic coding does not even

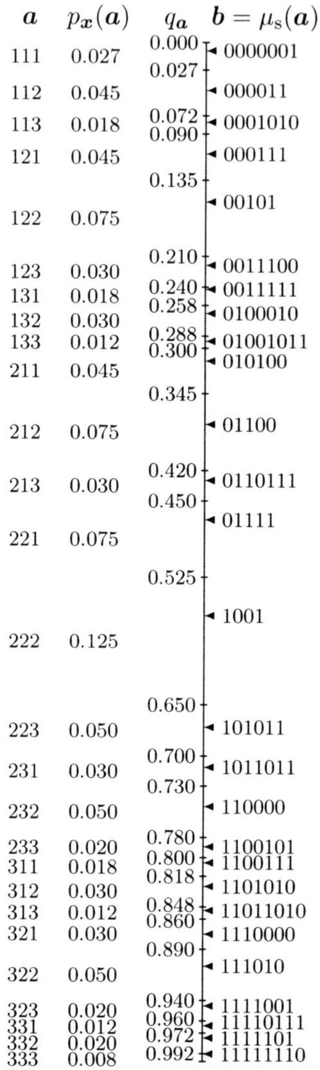

Figure 3.33 Information words, their PMD, and associated partition of $[0, 1)$ into intervals for arithmetic coding of 3-symbol words from the source in Example 3.4 F. The points N_a chosen according to the Elias code are shown with ◂, and the corresponding codewords are given.

need to know the source symbol PMD. It starts by assuming equally likely symbol values, and the PMD is updated during coding and decoding with the relative frequency of each value.

As regards the implementation of the encoder/decoder, observe that the lower end of the interval (3.123) can be written in terms of the PMD of x, by writing $\boldsymbol{a} = [a_1, \ldots, a_L]$, as

$$q_{\boldsymbol{a}} = \sum_{a_1' < a_1} p_x(a_1') + \sum_{a_2' < a_2} p_x(a_1) p_x(a_2') + \cdots \tag{3.131}$$

$$= \sum_{j=1}^{L} \prod_{i=1}^{j-1} p_x(a_i) \left(\sum_{a'_j < a_j} p_x(a'_j) \right) \tag{3.132}$$

and can be calculated recursively as the source symbols are input to the encoder from the prefixes of $\boldsymbol{a}$, as

$$q_{(\boldsymbol{a})_1^1} = \sum_{a'_1 < a_1} p_x(a'_1) \quad , \quad \cdots \quad , \quad q_{(\boldsymbol{a})_1^k} = q_{(\boldsymbol{a})_1^{k-1}} + \sum_{a'_k < a_k} p_x(a'_k) \tag{3.133}$$

and of course $q_{\boldsymbol{a}} = q_{(\boldsymbol{a})_1^L}$. Correspondingly the initial symbols of the codeword $\boldsymbol{b}$, which are the most significant digits of the binary representation of the interval midpoint, can be determined from the initial symbols of $\boldsymbol{a}$. Conversely, at the receiver, the initial symbols of $\boldsymbol{a}$ can be output from the decoder by processing the initial symbols of $\boldsymbol{b}$, without having to wait for reception of the whole word, thereby reducing the decoding latency.

The possibility of calculating intervals and output symbols while the input symbols are being received is illustrated in Fig. 3.34 for the source of Example 3.4 F and the sequence $\boldsymbol{a} = [122]$. First let us consider the encoder. Upon the input symbol $a_1 = 1$, the encoder knows that the interval $[q_{\boldsymbol{a}}, q_{\boldsymbol{a}} + p_{\boldsymbol{x}}(\boldsymbol{a}))$ is a subset of the interval $[q_1, q_2) = [0, 0.3)$. Then since all points in $[q_1, q_2) \subset [0, 1/2)$ have 0 as the first binary digit of their fractional part it determines $b_1 = 0$. After the input symbol $a_2 = 2$, the encoder knows that $[q_{\boldsymbol{a}}, q_{\boldsymbol{a}} + p_{\boldsymbol{x}}(\boldsymbol{a})) \subset [q_{12}, q_{13})$ and since $[q_{12}, q_{13}) = [0.09, 0.24) \subset [0, 1/4)$ it determines $(\boldsymbol{b})_1^2 = [00]$. Similarly, as the decoder receives the first two code symbols $(\boldsymbol{b})_1^2 = [00]$ it knows that $N_{\boldsymbol{a}}$ lies in the interval $[0, 1/4)$ and since $[0, 1/4) \subset [q_1, q_2)$ it determines that the first source symbol is $a_1 = 1$, and so on.

Lempel-Ziv algorithms For data storage applications, where the source message can be preprocessed before coding without real-time constraints, other compression methods are more commonly used, based on the construction of complex dictionaries and pointers, such as the *Lempel-Ziv* algorithms (LZ77, LZ80) included in the widespread *Zip* family of computer programs.

REFERENCES

1. N. Benvenuto and G. Cherubini, *Algorithms for communications systems and their applications*, Chichester, United Kingdom: Wiley, 2002.
2. R. E. Blahut, *Principles and practice of information theory*, Reading, MA: Addison-Wesley, 1987.
3. L.W. Couch II, *Digital and analog communication systems*. Upper Saddle River, NJ: Prentice-Hall, 1997.
4. T. M. Cover, J.A. Thomas, *Elements of information theory*, New York, NY: John Wiley & Sons, 1991.
5. R. Gallager, *Information theory and reliable communication* New York, NY: John Wiley & Sons, 1968.
6. R. V. L. Hartley, "Transmission of information", *The Bell System Technical Journal*, vol. 7, pp. 535–, July 1928.

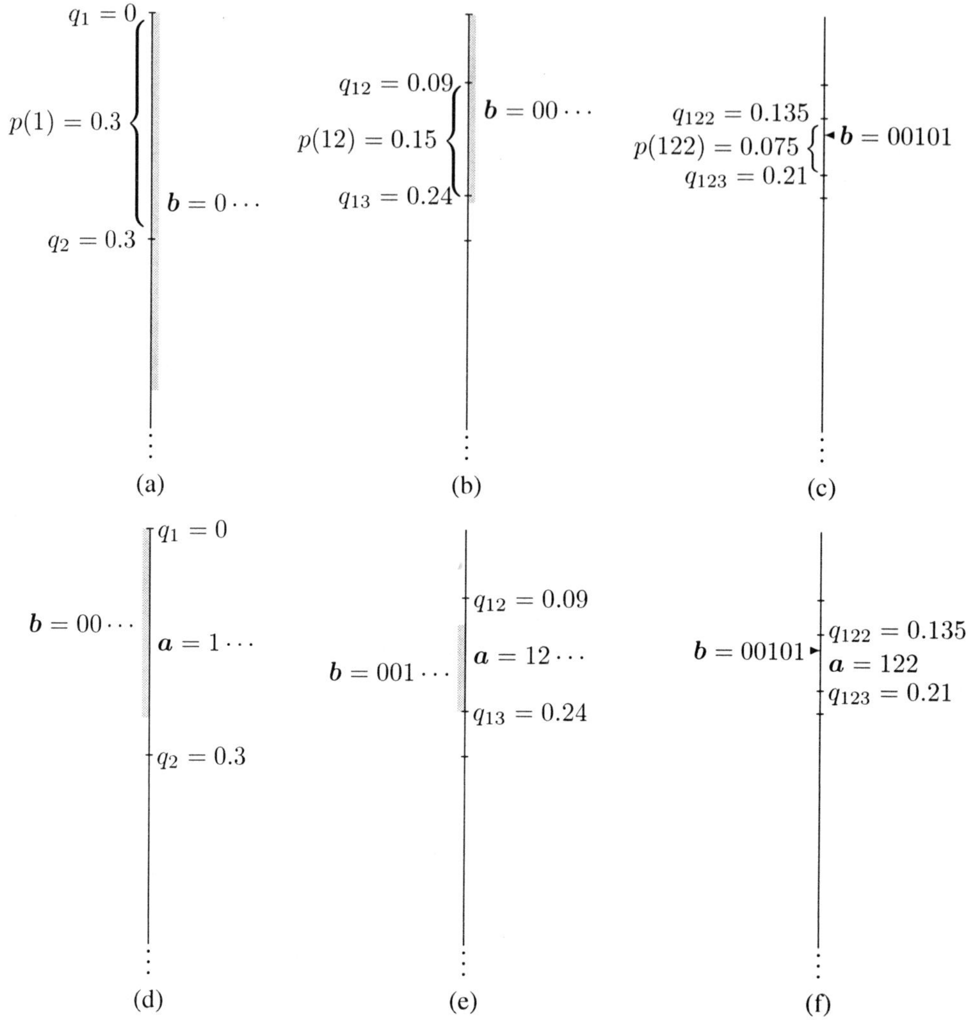

Figure 3.34 In (a)-(c), recursive calculation of intervals and codewords for arithmetic coding of the word $\boldsymbol{a} = [122]$ from the source in Example 3.4 F. In (d)-(f), the corresponding recursive decoding. All points in the gray-shaded intervals have a fractional binary representation beginning with the indicated codeword prefix $(\boldsymbol{b})_1^k$.

7. D. A. Huffman, "A method for the construction of minimum-redundancy codes", *Proceedings of the IRE*, vol. 40, pp. 1098–1102, September 1952.

8. F. Jelinek, *Probabilistic information theory*, New York, NY: McGraw-Hill, 1968.

9. J.G. Proakis and M. Salehi, *Communication systems engineering*. Englewood Cliffs, NJ: Prentice-Hall, 1994.

10. L. R. Rabiner and R.W. Schafer, *Digital processing of speech signals*. Englewood Cliffs, NJ: Prentice Hall, 1978.

11. M.S. Roden, *Analog and digital communication systems*. Upper Saddle River, NJ: Prentice-Hall, 1996.

12. C. E. Shannon, "A mathematical theory of communication", *The Bell System Technical Journal*, vol. 27, pp. 379–423, July 1948, pp. 623–656, October 1948.

13. C. E. Shannon, "Communication in the presence of noise", *Proceedings of the IRE*, vol. 37, pp. 10–21, 1949.

Problems

Uniform quantization

3.1 A signal has exponential bilateral probability density function with *mean* 1 Volt and variance $\sigma_a^2 = 2\,\text{V}^2$. Determine the *pre-processing* for this signal, the range and the number of bits of the uniform quantizer to guarantee a signal-to-quantization noise ratio of 42 dB. Assume $P_{\text{sat}} \leq 10^{-3}$.

3.2 The information signal $a(t)$ has uniform probability density function over the interval $(-5, +5)$ V. Determine the range of the uniform quantizer and the number of bits necessary to achieve a signal-to-quantization noise ratio of 40 dB.

3.3 Evaluate the statistical power of the saturation noise, $\mathtt{M}_{e_{\text{sat}}}$, using a 3 bit uniform quantizer with Gaussian input $a(t)$ with zero mean and variance σ_a^2. Assume $v_{\text{sat}} = 3\sigma_a$.

3.4 Find the number of bits necessary to achieve the signal-to-quantization noise ratio of 60 dB, using a uniform quantizer with a Gaussian input and imposing a saturation probability of 10^{-3}.

3.5 Consider a 4-bit uniform quantizer with dynamic range (0, 1) V. Assuming the input signal $a(t) = 20\cos(2\pi f_0 t)$:

- **a)** Determine the pre-processing needed to adapt the signal to the quantizer.
- **b)** Find the codewords corresponding to the input values -1.58 V and 5.22 V.
- **c)** Find the signal-to-quantization noise ratio Λ_q.

3.6 Evaluate the value of the signal-to-quantization noise ratio Λ_q for a signal with uniform probability density function in the interval $[-2, 2)$ V, using an uniform quantizer with 256 levels in the range $(-1, 1)$ V.

3.7 Derive the expression of the signal-to-quantization noise ratio, $(\Lambda_q)_{\text{dB}}$, as a function of $\mathtt{b}$ for the saw-tooth signal shown in figure

$$a(t) = \sum_{k=-\infty}^{+\infty} a_{\max}\,\text{saw}\left(\frac{t-kT}{T}\right)$$

where $\text{saw}(t) = t\ , -1 \leq t \leq 1$ and 0 otherwise.

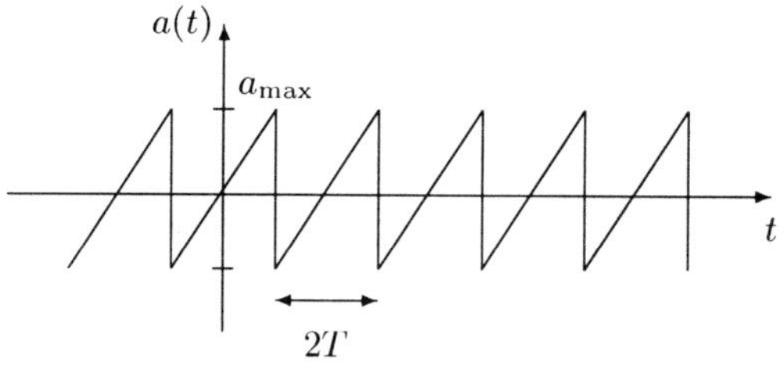

3.8 Find the number of bits necessary to achieve the signal-to-quantization noise ratio of 60 dB, using an uniform quantizer with a Laplacian input and imposing a saturation probability of 10^{-3}.

3.9 A musical signal $a(t)$ is described as a Gaussian signal having zero mean and power spectral density:

$$\mathcal{P}_a(f) = A\,\text{triang}\left(\frac{f}{20000}\right) \quad \text{V}^2/\text{Hz}.$$

The signal provides a power of 1 W when applied to a resistor of 100 Ω.
Determine:

a) The value of A.
b) The minimum sampling frequency of the signal $a(t)$.
c) The range $(-v_{sat}, v_{sat})$ of the uniform quantizer to obtain a saturation probability 10^{-6}.
d) The minimum number of bits for coding the quantization levels that guarantees a signal-to-quantization noise ratio of at least 100 dB.

ADC

3.10 Knowing that the sampling frequency in a CD player is 44.1 kHz and each sample is represented by a 16 bit word, determine the total number of bits corresponding to a recorded music song of 20 minutes.

3.11 Determine the characteristics of the uniform quantizer and the corresponding bit rate for an input signal with bandwidth $B = 10\,\text{kHz}$, uniform probability density function with zero mean and statistical power $\mathtt{M}_a = 2\,\text{V}^2$, imposing a signal-to-quantization noise ratio greater than 50 dB.

3.12 The information signal is modeled as a random process $a(t)$ with maximum amplitude A and correlation function

$$\mathtt{r}_a(\tau) = \frac{A^2}{2} e^{-|\tau|} \cos(2\pi f_0 \tau)$$

Assuming $A = 6\,\text{V}$ and $f_0 = 1\,\text{Hz}$, evaluate the following points.

a) Give a graphical representation of the correlation function $\mathtt{r}_a(\tau)$.
b) Evaluate the power spectral density $\mathcal{P}_a(f)$ and give its graphical representation.
c) Determine the number of quantization levels L needed to guarantee a value of Λ_q greater than 60 dB.
d) Assume that the signal is sampled (considering a conventional bandwidth corresponding to an attenuation of 24 dB of its PSD) and quantized by the quantizer considered at the previous point. Evaluate the corresponding bit rate R_b.

3.13 A signal $a(t)$ is modeled as a baseband stationary process with bandwidth 5 kHz and probability density function

$$p_a(u) = \frac{1}{2}\text{triang}\left(\frac{u}{2}\right) .$$

The signal is PCM encoded

a) What is the resulting Λ_q using a uniform quantizer with 32 levels?
b) What is the ADC bit rate R_b sampling at the minimum sampling frequency?
c) What is the highest Λ_q achievable for a binary channel with bit rate 80 kbit/s?
d) What is the highest Λ_q achievable, considering the same binary channel of the previous point, if a guard bandwidth of 1 kHz is required by the anti-aliasing filter?

Non-uniform quantization

3.14 Given the signal

$$a(t) = 20\cos(100\pi t) + 17\cos(500\pi t) ,$$

how many bits are necessary to achieve a signal-to-quantization noise ratio greater than 45 dB using an uniform quantizer? Using a standard μ-law *companding* technique, how many bits are necessary to achieve a signal-to-quantization noise ratio greater than 50 dB?

3.15 The signal $a(t)$ is obtained by filtering the periodic square wave $b(t)$ with period $T_p = 10^{-3}$ s and zero mean,

$$b(t) = \sum_{k=-\infty}^{+\infty} \left[\text{rect}\left(\frac{t}{T_p/2}\right) - \text{rect}\left(\frac{t - T_p/2}{T_p/2}\right) \right]$$

through an ideal lowpass filter with bandwidth 5.1 kHz. The signal $a(t)$ is applied to a quantizer with 6 bits. Evaluate the value of the signal-to-quantization noise ratio Λ_q when the quantizer has

a) Uniform characteristic.

b) Non-uniform characteristic with standard μ-law *companding*.

Hint: evaluate the Fourier series expansion of $a(t)$.

3.16 The signal $a(t)$ has probability density function

$$p_a(u) = \begin{cases} Ke^{-2|u|} & , -3 < u < 3 \\ 0 & , \text{otherwise} \end{cases}$$

a) Find the value of the signal-to-quantization noise ratio using an uniform quantizer with 3 bits.

b) Find the value of the signal-to-quantization noise ratio using a non-uniform quantizer with 3 bits and standard μ-law *companding*.

3.17 Consider an audio signal with bandwidth $B = 4$ kHz and uniformly distributed amplitude in $(-A, +A)$, with $A = 5$ V. The signal is PCM encoded with a uniform quantizer having $L = 2^{\mathsf{b}}$ levels.

a) Determine the relation between the signal to quantization noise ratio (in dB) and the number of bits b.

b) Discuss whether, with the same number of levels L, the system perfomance is improved if a non-uniform quantizer is used.

Information and entropy

3.18 Prove that

a) All functions of the type (3.59) satisfy properties P1–P4 on page 158.

b) Any continuous and differentiable function that satisfies properties P1–P4 must be of the type (3.59). Observe that this does not constitute a proof of the "only" part as continuity and differentiability were not among the axioms.

Hint: start by differentiating the relationship in P4 with respect to β, and then substitute $\beta = 1$.

3.19

a) Find the entropy of a geometric rv with

$$\mathcal{A}_x = \{0, 1, 2, \ldots\} \quad , \quad p_x(k) = (1 - p)p^k$$

where the parameter $p \in (0, 1)$.

b) Prove that the entropy of a binomial rv with parameters $n \in \mathbb{N}$ and $p \in (0, 1)$, having

$$\mathcal{A}_x = \{0, 1, \ldots, n\} \quad , \quad p_x(k) = \binom{n}{k}(1 - p)^{n-k}p^k \; ,$$

is bounded by

$$H(x) \leq n[p \log_{1/2} p + (1 - p) \log_{1/2}(1 - p)]$$

c) Calculate the values of the entropy and the upper bound above for a binomial rv with parameters $n = 8$ and $p = 1/2$.

For which of the above rvs is it possible to determine the efficiency?

3.20 Find the joint and conditional entropies for the pair of integer-valued random variables $[x, y]$ with

$$\mathcal{A}_{[x,y]} = \left\{[m,n] \in \mathbb{Z}^2 \; : \; 1 \le n \le m \le 4\right\} \quad , \quad p_{x,y}(m,n) = \frac{1}{4m}$$

What is the efficiency of the vector $[x, y]$?

3.21 Consider an infinite message $\{x_n\}$ with iid symbols and equally likely values in $\{0, 1\}$, and the message $\{y_n\}$ defined as $y_n = x_n x_{n-1}$.

Find the entropy of the rv y_n and the conditional entropy $H(y_n|x_n)$

3.22 Consider an unlimited binary message $\{x_n\}$ with iid symbols taking values in $\{0, 1\}$ with equal probability, and another message $\{y_n\}$ obtained from it as $y_n = x_n - x_{n-1}$.

a) Find the entropy of the rv y_0.

b) Find the entropy of the rve $[y_{-1}, y_0, y_1]$.

c) Find the entropy per symbol and the efficiency of the message $\{y_n\}$.

3.23 Prove the following statements

a) If x, y, z are discrete rvs,

$$H(x, y|z) \ge H(x|y, z)$$

b) If x is a discrete rv and $w = g(x)$ with g an arbitrary real function of a real variable,

$$H(w) \le H(x)$$

c) If x, y are discrete rvs and $u = x + y$,

$$H(u|x) = H(y|x)$$

Source coding

3.24 The Morse code used in telegraphy has the following coding map:

a	$\mu_s(a)$	a	$\mu_s(a)$	a	$\mu_s(a)$	a	$\mu_s(a)$	a	$\mu_s(a)$	a	$\mu_s(a)$
A	·-	B	-···	C	-·-·	D	-··	E	·	F	··-·
G	--·	H	····	I	··	J	·---	K	-·-	L	·-··
M	--	N	-·	O	---	P	·--·	Q	--·-	R	·-·
S	···	T	-	U	··-	V	···-	W	·--	X	-··-
Y	-·--	Z	--··		/						

a) While in the written version words are separated by a blank space, in Morse code they are separated by '/'. Thus, the phrase "*Morse code*" would be written as

M	O	R	S	E		C	O	D	E
--	---	·-·	···	·	/	-·-·	---	-··	·

We can view the written Morse code as a variable length code with $\mathcal{A}_y = \{\cdot, -, /\}$. Is it a prefix code? How is it decodable? Find its average length and efficiency assuming that letters are statistically independent with the following distribution (empirically derived from their relative frequency in the text of this chapter)

a	$p(a)$	a	$p(a)$	a	$p(a)$	a	$p(a)$	a	$p(a)$
A	0.0622	B	0.0328	C	0.0373	D	0.0284	E	0.0921
F	0.0199	G	0.0162	H	0.0352	I	0.0642	J	0.0011
K	0.0032	L	0.0322	M	0.0274	N	0.0566	O	0.0585
P	0.0233	Q	0.0047	R	0.0442	S	0.0524	T	0.0677
U	0.0173	V	0.0084	W	0.0128	X	0.0107	Y	0.0153
Z	0.0011		0.1748						

b) For transmission, the symbols '·' and '-' are represented by "signal on" intervals with duration T and $3T$, respectively. Consecutive symbols in the representation of a letter are separated by a "signal off" interval with duration T. Consecutive letters in a same word and consecutive words are separated by "signal off" intervals with duration $3T$ and $7T$, respectively. For example, the phrase "*Morse code*" would be transmitted as

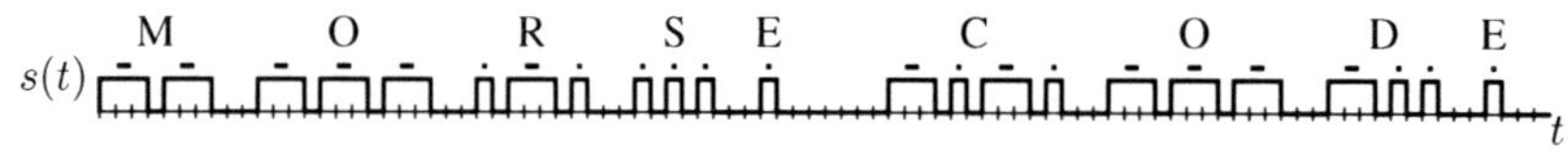

Find the average duration of the representation for a letter in the case $T = 0.12$ s (corresponding to 10 words per minute), the efficiency and the information rate of the message s.

3.25 A quaternary memoryless source has alphabet $\mathcal{A}_x = \{0, 1, 2, 3\}$ and PMD

$$p_x(0) = 1/3 \quad , \quad p_x(1) = 1/6 \quad , \quad p_x(2) = 1/12$$

and is coded symbol-by-symbol.

a) Find the efficiency of a binary code for it with fixed length words.

b) Find the efficiency of a Shannon-Fano binary code for it.

c) Find the efficiency of a Huffman binary code for it.

d) For what PMD would a binary variable length code with unit efficiency be possible?

3.26 Let x_n be a message with iid symbols, a quaternary alphabet and the following PMD

$$p_x(0) = 0.7 \quad , \quad p_x(1) = p_x(2) = p_x(3) = 0.1$$

a) Find a binary Huffman code with single symbols x_n as input and calculate its efficiency.

b) Find the efficiency of a ternary Shannon-Fano code with pairs $\boldsymbol{x} = [x_n, x_{n+1}]$ as input.

3.27 A stationary message $\{x_n\}$ with iid symbols and alphabet $\mathcal{A}_x = \{0, 1, 2\}$ has PMD $p_x(0) = 1/2$, $p_x(1) = 1/3$, $p_x(2) = 1/6$.

a) Construct a binary prefix code $\boldsymbol{y}$ with $\mathcal{A}_y = \{0, 1\}$, that is optimal for coding single symbols of x. Evaluate the code efficiency.

b) Consider the vector $\boldsymbol{x}_m = [x_{2m}, x_{2m+1}]$ made of 2 consecutive symbols taken from $\{x_n\}$ and find a binary code $\boldsymbol{y}'$ that is optimal for $\boldsymbol{x}_m$. Compare the efficiencies η_y and $\eta_{y'}$.

3.28 Design a source coding scheme for a memoryless source with alphabet $\mathcal{A}_x = \{0, 1, 2, 3, 4\}$ so that the efficiency of the coded message is at least 95% with equally likely values and at least 85% when $p_x(0) = 0.4, p_x(1) = 0.36, p_x(2) = 0.09, p_x(3) = 0.08, p_x(4) = 0.07$.

Give the input dictionary, code alphabet and coding map, and check that the code efficiency meets both requirements.

3.29 Choose a source coding scheme for a memoryless ternary source with symbol period $T_x = 10\,\mu s$ and PMD

$$p_x(0) = 0.8 \quad , \quad p_x(1) = p_x(2) = 0.1$$

so that it produces a binary output with symbol rate $F_y \leq 100\,\text{kHz}$. Justify your choice and explicitly provide the coding map.

3.30 A discrete time rp $x(nT_s)$ with iid symbols and Laplacian distribution

$$p_x(a) = \frac{1}{2V_0} e^{-|a|/V_0}$$

is quantized with an uniform quantizer with L levels and step size Δ. Let $x_q(nT_s)$ be the quantizer output rp.

a) Find alphabet, entropy and information rate of $x_q(nT_s)$, in the case $1/T_s = 8\,\text{kHz}$, $L = 6$ and $\Delta = V_0/2$.

b) Find an optimal binary prefix code for $x_q(nT_s)$, with the above values for L and Δ and evaluate its efficiency.

c) Show that when L is even and $\Delta = V_0 \ln 2$, there exists a prefix code with unit efficiency for $x_q(nT_s)$.

3.31 The discrete-time signal $x(nT)$ has iid samples with PDF

$$p_x(a) = \frac{3}{2}\frac{a^2}{V_1^3}\,\text{rect}\left(\frac{a}{2V_1}\right)$$

It is input to a uniform quantizer with $L = 6$ levels and $v_{\text{sat}} = V_1$. Find

a) the mean and variance of the quantization error

b) the average entropy per symbol and the efficiency of the quantized signal

c) the optimal source coding of the quantized signal with a single symbol dictionary, and the efficiency of the coded signal

3.32 In digital transmission systems with *pulse interval modulation* (PIM) having alphabet $\{0, \ldots, M-1\}$, we associate to the symbol value n the transmission of a particular waveform $h(t)$ with limited support $(0, T_b)$, preceded by a "mute" interval of length nT_b. For example, we have

digital message a_k: $2, 0, 3, \ldots$

transmitted signal $s(t)$:

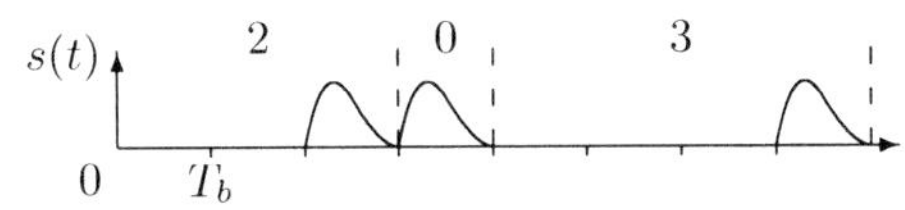

a) Show that the modulator can be decomposed into the cascade of a source encoder and an interpolate filter

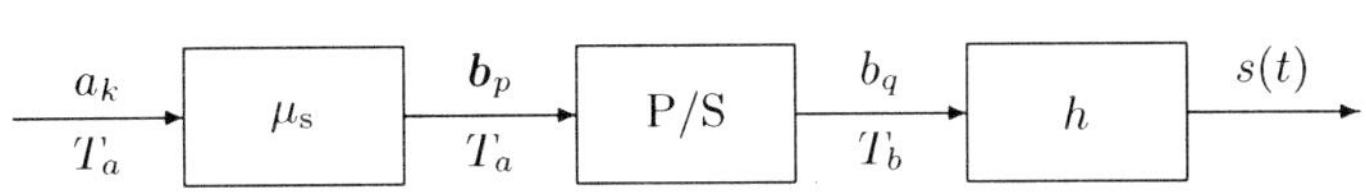

where the message $\{b_q\}$ has a binary alphabet $\mathcal{A}_b = \{0, 1\}$. Give the encoder map μ_s and state whether $\mathcal{D}_b$ is a prefix code.

b) Find the information rate of $\{b_q\}$ when $\{a_k\}$ is made of iid symbols with equally likely values, $M = 8$ and $T_b = 1$ ns.

c) Under the same hypotheses as in b) find the statistical power of the transmitted signal $s(t)$. Assume $h(t)$ has energy $E_h = \frac{3}{4}T_h V_0^2$, $V_0 = 2\,\text{mV}$ and $T_h = 0.2$ ns.

CHAPTER 4

CHARACTERIZATION OF TRANSMISSION MEDIA AND DEVICES

Roberto Corvaja

In communication systems, a signal bearing information is transferred from one location to another through a *transmission medium*. In practice the signals are electrical quantities, while the transmission medium can be any medium where electro-magnetic waves can propagate, such as an electrical cable, an optical fiber, the free space, or even water. With the term transmission medium often we may consider the overall system from the transmitter output to the receiver input, which may include antennas, filters and power amplifiers that allow to contrast the power losses due to propagation over long distances.

All of these media and devices may be described in different ways, depending on the scope of the investigation. Generally, they can be modeled as electrical systems, by using a two-port linear network formalism. This is a particularly useful approach allowing for a description of the transmission system both in terms of signal transformations, and thus of *bandwidth* and *statistical power*, and in terms of *electrical power*. Moreover, in this context a formulation of *noise* sources is readily available. These aspects are by far the most relevant for any communication system.

This chapter is devoted to introducing the two-port linear network model, with the primary purpose of providing to the reader a method for setting the power of a transmission system. Before getting into the description of two-port linear networks, we begin investigating the structure and properties of the basic element in an electrical circuit: the *two-terminal device*. The entire analysis is performed by assuming that the reader is familiar with the basics of circuit theory and electronics [1, 2].

Principles of Communications Networks and Systems N. Benvenuto, M. Zorzi

4.1 TWO-TERMINAL DEVICES

4.1.1 Electrical representation of a signal source

The model for an electrical signal source is an active two-terminal device. Its equivalent circuit (Thevenin equivalent) is illustrated in Fig. 4.1a, and consists of the series of a generator with open-circuit voltage $v_i(t)$ and of an impedance Z_S. The two-terminal device is connected to a load of impedance Z_L. We underline that in the frequency domain both impedances are functions of the frequency f, that is, we should write $Z_S(f)$ and $Z_L(f)$. However, to simplify the notation, often the frequency f is omitted. We also recall that an impedance $Z(f)$ exhibits a non negative resistive part, that is $\Re\left[Z(f)\right] = R(f) \geq 0$, and an Hermitian symmetry on f, that is $Z(f) = Z^*(-f)$.

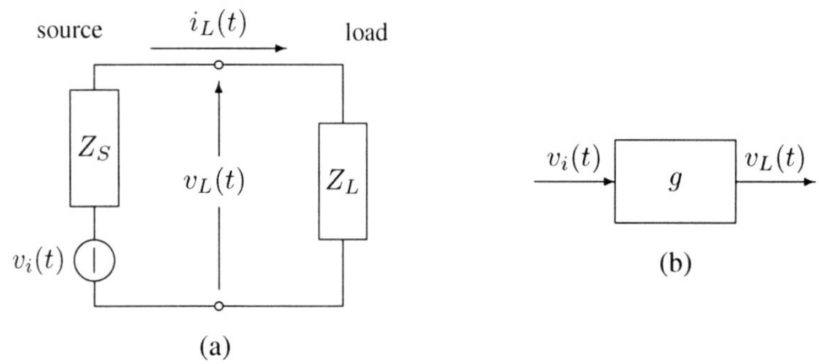

Figure 4.1 (a) Electrical model of an active two-terminal device connected to a load. (b) Equivalent linear system in terms of load voltage.

In the scheme of Fig. 4.1a, we can relate the voltage measured at the load, $v_L(t)$, to the generator signal given by the open voltage $v_i(t)$. The relation is straightforwardly expressed in the frequency domain as

$$\mathcal{V}_L(f) = \mathcal{V}_i(f)\,\mathcal{G}(f)\,, \qquad \mathcal{G}(f) = \frac{Z_L(f)}{Z_S(f) + Z_L(f)} \tag{4.1}$$

where $\mathcal{G}(f)$ is the frequency response of our system. In time domain the linear relation (4.1) gives the convolution $v_L(t) = (v_i * g)(t)$, where $g(t)$ is the impulse response of the system, which can be derived through the inverse Ftf of $\mathcal{G}(f)$. The equivalent time domain model is shown in Fig. 4.1b.

In terms of PSD, from (4.1) and (2.291), we obtain

$$\mathcal{P}_{v_L}(f) = \mathcal{P}_{v_i}(f)\,|\mathcal{G}(f)|^2 = \mathcal{P}_{v_i}(f)\frac{|Z_L(f)|^2}{|Z_S(f) + Z_L(f)|^2}\,. \tag{4.2}$$

4.1.2 Electrical power

The *average electrical power* transferred to the load due to generator voltage $v_i(t)$ is

$$\mathrm{P}_v = \lim_{u \to +\infty} \frac{1}{2u} \int_{-u}^{+u} v_L(t)\, i_L(t)\, dt\,, \tag{4.3}$$

where the current-voltage multiplication $v_L(t)\, i_L(t)$ plays the role of an *instantaneous electrical power* transferred to the load (see also definition (2.30)).

Power can be also related to statistical measures of electrical signals through ergodicity. If $v_L(t)$ and $i_L(t)$ in (4.3) are jointly wide sense stationary and ergodic rps, then from (2.235), (4.3) can be written in statistical terms as an expectation

$$\mathrm{P}_v = \mathrm{E}\left[v_L(t)\, i_L(t)\right]\ . \tag{4.4}$$

We notice that the meaning of the equivalence between (4.3) and (4.4) is that the integral in (4.3) converges (in some sense) to P_v for every realization of the couple of rps $v_L(t)$ and $i_L(t)$.

By exploiting the definition (2.258) of cross correlation $\mathtt{r}_{v_L i_L}$ and corresponding cross power spectral density $\mathcal{P}_{v_L i_L}$ as its Ftf, we can write

$$\mathrm{P}_v = \mathtt{r}_{v_L i_L}(0) = \int_{-\infty}^{+\infty} \mathcal{P}_{v_L i_L}(f)\, df\,. \tag{4.5}$$

This expression can be further simplified. In fact, both current and voltage are real-valued signals, which assures that the cross correlation $\mathtt{r}_{v_L i_L}(\tau)$ is a real-valued signal, hence from Tab. 2.1 its Ftf $\mathcal{P}_{v_L i_L}(f)$ is an Hermitian symmetric signal with even symmetric real part and odd symmetric imaginary part. So, the imaginary part does not contribute to the integral in (4.5) and we have

$$\mathrm{P}_v = \int_{-\infty}^{+\infty} \Re\left[\mathcal{P}_{v_L i_L}(f)\right]\, df\,. \tag{4.6}$$

Definition 4.1 *For the electrical model of Fig. 4.1, the function*

$$\mathrm{p}_v(f) = \Re\left[\mathcal{P}_{v_L i_L}(f)\right] \qquad \text{[W/Hz]} \tag{4.7}$$

is called power density *transfered to the load due to the generator voltage $v_i(t)$ and expresses the average power transfered to the load per unit frequency. The* average power *is thus derived as*

$$\mathrm{P}_v = \int_{-\infty}^{+\infty} \mathrm{p}_v(f)\, df \qquad \text{[W]}\,. \tag{4.8}$$

In other words, $\mathrm{p}_v(f)$ indicates how the power P_v is allocated in the frequency domain. This information is very useful in case we want to preserve throughout the system the power of a desired signal.

For practical purposes, it is important to derive the expression of the power density $\mathrm{p}_v(f)$ and of the average power P_v as a function of known quantities. We develop this derivation by considering the circuit of Fig. 4.1 for which the PSD of the generator voltage, $\mathcal{P}_{v_i}(f)$, is given. We preliminarily note that $\mathcal{P}_{v_L i_L}$ can be written in terms of $\mathcal{P}_{v_L}$ by exploiting (2.290) and the fact that $\mathcal{I}_L(f) = \mathcal{V}_L(f)/Z_L(f)$. We have

$$\mathcal{P}_{v_L i_L}(f) = \frac{\mathcal{P}_{v_L}(f)}{Z_L^*(f)} \tag{4.9}$$

From (4.2), it is

$$\mathcal{P}_{v_L i_L}(f) = \mathcal{P}_{v_i}(f) \frac{Z_L(f)}{|Z_S(f) + Z_L(f)|^2} . \tag{4.10}$$

Since $\mathcal{P}_{v_i}(f)$ is a real-valued quantity, (4.7) yields

$$\boxed{\mathtt{p}_v(f) = \mathcal{P}_{v_i}(f) \frac{R_L(f)}{|Z_S(f) + Z_L(f)|^2}} \tag{4.11}$$

where we implicitly considered $R_L = \Re[Z_L]$.

An alternative expression of (4.11) can be given if we know the PSD of the load voltage, $\mathcal{P}_{v_L}(f)$. From (4.9) and (4.7), it turns out $\mathtt{p}_v(f) = \Re[\mathcal{P}_{v_L}(f)/Z_L^*(f)]$, or

$$\boxed{\mathtt{p}_v(f) = \mathcal{P}_{v_L}(f) \frac{R_L(f)}{|Z_L(f)|^2}} \tag{4.12}$$

4.1.3 Measurement of electrical power

The generic power $\mathtt{P}_v$ is expressed in Watts (W). Also milli-Watts (1 mW= 10^{-3} W) or pico-Watts (1 pW= 10^{-12} W) scales are used, in which case we explicitly write $(\mathtt{P}_v)_{\text{mW}}$ and $(\mathtt{P}_v)_{\text{pW}}$ where sub-indexes indicate the unit of measure. Most of the times, a logarithmic scale is preferable. In this case we talk about dBW, dBm and dBrn, where

$$\begin{aligned}
(\mathtt{P}_v)_{\text{dBW}} &= 10 \log_{10}(\mathtt{P}_v)_{\text{W}} \\
(\mathtt{P}_v)_{\text{dBm}} &= 10 \log_{10}(\mathtt{P}_v)_{\text{mW}} = 30 + 10 \log_{10}(\mathtt{P}_v)_{\text{W}} \\
(\mathtt{P}_v)_{\text{dBrn}} &= 10 \log_{10}(\mathtt{P}_v)_{\text{pW}} = 120 + 10 \log_{10}(\mathtt{P}_v)_{\text{W}}
\end{aligned} \tag{4.13}$$

We should say that the most common unit of power measurement in communications is dBm. Incidentally, the average received power of a GSM mobile phone is around -70 dBm.

Correspondingly the power density $\mathtt{p}_v(f)$ is expressed in W/Hz, mW/Hz or pW/Hz. The logarithmic measure is available also in this case, and for example we have

$$(\mathtt{p}_v(f))_{\text{dBm/Hz}} = 10 \log_{10}(\mathtt{p}_v(f))_{\text{mW/Hz}} . \tag{4.14}$$

Notice that, when a logarithmic measure is given for $\mathtt{p}_v(f)$, a conversion to a linear scale is required for evaluating the average power $\mathtt{P}_v$ according to the integral (4.8).

Example 4.1 A A power of $\mathtt{P}_v = 0.5$ W can be equivalently expressed as $(\mathtt{P}_v)_{\text{mW}} = 500$ mW, $(\mathtt{P}_v)_{\text{dBW}} = -3$ dBW or $(\mathtt{P}_v)_{\text{dBm}} = 27$ dBm.

Example 4.1 B A power density of $(\mathtt{p}_v(f))_{\text{nW/Hz}} = 0.4$ nW/Hz can be equivalently expressed as $(\mathtt{p}_v(f))_{\text{pW/Hz}} = 400$ pW/Hz, $(\mathtt{p}_v(f))_{\text{dBW/Hz}} = -94$ dBW/Hz or $(\mathtt{p}_v(f))_{\text{dBm/Hz}} = -64$ dBm/Hz.

4.1.4 Load matching and available power

One of the targets of a communication system is to ensure the maximum transfer of electrical power to the user, that is to minimize the power dispersion through the transmission. In the device of Fig. 4.1, where the characteristics of the source are given (i.e. , $\mathcal{P}_{v_i}(f)$ and $Z_S(f)$ are given), this concept reduces to choosing the load impedance $Z_L(f)$ that maximizes the average power transferred to the load.

From inspection of (4.6), (4.7) and (4.11) we have

$$Z_{L,\text{opt}}(f) = \underset{Z_L(f)}{\text{argmax}} \int_{-\infty}^{+\infty} \mathcal{P}_{v_i}(\nu) \frac{R_L(\nu)}{|Z_S(\nu) + Z_L(\nu)|^2} \, d\nu \tag{4.15}$$

which, because of the linearity of integrals and of the fact that all terms are real-valued and positive, can be equivalently expressed for each frequency f as

$$Z_{L,\text{opt}}(f) = \underset{Z_L(f)}{\text{argmax}} \frac{R_L(f)}{|Z_S(f) + Z_L(f)|^2} \tag{4.16}$$

where $Z_L(f) = R_L(f) + jX_L(f)$ and $R_L(f) \geq 0$.

We have the important result

Theorem 4.1 *The maximum transfer of electrical power is obtained under the condition*

$$\boxed{Z_L(f) = Z_S^*(f)} \tag{4.17}$$

which is called load matching condition *for maximum transfer of power to the load.*

Proof We write $Z_S(f)$ as the sum of its real and imaginary components, $Z_S(f) = R_S(f) + jX_S(f)$, then expand the denominator term in (4.16) to obtain

$$Z_{L,\text{opt}}(f) = \underset{R_L(f), X_L(f)}{\text{argmax}} \frac{R_L(f)}{[R_S(f) + R_L(f)]^2 + [X_S(f) + X_L(f)]^2} \, . \tag{4.18}$$

For any choice of $R_L(f)$, the maximization of (4.18) with respect to $X_L(f)$ is obtained for $X_L(f) = -X_S(f)$. This choice sets the fraction to

$$\frac{R_L(f)}{[R_S(f) + R_L(f)]^2} \, . \tag{4.19}$$

Now, the maximum of (4.19) with respect to $R_L(f)$ is obtained by setting to zero the derivative with respect to $R_L(f)$, that is

$$\frac{1}{[R_S(f) + R_L(f)]^2} - 2 \frac{R_L(f)}{[R_S(f) + R_L(f)]^3} = \frac{R_S(f) - R_L(f)}{[R_S(f) + R_L(f)]^3} = 0$$

giving $R_L(f) = R_S(f)$. By collecting the results, we have that $Z_{L,\text{opt}}(f) = R_S(f) - jX_S(f)$.
□

Definition 4.2 *Under condition (4.17), the power density (4.12) is known as* available power density *and gives*

$$\boxed{\mathrm{p}_v(f) = \mathcal{P}_{v_L}(f)\,\frac{R_L(f)}{|Z_L(f)|^2}} \tag{4.20}$$

Furthermore, if impedances are resistive,

$$\mathrm{p}_v(f) = \frac{\mathcal{P}_{v_L}(f)}{R_L(f)} \quad (\text{if } Z_S(f) = R_S(f) = Z_L^*(f))\,. \tag{4.21}$$

In practical systems, the load matching condition (4.17) is always met to allow the maximum transfer of power. This implies that (4.17) is tacitly assumed unless otherwise stated, so that all the input/output relations we introduced should be particularized to the load matching case.

Remark From (4.20), under the assumption that $Z_S(f)$ is a constant within the band of v_i, it is possible to derive a straightforward relationship between the average electrical power at the load and the statistical power of the load voltage, that is

$$\mathrm{P}_v = \mathtt{M}_{v_L}\,\frac{R_L}{|Z_L|^2}\,. \tag{4.22}$$

In particular, (4.21) yields

$$\mathrm{P}_v = \frac{\mathtt{M}_{v_L}}{R_L} \quad (\text{if } Z_L(f) = R_L)\,. \tag{4.23}$$

Remark We note the difference between $\mathtt{M}_{v_L}$ and P_v in (4.22) or (4.23): the first refers to the average of the square of the load voltage $v_L(t)$, the latter instead is the the average of the instantaneous power $v_L(t)i_L(t)$.

Remark (On the distortion of the transmitted signal) An important requirement that is set in many practical systems is the absence of signal distortion. In particular, it is required that the useful signal at the load $v_L(t)$, be a replica, without distortion, of the source signal $v_i(t)$, that is

$$v_L(t) = A_0\, v_i(t - t_0)\,. \tag{4.24}$$

If the system were composed of only a two-terminal device closed on a load, from (4.1) the system frequency response is

$$\mathcal{G}(f) = \frac{Z_L(f)}{Z_L(f) + Z_S(f)}\,.$$

Now this system satisfies the Heaviside conditions of non distortion if $\mathcal{G}(f) = A_0\, e^{-j2\pi f t_0}$, $f \in \mathcal{B}$, for some $A_0 > 0$ and $t_0 > 0$. Equivalently we can write

$$Z_L(f)\,(1 - A_0\, e^{-j2\pi f t_0}) = Z_S(f)\,A_0\, e^{-j2\pi f t_0}\,, \qquad f \in \mathcal{B}\,.$$

We note that this is a strict request on impedances, which simplifies for $t_0 = 0$ giving the linear relation $Z_L(f) = K_0\, Z_S(f)$ with K_0 a real-valued constant. By adding the further constraint on load matching conditions, $Z_L(f) = Z_S^*(f)$, the only available solution to both constraints is to have purely resistive impedances, that is

$$Z_S(f) = R_S(f)\,, \qquad Z_L(f) = Z_S^*(f) = R_S(f)$$

which implies

$$v_L(t) = \tfrac{1}{2}\, v_i(t)\,.$$

4.1.5 Thermal noise

If the source shown in Fig. 4.1 were to model the output of a device we have that $v_i(t)$ contains, besides the useful signal $s_i(t)$, also a noise component $w_i(t)$. Indeed, practical systems are always affected by disturbances and *noise*. The sources of noise are numerous, for example: i) the electromagnetic coupling between adjacent systems; ii) noise radiated by the surrounding environment, and; iii) noise generated by the devices themselves.

Indeed, the most common source of noise is generated internally a conductor and is called *thermal noise*. Thermal noise is a phenomenon associated with Brownian or random motion of electrons in a conductor. As each electron carries a charge, its motion produces a short impulse of current. Actually, if we represent the motion of an electron within a conductor in a 2-dimensional plane, the typical behavior is represented in Fig. 4.2a where the sudden changes in direction are determined by random collisions with atoms at instants $\{\ldots, t_0, t_1, \ldots, t_k, \ldots\}$. Between two consecutive collisions, the electron produces a current that is proportional to the projection of the velocity onto the axis of the conductor (the horizontal axis in Fig. 4.2a). This behavior is illustrated in Fig. 4.2b.

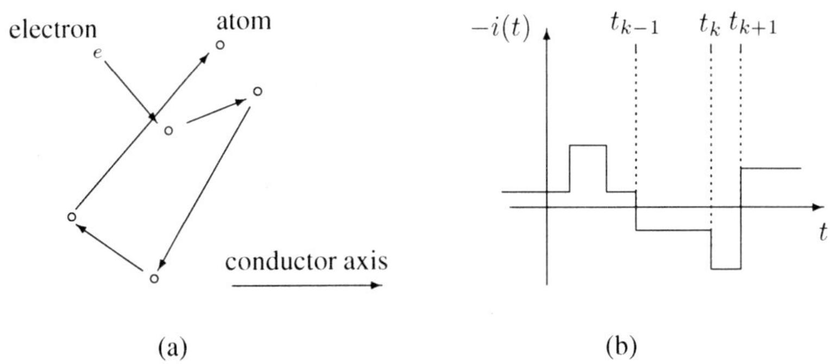

Figure 4.2 (a) Representation of electron motion; (b) current produced by the motion.

Although the average value of the current generated in this process is zero, so that there is no overall transport of charges, the large amount of electrons and collisions gives origin to a measurable alternating component whose statistical power may set the lower bound for the statistical power of useful signals. If a current flows through the conductor, an orderly motion (i.e. current) is superimposed on the disorderly motion of electrons, but the two motions do not interact with each other.

For a conductor of resistance R at an absolute temperature of T Kelvin the output open circuit voltage at the conductor terminals, $w_i(t)$, can be modeled as a WSS rp with PSD given by

$$\mathcal{P}_{w_i}(f) = 2\mathrm{kT}R\,\gamma(f)\,, \qquad \gamma(f) = \frac{hf}{\mathrm{kT}}\left(e^{\frac{hf}{\mathrm{kT}}} - 1\right)^{-1} \tag{4.25}$$

where

$$\mathrm{k} = 1.3805 \cdot 10^{-23}\ \mathrm{J/K} \qquad \text{(Boltzmann constant)}$$
$$h = 6.6262 \cdot 10^{-34}\ \mathrm{Js} \qquad \text{(Planck constant)}$$

Incidentally, for small values of $hf/(\mathrm{kT})$ the value of $\gamma(f)$ can be considered approximately constant and equal to 1. We have

$$\mathcal{P}_{w_i}(f) = 2\mathrm{kT}R \qquad \text{for} \quad f \ll \frac{\mathrm{kT}}{h}\,, \tag{4.26}$$

so, at the standard room temperature of $\mathrm{T} = \mathrm{T}_0 = 290\,\mathrm{K}$, we can set $\gamma(f) \simeq 1$ when $f \ll 6 \cdot 10^{12}$ Hz which is a situation widely met in practice.

The electrical model for a noisy resistance R is given in Fig. 4.3a, and consists of a noiseless resistance R having in series a generator with voltage $w_i(t)$ whose PSD is given by (4.25). Because at each instant the noise is due to the superposition of several current pulses, a suitable model for the amplitude of $w_i(t)$ is a Gaussian distribution with zero mean and PSD given by (4.25)

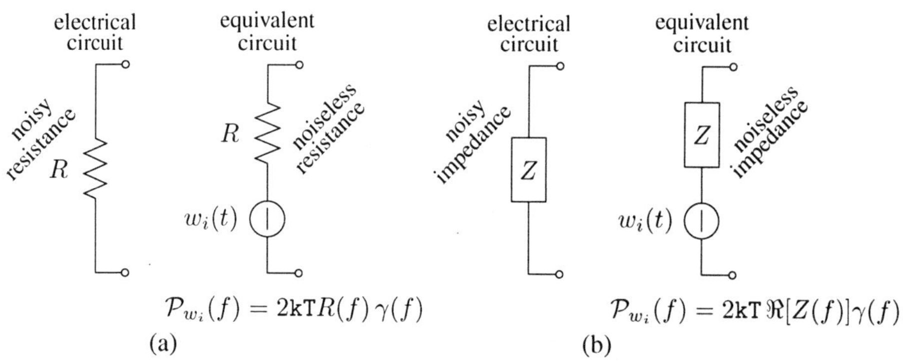

Figure 4.3 Electrical model for: (a) a noisy resistance; (b) a noisy impedance.

In the case of a (noisy) impedance $Z = R + jX$ at absolute temperature T, the model is illustrated in Fig. 4.3b and consists of a noiseless impedance Z in series with a voltage generator $w_i(t)$ whose PSD is

$$\mathcal{P}_{w_i}(f) = 2\mathrm{kT}\,\Re\left[Z(f)\right]\gamma(f) \simeq 2\mathrm{kT}\,R(f) \tag{4.27}$$

and depends on the resistive part of Z only.

So, for an impedance Z at temperature T, closed on a load (considered at a temperature of 0 K) with matched impedance, that is $Z_L = Z^*$, from (4.11) the available power density of thermal noise transferred to the load is

$$\boxed{\mathrm{p}_w(f) = \tfrac{1}{2}\,\mathrm{kT}\,\gamma(f) \simeq \tfrac{1}{2}\,\mathrm{kT}} \tag{4.28}$$

and depends only on temperature T.

4.1.6 Other sources of noise

Besides thermal noise, other sources of noise are present in practice. As an example, most devices are affected by *shot noise* caused by the fact that the current is carried by discrete charges (electrons).

Another well established noise is the so called *flicker noise* that is experienced in some devices at low frequencies and may be modeled as a noise current with PSD proportional to $1/|f|^2$.

In some applications it is also required to take into account the noise due to the discrete nature of the electron flow that prevents $v_i(t)$ from being a continuous function and forces it to be a piecewise constant function (as in Fig. 4.2b).

4.1.7 Noise temperature

Definition 4.3 *Consider the two terminal device of Fig. 4.4* in load matching conditions *where $w_i(t)$ is the open-circuit voltage due to noise contributes of the device. Let $\mathrm{p}_w(f)$ be the available noise power density at the load, due to $w_i(t)$. The* noise temperature *of the device is defined as*

$$\mathrm{T}(f) = \frac{\mathrm{p}_w(f)}{\mathrm{k}/2} \quad [\mathrm{K}]\,. \tag{4.29}$$

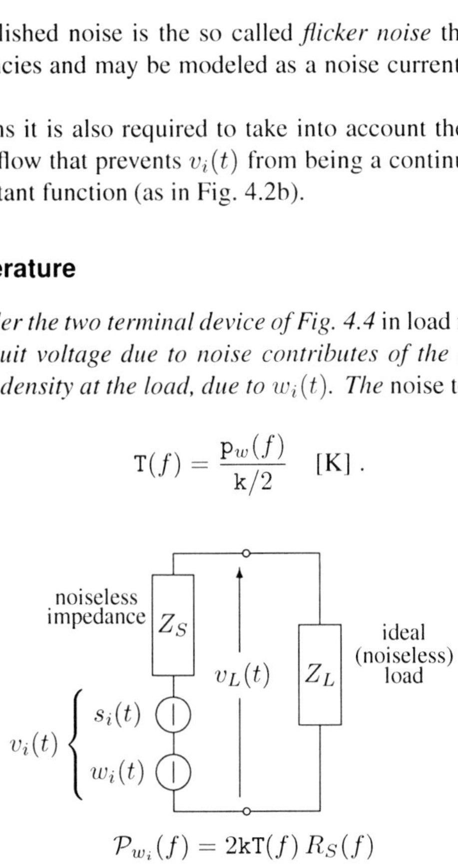

Figure 4.4 Equivalent scheme of a two-terminal device connected to a load where $s_i(t)$ is the useful signal part while $w_i(t)$ is the equivalent noise, including that due to Z_S. In the figure $v_L(t) = s_L(t) + w_L(t)$ where s_L is due to s_i and w_L is due to w_i.

Commonly, the device manufacturer does not provide the density $\mathrm{p}_w(f)$, but rather the noise temperature $\mathrm{T}(f)$. From the comparison of (4.28) and (4.29), for a fixed frequency f, (4.29) represents the temperature Z_S should have (as a thermal noise source) in order to produce the overall noise power density $\mathtt{p}_w(f)$. So, for a device at temperature $\mathtt{T}_\mathtt{t}$, we have

$$\mathtt{T}(f) \geq \mathtt{T}_\mathtt{t} \tag{4.30}$$

where equality is achieved when thermal noise is the *only* noise source (see (4.28)).

4.2 TWO-PORT NETWORKS

4.2.1 Reference model

From an electrical point of view, many components of a transmission system (e.g., cables, amplifiers, filters, etc.) may be interpreted as a cascade of *two-port linear networks*, that is as a cascade of devices which transfer a given source signal $v_i(t)$, with internal impedance $Z_S(f)$, to a load $Z_L(f)$, as illustrated in Fig. 4.5a.

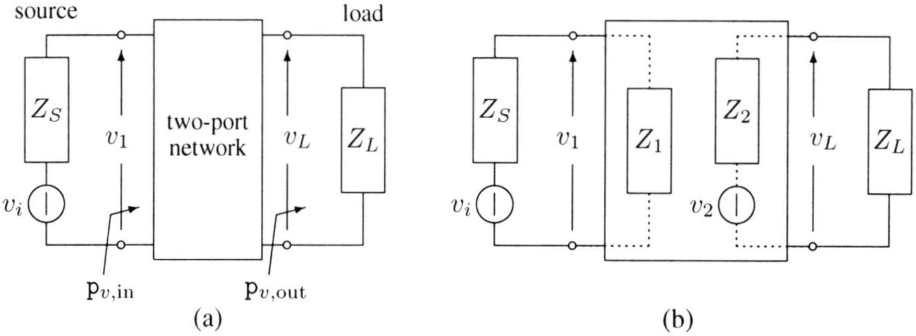

Figure 4.5 (a) Connection of a source to a load through a two-port linear network; (b) its equivalent electrical circuit.

The equivalent electrical circuit for a two-port linear network is also illustrated in Fig. 4.5b, where the network has input and output impedances $Z_1(f)$ and $Z_2(f)$, respectively. The frequency domain relation, linking the network input voltage $v_1(t)$ to the open circuit network output voltage $v_2(t)$, is

$$\mathcal{G}_2(f) = \frac{\mathcal{V}_2(f)}{\mathcal{V}_1(f)}, \tag{4.31}$$

which plays the role of *transfer characteristic of the network*.

The equivalent signal processing model is illustrated in Fig. 4.6 where the frequency response of g_1 is given by

$$\mathcal{G}_1(f) = \frac{\mathcal{V}_1(f)}{\mathcal{V}_i(f)} = \frac{Z_1(f)}{Z_S(f) + Z_1(f)} \tag{4.32}$$

and that of g_L by

$$\mathcal{G}_L(f) = \frac{\mathcal{V}_L(f)}{\mathcal{V}_2(f)} = \frac{Z_L(f)}{Z_2(f) + Z_L(f)}. \tag{4.33}$$

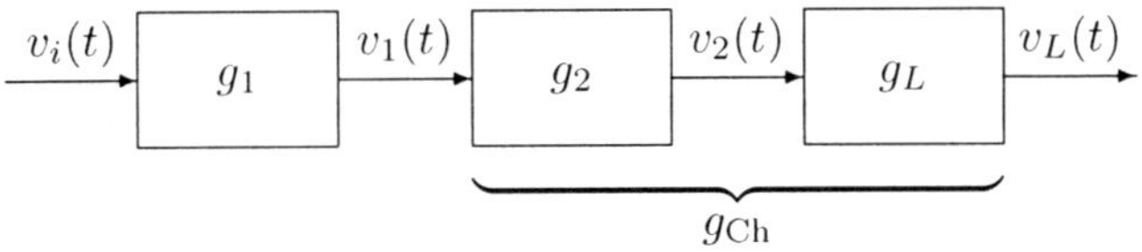

Figure 4.6 Signal processing model for a two-port network connected to a load.

It is also customary to relate the network input voltage $v_1(t)$ to the network output voltage (or load voltage) $v_L(t)$ through the filter with impulse response $g_{\rm Ch}$ which gives in the frequency domain,

$$\mathcal{G}_{\rm Ch}(f) = \frac{\mathcal{V}_L(f)}{\mathcal{V}_1(f)} = \mathcal{G}_2(f)\,\mathcal{G}_L(f)\,. \tag{4.34}$$

Remark A general description of a two-port network requires four-parameters models [1] to take into account the feedback from the load to the input port. Here, for simplicity, we limit to the above unilateral two-port network which assumes a control of the source and load impedances.

4.2.2 Network power gain and matched network

Recalling the two-port network model of Fig. 4.5a, from a transmission point of view, quantities of interest are the power densities at the input and at the output of the two-port network, $\mathtt{p}_{v,\rm in}(f)$ and $\mathtt{p}_{v,\rm out}(f)$, respectively. From these we can evaluate average powers.

In this context it is customary to introduce a function that relates the power density at the load with the power density at the input of the two-port network.

Definition 4.4 *The* network power gain *is defined as (see model of Fig. 4.7)*

$$\mathtt{g}(f) = \frac{\mathtt{p}_{v,\rm out}(f)}{\mathtt{p}_{v,\rm in}(f)} \tag{4.35}$$

which is a real valued and non negative function. The power gain is usually expressed in decibels, that is $(\mathtt{g}(f))_{\rm dB} = 10\,\log_{10}\mathtt{g}(f)$.

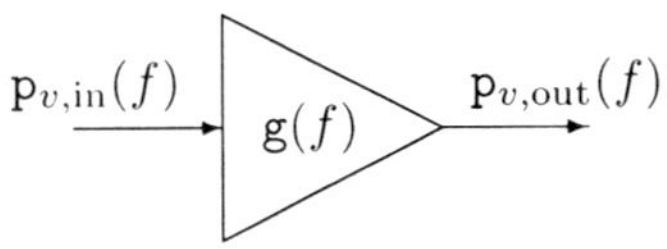

Figure 4.7 Relation between power densities in a two-port linear network.

For those values of f where $\mathtt{g}(f) < 1$ it is customary to speak of *attenuation of the network*

$$\mathtt{a}(f) = \frac{1}{\mathtt{g}(f)}\,, \tag{4.36}$$

which is as well expressed in decibels as $(\mathsf{a}(f))_{\text{dB}} = -(\mathsf{g}(f))_{\text{dB}}$.

Definition 4.5 *A two-port network is said to be* matched *if the conditions for maximum transfer of power are established at the source as well as at the load, that is if we have*

$$Z_1(f) = Z_S^*(f) \quad \textit{and} \quad Z_L(f) = Z_2^*(f)\ . \tag{4.37}$$

In this case the power densities (4.35) become available power densities, *and the power gain becomes an* available power gain.

Remark Usually in a device the condition for maximum transfer of power is imposed at the input with $Z_S = Z_1^*$, however, at the output often the condition is $Z_L = Z_2$, instead of $Z_L = Z_2^*$, to guarantee for the absence of reflections that would cause a further reduction in the power transferred to the load. This is especially true in transmission lines. Indeed the two conditions coincide when Z_L and Z_2 are real impedances. Furthermore, in the local loop often Z_2 is complex and both $Z_L \neq Z_2$ and $Z_L \neq Z_2^*$.

To summarize, there are many load situations in a transmission system and each should be considered separately. In this book we assume condition (4.37) because it allows a simple analysis of the noise power at the various points of a system. If (4.37) were not verified, we must simply replace the available gain with the specific gain (4.35).

4.2.3 Power gain in terms of electrical parameters

Starting from Fig. 4.5a and (4.12), it is

$$\begin{aligned} \mathsf{p}_{v,\text{in}}(f) &= \mathcal{P}_{v_1}(f)\,\frac{R_1(f)}{|Z_1(f)|^2} \\ \mathsf{p}_{v,\text{out}}(f) &= \mathcal{P}_{v_L}(f)\,\frac{R_L(f)}{|Z_L(f)|^2} \end{aligned} \tag{4.38}$$

with (see (4.34))

$$\mathcal{P}_{v_L}(f) = \mathcal{P}_{v_1}(f)\,|\mathcal{G}_{\text{Ch}}(f)|^2 \tag{4.39}$$

and from (4.35) it turns out

$$\mathsf{g}(f) = |\mathcal{G}_{\text{Ch}}(f)|^2\,\frac{|Z_1(f)|^2}{R_1(f)}\,\frac{R_L(f)}{|Z_L(f)|^2}\ . \tag{4.40}$$

This expression simplifies in the specific case of a *matched two-port network* (where (4.37) holds) having *equal input and output impedances*

$$Z_1(f) = Z_2(f)\ . \tag{4.41}$$

Finally, putting together (4.37) and (4.41), i.e. for

$$Z_1(f) = Z_2(f) = Z_S^*(f) = Z_L^*(f)\ , \tag{4.42}$$

then the available power gain can be related to $\mathcal{G}_{\text{Ch}}$ by

$$\boxed{\mathsf{g}(f) = |\mathcal{G}_{\text{Ch}}(f)|^2} \tag{4.43}$$

Incidentally, from (4.43) we note that the following equivalence holds

$$(\mathrm{g}(f))_{\mathrm{dB}} = 10\,\log_{10}\mathrm{g}(f) = 20\,\log_{10}|\mathcal{G}_{\mathrm{Ch}}(f)| = (|\mathcal{G}_{\mathrm{Ch}}(f)|)_{\mathrm{dB}} \tag{4.44}$$

that is, in dB the available power gain $\mathrm{g}(f)$ and the network frequency amplitude $|\mathcal{G}_{\mathrm{Ch}}(f)|$ perfectly correspond.

4.2.4 Noise temperature

The notion of noise and noise temperature of a two-terminal device can be extended to a two-port network. The situation is depicted in Fig. 4.8 where two independent noises are present: the noise introduced by the source, characterized by its noise temperature $\mathrm{T}_S(f)$, and the noise introduced by the two-port network, here thought as an amplifier (A).

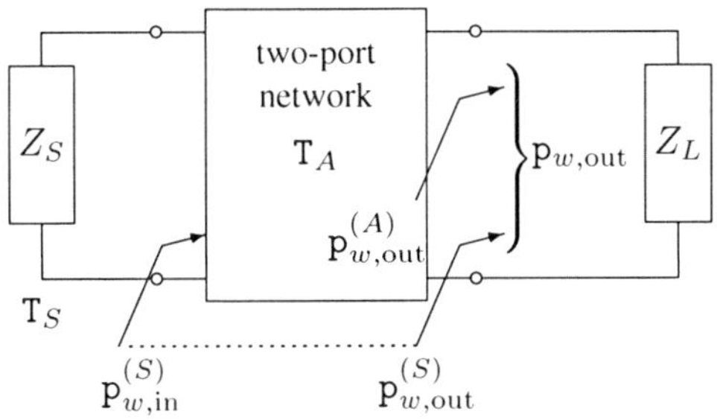

Figure 4.8 Sources of noise in a two-port network.

By assuming that the network is *matched*, the source noise provides the available power density at the two-port network input (recall (4.29))

$$\mathrm{p}_{w,\mathrm{in}}^{(S)}(f) = \tfrac{1}{2}\,\mathrm{k}\mathrm{T}_S(f) \tag{4.45}$$

and at the load the available power density is (recall (4.35))

$$\mathrm{p}_{w,\mathrm{out}}^{(S)}(f) = \mathrm{p}_{w,\mathrm{in}}^{(S)}(f)\,\mathrm{g}(f) = \tfrac{1}{2}\,\mathrm{k}\mathrm{T}_S(f)\,\mathrm{g}(f)\;. \tag{4.46}$$

Let $\mathrm{p}_{w,\mathrm{out}}^{(A)}(f)$ be the power density at the load due to the two-port network noise. Since the two noise sources are uncorrelated, the overall available noise power density at the load is thus

$$\mathrm{p}_{w,\mathrm{out}}(f) = \mathrm{p}_{w,\mathrm{out}}^{(S)}(f) + \mathrm{p}_{w,\mathrm{out}}^{(A)}(f)\;. \tag{4.47}$$

In this context, it is customary to introduce the following definition.

Definition 4.6 *The* noise temperature *of a matched two-port network is defined as*

$$\mathrm{T}_A(f) = \frac{\mathrm{p}_{w,\mathrm{out}}^{(A)}(f)}{\frac{1}{2}\,\mathrm{k}\mathrm{g}(f)}\;, \tag{4.48}$$

and represents the equivalent temperature that Z_S should have (as a thermal noise source) in order to produce the noise power density $\mathrm{p}_{w,\mathrm{out}}^{(A)}$ at the load.

Definition 4.7 *The* effective input noise temperature of the system, *consisting of a source with noise temperature $\mathrm{T}_S(f)$ connected to a matched two-port network with noise temperature $\mathrm{T}_A(f)$, is defined as*

$$\mathrm{T}_{\mathrm{eff,in}}(f) = \mathrm{T}_S(f) + \mathrm{T}_A(f) . \tag{4.49}$$

and denotes the equivalent temperature that the source should have in order to produce the overall noise power density of $\mathrm{p}_{w,\mathrm{out}}$ at the load.

The natural reference model we obtain is that of Fig. 4.9, where noises due to source and two-port network are now gathered together to obtain an input noise with effective input and output available power densities

$$\mathrm{p}_{w,\mathrm{in}}(f) = \tfrac{1}{2}\,\mathrm{k}\mathrm{T}_{\mathrm{eff,in}}(f) , \qquad \mathrm{p}_{w,\mathrm{out}}(f) = \mathrm{p}_{w,\mathrm{in}}(f)\,\mathrm{g}(f) = \tfrac{1}{2}\,\mathrm{k}\mathrm{T}_{\mathrm{eff,in}}(f)\,\mathrm{g}(f) . \tag{4.50}$$

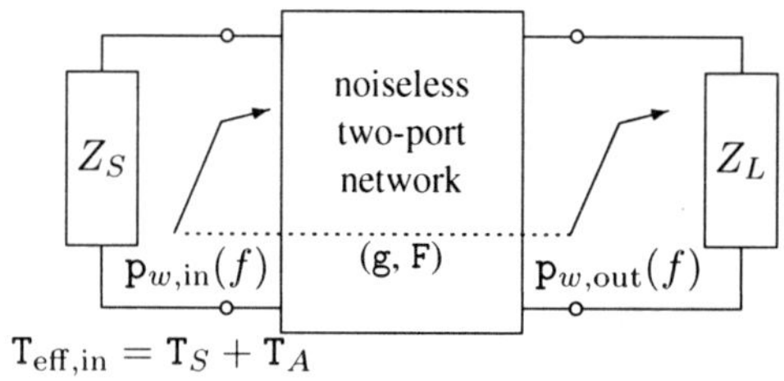

Figure 4.9 Equivalent model of a system consisting of a source connected to a matched two-port network with *all noises at the input*.

Remark Sometimes, it is useful to model *all* noises at the output of the two-port network and write

$$\mathrm{p}_{w,\mathrm{out}}(f) = \tfrac{1}{2}\,\mathrm{k}\mathrm{T}_{\mathrm{eff,out}}(f) , \tag{4.51}$$

where $\mathrm{T}_{\mathrm{eff,out}}(f)$ is the effective output noise temperature of the system (see (4.50))

$$\mathrm{T}_{\mathrm{eff,out}}(f) = \mathrm{T}_{\mathrm{eff,in}}(f)\,\mathrm{g}(f) , \tag{4.52}$$

i.e. the effective output temperature is related to the effective input temperature by the power gain.

4.2.5 Noise figure

Usually, the noise of a two-port network is not directly characterized by the noise temperature $\mathrm{T}_A(f)$, but through an equivalent parameter, the *noise figure* $\mathrm{F}(f)$.

Definition 4.8 *Let the source, connected to a two-port network, be constituted of an impedance* $Z_S(f)$ *set at room temperature* $T_0 = 290$ *K. In this peculiar situation, the* noise figure *of the two-port network is defined as the ratio*

$$F(f) = \frac{p_{w,\text{out}}(f)}{p^{(S)}_{w,\text{out}}(f)} = 1 + \frac{p^{(A)}_{w,\text{out}}(f)}{p^{(S)}_{w,\text{out}}(f)} \tag{4.53}$$

that is the ratio between the total *available power density of noise and that due to the* source only, *both measured at the load.*

Note that $F(f)$ is always greater than 1, and it is equal to 1 in the ideal case of a noiseless network. Moreover, $F(f)$ is a parameter of the network that does not depend on the noise temperature of the source device to which it is usually connected.

We now relate F to the noise temperature T_A defined in (4.48). From (4.46), the noise due to the source has an output available power density of $p^{(S)}_{w,\text{out}}(f) = \frac{1}{2}kT_0\, g(f)$. Hence, by recalling (4.48), equation (4.53) becomes

$$F(f) = 1 + \frac{T_A(f)}{T_0} \tag{4.54}$$

and so

$$\boxed{T_A(f) = T_0\,[F(f) - 1]} \tag{4.55}$$

The above relates the noise figure and noise temperature. Hence, from (4.48), the output available power density due to the network noise can be expressed as

$$p^{(A)}_{w,\text{out}}(f) = \tfrac{1}{2}kT_A(f)\, g(f) = \tfrac{1}{2}kT_0\,[F(f) - 1]\, g(f)\,.$$

From the above considerations we deduce that, to describe the noise of an active two-port linear network we must assign the gain g and the noise figure F (or, equivalently, the noise temperature T_A). In Tab. 4.1 typical values of noise temperature, noise figure and power gain are given for three devices: a MASER (microwave amplification by stimulated emission of radiation); a TWT (traveling wave tube) amplifier; a IC (integrated circuit) amplifier. These are common technologies that are used to build amplifiers. In the last column, the frequency range usually considered for the operations of each device is also given.

Table 4.1 Electrical parameters of three devices.

Device	Noise figure F	Noise temp. T_A	Gain g	Frequency band
Maser	0.16 dB	11 K	20 – 30 dB	6 GHz
TWT amplifier	2.7 dB	250 K	20 – 30 dB	3 GHz
IC amplifier	7.0 dB	1163 K	50 dB	$\leq$ 70 MHz

Passive networks For a *passive* network, that is a network composed only of passive elements (such as resistances, capacities and inductances), the noise figure $\mathtt{F}$ is in close relation to the network gain. When the *passive network* is at *room temperature* T_0, we have

$$\boxed{\mathtt{F}(f) = 1/\mathtt{g}(f) = \mathtt{a}(f)} \tag{4.56}$$

and, by (4.55),

$$\mathrm{T}_A(f) = \mathrm{T}_0[\mathtt{a}(f) - 1] \,. \tag{4.57}$$

This is due to the fact that the only source of noise in a passive network is thermal noise. Hence, only one parameter characterizes the passive network.

Let us evaluate the result (4.56). Now, being the network passive, the only source of noise is thermal noise. Moreover, being the network matched at the input and at the output, with respect to the model of Fig. 4.8 with $\mathrm{T}_S = \mathrm{T}_0$ we have

$$\mathtt{p}^{(S)}_{w,\mathrm{in}}(f) = \tfrac{1}{2}\mathtt{k}\mathrm{T}_0 \,, \qquad \mathtt{p}_{w,\mathrm{out}}(f) = \tfrac{1}{2}\mathtt{k}\mathrm{T}_0$$

for which we recall that the output thermal noise is the sum of the noise due to the source and that due to the two-port network. By then making explicit the output contribute of the source

$$\mathtt{p}^{(S)}_{w,\mathrm{out}}(f) = \mathtt{p}^{(S)}_{w,\mathrm{in}}(f)\,\mathtt{g}(f) = \tfrac{1}{2}\mathtt{k}\mathrm{T}_0\,\mathtt{g}(f) \,,$$

we obtain

$$\mathtt{F}(f) = \frac{\mathtt{p}_{w,\mathrm{out}}(f)}{\mathtt{p}^{(S)}_{w,\mathrm{out}}(f)} = \frac{1}{\mathtt{g}(f)} \,,$$

which proves (4.56).

The relation between the noise figure $\mathtt{F}(f)$ and the gain $\mathtt{g}(f)$ when the passive network is at generic temperature $\mathrm{T} \neq \mathrm{T}_0$ is illustrated in Problem 4.5.

4.2.6 Cascade of two-port networks

In many occasions, a transmission system is made of several two-port networks in cascade. It is thus of interest to model a cascade of two-port linear networks as a single two-port linear network with an equivalent gain and noise figure.

We begin with the simpler case of two linear networks in cascade, having as input a source with noise temperature $\mathrm{T}_S(f)$. We assume that all impedances are matched for the maximum transfer of power. As depicted in Fig. 4.10, there are three noise sources with available power densities of

$$\begin{aligned} \mathtt{p}^{(S)}_{w,\mathrm{in}}(f) = \tfrac{1}{2}\,\mathtt{k}\mathrm{T}_S(f) \,, \qquad & \mathtt{p}^{(A_1)}_{w,\mathrm{out}}(f) = \tfrac{1}{2}\,\mathtt{k}\mathrm{T}_{A_1}(f)\,\mathtt{g}_{A_1}(f) \\ & \mathtt{p}^{(A_2)}_{w,\mathrm{out}}(f) = \tfrac{1}{2}\,\mathtt{k}\mathrm{T}_{A_2}(f)\,\mathtt{g}_{A_2}(f) \end{aligned} \tag{4.58}$$

where the meaning of parameters is self-explaining.

We now wish to determine the parameters of the two-port linear network (as in Fig. 4.9) equivalent to the cascade of Fig. 4.10, that is we need to determine the overall power gain $\mathtt{g}(f)$,

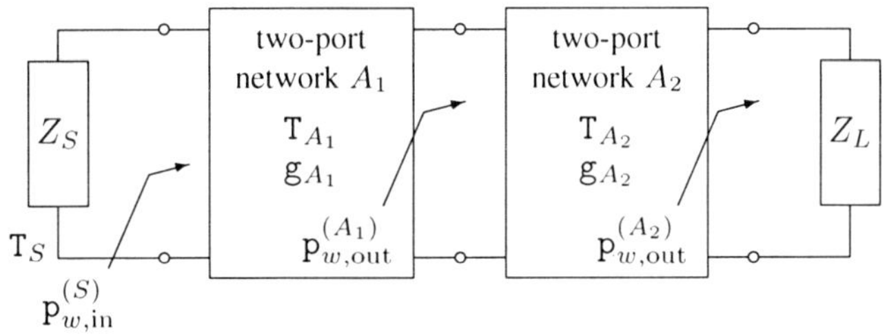

Figure 4.10 Sources of noise in a cascade of two-port networks driven by a source with noise temperature $T_S(f)$.

and the noise figure $F(f)$ or, equivalently, the noise temperature $T_A(f)$. For the power gain we immediately have

$$g(f) = \frac{p_{s,out}^{(A_2)}(f)}{p_{s,in}^{(A_1)}(f)} = \frac{p_{s,out}^{(A_1)}(f)}{p_{s,in}^{(A_1)}(f)} \frac{p_{s,out}^{(A_2)}(f)}{p_{s,in}^{(A_2)}(f)} = g_{A_1}(f)\, g_{A_2}(f)\,, \tag{4.59}$$

being $p_{s,out}^{(A_1)}(f) = p_{s,in}^{(A_2)}(f)$.

For the noise temperature, we can express the output power density of the noise due to the network cascade as

$$p_{w,out}^{(A)}(f) = p_{w,out}^{(A_1)}(f)\, g_{A_2}(f) + p_{w,out}^{(A_2)}(f)$$

and thus by applying (4.48) we obtain

$$T_A(f) = T_{A_1}(f) + \frac{T_{A_2}(f)}{g_{A_1}(f)}\,. \tag{4.60}$$

Finally, from (4.54) and after easy argumentation, the noise figure becomes

$$F(f) = F_{A_1}(f) + \frac{F_{A_2}(f) - 1}{g_{A_1}(f)}\,. \tag{4.61}$$

This result can be easily generalized to the cascade of N two-port linear networks with gains g_i, $i = 1, \ldots, N$, and noise figures F_i or, equivalently, noise temperatures T_i. For the overall gain we obtain

$$\boxed{g(f) = g_1(f)\, g_2(f) \, \ldots \, g_N(f)} \tag{4.62}$$

with its equivalent formulation in logarithmic scale given by

$$(g(f))_{dB} = (g_1(f))_{dB} + (g_2(f))_{dB} + \ldots + (g_N(f))_{dB}\,. \tag{4.63}$$

For the equivalent noise temperature we have

$$\boxed{T_A(f) = T_1(f) + \frac{T_2(f)}{g_1(f)} + \ldots + \frac{T_N(f)}{g_1(f) \, \ldots \, g_{N-1}(f)}} \tag{4.64}$$

The equivalent noise figure follows by use of relation (4.54). We have

$$\boxed{\mathtt{F}(f) = \mathtt{F}_1(f) + \frac{\mathtt{F}_2(f) - 1}{\mathtt{g}_1(f)} + \ldots + \frac{\mathtt{F}_N(f) - 1}{\mathtt{g}_1(f) \ldots \mathtt{g}_{N-1}(f)}} \tag{4.65}$$

This equation is known as *Friis equation* of the equivalent noise figure. We also observe that $\mathtt{F}$ strongly depends on the gain and noise parameters of the first stages. In particular, the smaller $\mathtt{F}_1$ and the larger $\mathtt{g}_1$, the more $\mathtt{F}$ will be reduced. Obviously, these specifications have an impact on the cost of the device.

Example 4.2 A Let us consider the block diagram of a typical configuration for a radio receiver, consisting of an antenna, a pre-amplifier and an amplifier, as illustrated in Fig. 4.11a. We assume that they are all matched. As shown in Fig. 4.11b, with respect to the noise $w_i^{(S)}(t)$, *the antenna is modeled as a resistance R with a noise temperature* $\mathtt{T}_S$. In this example we assume $\mathtt{T}_S = 500\,\text{K}$. Moreover, the pre-amplifier has a gain of 10 dB and a noise figure of 7 dB, and the amplifier has a gain of 30 dB and a noise temperature of 500 K. We want to identify the equivalent gain, noise figure and effective noise temperature of the system at the receiver input.

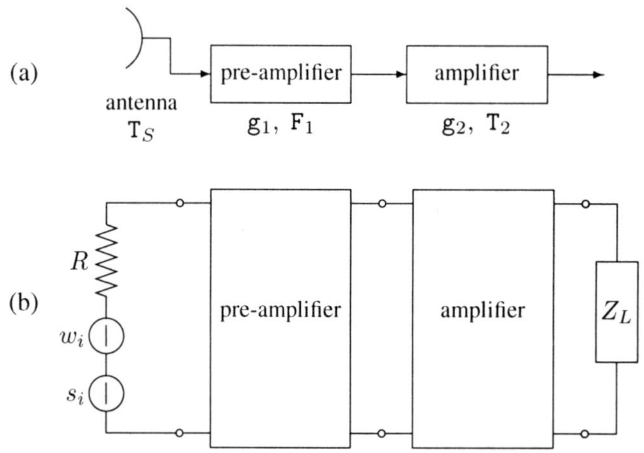

Figure 4.11 Radio receiver configuration with an antenna, a pre-amplifier and an amplifier: (a) block diagram; (b) two-port network electrical model.

The reference electrical model is thus that of Fig. 4.10, where the source is the antenna, the first network is the pre-amplifier, and the second network is the second amplifier. We have

$$\mathtt{T}_S = 500\,\text{K} \qquad (\mathtt{g}_1)_{\text{dB}} = 10\,\text{dB} \qquad (\mathtt{g}_2)_{\text{dB}} = 30\,\text{dB}$$
$$(\mathtt{F}_1)_{\text{dB}} = 7\,\text{dB} \qquad \mathtt{T}_2 = 500\,\text{K}\;.$$

By use of (4.62) and Friis equation (4.64), the amplifier cascade has

$$(\mathtt{g})_{\text{dB}} = (\mathtt{g}_1)_{\text{dB}} + (\mathtt{g}_2)_{\text{dB}} = 40\,\text{dB}$$
$$\mathtt{F} = \mathtt{F}_1 + \frac{\mathtt{F}_2 - 1}{\mathtt{g}_1} \quad \Longrightarrow \quad (\mathtt{F})_{\text{dB}} = 7.15\,\text{dB}$$

with $F_2 = 1 + T_2/T_0$, and

$$T_A = T_0\,(F-1) = 1213\,K\,.$$

Note that the dependency on f has been dropped, since all quantities are assumed constant over the band of interest. Overall, the effective noise temperature of the system at the receiver input is

$$T_{\text{eff,in}} = T_S + T_A = 1713\,K\,.$$

Example 4.2 B The idealized configuration of a transmission medium consisting of a very long cable where amplifiers are inserted at equally spaced points, to restore the signal power level, is illustrated in Fig. 4.12. Each section of the cable, which can be modeled as a passive network, has power attenuation a_C and noise figure $F_C = a_C$, while each amplifier has power gain $g_A = a_C$ and noise figure F_A. We want to determine the equivalent gain and noise figure of the cascade.

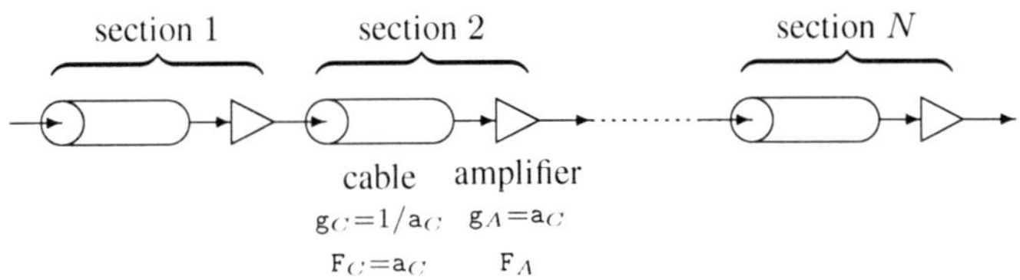

Figure 4.12 Transmission channel composed of N repeater sections.

We begin by addressing the problem for the cascade of a cable and an amplifier. This element will be referred to as *repeater section*, so that the transmission channel can be seen as the cascade of N repeater sections, each with gain

$$g_R = g_A\,g_C = g_A\,\frac{1}{a_C} = 1$$

and noise figure

$$F_R = F_C + \frac{F_A - 1}{g_C} = a_C\,F_A\,.$$

Therefore, the N sections have an equivalent unit gain, $g = 1$, and noise figure

$$F = F_R + \frac{F_R - 1}{g_R} + \ldots + \frac{F_R - 1}{g_R \cdots g_R} = N\,(F_R - 1) + 1 \simeq N\,F_R$$

so that the output noise power of N repeater sections is roughly N times the noise power introduced by an individual section. Finally, for the noise temperature of the cascade we have $T = N\,T_0\,(F_R - 1) \simeq N\,T_0 F_R$.

4.3 TRANSMISSION SYSTEM MODEL

4.3.1 Electrical model

From an electrical perspective a transmission system can be modeled as shown in Fig. 4.13, where we assume that all terminals are matched.

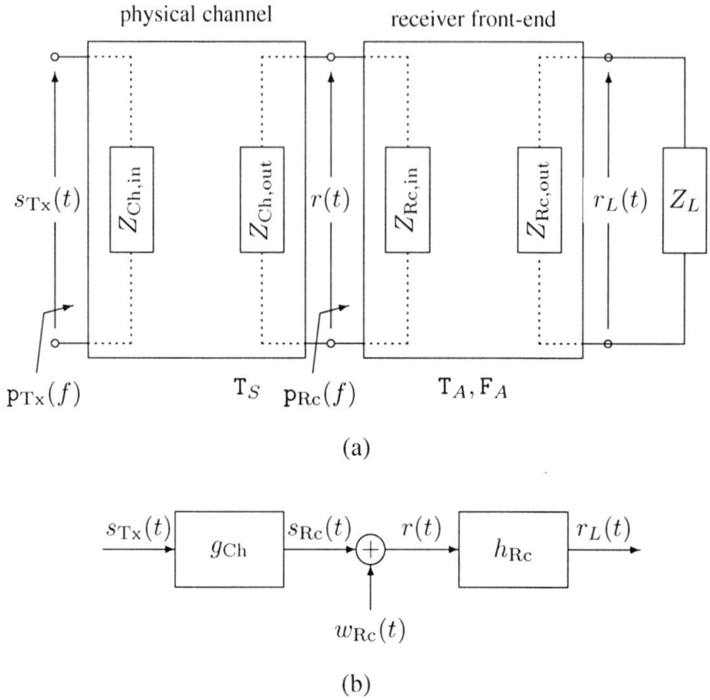

Figure 4.13 (a) Electrical model of a transmission system and (b) corresponding model in terms of signals.

The first two-port network in Fig. 4.13 models the transmission medium also denoted as physical channel (Ch) (a cable, a radio link inclusive of transmit and receive antennas, an optical fiber, an underwater link), which usually introduces a large attenuation. The second two-port network in the figure models the receiver front-end (Rc), which is usually constituted by the cascade of passband filters, to reject noise components that lie outside the band $\mathcal{B}$ of the useful signal, and amplifier, to re-establish suitable power levels.

In the electrical scheme of Fig. 4.13, the *transmitted signal* $s_{\text{Tx}}(t)$, of band $\mathcal{B}$ and bandwidth B, is the voltage across the input port of the physical channel. The *received signal* $r(t)$ is the voltage across the output port of the physical channel or, equivalently, at the input port of the receiver front-end. Noise sources are positioned at the physical channel output, i.e. at the receiver input. In particular, T_S is the noise temperature of all noises *seen* at the channel output, including the transmission medium noise and the transmitter noise. The receiver front-end is characterized by a noise figure F_A, or, equivalently, a noise temperature $\text{T}_A = \text{T}_0(\text{F}_A - 1)$. Thus, the channel (the source here) and receiver will produce an effective noise temperature $\text{T}_{\text{eff,in}} = \text{T}_S + \text{T}_A$. When the noise temperature T_S is not specified, it is common to assume $\text{T}_S = \text{T}_0$.

4.3.2 AWGN model

In the above context, the received signal $r(t)$ is the sum of a useful signal component, $s_{\rm Rc}(t)$, depending on $s_{\rm Tx}(t)$, and the overall effective noise $w_{\rm Rc}(t)$, which is modeled as Gaussian. This is illustrated in Fig. 4.13b, where

$$r(t) = s_{\rm Rc}(t) + w_{\rm Rc}(t)\,, \qquad s_{\rm Rc}(t) = (g_{\rm Ch} * s_{\rm Tx})(t) \tag{4.66}$$

and $g_{\rm Ch}(t)$ the channel impulse response. Moreover, the receiver front-end acts as a filter with impulse response $h_{\rm Rc}(t)$, to provide the signal $r_L(t)$.

The statistical characteristics of the noise $w_{\rm Rc}(t)$ can be evaluated from the effective noise temperature of the receiver, $\mathtt{T}_{\rm eff,Rc}(f) = \mathtt{T}_S(f) + \mathtt{T}_A(f)$, where the shorthand notation $\mathtt{T}_{\rm eff,Rc}$ is used in place of $\mathtt{T}_{\rm eff,in,Rc}$. Hence, from (4.50), the available noise power density at the receiver input is

$$\mathtt{p}_{w,\rm Rc}(f) = \tfrac{1}{2}\,\mathtt{k}\mathtt{T}_{\rm eff,Rc}(f)\,. \tag{4.67}$$

In turn, the PSD of the noise voltage can be obtained from the power density if we know the input receiver impedance $Z_{\rm Rc,in}(f)$ and we get (see also (4.12))

$$\mathcal{P}_{w_{\rm Rc}}(f) = \mathtt{p}_{w,\rm Rc}(f)\,\frac{|Z_{\rm Rc,in}(f)|^2}{R_{\rm Rc,in}(f)} = \tfrac{1}{2}\mathtt{k}\mathtt{T}_{\rm eff,Rc}(f)\,\frac{|Z_{\rm Rc,in}(f)|^2}{R_{\rm Rc,in}(f)}\,. \tag{4.68}$$

It is common to assume that $Z_{\rm Rc,in}(f)$ and $\mathtt{T}_{\rm eff,Rc}(f)$ are constant *over the band of interest* $\mathcal{B}$, so that the PSD of $w_{\rm Rc}(t)$ is constant over $\mathcal{B}$. In the technical literature it is also customary to denote the voltage noise PSD as $\frac{N_0}{2}$, that is

$$\boxed{\mathcal{P}_{w_{\rm Rc}}(f) = \frac{N_0}{2}} \tag{4.69}$$

In this case, the noise $w_{\rm Rc}(t)$ is said to be *white* (see Definition 2.35), and we reference to it as *additive white Gaussian noise* (AWGN).

4.3.3 Signal-to-noise ratio

In the context of communications, we are interested in measuring the relevance of noise with respect to the useful signal, for example across a given load. A common measure is given by the so called *signal-to-noise ratio* (SNR), as the ratio between the useful signal power and the noise power, which may take different forms.

Statistical SNR at the output By referencing to the two-port network of Fig. 4.13, the output signal r_L, voltage across the load, can be written as

$$r_L(t) = s_L(t) + w_L(t) \tag{4.70}$$

where s_L is the *useful signal* and w_L is the noise. In this context, the statistical SNR can be defined as the ratio between the statistical power of the useful signal s_L and the statistical power of the noise w_L, that is

$$\Lambda_{\mathtt{M}} = \frac{\mathtt{M}_{s_L}}{\mathtt{M}_{w_L}} = \frac{\mathrm{E}\left[|s_L(t)|^2\right]}{\mathrm{E}\left[|w_L(t)|^2\right]} = \frac{\displaystyle\int \mathcal{P}_{s_L}(f)\,df}{\displaystyle\int \mathcal{P}_{w_L}(f)\,df}\,. \tag{4.71}$$

Electrical SNR at the output In alternative to (4.71), the SNR can be defined as the ratio between electrical powers at the load, that is

$$\Lambda_{\mathtt{P}} = \frac{\mathtt{P}_{s,\text{out}}}{\mathtt{P}_{w,\text{out}}} = \frac{\displaystyle\int \mathtt{p}_{s,\text{out}}(f)\, df}{\displaystyle\int \mathtt{p}_{w,\text{out}}(f)\, df} . \tag{4.72}$$

Comparison In general, the two SNRs (4.71) and (4.72) do not coincide. The relation between the two can be made explicit by use of (4.12) which, in the present context, yields

$$\mathtt{p}_{s,\text{out}}(f) = \mathcal{P}_{s_L}(f) \frac{R_L(f)}{|Z_L(f)|^2} , \qquad \mathtt{p}_{w,\text{out}}(f) = \mathcal{P}_{w_L}(f) \frac{R_L(f)}{|Z_L(f)|^2} . \tag{4.73}$$

Hence, the SNRs become

$$\Lambda_{\mathtt{M}} = \frac{\displaystyle\int \mathcal{P}_{s_L}(f)\, df}{\displaystyle\int \mathcal{P}_{w_L}(f)\, df} \neq \Lambda_{\mathtt{P}} = \frac{\displaystyle\int \frac{R_L(f)}{|Z_L(f)|^2} \mathcal{P}_{s_L}(f)\, df}{\displaystyle\int \frac{R_L(f)}{|Z_L(f)|^2} \mathcal{P}_{w_L}(f)\, df} . \tag{4.74}$$

and the only practical condition where

$$\Lambda_{\mathtt{M}} = \Lambda_{\mathtt{P}} = \Lambda \tag{4.75}$$

is when $Z_L(f)$ is a constant within the bands of s_L and of w_L.

Example 4.3 A Let the received signal at the input of a matched front-end receiver have PSD

$$\mathcal{P}_{s_{\text{Rc}}}(f) = A \text{ rect}\left(\frac{f - f_0}{B}\right) + A \text{ rect}\left(\frac{f + f_0}{B}\right)$$

with $f_0 \gg B$, and let the front-end frequency response be an ideal bandpass filter

$$\mathcal{H}_{\text{Rc}}(f) = A_{\text{Rc}} \left[\text{rect}\left(\frac{f - f_0}{2B}\right) + \text{rect}\left(\frac{f + f_0}{2B}\right)\right] .$$

Let also the input and output impedances be $Z_{\text{Rc,in}} = Z_{\text{Rc,out}} = Z_L = R = 50\,\Omega$, and the noise temperature of the receiver front-end be $\mathtt{T}_A = 500$ K. We want to compute the value of A that guarantees an output SNR of at least 30 dB.

In this case, we evaluate the SNR $\Lambda_{\mathtt{M}}$ which, by the way, is equal to $\Lambda_{\mathtt{P}}$ because $Z_L(f) = R$ is a constant. To derive $\Lambda_{\mathtt{M}}$ we preliminarily need to express $\mathcal{P}_{s_L}(f)$ and $\mathcal{P}_{w_L}(f)$ in terms of the system parameters. For the useful signal, since s_L is the filtering of s_{Rc} by h_{Rc}, we have

$$\mathcal{P}_{s_L}(f) = \mathcal{P}_{s_{\text{Rc}}}(f)\, |\mathcal{H}_{\text{Rc}}(f)|^2 = A_{\text{Rc}}^2 \left[A \text{ rect}\left(\frac{f - f_0}{B}\right) + A \text{ rect}\left(\frac{f + f_0}{B}\right)\right] . \tag{4.76}$$

Similarly, for the noise

$$\mathcal{P}_{w_{\text{L}}}(f) = \mathcal{P}_{w_{\text{Rc}}}(f)\, |\mathcal{H}_{\text{Rc}}(f)|^2 , \qquad \text{with} \qquad \mathcal{P}_{w_{\text{Rc}}}(f) = \tfrac{1}{2}\, \mathtt{k}\mathtt{T}_{\text{eff,Rc}}(f) \frac{|Z_{\text{Rc,in}}(f)|^2}{R_{\text{Rc,in}}(f)} , \tag{4.77}$$

from (4.68).

Since in the present context the effective input noise temperature of the system is

$$\mathtt{T}_{\mathrm{eff,Rc}}(f) = \mathtt{T}_0 + \mathtt{T}_A \ ,$$

assuming that the source is at standard temperature $\mathtt{T}_0$, we obtain

$$\mathcal{P}_{w_L}(f) = A_{\mathrm{Rc}}^2 \left[A_1 \ \mathrm{rect}\left(\frac{f - f_0}{2B}\right) + A_1 \ \mathrm{rect}\left(\frac{f + f_0}{2B}\right)\right] \ , \qquad A_1 = \tfrac{1}{2}\mathtt{k}(\mathtt{T}_0 + \mathtt{T}_A)\, R \ .$$

Note that both the output signal and the output noise PSDs in (4.76) and (4.77) depend upon the amplitude A_{Rc} of the two-port network frequency response $\mathcal{H}_{\mathrm{Rc}}(f)$, which underlines the filtering action of the two-port network on both the input useful signal and the noise.

In conclusion, from (4.71) the output SNR is

$$\Lambda_{\mathtt{M}} = \frac{A\, 2B}{A_1\, 4B} = \frac{A}{\mathtt{k}(\mathtt{T}_0 + \mathtt{T}_A)\, R} \ ,$$

from which, by imposing $\Lambda_{\mathtt{M}} = 10^3$, we obtain

$$A = \Lambda_{\mathtt{M}}\, \mathtt{k}(\mathtt{T}_0 + \mathtt{T}_A)\, R = 5.45 \cdot 10^{-16} \ \mathrm{V^2/Hz}.$$

Bandpass network As seen in Example 4.3 A with reference to the model of Fig. 4.13, it is quite common that the receiver front-end acts as a bandpass filter which keeps undistorted the frequency components of the desired signal $s_{\mathrm{Rc}}(t)$, while attenuating the frequency components of the noise $w_{\mathrm{Rc}}(t)$ that lie outside the band $\mathcal{B}$ of $s_{\mathrm{Rc}}(t)$. In this situation, by assuming that $\mathcal{B}$ is also the *passband of the front-end filter* h_{Rc}, the SNRs (4.71) and (4.72) can be obtained by limiting the integration to the band $\mathcal{B}$ and become, respectively,

$$\Lambda_{\mathtt{M}} = \frac{\displaystyle\int_{\mathcal{B}} \mathcal{P}_{s_L}(f)\, df}{\displaystyle\int_{\mathcal{B}} \mathcal{P}_{w_L}(f)\, df} \ , \qquad \Lambda_{\mathtt{P}} = \frac{\displaystyle\int_{\mathcal{B}} \mathtt{p}_{s,\mathrm{out}}(f)\, df}{\displaystyle\int_{\mathcal{B}} \mathtt{p}_{w,\mathrm{out}}(f)\, df} \ . \tag{4.78}$$

Constant gain network Results drastically simplify when the system parameters are constant over the band of interest $\mathcal{B}$, of measure B, with the further assurance that statistical and electrical SNRs coincide. We assume that the *receiver front-end gain* $\mathtt{g}(f)$ is *constant* over $\mathcal{B}$, and zero outside. In this context, we have

$$\mathtt{P}_{s,\mathrm{out}} = 2\int_{\mathcal{B}} \mathtt{p}_{s,\mathrm{out}}(f)\, df = 2\int_{\mathcal{B}} \mathtt{p}_{\mathrm{Rc}}(f)\, \mathtt{g}(f)\, df = \mathtt{P}_{\mathrm{Rc}}\, \mathtt{g} \ , \tag{4.79}$$

where $\mathtt{P}_{\mathrm{Rc}}$ is the useful signal power at the receiver input. While, from (4.50)

$$\mathtt{P}_{w,\mathrm{out}} = 2\int_{\mathcal{B}} \mathtt{p}_{w,\mathrm{out}}(f)\, df = \mathtt{k}\mathtt{T}_{\mathrm{eff,Rc}}\, \mathtt{g}\, B \ , \tag{4.80}$$

where, we recall, $T_{\text{eff,Rc}}$ is the effective noise temperature of the system at the receiver input. Hence the output SNR becomes

$$\boxed{\Lambda_{\mathtt{P}} = \frac{P_{\text{Rc}}}{kT_{\text{eff,Rc}}\, B} = \Lambda_{\mathtt{M}}} \tag{4.81}$$

We stress the fact that the SNR *at the load* in Fig. 4.13a can be expressed in terms of the receiver input power and the receiver input noise power density. Moreover, it does not depend upon the receiver gain g, which amplifies both the useful signal and the noise.

4.3.4 Narrowband channel model and link budget

All the results seen so far are valid for a generic *broadband* signal $s_{\text{Tx}}(t)$, for which the system parameters (frequency response $\mathcal{G}_{\text{Ch}}(f)$, etc.) may be frequency varying over its band $\mathcal{B}$. However, the above results drastically simplify for *narrowband* signals.

Signal model For signals having a very small passband $\mathcal{B}$ positioned around the center frequency f_0, we can assume

$$\mathcal{G}_{\text{Ch}}(f) = \mathcal{G}_{\text{Ch}}(f_0) = \mathcal{G}_{\text{Ch}}\,, \qquad f \in \mathcal{B}\,. \tag{4.82}$$

Similar conditions are assumed for all system parameters.

Under (4.82), the channel satisfies the conditions for the absence of distortion. Specifically, by setting

$$\mathcal{C} = |\mathcal{G}_{\text{Ch}}|\,, \qquad t_0 = -\frac{\arg \mathcal{G}_{\text{Ch}}}{2\pi f_0} \tag{4.83}$$

we have

$$s_{\text{Rc}}(t) = \mathcal{C}\, s_{\text{Tx}}(t - t_0)\,. \tag{4.84}$$

The delay t_0 is not relevant to our analysis, since it has no effect on the statistical power. Thus, the scheme of Fig. 4.13b simplifies into that of Fig. 4.14.

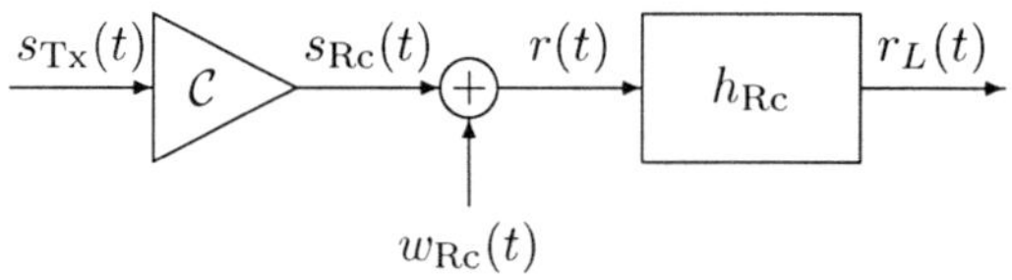

Figure 4.14 Narrowband transmission system model with AWGN.

In the narrowband channel model, the SNR measures can be expressed in terms of the transmitted signal.

Statistical powers From (4.84) it is $\mathtt{M}_{s_{\text{Rc}}} = \mathtt{M}_{s_{\text{Tx}}}\, \mathcal{C}^2$, and so the SNR (4.81) can also be written as (see (4.71) and assume a receiver front-end with a constant gain)

$$\boxed{\Lambda = \Lambda_{\mathtt{M}} = \frac{\mathtt{M}_{s_{\text{Tx}}}\, \mathcal{C}^2}{N_0 B}} \tag{4.85}$$

Electrical powers If g_{Ch} (a_{Ch}) is the transmission channel gain (attenuation), from Fig. 4.13a it is

$$P_{Rc} = P_{Tx}\, g_{Ch} = \frac{P_{Tx}}{a_{Ch}} \tag{4.86}$$

where P_{Tx} is the transmitted signal power, and the receiver output SNR (4.81) becomes

$$\boxed{\Lambda = \Lambda_P = \frac{P_{Tx}}{k T_{eff,Rc}\, B\, a_{Ch}}} \tag{4.87}$$

Equal impedance channel A further link between electrical and statistical measures is assured when the channel is a two-port network with equal input and output impedances, then (4.43) is valid and

$$\mathcal{C} = \sqrt{g_{Ch}} = \frac{1}{\sqrt{a_{Ch}}} \,. \tag{4.88}$$

This relation assures that

$$(\mathcal{C})_{dB} = 20\, \log_{10} \mathcal{C} = (g_{Ch})_{dB} = -(a_{Ch})_{dB} \tag{4.89}$$

and in fact $\mathcal{C}$ denotes a ratio of voltages (thus requiring a $20 \log_{10}$ dB scale).

Link budget The basic equation (4.87) is mostly used in logarithmic form. By expressing $T_{eff,Rc}$ as $T_0 \frac{T_{eff,Rc}}{T_0}$ in (4.87), we have

$$(\Lambda)_{dB} = (P_{Tx})_{dBm} - (a_{Ch})_{dB} - 10\, \log_{10}(10^9\, kT_0) - 10\, \log_{10}\left(\frac{T_{eff,Rc}}{T_0}\right) - 10\, \log_{10}(B)_{MHz} \tag{4.90}$$

or

$$\boxed{(\Lambda)_{dB} = (P_{Tx})_{dBm} - (a_{Ch})_{dB} + 114 - 10\, \log_{10}\left(\frac{T_{eff,Rc}}{T_0}\right) - 10\, \log_{10}(B)_{MHz}} \tag{4.91}$$

where B is expressed in MHz and $(P_{Tx})_{dBm} - (a_{Ch})_{dB}$ is the received power $(P_{Rc})_{dBm}$ in dBm. We also note that, *when* $T_S = T_0$, then the effective noise temperature at the input is proportional to the receiver noise figure F_A, that is from (4.49) $T_{eff,Rc} = T_0 + T_0\,(F_A - 1) = T_0\, F_A$. We thus have in (4.91)

$$\boxed{T_S = T_0 \qquad \Longrightarrow \qquad 10\, \log_{10}\left(\frac{T_{eff,in}}{T_0}\right) = (F_A)_{dB}} \tag{4.92}$$

It is very customary to set a reference minimum objective SNR Λ_{min} that the transmission system must guarantee, and to provide some of the the system parameters such as the transmitted signal power P_{Tx}, the signal bandwidth B or the receiver noise figure F_A. All remaining system parameters are then evaluated by use of (4.91). This operation is known as *link budget* analysis. This analysis, implicitly required by all the exercises proposed in this chapter, assumes a narrowband signal model, which is by far the simplest case. However, these same concepts can be applied to wide-band systems as well.

Example 4.3 B A station, receiving signals from a satellite, has an antenna gain $(\mathtt{g}_{\text{Ant}})_{\text{dB}} = 40\,\text{dB}$ and a noise temperature $\mathtt{T}_{\text{Ant}} = 60\,\text{K}$, that is, as seen in Example 4.2 A the antenna acts as a noisy resistor at a temperature of 60 K. The antenna feeds a pre-amplifier with a noise temperature $\mathtt{T}_{\text{Pre}} = 125\,\text{K}$ and a gain $(\mathtt{g}_{\text{Pre}})_{\text{dB}} = 20\,\text{dB}$. The pre-amplifier is followed by an amplifier with a noise figure $(\mathtt{F}_{\text{Amp}})_{\text{dB}} = 10\,\text{dB}$ and a gain $(\mathtt{g}_{\text{Amp}})_{\text{dB}} = 80\,\text{dB}$. The transmitted signal is narrowband with bandwidth $B = 1\,\text{MHz}$. The satellite has an antenna with a power gain of $(\mathtt{g}_{\text{Sat}})_{\text{dB}} = 6\,\text{dB}$ and the total attenuation due to the distance between transmitter and receiver (propagation) is $(\mathtt{a}_{\text{Prop}})_{\text{dB}} = 130\,\text{dB}$. The reference block diagram is given in Fig. 4.15. We want to find:

a) The average power of the overall noise at the receiver output.

b) The minimum power, in dBm, of the signal transmitted by the satellite to obtain a reference SNR of $(\Lambda)_{\text{dB}} = 20\,\text{dB}$ at the receiver output.

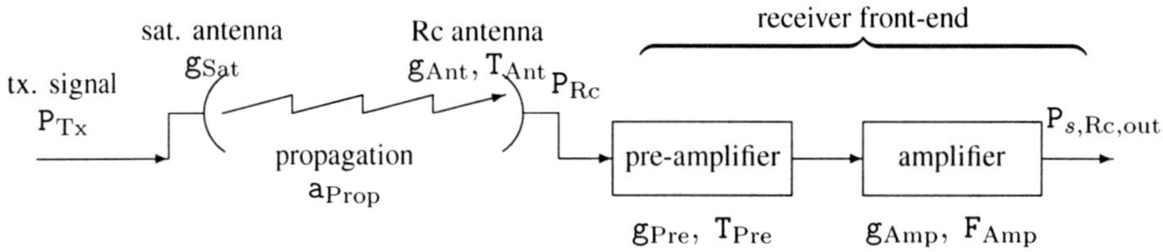

Figure 4.15 Satellite transmission system.

a) With respect to the noise, we note that the model proposed for the receive antenna is a noisy resistance. We thus set the noise temperature of the receiver source to $\mathtt{T}_S = \mathtt{T}_{\text{Ant}}$ and this measure includes noises introduced by the satellite antenna or by any propagation effect. Moreover, the input noise temperature of the receiver front-end is (see (4.64))

$$\mathtt{T}_A = \mathtt{T}_{\text{Pre}} + \frac{\mathtt{T}_0\,(\mathtt{F}_{\text{Amp}} - 1)}{\mathtt{g}_{\text{Pre}}} = 125 + 26 = 151\,\text{K}\,,$$

hence the receiver effective noise temperature is

$$\mathtt{T}_{\text{eff,Rc}} = \mathtt{T}_S + \mathtt{T}_A = \mathtt{T}_{\text{Ant}} + \mathtt{T}_A = 211\,\text{K}\,.$$

The receiver gain is instead $\mathtt{g}_{\text{Rc}} = \mathtt{g}_{\text{Pre}}\,\mathtt{g}_{\text{Amp}}$, giving $(\mathtt{g}_{\text{Rc}})_{\text{dB}} = 20 + 80 = 100\,\text{dB}$. In conclusion, from (4.80) and (4.91) the average noise power at the receiver output is

$$(\mathtt{P}_{w,\text{Rc,out}})_{\text{dBm}} = -114 - 1.38 + 0 + 3 + 100 = -12.38\,\text{dBm}$$

where $3 + 100$ accounts for the term $(2\mathtt{g}_{\text{Rc}})_{\text{dB}}$.

b) In Fig. 4.15 it is shown that the transmitted signal power, $\mathtt{P}_{\text{Tx}}$, is first amplified by the satellite antenna, then attenuated because of propagation, and finally amplified by receive antenna, pre-amplifier and amplifier. The overall channel gain is

$$\mathtt{g}_{\text{Ch,all}} = \mathtt{g}_{\text{Sat}}\,\frac{1}{\mathtt{a}_{\text{Prop}}}\,\mathtt{g}_{\text{Ant}} \quad\Longrightarrow\quad (\mathtt{g}_{\text{Ch,all}})_{\text{dB}} = 6 - 130 + 40 = -84\,\text{dB}\,.$$

So, the requested output SNR is

$$\frac{\mathtt{P}_{s,\text{Rc,out}}}{\mathtt{P}_{w,\text{Rc,out}}} \geq 100 \quad\Longrightarrow\quad (\mathtt{P}_{s,\text{Rc,out}})_{\text{dBm}} \geq (\mathtt{P}_{w,\text{Rc,out}})_{\text{dBm}} + 20 = 7.62\,\text{dB}\,.$$

By finally considering that $P_{s,Rc,out} = P_{Tx}\, g_{Ch,all}\, g_{Rc}$, in dBm we have

$$(P_{Tx})_{dBm} \geq (P_{s,Rc,out})_{dBm} - (g_{Ch,all})_{dB} - (g_{Rc})_{dB} = -8.38\,\text{dBm}\ .$$

4.4 TRANSMISSION MEDIA

The very general models presented so far can be particularized for specific transmission media, to identify their frequency response $\mathcal{G}_{Ch}(f)$ and power gain $g_{Ch}(f)$. In the following we review the basic characteristics of transmission lines, optical fibers, radio and underwater links.

4.4.1 Transmission lines and cables

An uniform transmission line consists of a two-conductor cable with a uniform cross section, that supports the propagation of electromagnetic waves [3, 4, 5]. Typical examples of transmission lines are *twisted pair* cables and *coaxial* cables.

A twisted-pair cable is the ordinary copper line connecting a customer premise to the telephone company. To reduce electromagnetic induction between pairs of wires, two insulated copper wires are twisted around each other (Fig. 4.16a). Twisted-pair is seldom enclosed in a shield that functions as a ground. This is known as *shielded twisted pair* (STP), while the ordinary wire is *unshielded twisted pair* (UTP).

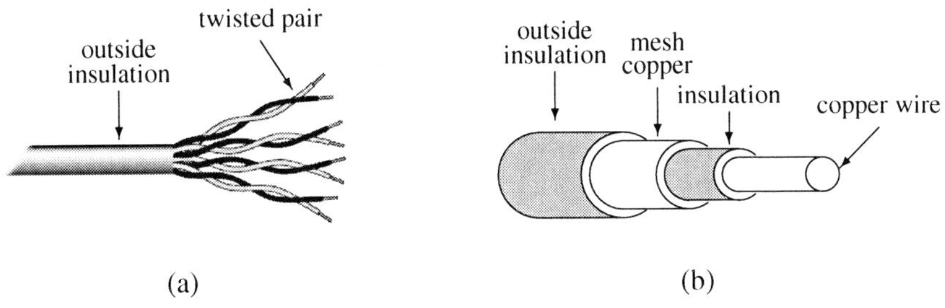

Figure 4.16 Transmission lines: (a) twisted pair; (b) coaxial cable.

A coaxial cable consists of a wire surrounded by insulation and a grounded shield (Fig. 4.16b). This configuration minimizes electrical and radio frequency interference. Coaxial cabling is primarily used by the cable television industry and is also widely used for computer networks. Although more expensive than standard telephone wire, it is much less susceptible to interference and can carry an higher data rate.

A widely used model for transmission lines is reported in Fig. 4.17, illustrating a uniform line segment of length dx where x denotes the distance from the origin. The parameters that determine the line characteristics are the *primary constants* r, ℓ, c and g which define, respectively, resistance, inductance, capacitance and conductance per unit length of the line.

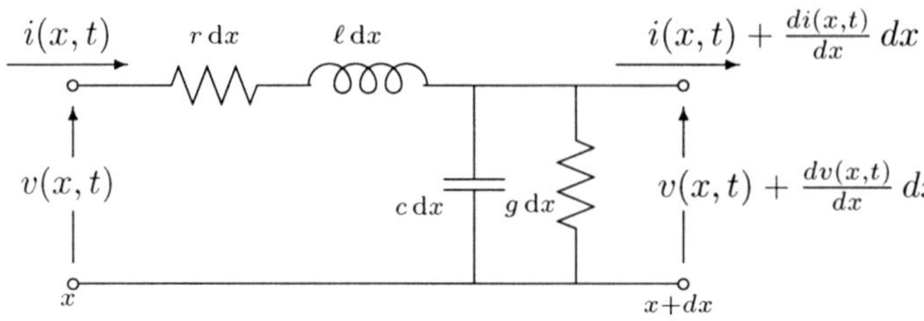

Figure 4.17 Transmission lines: line segment of infinitesimal length dx.

The input and output line impedances coincide, and are related to the *characteristic impedance* of the transmission line, given by

$$Z_0 = \sqrt{\frac{r + j2\pi f\ell}{g + j2\pi fc}} \tag{4.93}$$

A commonly used expression for the two-port network frequency response is

$$\mathcal{G}_{\text{Ch}}(f) = e^{-\gamma(f)\,d}\,, \qquad \gamma(f) = \alpha(f) + j\beta(f) \tag{4.94}$$

where d (for distance) is the length of the line and where α and β are the real and imaginary parts of γ, with α the *attenuation constant*, measured in neper per unit length, and β the *phase constant*, measured in radians per unit length. So, if the simplified relation (4.43) holds between available power gain and system frequency response, from (4.94) the power gain becomes

$$\mathtt{g}_{\text{Ch}}(f) = |\mathcal{G}_{\text{Ch}}(f)|^2 = e^{-2\alpha(f)\,d}\,. \tag{4.95}$$

All the above expressions are valid for both coaxial and twisted-pair cables insulated with plastic material, as long as we correctly choose the parameter values. Tab. 4.2 gives the characteristic quantities (experimentally measured [4]) for some telephone transmission lines characterized by their diameter, which is usually indicated by a parameter called *gauge*.

The most common model for γ takes into account the variations of r (resistance per unit length) with the frequency, due to the skin effect. With these assumptions, the attenuation and phase constants are of the form [3]

$$\begin{aligned} \alpha(f) &= K\,\sqrt{f} \qquad \text{[neper/m]}\,, \\ \beta(f) &= K\,\sqrt{f} + 2\pi f\,\sqrt{\ell c} \qquad \text{[rad/m]} \end{aligned} \tag{4.96}$$

where K is a further constant typical of the chosen line. Note that, both the attenuation constant $\alpha(f)$ and the phase constant $\beta(f)$ have a term that is proportional to the square root of frequency.

Using (4.96) in (4.94), the frequency response can also be written in the form

$$\boxed{\mathcal{G}_{\text{Ch}}(f) = \exp\left(-(1+j)\sqrt{f/f_c} - j2\pi f t_c\right)} \tag{4.97}$$

Table 4.2 Characteristic constants for some telephone lines.

gauge	diameter (mm)	frequency (kHz)	characteristic impedance Z_0 (Ω)	attenuation constant α (neper/km)	phase constant β (rad/km)	attenuation $\tilde{a}_{Ch} = 8.68\,\alpha$ (dB/km)
19	0.9119	1	$297 - j278$	0.09	0.09	0.78
		2	$217 - j190$	0.12	0.14	1.07
		3	$183 - j150$	0.15	0.18	1.27
22	0.6426	1	$414 - j401$	0.13	0.14	1.13
		2	$297 - j279$	0.18	0.19	1.57
		3	$247 - j224$	0.22	0.24	1.90
24	0.5105	1	$518 - j507$	0.16	0.17	1.43
		2	$370 - j355$	0.23	0.24	2.00
		3	$306 - j286$	0.28	0.30	2.42
26	0.4039	1	$654 - j645$	0.21	0.21	1.81
		2	$466 - j453$	0.29	0.30	2.55
		3	$383 - j367$	0.35	0.37	3.10

where $f_c = 1/(Kd)^2$ is the *characteristic frequency* and $t_c = d\sqrt{\ell c}$ is the *propagation delay*. From (4.97) we can derive the impulse response of the line as given by

$$g_{Ch}(t) = g_c(t - t_c)\,, \qquad g_c(t) = \frac{1}{2\pi\sqrt{f_c\,t^3}}\, e^{-1/(4\pi f_c t)}\, 1(t)\,. \tag{4.98}$$

The impulse response (4.98) is shown in Fig. 4.18 for some values of the product Kd and for $t_c = 0$. We note a large dispersion of $g_{Ch}(t)$ for increasing values of Kd, e.g., longer distances.

Incidentally, it is customary to provide the attenuation $\mathsf{a}_{Ch}(f) = 1/\mathsf{g}_{Ch}(f)$ as the parameter characterizing the line, taking also into account the fact that gain/attenuation and frequency response are closely related as in (4.95). Attenuation is provided either in dB or dB per unit length according to the expressions

$$\boxed{(\mathsf{a}_{Ch}(f))_{dB} = (\tilde{\mathsf{a}}_{Ch}(f))_{dB/km}\,(d)_{km}} \tag{4.99}$$

where from (4.95) the *specific attenuation* can be written as

$$(\tilde{\mathsf{a}}_{Ch}(f))_{dB/m} = 20\log_{10} e \cdot \alpha(f) = 8.68\,\alpha(f)\,. \tag{4.100}$$

Finally, in a linear scale

$$\mathsf{a}_{Ch}(f) = 10^{\frac{(\tilde{\mathsf{a}}_{Ch}(f))_{dB/km}\,(d)_{km}}{10}}\,. \tag{4.101}$$

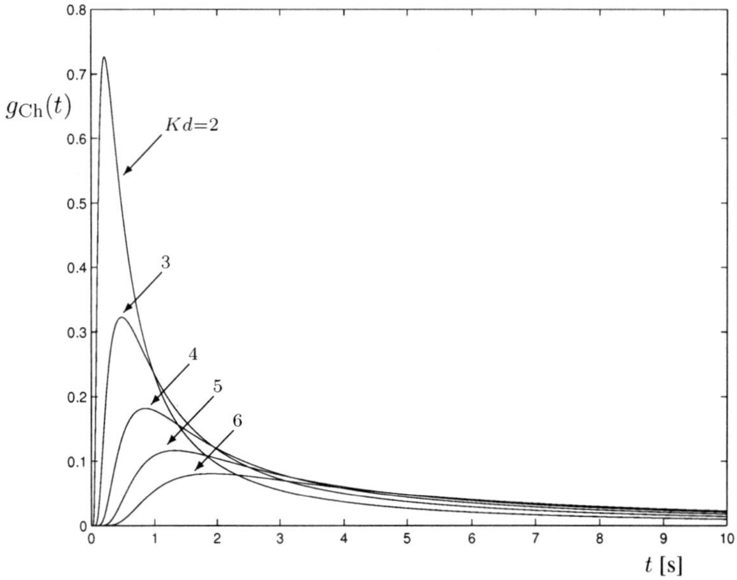

Figure 4.18 Impulse response of a matched transmission line for various values of Kd.

We note that, from (4.96), given the value of $\tilde{a}_{Ch}(f)$ for a certain frequency $f = f_1$, we can easily obtain the behavior at all frequencies by simply applying the relation

$$(\tilde{a}_{Ch}(f))_{dB/km} = (\tilde{a}_{Ch}(f_1))_{dB/km}\sqrt{f/f_1} \tag{4.102}$$

and similarly for $(a_{Ch}(f))_{dB}$.

Example 4.4 A A transmission system consists of a transmit amplifier, a cable and a receive amplifier, as illustrated in Fig. 4.19. The two amplifiers have gain $g_1 = g_2 = 20$ dB and noise figure $F_1 = F_2 = 10$ dB. The cable has an attenuation of 2 dB/km at the frequency $f_1 = 10$ kHz, while the transmitted signal is narrowband, with bandwidth $B = 10$ kHz, centered at the frequency $f_0 = 1$ MHz. We want to know what is the maximum length of the cable that guarantees a SNR at the receiver output greater than $\Lambda_{min} = 40$ dB for an average power before the transmit amplifier of $P_1 = 10$ mW.

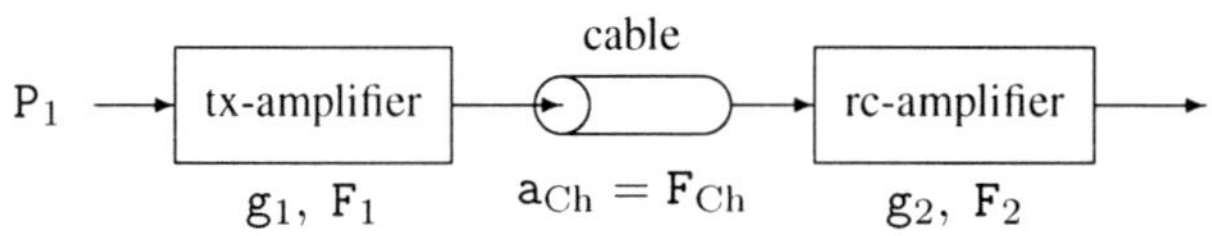

Figure 4.19 A simplified transmission system.

We first need to identify the cable attenuation a_{Ch} that guarantees the minimum required SNR. To do so we need to know the receiver effective noise temperature $T_{eff,Rc}$. Using the notation of Fig. 4.8, let

$T_{S,in}$ (T_S) be the equivalent input (output) noise temperature of the cascade transmit amplifier and cable. By exploiting Friis equation (4.64), we first determine $T_{S,in}$

$$T_{S,in} = T_0 + T_0 (F_1 - 1) + \frac{T_0 (a_{Ch} - 1)}{g_1} ,$$

where we exploited the fact that, being the transmission line a purely passive network, from (4.56) its noise figure coincides with the attenuation, that is $F_{Ch} = a_{Ch}$. Moreover, from (4.52)

$$T_S = \frac{g_1}{a_{Ch}} T_{S,in} ,$$

and the effective input noise temperature of the receiver is

$$T_{eff,Rc} = T_S + T_0 (F_2 - 1) = T_0 \left[\frac{g_1 F_1 - 1}{a_{Ch}} + F_2 \right] ,$$

which depends on a_{Ch}, to be determined. Also, the transmitted power is $(P_{Tx})_{dBm} = (P_1)_{dBm} + (g_1)_{dB} = 30\,\text{dBm}$. Now, being in the narrowband case we can exploit (4.91) to obtain the quantity

$$\begin{aligned} q &= 10 \log_{10} \left(\frac{T_{eff,Rc}}{T_0} a_{Ch} \right) \\ &= (P_{Tx})_{dBm} + 114 - 10 \log_{10}(B)_{MHz} - (\Lambda_{min})_{dB} \\ &= 30 + 114 + 20 - 40 = 124 . \end{aligned}$$

Since from the above expression

$$\frac{T_{eff,Rc}}{T_0} a_{Ch} = g_1 F_1 - 1 + F_2 a_{Ch} ,$$

we derive

$$a_{Ch} = \frac{1 + 10^{q/10} - g_1 F_1}{F_2} \simeq \frac{10^{q/10}}{F_2} \qquad \Longrightarrow \qquad (a_{Ch})_{dB} \simeq q - (F_2)_{dB} = 114\,\text{dB} .$$

Now, from (4.102), the specific attenuation of the cable is

$$(\tilde{a}_{Ch})_{dB/km} = (\tilde{a}_{Ch}(f_0))_{dB/km} = (\tilde{a}_{Ch}(f_1))_{dB/km} \sqrt{f_0/f_1} = 20\,\text{dB/km}$$

and (4.99) finally yields a maximum distance of

$$(d)_{km} = \frac{(a_{Ch})_{dB}}{(\tilde{a}_{Ch})_{dB/km}} = 5.7\,\text{km} .$$

Note that the transmit amplifier noise yields a negligible contribution to T_S and indeed $T_S \simeq T_0$ due to the cable noise.

Further insight A concluding aspect of transmission lines, not taken into account so far but of great relevance in practice, is interference. Interference is commonly referred to as *crosstalk*, and is determined by magnetic coupling and unbalanced capacitance between two adjacent transmission lines. Let us consider the two transmission lines of Fig. 4.20. The interference signals are called *near-end crosstalk* or NEXT, or *far-end crosstalk* or FEXT, depending on whether the receive side of the disturbed line is the same as the transmit side of the disturbing line, or the opposite side, respectively.

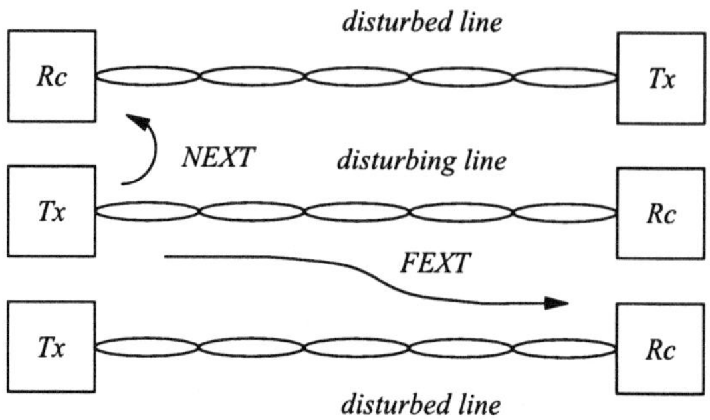

Figure 4.20 Transmission lines configuration for the study of crosstalk.

For example, the transmission characteristics of UTP cables used for data transmission over local area networks are divided into different categories according to the values of attenuation and NEXT, as reported in Tab. 4.3. Cables of category three (UTP-3) are commonly called *voice-grade*, those of categories four and five (UTP-4 and UTP-5) are *data-grade*. We note that the signal attenuation and the level of NEXT are substantially larger for UTP-3 cables than for UTP-4 and UTP-5 cables.

Table 4.3 Transmission characteristics for unshielded twisted pair (UTP) cables.

	signal attenuation at 16 MHz	NEXT attenuation at 16 MHz	characteristic impedance
UTP-3	13.15 dB/100 m	$\geq$ 23 dB	100 $\Omega \pm$ 15%
UTP-4	8.85 dB/100 m	$\geq$ 38 dB	100 $\Omega \pm$ 15%
UTP-5	8.20 dB/100 m	$\geq$ 44 dB	100 $\Omega \pm$ 15%

4.4.2 Power-line communications[1]

The term *power-line communications* (PLC) is used to indicate the transmission of data over existing electrical cables (power-line grid) superimposing a modulated higher frequency signal over the 50/60 Hz electrical signal.

The power-line grid, conventionally used for power distribution, has a very extensive infrastructure, connecting the power generation stations to a variety of users distributed over wide regions. Fig. 4.21 shows a typical European distribution grid, generally divided into three sections with different voltage levels: the high voltage, the medium voltage and the low voltage section. The high voltage lines, with voltages from 110 to 380 kV, cover distances of

[1]Paola Bisaglia, Simone Bois, and Eleonora Guerrini of Dora S.p.A., STMicroelectronics Group, and Daniele Veronesi of MGTech S.r.l., contributed to write this section.

tens of kilometers and connect the power generation stations to the distribution substations. The medium voltage lines, with voltages from 10 to 30 kV, cover distances of a few kilometers and connect the distribution stations to the low voltage transformers (aerial pole mounted or subterranean substations). Finally, the low voltage lines, with voltages from 200 to 400 V, cover distances of a few hundred meters and connect the low voltage transformers to individual users.

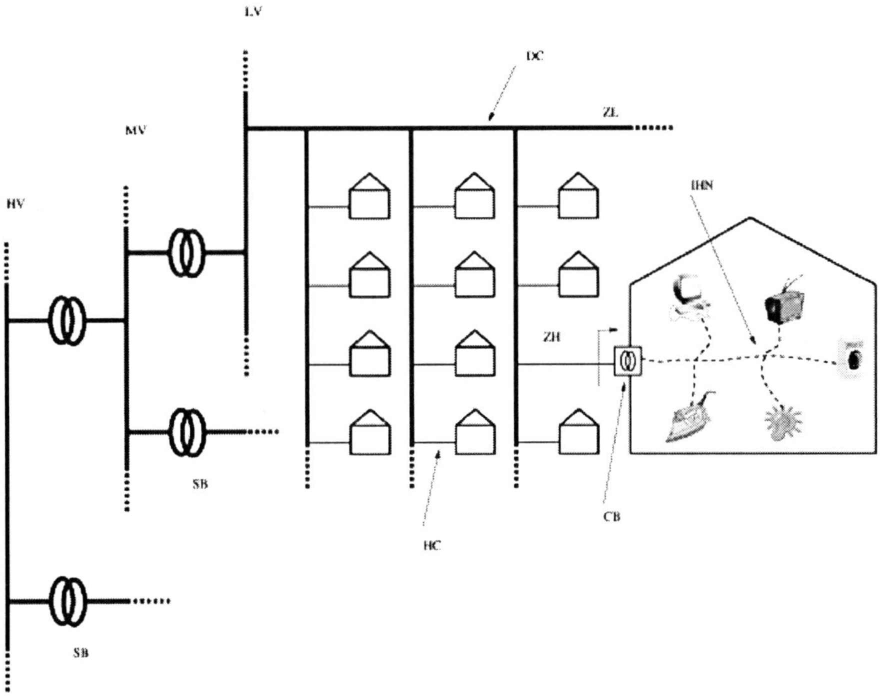

Figure 4.21 Typical topology of power distribution grid.

The idea of using electrical cables for low rate data transmission dates back many years ago. Indeed, in the 1920s power utilities in London started using power-lines to remotely control the grid. Since power-line has been devised for power transmission at 50/60 Hz, the use of this medium for data transmission at higher frequencies presents, however, several technical issues (e.g. high attenuation, different sources of noise), which make communication extremely difficult. Hence, for years the use of PLC has been limited to low rate data transmission such as signaling, remote monitoring of the electric grid and tele-metering. Recently, interest in medium to high data rate communications over low voltage lines has increased significantly. This growth has been fueled by several factors: the massive diffusion of Internet, technological advancements (i.e. very large scale integration, digital signal processing, error correction coding) and telecommunications market deregulation, first in the US and then in Europe and Asia.

The market for PLC is two-fold: to the home, referred as *access network* (or last mile access); and in the home, referred as *in-house network* or last inch access. Currently, the access network is only used for low rate data transmission such as automatic tele-metering and remote monitoring of the customer status parameters. However, in the near future, thanks to its ubiquity, the last mile power-line access could be the preferred medium for providing high rate broadband Internet connection to rural or remote areas where telephone and cable may not exist. The in-house network, thanks to its multiple sockets in each room, is used for providing a local area network to connect various electrical appliances allowing low rate home-automation (e.g. remote control and supervision of heating, air-conditioning and illumination, safety systems such as burglar or fire alarms). Furthermore, the most interesting opportunity for the in-house power-line network, is the broadband connection of PCs and their accessories (e.g. scanners, printers), sharing existing broadband Internet access and IP video distribution.

Low rate devices (up to 10 kbit/s), primarily used for tele-metering and home-automation, employ power-line in the range from 0 to 500 kHz. The use of these frequencies is restricted by the limitations imposed by the regulatory bodies, ensuring coexistence of various electro-magnetic devices in the same environment. Apart from band allocation, regulatory bodies also impose limits on the transmitted power and on out of band emissions. In particular, in Europe the European Committee for Electro technical Standardization (CENELEC) restricts the devices to operate in the frequency range from 3 to 148.5 kHz. This frequency range is further divided into several bands with different transmission purposes and restrictions. In North America regulatory rules, defined by the Federal Communications Commission (FCC), are less stringent in the use of the power-line from 0 to 500 kHz. The more significant products today available in the market for low rate data transmission over power-lines are: X-10, Intellon CEBus, Echelon LONWorks, Intelogis PLUG-IN, HomePlug CC, ST7538 and ST7540.

High rate devices (up to 150 Mbit/s), primarily used for broadband in-house distribution of data and multi-stream entertainment using existing power-line wiring, employ the unlicensed frequency band between 1 MHz and 30 MHz. In general, devices using this band need to mask certain frequency bands or to respect radiation emission limits in order to coexist with other communication systems using this portion of the spectrum, e.g. broadcast, citizen band and amateur radio frequencies. From a standardization point of view, several technologies for PLC compete among many consortiums worldwide: HomePlug, OPERA (a European-based consortium) and CEPCA (a Japanese-based consortium). The more significant products today available in the market for high rate data transmission over power-lines are: Inari IPL1201, DS2 DSS9101, Spidcom SPC200 and SPC300, Panasonic BL-PA100, HomePlug 1.0 and HomePlug AV (based on the IEEE P1901 standard).

Channel model In the literature, the power-line channel has been studied and modeled using two different approaches: i) *bottom-up*, or multiconductor transmission line, where the description of the network properties is based on its physical characteristics, i.e. cable geometry, losses, etc. [12, 13]; ii) *top-down* or multi-path, where a phenomenological or statistical description is given, based on on-field measurements [14, 15, 17]. For our purposes, the top-down approach is considered, since it is more general and it does not require a perfect knowledge of all line parameters.

Taking into consideration the network in Fig. 4.21, a multi-path power-line channel model, for both the low voltage access network and in-house network links, is now introduced. The link between the low voltage transformers and the users, referred as access network, consists of the distribution cable (bold line) and the branching house connection cables (thin line), both with a real valued characteristic impedance Z_L. The house connection cable ends at a house connection box, turning into the in-door wiring (dashed line), referred as in-house network. The house connection box and the following in-house network can be modeled as a complex terminal impedance $Z_H(f)$.

Along the path of the transmitted signal, each connection between cables yields a change of impedance and it causes a reflection in addition to the normal propagation. The same phenomenon occurs within the house, where several appliances with mismatched impedances are connected to the in-house network. In general, each impedance discontinuity, that occurs at every connection of two or more elements of the power-line network, give rise to both a reflection factor and a transmission factor. In definitive, due to branches and reflections, the signal not only propagates on the direct connection between a transmitter and a receiver, but also additional propagation paths or echoes have to be considered. The result is a *multi-path* signal propagation model [14, 15]. In this model, the low voltage power-line frequency response, for positive frequencies in the range from 500 kHz to 40 MHz, can be expressed as:

$$\mathcal{G}_{\mathrm{Ch}}(f) = \sum_{i=1}^{N} g_i \cdot A(f, d_i) \cdot e^{-j2\pi f \tau_i}, \qquad f > 0 \tag{4.103}$$

where

- N is the number of the propagation paths;
- g_i $(i = 1, 2, \ldots, N)$ is the weighting factor for the i-th path and is given by the product of all the reflection and the transmission factors along the path. In general it depends on the frequency. As all reflection and transmission factors are less or equal to 1, it is $|g_i| \leq 1$. The higher the number of reflections and transmissions is, the smaller the weighting factor $|g_i|$ is. Moreover, long paths have an higher attenuation and contribute less to the overall signal at the receiving point. Due to these considerations, it is reasonable to consider only the N more significant paths, instead of all the infinite paths that physically propagate in the network;
- $A(f, d_i)$ is the attenuation caused by the losses of the cable, and it increases with frequency f and with the length of the i-th path d_i. Based on physical considerations and extensive measurements, the attenuation can be expressed as

$$A(f, d_i) = e^{-(a_o + a_1 \cdot f^k) \cdot d_i} \tag{4.104}$$

  where a_0, a_1 and k are attenuation parameters that depend on the physical characteristics of the cables;
- τ_i is the time delay of the i-th path, expressed as

$$\tau_i = \frac{d_i \sqrt{\varepsilon_r}}{c} = \frac{d_i}{\nu_p} \tag{4.105}$$

  where $\nu_p = \frac{c}{\sqrt{\varepsilon_r}}$ is the phase velocity, ε_r is the dielectric constant of the insulating material of the cables and $c = 3 \cdot 10^8$ m/s is the speed of light in free-space propagation.

In (4.103) for $f < 0$, $\mathcal{G}_{\mathrm{Ch}}(f) = \mathcal{G}^*_{\mathrm{Ch}}(-f)$. Note that, if $A(f, d_i)$ is weakly frequency dependent, or better, assuming $g_i A(f, d_i)$ is a constant within the band of the transmitted signal, (4.103) can be transformed in the time domain. Let f_0 be the band center frequency of the useful signal and $h^{(a)}(t)$ be defined in the frequency domain as $H^{(a)}(f) = 2 \cdot 1(f)$. Then (4.103) yields a channel impulse response

$$g_{\mathrm{Ch}}(\tau) = \Re \left[\sum_{i=1}^{N} g_i A(f_0, d_i) \, h^{(a)}(\tau - \tau_i) \right] \tag{4.106}$$

where $g_i A(f_0, d_i)$ represents the gain of the i-th ray that arrives with delay τ_i.

To cover several real channel characteristics, different sets of the previously described parameters, detailed to access network or in-house network links, can be found in [14]. An example of typical values for an access network link is reported in Tab. 4.4 [16] while the respective frequency response $\mathcal{G}_{\mathrm{Ch}}(f)$ is depicted in Fig. 4.22. For comparison, a frequency response obtained by measurement is also reported. We can notice that the two curves have a very similar trend.

Table 4.4 Parameters of a four path channel model for an access power-line link [16].

Attenuation parameters

$k = 0.5$	$a_0 = 0$	$a_1 = 8 \cdot 10^{-6}$ s/m	$\varepsilon_r = 4$

Path parameters

i	g_i	d_i [m]	τ_i [μs]
1	0.4	150	1.0
2	-0.4	188	1.25
3	-0.8	264	1.76
4	-1.5	397	2.64

Besides multipath and attenuation, another critical aspect that makes the signal transmission over power-line particularly difficult is the noise. In general, five classes of noise can be defined [17]:

1. *Colored background noise*: the background noise is mainly caused by the summation of several noise sources with low power. The PSD is relatively low and it decreases with frequency. This noise can be considered stationary since it varies over time in terms of minutes or hours.

2. *Narrow-band noise*: it is mainly caused by the broadcast stations as well as radio services like amateur radio. It consists of several sinusoidal signals with modulated amplitudes. Their amplitudes vary over the daytime, becoming higher by night when the reflection properties of the atmosphere become stronger.

3. *Periodic impulsive noise asynchronous to the mains frequency*: it is mostly caused by switching power supplies. The spectrum is characterized by discrete lines with frequency spacing according to the repetition rate, which varies between 50 and 200 kHz. This noise is characterized by a very short time duration.

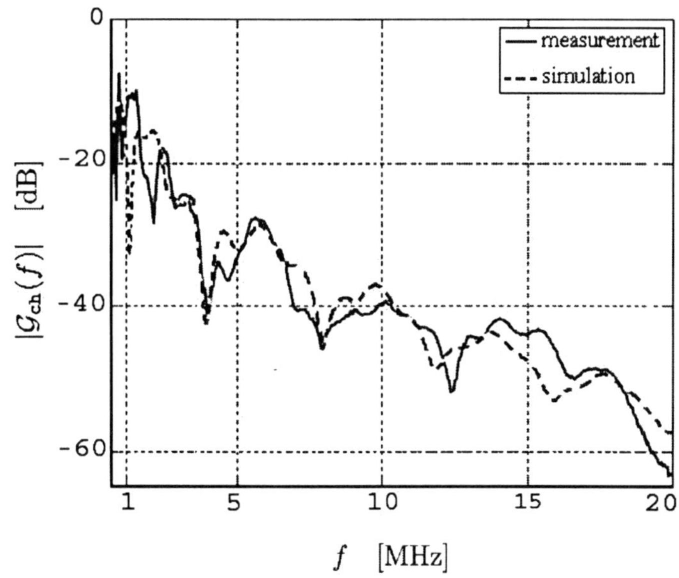

Figure 4.22 Example of an access power-line channel frequency response [16].

4. *Periodic impulsive noise synchronous to the mains frequency*: it is generally caused by power supplies operating synchronously with the main frequency, such as power converters connected to the mains supply. It consists of impulses of a repetition rate of 50 or 100 Hz with a short duration, in the order of μs.

5. *Asynchronous impulsive noise*: these impulses are caused by switching transients in the network. Their duration is of some μs up to a few ms with an arbitrary inter-arrival time. Their PSD can reach high values, making them the principal cause of error occurrences in the communication over power-line network.

4.4.3 Optical fibers

Transmission systems using light pulses that propagate over thin glass fibers were introduced in the 70's and have since then undergone a continuous development and experienced an increasing penetration, to the point that they now constitute a fundamental element of modern information highways. For in-depth study of optical fiber properties and of optical component characteristics we refer the reader to the vast literature existing on the subject [6, 7, 8]; in this section we limit ourselves to introducing some fundamental concepts.

The term *optical communications* is used to indicate the transmission of information by the propagation of electromagnetic fields at frequencies typically of the order of $10^{14} \div 10^{15}$ Hz, that are found in the optical band and are much higher than the frequencies of radio waves or microwaves; to identify a transmission band, the wavelength rather than the frequency is normally used. We recall that for electromagnetic wave propagation the relation $\lambda f = v$ holds, with v the speed of light in the medium, that can be expressed as $v = c/n$ with n the refractive index of the medium and $c = 3 \cdot 10^8$ m/s the speed of light in free-space propagation.

Although the propagation of electromagnetic fields in the atmosphere at these frequencies is also considered for transmission, the majority of optical communication systems employ as transmission medium an optical fiber. An optical fiber is a filament of transparent dielectric material, usually circular in cross section, that guides light. As shown in Fig. 4.23a, for the light to be guided by the fiber, the refractive index of the fiber core (n_1) must be slightly higher than that of the cladding (the material surrounding the core): that is, we require $n_1 > n_2$. In this case we talk of step-index (SI) fibers. To better compensate for light dispersion, fibers are often graded-index (GRIN), that is with a refractive index that decreases with distance from the radial axis, as shown in the example of Fig. 4.23b.

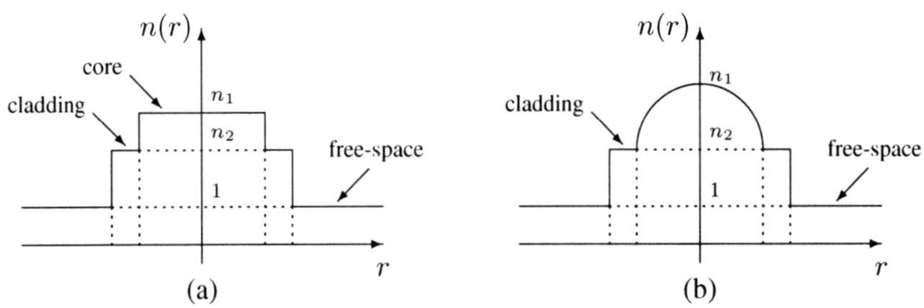

Figure 4.23 Radial behaviour of the refractive index in optical fibers: (a) step-index fiber, and (b) graded-index fiber.

Usually, fibers are made of glass with $n_1 \simeq 1.55$ or plastic (plastic optical fibers, POF) with $n_1 \simeq 1.35$. A wave with frequency $f = 3 \cdot 10^{14}$ Hz has therefore a wavelength $\lambda = c/f = 1\,\mu$m when propagating in free space, $\lambda_1 = \lambda/n_1 = 0.65\,\mu$m when propagating along a glass fibers, $\lambda_1 = 0.74\,\mu$m along a plastic fibers.

The signal attenuation as a function of the wavelength exhibits the behavior shown in Fig. 4.24 [9, 10]; we note that the useful interval for transmission is in the range from 800 to 1600 nm, that corresponds to a bandwidth of $2 \cdot 10^{14}$ Hz. Three regions (windows) are typically used for transmission, where the attenuation shows a minimum: the first window goes from 800 to 900 nm, the second from 1250 to 1350 nm, and the third from 1500 to 1600 nm. From Fig. 4.24 we immediately realize the enormous capacity of fiber transmission systems: for example, a system that uses only 1% of the $2 \cdot 10^{14}$ Hz bandwidth mentioned above, has an available bandwidth of $2 \cdot 10^{12}$ Hz, equivalent to that needed for the transmission of approximately 300000 television signals, each with a bandwidth of 6 MHz. To efficiently use the band, *multiplexing* techniques using optical devices have been developed, such as *wavelength-division multiplexing* (WDM) and *optical frequency-division multiplexing* (O-FDM).

A typical configuration for an optical communication system is presented in Fig. 4.25.

The signal $s_{\rm Tx}(t)$ that we wish to transmit is typically an electrical signal, that is converted into an optical signal by a proper light emitting source. This source is often a semiconductor *laser diode* (LD) or a *light-emitting diode* (LED), and is seldom followed by a *fiber optic coupler*. The conversion from an electrical signal to an electromagnetic field that propagates

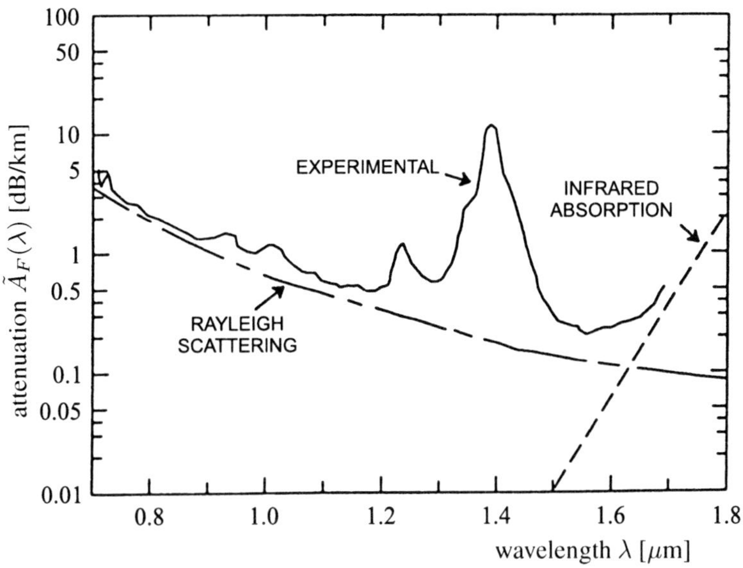

Figure 4.24 Attenuation curve as a function of wavelength for an optical fiber.

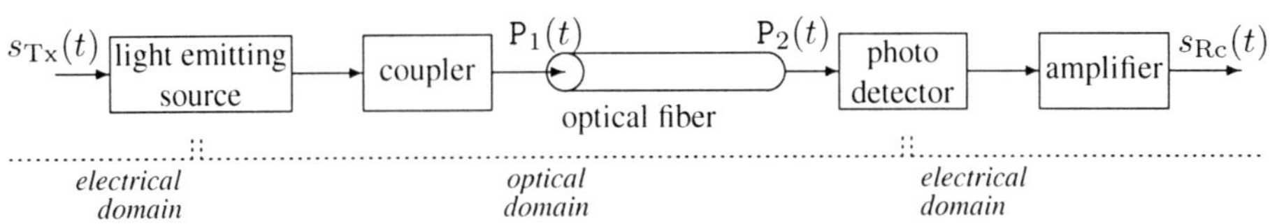

Figure 4.25 Optical transmission system.

along the fiber can be described in terms of the light signal power by the relation

$$P_1(t) = k_0 + k_1\, s_{Tx}(t) \tag{4.107}$$

where k_0 and k_1 are suitable constants that guarantee $P_1(t) \geq 0$. Therefore the transmitted signal can be seen as a replica of the modulation signal, provided that the linear relation (4.107) is satisfied. Note that $P_1(t)$ is the light power radiated through the fiber at a fixed wavelength λ typical of the chosen LD or LED.

The transmission along the optical fiber suffers from some dispersion with "spreading" of the transmitted waveforms $P_1(t)$. This phenomenon in turn causes interference between adjacent transmitted pulses and limits the available bandwidth of the transmission medium.

At the receiver, the more widely used photo detector devices are semiconductor photo diodes, which convert the optical signal into an electrical signal.

Apart from a constant additive term due to k_0 in (4.107), which can be neglected in practice, the overall frequency response of the system relating s_{Tx} and s_{Rc} is

$$\mathcal{G}_{\text{Ch}}(f) = A_F^{-1} \exp\left[-2\left(\pi f \sigma_F\right)^2\right] \exp(-j2\pi f t_F) \tag{4.108}$$

with A_F an attenuation, σ_F a dispersion coefficient (in seconds), and t_F the propagation delay. Thus the available power gain is given by

$$\mathtt{g}_{\text{Ch}}(f) = |\mathcal{G}_{\text{Ch}}(f)|^2 = A_F^{-2} \exp\left[-4\left(\pi f \sigma_F\right)^2\right] \tag{4.109}$$

if we assume that input and output impedances coincide as in (4.42). All parameters in (4.108) and (4.109) depend upon the fiber length d as described in the following.

Delay For the propagation delay we simply have

$$t_F = \frac{d}{v} \tag{4.110}$$

where $v = c/n$ is the speed of light in the fiber.

Attenuation For the attenuation in dB $(A_F)_{\text{dB}}$, a linear relation with the fiber length can be found, where we have

$$(A_F(\lambda))_{\text{dB}} = 20 \log_{10} A_F(\lambda) = (\tilde{A}_F(\lambda))_{\text{dB/km}}\ (d)_{\text{km}} \tag{4.111}$$

and depends upon the emitted wavelength λ. A typical behavior of $(\tilde{A}_F(\lambda))_{\text{dB/km}}$ was shown in Fig. 4.24.

Dispersion The dispersion coefficient σ_F depends both on the geometry and the material of the waveguide, and is directly proportional to the fiber length. We can write

$$\sigma_F = \tilde{\sigma}_F\, d\,, \qquad \tilde{\sigma}_F = (M_m + M_g)\, \Delta\lambda \tag{4.112}$$

where $\tilde{\sigma}_F$ is the dispersion per unit length, M_m is a dispersion coefficient of the material, M_g is a dispersion coefficient related to the geometry of the waveguide, and $\Delta\lambda$ denotes the spectral width of the light source. Laser diodes are characterized by a smaller $\Delta\lambda$ as compared to that of LEDs, and therefore lead to a lower dispersion.

The *total dispersion* $(M_m + M_g)$ has typical values about of 120 and 15 ps/(nm×km) at wavelengths of 850 and 1550 nm, respectively. For conventional fibers the total dispersion is minimum in the second window, with values near zero around the wavelength of 1300 nm.

Note that the bandwidth of the transmission medium is inversely proportional to σ_F. In Tab. 4.5 typical values of the bandwidth, normalized by the length of the optical fiber, are given for different types of fibers. Special fibers are also designed to compensate for the dispersion introduced by the material; because of the low attenuation and dispersion, these fibers are normally used in very long distance connections.

Table 4.5 Characteristic parameters of various types of optical fibers.

fiber	wavelength [nm]	source	bandwidth at $d = 1$ km [MHz]
multimode SI	850	LED	30
multimode GRIN	850	LD	500
multimode GRIN	1300	LD o LED	1000
monomode	1300	LD	> 10000
monomode	1550	LD	> 10000

Example 4.4 B An optical fiber connects Paris to Milan (approximately 700 km). It has a specific attenuation $(\tilde{A}_F(\lambda))_{\text{dB/km}} = 0.1$ dB/km and a dispersion coefficient $\tilde{\sigma}_F = 1$ ps/km. We assume that the photo-detector at the receiver can be modeled as a resistance at noise temperature $\mathtt{T}_S = 400$ K, and is followed by a matched impedance amplifier with noise figure $(\mathtt{F}_A)_{\text{dB}} = 5$ dB. The transmitted signal is approximately narrowband, with bandwidth $B = 100$ kHz centered at $f_0 = 1$ MHz. We want to know what transmitted power is required to assure an SNR of 20 dB at the receive amplifier output.

The optical fiber power gain is given by (4.109), or equivalently from (4.111) and (4.112)

$$(\mathtt{g}_{\text{Ch}})_{\text{dB}} = -(\tilde{A}_F)_{\text{dB/km}}\,(d)_{\text{km}} - 40\,(\log_{10} e)\left(\pi f_0\,(\tilde{\sigma}_F)_{\text{s/km}}\,(d)_{\text{km}}\right)^2 = -91\text{ dB}\,.$$

Now, by use of the effective receiver noise temperature

$$\mathtt{T}_{\text{eff,Rc}} = \mathtt{T}_S + \mathtt{T}_0(\mathtt{F}_A - 1) = 1027\text{ K}\,,$$

in the link budget relation (4.91) we derive

$$\begin{aligned}(\mathtt{P}_{\text{Tx}})_{\text{dBm}} &= (\Lambda)_{\text{dB}} - (\mathtt{g}_{\text{Ch}})_{\text{dB}} - 114 + 10\,\log_{10}\left(\frac{\mathtt{T}_{\text{eff,Rc}}}{\mathtt{T}_0}\right) + 10\,\log_{10}(B)_{\text{MHz}}\\ &= 20 + 91 - 114 + 5.5 - 10 = -7.5\,\text{dBm}\,.\end{aligned}$$

4.4.4 Radio links

The term *radio* is used to indicate the transmission of an electromagnetic field that propagates in free space. Some examples of radio transmission systems are: point-to-point terrestrial links [11]; mobile terrestrial communication systems [18, 19, 20, 21, 22]; earth-satellite links, with satellites employed as signal repeaters [23]; deep-space communication systems, with space probes at a large distance from earth.

Frequencies used for radio transmission are in the range from about 100 kHz to some tens of GHz. The choice of the carrier frequency depends on various factors, among which the dimensions of the transmit antenna play an important role. In fact, to achieve an efficient radiation of electromagnetic energy, one of the dimensions of the antenna must be at least equal to $\frac{1}{10}$ of the carrier wavelength. This means that an AM radio station, with carrier frequency $f_0 = 1\,\text{MHz}$ and wavelength $\lambda = \frac{c}{f_0} = 300\,\text{m}$, where we recall c is the speed of light in free space, requires an antenna of at least 30 m.

A radio wave usually propagates as a ground wave (or surface wave), via *reflection* and *scattering* in the atmosphere (or via tropospheric scattering), or as a direct wave. Recall that, if the atmosphere is non homogeneous (in terms of temperature, pressure, humidity, etc.), the electromagnetic propagation depends on the changes of the refraction index of the medium. In particular, this gives origin to the reflection of electromagnetic waves. We speak of diffusion or scattering phenomena if the molecules of the atmosphere absorb part of the power of the incident wave and then re-emit it in all directions. Obstacles such as mountains, buildings, etc., give also origin to signal reflection and/or diffusion. In any case, these are phenomena that may permit transmission between two points that are not in *line-of-sight* (LOS).

We will now consider the types of propagation associated with frequency bands. We have

- *Very low frequency* (VLF) for $f_0 < 0.3\,\text{MHz}$. The earth and the ionosphere form a waveguide for the electromagnetic waves. At these frequencies the signals propagate around the earth.

- *Medium frequency* (MF) for $0.3 < f_0 < 3\,\text{MHz}$. The waves propagate as ground waves up to a distance of 160 km.

- *High frequency* (HF) for $3 < f_0 < 30\,\text{MHz}$. The waves are reflected by the ionosphere at an altitude that may vary between 50 and 400 km.

- *Very high frequency* (VHF) for $30 < f_0 < 300\,\text{MHz}$. At these frequencies, the signal propagates through the ionosphere with small attenuation. Therefore these frequencies are adopted for satellite communications. They are also employed for line-of-sight transmissions, using high towers where the antennas are positioned to *cover* a wide area. The limit to the coverage is set by the earth curvature. If h is the height of the tower in meters, the range covered expressed in km is $r = 1.3\sqrt{h}$: for example, if $h = 100\,\text{m}$, coverage is up to about $r = 13\,\text{km}$. However, ionospheric and tropospheric scattering (at an altitude of 16 km or less) are present at frequencies in the range 30–60 MHz and 40–300 MHz, respectively, which cause the signal to propagate over long distances with large attenuation.

- *Ultra high frequency* (UHF) for $300\,\text{MHz} < f_0 < 3\,\text{GHz}$.

- *Super high frequency* (SHF) for $3 < f_0 < 30\,\text{GHz}$. At frequencies of about 10 GHz, atmospheric conditions play an important role in signal propagation. We note the following *absorption phenomena*, which cause additional signal attenuation: 1) due to oxygen: for $f_0 > 30\,\text{GHz}$, with peak attenuation at 60 GHz; 2) due to water vapor: for $f_0 > 20\,\text{GHz}$, with peak attenuation at around 20 GHz; 3) due to rain: for $f_0 > 10\,\text{GHz}$, assuming the diameter of the rain drops is of the order of the signal wavelength. We note that, if the antennas are not positioned high enough above the ground, the electromagnetic field propagates not only into the free space but also through ground waves.

- *Extremely high frequency* (EHF) for $f_0 > 30\,\text{GHz}$.

A brief illustration of the frequency bands associated with most common transmission systems is given in Fig. 4.26. Note that lower frequencies are devoted to radio broadcasting, VHF and UHF bands to the transmission of the analog television and the *digital video broadcasting* (DVB). UHF is also associated with cellular transmission and wireless *local area network* (LAN) technologies, while higher frequencies are used for satellite communications and also for deep space communications.

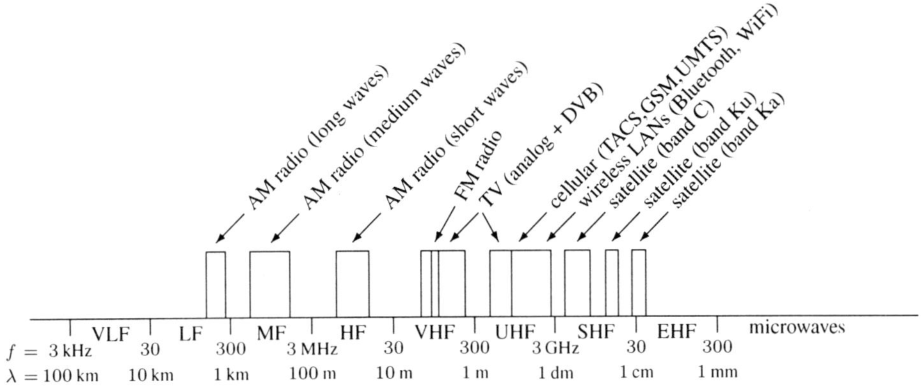

Figure 4.26 Overview of frequency bands and related transmission systems.

The propagation of electromagnetic waves through space should be studied using Maxwell equations with appropriate boundary conditions. Nevertheless, for our purposes a very simple model, which consists in approximating an electromagnetic wave as a ray (in the optical sense) is often adequate. This model is known as *narrowband radio channel model*, and it is in fact valid for transmit signals with a very narrow bandwidth B, centered around the carrier frequency f_0. This deterministic model is used to evaluate the power of the received signal when there are no obstacles between the transmitter and receiver, that is in the presence of line of sight. In this case we can think of only one wave that propagates from the transmitter to the receiver. This situation is typical of transmissions between satellites and terrestrial radio stations in the microwave frequency range ($3 < f_0 < 70$ GHz). In any case, we will not take into account the attenuation due to rain or other environmental factors such as signal reflections, nor the possibility that the antennas may not be correctly positioned: these conditions require more sophisticated channel models.

Narrowband channel model Let P_{Tx} be the power of the signal transmitted by an ideal *isotropic antenna*, which uniformly radiates in all directions in the free space. At a distance d from the antenna, the power density is

$$\Phi_0 = \frac{\mathrm{P}_{\mathrm{Tx}}}{4\pi d^2} \qquad [\mathrm{W/m^2}] , \tag{4.113}$$

where $4\pi d^2$ is the surface of a sphere of radius d that is uniformly illuminated by the antenna. We observe that the power density decreases with the square of the distance. On a logarithmic scale (dB) this is equivalent to a decrease of 20 dB-per-decade with the distance. In the case of a *directional antenna*, the power density is concentrated within a cone and is given by

$$\Phi = \Phi_0 \, \mathrm{g}_{\mathrm{Ant,Tx}} = \frac{\mathrm{P}_{\mathrm{Tx}} \, \mathrm{g}_{\mathrm{Ant,Tx}}}{4\pi d^2} , \tag{4.114}$$

where $g_{Ant,Tx}$ is the transmit antenna power gain. Obviously, $g_{Ant,Tx} = 1$ for an isotropic antenna and, usually, $g_{Ant,Tx} \gg 1$ for a directional antenna.

At the receive antenna, the *available power* in conditions of matched impedance is given by

$$P_{Rc} = \Phi \, A_{Ant,Rc} \, \eta_{Ant,Rc} \, , \tag{4.115}$$

where P_{Rc} is the received power, $A_{Ant,Rc}$ is the *effective area* of the receive antenna and $\eta_{Ant,Rc}$ is the *efficiency* of the receive antenna. The factor $\eta_{Ant,Rc} < 1$ takes into account the fact that the antenna does not capture all the incident radiation, because a part is reflected or lost. To conclude, the power of the received signal is given by

$$P_{Rc} = P_{Tx} \frac{A_{Ant,Rc}}{4\pi d^2} \, g_{Ant,Tx} \, \eta_{Ant,Rc} \, . \tag{4.116}$$

To obtain a simpler expression of (4.116), we recall that the antenna gain is usually expressed as [3]

$$g_{Ant} = \frac{4\pi \, A_{Ant}}{\lambda^2} \eta_{Ant} = \frac{4\pi \, A_{Ant} \, f_0^2}{c^2} \eta_{Ant} \, , \tag{4.117}$$

where A_{Ant} is the effective area of the antenna, $\lambda = c/f_0$ is the wavelength of the transmitted signal, f_0 is the carrier frequency and η_{Ant} is the efficiency factor. Equation (4.117) holds for the transmit as well as for the receive antenna. We also note that, because of the factor $A_{Ant} \, f_0^2$, working at higher frequencies presents the advantage of being able to use smaller antennas, for a given g_{Ant}. We also underline that, usually, $\eta \in [0.5, 0.6]$ for *parabolic* antennas, while $\eta \simeq 0.8$ for *horn* antennas.

Thus, observing (4.117) and (4.116), we get

$$g_{Ch} = \frac{P_{Rc}}{P_{Tx}} = g_{Ant,Tx} \, g_{Ant,Rc} \left(\frac{\lambda}{4\pi d} \right)^2 , \qquad \lambda = c/f_0 \tag{4.118}$$

which is known as the *Friis transmission equation* and is valid in conditions of maximum transfer of power. The term $(\lambda/(4\pi d))^2$ is called *free space path loss*. The *available attenuation* of the medium is thus

$$a_{Ch} = \frac{(4\pi d/\lambda)^2}{g_{Ant,Tx} \, g_{Ant,Rc}} \tag{4.119}$$

or, expressed in dB,

$$\boxed{(a_{Ch})_{dB} = 32.4 + 20 \log_{10}(d)_{km} + 20 \log_{10}(f_0)_{MHz} - (g_{Ant,Tx})_{dB} - (g_{Ant,Rc})_{dB}} \tag{4.120}$$

where $32.4 = 20 \log_{10}(4\pi/c)$, d is expressed in km, f_0 in MHz, and $g_{Ant,Tx}$ and $g_{Ant,Rc}$ in dB. For $g_{Ant,Tx} = g_{Ant,Rc} = 1$, the attenuation a_{Ch} coincides with the free space path loss.

When input and output system impedances coincide, from (4.88) and (4.84), we obtain the radio link frequency response

$$\mathcal{G}_{Ch}(f) = \sqrt{g_{Ch}} \, \exp(-j2\pi f t_0) \, , \tag{4.121}$$

where t_0 is the propagation delay, that can be typically expressed as $t_0 = d/c$ with c the speed of light.

Remark It is worthwhile to note that the attenuation in dB $(a_{Ch})_{dB}$, expressed by (4.120), increases with distance as $\log_{10} d$ and with frequency as $\log_{10} f_0$, whereas for metallic transmission lines the dependency on distance is linear (see (4.99)) and on frequency follows a square-root rule $\sqrt{f_0}$ (see (4.102)). So, radio transmission is a much more efficient approach than cable connection in terms of dissipated power over a covered distance. However, the bandwidth of a radio link is occupied over a vast geographical area (even with directive antennas) since electromagnetic waves are not confined within any medium. This is the reason why, historically, radio links have been mainly used for radio and television broadcasting purposes.

Example 4.4 C A local radio station transmits over a bandwidth $B = 60\,\text{kHz}$ centered around the frequency $f_0 = 90\,\text{MHz}$ with an average power $P_{Tx} = 0.1\,\text{W}$. The receiver consists of an amplifier with gain $(g_A)_{dB} = 20\,\text{dB}$ and noise figure $(F_A)_{dB} = 7\,\text{dB}$. What is the radius of the covered area if the minimum SNR at the receiver output that guarantees adequate reception is 40 dB?

The received signal power is $P_{Rc} = P_{Tx}/a_{Ch}$ where a_{Ch} is given by (4.120) with $(g_{Ant,Tx})_{dB} = (g_{Ant,Rc})_{dB} = 0$. The effective noise temperature at the amplifier input is

$$T_{eff,Rc} = T_0 + T_0(F_A - 1) = T_0 F_A$$

where we assumed $T_S = T_0$. Hence, the SNR at the receiver output is given by (4.90), that is

$$(\Lambda)_{dB} = 20 - (a_{Ch})_{dB} + 114 - 7 + 12.2 = 139.2 - (a_{Ch})_{dB} \,.$$

Since we require $(\Lambda)_{dB} \geq (\Lambda_{min})_{dB} = 40\,\text{dB}$, we have

$$(a_{Ch})_{dB} \leq 139.2 - (\Lambda_{min})_{dB} = 99.2\,\text{dB} \,.$$

So, from (4.120) the covering radius is

$$20 \log_{10}(d)_{km} \leq 99.2 - 32.4 - 39.1 = 27.7 \quad \Longrightarrow \quad d \leq 24.26\,\text{km} \,.$$

Example 4.4 D We want to design a radio transmission system between two stations at distance $d = 3\,\text{km}$ using a narrowband signal with center frequency $f_0 = 500\,\text{MHz}$ and transmitted power $(P_{Tx})_{dBm} = -20\,\text{dBm}$. The system must be such that: 1) the received power is at least $-85\,\text{dBm}$; 2) the radiated power is at most $50\,\mu\text{W/st}$ (st = steradians, the measure for solid angles) in the LOS direction; 3) the effective area of the receive antenna is minimized (assume $\eta_{Ant,Rc} = 1$). What are the required effective area of the receive antenna $A_{Ant,Rc}$ and transmit antenna gain $g_{Ant,Tx}$?

Now, from (4.117) and (4.118) we have

$$P_{Rc} = P_{Tx}\, g_{Ant,Tx} \left(\frac{4\pi\, A_{Ant,Rc}\, f_0^2}{c^2} \right) \left(\frac{c}{4\pi d\, f_0} \right)^2 = P_{Tx}\, g_{Ant,Tx} \frac{A_{Ant,Rc}}{4\pi\, d^2} \geq P_{min}$$

where $(P_{min})_{dB} = -85\,\text{dBm}$. In logarithmic scale we have

$$(g_{Ant,Tx})_{dB} + 10\, \log_{10}(A_{Ant,Rc})_{m^2} \geq (P_{min})_{dBm} + 10\, \log_{10}(4\pi\, d^2) - (P_{Tx})_{dBm} = 15.5 \,.$$

The further constraint given by 2) implies a limitation in the emitted power along the LOS direction. Being 4π st the maximum solid angle, the limitation reads as (compare with (4.114))

$$\frac{P_{Tx}\, g_{Ant,Tx}}{4\pi} \leq \tilde{P}_{max} = 50\,\mu\text{W/st}$$

which in logarithmic scale yields

$$(\mathtt{g}_{\mathrm{Ant,Tx}})_{\mathrm{dB}} \leq 10\,\log_{10} 4\pi\,(\tilde{\mathtt{P}}_{\mathrm{max}})_{\mathrm{mW/st}} - (\mathtt{P}_{\mathrm{Tx}})_{\mathrm{dBm}} = -2 + 20 = 18\,\mathrm{dB}\ .$$

The minimum effective area is thus obtained by using the maximum allowable gain $(\mathtt{g}_{\mathrm{Ant,Tx}})_{\mathrm{dB}} = 18$ dB, and we have

$$10\,\log_{10}(\mathtt{A}_{\mathrm{Ant,Rc}})_{\mathrm{m}^2} = 15.5 - 18 \qquad \Longrightarrow \qquad \mathtt{A}_{\mathrm{Ant,Rc}} = 0.56\,\mathrm{m}^2\ .$$

Wideband channel model In general, radio propagation is described by a model which includes, besides attenuation, also distortion on the transmitted signal. In fact the channel frequency response $\mathcal{G}_{\mathrm{Ch}}(f)$ cannot be considered constant but over very narrow bands, hence we talk of a frequency selective channel.

In practice, in many radio systems, and especially in wireless indoor and mobile radio communications, propagation takes place by means of several different paths, characterized each by a different delay and a different attenuation. This gives a channel impulse response of the type (4.106), where the distinctive parameters of the model are set by the propagation environment, such as indoor, outdoor, urban, vehicular, etc.

4.4.5 Underwater acoustic propagation[2]

Historical Notes In this section we give a high level overview of the way sound propagates in water, a topic that is currently gaining importance within the telecommunications community. The knowledge that sound may in fact propagate beneath the sea surface is due to Leonardo da Vinci, who wrote in 1490 that "If you cause your ship to stop, and place the head of a long tube in the water, and place the outer extremity to your ear, you will hear ships at a great distance from you." This first example of *passive sonar* does not give any indication of the direction of the target and suffers from acoustical mismatch between the water and the air; however, it is very similar to the systems used in World War I, whereby a second tube was added so that a listener could infer the bearing of a target. Since Leonardo's statement, a number of discoveries have more or less directly contributed to the development of underwater sound systems. The first accurate measurement of the propagation speed of sound in water is due to Daniel Colladon, a Swiss physicist, and Charles Sturm, a French mathematician. In 1827, they inferred the speed of sound by measuring the difference between the time of arrival of a flash of light and the striking of an underwater bell. Other discoveries, such as magnetostriction[3] by James Joule in the 1840s, and piezoelectricity[4] by Pierre and Jacques Curie in 1880, laid the foundation for the construction of modern electroacoustic transducers and hydrophones.

The first practical application of underwater sound came, however, at the and of the eighteenth century, when an underwater bell-based ranging system was invented: by timing the difference between the sound of the bell and the sound of a simultaneously rang horn, a ship could determine its distance from the ship or platform where both sound sources were

[2]Paolo Casari of DEI, University of Padova, contributed to write this section.
[3]The change of shape in a substance induced by a magnetic field.
[4]The ability of certain crystals to develop electric charges across certain pair of faces when posed under stress.

installed. In 1912, Richardson filed a patent with the British Patent Office which involved using airborne sound for echo ranging; a month later, he also filed a patent for the underwater counterpart of his system, which is among the ancestors of recent sonars. Unfortunately, the "Titanic" passenger ship could not benefit from this technology, as it collided with an iceberg five days before the first patent application was filed. The first to come out with a prototype implementing Richardson's ideas was Fessenden, in the United States, who invented a similar system, actually working at lower acoustic frequency than Richardson's. This, as we will see, allows signals to travel longer distances. Before 1914, Fessenden demonstrated that his sonar could detect an iceberg two miles away, which favored the installation of his system on US submarines during World War I. The outbreak of the war gave a big boost to underwater acoustics research, until systems using sound echoes allowed to automatically detect steel plates at a distance of more than one kilometer. By the time the World War II begun, large-scale production of relatively cheap sonar equipment had started, and submarines had been outfitted with this technology.

One of the most notable discoveries in underwater acoustics, however, came in the years between the two World Wars, as at-sea experiments on echo ranging showed unpredictably changing performance. Good echoes were often obtained in the morning, while poor if any echoes were received in the afternoon. This "afternoon effect" as it was named by Stephenson, was discovered to be originated by slight temperature gradients, which were able to refract sound waves, curving their trajectory and creating "shadow zones" where acoustic signals did not propagate at all. Unveiling such a peculiar aspect of underwater propagation has represented a key outbreak among other discoveries in this field.

Since the end of the Wars, the exploitation of underwater sound in civilian applications has become widespread, and many instruments for, e.g., navigation, beaconing, fish finding, diver aiding, wave height measurement, telemetry and communication are commercially available as of today.

Channel model The use of sound pressure waves for digital underwater communications poses a number of challenges. A first difficulty comes from the propagation speed of sound waves, approximately equal to 1500 m/s, which is five orders of magnitude smaller than the speed of radio waves in the air. Actually it depends on depth (i.e., pressure), temperature, and local salinity level. Moreover, sound attenuation phenomena are different than those experienced by radio waves, and critically depend on the frequency of the signal being transmitted, so that, e.g., lower-frequency signals are less attenuated and can travel longer distances. Also, shallow waters act as waveguides, and cause sound to propagate in a cylindrical rather than spherical geometry. A third macroscopic difference is in the sources of noise that affect communications, whose PSD is frequency-dependent, or in other words, the channel noise is colored. Last, the complicated refraction patterns of sound in water cause very strong multipath phenomena and the constant movement of the transmitter and receiver equipment (e.g., due to marine currents) originated relevant Doppler shifts with respect to the carrier frequency used for communication. These challenges have fostered a number of contributions in the field of signal processing [24], which allowed the setup of reliable communication links between two nodes, thereby releasing the first impulse to research in the field of underwater acoustic networks [25].[5] In this section we will review the fundamental equations that allow an approximated link budget.

The system model is depicted in Fig. 4.27.

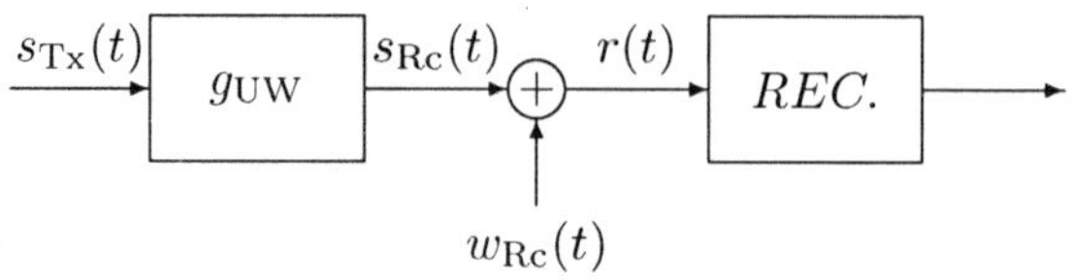

Figure 4.27 Model of the underwater transmission system.

The transmitted signal $s_{\rm Tx}(t)$ is an acoustic waveform, characterized by a statistical power $\mathtt{M}_{s_{\rm Tx}} = E[s_{\rm Tx}^2(t)]$ represented by a pressure, measured in Pascal (Pa). In many practical cases the transmitted signal is assumed to be a sinusoidal acoustic tone of frequency f_0. The underwater link has length d. In order to translate acoustic power into electrical power the following relation is applied [26]:

$$(\mathtt{P}_{\rm Tx})_{\rm dBm} = (\mathtt{M}_{s_{\rm Tx}})_{{\rm dB}\mu{\rm Pa}} - 17.2 - 10\log_{10}(\eta)\ , \tag{4.122}$$

where -17.2 is the conversion factor from acoustic power in dBμPa to electrical power, and η is the overall efficiency of the electronic circuitry (power amplifier and transducer). A typical value for this parameter is $\eta = 0.25$.

The attenuation $\mathtt{a}_{\rm Ch}(d, f)$ that a tone of frequency f_0 incurs can be expressed as the product of two components as [26]

$$\mathtt{a}_{\rm Ch}(d, f_0) = d^k \cdot [\alpha(f_0)]^d\ , \tag{4.123}$$

where the term d^k is called the *spreading loss* and depends only on the geometry of the propagation, whereas the term $[\alpha(f_0)]^d$ is called the *absorption loss*, and models the conversion of acoustic pressure into heat due to the resonance with certain ions present in the water. The factor k is called the *spreading coefficient*, and is the counterpart of the path loss exponent in terrestrial radio; it is used to define the geometry of the propagation (*i.e.*, $k = 1$ is cylindrical, $k = 2$ is spherical, and $k = 1.5$ is a practical value used to represent mixed propagation conditions). The *absorption coefficient* $\alpha(f)$ has been empirically derived by Thorp and can be expressed as [26]

$$10\log_{10}\alpha(f) = 0.11\frac{f^2}{1+f^2} + 44\frac{f^2}{4100+f^2} + 2.75\cdot 10^{-4} f^2 + 0.003\ . \tag{4.124}$$

(4.124) returns $\alpha(f)$ in dB/km for f in kHz. Fig. 4.28 plots $\alpha(f)$ in a broad range of frequencies. The absorption coefficient is the major factor that limits the power received at a given distance as it increases very rapidly with frequency. Its presence is the most significant difference with respect to radio attenuation (see (4.120)). Note that (4.123) describes the attenuation on a single, unobstructed propagation path. Therefore, if a tone of frequency f_0 and power $\mathtt{M}_{s_{\rm Tx}}$ is transmitted over a distance d, the received signal power will be $\mathtt{M}_{s_{\rm Tx}}/\mathtt{a}_{\rm Ch}(d, f_0)$.

[5]It is worth noting that optic as well as electromagnetic waves have also been considered for digital underwater communications: however, optic systems bear a very low range and require to keep the transmitter and the receiver

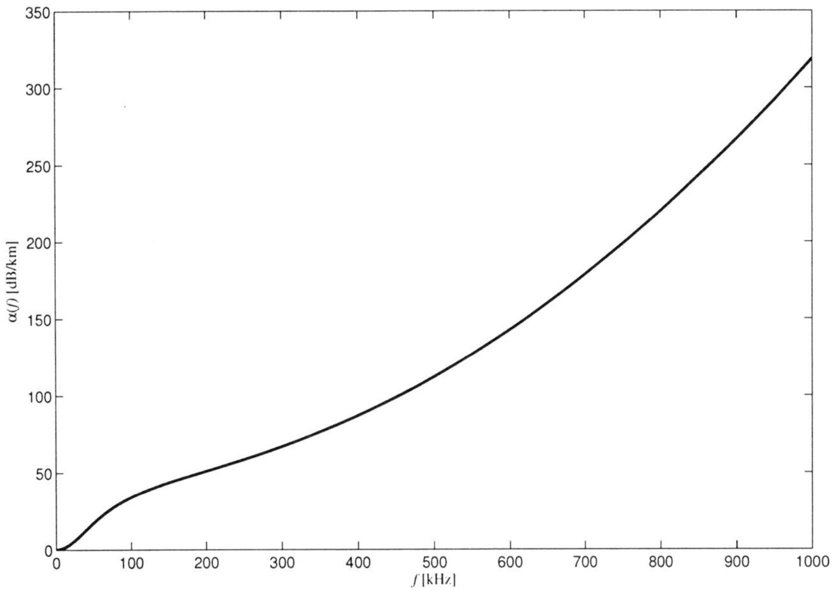

Figure 4.28 Behavior of the absorption coefficient $a(f)$.

The effective noise $w_{\mathrm{Ch}}(t)$ in the ocean can be modeled as the superposition of four components: turbulence, shipping, waves and thermal noise. Each component is assumed to have a Gaussian statistic and is described by a continuous suitable PSD. By using the subscript t for turbulence, s for shipping, w for waves, and th for thermal noise (at the receiver input), for each noise component the following formulae return the corresponding PSD in dBμPa/Hz,

$$\begin{aligned}
10\log_{10}\mathcal{P}_{w_t}(f) &= 17 - 30\log_{10}(f)\\
10\log_{10}\mathcal{P}_{w_s}(f) &= 40 + 20(S-0.5) + 26\log_{10}(f) - 60\log_{10}(f+0.03)\\
10\log_{10}\mathcal{P}_{w_w}(f) &= 50 + 7.5\sqrt{W} + 20\log_{10}(f) - 40\log_{10}(f+0.4)\\
10\log_{10}\mathcal{P}_{w_{th}}(f) &= -15 + 20\log_{10}(f)\,.
\end{aligned} \tag{4.125}$$

meaning that turning each equation to linear scale yields a PSD in μPa/Hz. In (4.125), S is the so called *shipping factor*, representing the intensity of shipping activities on the surface of the water, and takes values ranging between 0 and 1, while W is the wind speed in $\mathrm{m/s}$. The total noise PSD is finally expressed as $\mathcal{P}_{w_{\mathrm{Rc}}}(f) = \mathcal{P}_{w_t}(f) + \mathcal{P}_{w_s}(f) + \mathcal{P}_{w_w}(f) + \mathcal{P}_{w_{th}}(f)$. Fig. 4.29 plots $\mathcal{P}_{w_{\mathrm{Rc}}}(f)$ for different values of the shipping factor ($S = 0$, 0.5, and 1) and of the wind speed ($W = 0$ and 10 m/s). From this figure, we observe that each noise component impacts the PSD at different frequencies. In particular, turbulence influences the very low frequency region, for $f < 10$ Hz. Shipping noise has greater importance in the 10 Hz–100 Hz

aligned for best performance; electromagnetic waves, instead, achieve a wider coverage range, but can be effectively used only at very low frequencies, which greatly reduces the bandwidth of the system, and thus the system bit rate.

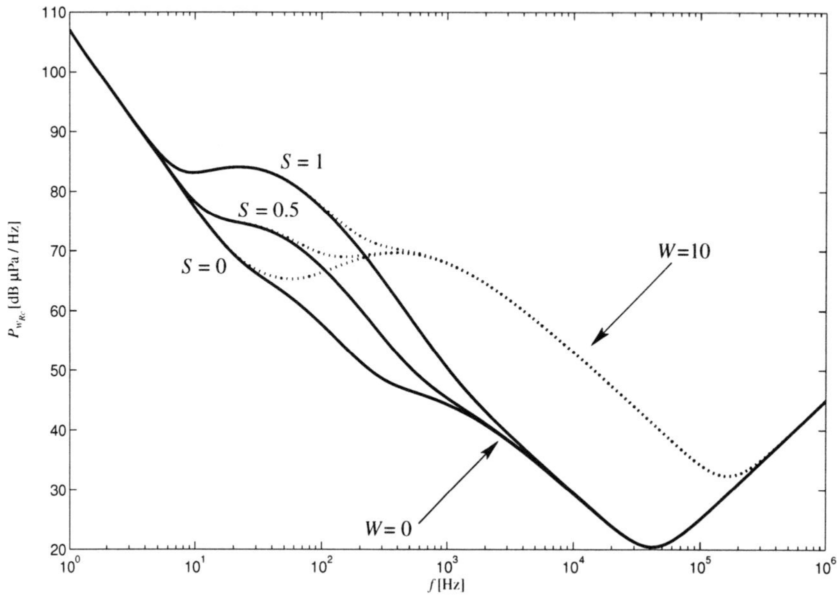

Figure 4.29 Noise power spectral density against f, for different values of the shipping factor S and wind speed W.

region, whereas wind impacts the frequency portion from 100 Hz to 100 kHz. Finally thermal noise is the dominant component for $f > 100\,\text{kHz}$.

By using the attenuation and noise equations, we can figure out the reference signal-to-noise ratio Γ associated to a tone of frequency f_0 and power $\mathtt{M}_{s\text{Tx}}$ received at a distance d from the source (assume a system bandwidth B):

$$\Gamma(d, f_0) = \frac{\mathtt{M}_{s\text{Tx}}}{\mathcal{P}_{w\text{Rc}}(f_0) 2B\, \mathtt{a}_{\text{Ch}}(d, f_0)} \tag{4.126}$$

where the noise PSD $\mathcal{P}_{w\text{Rc}}(f)$ is assumed constant within the bandwidth B around f_0. In (4.126), the factor $1/[\mathtt{a}_{\text{Ch}}(d, f)\mathcal{P}_{w\text{Rc}}(f)]$ is the frequency-dependent term. Since $\mathtt{a}_{\text{Ch}}(d, f)$ increases with frequency while $\mathcal{P}_{w\text{Rc}}(f)$ decreases at least to a certain point, the product between the two has a maximum for some frequency f_{opt}. This maximum yields minimal combined attenuation and noise effects, hence it is the best frequency to use for transmission. In order to depict this phenomenon, Fig. 4.30 shows a number of concave gray lines that represent the factor $[\mathtt{a}_{\text{Ch}}(d, f)\mathcal{P}_{w\text{Rc}}(f)]^{-1}$ as a function of frequency with d as a parameter varying from 10 m to 100 km.

The first observation is that each curve exhibits a maximum, showing the existence of an optimal frequency $f_{\text{opt}}(d)$ to use for transmission for each link distance d. We note that $f_{\text{opt}}(d)$ decreases for increasing distances, explaining why Fessenden's low frequency echo ranging system introduced in the first part of this section performed better than Richardson's. Another important observation can be made by defining the acoustic bandwidth $B_{\text{Ch}}(d)$ around $f_{\text{opt}}(d)$ according to a -3 dB definition on the Γ value, i.e., let f_1 and f_2, with $f_1 < f_{\text{opt}}(d)$ and

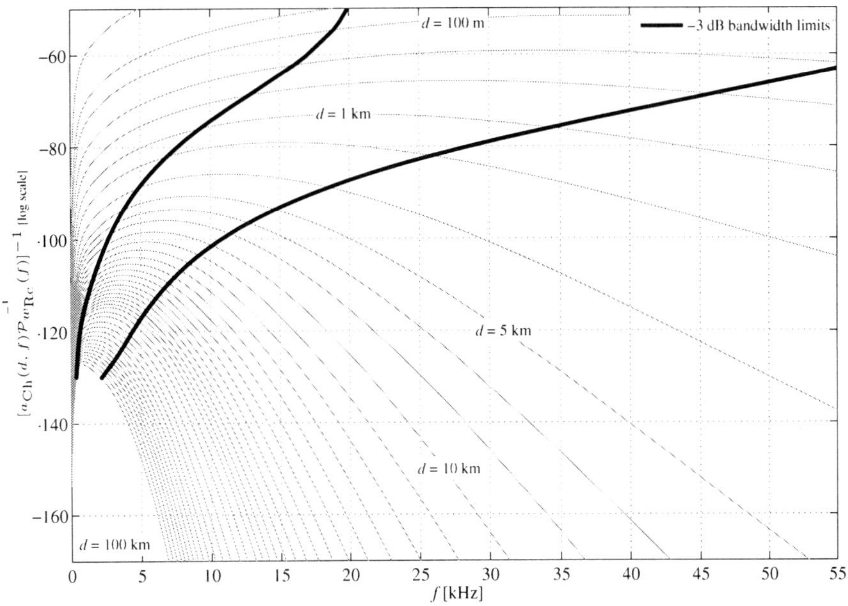

Figure 4.30 Frequency-dependent part of Γ for an acoustic tone transmitted underwater. Grey lines represent the $[a_{Ch}(d,f)\mathcal{P}_{w_{Rc}}(f)]^{-1}$ factor for different distances. The two black lines represent the lower and upper limit of the available channel band.

$f_2 > f_{pt}(d)$, such that $\Gamma(d, f_2) = \Gamma(d, f_1) = \frac{1}{2}\Gamma(d, f_{opt}(d))$, it is $B_{Ch}(d) = f_2 - f_1$. Values of f_1 and f_2 which define the channel band, for various values of d, are shown in Fig. 4.30 by means of two black lines. Notably, the available bandwidth shrinks for increasing distances. This is different from what happens in radio and must be carefully accounted for in the design of network protocols. For example, performing a few long-range hops in a multi-hop path would require to transmit at very high power (because of the large attenuation) and at low bit rate (because of the small bandwidth), thus increasing the energy consumption considerably.

Example 4.4 E Based on the above equations, we infer that the acoustic power that needs to be transmitted in order to meet a prescribed SNR at the receiver depends on the distance. To fix ideas, assume that we require a reference SNR at the receiver input Γ of 20 dB. Let us also assume that our system can perfectly adapt the transmitted signal band to the optimal channel band of Fig. 4.30. Moreover, we assume the band of the transmitted signal contained within that of the channel, hence we can approximate Γ as a constant and equal to $\Gamma(d, f_{opt}(d))$ which we denote simply as $\Gamma(d)$. Finally, assume that $d = 1$ km.

The first step is to find the central optimal transmit frequency $f_{opt}(d)$, for the given distance $d = 1$ km, found by combining (4.124)–(4.126) (see Fig. 4.30) to achieve the maximum SNR, giving $f_{opt}(d) = 20.164$ kHz. Next, from Fig. 4.30 the 3 dB bandwidth that the transmit signal can occupy is $B = B_{Ch} = 25.113$ kHz. For this bandwidth B, the statistical power $\mathtt{M}_{s_{Tx}}$ of the transmitted acoustic signal is otained by inverting (4.126), and gives $\mathtt{M}_{s_{Tx}} = 5.02 \cdot 10^{10}$ μPa, or equivalently 107 dBμPa. Lastly, the transmit electrical power $\mathtt{P}_{Tx}(d)$ is derived by (4.122), and yields $\mathtt{P}_{Tx}(d) = 1.3\,\mu$W.

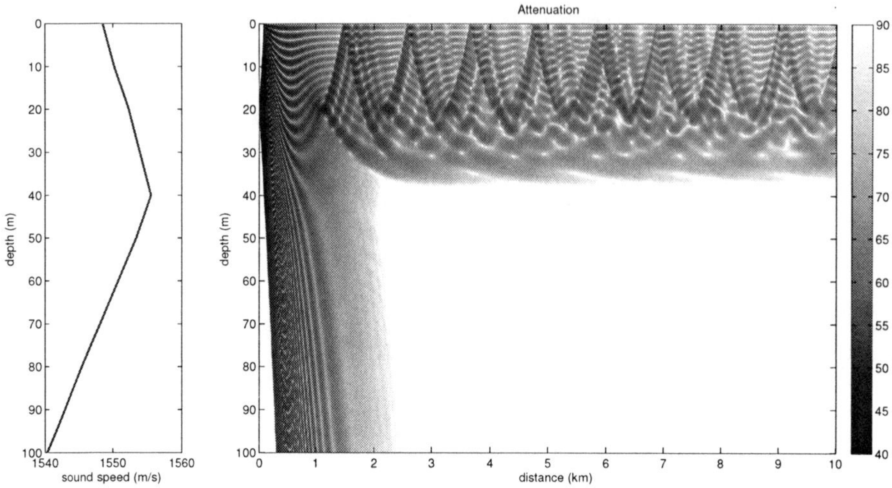

Figure 4.31 Shallow water channel: (left) sound speed profile and (right) sound attenuation profile showing the presence of a "shadow zone" after 2500 m and below 38 m of depth.

A few details on underwater propagation While the above equations allow to set up a coarse link budget, underwater propagation is more complicated than what expressed by the model considered so far. It is outside the scope of this section to delve into the details of this matter, but it is worth mentioning how the main physical features of water impact on the propagation of acoustic pressure waves. Let us recall that the speed of sound is approximately equal to 1500 m/s but actually depends on temperature, depth and water salinity. A typical procedure involves sampling these three quantities at various depths, and then calculate the speed of sound as a function of depth. This function is called the sound speed profile (SSP). The SSP influences the way sound waves are refracted (i.e., curved) while propagating, and is thus a factor of primary importance. Different SSPs yield different propagation profiles, which in turn may give rise to strange phenomena such as the "afternoon effect" described in the first part of this section, whereby the intensity of the received acoustic signal dramatically varies as a function of the time of the day.

Figs. 4.31 and 4.31 refer to, respectively, a shallow-water scenario and a deep water scenario. Each picture contains on the left a plot of the sound speed profile, with in ordinate the depth and in abscissa the corresponding sound speed. Instead, the plots on the right side depict, by a gray shade, the attenuation (in dB) incurred by the transmitted signal as a function of distance from the transmitter (abscissa) and depth (ordinate). Hence, a darker gray represents lower attenuation: white portions are hence regions where no signal is actually received. Note that, the transmitter is placed at a depth of 20 m in Fig. 4.31 and of 1000 m in Fig. 4.32, and it emits sounds toward the right.

Fig. 4.31 highlights a phenomenon known as "shadow zone" and, due to the way sound is refracted, no signal reaches the area below 38 m of depth, after a distance of roughly

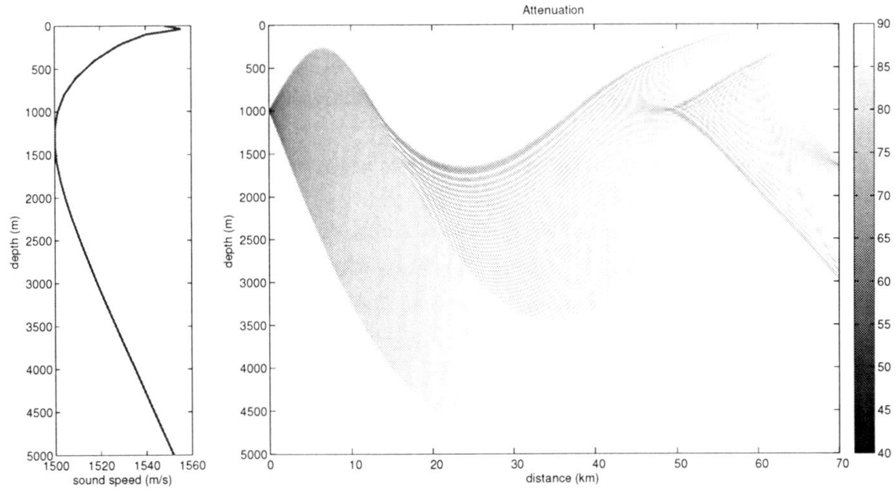

Figure 4.32 Deep water channel: (left) sound speed profile and (right) sound attenuation profile showing the presence of a "convergence zone" at a distance of 50000 m and a depth of 1000 m.

2500 m. The propagation pattern is also complicated by surface reflections, which contribute to originating multiple copies of the transmitted signal, which in turn superimpose differently in different locations of space, causing attenuation to vary with respect to the average value provided by (4.123).

Fig. 4.32 reports an example of sound propagation in deep water (a graph of this type was plotted for the first time during World War II), and shows a phenomenon known as "deep sound axis." In this figure, the SSP represents the conditions of water in spring, at around 43° of latitude, in a point mid-way between Newfoundland and Great Britain [26]. In this case, we see that after having been transmitted, the sound waves get trapped in the water and never reach the surface again. Furthermore, we observe the presence of a *convergence zone* at a distance of 50000 m and a depth of 1000 m, i.e., a zone where different propagation paths converge to yield locally higher signal power.

Figs. 4.31 and 4.32 give a clue of how complicated underwater propagation actually is: indeed the most accurate reproduction of sound propagation can only be obtained by solving the wave equations, e.g., by ray tracing (as done in figures). Moreover, the problem of summarizing all propagation effects into statistical phenomena such as fading is still an open issue. In fact, very large number of factors affect the way sound propagates, for which no exhaustive statistical models currently exist. However, the general link budget equations reported earlier in this section can help understand at least the average behavior of underwater communications, and have been used to date in a wealth of papers on underwater networking.

REFERENCES

1. L. O. Chua, C. A. Desoer, E. S. Kuh, *Linear and nonlinear circuits*, McGraw - Hill International Edition 1987.

2. A. Agarwal, and J. H. Lang, *Foundations of analog and digital electronic circuits.* Elsevier, 2005.

3. G. C. Someda, *Electromagnetic waves.* London: Chapman & Hall, 1998.

4. M. of the Technical Staff, *Transmission systems for communications.* Winston, NC: Bell Telephone Laboratories, 5th ed., 1982.

5. S. Ramo, J.R. Whinnery, and T. Van Duzer, *Fields and waves in communication electronics.* New York: John Wiley & Sons, 1965.

6. R. J Hoss, *Fiber optic communications.* Englewood Cliffs, NJ: Prentice Hall, 1990.

7. L. B. Jeunhomme, *Single-mode fiber optics.* New York: Marcel Dekker, 2nd ed., 1990.

8. J. C. Palais, *Fiber optic communications.* Englewood Cliffs, NJ: Prentice Hall, 3rd ed., 1992.

9. T. Li, "Structures, parameters, and transmission properties of optical fibers," *Proc. IEEE*, pp. 1175–1180, Oct. 1980.

10. T. Miya, Y. Terunuma, T. Hosaka, and T. Miyashita, "Ultimate low-loss single-mode fiber at 1.55 μm," *Proc. IEE Electronics Letters*, vol. 15, pp. 106-108, Feb. 1979.

11. D. G. Messerschmitt, and E.A. Lee, *Digital communication.* Boston, MA: Kluwer Academic Publishers, 2nd ed., 1994.

12. T. Banwell and S. Galli, "A novel approach to the modeling of the indoor power line channel part I: circuit analysis and companion model," *IEEE Trans. on Power Delivery*, vol. 20, pp. 655-663, Apr. 2005.

13. S. Galli, and T. Banwell, "A novel approach to the modeling of the indoor power line channel-Part II: transfer function and its properties," *IEEE Trans. on Power Delivery*, vol. 20, pp. 1869-1878, July 2005.

14. M. Babic, M. Hagenau, K. Dostert and J. Bausch, "Theoretical postulation of PLC channel model," *Tech. Rep. OPERA,* March 2005.

15. H. Philipps, "Modelling of powerline communication channels," *ISPLC 1999*, pp. 14-21, Lancaster UK, March 1999.

16. P.J. Langfeld, M. Zimmermann and K. Dostert, "Power line communication system design strategies for local loop access," *Proceedings of the Workshop Kommunikationstechnik, Technical Report ITUU-TR-1999/02*, pp. 21-26, July 1999.

17. M. Zimmermann and K. Dostert, "Analysis and modeling of impulsive noise in broad-band powerline communications," *IEEE Trans. on Electr. Compat.*, vol. 44, pp. 249-258, Feb. 2002.

18. K. Feher, *Wireless digital communications.* Upper Saddle River, NJ: Prentice Hall, 1995.

19. T. S. Rappaport, *Wireless communications: principles and practice.* Englewood Cliffs, NJ: Prentice Hall, 1996.

20. K. Pahlavan, and A.H. Levesque, *Wireless information networks.* New York: John Wiley & Sons, 1995.

21. W. C. Jakes, *Microwave mobile communications.* New York: IEEE Press, 1993.

22. G. L. Stuber, *Principles of mobile communications.* Norwell, MA: Kluwer Academic Publishers, 1996.

23. J. J. Spilker, *Digital communications by satellite.* Englewood Cliffs, NJ: Prentice Hall, 1977.

24. M. Stojanovic, "Recent advances in high-speed underwater acoustic communications", *IEEE Journal of Oceanic Engineering*, Vol. 21, No, 2, pp. 125–136, Apr. 1996.

25. E. M. Sozer, M. Stojanovic, and J. G. Proakis, "Underwater acoustic networks", *IEEE Journal of Oceanic Engineering*, Vol. 25, No, 1, pp. 72–83, Jan. 2000.

26. R. Urick, *Principles of Underwater Sound*, New York: McGraw - Hill, 1983.

Problems

Two-terminal devices

4.1 Calculate the following:

a) $(\mathrm{P}_1)_{\mathrm{mW}} = 0.10\,\mathrm{mW}$ in dBm.
b) $(\mathrm{P}_2)_{\mathrm{dBm}} = 40\,\mathrm{dBm}$ in W.
c) $(\mathrm{P}_3)_{\mathrm{dBm}} = -30\,\mathrm{dBm}$ in dBrn.
d) $(\mathrm{P}_4)_{\mathrm{dBm}} = -40\,\mathrm{dBm}$ in dBW.
e) $(\mathrm{P}_5)_{\mathrm{pW}} = 30\,\mathrm{pW}$ in dBrn.

4.2 Let a matched two-terminal device have open circuit source voltage

$$v_i(t) = A\left[1 + \cos(2\pi f_c t)\right] \cos(2\pi f_0 t)$$

with $f_0 = 10\, f_c$. Let also the device be closed on a resistive load $R_L = 100\,\Omega$. What is the value of A in order to guarantee a maximum dissipated power at the load $(\mathrm{P}_{\max})_{\mathrm{dBm}} = 5\,\mathrm{dBm}$?

4.3 A power $(\mathrm{P}_v)_{\mathrm{dBm}} = -16\,\mathrm{dBm}$ is measured across a resistive load $R_L = 500\,\Omega$ for the signal $v_L(t) = V_0 \sin(2\pi f_0 t)$ with frequency $f_0 = 500\,\mathrm{Hz}$. Express the signal amplitude V_0 in V and dBm, where $(V_0)_{\mathrm{dBm}} = 20\,\log_{10}(V_0)_{\mathrm{mV}}$.

4.4 Let the source PSD of a two-terminal device, closed on a load, be

$$\mathcal{P}_{v_i}(f) = V_0 \left|\frac{f}{B}\right| \mathrm{rect}\left(\frac{f}{2B}\right) ,$$

with $(V_0)_{\mathrm{V}^2/\mathrm{kHz}} = 2\,\mathrm{V}^2/\mathrm{kHz}$ and $B = 750\,\mathrm{Hz}$. Knowing that the source impedance is $Z_S = 100 + j50$ and that the system is matched, what is the average power P_v and the power density $\mathrm{p}_v(f)$ at the load?

Two-port networks

4.5 Determine the noise figure of a passive network at temperature $\mathrm{T}_1 \neq \mathrm{T}_0$.

4.6 An antenna for radio transmission is pointed to a direction where the equivalent noise temperature of the sky is 50 K. The antenna output is feeding an amplifier with bandwidth $B = 5\,\mathrm{MHz}$, gain $(\mathrm{g}_\mathrm{A})_{\mathrm{dB}} = 40\,\mathrm{dB}$, and noise figure $(\mathrm{F}_\mathrm{A})_{\mathrm{dB}} = 5\,\mathrm{dB}$. The receiver is matched and the load at the output is purely resistive with $R_L = 500\,\Omega$. Determine:

a) The amplifier noise temperature and the effective input temperature of the receiver.
b) The average noise power at the system output, both in mW and dBm.
c) The standard deviation of the noise voltage across the load.

4.7 A two-port linear passband network has a bandwidth $B = 50\,\mathrm{kHz}$. The network has a 6 dB noise figure, it has equal input and output impedances $Z_1 = Z_2 = 50\,\Omega$ (purely resistive), and it is perfectly matched. Calculate the input signal power (in dBm) and the corresponding source *statistical* power (in V^2) that guarantees an output SNR of 40 dB.

4.8 Consider a receiver consisting on the cascade of: 1) an antenna with noise temperature $\mathrm{T}_a = 50\,\mathrm{K}$; 2) an amplifier, directly connected to the antenna, with gain $(\mathrm{g}_1)_{\mathrm{dB}} = 20\,\mathrm{dB}$ and noise temperature $\mathrm{T}_1 = 40\,\mathrm{K}$; 3) a second amplifier with gain $(\mathrm{g}_2)_{\mathrm{dB}} = 30\,\mathrm{dB}$ and noise figure $(\mathrm{F}_2)_{\mathrm{dB}} = 6\,\mathrm{dB}$; 4) a device consisting of a converter to an intermediate frequency and of a series of amplifiers, with a total noise figure $(\mathrm{F}_3)_{\mathrm{dB}} = 15\,\mathrm{dB}$. Assume each device be matched. Determine:

a) The noise figure of the receiver.

b) The average power (in dBm) of the input signal, $P_{s,in}$, to guarantee an output SNR of 30 dB (assume the system bandwidth $B = 100$ kHz).

c) How do the above results change in case the second amplifier at item 3) is removed from the receiver?

4.9 If we want to minimize the noise power at the system output, is it more convenient to use a cascade of N *passive* two-port networks each at temperature T_1 and gain g_1, or a unique passive two-port network at temperature T_1 and gain $(g_1)^N$?

4.10 If our aim is to minimize the noise power at the system output, is it more convenient to use a single amplifier with noise figure F and gain g, or the cascade of N amplifiers each with noise figure F and total (of the cascade) gain g?

Transmission lines and cables

4.11 A coaxial cable of length $d = 10$ m has an attenuation of 5.3 dB/km at the frequency $f_1 = 1$ MHz. Determine the cable constant K (see (4.96)) and the attenuation for a narrowband signal centered at $f_0 = 20$ MHz.

4.12 Consider a transmission line at temperature $T = 315$ K, with characteristic impedance Z_0 and power attenuation $(a_{Ch})_{dB} = 20$ dB, in cascade with an amplifier, having noise figure $(F_A)_{dB} = 7$ dB, power gain $(g_A)_{dB} = (a_{Ch})_{dB}$, and impedances $Z_1 = Z_2 = Z_0^*$ matched to the line. Determine:

a) The noise figure of: 1) the cascade line-amplifier, 2) the cascade amplifier-line.

b) The input signal power $P_{s,in}$ required in the two cited cases to guarantee an SNR of 40 dB at the output of the system (assume the noise temperature of the source equal to the standard temperature T_0 and the signal bandwidth $B = 10$ kHz).

4.13 A twisted-pair telephone cable with characteristic impedance $Z_0 = 300\,\Omega$, is connected to a load $Z_L = 300\,\Omega$. The telephone line is $d = 200$ km long and exhibits a specific attenuation of $(\tilde{a}_{Ch})_{dB/km} = 2$ dB/km. Determine:

a) The average power at the load with a transmitted power $(P_{Tx})_{dBm} = 10$ dBm.

b) The number N of repeaters (uniformly placed along the line, with gain $(g_A)_{dB} = 20$ dB and noise figure $(F_A)_{dB} = 6$ dB) that guarantee a minimum power at the load of 10 dBm.

c) The noise figure of the system and the SNR at the load for the situation depicted in b) and for a signal bandwidth $B = 10$ kHz.

4.14 A transmission line of length $d = 20$ km has a constant characteristic impedance $Z_0 = 100\,\Omega$, and an attenuation of 1 dB/km at the frequency $f_0 = 10$ MHz. By assuming matched conditions, determine:

a) The amplitude of the line frequency response.

b) The received power, by assuming that the transmitted signal has a band $[f_0, f_1]$, $f_1 = 40$ MHz, and a power $P_{Tx} = 1$ mW uniformly distributed over the band. Note that the narrowband assumption is not valid in this case.

c) The SNR at the output of a constant gain receive amplifier with noise figure $(F_{Rc})_{dB} = 8$ dB and suitable band (to be specified).

Optical fibers

4.15 An optical fiber is transmitting in the 3rd window ($\lambda = 1.55\,\mu$m) a signal with bandwidth $B = 30$ kHz centered at $f_0 = 10$ MHz. The fiber has specific attenuation, delay and dispersion coefficients per unit length respectively of $(\tilde{A}_F)_{\text{dB/km}} = 0.1$ dB/km, $\tilde{t}_F = 5\,\mu$s/km and $\tilde{\sigma}_F = 1$ ns/km. Determine the receive amplifier noise figure to assure a SNR of 20 dB at the system output if the fiber link is 70 km long and the transmitted power is 1 dBm.

4.16 We want to design a fiber optical system 6000 km long connecting New York to Paris (and running under the Atlantic ocean). The fiber has a specific attenuation of 0.005 dB/km, a dispersion coefficient of 0.01 ns/km, while the transmitted signal has bandwidth $B = 100$ kHz centered at $f_0 = 1$ MHz. Let the required output SNR be 50 dB and let the noise figure of the receiver be 5 dB. What is the required transmitted power to satisfy the SNR constraint?

Radio links

4.17 A narrowband radio transmission system has a transmit antenna with constant gain $(\mathtt{g}_{\text{Ant,Tx}})_{\text{dB}} = 10$ dB and a receive antenna with constant gain $(\mathtt{g}_{\text{Ant,Rc}})_{\text{dB}} = 7$ dB. The receive antenna is modeled as a noisy resistor with noise temperature $\mathtt{T}_S = 120$ K, and is followed by a pre-amplifier with gain $(\mathtt{g}_A)_{\text{dB}} = 20$ dB and noise figure $(\mathtt{F}_A)_{\text{dB}} = 7$ dB. Let the receiver be matched, and let the input signal have bandwidth $B = 5$ MHz and center frequency $f_0 = 500$ MHz. Determine:

a) The power density of the noise, in dBm/Hz, at the pre-amplifier output.

b) The required transmitted power for guaranteeing an output SNR of 20 dB at a distance of $L = 300$ m.

c) The maximum distance that guarantees an output SNR of 20 dB with a transmitted power of $(\mathtt{P}_{\text{Tx}})_{\text{dBm}} = 10$ dBm.

4.18 Consider a radio link of length $d = 100$ km. The transmit and receive antenna gains are, respectively, $(\mathtt{g}_{\text{Ant,Tx}})_{\text{dB}} = 3$ dB and $(\mathtt{g}_{\text{Ant,Rc}})_{\text{dB}} = 5$ dB, and the noise temperature of the receive antenna is $\mathtt{T}_S = 250$ K. The transmitted signal has bandwidth $B = 4$ kHz centered at frequency $f_0 = 1$ GHz. The output of the receive antenna feeds a receiver with noise figure $(\mathtt{F}_R)_{\text{dB}} = 6$ dB.

a) Determine the required transmitted power (in dBm) that guarantees an output SNR of 30 dB.

b) By considering that all parameters remain unchanged (including SNR and transmitted power) what noise figure $\mathtt{F}_R$ is required if the length of the link is increased to $d = 150$ km?

4.19 A receive antenna, with gain $(\mathtt{g}_{\text{Ant,Rc}})_{\text{dB}} = 8$ dB and noise temperature $\mathtt{T}_S = 120$ K, feeds a coaxial cable. The cable line has length $d_C = 10$ km and a specific attenuation $(\tilde{\mathtt{a}}_C)_{\text{dB/km}} = 2$ dB/km at the frequency of 1 MHz. The coaxial cable is followed by an amplifier with gain $(\mathtt{g}_A)_{\text{dB}} = 20$ dB and noise figure $(\mathtt{F}_A)_{\text{dB}} = 5$ dB. Let the system be matched. Determine:

a) The amplifier output power if the received power at the antenna input is $(\mathtt{P}_{\text{Ant,Rc}})_{\text{dBm}} = 10$ dBm and the transmitted signal is narrowband with bandwidth $B = 5$ MHz centered at $f_0 = 100$ MHz.

b) The noise power at the system output.

c) How does the noise power change if the cascade cable-amplifier is substituted by two coaxial cables each of length 5 km (and with the same characteristics of the 10 km cable) and each followed by a repeater with gain $(\mathtt{g}_A)_{\text{dB}} = 10$ dB and noise figure $(\mathtt{F}_A)_{\text{dB}} = 6$ dB?

4.20 In a GSM network the transmitted signal is narrowband with bandwidth $B = 200$ kHz centered at frequency $f_0 = 900$ MHz. Each base station has an isotropic antenna that radiates the signal over a *cell*, that is a circular region with radius $d = 3.5$ km. Determine:

a) The required receive antenna gain to obtain a *maximum* attenuation of 90 dB over the entire cell.

b) The required receive antenna gain to obtain an *average* attenuation of 90 dB over the entire cell, by assuming that receivers are uniformly distributed over the cell.

Review problems

4.21 An antenna with gain 7 dB and noise temperature 120 K is connected to a coaxial cable. The coaxial cable is long 10 km and has a specific attenuation of 2 dB/km at the frequency 1 MHz. The cable is followed by an amplifier with gain 20 dB and noise figure 5 dB. Assume the impedance matching conditions between al the system devices.

The signal, with bandwidth 10 MHz is centered at the frequency of 100 MHz.

a) What is the available power (in dBm) at the amplifier output if at the *antenna input* we have a useful signal with power $\mathtt{P}_{\mathrm{i,Ant}} = 10\,\mathrm{dBm}$?

b) What is the noise power (in dBm) at the amplifier output?

c) What is the noise power (in dBm) at the output if the cascade cable + amplifier is replaced by two cables of 5 km (with the same characteristics of the previous cable of 10 km), each followed by a repeater with gain 10 dB and noise figure 6 dB?

4.22 Consider $s_{\mathrm{Tx}}(t)$ the signal at the input of the channel. $s_{\mathrm{Tx}}(t)$ has a Gaussian amplitude probability density function with zero mean and power density given by

$$\mathtt{p}_{\mathrm{Tx}}(f) = \frac{10^{-2}}{2B}\,\mathrm{rect}\left(\frac{f}{2B}\right)\ [\mathrm{W/Hz}], \qquad B = 1\,\mathrm{MHz}\,.$$

The channel is characterized by the frequency response

$$\mathcal{G}_{\mathrm{Ch}}(f) = \mathcal{G}_0\,\mathrm{rect}\left(\frac{f}{2B_{\mathrm{Ch}}}\right)$$

with $\mathcal{G}_0 = 0.1$, $B_{\mathrm{Ch}} = 1\,\mathrm{GHz}$ and has output impedance $100\,\Omega$ and noise temperature $\mathtt{T}_S = \mathtt{T}_0$. At the channel output we have an amplifier with input impedance $100\,\Omega$, gain 50 dB and noise figure 6 dB.

a) What is the probability that the signal at the amplifier input port takes values in the interval $[-0.5, 0.5]$ V?

b) What is the reference SNR (in dB) at the receiver input?

c) Consider now the channel made by a radio link with carrier 1 GHz, length 50 km, transmit and receive antennas with gains 12 dB e 15 dB, respectively, and noise temperature of the receive antenna of 250 K. Determine the reference SNR (in dB) for this system.

4.23 Consider the signal $a(t)$ with PSD

$$\mathcal{P}_a(f) = A\,\mathrm{rect}\left(\frac{f}{2B_a}\right)$$

where $B_a = 4\,\mathrm{kHz}$, $A = 10^{-9}\,\mathrm{V^2/Hz}$. The signal is centered at the frequency 10 kHz (see Example) and transmitted through a channel with frequency response

$$\mathcal{H}_{\mathrm{Ch}}(f) = \frac{1}{1 + j2\pi f T_{\mathrm{Ch}}}$$

with $T_{\mathrm{Ch}} = 1\,\mathrm{ms}$.

At the receiver we have a first stage made by an antenna (modeled by a $100\,\Omega$ resistance with noise temperature 200 K), followed by an amplifier with noise temperature 500 K and gain 30 dB. Assuming a narrowband channel model, determine:

a) The channel attenuation (in dB).

b) The noise PSD at the output of the amplifier.

c) The reference SNR (in dB) at the receiver input.

d) The output SNR (in dB).

4.24 Consider a radio transmission system (narrowband channel model) where the transmit antenna has a gain of 10 dB and the receive antenna of 7 dB. The receive antenna is modeled by a resistance with noise temperature 120 K. In cascade with the antenna we have an amplifier with gain 20 dB and noise figure 7 dB. Consider the receiver matched for the maximum power transfer.

The carrier frequency is 500 MHz and the information signal has a bandwidth of 5 MHz.

a) Determine the noise power density (in dBm/Hz) at the receive amplifier output.

b) In the case the distance between transmitter and receiver is 300 m, determine the transmit power (in dBm) required in order to get an output SNR of 20 dB.

c) Determine the maximum distance to guarantee an output SNR of 20 dB if the transmit power is 10 dBm.

4.25 Consider the signal $a(t) = V\cos(2\pi f_a t + \phi_a)$, with $f_a = 4$ kHz and $V = 1$ mV. The signal is shifted in frequency around the center frequency of 8 MHz and sent through a channel made by two sections: Section 1 is represented by a copper pair follwed by an amplificier restoring the signal level. The cable impedance is 100 Ω, the length is 5 km, and the specific attenuation at 2 MHz is 2 dB/km and it is at room temerature 290 K. The amplifier has noise figure F_1 equal to 10 dB. Section 2 is represented by a link with the same cable, with length 10 km, followed by the amplifier–receiver with gain 12 dB and noise figure $\mathsf{F}_2 = 6$ dB.

a) Determine the effective noise temperature at the input of the first cable, due *only* to Section 1.

b) Determine the effective noise temperature at the input of the first cable, due to both Sections.

c) Determine the output SNR assuming the source temperature $\mathsf{T}_S = 290$ K.

d) Exchanging the value of only the noise figure of the two amplifiers, does the output SNR get better, worse, or remain unchanged? Why?

4.26 A transmission system is characterized by the following devices

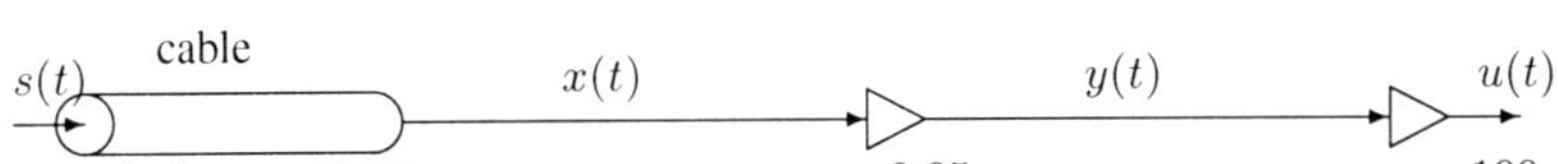

$(\mathsf{a}_c)_{\rm dB} = 40$ dB	$\mathsf{g}_2 = 0.05$	$\mathsf{g}_3 = 100$
$Z_{\rm out} = R_1 = 50\Omega$	$\mathsf{F}_2 = 10$ dB	$\mathsf{F}_3 = 10$ dB
	$Z_{\rm in} = R_1 = 50\Omega$	$Z_{\rm in} = Z_{\rm out} = R_2 = 1{\rm k}\Omega$
	$Z_{\rm out} = R_2 = 1{\rm k}\Omega$	

a) Determine the analytical relation between the signals $u(t)$ and $s(t)$.

b) Evaluate the noise figure of the two-port network with input $s(t)$ and output $u(t)$.

c) Evaluate the noise PSD at the system output.

d) Given the signal

$$s(t) = s_0 \cos(2\pi\, 1000\, t) + s_1 \sin(2\pi\, 100000\, t) - s_2 \cos(2\pi\, 999000\, t)$$

with $s_0 = 0.1$ V, $s_1 = 0.07$ V, $s_2 = 0.7$ V, evaluate the power of $u(t)$.

e) Evaluate the system output SNR (on the electrical powers) assuming the same input signal of the previous point and a system bandwidth of 1 MHz.

4.27 Assume a transmission over a coaxial cable of a baseband information signal $a(t)$ which is then shifted to the center frequency $f_0 = 1$ MHz. The cable acts as a filter with impulse response

$$g_{\rm Ch} = G\,\delta(t - t_0)\,,$$

where $t_0 = (1.6)^{-1}10^{-6}$ s and G is a parameter which depends on the transmission center frequency. The cable attenuation at 4 MHz is 6 dB.

a) Determine the value of G for this transmission.

b) Assuming the information signal $a(t) = \cos(2\pi f_a t)$ V con $f_a = 0.2$ MHz, determine the expression of the received signal.

c) Determine if the channel is distorting the signal of point b), justifying the answer.

d) Determine the system output SNR for an information signal with bandwidth 10 kHz, statistical transmit power 4 V^2 and white noise at the receiver input with PSD $1 \cdot 10^{-5}$ V^2/Hz.

CHAPTER 5

DIGITAL MODULATION SYSTEMS

Tomaso Erseghe

The term *modulation* denotes the process of transforming the information generated by a source into a signal that is suitable for transmission over a physical channel, in order to convey it to a *receiver*. When the information is represented by a sequence of bits, $\{b_\ell\}$, we talk of *digital modulation*. The binary sequence detected by the receiver, $\{\hat{b}_\ell\}$, may well be affected by errors, i.e. $\hat{b}_\ell \neq b_\ell$ for some values of ℓ, due to distortion and noise introduced by the transmission medium. In this chapter we present a survey of the main modulation techniques used in modern digital communication systems (*e.g.*, PAM, PSK, QAM, etc.). The performance of each modulation-demodulation method is evaluated with reference to the *bit error probability*, $P_{\text{bit}} = \text{P}[\hat{b}_\ell \neq b_\ell]$, and a comparison of the various approaches is given.

5.1 INTRODUCTION

In the case of *digital transmission*, the information is represented by a sequence of bits $\{b_\ell\}$ of period T_b, as depicted in Fig. 5.1.

The transformation of a digital sequence into a continuous time signal $s_{\text{Tx}}(t)$ is called *digital modulation*, and the device that performs the transformation is called *digital modulator*. A modulator may employ a set of $M = 2$ waveforms to generate the transmitted signal $s_{\text{Tx}}(t)$,

Principles of Communications Networks and Systems N. Benvenuto, M. Zorzi

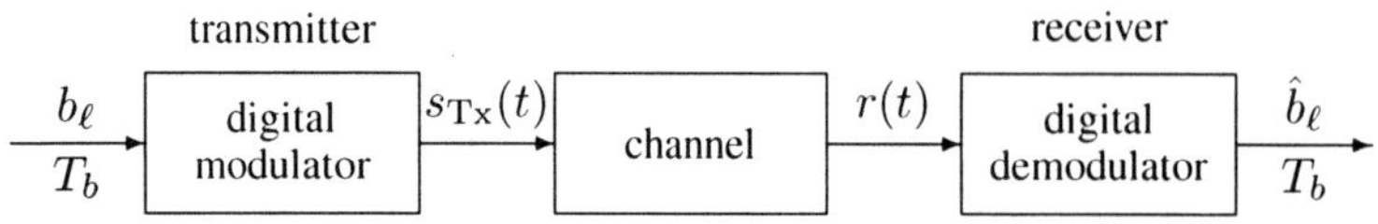

Figure 5.1 Digital transmission scheme.

in which case we talk of *binary modulation*, or, in general, $M > 2$ waveforms, in which case we talk of *M-ary modulation*.

The channel was discussed in Chapter 4, and modifies the transmitted signal by introducing distortion, interference, and noise. In general, it is assumed that it may introduce both distortion and additive white Gaussian noise (AWGN), so that the received signal is of the form

$$r(t) = s_{\text{Rc}}(t) + w_{\text{Rc}}(t)\ , \qquad s_{\text{Rc}}(t) = (s_{\text{Tx}} * g_{\text{Ch}})(t) \tag{5.1}$$

where from (4.66) we recall that g_{Ch} is the impulse response of the channel, and w_{Rc} is a real-valued additive white Gaussian *noise* (AWGN) having *zero mean and constant power spectral density* $\mathcal{P}_{w_{\text{Rc}}}(f) = N_0/2$.

Finally, the task of the digital demodulator (receiver) in Fig. 5.1 is to detect the transmitted bits, based on the received signal $r(t)$.

5.2 DIGITAL MODULATION THEORY FOR AN AWGN CHANNEL

5.2.1 Transmission of a single pulse

We consider first the transmission of an isolated pulse.

With reference to the system illustrated in Fig. 5.2, a_0 is the transmitted symbol, modeled as a discrete rv with values in $\{1, \ldots, M\}$ and associated probabilities $p_n = \text{P}[a_0 = n]$, also called *a priori probabilities*. The transmission system aims at transferring the value of a_0 to the receiver through the channel. In particular, the digital modulator, defined by a given set of M real-valued waveforms $s_n(t)$, $n = 1, \ldots, M$, selects the waveform to transmit in accordance to the value of a_0, that is if the selected symbol value is $a_0 = n$, then the transmitted signal is $s_{a_0}(t) = s_n(t)$.

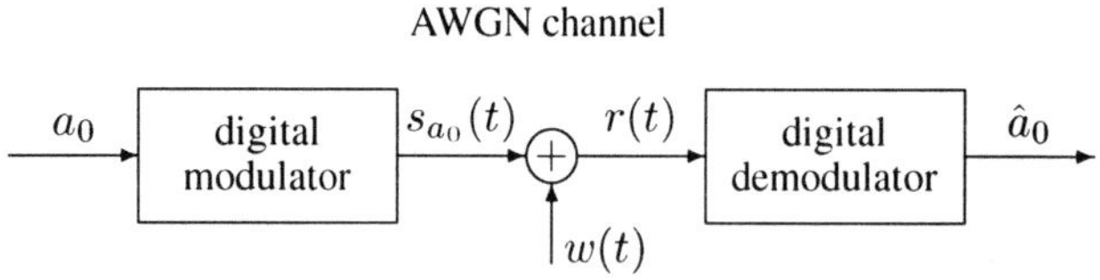

Figure 5.2 System model for digital transmission (isolated pulse).

For analytical tractability, at first the channel impulse response (5.1) is considered ideal, that is $g_{\text{Ch}}(t) = \delta(t)$, and

$$r(t) = s_{\text{Tx}}(t) + w(t)\ , \tag{5.2}$$

where $w(t) = w_{\mathrm{Rc}}(t)$ for notation simplicity.

Assuming that the waveform with index n is transmitted, that is $a_0 = n$, the received, or observed, signal is given by

$$r(t) = s_n(t) + w(t) \ . \tag{5.3}$$

The receiver, based on $r(t)$, must decide which among the M hypotheses

$$H_n \ : \ r(t) = s_n(t) + w(t) \ , \qquad n = 1, 2, \ldots, M \tag{5.4}$$

is the most likely, and correspondingly must select the detected value $\hat{a}_0$ [3]. To do so, it is convenient to represent signals using the vector notation introduced in Section 2.5.

Let $\{\phi_i(t)\}$, $i = 1, \ldots, I$, be a complete orthonormal basis for the M waveforms $\{s_n(t)\}$, $n = 1, \ldots, M$, and let $\boldsymbol{s}_n$ be the vector representation of $s_n(t)$,

$$s_n(t) \qquad \Longleftrightarrow \qquad \boldsymbol{s}_n = [s_{n,1}, s_{n,2}, \ldots, s_{n,I}] \tag{5.5}$$

where

$$s_n(t) = \sum_{i=1}^{I} s_{n,i}\, \phi_i(t) \ , \qquad s_{n,i} = \langle s_n(t), \phi_i(t) \rangle = \int_{-\infty}^{+\infty} s_n(t) \phi_i^*(t)\, dt \ . \tag{5.6}$$

The basis of I functions may *not be complete* for the representation of the noise signal $w(t)$. In any case, we can express the noise as

$$w(t) = w_\phi(t) + w_\perp(t) \tag{5.7}$$

where

$$w_\phi(t) = \sum_{i=1}^{I} w_i\, \phi_i(t) \ , \qquad w_i = \langle w(t), \phi_i(t) \rangle = \int_{-\infty}^{+\infty} w(t) \phi_i^*(t)\, dt \tag{5.8}$$

is the projection of $w(t)$ onto the space V spanned by the basis $\{\phi_i(t)\}$, while $w_\perp(t)$ is the noise component orthogonal to V. In other words, $w_\perp(t)$ is the error due to the vector representation, and it lies outside of the useful signal space. As we will state later on using the *theorem of irrelevance*, $w_\perp(t)$ can be neglected because it is irrelevant for detection. In any case, the vector representation of the component of the noise that lies in the span of $\{\phi_i(t)\}$ is given by

$$\boldsymbol{w} = [w_1, w_2, \ldots, w_I] = \boldsymbol{w}_\phi \ . \tag{5.9}$$

Similarly, the received signal $r(t)$ can also be expressed in vector form as

$$\boldsymbol{r} = [r_1, r_2, \ldots, r_I] \ , \qquad r_i = \langle r(t), \phi_i(t) \rangle \tag{5.10}$$

which, from (5.3), we can also write as

$$\boldsymbol{r} = \boldsymbol{s}_{a_0} + \boldsymbol{w}_\phi = \boldsymbol{s}_{a_0} + \boldsymbol{w} \tag{5.11}$$

to underline the vector dependence on both the useful signal and the noise. The components of the vector $\boldsymbol{r}$ are called *sufficient statistics* in the sense that they allow optimum detection. In other words, no information is lost in considering the vector $\boldsymbol{r}$ instead of the received signal $r(t)$, $t \in \mathbb{R}$.

From (5.11), we therefore get a vector formulation equivalent to (5.4), and must decide among the M hypotheses

$$H_n \; : \boldsymbol{r} = \boldsymbol{s}_n + \boldsymbol{w} \; , \qquad n = 1, 2, \ldots, M \; . \tag{5.12}$$

The framework for taking a decision is to realize that even repetitive transmissions of the same waveform yield different observed values $\boldsymbol{r}$ because of the noise which is random. Hence, a decision criterion must be based on a statistical interpretation of (5.12), that is on repetitive trials of the experiment of Fig. 5.2.

5.2.2 Optimum detection

The theory for designing the decision criterion is based on three elements:

1. *Subdivide the space* $\mathbb{R}^I$ of the received vector $\boldsymbol{r}$ into M non overlapping regions R_n, each one associated to one of the possible outcomes of a_0. Mathematically speaking, the regions R_n must identify a *partition* of the space $\mathbb{R}^I$, that is we must have

$$\bigcup_{n=1}^{M} R_n = \mathbb{R}^I \qquad \text{and} \qquad R_m \cap R_n = \emptyset \; , \quad m \neq n \; . \tag{5.13}$$

2. *Adopt a decision rule*: the hypothesis H_m (and so $\hat{a}_0 = m$) is chosen if the received vector $\boldsymbol{r}$ belongs to R_m, that is

$$\text{if } \boldsymbol{r} \in R_m \text{ then choose } H_m \text{ and } \hat{a}_0 = m \; . \tag{5.14}$$

The receiver block diagram is reported in Fig. 5.3, where the decision rule is realized by the *detector*, whose input is often denoted as *decision point*.

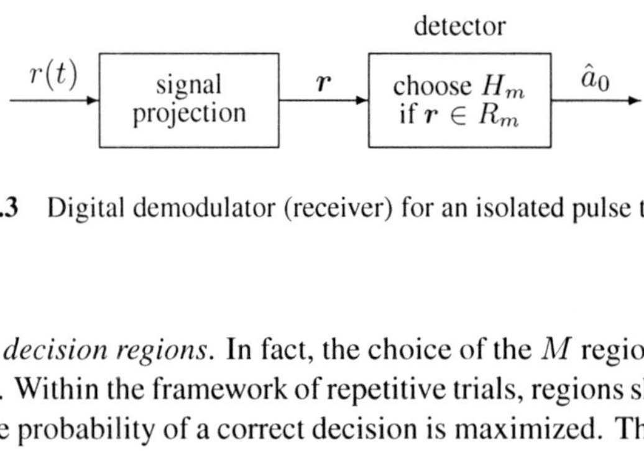

Figure 5.3 Digital demodulator (receiver) for an isolated pulse transmission.

3. *Optimize the decision regions.* In fact, the choice of the M regions $\{R_n\}$ affects system performance. Within the framework of repetitive trials, regions should be chosen in such a way that the probability of a correct decision is maximized. The *probability of correct decision* is defined as

$$\mathrm{P}\left[\mathrm{C}\right] = \mathrm{P}\left[\hat{a}_0 = a_0\right] \tag{5.15}$$

and it is the probability that the detected symbol coincides with the transmitted symbol. Hence, we determine decision regions as

$$\{R_{n,\text{opt}}\} = \underset{\{R_n\}}{\operatorname{argmax}} \, \mathrm{P}\left[\mathrm{C}\right] \; . \tag{5.16}$$

Equivalently, because P [C] is usually a number very close to 1, it can be more convenient to express performance in terms of the *error probability*

$$\mathrm{P}[\mathrm{E}] = \mathrm{P}[\hat{a}_0 \neq a_0] = 1 - \mathrm{P}[\mathrm{C}] \ , \tag{5.17}$$

in which case the optimum solution will be to choose regions is such a way that the probability of error is minimized, that is

$$\{R_{n,\mathrm{opt}}\} = \underset{\{R_n\}}{\mathrm{argmin}}\, \mathrm{P}[\mathrm{E}] \ . \tag{5.18}$$

Operatively, from (2.180) and (5.14) the probability of correct decision can be written as

$$\mathrm{P}[\mathrm{C}] = \sum_{n=1}^{M} \mathrm{P}[\hat{a}_0 = n | a_0 = n]\, \mathrm{P}[a_0 = n] = \sum_{n=1}^{M} \mathrm{P}[\boldsymbol{r} \in R_n | a_0 = n]\, p_n \tag{5.19}$$

while for the probability of error we have

$$\mathrm{P}[\mathrm{E}] = \sum_{n=1}^{M} \mathrm{P}[\boldsymbol{r} \notin R_n | a_0 = n]\, p_n \ , \tag{5.20}$$

It turns out that often it is easier to evaluate P [C] from (5.19), and then use (5.17) to obtain P [E].

5.2.3 Statistical characterization of random vectors

In order to proceed in evaluating (5.16) or (5.18), it is now important to statistically characterize the rvs $\{w_i\}$ and $\{r_i\}$, and so the rves $\boldsymbol{w}$ and $\boldsymbol{r}$. We begin with inspection of the statistical properties of noise.

We immediately note that $\{w_i\}$, $i = 1, \ldots, I$, are jointly Gaussian rvs, as they are linear transformations of the Gaussian rp $w(t)$ (see (5.8)). In addition, for the mean of w_i we find

$$\mathrm{E}[w_i] = \mathrm{E}\left[\int_{-\infty}^{+\infty} w(t)\phi_i^*(t)\, dt\right] = \int_{-\infty}^{+\infty} \mathrm{E}[w(t)]\, \phi_i^*(t)\, dt = 0 \tag{5.21}$$

since $w(t)$ has zero mean. Concerning the correlation, we need to evaluate the following expectation:

$$\begin{aligned}\mathrm{E}[w_i w_j^*] &= \mathrm{E}\left[\int_{-\infty}^{+\infty} w(t)\phi_i^*(t)\, dt \cdot \int_{-\infty}^{+\infty} w^*(u)\phi_j(u)\, du\right] \\ &= \int_{-\infty}^{+\infty} \left(\int_{-\infty}^{+\infty} \mathrm{E}[w(t)w^*(u)]\, \phi_j(u)\phi_i^*(t)\, du\right) dt \ .\end{aligned}$$

Here the noise autocorrelation function is

$$\mathrm{E}[w(t)w^*(u)] = \mathtt{r}_w(t-u) = \frac{N_0}{2}\,\delta(t-u)$$

since $w(t)$ has PSD $\mathcal{P}_w(f) = N_0/2$. By substitution we finally have

$$\begin{aligned} \mathrm{E}\left[w_i w_j^*\right] &= \int_{-\infty}^{+\infty} \left(\int_{-\infty}^{+\infty} \frac{N_0}{2}\, \delta(t-u)\, \phi_j(u) \phi_i^*(t)\, du \right) dt \\ &= \frac{N_0}{2} \int_{-\infty}^{+\infty} \phi_j(t)\phi_i^*(t)\, dt \\ &= \frac{N_0}{2}\, \delta_{i-j} \end{aligned} \tag{5.22}$$

because of the orthogonality of the basis $\{\phi_i(t)\}$. Hence the noise components $\{w_i\}$ are uncorrelated. As $\{w_i\}$ are jointly Gaussian uncorrelated rvs with zero mean, then from Proposition 2.17 they are statistically independent with equal variance given by

$$\sigma_I^2 = \frac{N_0}{2} \ . \tag{5.23}$$

By summarizing the results, the noise components $\{w_i\}$, $i = 1, \ldots, I$, are iid rvs with probability density functions given by (see Example 2.6 A)

$$p_{w_i}(u_i) = \frac{1}{\sqrt{2\pi\sigma_I^2}} \exp\left(-\frac{1}{2}\frac{u_i^2}{\sigma_I^2}\right) = \frac{1}{\sqrt{\pi N_0}} \exp\left(-\frac{u_i^2}{N_0}\right) . \tag{5.24}$$

So, the probability density function of the noise rve $\boldsymbol{w}$,

$$p_{\boldsymbol{w}}(\boldsymbol{u}) = \prod_{i=1}^{I} p_{w_i}(u_i) \ , \tag{5.25}$$

may be expanded as

$$p_{\boldsymbol{w}}(\boldsymbol{u}) = \frac{1}{(2\pi\sigma_I^2)^{I/2}} \exp\left(-\frac{1}{2}\frac{\|\boldsymbol{u}\|^2}{\sigma_I^2}\right) = \frac{1}{(\pi N_0)^{I/2}} \exp\left(-\frac{\|\boldsymbol{u}\|^2}{N_0}\right) . \tag{5.26}$$

Concerning the transmit signal $s_{a_0}(t)$ or equivalently $\boldsymbol{s}_{a_0}$ in (5.11), statistically it is described by the *a priori probabilities* $\{\mathrm{P}\left[a_0 = n\right] = p_n\}$, $n = 1, \ldots, M$.

A quantity of interest is the *average* (in a statistical sense) signalling energy which can be written as

$$E_s = \sum_{n=1}^{M} \|\boldsymbol{s}_n\|^2 p_n \ . \tag{5.27}$$

Occasionally (5.27) is also used as a measure of the compactness of the constellation.

We finally concentrate on $\boldsymbol{r}$ which, according to (5.11), is a sum of the noise $\boldsymbol{w}$ and the useful signal $\boldsymbol{s}_{a_0}$. In general, the probability density function $p_{\boldsymbol{r}}$ is rather complicated to evaluate. However, if we condition $\boldsymbol{r}$ to a given transmitted symbol value $a_0 = n$, things become very simple. In fact, under the condition that $a_0 = n$, the received vector $\boldsymbol{r} = \boldsymbol{s}_n + \boldsymbol{w}$ can be seen as the zero mean Gaussian rve $\boldsymbol{w}$ to which the constant component $\boldsymbol{s}_n$ has been added. In other terms, *given* $a_0 = n$, $\boldsymbol{r}$ turns out to be a Gaussian rve with mean $\boldsymbol{s}_n$. In mathematical terms, this can be expressed in terms of the conditional probability density function

$$p_{\boldsymbol{r}|a_0}(\boldsymbol{\rho}|n) = p_{\boldsymbol{w}}(\boldsymbol{\rho} - \boldsymbol{s}_n) = \frac{1}{(\pi N_0)^{I/2}} \exp\left(-\frac{\|\boldsymbol{\rho} - \boldsymbol{s}_n\|^2}{N_0}\right) . \tag{5.28}$$

As a consequence of (5.28), and by exploiting the *total probability theorem* (Theorem 2.12), we could also express the probability density function of $\boldsymbol{r}$ as

$$p_{\boldsymbol{r}}(\boldsymbol{\rho}) = \sum_{n=1}^{M} p_{\boldsymbol{r}|a_0}(\boldsymbol{\rho}|n)\, p_n \,. \tag{5.29}$$

Remark The theory outlined in this section implies that $s_n(t)$, $w(t)$ and $r(t)$ are all real-valued signals. The theory can be extended to the case of complex-valued signals by carefully considering the probability density function for complex-valued Gaussian noise (see Section 2.6.5).

5.2.4 Optimum decision regions

The statistical description of $\boldsymbol{r}$ allows us to derive a meaningful expression for the optimum decision regions (5.16). For the sake of clarity we first present the result for the one-dimensional case $I = 1$, in which the waveform vector $\boldsymbol{s}_n$ $(n = 1, \ldots, M)$ reduces to a scalar s_n, and the noise vector $\boldsymbol{w}$ to a Gaussian rv w with zero mean and variance σ_I^2.

According to the optimal criterion (5.16) we wish to evaluate what choice of the regions $\{R_n\}$ maximizes the probability of correct decision. Recalling (5.19) in the one-dimensional case, the probability of correct decision becomes

$$\mathrm{P}\left[\mathrm{C}\right] = \sum_{n=1}^{M} \mathrm{P}\left[r \in R_n | a_0 = n\right]\, p_n \tag{5.30}$$

with $r = s_{a_0} + w$ the received sample and $\{R_n\}$, $n = 1, \ldots, M$, the one-dimensional decision regions constituting a partition of $\mathbb{R}$. By using the conditional probability density function (5.28) we further have

$$\mathrm{P}\left[\mathrm{C}\right] = \sum_{n=1}^{M} \int_{R_n} p_{r|a_0}(\rho|n)\, p_n \, d\rho \,. \tag{5.31}$$

It is now convenient to introduce an indicator function for the region R_n, that is

$$\mu_n(\rho) = \begin{cases} 1 & , \ \rho \in R_n \\ 0 & , \ \text{elsewhere} \end{cases} \tag{5.32}$$

so that (5.31) can be rewritten as

$$\begin{aligned} \mathrm{P}\left[\mathrm{C}\right] &= \int_{-\infty}^{+\infty} \sum_{n=1}^{M} \mu_n(\rho)\, p_{r|a_0}(\rho|n)\, p_n \, du \\ &= \int_{-\infty}^{+\infty} \sum_{n=1}^{M} \mu_n(\rho)\, D(\rho; n)\, du \end{aligned} \tag{5.33}$$

where we used the notation

$$D(\rho; n) = p_{r|a_0}(\rho|n)\, p_n \,. \tag{5.34}$$

The meaning of (5.33) is depicted in Fig. 5.4, where $M = 3$. The three gray areas indicate the integral (5.33) over each of the regions R_1, R_2 and R_3, and so the total gray area is the probability P [C] of correct decision. We immediately note that the regions of Fig. 5.4 are not optimal, and in fact we would obtain a bigger probability P [C] (bigger gray area) by moving to the right the threshold separating R_1 from R_2. Similarly, the threshold separating R_2 from R_3 should be moved to the left.

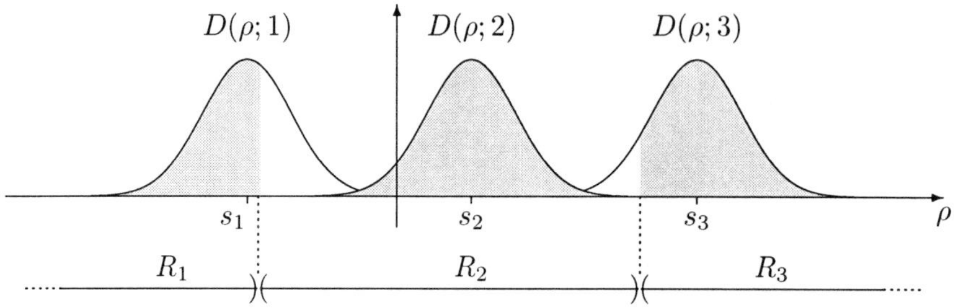

Figure 5.4 Illustration of the relation between the probability of correct decision P [C] (gray area) and the decision regions.

In general, since the integrand function in (5.33) consists of M terms, $\{\mu_n(\rho)\, D(\rho; n)\}$, $n = 1, \ldots, M$, and since the M regions are non-overlapping, for each value of ρ only one of the terms is selected. Therefore the maximum value of the integrand function is obtained if for each value of ρ we select among M terms the one that yields the maximum value of $D(\rho; n)$. Thus we have identified the following optimum decision regions

$$R_m = \left\{ \rho \,\middle|\, m = \underset{n \in \{1,\ldots,M\}}{\operatorname{argmax}} \; D(\rho; n) \right\}, \qquad m = 1, \ldots, M . \tag{5.35}$$

For the example of Fig. 5.4 the optimum decision regions are shown in Fig. 5.5.

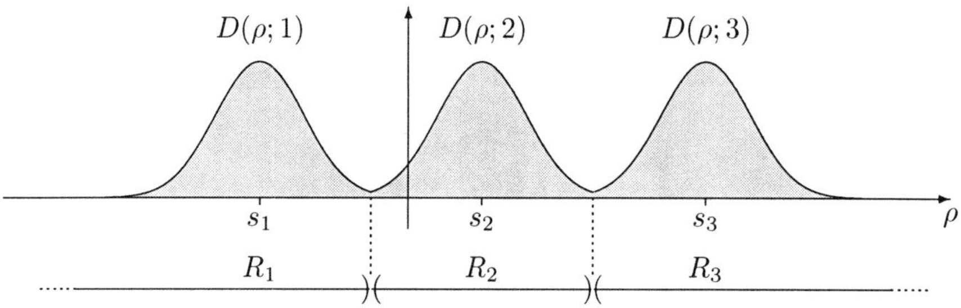

Figure 5.5 Optimum decision regions.

Once the optimum decision regions have been determined, the probability of correct decision can be evaluated by (5.31).

Example 5.2 A Let a digital transmission system have signal space dimension $I = 1$, constellation dimension $M = 4$, and waveforms with vector representation $s_1 = -3V_0$, $s_2 = -V_0$, $s_3 = V_0$, $s_4 = 3V_0$. Let also the a priori probabilities be $p_1 = p_4 = 1/8$ and $p_2 = p_3$, and the Gaussian noise PSD be $N_0/2 = V_0^2/\ln 3$. We want to determine the optimum decision regions and the maximum achievable probability of correct decision.

To do so, we first note that $p_2 = p_3 = 3/8$, which implies that the functions

$$D(\rho; n) = p_n \frac{1}{\sqrt{\pi N_0}} \exp\left(-\frac{(\rho - s_n)^2}{N_0}\right)$$

are as depicted in Fig. 5.6. Because of the Gaussian-shape curves, it is obvious that the threshold separating R_2 and R_3 is $v_2 = 0$. From the figure it is less obvious what the threshold v_1 separating R_1 from R_2, and what the threshold v_3 separating R_3 from R_4 should be. Nonetheless, we can immediately say that $v_3 = -v_1$.

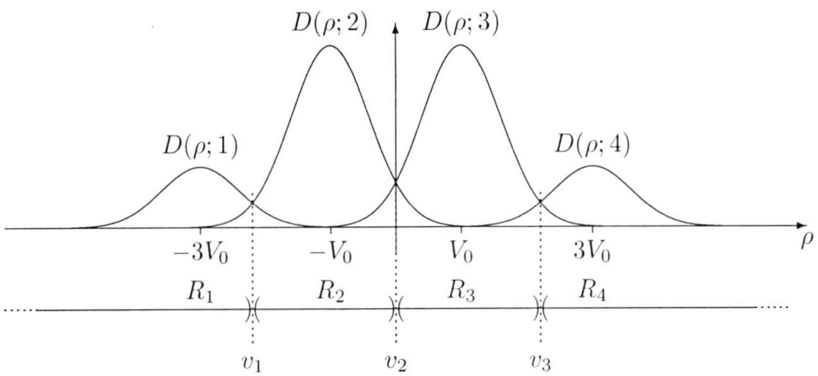

Figure 5.6 Optimum decision regions.

We then need to evaluate v_3, which is that value such that $D(v_3; 3) = D(v_3; 4)$, that is

$$\frac{3}{8} \frac{1}{\sqrt{\pi N_0}} \exp\left(-\frac{(v_3 - V_0)^2}{N_0}\right) = \frac{1}{8} \frac{1}{\sqrt{\pi N_0}} \exp\left(-\frac{(v_3 - 3V_0)^2}{N_0}\right).$$

After taking the logarithm and after some straightforward rearrangements, this gives

$$\ln 3 - \frac{(v_3 - V_0)^2}{N_0} = -\frac{(v_3 - 3V_0)^2}{N_0} \qquad \Longrightarrow \qquad v_3 = 2V_0 + \frac{N_0 \ln 3}{4V_0} = \tfrac{5}{2}V_0 .$$

So, the optimum decision regions are

$$R_1 = \left(-\infty, -\tfrac{5}{2}V_0\right) \quad R_2 = \left[-\tfrac{5}{2}V_0, 0\right) \quad R_3 = \left[0, \tfrac{5}{2}V_0\right) \quad R_4 = \left(\tfrac{5}{2}V_0, +\infty\right) .$$

Now, from (5.31) the maximum achievable probability of correct decision is

$$\begin{aligned} \mathrm{P}\,[\mathrm{C}] &= \int_{R_1} D(\rho;1)\,d\rho + \int_{R_2} D(\rho;2)\,d\rho + \int_{R_3} D(\rho;3)\,d\rho + \int_{R_4} D(\rho;4)\,d\rho \\ &= 2\left[\int_{R_3} D(\rho;3)\,d\rho + \int_{R_4} D(\rho;4)\,d\rho\right] \end{aligned}$$

because of the symmetry in Fig. 5.6. Finally, from (5.34) we obtain

$$\begin{aligned} \mathrm{P}\,[\mathrm{C}] &= \frac{3}{4}\int_0^{\frac{5}{2}V_0} p_w(\rho - V_0)\,d\rho + \frac{1}{4}\int_{\frac{5}{2}V_0}^{+\infty} p_w(\rho - 3V_0)\,d\rho \\ &= \frac{3}{4}\int_{-V_0}^{\frac{3}{2}V_0} p_w(u)\,du + \frac{1}{4}\int_{-\frac{1}{2}V_0}^{+\infty} p_w(u)\,du\,. \end{aligned}$$

The result is finally achieved by considering that w is Gaussian with variance $\sigma_I^2 = N_0/2$, so that in terms of the Q function (2.148), we have

$$\begin{aligned} \mathrm{P}\,[\mathrm{C}] &= \frac{3}{4}\left[\mathrm{Q}\left(\frac{-V_0}{\sigma_I}\right) - \mathrm{Q}\left(\frac{\frac{3}{2}V_0}{\sigma_I}\right)\right] + \frac{1}{4}\left[\mathrm{Q}\left(\frac{-\frac{1}{2}V_0}{\sigma_I}\right)\right] \\ &= \frac{3}{4}\left[1 - \mathrm{Q}\left(\frac{V_0}{\sigma_I}\right) - \mathrm{Q}\left(\frac{\frac{3}{2}V_0}{\sigma_I}\right)\right] + \frac{1}{4}\left[1 - \mathrm{Q}\left(\frac{\frac{1}{2}V_0}{\sigma_I}\right)\right] \\ &= 1 - \frac{3}{4}\mathrm{Q}\left(\frac{V_0}{\sigma_I}\right) - \frac{3}{4}\mathrm{Q}\left(\frac{\frac{3}{2}V_0}{\sigma_I}\right) - \frac{1}{4}\mathrm{Q}\left(\frac{\frac{1}{2}V_0}{\sigma_I}\right) \simeq 0.88\,. \end{aligned}$$

The theory presented so far can be easily extended to the multidimensional case where $I > 1$. In particular, the received sample r is substituted with the received vector $\boldsymbol{r}$, and the probability of correct decision (5.31) becomes

$$\mathrm{P}\,[\mathrm{C}] = \sum_{n=1}^{M} \mathrm{P}\,[\boldsymbol{r} \in R_n | a_0 = n]\; p_n = \sum_{n=1}^{M} \int_{R_n} p_{\boldsymbol{r}|a_0}(\boldsymbol{\rho}|n)\, p_n\, d\boldsymbol{\rho}\,. \tag{5.36}$$

Again, (5.36) expressed through the indicator functions becomes

$$\mathrm{P}\,[\mathrm{C}] = \int_{\mathbb{R}^I} \sum_{n=1}^{M} \mu_n(\boldsymbol{\rho})\, D(\boldsymbol{\rho}; n)\, d\boldsymbol{\rho} \tag{5.37}$$

where

$$D(\boldsymbol{\rho}; n) = p_{\boldsymbol{r}|a_0}(\boldsymbol{\rho}|n)\, p_n\,, \qquad \mu_n(\boldsymbol{\rho}) = \begin{cases} 1 & , \ \boldsymbol{\rho} \in R_n \\ 0 & , \ \text{elsewhere} \end{cases} \tag{5.38}$$

Also the procedure leading to the optimal decision regions is equivalent, and we thus have

$$R_m = \left\{\boldsymbol{\rho} \,\middle|\, m = \underset{n\in\{1,\ldots,M\}}{\operatorname{argmax}}\; D(\boldsymbol{\rho}; n)\right\} \tag{5.39}$$

which should be compared with (5.35)

Remark In all the above examples, the decision regions are compact subsets of $\mathbb{R}^I$. This is due to the fact that we are considering additive Gaussian noise.

However, in communications there exist situations where noise (or interference) is not Gaussian, and it may even depend on the transmitted signal. Such a situation is depicted in Fig. 5.7 for the one-dimensional case $I = 1$ and for $M = 3$. If we denote with v_1, v_2, and v_3 the intersection points of the various functions as illustrated in Fig. 5.7, it is easy to verify that

$$R_1 = (-\infty, v_1] \qquad R_2 = (v_2, v_3] \qquad R_3 = (v_1, v_2] \cup (v_3, +\infty) \, .$$

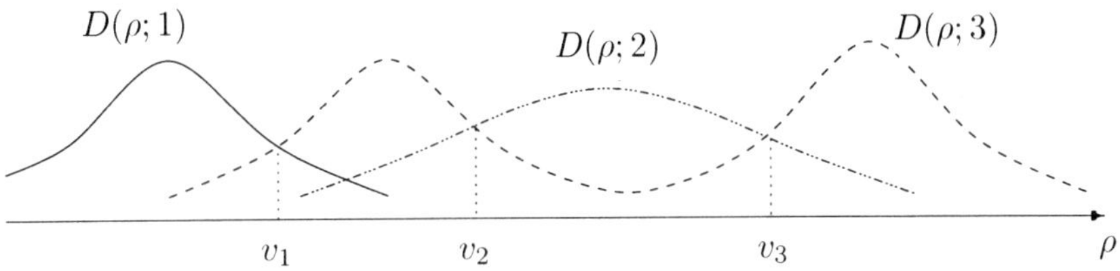

Figure 5.7 Illustration of optimum decision regions in a non-AWGN channel.

5.2.5 Maximum a posteriori criterion

Determination of optimum decision regions by (5.39) gives us the possibility to reinterpret the decision rule (5.14). For the sake of clarity, in the following we make a distinction between $\boldsymbol{r}$, the rve obtained from the projection of the rp $r(t)$ onto the waveform basis, and $\boldsymbol{\rho}$, the realization of $\boldsymbol{r}$. This is to avoid expressions like $p_{\boldsymbol{r}|a_0}(\boldsymbol{r}|n)$, whose meaning may be unclear.

Now, according to (5.14), the hypothesis H_m, that is $\hat{a}_0 = m$, is chosen if the received vector $\boldsymbol{\rho}$ belongs to R_m. Now, because of (5.39), to say that $\boldsymbol{\rho}$ belongs to R_m is equivalent to writing

$$\boldsymbol{\rho} \in R_m \qquad \Longleftrightarrow \qquad m = \underset{n \in \{1,\dots,M\}}{\operatorname{argmax}} \; D(\boldsymbol{\rho}; n) \tag{5.40}$$

so that the decision rule (5.14) can be restated as

$$\boxed{\hat{a}_0 = \underset{n \in \{1,\dots,M\}}{\operatorname{argmax}} \; D(\boldsymbol{\rho}; n)} \tag{5.41}$$

This criterion is commonly termed as *maximum a posteriori probability (MAP) criterion* since (5.41) is equivalent to the formulation

$$\hat{a}_0 = \underset{n \in \{1,\dots,M\}}{\operatorname{argmax}} \; \mathrm{P}\left[a_0 = n \,|\, \boldsymbol{r} = \boldsymbol{\rho}\right] \tag{5.42}$$

where the conditional probabilities $\mathrm{P}[a_0 = n \mid \boldsymbol{r} = \boldsymbol{\rho}]$ are called *a posteriori probabilities* and indicate the probability that a_0 was transmitted based on the (*a posteriori*) observation of $\boldsymbol{\rho}$. The equivalence between (5.42) and (5.41) can be proved by using Bayes's rule (Proposition 2.13) in the conditional probability (5.42) to obtain

$$\mathrm{P}[a_0 = n \mid \boldsymbol{r} = \boldsymbol{\rho}] = \frac{\mathrm{P}[a_0 = n,\ \boldsymbol{r} = \boldsymbol{\rho}]}{p_{\boldsymbol{r}}(\boldsymbol{\rho})} = \frac{p_{\boldsymbol{r}|a_0}(\boldsymbol{\rho}|n)\, p_n}{p_{\boldsymbol{r}}(\boldsymbol{\rho})} = \frac{D(\boldsymbol{\rho}; n)}{p_{\boldsymbol{r}}(\boldsymbol{\rho})}$$

and by observing that the denominator $p_{\boldsymbol{r}}(\boldsymbol{\rho})$ is irrelevant in the decision rule since it does not depend on n.

5.2.6 Maximum likelihood criterion

A very common situation is when the transmitted symbol assumes equally likely values, that is

$$p_n = \frac{1}{M}, \qquad n = 1, 2, \ldots, M \,. \tag{5.43}$$

This assumption is equivalent to have no *a priori* knowledge on what a_0 is more likely to have been transmitted. Since all p_n are equal to $1/M$, the decision rule (5.41) can be restated as

$$\boxed{\hat{a}_0 = \operatorname*{argmax}_{n \in \{1,\ldots,M\}} p_{\boldsymbol{r}|a_0}(\boldsymbol{\rho}|n)} \tag{5.44}$$

where we have simplified the common factor $1/M$. The criterion (5.44) is known as *maximum likelihood (ML) criterion* in that it selects the value n which maximizes the conditional probability that $\boldsymbol{r} = \boldsymbol{\rho}$ is observed given $a_0 = n$.

5.2.7 Minimum distance criterion

A simplification of the ML criterion which leads to a very simple decision rule is obtained when dealing with AWGN channels, in which case the conditional probability $p_{\boldsymbol{r}|a_0}(\boldsymbol{\rho}|n)$ has a Gaussian shape, as given in (5.28). By discarding the constant factor $(\pi N_0)^{-I/2}$ from (5.28), we get

$$\hat{a}_0 = \operatorname*{argmax}_{n \in \{1,\ldots,M\}} \exp\left(-\frac{\|\boldsymbol{\rho} - \boldsymbol{s}_n\|^2}{N_0}\right).$$

Taking the logarithm, which is a monotonic function, we then obtain

$$\hat{a}_0 = \operatorname*{argmax}_{n \in \{1,\ldots,M\}} -\frac{\|\boldsymbol{\rho} - \boldsymbol{s}_n\|^2}{N_0}$$

where the negative constant $-1/N_0$ can be discarded and $\|\boldsymbol{\rho} - \boldsymbol{s}_n\|^2 = d^2(\boldsymbol{\rho}, \boldsymbol{s}_n)$. By taking the squared root, we finally obtain

$$\boxed{\hat{a}_0 = \operatorname*{argmin}_{n \in \{1,\ldots,M\}} d(\boldsymbol{\rho}, \boldsymbol{s}_n)} \tag{5.45}$$

The decision rule expressed by (5.45) is the *minimum distance criterion* and yields the signal vector $\boldsymbol{s}_n$ that is closest to the received signal vector $\boldsymbol{\rho}$. Moreover, the optimum decision regions $\{R_m\}$, $m = 1, \ldots, M$, are also easily determined as

$$R_m = \left\{ \boldsymbol{\rho} \,\middle|\, m = \underset{n \in \{1,\ldots,M\}}{\text{argmin}} \; d(\boldsymbol{\rho}, \boldsymbol{s}_n) \right\} . \tag{5.46}$$

Remark Be aware of the fact that the minimum distance criterion (5.45) is equivalent to the ML criterion *only* for AWGN channels. If the noise introduced by the channel is not Gaussian, then we must resort to the ML criterion (5.44) or, if the symbol values are not equally likely, to the MAP criterion (5.41).

Definition 5.1 *The minimum distance regions (5.46) are often called* Voronoi regions *or, when they have finite measure,* Voronoi cells.

Example 5.2 B We solve Example 5.2 A in the case the symbol values are equally likely. This situation is depicted in Fig. 5.8.

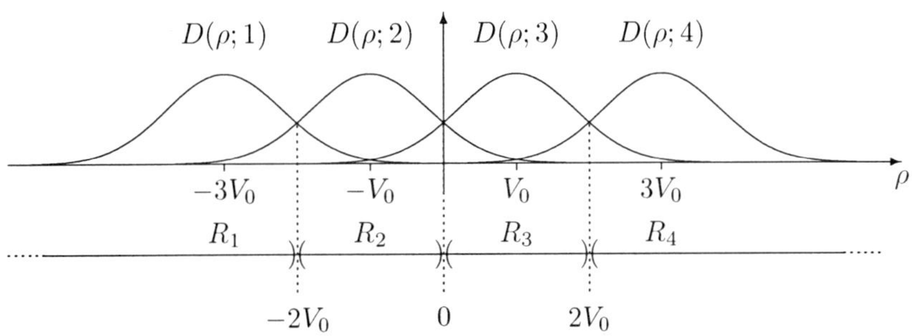

Figure 5.8 Optimum decision regions with equally likely symbol values.

The decision regions can be immediately identified, through the minimum distance criterion (5.46), as

$$R_1 = (-\infty, -2V_0) \quad R_2 = [-2V_0, 0) \quad R_3 = [0, 2V_0) \quad R_4 = (2V_0, +\infty) \ ,$$

and the probability of correct decision (5.37) becomes

$$\begin{aligned}
\mathrm{P}\left[\mathrm{C}\right] &= 2\left[\int_{R_3} D(\rho;3)\, d\rho + \int_{R_4} D(\rho;4)\, d\rho\right] \\
&= \tfrac{1}{2}\int_0^{2V_0} p_w(\rho - V_0)\, d\rho + \tfrac{1}{2}\int_{2V_0}^{+\infty} p_w(\rho - 3V_0)\, d\rho \\
&= \tfrac{1}{2}\int_{-V_0}^{V_0} p_w(u)\, du + \tfrac{1}{2}\int_{-V_0}^{+\infty} p_w(u)\, du \\
&= 1 - \tfrac{3}{2}\,\mathrm{Q}\left(\frac{V_0}{\sigma_I}\right) = 0.90 \ .
\end{aligned}$$

Example 5.2 C We derive the two-dimensional optimum decision regions for the transmitted waveforms of Examples 2.5 D and 2.5 E (see also Example 2.5 F). Constellations and the relative two-dimensional decision regions are shown in Fig. 5.9 when considering the minimum distance criterion.

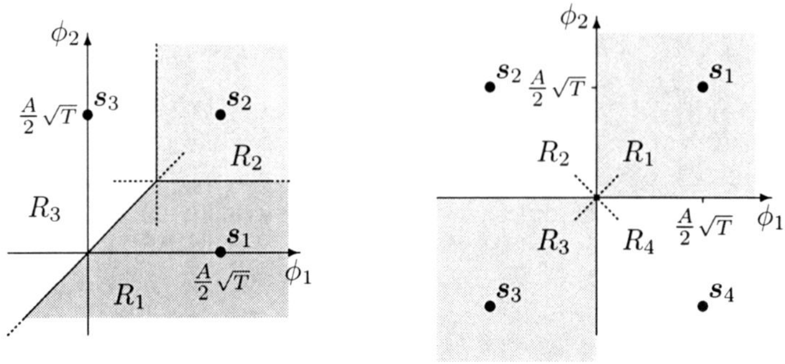

(a) (b)

Figure 5.9 Two-dimensional decision regions based upon the minimum distance criterion: (a) Decision regions of Example 2.5 D; (b) Decision regions of Example 2.5 E.

The procedure for identifying the optimum decision regions is as follows: for any pair of vectors $\boldsymbol{s}_m, \boldsymbol{s}_n$, to draw the straight line of points that are equidistant from $\boldsymbol{s}_m$ and $\boldsymbol{s}_n$; this straight line defines the boundary between R_m and R_n. The decision region associated with each vector $\boldsymbol{s}_m$ is then given by the intersection of half-planes as illustrated in Fig. 5.9.

5.2.8 Implementation of minimum distance receivers

By considering the minimum distance criterion, the typical receiver implementation for an M-ary transmission is illustrated in Fig. 5.10, which we call *implementation type I*, that consists of three fundamental blocks. The first block is the filter bank projecting the received signal $r(t)$ onto the basis, to yield the vector $\boldsymbol{r}$ of length I. The projection $r_i = \langle r(t), \phi_i(t) \rangle$ is achieved through a filter with impulse response

$$\psi_i(t) = \phi_i^*(t_0 - t) \,, \qquad i = 1, \ldots, I \tag{5.47}$$

where t_0 is usually chosen to guarantee a causal filter. The filter output is then sampled at instant t_0, to obtain

$$r_i = (r * \psi_i)(t_0) = \int r(u)\, \psi_i(t_0 - u)\, du = \int r(u)\, \phi_i^*(u)\, du \,. \tag{5.48}$$

The second block computes the M distances

$$D_n = d(\boldsymbol{r}, \boldsymbol{s}_n) \,, \qquad n = 1, \ldots, M \,. \tag{5.49}$$

The final block is the decision block that selects the output $\hat{a}_0$ through a minimum distance criterion

$$\hat{a}_0 = \underset{n \in \{1, \ldots, M\}}{\operatorname{argmin}} \; D_n \,. \tag{5.50}$$

In many applications, instead of using steps (5.49) and (5.50), it is simpler to apply the decision criterion (5.14) based on the decision regions.

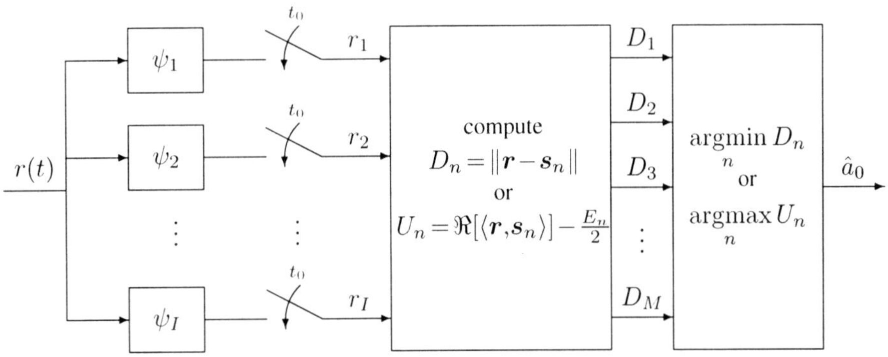

Figure 5.10 Receiver implementation based on the minimum distance criterion. *Implementation type I*, where $\psi_i(t) = \phi_i^*(t_0 - t)$.

An alternative approach to (5.50) can be pursued by expanding the expression of the distance (5.49) by (2.124),

$$\begin{aligned}\hat{a}_0 &= \underset{n\in\{1,\ldots,M\}}{\text{argmin}} \ \|\boldsymbol{r} - \boldsymbol{s}_n\|^2 \\ &= \underset{n\in\{1,\ldots,M\}}{\text{argmin}} \ \|\boldsymbol{r}\|^2 - 2\Re[\langle \boldsymbol{r}, \boldsymbol{s}_n\rangle] + \|\boldsymbol{s}_n\|^2 \ ,\end{aligned} \tag{5.51}$$

to note that $\|\boldsymbol{r}\|^2$ can be discarded from the decision rule since it does not depend on n. By further multiplication by $-\frac{1}{2}$, we obtain the following criterion, equivalent to (5.50),

$$\hat{a}_0 = \underset{n\in\{1,\ldots,M\}}{\text{argmax}} \ U_n \ , \qquad U_n = \Re[\langle \boldsymbol{r}, \boldsymbol{s}_n\rangle] - \tfrac{1}{2}E_n \tag{5.52}$$

where E_n is the energy of $\boldsymbol{s}_n$. This interpretation leads to the second decision criterion of Fig. 5.10 or to the *implementation type II* receiver of Fig. 5.11. Note that, equivalently to what we saw in (5.48), U_n is obtained by convolving the received signal $r(t)$ by the waveforms

$$\xi_n(t) = s_n^*(t_0 - t) \ , \qquad n = 1, \ldots, M$$

and then sampling the output at the instant t_0, to obtain the projection $\langle r(t), s_n(t)\rangle$. Note also that, when the waveforms have equal energy, $E_s = E_n$, $n = 1, \ldots, M$, then the term $-\frac{1}{2}E_n$ in (5.52) can be dropped, thus further simplifying the receiver implementation.

Remark We generally have $I < M$, so that the *implementation type I* receiver is more convenient since it requires a bank of I ($< M$) filters only. However, when $I = M$ or I and M are very close numbers, the *implementation type II* receiver may be simpler to formalize since it does not require the identification of an orthonormal basis. An example follows.

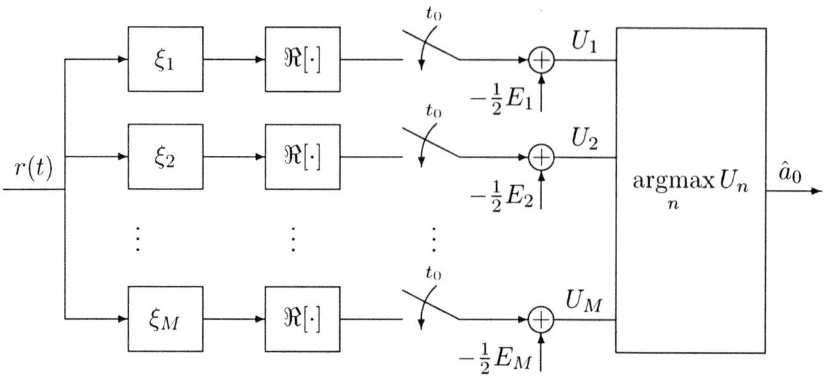

Figure 5.11 Receiver implementation based on the minimum distance criterion. *Implementation type II*, where $\xi_n(t) = s_n^*(t_0 - t)$.

Example 5.2 D (4-PPM) A *quaternary pulse position modulation* (quaternary PPM or 4-PPM) waveform set is given by

$$s_n(t) = \mathrm{rect}\left(\frac{t - nT_0}{2T_0}\right), \qquad n = 1, 2, 3, 4$$

where $T = 5T_0$, and is illustrated in Fig. 5.12a.

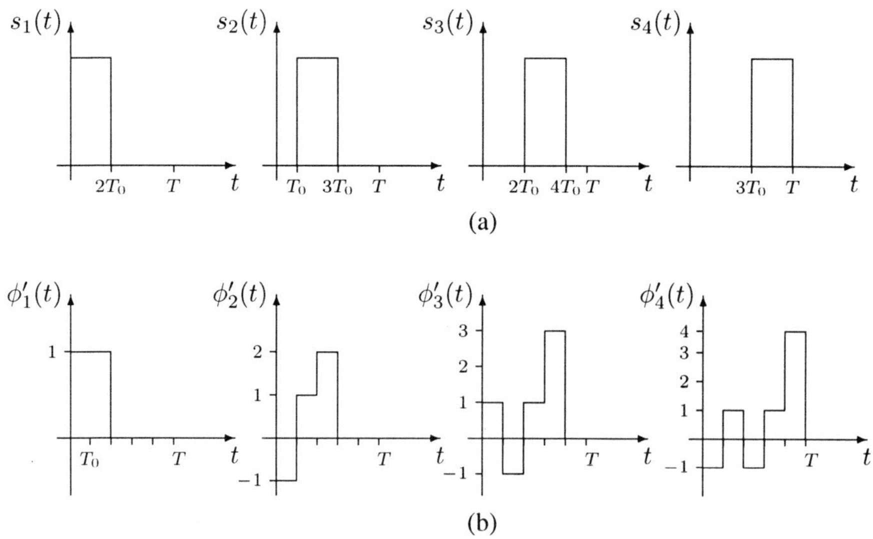

Figure 5.12 (a) 4-PPM waveform set and (b) orthogonal basis.

The waveforms have equal energy $E_s = E_1 = E_2 = E_3 = E_4 = 2T_0$, and an orthonormal basis can be identified starting from

$$\begin{aligned}
\phi_1'(t) &= s_1(t)\\
\phi_2'(t) &= -s_1(t) + 2s_2(t)\\
\phi_3'(t) &= s_1(t) - 2s_2(t) + 3s_3(t)\\
\phi_4'(t) &= -s_1(t) + 2s_2(t) - 3s_3(t) + 4s_4(t)
\end{aligned}$$

illustrated in Fig. 5.12b. We let the reader verify that the basis of Fig. 5.12b is orthogonal. Now, the *implementation type I* receiver can be implemented starting from the orthonormal set $\{\phi_i(t)\}$, or we could implement an *implementation type II* receiver starting from the waveforms $\{s_n(t)\}$ where the summations by $-\frac{1}{2}E_n$ can be avoided since all waveforms have equal energies. The computational complexity of the two receivers may seem equivalent, since $I = M$. However, *implementation type II* receiver is much easier to implement since its filters are simple integrators over a specified interval, while the *implementation type I* receive filters have much more complex shapes. So, in this case, the *implementation type II* receiver is the best choice for implementation.

5.2.9 The theorem of irrelevance

In the formalization of the optimum decision process, we have so far assumed that the noise component $w_\perp(t)$, orthogonal to the waveform space spanned by the orthonormal basis $\{\phi_i(t)\}$, $i = 1, \ldots, I$, is irrelevant to the decision process, that is

$$r(t) = \underbrace{s_{a_0}(t) + w_\phi(t)}_{\text{sufficient statistic signal}} + \underbrace{w_\perp(t)}_{\text{irrelevant signal}} . \tag{5.53}$$

In such a situation, the projection of interest was $\boldsymbol{r} = \boldsymbol{s}_{a_0} + \boldsymbol{w}_\phi$.

We now prove that the assumption made on the irrelevancy of $w_\perp(t)$ is true. Firstly, we introduce a very general result given by the *theorem of irrelevance*.

Theorem 5.1 (Theorem of irrelevance) *Let the received vector $\boldsymbol{r}$ be split into two parts, $\boldsymbol{r} = [\boldsymbol{r}_1, \boldsymbol{r}_2]$. If the conditional probability density function $p_{\boldsymbol{r}_2|\boldsymbol{r}_1,a_0}(\boldsymbol{\rho}_2|\boldsymbol{\rho}_1, n)$ does not depend on the particular value n assumed by a_0, that is if we can write*

$$p_{\boldsymbol{r}_2|\boldsymbol{r}_1,a_0}(\boldsymbol{\rho}_2|\boldsymbol{\rho}_1, n) = p_{\boldsymbol{r}_2|\boldsymbol{r}_1}(\boldsymbol{\rho}_2|\boldsymbol{\rho}_1) \tag{5.54}$$

then the optimum receiver can disregard the component $\boldsymbol{r}_2$ and base its decision only on the component $\boldsymbol{r}_1$.

Proof By recalling the property of conditional probability it is

$$p_{\boldsymbol{r}|a_0}(\boldsymbol{\rho}|n) = p_{\boldsymbol{r}_1,\boldsymbol{r}_2|a_0}(\boldsymbol{\rho}_1, \boldsymbol{\rho}_2|n) = p_{\boldsymbol{r}_2|\boldsymbol{r}_1,a_0}(\boldsymbol{\rho}_2|\boldsymbol{\rho}_1, n)\, p_{\boldsymbol{r}_1|a_0}(\boldsymbol{\rho}_1|n)$$

hence, by making use of hypothesis (5.54), we have

$$p_{\boldsymbol{r}|a_0}(\boldsymbol{\rho}|n) = p_{\boldsymbol{r}_2|\boldsymbol{r}_1}(\boldsymbol{\rho}_2|\boldsymbol{\rho}_1)\, p_{\boldsymbol{r}_1|a_0}(\boldsymbol{\rho}_1|n) .$$

This expression can now be used in the MAP decision criterion (5.41) to give

$$\begin{aligned}\hat{a}_0 &= \underset{n\in\{1,\ldots,M\}}{\text{argmax}}\ p_{\boldsymbol{r}_2|\boldsymbol{r}_1}(\boldsymbol{\rho}_2|\boldsymbol{\rho}_1)\, p_{\boldsymbol{r}_1|a_0}(\boldsymbol{\rho}_1|n)\, p_n \\ &= \underset{n\in\{1,\ldots,M\}}{\text{argmax}}\ p_{\boldsymbol{r}_1|a_0}(\boldsymbol{\rho}_1|n)\, p_n\end{aligned}$$

because $p_{\boldsymbol{r}_2|\boldsymbol{r}_1}(\boldsymbol{\rho}_2|\boldsymbol{\rho}_1)$ does not depend on n. So, the theorem is proved. □

Let V be the space spanned by the original waveforms $\{s_n(t)\}, n = 1, \ldots, M$, of dimension I and let $\{\phi_i(t)\}$, $i = 1, \ldots, I$, be an orthonormal basis for V. Use of this basis leads to the results seen so far, with a received signal projection of the form (see (5.11))

$$\boldsymbol{r} = \boldsymbol{s}_{a_0} + \boldsymbol{w}_\phi \ . \tag{5.55}$$

Let now $\{\psi_i(t)\}$, $i = 1, \ldots, I'$, with $I' > I$, be a new orthonormal basis spanning a space V' that is larger than V, that is $V \subset V'$, with $\psi_i(t) = \phi_i(t)$ for $i = 1, \ldots, I$. With this extended basis, the resulting received vector can be written as

$$\boldsymbol{r}' = [\boldsymbol{r}'_1,\ \boldsymbol{r}'_2] = [\boldsymbol{s}_{a_0} + \boldsymbol{w}_\phi,\ \boldsymbol{w}_\perp] \tag{5.56}$$

where $\boldsymbol{r}'_1 = \boldsymbol{r} = \boldsymbol{s}_{a_0} + \boldsymbol{w}_\phi$ is the former received vector, and $\boldsymbol{r}'_2 = \boldsymbol{w}_\perp$ is the newly introduced vector component which solely depends on the noise $w_\perp(t)$ orthogonal to V and which is independent of a_0. Therefore we have

$$p_{\boldsymbol{r}'_2|\boldsymbol{r}'_1,a_0}(\boldsymbol{\rho}_2|\boldsymbol{\rho}_1, n) = p_{\boldsymbol{r}'_2|\boldsymbol{r}'_1}(\boldsymbol{\rho}_2|\boldsymbol{\rho}_1) \tag{5.57}$$

which assures that we can apply Theorem 5.1, and thus proves the irrelevancy of any component due to $w_\perp(t)$.

Incidentally, for a Gaussian noise, (5.57) can be further simplified since $\boldsymbol{w}_\perp$ is also independent of $\boldsymbol{w}_\phi$, and so $\boldsymbol{r}'_2$ is also independent of $\boldsymbol{r}'_1$, that is we can write

$$p_{\boldsymbol{r}'_2|\boldsymbol{r}'_1,a_0}(\boldsymbol{\rho}_2|\boldsymbol{\rho}_1, n) = p_{\boldsymbol{r}'_2}(\boldsymbol{\rho}_2) \ . \tag{5.58}$$

5.3 BINARY MODULATION

5.3.1 Error probability

The basic form of digital transmission is *binary transmission* with $M = 2$, where the modulation employs only two waveforms. If we let $s_1(t)$ and $s_2(t)$ be the reference waveforms, then the waveform basis can be identified through a standard Gram-Schmidt procedure. The resulting constellation is illustrated in Fig. 5.13 where coordinates are functions of ρ, the *correlation coefficient* between $s_1(t)$ and $s_2(t)$ which is defined as

$$\rho = \frac{\langle s_1(t), s_2(t)\rangle}{\sqrt{E_1}\,\sqrt{E_2}} \ , \tag{5.59}$$

with $|\rho| < 1$ because of the Cauchy-Schwartz inequality (2.94). Note also that here ρ is a real valued quantity. The waveform representation of Fig. 5.13 is proved in the following example.

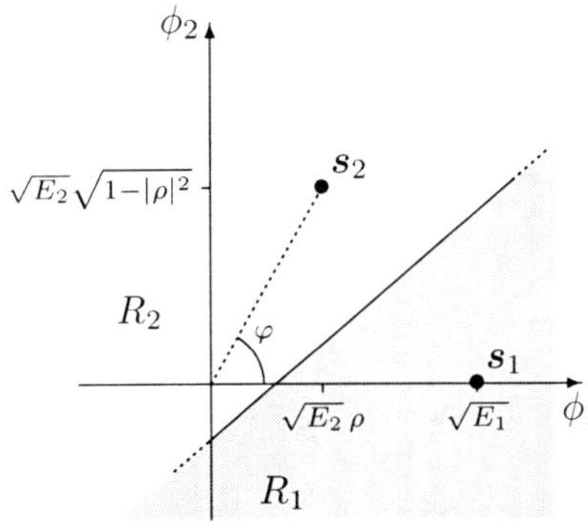

Figure 5.13 Binary constellation and minimum distance decision regions.

Example 5.3 A We want to identify an orthonormal basis and the corresponding signal constellation for a binary transmission employing general waveforms $s_1(t)$ and $s_2(t)$.

Starting from $\phi_1'(t) = s_1(t)$, the first signal of the basis is

$$\phi_1(t) = \frac{s_1(t)}{\sqrt{E_1}} . \tag{5.60}$$

The second signal of the basis can be obtained, as in Fig. 2.29, starting from the signal

$$\phi_2'(t) = s_2(t) - c\,\phi_1(t) , \qquad c = \langle s_2(t), \phi_1(t) \rangle . \tag{5.61}$$

The main difficulty here is to determine the energy of $\phi_2'(t)$,

$$\begin{aligned} E_2' &= \langle s_2(t) - c\,\phi_1(t), s_2(t) - c\,\phi_1(t) \rangle \\ &= \langle s_2(t), s_2(t) \rangle - c^* \langle s_2(t), \phi_1(t) \rangle - c \langle \phi_1(t), s_2(t) \rangle + |c|^2 \langle \phi_1(t), \phi_1(t) \rangle \\ &= E_2 - c^* c - c\,c^* + |c|^2 = E_2 - |c|^2 . \end{aligned} \tag{5.62}$$

Hence, from (5.61) and (5.62), we have

$$\phi_2(t) = \frac{s_2(t) - c\,\phi_1(t)}{\sqrt{E_2 - |c|^2}} . \tag{5.63}$$

The vector representation of the waveforms is thus

$$\boldsymbol{s}_1 = \left[\sqrt{E_1}, 0\right] , \qquad \boldsymbol{s}_2 = \left[c, \sqrt{E_2 - |c|^2}\right] . \tag{5.64}$$

With the notation (5.59) in mind we further have $c = \sqrt{E_2}\,\rho$, and the vector representation (5.64) can be rewritten as

$$\boldsymbol{s}_1 = \left[\sqrt{E_1}, 0\right] , \qquad \boldsymbol{s}_2 = \left[\sqrt{E_2}\,\rho, \sqrt{E_2}\sqrt{1 - |\rho|^2}\right] . \tag{5.65}$$

The reference constellation is shown in Fig. 5.13 together with the minimum distance decision regions R_1 and R_2. By observing Fig. 5.13, where φ is the phase of $\boldsymbol{s}_2$, we can also write $\rho = \cos(\varphi)$ and so $\boldsymbol{s}_2 = \sqrt{E_2}\,[\cos(\varphi), \sin(\varphi)]$.

In evaluating the probability of correct reception, $\mathrm{P}[\mathrm{C}]$, we assume that waveforms (i.e. symbol values) are equally likely, $p_1 = p_2 = \frac{1}{2}$, so that the optimum regions are minimum distance regions, as illustrated in Fig. 5.13. The derivation of $\mathrm{P}[\mathrm{C}]$, outlined in the following example, makes use of the distance d between the reference waveforms. From (2.123) and the definition of ρ in (5.59), we have

$$d = d(s_1(t), s_2(t)) = \sqrt{E_1 + E_2 - 2\rho\sqrt{E_1 E_2}} \tag{5.66}$$

which, for *equal energy* waveforms, becomes

$$d = d(s_1(t), s_2(t)) = \sqrt{2E_s(1-\rho)}\,, \qquad E_s = E_1 = E_2\,. \tag{5.67}$$

Example 5.3 B We want to evaluate the probability of correct reception $\mathrm{P}[\mathrm{C}]$ of a binary transmission with constellation $\{\boldsymbol{s}_1, \boldsymbol{s}_2\}$. We start by rewriting (5.19) in a binary context, for which we have

$$\mathrm{P}[\mathrm{C}] = \tfrac{1}{2}\,\mathrm{P}[\boldsymbol{r} \in R_1 | a_0 = 1] + \tfrac{1}{2}\,\mathrm{P}[\boldsymbol{r} \in R_2 | a_0 = 2]\ . \tag{5.68}$$

We first concentrate on evaluating the first term of (5.68). Now, because of the minimum distance regions, $\mathrm{P}[\boldsymbol{r} \in R_1 | a_0 = 1]$ is the probability that $\boldsymbol{r}$ is closer to $\boldsymbol{s}_1$ than to $\boldsymbol{s}_2$ given that the transmitted symbol is $a_0 = 1$. So, we can write

$$\mathrm{P}[\boldsymbol{r} \in R_1 | a_0 = 1] = \mathrm{P}\left[d(\boldsymbol{r}, \boldsymbol{s}_1) < d(\boldsymbol{r}, \boldsymbol{s}_2) \middle| a_0 = 1\right] \tag{5.69}$$

and by expressing $\boldsymbol{r}$ as the sum $\boldsymbol{s}_1 + \boldsymbol{w}$, as in (5.11), we have

$$\begin{aligned}\mathrm{P}[\boldsymbol{r} \in R_1 | a_0 = 1] &= \mathrm{P}\left[d(\boldsymbol{w}+\boldsymbol{s}_1, \boldsymbol{s}_1) < d(\boldsymbol{w}+\boldsymbol{s}_1, \boldsymbol{s}_2) \middle| a_0 = 1\right] \\ &= \mathrm{P}\left[d(\boldsymbol{w}+\boldsymbol{s}_1, \boldsymbol{s}_1) < d(\boldsymbol{w}+\boldsymbol{s}_1, \boldsymbol{s}_2)\right]\end{aligned} \tag{5.70}$$

where the condition has been dropped since it has been already taken into account in the expansion of $\boldsymbol{r}$. The result is now obtained by use of the properties of norms and distances. By extending our reasoning to square distances, we have

$$\begin{aligned}\mathrm{P}[\boldsymbol{r} \in R_1 | a_0 = 1] &= \mathrm{P}\left[d^2(\boldsymbol{w}, \boldsymbol{0}) < d^2(\boldsymbol{w}, \boldsymbol{s}_2 - \boldsymbol{s}_1)\right] \\ &= \mathrm{P}\left[\|\boldsymbol{w}\|^2 < \|\boldsymbol{w}\|^2 + \|\boldsymbol{s}_2 - \boldsymbol{s}_1\|^2 - 2\langle \boldsymbol{w}, \boldsymbol{s}_2 - \boldsymbol{s}_1\rangle\right] \\ &= \mathrm{P}\left[\langle \boldsymbol{w}, \boldsymbol{s}_2 - \boldsymbol{s}_1\rangle < \tfrac{1}{2}\|\boldsymbol{s}_2 - \boldsymbol{s}_1\|^2\right]\end{aligned} \tag{5.71}$$

In (5.71), $\|\boldsymbol{s}_2 - \boldsymbol{s}_1\|^2 = d^2$ (see (5.66)), and

$$w_0 = \langle \boldsymbol{w}, \boldsymbol{s}_2 - \boldsymbol{s}_1\rangle = w_1\,(s_{2,1} - s_{1,1}) + w_2\,(s_{2,2} - s_{1,2}) \tag{5.72}$$

is the weighted sum of Gaussian noises w_1 and w_2, both with variance σ_I, so it is itself a Gaussian noise. By recalling the statistical independence (5.25) between w_1 and w_2, the variance of w_0 becomes

$$\begin{aligned}\mathrm{E}\left[w_0^2\right] &= \mathrm{E}\left[w_1^2\right](s_{2,1} - s_{1,1})^2 + \mathrm{E}\left[w_2^2\right](s_{2,2} - s_{1,2})^2 \\ &= \sigma_I^2\,(s_{2,1} - s_{1,1})^2 + \sigma_I^2\,(s_{2,2} - s_{1,2})^2 = \sigma_I^2\, d^2\ .\end{aligned} \tag{5.73}$$

Hence, (5.71) becomes

$$\mathrm{P}\left[\boldsymbol{r} \in R_1 | a_0 = 1\right] = \mathrm{P}\left[w_0 < \tfrac{1}{2}\,d^2\right] = 1 - \mathrm{Q}\left(\frac{\frac{1}{2}\,d^2}{\sigma_I\,d}\right) = 1 - \mathrm{Q}\left(\frac{d}{2\sigma_I}\right) . \tag{5.74}$$

Since the result (5.71) holds for the second term of (5.68) as well, as we let the reader verify, then the probability of correct decision is $\mathrm{P}\left[\mathrm{C}\right] = 1 - \mathrm{Q}\left(d/(2\sigma_I)\right)$.

So, from Example 5.3 B, the probability of correct decision $\mathrm{P}\left[\mathrm{C}\right]$ and the probability of error $\mathrm{P}\left[\mathrm{E}\right]$ are, respectively,

$$\mathrm{P}\left[\mathrm{C}\right] = 1 - \mathrm{Q}\left(\frac{d}{2\sigma_I}\right) , \qquad \mathrm{P}\left[\mathrm{E}\right] = \mathrm{Q}\left(\frac{d}{2\sigma_I}\right) , \tag{5.75}$$

which depend only on the ratio between the distance of the waveforms (at the decision point) and the standard deviation (per dimension) of the noise. Note also that, in the context of binary transmission, $\mathrm{P}\left[\mathrm{E}\right]$ is the *bit error probability*, that is

$$P_{\mathrm{bit}} = \mathrm{P}\left[\mathrm{E}\right] . \tag{5.76}$$

Remark For a given noise level σ_I, the bit error probability is lowered by increasing the distance between constellation points. For signals having equal energies where (5.67) holds, the bit error probability becomes

$$P_{\mathrm{bit}} = \mathrm{Q}\left(\sqrt{\frac{E_s\,(1-\rho)}{2\sigma_I^2}}\right) . \tag{5.77}$$

Now P_{bit} is minimized by increasing the average waveform energy E_s or by minimizing the correlation coefficient ρ. Since $|\rho| < 1$, the minimum with respect to ρ is achieved for $\rho = -1$.

Example 5.3 C (2-FSK) The modulation technique of *binary frequency shift keying* (binary FSK or 2-FSK) employs the waveforms

$$s_1(t) = \begin{cases} A\cos(2\pi(f_0 - f_d)t + \varphi_0) & , \ 0 < t < T \\ 0 & , \ \text{otherwise} \end{cases}$$

$$s_2(t) = \begin{cases} A\cos(2\pi(f_0 + f_d)t + \varphi_0) & , \ 0 < t < T \\ 0 & , \ \text{otherwise} \end{cases}$$

where φ_0 is an arbitrary phase. Note that the waveforms are two "windowed" sinusoidal functions, one with frequency $f_0 - f_d$ and the other with frequency $f_0 + f_d$ (Fig. 5.14). We call f_0 the *carrier frequency* and f_d the *frequency deviation*. We also assume that $f_0 \pm f_d \gg 1/T$, and so $f_0 \gg 1/T$, as it is common in applications. We want to determine the bit error probability P_{bit} of 2-FSK.

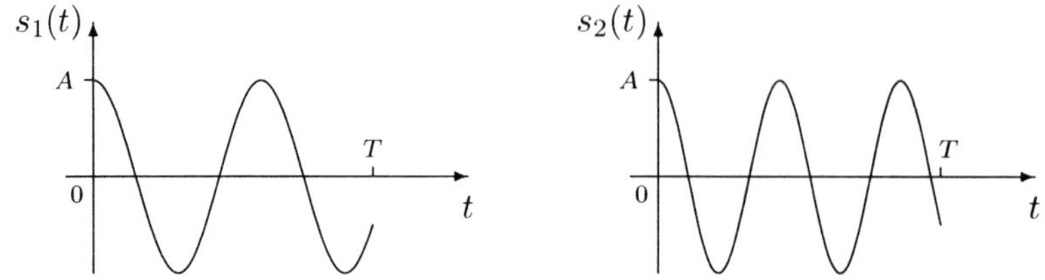

Figure 5.14 2-FSK waveform set for $f_0 = 2/T$, $f_d = 1/(3T)$ and $\varphi_0 = 0$.

From the above discussion, the main parameters of interest are the two energies, E_1 and E_2, and the correlation coefficient ρ, which we now evaluate. For the energy of $s_1(t)$ we have

$$\begin{aligned} E_1 &= \int_0^T A^2 \cos^2(2\pi(f_0 - f_d)t + \varphi_0)\, dt \\ &= \tfrac{1}{2}A^2 \int_0^T \Big[1 + \cos(2\pi\, 2(f_0 - f_d)t + 2\varphi_0)\Big]\, dt \\ &= \tfrac{1}{2}A^2 T \left[1 + \frac{\sin(2\pi\, 2(f_0 - f_d)T + 2\varphi_0) - \sin(2\varphi_0)}{2\pi\, 2(f_0 - f_d)T}\right] \simeq \tfrac{1}{2}A^2 T \end{aligned} \tag{5.78}$$

where the last equivalence is assured because $f_0 - f_d \gg 1/T$. Similarly, for the energy of $s_2(t)$ we have

$$E_2 = \tfrac{1}{2}A^2 T \left[1 + \frac{\sin(2\pi\, 2(f_0 + f_d)T + 2\varphi_0) - \sin(2\varphi_0)}{2\pi\, 2(f_0 + f_d)T}\right] \simeq \tfrac{1}{2}A^2 T\,. \tag{5.79}$$

Finally, for the correlation $\langle s_1(t), s_2(t)\rangle$, we have

$$\begin{aligned} \langle s_1(t), s_2(t)\rangle &= \int_0^T A^2 \cos(2\pi(f_0 - f_d)t + \varphi_0)\cos(2\pi(f_0 + f_d)t + \varphi_0)\, dt \\ &= \tfrac{1}{2}A^2 \int_0^T \Big[\cos(2\pi\, 2f_0 t + 2\varphi_0) + \cos(2\pi\, 2f_d t)\Big]\, dt \\ &= \tfrac{1}{2}A^2 T \left[\frac{\sin(2\pi\, 2f_0 T + 2\varphi_0) - \sin(2\varphi_0)}{2\pi\, 2f_0 T} + \mathrm{sinc}(4f_d T)\right] \end{aligned} \tag{5.80}$$

where we have used the trigonometric equivalence $2\cos\alpha\cos\beta = \cos(\alpha+\beta) + \cos(\alpha-\beta)$.

By then considering that $f_0 \gg 1/T$ also holds, we have

$$\rho \simeq \mathrm{sinc}(4f_d T)\,, \qquad E_1 \simeq E_2 \simeq \tfrac{1}{2}A^2 T\,, \tag{5.81}$$

so that (5.67) yields the distance

$$d = \sqrt{2E_1(1-\rho)} = A\sqrt{T[1 - \mathrm{sinc}(4f_d T)]}\,, \tag{5.82}$$

and the bit error probability is, from (5.77),

$$P_{\text{bit}} = \mathrm{Q}\left(\sqrt{\frac{E_1(1-\rho)}{2\sigma_I^2}}\right) = \mathrm{Q}\left(\sqrt{\frac{A^2 T\,[1 - \mathrm{sinc}(4f_d T)]}{4\sigma_I^2}}\right)\,. \tag{5.83}$$

The minimum P_{bit} is found either by maximizing d, or by minimizing ρ. The prospect for the distance is illustrated in Fig. 5.15, where the maximum value of d is achieved somewhere between $f_d = 1/(4T)$ and $f_d = 1/(2T)$. This position must be identified numerically, giving $f_d \simeq 1.43/(4T)$, a correlation coefficient $\rho_{\min} \simeq -0.217$, and a maximum distance $d \simeq 1.1\, A\sqrt{T}$.

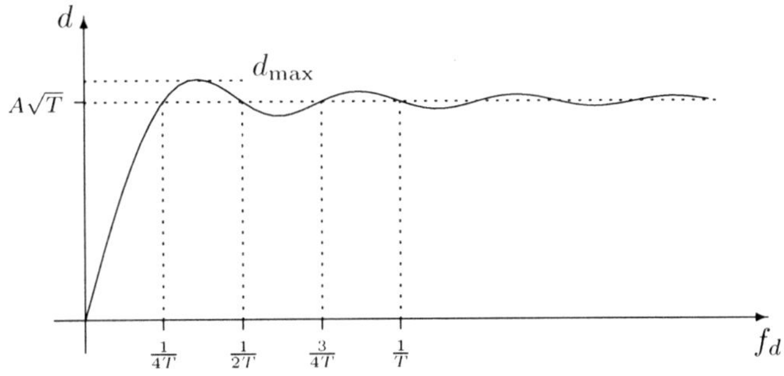

Figure 5.15 Distance as a function of f_d in 2-FSK.

5.3.2 Antipodal and orthogonal signals

We now examine in detail two peculiar classes of binary signaling, namely the class of *antipodal* signals (for which $\rho = -1$) and the class of *orthogonal* signals (for which $\rho = 0$).

Antipodal signals A binary waveform set is said to be *antipodal* when

$$s_2(t) = -s_1(t) \,. \tag{5.84}$$

Note that (5.84) is a binary waveform set with equal energies, $E_1 = E_2 = E_s$, and the correlation coefficient is

$$\rho = \frac{\langle s_1(t), s_2(t)\rangle}{\sqrt{E_1 E_2}} = -\frac{\|s_1(t)\|^2}{E_1} = -1 \,. \tag{5.85}$$

Since waveforms are linearly dependent, the basis has only one element, $I = 1$, with

$$\phi_1(t) = \frac{s_1(t)}{\sqrt{E_1}} \tag{5.86}$$

so that the vector representation of waveforms is

$$\mathbf{s}_1 = \left[\sqrt{E_s}\right], \qquad \mathbf{s}_2 = \left[-\sqrt{E_s}\right] \tag{5.87}$$

as represented in Fig. 5.16a.

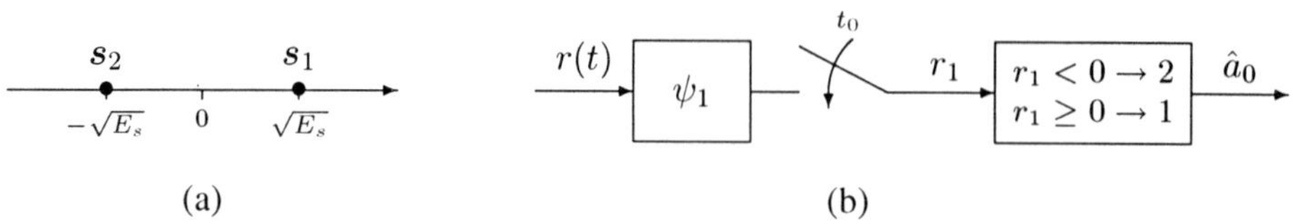

Figure 5.16 *Antipodal* transmission: (a) signal constellation; (b) receiver implementation, where $\psi_1(t) = \phi_1^*(t_0 - t) = \frac{s_1^*(t_0-t)}{\sqrt{E_1}}$.

The distance between the two waveforms is $d = 2\sqrt{E_s}$ and the minimum distance decision regions are

$$R_2 = (-\infty; 0)\ , \qquad R_1 = [0, +\infty) \tag{5.88}$$

leading to the receiver implementation illustrated in Fig. 5.16b where the decision rule is uniquely based upon the sign of r_1. Finally, from (5.77) the bit error probability is

$$P_{\text{bit}} = \mathrm{Q}\left(\sqrt{\frac{E_s}{\sigma_I^2}}\right) \tag{5.89}$$

since we are dealing with equal energy waveforms with $\rho = -1$.

Example 5.3 D (BPSK) A modulation technique with antipodal waveforms is *binary phase shift keying* (2-PSK or BPSK), where

$$s_1(t) = \begin{cases} A\cos(2\pi f_0 t) & , \ 0 < t < T \\ 0 & , \ \text{otherwise} \end{cases}$$

$$s_2(t) = \begin{cases} A\cos(2\pi f_0 t + \pi) & , \ 0 < t < T \\ 0 & , \ \text{otherwise} \end{cases}$$

In this situation the waveform energy is

$$E_s = \int_0^T A^2 \cos^2(2\pi f_0 t)\, dt = \tfrac{1}{2}A^2 \int_0^T [1 + \cos(2\pi\, 2 f_0 t)] = \tfrac{1}{2}A^2 T \left[1 + \mathrm{sinc}(4 f_0 T)\right]$$

and so the bit error probability is, from (5.89),

$$P_{\text{bit}} = \mathrm{Q}\left(\sqrt{\frac{A^2 T}{2\sigma_I^2}\left[1 + \mathrm{sinc}(4 f_0 T)\right]}\right) .$$

Orthogonal signals A binary waveform set is said to be *orthogonal* when s_1 and s_2 are orthogonal, that is when $\rho = 0$ and the basis dimension is $I = 2$. We also assume to be dealing with equal energy waveforms, $E_s = E_1 = E_2$, in which case the orthonormal basis turns out to be

$$\phi_1(t) = \frac{s_1(t)}{\sqrt{E_s}}\ , \qquad \phi_2(t) = \frac{s_2(t)}{\sqrt{E_s}} \tag{5.90}$$

and the vector representation of waveforms is

$$s_1 = \left[\sqrt{E_s},\ 0\right], \qquad s_2 = \left[0,\ \sqrt{E_s}\right] \tag{5.91}$$

as represented in Fig. 5.17a. As the two vectors s_1 and s_2 are orthogonal, their distance is $d = \sqrt{2E_s}$. This distance is reduced by a factor of $\sqrt{2}$ as compared to the case of antipodal waveforms for the same value of E_s.

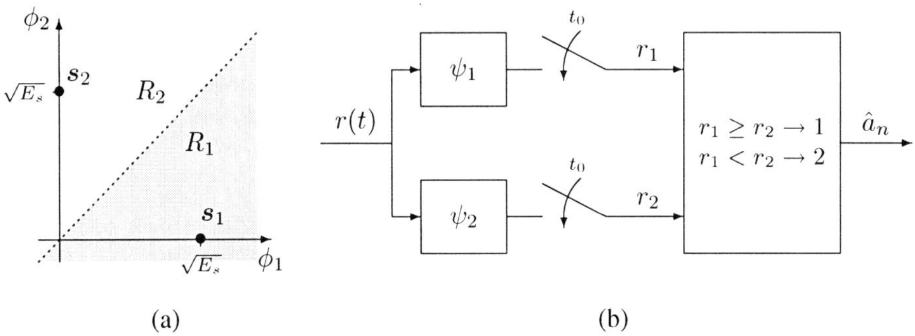

Figure 5.17 Orthogonal transmission: (a) signal constellation; (b) receiver implementation, where $\psi_i(t) = \phi_i^*(t_0 - t) = \frac{s_i^*(t_0-t)}{\sqrt{E_s}}$.

As illustrated in Fig. 5.17a, the boundary between the minimum distance decision regions is the line $\phi_2 = \phi_1$, that is the regions are

$$R_1 = \left\{(x,y) \,\middle|\, x \geq y\right\} \qquad R_2 = \left\{(x,y) \,\middle|\, x < y\right\} \tag{5.92}$$

so that the optimum receiver is illustrated in Fig. 5.17b where the decision rule is uniquely based upon the comparison between r_1 and r_2 (that is on the sign of $r_1 - r_2$). Finally, since $\rho = 0$, for the bit error probability we have (see (5.77))

$$P_{\text{bit}} = \mathrm{Q}\left(\sqrt{\frac{E_s}{2\sigma_I^2}}\right) . \tag{5.93}$$

Example 5.3 E (2-MSK) The 2-FSK waveform set of Example 5.3 C is orthogonal under specific values of the frequency deviation f_d. By recalling that the correlation coefficient is $\rho = \mathrm{sinc}(4f_dT)$, an orthogonal waveform set is obtained for $f_d = 1/(4T)$ and in this case we speak of *minimum shift keying* (MSK). There are other values of f_d ($f_d = k/(4T)$) that yield $\rho = 0$, however they imply larger f_d, with the consequent requirement of a larger channel bandwidth. In any case, the bit error probability becomes

$$P_{\text{bit}} = \mathrm{Q}\left(\sqrt{\frac{A^2T}{4\sigma_I^2}}\right) .$$

Example 5.3 F The two waveforms

$$s_1(t) = \begin{cases} \cos(2\pi f_0 t) & , \; 0 < t < T \\ 0 & , \; \text{elsewhere} \end{cases}$$

$$s_2(t) = \begin{cases} \sin(2\pi f_0 t) & , \; 0 < t < T \\ 0 & , \; \text{elsewhere} \end{cases}$$

constitute, under the condition $f_0 \gg 1/T$, an orthogonal waveform set with equal energies $E_s = \frac{1}{2}A^2T$, as proved in Example 2.5 D.

Fig. 5.18 shows the behavior of P_{bit}, in logarithmic scale, versus $E_s/N_0 = E_s/(2\sigma_I^2)$ (by (5.23)), in dB, for both antipodal (5.89) and orthogonal (5.93) signaling.

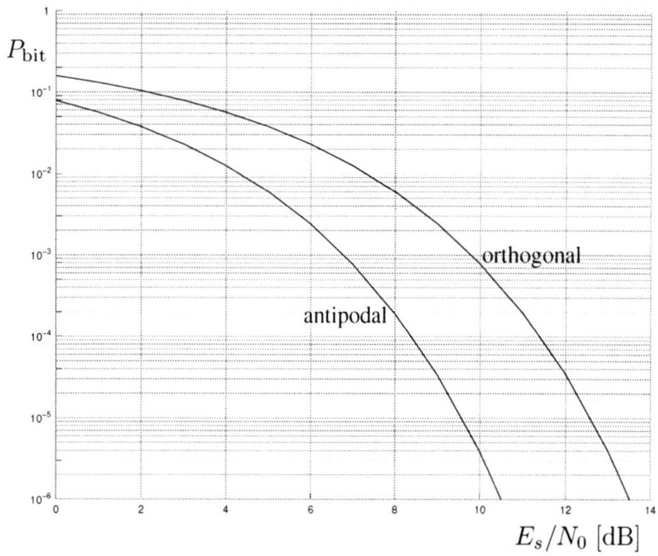

Figure 5.18 Bit error probability as a function of E_s/N_0 for binary antipodal and orthogonal signalling.

We note that, for a given P_{bit}, we have a loss of 3 dB in E_s/N_0 for the orthogonal signaling scheme as compared to the antipodal scheme, and in fact the two curves are horizontally separated by a constant 3 dB gap. Note that this 3 dB difference is due to the factor 2 in the antipodal bit error probability (5.89), since $10 \log_{10} 2 \simeq 3$ dB, and assures that antipodal modulation performs better than orthogonal modulation.

5.3.3 Single filter receivers

As we have seen, the dimension of binary signaling is in general $I = 2$ and the receiver requires a bank of two filters. However, in some specific situations where $s_1(t)$ and $s_2(t)$ are linearly dependent, the signaling dimension is $I = 1$ and the receiver requires one filter only, as illustrated in Fig. 5.16 for antipodal signals.

It is easy to see that, with binary signaling, it is always possible to implement an optimum receiver by using only one receive filter. For example, for the orthogonal transmission of Fig. 5.17 where the decision is based on the comparison between r_1 and r_2, an equivalent detector working on the difference $q = r_1 - r_2$ could be designed. Now, q can be obtained by use of a single filter with impulse response $\psi_1(t) - \psi_2(t)$, as depicted in Fig. 5.19.

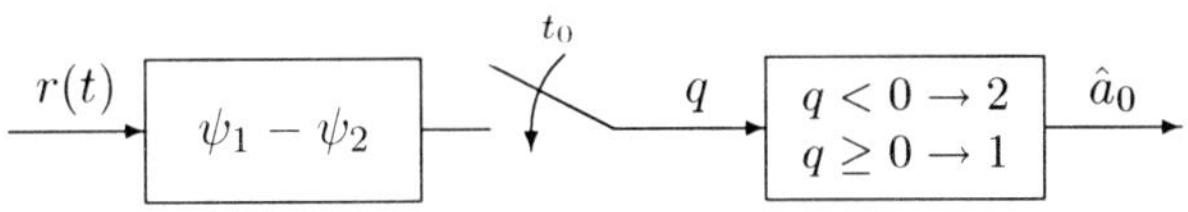

Figure 5.19 Receiver implementation for *orthogonal signaling* using a single filter, where $\psi_i(t) = \phi_i^*(t_0 - t) = \frac{s_i^*(t_0-t)}{\sqrt{E_i}}$, $i = 1, 2$.

In the general case, the derivation of the single filter receiver is obtained by carefully choosing the waveform basis $\phi_1(t)$, $\phi_2(t)$. In particular, instead of following the standard Gram-Schmidt procedure, the first signal of the basis must be chosen proportional to the difference between $s_1(t)$ and $s_2(t)$, that is $\phi_1'(t) = s_1(t) - s_2(t)$ with $E_1' = d^2$. In this situation, the first element of the basis is

$$\phi_1(t) = \frac{s_1(t) - s_2(t)}{d} \tag{5.94}$$

while the second element of the basis must be derived either from $s_1(t)$ or from $s_2(t)$ by the standard Gram-Schmidt procedure. Taking $s_1(t)$ as our reference, we have

$$\phi_2'(t) = s_1(t) - c\,\phi_1(t)\,, \qquad c = \langle s_1(t), \phi_1(t)\rangle = \frac{E_1 - \sqrt{E_1 E_2}\,\rho}{d} \tag{5.95}$$

and letting

$$E_2' = \|\phi_2'(t)\|^2\,, \tag{5.96}$$

it is

$$\phi_2(t) = \frac{\phi_2'(t)}{\sqrt{E_2'}} = \frac{s_1(t) - c\,\phi_1(t)}{\sqrt{E_2'}}\,. \tag{5.97}$$

Now, from (5.94) and (5.97), the vector representation of waveforms is

$$\boldsymbol{s}_1 = \left[c,\ \sqrt{E_2'}\right], \qquad \boldsymbol{s}_2 = \left[c - d,\ \sqrt{E_2'}\right] \tag{5.98}$$

as shown in Fig. 5.20a where the threshold v is given by

$$v = c - \tfrac{1}{2}d = \frac{2(E_1 - \sqrt{E_1 E_2}\,\rho) - d^2}{2d} = \frac{E_1 - E_2}{2d}\,. \tag{5.99}$$

We also explicit the minimum distance decision regions

$$R_1 = \left\{(x, y)\,\middle|\, x \geq v\right\} \qquad R_2 = \left\{(x, y)\,\middle|\, x < v\right\}. \tag{5.100}$$

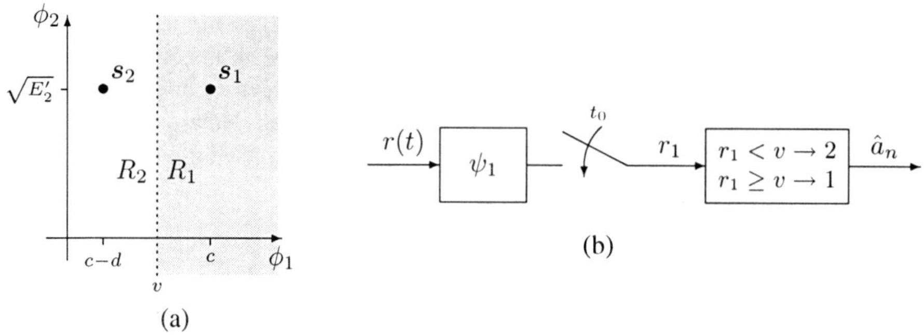

Figure 5.20 Receiver implementation for *binary signaling* using a single filter. Here the basis signal $\phi_1(t)$ is proportional to $s_1(t) - s_2(t)$: (a) signal constellation; (b) receiver with threshold $v = \frac{E_1 - E_2}{2d}$ and $\psi_1(t) = \phi_1^*(t_0 - t) = \frac{1}{d}\left[s_1^*(t_0 - t) - s_2^*(t_0 - t)\right]$.

Because of the shape of R_1 and R_2, the optimal decision rule is

$$\begin{aligned} &\text{if } \boldsymbol{r} \in R_1, \text{ that is if } r_1 \geq v, \text{ then decide } \hat{a}_0 = 1 \\ &\text{if } \boldsymbol{r} \in R_2, \text{ that is if } r_1 < v, \text{ then decide } \hat{a}_0 = 2. \end{aligned} \tag{5.101}$$

Note that the value r_2 is not used. The resulting receiver scheme is presented in Fig. 5.20b.

Example 5.3 G We let the reader verify the general result (5.75), $P_{\text{bit}} = Q\left(\frac{d}{2\sigma_I}\right)$, for a binary transmission of equally likely waveforms, by using the above signal basis and decision rule (5.101).

Example 5.3 H (2-PPM) Let the binary waveform set be 2-PPM

$$s_1(t) = A \operatorname{rect}\left(\frac{t - \frac{1}{2}T_0}{T_0}\right) \qquad s_2(t) = s_1(t - T_0)\,.$$

We immediately note that waveforms are orthogonal, $\rho = 0$, with equal energy, $E_s = E_1 = E_2 = A^2 T_0$, so that their distance is $d = \sqrt{2E_s}$. Now, the filter $\psi_1(t) = \phi_1^*(t_0 - t)$ of Fig. 5.20b can be designed by use of (5.94) and by setting $t_0 = 2T_0$ to guarantee causality. So, we have

$$\psi_1(t) = \frac{s_1^*(2T_0 - t) - s_2^*(2T_0 - t)}{A\sqrt{2T_0}} = \begin{cases} -\dfrac{1}{\sqrt{2T_0}} & , \ 0 < t < T_0 \\ +\dfrac{1}{\sqrt{2T_0}} & , \ T_0 < t < 2T_0 \\ 0 & , \ \text{otherwise} \end{cases}$$

while from (5.99) the threshold value is $v = 0$.

Remark An alternative receiver, equivalent to that of Fig. 5.20b, is illustrated in Fig. 5.21 where the filter shape is now

$$\xi_1(t) = d\psi_1(t) = s_1^*(t_0 - t) - s_2^*(t_0 - t)$$

and the threshold level in the decision point is set to 0 through a shift by $\frac{1}{2}(E_2 - E_1)$ of r_1, so that the resulting detector is a simple sign discriminator. This is quite a common approach to binary receivers, which is further simplified when the waveforms have equal energies, in which case the shift value is 0.

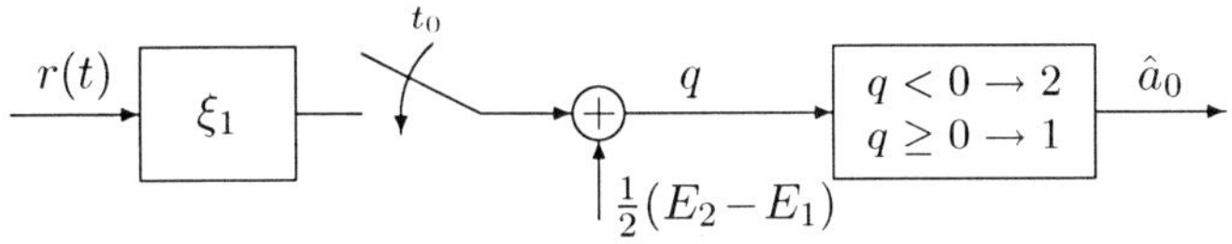

Figure 5.21 Receiver implementation for *binary signaling* using a single filter with $\xi_1(t) = s_1^*(t_0 - t) - s_2^*(t_0 - t)$.

Note also that the scheme of Fig. 5.21 can be easily derived from the *implementation type II* receiver of Fig. 5.11 when $M = 2$. In fact q in Fig. 5.21 is equivalent to the difference $U_1 - U_2$ of Fig. 5.11.

5.4 M-ARY MODULATION

5.4.1 Bounds on the error probability

Unlike the binary case, for M-ary modulations with $M > 2$ a general expression for orthonormal bases and waveform constellations cannot be found. So, we assume that we are dealing with a generic orthonormal basis $\{\phi_i(t)\}$, $i = 1, \ldots, I$, and that the resulting waveform constellation is $\{\boldsymbol{s}_n\}$, $n = 1, \ldots, M$. We also assume that symbol values are equally likely, $p_n = 1/M$, so that the optimum decision criterion is that associated to minimum distance regions.

In this context, a particular role is played by the relative distances

$$d_{m,n} = d(\boldsymbol{s}_m, \boldsymbol{s}_n) \tag{5.102}$$

and the *minimum distance*

$$d_{\min} = \min_{\substack{m,n \\ m \neq n}} d_{m,n} \ , \tag{5.103}$$

which allows us to obtain upper and lower bounds to the probability of error (5.20) by simple criteria that exploit the closed form error probability expression of the binary case.

Upper bound The upper bound derivation relies on (5.20) where the conditional probability of error, $\mathrm{P}[\boldsymbol{r} \notin R_n | a_0 = n]$, can be rewritten by taking into account the minimum distance criterion. In particular, $\mathrm{P}[\boldsymbol{r} \notin R_n | a_0 = n]$ expresses the probability that $\boldsymbol{s}_n$ is not the closest constellation point to $\boldsymbol{r}$, and this is the probability that (at least) one of the following events is verified:

$$\begin{aligned} \mathcal{E}_{m,n} &= \left\{ \boldsymbol{r} \text{ is closer to } \boldsymbol{s}_m \text{ than } \boldsymbol{s}_n \middle| a_0 = n \right\} \\ &= \left\{ d(\boldsymbol{r}, \boldsymbol{s}_n) > d(\boldsymbol{r}, \boldsymbol{s}_m) \middle| a_0 = n \right\} \ , \qquad m \neq n \ . \end{aligned} \tag{5.104}$$

We stress the fact that $\mathcal{E}_{m,n}$ concerns the event that the received signal vector $\boldsymbol{r}$ is closer to $\boldsymbol{s}_m$ than to $\boldsymbol{s}_n$, i.e. we are in the presence of *binary detection*. So the conditional probability of error can be expressed by the *union* of the events in (5.104), that is by

$$\mathrm{P}\left[\boldsymbol{r} \notin R_n | a_0 = n\right] = \mathrm{P}\left[\bigcup_{\substack{m=1\\m\neq n}}^{M} \mathcal{E}_{m,n}\right] . \tag{5.105}$$

Therefore, by applying the *union bound* [5], we obtain

$$\mathrm{P}\left[\boldsymbol{r} \notin R_n | a_0 = n\right] \leq \sum_{\substack{m=1\\m\neq n}}^{M} \mathrm{P}\left[\mathcal{E}_{m,n}\right] \tag{5.106}$$

where equality holds only if the events $\{\mathcal{E}_{m,n}\}$ are disjoint. As a matter of fact, the upper bound in (5.106) can be evaluated in closed form by exploiting (5.75) of binary modulations. We have

$$\mathrm{P}\left[\mathcal{E}_{m,n}\right] = \mathrm{Q}\left(\frac{d_{m,n}}{2\sigma_I}\right) . \tag{5.107}$$

So, by combining (5.20) with (5.106) and (5.107), the bound on error probability becomes

$$\boxed{\mathrm{P}\left[\mathrm{E}\right] \leq \frac{1}{M} \sum_{n=1}^{M} \sum_{\substack{m=1\\m\neq n}}^{M} \mathrm{Q}\left(\frac{d_{m,n}}{2\sigma_I}\right)} \tag{5.108}$$

where we have taken into account that symbol values are equally likely, $p_n = 1/M$. Note that, when the noise power σ_I^2 is sufficiently small, then (5.108) is quite a strict bound. The upper bound (5.108) can be further simplified by using the minimum distance (5.103) to obtain the looser bound

$$\mathrm{P}\left[\mathrm{E}\right] \leq (M-1)\, \mathrm{Q}\left(\frac{d_{\min}}{2\sigma_I}\right) . \tag{5.109}$$

Further insight The derivation of (5.108) and (5.109) by use of the union bound is by far the most used in the literature. Here we present an alternative derivation. The idea is to start from the conditional probability of error, $\mathrm{P}\left[\boldsymbol{r} \notin R_n | a_0 = n\right]$, and to express the condition of not belonging to the region R_n as the condition of belonging to one of the regions $\{R_m\}$, with $m \neq n$. So, we have

$$\left\{\boldsymbol{r} \notin R_n | a_0 = n\right\} = \bigcup_{\substack{m=1\\m\neq n}}^{M} \left\{\boldsymbol{r} \in R_m | a_0 = n\right\} \tag{5.110}$$

where events are now disjoint (because decision regions are disjoint), and so

$$\mathrm{P}\left[\boldsymbol{r} \notin R_n | a_0 = n\right] = \sum_{\substack{m=1\\m\neq n}}^{M} \mathrm{P}\left[\boldsymbol{r} \in R_m | a_0 = n\right] . \tag{5.111}$$

We can now upper bound the terms in (5.111). We have the inequality

$$\mathrm{P}\left[\boldsymbol{r} \in R_m | a_0 = n\right] \leq \mathrm{P}\left[\boldsymbol{r} \text{ is closer to } \boldsymbol{s}_m \text{ than } \boldsymbol{s}_n \middle| a_0 = n\right] = \mathrm{P}\left[\mathcal{E}_{m,n}\right] . \tag{5.112}$$

The pictorial interpretation of this inequality is shown in Fig. 5.22 for $M = 3$ and $I = 2$, and by assuming $n = 3$ and $m = 1$.

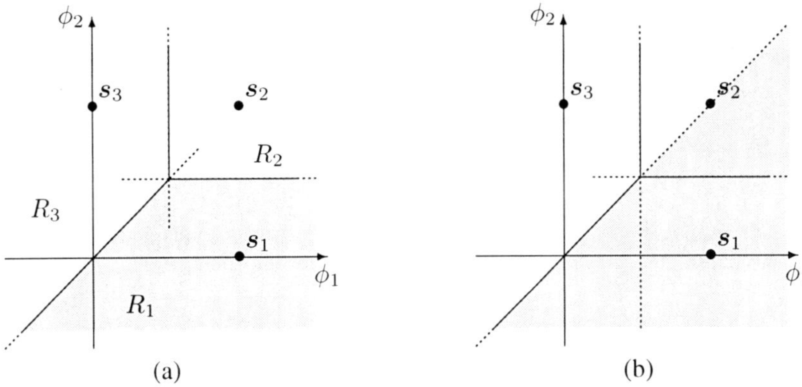

Figure 5.22 Pictorial interpretation of the upper bound (5.112) for $n = 3$ and $m = 1$: (a) region $R_m = R_1$; (b) region associated to the event $\mathcal{E}_{m,n} = \mathcal{E}_{1,3}$.

Specifically, $\mathrm{P}\left[\boldsymbol{r} \in R_1 | a_0 = 3\right]$ is the probability that $\boldsymbol{r}$ belongs to the gray region of Fig. 5.22a. Similarly, the probability $\mathrm{P}\left[\mathcal{E}_{1,3}\right]$ is the probability that $\boldsymbol{r}$ is closer to $\boldsymbol{s}_1$ than $\boldsymbol{s}_3$, that is the probability that $\boldsymbol{r}$ belongs to the gray region of Fig. 5.22b. Since the gray region of Fig. 5.22b contains that of Fig. 5.22a, the upper bound (5.112) is assured.

The combination of (5.20) with (5.111) and (5.112) finally leads to the bound (5.108).

Example 5.4 A The evaluation of the upper bound to the error probability for the 4-PPM waveform set of Example 5.2 D can be obtained by first evaluating the waveform distances. From (5.102) and (5.103), we have

$$d_{\min} = d_{1,2} = d_{2,3} = d_{3,4} = \sqrt{2T_0} \,, \qquad d_{1,3} = d_{1,4} = d_{2,4} = 2\sqrt{T_0}$$

so that the upper bounds (5.108) and (5.109) give

$$\mathrm{P}\left[\mathrm{E}\right] \leq \tfrac{3}{2}\,\mathrm{Q}\left(\frac{\sqrt{2T_0}}{2\sigma_I}\right) + \tfrac{3}{2}\,\mathrm{Q}\left(\frac{2\sqrt{T_0}}{2\sigma_I}\right) \leq 3\,\mathrm{Q}\left(\frac{\sqrt{2T_0}}{2\sigma_I}\right) .$$

Lower bound The derivation of the lower bound is rather straightforward by exploiting a pictorial approach. The starting point is again (5.20) where the conditional probability of error, $\mathrm{P}\left[\boldsymbol{r} \notin R_n | a_0 = n\right]$, can be lower bounded by

$$\mathrm{P}\left[\boldsymbol{r} \notin R_n | a_0 = n\right] \geq \mathrm{P}\left[\boldsymbol{r} \text{ is closer to } \boldsymbol{s}_m \text{ than } \boldsymbol{s}_n \middle| a_0 = n\right] = \mathrm{P}\left[\mathcal{E}_{m,n}\right] \tag{5.113}$$

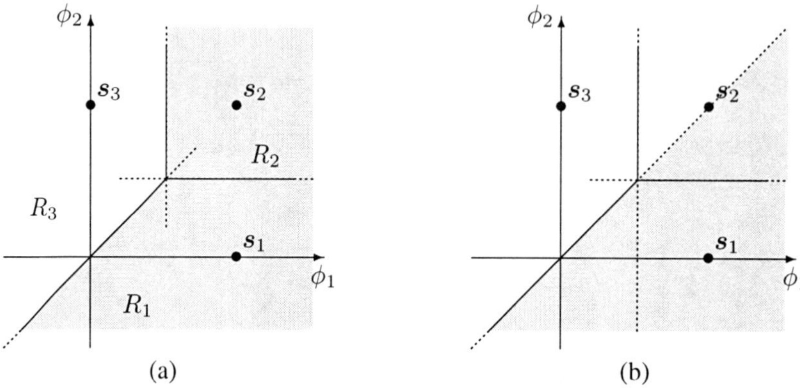

Figure 5.23 Pictorial interpretation of the lower bound to the error probability under the condition that $a_0 = n = 3$ was transmitted; gray areas indicate areas over which integration of $p_{\boldsymbol{r}_n|a_n}$ is performed.

for any $m \neq n$. The graphical interpretation of this result is given in Fig. 5.23 for $M = 3$ and $I = 2$, and by assuming $n = 3$ and $m = 1$.

Specifically, $\mathrm{P}\left[\boldsymbol{r} \notin R_3 | a_0 = 3\right]$ is the probability that $\boldsymbol{r}$ belongs to the gray region of Fig. 5.23a. Similarly, the probability $\mathrm{P}\left[\mathcal{E}_{1,3}\right]$ is the probability that $\boldsymbol{r}$ is closer to $\boldsymbol{s}_1$ than to $\boldsymbol{s}_3$, that is the probability that $\boldsymbol{r}$ belongs to the gray region of Fig. 5.23b. Since the gray region of Fig. 5.23b is contained in that of Fig. 5.23a, the lower bound (5.113) is assured.

We note that (5.113) is valid for any $m \neq n$. To assure that the lower bound is as tight as possible, it is convenient to select in (5.113) the value of $m \neq n$ with corresponding waveform vector at the minimum distance from $\boldsymbol{s}_n$. By defining

$$d_{\min,n} = \min_{\substack{m \in \{1,\ldots,M\} \\ m \neq n}} d_{m,n} \tag{5.114}$$

the lower bound is finally obtained by combining (5.20) with (5.107) and (5.113). We obtain

$$\boxed{\mathrm{P}\left[\mathrm{E}\right] \geq \frac{1}{M} \sum_{n=1}^{M} \mathrm{Q}\left(\frac{d_{\min,n}}{2\sigma_I}\right) .} \tag{5.115}$$

As for the upper bound, the lower bound (5.115) can be further simplified by selecting only those distances $\{d_{\min,n}\}$ which are equal to the minimum constellation distance $d_{\min}$. By assuming that these are in number of $N_{\min}$, we have

$$\mathrm{P}\left[\mathrm{E}\right] \geq \frac{N_{\min}}{M} \mathrm{Q}\left(\frac{d_{\min}}{2\sigma_I}\right) . \tag{5.116}$$

Example 5.4 B The evaluation of the lower bound to the error probability for the 4-PPM waveform set of Example 5.2 D can be obtained by first evaluating minimum distances $\{d_{\min,n}\}$ (see also Example 5.4 A). We have

$$d_{\min,1} = d_{\min,2} = d_{\min,3} = d_{\min,4} = \sqrt{2T_0} ,$$

so that

$$\mathrm{P}[\mathrm{E}] \geq \mathrm{Q}\left(\sqrt{\frac{T_0}{N_0}}\right) .$$

We note that in this case we have $d_{\min,n} = d_{\min}$, $n = 1, \ldots, 4$, and so the bound cannot be further simplified.

Example 5.4 C For the waveform constellation of Example 5.2 C, where $M = 3$ and $I = 2$, the distances are

$$d_{1,2} = d_{2,3} = \frac{A}{2}\sqrt{T} , \qquad d_{1,3} = \frac{A}{\sqrt{2}}\sqrt{T}$$

with

$$d_{\min} = d_{\min,1} = d_{\min,2} = d_{\min,3} = \frac{A}{2}\sqrt{T}$$

so that the bounds on the error probability become

$$\mathrm{Q}\left(\sqrt{\frac{A^2T}{8N_0}}\right) \leq \mathrm{P}[\mathrm{E}] \leq \tfrac{4}{3}\mathrm{Q}\left(\sqrt{\frac{A^2T}{8N_0}}\right) + \tfrac{2}{3}\mathrm{Q}\left(\sqrt{\frac{A^2T}{4N_0}}\right) \leq 2\mathrm{Q}\left(\sqrt{\frac{A^2T}{8N_0}}\right) .$$

5.4.2 Orthogonal and bi-orthogonal modulations

As for the case of binary signaling, we now introduce two classes of M-ary signaling that are of interest. These are the classes of *orthogonal* and of *bi-orthogonal* modulations.

Orthogonal modulation An M-ary waveform set is said to be *orthogonal* when the waveforms are pairwise orthogonal, that is

$$\langle s_m(t), s_n(t)\rangle = 0 , \quad n \neq m , \tag{5.117}$$

in which case an orthonormal basis has dimension $I = M$, and it is very simply derived as

$$\phi_i(t) = \frac{s_i(t)}{\sqrt{E_i}} , \quad i = 1, \ldots, M , \tag{5.118}$$

and the waveform vector representation can be written as

$$\boldsymbol{s}_n = \sqrt{E_n}\,\boldsymbol{e}_n , \qquad \boldsymbol{e}_n = [\underbrace{0, 0, \ldots, 0}_{n-1}, 1, \underbrace{0, \ldots, 0}_{M-n}] \tag{5.119}$$

where $\boldsymbol{e}_n$ is a vector with a single 1 in position n, and the rest set to 0. Typically waveforms have equal energy $E_s = E_n$, $n = 1, \ldots, M$, and so they are all equally spaced

$$d_{m,n} = \sqrt{2E_s} , \qquad n \neq m \tag{5.120}$$

in which case (5.120) also expresses the minimum distance,

$$d_{\min} = \sqrt{2E_s} . \tag{5.121}$$

By assuming that (5.120) and (5.121) are verified, the minimum distance regions (5.46) assume a particularly simple expression. Since, according to (5.46), R_n is the region of points $\boldsymbol{\rho}$ that are closer to $\boldsymbol{s}_n$ than to any other point $\boldsymbol{s}_m$, with $m \neq n$, for expressing R_n we are interested in knowing when the condition

$$\left\{ d(\boldsymbol{\rho}, \boldsymbol{s}_n) < d(\boldsymbol{\rho}, \boldsymbol{s}_m) \right\} \tag{5.122}$$

is verified. By expanding (5.122), with $\boldsymbol{\rho} = [\rho_1, \ldots, \rho_M]$ we have

$$\rho_1^2 + \ldots + (\rho_n - \sqrt{E_s})^2 + \ldots + \rho_M^2 < \rho_1^2 + \ldots + (\rho_m - \sqrt{E_s})^2 + \ldots + \rho_M^2$$

which is equivalent to writing

$$\rho_n > \rho_m$$

so that the region R_n becomes

$$R_n = \left\{ \boldsymbol{\rho} \,\middle|\, \rho_1 < \rho_n, \rho_2 < \rho_n, \ldots, \rho_{n-1} < \rho_n, \rho_{n+1} < \rho_n, \ldots, \rho_M < \rho_n \right\} . \tag{5.123}$$

This result is in perfect accordance with the *implementation type II* receiver structure of Fig. 5.11 where in this case $U_n = \sqrt{E_s}\,\rho_n - \frac{1}{2}E_s$, so that selecting the biggest U_n is equivalent to selecting the biggest ρ_n, as in (5.123).

In any case, the probability of correct decision can be evaluated from (5.30) where the conditional probability $\mathrm{P}\left[\boldsymbol{r} \in R_n | a_0 = n\right]$ can be expressed by use of (5.28) as

$$\mathrm{P}\left[\boldsymbol{r} \in R_n | a_0 = n\right] = \int_{R_n} p_{\boldsymbol{w}}(\boldsymbol{\rho} - \boldsymbol{s}_n)\, d\boldsymbol{\rho}$$

which, by exploiting (5.123) and (5.25), can be further expressed as

$$\begin{aligned} \mathrm{P}\left[\boldsymbol{r} \in R_n | a_0 = n\right] &= \int_{-\infty}^{+\infty} p_{w_n}(\rho_n - \sqrt{E_s}) \left[\prod_{\substack{m=1 \\ m \neq n}}^{M} \int_{-\infty}^{\rho_n} p_{w_m}(\rho_m)\, d\rho_m \right] d\rho_n \\ &= \int_{-\infty}^{+\infty} \frac{1}{\sqrt{2\pi\sigma_I^2}} e^{-\frac{\left(\rho_n - \sqrt{E_s}\right)^2}{2\sigma_I^2}} \left[1 - \mathrm{Q}\left(\frac{\rho_n}{\sigma_I}\right)\right]^{M-1} d\rho_n \, . \end{aligned} \tag{5.124}$$

By considering the existing symmetry among waveforms, we note that $\mathrm{P}\left[\boldsymbol{r} \in R_n | a_0 = n\right]$ is independent on the chosen value of n, so that $\mathrm{P}\left[\mathrm{C}\right] = \mathrm{P}\left[\boldsymbol{r} \in R_n | a_0 = n\right]$ for any n. Anyway, this can be seen from (5.124) as well. So, by rewriting (5.124) in a more standard form and by recalling (5.24), for the symbol error probability we obtain

$$P_e = \mathrm{P}\left[\mathrm{E}\right] = 1 - \int_{-\infty}^{+\infty} \frac{1}{\sqrt{2\pi}} e^{-\frac{1}{2}\left(x - \sqrt{E_s/\sigma_I^2}\right)^2} \left[1 - \mathrm{Q}\left(x\right)\right]^{M-1} dx \tag{5.125}$$

which needs to be evaluated through numerical integration, and is illustrated in Fig. 5.24 as a function of E_s/N_0 where $N_0/2 = \sigma_I^2$, as we know from (5.23).

In the figure we also show the upper (dashed lines) and lower (dash-dot lines) bounds given by (5.109) and (5.116), respectively,

$$\mathrm{Q}\left(\sqrt{\frac{E_s}{2\sigma_I^2}}\right) \leq P_e \leq (M-1)\,\mathrm{Q}\left(\sqrt{\frac{E_s}{2\sigma_I^2}}\right) . \tag{5.126}$$

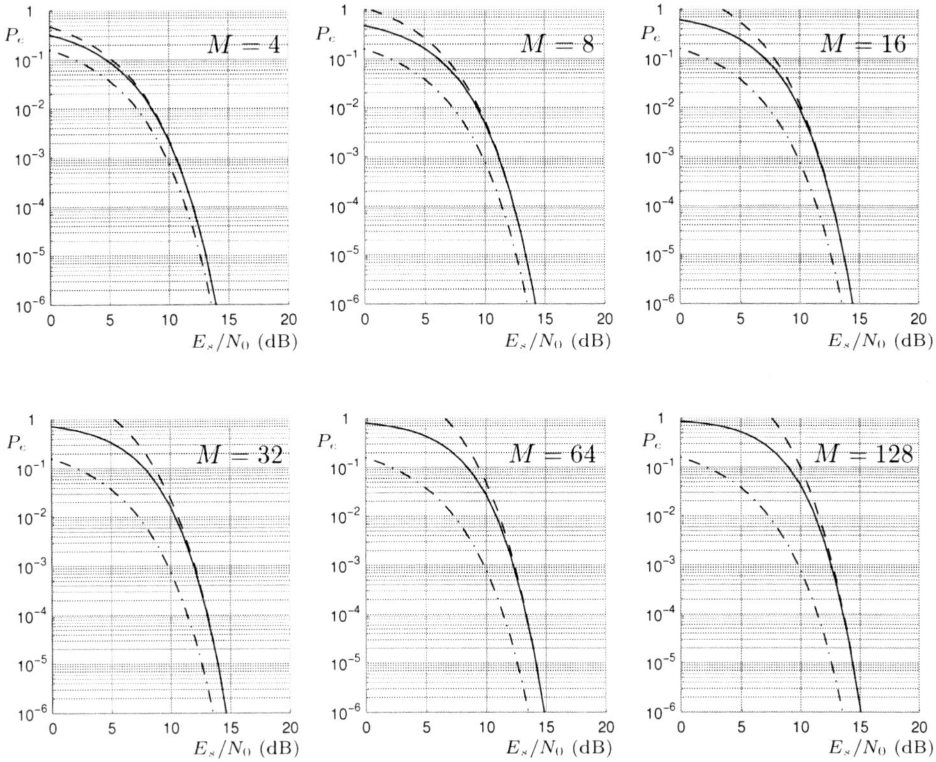

Figure 5.24 Symbol error probability P_e of M-ary orthogonal signalling as a function of E_s/N_0: error probability (5.125) in solid lines; from (5.126) upper bound in dashed lines; lower bound in dash-dot lines.

Note that the upper bound looks quite tight for high values of E_s/N_0, while the lower bound is rather loose as soon as M gets bigger than 4.

Bi-orthogonal modulation The elements of a set of M bi-orthogonal waveforms are $\frac{M}{2}$ orthogonal waveforms and their antipodal waveforms. In a sense, bi-orthogonal waveforms are the extension to $M > 2$ of the antipodal concept seen for binary transmission. For example, 4-PSK is a bi-orthogonal signaling scheme.

We assume that the bi-orthogonal waveform set is built from the $M/2$ orthogonal waveforms $s_1(t), \ldots, s_{M/2}(t)$ and that

$$s_{n+M/2}(t) = -s_n(t)\,, \quad n = 1, \ldots, M/2\,. \tag{5.127}$$

We also assume that waveforms have equal energy, that is $E_n = E_s$, $n = 1, \ldots, M$. Here, the orthonormal basis of $I = M/2$ waveforms is

$$\phi_i(t) = \frac{s_i(t)}{\sqrt{E_s}}\,, \quad i = 1, \ldots, M/2 \tag{5.128}$$

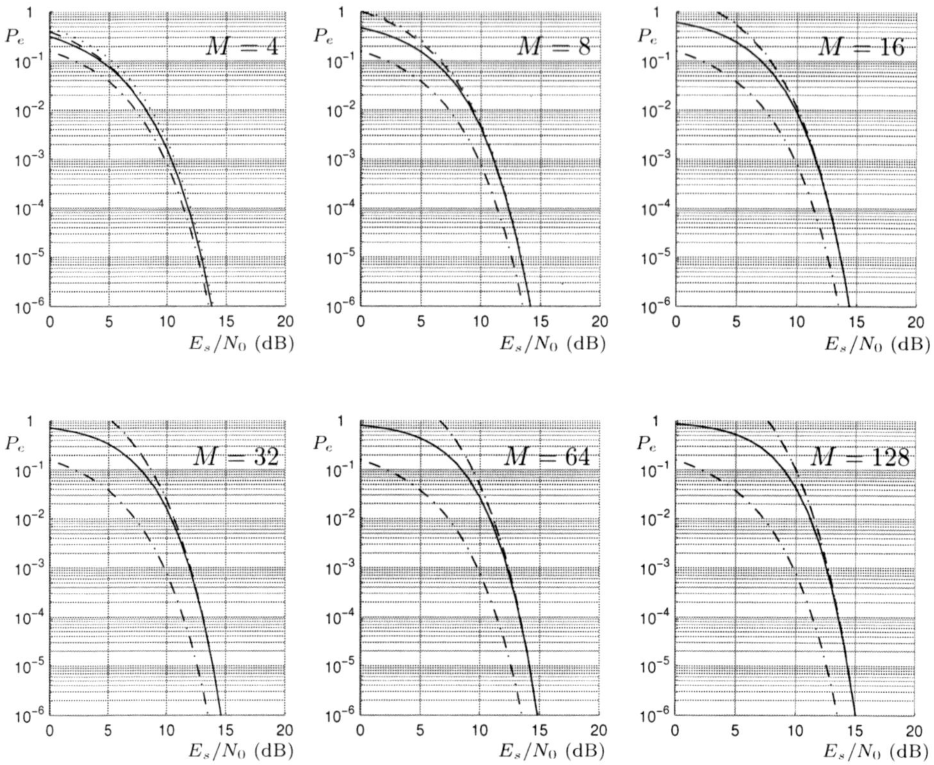

Figure 5.25 Symbol error probability P_e of M-ary bi-orthogonal signaling as a function of E_s/N_0: error probability (5.132) in solid lines; from (5.130), upper bounds in dashed and dotted lines; lower bound in dash-dot lines.

and the vector representation of waveforms is

$$\boldsymbol{s}_n = \begin{cases} +\sqrt{E_s}\,\boldsymbol{e}_n & , \; 1 \le n \le M/2 \\ -\sqrt{E_s}\,\boldsymbol{e}_{n-M/2} & , \; \frac{M}{2} < n \le M \, . \end{cases} \tag{5.129}$$

Note that in this case each waveform has $M-2$ waveforms at distance $d_\perp = \sqrt{2E_s}$, that is those lying on an orthogonal space, and one waveform at distance $d_{\text{ant}} = 2\sqrt{E_s}$, which is the antipodal counterpart. In addition $d_{\text{min}} = d_\perp$. So, the upper and lower bounds to the symbol error probability (given by (5.108), (5.109) and (5.116) respectively) become

$$\mathrm{Q}\left(\sqrt{\frac{E_s}{2\sigma_I^2}}\right) \le P_e \le (M-2)\,\mathrm{Q}\left(\sqrt{\frac{E_s}{2\sigma_I^2}}\right) + \mathrm{Q}\left(\sqrt{\frac{E_s}{\sigma_I^2}}\right) \le (M-1)\,\mathrm{Q}\left(\sqrt{\frac{E_s}{2\sigma_I^2}}\right) \, . \tag{5.130}$$

The derivation of the exact error probability is similar to the orthogonal case and requires the identification of minimum distance decision regions. We let the reader verify that the decision regions are

$$\begin{aligned} R_n &= \left\{ \boldsymbol{\rho} \,\middle|\, \rho_n > 0 \text{ and } |\rho_n| > |\rho_m|, \text{ for all } m \neq n \right\} \quad , \qquad 1 \le n \le \tfrac{M}{2} \\ R_{n+M/2} &= \left\{ \boldsymbol{\rho} \,\middle|\, \rho_n \le 0 \text{ and } |\rho_n| > |\rho_m|, \text{ for all } m \neq n \right\} \quad , \qquad 1 \le n \le \tfrac{M}{2} \, , \end{aligned} \tag{5.131}$$

respectively for the first $M/2$ signals, and for their antipodal counterparts. From this result we obtain that the symbol error probability is

$$P_e = 1 - \int_{-\infty}^{+\infty} \frac{1}{\sqrt{2\pi}} e^{-\frac{1}{2}\left(x - \sqrt{E_s/\sigma_I^2}\right)^2} \left[1 - 2\,\mathrm{Q}\,(x)\right]^{M/2-1} \, dx \tag{5.132}$$

which needs to be evaluated through numerical integration, and is illustrated in Fig. 5.25 as a function of E_s/N_0. In the figure we also show the upper and lower bounds (5.130). Note that the two upper bounds are very close, while the lower bound is rather loose.

5.5 THE DIGITAL MODULATION SYSTEM

5.5.1 System overview

The transmission model of Fig. 5.2 was based on the assumption that a single symbol had to be transmitted. However, in a digital communication system we wish to transmit a sequence of symbols. More commonly, we wish to transmit a *binary sequence*

$$\{\ldots, b_{-1}, b_0, b_1, b_2, \ldots, b_\ell, \ldots\}$$

where each symbol b_ℓ belongs to a binary alphabet such as $\{0, 1\}$ or $\{-1, 1\}$ or $\{1, 2\}$. The way this transmission takes place is illustrated in Fig. 5.26, where the binary sequence $\{b_\ell\}$ is first mapped into a symbol sequence $\{\ldots, a_{-1}, a_0, a_1, a_2, \ldots, a_k, \ldots\}$ with symbols belonging to an M-ary alphabet. Next an M-ary modulator such as that of Fig. 5.2 is used.

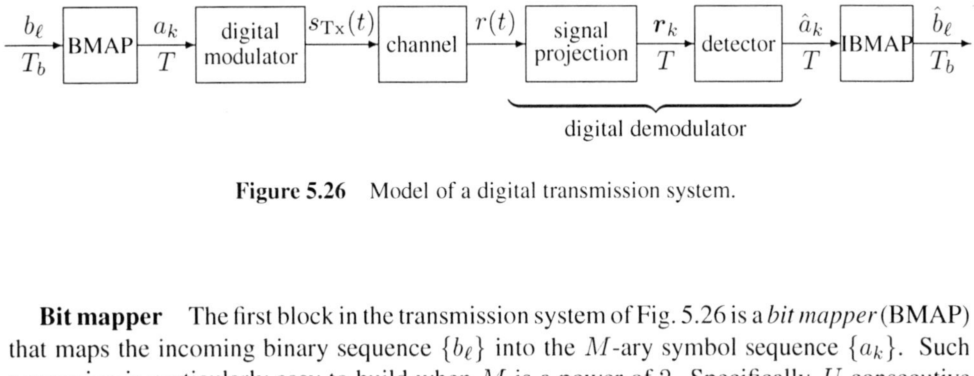

Figure 5.26 Model of a digital transmission system.

Bit mapper The first block in the transmission system of Fig. 5.26 is a *bit mapper* (BMAP) that maps the incoming binary sequence $\{b_\ell\}$ into the M-ary symbol sequence $\{a_k\}$. Such a mapping is particularly easy to build when M is a power of 2. Specifically, U consecutive bits are grouped to form a binary word $\boldsymbol{b}_k$ of length $U = \log_2(M)$, which is then one-to-one mapped into the symbol a_k, that is

$$\boldsymbol{b}_k = [b_{kU}, b_{kU+1}, \ldots, b_{kU+U-1}] \quad \overset{\text{BMAP}}{\Longrightarrow} \quad a_k = \mathrm{BMAP}(\boldsymbol{b}_k) \, . \tag{5.133}$$

When $b_\ell \in \{0, 1\}$ one possible mapping is to choose $\boldsymbol{b}_k$ as the binary representation of a_k. However, other, more suitable one-to-one mappings between $\boldsymbol{b}_k$ and a_k are used in practice. In any case, by letting T_b be the quantum or *bit period* of the binary sequence $\{b_\ell\}$, and T be the quantum or *symbol period* of the M-ary sequence $\{a_k\}$, we have

$$T = T_b \log_2 M \ . \tag{5.134}$$

By further introducing the *bit rate* $R_b = 1/T_b$ and the *symbol rate* $F = 1/T$, we also have

$$F = \frac{R_b}{\log_2 M} \ . \tag{5.135}$$

Incidentally, the symbol rate F is measured in baud.

Digital modulation In the digital modulator of Fig. 5.26 the transmission of a waveform is repeated every *symbol period* T. In particular, if at instant kT the symbol $a_k = n$ is selected, the modulator transmits the waveform $s_n(t)$ shifted in time of kT, that is we have

$$a_k = n \longrightarrow s_{a_k}(t - kT) = s_n(t - kT) \ . \tag{5.136}$$

Therefore the digital modulator generates the signal

$$s_{\text{Tx}}(t) = \sum_{k=-\infty}^{+\infty} s_{a_k}(t - kT) \ , \tag{5.137}$$

which is sent over the channel.

Channel The received signal $r(t)$ is related to $s_{\text{Rc}}(t)$ and the noise according to (5.1). The received waveforms are given by $\{\tilde{s}_n(t) = (s_n * g_{\text{Ch}})(t)\}$, which determine a new orthonormal basis $\tilde{\phi}_i(t)$. In the simplified context of narrowband transmission, the channel frequency response is flat in the band of s_{Tx} and the received signal is (see Fig. 4.14)

$$s_{\text{Rc}}(t) = \mathcal{C}\, s_{\text{Tx}}(t) \quad , \qquad r(t) = \mathcal{C}\, s_{\text{Tx}}(t) + w_{\text{Rc}}(t) \ , \tag{5.138}$$

where $\mathcal{C}$ is the channel gain. Since $\tilde{s}_n(t) = \mathcal{C}\, s_n(t)$, the received waveform set is simply the transmitted waveform set scaled by $\mathcal{C}$ and both have the same reference basis, that is $\tilde{\phi}_i(t) = \phi_i(t)$.

Received power If E_s is the average signaling energy (5.27) at the receiver input, then the received average statistical power can be expressed as

$$\mathtt{M}_{s_{\text{Rc}}} = \frac{E_s}{T} \ . \tag{5.139}$$

Signal projection The demodulator in Fig. 5.26 is the standard *implementation type I* or *II* receiver of Fig. 5.10 and Fig. 5.11, respectively, with the unique variation of being able to cope with a sequence of symbols rather than with a single symbol. So, for example when using an *implementation type I* receiver and $\{\phi_i(t)\}$, $i = 1, \ldots, I$, is the waveform basis, the decision rule relative to the symbol a_k is based on the projection vector $\boldsymbol{r}_k = [r_{k,1}, r_{k,2}, \ldots, r_{k,I}]$ where

$$r_{k,i} = \langle r(t), \phi_i(t - kT) \rangle \ . \tag{5.140}$$

which is in agreement with (5.137). When the channel distorts waveforms, then a new base $\tilde{\phi}_i(t)$ should be used in place of $\phi_i(t)$.

Detector The second operation performed by the demodulator is *detection*, i.e. to decide on $\hat{a}_k$ based on the observation $\boldsymbol{\rho}_k$ of the rve $\boldsymbol{r}_k$. Depending on the chosen approach (MAP, ML, minimum distance), we have the following decision rules (see (5.41), (5.44), (5.45))

$$\hat{a}_k = \begin{cases} \underset{n\in\{1,\ldots,M\}}{\text{argmax}} \ p_{\boldsymbol{w}}(\boldsymbol{\rho}_k - \tilde{\boldsymbol{s}}_n)\, p_n & , \ \text{MAP} \\ \underset{n\in\{1,\ldots,M\}}{\text{argmax}} \ p_{\boldsymbol{w}}(\boldsymbol{\rho}_k - \tilde{\boldsymbol{s}}_n) & , \ \text{ML} \\ \underset{n\in\{1,\ldots,M\}}{\text{argmin}} \ \|\boldsymbol{\rho}_k - \tilde{\boldsymbol{s}}_n\|^2 & , \ \text{minimum distance} \end{cases} \tag{5.141}$$

where $\{\tilde{\boldsymbol{s}}_n\}$ is now the constellation of the received waveform set. With narrowband transmission, the link between the transmitted and the received constellation is simply a scale factor $\mathcal{C}$, that is $\tilde{\boldsymbol{s}}_n = \mathcal{C}\, \boldsymbol{s}_n$.

Inverse bit mapper The final block in the transmission system is the *inverse bit mapper* (IBMAP) which, using the inverse of (5.133) builds the detected binary sequence $\{\hat{b}_\ell\}$ starting from the detected symbol sequence $\{\hat{a}_k\}$.

Example 5.5 A (4-PPM) Let a digital transmission be based upon a *quaternary pulse position modulation* (quaternary PPM or 4-PPM) waveform set $s_n(t) = \text{rect}\left(t/T_0 - n - \frac{1}{2}\right)$, $n = 0, 1, 2, 3$, where $T = 4T_0$. The reference orthonormal basis is thus $\phi_i = s_i(t)/\sqrt{T_0}$, $i = 0, 1, 2, 3$ with dimension $I = 4$. The signals of Fig. 5.26 are shown in Fig. 5.27 and Fig. 5.28. Note that bits $b_\ell \in \{0, 1\}$ are arranged in groups of two to extract the quaternary symbols $a_k \in \{0, 1, 2, 3\}$ using the trivial map $a_k = b_{2k} + 2b_{2k+1}$. Then, the PPM signal $s_{\text{Tx}}(t)$ is built. Note in Fig. 5.27 how AWGN corrupts the signal. At the receiver, after projection of the received signal we obtain the vector $\boldsymbol{r}_k = [r_{k,0}, r_{k,1}, r_{k,2}, r_{k,3}]$, whose components are illustrated in the first four plots of Fig. 5.28. Being PPM an orthogonal modulation, the detector selects the index corresponding to the maximum value among the four components of $\boldsymbol{r}_k$. This is the operation performed to extract $\widehat{a}_k$ in Fig. 5.28. Then $\widehat{b}_\ell$ is obtained by using the inverse bit map. Note that, in our specific case, only one error occurs on the detected symbol sequence, as $\widehat{a}_4 = 2 \neq a_4 = 1$. Correspondingly, this yields two errors on the detected binary sequence $\widehat{b}_\ell$, namely $\widehat{b}_8 = 0 \neq b_8 = 1$ and $\widehat{b}_9 = 1 \neq b_9 = 0$. Note that a symbol error is evidenced in Fig. 5.28 by a cross, $\times$.

Further insight We note that, the decision criteria (5.141) are optimum only if the received vector can be written as

$$\boldsymbol{r}_k = \tilde{\boldsymbol{s}}_{a_k} + \boldsymbol{w}_k \ , \tag{5.142}$$

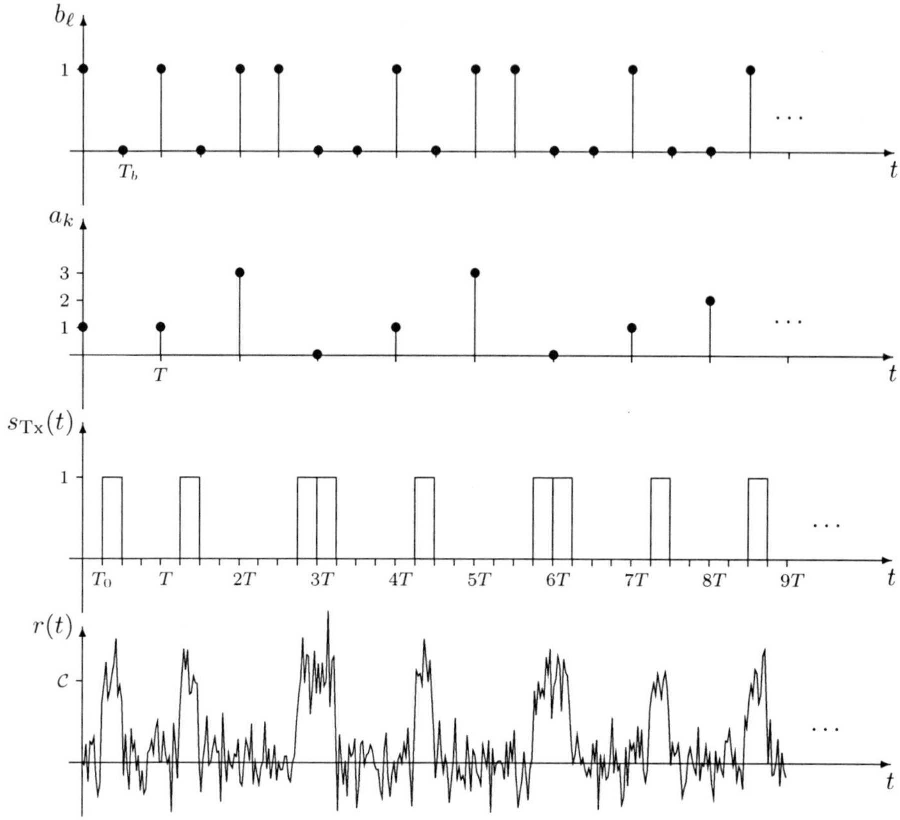

Figure 5.27 4-PPM digital transmission: signals at transmission and after the AWGN channel.

that is as the vector representation related to the transmitted symbol a_k *plus* noise (not necessarily Gaussian). This is certainly true when the waveforms $\{\tilde{s}_n(t)\}$ are confined within a symbol period T, that is

$$\tilde{s}_n(t) = 0 \,, \qquad t < 0,\, t \geq T \tag{5.143}$$

and the channel is narrowband. In this case both the received waveforms and the reference basis will be confined within a symbol period T and (5.140) will only collect the contribution due to a_k.

When the channel is wideband, or the transmitted waveforms do not satisfy (5.143), there is superpositions between waveforms transmitted at successive instants which *may* cause $\boldsymbol{r}_k$ to be written as

$$\boldsymbol{r}_k = \tilde{\boldsymbol{s}}_{a_k} + \boldsymbol{w}_k + \underbrace{f(\ldots, a_{k-2}, a_{k-1}, a_{k+1}, a_{k+2}, \ldots)}_{\text{intersymbol interference}} \tag{5.144}$$

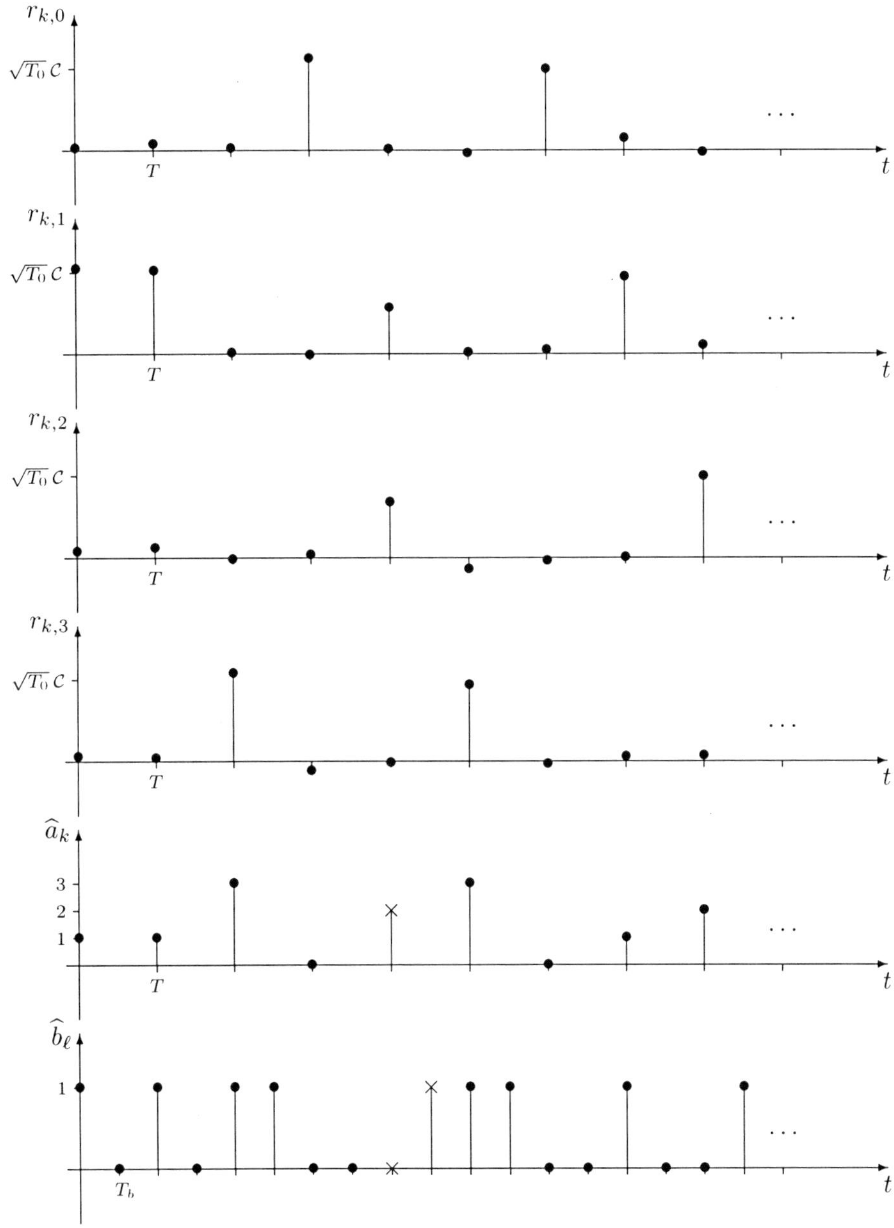

Figure 5.28 4-PPM digital transmission: signals extracted at reception; a cross, ×, evidences a symbl error.

where the contribution due to the symbols $\{\ldots, a_{k-2}, a_{k-1}, a_{k+1}, a_{k+2}, \ldots\}$ is known as *intersymbol interference* (ISI), an undesired form of disturbance that we do not take into account in this book, but which needs to be carefully considered in practical communication systems.

In narrowband channels, the general condition for the absence of ISI is

$$\langle \phi_i(t), \phi_\ell(t - kT) \rangle = \begin{cases} 1 & , \ i = \ell \ , \quad k = 0 \\ 0 & , \ \text{otherwise} \end{cases} \tag{5.145}$$

konown as *Nyquist criterion*, and expressing both the absence of ISI (for $k \neq 0$) and the orthogonality between waveforms (for $k = 0$). The criterion can be extended to any channel by simply replacing ϕ_i with $\tilde{\phi}_i$.

5.5.2 Front-end receiver implementation

In agreement with the implementation type I receiver of Fig. 5.10, the inner product (5.140) can be implemented in different ways, depending on the structure of the waveform basis $\{\phi_i(t)\}$, $i = 1, \ldots, I$.

For a generic basis, the implementation of a single branch of the receive filter bank is illustrated in Fig. 5.29a, through the cascade of a filter and a sampler. The filter has impulse response $\psi_i(t)$ as in (5.47) where t_0 is such to guarantee filter causality. The sampler collects the filter output at instants $t_0 + kT$, so that

$$(\psi_i * r)(t_0 + kT) = \int \psi_i(t_0 + kT - t)\, r(t)\, dt = \int \phi_i^*(t - kT)\, r(t)\, dt = r_{k,i} \tag{5.146}$$

as in (5.140).

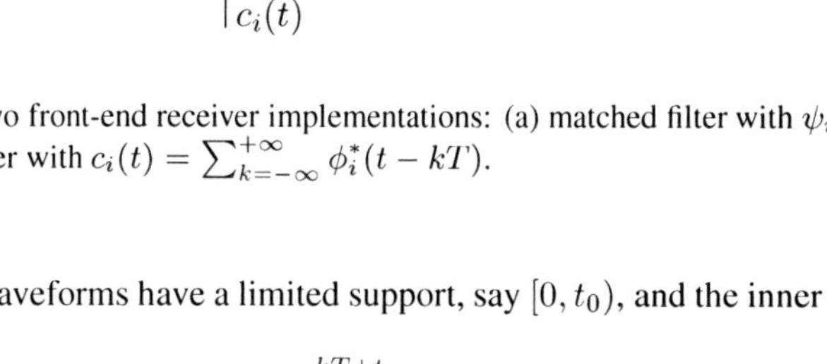

Figure 5.29 Two front-end receiver implementations: (a) matched filter with $\psi_i(t) = \phi_i^*(t_0 - t)$; (b) correlation receiver with $c_i(t) = \sum_{k=-\infty}^{+\infty} \phi_i^*(t - kT)$.

In general, waveforms have a limited support, say $[0, t_0)$, and the inner product (5.140) can be written as

$$r_{k,i} = \int_{kT}^{kT+t_0} r(t)\, \phi_i^*(t - kT)\, dt \ . \tag{5.147}$$

Since ψ_i is matched to ϕ_i, the filter bank is also called matched filter bank.

As an alternative, the receiver may be implemented as shown in Fig. 5.29b, where the received signal $r(t)$ is first multiplied by the periodic repetition of each waveform of the orthonormal basis, or more generally by its complex conjugate version

$$c_i(t) = \sum_{k=-\infty}^{+\infty} \phi_i^*(t - kT) , \tag{5.148}$$

and the result is repeatedly integrated every symbol period T, according to (5.147). This scheme is also known as *correlation receiver*.

5.5.3 The binary channel

It is useful to interpret the whole system of Fig. 5.26 as a *binary channel*, as reported in Fig. 5.30, with input the binary sequence $\{b_\ell\}$ and output the binary sequence $\{\hat{b}_\ell\}$.

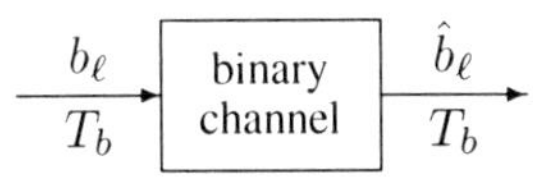

Figure 5.30 Binary channel model.

In this context, the binary channel, besides its rate $R_b = 1/T_b$, is specified (in statistical terms) by the so called *transition probabilities*

$$p_{m|n} = \mathrm{P}\left[\hat{b}_\ell = m \mid b_\ell = n\right] , \qquad m, n = 1, 2 \tag{5.149}$$

that is the probability that $\hat{b}_\ell = m$ is detected under the condition that $b_\ell = n$ was transmitted. The typical diagram representation of transition probabilities is that given in Fig. 5.31a, where we note that only two quantities must be defined since

$$p_{1|1} + p_{2|1} = 1 , \qquad p_{1|2} + p_{2|2} = 1 \tag{5.150}$$

by the normalization condition (analogous to (2.177)). Incidentally the transition probabilities $p_{m|n}$, $m \neq n$ are called *incorrect transition probabilities*, while the transition probabilities $p_{n|n}$ are called *correct transition probabilities*.

With the notation (5.149), the bit error probability of the binary channel can be written as (see the total probability Theorem 2.12)

$$\begin{aligned} P_{\text{bit}} = \mathrm{P}\left[\hat{b}_\ell \neq b_\ell\right] &= \sum_{n=1}^{2} \mathrm{P}\left[\hat{b}_\ell \neq n | b_\ell = n\right] \mathrm{P}\left[b_\ell = n\right] \\ &= \sum_{n=1}^{2} \sum_{\substack{m=1 \\ m \neq n}}^{2} p_{m|n} p_n \\ &= p_{2|1}\, p_1 + p_{1|2}\, p_2 , \end{aligned} \tag{5.151}$$

where $p_n = \mathrm{P}\left[b_\ell = n\right]$.

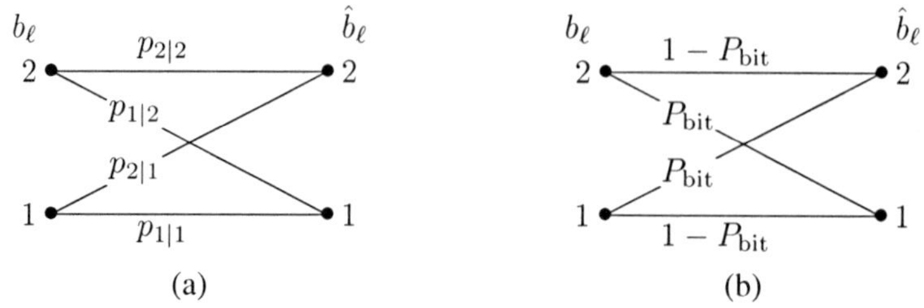

Figure 5.31 Transitions and transition probabilities of the binary digital channel: (a) general case; (b) binary symmetric channel.

When the two incorrect transition probabilities are equal, that is when $p_{2|1} = p_{1|2}$, then (5.151) simplifies to

$$P_{\text{bit}} = p_{2|1}(p_1 + p_2) = p_{2|1} = p_{1|2} \tag{5.152}$$

and the binary channel is uniquely identified by the bit error probability P_{bit}. In this situation, illustrated in Fig. 5.31b, we talk of a *binary symmetric channel.* Note also that, in a binary symmetric channel, the correct transition probabilities are equal, that is $p_{1|1} = p_{2|2} = 1 - P_{\text{bit}}$.

5.5.4 The inner numerical channel

The binary channel model of Fig. 5.30 compacts the entire transmission system of Fig. 5.26 into a unique transformation, mapping binary digits into binary digits. A similar transformation can be identified in the scheme of Fig. 5.26, mapping the transmitted symbol sequence $\{a_k\}$ into the received symbol sequence $\{\hat{a}_k\}$. This transformation is commonly referred to as *numerical channel* (M-ary channel), and leads to the model of Fig. 5.32.

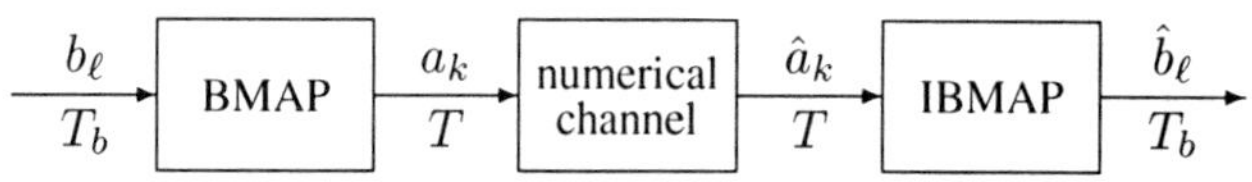

Figure 5.32 Model of a digital transmission system with inner numerical channel model.

Similarly to the binary channel case, besides its symbol rate $1/T$, the numerical channel is described by the *transition probabilities*

$$p_{m|n} = \mathrm{P}\left[\hat{a}_k = m | a_k = n\right] \,, \qquad m, n = 1, \ldots, M \,, \tag{5.153}$$

that is the probability that $\hat{a}_k = m$ is chosen under the condition that $a_k = n$ was transmitted. The typical diagram representation of transition probabilities is given in Fig. 5.33 for $M = 4$, where the probability $p_{m|n}$ is associated to the branch linking $a_k = n$ to $\hat{a}_k = m$.

By referring to the decision rule (5.14), the expression for the transition probabilities is

$$p_{m|n} = \mathrm{P}\left[\boldsymbol{r}_k \in R_m | a_k = n\right] \,, \tag{5.154}$$

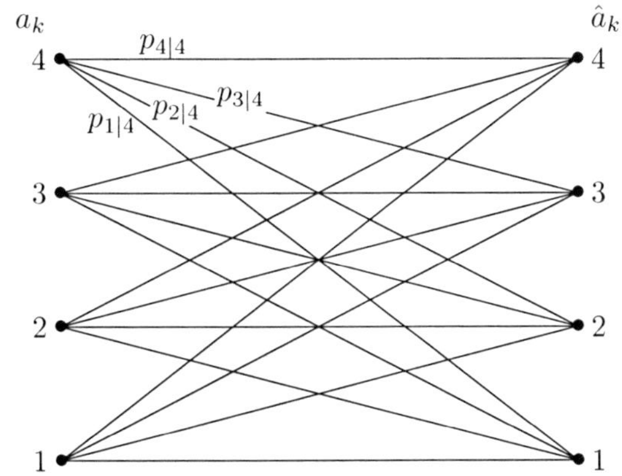

Figure 5.33 Transitions and transition probabilities of the M-ary numerical channel for $M = 4$.

which, by use of (5.28), can be further expressed as

$$p_{m|n} = \int_{R_m} p_{\boldsymbol{r}_k|a_k}(\boldsymbol{\rho}|n)\, d\boldsymbol{\rho} = \int_{R_m} p_{\boldsymbol{w}}(\boldsymbol{\rho} - \boldsymbol{s}_n)\, d\boldsymbol{\rho} \,. \tag{5.155}$$

which is an integral that is hardly solvable in closed form. Also recall the following total probability equivalence:

$$\sum_{m=1}^{M} p_{m|n} = 1 \,. \tag{5.156}$$

With the above notation, the probability of correct decision (5.19) can be written as

$$\mathrm{P}\left[\mathrm{C}\right] = \sum_{n=1}^{M} p_{n|n}\, p_n \tag{5.157}$$

and the probability of error (5.20) as

$$P_e = \sum_{n=1}^{M} \left(1 - p_{n|n}\right) p_n = \sum_{n=1}^{M} \sum_{\substack{m=1 \\ m \neq n}}^{M} p_{m|n}\, p_n \,. \tag{5.158}$$

In addition, it is possible to relate the transition probabilities of the binary channel of Fig. 5.30 to the transition probabilities of the inner numerical channel of Fig. 5.32. This relation is in general quite difficult to express, but can take simple forms in specific cases, for example, when the modulation size is $M = 2^U$. In this situation (5.133) holds true and the binary word $\boldsymbol{b}_k$ of length $U = \log_2 M$ is mapped into the detected binary word $\hat{\boldsymbol{b}}_k$ through the steps

$$\boldsymbol{b}_k \xrightarrow{\text{BMAP}} a_k \xrightarrow{\text{numerical channel}} \hat{a}_k \xrightarrow{\text{IBMAP}} \hat{\boldsymbol{b}}_k \,. \tag{5.159}$$

We introduce the concept of *Hamming distance* between the two binary words $\boldsymbol{c} = [c_1, \ldots, c_U]$ and $\hat{\boldsymbol{c}} = [\hat{c}_1, \ldots, \hat{c}_U]$ (see also Chapter 6), that is the number of binary digits in which $\boldsymbol{c}$ and $\hat{\boldsymbol{c}}$

differ. This distance can be expressed as

$$d_{\rm H}(\boldsymbol{c}, \hat{\boldsymbol{c}}) = \sum_{i=1}^{U} (1 - \delta_{c_i - \hat{c}_i}) \ . \tag{5.160}$$

where δ_n is the Kronecker delta. By this notation, the average number of wrong bits per transmitted binary word is

$$\mathrm{E}\left[N_{\rm wrong}\right] = \sum_{n=1}^{M} \sum_{\substack{m=1 \\ m \neq n}}^{M} d_{\rm H}\Big(\underbrace{\mathrm{IBMAP}(m)}_{\hat{\boldsymbol{b}}_k}, \underbrace{\mathrm{IBMAP}(n)}_{\boldsymbol{b}_k}\Big) \ \mathrm{P}\left[\hat{a}_k = m, \, a_k = n\right] \ . \tag{5.161}$$

By assuming that errors are uniformly distributed in the detected binary word, the bit error probability can be derived by dividing $\mathrm{E}\left[N_{\rm wrong}\right]$ by the vector length U, to obtain

$$P_{\rm bit} = \frac{\mathrm{E}\left[N_{\rm wrong}\right]}{\log_2 M} = \sum_{n=1}^{M} \sum_{\substack{m=1 \\ m \neq n}}^{M} \frac{d_{\rm H}\Big(\mathrm{IBMAP}(m), \mathrm{IBMAP}(n)\Big)}{\log_2 M} \, p_{m|n} \, p_n \ , \tag{5.162}$$

using the expression of the transition probabilities (5.154).

Situations leading to much simpler expressions of $P_{\rm bit}$ will be considered in specific constellations.

Definition 5.2 *The above numerical channel is called memoryless if, for every choice of N different instants $k_1, k_2, \ldots, k_N$, the following relation holds*

$$\begin{aligned} &\mathrm{P}\left[\hat{a}_{k_1} \neq a_{k_1}, \, \hat{a}_{k_2} \neq a_{k_2}, \, \ldots, \, \hat{a}_{k_N} \neq a_{k_N}\right] \\ &\qquad = \mathrm{P}\left[\hat{a}_{k_1} \neq a_{k_1}\right] \mathrm{P}\left[\hat{a}_{k_2} \neq a_{k_2}\right] \ldots \mathrm{P}\left[\hat{a}_{k_N} \neq a_{k_N}\right] \ . \end{aligned} \tag{5.163}$$

This is verified for example when all these conditions hold:

1. *The noise sample vectors $\{\boldsymbol{w}_k\}$ are statistically independent.*
2. *There is no ISI at the decision point.*
3. *Symbols $\{a_k\}$ are uncorrelated.*

5.5.5 Realistic receiver structure

Because of the introductory nature of this book, many aspects that are found in realistic digital transmission systems have been intentionally omitted. A reasonable receiver configuration is depicted in Fig. 5.34, where *demodulation* is only one of many building blocks, and these are needed because many of the assumptions on which our derivation was based so far do not apply in practice.

Here, without the intention of being exhaustive, we give a brief excursus on the building blocks of Fig. 5.34.

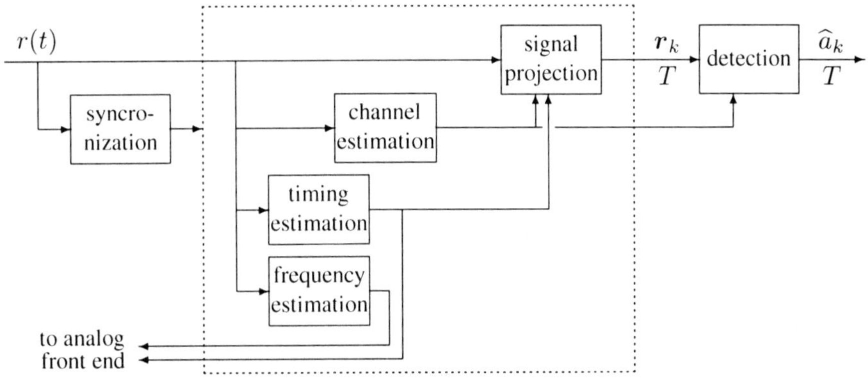

Figure 5.34 Building blocks of a digital receiver.

Rough signal synchronization. As the digital receiver is turned on, the very first operation it must perform is to detect the arrival time of the signal it wishes to demodulate. To this end, every digital signal has, alongside the transmitted data stream, some auxiliary information whose unique function is to provide an efficient mean for signal synchronization. Two different situations can be identified. When transmission is *packet based*, known data patterns (the *preamble*) are pre-appended to the useful data. When, instead, the transmission is *continuous*, e.g. in TV broadcasting, synchronization data patterns are periodically inserted in the data stream. In both cases, once the signal position has been detected, the entire receiver is turned on.

Timing and frequency estimation. Although we have so far assumed that the timing phase t_0, the sampling period T (or the symbol rate $F = 1/T$), and the carrier frequency f_0 (in passband transmissions) are perfectly known at reception, this is not the case in practice. Indeed only their rough value is known, and the precise value must be evaluated otherwise demodulation can be severely affected.

A rough estimate of t_0 is usually known thanks to the signal synchronization module, but a fine evaluation helps in enhancing the system performance.

A rough estimate of F and f_0 is given by their nominal values as defined by standards. However, minimal differences between the clock frequencies at transmission and reception, namely $\Delta F = F_{\mathrm{Tx}} - F_{\mathrm{Rc}}$ and $\Delta f_0 = f_{0,\mathrm{Tx}} - f_{0,\mathrm{Rc}}$, may drastically affect system performance. Note that, frequency errors must be very small, and clock differences are in fact measured in *parts per million* (ppm), that is $\Delta f_0 / f_0 \simeq 10^{-6} = 1$ ppm. For example, the impairment given by a carrier frequency offset Δf_0 in the binary modulation of Example 5.3 F (see also later QAM modulation in Section 5.6.2) yields a received signal of the form

$$r_k = \mathcal{C} \begin{bmatrix} \cos(2\pi\Delta f_0 kT) & 0 \\ 0 & \sin(2\pi\Delta f_0 kT) \end{bmatrix} s_{a_k} + w_k \tag{5.164}$$

whose effect is to induce a rotation of the constellation which linearly increases with time kT. The rotation is induced also for very small Δf_0, as long as one waits a sufficient long time.

Instead, the impairment given by a sampling frequency offset yields a drift in the timing phase t_0, in fact

$$t_0 + kT_{\rm Rc} = t_0 + kT_{\rm Tx} + k\left(\frac{1}{F_{\rm Rc}} - \frac{1}{F_{\rm Tx}}\right) \simeq \left(t_0 + k\,\frac{\Delta F}{F_{\rm Rc}} T_{\rm Tx}\right) + kT_{\rm Tx}\,. \tag{5.165}$$

Note that, also this drift linearly increases as the time $kT_{\rm Tx}$ elapses.

The result is that, in both situations (5.164) and (5.165), the frequency errors must be continuously estimated and corrected. In Fig. 5.34 this is achieved by controlling the clock sources in the analog front end in a *feedback* fashion, even if in these days *feedforward* options that correct the digital signal $\boldsymbol{r}_k$ are more common.

Channel estimation. The last block presented in Fig. 5.34 performs channel estimation, whose aim is to estimate the channel impulse response $g_{\rm Ch}(t)$, or, equivalently, the received and distorted waveforms $\tilde{s}_n(t)$. This information is then used by the detector. Note that channel estimation is also needed in the case of narrowband channels since the presence of a channel gain $\mathcal{C}$ affects the shape of decision regions.

5.6 EXAMPLES OF DIGITAL MODULATIONS

5.6.1 Pulse amplitude modulation (PAM)

Pulse amplitude modulation (PAM), also called *amplitude shift keying* (ASK), is the simplest example of multilevel baseband signaling, i.e. M may take values larger than 2. Let $h_{\rm Tx}$ be a real-valued finite-energy pulse. The waveform set consists of

$$s_n(t) = \alpha_n h_{Tx}(t)\ , \qquad \alpha_n = 2n - 1 - M \tag{5.166}$$

where $n = 1, 2, \ldots, M$. In other words, PAM waveforms are obtained by modulating in amplitude the pulse shape $h_{\rm Tx}$, and the waveform set has dimension $I = 1$. Often the scheme is denoted as M-PAM.

Remark (on notation) From (5.166) there is a one-to-one relationship between the waveform index n and the amplitude α_n. Hence for PAM we can identify the waveforms of the waveform set by α_n rather than n. Correspondingly, for PAM a_k will assume values in the alphabet

$$\mathcal{A} = \{\alpha_1, \ldots, \alpha_M\} = \{-(M-1), -(M-3), \ldots, (M-3), (M-1)\}\,. \tag{5.167}$$

rather than in the set $\{1, 2, \ldots, M\}$ as was done in the previous sections.

Orthonormal basis By defining $E_h = \|h_{Tx}(t)\|^2$ the energy of $h_{Tx}(t)$, we immediately have that the orthonormal basis for the waveform set is given by the signal

$$\phi_1(t) = \frac{h_{Tx}(t)}{\sqrt{E_h}}\,. \tag{5.168}$$

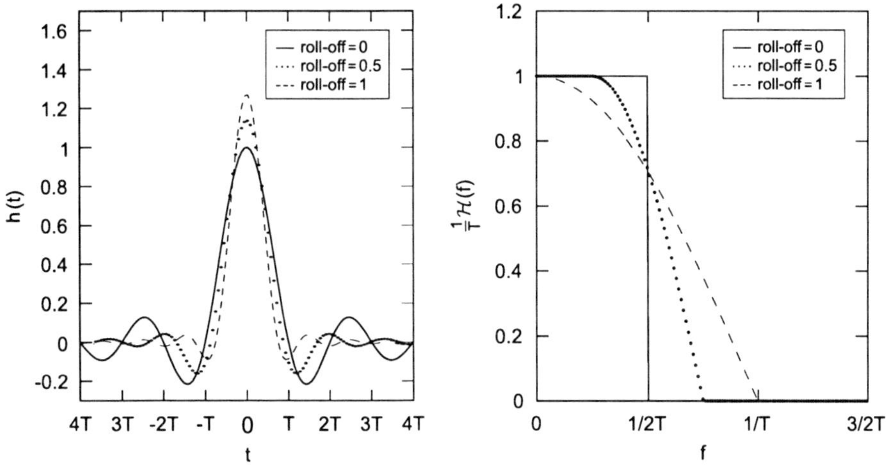

Figure 5.35 Time and frequency plots of *square root raised cosine* (rrcos) pulses for three values of the roll-off factor ρ.

Filter shape In order to fulfill condition (5.145) for the absence of ISI , $h_{\text{Tx}}(t)$ can be chosen to be confined in the interval $[0, T)$. As an alternative solution $h_{\text{Tx}}(t) = \text{sinc}(t/T)$ is also a valid choice, but of difficult realization with practical circuitry. A widely used choice in telecommunications is the so called *square root raised cosine* pulse, namely a pulse whose frequency response is

$$\mathcal{H}_{\text{Tx}}(f) = T\sqrt{\text{rcos}(fT, \rho)} \tag{5.169}$$

with $\text{rcos}(f, \rho)$ defined in (2.2). Namely $\mathcal{H}_{\text{Tx}}$ is a pulse obtained by connecting the level 1, for $|f| \leq \frac{1}{2T}(1-\rho)$, with level 0, for $|f| \geq \frac{1}{2T}(1+\rho)$, through a sinusoidal shape, where $0 < \rho \leq 1$ is a filter parameter called *roll off factor* setting the transition width. In the time domain, (5.169) gives $h_{\text{Tx}}(t) = \text{irrcos}(t/T, \rho)$ where $\text{irrcos}(x, \rho)$ is defined in (2.5). Plots of $\mathcal{H}_{\text{Tx}}(f)$ and $h_{\text{Tx}}(t)$ are shown in Fig. 5.35 for some values of ρ. Note that the pulse energy is given by

$$E_h = \int_{-\infty}^{+\infty} T^2 \text{rcos}(fT, \rho) df = T\ . \tag{5.170}$$

Constellation The vector representation of waveforms immediately follows from (5.168) and (5.166) to give

$$s_n = [\alpha_n \sqrt{E_h}]\ , \tag{5.171}$$

i.e. $\boldsymbol{s}_n$ is proportional to α_n. The energy of $s_n(t)$ can be written as $E_n = \|\boldsymbol{s}_n\|^2 = \alpha_n^2 E_h$ and the average signaling energy can be derived by the application of (5.27), giving[1]

$$E_s = \frac{E_h}{M}\sum_{n=1}^{M}(2n-1-M)^2 = \frac{M^2-1}{3}E_h \tag{5.172}$$

where we have implicitly considered that waveforms are equally likely. The PAM waveform constellation is shown in Fig. 5.36 for $M = 8$, from which we find that the minimum distance is $d_{\min} = 2\sqrt{E_h}$.

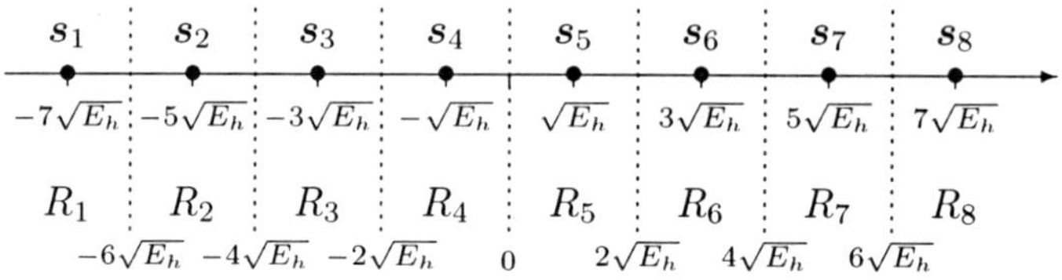

Figure 5.36 PAM transmission with $M = 8$: waveform constellation.

Decision regions In Fig. 5.36 we also note that the minimum distance decision regions are identified by $M-1$ thresholds placed at $(2n-M)\sqrt{E_h}$, for $n = 1, \ldots, M-1$. The decision regions associated to the various waveforms $\{s_n(t)\}$ are thus

$$\begin{aligned}
R_1 &= \left\{\rho \,\middle|\, \rho \leq (2-M)\sqrt{E_h}\right\} \\
R_n &= \left\{\rho \,\middle|\, (2n-2-M)\sqrt{E_h} < \rho \leq (2n-M)\sqrt{E_h}\right\}, \qquad n = 2, \ldots, M-1 \\
R_M &= \left\{\rho \,\middle|\, \rho > (M-2)\sqrt{E_h}\right\}.
\end{aligned} \tag{5.173}$$

A more useful expression is

$$\begin{aligned}
R_1 &= \left\{s_1 + e \,\middle|\, e \leq \sqrt{E_h}\right\} \\
R_n &= \left\{s_n + e \,\middle|\, |e| \leq \sqrt{E_h}\right\}. \qquad n = 2, \ldots, M-1 \\
R_M &= \left\{s_M + e \,\middle|\, e \geq -\sqrt{E_h}\right\}.
\end{aligned} \tag{5.174}$$

We note that the decision regions have a simple structure, hence the detector based on the criterion (5.14), rather than (5.51), is used.

[1] A few useful expressions are:

$$\sum_{i=1}^{M} i = \frac{M(M+1)}{2}, \qquad \sum_{i=1}^{M} i^2 = \frac{M(M+1)(2M+1)}{6}, \qquad \sum_{i=1}^{M} i^3 = \left(M\frac{M+1}{2}\right)^2.$$

System implementation The system implementation is shown in Fig. 5.37. From the binary input sequence $\{b_\ell\}$, the transmitter extracts the symbol sequence $\{a_k\}$, where $a_k \in \mathcal{A}$. The modulation is then performed by an interpolate filter with output

$$s_{\text{Tx}}(t) = \sum_{k=-\infty}^{+\infty} a_k \, h_{\text{Tx}}(t - kT) \quad , \qquad a_k \in \mathcal{A} \, . \tag{5.175}$$

The receiver is the conventional *implementation type I* receiver of Fig. 5.10 with decision rule (5.14) where the decision regions are given in (5.173).

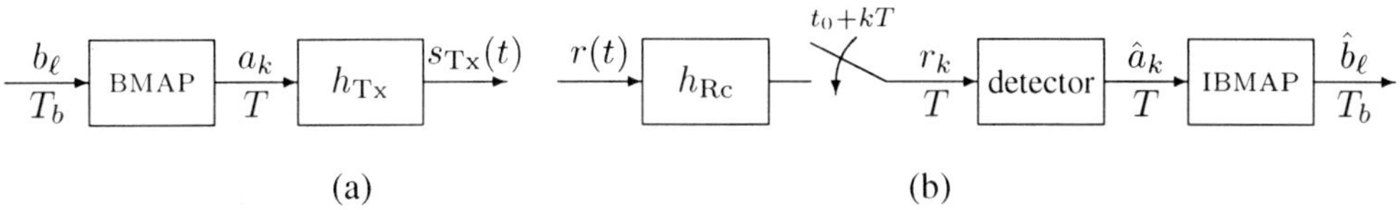

Figure 5.37 PAM system implementation: (a) modulator (transmitter); (b) demodulator (receiver). Here $h_{\text{Rc}}(t) = h^*_{\text{Tx}}(t_0 - t)/\sqrt{E_h}$ and $a_k \in \mathcal{A}$ (see (5.167)).

Moreover, for an ideal channel with $g_{\text{Ch}}(t) = \delta(t)$, from (5.11) and (5.171) at the detection point we have

$$r_k = s_{a_k} + w_k = a_k \sqrt{E_h} + w_k \quad , \qquad a_k \in \mathcal{A} \, , \; w_k \sim \mathcal{N}(0, \sigma_I^2) \, . \tag{5.176}$$

Symbol error probability The symbol error probability can now be evaluated by taking into account that the cases $a_k = \alpha_1$ and $a_k = \alpha_M$ must be treated separately because of the peculiar form of the corresponding decision regions. We first evaluate the correct transition probabilities $\{p_{n|n}\}$, $n = 1, \ldots, M$, defined in (5.154). For the case $a_k = \alpha_1$, we have (see (5.176) and (5.174))

$$\begin{aligned} p_{1|1} &= \text{P}\left[r_k \in R_1 | a_k = \alpha_1\right] = \text{P}\left[w_k + s_1 \in R_1\right] \\ &= \text{P}\left[w_k < \sqrt{E_h}\right] = 1 - \text{Q}\left(\sqrt{\frac{2E_h}{N_0}}\right) \, . \end{aligned} \tag{5.177}$$

The same result holds for $p_{M|M}$ (that is $p_{M|M} = p_{1|1}$) as is intuitive from the symmetry of the system. For the case $1 < n < M$, we have

$$\begin{aligned} p_{n|n} &= \text{P}\left[r_k \in R_n | a_k = \alpha_n\right] = \text{P}\left[w_k + s_n \in R_n\right] \\ &= \text{P}\left[-\sqrt{E_h} < w_k \le \sqrt{E_h}\right] = 1 - 2\,\text{Q}\left(\sqrt{\frac{2E_h}{N_0}}\right) \end{aligned} \tag{5.178}$$

so that, from (5.157) the probability of correct decision is

$$\text{P}\left[\text{C}\right] = \frac{1}{M} \sum_{n=1}^{M} p_{n|n} = 1 - 2\frac{M-1}{M} \, \text{Q}\left(\sqrt{\frac{2E_h}{N_0}}\right) \tag{5.179}$$

and the symbol error probability becomes

$$P_e = 1 - \mathrm{P}[\mathrm{C}] = 2\left(1 - \frac{1}{M}\right) \mathrm{Q}\left(\sqrt{\frac{2E_h}{N_0}}\right) , \tag{5.180}$$

By exploiting the relation (5.172), the error probability can be expressed in a more convenient form as

$$\boxed{P_e = 2\left(1 - \frac{1}{M}\right) \mathrm{Q}\left(\sqrt{\frac{6}{M^2-1}\frac{E_s}{N_0}}\right)} \tag{5.181}$$

The behavior of P_e as a function of E_s/N_0 is shown in Fig. 5.38.

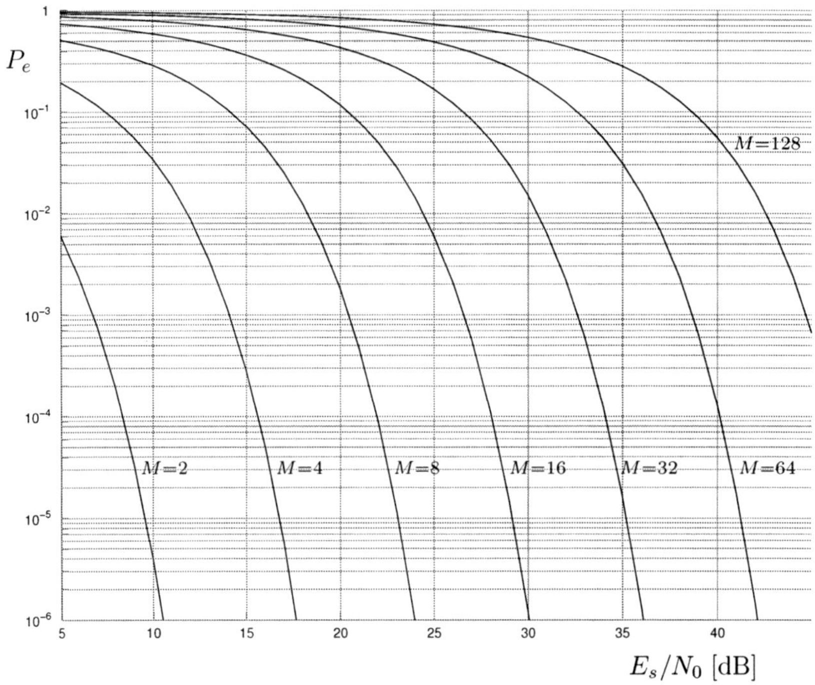

Figure 5.38 Symbol error probability P_e as a function of E_s/N_0 for M-PAM.

Example 5.6 A In PAM the transition probabilities can be expressed in closed form. As an example we determine the transition probability $p_{M|1}$. It is

$$\begin{aligned} p_{M|1} &= \mathrm{P}[r_k \in R_M | a_k = \alpha_1] = \mathrm{P}\left[w_k + (1-M)\sqrt{E_h} > (M-2)\sqrt{E_h}\right] \\ &= \mathrm{P}\left[w_k > (2M-3)\sqrt{E_h}\right] = \mathrm{Q}\left(\sqrt{\frac{2\,(2M-3)^2\,E_h}{N_0}}\right) = \mathrm{Q}\left(\sqrt{\frac{24\,(M-\frac{3}{2})^2}{(M^2-1)}\frac{E_s}{N_0}}\right) . \end{aligned}$$

Bit error probability In general, the derivation of the bit error probability should be approached by evaluating (5.162). However, some simplified results can be obtained when the BMAP is a Gray encoder (see below), that is an encoder where adjacent constellation points are associated with binary words $\boldsymbol{b}_k$ in (5.133) that differ in only one bit. An example of Gray bit mapping for $M = 8$ is illustrated in Fig. 5.39.

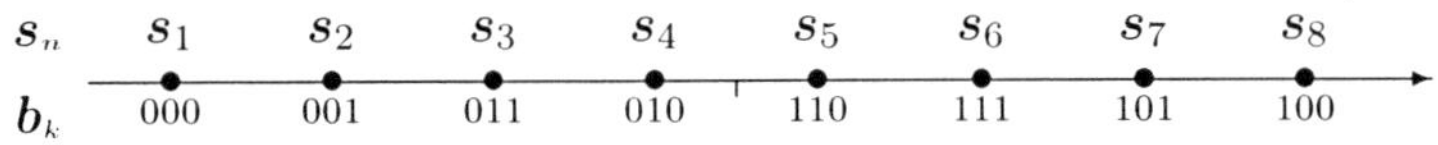

Figure 5.39 Gray bit mapping for M-PAM with $M = 8$.

Assuming that Gray coding is used, and that the signal-to-noise ratio E_s/N_0 is sufficiently large, if an error event occurs, it is very likely that one of the symbols at minimum distance from the transmitted symbol is detected. Thus, with very high probability only one bit of the $\log_2 M$ bits associated with the transmitted signal is incorrectly recovered. So, for the bit error probability we can write

$$\boxed{P_{\text{bit}} \simeq \frac{P_e}{\log_2 M}, \qquad \frac{E_s}{N_0} \gg 1} \tag{5.182}$$

where P_e is given in (5.181).

Gray coding A Gray coding procedure can be easily determined for a list of 2^n binary words of n bits. The case for $n = 1$ is immediate. We have two words with possible values given by

$$\begin{matrix} 0 \\ 1 \end{matrix} \tag{5.183}$$

and they differ by one bit.

The list for $n = 2$ is built by considering first the list of 2 words that are obtained by appending a 0 in front of the word list given by (5.183), that is

$$\begin{matrix} 0\,0 \\ 0\,1 \end{matrix}$$

Then, the remaining two words are obtained by inverting the order of the words in (5.183) and appending a 1 in front, to obtain

$$\begin{matrix} 1\,1 \\ 1\,0 \end{matrix}$$

The final result is the following list of words

$$\begin{matrix} 0\,0 \\ 0\,1 \\ 1\,1 \\ 1\,0 \end{matrix} \tag{5.184}$$

which, again, is a list where adjacent words differ by one bit.

For $n = 3$ we follow the same procedure. The first 4 words are obtained by appending a 0 in front of (5.184), while the other 4 words are obtained by appending a 1 to the list (5.184) taken in reverse order. We obtain

$$\begin{array}{c} 000 \\ 001 \\ 011 \\ 010 \\ 110 \\ 111 \\ 101 \\ 100 \end{array} \tag{5.185}$$

which is the sequence used in Fig. 5.39. The result can be extended to any n by induction. By induction it is just as easy to prove that two adjacent words in each list differ by one bit.

5.6.2 Quadrature amplitude modulation (QAM)

Quadrature amplitude modulation (QAM) is an example of passband modulation. Let

$$\alpha_n = \alpha_{n,I} + j\,\alpha_{n,Q}\ , \qquad n = 1, 2, \ldots, M \tag{5.186}$$

be a set of complex numbers with real and imaginary part, respectively, $\alpha_{n,I}$ and $\alpha_{n,Q}$. Here the transmitted symbol a_k will assume complex values in the alphabet

$$\mathcal{A} = \{\alpha_1, \ldots, \alpha_M\}\ . \tag{5.187}$$

Moreover, let h_{Tx} be a real-valued pulse with bandwidth B_h and energy E_h. The waveforms of the waveform set are expressed as

$$s_n(t) = \alpha_{n,I}\, h_{\text{Tx}}(t)\, \cos(2\pi f_0 t) - \alpha_{n,Q}\, h_{\text{Tx}}(t)\, \sin(2\pi f_0 t) \tag{5.188}$$

where $\cos(2\pi f_0 t)$ and $\sin(2\pi f_0 t)$ are two carriers in quadrature, with frequency f_0. From expression (5.188) the transmitted waveform is obtained by modulating in amplitude two carriers in quadrature. From (5.188) we also have

$$s_n(t) = \text{Re}\left[\alpha_n\, h_{\text{Tx}}(t)\, e^{j2\pi f_0 t}\right] = |\alpha_n|\, h_{\text{Tx}}(t)\, \cos(2\pi f_0 t + \angle\alpha_n)\ , \tag{5.189}$$

which suggests that the transmitted waveforms are obtained by varying both the amplitude and the phase of the carrier, hence the alternative name *amplitude modulation-phase modulation* (AM-PM). Incidentally, the presence of sinusoids with frequency f_0 in both (5.188) and (5.189) evidences that the considered signals are passband signals, and that QAM is a passband modulation.

Orthonormal basis From (5.188) the waveforms are obtained as linear combinations of $v_1(t) = h_{\text{Tx}}(t)\cos(2\pi f_0 t)$ and $v_2(t) = h_{\text{Tx}}(t)\sin(2\pi f_0 t)$. Under the fairly general condition

$$f_0 > B_h\ , \tag{5.190}$$

it can be proved that v_1 and v_2 are orthogonal signals, and that an orthonormal basis is given by

$$\begin{aligned}\phi_1(t) &= +\sqrt{\frac{2}{E_h}}\, h_{\mathrm{Tx}}(t)\, \cos(2\pi f_0 t)\\ \phi_2(t) &= -\sqrt{\frac{2}{E_h}}\, h_{\mathrm{Tx}}(t)\, \sin(2\pi f_0 t)\end{aligned} \tag{5.191}$$

where the sign $-$ in ϕ_2 was chosen for later convenience. Hence the QAM waveform set dimension is $I = 2$. The result (5.191) is proved in the following example.

Example 5.6 B We want to identify a basis for the signal set

$$v_1(t) = h_{\mathrm{Tx}}(t)\, \cos(2\pi f_0 t)\ , \qquad v_2(t) = h_{\mathrm{Tx}}(t)\, \sin(2\pi f_0 t)$$

under condition (5.190). We first verify that the two signals are orthogonal, that is

$$\langle v_1(t), v_2(t)\rangle = \int v_1(t)\, v_2^*(t)\, dt = 0\ .$$

This result is more evident if we evaluate the inner product in the frequency domain, where from Parseval's theorem (see Tab. 2.1) we have

$$\langle v_1(t), v_2(t)\rangle = \int \mathcal{V}_1(f)\, \mathcal{V}_2^*(f)\, df\ .$$

Now, since

$$\begin{aligned}\mathcal{V}_1(f) &= \tfrac{1}{2}\,\mathcal{H}_{\mathrm{Tx}}(f-f_0) + \tfrac{1}{2}\,\mathcal{H}_{\mathrm{Tx}}(f+f_0)\\ \mathcal{V}_2(f) &= -j\tfrac{1}{2}\,\mathcal{H}_{\mathrm{Tx}}(f-f_0) + j\tfrac{1}{2}\,\mathcal{H}_{\mathrm{Tx}}(f+f_0)\end{aligned}$$

by substitution we obtain

$$\begin{aligned}\langle v_1(t), v_2(t)\rangle &= j\tfrac{1}{4}\int |\mathcal{H}_{\mathrm{Tx}}(f-f_0)|^2\, df - j\tfrac{1}{4}\int \mathcal{H}_{\mathrm{Tx}}(f-f_0)\,\mathcal{H}_{\mathrm{Tx}}^*(f+f_0)\, df\\ &\quad + j\tfrac{1}{4}\int \mathcal{H}_{\mathrm{Tx}}(f+f_0)\,\mathcal{H}_{\mathrm{Tx}}^*(f-f_0)\, df - j\tfrac{1}{4}\int |\mathcal{H}_{\mathrm{Tx}}(f+f_0)|^2\, df\\ &= j\tfrac{1}{4}E_h - \tfrac{1}{2}\int \mathrm{Im}\left[\mathcal{H}_{\mathrm{Tx}}(f+f_0)\,\mathcal{H}_{\mathrm{Tx}}^*(f-f_0)\right]\, df - j\tfrac{1}{4}E_h\end{aligned}$$

where the cross product $\mathcal{H}_{\mathrm{Tx}}(f-f_0)\,\mathcal{H}_{\mathrm{Tx}}^*(f+f_0)$ is zero because of the hypothesis (5.190). So, also the inner product is zero, and the signals are orthogonal.

The identification of the orthonormal basis is finally obtained by evaluating the energy of both $v_1(t)$ and $v_2(t)$. Again, it is convenient to develop the derivation in the frequency domain, obtaining (with simplifications identical to the one seen with the inner product)

$$E_1 = \|v_1(t)\|^2 = \int |\mathcal{V}_1(f)|^2\, df = \tfrac{1}{2}E_h$$

and, similarly, $E_2 = \|v_2(t)\|^2 = \frac{1}{2}E_h$. So, (5.191) is an orthonormal basis.

The filter shapes used in practice for $h_{\rm Tx}(t)$ are the same that were introduced for PAM, namely pulses confined in $[0, T)$, sinc functions or inverses of squared root raised cosine pulses (see Section 5.6.1).

Constellation With the choice (5.191) for the orthonormal basis, the vector representation for the waveforms (5.188) is

$$\boldsymbol{s}_n = \sqrt{\frac{E_h}{2}}\,[\alpha_{n,I}\,,\,\alpha_{n,Q}\,]\;. \tag{5.192}$$

So, except for the factor $\sqrt{\frac{E_h}{2}}$, in QAM *the vector $\boldsymbol{s}_n$ coincides with α_n*. Moreover, the energy of s_n can be written as

$$E_n = \|s_n\|^2 = |\alpha_n|^2\,\frac{E_h}{2}\;, \tag{5.193}$$

and the average signaling energy is

$$E_s = \frac{1}{M}\sum_{n=1}^{M}|\alpha_n|^2\,\frac{E_h}{2}\;. \tag{5.194}$$

Some of the typical constellations used in QAM systems are shown in Fig. 5.40.

In the following, we will concentrate on *rectangular constellations* with $M = L^2$ points and

$$\alpha_{n,I}, \alpha_{n,Q} \in \Big\{-(L-1), -(L-3), \ldots, (L-3), (L-1)\Big\}\;. \tag{5.195}$$

In this context it is important to observe that the minimum distance between the constellation points is

$$d_{\min} = 2\sqrt{\frac{E_h}{2}} = \sqrt{2E_h}\;. \tag{5.196}$$

Moreover, for rectangular constellations, by comparison with (5.172) we obtain

$$E_s = \frac{E_h}{2M}\sum_{\ell,k=1}^{L}\Big((2\ell-L-1)^2+(2k-L-1)^2\Big) = \frac{M-1}{3}\,E_h\;. \tag{5.197}$$

System implementation The implementation of a QAM system is shown in Fig. 5.41.

In the transmitter of Fig. 5.41a, the binary stream $\{b_\ell\}$ is mapped into the complex-valued symbol sequence $\{a_k\}$, where $a_k \in \mathcal{A}$. From this we can extract two parallel symbol sequences

$$a_{k,I} = \Re[a_k]\;, \qquad a_{k,Q} = \Im[a_k]\;, \tag{5.198}$$

which are separately interpolated by the filter $h_{\rm Tx}$. Then the in-phase component is modulated by $\cos(2\pi f_0 t)$, while the quadrature component is modulated by $-\sin(2\pi f_0 t)$. So, the transmitted signal is

$$s_{\rm Tx}(t) = \sum_{k=-\infty}^{+\infty} a_{k,I}\,h_{\rm Tx}(t-kT)\,\cos(2\pi f_0 t) - \sum_{k=-\infty}^{+\infty} a_{k,Q}\,h_{\rm Tx}(t-kT)\,\sin(2\pi f_0 t)\;, \tag{5.199}$$

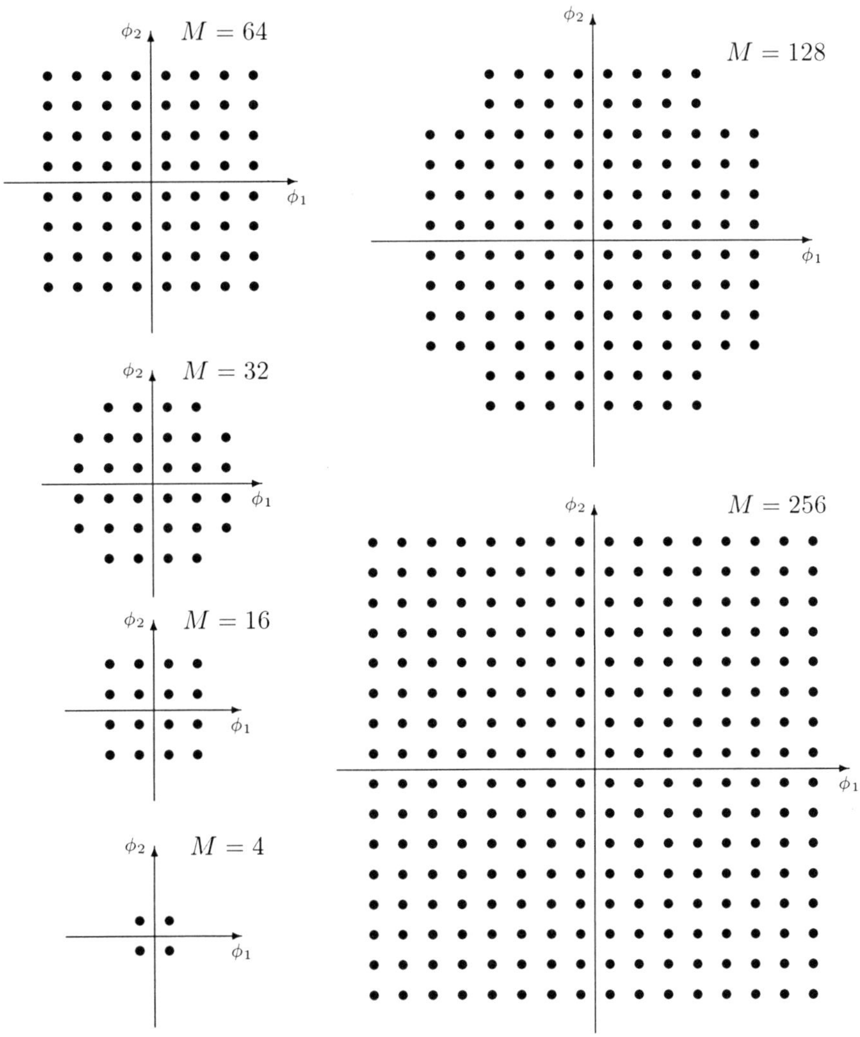

Figure 5.40 Some QAM constellations.

which evidently is a passband signal. The figure also shows that the two carriers can be obtained from a single oscillator by means of a $\pi/2$ phase shifter (Hilbert filter). An alternative expression to (5.199) is

$$s_{\mathrm{Tx}}(t) = \Re\left[\sum_{k=-\infty}^{+\infty} a_k\, h_{\mathrm{Tx}}(t-kT)e^{j2\pi f_0 t}\right] . \tag{5.200}$$

In the receiver of Fig. 5.41b, a carrier recovery system extracts the carrier $\cos(2\pi f_0 t)$ from the received signal $r(t)$, while the carrier $-\sin(2\pi f_0 t)$ is again obtained by means of a $\pi/2$

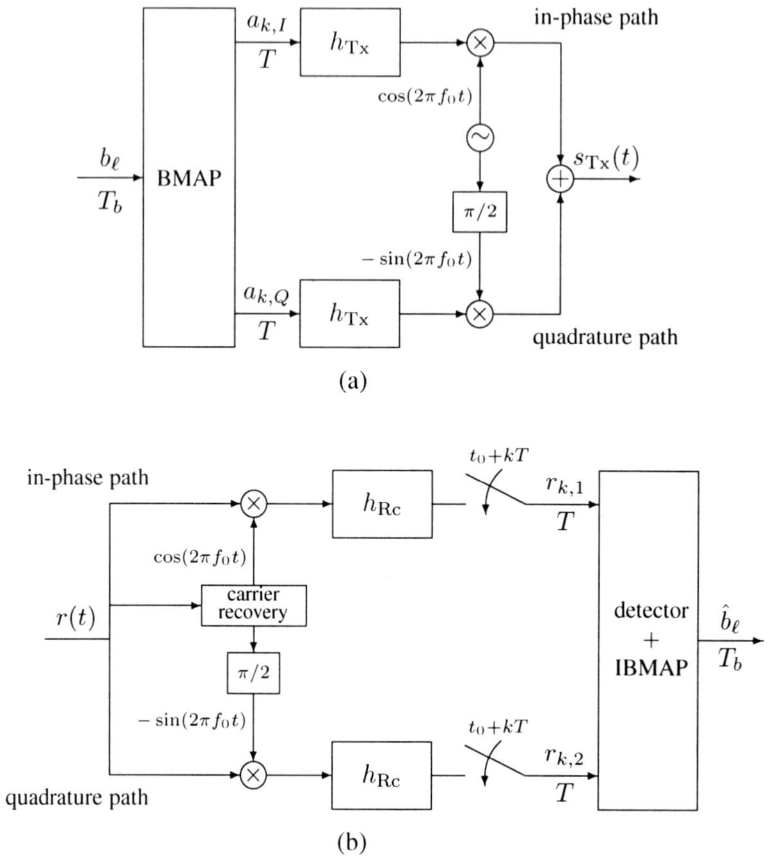

Figure 5.41 QAM system implementation: (a) modulator (transmitter); (b) demodulator (receiver). Here $h_{Rc}(t) = h^*_{Tx}(t_0 - t)/\sqrt{E_h/2}$ and $a_k \in \mathcal{A}$ (see (5.187)).

phase-shifter. Both the in-phase and the quadrature signal components are then filtered by

$$h_{Rc}(t) = \frac{h^*_{Tx}(t_0 - t)}{\sqrt{E_h/2}} , \tag{5.201}$$

and successively sampled. Note that the projection of $r(t)$ on the orthonormal basis (5.191) is obtained by the cascade of a correlator (by the carrier) and a filter matched to h_{Tx}. Let $\boldsymbol{r}_k = [r_{k,1} , r_{k,2}]$, $\boldsymbol{a}_k = [a_{k,I} , a_{k,Q}]$ and $\boldsymbol{w}_k = [w_{k,1} , w_{k,2}]$, where $w_{k,1} , w_{k,2} \sim \mathcal{N}(0, \sigma_I^2 = \frac{N_0}{2})$. For an ideal channel with $g_{Ch}(t) = \delta(t)$, at the decision point we have (see (5.11) and (5.192))

$$\boldsymbol{r}_k = \boldsymbol{s}_{a_k} + \boldsymbol{w}_k = \boldsymbol{a}_k \sqrt{\frac{E_h}{2}} + \boldsymbol{w}_k . \tag{5.202}$$

Further insight By comparison of (5.199) and (5.188) it is seen that indeed $s_{Tx}(t)$ does not satisfy property (5.137), due to the fact that the transmit carrier phase is not set to zero at each symbol period. However, because the carrier at the receiver is coherent with the transmit carrier, the equivalent relationship (5.202) still holds.

Decision regions The final receiver block is the decision block. Under a minimum distance criterion, the decision regions are as those of Fig. 5.42 in the specific case $M = 16$. In particular, three different situations can be identified, which are here named A, B, and C.

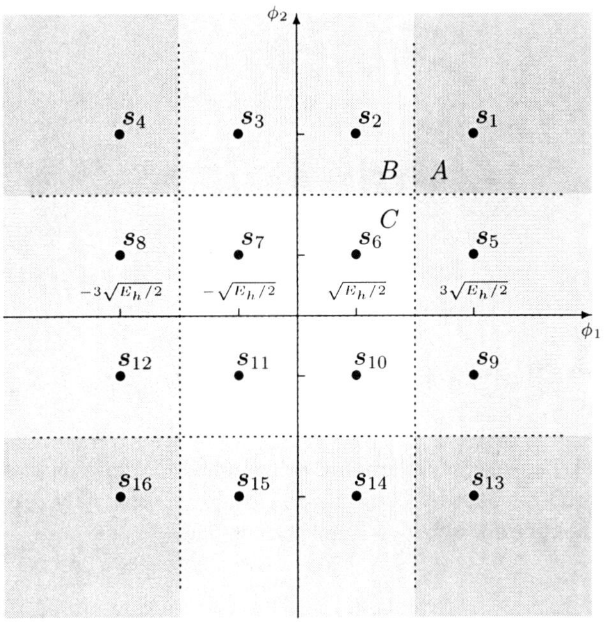

Figure 5.42 QAM decision regions for a rectangular constellation with $M = 16$.

A. These are the $N_A = 4$ dark gray regions at the corners, namely R_1, R_4, R_{13}, and R_{16}. Letting $\boldsymbol{e} = [e_1, e_2]$, they can be defined as follows:

$$\begin{aligned}
R_1 &= \left\{ \boldsymbol{s}_1 + \boldsymbol{e} \,\middle|\, e_1 > -\sqrt{E_h/2},\, e_2 > -\sqrt{E_h/2} \right\} \\
R_4 &= \left\{ \boldsymbol{s}_4 + \boldsymbol{e} \,\middle|\, e_1 \leq \sqrt{E_h/2},\, e_2 > -\sqrt{E_h/2} \right\} \\
R_{13} &= \left\{ \boldsymbol{s}_{13} + \boldsymbol{e} \,\middle|\, e_1 > -\sqrt{E_h/2},\, e_2 \leq \sqrt{E_h/2} \right\} \\
R_{16} &= \left\{ \boldsymbol{s}_{16} + \boldsymbol{e} \,\middle|\, e_1 \leq \sqrt{E_h/2},\, e_2 \leq \sqrt{E_h/2} \right\}
\end{aligned} \tag{5.203}$$

B. These are the light gray regions at the boundaries, and are $N_B = 4(L-2) = 8$ in number. We have

$$\begin{aligned}
R_2 &= \left\{ \boldsymbol{s}_2 + \boldsymbol{e} \,\middle|\, |e_1| \le \sqrt{E_h/2},\ e_2 > -\sqrt{E_h/2} \right\} \\
R_3 &= \left\{ \boldsymbol{s}_3 + \boldsymbol{e} \,\middle|\, |e_1| \le \sqrt{E_h/2},\ e_2 > -\sqrt{E_h/2} \right\} \\
R_5 &= \left\{ \boldsymbol{s}_5 + \boldsymbol{e} \,\middle|\, e_1 > -\sqrt{E_h/2},\ |e_2| \le \sqrt{E_h/2} \right\} \\
R_8 &= \left\{ \boldsymbol{s}_8 + \boldsymbol{e} \,\middle|\, e_1 \le \sqrt{E_h/2},\ |e_2| \le \sqrt{E_h/2} \right\} \\
R_9 &= \left\{ \boldsymbol{s}_9 + \boldsymbol{e} \,\middle|\, e_1 > -\sqrt{E_h/2},\ |e_2| \le \sqrt{E_h/2} \right\} \\
R_{12} &= \left\{ \boldsymbol{s}_{12} + \boldsymbol{e} \,\middle|\, e_1 \le \sqrt{E_h/2},\ |e_2| \le \sqrt{E_h/2} \right\} \\
R_{14} &= \left\{ \boldsymbol{s}_{14} + \boldsymbol{e} \,\middle|\, |e_1| \le \sqrt{E_h/2},\ e_2 \le \sqrt{E_h/2} \right\} \\
R_{15} &= \left\{ \boldsymbol{s}_{15} + \boldsymbol{e} \,\middle|\, |e_1| \le \sqrt{E_h/2},\ e_2 \le \sqrt{E_h/2} \right\}
\end{aligned} \tag{5.204}$$

C. These are the square white regions in the center, and are $N_C = (L-2)^2 = 4$ in number. These can be simply defined as

$$R_n = \left\{ \boldsymbol{s}_n + \boldsymbol{e} \,\middle|\, |e_1| \le \sqrt{E_h/2},\ |e_2| \le \sqrt{E_h/2} \right\}, \qquad n = 6, 7, 10, 11\,. \tag{5.205}$$

Note that, because all the decision regions are rectangular, detection on the in-phase and in-quadrature branches can be made by observing $r_{k,1}$ and $r_{k,2}$ separately. This is not the case when $M \neq L^2$, or when the constellation is not rectangular.

Symbol error probability As for PAM, also with rectangular QAM constellations the transition and error probabilities can be evaluated in closed form. In particular, concerning the correct transition probability $p_{n|n}$, corresponding to the transmission of $s_n(t)$, i.e. $a_k = \alpha_n$, we have the following three cases (which correspond to the three decision regions A, B, and C):

A. For constellation points belonging to A regions, the correct transition probabilities are all equal, so we evaluate the one corresponding to $n = 1$. We have

$$\begin{aligned}
p_{n|n,A} &= \mathrm{P}\left[\boldsymbol{w}_k + \boldsymbol{s}_1 \in R_1\right] \\
&= \mathrm{P}\left[w_{k,1} > -\sqrt{E_h/2},\ w_{k,2} > -\sqrt{E_h/2}\right] \\
&= \left(1 - \mathrm{Q}\left(\sqrt{\frac{E_h}{N_0}}\right)\right)^2
\end{aligned} \tag{5.206}$$

since $w_{k,1}$ and $w_{k,2}$ are statistically independent Gaussian noises.

B. For the case of B regions, we evaluate the correct transition probability for $n = 2$. We have

$$\begin{aligned} p_{n|n,B} &= \mathrm{P}\left[\boldsymbol{w}_k + \boldsymbol{s}_2 \in R_2\right] \\ &= \mathrm{P}\left[|w_{k,1}| \leq \sqrt{E_h/2},\ w_{k,2} > -\sqrt{E_h/2}\right] \\ &= \left(1 - 2\,\mathrm{Q}\left(\sqrt{\frac{E_h}{N_0}}\right)\right)\left(1 - \mathrm{Q}\left(\sqrt{\frac{E_h}{N_0}}\right)\right) . \end{aligned} \tag{5.207}$$

C. Finally, for C regions we evaluate the correct transition probability for $n = 6$, and obtain

$$\begin{aligned} p_{n|n,C} &= \mathrm{P}\left[\boldsymbol{w}_k + \boldsymbol{s}_6 \in R_6\right] \\ &= \mathrm{P}\left[|w_{k,1}| \leq \sqrt{E_h/2},\ |w_{k,2}| \leq \sqrt{E_h/2}\right] \\ &= \left(1 - 2\,\mathrm{Q}\left(\sqrt{\frac{E_h}{N_0}}\right)\right)^2 . \end{aligned} \tag{5.208}$$

By collecting the results, from (5.157) the probability of correct decision becomes

$$\mathrm{P}\left[\mathrm{C}\right] = \frac{1}{M}\left(N_A\, p_{n|n,A} + N_B\, p_{n|n,B} + N_C\, p_{n|n,C}\right)$$

which, after some rearrangement, can be written as

$$\mathrm{P}\left[\mathrm{C}\right] = \left(1 - 2\,\frac{L-1}{L}\,\mathrm{Q}\left(\sqrt{\frac{E_h}{N_0}}\right)\right)^2 , \tag{5.209}$$

so that the probability of error, $P_e = 1 - \mathrm{P}\left[\mathrm{C}\right]$, is

$$P_e = 4\frac{L-1}{L}\,\mathrm{Q}\left(\sqrt{\frac{E_h}{N_0}}\right) - \left(2\,\frac{L-1}{L}\,\mathrm{Q}\left(\sqrt{\frac{E_h}{N_0}}\right)\right)^2 . \tag{5.210}$$

In general, it is customary to express the error probability in terms of M and the signal-to-noise ratio E_s/N_0 (see also (5.197)) and, moreover, to neglect the square term. We thus obtain the expression

$$\boxed{P_e \simeq 4\left(1 - \frac{1}{\sqrt{M}}\right)\mathrm{Q}\left(\sqrt{\frac{3}{M-1}\,\frac{E_s}{N_0}}\right)} \tag{5.211}$$

Symbol error probability curves are shown in Fig. 5.43 as a function of E_s/N_0.

In the figure the behavior of the exact expression (5.210) is shown in dashed lines, while the approximate expression (5.211) is shown in solid lines. We note a very good coincidence between the two expressions, except for very high values of error probability.

Example 5.6 C As in PAM, also in QAM systems the transition probabilities can be evaluated in closed form. By considering the QAM system of Fig. 5.42, we evaluate the transition probability $p_{10|5}$. Since

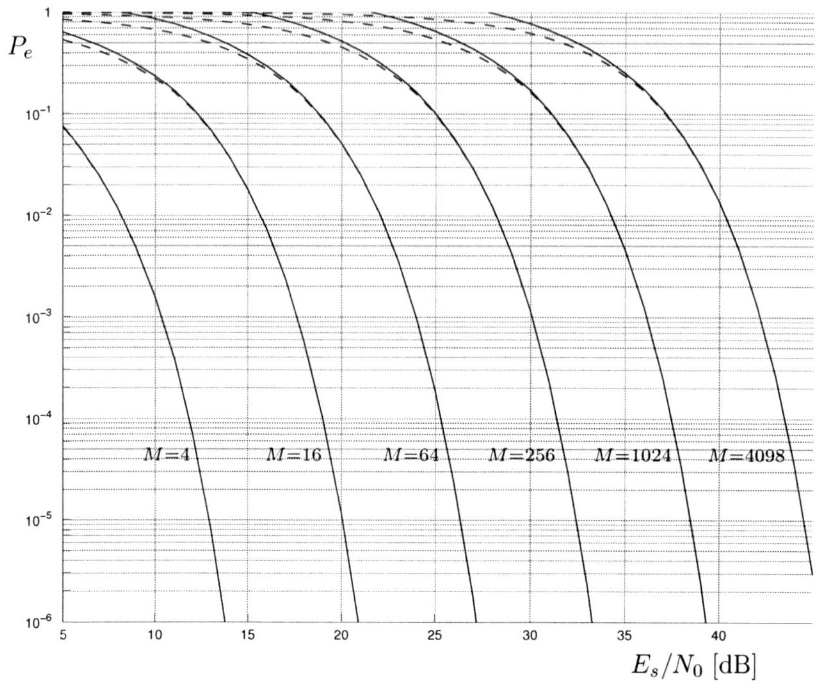

Figure 5.43 Symbol error probability P_e as a function of E_s/N_0 for M-QAM: approximated error probability (5.211) in solid lines, and exact error probability (5.210) in dashed lines.

R_{10} belongs to class C regions, from (5.202) we have

$$\begin{aligned} p_{10|5} &= \mathrm{P}\left[\boldsymbol{w}_k + \boldsymbol{s}_5 \in R_{10}\right] \\ &= \mathrm{P}\left[\left|w_{k,1} + 3\sqrt{E_h/2} - \sqrt{E_h/2}\right| \le \sqrt{E_h/2},\ \left|w_{k,2} + \sqrt{E_h/2} + \sqrt{E_h/2}\right| \le \sqrt{E_h/2}\right] \\ &= \mathrm{P}\left[-3\sqrt{E_h/2} \le w_{k,1} \le -\sqrt{E_h/2},\ -3\sqrt{E_h/2} \le w_{k,2} \le -\sqrt{E_h/2}\right] \\ &= \left(\mathrm{Q}\left(\sqrt{\frac{E_h}{N_0}}\right) - \mathrm{Q}\left(\sqrt{\frac{9E_h}{N_0}}\right)\right)^2 . \end{aligned}$$

Bit error probability As for the case of PAM, an approximation of the bit error probability can be simply derived from the symbol error probability by assuming that a Gray bit mapping (where constellation points at minimum distance have a binary representation that differs by one bit) is used. In this case, (5.182) holds also for QAM where P_e is given in (5.211).

For a rectangular QAM constellation with $M = L^2$, where detection is taken separately on the two branches, we can assume that the binary word $\boldsymbol{b}_k$ of (5.133) is first split into two halves, $\boldsymbol{b}_{k,\mathrm{I}}$ and $\boldsymbol{b}_{k,\mathrm{Q}}$, each of length $\frac{1}{2}U = \frac{1}{2}\log_2(M)$, and that Gray BMAP is applied separately to the in-phase and the quadrature branches. The resulting map and the component encoders/decoders are shown in Fig. 5.44 for the case $M = 16$. The reader can verify that constellation points at minimum distance have binary representations that differ by one bit only.

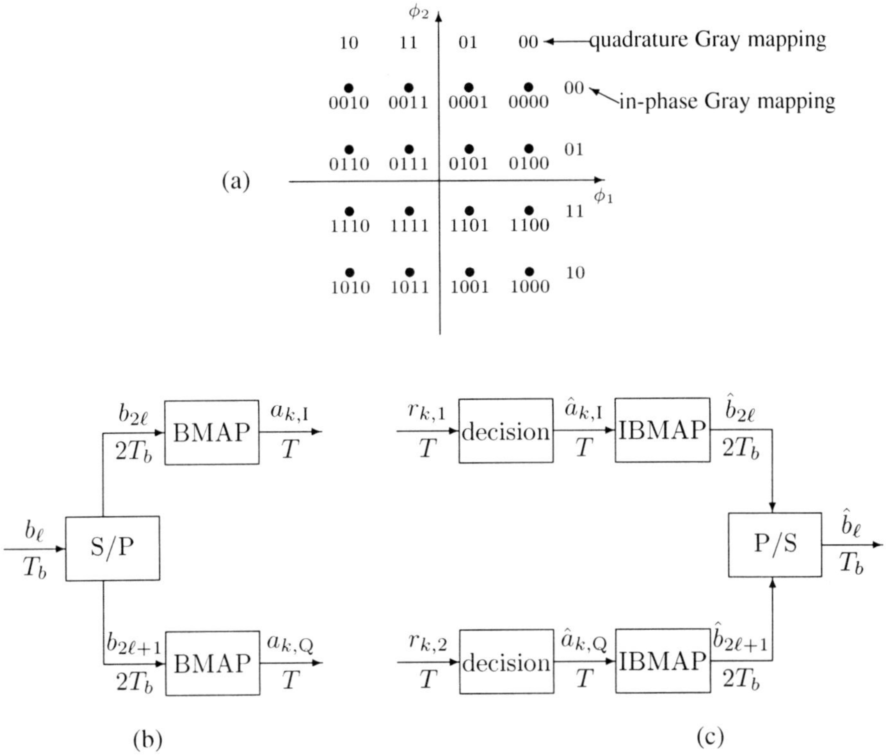

Figure 5.44 QAM modulation: (a) Gray BMAP for $M = 16$; (b) encoder; (c) decoder. S/P and P/S stand for serial to parallel and parallel to serial transformations, respectively.

Further insight QAM is employed in many existing communication systems. For example, the *Digital Video Broadcasting* (DVB) standard [13], for distributing the digital cable television (DVB-Cable), uses QAM constellations with $M = 16, 32, 64, 128$ or 256 points [6]. Similarly, QAM with $M = 16$ or 64 is used for the broadcasting of the digital terrestrial television signal (DVB-Terrestrial) [8].

5.6.3 Phase shift keying (PSK)

Phase shift keying (PSK) is another example of passband modulation. Let h_{Tx} be a real-valued finite-energy baseband pulse. Letting[2]

$$\varphi_n = \frac{\pi}{M}(2n-1)\,, \qquad n = 1, \ldots, M\,, \tag{5.212}$$

the generic transmitted waveform is given by

$$s_n(t) = h_{\mathrm{Tx}}(t)\cos(2\pi f_0 t + \varphi_n) \tag{5.213}$$

that is, signals are obtained by choosing one of the M possible values of the phase of a sinusoidal signal with frequency f_0, modulated by h_{Tx}. An alternative expression to (5.213) is given by

$$s_n(t) = \mathrm{Re}\left[h_{\mathrm{Tx}}(t)e^{j(2\pi f_0 t + \varphi_n)}\right] \tag{5.214}$$

or by

$$s_n(t) = \cos(\varphi_n)\, h_{\mathrm{Tx}}(t)\cos(2\pi f_0 t) - \sin(\varphi_n)\, h_{\mathrm{Tx}}(t)\sin(2\pi f_0 t)\,. \tag{5.215}$$

Orthonormal basis By relating (5.213) and (5.214) to (5.189), and (5.215) to (5.188), we immediately recognize PSK as a peculiar form of QAM modulation where $\alpha_n = e^{j\varphi_n}$. In particular, the PSK orthonormal basis is the QAM orthonormal basis (5.191) under the general condition that $f_0 > B_h$, with B_h the bandwidth of $h_{\mathrm{Tx}}(t)$.

Constellation As a consequence, the vector representation of PSK waveforms (5.213) is

$$\boldsymbol{s}_n = \sqrt{\frac{E_h}{2}}\,[\cos(\varphi_n)\,, \sin(\varphi_n)] \tag{5.216}$$

all with the same energy, hence

$$E_s = E_n = \frac{E_h}{2}\,. \tag{5.217}$$

An example of PSK constellation with $M = 8$ is illustrated in Fig. 5.45. Note that the waveform constellation lies on a circle and the various vectors differ in the phase φ_n. From the figure we also note that the minimum distance between constellation points is

$$d_{\min} = 2\sqrt{\frac{E_h}{2}}\sin\left(\frac{\pi}{M}\right) = \sqrt{2E_h}\sin\left(\frac{\pi}{M}\right)\,. \tag{5.218}$$

Decision regions From the example of Fig. 5.45 we note that PSK decision regions (minimum distance) are angular sectors with angle $2\pi/M$. For $M = 2, 4, 8$, simple decision rules can be defined. For $M > 8$ detection can be made by observing the phase of the received vector $\boldsymbol{r}_k$. In the complex plane, the general definition for decision regions is thus ($\rho = \rho_1 + j\rho_2$)

$$R_n = \left\{\rho = |\rho|\, e^{j(\varphi_n + \psi)}\,\middle|\, -\tfrac{\pi}{M} \le \psi < \tfrac{\pi}{M}\right\}. \tag{5.219}$$

[2] A more general definition of φ_n is given by $\varphi_n = (\pi/M)(2n-1) + \varphi_0$, where φ_0 is a constant phase.

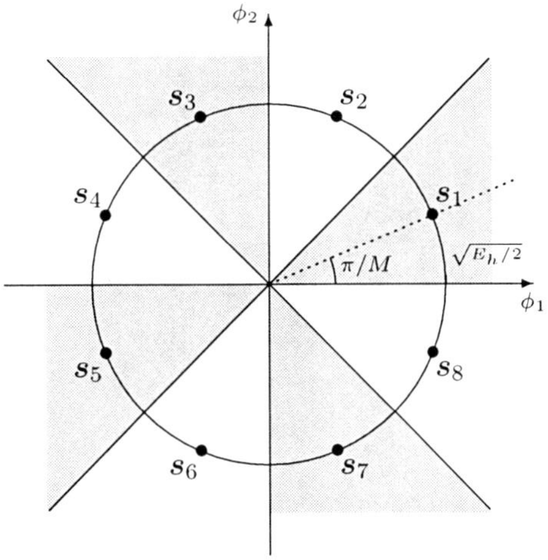

Figure 5.45 PSK constellation and decision regions for $M = 8$.

System implementation The general transmitter and receiver implementation for PSK is that of QAM illustrated in Fig. 5.41, where $\alpha_n = e^{j\varphi_n}$, and thus

$$\begin{aligned} a_{k,I} &\in \left\{\alpha_{n,I} = \cos(\varphi_n) = \cos\left(\tfrac{\pi}{M}(2n-1)\right), \quad n = 1, \ldots, M\right\} \\ a_{k,Q} &\in \left\{\alpha_{n,Q} = \sin(\varphi_n) = \sin\left(\tfrac{\pi}{M}(2n-1)\right), \quad n = 1, \ldots, M\right\} . \end{aligned} \tag{5.220}$$

Moreover, the detector implements the decision rule

$$\hat{a}_k = m \quad \text{if} \quad \frac{2\pi}{M}(m-1) < \arg\left(r_{k,1} + j\, r_{k,2}\right) \le \frac{2\pi}{M}m . \tag{5.221}$$

Symbol error probability The evaluation of the symbol error probability is pretty cumbersome for a generic value of M, but can be straightforwardly identified when $M = 2$ or $M = 4$. We thus treat these two cases separately.

Binary PSK (BPSK) For $M = 2$ we get $\varphi_1 = \varphi_0$ and $\varphi_2 = \pi + \varphi_0$, where φ_0 is an arbitrary phase. Then $I = 1$, and a basis is given by the pulse

$$\phi_1(t) = \sqrt{\frac{2}{E_h}}\, h_{Tx}(t)\cos(2\pi f_0 t + \varphi_0) \tag{5.222}$$

and the waveform vector representation is

$$s_1 = \left[\sqrt{\frac{E_h}{2}}\right], \qquad s_2 = \left[-\sqrt{\frac{E_h}{2}}\right] . \tag{5.223}$$

The signal constellation is illustrated in Fig. 5.46a.

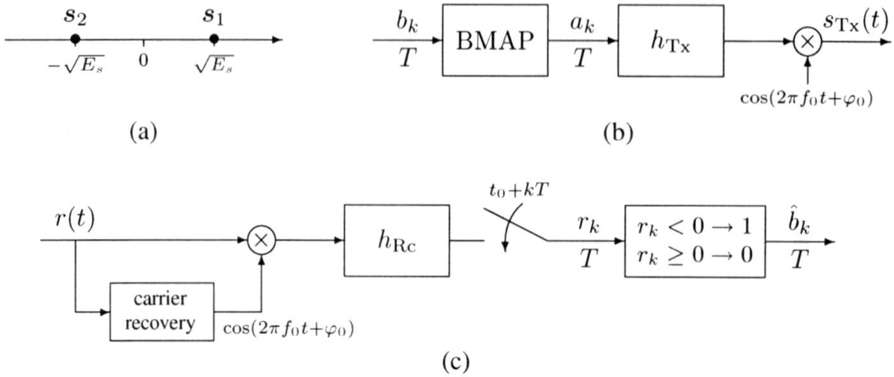

Figure 5.46 Binary PSK: (a) waveform constellation; (b) modulator; (c) demodulator. Here, $h_{\mathrm{Rc}}(t) = h^*_{\mathrm{Tx}}(t_0 - t)/\sqrt{E_h/2}$ and $a_k \in \{-1, 1\}$.

The transmitter and the receiver for BPSK are shown in Fig. 5.46b and c, respectively, and have a very simple implementation. At the transmitter the bit mapper maps '0' in '-1' and '1' in '$+1$' to generate the stream $\{a_k\}$ that is first interpolated by the filter h_{Tx} and then modulated by the carrier $\cos(2\pi f_0 t + \varphi_0)$. At the receiver, $r(t)$ is immediately demodulated by the carrier $\cos(2\pi f_0 t + \varphi_0)$, which, in turn, is obtained from $r(t)$ through a carrier recovery mechanism. The resulting signal is then filtered by the receive filter (5.201) matched to h_{Tx}, and successively sampled to obtain r_k. Finally, the detector implements a "sign" function to detect the binary data $\hat{b}_k$.

Apart from specific implementation aspects BPSK is an antipodal modulation, for which (5.89) holds, that is

$$\boxed{P_{\mathrm{bit}} = P_e = \mathrm{Q}\left(\sqrt{\frac{2E_s}{N_0}}\right), \qquad M = 2\,.} \tag{5.224}$$

Quaternary PSK (QPSK) PSK for $M = 4$ is usually called *quadrature* PSK (QPSK). The dimension of the basis is $I = 2$, for which the general implementation model of Fig. 5.41 holds. The QPSK waveform constellation is shown in Fig. 5.47 together with the respective decision regions. Note that the decision regions coincide with the four quadrants, and thus decision can be accomplished by simply testing the signs of $r_{k,1}$ and of $r_{k,2}$.

Also for QPSK, the symbol error probability is easy to determine since QPSK is a form of quaternary QAM for which (5.211) holds, that is

$$P_e \simeq 2\,\mathrm{Q}\left(\sqrt{\frac{E_s}{N_0}}\right). \tag{5.225}$$

In addition, we could also determine the exact error probability from (5.210), where we must take into account the different constellations in 4-QAM and QPSK, the difference being a

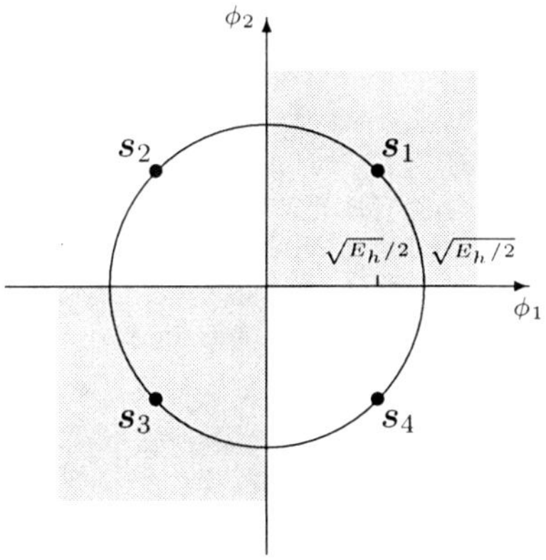

Figure 5.47 QPSK constellation and decision regions.

scaling factor of $\sqrt{2}$. In conclusion, we have

$$P_e = 2\,\mathrm{Q}\left(\sqrt{\frac{E_s}{N_0}}\right) - Q^2\left(\sqrt{\frac{E_s}{N_0}}\right) . \tag{5.226}$$

***M*-PSK** With equally likely signals, by exploiting the symmetry of the signaling scheme, the symbol error probability is

$$P_e = \mathrm{P}\left[\hat{a}_k \neq \alpha_1 | a_k = \alpha_1\right] = \mathrm{P}\left[\boldsymbol{r}_k \notin R_1 | a_k = \alpha_1\right] = 1 - \mathrm{P}\left[\boldsymbol{r}_k \in R_1 | a_k = \alpha_1\right] \tag{5.227}$$

where R_1 identifies the angular sector $[0, 2\pi/M)$. For $a_k = \alpha_1$ we have $\boldsymbol{r}_k = \boldsymbol{w} + \boldsymbol{s}_1$ with $\boldsymbol{s}_1 = \sqrt{E_s}\,[\cos(\varphi_1),\, \sin(\varphi_1)]$ and $\varphi_1 = \pi/M$, so that from (5.26) we obtain

$$\begin{aligned} P_e &= 1 - \int_{R_1} p_{\boldsymbol{w}}(\boldsymbol{\rho} - \boldsymbol{s}_1)\, d\boldsymbol{\rho} \\ &= 1 - \frac{1}{\pi N_0} \int\!\!\int_{R_1} \exp\left(- \frac{(\rho_1 - \sqrt{E_s}\cos(\frac{\pi}{M}))^2 + (\rho_2 - \sqrt{E_s}\sin(\frac{\pi}{M}))^2}{N_0} \right) d\rho_1 d\rho_2 \end{aligned} \tag{5.228}$$

A more useful expression can be obtained by using polar coordinates $[z, \psi]$ whose relation with Cartesian coordinates is $\rho_1 + j\rho_2 = z\,e^{j\psi}$. Since $d\rho_1\, d\rho_2 = z\, dz\, d\psi$, equation (5.228) becomes

$$P_e = 1 - \frac{1}{\pi N_0} \int_0^{2\pi/M} \int_0^{+\infty} z\, \exp\left(- \frac{z^2 + E_s - 2\sqrt{E_s}\,\cos(\psi + \frac{\pi}{M})}{N_0} \right) dz\, d\psi \tag{5.229}$$

Now (5.229) can be further simplified to give

$$P_e = 1 - \int_{-\pi/M}^{+\pi/M} p_\vartheta(\psi)\, d\psi \tag{5.230}$$

where $p_\vartheta(\psi)$ is the PDF of the phase error ϑ given $a_k = \alpha_n$,

$$p_\vartheta(\psi) = \frac{e^{-E_s/N_0}}{2\pi} + \sqrt{\frac{E_s}{N_0\,\pi}}\,\cos\psi\, e^{-\frac{E_s}{N_0}\sin^2\psi}\left[1 - \mathrm{Q}\left(\sqrt{\frac{2E_s}{N_0}}\cos\psi\right)\right] . \qquad (5.231)$$

No further simplifications are possible and we must resort to numerical evaluation in order to solve (5.230). If $E_s/N_0 \gg 1$, we can make use of the approximation (2.A.4) for the Gaussian cumulative distribution function, so that (5.231) becomes

$$p_\vartheta(\psi) \simeq \sqrt{\frac{E_s}{\pi N_0}}\,\cos\psi\, e^{-\frac{E_s}{N_0}\sin^2\psi}$$

and the approximation on error probability yields

$$\boxed{P_e \simeq 2\,\mathrm{Q}\left(\sqrt{\frac{2E_s}{N_0}}\sin\frac{\pi}{M}\right) , \qquad \frac{E_s}{N_0} \gg 1 , \qquad M \geq 4 .} \qquad (5.232)$$

Note that, for $M = 4$ the two approximations (5.232) and (5.225) coincide. Error probability curves are shown in Fig. 5.48 as functions of E_s/N_0. In the figure the behavior of the exact error probability (5.230) is shown in dashed lines. We note the closeness between the two expressions.

Bit error probability Gray bit mapping can be also pursued with PSK constellations by following the natural order in the clockwise (or counterclockwise) direction. An example is given in Fig. 5.49 for $M = 4$ and $M = 8$. Again, the bit error probability with Gray bit mapping is given by (5.182) where P_e is now given by the expressions (5.224) or (5.232).

Further insight (Differential PSK, DPSK) With PSK, the receiver often recovers the carrier signal, except for a phase offset φ_{off}. In particular, with reference to the scheme of Fig. 5.41b, the reconstructed carrier is $\cos(2\pi f_0 t - \varphi_{\text{off}} + \varphi_0)$ instead of $\cos(2\pi f_0 t + \varphi_0)$. In this case, $\boldsymbol{s}_n$ coincides with $\sqrt{E_s}\,[\cos(\varphi_n + \varphi_{\text{off}}), \sin(\varphi_n + \varphi_{\text{off}})]$, where φ_n is given by (5.212). Consequently, it is as if at the receiver the constellation were rotated by φ_{off}.

To prevent this problem there are two strategies. By a *coherent method* the receiver estimates φ_{off} from the received signal, and considers the received constellation rotated by $-\varphi_{\text{off}}$ to solve the problem. Alternatively, the receiver can avoid estimating φ_{off} by using a *differential (non coherent) method*, where the phase of the transmitted complex symbol a_k in Fig. 5.41a is obtained in an incremental fashion. If the selected phase (associated to the information bits) to transmit at the instant kT is φ_n, we do not choose $a_k = e^{j\varphi_n}$ as in PSK, but we form $a_k = a_{k-1}\, e^{j\varphi_n}$, that is

$$\arg(a_k) = \arg(a_{k-1}) + \varphi_n . \qquad (5.233)$$

In such a way, the information of φ_n at the receiver can be estimated by simply observing the difference $\arg(\hat{a}_k) - \arg(\hat{a}_{k-1})$. So, the receiver will detect the transmitted information using the difference between the phases of samples at successive sampling instants, $\boldsymbol{r}_k$ and $\boldsymbol{r}_{k-1}$. In general, DPSK gives lower performance with respect to PSK, especially for $M \geq 4$, because both the current sample and the previous sample are corrupted by noise.

PSK, especially in its differential form, is widely used in satellite communications. This is due to the fact that satellite transmission undergoes very sudden changes in signal attenuation,

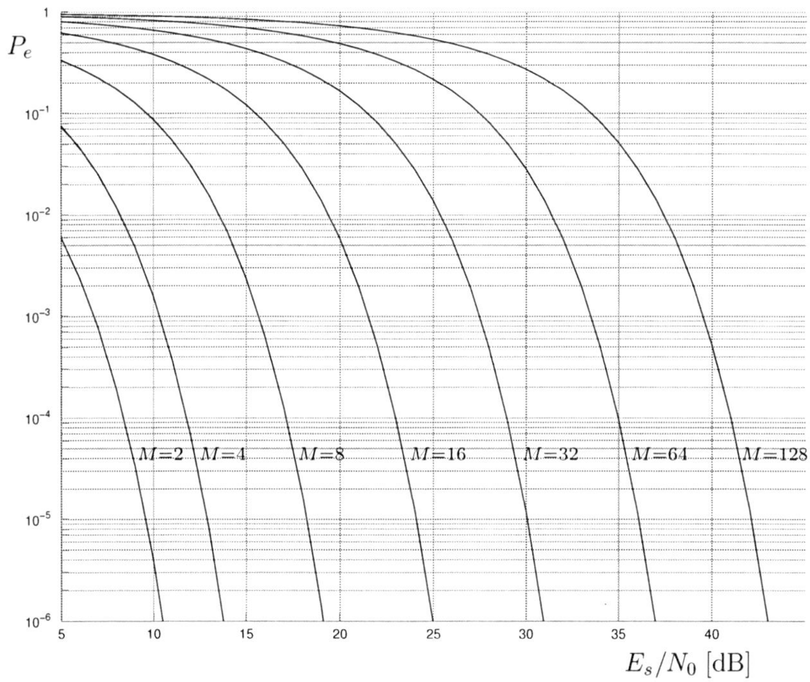

Figure 5.48 Symbol error probability P_e as a function of E_s/N_0 for M-PSK: approximate error probability (5.224) and (5.232) in solid lines, and exact error probability (5.230) in dashed lines. The two sets of lines overlap.

which do not affect the performance of a DPSK receiver as would be the case for a QAM receiver. As an application example we recall that 8-PSK is used in television broadcasting (DVB-Satellite) [7].

5.6.4 Frequency shift keying (FSK)

Frequency shift keying (FSK) is a widely used form of orthogonal modulation. There are two important categories of FSK signaling, coherent FSK and noncoherent FSK, which we now review in detail.

Noncoherent FSK In *noncoherent* FSK, the waveforms are

$$s_n(t) = \begin{cases} A \sin(2\pi f_n t + \varphi_n) & , \ 0 \leq t < T \\ 0 & , \ \text{otherwise} \end{cases} , \qquad n = 1, \ldots, M , \tag{5.234}$$

each with a different carrier frequency f_n and a different phase offset φ_n. The carrier frequencies f_n are chosen in such a way to obtain orthogonality between waveforms. A widely used

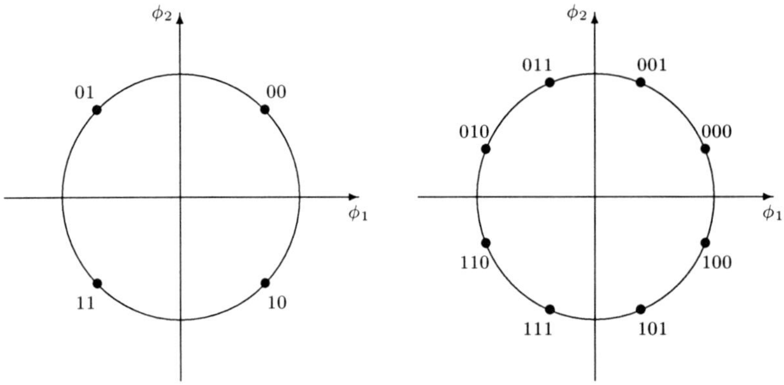

Figure 5.49 Gray bit mapping with two PSK constellations.

setting that assures orthogonality is

$$f_n = f_0 + \frac{n}{T}, \qquad f_0 \gg \frac{1}{T} \tag{5.235}$$

as illustrated in the following example.

Example 5.6 D We want to identify the conditions on f_n that guarantee orthogonality between FSK waveforms (5.234). To do so, we expand the inner product between waveforms as

$$\langle s_m(t), s_n(t)\rangle = A^2 \int_0^T \sin(2\pi f_m t + \varphi_m)\sin(2\pi f_n t + \varphi_n)\, dt$$

$$= \frac{A^2}{2}\int_0^T \cos[2\pi(f_m - f_n)t + (\varphi_m - \varphi_n)] - \frac{A^2}{2}\int_0^T \cos[2\pi(f_m + f_n)t + (\varphi_m + \varphi_n)]\, dt$$

which becomes

$$\begin{aligned}\langle s_m(t), s_n(t)\rangle = {} & \frac{A^2}{2}\,\frac{\sin[2\pi(f_m - f_n)T + (\varphi_m - \varphi_n)] - \sin(\varphi_m - \varphi_n)}{2\pi(f_m - f_n)} + \\ & - \frac{A^2}{2}\,\frac{\sin[2\pi(f_m + f_n)T + (\varphi_m + \varphi_n)] - \sin(\varphi_m + \varphi_n)}{2\pi(f_m + f_n)}\,.\end{aligned} \tag{5.236}$$

So, the condition on orthogonality requires that both terms in (5.236) vanish. For the first term, the phase $2\pi(f_m - f_n)T$ must be an integer multiple of 2π, that is

$$f_m - f_n = \frac{k}{T} \quad (k \text{ integer}) \tag{5.237}$$

with a minimum spacing of $\frac{1}{T}$ between adjacent carrier frequencies. For the second term we have

$$f_m + f_n = \frac{k}{T} \quad (k \text{ integer}) \qquad \text{or} \qquad f_n \gg \frac{1}{T} \tag{5.238}$$

where the second condition is by far the most widely used. Note that settings (5.235) satisfy both (5.237) and (5.238).

Coherent FSK In *coherent* FSK, the signaling waveforms are instead

$$s_n(t) = \begin{cases} A \sin(2\pi f_n t + \varphi) & , 0 \le t < T \\ 0 & , \text{otherwise} \end{cases} , \qquad n = 1, \ldots, M , \tag{5.239}$$

each with a different carrier frequency f_n but with synchronized phases $\varphi_n = \varphi$. In this context, a possible setting that guarantees orthogonality between waveforms is

$$f_n = f_0 + \frac{n}{2T} , \qquad f_0 \gg \frac{1}{T} , \tag{5.240}$$

as we prove in the example below. Note that the frequency span of coherent FSK transmission is reduced by 50% with respect to that of noncoherent FSK, at the cost of added complexity to maintain synchronization between carrier phases.

Example 5.6 E We want to identify the conditions on f_n that guarantee orthogonality between FSK waveforms (5.239). To do so, we start from the results of Example 5.6 D. Specifically, in the present context where $\varphi_n = \varphi$, equation (5.236) can be written as

$$\langle s_m(t), s_n(t) \rangle = \frac{A^2 T}{2} \text{sinc}[2(f_m - f_n)T] - \frac{A^2}{2} \frac{\sin[2\pi(f_m + f_n)T + 2\varphi] - \sin(2\varphi)}{2\pi(f_m + f_n)} \tag{5.241}$$

so that (5.237) must be substituted by the condition

$$f_m - f_n = \frac{k}{2T} \quad (k \text{ integer}) , \tag{5.242}$$

while (5.238) is still valid. In this case, the minimum spacing between adjacent carrier frequencies is $1/(2T)$. Note also that the settings (5.240) satisfy both (5.242) and (5.238).

System implementation An implementation of the coherent FSK system is given in Fig. 5.50 when $f_0 \gg 1/T$, so that all the waveforms have equal energy and

$$E_s = E_n = \frac{A^2 T}{2} . \tag{5.243}$$

The transmitter implementation is given in Fig. 5.50a where one from M carriers is selected every symbol period T according to the value of a_k. The receiver is illustrated in Fig. 5.50b, showing the *implementation type II* approach of Fig. 5.11, where 1) we set $t_0 = T$, 2) the additive terms due to the energy have been neglected because of (5.243), and 3) the front-end has been implemented through the correlation approach of Fig. 5.29b. Last, the detector selects the signal corresponding to the branch with the highest sample value.

Symbol error probability When $f_0 \gg \frac{1}{T}$ and (5.243) is satisfied, FSK is a form of orthogonal transmission with equal energy signals for which (5.125) holds for the symbol error probability P_e, together with bounds (5.126) (see also Fig. 5.24).

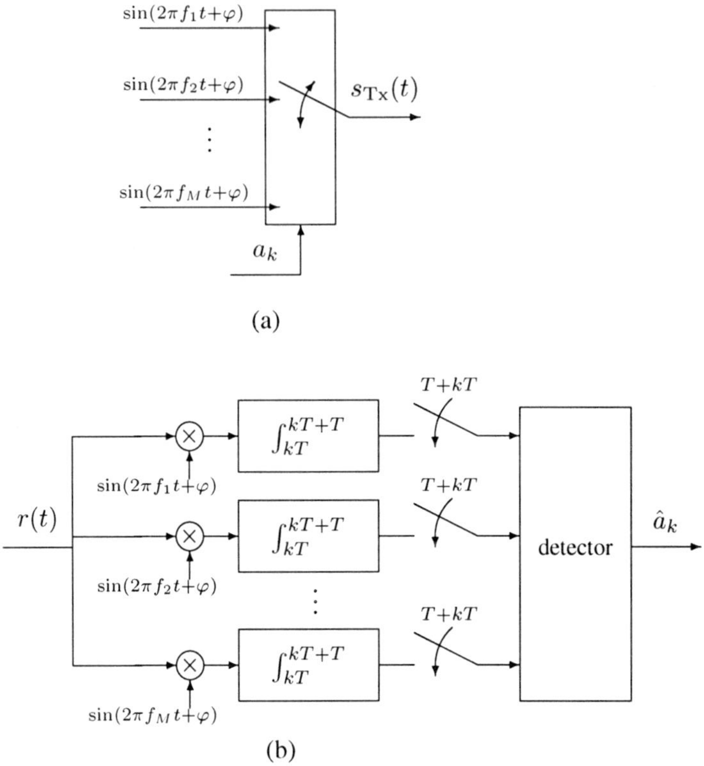

Figure 5.50 M-FSK: (a) modulator; (b) demodulator.

Bit error probability The calculation of the bit error probability with an orthogonal transmission is very different from the one seen with PAM and QAM. In particular, because of the symmetry of an orthogonal transmission the following equivalences hold for transition probabilities (5.153), that is

$$p_{n|n} = 1 - P_e \,, \qquad p_{m|n} = \frac{P_e}{M-1} \,, \qquad m \neq n \,. \tag{5.244}$$

In other words, (5.244) states that the probabilities of incorrect transition are equally likely. So, the relation (5.162) can now be written as

$$P_{\text{bit}} = \frac{P_e \, \overline{d}}{M-1} \,, \qquad \overline{d} = \sum_{\substack{m=1 \\ m \neq n}}^{M} \frac{d_{\text{H}}\Big(\text{IBMAP}(m), \text{IBMAP}(n)\Big)}{\log_2 M}$$

with $\overline{d}$ the average Hamming distance between binary words, giving $\overline{d} = M/2$. So, we finally obtain

$$\boxed{P_{\text{bit}} = \frac{M/2}{M-1} \, P_e \simeq \tfrac{1}{2} \, P_e} \tag{5.245}$$

independently of the chosen BMAP function. Evidently, the approximation in (5.245) is valid for sufficiently large M.

Further insight FSK is a modulation that has been widely used as signaling technique in low-cost modem applications, because of its reduced implementation complexity.

Nowadays, a hybrid form of binary FSK is adopted in the Bluetooth standard [9], [10], targeted at low-cost and low-range radio transmissions (see also Section 5.8.2). In Bluetooth, the transmitted waveform is an arch of cosine modulated by a sinusoid at frequency f_n, as opposed to the modulated rectangular waveform of (5.234) and (5.239). The frequency f_n at time kT, i.e. symbol a_k, is chosen according to two parameters: the value of the incoming binary symbol b_k (we implicitly consider that $T = T_b$) and the value of a pseudo-random generated codeword $\{c_k\}$ with alphabet $\{1, 2, \dots, M/2\}$ to guarantee a possible choice between M frequencies. We can thus write

$$a_k = f(b_k, c_k) \in \{1, 2, \dots, M\} .$$

Because of the jumping of the carrier frequency, the modulation system is commonly known as *frequency hopping* (FH) code, and the modulation approach as FH technique. By associating a different codeword $\{c_k\}$ to different users, FH allows the simultaneous transmission of multiple communications, hence it is a *multiple user* communication technique.

5.6.5 Code division modulation

Another important form of orthogonal modulation is given by *code division modulation*, where waveforms are built starting from binary codewords. A typical example is the signal set built starting from Walsh codes.

Walsh codes Walsh codes are formed by binary codewords with values $\{1, -1\}$ and length $N = 2^n$. Codewords are N in number. For $N = 2^0 = 1$ the codeword is the unique vector $\boldsymbol{C}_0 = [1]$. For $N = 2^1 = 2$ the two Walsh codewords are the rows of matrix

$$\boldsymbol{C}_1 = \begin{bmatrix} 1 & 1 \\ 1 & -1 \end{bmatrix} \tag{5.246}$$

For higher values of N, the $N = 2^n$ codewords are the rows of matrix $\boldsymbol{C}_n$ that is derived from $\boldsymbol{C}_{n-1}$ by the recursion

$$\boldsymbol{C}_n = \begin{bmatrix} \boldsymbol{C}_{n-1} & \boldsymbol{C}_{n-1} \\ \boldsymbol{C}_{n-1} & -\boldsymbol{C}_{n-1} \end{bmatrix} \tag{5.247}$$

so that, we have

$$\boldsymbol{C}_2 = \begin{bmatrix} 1 & 1 & 1 & 1 \\ 1 & -1 & 1 & -1 \\ 1 & 1 & -1 & -1 \\ 1 & -1 & -1 & 1 \end{bmatrix} \tag{5.248}$$

and

$$C_3 = \begin{bmatrix} 1 & 1 & 1 & 1 & 1 & 1 & 1 & 1 \\ 1 & -1 & 1 & -1 & 1 & -1 & 1 & -1 \\ 1 & 1 & -1 & -1 & 1 & 1 & -1 & -1 \\ 1 & -1 & -1 & 1 & 1 & -1 & -1 & 1 \\ 1 & 1 & 1 & 1 & -1 & -1 & -1 & -1 \\ 1 & -1 & 1 & -1 & -1 & 1 & -1 & 1 \\ 1 & 1 & -1 & -1 & -1 & -1 & 1 & 1 \\ 1 & -1 & -1 & 1 & -1 & 1 & 1 & -1 \end{bmatrix} \tag{5.249}$$

Walsh codes are closely related to the so-called Hadamard matrices, which are matrices $\boldsymbol{C}_n$ where each -1 is replaced throughout by a 0.

Moreover, Walsh codes have the characteristic property that the codewords are pairwise orthogonal, as the reader can verify by induction.

Signal set and orthonormal basis Let $M \leq N$, the waveform set of code division modulation is derived from the $N = 2^n$ codewords of the Walsh matrix $\boldsymbol{C}_n$ with entries $c_{n,i}$ (entry at row n and column i) as

$$s_n(t) = A \sum_{i=1}^{N} c_{n,i} \, \mathrm{w}_{T_c}(t - (i-1)T_c) \,, \qquad n = 1, \ldots, M \tag{5.250}$$

where

$$\mathrm{w}_{T_c}(t) = \begin{cases} 1 & , \; 0 \leq t < T_c \\ 0 & , \; \text{otherwise} \end{cases} \qquad T_c = \frac{T}{N} \tag{5.251}$$

is a rectangular pulse with extension $T_c = T/N$. Waveforms (5.250) are illustrated for $M = N = 8$ in Fig. 5.51.

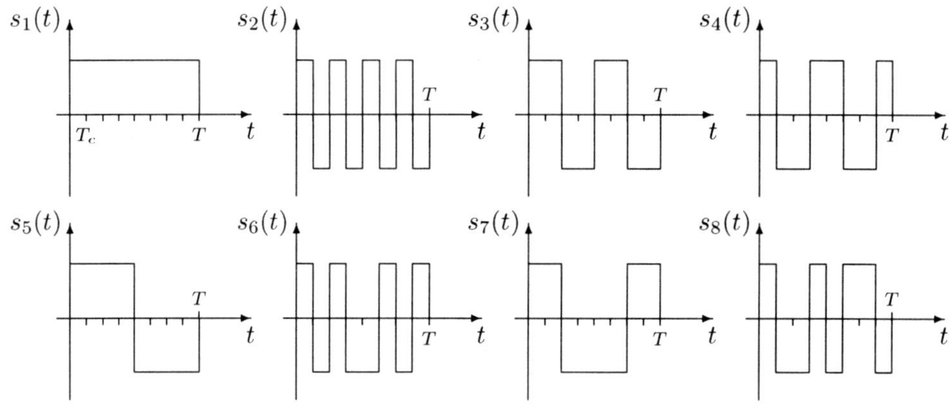

Figure 5.51 Orthogonal signal set obtained from a Walsh code of length $N = M = 8$.

We note that, because the vector set of $\boldsymbol{C}_n$ is orthogonal, the waveforms (5.250) form an orthogonal set with basis

$$\phi_i(t) = \frac{\mathrm{w}_{T_c}(t - (i-1)T_c)}{\sqrt{T_c}} \,, \qquad i = 1, \ldots, N \,. \tag{5.252}$$

The corresponding vector representation of waveforms (5.250) is given by the rows of $A\sqrt{T_c}\,C_n$.

System implementation Code division modulation is a form of orthogonal modulation where waveforms have equal energy and

$$E_s = E_n = A^2 T \ . \tag{5.253}$$

A simple receiver implementation is the *implementation type I* receiver of Fig. 5.10. By exploiting the fact that the reference basis (5.252) is a shifted version of $\phi_1(t)$, that is $\phi_i(t) = \phi_1(t-(i-1)T_c)$, $i = 2, \dots, N$, the receiver frontend can be implemented by means of only one filter as shown in Fig. 5.52.

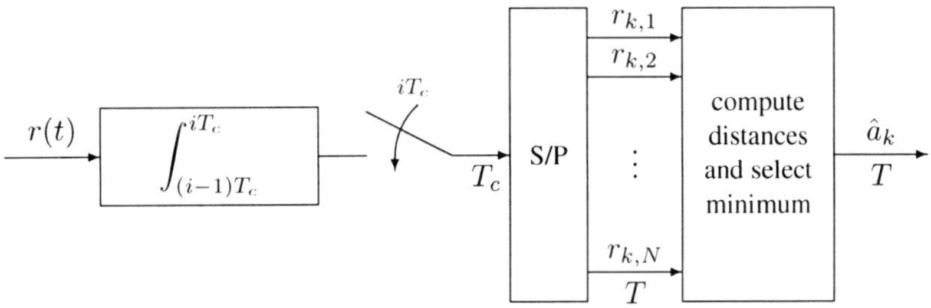

Figure 5.52 Code division modulation: receiver implementation. The "compute distances and select minimum" block is as described in Fig. 5.10.

Error probability Symbol and bit error probabilities of a code division modulation system behave as in the case of M-FSK, since they both rely upon orthogonal signaling with equal-energy waveforms. So, the reference expression for P_e is given by (5.125) with the bounds (5.126). For the bit error probability, $P_{\rm bit}$, we take (5.245) as reference.

Further insight Code division modulation is widely employed in modern communications under the name of *code division multiple access* (CDMA), for example in the *Universal Mobile Telecommunications System* (UMTS) standard [11], [12]. CDMA is an hybrid modulation where a bi-orthogonal extension of the code division modulation signal set is used, and where different codewords are associated to different users to guarantee the orthogonality between simultaneous transmission, thus obtaining a *multiple user* communication (see also later Section 5.8.2). Since different codewords are associated with different users, in CDMA each user is effectively exploiting a binary antipodal transmission scheme. Extensions to higher modulation formats as QPSK are also used.

5.7 COMPARISON OF DIGITAL MODULATION SYSTEMS

5.7.1 Reference bandwidth and link budget

A fair comparison between different communication schemes can be obtained by investigating their bit error rate performance in conjunction with the two parameters of major interest in a communication system: required bandwidth and power. We now clarify these two aspects.

Bandwidth and spectral efficiency For each modulation scheme it is customary to define a reference bandwidth, called minimum bandwidth $B_{\min}$, for which transmission can be performed in the absence of intersymbol interference at the receiver decision point. The conventional approach is to define

$$B_{\min} = \begin{cases} \dfrac{1}{2T} & , \text{ for baseband signals} \\ \dfrac{1}{T} & , \text{ for passband signals} \end{cases} \tag{5.254}$$

although this definition may not fit all modulation schemes (e.g. orthogonal and biorthogonal modulations), as can be seen from the detailed prospect given in Tab. 5.1. Incidentally, the pulse shape to be used for obtaining the reference $B_{\min}$ is (see also Section 5.6.1)

$$h_{\text{Tx}}(t) = A \,\text{sinc}\left(\frac{t}{T}\right) , \tag{5.255}$$

with A a suitable scaling factor.

Besides the definition of minimum bandwidth it is common to introduce the *spectral efficiency*

$$\nu = \frac{R_b}{2B_{\min}} = \frac{\log_2(M)}{T\,2B_{\min}} , \tag{5.256}$$

using (5.134). In fact ν measures how many bits per unit of time are sent over a channel with the conventional bandwidth $B_{\min}$ and it is expressed in bit/s/Hz. In other words, ν is a measure of how effectively the communication system is exploiting the required bandwidth $B_{\min}$. We note also that the spectral efficiency is a normalized measure that does not require the knowledge of T.

Power and reference SNR Here the reference SNR Γ is defined as the ratio between $\mathtt{M}_{s_{\text{Rc}}}$, the received statistical power, and the product of the noise PSD $N_0/2$ with $2B_{\min}$, i.e.

$$\Gamma = \frac{\mathtt{M}_{s_{\text{Rc}}}}{(N_0/2)\,2B_{\min}} , \tag{5.257}$$

or equivalently, from (5.139)

$$\Gamma = \frac{E_s}{N_0(TB_{\min})} , \tag{5.258}$$

where E_s denotes the average energy of the received waveform set. Moreover, if $E_{s_{\text{Tx}}}$ denotes the average energy of the transmitted waveform set, we have

$$\Gamma = \frac{E_{s_{\text{Tx}}}}{N_0(TB_{\min})\mathtt{a}_{\text{Ch}}} , \tag{5.259}$$

Table 5.1 Comparison of various modulation schemes in terms of minimum bandwidth and performance.

modulation	min. bandwidth $B_{\min}$	approximated P_{bit}	reference SNR Γ
M-PAM	$\frac{1}{2T}$	$\frac{2\left(1-\frac{1}{M}\right)}{\log_2 M}\,\mathrm{Q}\left(\sqrt{\frac{3\Gamma}{M^2-1}}\right)$	$\frac{2E_s}{N_0}$
M-QAM ($M=L^2$)	$\frac{1}{T}$	$\frac{4\left(1-\frac{1}{\sqrt{M}}\right)}{\log_2 M}\,\mathrm{Q}\left(\sqrt{\frac{3\Gamma}{M-1}}\right)$	$\frac{E_s}{N_0}$
M-PSK	$\frac{1}{T}$	$\mathrm{Q}\left(\sqrt{2\Gamma}\right)$, $M=2$; $\frac{2}{\log_2 M}\,\mathrm{Q}\left(\sqrt{2\Gamma\sin^2\frac{\pi}{M}}\right)$, $M>2$	$\frac{E_s}{N_0}$
orthogonal (coherent)	$\frac{M}{2T}$	$\frac{M-1}{2}\,\mathrm{Q}\left(\sqrt{\frac{M}{2}\Gamma}\right)$	$\frac{2E_s}{MN_0}$
bi-orthogonal (coherent)	$\frac{M}{4T}$	$\frac{M}{2(M-1)}\left[(M-2)\,\mathrm{Q}\left(\sqrt{\frac{M}{4}\Gamma}\right)+\mathrm{Q}\left(\sqrt{\frac{M}{2}\Gamma}\right)\right]$	$\frac{4E_s}{MN_0}$

where $\mathtt{a}_{\mathrm{Ch}}$ is the channel power attenuation. Expressions of Γ as a function of E_s/N_0 are reported in Tab. 5.1 for the various modulation schemes.

As $\mathtt{M}_{s_{\mathrm{Tx}}} = E_{s_{\mathrm{Tx}}}/T$, we also have

$$\Gamma = \frac{\mathtt{M}_{s_{\mathrm{Tx}}}}{N_0 B_{\min}\mathtt{a}_{\mathrm{Ch}}} . \tag{5.260}$$

Furthermore, expressions of Γ in terms of electrical powers are (see also (4.87))

$$\Gamma = \frac{\mathtt{P}_{\mathrm{Rc}}}{\mathtt{k}\mathtt{T}_{\mathrm{eff,Rc}} B_{\min}} . \tag{5.261}$$

and

$$\Gamma = \frac{\mathtt{P}_{\mathrm{Tx}}}{\mathtt{k}\mathtt{T}_{\mathrm{eff,Rc}} B_{\min}\mathtt{a}_{\mathrm{Ch}}} , \tag{5.262}$$

which in logarithmic form yields

$$(\Gamma)_{\mathrm{dB}} = (\mathtt{P}_{\mathrm{Tx}})_{\mathrm{dBm}} - (\mathtt{a}_{\mathrm{Ch}})_{\mathrm{dB}} + 114 - 10\log_{10}\left(\frac{\mathtt{T}_{\mathrm{eff,Rc}}}{\mathtt{T}_0}\right) - 10\log_{10}(B_{\min})_{\mathrm{MHz}} . \tag{5.263}$$

Typically, for a given modulation scheme with a target bit error probability, we first derive the required E_s/N_0. This requirement is then expressed in terms of Γ by (5.258). Last, we may use (5.263) to derive for example $\mathtt{P}_{\mathrm{Tx}}$ given Γ and all system parameters. This is the so called "link budget" evaluation.

5.7.2 Comparison in terms of performance, bandwidth and spectral efficiency

A comparison between the different modulation systems of Tab. 5.1 is reported in Fig. 5.53 in terms of the spectral efficiency ν and of the reference SNR Γ, for some values of the waveform cardinality M. The points in figure refer to a common value of $P_{\rm bit}$ that was chosen equal to 10^{-6}, that is each modulation is represented by the point at coordinate (ν, Γ) where ν is given by (5.256) and where Γ is the reference SNR that yields $P_{\rm bit}(\Gamma) = 10^{-6}$ according to the prospect of Tab. 5.1.

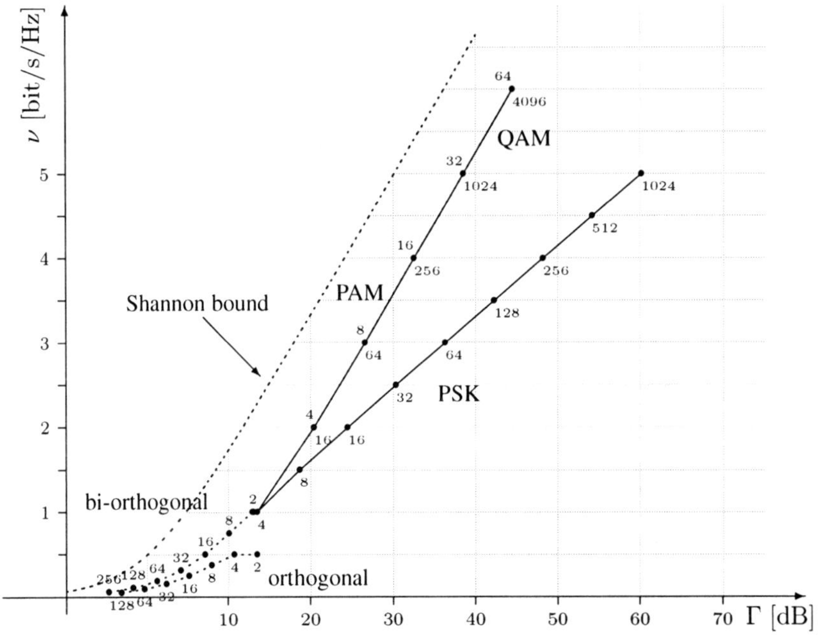

Figure 5.53 Spectral efficiency ν versus required reference SNR Γ at $P_{\rm bit} = 10^{-6}$ for various modulation systems.

The figure also shows the Shannon bound [1], [4] (see also Theorem 6.17)

$$\nu \leq \tfrac{1}{2}\log_2(1+\Gamma) \tag{5.264}$$

which expresses the maximum spectral efficiency ν that can be achieved for a given SNR Γ. Note in figure that the bound (5.264) sets the available region of practical interest.

From Fig. 5.53 we observe the following:

1. PAM and QAM systems behave in an equivalent way; in particular they can reach high spectral efficiencies ν at the cost of using higher power (high values of Γ).

2. PSK is less efficient than PAM or QAM in that, for the same Γ, it can reach lower spectral efficiencies; in any case, PSK systems are used for other reasons (see comments at page 318).

3. Orthogonal and bi-orthogonal modulations are not very efficient in terms of ν because they require a large bandwidth. However they require also very low values of Γ to guarantee a very low bit error probability; for this reason they are targeted to lower-bit-rate applications requiring low power dissipation and low implementation complexity.

4. In any case, bi-orthogonal modulations are twice as efficient as orthogonal modulations in terms of bandwidth.

So, PAM, QAM, and PSK are bandwidth-efficient modulation methods as they cover the region for $\nu > 1$, or $B_{\min} < 1/(2T_b)$, as illustrated in Fig. 5.53. For the same bit rate, with these modulations we can increase the number M of transmit waveforms, and hence lower the required bandwidth (higher ν) by using a higher power (higher Γ). Conversely, by limiting the power (smaller M) a larger bandwith is required. So, the bandwidth can be traded off with the power.

Orthogonal and biorthogonal modulations are not very efficient in bandwidth ($\nu < 1$), however they require much lower values of Γ. As illustrated in Fig. 5.53, biorthogonal modulation has the same performance as orthogonal modulation, but requires half the bandwidth; in this region, by increasing the bandwidth it is possible to decrease Γ. However, a slight decrease in Γ may determine a large increase of bandwidth. The P_{bit} of orthogonal or biorthogonal modulation is almost independent of M and mainly depends on the signaling energy E_s and on the noise PSD $N_0/2$.

In any case, in addition to the required power and bandwidth, the choice of a modulation scheme is based on the channel characteristics and on the cost of the implementation.

5.8 ADVANCED DIGITAL MODULATION TECHNIQUES

In this section, we briefly review the two most relevant digital modulation techniques which are used in modern communication systems, namely OFDM (*ortogonal frequency division multiplexing*) and *spread spectrum* approaches. Both are extensions of the basic modulations seen so far.

5.8.1 Orthogonal frequency division multiplexing

OFDM is one of the most widely used modulation techniques in communication systems providing high bit rates in dispersive channels. It is used in the *digital video broadcasting* (DVB) of television programs [13, 8], in the DSL (*digital subscriber line*) connection of home appliances, in high-rate wireless domotic applications (e.g., in the wireless connection between a DVD source and a TV), in data transmission over power lines [17], etc..

The idea behind ODFM is simple and it is illustrated in Fig. 5.54. For a wideband channel, the reference model is that of (5.1) where $g_{\text{Ch}}(t)$ is dispersive and causes ISI. Specifically, the channel frequency response shows selectivity, as shown in Fig. 5.54a where amplitude experiences wide fluctuations. In this case we talk of a *frequency selective channel*, in the sense that some frequencies suffer deeper attenuation than others.

Now, the modulation approaches seen so far (e.g., PAM and QAM) can still be used in such a channel, but they require a suitable filter, either at transmission or at reception, to equalize the channel dispersion. However, complexity of equalizer design and of filtering itself is quite high. An alternative approach would be to partition the frequency domain in small frequency bands,

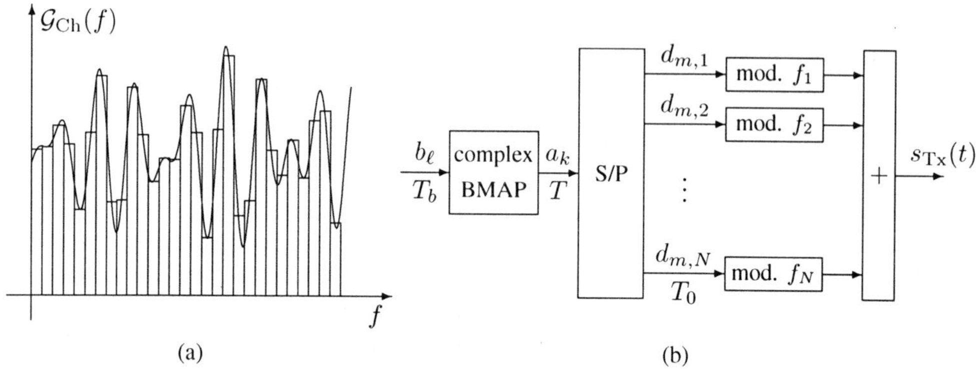

Figure 5.54 OFDM principle: (a) frequency domain narrowband partition and (b) bank of digital modulators.

of width $B_0 = 1/T_0$ and center frequencies f_i, with B_0 so small to assure a quasi-constant gain $\mathcal{C}_i \simeq \mathcal{G}_{\rm Ch}(f_i)$ over each subband. Then, transmission in each subband

$$\mathcal{B}_i = (f_i - \tfrac{1}{2}B_0; f_i + \tfrac{1}{2}B_0) \ , \qquad B_0 = \frac{1}{T_0} \tag{5.265}$$

can be accomplished through a standard narrowband technique, as shown in Fig. 5.54b.

In Fig. 5.54b, the bit sequence $\{b_\ell\}$ needs first to be mapped into a complex valued symbol sequence $a_k = a_{k,\rm I} + j\, a_{k,\rm Q}$, typically with QAM constellation. Then the symbol sequence is converted through a serial-to-parallel conversion to a symbol vector $\boldsymbol{d}_m$ whose N components are modulated over separate frequency bands through a bank of modulators. Note that $T_0 = NT \gg T$ is the symbol period at which the interpolate filters $h_{\rm Tx}$ of the digital modulators bank operate. When interpolate filters are chosen limited in time, that is when $h_{\rm Tx}$ has a rectangular shape, the OFDM system is known as *discrete multi tone* (DMT), while when the interpolate filters are chosen limited in frequency then it is known as *frequency multi-tone* (FMT). Note also that, assuming QAM constellations, the waveform set dimension is $I = 2N$, two for each subband $\mathcal{B}_i$. Also, N is typically a big number, however, due to implementation considerations, not all subbands are active transmitting data. Tab. 5.2 summarizes parameters for some of the more relevant standards employing OFDM.

It is evident that, in principle, the modulation scheme of Fig. 5.54b (and also the corresponding demodulation illustrated in Fig. 5.55) may be complex to implement, as it requires a non negligible number of parallel modulators (demodulators). However, techniques based on the computational efficiency of the fast Fourier transform (FFT) exist to drastically reduce the complexity burden. Among other benefits, this is the reason for the success of OFDM.

According to (5.202) and to the fact that gain $\mathcal{C}_i$ is experienced at subband $\mathcal{B}_i$, the signal at output of each demodulator (i.e. at the decision point) in Fig. 5.55 are of the form

$$r_{m,i} = \mathcal{C}_i \sqrt{\frac{E_h}{2}}\, d_{m,i} + w_{m,i} \tag{5.266}$$

Table 5.2 Some of the standards using OFDM.

standard (issued by, year, nation)	application	number of active subbands	badwidth B (MHz)	bit rate R_b (Mbit/s)
DAB (ETSI, 1995, Europe)	digital radio broadcasting	1536 768 384 192	1.712	from 0.576 to 1.152
DVB-T (ETSI, 1997, Europe)	digital video broadcasting (terrestrial)	6817 1705	8 7 6	from 4.98 to 31.67
DVB-H (ETSI, 2004, Europe)	digital video broadcasting (hand-held devices)	6817 3409 1705	8 7 6 5	from 3.7 to 23.8
ADSL (ANSI, 1995, USA) (ETSI, 2001, Europe)	digital transmission over copper telephone lines	224 31	1.10	from 1.3 to 12

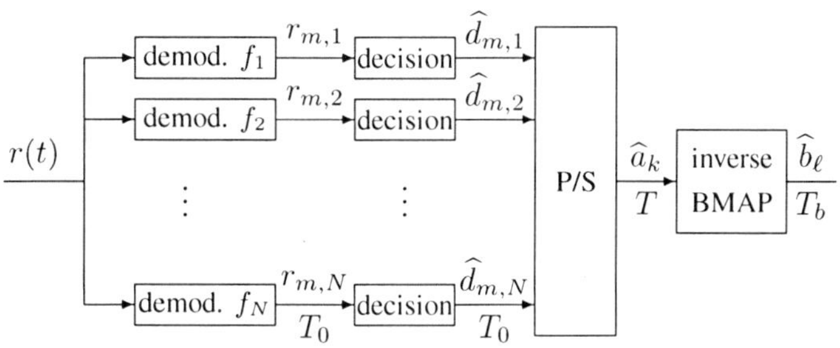

Figure 5.55 OFDM demodulation principle

where $w_{m,i}$ are uncorrelated, complex AWGN components with the usual $N_0/2$ variance in both the real and the imaginary part. Then different SNRs are experienced over different subbands. This is the reason why, the best out of OFDM can be obtained by two approaches that take into account this variability. These are:

1. *Power loading* Signal power over each subband is assigned depending on the channel attenuation. Within an overall transmitted power budget, a higher (lower) power is assigned to a subband with low (high) attenuation.

2. *Bit loading* Correspondingly, a higher number of bits is sent on those subbands with higher gain C_i, or, equivalently, higher constellation cardinalities are chosen for the best subbands (i.e., channels with an higher SNR). On the contrary, subbands with severe

attenuation are assigned a very low number of bits (low constellation cardinality), or they can even be silenced, that is they are not used for data transmission.

5.8.2 Spread spectrum techniques

While OFDM is targeted to high-rate applications, *spread spectrum* modulation is commonly adopted for low-rate and very-low-power wireless applications. Notable standards employing the spread spectrum modulation are 3rd generation UMTS (*universal mobile telecommunication standard*) cellular phones [11], [12], United States GPS and European Galileo satellite navigation systems [2, 15], wireless network based upon Bluetooth [9], [10], ZigBee and WiFi [16].

The peculiarity of spread spectrum modulation is to exploit a bandwidth which is much larger than the minimum bandwidth required for transmitting a digital stream with symbol period T, that is $B \gg B_{\min} = 1/T$. Although at a first glance this may seem a waste of bandwidth resources, it assures many important features, namely:

1. *Low probability of intercept* Spread spectrum signals do cover large bandwidths with low power emissions in such a way that their power spectrum level may lay below the noise level. When the modulation settings are not perfectly known at the receiver, the signal can hardly be demodulated, or even detected, this being the main reason why spread spectrum techniques were initially developed for military applications.

2. *Jam resistance* Spread spectrum signals are very robust against noise and intentional disturbances.

3. *Robustness to ISI* Using a suitable receiver (rake, or even an equalizer) provides robustness to ISI also in severe dispersive channels.

4. *Multiple access* It provides a multiple user access, that is it allows that more than a single communication can coexist in time and space, that is in the same instant and within the same physical region (i.e., a room).

5. *Localization capabilities* Because of the large bandwidths employed, spread spectrum signals allow to accurately estimate arrival times with a temporal precision which is approximately the inverse of the bandwidth B. In turn, by exploiting the speed of light $c = d/t$, they allow for precise distance measure, hence for precise object localization.

The typical modulations of spread spectrum are three: direct sequence, frequency hopping and time hopping. We briefly review them in the following.

Direct sequence. Direct sequence (DS) is obtained by using the waveform typical of code division modulation of Section 5.6.5. Specifically, each user u is assigned a specific codeword $\boldsymbol{C}_u = [c_{u,1}, \ldots, c_{u,N}]$, with N the code length, which is then employed for PAM (usually binary) and QAM (usually QPSK) transmission using the pulse shape (see (5.250))

$$h_{u,\mathrm{Tx}}(t) = \sum_{k=1}^{N} c_{u,k}\, \mathrm{w}_{T_c}(t-(k-1)T_c)\ , \qquad T_c = T/N\ . \tag{5.267}$$

Note that the reference bandwidth for the pulse in (5.267) is $1/T_c = N/T = NB_{\min}$.

DS signals belonging to different users are orthogonal if transmission is perfectly synchronized and the channel is ideal, a situation which can hardly be experienced in practice. In more reasonable hypotheses with a dispersive channel and absence of network synchronization, orthogonality is only approximate, and some form of disturbance due to the multiuser access is experienced. In common situations, this disturbance can be simply approximated as a further AWGN noise deteriorating the useful signal.

DS is used in GPS and Galileo satellite navigation systems [2, 15], CDMA cellular phones [11, 12], IEEE 802.11 and 802.11b Wi-Fi standards [16], and in the IEEE 802.15.4 ZigBee standard [16].

Frequency hopping. The idea behind frequency hopping (FH) is illustrated in Fig. 5.56. The available bandwidth B is partitioned in N slices ($N = 8$ in figure) with center frequencies f_i, $i = 1, \ldots, N$. At each symbol period kT, a different subband is chosen for transmission according to a pseudo-randomly generated code $c_{u,k}$, $0 < c_{u,k} \leq N$, associated with user u, so that $f_{c_{u,k}}$ is the selected frequency. In general, we may transmit more symbols for a given center frequency; this will be called *slow FH*. Conversely, we could change center frequency more times within a symbol period (*fast FH*).

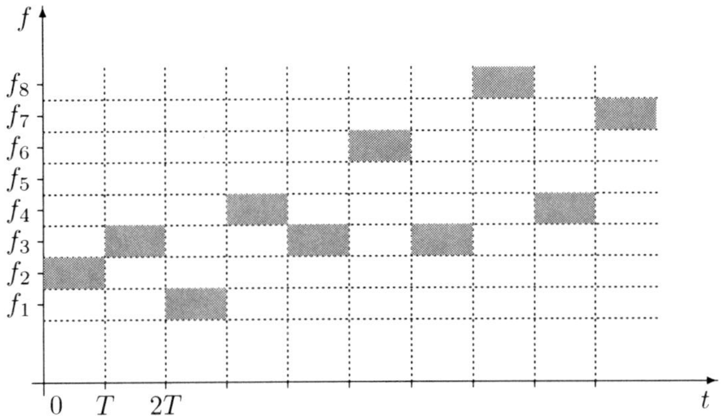

Figure 5.56 Exemplification of the frequency hopping paradigm with code $\{2, 3, 1, 4, 3, 6, 3, 8, 4, 7\}$ and $N = 8$ subbands.

A common modulation approach with FH is non-coherent binary FSK (with some modifications, this is the approach used by Bluetooth [9], [10]), in which case the transmitted signal is of the form (see (5.234))

$$s_{u,\text{Tx}}(t) = A \sin(2\pi\,(a_k \Delta F + f_{c_{u,k}})\,t + \varphi_k)\,, \quad \text{for } kT \leq t < (k+1)T \qquad (5.268)$$

where $\Delta F = 1/(2T)$ and $a_k = \pm 1$, with frequencies $f_{c_{u,k}}$ separated by multiples of $1/T$ to guarantee orthogonality. This form of FH is the first spread spectrum system ever implemented. Other forms of modulations, such as PAM or QAM, are also possible in conjunction with FH.

In multiuser FH, orthogonality between users is assured by carefully constructing the random codes $c_{u,k}$ in such a way that the probability of transmitting in the same instant over the same frequency is limited. This is better achieved if N is a large number.

Time hopping. Time hopping (TH) is a modification of the FH idea where time, instead of frequency, is partitioned in N sub-slots. Specifically, as shown in Fig. 5.57, each symbol period T is divided in N intervals of length $T_s = T/N$. At each symbol period kT, a pseudo random code $c_{u,k}$, $0 \le c_{u,k} < N$, associated with user u selects which of the N sub-slots is used for transmission. Modulation (typically binary PPM) is then applied inside the chosen sub-slot.

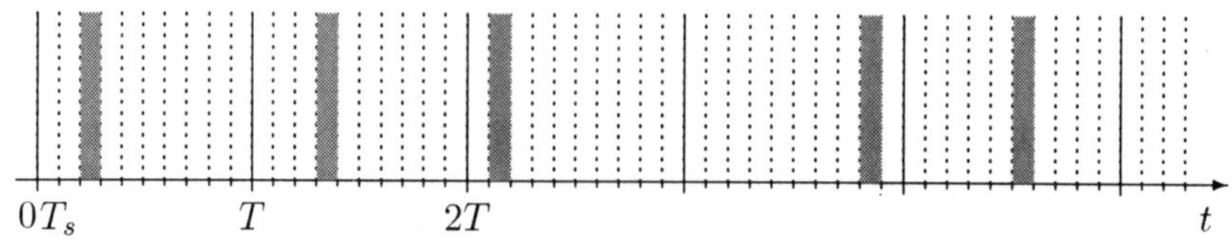

Figure 5.57 Exemplification of the TH access, with code sequence $\{3, 4, 2, 9, 6\}$ and $N = 10$.

A typical implementation of TH is in connection with binary PPM, in which case the transmitted signal is

$$s_{u,\mathrm{Tx}}(t) = \sum_{k=-\infty}^{+\infty} h_{\mathrm{Tx}}(t - kT - c_{u,k}T_s - a_k\, T_{\mathrm{PPM}}) \tag{5.269}$$

where $T_{\mathrm{PPM}} = T_s/2$ is the chosen PPM displacement (half a sub-slot) and $a_k \in \{0, 1\}$. Note that, the reference width for h_{Tx} is T_s, so that, also in this case, the reference bandwidth is $1/T_s = N/T$, N times larger than the minimum bandwidth. In multiuser TH, the orthogonality between users is assured by using suitable TH codes $c_{u,k}$ in such a way to minimize collisions between pulses belonging to different users.

5.9 DIGITAL TRANSMISSION OF ANALOG SIGNALS

As previously outlined, the objective of a communication system is to transmit besides data also analog information signals (e.g., voice or video). In modern communication systems, the transmission takes place as in the composite transmission system of Fig. 5.58 by converting the analog information signal into a sequence of binary digits (bits) using an ADC, as those shown in Chapter 3. The resulting bit stream $\{b_\ell\}$ is digitally transmitted, and the signal $\tilde{a}(t)$ is then reconstructed from the received bits by a DAC.

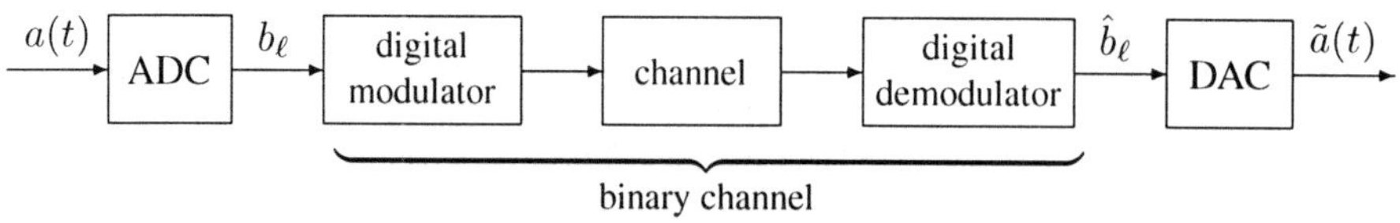

Figure 5.58 Digital transmission of an analog information signal.

Although the cascade of ADC and DAC blocks was fully investigated in Chapter 3, there the effect of transmission errors $\widehat{b}_\ell \neq b_\ell$ was not taken into account. Here we analyze this problem.

5.9.1 Transmission through a binary channel

We review the building blocks of Fig. 5.58, where ADC and DAC follow the notation previously introduced in Section 3.1.1.

ADC Here the quantizer is assumed uniform in the range $(-v_{\text{sat}}, v_{\text{sat}})$. The IBMAP of the ADC performs the following function:

$$a_q(nT_s) = Q_i \longrightarrow \boldsymbol{c}(nT_s) = \left[c_{\mathtt{b}-1}, \ldots, c_0\right] , \tag{5.270}$$

where, in order to simplify the analysis, we assume the rule ($L = 2^{\mathtt{b}}$)

$$\begin{cases} i = \displaystyle\sum_{j=0}^{\mathtt{b}-1} c_j 2^j , \\ Q_i = -v_{\text{sat}} + (i + \frac{1}{2})\Delta . \end{cases} \tag{5.271}$$

In other words, if Q_i is the level, the transmitted word is the binary representation of i. Recall that the binary sequence $\{b_\ell\}$ is obtained by P/S conversion of words $\boldsymbol{c}(nT_s)$.

Binary channel The model we consider is the simplest, considering a *memoryless binary symmetric channel* (see also Fig. 5.31) characterized by the transition probabilities

$$\begin{aligned} \mathrm{P}\left[\hat{b}_\ell = 1 | b_\ell = 0\right] &= \mathrm{P}\left[\hat{b}_\ell = 0 | b_\ell = 1\right] = P_{\text{bit}} \\ \mathrm{P}\left[\hat{b}_\ell = 0 | b_\ell = 0\right] &= \mathrm{P}\left[\hat{b}_\ell = 1 | b_\ell = 1\right] = 1 - P_{\text{bit}} \end{aligned} \tag{5.272}$$

Moreover, we consider errors to be iid. Depending on the chosen modulation, P_{bit} may assume one of the expressions in Tab. 5.1.

DAC Let $\tilde{\boldsymbol{c}}(nT_s) = [\tilde{c}_{\mathtt{b}-1}, \ldots, \tilde{c}_0]$ be the received word. The BMAP of the DAC performs the inverse operation of (5.270):

$$\tilde{\boldsymbol{c}}(nT_s) \rightarrow \tilde{a}_q(nT_s) = Q_{\tilde{i}} , \tag{5.273}$$

where

$$\begin{cases} \tilde{i} = \displaystyle\sum_{j=0}^{\mathtt{b}-1} \tilde{c}_j 2^j , \\ Q_{\tilde{i}} = -v_{\text{sat}} + (\tilde{i} + \frac{1}{2})\Delta . \end{cases} \tag{5.274}$$

Remark We incidentally note that, according to the binary channel model, the error probability on the word bits is given by $\mathrm{P}\left[\tilde{c}_j \neq c_j\right] = P_{\text{bit}}$. It is then useful to evaluate the error probability of words composed of b bit,

$$\begin{aligned} P_{e,c} &= \mathrm{P}\left[\tilde{\mathbf{c}}(nT_s) \neq \mathbf{c}(nT_s)\right] \\ &= 1 - \mathrm{P}\left[\tilde{\mathbf{c}}(nT_s) = \mathbf{c}(nT_s)\right] \\ &= 1 - (1 - P_{\mathrm{bit}})^{\mathsf{b}} , \end{aligned} \tag{5.275}$$

assuming errors are iid. On the other hand, if $P_{\mathrm{bit}} \ll 1$ it follows that $(1 - P_{\mathrm{bit}})^{\mathsf{b}} \simeq 1 - \mathsf{b}P_{\mathrm{bit}}$ and

$$P_{e,c} \simeq \mathsf{b}P_{\mathrm{bit}} . \tag{5.276}$$

Moreover, given the one-to-one map between words and quantizer levels, using (5.276) it follows that $\mathrm{P}\left[\tilde{a}_q(nT_s) \neq a_q(nT_s)\right] = P_{e,c} \simeq \mathsf{b}P_{\mathrm{bit}}$.

5.9.2 Evaluation of the overall SNR

Performance of the overall system will be evaluated in terms of the signal-to-noise ratio at the output of Fig. 5.58, where we can write $\tilde{a}(t) = a(t) + (\tilde{a}(t) - a(t))$ with $\tilde{a}(t) - a(t)$ the overall system error. Hence, we write

$$\Lambda_{\mathrm{PCM}} = \frac{\mathrm{E}\left[a^2(t)\right]}{\mathrm{E}\left[(\tilde{a}(t) - a(t))^2\right]} = \frac{\mathrm{E}\left[a^2(nT_s)\right]}{\mathrm{E}\left[(\tilde{a}(nT_s) - a(nT_s))^2\right]} , \tag{5.277}$$

using the fact of Section 2.8.3 that the statistical power of the output signal of an interpolate filter in a DAC is equal to the statistical power of the input samples, assuming they are stationary.

Now, in Fig. 5.58 the reconstructed analog signal $\tilde{a}(t)$ is different from the information signal $a(t)$ for two reasons: 1) presence of the quantization error in the ADC; 2) errors on the detection of the binary sequence at the output of the binary channel. The quantizer introduces an error e_q such that

$$a_q(nT_s) = a(nT_s) + e_q(nT_s) , \tag{5.278}$$

and the binary channel reconstructs a_q with a certain error e_{Bc},

$$\tilde{a}_q(nT_s) = a_q(nT_s) + e_{\mathrm{Bc}}(nT_s) . \tag{5.279}$$

Therefore the overall relation is

$$\tilde{a}_q(nT_s) = a(nT_s) + e_q(nT_s) + e_{\mathrm{Bc}}(nT_s) , \tag{5.280}$$

where the two errors are assumed uncorrelated, as they are ascribed to different phenomena.

In particular, assuming the quantization noise uniform, from (3.26) it follows

$$\mathtt{M}_{e_q} = \frac{\Delta^2}{12} = \frac{v_{\mathrm{sat}}^2}{3 \cdot 2^{2\mathsf{b}}} . \tag{5.281}$$

The computation of $\mathtt{M}_{e_{\mathrm{Bc}}}$ is somewhat more difficult. Hereafter the dependence of the signals on the time nT_s, will be sometimes omitted. Firstly, from (5.279), using (5.270), (5.271) and

(5.273), (5.274), we have

$$\begin{aligned} e_{\mathrm{Bc}}(nT_s) &= \tilde{a}_q(nT_s) - a_q(nT_s) \\ &= -v_{\mathrm{sat}} + \left(\tilde{i} + \frac{1}{2}\right)\Delta - \left[-v_{\mathrm{sat}} + \left(i + \frac{1}{2}\right)\Delta\right] \\ &= \Delta \sum_{j=0}^{\mathtt{b}-1} \left(\tilde{c}_j - c_j\right) 2^j \,. \end{aligned} \tag{5.282}$$

Let the error on the j-th transmitted bit be

$$\varepsilon_j = \tilde{c}_j - c_j \,, \tag{5.283}$$

then (5.282) becomes

$$e_{\mathrm{Bc}}(nT_s) = \Delta \sum_{j=0}^{\mathtt{b}-1} \varepsilon_j \, 2^j \,. \tag{5.284}$$

We note that $\varepsilon_j \in \{-1, 0, 1\}$, with probabilities given by

$$\begin{cases} \mathrm{P}\left[\varepsilon_j = 1\right] = \mathrm{P}\left[\tilde{c}_j = 1, c_j = 0\right] = \frac{1}{2} P_{\mathrm{bit}} \\ \mathrm{P}\left[\varepsilon_j = -1\right] = \mathrm{P}\left[\tilde{c}_j = 0, c_j = 1\right] = \frac{1}{2} P_{\mathrm{bit}} \\ \mathrm{P}\left[\varepsilon_j = 0\right] = \mathrm{P}\left[\tilde{c}_j = c_j\right] = 1 - \mathrm{P}\left[\tilde{c}_j \neq c_j\right] = 1 - P_{\mathrm{bit}} \end{cases} \tag{5.285}$$

Then the error on each bit has zero mean

$$\mathrm{E}\left[\varepsilon_j\right] = 0 \,, \tag{5.286}$$

and statistical power

$$\begin{aligned} \mathrm{E}\left[\varepsilon_j^2\right] &= 1 \; \mathrm{P}\left[\varepsilon_j \neq 0\right] + 0 \; \mathrm{P}\left[\varepsilon_j = 0\right] \\ &= \mathrm{P}\left[\tilde{c}_j \neq c_j\right] \\ &= P_{\mathrm{bit}} \,. \end{aligned} \tag{5.287}$$

Moreover, since we consider a memoryless binary channel, errors on different bits are uncorrelated

$$\mathrm{E}\left[\varepsilon_i \, \varepsilon_j\right] = \begin{cases} \mathrm{E}\left[\varepsilon_j^2\right] = P_{\mathrm{bit}} & , \text{ for } i = j \,, \\ 0 & , \text{ for } i \neq j \,. \end{cases} \tag{5.288}$$

Hence from (5.284)

$$\mathtt{M}_{e_{\mathrm{Bc}}} = \mathrm{E}\left[e_{\mathrm{Bc}}^2(nT_s)\right] = \Delta^2 \sum_{i=0}^{\mathtt{b}-1} \sum_{j=0}^{\mathtt{b}-1} \mathrm{E}\left[\varepsilon_i \, \varepsilon_j\right] 2^i 2^j \,, \tag{5.289}$$

and recalling (5.288) we get

$$\mathtt{M}_{e_{\mathrm{Bc}}} = \Delta^2 P_{\mathrm{bit}} \sum_{j=0}^{\mathtt{b}-1} 2^{2j} = \Delta^2 P_{\mathrm{bit}} \frac{2^{2\mathtt{b}} - 1}{3} \,. \tag{5.290}$$

Consequently, from (5.277) and (5.280) the output SNR is given by

$$\Lambda_{\rm PCM} = \frac{\mathtt{M}_a}{\mathrm{E}\left[(e_q(nT_s) - e_{\rm Bc}(nT_s))^2\right]} = \frac{\mathtt{M}_a}{\mathtt{M}_{e_q} + \mathtt{M}_{e_{\rm Bc}}}, \tag{5.291}$$

being the two errors uncorrelated. Using (5.281) and (5.290), and for a signal-to-quantization noise ratio $\Lambda_q = \mathtt{M}_a/(\Delta^2/12)$ (see (3.31)), we get

$$\boxed{\Lambda_{\rm PCM} = \frac{\Lambda_q}{1 + 4\,(2^{2\mathsf{b}} - 1)\,P_{\rm bit}}} \tag{5.292}$$

We note that usually $P_{\rm bit}$ is such that $P_{\rm bit} 2^{2\mathsf{b}} \ll 1$ and it results $\Lambda_{\rm PCM} \simeq \Lambda_q$, that is the output noise is mainly due to the quantization noise.

Remark We observe that in the general case of non-uniform quantization there are no simple expressions similar to (5.290) and (5.292); however, the above observations remain valid.

Example 5.9 A For a signal $a \sim \mathcal{U}(-v_{\rm sat}, v_{\rm sat}]$ whereby $\Lambda_q = 2^{2\mathsf{b}}$, (5.292) is represented in Fig. 5.59 for various values of b. For $P_{\rm bit} < 1/(4 \cdot 2^{2\mathsf{b}})$ the output signal is corrupted mainly by the quantization noise, whereas for $P_{\rm bit} > 1/(4 \cdot 2^{2\mathsf{b}})$ the output is affected mainly by errors introduced by the binary channel. For example, for $P_{\rm bit} = 10^{-4}$, going from $\mathsf{b} = 6$ to $\mathsf{b} = 8$ bits per sample yields an increment of $\Lambda_{\rm PCM}$ of only 2 dB.

5.9.3 Digital versus analog transmission

To understand the importance of digital modulation, it is interesting to compare the performance of the scheme of Fig. 5.58 with alternative analog based schemes. This will clarify the reason why modern communication schemes are all digital. To this end, we first need to introduce some concepts on analog transmission.

Analog transmission The model of an analog modulation system is presented in Fig. 5.60, where $a(t)$ is the *information signal or modulating signal*, while $s_{\rm Tx}(t)$ is the *modulated signal*, obtained from $a(t)$. In the following we will consider a real valued, modulating signal $a(t)$, with bandwidth B, as shown in Fig. 5.61.

The modulated signal is given by the multiplication, realized by the *mixer*, between the information signal $a(t)$ and the carrier,

$$s_{\rm Tx}(t) = a(t)\, \cos(2\pi f_0 t)\,. \tag{5.293}$$

In the frequency domain we get

$$\mathcal{S}_{\rm Tx}(f) = \tfrac{1}{2}\,\mathcal{A}(f - f_0) + \tfrac{1}{2}\,\mathcal{A}(f + f_0)\,. \tag{5.294}$$

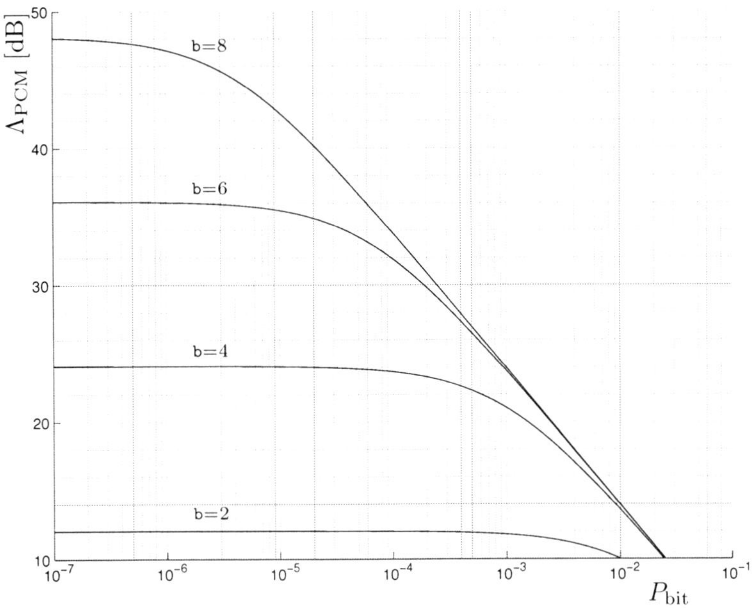

Figure 5.59 Signal-to-noise ratio of a PCM system as a function of P_{bit}.

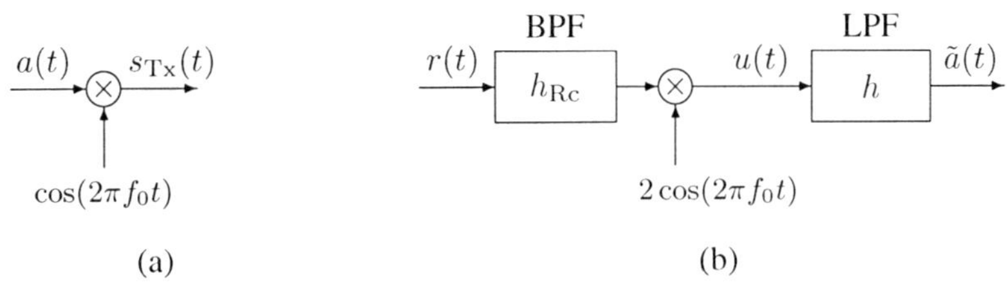

Figure 5.60 Analog modulation scheme: (a) modulator (transmitter); (b) demodulator (receiver).

This transformation is illustrated in Fig. 5.61. Note that modulating $a(t)$ with the carrier, apart for the factor $\frac{1}{2}$, achieves the objective of shifting the Ftf of the information signal around the carrier frequency f_0. A first consequence of this operation is that the modulated signal s_{Tx} has a bandwidth $B_s = 2B$, double of that of a. This explains the terminology with which the scheme of Fig. 5.60 is referred to, namely *double side band* (DSB) modulation.

As shown in Fig. 5.60b, the demodulator consists of a bandpass filter, a mixer and a lowpass filter. The purpose of the bandpass filter h_{Rc} is to remove the noise frequency components outside the band of the useful signal part s_{Rc} (see (5.1)), so that ideally its frequency response is given by

$$\mathcal{H}_{\text{Rc}}(f) = \text{rect}\left(\frac{f - f_0}{2B}\right) + \text{rect}\left(\frac{f + f_0}{2B}\right) . \tag{5.295}$$

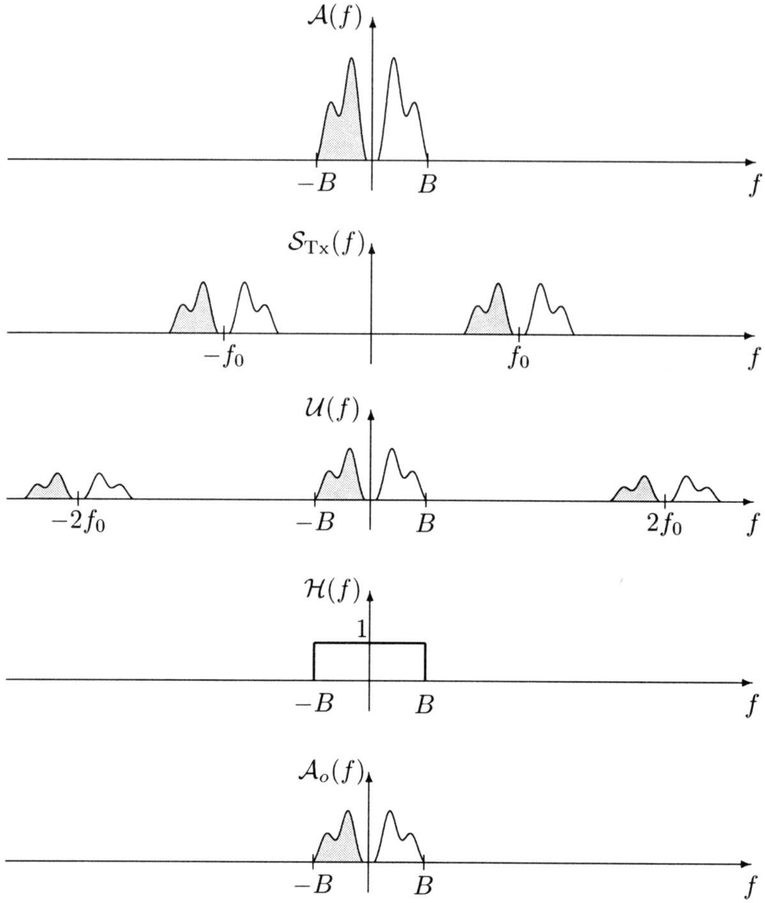

Figure 5.61 Frequency domain representation of the signals in a DSB system.

Then, the mixer multiplies the received signal $r(t)$ (after bandpass filtering) by the carrier $2\cos(2\pi f_0 t)$, that should have the same frequency of the carrier of the received signal. The *mixer* output is then filtered by a lowpass filter having the same bandwidth B of the information signal. We denote the output signal as $\tilde{a}(t)$. The various transformations introduced by the demodulator are illustrated in the frequency domain in Fig. 5.61 in the idealized case $r(t) = s_{Tx}(t)$, and discussed in the following.

For $r(t) = s_{Tx}(t)$, the receive BPF does not alter the received signal, and the output of the mixer is given by

$$\begin{aligned} u(t) &= 2\, s_{Tx}(t)\, \cos(2\pi f_0 t) \\ &= 2\, a(t)\, \cos(2\pi f_0 t)\, \cos(2\pi f_0 t) \\ &= a(t) \left[1 + \cos(2\pi 2 f_0 t)\right], \end{aligned} \tag{5.296}$$

where we used standard trigonometric identities. In (5.296) we see one term proportional to the information signal a, while $a(t)\,\cos(2\pi 2 f_0 t)$ represents in the frequency domain the shift of $\mathcal{A}(f)$ around the frequencies $\pm 2 f_0$. The information signal $a(t)$ can be recovered by using a lowpass filter h with a constant frequency response, here assumed of unit gain, over the *pass band* $(0, B)$ of $a(t)$, to avoid distortion of the first term in (5.296), and zero gain over the *stop band* $(2f_0 - B, 2f_0 + B)$, to attenuate the second term in (5.296). At the filter output we thus have $\tilde{a}(t) = a(t)$, so that the signal is perfectly recovered in ideal channel conditions.

When the channel model encompasses AWGN and attenuation (but no distortion) as in (5.138), then it can be observed that, thanks to the linearity of the modulation in Fig. 5.60, we have

$$\tilde{a}(t) = \mathcal{C}\, a(t) + \tilde{w}(t) \tag{5.297}$$

where $\tilde{w}(t)$ is a Gaussian noise, generated by $w_{\mathrm{Rc}}(t)$, whose statistical power can be proved to be (see [14])

$$\mathtt{M}_{\tilde{w}} = 2\, N_0\, B\ . \tag{5.298}$$

Hence, the output SNR is

$$\Lambda_{\mathrm{DSB}} = \frac{\mathcal{C}^2\, \mathtt{M}_a}{2\, N_0\, B} = \frac{\mathtt{M}_{s_{\mathrm{Rc}}}}{N_0\, B} \tag{5.299}$$

where we exploited the relations $\mathtt{M}_{s_{\mathrm{Rc}}} = \mathcal{C}^2\, \mathtt{M}_{s_{\mathrm{Tx}}}$ and $\mathtt{M}_{s_{\mathrm{Tx}}} = \frac{1}{2}\, \mathtt{M}_a$ (see Example 2.7 F). Other modulation methods can be found in [14].

Digital transmission If, instead, we wish to digitally transmit the signal $a(t)$, the overall performance is given by (5.292) where, as seen in Tab. 5.1, the value of P_{bit} is determined by the type of modulation and by the value of the reference SNR Γ.

To obtain an expression of P_{bit}, it is necessary to specify the type of modulator. Let us consider an M-PAM system, where (see Tab. 5.1)

$$P_{\mathrm{bit}} = \frac{2(M-1)}{M \log_2 M}\, Q\left(\sqrt{\frac{3}{M^2 - 1}\Gamma}\right)\ , \tag{5.300}$$

and where (see (5.257))

$$\Gamma = \frac{\mathtt{M}_{s_{\mathrm{Rc}}}}{N_0\, B_{\mathrm{min}}}\ . \tag{5.301}$$

Moreover, from (5.134) and the PCM coding bit rate $R_b = \mathtt{b}\, 2B$, the symbol period T can be written as

$$\begin{aligned} T &= T_b \log_2 M \\ &= \frac{\log_2 M}{R_b} \\ &= \frac{\log_2 M}{\mathtt{b} 2B} \end{aligned} \tag{5.302}$$

so that the minimum bandwidth is given by (see Tab. 5.1)

$$B_{\mathrm{min}} = \frac{1}{2T} = \frac{\mathtt{b}}{\log_2 M} B\ . \tag{5.303}$$

Hence the digital transmission of an analog signal may require a considerable expansion of the required bandwidth, if M is small. Obviously, using a more efficient digital representation of waveforms and/or a modulator with higher spectral efficiency, for example by resorting to a multilevel format, $B_{\min}$ may result very close to B or even smaller.

Using (5.303) in (5.301), and comparing the result with (5.299) we obtain

$$\Gamma = \frac{\log_2 M}{\mathsf{b}} \Lambda_{\rm DSB} \, , \tag{5.304}$$

which gives us a mean to compare the performance of analog and digital transmissions. To simplify the expression of $\Lambda_{\rm PCM}$ we assume 1) 2-PAM, i.e. $M = 2$, as modulation format, and 2) linear PCM encoding with uniform input signal for which $\Lambda_q = 2^{2\mathsf{b}}$. The expression of $\Lambda_{\rm PCM}$ in (5.292) becomes

$$\Lambda_{\rm PCM} = \frac{2^{2\mathsf{b}}}{1 + 4(2^{2\mathsf{b}} - 1) Q\left(\sqrt{\frac{\Lambda_{\rm DSB}}{\mathsf{b}}}\right)} \, , \tag{5.305}$$

and is plotted in Fig. 5.62 as a function of $\Lambda_{\rm DSB}$.

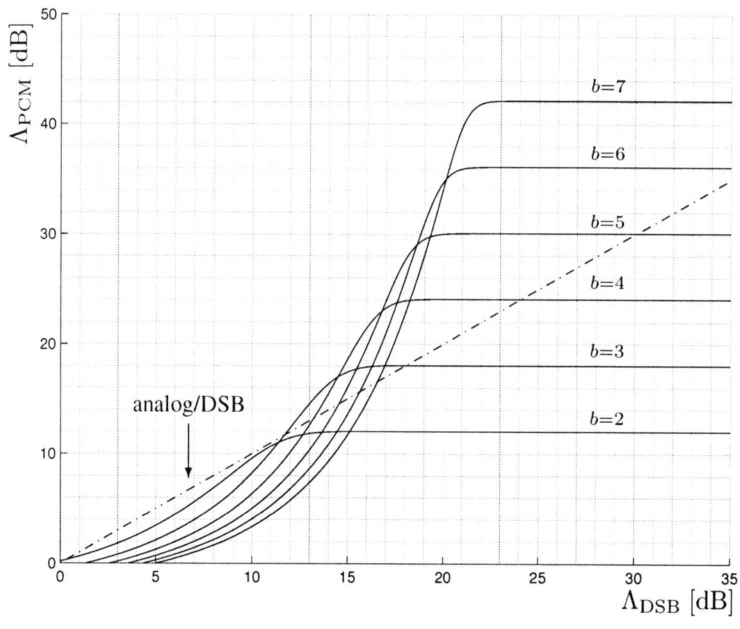

Figure 5.62 $\Lambda_{\rm PCM}$ as a function of $\Lambda_{\rm DSB}$ when 2-PAM is employed in the digital transmission: the parameter b denotes the number of bits of the linear PCM encoder.

We note that in the region of interest, for $(\Lambda_{\rm DSB})_{\rm dB} > 20\,{\rm dB}$, $\Lambda_{\rm PCM}$ is typically higher than $\Lambda_{\rm DSB}$, as long as a sufficiently large number of bits in the PCM representation is used and high values of $\Lambda_{\rm DSB}$ are considered. However, as we mentioned above, the PCM system may be penalized by the increment of the required channel bandwidth. In general, the advantage of digital transmission is even more evident using coding, but this aspect will be investigated in Chapter 6.

5.9.4 Digital transmission over long distances: analog versus regenerative repeaters

An analogous comparison between the performance of analog and digital approaches can be observed in long distance transmission, where it is necessary to place repeaters along the transmission line, otherwise the signal would be too attenuated and hence corrupted by the noise to allow correct detection at the receiver. Here we focus only on digital transmission and report a comparison between two approaches: analog and regenerative repeaters. Indeed, it can be seen when using regenerative repeaters that, for a given overall SNR, the transmitted power required in a digital transmission is a fraction of that required in an analog transmission.

Analog repeaters Here the solution is to place *analog repeaters* constituted of amplifiers and filters to restore the signal level and eliminate the noise outside the band of the useful signal. The scheme is presented in Fig. 5.63a.

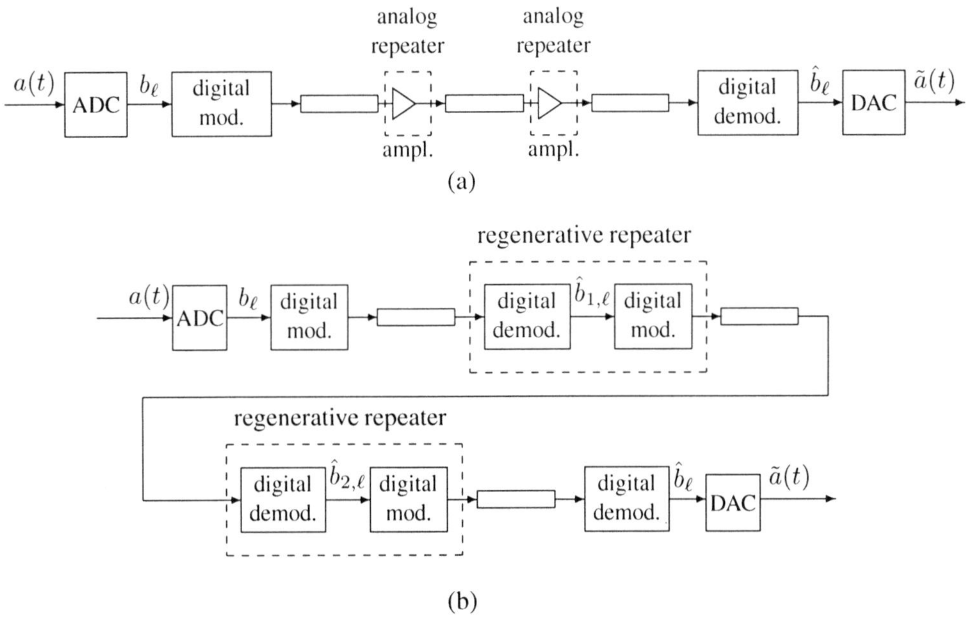

Figure 5.63 Digital transmission system with (a) analog repeaters; (b) regenerative repeaters.

The cascade of amplifiers along a transmission line, however, deteriorates the signal-to-noise ratio. As in Example 4.2 B we consider N repeater sections, *each* characterized by a noise figure $\mathsf{F}_R = \mathsf{a}_{\mathrm{Ch}}\mathsf{F}_A$ and by a *reference SNR* Γ. At the end of N analog repeater sections, the overall (including the first transmission link) noise figure is equal to $\mathsf{F} = N\mathsf{F}_R = N\mathsf{a}_{\mathrm{Ch}}\mathsf{F}_A$ (see (4.65)), and the value of the reference SNR is reduced to Γ/N due to the fact that the overall noise figure has increased by a factor of N. Hence, in a system with analog repeaters, the noise builds up repeater after repeater and the overall SNR worsens as the number of repeaters increases. Moreover, it must be remembered that in practical systems, possible distortion experienced by the desired signal through the various channels and amplifiers also accumulates. In conclusion, the performance of a M-PAM system becomes (compare with (5.300))

$$N \text{ analog repeaters :} \qquad P_{\text{bit},N} = \frac{2(M-1)}{M \log_2 M} Q\left(\sqrt{\frac{3}{M^2-1}\frac{\Gamma}{N}}\right) . \qquad (5.306)$$

Digital repeaters As an alternative to the simple amplification of the received signal, we can resort to the *regeneration* of the digital signal. In each repeater, from the received signal $r(t)$, the digital message $\{b_\ell\}$ is first reconstructed, and then re-transmitted by a modulator, as illustrated in Fig. 5.63b. The basic operations performed by a digital regenerative repeater are shown in greater detail in Fig. 5.64.

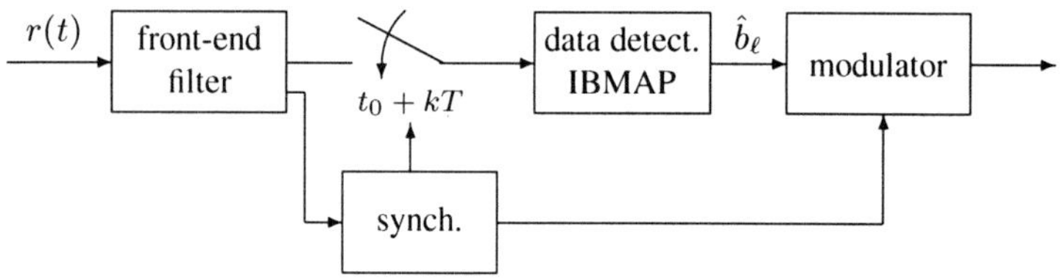

Figure 5.64 Regenerative repeater.

Each *regenerative section*, from a binary stream to the next, can be modeled by a memoryless binary symmetric channel, with error probability P_{bit}, and the errors on the different links are considered statistically independent. Neglecting the probability that a bit undergoes more than one error along the various repeaters, the bit error probability at the output of N regenerative repeaters is equal to

$$P_{\text{bit},N} \simeq 1 - (1 - P_{\text{bit}})^N \simeq N P_{\text{bit}} \,, \qquad (5.307)$$

where the last approximation holds if $P_{\text{bit}} \ll 1$. Hence, from (5.300), the performance with M-PAM is

$$N \text{ regenerative repeaters :} \qquad P_{\text{bit},N} = \frac{2(M-1)}{M \log_2 M} N\, Q\left(\sqrt{\frac{3}{M^2-1}\Gamma}\right) . \qquad (5.308)$$

The comparison between (5.306) and (5.308) is shown in Fig. 5.65. Even if a regenerative repeater is much more complex than an analog repeater, for a given overall P_{bit}, regeneration allows a significant saving in the power of the transmitted signal. Note that (5.308) is based on the assumption that $P_{\text{bit}} \ll 1$ hence for Γ sufficiently high.

REFERENCES

1. R.G. Gallager, *Information theory and reliable communications.* New York: John Wiley & Sons, 1968.
2. Alfred Leick, *GPS satellite surveying*, John Wiley & Sons, New York: 1990.

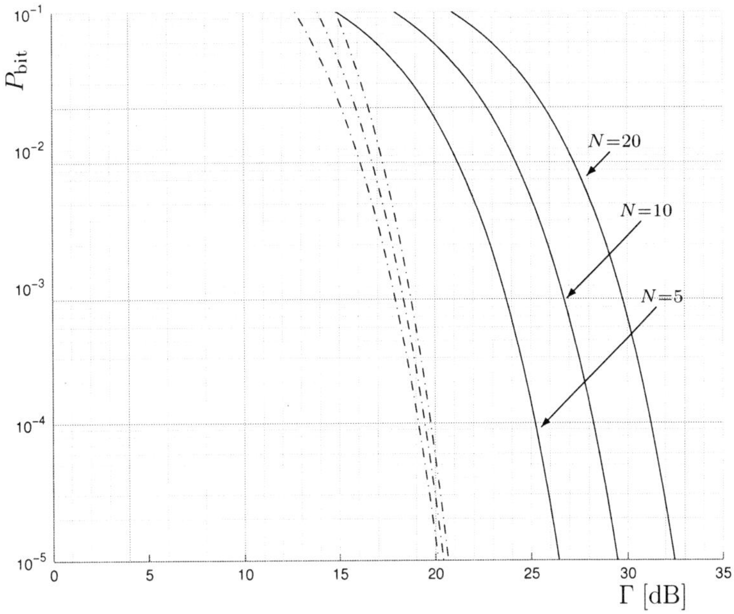

Figure 5.65 Bit error probability of digital transmission system using 4-PAM with analog (solid lines) and regenerative (dash-dot lines) repeaters, for three values of the number of repeaters N.

3. J.M. Wozencraft and I.M. Jacobs, *Principles of communication engineering*, New York: Waveland Press, reprint edition, 1990.
4. T.M. Cover and J. Thomas, *Elements of information theory*. New York: John Wiley & Sons, 1991.
5. A. Papoulis, *Probability, random variables and stochastic processes*. New York, NY: McGraw-Hill, 3rd ed., 1991.
6. DVB Project, ETSI, *"Digital broadcasting system for television, sound and data services; framing structure, channel coding and modulation for cable systems"*, EN 300 429 DVB-C standard, 1993.
7. DVB Project, ETSI, *Digital broadcasting system for television, sound and data services; framing structure, channel coding and modulation for 11/12 GHz satellite services"*, EN 300 421 DVB-S standard, 1993.
8. DVB Project, ETSI, *"Digital broadcasting system for television, sound and data services; framing structure, channel coding and modulation for digital terrestrial television"*, ETS 300 744 DVB-T standard, 1996.
9. J.C. Haartsen, "The Bluetooth radio system," *IEEE Personal Communications*, Vol. 7, No. 1, pp.28-36, Feb. 2000.
10. Bluetooth Special Interest Group (SIG), *Specification of Bluetooth system, ver. 1.1*, February 22, 2001, available online at www.bluetooth.org.
11. H. Holma and A. Toskala, *WCDMA for UMTS: radio access for third generation mobile communications*, Chichester, UK: John Wiley & Sons, 2002.
12. 3GPP IS-2002.2, *Physical layer standard for CDMA2000 spread spectrum systems*, June 2002.

13. U. Reimers, *Digital video broadcasting*. New York: Springer, 2004.

14. N. Benvenuto, R. Corvaja, T. Erseghe, N. Laurenti, *Communication Systems: fundamentals and design methods*, Chichester, UK: John Wiley & Son, 2006.

15. Guochang Xu, *GPS: Theory, Algorithms and Applications*, Springer, 2nd ed., 2007.

16. H. Labiod, H. Afifi, C. De Santis, *Wi-Fi, Bluetooth, Zigbee and WiMax*, Springer, 2007.

17. S. Galli, O. Logvinov, "Recent Developments in the Standardization of Power Line Communications within the IEEE," *IEEE Communications Magazine*, Vol. 46, No. 7, pp. 64-71, July 2008.

Problems

Digital modulation theory

5.1 Consider the two constellations in figure, where the minimum distance between waveforms is A. Which is the average energy in the two cases? Which constellation requires less energy?

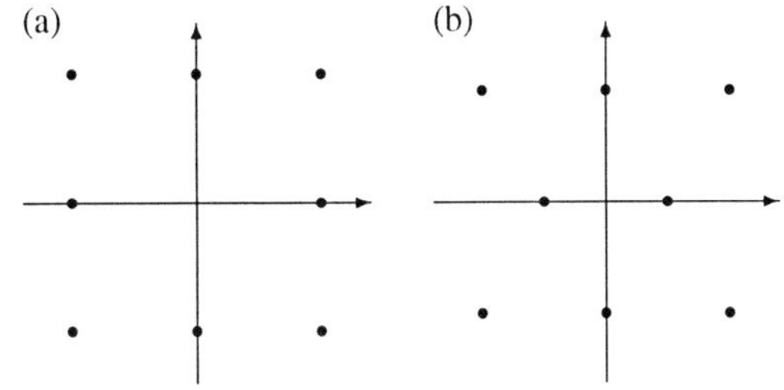

5.2 A transmission system employs 4 waveforms with probabilities $p_1 = p_4 = \frac{1}{8}$ and $p_2 = p_3$. The waveform space has dimension $I = 1$ and constellation

$$s_1 = -3V_0\ , \qquad s_2 = -V_0\ , \qquad s_3 = V_0\ , \qquad s_4 = 3V_0\ .$$

In the presence of AWGN with PSD $\frac{N_0}{2} = \frac{1}{4}V_0^2$, evaluate:

a) The error probability with minimum distance decision regions.

b) The error probability with optimum decision regions.

5.3 A transmission system employs 3 waveforms with probabilities $p_1 = p_3 = \frac{1}{4}$. The signal space has dimension $I = 1$ and constellation

$$s_1 = -V_0\ , \qquad s_2 = 0\ , \qquad s_3 = V_0\ .$$

In the presence of AWGN with PSD $\frac{N_0}{2} = \frac{2}{9}V_0^2$, evaluate:

a) The error probability with decision region thresholds at $\pm V_0/3$.

b) The error probability with optimum decision regions.

5.4 A binary transmission system with dimension $I = 1$ has constellation $s_1 = -A$ and $s_2 = +A$. Let the receiver employ two decision regions $R_1 = (-\infty, v)$ and $R_2 = [v, +\infty)$ with threshold $v = A/5$.

a) Evaluate the probability of error when $p_1 = p_2$ and the additive Gaussian noise has standard deviation $\sigma_I = A/5$.

b) Evaluate the probabilities p_1 and p_2 which guarantee that the chosen regions are optimum (consider again $\sigma_I = A/5$).

5.5 A digital transmission system employs the waveforms

$$s_1(t) = A\,e^{-|t|/T}\ \text{rect}\left(\frac{t - \frac{1}{4}T}{T}\right)\ , \qquad s_2(t) = -s_1(t)$$

with equal probability $p_1 = p_2$. Evaluate the error probability when the noise PSD is $\frac{N_0}{2} = A^2T/8$ and the threshold is 0.

5.6 A digital transmission system employs two linearly dependent waveforms, with constellation $s_1 = V_0$ and $s_2 = -V_0$, and probabilities $p_1 = \frac{2}{5}$ and $p_2 = \frac{3}{5}$. Let the noise at the decision point be independent of the useful signal, have standard deviation $\sigma_I = V_0/8$, and a Laplacian PDF

$$p_w(u) = \frac{1}{\sqrt{2\,\sigma_I^2}} \exp\Big(- \sqrt{2}\,|u/\sigma_I| \Big) .$$

Determine the optimum detection rule and the corresponding error probability.

5.7 Let a digital transmission system employ, with equal probability, two linearly dependent waveforms with projections $s_1 = V_0$ and $s_2 = 0$, respectively. Let the sample at the decision point have an additive Gaussian noise term with a standard deviation which is dependent on the transmitted signal according to the rule

$$\sigma_I = \begin{cases} 4\sigma & , a_0 = 1 \\ \sigma & , a_0 = 2 \end{cases} \qquad \sigma = 5 \cdot 10^{-2}\, V_0 \,.$$

Determine the optimum decision rule and the corresponding error probability.

Binary modulation

5.8 **(On-off keying, OOK)** A binary transmission system employs the two waveforms

$$s_1(t) = 0 \,, \qquad s_2(t) = A, \quad 0 \le t < T$$

for transmission over an AWGN channel with PSD $\frac{N_0}{2}$. By assuming that the two waveforms are equally likely, determine:

- **a)** The optimum receiver.
- **b)** The corresponding error probability as a function of E_s/N_0, where E_s is the average signaling energy.
- **c)** Compare the system performance with that of antipodal modulation.

5.9 An antipodal transmission system employs the waveform

$$s_1(t) = A \left[\mathrm{rect}\left(\frac{t - T_1}{2T_1} \right) + \mathrm{rect}\left(\frac{t - 5T_1}{2T_1} \right) \right] .$$

The channel introduces an AWGN with PSD $\frac{N_0}{2}$. Determine:

- **a)** The optimum receiver by assuming equally likely waveforms.
- **b)** The error probability as a function of A, T_1 and N_0.

5.10 **(Manchester encoding)** The two waveforms employed are

$$s_1(t) = \mathrm{rect}\left(\frac{t - \frac{1}{2}T}{T} \right) \mathrm{sgn}(t - \tfrac{1}{2}T) \,, \qquad s_2(t) = -s_1(t) \,.$$

By assuming that they are equally likely, and that the channel introduces AWGN with PSD $\frac{N_0}{2}$, determine the optimum receiver and the corresponding error probability expression.

5.11 An antipodal binary modulation employs, with equal probability, the waveforms

$$s_1(t) = A\ \mathrm{rect}\left(\frac{t - \frac{1}{2}T}{T} \right) , \qquad s_2(t) = -s_1(t) \,.$$

By assuming an AWGN channel with PSD $\frac{N_0}{2}$ determine:

- **a)** The expression for the SNR at the input of the detector (that is the SNR based on $\boldsymbol{r}$) by assuming that an optimum receive filter ψ_1 is used.

b) The expression for the SNR *loss* by assuming that the receive filter

$$\tilde{\psi}_1(t) = \sqrt{\tfrac{2}{\tau}}\, e^{-t/\tau}\, 1(t)$$

is used in place of the optimum one.

5.12 **(Pulse position modulation, PPM)** A binary transmission system employs the two waveforms

$$s_1(t) = \text{triang}(t-1)\ , \qquad s_2(t) = s_1(t-1)$$

with equal probability. By assuming that the channel is AWGN with PSD $\frac{N_0}{2} = 0.03$, determine:

a) The optimum receiver and the corresponding error probability.

b) An alternative waveform set (with the same average signaling energy) that guarantees an higher value of P_{bit}.

c) An alternative waveform set (with the same average signaling energy) that guarantees a smaller value of P_{bit}.

5.13 Consider a binary transmission system in AWGN noise with PSD $\frac{N_0}{2}$. Let the two employed waveforms be non equally likely, that is $p_1 = p$ and $p_2 = 1-p$. By assuming that the system has dimension $I = 1$ and waveform constellation s_1 and s_2, evaluate the optimum decision regions and the corresponding error probability.

5.14 **(Pulse amplitude modulation, PAM)** A binary antipodal signaling scheme employs two waveforms with probabilities $p_1 = 0.2$ and $p_2 = 0.8$. Determine the optimum receiver and the resulting error probability when $E_s/N_0 = 10$. Re-evaluate the result when $p_1 = 0.5$.

5.15 A binary antipodal transmission scheme employs the waveforms

$$s_1(t) = A\ \text{triang}\left(\frac{t - \frac{1}{2}T}{\frac{1}{2}T}\right)\ , \qquad s_2(t) = -s_1(t)\ .$$

By assuming that the channel is AWGN with PSD $\frac{N_0}{2} = \frac{1}{64}A^2T$, and the waveform probabilities are $p_1 = \frac{1}{3}$ and $p_2 = \frac{2}{3}$, determine the optimum detector and the corresponding error probability.

M-ary modulation

5.16 A digital transmission system employs the waveforms

$$s_1(t) = +\cos(2\pi f_0 t)\, h_{\text{Tx}}(t)\ , \quad s_2(t) = 2\sin(2\pi f_0 t)\, h_{\text{Tx}}(t)\ ,$$
$$s_3(t) = -\sin(2\pi f_0 t)\, h_{\text{Tx}}(t)\ ,$$

where $h_{\text{Tx}}(t)$ is a rectangular pulse of unit amplitude and duration $T = 1\,\mu$s, and $f_0 = 213/T$. The channel noise is AWGN with PSD $\frac{N_0}{2} = \frac{1}{2}\,10^{-7}$ V^2/Hz. Determine:

a) An orthonormal basis for the waveform set, the corresponding constellation and decision regions.

b) An upper bound to the error probability by using the minimum distance between constellation points.

5.17 Consider the two waveform sets

$$\left\{s_n(t) = A\,(2n-5)\,g(t)\right\} \qquad \text{and} \qquad \left\{v_n(t) = B\,g(t - nT)\right\}$$

where $n = 1, 2, 3, 4$ and $g(t) = \text{rect}(t/T)$. Derive bounds for the error probability and determine which transmission system is more efficient under the constraint of equal average signaling energy.

5.18 Consider the constellations of Problem 5.1. Let the average energy of the two systems be equal. Determine lower and upper bounds for the error probabilities as a function of E_s/N_0. Which of the two constellations is more robust to noise?

Digital modulation systems

5.19 A binary antipodal signaling is used over an AWGN channel with PSD $\frac{N_0}{2} = 10^{-10}\,\text{V}^2/\text{Hz}$. By assuming rectangular pulses, determine:

- **a)** The signal amplitude A required to meet a $P_{\text{bit}} = 10^{-6}$ when the bit rate is $R_b = 100\,\text{kbit/s}$.
- **b)** The loss in P_{bit} if the transmitter, due to energy constraints, is able to deliver only 80% of the amplitude A.

5.20 A digital transmission system employs a binary antipodal modulation with a rectangular pulse of duration T and amplitude A [Volts]. Let the bit rate be $R_b = 100\,\text{kbit/s}$ and assume the channel attenuates by a factor $C = \frac{1}{10}$. Let also the noise at the receiver input have PSD $\frac{N_0}{2} = 10^{-2}\,\text{V}^2/\text{Hz}$. By assuming equally likely waveforms, determine the value of A which guarantees an error probability $P_{\text{bit}} = 10^{-6}$.

5.21 A binary PSK system employs two 'windowed' sinusoidal waveforms with maximum amplitude $A = 1\,\text{V}$ and frequency $f_0 \gg 1/T$, with T the symbol period. Let the noise at the receiver input have PSD $\frac{N_0}{2} = 10^{-10}\,\text{V}^2/\text{Hz}$. Determine the maximum channel power attenuation that can be tolerated to guarantee a bit error probability of $P_{\text{bit}} = 10^{-6}$ when the bit rate is $R_b = 10\,\text{kbit/s}$, $R_b = 100\,\text{kbit/s}$, and $R_b = 1\,\text{Mbit/s}$, respectively.

5.22 Consider a digital communication system employing the waveforms

$$s_1(t) = -4\,h_{\text{Tx}}(t)\,, \quad s_2(t) = -3\,h_{\text{Tx}}(t)\,, \quad s_3(t) = 3\,h_{\text{Tx}}(t)\,, \quad s_4(t) = 4\,h_{\text{Tx}}(t)\,,$$

where $h_{\text{Tx}}(t)$ is a rectangular pulse with support $[0, T)$, $T = 1\,\text{ms}$ and amplitude $A = 1\,\text{V}$. Assume that the channel introduces a 20 dB attenuation and an AWGN noise with PSD $\frac{N_0}{2} = 10^{-3}\,\text{V}^2/\text{Hz}$. Determine:

- **a)** The system block diagram, the waveform constellation at the decision point and the corresponding decision regions.
- **b)** The system error probability.
- **c)** The system error probability in the presence of an additional periodic disturbance at the receiver input given by

$$d(t) = 2\,\sin(2\pi f_0 t + \varphi_0)\,, \qquad f_0 = \frac{4}{T}\,, \quad \varphi_0 = \frac{\pi}{6}\,.$$

5.23 A digital transmission system has the constellation

$$s_1 = (A, A)\,, \qquad s_2 = (-A, A)\,, \qquad s_3 = (A, -A)\,,$$

with $A = 50$. By assuming equally likely waveforms, and a Gaussian noise variance $\sigma_I^2 = 100$, determine:

- **a)** The optimum decision regions.
- **b)** The tighest upper bound on error probability.
- **c)** The error probability if s_1 is not used.

5.24 A binary PPM transmission system employs the waveforms

$$s_1(t) = A \operatorname{rect}\left(\frac{2t}{T_0} - \frac{1}{2}\right), \qquad s_2(t) = s_1(t - \tfrac{1}{2}T_0)$$

with $T_0 = 2\,\mu$s and $A = 2$ V, for transmission over an AWGN channel. Let the PSD of noise at the receiver input be $\frac{N_0}{2} = 10^{-8}$ V^2/Hz. Assume that the channel has impulse response

$$g_{\text{Ch}}(t) = \tfrac{1}{2}\,\delta(t) + \tfrac{1}{4}\,\delta(t - \tfrac{1}{2}T_0)\ .$$

Determine:

a) The symbol period T which yields the maximum value of the bit rate and at the same time guarantees the absence of ISI at the decision point (use this value throughout the problem). Which is the resulting bit rate R_b?

b) The optimum receiver and the corresponding error probability.

c) The bit error probability when the receive filter and the detection thresholds have been chosen based on the transmitted waveforms.

5.25 A transmission system employs a quaternary amplitude modulation with waveforms

$$s_1(t) = A\,h_{\text{Tx}}(t)\ , \qquad s_2(t) = -A\,h_{\text{Tx}}(t)\ , \qquad s_3(t) = 2A\,h_{\text{Tx}}(t)\ , \qquad s_4(t) = -2A\,h_{\text{Tx}}(t)\ ,$$

where $A = 1$ V and $h_{\text{Tx}}(t) = \operatorname{rect}(t/T)$ is a rectangular pulse with duration $T = 16$ ns. By considering a channel with AWGN noise having PSD $\frac{N_0}{2} = 10^{-9}$ V^2/Hz, determine:

a) The incorrect transition probabilities of the inner numerical channel by assuming an optimum receiver.

b) The bit error probability when the bit mapping is

$$\{0,0\} \to 1\ , \quad \{1,0\} \to 2\ , \quad \{0,1\} \to 3\ , \quad \{1,1\} \to 4\ .$$

c) The bit error probability when the bit mapping is

$$\{0,0\} \to 1\ , \quad \{1,0\} \to 2\ , \quad \{1,1\} \to 3\ , \quad \{0,1\} \to 4\ .$$

Which of the two bit mappings is more appropriate?

Examples of digital modulations

5.26 Consider a 4-QAM system with constellation

$$s_1 = (A, B)\ , \quad s_2 = (-A, B)\ , \quad s_3 = (-A, -B)\ , \quad s_4 = (A, -B)\ ,$$

where $A^2 = (2.3)^2$ V^2/Hz e $B^2 = 25$ V^2/Hz.

a) Illustrate the most efficient ML receiver architecture for the system.

b) Determine both the conditional probability of correct decision $\mathrm{P}\,[\mathrm{C}|a_0 = 1]$ in closed form and the probability of correct decision $\mathrm{P}\,[\mathrm{C}]$.

c) What is the value of P_{bit} if the noise variance is $\sigma_I^2 = 1$ V^2/Hz?

5.27 Consider a QAM system over an AWGN channel. Let the maximum symbol rate be $\frac{1}{T} =$ 2000 Baud. Determine:

a) The required modulation cardinality, M, and ratio E_s/N_0, in dB, to guarantee a bit error probability of $P_{\text{bit}} = 10^{-5}$ with a bit rate $R_b = 4000$ bit/s.

b) Re-determine a) when $R_b = 8$ kbit/s and $R_b = 16$ kbit/s and draw some conclusions.

5.28 Consider the two constellations of figure where the minimum distance between constellation points is A.

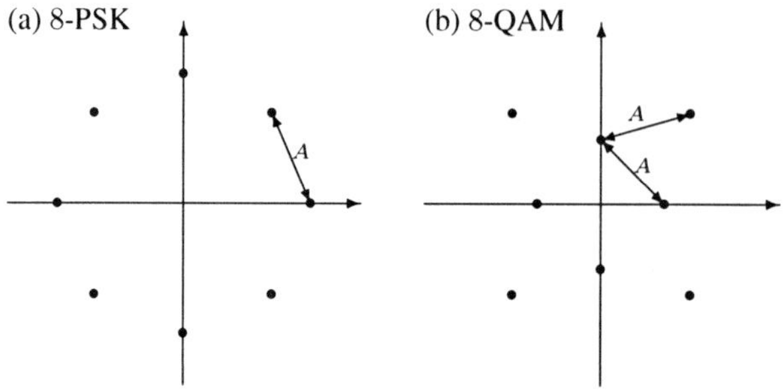

a) Is it possible to devise a Gray bit mapping for the above 8-QAM constellation?
b) Determine the required symbol rate if the system bit rate is $R_b = 60$ Mbit/s.
c) Compare the SNR E_s/N_0 required for obtaining the same bit error probability in a 8-QAM and a 8-PSK system (assume that the expression in Tab. 5.1 is valid for 8-QAM).
d) Which of the constellations in figure is more robust to phase errors, i.e. errors in the phase of the received vector $\boldsymbol{r}$?

Comparison of digital modulations and link budget

5.29 A binary communication system utilizes the waveform set

$$s_1(t) = 2\,\mathrm{rect}\,(t/T) \;\; [\mathrm{V}] \quad (T = 1\,\mu\mathrm{s})$$
$$s_2(t) = s_1(t)\,\mathrm{sgn}(t)\ .$$

At the output of the digital modulator there is a transmission line having impedance 100 Ω, length 13 km, specific attenuation 6 dB/km. The line is connected to the optimum receiver, having noise figure 10 dB. Determine:

a) The maximum amplitude of the signal at the receiver input.
b) The impulse responses of the filters of the optimum receiver.
c) The signal-to-noise ratio (in dB) at the decision point.
d) The statistical power of each component of the noise (in dB), at the decision point.

5.30 Determine the maximum bit rate R_b that can be transmitted over a bandpass telephone channel with bandwidth $B_{\mathrm{Ch}} = 5$ kHz and the following modulation schemes (assume $B_{\min} = B_{\mathrm{Ch}}$):

a) binary PAM;
b) 4-PAM;
c) 16-QAM;
d) non-coherent binary FSK;
e) coherent quaternary FSK;
f) non-coherent 8-FSK.

5.31 A space probe is sending binary data from a distance of 10^5 km using BPSK. The transmitted power of 40 dBm is sent over a narrowband channel centered at $f_0 = 1$ GHz. The transmit and receive antenna gains are, respectively, 20 and 40 dB, while the effective noise temperature at the receiver input is $\mathrm{T}_{\mathrm{eff,Rc}} = 300$ K.

a) Determine the signal power at the receiver input.

b) Determine what is the maximum allowable bit rate R_b by assuming a bit error probability $P_{\rm bit} = 10^{-6}$.

c) What is the answer to **b)** if QPSK is used?

5.32 A transmission line of bandwidth 1200 Hz and of length $L = 1000\,$km, where amplifiers are uniformly placed along the line every 50 km, is employed for data transmission. The modulation format is binary PAM. Let the line be modeled as an ideal channel with constant attenuation of 1 dB/km. Let the amplifiers along the line have gain $\mathtt{g}_A = 50\,$dB and noise figure $\mathtt{F}_A = 7\,$dB. Finally, let the noise figure of the receive amplifier be $\mathtt{F}_{\rm Rc} = 5\,$dB. Determine:

a) The maximum bit rate that guarantees the absence of intersymbol interference.

b) The required transmitted power to guarantee a bit error probability of 10^{-6}.

5.33 A narrowband communication system employs the two waveforms

$$s_1(t) = V_0 \operatorname{rect}\left(\frac{t - \frac{1}{2}T}{T}\right), \qquad s_2(t) = -s_1(t)\operatorname{sgn}(t - \tfrac{1}{2}T)$$

where V_0 is measured in Volt and $T = 1\,\mu$s. The transmission line has output impedance $100\,\Omega$, length 15 km, and specific attenuation 6 dB/km. The line is connected to a receiver with noise figure $(\mathtt{F}_{\rm Rc})_{\rm dB} = 13\,$dB. Determine:

a) The maximum amplitude of the signal at the receiver input by assuming $V_0 = 2\,$V.

b) The matched filter impulse response of the optimum receiver.

c) The required transmitted power to guarantee a bit error probability $P_{\rm bit} = 10^{-6}$.

d) The required value of V_0 to guarantee a bit error probability $P_{\rm bit} = 10^{-6}$.

Transmission through a binary channel

5.34 Evaluate the accuracy of the approximation (5.276) on the error probability of a word with $\mathtt{b}$ bits, $P_{e,\mathrm{c}}$, by plotting the exact error probability and expression (5.276) as a function of $P_{\rm bit}$ for $\mathtt{b} = 8, 12, 16$.

5.35 For an input signal with uniform amplitude, a linear PCM transmission system must guarantee a SNR $\Lambda_{\rm PCM}$ greater than 40 dB. Find the number of bits required, knowing that the binary channel bit error probability is $P_{\rm bit} = 10^{-5}$. Repeat with $P_{\rm bit} = 10^{-3}$.

5.36 A PCM transmission system has the following characteristics:

- Gaussian input signal $a(t)$ with zero mean and standard deviation $\sigma_a = 2\,$V;
- uniform quantizer with load factor $1/k_f = 4$;
- required signal-to-quantization noise ratio greater than 40 dB.

Find the probability of having a wrong *sample* at the receiver, knowing that the binary channel is characterized by the bit error probability $P_{\rm bit} = 10^{-6}$.

5.37 For an input signal with uniform amplitude, a linear PCM transmission system must guarantee a SNR $\Lambda_{\rm PCM}$ greater than 55 dB. Determine the number of bits required, knowing that the digital modulation is 4-PAM and the reference SNR on the link is $\Gamma = 21\,$dB.

5.38 A linear PCM transmission system has as input a Gaussian signal with autocorrelation function

$$\mathtt{r}_a(\tau) = A \operatorname{sinc}^2\left(\frac{\tau}{T_a}\right),$$

where $T_a = 0.2\,\mu$s. The digital transmitter employs M-QAM and the channel available bandwidth is 16 MHz. Determine:

a) The number of bits of the uniform quantizer to guarantee a signal-to-quantization noise ratio Λ_q greater than 45 dB. Assume a saturation probability of 10^{-3}.

b) The ADC output bit rate.

c) The symbol rate and the cardinality M of the modulator constellation.

d) The global SNR Λ_{PCM} (in dB), knowing that the received signal power is $\mathtt{P}_{\text{Rc}} = -60\,\text{dBm}$, the receiver noise figure is $\mathtt{F} = 15\,\text{dB}$ and the channel (source) has a noise temperature $\mathtt{T}_s = 400\,\text{K}$.

5.39 Consider a linear PCM transmission system with $N = 4$ *regenerative* repeaters, each characterized by a bit error probability $P_{\text{bit}} = 10^{-5}$. Determine the value of Λ_{PCM}, knowing that the uniform quantizer with 9 bits achieves the signal-to-quantization noise ratio $(\Lambda_q)_{\text{dB}} = 50\,\text{dB}$.

5.40 Consider a linear PCM transmission system with N repeaters, each using 16-QAM with a reference SNR $\Gamma = 20\,\text{dB}$. The information input signal is uniform and the ADC employs an uniform quantizer with 256 levels.

a) Determine the minimum number of repeaters N for which it is more convenient using regenerative repeaters than analog repeaters.

b) Repeat using 2-PAM.

5.41 Consider two uniform quantizers, optimized for two different inputs: i) a Gaussian signal having zero mean and unit statistical power (let the saturation probability be $1.9\ 10^{-1}$), ii) a sinusoidal signal with zero mean and unit statistical power (let the saturation probability be zero).

a) Using 8 bits per sample, determine the ratio (in dB) between the step size of the two quantizers.

b) Determine the relation between the signal-to-quantization noise ratio (in dB) of the two quantizers for the same generic number $\mathtt{b}$ of bits per sample. Assume no saturation noise.

c) Imposing the same signal-to-quantization noise ratio of 42 dB, determine the integer difference between the numbers of bits per sample required by the two quantizers.

d) Suppose that the Gaussian signal is obtained by sampling a voice signal for telephone applications, what is the bit rate at the output of the quantizer, requiring a signal-to-quantization noise ratio of 42 dB?

5.42 Consider the waveforms

$$s_1(t) = \text{triang}\left(\frac{t - T/2}{T/2}\right), \qquad s_2(t) = -s_1(t).$$

with $T = 2\ \mu\text{s}$, used for a binary transmission on a cable with specific attenuation $(\tilde{a}_{\text{Ch}})_{\text{dB/km}} = 5\,\text{dB/km}$, with a length of 24.89 km. The cable terminates on an impedence of $10\,\Omega$. Let $\mathtt{F}_A = 8\,\text{dB}$ be the noise figure of the receiver amplifier.

a) Determine whether $\{s_1(t), s_2(t)\}$ is a complete orthonormal basis for the signalling.

b) Compute the maximum number of sections cable-analog repeater that ensure a global error probability $P_{\text{bit}} < 10^{-3}$.

c) Compute the maximum number of sections cable-regenerative repeater that ensure a global error probability $P_{\text{bit}} < 10^{-3}$.

5.43 A PCM transmission system makes use of M-PAM to transmit a signal $a(t)$ with bandwidth 4 kHz.

a) Assume $M = 4$, a reference signal-to-noise ratio at the receiver input $(\Gamma)_{\text{dB}} = 20.25\,\text{dB}$, $\mathtt{b} = 4$ bits for quantization and a signal-to-quantization noise ratio $(\Lambda_q)_{\text{dB}} = 24\,\text{dB}$. Determine the signal-to-noise ratio, in dB, relative to the PCM transmission of $a(t)$.

b) If the systems makes use of 100 identical regenerative repeaters with M-PAM ($M = 4$) and each section is characterized by the same value of Γ as in the previous point, compute the signal-to-noise ratio, in dB, relative to the PCM transmission of $a(t)$.

c) Assuming that PCM uses 16 level quantization, discuss if there exists a PAM format that allows to use the channel with a minimum bandwidth of 2 kHz.

Review Problems

5.44 A 16-QAM transmission system uses the transmission pulse

$$h_{\text{Tx}}(t) = g_0 \text{ rect}\left(\frac{t}{T}\right)$$

con $T = 1\,\mu$s, $g_0 = 1$ V. The carrier has frequency $f_0 = 40$ MHz. The modulated signal is transmitted over a cable with specific attenuation of 10 dB/km. At the receiver the AWGN has PSD $N_0/2 = 2 \cdot 10^{-18}\ \text{V}^2/\text{Hz}$.

a) Find the statistical power of the transmitted signal (in V^2).

b) Draw the optimal receiver scheme, specifying all its parameters.

c) Determine the receive reference signal/noise ratio in order to get $P_e = 10^{-5}$.

d) Find the maximum cable length in order to obtain the error probability of point c).

5.45 Consider a 50 km radio link, using a carrier at 900 MHz and transmit and receive antennas with gain 6 and 10 dB, respectively. The input and output antenna impedance is 100 Ω. We want to transmit a 2 Mbit/s bit rate using a QPSK system, with transmit impulse h_{Tx} given by

$$h_{\text{Tx}}(t) = A \text{ rect}\left[\frac{t - (d + 0.5)T}{dT}\right]$$

with $d = 0.8$, where T is the symbol period. The receive antenna has noise temperature T_0. Assume that also the receiver has noise temperature T_0.

a) Determine the value of the symbol period.

b) Draw the impulse $h_{\text{Tx}}(t)$ as a function of the time, assuming $A = 1$.

c) Consider one of the possible transmit signalling waveforms. Give the waveform expression in time and frequency domain. Draw the corresponding graphs (consider only the amplitude for the frequency case). For this item only, normalize the waveform to unit energy.

d) Using one of the definitions of practical bandwidth, evaluate the bandwidth of the signal of the previous point. Specify the bandwidth definition used.

e) Using the optimal receiver, determine the sampling phase t_0 and the expression of the samples and of the noise at the decision point, as a function of A, giving their statistical description.

f) Determine the minimum value of the reference SNR Γ in order to achieve an error probability lower than 10^{-9}.

g) Determine the transmit power (in dBm) in order to meet the requirements of the previous point. Determine the corresponding value of A.

h) Determine how much we can increase the link distance if we assume that the receiver does not introduce any noise, with the same conditions of the point f).

5.46 Consider the binary signalling shown in the figure.

a) Represent the signalling constellation.

b) Represent the block scheme of the ML receiver, specifying the decision regions.

c) Determine the error probability, assuming AWGN with PSD $N_0/2 = 10^{-8}\ \text{V}^2/\text{Hz}$.

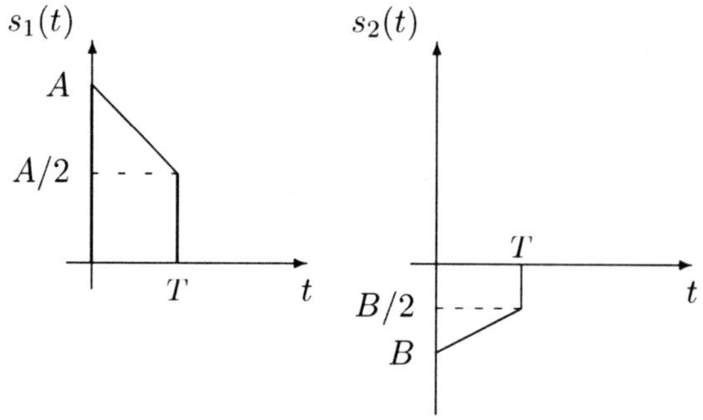

d) Considering now a variable value of B and imposing the same energy of $s_2(t)$ and $s_1(t)$, determine the value of B that minimize the error probability.

5.47 Consider a digital transmission system with binary signalling (in Volt)

$$s_{\mathrm{Tx},1}(t) = 0.5\,\mathrm{rect}\,(t - 0.5)\ , \qquad s_{\mathrm{Tx},2}(t) = -\mathrm{rect}\,(t - 0.5)\ ,$$

and channel with impulse response

$$g_{\mathrm{Ch}}(t) = \frac{1}{2}\delta(t) - \frac{1}{2}\delta(t - 1.2)$$

and AWGN with PSD $N_0/2 = 1.4 \cdot 10^{-1}$ V^2/Hz.

a) Determine the signalling waveforms *at the channel output*. Give the analytic expressions and draw the graphs.

b) Determine the minimum symbol period of the transmission system in order to avoid interference between subsequent pulses at the receiver.

c) Determine an optimal receiver block scheme: receive filter (draw the impulse response); optimal sampling phase value; decision element (give the decision rule).

d) Determine the system error probability.

5.48 Consideri a digital trasnmission system characterized by the constellation of the figure with $A^2 = (2.3)^2$ V^2/Hz and $B^2 = 25$ V^2/Hz.

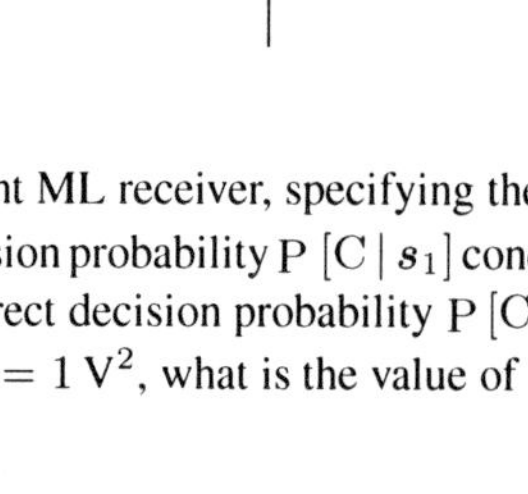

a) Represent the most efficient ML receiver, specifying the decision regions.

b) Determine the correct decision probability $\mathrm{P}\,[\mathrm{C} \mid s_1]$ conditioned by transmission of the symbol 1. Determine the total correct decision probability $\mathrm{P}\,[\mathrm{C}]$.

c) If the noise variance is $\sigma_I^2 = 1$ V^2, what is the value of P_{bit}?

5.49 A ternary digital transmission system employs the following waveforms:

$$\begin{aligned} s_1(t) &= S_0 \operatorname{rect}\left(\frac{t}{T}\right) \\ s_2(t) &= S_0 \operatorname{rect}\left(\frac{t}{T}\right) \operatorname{sign}(t) \\ s_3(t) &= -S_0 \operatorname{rect}\left(\frac{t}{T}\right) \end{aligned}$$

where $S_0 = 1$ V and the symbol period is $T = 0.1$ ms.
Determine:

a) An orthonormal base for the signalling.
b) The simplest optimal receiver scheme.
c) The optimal decision regions for AWGN.
d) The reference SNR in dB at the receiver input to obtain $P_e \leq 10^{-7}$. Use a suitable bound for the error probability.

5.50 A digital modulation scheme uses the following six waveforms

$$s_n(t) = A \operatorname{rect}\left(\frac{t - t_n}{T_0}\right) \quad , \quad n = 1, \ldots, 6$$

with $A = 2$ and $t_n = T_0/2 + (n-1)2T_0$. Assume $T_0 = \frac{T}{12}$, with $T = 1$ s symbol period. the channel is AWGN with PSD $\frac{N_0}{2} = 0.01$ V^2/Hz.

a) Draw the six waveforms of the signalling.
b) Determine the optimal receiver with all its parameters.
c) Simplify the receiver scheme by using a single receive filter.
d) Determine the system error probability.

5.51 A 16-QAM digital transmission system employs the transmit pulse $h_{Tx}(t) = V_0 \operatorname{rect}\left(\frac{2t}{T}\right)$ with $V_0 = 1$ V and symbol period $T = 1\,\mu$s.

a) Evaluate the statistical power of the transmitted signal.
b) Determine the *causal* receive filter of the optimum receiver and the corresponding sampling phase, assuming a channel with attenuation of 50 dB, constant over all the signal band.
c) Assuming AWGN with PSD $\frac{N_0}{2} = 2 \cdot 10^{-13}$ V^2/Hz, determine the variance per component of the noise at the decision point.
d) Evaluate the symbol error probability.

5.52 A digital transmission system with equally probable symbols employs the following waveforms:

$$s_1(t) = V_0 \operatorname{rect}\left(\frac{2t}{T_s}\right) \quad , \quad s_2(t) = V_1 \left(1 - \frac{|t|}{T_s}\right) \operatorname{rect}\left(\frac{2t}{T_s}\right)$$

with $V_0 = 10$ mV, $V_1 = 5$ mV and $T_s = 50$ ns.
The modulated signal is applied to a cable with impedance 100 Ω and attenuation 40 dB. The receiver noise figure is 23 dB.

a) Evaluate the minimum symbol period T to avoid interference between consecutive symbols.
b) Draw the optimum receiver scheme with lowest complexity, specifying all its parameters.
c) Evaluate the effective noise PSD at the receiver input port.
d) Evaluate the system error probability.

5.53 Consider a ternary PAM signalling with alphabet $\{-3, 0, 2\}$ with symbol probabilities $\left\{\frac{1}{4}, \frac{1}{2}, \frac{1}{4}\right\}$. The (voltage) transmit pulse is

$$h_{\mathrm{Tx}}(t) = 10 \,\mathrm{triang}\left(\frac{t - \tau_0}{5}\right).$$

a) Determine the minimum symbol period T to avoid interference between successive transmit pulses. Determine also the minimum value of the sampling phase τ_0 in order to have a causal transmit pulse. Draw the transmit pulse.

b) Determine the expression of the modulated signal $s_{\mathrm{Tx}}(t)$ for the following five-symbol input sequence:

$$a_0 = 2, \quad a_1 = 0, \quad a_2 = 2, \quad a_3 = -3, \quad a_4 = 2.$$

c) Draw the modulated signal of point 2).

d) For a channel with attenuation 40 dB and AWGN power that gives the reference SNR $(\Gamma)_{\mathrm{dB}} = 10\,\mathrm{dB}$, determine the noise variance in dBmV^2/Hz at the decision point.

e) Draw the constellation at the decision point, specifying the values of the waveform vectors.

f) Determine the system error probability assuming symmetrical decision thresholds (with the same amplitude and opposite sign) with amplitude equal to half the constellation value representing the signal s_2.

5.54 Consider a QPSK transmission system where the baseband transmit pulse h_{Tx} has bandwidth $\frac{0.75}{T}$ ($T = 1$ ms is the symbol period) and the carrier frequency is 100 MHz. The modulated signal is amplified and sent through and isotropic antenna (unit gain) to the radio channel.

The transmit amplifier has an available power of 2 mW and a resistive output impedance of 10 Ω. The antenna has input impedance given by the series of a resistance $R = 10\,\Omega$ and an inductance $L = 31.8$ nH. All the input power is then radiated.

At the receiver the isotropic receive antenna has noise temperature 300 K and is matched for the maximum power transfer to the QPSK demodulator. The demodulator noise figure is 4 dB.

a) Represent in a graph the frequency allocation of the waveforms of the digital signalling and evaluate the channel bandwidth required for the transmission.

b) Represent the electrical scheme of the connection amplifier-antenna at the transmitter. Evaluate the radiated power in dBm, assuming a constant antenna impedance over the useful band.

c) Determine the maximum distance for the receiver to guarantee a symbol error probability lower than $2 \cdot 10^{-6}$.

d) Given the same transmit power and symbol period, how could one modify the constellation to achieve a distance greater than the one of point c), keeping the same system performance? What is the price to pay for this change?

5.55 Consider the following digital segnalling, null outside the interval $0 < t < T$ (T is the symbol period):

$$\begin{aligned} s_{1,2,3,4}(t) &= \pm A\cos(2\pi f_0 t + \varphi_0) \pm A\sin(2\pi f_0 t + \varphi_0) \\ s_{5,6,7,8}(t) &= \pm 3A\cos(2\pi f_0 t + \varphi_0) \pm A\sin(2\pi f_0 t + \varphi_0) \end{aligned}$$

where $A = \frac{1}{\sqrt{2}}$, $T = 4$ s, $f_0 = 1$ MHz and $\varphi_0 = \frac{\pi}{10}$.

The following three questions are referred to the transmission of the previous signalling over a channel with impulse response

$$g_{\mathrm{Ch}}(t) = \delta(t)$$

and AWGN with PSD $\frac{N_0}{2} = 3.8 \cdot 10^{-2}$ V^2/Hz.

a) Determine a suitable base and draw the constellation at the decision point, specifying the signal coordinates.

b) Determine the decision rule and the decision regions. Write a pseudo-code (by using IF, THEN, ELSE, etc.) implementing the decision rule with input the components of $\boldsymbol{r}$ (received signal projection) and output $\hat{a}_0 \in \{1, 2, \ldots, 8\}$.

c) Determine the system error probability.

d) Suppose now that the channel frequency response is

$$\mathcal{G}_{\mathrm{Ch}}(f) = \begin{cases} e^{j\frac{\pi}{4}} & , \quad f > 0 \\ e^{-j\frac{\pi}{4}} & , \quad f < 0 \end{cases}$$

while the noise is still AWGN with PSD $\frac{N_0}{2} = 3.8 \cdot 10^{-2}$ V^2/Hz. Draw the new constellation at the decision point using the same base of point a).

e) In order to simplify the decision regions, determine a new orthonormal base for the new signalling at the channel output. Draw the corresponding constellation at the decision point.

f) Determine the system error probability of the previous point.

5.56 Consider three bit streams, with bit rate, respectively

$$R_{b,1} = 60 \quad \text{kbit/s}\,, \qquad R_{b,2} = 60 \quad \text{kbit/s}\,, \qquad R_{b,3} = 20 \quad \text{kbit/s}\,.$$

These messages are multiplex by TDM into a single binary stream.

a) Design a multiplexing scheme. In particular give an example of the multiplexed frame, identifying the three elementary stream bits.

b) Determine the overall bit rate.

c) The overall bit stream is modulated by 2-PAM. The PAM signal is transmitted with a power of 12 dBm around a central frequency of 2 MHz, over a 45 km cable with specific attenuation 2 dB/km at 1 MHz and impedance 10 Ω. The receiver noise figure is 5 dB. Design the optimum riceiver.

d) Determine the system error probability of point c).

e) Using now 8-PAM (instead of 2-PAM), determine the system error probability. Give only its expression.

CHAPTER 6

CHANNEL CODING AND CAPACITY

Nicola Laurenti

In the previous chapters we have examined several methods to transmit digital signals through a channel without wondering how much information can be effectively carried by the channel, and how it can be robustly encoded into the signal for reliable transmission. In this chapter, based upon the fundamentals of Information Theory introduced in Chapter 3, we introduce the concept of channel capacity, and some techniques of channel coding.

6.1 PRINCIPLES OF CHANNEL CODING

6.1.1 The purpose of channel coding

Channel coding is the technique to transform a digital information message $\{b_\ell\}$ into another message $\{c_\ell\}$ (hence the term *coding*), which is more robust to the errors that the channel may introduce. The robustness of the coded message is obtained at the price of some redundancy, which is added on purpose in the form of additional symbols (*parity symbols*) uniquely determined by the information message. Therefore the transmitted message is one of a set $\mathcal{C}$ of possibly transmitted messages, which is a subset of a larger set $\tilde{\mathcal{C}}$ of possibly received messages. When the channel corrupts the transmitted message, it will produce a message $\{\tilde{c}_\ell\}$ that is in $\tilde{\mathcal{C}}$ but may (with a high probability) not be in $\mathcal{C}$, as is illustrated in Fig. 6.1. This

Principles of Communications Networks and Systems N. Benvenuto, M. Zorzi

allows the receiver to recognize the message as invalid, and in some cases even to correct it, thus recovering the information message.

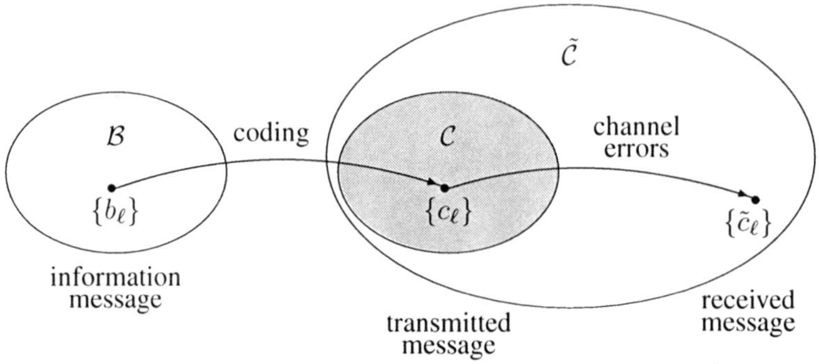

Figure 6.1 Conceptual illustration of the channel coding principle.

The reader may wonder what is the purpose of removing redundancy through source coding as we saw in Chapter 3, if it is then necessary to add some redundancy into the message for transmission. The answer is that while the initial redundancy is given by the statistics of the source, the redundancy introduced by channel coding is matched to the channel characteristics and the required level of robustness. More importantly, the latter is properly designed to help the receiver retrieve the information message with higher reliability. Hence the two coding methods are often separately designed.

There are two main ways that the receiver can deal with invalid messages:

Automatic retransmission query or automatic repeat request The automatic repeat request (ARQ) scheme consists in requesting the retransmission of the entire message or a section of it upon detection of an error.

In principle ARQ is similar to what we would do when speaking on the phone with somebody, and we cannot understand what he/she is saying: we ask him/her to repeat.

Forward error correction It consists in replacing the invalid received message with a possible transmitted message that is in some sense "close" to it. This may only partially correct the channel errors.

Somehow, forward error correction (FEC) is similar to what we would do when reading a typographical error in a newspaper: we mentally replace the incorrect word with a close word that makes sense and continue reading, instead of asking for the newspaper to be reprinted. While in this case we would make use of the intrinsic redundancy of natural language, the receiver of a digital transmission system makes use of the redundancy introduced by the channel encoder.

The choice of whether using FEC or ARQ techniques in designing a system should take into account different aspects, such as the following:

- The ARQ requires the presence of a reverse link, over which the receiver can communicate its request to the transmitter, and this may represent a strong constraint for some applications. On the other side, FEC can be implemented over simplex links.

- With ARQ, the delay implied by repeated retransmissions (and retransmission requests) in the reception of a message can be intolerable for some applications, as in telephony. Moreover, frequent retransmissions increase the link traffic, and thus lower the system efficiency.

- Transmission systems based on FEC are less reliable, since in the case of wrong decoding, i.e. decoding the received message to an information message different from the actually transmitted one, we increase the amount of errors, instead of reducing it.

For the above reasons, ARQ is more commonly employed in transmission of data, where delay constraints are not so stringent, e.g. transmission control protocol (TCP). On the other hand for broadcast transmissions or real-time applications, e.g. digital video broadcasting (DVB) and GSM, the FEC technique is preferred.

Hybrid schemes are also possible, where only the most likely (and hence frequent) errors are forward corrected, whereas in the remaining cases retransmission is requested. Such schemes operate a trade-off between the reliability of ARQ and the efficiency of FEC techniques.

The effectiveness of a coding technique can be expressed in terms of the *coding gain*, $g_c(P_{\text{bit}})$, given by the reduction in the reference SNR Γ that is required to achieve a certain bit error probability P_{bit}, with respect to uncoded transmission. The redundancy introduced by the coding is expressed via the *code rate*, given by the ratio between the number of input information symbols and the number of coded symbols that are transmitted in a given time interval.

For further study of error correcting codes we refer the reader to [5, 13].

6.1.2 Binary block codes

Although many different structures for channel coding are possible, we focus on *block coding* to introduce the principles of channel coding, while keeping the notation simple. Moreover we consider binary codes since these are used in nearly all cases, and even when nonbinary codes are employed, encoder and decoder are preceded (and followed) by a binary-to-M-ary (M-ary-to-binary) conversion, so that the cascade can be seen as a binary block encoder/decoder.

As illustrated in Fig. 6.2 the incoming sequence of binary symbols $\{b_\ell\}$ at rate $1/T_b$ (which we call *information message*) is segmented into separate blocks of k symbols named *information words* $\boldsymbol{b}_m$, one every $T_w = kT_b$ seconds. Each information word is then associated to a *codeword* of n binary symbols via the map

$$\mu_c : \mathcal{A}^k \mapsto \mathcal{A}^n \quad , \quad \mathcal{A} = \{0, 1\} \tag{6.1}$$

The set of possible codewords, that is the range of μ_c, $\mathcal{C} = \mu_c\left(\mathcal{A}^k\right)$ is named a (n, k) *block code*. The codeword symbols $\{c_\ell\}$ are then sequentially transmitted through a binary channel at rate $1/T_c$, with $T_c = T_w/n = kT_b/n$. At the channel output the *received symbols* $\{\tilde{c}_\ell\}$ will in general differ from the corresponding *transmitted symbols* $\{c_\ell\}$ due to the errors introduced by the channel. At the receiver, the sequence $\{\tilde{c}_\ell\}$ is segmented into n-symbol words $\{\tilde{\boldsymbol{c}}_m\}$. According to whether a FEC or an ARQ scheme is employed, the decoder performs different operations. In the ARQ scheme the decoder checks whether $\tilde{\boldsymbol{c}}_m \in \mathcal{C}$ and in this case yields $\hat{\boldsymbol{b}}_m = \mu_c^{-1}(\tilde{\boldsymbol{c}}_m)$, otherwise the receiver asks for retransmission of the message. In the FEC scheme, the decoder *detects* a suitable codeword $\hat{\boldsymbol{c}}_m \in \mathcal{C}$ from $\tilde{\boldsymbol{c}}_m$, then yields $\hat{\boldsymbol{b}}_m = \mu_c^{-1}(\hat{\boldsymbol{c}}_m)$

as detection of the information word $\boldsymbol{b}_m$. Therefore, the decoder can be seen as a map

$$\mu_{\rm d} \, : \, \mathcal{A}^n \mapsto \mathcal{A}^k \tag{6.2}$$

The parameter k/n is the *code rate*.

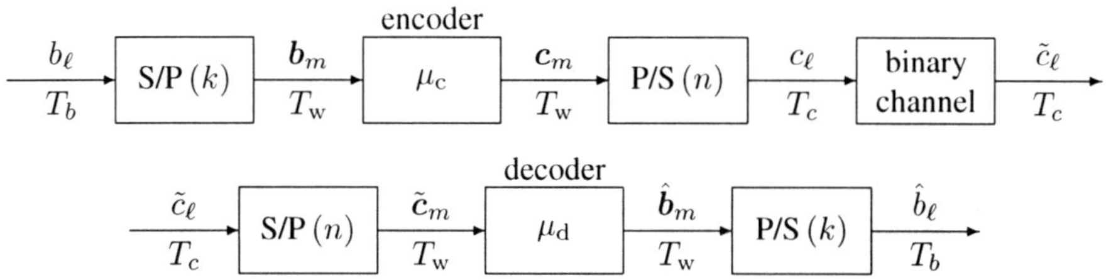

Figure 6.2 Block diagram of a system with binary (n, k) block channel coding.

A coding map, and the corresponding code $\mathcal{C}$ are said to be *systematic* if the information word $\boldsymbol{b}_m$ is the prefix of the codeword $\boldsymbol{c}_m$. In this case, the inverse map $\mu_{\rm c}^{-1}$ can be easily implemented by taking the k-symbol prefix of the detected codeword, $\hat{\boldsymbol{b}}_m = \mu_{\rm c}^{-1}(\hat{\boldsymbol{c}}_m) = (\hat{\boldsymbol{c}}_m)_1^k$.

6.1.3 Decoding criteria. Minimum distance decoding

With block coding, the possibility to detect and correct errors introduced by the channel lies in the fact that not all possibly received words $\tilde{\boldsymbol{c}}_m$ are codewords. Indeed, an error will be detected if and only if the corrupted word $\tilde{\boldsymbol{c}}_m$ does not lie in the code $\mathcal{C}$.

As general decoding creterion we outline the rule based on the decoding regions. Consider the space $\mathcal{A}^n$ of possible received words $\tilde{\boldsymbol{c}}_m = \boldsymbol{\xi} \in \mathcal{A}^n$. We partition it into 2^k disjoint sets $\{R_{\boldsymbol{\beta}_i}, i = 1, \ldots, 2^k\}$, called *decoding regions*, one for each information word $\boldsymbol{b}_m = \boldsymbol{\beta}_i$, as illustrated in Fig. 6.3.

Hence the decondig rule is

$$\text{if} \quad \tilde{\boldsymbol{c}}_m \in R_{\boldsymbol{\beta}_i} \quad \text{then} \quad \hat{\boldsymbol{b}}_m = \mu_{\rm d}(\tilde{\boldsymbol{c}}_m) = \boldsymbol{\beta}_i \ . \tag{6.3}$$

For later use, we recall that the Hamming distance $d_{\rm H}(\boldsymbol{\xi}_1, \boldsymbol{\xi}_2)$ between two words $\boldsymbol{\xi}_1, \boldsymbol{\xi}_2 \in \mathcal{A}^n$ is the number of symbols in which $\boldsymbol{\xi}_1$ and $\boldsymbol{\xi}_2$ differ, as given in (5.160).

We now introduce one of the most important parameters of a code.

Definition 6.1 *The* minimum Hamming distance of a code $\mathcal{C}$ *is defined as*

$$d_{\min} = \min_{\substack{\boldsymbol{\gamma}_1, \boldsymbol{\gamma}_2 \in \mathcal{C} \\ \boldsymbol{\gamma}_1 \neq \boldsymbol{\gamma}_2}} d_{\rm H}(\boldsymbol{\gamma}_1, \boldsymbol{\gamma}_2) \ . \tag{6.4}$$

Remark Note that if the binary channel introduces t symbol errors into the codeword $\boldsymbol{c}_m$ it results $d_{\rm H}(\tilde{\boldsymbol{c}}_m, \boldsymbol{c}_m) = t$.

Hence we have the following.

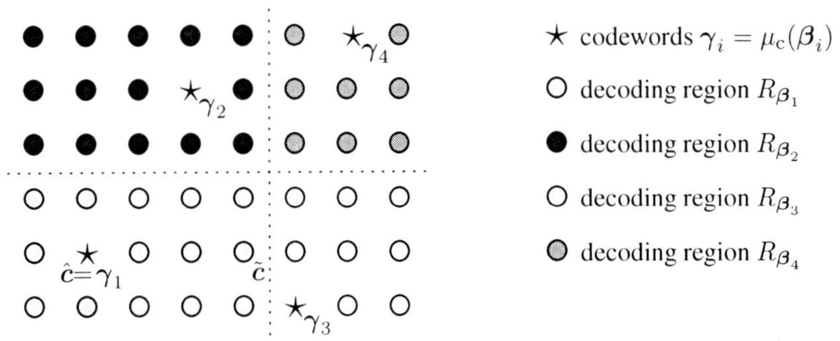

Figure 6.3 Illustrative representation of the partitioning of $\mathcal{A}^n$ into decoding regions, and corresponding decoding rule.

Proposition 6.1 *A binary block code with minimum Hamming distance* $d_{\min}$ *can* detect *all patterns of* $(d_{\min} - 1)$ *or fewer errors.*

Proof If $0 < t < d_{\min}$, $\tilde{\boldsymbol{c}}_m$ is not a codeword and the decoder is able to *detect* it, for example by comparison, even if it may not be able to determine the transmitted codeword. In other words, if $0 < t < d_{\min}$, the decoder is only able to determine wheather $\tilde{\boldsymbol{c}}_m$ is a codeword. If $t \geq d_{\min}$ nothing can be said in general because $\tilde{\boldsymbol{c}}_m$ may be another codeword. □

Error probabilities The performance of a block code over a binary channel can be measured in terms of the *word error probability*

$$\mathrm{P}\left[\hat{\boldsymbol{b}}_m \neq \boldsymbol{b}_m\right] \tag{6.5}$$

or the *symbol error probability*

$$\mathrm{P}\left[\hat{b}_\ell \neq b_\ell\right] \tag{6.6}$$

The word error probability can in principle be calculated as

$$\begin{aligned}
\mathrm{P}\left[\hat{\boldsymbol{b}}_m \neq \boldsymbol{b}_m\right] &= 1 - \mathrm{P}\left[\hat{\boldsymbol{b}}_m = \boldsymbol{b}_m\right] \\
&= 1 - \sum_{\boldsymbol{\beta}\in\mathcal{A}^k} \mathrm{P}\left[\hat{\boldsymbol{b}}_m = \boldsymbol{\beta} \mid \boldsymbol{b}_m = \boldsymbol{\beta}\right] p_{\boldsymbol{b}_m}(\boldsymbol{\beta}) \\
&= 1 - \sum_{\boldsymbol{\beta}\in\mathcal{A}^k} \mathrm{P}\left[\tilde{\boldsymbol{c}}_m \in R_{\boldsymbol{\beta}} \,\middle|\, \boldsymbol{b}_m = \boldsymbol{\beta}\right] p_{\boldsymbol{b}_m}(\boldsymbol{\beta}) \\
&= 1 - \sum_{\boldsymbol{\beta}\in\mathcal{A}^k} p_{\boldsymbol{b}_m}(\boldsymbol{\beta}) \sum_{\boldsymbol{\xi}\in R_{\boldsymbol{\beta}}} p_{\tilde{\boldsymbol{c}}_m|\boldsymbol{b}_m}(\boldsymbol{\xi}|\boldsymbol{\beta}) \\
&= 1 - \sum_{\boldsymbol{\beta}\in\mathcal{A}^k} p_{\boldsymbol{b}_m}(\boldsymbol{\beta}) \sum_{\boldsymbol{\xi}\in R_{\boldsymbol{\beta}}} p_{\tilde{\boldsymbol{c}}_m|\boldsymbol{c}_m}(\boldsymbol{\xi}|\mu_c(\boldsymbol{\beta})) \,.
\end{aligned} \tag{6.7}$$

In (6.7) the word error probability is expressed in terms of the *a priori* PMD $p_{\boldsymbol{b}_m}(\boldsymbol{\beta})$ of the information word $\boldsymbol{b}_m$ and the transition probabilities between input and output words $\boldsymbol{c}_m$ and $\tilde{\boldsymbol{c}}_m$, $p_{\tilde{\boldsymbol{c}}_m|\boldsymbol{c}_m}(\boldsymbol{\xi}|\boldsymbol{\gamma})$ of the binary channel. In the case of a memoryless binary symmetric channel (BSC) with error probability P_{bit} the word transition probability has a simple expression. Let $\boldsymbol{c} = [c_0, c_1, \ldots, c_{n-1}]$ and $\tilde{\boldsymbol{c}} = [\tilde{c}_0, \tilde{c}_1, \ldots, \tilde{c}_{n-1}]$, dropping the dependence on m for the ease of notation. Since the channel is memoryless we write

$$p_{\tilde{\boldsymbol{c}}|\boldsymbol{c}}(\boldsymbol{\xi}|\boldsymbol{\gamma}) = \prod_{i=0}^{n-1} p_{\tilde{c}_i|c_i}(\xi_i|\gamma_i) \tag{6.8}$$

where by the channel symmetry

$$p_{\tilde{c}_i|c_i}(\xi_i|\gamma_i) = \begin{cases} P_{\text{bit}} & , \quad \xi_i \neq \gamma_i \\ 1 - P_{\text{bit}} & , \quad \xi_i = \gamma_i \end{cases} \tag{6.9}$$

Thus in the product (6.8) there will be $d_{\text{H}}(\boldsymbol{\xi}, \boldsymbol{\gamma})$ factors equal to P_{bit} and $n - d_{\text{H}}(\boldsymbol{\xi}, \boldsymbol{\gamma})$ factors equal to $1 - P_{\text{bit}}$ yielding

$$p_{\tilde{\boldsymbol{c}}|\boldsymbol{c}}(\boldsymbol{\xi}|\boldsymbol{\gamma}) = P_{\text{bit}}^{d_{\text{H}}(\boldsymbol{\xi},\boldsymbol{\gamma})} (1 - P_{\text{bit}})^{n - d_{\text{H}}(\boldsymbol{\xi},\boldsymbol{\gamma})} = (1 - P_{\text{bit}})^n \left(\frac{P_{\text{bit}}}{1 - P_{\text{bit}}} \right)^{d_{\text{H}}(\boldsymbol{\xi},\boldsymbol{\gamma})} \tag{6.10}$$

For the symbol error probability (6.6) the calculation is more cumbersome. When the decoded word $\hat{\boldsymbol{b}}_m$ does not coincide with the transmitted word $\boldsymbol{b}_m$, the number of incorrect information symbols is $d_{\text{H}}(\boldsymbol{b}_m, \hat{\boldsymbol{b}}_m)$ out of the k symbols in $\hat{\boldsymbol{b}}_m$. Similarly to the derivation in Section 5.5.4, we can calculate the symbol error probability as

$$\begin{aligned}
\text{P}\left[\hat{b}_\ell \neq b_\ell\right] &= \sum_{\boldsymbol{\beta},\hat{\boldsymbol{\beta}} \in \mathcal{A}^k} \text{P}\left[\hat{b}_\ell \neq b_\ell | \boldsymbol{b}_m = \boldsymbol{\beta}, \hat{\boldsymbol{b}}_m = \hat{\boldsymbol{\beta}}\right] p_{\boldsymbol{b}_m,\hat{\boldsymbol{b}}_m}(\boldsymbol{\beta}, \hat{\boldsymbol{\beta}}) \\
&= \sum_{\boldsymbol{\beta},\hat{\boldsymbol{\beta}} \in \mathcal{A}^k} \frac{d_{\text{H}}(\boldsymbol{\beta}, \hat{\boldsymbol{\beta}})}{k} p_{\hat{\boldsymbol{b}}_m|\boldsymbol{b}_m}(\hat{\boldsymbol{\beta}}|\boldsymbol{\beta}) p_{\boldsymbol{b}_m}(\boldsymbol{\beta}) \\
&= \sum_{\boldsymbol{\beta},\hat{\boldsymbol{\beta}} \in \mathcal{A}^k} p_{\boldsymbol{b}_m}(\boldsymbol{\beta}) \frac{d_{\text{H}}(\boldsymbol{\beta}, \hat{\boldsymbol{\beta}})}{k} \sum_{\boldsymbol{\xi} \in R_{\hat{\boldsymbol{\beta}}}} p_{\tilde{\boldsymbol{c}}_m|\boldsymbol{b}_m}(\boldsymbol{\xi}|\boldsymbol{\beta}) \\
&= \sum_{\boldsymbol{\beta},\hat{\boldsymbol{\beta}} \in \mathcal{A}^k} p_{\boldsymbol{b}_m}(\boldsymbol{\beta}) \frac{d_{\text{H}}(\boldsymbol{\beta}, \hat{\boldsymbol{\beta}})}{k} \sum_{\boldsymbol{\xi} \in R_{\hat{\boldsymbol{\beta}}}} p_{\tilde{\boldsymbol{c}}_m|\boldsymbol{c}_m}(\boldsymbol{\xi}|\mu_{\text{c}}(\boldsymbol{\beta}))
\end{aligned} \tag{6.11}$$

The decoding map μ_{d} can be chosen in different ways, each corresponding to a different *decoding criterion*. In drawing a parallel with the discussion in Sections 5.2.5–5.2.7, we formulate, assuming $\tilde{\boldsymbol{c}}_m = \boldsymbol{\xi}$, the following criteria.

Maximum *a posteriori* (MAP) decoding Analogously to (5.42) we get

$$\hat{\boldsymbol{b}}_m = \underset{\boldsymbol{\beta} \in \mathcal{A}^k}{\text{argmax}}\, p_{\boldsymbol{b}_m|\tilde{\boldsymbol{c}}_m}(\boldsymbol{\beta}|\boldsymbol{\xi}) \tag{6.12}$$

Alternatively, as the coding map μ_{c} is one-to-one, (6.12) can be expressed as

$$\hat{\boldsymbol{b}}_m = \mu_{\text{c}}^{-1}(\hat{\boldsymbol{c}}_m) \quad , \quad \text{with} \;\; \hat{\boldsymbol{c}}_m = \underset{\boldsymbol{\gamma} \in \mathcal{C}}{\text{argmax}}\, p_{\boldsymbol{c}_m|\tilde{\boldsymbol{c}}_m}(\boldsymbol{\gamma}|\boldsymbol{\xi}) \tag{6.13}$$

where $\hat{c}_m$ is the most likely transmitted codeword given that $\tilde{c}_m = \boldsymbol{\xi}$ is received.

Following from (6.7), the maximum a posteriori (MAP) criterion is optimal in the sense that it yields the lowest word error probability $\mathrm{P}[\hat{\boldsymbol{b}}_m \neq \boldsymbol{b}_m]$.

Maximum likelihood (ML) decoding Similarly to (5.44) we write

$$\hat{\boldsymbol{b}}_m = \underset{\boldsymbol{\beta} \in \mathcal{A}^k}{\operatorname{argmax}} \, p_{\tilde{\boldsymbol{c}}_m | \boldsymbol{b}_m}(\boldsymbol{\xi} | \boldsymbol{\beta}) \tag{6.14}$$

or equivalently

$$\hat{\boldsymbol{b}}_m = \mu_c^{-1}(\hat{\boldsymbol{c}}_m) \quad , \quad \text{with} \quad \hat{\boldsymbol{c}}_m = \underset{\boldsymbol{\gamma} \in \mathcal{C}}{\operatorname{argmax}} \, p_{\tilde{\boldsymbol{c}}_m | \boldsymbol{c}_m}(\boldsymbol{\xi} | \boldsymbol{\gamma}) \tag{6.15}$$

In general, the maximum likelihood (ML) criterion is suboptimal with respect to the MAP, but it has the advantage of being more easily implementable, since it only requires the knowledge of the transition probabilities of the binary channel $p_{\tilde{\boldsymbol{c}}_m | \boldsymbol{c}_m}$, and not of the *a priori* distribution of the information words $p_{\boldsymbol{b}_m}$.

The ML criterion coincides with the MAP, and is therefore optimal, in the case the 2^k possible information words (and so the codewords) are equally likely, since in this case, by Bayes rule (2.181) we have

$$p_{\tilde{\boldsymbol{c}}_m | \boldsymbol{b}_m}(\boldsymbol{\xi} | \boldsymbol{\beta}) = p_{\boldsymbol{b}_m | \tilde{\boldsymbol{c}}_m}(\boldsymbol{\beta} | \boldsymbol{\xi}) \, p_{\tilde{\boldsymbol{c}}_m}(\boldsymbol{\xi}) \, 2^k \tag{6.16}$$

so the conditional probabilities in (6.12) and (6.14) coincide, apart from a factor that is independent of $\boldsymbol{\beta}$. The hypothesis of equally likely information words is indeed quite reasonable. As a matter of fact, it is implied by assuming that the information symbols $\{b_\ell\}$ are iid with equally likely values. This is the case if the message $\{b_\ell\}$ is the output of an effective source coder and thus has a low redundancy. Moreover, usually the symbols of the information message are *scrambled*, i.e. rearranged in a pseudo random fashion, before coding.

Minimum Hamming distance decoding Corresponding to (5.45)

$$\hat{\boldsymbol{b}}_m = \mu_c^{-1}(\hat{\boldsymbol{c}}_m) \quad , \quad \text{with} \quad \hat{\boldsymbol{c}}_m = \underset{\boldsymbol{\gamma} \in \mathcal{C}}{\operatorname{argmin}} \, d_H(\boldsymbol{\xi}, \boldsymbol{\gamma}) \tag{6.17}$$

The minimum distance criterion chooses as $\hat{\boldsymbol{c}}_m$ the closest codeword to the received word $\tilde{\boldsymbol{c}}_m$, in terms of the Hamming distance. It has, in principle, a simple implementation since it does not even require the channel statistical description. It coincides with the ML in the case of a symmetric memoryless channel as we prove in the following proposition, thus it is optimal in the case of iid information simbols with equally likely values *and* a symmetric memoryless channel.

Proposition 6.2 *If the binary channel is memoryless and symmetric with $P_{\text{bit}} < 1/2$ the ML and minimum distance decoding criteria coincide.*

Proof We show that in the given hypotheses, the values of $\hat{\boldsymbol{c}}_m$ obtained in (6.15) and in (6.17) coincide.

Since the channel is memoryless and symmetric the conditional probability in (6.15) is given by (6.10). If $P_{\text{bit}} < 1/2$, we have $P_{\text{bit}}/(1 - P_{\text{bit}}) < 1$ so that $p_{\tilde{c}|c}(\boldsymbol{\xi}|\boldsymbol{\gamma})$ is a decreasing function of $d_H(\boldsymbol{\xi}, \boldsymbol{\gamma})$ and its maximum over $\boldsymbol{\gamma}$ is attained at the minimum of $d_H(\boldsymbol{\xi}, \boldsymbol{\gamma})$. □

An illustration of the minimum distance decoding criterion, applied to the code representation of Fig. 6.3, is given in Fig. 6.4.

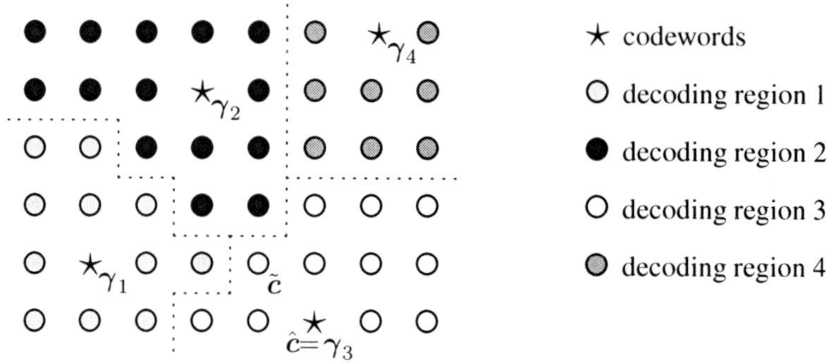

Figure 6.4 Illustration of the minimum Hamming distance decoding criterion.

Example 6.1 A Consider the following $(5, 3)$ systematic binary code

$$\mu_c(000) = 00010 \ , \quad \mu_c(001) = 00101 \ , \quad \mu_c(010) = 01000 \ , \quad \mu_c(011) = 01101$$

$$\mu_c(100) = 10001 \ , \quad \mu_c(101) = 10101 \ , \quad \mu_c(110) = 11010 \ , \quad \mu_c(111) = 11110$$

and assume we use minimum Hamming distance decoding. Then, if the received word is $\tilde{\boldsymbol{c}} = [10011]$, the closest codeword is $\hat{\boldsymbol{c}} = [10001] = \mu_c(100)$ with $d_H(\tilde{\boldsymbol{c}}, \hat{\boldsymbol{c}}) = 1$. Hence the decoded word is $\hat{\boldsymbol{b}} = \mu_c^{-1}(\hat{\boldsymbol{c}}) = [100]$.

On the other hand, if the received word is $\tilde{\boldsymbol{c}} = [01010]$ the three codewords $[00010] = \mu_c(000)$, $[01000] = \mu_c(010)$ and $[11010] = \mu_c(110)$ all have Hamming distance 1 from $\tilde{\boldsymbol{c}}$, so $\tilde{\boldsymbol{c}}$ can be equivalently decoded either as $\hat{b} = [000]$, or $\hat{b} = [010]$ or $\hat{b} = [110]$.

The Hamming distances of all possible received words in $\mathcal{A}^n$ from all codewords can be computed before transmission and the closest codeword stored in a decoding table as in Tab. 6.1. However, it should be noted that the dimension of the table grows exponentially with the codeword length n, and this decoding method may not be viable.

The following proposition states how many errors can be corrected in a single word with minimum Hamming distance decoding.

Proposition 6.3 *Let $\mathcal{C}$ be a binary block code with minimum Hamming distance d_{min}. Let $\boldsymbol{c}_m \in \mathcal{C}$ and $\tilde{\boldsymbol{c}}_m \in \mathcal{A}^n$ be the transmitted codeword and received word, respectively. If the*

received word $\tilde{c}$					closest codeword $\hat{c}$	decoded word $\hat{b}$
	00000	00011	00010	00110	00010	000
	00001	00100	00101	00111	00101	001
		01000	01001	11000	01000	010
		01101	01100	01111	01101	011
	10000	10011	10001	11001	10001	100
	10100	10111	10101	11101	10101	101
01010	10010	11010	11011	01011	11010	110
01110	10110	11110	11100	11111	11110	111

Table 6.1 Decoding table for the code in Example 6.1 A.

number of errors, $t = d_H(\tilde{\mathbf{c}}_m, \mathbf{c}_m)$, introduced by the binary channel in the word $\tilde{\mathbf{c}}_m$ satisfies

$$t < \frac{d_{\min}}{2} \tag{6.18}$$

then, regardless of the error positions, the received word will be correctly detected as $\hat{\mathbf{c}}_m = \mathbf{c}_m$.

Proof For any other codeword $\boldsymbol{\gamma} \in \mathcal{C}, \boldsymbol{\gamma} \neq \mathbf{c}_m$, it is

$$d_H(\tilde{\mathbf{c}}_m, \boldsymbol{\gamma}) \geq d_H(\mathbf{c}_m, \boldsymbol{\gamma}) - d_H(\tilde{\mathbf{c}}_m, \mathbf{c}_m) \geq d_{\min} - \frac{d_{\min}}{2} = \frac{d_{\min}}{2}$$

while

$$d_H(\tilde{\mathbf{c}}_m, \mathbf{c}_m) < \frac{d_{\min}}{2} ,$$

so that $\mathbf{c}_m$ is indeed the closest codeword to $\tilde{\mathbf{c}}_m$. □

So we say that the code $\mathcal{C}$ can *correct* up to

$$t = \begin{cases} d_{\min}/2 - 1 & , \ d_{\min} \text{ even} \\ (d_{\min} - 1)/2 & , \ d_{\min} \text{ odd} \end{cases} \tag{6.19}$$

errors with minimum distance decoding.

The reader may wonder whether the possibility to correct a higher number of errors in a code comes at the price of a lower code rate (and hence higher redundancy). The following proposition answers this question

Proposition 6.4 (Hamming bound) *For every code of length n that can correct up to t errors with minimum distance decoding, the code rate is upper bounded by*

$$\frac{k}{n} \leq 1 - \frac{1}{n} \log_2 \left(\sum_{r=0}^{t} \binom{n}{r} \right) \tag{6.20}$$

Proof Since the code can correct up to t errors, each decoding region $\mathcal{R}_\beta$ contains at least all the words at Hamming distance r from $\gamma = \mu_c(\beta)$, for $r = 0, \ldots, t$. Since the number of distinct words that have distance r from γ is given by the binomial coefficient $\binom{n}{r}$ (as the distinct ways of flipping r symbols out of n in γ), the cardinality of $\mathcal{R}_\beta$ is lower bounded by

$$|\mathcal{R}_\beta| \geq \sum_{r=0}^{t} \binom{n}{r} \tag{6.21}$$

By summing over the 2^k disjoint decoding regions, we obtain a lower bound on the cardinality of the whole space of $\tilde{c}_m$, $\mathbb{F}^n$

$$|\mathbb{F}^n| = \sum_{\beta \in \mathbb{F}^k} |\mathcal{R}_\beta| \geq \sum_{\beta \in \mathbb{F}^k} \sum_{r=0}^{t} \binom{n}{r} \tag{6.22}$$

and since the term in the sum do not depend on β,

$$2^n \geq 2^k \sum_{r=0}^{t} \binom{n}{r} \tag{6.23}$$

Then by taking the base-2 logarithm of both sides we get

$$n \geq k + \log_2 \left(\sum_{r=0}^{t} \binom{n}{r} \right) \tag{6.24}$$

and hence the statement (6.20) □

6.2 LINEAR BLOCK CODES

6.2.1 Construction of linear codes

A block code as a linear space In order to introduce *linear* block codes we must first endow the set of n-symbol words with the structure of a linear space. The proper way to do this is to make the binary alphabet $\mathcal{A}$ a field. This can be done through the sum and product operations modulo 2,

$$\begin{array}{c|cc} + & 0 & 1 \\ \hline 0 & 0 & 1 \\ 1 & 1 & 0 \end{array} \, , \qquad \begin{array}{c|cc} \cdot & 0 & 1 \\ \hline 0 & 0 & 0 \\ 1 & 0 & 1 \end{array} \tag{6.25}$$

To underline that we have made $\mathcal{A}$ a field by introducing the proper sum and product operations we will denote the field $(\mathcal{A}, +, \cdot)$ with the symbol $\mathbb{F}$. Then $V = \mathbb{F}^n$ is seen to be a linear space over the field $\mathbb{F}$ itself, since it satisfies the properties in Definition 2.8.

Example 6.2 A In $\{0, 1\}^4$, given the two words $\gamma_1 = [1010]$, $\gamma_2 = [0011]$, their sum is $\gamma_3 = [1001]$.

Observe that the sum of any word γ with itself is the all-zero word, $\gamma + \gamma = \mathbf{0}$, and so we write $\gamma = -\gamma$.

Definition 6.2 *We define the* Hamming weight *of a binary word, as the function*

$$\|\cdot\|_{\mathrm{H}} : \mathbb{F}^n \mapsto \mathbb{N} \quad , \quad \|\boldsymbol{x}\|_{\mathrm{H}} = \textit{number of nonzero symbols in } \boldsymbol{x} \tag{6.26}$$

The Hamming weight has the properties of a norm:

i) The all-zero word has zero weight, $\|\mathbf{0}\|_{\mathrm{H}} = 0$.

ii) All other words have positive weight, $\|\gamma\|_{\mathrm{H}} > 0$, for any $\gamma \neq \mathbf{0}$.

iii) The triangular inequality holds, $\|\gamma_1 + \gamma_2\|_{\mathrm{H}} \leq \|\gamma_1\|_{\mathrm{H}} + \|\gamma_2\|_{\mathrm{H}}$, for any $\gamma_1, \gamma_2 \in \mathbb{F}^n$

We can now express the Hamming distance between two words as the Hamming weight of their difference

$$d_{\mathrm{H}}(\gamma_1, \gamma_2) = \|\gamma_1 - \gamma_2\|_{\mathrm{H}} \tag{6.27}$$

which is the analogous of writing the Euclidean distance between two vectors as the norm of their difference (2.120), or the distance between two signals as the square root of the energy of the difference signal (2.121).

Further insight Unlike the Euclidean norm seen in Chapter 2, the Hamming weight can not be derived from an inner product. This fact has the important consequence that we cannot use the notion of projection (2.132), or the orthogonality condition (2.134), to find the closest vector, belonging to a subspace $V' \subset V$, to a given vector in terms of the Hamming distance.

We are now ready to give the following definition.

Definition 6.3 *A* (n, k) *block code* $\mathcal{C}$ *is said to be* linear *if it is a linear subspace of* $\mathbb{F}^n$, *that is if*

$$\textit{for all } \gamma_1, \gamma_2 \in \mathcal{C} \quad , \quad \gamma_1 + \gamma_2 = \gamma_1 - \gamma_2 \in \mathcal{C} \tag{6.28}$$

Since we require the coding map μ_{c} to be one-to-one, and $\mathcal{C}$ must have 2^k codewords, its dimension as a linear space is $\dim \mathcal{C} = k$.

An important parameter of a linear code is its *minimum weight* $w_{\min}$, which is defined as the minimum among the Hamming weights of its nonzero codewords

$$w_{\min} = \min_{\substack{\gamma \in \mathcal{C} \\ \gamma \neq \mathbf{0}}} \|\gamma\|_{\mathrm{H}} \ . \tag{6.29}$$

Proposition 6.5 *In a linear code, the minimum Hamming distance and the minimum Hamming weight coincide.*

Proof From (6.4), (6.28) and (6.29), we have

$$d_{\min} = \min_{\substack{\gamma_1, \gamma_2 \in \mathcal{C} \\ \gamma_1 \neq \gamma_2}} d_{\mathrm{H}}(\gamma_1, \gamma_2) = \min_{\substack{\gamma_1, \gamma_2 \in \mathcal{C} \\ \gamma_1 \neq \gamma_2}} \|\gamma_1 - \gamma_2\|_{\mathrm{H}} = \min_{\substack{\gamma_3 \in \mathcal{C} \\ \gamma_3 \neq \mathbf{0}}} \|\gamma_3\|_{\mathrm{H}} = w_{\min} \tag{6.30}$$

□

Generating matrix Any matrix $G \in \mathbb{F}^{n\times k}$ whose *columns span the linear code* $\mathcal{C}$ is said to be a *generating matrix* for $\mathcal{C}$. Since G must have rank $\dim \mathcal{C} = k$, its columns must be linearly independent codewords. Such a matrix is not unique, as any other matrix obtained from it with elementary column operations will have the same range space and hence be a generating matrix for $\mathcal{C}$, too.

A systematic code $\mathcal{C}$ has k codewords having as prefix the words (of length k) $\boldsymbol{\beta}_1 = [100\cdots0]$, $\boldsymbol{\beta}_2 = [010\cdots0]$, ..., $\boldsymbol{\beta}_k = [000\cdots1]$, respectively. By observing that the corresponding codewords $\gamma_1, \gamma_2, \ldots, \gamma_k$ are linearly independent we can use them as columns $\boldsymbol{g}_i = \gamma_i$ to build a generating matrix of the type

$$G = \left[\begin{array}{c} I_k \\ \hline A \end{array}\right] \tag{6.31}$$

with I_k the $k \times k$ identity matrix and A a suitable $(n-k) \times k$ matrix with entries in $\mathbb{F}$.

Example 6.2 B The systematic code known as $(7, 4)$ Hamming code

$$\begin{aligned}\mathcal{C} = \{\, & 0000000, 1000110, 0100101, 0010011, 0001111, 1100011, 1010101, 0110110,\\ & 1110000, 1001001, 0101010, 1101100, 0011100, 1011010, 0111001, 1111111\}\end{aligned}$$

has the following generating matrix

$$G = \begin{bmatrix} 1 & 0 & 0 & 0 \\ 0 & 1 & 0 & 0 \\ 0 & 0 & 1 & 0 \\ 0 & 0 & 0 & 1 \\ 1 & 1 & 0 & 1 \\ 1 & 0 & 1 & 1 \\ 0 & 1 & 1 & 1 \end{bmatrix}$$

A possible coding map for a code with generating matrix G is the linear map

$$c = \mu_c(b) = Gb\,. \tag{6.32}$$

The k columns $g_1, \ldots, g_k$ of the matrix G will then be the codewords corresponding to the standard basis for $\mathbb{F}^k$, that is

$$g_1 = \mu_c(100\cdots0)\;,\quad g_2 = \mu_c(010\cdots0)\;,\quad \cdots\;,\quad g_k = \mu_c(000\cdots1) \tag{6.33}$$

In particular, if $\mathcal{C}$ is systematic, we can use the generating matrix (6.31) in (6.32) and obtain a systematic encoding map.

Consider a matrix $G' \in \mathbb{F}^{n\times k}$, obtained from a generating matrix G for a code $\mathcal{C}$ by a *row permutation* operation. Then G' is a generating matrix for a new code $\mathcal{C}'$, where each word $\gamma' \in \mathcal{C}'$ can be obtained from $\gamma \in \mathcal{C}$ by a permutation of its symbols. We call the codes $\mathcal{C}$ and $\mathcal{C}'$ *equivalent* since their performance over a memoryless binary channel are identical.

In particular, since any matrix with rank k can be transformed into a matrix of the form (6.31) by elementary column operations and row permutations, we can state that any linear code is equivalent to a systematic code. Then, by writing the systematic code generating matrix as in (6.31), the choice of $\boldsymbol{A}$ uniquely determines the properties of the code $\mathcal{C}$ (e.g. its minimum distance).

We give now another result on the minimum distance of a linear code.

Proposition 6.6 (Singleton bound) *For a (n, k) linear block code the minimum Hamming distance is upper bounded by*

$$d_{\min} \leq n - k + 1 \tag{6.34}$$

Proof By Proposition 6.5, it is sufficient to show that there certainly exist in $\mathcal{C}$ one codeword γ with weight $\|\gamma\|_{\mathrm{H}} \leq n - k + 1$. Moreover, since any linear code $\mathcal{C}$ is equivalent to a systematic one, we only need to prove the proposition for a systematic code. Indeed, in a systematic code, the codeword γ having $[10 \cdots 0]$ as its k-symbol prefix has at least $k - 1$ zeros, and therefore $\|\gamma\|_{\mathrm{H}} \leq n - k + 1$. □

From Proposition 6.1 and (6.19), the bound (6.34) also represents a constraint on the error detection and correction capabilities of a linear code with given codeword and information word lengths (n, k). Codes for which the bound (6.34) holds as equality, and hence can detect up to $n - k$ errors and correct up to $\lfloor (n - k)/2 \rfloor$ errors, are called *maximum distance* codes. A class of maximum distance codes is given by Reed-Solomon codes.

6.2.2 Decoding of linear codes

Parity check matrix A *parity check matrix* for a (n, k) linear code $\mathcal{C}$ is any matrix $\boldsymbol{H}$ with ℓ rows and n columns having the property that its null space coincides with $\mathcal{C}$, that is

$$\boldsymbol{H}\gamma = \mathbf{0} \quad \text{if and only if} \quad \gamma \in \mathcal{C} \tag{6.35}$$

Its name is justified by the fact that in order to determine whether a received word $\tilde{c}$ is a codeword or not, it is sufficient to multiply it by $\boldsymbol{H}$ and check whether the resulting word of length ℓ is the all-zero word. This procedure is certainly more efficient than having to browse through the 2^k words of a nonlinear code.

The following proposition gives a characterization of the parity check matrix in terms of the generating matrix.

Proposition 6.7 *A matrix $\boldsymbol{H} \in \mathbb{F}^{\ell \times n}$ is a parity check matrix for a linear code $\mathcal{C}$ with generating matrix $\boldsymbol{G} \in \mathbb{F}^{n \times k}$ if and only if*

$$\boldsymbol{H}\boldsymbol{G} = \mathbf{0} \quad \textit{and} \quad \operatorname{rank}(\boldsymbol{H}) = n - k \tag{6.36}$$

Proof We first prove that if (6.36) is satisfied then $\mathcal{C}$ is the null space of $\boldsymbol{H}$. Let γ be any codeword in $\mathcal{C}$: then, by (6.32), we can write $\gamma = \boldsymbol{G}\beta$ for a suitable $\beta \in \mathbb{F}^k$. Therefore by (6.36) it must be

$$\boldsymbol{H}\gamma = \boldsymbol{H}\boldsymbol{G}\beta = \mathbf{0} \cdot \beta = \mathbf{0}$$

Thus $\mathcal{C}$ is a subspace of the null space of $\boldsymbol{H}$, but since the null space of $\boldsymbol{H}$ has dimension $n - \operatorname{rank}(\boldsymbol{H}) = k = \dim \mathcal{C}$, the two spaces must coincide.

Conversely, if $\boldsymbol{H}$ is a parity check matrix for $\mathcal{C}$, since the columns of $\boldsymbol{G}$ are in $\mathcal{C}$, we have that the ith column of the matrix product $\boldsymbol{HG}$ is given by

$$\boldsymbol{H}\boldsymbol{g}_i = \boldsymbol{0} \quad , \quad i = 1, \ldots, k$$

so $\boldsymbol{HG}$ is the null matrix. Moreover, as $\mathcal{C}$ is the null space of $\boldsymbol{H}$, it must be

$$\text{rank}(\boldsymbol{H}) = n - \dim \mathcal{C} = n - k \ .$$

□

From the above result it follows that the number of rows in a parity check matrix must be $\ell \geq n - k$. As it is customary, from now on we will consider parity check matrices with minimum size $\ell = n - k$, unless it is necessary to address parity check matrices with a particular structure as in Section 6.5.3.

In particular, for a systematic linear code a parity check matrix is immediately found from the generating matrix.

Proposition 6.8 *Let $\boldsymbol{G}$ be the generating matrix of a systematic linear code $\mathcal{C}$, partitioned as in (6.31). Then the matrix*

$$\boldsymbol{H} = \left[\ -\boldsymbol{A} \mid \boldsymbol{I}_{n-k} \ \right] \tag{6.37}$$

is a parity check matrix for $\mathcal{C}$.

Proof We first observe that $\text{rank}(\boldsymbol{H}) = n - k$. Moreover, by writing

$$\boldsymbol{HG} = -\boldsymbol{A}\boldsymbol{I}_k + \boldsymbol{I}_{n-k}\boldsymbol{A} = -\boldsymbol{A} + \boldsymbol{A} = \boldsymbol{0}$$

(recall that in the binary field $\boldsymbol{A} = -\boldsymbol{A}$), we have proved that $\boldsymbol{H}$ satisfies (6.36), hence by Proposition 6.7 it is a parity check matrix for $\mathcal{C}$. □

Example 6.2 C For the $(7, 4)$ Hamming code whose systematic generating matrix was given in Example 6.2 B, we have

$$\boldsymbol{A} = \begin{bmatrix} 1 & 1 & 0 & 1 \\ 1 & 0 & 1 & 1 \\ 0 & 1 & 1 & 1 \end{bmatrix}$$

so we can write the parity check matrix

$$\boldsymbol{H} = \begin{bmatrix} 1 & 1 & 0 & 1 & 1 & 0 & 0 \\ 1 & 0 & 1 & 1 & 0 & 1 & 0 \\ 0 & 1 & 1 & 1 & 0 & 0 & 1 \end{bmatrix}$$

We let the reader verify that $\boldsymbol{H}\boldsymbol{g}_i = [000]^{\mathrm{T}}$ for each column $\boldsymbol{g}_i$ of $\boldsymbol{G}$.

Syndrome decoding Besides error detection, the parity check matrix of a linear code also allows a more efficient algorithm of error correction through an efficient implementation of minimum Hamming distance decoding, named *syndrome decoding*. Let $\boldsymbol{\xi} \in \mathbb{F}^n$ be a possible received word and multiply it by the parity check matrix $\boldsymbol{H}$. The result $\varsigma = \boldsymbol{H}\boldsymbol{\xi}$ is a word of $n-k$ symbols known as the *syndrome* of $\boldsymbol{\xi}$. The correspondence between the received word and its syndrome is not one-to-one, as other words will have the same syndrome. For example we have seen in (6.35) that all codewords $\boldsymbol{\gamma} \in \mathcal{C}$ have the all-zero syndrome.

Proposition 6.9 *Two words $\boldsymbol{\xi}_1, \boldsymbol{\xi}_2$ have the same syndrome if and only if their difference $\boldsymbol{\gamma} = \boldsymbol{\xi}_1 - \boldsymbol{\xi}_2$ is a codeword.*

Proof Let $\boldsymbol{\xi}_1$ and $\boldsymbol{\xi}_2$ have the same syndrome ς. Then, we have

$$\boldsymbol{H}\boldsymbol{\gamma} = \boldsymbol{H}\boldsymbol{\xi}_1 - \boldsymbol{H}\boldsymbol{\xi}_2 = \varsigma - \varsigma = 0$$

so that $\boldsymbol{\gamma}$ is a codeword by (6.35). Conversely, if $\boldsymbol{\gamma}$ is a codeword, we get

$$\boldsymbol{H}\boldsymbol{\xi}_1 = \boldsymbol{H}(\boldsymbol{\xi}_2 + \boldsymbol{\gamma}) = \boldsymbol{H}\boldsymbol{\xi}_2 + \boldsymbol{H}\boldsymbol{\gamma} = \boldsymbol{H}\boldsymbol{\xi}_2 + \boldsymbol{0} = \boldsymbol{H}\boldsymbol{\xi}_2$$

□

Therefore if a certain word $\boldsymbol{\xi}$ has syndrome ς, the set of all possible received words that share the same syndrome ς can be written as $\mathcal{C} + \boldsymbol{\xi}$, i.e. it is a shift of the code space $\mathcal{C}$, named *coset*. The coset $\mathcal{C} + \boldsymbol{\xi}$ is also the set of possible error words $\boldsymbol{e} = \tilde{\boldsymbol{c}} - \boldsymbol{c}$ introduced by the channel, in correspondence of the various codewords $\boldsymbol{\gamma} \in \mathcal{C}$, that yield the received word $\tilde{\boldsymbol{c}} = \boldsymbol{\xi}$. Within the coset, we can find the minimum weight error word (*coset leader*) as

$$\boldsymbol{\varepsilon}_{\min}(\varsigma) = \underset{\boldsymbol{\varepsilon}\,:\,\boldsymbol{H}\boldsymbol{\varepsilon}=\varsigma}{\operatorname{argmin}} \|\boldsymbol{\varepsilon}\|_{\mathrm{H}} \tag{6.38}$$

Proposition 6.10 *The minimum Hamming distance decoding of the received word $\boldsymbol{\xi}$ can be obtained as*

$$\hat{\boldsymbol{b}}_m = \mu_{\mathrm{c}}^{-1}(\hat{\boldsymbol{c}}_m) \quad , \quad \textit{with} \quad \hat{\boldsymbol{c}}_m = \boldsymbol{\xi} - \boldsymbol{\varepsilon}_{\min}(\boldsymbol{H}\boldsymbol{\xi}) \tag{6.39}$$

Proof From (6.17) we have

$$\hat{\boldsymbol{c}}_m = \underset{\boldsymbol{\gamma}\in\mathcal{C}}{\operatorname{argmin}}\, d_{\mathrm{H}}(\boldsymbol{\xi}, \boldsymbol{\gamma}) = \underset{\boldsymbol{\gamma}\in\mathcal{C}}{\operatorname{argmin}} \|\boldsymbol{\xi} - \boldsymbol{\gamma}\|_{\mathrm{H}} = \boldsymbol{\xi} - \underset{\boldsymbol{\varepsilon}\in\mathcal{C}+\boldsymbol{\xi}}{\operatorname{argmin}} \|\boldsymbol{\varepsilon}\|_{\mathrm{H}} = \boldsymbol{\xi} - \boldsymbol{\varepsilon}_{\min}(\boldsymbol{H}\boldsymbol{\xi})$$

□

Remark From (6.39) the coset leaders are all and only the possible error words that the decoder can correct.

The syndrome decoding procedure, starting from the received word $\tilde{\boldsymbol{c}}_m = \boldsymbol{\xi}$, is as follows:

1. Calculate the syndrome of $\boldsymbol{\xi}$ as $\varsigma = \boldsymbol{H}\boldsymbol{\xi}$.

2. Find the corresponding coset leader $\boldsymbol{\varepsilon}_{\min}(\varsigma)$ (can be done via a lookup table indexed by the syndrome).

3. Find the closest codeword $\hat{c}_m = \xi - \varepsilon_{\min}(\varsigma)$.

4. Obtain the corresponding information word by the inverse coding map $\hat{b}_m = \mu_c^{-1}(\hat{c}_m)$. If the code is systematic, simply strip the last $n - k$ symbols from $\hat{c}_m$.

Observe that, with respect to the direct method of Example 6.1 A, the above procedure can save a lot of space in storing the lookup table since there are 2^{n-k} possible syndromes versus 2^n possible received words in a decoding table. This is accoplished at the price of some additional processing, i.e. the syndrome calculation.

Example 6.2 D Consider the $(4, 2)$ systematic linear code

$$\mathcal{C} = \{0000, 1010, 0101, 1111\}$$

It has a generating matrix and a parity check matrix given by

$$G = \left[\begin{array}{c} I_2 \\ \hline I_2 \end{array}\right], \quad H = \left[\begin{array}{c|c} I_2 & I_2 \end{array}\right]$$

The partitioning of $\mathbb{F}^4$ into cosets of $\mathcal{C}$ and the corresponding syndromes are

coset	coset leader $\varepsilon_{\min}(\varsigma)$	syndrome ς
$\mathcal{C} = \{0000, 1010, 0101, 1111\}$	0000	00
$\mathcal{C} + 0001 = \{0001, 1011, 0100, 1110\}$	0001	01
$\mathcal{C} + 0010 = \{0010, 1000, 0111, 1101\}$	0010	10
$\mathcal{C} + 0011 = \{0011, 1001, 0110, 1100\}$	0011	11

then if we receive the word $\xi = [1011]$ that has syndrome $\varsigma = [01]$, and lookup its coset leader $\varepsilon_{\min}(\varsigma) = [0101]$ we obtain the closest codeword $\hat{c}_m = \xi - \varepsilon_{\min}(\varsigma) = [1010]$.

Example 6.2 E Consider again the $(7, 4)$ Hamming code, with the parity check matrix given in Example 6.2 C.

The syndrome lookup table can be built as follows. Consider the words in $\{0, 1\}^n$, one word at a time, starting from the all-zero word and in increasing order of weight, and calculate their syndromes. If the syndrome ς of the word ξ is not already in the table, this means that no word with lower weight than ξ has the same syndrome ς, so that ξ is the coset leader for ς. Insert ς and the corresponding value of $\varepsilon_{\min}(\varsigma) = \xi$ in the table, and continue until all 2^{n-k} syndromes are found. Thus we get the table for this code

syndrome ς	coset leader $\varepsilon_{\min}(\varsigma)$
000	0000000
110	1000000
101	0100000
011	0010000
111	0001000
100	0000100
010	0000010
001	0000001

Upon reception of the word

$$\tilde{c} = [1000100]$$

we calculate its syndrome (sum of the 1st and 5th columns of $\boldsymbol{H}$)

$$\varsigma = [010] \ ,$$

lookup the coset leader

$$\varepsilon_{\min}(\varsigma) = [0000010]$$

and obtain the detected codeword

$$\hat{c} = \tilde{c} - \varepsilon_{\min}(\varsigma) = [1000110]$$

then decode to the information word

$$\hat{b} = [1000]$$

6.2.3 Cyclic codes

Given a word $\boldsymbol{\gamma} = [\gamma_0, \ldots, \gamma_{n-1}]$ we call *cyclic shift* of $\boldsymbol{\gamma}$ by r symbols forward the word $\boldsymbol{\gamma}' = [\gamma_{n-r}, \ldots, \gamma_{n-1}, \gamma_0, \ldots, \gamma_{n-r-1}]$ so that

$$\gamma_i = \gamma'_{i+r \bmod n} \tag{6.40}$$

.

Moreover we define the *cyclic convolution* between two n-symbol words $\boldsymbol{\gamma}' = [\gamma'_0, \ldots, \gamma'_{n-1}]$ and $\boldsymbol{\gamma}'' = [\gamma''_0, \ldots, \gamma''_{n-1}]$ the word $\boldsymbol{\gamma} = \boldsymbol{\gamma}' * \boldsymbol{\gamma}''$ whose symbols are

$$\gamma_i = \sum_{j=0}^{n-1} \gamma'_j \gamma''_{i-j \bmod n} \ , \quad i = 0, \ldots, n-1 \tag{6.41}$$

Definition 6.4 *A linear block code is said to be* cyclic *if all the cyclic shifts of each codeword are codewords themselves.*

Example 6.2 F The $(4, 2)$ binary code

$$C = \{0000, 1010, 0101, 1111\}$$

is cyclic. As a matter of fact, all cyclic shifts of the words $[0000]$ and $[1111]$ coincide with the original word, whereas the words $[0101]$ and $[1010]$ are cyclic shifts of each other.

The following proposition that we state without proof (see for example [5]) allows to characterize a cyclic code by a single codeword.

Proposition 6.11 *In a (n, k) cyclic code C no codeword, apart from the all-zero word, has more than $k-1$ trailing zeros. Moreover, there is a unique codeword $\boldsymbol{g} = [g_0, \ldots, g_{n-k}, 0, \ldots, 0]$ with $k-1$ trailing zeros and $g_{n-k} = 1$.*

As a matter of fact, we can easily build a generating matrix for the code.

Proposition 6.12 *Let C be a (n, k) cyclic code and $\boldsymbol{g}$ be as in Proposition 6.11. Then the circulant matrix*

$$\boldsymbol{G} = \begin{bmatrix} \boldsymbol{g} & \boldsymbol{g}_1 & \cdots & \boldsymbol{g}_{k-1} \end{bmatrix} = \begin{bmatrix} g_0 & \cdots & 0 \\ g_1 & \ddots & \vdots \\ \vdots & \ddots & g_0 \\ 1 & \ddots & g_1 \\ 0 & \ddots & \vdots \\ \vdots & \ddots & 1 \end{bmatrix} \tag{6.42}$$

where $\boldsymbol{g}_i$ is the cyclic shift of $\boldsymbol{g}$ by i symbols forward, is a generating matrix for C.

Proof It suffices to observe that the k columns of $\boldsymbol{G}$ are codewords of C, and are linearly independent. Hence they span C. □

Based on Proposition 6.12 we can build the coding map as a cyclic convolution since

$$\mathbf{c} = \boldsymbol{G}\boldsymbol{b} = [\boldsymbol{b} \; \underbrace{0 \ldots 0}_{n-k \text{ zeros}}] * \boldsymbol{g} \tag{6.43}$$

Hence $\boldsymbol{g}$ is the *generating word* for the code C, and it is often represented via the polynomial

$$g(x) = \sum_{i=0}^{n-k} g_i x^i \tag{6.44}$$

called the *generating polynomial*. The above result allows an efficient implementation of the encoder in the form of a shift register.

More importantly, there exists an efficient implementation of the syndrome decoding for cyclic codes, which is based on the following result.

Proposition 6.13 *Let g be the generating word of a cyclic code $\mathcal{C}$, and let $\boldsymbol{h}$ be a word of $n-k+1$ symbols such that*

$$[\boldsymbol{g}, \mathbf{0}] * [\boldsymbol{h}, \mathbf{0}] = \mathbf{0} \,. \tag{6.45}$$

Then the circulant matrix

$$\boldsymbol{H} = \begin{bmatrix} h_{n-k} & \cdots & h_0 & 0 & \cdots \\ \cdots & \ddots & \ddots & \ddots & \ddots \\ 0 & \cdots & h_{n-k} & \cdots & h_0 \end{bmatrix} \tag{6.46}$$

is a parity check matrix for $\mathcal{C}$.

Proof We observe that $\boldsymbol{H}$ has rank $n-k$ and that due to (6.45) all the products between rows of $\boldsymbol{H}$ and columns of $\boldsymbol{G}$ given in (6.42) yield 0. Hence, by Proposition 6.7, the statement is proved. □

The multiplication of the received word $\tilde{c}$ by the matrix (6.46) (which is the core of syndrome decoding) is equivalent to a cyclic convolution between $\tilde{c}$ and $\boldsymbol{h}$, and can be performed with computationally efficient techniques. Beside allowing for efficient encoding and decoding, cyclic codes also have an intrinsic capability in detecting burst errors.

Proposition 6.14 *In a (n, k) cyclic code there are no codewords with k consecutive zeros, apart from the all zero word.*

Proof The proof is by contradiction. Assume there is a codeword $\gamma \neq \mathbf{0}$ with k consecutive zeros and let γ' be the cyclic shift of γ with k trailing zeros. Since the code is cyclic, γ' must be a codeword too but this is not the case by Proposition 6.11. □

6.2.4 Specific classes of linear block codes

Hamming codes

Definition 6.5 *A (n, k)* Hamming code *is a binary linear code for which a parity check matrix $\boldsymbol{H}$ can be constructed with all the nonzero columns of length $n-k$.*

As the number of $(n-k)$-bit words is 2^{n-k}, and we exclude the all-zero word, the number of columns in the parity check matrix $\boldsymbol{H}$ is $2^{n-k}-1$. Hence, there exists a (n, k) Hamming code if and only if

$$n = 2^{n-k} - 1$$

Parameters for the shortest Hamming codes are given in Tab. 6.2.

Proposition 6.15 *The minimum distance of a Hamming code is $d_{\min} = 3$*

Proof By Proposition 6.5, as Hamming codes are linear, it is sufficient to show that $w_{\min} = 3$. To this aim, we must prove that there are no codewords of weight 2 and there is at least one codeword of weight 3.

Let $\boldsymbol{\xi} \in \mathbb{F}^n$ be such that $\|\boldsymbol{\xi}\|_{\mathrm{H}} = 2$, and let its two 1's be at positions i and j. Then, with $\boldsymbol{h}_i$ indicating the ith column of a parity check matrix $\boldsymbol{H}$, we have $\boldsymbol{H}\boldsymbol{\xi} = \boldsymbol{h}_i + \boldsymbol{h}_j \neq \mathbf{0}$, as $\boldsymbol{h}_i \neq \boldsymbol{h}_j$. So, $\boldsymbol{\xi}$ cannot be a codeword.

Now let $\boldsymbol{h}_1, \boldsymbol{h}_2$ be the first two columns of $\boldsymbol{H}$. As $\boldsymbol{h}_1 + \boldsymbol{h}_2 \neq 0$ there must be a column $\boldsymbol{h}_i$, $i \neq 1, 2$ such that $\boldsymbol{h}_i = \boldsymbol{h}_1 + \boldsymbol{h}_2$. Let $\gamma \in \mathbb{F}^n$ be the word with 1's at positions $1, 2, i$ and 0's elsewhere. It has $\|\gamma\|_{\mathrm{H}} = 3$, and it is a codeword since $\boldsymbol{H}\gamma = \boldsymbol{h}_1 + \boldsymbol{h}_2 + \boldsymbol{h}_i = \mathbf{0}$ □

From the above proposition we see that Hamming codes can detect at most 2 errors and can only correct 1 error. A welcome property of all Hamming codes is that they achieve the Hamming bound (6.20). In fact, as $t = 1$ we get

$$\sum_{r=0}^{t} \binom{n}{r} = \binom{n}{0} + \binom{n}{1} = 1 + n = 2^{n-k}$$

so that (6.23) is satisfied with equality. Thus, the (n, k) Hamming code has the highest rate among all the single-error correcting codes of the same length n. Furthermore, we see that the decoding regions can be convenently described as "spheres" of radius 1

$$\mathcal{R}_{\boldsymbol{\beta}} = \{\boldsymbol{\xi} \in \mathbb{F}^n \,:\, d_{\mathrm{H}}(\boldsymbol{\xi}, \mu_{\mathrm{c}}(\boldsymbol{\beta})) \leq 1\}$$

which, by combining (6.7) and (6.10), allows to compactly express the word error probability on a BSC as

$$\mathrm{P}\left[\boldsymbol{b}_m \neq \boldsymbol{b}_m\right] = 1 - (1 - P_{\mathrm{bit}})^{n-1}[1 - (n-1)P_{\mathrm{bit}}] \,. \tag{6.47}$$

For $P_{\mathrm{bit}} \ll 1$ the above expression can be conveniently approximated as $\mathrm{P}\left[\boldsymbol{b}_m \neq \boldsymbol{b}_m\right] \simeq \binom{n}{2} P_{\mathrm{bit}}^2$

Reed-Muller codes These codes can be built, starting with arbitrary integer parameters $m \geq r \geq 0$, and obtaining the length of information words and codewords as

$$k = \sum_{i=0}^{r} \binom{m}{i} \quad , \quad n = 2^m \,. \tag{6.48}$$

They do not exhibit particularly good distance properties, in general being

$$d_{\min} = 2^{m-r} \tag{6.49}$$

However they can be efficiently decoded via a recursive majority rule algorithm.

Reed-Solomon (RS) codes These codes are non binary BCH codes (typically the cardinality of the symbol alphabet is 2^m, with $m \geq 3$) that exhibit the maximum Hamming distance, i.e. they achieve the Singleton bound (6.34). This means that no other block code with the same lengths (n, k) and symbol cardinality can perform better than the corresponding RS code. However, known RS codes do not have very long words.

Bose-Chauduri-Hocquenghem (BCH) codes This large class of cyclic codes have good distance properties, and have a strong algebraic structure that allows efficient implementations of encoding and syndrome decoding. Namely, decoding is usually performed via the efficient Berlekamp-Massey and Forney algorithms [5].

Examples of code parameters from the above classes are given in Tab. 6.2. For the same codes the trade-off between code rate k/n and the ratio t_c/n between the maximum number of correctable errors and the code length is self-evident from Fig. 6.5.

code class	n	k	$d_{\min}$	t_c	n	k	$d_{\min}$	t_c
Hamming	7	4	3	1	15	11	3	1
	31	26	3	1	63	57	3	1
	127	120	3	1	255	247	3	1
Reed-Muller	8	4	4	1	16	11	4	1
	16	5	8	3	32	26	4	1
	32	16	8	3	32	6	16	7
	64	57	4	1	64	42	8	3
	64	22	16	7	64	7	32	15
BCH	15	7	5	2	15	5	7	3
	31	21	5	2	31	16	7	3
	31	11	11	5	31	6	15	7
	63	51	5	2	63	45	7	3
	63	39	9	4	63	36	11	5
	63	30	13	6	63	24	15	7
	63	18	21	10	63	16	23	11
	63	10	27	13	63	7	31	15

Table 6.2 Parameters of some binary block codes. Single-error correcting binary BCH codes are not given as they coincide with Hamming codes of the same length.

Alterations of code parameters Being strongly structured, codes belonging to the above classes do not exist for all possible values of word lengths n and k. However one can still build codes with good properties and arbitrary lengths, starting from a structured code and applying one of the following techniques.

Expanding/puncturing i.e. increasing/reducing the code length n while keeping the same information word length k, by adding/removing some parity symbols.

Lengthening/shortening i.e. increasing/reducing the lengths n and k of both the code and information words, while keeping the same number $n - k$ of parity symbols.

Augmenting/expurgating i.e. increasing/reducing the length k of the information words while keeping the same codeword length n, and correspondingly reducing/increasing the number of parity symbols.

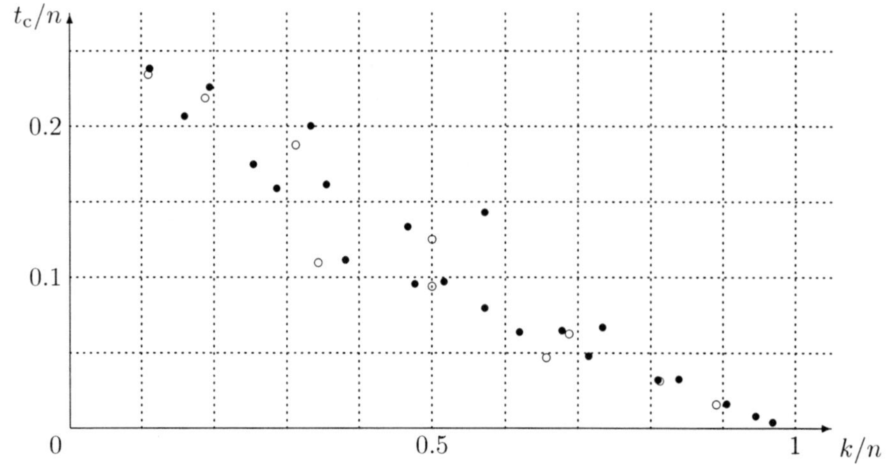

Figure 6.5 Error correction rate t_c/n versus code rate k/n for BCH (dots) and Reed-Muller (circles) codes, with parameters given in Tab. 6.2.

Puncturing, lengthening and augmenting lead to an increase of the code rate, while the complementary operations decrease it. On the other hand, expanding and augmenting a code allow to increase the minimum Hamming distance of the code, while puncturing and augmenting decrease it.

6.2.5 Performance of linear codes

Although the parameters $d_{\min}$ and t_c can give valuable information on a code, it is more appropriate to evaluate the code performance in terms of its symbol error probability (6.8), for example when the channel is a memoryless BSC. Fig. 6.6 shows the symbol error probability of a $(7, 4)$ and a $(15, 11)$ Hamming codes, depending on the error probability P_{bit} of the BSC. Observe that for low P_{bit} the symbol error probability follows a power law in which the exponent (the slope of the straight line in the logarithmic plot) is essentially determined by t_c. Also, note that among codes with the same error correcting capability, longer codes have a higher symbol error probability which compensates for their higher rate.

On the other hand, it is also appopriate to keep the rate of information symbols fixed and compare different codes in terms of the symbol error probability. In this case the coded messages will in general have different symbol rates, and so different modulation systems and bandwidths must be employed. If modulation is done over an AWGN channel as in Chapter 5, for instance with a binary pulse amplitude modulation (PAM), it is customary to compare different codes under the same noise PSD N_0 and average energy *per information bit E_b*, defined as

$$E_b = \frac{n}{k \log_2 M} E_s \,, \tag{6.50}$$

where E_s is the energy per modulated symbol and M the modulation cardinality, introduced in Chapter 5. In Fig. 6.7 we compare the performance of the same codes seen in Fig. 6.6 with the uncoded system, when using a binary PAM of the proper symbol rate and bandwidth for each

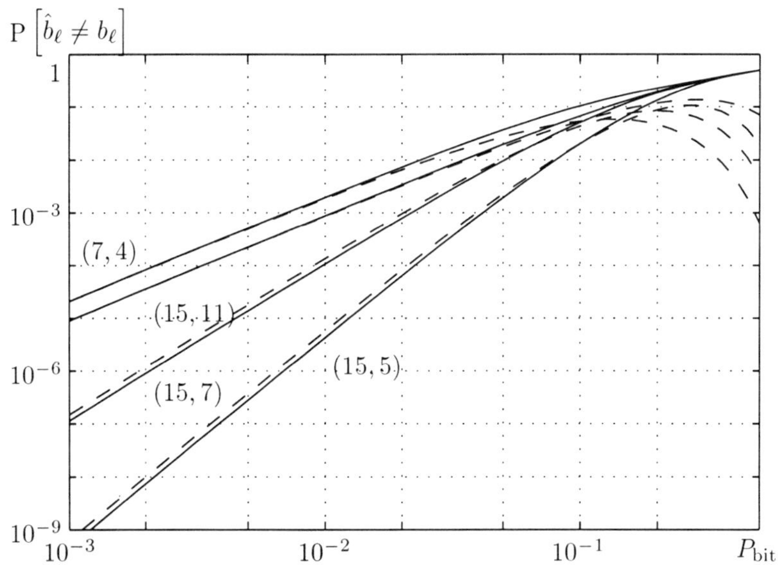

Figure 6.6 Performance comparison among some BCH/Hamming codes in terms of the symbol error probability over a BSC channel with error probability P_{bit}. The dashed lines plot the approximate expression (6.51).

code and the same ratio E_b/N_0. The symbol error probabilities are now much closer among the codes, and it can be seen that at higher SNR the (15,11) Hamming code performs best of all, exhibiting a coding gain of 1.4 dB at a symbol error probability 10^{-6}.

We note that apart from some particular cases (e.g. the calculation of the word error probability (6.2) for Hamming codes) an exact evaluation of (6.7) and (6.8) is not practical, even for codes with moderate length. However, a simple approximation can be found, by assuming the following simplifications:

1. we neglect the probability of transition $p_{\tilde{\mathbf{c}}|\mathbf{c}}(\boldsymbol{\xi}|\boldsymbol{\gamma})$ to words $\boldsymbol{\xi}$ with higher Hamming distances from the transmitted codeword $\boldsymbol{\gamma}$, thereby considering only those at distance $t_c + 1$ (where t is the number of correctable errors);

2. we consider that all words at distance $t_c + 1$ lie outside the correct decoding region (this holds true for Hamming codes, but in general some words at distance higher than t_c may still be included in the correct region)

3. we consider that decoding to codewords at distance close to $d_{\min}$ from the trasmitted one is much more likely than to more distant codewords;

4. we assume that the fraction of incorrect bits remains approximately the same going form codewords to the information words.

By combining assumptions 1 and 2 we get for any $\boldsymbol{\beta} \in \mathbb{F}^k$

$$\sum_{\hat{\boldsymbol{\beta}} \in \mathcal{A}^k} \sum_{\boldsymbol{\xi} \in R_{\hat{\boldsymbol{\beta}}}} p_{\tilde{\mathbf{c}}_m|\mathbf{c}_m}(\boldsymbol{\xi}|\mu_c(\boldsymbol{\beta})) \simeq \binom{n}{t_c + 1} P_{\text{bit}}^{t_c+1}(1 - P_{\text{bit}})^{n-t_c-1}$$

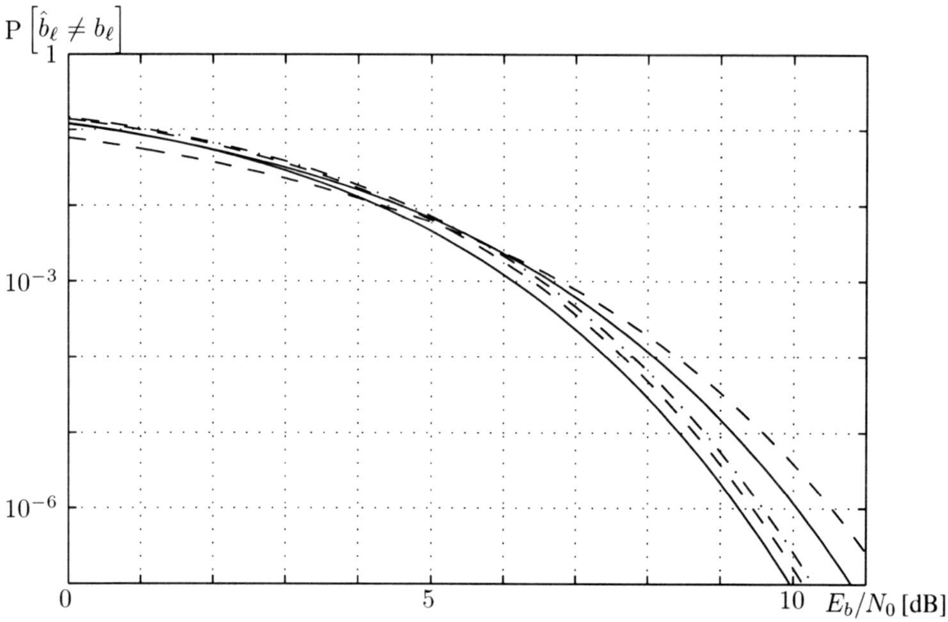

Figure 6.7 Performance comparison among some BCH/Hamming codes in terms of the symbol error probability with a binary PAM system over an AWGN channel with PSD $N_0/2$. The two solid lines represents two Hamming codes: (15,11) the lower, (7,4) the higher. The higher dashed line represents the uncoded system, the lower dashed line and the dash-dotted line represent the (15,5) and (15,7) BCH codes respectively. The coding gain can be seen as the dB difference between the crossing point with a given error probability between the coded and uncoded system

while assumptions 3 and 4 yield

$$\frac{d_\mathrm{H}(\boldsymbol{\beta}, \hat{\boldsymbol{\beta}})}{k} \simeq \frac{d_\mathrm{min}}{n}$$

so that we can simplify (6.8) to the expression

$$\mathrm{P}\left[\hat{b}_\ell \neq b_\ell\right] \simeq \frac{d_\mathrm{min}}{n}\binom{n}{t_\mathrm{c}+1} P_\mathrm{bit}^{t_\mathrm{c}+1}(1-P_\mathrm{bit})^{n-t_\mathrm{c}-1} \tag{6.51}$$

Fig. 6.6 shows that the above approximation is quite accurate for small P_bit.

6.3 CONVOLUTIONAL CODES

6.3.1 Construction and properties

Beside block codes, a class of codes frequently used in applications is that of *convolutional codes*. In this case, the reference scheme for the system is again that of Fig. 6.2, with the map μ_c replaced by a $n \times k$ matrix filter $\boldsymbol{g}(hT_\mathrm{w})$ with finite support, hence the name "convolutional".

The input-output relation for the encoder is therefore

$$c_m = \sum_{h=0}^{\nu} g(hT_w) b_{m-h} \tag{6.52}$$

where it is understood that the sum and product are those defined in (6.25) for the binary field $\mathbb{F}$. Hence, convolutional codes are linear in the same sense as of Definition 6.3. Typically, n and k are chosen as small integers and the code rate is still given by the ratio k/n. From (6.52) we can see that each output word c_m depends on the current information word b_m and the previous ν information words. The parameter ν is then called the *constraint length* of the code.

The convolutional encoder can also be represented as a finite state sequential machine, where the current state at time mT_w is given by the last ν input words

$$s_m = [b_{m-1}, \ldots, b_{m-\nu}] \tag{6.53}$$

and is updated accordingly. We then have

$$\begin{cases} s_{m+1} = u(b_m, s_m) \\ c_m = v(b_m, s_m) \end{cases} \tag{6.54}$$

with the update and output functions given by

$$u(\beta, [\sigma_1, \ldots, \sigma_\nu]) = [\beta, \sigma_1, \ldots, \sigma_{\nu-1}] \tag{6.55}$$

$$v(\beta, [\sigma_1, \ldots, \sigma_\nu]) = g(0)\beta + \sum_{h=1}^{\nu} g(hT_w)\sigma_h \quad , \tag{6.56}$$

respectively. As the state space is $\mathbb{F}^{k\nu}$, its cardinality (i.e. the number of possible states of the sequential machine) is $2^{k\nu}$. An example of the state transition diagram for a $(3, 1)$ convolutional code with constraint length $\nu = 3$ is shown in Fig. 6.8, where state transitions are labeled with the corresponding input and output words.

Another useful graphical representation of a convolutional code is through its *trellis diagram*, shown in Fig. 6.9. In such a diagram the $2^{k\nu}$ possible states are repeated for each word interval T_w, thus adding a time axis to the state diagram. Possible transitions are then indicated by arrows linking states at adjacent word intervals.

The role that minimum Hamming distance plays for block codes is played for a convolutional code by its *free distance* d_f, which can be defined as the minimum number of different symbols between any two distinct infinitely long coded sequences. More formally,

$$d_f = \lim_{N \to \infty} d_{\min,N} \tag{6.57}$$

where $d_{\min,N}$ is the minimum Hamming distance between two coded sequences of N words that differ on the first word[1].

[1]It is straightforward to see that if $N < N'$, then $d_{\min,N} \leq d_{\min,N'}$, and that for any N, we have $d_{\min,N} \leq n(\nu + 1)$. Therefore the sequence $\{d_{\min,N}\}$ is monotonically increasing and bounded above, so the limit in (6.57) always exists and is finite.

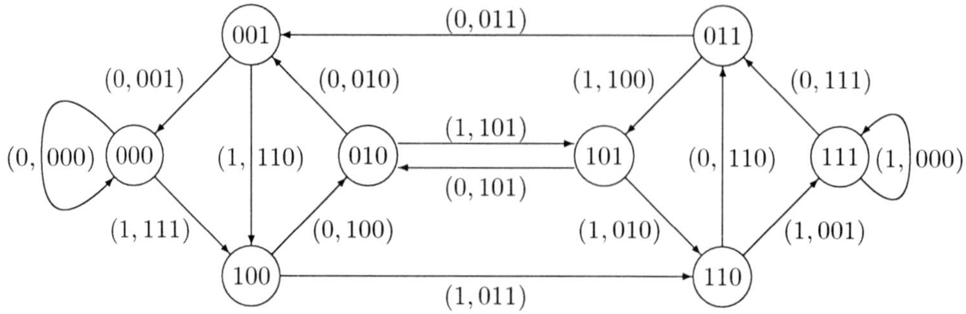

Figure 6.8 State transition diagram for a convolutional encoder with $k = 1$ (i.e. input words are made of a single binary symbol), $n = 3$, $\nu = 3$, and filter matrices $\boldsymbol{g}(0) = [111]^{\mathrm{T}}$, $\boldsymbol{g}(T_{\mathrm{w}}) = [100]^{\mathrm{T}}$, $\boldsymbol{g}(2T_{\mathrm{w}}) = [010]^{\mathrm{T}}$, $\boldsymbol{g}(3T_{\mathrm{w}}) = [001]^{\mathrm{T}}$. There are $2^{kH} = 8$ states and from every state there are $2^k = 2$ possible transitions, one for each possible value on the input symbol β. Each state transition is labeled with the corresponding input symbol and output word (β, γ).

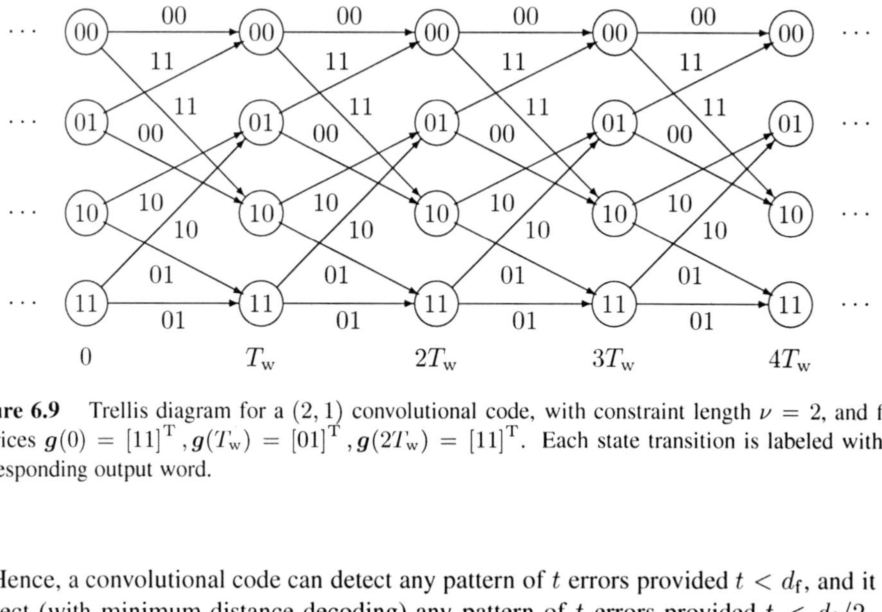

Figure 6.9 Trellis diagram for a $(2, 1)$ convolutional code, with constraint length $\nu = 2$, and filter matrices $\boldsymbol{g}(0) = [11]^{\mathrm{T}}, \boldsymbol{g}(T_{\mathrm{w}}) = [01]^{\mathrm{T}}, \boldsymbol{g}(2T_{\mathrm{w}}) = [11]^{\mathrm{T}}$. Each state transition is labeled with the corresponding output word.

Hence, a convolutional code can detect any pattern of t errors provided $t < d_{\mathrm{f}}$, and it can correct (with minimum distance decoding) any pattern of t errors provided $t < d_{\mathrm{f}}/2$. As convolutional codes are linear, results analogous to Propositions 6.5 and 6.6 hold. However, since not every convolutional code is equivalent to a systematic one, the convolutional version of the Singleton bound (6.34) only holds for systematic codes. It includes the constraint length ν and yields

$$d_{\mathrm{f}} \leq (n - k)(\nu + 1) + 1 \tag{6.58}$$

Example 6.3 A Consider a $(3, 2)$ convolutional code with constraint length $\nu = 2$ and filter matrices

$$g(0) = \begin{bmatrix} 1 & 0 \\ 0 & 1 \\ 1 & 1 \end{bmatrix} \quad , \quad g(T_\mathrm{w}) = \begin{bmatrix} 1 & 1 \\ 0 & 0 \\ 1 & 0 \end{bmatrix} \quad , \quad g(2T_\mathrm{w}) = \begin{bmatrix} 1 & 1 \\ 1 & 0 \\ 0 & 1 \end{bmatrix}$$

The possible encoder states are all the 2×2 (that is, $k \times \nu$) binary matrices

$$\begin{bmatrix} 0 & 0 \\ 0 & 0 \end{bmatrix}, \begin{bmatrix} 0 & 0 \\ 0 & 1 \end{bmatrix}, \ldots, \begin{bmatrix} 1 & 1 \\ 1 & 1 \end{bmatrix}$$

Then, for example, the sequence of information words

$$\boldsymbol{b}_0 = \begin{bmatrix} 1 \\ 0 \end{bmatrix} \quad , \quad \boldsymbol{b}_1 = \begin{bmatrix} 1 \\ 0 \end{bmatrix} \quad , \quad \boldsymbol{b}_2 = \begin{bmatrix} 0 \\ 1 \end{bmatrix} \quad , \quad \boldsymbol{b}_3 = \begin{bmatrix} 1 \\ 1 \end{bmatrix} \quad , \quad \cdots$$

yields the state sequence

$$\boldsymbol{s}_0 = \begin{bmatrix} 0 & 0 \\ 0 & 0 \end{bmatrix} , \ \boldsymbol{s}_1 = \begin{bmatrix} 1 & 0 \\ 0 & 0 \end{bmatrix} , \ \boldsymbol{s}_2 = \begin{bmatrix} 1 & 1 \\ 0 & 0 \end{bmatrix} , \ \boldsymbol{s}_3 = \begin{bmatrix} 0 & 1 \\ 1 & 0 \end{bmatrix} , \ \boldsymbol{s}_4 = \begin{bmatrix} 1 & 0 \\ 1 & 1 \end{bmatrix} , \ \cdots$$

and the output sequence

$$\boldsymbol{c}_0 = \begin{bmatrix} 1 \\ 0 \\ 1 \end{bmatrix} \quad , \quad \boldsymbol{c}_1 = \begin{bmatrix} 0 \\ 0 \\ 0 \end{bmatrix} \quad , \quad \boldsymbol{c}_2 = \begin{bmatrix} 1 \\ 1 \\ 0 \end{bmatrix} \quad , \quad \boldsymbol{c}_3 = \begin{bmatrix} 0 \\ 1 \\ 0 \end{bmatrix} \quad , \quad \cdots$$

Example 6.3 B The ECMA-368 standard for Ultra Wideband data transmission (aka WiMedia) from 55 Mbit/s to 480 Mbit/s, uses a base (3,1) convolutional code with constraint length $\nu = 6$. The filter matrices are

$$g(0) = \begin{bmatrix} 1 \\ 1 \\ 1 \end{bmatrix} \quad , \quad g(T_\mathrm{w}) = \begin{bmatrix} 0 \\ 1 \\ 1 \end{bmatrix} \quad , \quad g(2T_\mathrm{w}) = \begin{bmatrix} 1 \\ 1 \\ 1 \end{bmatrix} \quad , \quad g(3T_\mathrm{w}) = \begin{bmatrix} 1 \\ 0 \\ 1 \end{bmatrix} \quad ,$$

$$g(4T_\mathrm{w}) = \begin{bmatrix} 0 \\ 1 \\ 0 \end{bmatrix} \quad , \quad g(5T_\mathrm{w}) = \begin{bmatrix} 1 \\ 0 \\ 0 \end{bmatrix} \quad , \quad g(6T_\mathrm{w}) = \begin{bmatrix} 1 \\ 1 \\ 1 \end{bmatrix} \quad .$$

An alternative specification of a convolutional code is often given in terms of the nk vectors of length $\nu + 1$, each one representing the impulse response of a single filter.

$$\tilde{g}_{ij} = [g_{ij}(0), g_{ij}(T_\mathrm{w}), \ldots, g_{ij}(\nu T_\mathrm{w})] \quad , \quad i = 1, \ldots, n \quad , \quad j = 1, \ldots, k$$

For the sake of compactness, this is usually done via octal representation. In the case of the ECMA-368 code we get

$$\tilde{g}_{11} = [1011011] = 133_8 \quad , \quad \tilde{g}_{21} = [1110101] = 165_8 \quad , \quad \tilde{g}_{31} = [1111001] = 171_8$$

Example 6.3 C The free distance of a code can also be determined, either from its state transition diagram or its trellis, as the minimum total Hamming weight among the paths that depart from the 0 state and merge back into it.

For example in the code shown in Fig. 6.8 such a path is easily identified by inspection as

$$000 \rightarrow 100 \rightarrow 010 \rightarrow 001 \rightarrow 000$$

which corresponds to the codeword sequence

$$111, 100, 010, 001$$

yielding a total Hamming weight if $d_{\mathrm{f}} = 6$.

Analogously, for the code whose trellis diagram is illustrated in Fig. 6.9 the path of minimum Hamming weight is

$$00 \rightarrow 10 \rightarrow 01 \rightarrow 00$$

with codeword sequence

$$11, 10, 11$$

and $d_{\mathrm{f}} = 5$.

6.3.2 Decoding of convolutional codes and the Viterbi algorithm

In decoding a convolutional code, unlike the block coding case, since each transmitted word also depends on previous information words, the received words can not be separately decoded from each other. Indeed, they must be *jointly* decoded to obtain the corresponding sequence of information words, and this operation is called *sequence decoding*.

Given a finite-length received sequence of L words, the convolutional decoder is therefore a map

$$\mu_{\mathrm{d}} : \mathbb{F}^{nL} \mapsto \mathbb{F}^{kL} \quad , \quad [\hat{\boldsymbol{b}}_1, \ldots, \hat{\boldsymbol{b}}_L] = \mu_{\mathrm{d}}(\tilde{\boldsymbol{c}}_1, \ldots, \tilde{\boldsymbol{c}}_L)$$

Observe, however, that since the encoder has a finite memory, a suitable initialization and termination is usually applied, for example by letting $\boldsymbol{b}_{-\nu} = \ldots = \boldsymbol{b}_{-1} = \boldsymbol{b}_{L+1} = \ldots = \boldsymbol{b}_{L+\nu} = \boldsymbol{0}$.

Given the received sequence $[\tilde{\boldsymbol{c}}_1, \ldots, \tilde{\boldsymbol{c}}_L] = [\boldsymbol{\xi}_1, \ldots, \boldsymbol{\xi}_L]$, the same criteria outlined in Section 6.1.3 can be applied to sequence decoding yielding the following maps

MAP decoding

$$[\hat{\boldsymbol{b}}_1, \ldots, \hat{\boldsymbol{b}}_L] = \underset{\boldsymbol{\beta}_1, \ldots, \boldsymbol{\beta}_L \in \mathbb{F}^k}{\operatorname{argmax}} \; p_{\boldsymbol{b}_1, \ldots, \boldsymbol{b}_L | \tilde{\boldsymbol{c}}_1, \ldots, \tilde{\boldsymbol{c}}_L}(\boldsymbol{\beta}_1, \ldots, \boldsymbol{\beta}_L | \boldsymbol{\xi}_1, \ldots, \boldsymbol{\xi}_L) \tag{6.59}$$

ML decoding

$$[\hat{\boldsymbol{b}}_1, \ldots, \hat{\boldsymbol{b}}_L] = \underset{\boldsymbol{\beta}_1, \ldots, \boldsymbol{\beta}_L \in \mathbb{F}^k}{\operatorname{argmax}} \; p_{\tilde{\boldsymbol{c}}_1, \ldots, \tilde{\boldsymbol{c}}_L | \boldsymbol{b}_1, \ldots, \boldsymbol{b}_L}(\boldsymbol{\xi}_1, \ldots, \boldsymbol{\xi}_L | \boldsymbol{\beta}_1, \ldots, \boldsymbol{\beta}_L) \tag{6.60}$$

minimum distance decoding

$$[\hat{\boldsymbol{b}}_1, \ldots, \hat{\boldsymbol{b}}_L] = \underset{\boldsymbol{\beta}_1, \ldots, \boldsymbol{\beta}_L \in \mathbb{F}^k}{\operatorname{argmin}} \sum_{m=1}^{L} d_{\mathrm{H}} \left(\boldsymbol{\xi}_m, \sum_{h=0}^{\nu} \boldsymbol{g}(hT_{\mathrm{w}}) \boldsymbol{\beta}_{m-h} \right) \tag{6.61}$$

From the above formulation it is clear that a straightforward implementation of the decoding criteria requires maximization over 2^{kL} possible sequences, a complexity that incresaes exponentially with the sequence length L, making it unviable for large L.

The *Viterbi algorithm* [17, 8, 5] allows to efficiently accomplish ML sequence decoding in the case of a memoryless channel, with a complexity that is only linear in the sequence length. It is based on the following result

Proposition 6.16 *In a memoryless channel the maximum of the likelihood function in (6.60) can be written as*

$$\max_{\boldsymbol{\beta}_1,\ldots,\boldsymbol{\beta}_L \in \mathbb{F}^k} p_{\tilde{\mathbf{c}}_1,\ldots,\tilde{\mathbf{c}}_L|\boldsymbol{b}_1,\ldots,\boldsymbol{b}_L}(\boldsymbol{\xi}_1,\ldots,\boldsymbol{\xi}_L|\boldsymbol{\beta}_1,\ldots,\boldsymbol{\beta}_L) = \exp\left[-\min_{\boldsymbol{\sigma}\in\mathbb{F}^{k\nu}} C_L(\boldsymbol{\sigma})\right] \tag{6.62}$$

where $C_L(\boldsymbol{\sigma})$ is a cumulative state cost function defined recursively as

$$C_1(\boldsymbol{\sigma}) = 0 \tag{6.63}$$

$$C_{m+1}(\boldsymbol{\sigma}) = \min_{(\boldsymbol{\beta},\boldsymbol{\sigma}')\in u^{-1}(\boldsymbol{\sigma})} [C_m(\boldsymbol{\sigma}') + \Delta_C(\boldsymbol{\xi}_m, v(\boldsymbol{\sigma}_1,\boldsymbol{\sigma}'))] \tag{6.64}$$

$$\Delta_C(\boldsymbol{\xi},\boldsymbol{\gamma}) = -\ln p_{\tilde{\mathbf{c}}|\mathbf{c}}(\boldsymbol{\xi}|\boldsymbol{\gamma}) \tag{6.65}$$

Proof By iteratively applying the definition of conditional PMD (2.176) we get

$$\begin{aligned} & p_{\tilde{\mathbf{c}}_1,\ldots,\tilde{\mathbf{c}}_L|\boldsymbol{b}_1,\ldots,\boldsymbol{b}_L}(\boldsymbol{\xi}_1,\ldots,\boldsymbol{\xi}_L|\boldsymbol{\beta}_1,\ldots,\boldsymbol{\beta}_L) \\ &= \prod_{m=1}^{L} p_{\tilde{\mathbf{c}}_m|\tilde{\mathbf{c}}_1,\ldots,\tilde{\mathbf{c}}_{m-1},\boldsymbol{b}_1,\ldots,\boldsymbol{b}_L}(\boldsymbol{\xi}_m|\boldsymbol{\xi}_1,\ldots,\boldsymbol{\xi}_{m-1},\boldsymbol{\beta}_1,\ldots,\boldsymbol{\beta}_L) \\ &= \prod_{m=1}^{L} p_{\tilde{\mathbf{c}}_m|\mathbf{c}_m}\left(\boldsymbol{\xi}_m \middle| \sum_{h=0}^{\nu} \boldsymbol{g}(hT_{\mathrm{w}})\boldsymbol{\beta}_{m-h}\right) \end{aligned}$$

where the latter step is justified by the memoryless property of the channel.[2] Then by taking the logarithm of both sides we get

$$-\ln p_{\tilde{\mathbf{c}}_1,\ldots,\tilde{\mathbf{c}}_L|\boldsymbol{b}_1,\ldots,\boldsymbol{b}_L}(\boldsymbol{\xi}_1,\ldots,\boldsymbol{\xi}_L|\boldsymbol{\beta}_1,\ldots,\boldsymbol{\beta}_L) = \sum_{m=1}^{L} \Delta_C\left(\boldsymbol{\xi}_m, \sum_{h=0}^{\nu} \boldsymbol{g}(hT_{\mathrm{w}})\boldsymbol{\beta}_{m-h}\right)$$

Observe that the sum on the right-hand side can be evaluated recursively in terms of the state vector that comprises ν consecutive information words, and the statement is proved. □

The Viterbi algorithm for ML decoding of a sequence of L received words $\tilde{\mathbf{c}}_1,\ldots,\tilde{\mathbf{c}}_m$ over a memoryless channel can therefore be formulated as follows

[2] A sufficient condition for Proposition 6.16 to hold is that the channel be memoryless *at the word level*, that is

$$p_{\tilde{\mathbf{c}}_1,\ldots,\tilde{\mathbf{c}}_L|\mathbf{c}_1,\ldots,\mathbf{c}_L}(\boldsymbol{\xi}_1,\ldots,\boldsymbol{\xi}_L|\boldsymbol{\gamma}_1,\ldots,\boldsymbol{\gamma}_L) = \prod_{m=1}^{L} p_{\tilde{\mathbf{c}}|\mathbf{c}}(\boldsymbol{\xi}_m|\boldsymbol{\gamma}_m)$$

However it is customary to set forth the stronger yet simpler condition that it be memoryless at the symbol level (as stated by (5.163)).

initialization Let the starting cost $C_1(\boldsymbol{\sigma}) = 0$, for all $\boldsymbol{\sigma} \in \mathbb{F}^{k\nu}$ and the

iteration step For $m = 1, 2, \dots, L$:

- for all $\boldsymbol{\sigma} \in \mathbb{F}^{k\nu}$,
  - let $A_m(\boldsymbol{\sigma}) = \arg\min_{\boldsymbol{\sigma}' \in \mathbb{F}^{k\nu}} C_m(\boldsymbol{\sigma}') + \Delta_C(\tilde{c}_m, v(\boldsymbol{\sigma}_1, \boldsymbol{\sigma}'))$, with ;
  - let $C_{m+1}(\boldsymbol{\sigma}) = \arg\min_{\boldsymbol{\sigma}' \in \mathbb{F}^{k\nu}} C_m(\boldsymbol{\sigma}') + \Delta_C(\tilde{c}_m, v(\boldsymbol{\sigma}_1, \boldsymbol{\sigma}'))$;

termination Choose $\hat{\boldsymbol{s}}_{L+1} = \arg\min_{\boldsymbol{\sigma} \in \mathbb{F}^{k\nu}} C_{L+1}(\boldsymbol{\sigma})$

backtracing For $m = L, L-1, \dots, 1$:

- let $\hat{\boldsymbol{b}}_m$ be the first column of $\boldsymbol{s}_{m+1}$;
- let $\hat{\boldsymbol{s}}_m = A_m(\boldsymbol{s}_{m+1})$;

Observe the both the computational complexity and memory space requirement of the algorithm is determined by the iteration step, and is proportional to $2^{k\nu}L$, thus being linear in the sequence length and exponential in the code dimension and the constraint length.

When the memoryless channel is a BSC the incremental cost function can be written as

$$\Delta_C(\boldsymbol{\xi}, \boldsymbol{\gamma}) = d_{\text{H}}(\boldsymbol{\xi}, \boldsymbol{\gamma}) \ln\left(\frac{P_{\text{bit}}}{1 - P_{\text{bit}}}\right) + n \ln(1 - P_{\text{bit}})$$

and by removing the multiplicative and additive constants, it can be replaced by just the Hamming distance.

On the other hand, the Viterbi algorithm lends itself to a wider class of applications, and unlike other algebraic decoding techniques can also be used for soft decoding (see Section 6.5) of convolutional codes over AWGN channels, by using the Euclidean distance (or other appropriate metrics) in the cost definition.

Example 6.3 D The implementation of the Viterbi algorithm over the code of the trellis diagram in Fig. 6.9 and with the following received sequence

$$\tilde{c}_1 = 01 \quad , \quad \tilde{c}_2 = 10 \quad , \quad \tilde{c}_3 = 01 \quad , \quad \tilde{c}_4 = 11 \; .$$

After initialization, in the first iteration step we get the incremental costs

$$\Delta_C(01, 00) = d_{\text{H}}(01, 00) = 1 \quad , \quad \Delta_C(01, 01) = 0 \quad , \quad \Delta_C(01, 10) = 2 \quad , \quad \Delta_C(01, 11) = 1$$

that determine the costs

$$C_2(00) = \min\{\Delta_C(01, 00), \Delta_C(01, 11)\} = 1 \quad , \quad C_2(01) = 2 \quad , \quad C_2(10) = 1 \quad , \quad C_2(11) = 0$$

and the corresponding ancestors (any one in the pair is fine, here)

$$A_1(00) \in \{00, 01\} \quad , \quad A_1(01) \in \{10, 11\} \quad , \quad A_1(10) \in \{00, 01\} \quad , \quad A_1(11) \in \{10, 11\}$$

At the second step the costs are

$$C_3(00) = \min\{C_2(00) + \Delta_C(10, 00), C_2(01) + \Delta_C(10, 11)\} = \min\{1 + 1, 2 + 1\} = 2$$

$$C_3(01) = 0 \quad , \quad C_3(10) = 2 \quad , \quad C_3(11) = 2$$

and the ancestors

$$A_2(00) = 00, A_2(01) = 11 \quad , \quad A_2(10) = 00 \quad , \quad A_2(11) = 11$$

and so on with the following steps until we find

$$C_5(00) = 3 \quad , \quad C_5(01) = 2 \quad , \quad C_5(10) = 1 \quad , \quad C_3(11) = 2 \; .$$

The cumulative and incremental costs for all states and transitions at all steps are shown on the trellis in Fig. 6.10. Then by backtracing we find the ML decoded sequence

$$\hat{b}_4 = 1 \quad , \quad \hat{b}_3 = 0 \quad , \quad \hat{b}_2 = 0 \quad , \quad \hat{b}_1 = 1$$

Example 6.3 E Convolutional codes are often used together with block codes. In the DVB standard, a pair of serially concatenated codes is used. The base inner code is a $(2, 1)$ convolutional code with constraint length $\nu = 6$. However by proper puncturing, its rate can be varied from the original 1/2 to 2/3, 3/4, 5/6 or 7/8. The outer code is a code with symbol (byte) cardinality $2^8 = 256$, information words (packets) of length $k = 188$, and code words of length $n = 204$, that is obtained by shortening a $(255, 239)$ Reed-Solomon code.

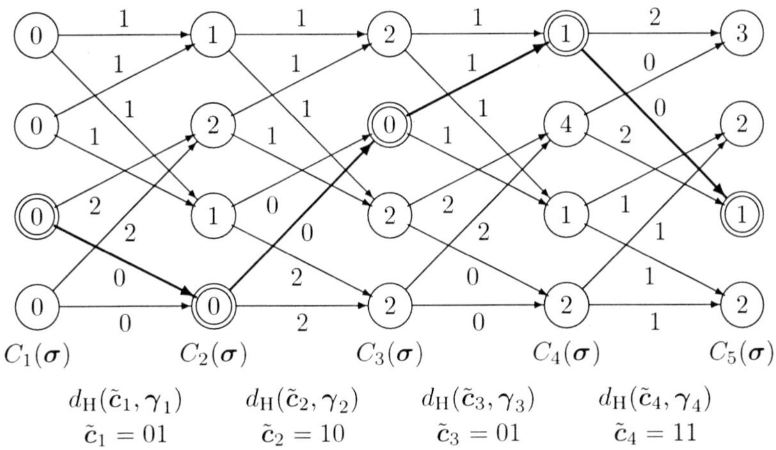

Figure 6.10 Illustration of the Viterbi decoding algorithm on the trellis diagram of Fig. 6.9. The received word sequence is shown below the trellis. Each transition is labeled with the incremental cost of the corresponding output word, each state is labeled with the corresponding cumulative minimum cost. The path with minimum total cost is emphasized.

6.4 CHANNEL CAPACITY

6.4.1 Capacity of a numerical channel

Mutual information We start by introducing the following definition

Definition 6.6 *The* mutual information *between two rvs x and y is defined as*

$$I(x, y) = H(x) + H(y) - H(x, y) \, . \tag{6.66}$$

The meaning of its name is clear by combining (6.66) with (3.79), obtaining

$$I(x, y) = H(x) - H(x|y) \, . \tag{6.67}$$

Here $I(x, y)$ is the difference between the *a priori* uncertainty on x, and the uncertainty on x once we know y. From (6.66) we see that $I(x, y)$ is symmetrical, with $I(x, y) = I(y, x)$.

By combining (6.66) with (3.73) we obtain the following bounds for the mutual information

$$0 \leq I(x, y) \leq \min \{H(x), H(y)\} \tag{6.68}$$

with the lower bound corresponding to x and y being statistically independent, and the upper bound corresponding to one rv being a.s. a function of the other.

Similarly to entropy, the notion of mutual information can be extended to pairs of vectors or messages. In particular, for a pair of messages we can define their *mutual information per symbol* by replacing the entropies in (6.66) with average entropies per symbol as defined in (3.87) and (3.89), to obtain

$$I_s(x, y) = H_s(x) + H_s(y) - H_s(x, y) \tag{6.69}$$

Channel information rate and capacity For a M-ary numerical channel $\mathcal{G}$, with symbol rate $F = 1/T$, M-ary input message $\{c_\ell\}$ and M-ary output message $\{\tilde{c}_\ell\}$, we define its *information rate*[3] as

$$R(\mathcal{G}, c) = F \, I_s(c, \tilde{c}) \qquad \text{[bit/s]} \tag{6.70}$$

where I_s is the mutual information per symbol (6.69). Observe that the information rate is a function of both the channel input and the channel output through the joint statistical description of the two messages. On the other hand, we can see it as a function of both the channel input and the channel itself, as the joint statistical description of input and output can be derived from that of the input and the channel.

Definition 6.7 *The* capacity *of a numerical channel is defined as the maximum of its information rate over all possible statistical description of the input message $\{c_\ell\}$*

$$C = \max_{\text{statistics of } c} R(\mathcal{G}, c) \qquad \textit{[bit/s]} \tag{6.71}$$

It is also common to consider the capacity per symbol which is defined as

$$C_s = \max_{\text{statistics of } c} I_s(c, \tilde{c}) \qquad \textit{[bit], [bit/symbol] or [bit/s/Hz]} \tag{6.72}$$

[3]Although they have the same name, the reader should not confuse the concept of information rate of a channel with the information rate of a message that was defined in (3.90).

and is equivalently measured in bit, bit/symbol or bit/s/Hz. The two are related by $C = FC_s$.

However the meaning of channel capacity is more important that what (6.76) would suggest. The Shannon theorem on channel coding and its converse, which are stated below without proof, allow to consider the capacity of a channel as the upper bound of the nominal information rate that can be transmitted with an arbitrary degree of reliability through appropriate channel coding.

Theorem 6.17 (Shannon theorem on channel coding) *Consider a memoryless M-ary channel with capacity C given by (6.76), and let $\{b_\ell\}$ be a message with nominal information rate $R < C$. Then for any $\delta > 0$ and for any sufficiently large n there exist*

i) an information dictionary $\mathcal{D}_{\boldsymbol{b}}$ of $N = \lceil 2^{RnT} \rceil$ words

ii) a coding map with codewords of length n, $\mu_c : \mathcal{D}_{\boldsymbol{b}} \mapsto \mathcal{A}^n$

iii) a decoding rule $\mu_d : \mathcal{A}^n \mapsto \mathcal{D}_{\boldsymbol{b}}$, with $\hat{\boldsymbol{b}} = \mu_d(\tilde{\boldsymbol{c}})$

such that the word error probability satisfies

$$\mathrm{P}\left[\hat{\boldsymbol{b}} \neq \boldsymbol{b}\right] < \delta \tag{6.73}$$

Theorem 6.18 (Converse of Theorem 6.17) *Consider a memoryless M-ary channel with capacity C given by (6.76), and let $\{b_\ell\}$ be a message with information rate $R > C$. Then there exists $\delta > 0$ such that, for any code and decoding strategy, the symbol error probability is bounded below*

$$\mathrm{P}\left[\hat{b}_\ell \neq b_\ell\right] > \delta \tag{6.74}$$

Observe that Theorem 6.17 only proves that a block coding method yielding arbitrarily small word error probabilities exists, but does not give any indication on how to find or build such a code. For this reason the Shannon theorem was long regarded to as a merely theoretical result, far from practical viability.

More recently, however, with the discovery of coding schemes that yield performance surprisingly close to the Shannon bound, such as concatenated codes with iterative soft decoding (the so called *turbo codes*) and *low density parity check* codes, that we will briefly outline in Section 6.5 , the result by Shannon is put into a new perspective as an attainable limit.

Capacity of a memoryless channel If the *channel is memoryless* (see Section 5.5.4) we can consider the input message to be made of iid symbols without loss of generality. Thus, the output symbols are also iid and the mutual information equals that of the pair of rvs $(c_\ell, \tilde{c}_\ell)$, so that $I_s(c, \tilde{c}) = I(c, \tilde{c})$, so the information rate can be written as

$$R(\mathcal{G}, c) = FI(c, \tilde{c}) \; . \tag{6.75}$$

Hence if $p_c(\cdot)$ is the PMD of c_ℓ, we can write

$$C = F \max_{p_c(\cdot)} I(c, \tilde{c}) \; . \tag{6.76}$$

From (6.67) we can express the mutual information in (6.76) in terms of the PMD of the input message and the channel transition probabilities

$$I(c, \tilde{c}) = H(\tilde{c}) - H(\tilde{c}|c) = \sum_{\xi,\gamma} p_{\tilde{c},c}(\xi, \gamma) \log_2 \left(\frac{p_{\tilde{c}|c}(\xi|\gamma)}{p_{\tilde{c}}(\xi)} \right) \tag{6.77}$$

$$= \sum_{\xi,\gamma} p_c(\gamma) p_{\tilde{c}|c}(\xi|\gamma) \log_2 \left(\frac{p_{\tilde{c}|c}(\xi|\gamma')}{\sum_{\gamma'} p_c(\gamma') p_{\tilde{c}|c}(\xi|\gamma')} . \right) \tag{6.78}$$

Hence, the maximization in (6.76) corresponds to the problem of finding the maximum of the nonlinear function (6.78) with respect to the M real-valued variables

$$p_c(\gamma_1), \ldots, p_c(\gamma_M) \tag{6.79}$$

under the constraints

$$p_c(\gamma) \geq 0 \quad , \quad \text{for all } \gamma \quad , \quad \sum_{\gamma} p_c(\gamma) = 1 \tag{6.80}$$

and with the M^2 transition probabilities $\{p_{\tilde{c}|c}(\xi|\gamma)\}$ regarded as parameters.

In this chapter we will derive (6.76) for a few simple cases, based upon symmetry considerations. In general, while standard optimization methods can effectively be used for small M, an efficient iterative technique for the numerical maximization of (6.78) is the Arimoto-Blahut algorithm, introduced in [1] and [4], and also described in [7, §10.8].

From the results in Sections 5.5.3-5.5.4, the transition probabilities of the channel can be conveniently arranged into an $M \times M$ *transition probability matrix* $\boldsymbol{P}$, where the entry in the ith row and jth column is the probability

$$p_{\tilde{c}|c}(\xi_i|\gamma_j) \quad , \quad i, j = 1, \ldots, M .$$

The matrix $\boldsymbol{P}$ hence completely describes the behavior of a memoryless channel.

Example 6.4 A A *perfect channel* is a memoryless M-ary channel with its transition probability matrix $\boldsymbol{P}$ having only one nonzero entry (equal to 1) in each row and column. It has the transition probabilities diagram shown in Fig. 6.11.

$$\boldsymbol{P} = \begin{bmatrix} 1 & 0 & 0 & 0 & 0 \\ 0 & 0 & 0 & 1 & 0 \\ 0 & 0 & 1 & 0 & 0 \\ 0 & 0 & 0 & 0 & 1 \\ 0 & 1 & 0 & 0 & 0 \end{bmatrix}$$

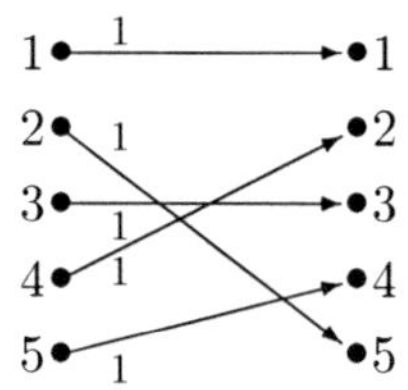

Figure 6.11 Transition probability matrix and corresponding diagram for a perfect channel with $M = 5$.

Such a channel does not introduce any error, it merely makes a permutation of the input symbol values. Since the output $\tilde{c}$ is a.s. a function of the input c, by (6.68) their mutual information is

$$I(\tilde{c}, c) = H(\tilde{c}) \ .$$

By Proposition 3.3, the maximum value of $H(\tilde{c})$ is $\log_2 M$ bit, obtained with all the values of $\tilde{c}$ equally likely. Observe that this corresponds to all values of c being equally likely as well. Therefore $\log_2 M$ bit represent the capacity per symbol of the perfect M-ary channel, and its capacity is

$$C = F \log_2 M \ .$$

Example 6.4 B A numerical channel in which each possible input value is mapped with the same probability to any of the output values is called a *useless channel*. Its probability transition matrix has all entries equal to $1/M$, and by applying the total probability theorem (2.180) and the normalization condition (2.141) we get

$$p_{\tilde{c}}(\xi) = \sum_{\gamma \in \mathcal{A}_c} p_{\tilde{c}|c}(\xi|\gamma) p_c(\gamma) = \frac{1}{M} \sum_{\gamma \in \mathcal{A}_c} p_c(\gamma) = \frac{1}{M}$$

that is

$$p_{\tilde{c}}(\xi) = p_{\tilde{c}|c}(\xi|\gamma) \quad , \quad \text{for all } \xi \in \mathcal{A}_{\tilde{c}}, \gamma \in \mathcal{A}_c \ .$$

Hence, the output symbol $\tilde{c}$ is independent of the input c, and by (6.68) we get $I(\tilde{c}, c) = 0$, so that this channel has null capacity, which justifies the term "useless".

Example 6.4 C The BSC with error probability P_{bit} has transition probability matrix

$$\boldsymbol{P} = \begin{bmatrix} 1 - P_{\text{bit}} & P_{\text{bit}} \\ P_{\text{bit}} & 1 - P_{\text{bit}} \end{bmatrix}$$

and its transition probability diagram was shown in Fig. 5.31b.

First, we derive its capacity via an intuitive approach based on the channel symmetry that also gives some insight on the result. We observe that the conditional information can only take two possible values

$$i_{\tilde{c}|c}(\tilde{c}|c) = \begin{cases} -\log_2 P_{\text{bit}} & , \ \text{for } \tilde{c} \neq c \\ -\log_2(1 - P_{\text{bit}}) & , \ \text{for } \tilde{c} = c \end{cases}$$

We can then regard $i_{\tilde{c}|c}(\tilde{c}|c)$ as a binary rv with alphabet $\{-\log_2 P_{\text{bit}}, -\log_2(1 - P_{\text{bit}})\}$ and PMD

$$\mathrm{P}\left[i_{\tilde{c}|c}(\tilde{c}|c) = -\log_2 P_{\text{bit}}\right] = \mathrm{P}\left[\tilde{c} \neq c\right] = P_{\text{bit}}$$
$$\mathrm{P}\left[i_{\tilde{c}|c}(\tilde{c}|c) = -\log_2(1 - P_{\text{bit}})\right] = \mathrm{P}\left[\tilde{c} = c\right] = 1 - P_{\text{bit}} \ .$$

Hence we get

$$H(\tilde{c}|c) = \mathrm{E}\left[i_{\tilde{c}|c}(\tilde{c}|c)\right] = -P_{\text{bit}} \log_2 P_{\text{bit}} - (1 - P_{\text{bit}}) \log_2(1 - P_{\text{bit}})$$

regardless of the input distribution. On the other hand, $\tilde{c}$ is binary and the maximum of its entropy is obtained with equally likely symbol values, that is $\max H(\tilde{c}) = 1$, and this is actually attainable with equally likely input symbols. Thus we get the capacity per symbol

$$\begin{aligned} C_s &= \max_{p_c(\cdot)} H(\tilde{c}) - H(\tilde{c}|c) \\ &= 1 + P_{\text{bit}} \log_2 P_{\text{bit}} + (1 - P_{\text{bit}}) \log_2(1 - P_{\text{bit}}) \end{aligned} \tag{6.81}$$

A more formal derivation follows. To solve the maximization of (6.78) we express the input PMD as

$$p_c(0) = q \quad , \quad p_c(1) = 1 - q$$

and obtain the output PMD as

$$p_{\tilde{c}}(0) = (1 - P_{\text{bit}})q + P_{\text{bit}}(1 - q) \quad , \quad p_{\tilde{c}}(1) = (1 - q)(1 - P_{\text{bit}}) + P_{\text{bit}}q$$

Hence we get

$$\begin{aligned} H(\tilde{c}) \quad = \quad & -(P_{\text{bit}} + q - 2qP_{\text{bit}}) \log_2(P_{\text{bit}} + q - 2qP_{\text{bit}}) \\ & -(1 + 2qP_{\text{bit}} - q - P_{\text{bit}}) \log_2(1 + 2qP_{\text{bit}} - q - P_{\text{bit}}) \end{aligned} \tag{6.82}$$

and

$$H(\tilde{c}|c) = -P_{\text{bit}} \log_2 P_{\text{bit}} - (1 - P_{\text{bit}}) \log_2(1 - P_{\text{bit}})$$

Observe that $H(\tilde{c}|c)$ does not depend on q, hence we only need to seek the maximum of the function $H(\tilde{c})$ with respect to $q \in [0, 1]$. The maximum can not be attained at the extrema, since $q = 0$ and $q = 1$ yield $H(c) = 0$ and hence by (6.68), $I(\tilde{c}, c) = 0$. Therefore we seek a maximum in the open interval $(0, 1)$ by imposing

$$\frac{d}{dq} H(\tilde{c}) = 0$$

and obtain the equation

$$\log_2(P_{\text{bit}} + q - 2qP_{\text{bit}}) = \log_2(1 - P_{\text{bit}} - q + 2qP_{\text{bit}})$$

which is solved by $q = 1/2$. Hence we can say that the source that allows to achieve the capacity (i.e. gives the maximum information rate) through a memoryless BSC is a memoryless symmetric source, that is with $p_c(0) = p_c(1) = 1/2$. This result is not surprising, since the function $I(\tilde{c}, c)$ is symmetric with respect to the variables $p_c(0)$ and $p_c(1)$. Substituting $q = 1/2$ in (6.82) yields back $H(\tilde{c}) = 1$ and hence the expression (6.81) for the capacity per symbol C_s, which is illustrated in Fig. 6.12 as a function of P_{bit}.

Observe that for $P_{\text{bit}} = 0$ or 1, the channel has unit capacity per symbol, yielding the same result as in Example 6.4 A, since in this case the channel is deterministic and perfect. On the contrary, when $P_{\text{bit}} = 1/2$ the BSC has null capacity: as a matter of fact, since each input symbol is transformed into a 0 or a 1 with equal probability, the channel is useless as in Example 6.4 B.

With a closer look at the low P_{bit} region, as in Fig. 6.12b, we see that for values of P_{bit} up to 10^{-2} the BSC still retains more than 90% of the capacity of the perfect channel, whereas for higher error rates its capacity drops and nearly halves its value at $P_{\text{bit}} = 1/10$.

Example 6.4 D The M-ary symmetric channel with symbol error probability P_e has all correct transition probabilities equal to $1 - P_e$ and all incorrect transition probabilities equal to $P_e/(M - 1)$

$$\boldsymbol{P} = \begin{bmatrix} 1 - P_e & \frac{P_e}{M-1} & \cdots & \frac{P_e}{M-1} \\ \frac{P_e}{M-1} & 1 - P_e & \cdots & \frac{P_e}{M-1} \\ \vdots & \vdots & \ddots & \vdots \\ \frac{P_e}{M-1} & \frac{P_e}{M-1} & \cdots & 1 - P_e \end{bmatrix}$$

Then, the mutual information (6.78) turns out to be a symmetric function of the M variables $p_c(1), \ldots, p_c(M)$.

As in Example 6.4 C, the conditional information takes only two possible values

$$i_{\tilde{c}|c}(\tilde{c}|c) = \begin{cases} -\log_2 \frac{P_e}{M-1} & , \text{ for } \tilde{c} \neq c \\ -\log_2(1 - P_e) & , \text{ for } \tilde{c} = c \end{cases}$$

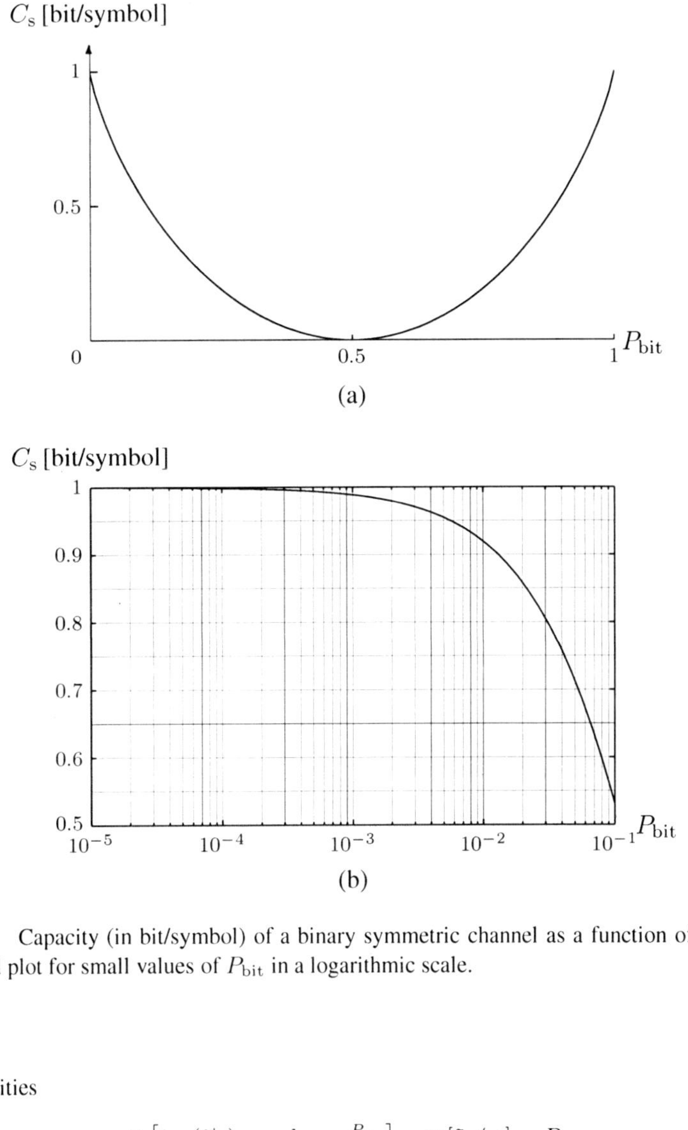

Figure 6.12 Capacity (in bit/symbol) of a binary symmetric channel as a function of $P_{\rm bit}$. In (b) a more detailed plot for small values of $P_{\rm bit}$ in a logarithmic scale.

with probabilities

$$\mathrm{P}\left[i_{\tilde{c}|c}(\tilde{c}|c) = -\log_2 \tfrac{P_e}{M-1}\right] = \mathrm{P}\left[\tilde{c} \neq c\right] = P_e$$
$$\mathrm{P}\left[i_{\tilde{c}|c}(\tilde{c}|c) = -\log_2(1-P_e)\right] = \mathrm{P}\left[\tilde{c} = c\right] = 1 - P_e$$

that do not depend on the PMD of c. Hence, we get

$$H(\tilde{c}|c) = \mathrm{E}\left[i_{\tilde{c}|c}(\tilde{c}|c)\right] = -P_e \log_2 \frac{P_e}{M-1} - (1-P_e)\log_2(1-P_e) \quad .$$

On the other hand, the maximum entropy for $\tilde{c}$ is obtained with equally likely symbol values, which by symmetry are given by equally likely values for c, so that $\max_{p_c(\cdot)} H(\tilde{c}) = \log_2 M$. Thus we get the capacity per symbol

$$C_s = \log_2 M + P_e \log_2 \frac{P_e}{M-1} + (1-P_e)\log_2(1-P_e) \quad .$$

We show a plot of C_s versus P_e for this channel in Fig. 6.13. Observe that for $P_e = 1 - 1/M$ the channel becomes useless

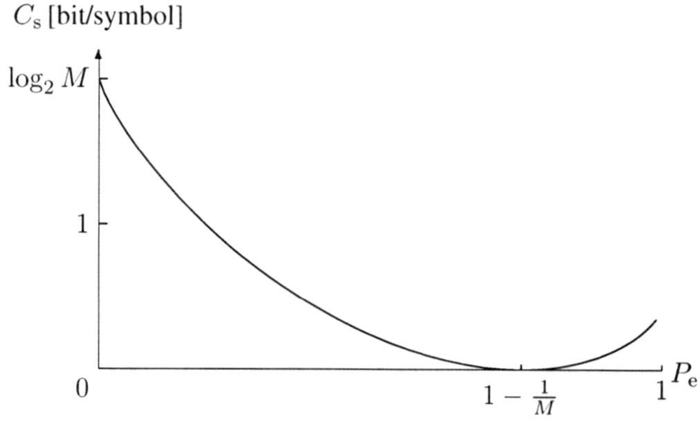

Figure 6.13 Capacity (in bit/symbol) of an M-ary symmetric channel as a function of P_e, for $M = 4$.

6.4.2 Capacity of the AWGN channel

Given the importance of the AWGN channel model introduced in Fig. 4.14 it is of interest to introduce the capacity of such a channel. This poses the problem of how to define capacity for analog channels, starting from that of numerical channels. It requires two distinct steps: i) moving from discrete amplitudes to continuous amplitudes and ii) from discrete-time to continuous time. We begin by extending the definition of mutual information to continuous rvs.

Mutual information between continuous rvs We shall derive the definition of mutual information between continuous rvs from that between discrete rvs introduced in Chapter 3 through the following device.

Let x, y be two continuous rvs and Δ_x, Δ_y two fixed positive quantities homogeneous with x and y, respectively. Then we define the two discrete rvs x_d, y_d as

$$\begin{array}{ll} x_d = i\Delta_x & \text{for } \Delta_x(i - \frac{1}{2}) \le x < \Delta_x(i + \frac{1}{2}) \\ y_d = j\Delta_y & \text{for } \Delta_y(j - \frac{1}{2}) \le y < \Delta_y(j + \frac{1}{2}) \end{array} \quad , \quad i, j \in \mathbb{Z} \,. \tag{6.83}$$

Observe that we can read x_d as the output of a uniform quantizer with step size Δ_x and ideally infinite levels, with x at its input. Similarly for y_d.

As $|x_d - x| \le \Delta_x/2$, if we let $\Delta_x \to 0$, the discrete rv x_d will uniformly converge to the continuous rv x. Similarly for y_d and y. We can therefore define the mutual information between x and y by continuity.

Definition 6.8 *We define the mutual information between the continuous rvs x and y as*

$$I(x, y) = \lim_{\Delta_x, \Delta_y \to 0} I(x_{\rm d}, y_{\rm d}) \tag{6.84}$$

where $x_{\rm d}, y_{\rm d}$ are defined in (6.83), and providing the limit exists.

Despite being quite reasonable and intuitive, the above definition does not lend itself to a practical calculation of $I(x, y)$. To this aim we give the following proposition, directly relating the mutual information to the joint statistical description of x and y.

Proposition 6.19 *The mutual information between continuous rvs x and y can be calculated as*

$$I(x, y) = \mathrm{E}\left[\log_2 \frac{p_{xy}(x, y)}{p_x(x)p_y(y)}\right] \tag{6.85}$$

$$= \mathrm{E}\left[\log_2 \frac{p_{y|x}(y|x)}{p_y(y)}\right] \tag{6.86}$$

Proof We give the proof assuming that the joint PDF $p_{xy}(a, b)$ is continuous (and so are the marginal PDFs $p_x(a), p_y(b)$), the other cases bearing only formal complications but no further insight. In fact, due to the continuity of the PDFs we can make use of the Mean Value Theorem from integral calculus and write the general term of the PMD of $x_{\rm d}$ as

$$p_{x_{\rm d}}(i\Delta_x) = \int_{(i-1/2)\Delta_x}^{(i+1/2)\Delta_x} p_x(a)\, da = \Delta_x p_x(a'_i) \tag{6.87}$$

with a'_i some point in the integration interval $[(i - 1/2)\Delta_x, (i + 1/2)\Delta_x]$, and analogously

$$p_{y_{\rm d}}(j\Delta_y) = \Delta_y p_y(b'_j) \quad , \quad p_{x_{\rm d}, y_{\rm d}}(i\Delta_x, j\Delta_y) = \Delta_x \Delta_y p_{x,y}(a''_i, b''_j) \tag{6.88}$$

Following the definition (6.66) we can express the mutual information between $x_{\rm d}$ and $y_{\rm d}$ as

$$I(x_{\rm d}, y_{\rm d}) = \mathrm{E}\left[\log_2 \frac{p_{x_{\rm d}, y_{\rm d}}(x_{\rm d}, y_{\rm d})}{p_{x_{\rm d}}(x_{\rm d})p_{y_{\rm d}}(y_{\rm d})}\right] \tag{6.89}$$

$$= \sum_{i,j} p_{x_{\rm d}, y_{\rm d}}(i\Delta_x, j\Delta_y) \log_2 \frac{p_{x_{\rm d}, y_{\rm d}}(i\Delta_x, j\Delta_y)}{p_{x_{\rm d}}(i\Delta_x)p_{y_{\rm d}}(j\Delta_y)} \tag{6.90}$$

$$= \sum_{i,j} \Delta_x \Delta_y \, p_{x,y}(a''_i, b''_j) \log_2 \frac{p_{x,y}(a''_i, b''_j)}{p_x(a'_i)p_y(b'_j)} \,. \tag{6.91}$$

As we let $\Delta_x, \Delta_y \to 0$, the Riemann sum in (6.91) converges to the integral

$$\lim_{\Delta_x, \Delta_y \to 0} I(x_{\rm d}, y_{\rm d}) = \int_{-\infty}^{+\infty} \int_{-\infty}^{+\infty} p_{x,y}(a, b) \log_2 \frac{p_{x,y}(a, b)}{p_x(a)p_y(b)} \, da \, db \tag{6.92}$$

thereby proving (6.85). Eventually, (6.86) follows by the definition of conditional PDF.[4] □

[4]Many authors [6, 7] use a similar approach to define entropy for continuous rvs, and build mutual information from it, as we did in the discrete case. However, it can be easily observed that, as $\Delta_x \to 0$,

Example 6.4 E Let x, y be jointly Gaussian rvs with means $\mathtt{m}_x, \mathtt{m}_y$, variances σ_x^2, σ_y^2, respectively, and covariance $\mathtt{k}_{xy}$. By rewriting (6.85) as

$$I(x,y) = -\frac{1}{\ln 2}\,\mathrm{E}\left[\ln \frac{p_{xy}(x,y)}{p_x(x)p_y(y)}\right]$$

and making use of the Gaussian PDFs given in Example 2.6 A and (2.192) so that

$$\ln \frac{p_{xy}(a,b)}{p_x(a)p_y(b)} = \ln \sqrt{\frac{1}{1-k_{xy}^2/(\sigma_x^2\sigma_y^2)}} - \frac{(a-\mathtt{m}_x)^2\sigma_y^2 + (b-\mathtt{m}_y)^2\sigma_x^2 - 2(a-\mathtt{m}_x)(b-\mathtt{m}_y)\mathtt{k}_{xy}}{2\sigma_x^2\sigma_y^2 - 2\mathtt{k}_{xy}^2} + \frac{(a-\mathtt{m}_x)^2}{2\sigma_x^2} + \frac{(b-\mathtt{m}_y)^2}{2\sigma_x^2}$$

we have

$$\mathrm{E}\left[\ln \frac{p_{xy}(x,y)}{p_x(x)p_y(y)}\right] = \ln \sqrt{\frac{1}{1-k_{xy}^2/(\sigma_x^2\sigma_y^2)}} - \frac{\mathrm{E}\left[(x-\mathtt{m}_x)^2\right]\sigma_y^2 + \mathrm{E}\left[(y-\mathtt{m}_y)^2\right]\sigma_x^2 - 2\,\mathrm{E}\left[(x-\mathtt{m}_x)(y-\mathtt{m}_y)\right]\mathtt{k}_{xy}}{2\sigma_x^2\sigma_y^2 - 2\mathtt{k}_{xy}^2} + \frac{1}{2\sigma_x^2}\mathrm{E}\left[(x-\mathtt{m}_x)^2\right] + \frac{1}{2\sigma_x^2}\mathrm{E}\left[(y-\mathtt{m}_y)^2\right]$$

As $\mathrm{E}\left[(x-\mathtt{m}_x)^2\right] = \sigma_x^2$, $\mathrm{E}\left[(y-\mathtt{m}_y)^2\right] = \sigma_y^2$, and $\mathrm{E}\left[(x-\mathtt{m}_x)(y-\mathtt{m}_y)\right] = \mathtt{k}_{xy}$, the last three terms in the sum cancel out, leaving only the first logarithm and yielding

$$I(x,y) = -\frac{1}{2}\log_2\left(1 - \frac{k_{xy}^2}{\sigma_x^2\sigma_y^2}\right) \tag{6.94}$$

Capacity of the discrete-time AWGN channel The discrete-time AWGN channel model is illustrated in Fig. 6.14, where T is the symbol period $\{s_n\}$ is the input process, $\{w_n\}$ is the additive noise, which is assumed to be Gaussian with iid samples having zero mean and

$H(x_\mathrm{d}) = -\sum_i \Delta_x p_x(a_i') \log_2(\Delta_x p_x(a_i'))$ diverges to ∞, and hence it is necessary to define a *differential entropy*

$$H(x) = \lim_{\Delta_x \to 0} \left[H(x_\mathrm{d}) + \log_2 \Delta_x\right] = -\int_{-\infty}^{+\infty} p_x(a)\log_2 p_x(a)\,\mathrm{d}a \tag{6.93}$$

In this book continuous rvs often denote physical quantities (usually a voltage, in the additive channel model) and have a physical dimension attached, and so do the quantization step size Δ_x and the PDF $p_x(a)$. Therefore, we prefer to avoid the definition of differential entropy (6.93), since the logarithms would not be well defined, and instead use that of mutual information (6.85)-(6.86) which involves the ratio of PDFs. See also the discussion in [6].

variance σ_w^2, and the output process $\{r_n\}$ is given by

$$r_n = \mathcal{C}s_n + w_n \tag{6.95}$$

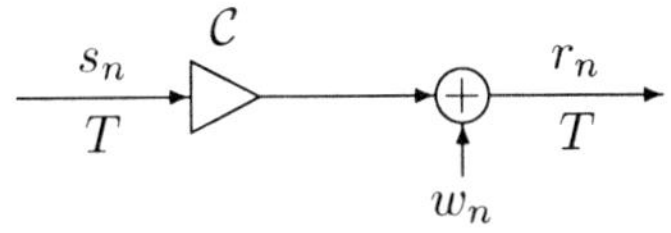

Figure 6.14 Illustration of the discrete-time AWGN channel.

It is a memoryless channel as, in fact, for every N the conditional PDF of the vector of output samples $\boldsymbol{r} = [r_{-N}, \ldots, r_N]$ given the corresponding vector of input samples $\boldsymbol{s} = [s_{-N}, \ldots, s_N]$ can be factored as

$$p_{\boldsymbol{r}|\boldsymbol{s}}(\boldsymbol{b}|\boldsymbol{a}) = p_{\boldsymbol{w}}(\boldsymbol{b} - \mathcal{C}\boldsymbol{a}) = \prod_{n=-N}^{N} p_w(b_n - \mathcal{C}a_n) = \prod_{n=-N}^{N} p_{r|s}(b_n|a_n) \tag{6.96}$$

where $\boldsymbol{w} = [w_{-N}, \ldots, w_N]$ and the marginal PDFs on the right-hand side do not depend on the time index n. Hence, based on (6.72) and the discussion leading to (6.75), we assume that the input $\{s_n\}$ has iid samples, i.e. it is produced by a memoryless source. Then, we derive the mutual information per symbol $I_s(s, r)$ between input and output, which coincides with the mutual information $I(s, r)$ between each pair of rvs (s_n, r_n). In fact, we can write (6.86) as

$$\begin{aligned} I(s,r) &= \mathrm{E}\left[\log_2 \frac{p_w(r_n - \mathcal{C}s_n)}{p_r(r_n)}\right] = \mathrm{E}\left[-\log_2 \frac{p_r(r_n)}{p_w(w_n)}\right] \\ &= -\,\mathrm{E}\left[\log_2 \frac{\sqrt{2\pi}\sigma_w p_r(r_n)}{e^{-w_n^2/(2\sigma_w^2)}}\right] \end{aligned}$$

We can then use properties of the logarithm and the linearity of expectation to get

$$I(s,r) = -\,\mathrm{E}\left[\log_2\left(\sigma_w p_r(r_n)\right)\right] - \frac{1}{2}\log_2(2\pi) - \frac{\mathrm{E}\left[w_n^2\right]}{2\sigma_w^2}\log_2 e \tag{6.97}$$

$$= -\,\mathrm{E}\left[\log_2\left(\sigma_w p_r(r_n)\right)\right] - \frac{1}{2}\log_2 2\pi e \tag{6.98}$$

In particular, if the transmitted signal $\{s_n\}$ is Gaussian with zero mean and variance σ_s^2, the received signal $\{r_n\}$ will itself be Gaussian, with zero mean. More than that, $\{r_n\}$ is also jointly Gaussian with $\{s_n\}$: from (6.95) the variance of $\{r_n\}$ and the covariance between s_n and r_n are, respectively, given by

$$\sigma_r^2 = \mathcal{C}^2\sigma_s^2 + \sigma_w^2 \quad , \quad \mathtt{k}_{rs} = \mathcal{C}\sigma_s^2\,. \tag{6.99}$$

We can thus apply (6.94) yielding

$$I(s,r) = -\frac{1}{2}\log_2\left(1 - \frac{\mathcal{C}^2\sigma_s^2}{\mathcal{C}^2\sigma_s^2 + \sigma_w^2}\right) = \frac{1}{2}\log_2\left(1 + \frac{\mathcal{C}^2\sigma_s^2}{\sigma_w^2}\right) \tag{6.100}$$

and we get the information rate of the AWGN channel with memoryless zero-mean Gaussian input

$$\boxed{R(\mathcal{G}, s) = \frac{1}{T} I(s, r) = \frac{1}{2T} \log_2 \left(1 + \frac{C^2 \sigma_s^2}{\sigma_w^2}\right)} \tag{6.101}$$

Two observations are appropriate at this point. First of all, we observe that the information rate (6.101) is an increasing function of the channel SNR $\Gamma = C^2\sigma_s^2/\sigma_w^2$ so that a higher transmit power allows a better exploitation of the channel. Secondly, we point out that $I(r, s)$ is unbounded for $\sigma_s^2 \to \infty$, hence it makes no sense to search for a maximum of the information rate (and speak of capacity) unless we pose a constraint on the transmitted signal power. It turns out, however, that for a given maximum input power $\mathtt{M}_{s,\max}$, the information rate is upper bounded, as we show in the following, and that it is therefore possible to define a channel capacity depending on $\mathtt{M}_{s,\max}$.

Proposition 6.20 *The mutual information (6.98) with an arbitrary memoryless input with statistical power $\mathtt{M}_s$ is upper bounded by that of a zero-mean Gaussian input with the same power,i.e.*

$$I(s, r) \le \frac{1}{2} \log_2 \left(1 + \frac{C^2 \mathtt{M}_s}{\sigma_w^2}\right) \tag{6.102}$$

The equality in (6.102) only holds for the zero-mean Gaussian case.

Proof Denote with $\{\tilde{s}_n\}$ a Gaussian, zero-mean process with iid rvs and variance $\sigma_s^2 = \mathtt{M}_s$ and let $\tilde{r}_n = \tilde{s}_n + w_n$ be the corresponding output of the AWGN channel with $\tilde{s}_n$ as input. As $I(\tilde{s}, \tilde{r})$ is given by the right-hand side of (6.102), we shall prove that $I(s, r) \le I(\tilde{s}, \tilde{r})$ for any possible input with the same statistical power. Since (6.98) holds for both cases (general input and zero-mean Gaussian input), by writing

$$I(s, r) - I(\tilde{s}, \tilde{r}) = \mathrm{E}\left[\log_2\left(\sigma_w p_{\tilde{r}}(\tilde{r}_n)\right)\right] - \mathrm{E}\left[\log_2\left(\sigma_w p_r(r_n)\right)\right]$$

and observing that, since $\{s_n\}$ and $\{w_n\}$ are uncorrelated, $\mathtt{M}_{\tilde{r}} = C\mathtt{M}_s + \sigma_w^2 = \mathtt{M}_r$,

$$\begin{aligned}
\mathrm{E}\left[\log_2\left(\sigma_w p_{\tilde{r}}(\tilde{r}_n)\right)\right] &= \log_2 \frac{\sigma_w}{\sqrt{2\pi \mathtt{M}_{\tilde{r}}}} + \frac{1}{2}\log_2 e \\
&= \log_2 \frac{\sigma_w}{\sqrt{2\pi \mathtt{M}_{\tilde{r}}}} + \frac{\mathtt{M}_r}{2\mathtt{M}_{\tilde{r}}}\log_2 e \\
&= \mathrm{E}\left[\log_2\left(\sigma_w p_{\tilde{r}}(r_n)\right)\right]
\end{aligned}$$

we have

$$\begin{aligned}
I(s, r) - I(\tilde{s}, \tilde{r}) &= \mathrm{E}\left[\log_2\left(\sigma_w p_{\tilde{r}}(r_n)\right)\right] - \mathrm{E}\left[\log_2\left(\sigma_w p_r(r_n)\right)\right] \\
&= \mathrm{E}\left[\log_2 \frac{p_{\tilde{r}}(r_n)}{p_r(r_n)}\right] .
\end{aligned}$$

Then, by Jensen's inequality (3.66)

$$\begin{aligned}
I(s, r) - I(\tilde{s}, \tilde{r}) &\le \log_2\left(\mathrm{E}\left[\frac{p_{\tilde{r}}(r_n)}{p_r(r_n)}\right]\right) \\
&= \log_2\left(\int_{\mathcal{S}_r} \frac{p_{\tilde{r}}(a)}{p_r(a)} p_r(a)\, \mathrm{d}a\right)
\end{aligned}$$

where $\mathcal{S}_r$ is the support of p_r, and since $p_{\tilde{r}}(a) \geq 0$ we have

$$\begin{aligned} I(s,r) - I(\tilde{s},\tilde{r}) &\leq \log_2\left(\int_{\mathcal{S}_r} p_{\tilde{r}}(a)\, \mathrm{d}a\right) \\ &\leq \log_2\left(\int_{-\infty}^{+\infty} p_{\tilde{r}}(a)\, \mathrm{d}a\right) \\ &= \log_2 1 = 0\,. \end{aligned}$$

We have thus proved the inequality (6.102). Now observe that equality holds if and only if $p_{\tilde{r}}$ has the same support as p_r and $p_{\tilde{r}}(\tilde{r}_n)/p_r(r_n)$ is almost surely constant. This is equivalent to have $p_r(a) = p_{\tilde{r}}(a)$ for all a, that is r_n itself Gaussian with zero mean. Eventually, $s_n = r_n - w_n$ must be a zero-mean Gaussian rvs, too. □

We are now ready to state the following theorem.

Theorem 6.21 (Capacity of the discrete-time AWGN channel) *The capacity of a discrete-time AWGN channel with bounded input power* $\mathtt{M}_s \leq M_{s,\max}$ *is given by*

$$C = \frac{1}{2T}\log_2\left(1 + \frac{\mathcal{C}^2 M_{s,\max}}{\sigma_w^2}\right) \tag{6.103}$$

It can be only attained with a zero-mean iid Gaussian process with variance $\sigma_s^2 = M_{s,\max}$.

Proof We start by defining the capacity under the input power bound as

$$C = \max_{\mathtt{M}_s \leq \mathtt{M}_{s,\max}} \frac{1}{T} I(s,r)$$

then we observe from Proposition 6.20 that for any $\mathtt{M}_s \leq \mathtt{M}_{s,\max}$ we have

$$I(s,r) \leq \frac{1}{2}\log_2\left(1 + \frac{\mathcal{C}^2\mathtt{M}_s}{\sigma_w^2}\right) \leq \frac{1}{2}\log_2\left(1 + \frac{\mathcal{C}^2\mathtt{M}_{s,\max}}{\sigma_w^2}\right) \tag{6.104}$$

On the other hand, the right-hand side of (6.104) is attained only with a Gaussian input process that has the maximum allowed power. □

Capacity of the bandlimited continuous-time AWGN channel Consider the model of Fig. 4.14 for a continuous time bandlimited channel with constant power gain $\mathtt{g}_{\mathrm{Ch}}$ and attenuation $\mathtt{a}_{\mathrm{Ch}} = 1/\mathtt{g}_{\mathrm{Ch}}$ over its band $\mathcal{B} = [0, B)$, and AWGN with PSD $N_0/2$. By antialiasing and sampling it can be shown equivalent to the discrete-time AWGN channel of the previous section, with $\sigma_w^2 = N_0 B$, $\mathcal{C}^2 = \mathtt{g}_{\mathrm{Ch}}$ and $\mathtt{M}_s = \mathtt{M}_{s_{\mathrm{Tx}}}$. Hence we can show [15] that in this case capacity is obtained s_{Tx} has Gaussian iid samples with mximum power, that is when it is a stationary Gaussian rp with constant PSD over $\mathcal{B}$

$$\mathcal{P}_{s_{\mathrm{Tx}}}(f) = \mathcal{E}_0 \operatorname{rect}\left(\frac{f}{2B}\right)\,. \tag{6.105}$$

The capacity value is given by

$$C = B\log_2(1 + \Gamma) \quad \text{[bit/s]} \tag{6.106}$$

where $\Gamma = \mathtt{M}_{s\mathrm{Tx}}/(N_0 B \mathsf{a}_{\mathrm{Ch}})$ is the reference SNR at the receiver input. From (6.105) it is also

$$\Gamma = \frac{2\mathcal{E}_0}{N_0 \mathsf{a}_{\mathrm{Ch}}} . \tag{6.107}$$

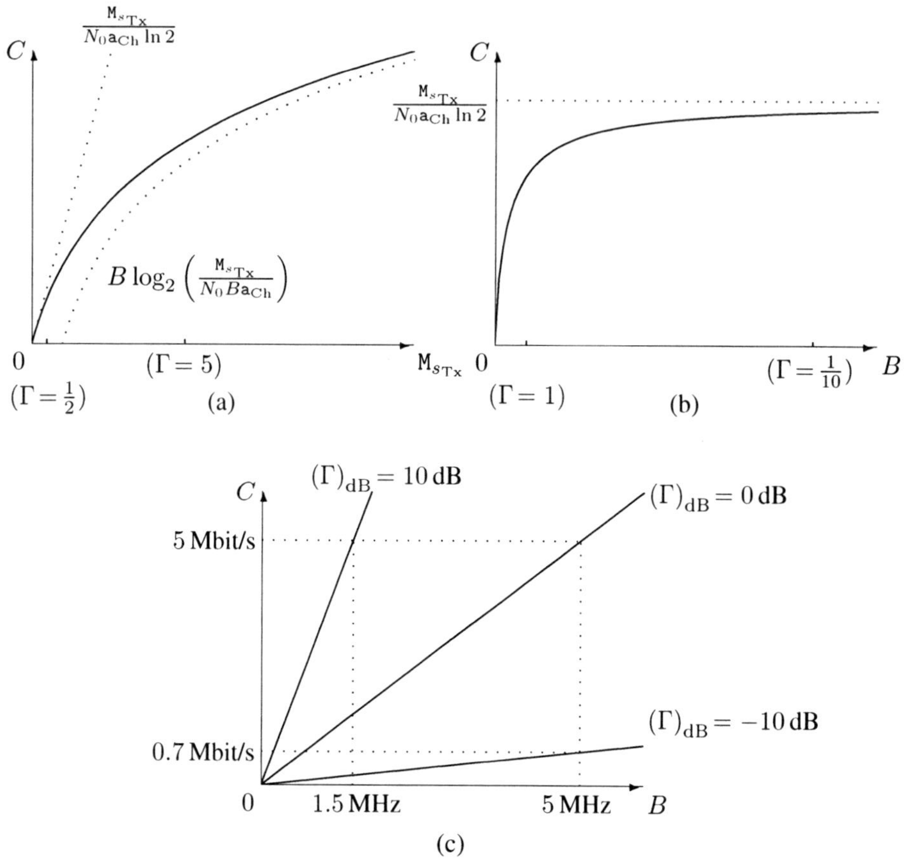

Figure 6.15 Illustration of the capacity for a bandlimited AWGN channel. (a) capacity vs. transmit power for bandwidth-constrained systems; (b) capacity vs. bandwidth for power-constrained systems; (c) capacity vs. bandwidth for PSD-constrained systems.

Equation (6.106) was introduced in Chapter 5 as (5.264), and $C/(2B)$ was shown in Fig. 5.53 as an upper bound to the spectral efficency of modulation systems for a given bit error probability. Thanks to the Shannon theorem on channel coding, we now know that its iportance is more general, since such bound holds for asymptotically vanishing error probability.

We illustrate (6.106) in Fig. 6.15. Namely, Fig. 6.15a shows a plot of C versus $\mathtt{M}_{s\mathrm{Tx}}$ for fixed B, thus representing how the capacity of bandwidth-constrained systems can increase by increasing the transmitted power. Observe that for a low power (corresponding to $\Gamma < 1$) the increase in capacity is nearly linear with power, as (6.106) is asymptotic to

$$C \sim \frac{\mathtt{M}_{s\mathrm{Tx}}}{N_0 \mathsf{a}_{\mathrm{Ch}} \ln 2} \simeq 1.44 \Gamma B \quad , \qquad \text{for } \Gamma \to 0 . \tag{6.108}$$

On the other hand in the high power regime (that is $\Gamma \gg 1$) equation (6.106) is asymptotic to

$$C \sim B \log_2 \Gamma \quad , \qquad \text{for } \Gamma \to \infty \, . \tag{6.109}$$

so the increase in capacity with power follows an approximate $\log_2$ rule. For example, by doubling the transmitted power C_s will only increase by 1 bit.

In Fig. 6.15b we plot C versus B for fixed $\mathtt{M}_{s\mathrm{Tx}}$ and hence show how the capacity of power-constrained systems can be increased by enlarging their bandwidth. Observe that while for low bandwidths (corresponding to $\Gamma > 1$) the capacity rapidly increases with B, as the bandwidth grows further (with corresponding increase of the noise power so that $\Gamma \ll 1$) the capacity approaches the limit value of $\mathtt{M}_{s\mathrm{Tx}}/(N_0 \mathtt{a}_{\mathrm{Ch}} \ln 2)$, which represents an upper bound for power-constrained systems.

On the contrary, if we keep the PSD value $\mathcal{E}_0$ fixed and increase the bandwidth B, the capacity increases linearly according to (6.106). So, under a constraint on the maximum PSD we can obtain an arbitrarily large system capacity, by proportionally increasing the bandwidth (and the transmit power with it). While constraints on power are usually set by budget issues (power consumption, battery life, chip area, etc.), constraints on the transmit PSD are set by regulatory bodies, especially with regards to wireless systems, in order to limit electromagnetic pollution and interference between systems. Fig. 6.15c shows a possible way to reach a good capacity within a PSD constraint by using a suitable bandwidth, which can be very large when Γ is small. This principle is exploited by ultra wide band (UWB) systems that transmit with very low power densities (below 10^{-7} W/Hz) and a bandwidth of several GHz in the unlicensed spectral range [3 GHz, 10 GHz]. In this way, they do not require a specific band to be reserved for their transmissions, as their low signal level is seen as wideband noise by other systems.

Example 6.4 F Consider a binary transmission system over an AWGN channel with unit gain as described in Section 5.3. Assume it employs antipodal waveforms with minimum bandwidth, $B = \frac{1}{2T}$. From (5.89) we can write the symbol error probability as

$$P_{\mathrm{bit}} = \mathrm{Q}\left(\sqrt{\Gamma}\right) \tag{6.110}$$

Thanks to the absence of ISI we can model the cascade of modulator/AWGN channel/demodulator as a memoryless BSC, with capacity in bit/symbol given by (6.81) where P_{bit} is given in (6.110).

In Fig. 6.16 we show the capacity in bit/symbol of the binary system as a function of Γ, together with the capacity in bit/symbol $C_s = TC$ of the inner analog AWGN channel, with C given by (6.106). It is understood that the capacity of the BSC can not be greater than the AWGN channel capacity, which then represents an upper bound. Moreover, the binary system capacity is also upper bounded by the capacity of the perfect binary channel, given by $\log_2 M = 1$.

Observe that for $\Gamma > 5\,\mathrm{dB}$, the distance between the two capacity curves becomes quite large, and hence a binary system (even with antipodal modulation and minimum bandwidth) is very inefficient in exploiting the capacity of the AWGN channel at high SNR. Indeed higher order modulation schemes are preferred for high SNR, as we discussed in Chapter 5.

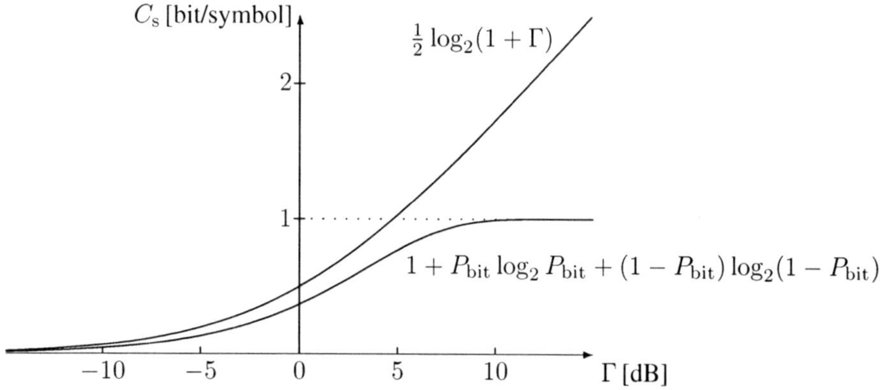

Figure 6.16 Capacity C_s in bit/symbol versus reference SNR Γ for the binary system of Example 6.4 F, compared with the corresponding capacity $B\log_2(1+\Gamma)$ of the bandlimited AWGN channel.

6.5 CODES THAT APPROACH CAPACITY

6.5.1 Soft decoding

A first step toward approaching capacity in the AWGN channel is to replace minimum Hamming distance decoding of the received binary sequence $\{\tilde{c}_\ell\}$ with joint ML detection and decoding from the actual projection vectors $\{\boldsymbol{r}_\ell\}$ introduced in Section 5.5. For block codes, soft decoding is thus based on the *Euclidean distance* between $[\boldsymbol{r}_1, \ldots, \boldsymbol{r}_h]$ and the received constellation symbols $[\tilde{\boldsymbol{s}}_{a_1}, \ldots, \tilde{\boldsymbol{s}}_{a_h}]$ that correspond to a word of n binary symbols, with $h = n/\log_2 M$ and M the constellation cardinality. Similarly, for convolutional codes, the Euclidean distance is considered between sequences of received constellation symbols and corresponding projection vectors.

The system performance in terms of symbol and word error probability increases, as the joint detection and decoding process makes also use of information that is lost in separate detection. More generally, soft decoding can be implemented by associating a continuous-valued *reliability measure* to each detected binary symbol based on the projection vectors, and then decoding to the information word or sequence that exhibits the highest total reliability.

Example 6.5 A Consider that a binary block code is used in a system with binary PAM transmission over an AWGN channel with unit gain. Then, each transmitted symbol coincides with one coded binary symbol, that is $a_\ell = 2c_\ell - 1$ and the corresponding receive sample is

$$r_\ell = \sqrt{E_h}(2c_\ell - 1) + w_\ell$$

We can then write the channel conditional PDF for received samples as

$$p_{r|c}(\rho|\gamma) = \frac{1}{\sqrt{\pi N_0}} e^{-[\rho - \sqrt{E_h}(2\gamma-1)]^2/N_0} \tag{6.111}$$

whereas for words, with $\boldsymbol{r} = [r_1, \ldots, r_n]$ we have

$$p_{\boldsymbol{r}|\boldsymbol{b}}(\boldsymbol{\rho}|\boldsymbol{\beta}) = \frac{1}{(\pi N_0)^{n/2}} e^{-\|\rho - \sqrt{E_h}[2\mu_c(\beta)-1]\|^2/N_0} \tag{6.112}$$

so that ML joint detecton and decoding can be written in terms of the Euclidean distance as

$$\mu_d'(\boldsymbol{\rho}) = \underset{\boldsymbol{\beta}\in\mathbb{F}^k}{\operatorname{argmin}} \|\boldsymbol{\rho} - \sqrt{E_h}[2\mu_c(\boldsymbol{\beta}) - 1]\| \tag{6.113}$$

Assume for example a $(7, 4)$ Hamming code, and let us pick the transmitted information word to be $\boldsymbol{b} = [1010]$. The corresponding codeword is $\boldsymbol{c} = [1010101]$, which yields the PAM symbol sequence $\boldsymbol{a} = [1, -1, 1, -1, 1, -1, 1]$ and the received constellation sequence $\boldsymbol{s} = [s_{c_1}, \ldots, s_{c_7}] = \sqrt{E_h}\,[1, -1, 1, -1, 1, -1, 1]$. Now, suppose the received sequence, at the demodulator output, is

$$\boldsymbol{r} = \boldsymbol{s} + \boldsymbol{w} = [r_1, \ldots, r_7] = \sqrt{E_h}\,[1, -1.1, 0.9, -0.8, -0.1, 0.2, 1.1]\ . \tag{6.114}$$

With symbol by symbol ML detection we have

$$\hat{\boldsymbol{a}} = [1, -1, 1, -1, -1, 1, 1] \quad , \quad \tilde{c} = [1010011]$$

and with minimum Hamming distance decoding

$$\hat{c} = [0010011] \quad , \quad \hat{\boldsymbol{b}} = [0010011]$$

since the incorrect codeword $\hat{c}$ is the closest to $\tilde{c}$ in terms of Hamming distance $d_H(\hat{c}, \tilde{c}) = 1$. On the contrary, with soft decoding, since the minimum Euclidean distance between symbol sequences corresponding to two distinct codewords is $d_{E,\min} = \sqrt{3 \cdot 4E_h} = 2\sqrt{3E_h}$, the ML decoding of $\boldsymbol{r}$ given in (6.114) yields the correct $\boldsymbol{s}$, since their Euclidean distance is $\|\boldsymbol{r} - \boldsymbol{s}\| = \sqrt{68E_h}/5 \simeq 1.64\sqrt{E_h} < d_{E,\min}/2$.

In some cases soft decoding might yield an incorrect decoding, but its average performance with equally likely information words in an AWGN channel is optimal, as can be expected from a ML criterion. In fact, for the system in this example the word error probability for hard decoding can be calculated as in (6.47), with $P_{\rm bit} = \mathrm{Q}\left(\sqrt{2E_h/N_0}\right)$, as given by (5.89). On the other hand, for the same probability in the soft decoding case we can derive an upper bound analogous to (5.108) and a lower bound analogous to (5.115), by observing that the Euclidean distance between symbol sequences corresponding to two distinct codewords $\boldsymbol{\gamma}, \boldsymbol{\gamma}'$ is

$$d_E\left((2\boldsymbol{\gamma} - 1)\sqrt{E_h}, (2\boldsymbol{\gamma}' - 1)\sqrt{E_h}\right) = \left\|2\sqrt{E_h}(\boldsymbol{\gamma} - \boldsymbol{\gamma}')\right\| = 2\sqrt{d_H(\boldsymbol{\gamma} - \boldsymbol{\gamma}')E_h} \tag{6.115}$$

and obtain

$$\mathrm{Q}\left(\sqrt{\frac{2E_h w_{\min}}{N_0}}\right) \leq \mathrm{P}\left[\hat{\boldsymbol{b}} \neq \boldsymbol{b}\right] \leq \sum_{\substack{\boldsymbol{\gamma}\in\mathcal{C}\\ \boldsymbol{\gamma}\neq\boldsymbol{0}}} \mathrm{Q}\left(\sqrt{\frac{2E_h\,\|\boldsymbol{\gamma}\|_H}{N_0}}\right) \tag{6.116}$$

Fig. 6.17 compares the performance of soft and hard decoding in terms of word and symbol error probabilities for the system described in this example, showing the gain in SNR for soft decoding.

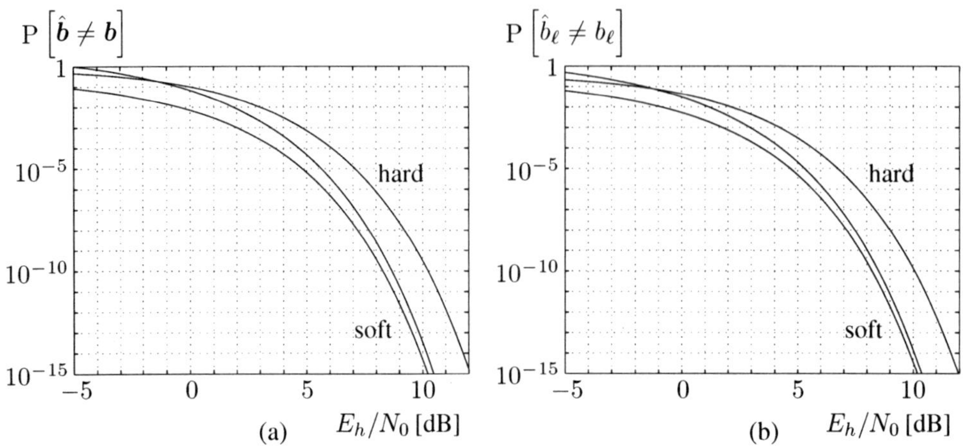

Figure 6.17 Performance comparison between soft and hard decoding for the system in Example 6.5 A that employs a $(7, 4)$ Hamming code and binary PAM over an AWGN channel. (a) word error probability; (b) symbol error probability. The two curves labeled "soft" refer to the lower and upper bounds in (6.116). Observe that they exhibit a gain of approximately 1.6 dB over hard decoding at higher SNR

6.5.2 Concatenated codes

As we have seen in Section 6.4, the key to improve code performance and hence approach the channel capacity, is to use very long codewords (in block codes) or large constraint lengths (in convolutional codes). However this would also increase the complexity of encoders and decoders, beyond practical feasibility.

A possible solution is illustrated in the following. Two encoders for short codes are concatenated with the insertion of a symbol interleaver that merely permutes the incoming symbol sequence. The maximum shift that a symbol undergoes in the interleaving process is called the *interleaver depth* $L_{\rm i}$ and it is a key parameter. In fact it creates correlation between distant symbols, thereby making the whole scheme somehow equivalent to having much longer codewords.

In the serial concatenation scheme, illustrated in Fig. 6.18, the output of the outer encoder is fed through interleaving as input to the inner encoder. The overall coder rate k/n is therefore the product of the rates for the constituent codes. At the decoder, the reverse process is carried out where the output of the inner decoder is deinterleaved (the inverse permutation with respect to interleaving) to build the input of the outer decoder.

However this scheme would not perform so well if the two decoding stages were kept separate. What allows to approach the performance of very long codes is the fact that soft information on the symbols that are dcoded by the outer decoder is fed back (through interleaving) to the inner decoder to improve its decoding, and so on. A few iterations are sufficient to achieve very good performance.

In the parallel concatenation scheme, shown in Fig. 6.19, the same information sequence is input to two distinct systematic encoders, except that it is interleaved before the second encoder. Then only the parity symbols from the second encoder are appended to the output of the first encoder.

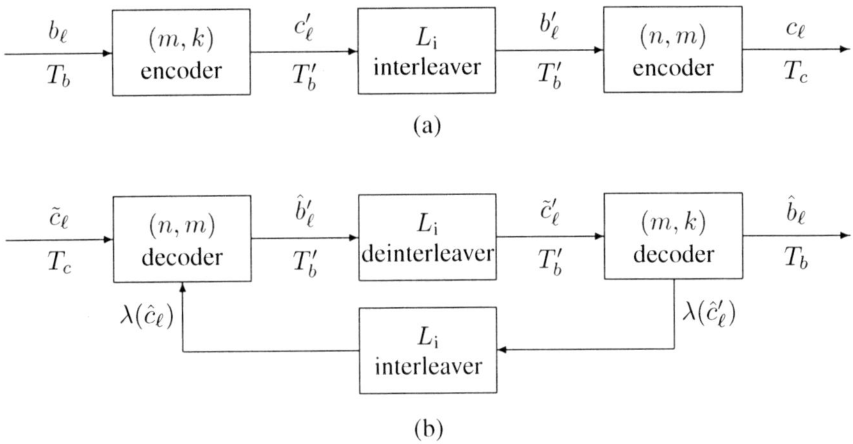

Figure 6.18 Block diagram of (a) the encoder and (b) the decoder for a pair of serially concatenated codes. The (n, m) code is called the *inner code*, while the (m, k) is the *outer code*.

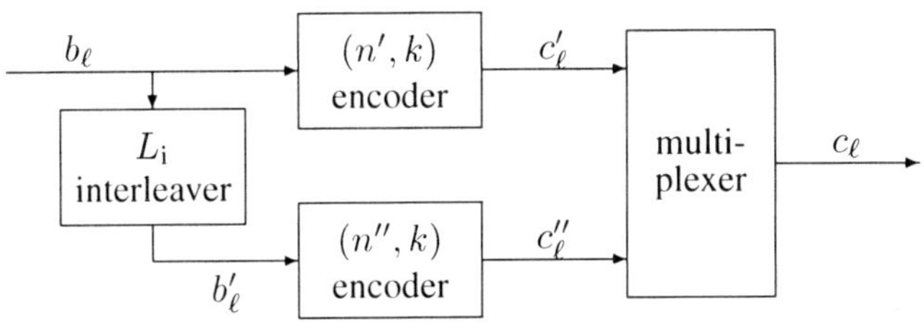

Figure 6.19 Block diagram of the encoder for a pair of parallel concatenated codes.

6.5.3 Low density parity check codes

Another way of allowing long codewords in block codes, while keeping the complexity of decoding low, is to construct codes with a sparse parity check matrix [9, 14], i.e. such that the number of nonzero elements in $\boldsymbol{H}$ is much lower than $n\ell$. In this way, the decoding procedure via syndrome calculation, or better its soft decoding counterpart, can be performed with fewer arithmetic operations.

Definition 6.9 *A* regular (n, k) *binary low density parity check (LDPC) code with column weight* λ *and row weight* ρ *is a* (n, k) *block code that has a* $\ell \times n$ *parity check matrix* $\boldsymbol{H}$ *in which the rows* $\boldsymbol{h}_i$ *and the columns* $\boldsymbol{h}'_j$ *satisfy*

$$\|\boldsymbol{h}_i\|_{\mathrm{H}} = \rho \ , \quad i = 1, \ldots, \ell \tag{6.117}$$

$$\left\|\boldsymbol{h}'_j\right\|_{\mathrm{H}} = \lambda \ , \quad j = 1, \ldots, n \tag{6.118}$$

Observe that since $\boldsymbol{H}$ is not guaranteed to have full row rank, the code dimension k may be higher than $n - \ell$. A few general properties of regular LDPC codes can nevertheless be deduced from the definition above, such as a lower bound on minimum distance.

Proposition 6.22 *In a regular binary LDPC code with column weight* λ*, the minimum distance satisfies*

$$d_{\min} \geq \lambda + 1$$

Example 6.5 B *Gilbert codes.* A regular Gilbert code is defined from parameters m and r with $m \geq r \geq 2$ as having the following parity check matrix

$$\boldsymbol{H} = \left[\begin{array}{c|c|c|c} \boldsymbol{I}_m & \boldsymbol{I}_m & \cdots & \boldsymbol{I}_m \\ \hline \boldsymbol{V}_m^0 & \boldsymbol{V}_m^1 & \cdots & \boldsymbol{V}_m^{r-1} \end{array}\right]$$

where $\boldsymbol{V}_m^q$ is the $m \times m$ matrix

$$\boldsymbol{V}_m^0 = \boldsymbol{I}_m \;, \quad \boldsymbol{V}_m^1 = \left[\begin{array}{c|c} \mathbf{0} & 1 \\ \hline \boldsymbol{I}_{m-1} & \mathbf{0} \end{array}\right] \;, \quad \ldots \;, \quad \boldsymbol{V}_m^q = \left[\begin{array}{c|c} \mathbf{0} & \boldsymbol{I}_q \\ \hline \boldsymbol{I}_{m-q} & \mathbf{0} \end{array}\right]$$

It is a regular binary LDPC code with $n = mr$, and weights $\rho = r$, $\lambda = 2$. Moreover, since its $2m$ rows sum to the all-zero row, while the first $2m - 1$ are linearly independent, we have $\text{rank}(\boldsymbol{H}) = 2m - 1$ and hence the code dimension is $k = (r-2)m + 1$.

It is easy to see that the minimum distance of a Gilbert code is $d_{\min} = 4$, since any 3 columns of $\boldsymbol{H}$ are linearly independent while 4 linearly dependent columns can be found (see Problem 6.14).

REFERENCES

1. S. Arimoto, "An algorithm for computing the capacity of arbitrary discrete memoryless channels," *IEEE Transactions on Information Theory*, vol. 18, 1972, pp. 14-20.
2. S. Benedetto, G. Montorsi, "Design of parallel concatenated convolutional codes," *IEEE Transactions on Communications*, vol. 44, n. 5, pp. 591 - 600, May 1996.
3. C. Berrou, A. Glavieux, "Near optimum error correcting coding and decoding: turbo-codes," *IEEE Transactions on Communications*, vol. 44, n. 10, pp. 1261-1271, October 1996.
4. R. E. Blahut, "Computation of channel capacity and rate-distortion functions," *IEEE Transactions on Information Theory*, vol. 18, 1972, pp. 460-473.
5. R. E. Blahut, *Theory and practice of error control codes*, Reading, MA: Addison-Wesley, 1983.
6. R. E. Blahut, *Principles and practice of information theory*, Reading, MA: Addison-Wesley, 1987.
7. T. M. Cover, J.A. Thomas, *Elements of information theory*, New York, NY: John Wiley & Sons, 1991.
8. G. D. Forney, Jr., "The Viterbi algorithm", *Proceedings of the IEEE*, vol. 61, pp. 268–278, March 1973.
9. R. G. Gallager, *Low-Density Parity-Check Codes*, Cambridge, MA: 1963.
10. R. G. Gallager, *Information theory and reliable communication*, New York, NY: John Wiley & Sons, 1968.

11. R. V. L. Hartley, "Transmission of information", *The Bell System Technical Journal*, vol. 7, pp. 535–, July 1928.

12. F. Jelinek, *Probabilistic information theory*, New York, NY: McGraw-Hill, 1968.

13. S. Lin, D. J. Costello Jr., *Error control coding*, Englewood Cliffs, NJ: Prentice-Hall, 1983.

14. D. MacKay, "Good error-correcting codes based on very sparse matrices", *IEEE Transactions on Information Theory*, vol. 45, pp. 399–431, March 1999.

15. C. E. Shannon, "A mathematical theory of communication", *The Bell System Technical Journal*, vol. 27, pp. 379–423, July 1948, pp. 623–656, October 1948.

16. C. E. Shannon, "Communication in the presence of noise", *Proceedings of the IRE*, vol. 37, pp. 10–21, 1949.

17. A. Viterbi, "Error bounds for convolutional codes and an asymptotically optimum decoding algorithm", *IEEE Transactions on Information Theory*, vol. 13, pp. 260–269, April 1967.

Problems

Principles of channel coding

6.1 Consider a block binary code with encoding map

$$\mu_c(00) = 0000 \quad , \quad \mu_c(01) = 0111 \quad , \quad \mu_c(10) = 1011 \quad , \quad \mu_c(11) = 1110$$

a) Find the maximum number of errors that can be detected in any received word

b) Find the information efficiency of the code, assuming the information symbols are the output of a memoryless source with PMD

$$p_b(0) = 1/4 \quad , \quad p_b(1) = 3/4$$

c) Calculate the word decoding error probability, conditioned on the transmitted word $\boldsymbol{b} = [00]$, for a memoryless BSC channel with $P_{\text{bit}} = 0.1$ and minimum distance decoding.

6.2 A ternary memoryless channel has input and output alphabet $\mathcal{A}_c = \mathcal{A}_{\tilde{c}} = \{1, 2, 3\}$, symbol period $T_c = 100$ ns and transition probabilities

$$p_{\tilde{c}|c}(\xi|\gamma) = \begin{cases} 1/\gamma & , \ \xi \leq \gamma \\ 0 & , \ \xi > \gamma \end{cases}$$

a) Find the probability $\mathrm{P}\left[\tilde{\boldsymbol{c}} \neq \boldsymbol{c}\right]$ for 3-symbol words when the input message $\{c_\ell\}$ has iid symbols with equally likely values.

b) Show the MAP decoding region for the codeword $[222]$ in the code $\mathcal{C} = \{111, 222, 333\}$ with equally likely codewords.

Linear block codes

6.3 Consider the binary block code $\mathcal{C} = \{000000, 111100, 101011, 010111\}$

a) Show that it is a linear code and find the information word length.

b) Find its generating and parity check matrices in systematic form.

c) How many errors can this code detect? How many errors can it correct with ML decoding over a memoryless symmetric channel? Decode the received word $\tilde{\boldsymbol{c}} = [011110]$.

6.4 Consider the following binary block code

$$\mathcal{C} = \{0000, 1010, 0110, 1100\}$$

a) State whether it is a linear code.

b) Find how many errors it can detect.

c) Find a parity check matrix for it.

d) Find the redundancy of the codewords when the input information words have the following probability mass distribution

$$p_b(00) = 1/2, p_b(11) = 1/4, p_b(10) = 1/8, p_b(01) = 1/8 \ .$$

6.5 A linear binary block code has a parity check matrix

$$H = \begin{bmatrix} 1 & 1 & 1 & 0 & 0 & 0 & 0 \\ 1 & 1 & 0 & 1 & 0 & 0 & 0 \\ 1 & 1 & 0 & 0 & 1 & 0 & 0 \\ 1 & 0 & 0 & 0 & 0 & 1 & 0 \\ 0 & 1 & 0 & 0 & 0 & 0 & 1 \end{bmatrix}$$

a) How many errors can it correct with minimum Hamming distance decoding?

b) If the binary channel is memoryless but not symmetric, with transition probabilities $p(1|0) = 1/10$ and $p(0|1) = 1/100$, what is the ML decoding of the received word $\tilde{\mathbf{c}} = [1011000]$?

6.6 In the following system

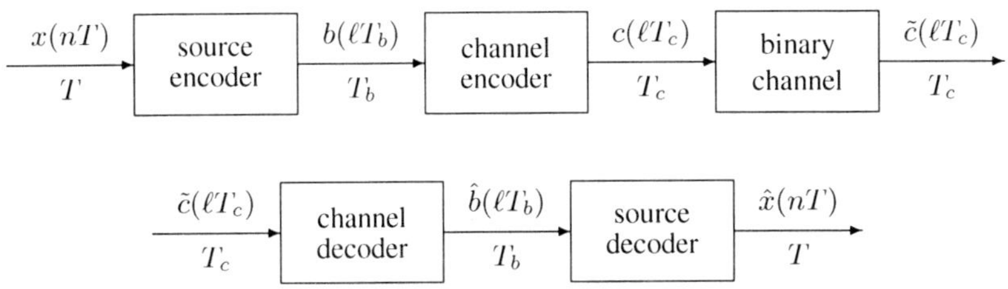

- the digital message $\{x(nT)\}$ is made of iid symbols with period $T = 1\,\mu$s, alphabet $\mathcal{A}_x = \{0, 1, 2, \ldots, 9\}$ and PMD

$$p_x(0) = p_x(1) = 1/4 \quad , \quad p_x(2) = \ldots = p_x(9) = 1/16$$

- the source coding scheme uses a binary prefix code with efficiency $\eta = 1$.
- the channel coding uses a (3,2) linear systematic code with generating matrix

$$G = \begin{bmatrix} & \boldsymbol{I}_2 & \\ 1 & & 1 \end{bmatrix}$$

- the binary channel is memoryless and symmetric with $P_{\text{bit}} = 1/10$

a) Write the source coding map $\mu_{\text{s}}(\cdot)$ and the channel decoding map $\mu_{\text{d}}(\cdot)$

b) Find the bit period T_c required for the binary channel

c) Find the probability that $\text{P}\left[\hat{x}(nT) = 1 | x(nT) = 0\right]$

6.7 A (n, k) Hamming code $\mathcal{C}$ can be *expanded* by appending one parity symbol at the end of each codeword so that it has even Hamming weight. Write the rule to determine c_{n+1} from $c_1, \ldots, c_n$.

a) Show that the expanded $(n+1, k)$ code $\mathcal{C}'$ is linear.

b) Write generating and parity check matrices for $\mathcal{C}'$ starting from $\mathcal{C}$ being the $(7, 4)$ Hamming code.

c) Find the minimum distance of $\mathcal{C}'$. How many errors can $\mathcal{C}'$ detect?

Channel capacity

6.8 Consider the ternary channel and input signal described in Problem 6.2. Find the information rate through the channel.

6.9 A digital message $x(nT)$ with $T = 1\ \mu$s, has iid symbols with PMD

$$p_x(1) = p_x(-1) = \frac{1}{3} \quad , \quad p_x(2) = p_x(-2) = p_x(0) = \frac{1}{9}$$

a) Find the message efficiency.

b) Find a source coding scheme in which the output message y_n has a ternary alphabet and unit efficiency. What is the symbol period for y_n?

c) Is it possible to find a channel coding scheme for y_n that will allow transmission over a BSC with symbol period $T_b = 400$ ns and bit error probability $P_{\text{bit}} = 10^{-2}$, with an overall error probability $\text{P}[\hat{y}_n \neq y_n] < 10^{-7}$?

6.10 In a binary modulation system without intersymbol interference the waveform space has dimension $I = 1$ and waveforms are represented by the constellation points $s_0 = 0$ e $s_1 = 1$. Assume that the noise component at the decision point has exponential PDF

$$p_w(b) = \begin{cases} 2e^{-2b} & , \quad b \geq 0 \\ 0 & , \quad b < 0 \end{cases}$$

and the input message $\{a_k\}$, $a_k \in \{0, 1\}$, is composed of statistically independent symbols with $p_a(0) = 1/4$.

a) Find the error probability of the system with the threshold $v = \ln 2$.

b) Find the threshold v_{opt} that minimizes the error probability.

c) Find the mutual information between a_k and $\hat{a}_k$ with threshold v.

d) Find the capacity (in bit/symbol) of the binary channel with threshold v.

6.11 Consider a QPSK digital modulation system over an AWGN channel as a quaternary channel with a_n the transmitted symbols and $\hat{a}_n$ the detected symbol, as described in Chapter 5.

a) Assuming that the transmitted symbols are iid with equally likely values, express the entropies $H(\hat{a})$, $H(\hat{a}|a)$, $H(a|\hat{a})$ and the mutual information $I(a, \hat{a})$, as functions of the ratio E_s/N_0. Find their values for $(E_s/N_0)_{\text{dB}} = 0$ dB, 10 dB, 20 dB.

b) Calculate the capacity of the quaternary channel when the symbol period is $T = 1$ ms, and the transmit pulse is $h_{\text{Tx}}(t) = A\,\text{sinc}(t/T)$, with amplitude $A = 2$ mV, while the noise PSD is $N_0/2 = 10^{-10}\ \text{V}^2/\text{Hz}$.

6.12 In orthogonal frequency division multiplexing (OFDM) communication systems such as those used for digital subscriber lines (ADSL, HDSL, VDSL), the channel can be thought of as the parallel of N independent *subchannels*. Each subchannel m is AWGN with its own bandwidth B_i, power gain $\mathtt{g}_i$ and noise PSD $N_i/2$, and has a capacity C_i that depends on its transmit power $\mathtt{M}_i$. A possible aim in designing such systems is to distribute the total transmit power $\mathtt{M}_{\text{Tx}}$ among the subchannels so that the total channel capacity is maximized, that is

$$\max_{\{\mathtt{M}_i \geq 0\}} \sum_{i=1}^{N} C_i \quad , \quad \text{under the constraint} \quad \sum_{i=1}^{N} \mathtt{M}_i \leq \mathtt{M}_{\text{Tx}}$$

a) For $N = 2$ subchannels find the optimal values of $\mathtt{M}_1$ and $\mathtt{M}_2$ as a function of the parameters $\{B_i, \mathtt{g}_i, N_i\}$ and total power $\mathtt{M}_{\text{Tx}}$. Plot the capacity C as a function of $\mathtt{M}_1$ for $B_1 = 4$ MHz, $B_2 = 1$ MHz, $\mathtt{g}_1 = -40$ dB, $\mathtt{g}_2 = -45$ dB, $N_1 = N_2 = 10^{-10}\ \text{V}^2/\text{Hz}$, $\mathtt{M}_{\text{Tx}} = 1\ \text{V}^2$. Repeat the plot in the case the gains are exchanged, $\mathtt{g}_1 = -45$ dB, $\mathtt{g}_2 = -40$ dB.

b) For general N, prove that if all subchannels have the same bandwidth $B_i = B$ and

$$\frac{N_i}{\mathtt{g}_i} \le \mathcal{P}_0 \ , \quad \text{with } \mathcal{P}_0 = \frac{1}{N}\left(\frac{\mathtt{M}_{\mathrm{Tx}}}{B} + \sum_{i=1}^{N}\frac{N_i}{\mathtt{g}_i}\right)$$

the optimal solution is

$$\mathtt{M}_i = B\left(\mathcal{P}_0 - \frac{N_i}{\mathtt{g}_i}\right)$$

Codes that approach capacity

6.13 Consider the following binary transmission system including channel coding and digital modulation over an AWGN channel

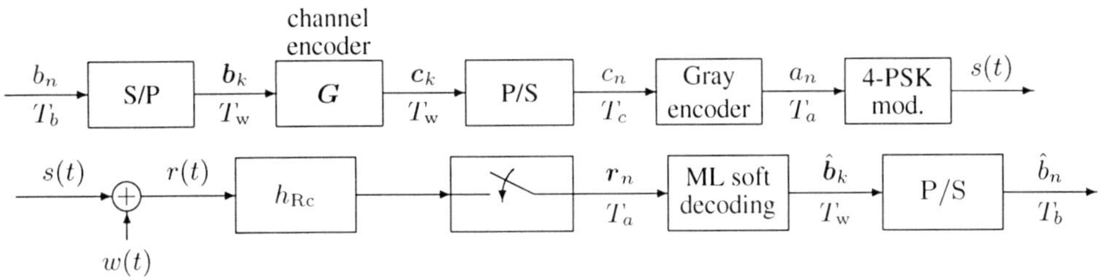

Given that

- the 4-PSK modulator has the waveform

$$h_{\mathrm{tx}}(t) = A\,\mathrm{rect}(2t/T_a)$$

 with $A = 10\,\mathrm{mV}$ and $T_a = 1\,\mu\mathrm{s}$
- $w(t)$ is a WGN with PSD $N_0/2 = 2 \cdot 10^{-12}\ \mathrm{V^2/Hz}$
- the channel coding is a linear block code with generating matrix

$$\boldsymbol{G} = \begin{bmatrix} 1 & 0 & 0 \\ 0 & 1 & 0 \\ 0 & 0 & 1 \\ 1 & 1 & 0 \\ 1 & 0 & 1 \\ 0 & 1 & 1 \end{bmatrix}$$

a) Find the bit rate R_b of the system, and the bit error probability P_{bit} for the digital modulation system (from c_n to $\tilde{c}_n$).

b) Show that the minimum distance decoding of $\tilde{\boldsymbol{c}}_k = [100100]$ is $\hat{\boldsymbol{b}}_k = [100]$.

c) Find minimum Hamming distance of the code, and the minimum Euclidean distance between the sequences of constellation symbols corresponding to two distinct codewords.

d) Decode the received sequence

$$\boldsymbol{r}(0) = (0.02\sqrt{E_h}, -0.35\sqrt{E_h})$$
$$\boldsymbol{r}(T_a) = (0.47\sqrt{E_h}, -0.02\sqrt{E_h})$$
$$\boldsymbol{r}(2T_a) = (-0.51\sqrt{E_h}, -0.59\sqrt{E_h}$$

to the corresponding detected information sequence $\left\{\hat{b}_\ell\right\}$.

e) Compare the above result to that obtained with separate ML detection and hard decoding, as illustrated below

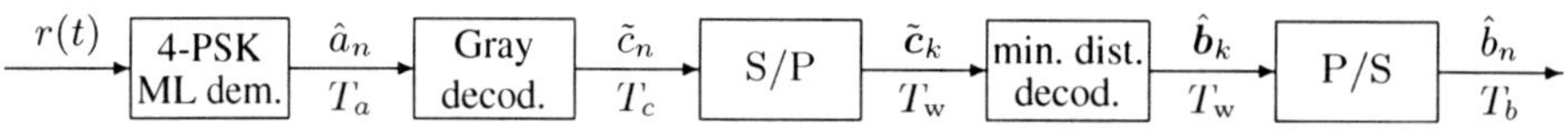

6.14 Show that in the low density parity check matrix of any Gilbert code with $m \geq 2, \ell \geq 3$ there are 4 linearly dependent columns, and hence the code minimum distance is 4.

Hint: Observe that if we write the column index as $j = hm + k$, with $h = 0, \ldots, \ell - 1$, and $k = 1, \ldots, m$, the two ones of the jth column lie in rows k and $m + [(k + h) \bmod m]$

CHAPTER 7

MARKOV CHAINS THEORY

Andrea Zanella

This chapter introduces some basic mathematical tools that are widely used in the performance analysis of communication systems. Generally, an accurate performance evaluation of modern communications systems requires sophisticated computer simulations or extensive and expansive testbeds. However, a suitable mathematical model of the system often makes it possible to investigate in a systematic way the cause-effect relations that govern the system behavior, thus shedding light on the most significant tradeoffs between different performance indices. Furthermore, a mathematical model makes possible to investigate the system behavior in limiting scenarios, such as in the presence of traffic overload or breakdown of parts of the system, that could not be observed in reality nor reproduced in simulations for different reasons (economic, service-continuity, complexity, and so on). The comprehension of the fundamental mathematical tools for modeling and analyzing a telecommunication system is, hence, essential for a system designer. To this purpose, this chapter describes the elementary theory of both discrete-time and continuous-time *Markov chains* and of *birth-death processes*, which are a special case of Markov chains that play a pivotal role in the queueing theory, presented in Chapter 8. It might be worth remarking that this chapter is not intended to offer a complete and detailed coverage of these subjects, which would require an entire book. Rather, the aim is to provide the principles and the fundamental results that will be used in the analysis developed in later chapters.

Principles of Communications Networks and Systems N. Benvenuto, M. Zorzi

7.1 INTRODUCTION

The purpose of this chapter is to provide the theoretical background required to develop the queueing theory in Chapter 8 and the performance analysis of network protocols in Chapter 9. More specifically, this chapter focuses on *discrete-time Markov chains, continuous-time Markov chains* and *birth-death processes.*

A Markov chain (MC) is a stochastic process characterized by the *Markov* property, which makes it possible to separate the statistical description of the future evolution of the process from its past, provided that the present is known. This concept, which will be mathematically formalized at the beginning of the next section, is the starting point upon which most of the theory presented in this and the next chapter is based.

A MC can either be discrete-time or continuous-time. Following the classical approach [2, 3, 4], we will first develop the theory of discrete-time MCs, which are particularly suitable for modeling a number of communication systems and protocols that operate at *discrete intervals of time.* Examples of applications of discrete-time MC are discussed in Chapter 9, where the theory is largely used to determine and compare the performance of different medium-access control protocols and packet-retransmission schemes. The discussion of discrete-time MCs paves the way for the development of the theory of continuous-time MCs, in Section 7.3, which is slightly more intricate than the discrete counterpart. The study of continuous-time MCs leads directly into Section 7.4, where we discuss birth-death processes, which provide the building blocks of queueing theory, developed in Chapter 8.

7.2 DISCRETE-TIME MARKOV CHAINS

The first study on MCs dates back to 1907, when the Russian mathematician Andrei Andreyevich Markov defined and investigated the properties of this specific type of random process [1]. Since then, the literature on the subject has kept growing in size and depth, producing a number of interesting results that find application in several different fields. In particular, a branch of MC theory has originated the so-called queueing theory, which will be presented in Chapter 8.

7.2.1 Definition of discrete-time MC

Let $X(nT)$, with $n \in \{0, 1, 2, \ldots\}$, be a discrete-time random process (rp) that takes values in a countable *state space* $\mathcal{S}$. For ease of notation, in the following we will use X_n in place of $X(nT)$ to denote the random process at the nth time instant. Furthermore, unless otherwise stated, the state space $\mathcal{S}$ will be hereinafter labeled by the nonnegative integers, that is to say, $\mathcal{S} = \{0, 1, 2, \ldots\}$. Hence, X_n is also discrete valued.

Definition 7.1 *A discrete-time rp* $\{X_n, n \geq 0\}$ *is a MC if it possesses the* Markov property, *i.e., such that for any integer* k *and for any ordered sequence of time instants* $n_0 < n_1 < \ldots < n_k$ *and for any set* $\{\sigma_0, \sigma_1, \ldots, \sigma_k\}$, *with* $\sigma_i \in \mathcal{S}$, *it holds*

$$\mathrm{P}\left[X_{n_k} = \sigma_k | X_{n_{k-1}} = \sigma_{k-1}, \ldots, X_{n_0} = \sigma_0\right] = \mathrm{P}\left[X_{n_k} = \sigma_k | X_{n_{k-1}} = \sigma_{k-1}\right] . \quad (7.1)$$

The Markov property lends itself to a fascinating interpretation. Consider the instant n_{k-1} as the "present" (or current) time, so that the previous instants $n_0, n_1, \ldots, n_{k-2}$ represent the past and the successive instant n_k is the "future". Then, (7.1) states that the knowledge of the past history of the process does not help refining the prediction of the future evolution of the process from the current state σ_{k-1}. In other words, the way in which the previous history of the process affects its future evolution is completely summarized by the present state of the process. The Markov property, then, makes it possible to *forget* the past history of the process, given that the present state is known: the statistical evolution of the process from a given state is independent of the *path* followed by the process to reach that state.

Further insight Definition 7.1 refers to a generic ordered sequence of time instants $n_0 < n_1 < \ldots < n_k$. However, an equivalent definition can be given by considering a sequence of *consecutive* time instants, $n_0, n_0 + 1, \ldots, n_0 + k$, for which the Markov property can be expressed as follows:

$$\mathrm{P}\left[X_{n_0+k} = \sigma_k | X_{n_0+k-1} = \sigma_{k-1}, \ldots, X_{n_0} = \sigma_0\right] = \mathrm{P}\left[X_{n_0+k} = \sigma_k | X_{n_0+k-1} = \sigma_{k-1}\right] \tag{7.2}$$

where $\sigma_0, \ldots, \sigma_k \in \mathcal{S}$. Although this second formulation of the Markov property may appear to be less restrictive than the previous one, for (7.2) can actually be considered as a particular case of (7.1), it is possible to prove that they are completely equivalent (see Problem 7.9). We observe that, according to (7.2), a discrete-time discrete-state process is a MC when the probability that the rp assumes a certain state at the next instant given the state at the present instant does not depend upon states assumed at previous instants.

Many MCs that we shall encounter actually have infinite state space, $\mathcal{S} = \{0, 1, 2 \ldots\}$. For simplicity, however, we develop the theory for MCs with finite state space $\mathcal{S} = \{0, 1, 2, \ldots, N\}$. Nonetheless, most of the concepts and results discussed in the following also apply to MCs with infinite states. As a matter of fact, it is possible to develop an almost-parallel theory for processes with infinite state space ($N = \infty$) that, however, would require an extra effort to formally justify some mathematical operations, which are instead straightforward when dealing with finite state spaces.

Example 7.2 A Let us consider a dice game. X_n denotes the number on the face up of the dice at the nth roll. X_n is a discrete-time rp, with state space $\mathcal{S} = \{1, 2, \ldots, 6\}$. Assuming that successive rolls give independent outcomes, the rp X_n is a sequence of independent and identically distributed (iid) random variables. Hence, the conditional probability $\mathrm{P}\left[X_n = \sigma_n | X_{n-1} = \sigma_{n-1}, \ldots, X_0 = \sigma_0\right]$ is equal to $\mathrm{P}\left[X_n = \sigma_n\right]$ and it does not depend on the past outcomes (not even on the present outcome, in this specific case). Therefore, the rp exhibits the Markov property.

Remark The property of statistical independence of X_n is *stronger* than the Markovian property, in the sense that a rp that assumes independent states at different time instants exhibits the Markov property, whereas the states assumed by a MC at different time instants are not necessarily independent.

Example 7.2 B Consider the counting rp $Y_n = \sum_{j=0}^{n} X_j$, with $\{X_j\}$ iid (e.g., see Example 7.2 A). This process does not evolve with independent states, as it can be immediately realized by rewriting Y_n as $Y_n = Y_{n-1} + X_n$ from which it clearly appears that states assumed by the rp in successive time instants are correlated. Indeed, it can be proved that Y_n is a MC. In fact, for any instant n and for any set $\{\sigma_0, \sigma_1, \ldots, \sigma_n\}$, with $\sigma_j \in \mathcal{S}$, we have

$$\begin{aligned} \mathrm{P}\left[Y_n = \sigma_n | Y_{n-1} = \sigma_{n-1}, \ldots, Y_0 = \sigma_0\right] &= \mathrm{P}\left[Y_{n-1} + X_n = \sigma_n | Y_{n-1} = \sigma_{n-1}, \ldots, Y_0 = \sigma_0\right] \\ &= \mathrm{P}\left[X_n = \sigma_n - \sigma_{n-1} | Y_{n-1} = \sigma_{n-1}, \ldots, Y_0 = \sigma_0\right] = \mathrm{P}\left[X_n = \sigma_n - \sigma_{n-1}\right] \end{aligned} \tag{7.3}$$

where the last step follows from the independence of X_n and any Y_j with $j < n$. Since the right-most probability in (7.3) depends only upon σ_{n-1} and σ_n, we can conclude that, given the state at step $n-1$, e.g., $Y_{n-1} = \sigma_{n-1}$, the rp state before step $n-1$ is irrelevant to determine the probability of $Y_n = \sigma_n$. In other words, the rp exhibits the Markov property and, hence, it is a MC.

Example 7.2 C Node A transmits packets to node B through a noisy channel that can introduce errors in the received packet. Each packet is transmitted and retransmitted until correctly received by node B. Let X_n be a binary rv indicating the success or failure of a transmission attempt, that is to say

$$X_n = \chi\left\{\text{the } n\text{th transmission attempt is successful}\right\} ,$$

where $\chi\{\cdot\}$ is the indicator function defined in (2.163). Assume that each transmission attempt has a probability p of success and $1-p$ of failure, independently of the other attempts. Then, X_n is a rp with independent states and, as such, it possesses the Markov property (see Example 7.2 A). Furthermore, the process Y_n that counts the number of packets successfully delivered to node B after n transmissions by node A is also a MC, as it can be easily realized by following the same rationale of Example 7.2 B.

Example 7.2 D Let X_0 denote the number of students enrolled in the Telecommunications Systems course in a given year. Students keep taking the exam at every session, until they succeed. The probability that a student passes the nth exam session is p_n. Denoting by X_n the number of students that have not passed the test after n exam sessions, we can write the following equation

$$X_{n+1} = X_n - y_{n+1}(X_n)$$

where $y_{n+1}(X_n)$ denotes the number of students that pass the $(n+1)$th test, given that X_n students took the test. Note, that X_n is a rv. The conditional probability mass distribution (PMD) of $y_{n+1}(X_n)$, given that $X_n = i$, is binomial of index i and parameter p_n, namely

$$\mathrm{P}\left[y_{n+1}(X_n) = j | X_n = i\right] = \binom{i}{j} p_n^j (1-p_n)^{i-j} , \qquad j = 0, 1, \ldots, i .$$

Hence, we have

$$\begin{aligned} \mathrm{P}\left[X_{n+1} = j | X_n = i\right] &= \mathrm{P}\left[X_n - y_{n+1}(X_n) = j | X_n = i\right] \\ &= \mathrm{P}\left[y_{n+1}(X_n) = i - j | X_n = i\right] = \binom{i}{i-j} p_n^{i-j} (1-p_n)^j . \end{aligned} \tag{7.4}$$

Since the conditional probability of X_{n+1} given X_n does not depend on any past values of the process, then X_n exhibits the Markov property in the form (7.2) and, consequently, it is a MC.

Example 7.2 E Node A transmits a data packet every T seconds to node B, through a wireless link. In general, the signal to noise ratio at the receiver is extremely good and the probability of error is practically negligible. After a while, however, a fire alarm system starts broadcasting a radio signal at regular intervals, producing destructive interference spikes every $4T$ seconds. As a consequence, one packet every four is not correctly received by B. Let $X_n = 1$ if the nth packet is successfully received and zero otherwise. It is easy to realize that the process X_n does not satisfy the Markov condition (7.1). To prove this statement, it suffices to find a case that violates the Markov property. Let us divide the sequence of transmissions in blocks of four, in such a way that the last transmission of each block is lost because of interference from the fire alarm. Now, observing that $X_{n-1} = 1$ we only know that the $(n-1)$th packet is correctly received, so that it cannot be the last of such a block. The next transmission can either be the second, the third or the forth of that block, and only in the latter case, the transmission would be unsuccessful. Then, the probability that $X_n = 1$, given that $X_{n-1} = 1$, is equal to

$$\mathrm{P}[X_n = 1 | X_{n-1} = 1] = \frac{2}{3}.$$

Now suppose that we have some knowledge of the past history of the process as, for example, that $X_{n-2} = 0$. This information tells us that the $(n-2)$th packet was the last of a block. Therefore, we know that the successive three transmissions will be successful, thus we can conclude that

$$\mathrm{P}[X_n = 1 | X_{n-1} = 1, X_{n-2} = 0] = 1.$$

Since the knowledge of the past of the process makes it possible to refine the statistic of the future of the process, we conclude that X_n does not possess the Markov property and, hence, it is not a MC.

7.2.2 Transition probabilities of discrete-time MC

In MC theory it is customary to refer to the discrete time instants as "steps". Accordingly, the MC is said to be in state $j \in \mathcal{S}$ at the nth step when $X_n = j$. Note that, the term *step* is also used to indicate the passing of a time unit, so that we can say that the MC moves from state i to state j in k steps when $X_n = i$ and $X_{n+k} = j$. The conditional probabilities $\mathrm{P}[X_{n+k} = j | X_n = i]$ are called *state transition probabilities in k steps* of the MC. When the transition probabilities depend only on the time lag k and not on the reference step n, they are said to be *stationary* and the related MC is said to be *homogeneous*. The transition probability from state i to state j in k steps of a homogeneous MC is denoted by

$$P_{i,j}(k) = \mathrm{P}[X_{n+k} = j | X_n = i] \tag{7.5}$$

In the sequel we confine ourselves to homogeneous MCs only.

Example 7.2 F Let us further elaborate on Example 7.2 D. If the probability p_n for a student to pass the exam decreases at every session, for example $p_n = 1/(n+1)$, then the transition probabilities $\mathrm{P}[X_{n+k} = j | X_n = i]$ given by (7.4) actually depend on the reference time instant n and not only on the time difference k. In this case the MC is *non-homogeneous*. Conversely, if the probability p_n does not depend on n, e.g., $p_n = 1/3$, then the transition probabilities become stationary. In this case, the number of students that pass an exam session only depends on the number of students that take that exam and not on the number of sessions that have been already offered. Accordingly, the MC is homogeneous.

A very convenient representation of the transition probabilities in k steps is provided by the so-called *transition matrix in k steps* $\boldsymbol{P}(k)$, which is a square matrix of size $(N+1) \times (N+1)$, having in the ith row and jth column the transition probability from state i to state j in k steps:

$$\boldsymbol{P}(k) = \begin{bmatrix} P_{0,0}(k) & P_{0,1}(k) & \dots & P_{0,j}(k) & \dots & P_{0,N-1}(k) & P_{0,N}(k) \\ P_{1,0}(k) & P_{1,1}(k) & \dots & P_{1,j}(k) & \dots & P_{1,N-1}(k) & P_{1,N}(k) \\ \vdots & \vdots & \vdots & \vdots & \vdots & \vdots & \vdots \\ P_{i,0}(k) & P_{i,1}(k) & \dots & P_{i,j}(k) & \dots & P_{i,N-1}(k) & P_{i,N}(k) \\ \vdots & \vdots & \vdots & \vdots & \vdots & \vdots & \vdots \\ P_{N,0}(k) & P_{N,1}(k) & \dots & P_{N,j}(k) & \dots & P_{N,N-1}(k) & P_{N,N}(k) \end{bmatrix} . \tag{7.6}$$

By definition,

$$\boldsymbol{P}(0) = \boldsymbol{I}, \tag{7.7}$$

where $\boldsymbol{I}$ is the identity matrix. An interpretation of (7.7) is that in zero time the process remains in its current state with probability one.

One step transition probabilities. The transition probabilities in one step $P_{i,j}(1)$ hold a special role in MC theory and, for this reason, they are simply referred to as *transition probabilities* and denoted by

$$P_{i,j} = \mathrm{P}\left[X_{n+1} = j | X_n = i\right] \tag{7.8}$$

thus omitting the time different that is implicitly assumed equal to 1. Analogously, the transition matrix in one step is simply said *transition matrix* and denoted by

$$\boldsymbol{P} = \begin{bmatrix} P_{0,0} & P_{0,1} & \dots & P_{0,j} & \dots & P_{0,N-1} & P_{0,N} \\ P_{1,0} & P_{1,1} & \dots & P_{1,j} & \dots & P_{1,N-1} & P_{1,N} \\ \vdots & \vdots & \vdots & \vdots & \vdots & \vdots & \vdots \\ P_{i,0} & P_{i,1} & \dots & P_{i,j} & \dots & P_{i,N-1} & P_{i,N} \\ \vdots & \vdots & \vdots & \vdots & \vdots & \vdots & \vdots \\ P_{N,0} & P_{N,1} & \dots & P_{N,j} & \dots & P_{N,N-1} & P_{N,N} \end{bmatrix} \begin{matrix} \\ \\ \\ \leftarrow [\boldsymbol{P}]_{(i,:)} \\ \\ \\ \end{matrix}$$

$$[\boldsymbol{P}]_{(i,j)} \qquad [\boldsymbol{P}]_{(:,j)} \tag{7.9}$$

In (7.9) we have also exemplified the use of a shorthand notation for referencing to single elements, rows and columns of a generic matrix $\boldsymbol{A}$.[1]

We note that the ith row of the transition matrix $\boldsymbol{P}$ collects all the possible transition probabilities from state i to any other state $j \in \mathcal{S}$. Adding up all these terms we hence obtain

$$\sum_{j=0}^{N} P_{i,j} = \sum_{j=0}^{N} \mathrm{P}\left[X_{n+1} = j | X_n = i\right] = \mathrm{P}\left[X_{n+1} \in \mathcal{S} | X_n = i\right] .$$

[1]The notation $[\boldsymbol{A}]_{(i,j)}$ refers to the element in the ith row and jth column of $\boldsymbol{A}$, whereas $[\boldsymbol{A}]_{(i,:)}$ and $[\boldsymbol{A}]_{(:,j)}$ denote the ith row and jth column vectors of the matrix $\boldsymbol{A}$, respectively.

Since the event $\{X_n \in \mathcal{S}\}$ has probability one for any $n \geq 0$, we conclude that the elements of each row of $\boldsymbol{P}$ add up to one, i.e.,

$$\sum_{j=0}^{N} P_{i,j} = 1 \,. \tag{7.10}$$

Denoting by $\mathbf{1} = [1, 1, \ldots, 1]^{\mathrm{T}}$ the column vector with all unit elements, (7.10) can be written as

$$[\boldsymbol{P}]_{(i,:)}\, \mathbf{1} = 1$$

and, considering that (7.10) applies to any row of the transition matrix $\boldsymbol{P}$, we get

$$\boldsymbol{P}\, \mathbf{1} = \mathbf{1} \,. \tag{7.11}$$

This result holds also for any transition matrix in k steps $\boldsymbol{P}(k)$, i.e., we have

$$\boldsymbol{P}(k)\, \mathbf{1} = \mathbf{1} \,, \qquad k = 0, 1, 2, \ldots \,. \tag{7.12}$$

A matrix with this property, i.e., with non-negative elements that add up to one on each row, is called *stochastic matrix*.

7.2.3 Sojourn times of discrete-time MC

The *sojourn time* s_j in a state j is defined as the number of consecutive steps that a rp spends in state j before moving to another state. At every step, the probability that the rp makes a self-transition from state j to itself is $P_{j,j}$, whereas the probability of leaving state j is $1 - P_{j,j}$, irrespective of the time already spent in state j because of the Markov property. Therefore, the number of steps that the process spends in state j before moving to any other state is a random variable (rv) with geometric distribution with parameter $1 - P_{j,j}$. The following proposition formalizes this intuitive result.

Proposition 7.1 *The sojourn times s_j of a discrete-time MC in state $j \in \mathcal{S}$ are statistically iid rvs with geometric distribution with parameter $1 - P_{j,j}$, i.e., with probability mass distribution (PMD) given by*

$$p_{s_j}(k) = (1 - P_{j,j})\, P_{j,j}^{k-1} \,, \tag{7.13}$$

and mean

$$\mathrm{m}_{s_j} = \frac{1}{1 - P_{j,j}} \,. \tag{7.14}$$

Proof The sojourn time s_j accounts for the number of *consecutive* time instants that the MC spends in state j: it begins when the MC steps into j from another state and ends when the MC leaves state j for a different state. Therefore, for any integer k, we have $s_j > k$ if, once entered in state j from another state, the MC remains for *at least* k consecutive steps in state j. The probability of such an event is given by

$$\mathrm{P}\left[s_j > k\right] = \mathrm{P}\left[X_{n+k} = j, X_{n+k-1} = j, \ldots, X_{n+1} = j | X_n = j, X_{n-1} \neq j\right] \tag{7.15}$$

where the condition $\{X_n = j, X_{n-1} \neq j\}$ indicates that state j was entered exactly at the nth step. Due to the Markov property, however, the condition $X_{n-1} \neq j$ is irrelevant and can be neglected, so that (7.15) becomes

$$\mathrm{P}[s_j > k] = \mathrm{P}[X_{n+k} = j, X_{n+k-1} = j, \ldots, X_{n+1} = j | X_n = j] \,.$$

Applying properties of the conditional probability, we obtain

$$\begin{aligned}\mathrm{P}[s_j > k] = &\mathrm{P}[X_{n+k} = j | X_{n+k-1} = j, \ldots, X_{n+1} = j, X_n = j] \\ &\mathrm{P}[X_{n+k-1} = j, X_{n+k-2} = j, \ldots, X_{n+1} = j | X_n = j]\end{aligned}$$

and, invoking again the Markov property on the first conditional probability, we get

$$\begin{aligned}\mathrm{P}[s_j > k] &= \mathrm{P}[X_{n+k} = j | X_{n+k-1} = j] \\ &\quad \mathrm{P}[X_{n+k-1} = j, X_{n+k-2} = j, \ldots, X_{n+1} = j, X_n = j] \\ &= P_{j,j}\,\mathrm{P}[X_{n+k-1} = j, X_{n+k-2} = j, \ldots, X_{n+1} = j, X_n = j] \,.\end{aligned}$$

Repeating recursively these steps, we finally obtain

$$\mathrm{P}[s_j > k] = P_{j,j}^k \,,$$

which is the complementary cumulative distribution function of a geometric rv with parameter $1 - P_{j,j}$ (see Example 2.6 G).

The independence of the different sojourn times follows directly from the Markov property. To prove this assert, let us consider two distinct visits of the MC to the same state j. Let $s_{j,1}$ and $s_{j,2}$ denote the sojourn times of the first and second visit, respectively. The two rvs are independent if and only if for any pair of non-negative integers k_1 and k_2 it holds

$$\mathrm{P}[s_{j,2} = k_2 | s_{j,1} = k_1] = \mathrm{P}[s_{j,2} = k_2] \,. \tag{7.16}$$

Thanks to the homogeneity of the MC, we can assume that the first visit to state j begins at step $n_1 = 0$ and ends at step $k_1 - 1$. The other visit will, hence, starts at a successive step $n_2 \geq k_1$. Accordingly, the left-hand side of (7.16) can be written as

$$\begin{aligned}&\mathrm{P}[s_{j,2} = k_2 | s_{j,1} = k_1] = \\ &\mathrm{P}[X_{n_2+k_2}{\neq}j, X_{n_2+k_2-1}{=}j, \ldots, X_{n_2+1}{=}j | X_{n_2}{=}j, X_{k_1}{\neq}j, X_{k_1-1}{=}j, \ldots, X_0{=}j]\end{aligned}$$

Applying the Markov property, we have

$$\begin{aligned}&\mathrm{P}[s_{j,2} = k_2 | s_{j,1} = k_1] = \\ &= \mathrm{P}[X_{n_2+k_2} \neq j, X_{n_2+k_2-1} = j, \ldots, X_{n_2+1} = j | X_{n_2} = j] = P_{j,j}^{k_2}(1 - P_{j,j}) \,,\end{aligned}$$

that equals $\mathrm{P}[s_{j,2} = k_2]$, thus proving (7.16). The same reasoning can be applied to prove the independence of sojourn times in different states. □

Remark Sojourn times are always statistically independent and geometrically distributed. While sojourn times in a same state $j \in \mathcal{S}$ are identically distributed, the sojourn times in different states i and j are identically distributed only if $P_{i,i} = P_{j,j}$.

7.2.4 Chapman-Kolmogorov equations for discrete-time MC

The (one-step) transition matrix $\boldsymbol{P}$ plays a major role in the MC analysis. In fact, $\boldsymbol{P}$ has the noticeable property of "generating" any other transition matrix in k steps $\boldsymbol{P}(k)$, as stated by the following proposition.

Proposition 7.2 (Chapman-Kolmogorov equations) *For any pair of positive integers n, k, it holds*

$$P_{i,j}(n+k) = \sum_{s=0}^{N} P_{i,s}(n) P_{s,j}(k)\,, \quad i,j \in \mathcal{S}\,, \tag{7.17}$$

which can be expressed in matrix form as

$$\boldsymbol{P}(n+k) = \boldsymbol{P}(n)\boldsymbol{P}(k) \tag{7.18}$$

or, equivalently, as

$$\boxed{\boldsymbol{P}(k) = \boldsymbol{P}^k\,.} \tag{7.19}$$

Equation (7.17) and its equivalent matrix formulations are known as the Chapman-Kolmogorov equations.

Proof Applying the Total Probability Theorem 2.12, the transition probability $P_{i,j}(n+k)$ can be written as

$$P_{i,j}(n+k) = \mathrm{P}\left[X_{n+k} = j | X_0 = i\right] = \sum_{s=0}^{N} \mathrm{P}\left[X_{n+k} = j, X_n = s | X_0 = i\right]$$

and, using properties of the conditional probability, we obtain

$$P_{i,j}(n+k) = \sum_{s=0}^{N} \mathrm{P}\left[X_{n+k} = j | X_n = s, X_0 = i\right] \mathrm{P}\left[X_n = s | X_0 = i\right]\,. \tag{7.20}$$

Since X_n is a MC, the condition $X_0 = i$ of the first conditional probability is not relevant, thus (7.20) becomes

$$P_{i,j}(n+k) = \sum_{s=0}^{N} \mathrm{P}\left[X_{n+k} = j | X_n = s\right] \mathrm{P}\left[X_n = s | X_0 = i\right]$$

that, for the homogeneity of the MC, can be rewritten as

$$P_{i,j}(n+k) = \sum_{s=0}^{N} P_{s,j}(k) P_{i,s}(n)\,. \tag{7.21}$$

Note that each term in the form $P_{s,j}(k)P_{i,s}(n)$ on the right-hand side of (7.21) accounts for the probability that the process starts from state i at step zero, then passes through state s at step n and ends up in state j after k additional steps. In matrix notation, (7.21) becomes

$$\left[\boldsymbol{P}(n+k)\right]_{(i,j)} = \left[\boldsymbol{P}(n)\right]_{(i,:)} \left[\boldsymbol{P}(k)\right]_{(:,j)} \tag{7.22}$$

which remarks that the (i, j)th element of $\boldsymbol{P}(n+k)$ is given by the inner product of the ith row of $\boldsymbol{P}(n)$ and the jth column of $\boldsymbol{P}(k)$. Since the result holds for any $(i, j) \in \mathcal{S} \times \mathcal{S}$, we find (7.18). Finally, applying recursively (7.18) we can write $\boldsymbol{P}(k)$ as

$$\boldsymbol{P}(k) = \boldsymbol{P}\,\boldsymbol{P}(k-1) = \boldsymbol{P}\,\boldsymbol{P}\,\boldsymbol{P}(k-2) = \ldots = \boldsymbol{P}^k$$

for any $k > 0$, which concludes the proof. □

Remark The transition matrix $\boldsymbol{P}$ generates the transition probabilities $P_{i,j}(k)$ for any $k > 0$ and any pair of states $i, j \in \mathcal{S}$.

7.2.5 Transition diagram of discrete-time MC

It is often useful to represent the transition matrix $\boldsymbol{P}$ by a graph, named *transition diagram*. The nodes of the graph are in one-to-one correspondence with the states of the state-space $\mathcal{S}$ of the MC. In most cases, nodes are labelled after the associated states, so that node associated to state i is generally labelled as "node i". The arcs in the transition diagram are in one-to-one correspondence with the positive elements of $\boldsymbol{P}$. More specifically, for each element $P_{i,j} > 0$ of $\boldsymbol{P}$, there is an arc departing from node i and entering into node j, which is labelled (or weighted) as $P_{i,j}$. Note, that i is connected to j only if the state of the MC can change from i to j in a single step with non-zero probability. Therefore, due to the condition (7.11), the sum of the weights of outgoing arcs from any state is always equal to one. Fig. 7.1 gives an example of transition diagram associated to the transition matrix $\boldsymbol{P}$ reported in the same figure.

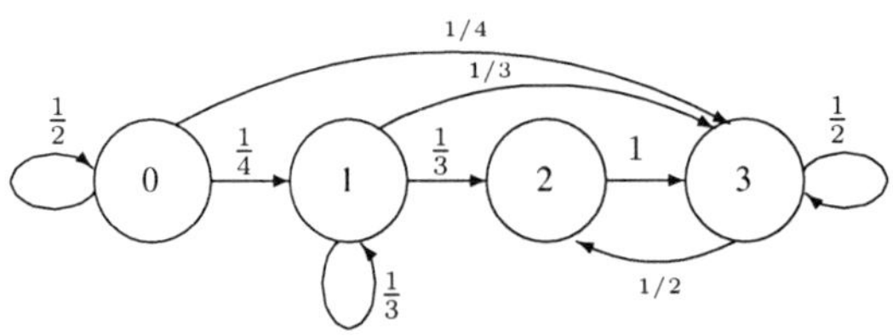

$$\boldsymbol{P} = \begin{bmatrix} 1/2 & 1/4 & 0 & 1/4 \\ 0 & 1/3 & 1/3 & 1/3 \\ 0 & 0 & 0 & 1 \\ 0 & 0 & 1/2 & 1/2 \end{bmatrix}$$

Figure 7.1 Example of transition diagram for the MC with transition matrix $\boldsymbol{P}$ shown on the right of the figure.

The graphical representation makes sometimes easier the analysis of the MC's time evolution, which can be associated to a journey of an imaginary "particle" that moves along the transition diagram. The particle starts its trip from the node associated to the initial state of the process and, at each step, follows an arc corresponding to a state-transition performed by the process, thus entering the node associated to the next state of the process, an so on. Multiplying the labels of all the arcs crossed by the particle along its path, we obtain the probability of the MC realization associated to that path.

7.2.6 State probability of discrete-time MC

So far we have considered only the transition probabilities of the MC, which are the *conditional* probabilities that the process moves from the current state into another state. We now turn our attention to the *absolute* probability that the MC is in a given state $j \in \mathcal{S}$ at the nth step, that is

$$p_j(n) = \mathrm{P}\left[X_n = j\right], \qquad j \in \mathcal{S}. \tag{7.23}$$

Note that $p_j(n)$ for $j \in \mathcal{S}$ is just the PMD of the discrete rv X_n, which can be conveniently represented through the *state probability vector*:[2]

$$\boldsymbol{p}(n) = [p_0(n), p_1(n), \ldots, p_N(n)] \tag{7.24}$$

where

$$\boldsymbol{p}(n)\,\mathbf{1} = \sum_{j=0}^{N} p_j(n) = 1 \tag{7.25}$$

because of the normalization condition. Furthermore, for any $j \in \mathcal{S}$, the Total Probability Theorem 2.12 gives

$$p_j(n) = \mathrm{P}\left[X_n = j\right] = \sum_{i=0}^{N} \mathrm{P}\left[X_n = j | X_0 = i\right] \mathrm{P}\left[X_0 = i\right]$$

or,

$$p_j(n) = \sum_{i=0}^{N} p_i(0) P_{i,j}(n). \tag{7.26}$$

In matrix notation, (7.26) becomes

$$\boldsymbol{p}(n) = \boldsymbol{p}(0)\boldsymbol{P}(n) \tag{7.27}$$

according to which, the state probability vector at the nth step can be obtained from the probability vector $\boldsymbol{p}(0)$ of the initial state and the transition matrix in n steps $\boldsymbol{P}(n)$. Now, using the Chapman-Kolmogorov equation (7.17), equation (7.27) yields

$$\boxed{\boldsymbol{p}(n) = \boldsymbol{p}(0)\boldsymbol{P}^n\,.} \tag{7.28}$$

Remark The state probability vector of a discrete time MC is entirely defined by the transition matrix $\boldsymbol{P}$ and the initial state probability vector $\boldsymbol{p}(0)$. In other words, the knowledge of $\boldsymbol{P}$ and $\boldsymbol{p}(0)$ does permit to determine the *state probability vector* of the MC at any time instant n. Hence, the MC is, at any given time instant n, fully described by $\boldsymbol{P}$ and $\boldsymbol{p}(0)$: however, its actual time evolution remains uncertain.

Example 7.2 G Consider a given facility, for instance a parking lot or a table in a restaurant, that can be in either of two states: idle (state 0) or busy (state 1). Suppose that the state of the facility is regularly checked (for instance, every day at the same hour) and let X_n denote the state of the facility at the nth check. Suppose that X_n is a MC with state transitions as in Fig. 7.2.

[2]Note that, in order to maintain consistency with most of the literature on the subject, the state probability vectors are defined as *row vectors*, conversely to any other vector encountered in this book.

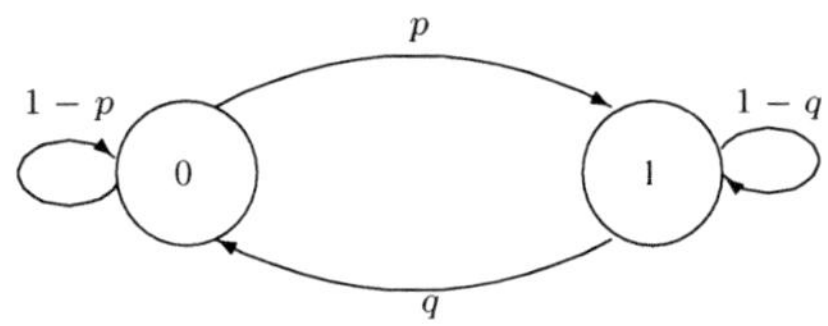

$$P = \begin{bmatrix} 1-p & p \\ q & 1-q \end{bmatrix}$$

Figure 7.2 Transition diagram and corresponding transition matrix of a two state Markov chain, also known as discrete-time Gilbert model.

We are interested in determining the state probability vector $\boldsymbol{p}(n) = [p_0(n), p_1(n)]$ at generic time instant n. To begin with, we observe that for any n we have $p_0(n) = 1 - p_1(n)$, hence it suffices to compute $p_0(n)$. Applying (7.28), we have

$$p_0(n+1) = P_{1,0}p_1(n) + P_{0,0}p_0(n) \tag{7.29}$$

which basically states that the MC can be in state zero at time $n+1$ only if it was in state one at time n and made a transition into state zero or it was already in state zero time n and made a self-transition to the same state. Replacing the transition probabilities with their values, we get

$$\begin{aligned} p_0(n+1) &= qp_1(n) + (1-p)p_0(n) \\ &= q(1-p_0(n)) + (1-p)p_0(n) \\ &= q + (1-p-q)p_0(n)\,. \end{aligned} \tag{7.30}$$

Iterating back this recursion to $n = 0$ we get

$$\begin{aligned} p_0(n) &= q + (1-p-q)p_0(n-1) \\ &= q + (1-p-q)(q + (1-p-q)p_0(n-2)) \\ &\vdots \\ &= (1-p-q)^n p_0(0) + q\sum_{k=0}^{n-1}(1-p-q)^k\,. \end{aligned} \tag{7.31}$$

and, recalling that $\sum_{k=0}^{n-1} x^k = \frac{1-x^n}{1-x}$ for $x \neq 1$, we obtain

$$p_0(n) = \frac{q}{p+q} + (1-p-q)^n\left(p_0(0) - \frac{q}{p+q}\right) \tag{7.32}$$

which is the result we were looking for. The complementary probability $p_1(n) = 1 - p_0(n)$ is

$$p_1(n) = \frac{p}{p+q} + (1-p-q)^n\left(p_1(0) - \frac{p}{p+q}\right)\,. \tag{7.33}$$

The time-dependent behavior of the state probabilities is plotted in Fig. 7.3 for $p = 1/20$ and $q = 1/10$. The figure reports two sets of curves, the solid lines have been obtained by setting the initial state to zero, i.e., assuming $p_0(0) = 1$ and $p_1(0) = 0$, whereas the dashed lines refer to the

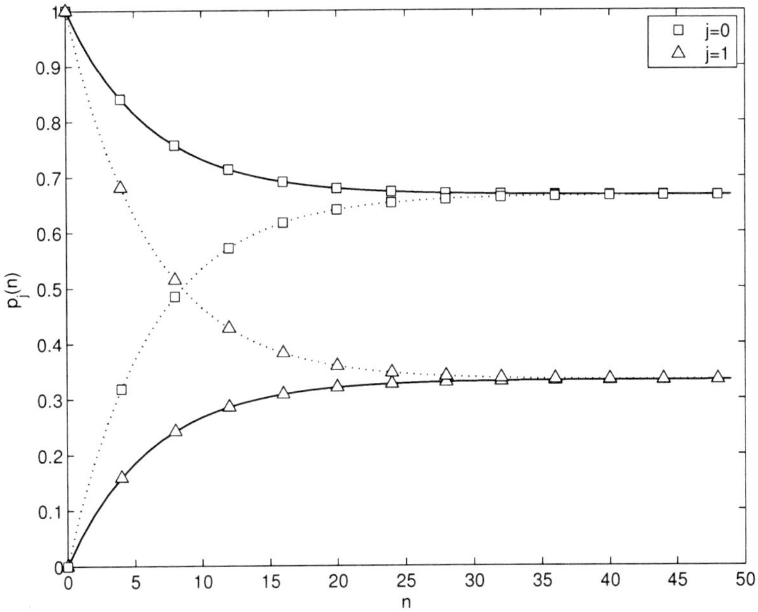

Figure 7.3 State probabilities $p_0(n)$ (squares) and $p_1(n)$ (triangles) as a function of time instant n for the MC of Fig. 7.2, with $p = 1/20$ and $q = 1/10$ (see Example 7.2 G). Solid and dotted lines refer to the time evolution of the state probabilities when starting from $X_0 = 0$ and $X_0 = 1$, respectively.

complementary case, i.e., $p_0(0) = 0$ and $p_1(0) = 1$. We observe that in both cases the probability of each state tends to the same asymptotic value, irrespective of the initial state distribution. As a matter of fact, by letting n approach infinity in (7.32) and (7.33), we easily realize that asymptotic values are equal to

$$\lim_{n\to\infty} p_0(n) = \frac{q}{p+q} \quad \text{and} \quad \lim_{n\to\infty} p_1(n) = \frac{p}{p+q} \tag{7.34}$$

and do not depend on the initial state probability vector $\boldsymbol{p}(0)$.

Further insight The two-state model considered in Example 7.2 G is also known as *Gilbert model* since it was first proposed by E. N. Gilbert in [8] as a model to analyze the capacity of a binary transmission channel with error bursts. In that case, the Markov process was used to model the radio channel condition that could either be *Good* or *Bad*. In Good state, the channel was supposed to be error-free, whereas in Bad state the error probability was supposed to be equal to $1/2$. The transition step period was equal to the bit period. Gilbert assumed that the sojourn times in the two states could be approximated by geometric random variables with parameters p and q, respectively, thus permitting the analysis of the system through MC theory. The same model can be applied to any system that admits a binary classification of the states, provided that the state evolution exhibits the Markov property.

Example 7.2 H Consider the MC represented in Fig. 7.4. Despite the rather simple structure of this MC, closed-form evaluation of the time-dependent state probability vector $\boldsymbol{p}(n)$ is very complex. However, numerical computation by (7.28) is a trivial matter. In Fig. 7.5 we report the state probabilities $p_j(n)$ as a function of n, for states $j = 0$ (square) and $j = 5$ (triangle). The probabilities of the other states are

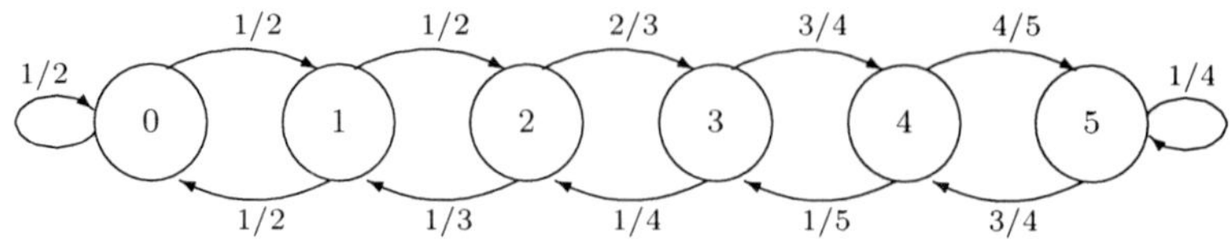

Figure 7.4 Transition diagram of the MC in Example 7.2 H.

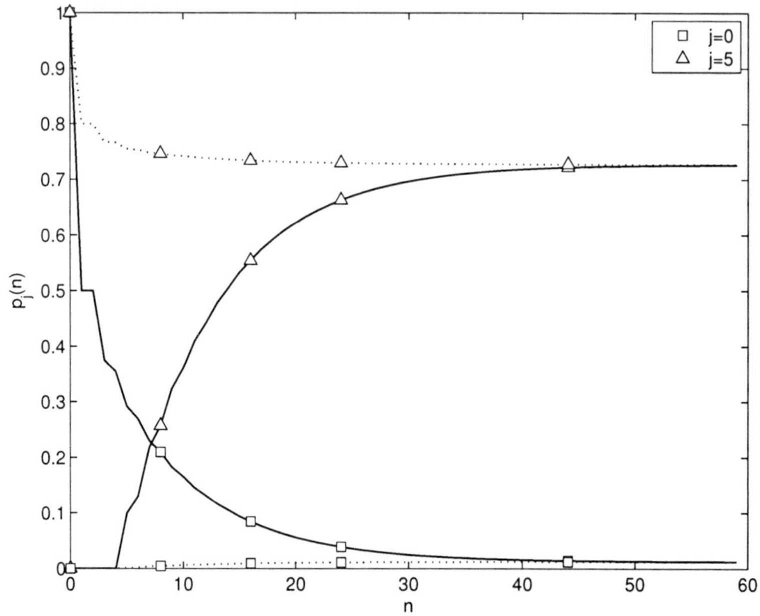

Figure 7.5 State probabilities $p_0(n)$ and $p_5(n)$ versus n for the MC of Fig. 7.4 (see Example 7.2 H). Solid and dotted lines refer to the time evolution of the state probabilities when starting from $X_0 = 0$ and $X_0 = 5$, respectively.

not reported to reduce clutter in the figure. The graph contains two sets of curves, plotted with solid and dashed lines, that correspond to two different choices of the initial state probability vector $\boldsymbol{p}(0)$. The solid lines have been obtained by assuming that the process starts in state 0, i.e., $\boldsymbol{p}(0) = [1, 0, 0, 0, 0, 0]$, whereas the dotted lines refer to the case in which the initial state is $i = 5$, i.e., $\boldsymbol{p}(0) = [0, 0, 0, 0, 0, 1]$. We observe that the curves present an asymptote as n goes to infinity, so that after a certain number of steps the state probability vector tends to become time-invariant. Furthermore, we note that the asymptotic values are the same for both solid and dashed lines, i.e., the state probabilities tend to converge to an asymptotic value that does not depend on the initial state probability vector. This interesting property is actually typical of a large family of MCs, and will be thoroughly investigated in the following sections.

7.2.7 Classification of discrete-time Markov chains

The statistical properties of a MC are determined by the structure of its transition matrix $\boldsymbol{P}$ or, equivalently, of the associated transition diagram. In this section, we classify the MC according to property of its states and structure of the transition matrix $\boldsymbol{P}$.

Accessible and communicating states.

Definition 7.2
State i is said to access *(or* reach*) state j if $P_{i,j}(k) > 0$ for some $k \geq 0$. This relation is denoted as $i \rightarrow j$.*

States i and j are said to communicate *when they are mutually accessible, i.e., $i \rightarrow j$ and $j \rightarrow i$. This relation is denoted as $i \leftrightarrow j$.*

According to this definition, state i can access state j if the transition diagram contains a path from i to j. In fact, if $P_{i,j}(k) > 0$, by applying the Chapman-Kolmogorov equation we have

$$P_{i,j}(k) = \sum_{s=0}^{N} P_{i,s} P_{s,j}(k-1) > 0\,,$$

thus there exists at least one state $s_1 \in \mathcal{S}$ such that $P_{i,s_1} P_{s_1,j}(k-1) > 0$. Repeating the reasoning recursively, we can find at least one sequence of states $\{i, s_1, s_2, \ldots, s_{k-1}, j\}$, $s_h \in \mathcal{S}$, such that

$$P_{i,j}(k) \geq P_{i,s_1} P_{s_1,s_2} \cdots P_{s_{k-2},s_{k-1}} P_{s_{k-1},j} > 0\,.$$

The sequence of states $\{i, s_1, s_2, \ldots, s_{k-1}, j\}$ identifies a directed path from i to j in the transition diagram of the MC, which offers a "graphical" interpretation of the accessibility relation.

It is then immediate to realize that i and j communicate if there exists (at least) one closed path that includes both i and j.

Further insight In a transition diagram with a finite number of nodes, numbered from 0 to N, any loop-free path[3] contains N or less arcs. Hence, if a state i can access another state j there exists at least one loop-free path from i to j with no more than N arcs. In other words, $i \rightarrow j$ if and only if $P_{i,j}(k) > 0$ for some $k \leq N$.

Example 7.2 I With reference to the MC represented in Fig. 7.1, we can see that each state is accessible from state zero, whereas state zero is not accessible from any state except itself. State one is accessible from states zero and one, but not from any other state. States two and three are mutually reachable and, hence, they communicate.

[3] A loop-free path is a trajectory in the transition diagram that does not touch any node more than once.

Communication classes.

Proposition 7.3 *Communication is an equivalence relation, i.e., it is*

(i) reflexive: $i \leftrightarrow i$*;*

(ii) symmetric: if $i \leftrightarrow j$*, then* $j \leftrightarrow i$*;*

(iii) transitive: if $i \leftrightarrow \sigma$ *and* $\sigma \leftrightarrow j$*, then* $i \leftrightarrow j$*.*

Proof The reflexive and symmetric properties follow trivially from the definition of the communication relation and from the assumption $\boldsymbol{P}(0) = \boldsymbol{I}$. It remains to prove (iii). Since $i \leftrightarrow \sigma$ and $\sigma \leftrightarrow j$ then $P_{i,\sigma}(n) > 0$ and $P_{\sigma,j}(k) > 0$ for some $n, k > 0$. Now, applying the Chapman-Kolmogorov equation (7.21) we have

$$P_{i,j}(n+k) = \sum_{s=0}^{N} P_{i,s}(n)P_{s,j}(k) \geq P_{i,\sigma}(n)P_{\sigma,j}(k) > 0 \tag{7.35}$$

which proves that j is accessible from i. Following the same rationale, we can also prove that i is accessible from j. □

On the basis of the previous proposition, the relation of communication yields a partition of the states of the MC in *classes* of equivalence, in such a way that a class collects all and only the states that communicate with one another. Formally, we have the following definition.

Definition 7.3 *A subset* $\mathcal{S}_c \subseteq \mathcal{S}$ *is a* communication class *if all its states communicate one another and do not communicate with any other state.*

Note that, in general, there may exist arcs from a communication class to another class. However, these transitions do not admit return, otherwise we would have communication between states of different classes, which contradicts the definition of communication class. Therefore, with reference to the transition diagram, a set of states is a communication class if removing all the inbound and outbound arcs, we obtain a strongly connected graph, i.e., a graph in which there is a path between any pair of distinct nodes in the graph. Conversely, by including in the set any other group of nodes, together with their connecting arcs, the resulting graph will not be strongly connected anymore.

The state space $\mathcal{S}$ of a MC can, hence, be partitioned into $\mathcal{S}_1, \ldots, \mathcal{S}_C$, disjoint subsets, each corresponding to a different communication class. A communication class is *closed* when it is not possible to leave the states of the class once entered, i.e., there are no outbound arcs from the states of the class towards other classes. Conversely, a communication class is *open* when it is possible to leave the states of that class to states of other classes.

Further insight Let partition the state space $\mathcal{S}$ in the subsets $\mathcal{S}_1, \ldots, \mathcal{S}_C$, each corresponding to a distinct communication class. Let us enumerate the states sequentially, so that $\mathcal{S}_1 = \{0, 1, \ldots, k_1 - 1\}$, $\mathcal{S}_2 = \{k_1, k_1 + 1, \ldots, k_2 - 1\}$, and so on, till $\mathcal{S}_C = \{k_{C-1} + 1, \ldots, N\}$. The transition probabilities among the states of the communication class S_j, $j = 1, 2, \ldots, C$ will hence occupy the diagonal squared block of the transition matrix $\boldsymbol{P}$, from row k_{j-1} to k_j

and columns k_{j-1} to k_j. The communication class S_j is closed if all the *other* elements in the *rows* from k_{j-1} to k_j are zero. On the contrary case, i.e., if there exist non-zero elements in the rows from k_{j-1} to k_j, excluded those in the columns from k_{j-1} to k_j, the class S_j is open.

Remark If the MC enters a closed communication class S_c at some time instant n, from that instant onward the state of the MC will only take values in that closed class, that is to say

$$X_n = \sigma \text{ for some } \sigma \in S_c \Rightarrow X_{n+k} \in S_c, \text{ for } k = 0, 1, 2, \ldots .$$

In some sense, once entered a closed communication class, the MC remains *entrapped* into the states of that class forever. Conversely, states of an open class are predestined to be abandoned by the MC in the long run.

This simple observation motivates the following definition.

Definition 7.4 *A MC is called* irreducible *if there is only one communicaiton class, that is, if all states communicate with each other, forming a single closed class. Otherwise, the MC is called* reducible. *In this case, the MC contains several (closed or open) communication classes.*

Example 7.2 J Consider the MC with state-transition diagram and transition matrix $\boldsymbol{P}$ shown in Fig. 7.6. We can observe that the MC contains three communication classes, namely $\mathcal{S}_1 = \{0\}$, $\mathcal{S}_2 = \{1, 2\}$ and

$$\boldsymbol{P} = \begin{bmatrix} 0 & 1/2 & 1/4 & 1/4 & 0 \\ 0 & 1/3 & 2/3 & 0 & 0 \\ 0 & 1/4 & 1/4 & 1/2 & 0 \\ 0 & 0 & 0 & 1/2 & 1/2 \\ 0 & 0 & 0 & 2/3 & 1/3 \end{bmatrix}$$

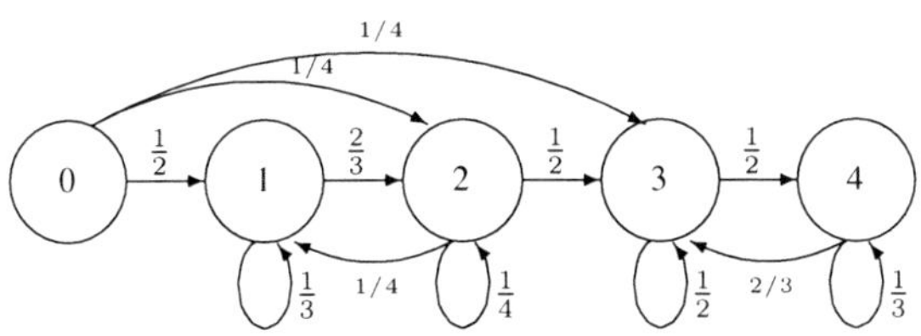

Figure 7.6 Example of reducible MC.

$\mathcal{S}_3 = \{3, 4\}$. Classes $\mathcal{S}_1$ and $\mathcal{S}_2$ are open because there are outbound transitions from these classes to other classes. Class $\mathcal{S}_3$ is closed, for all transitions originating from states belonging to $\mathcal{S}_3$ also terminate into states of $\mathcal{S}_3$. Since the MC contains multiple communication classes it is reducible.

Example 7.2 K The MC in Fig. 7.7 is irreducible, since all its states do communicate with one another. Inspecting the transition diagram, we note that it is always possible to find a path between any pair of states, i.e., the graph is strongly connected. We can also observe that it is possible to find a loop-free path that touches all the states exactly once before closing on itself.

$$\boldsymbol{P} = \begin{bmatrix} 0 & 1/2 & 1/4 & 1/4 & 0 \\ 0 & 1/3 & 2/3 & 0 & 0 \\ 0 & 1/4 & 1/4 & 1/2 & 0 \\ 0 & 0 & 0 & 1/2 & 1/2 \\ 2/3 & 0 & 0 & 0 & 1/3 \end{bmatrix}$$

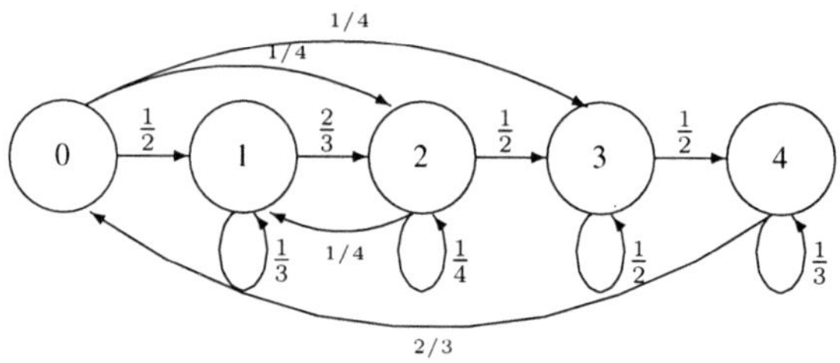

Figure 7.7 Example of irreducible MC.

Recurrent and transient states.
For any state j, we define $f_{j,j}(k)$ to be the probability that starting in j at step n, the next transition into j occurs at step $n+k$. Formally,

$$\begin{aligned} f_{j,j}(0) &= 0\,; \\ f_{j,j}(k) &= \mathrm{P}\left[X_{n+k} = j, X_{n+k-1} \neq j, \ldots, X_{n+1} \neq j | X_n = j\right]. \end{aligned} \tag{7.36}$$

The *hitting probability* of state j, denoted by $f_{j,j}$, is the probability of ever making a transition into state j, given that the MC starts in state j, i.e.,

$$f_{j,j} = \mathrm{P}\left[\bigcup_{k=1}^{\infty}\{X_{n+k} = j\} \middle| X_n = j\right] = \sum_{k=1}^{\infty} f_{j,j}(k)\,. \tag{7.37}$$

Definition 7.5 *State j is said to be* recurrent *if $f_{j,j} = 1$, and* transient *if $f_{j,j} < 1$.*

Hence, state j is recurrent if, when starting at j, the MC will eventually return to j with probability one. Conversely, state j is transient if, starting at j, there is a positive probability that the MC will never return to j in the future. Another interpretation of the recurrence and transiency properties is provided by the following propositions.

Proposition 7.4 *Let v_j denote the overall number of times a MC visits state j during its time-evolution. The conditional expectation of v_j, given that the process starts at j, is given by*

$$\mathrm{E}\left[v_j | X_0 = j\right] = \sum_{n=0}^{\infty} P_{j,j}(n) = \frac{1}{1 - f_{j,j}} . \tag{7.38}$$

Proof To prove the first part of the proposition, we express the number of visits to state j as

$$v_j = \sum_{n=0}^{\infty} \chi \left\{X_n = j\right\} , \tag{7.39}$$

where we used the indicator function (2.163). Taking the conditional expectation of both sides of (7.39), given $X_0 = j$, we then have

$$\begin{aligned} \mathrm{E}\left[v_j | X_0 = j\right] &= \mathrm{E}\left[\sum_{n=0}^{\infty} \chi\left\{X_n = j\right\} | X_0 = j\right] = \sum_{n=0}^{\infty} \mathrm{E}\left[\chi\left\{X_n = j\right\} | X_0 = j\right] \\ &= \sum_{n=0}^{\infty} \mathrm{P}\left[X_n = j | X_0 = j\right] = \sum_{n=0}^{\infty} P_{j,j}(n) . \end{aligned} \tag{7.40}$$

Now, let T_j denote the time of the *first* return into j. For any positive integer n, the Total Probability Theorem 2.12 yields

$$\begin{aligned} P_{j,j}(n) &= \sum_{k=1}^{n} \mathrm{P}\left[X_n = j, T_j = k | X_0 = j\right] \\ &= \sum_{k=1}^{n} \mathrm{P}\left[X_n = j | T_j = k, X_0 = j\right] \mathrm{P}\left[T_j = k | X_0 = j\right] . \end{aligned} \tag{7.41}$$

Since the condition $T_j = k$ means that the process is in state j at step k for the first time, we can write

$$\begin{aligned} \mathrm{P}\left[X_n = j | T_j = k, X_0 = j\right] &= \mathrm{P}\left[X_n = j | X_k = j, X_{k-1} \neq j, \ldots, X_1 \neq j, X_0 = j\right] \\ &= P_{j,j}(n - k) \end{aligned}$$

where the last step follows from the Markov property. Furthermore, the conditional probability $\mathrm{P}\left[T_j = k | X_0 = j\right]$ corresponds to the probability $f_{j,j}(k)$ of returning in j in k steps, given that the process starts at j. Hence, (7.41) can be written as

$$P_{j,j}(n) = \sum_{k=1}^{n} P_{j,j}(n - k) f_{j,j}(k) . \tag{7.42}$$

Summing both terms from $n = 1$ to infinity, we get

$$\begin{aligned}\sum_{n=1}^{\infty} P_{j,j}(n) &= \sum_{n=1}^{\infty}\sum_{k=1}^{n} P_{j,j}(n-k) f_{j,j}(k) \\ &= \sum_{k=1}^{\infty}\sum_{n=k}^{\infty} P_{j,j}(n-k) f_{j,j}(k) \\ &= \sum_{k=1}^{\infty} f_{j,j}(k) \sum_{r=0}^{\infty} P_{j,j}(r)\end{aligned} \tag{7.43}$$

where the second row follows from a simple change of the summations order, whereas the third row is obtained by changing the index of the inner sum to $r = n - k$ and adjusting the limits of the sum accordingly. At this point the two originally nested sums are decoupled and we can now replace the first sum with the hitting probability $f_{j,j}$, obtaining

$$\sum_{n=0}^{\infty} P_{j,j}(n) - P_{j,j}(0) = f_{j,j} \sum_{r=0}^{\infty} P_{j,j}(r) \,. \tag{7.44}$$

Collecting the terms $\sum_{n=0}^{\infty} P_{j,j}(n)$ in (7.44) and recalling that $P_{j,j}(0) = 1$, we finally obtain

$$\sum_{n=0}^{\infty} P_{j,j}(n) = \frac{1}{1 - f_{j,j}} \tag{7.45}$$

which concludes the proof. □

Proposition 7.5 *State j is recurrent if and only if*

$$\sum_{n=0}^{\infty} P_{j,j}(n) = \infty \,. \tag{7.46}$$

Accordingly, state j is transient if and only if

$$\sum_{n=0}^{\infty} P_{j,j}(n) < \infty \,. \tag{7.47}$$

Proof The proof follows directly from Proposition 7.4 and Definition 7.5. □

According to Propositions 7.4 and 7.5, when a MC starts from a recurrent state j, then it returns to state j infinitely many times during its time evolution. Conversely, if state j is transient, then the average number of visits to state j is finite. Therefore, Proposition 7.5 provides a method to verify whether a state is recurrent or transient. The next proposition shows that recurrence and transiency are class properties. Therefore, to figure out the recurrence/transience for a whole communicating class, it suffices to check one of its members.

Proposition 7.6 *Recurrence and transiency are* class properties, *i.e., the states of the same communication class are either all recurrent or all transient.*

Proof Consider two states i, j belonging to the same communication class and suppose that i is recurrent. Since $i \leftrightarrow j$, then there exist two positive integers n, m such that $P_{j,i}(n) > 0$ and $P_{i,j}(m) > 0$. Now, for any integer $k \geq 0$, from (7.17) we have

$$P_{j,j}(n+k+m) = \sum_{s_1=0}^{N} \sum_{s_2=0}^{N} P_{j,s_1}(n) P_{s_1,s_2}(k) P_{s_2,j}(m) \geq P_{j,i}(n) P_{i,i}(k) P_{i,j}(m) \quad (7.48)$$

where the inequality follows from the fact that we extract only the term with $s_1 = i$ and $s_2 = i$ from the sum that contains only non-negative terms. From (7.48) we thus obtain

$$\sum_{k=0}^{\infty} P_{j,j}(n+k+m) \geq P_{j,i}(n) P_{i,j}(m) \sum_{k=0}^{\infty} P_{i,i}(k) = \infty \quad (7.49)$$

where the last step follows from Proposition 7.5 being i recurrent. We finally have

$$\sum_{k=0}^{\infty} P_{j,j}(k) \geq \sum_{k=n+m}^{\infty} P_{j,j}(k) \geq \infty ,$$

where the second inequality follows from (7.49). Hence, according to Proposition 7.5, also state j is recurrent. Hence, if state i is recurrent, all states communicating with i are also recurrent. Conversely, if state i is transient, all its communicating states must as well be transient. □

Further insight We note that states of an open class are all transient, since they will be visited, on average, a finite number of times. Conversely, states of a closed class with a *finite number* of states are necessarily recurrent. In fact, being the number of states in the closed class limited, at least one state will be visited infinitely many times during the time-evolution of the MC, so that, according to Proposition 7.5, it is recurrent. In this case, however, all its communicating states are also recurrent, as stated by Proposition 7.6. Note, that this conclusion does not hold in general for closed classes with infinite states. A notable example is provided by the bidirectional random walk discussed in Example 7.2 M, whose states can either be (all) recurrent or (all) transient, depending on the parameters that define the process.

Now, considering a recurrent state j, we define the *mean recurrence time* τ_j as the average number of steps required to return in j when starting from j. In formula, we have

$$\tau_j = \sum_{k=1}^{\infty} k f_{j,j}(k) . \quad (7.50)$$

The mean recurrence time allows us to further specify the recurrence property.

Definition 7.6 *A recurrent state j is said to be* positive recurrent *if $\tau_j < \infty$ and* null recurrent *if $\tau_j = \infty$.*

It is possible to prove that positive and null recurrence are both class properties, that is to say, if a state is positive (null) recurrent, then all its communicating states are also positive (null) recurrent. Moreover, it is possible to show that the states of a closed class with a finite number of states are all positive recurrent (see [7]).

Example 7.2 L Consider a sequence y_k of independent Bernoulli trials, having probability of success p and probability of failure $q = 1 - p$. Let X_n be the counter process of consecutive successes up to the nth trial, defined as

$$X_n = \begin{cases} X_n + 1 & \text{if} \quad y_n \text{ is a success ;} \\ 0 & \text{if} \quad y_n \text{ is a failure .} \end{cases} \tag{7.51}$$

It is easy to realize that X_n is a MC with transition diagram and transition matrix as reported in Fig. 7.8.

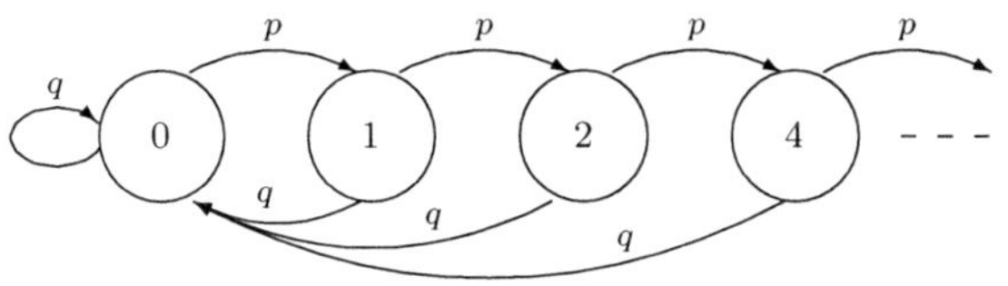

$$P = \begin{bmatrix} q & p & 0 & 0 & \cdots \\ q & 0 & p & 0 & \cdots \\ q & 0 & 0 & p & \cdots \\ \vdots & \vdots & \vdots & \vdots & \ddots \end{bmatrix}$$

Figure 7.8 Transition diagram and transition matrix of the process X_n of Example 7.2 L.

Inspecting the transition diagram we realize that the probability of first return to state zero in k steps is given by

$$f_{0,0}(k) = p^{k-1}q \qquad k \geq 1 . \tag{7.52}$$

For $p, q > 0$, the hitting probability of state zero is given by

$$f_{0,0} = \sum_{k=1}^{\infty} f_{0,0}(k) = \frac{q}{1-p} = 1 \tag{7.53}$$

and the mean recurrence time τ_0 is equal to

$$\tau_0 = \sum_{k=1}^{\infty} k f_{0,0}(k) = \sum_{k=1}^{\infty} kp^{k-1}q = \frac{1}{1-p} < \infty . \tag{7.54}$$

State zero is hence positive recurrent. Since the MC is irreducible, then according to Proposition 7.6 we can conclude that all states are positive recurrent.

Example 7.2 M The MC represented in Fig. 7.9 is an example of *bidirectional random walk* (also known as unrestricted or bilateral random walk).

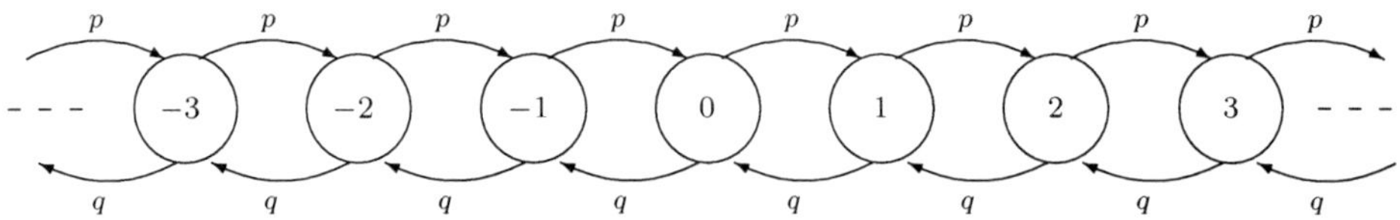

Figure 7.9 Transition diagram of a bidirectional random walk (see Example 7.2 M).

In general, a random walk is a Markov process in which transitions can occur only between adjacent states, so that in a single step the process can either remain in the current state, or move to the next or to the preceding state (if any). Transitions between non-adjacent states are not allowed. In the specific case considered in this example, forward and backward transition probabilities are equal to p and $q = 1 - p$, respectively, whereas self-transitions are not permitted. The state space of the process is unlimited in both directions, extending from $-\infty$ to $+\infty$. Observe that, except for $p = 0$ or $p = 1$, the MC is irreducible, so that all its states are communicating. The random walk is said to be symmetric if $p = q = 1/2$ and asymmetric otherwise. Many phenomena can be modeled quite faithfully using a random walk: the motion of gas molecules in a diffusion process, the trajectory of a ball played by a large group of kids, the stock value of a particular share. Assuming that the process starts in state zero, the events of particular interest are the visits to the origin (state zero), since the process starts all over again from that point onward. We are hence interested in determining the transition probabilities $P_{0,0}(n)$. First of all, we notice that the process may return to the origin only after an even number of steps, so that we have $P_{0,0}(2n+1) = 0$ for any integer n. Furthermore, the probability that the process returns to the origin after $2n$ steps is given by the probability that the process makes exactly n transitions in forward direction and n transitions in backward directions, in any order. Since forward and backward transitions occur with probability p and q, respectively, independently of the past transitions, we have

$$P_{0,0}(2n) = \binom{2n}{n}(pq)^n = \frac{(2n)!}{n!n!}(pq)^n \tag{7.55}$$

where the binomial coefficient represents the number of distinct paths of $2n$ steps that start and end in the origin. Replacing the factorial terms in (7.55) with the Stirling's approximation

$$k! \simeq k^{k+1/2}e^{-k}\sqrt{2\pi} \tag{7.56}$$

we obtain

$$P_{0,0}(2n) \simeq \frac{n^{2n}(pq)^n}{\sqrt{n\pi}} = \frac{(4pq)^n}{\sqrt{\pi n}}\,. \tag{7.57}$$

Since $pq \leq 1/4$, with equality holding if and only if $p = q = 1/2$, then $\sum_{n=0}^{\infty} P_{0,0}(n) = \infty$ if and only if $p = 1/2$. Therefore, according to Proposition 7.5, state zero is recurrent if and only if the random walk is symmetric, i.e., $p = q = 1/2$. For any other value of p and q, state zero is transient, as well as all the other states of the MC. For the exact expression of $\sum_{n=0}^{\infty} P_{0,0}(n)$ we refer to Problem 7.23.

When states are recurrent, it is possible to derive the mean recurrence times. Referring to Problem 7.23 for the (non-trivial) derivation, here we report only the final expression of the mean recurrence time of state zero, which is equal to

$$\tau_0 = \frac{4pq}{\sqrt{1-4pq}}\,. \tag{7.58}$$

We note that for $p = q = 1/2$, (7.58) gives $\tau_0 = \infty$. Hence, we conclude that the process is either transient ($pq < 1/4$), or null recurrent ($pq = 1/4$). In gambling terminology, this means that in a fair game ($p = q$) with infinite resources on both sides, sooner or later one should be able to recover all losses, since the return to the break even is certain. Unfortunately, the expected number of trials to reach the break even point is infinite. Therefore, even in a fair game (and games are rarely fair to the gambler), a player with finite resources may never arrive to balance his losses, let alone to realize a positive net gain!

◦

Periodic and aperiodic states.

Definition 7.7 *The period d_j of state j is the greatest common divisor of all integers k for which $P_{j,j}(k) > 0$. If $d_j > 1$ then state j is said to be* periodic *of period d_j, whereas if $d_j = 1$ the state is* aperiodic.

Note that if state j is periodic with period $d_j > 1$, then $P_{j,j}(\ell d_j) > 0$ for *some* but not necessarily *all* integers ℓ. In particular, it is not certain that the process is able to return into state j in exactly d_j steps. Instead, we can certainly claim that the process will *not* return to state j in a number of steps that is not multiple of d_j, i.e., $P_{j,j}(m) = 0$ for any integer m that is not divisible by d_j. In the transition diagram, state j is periodic with periodicity $d_j > 1$ when any closed path passing through j contains a number of arcs multiple of d_j or, analogously, any path originating from j cannot return in j in a number of arcs that is not multiple of d_j.

Further insight Observe that, $P_{j,j} > 0$ is a sufficient (not necessary) condition for state j to be aperiodic. In fact, when $P_{j,j} > 0$ then the MC can return into state j in a single step, so that the greatest common divisor of all k for which $P_{j,j}(k) > 0$ is necessarily one.

Proposition 7.7 *Periodicity is a class property, i.e., communicating states have equal periodicity.*

Proof Let us consider a pair of communicating states i, j. Since $i \leftrightarrow j$, then $P_{j,i}(n) > 0$ and $P_{i,j}(m) > 0$ for some n and m. Applying (7.17), we can write

$$P_{j,j}(n+m) = \sum_{s=0}^{N} P_{j,s}(m) P_{s,j}(n) \geq P_{j,i}(n) P_{i,j}(m) > 0\,. \tag{7.59}$$

Analogously, for any $k > 0$ for which $P_{i,i}(k) > 0$, from (7.48) we have

$$P_{j,j}(n+k+m) \geq P_{j,i}(n) P_{i,i}(k) P_{i,j}(m) > 0\,.$$

Therefore, d_j divides both $n+m$ and $n+k+m$, thus it is a divisor of k as well. Since d_i is the greatest common divisor of all k such that $P_{i,i}(k) > 0$ and d_j divides any such k, then d_j also divides d_i. Applying a similar argument, we can prove that d_i divides d_j as well. Since d_i and d_j divide each other, we conclude that $d_i = d_j$. □

Example 7.2 N Observe the MC represented on the left-hand side of Fig. 7.10. We see that, starting from state zero, an imaginary particle can return to zero by visiting the states $0 \to 1 \to 2 \to 3 \to 4 \to 5$ and 0, in sequence, or by visiting the states $0 \to 2 \to 3 \to 4 \to 5$ and 0. Since the first path counts six transitions, whereas the second path counts only five transitions, then the period of the state zero is $d_0 = 1$ and the MC is aperiodic. The same conclusion can be reached by considering any other state. For instance, we can find a closed path that starts and ends in state one with six transitions and another with 11

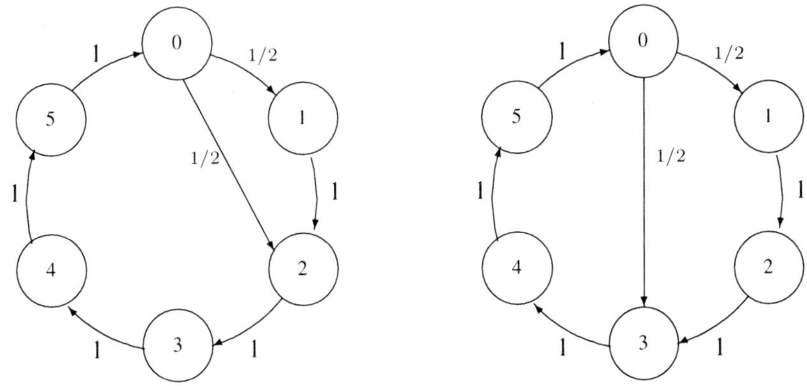

Figure 7.10 Example of irreducible aperiodic (left) and periodic (right) MC.

transitions, so that $d_1 = 1$. Conversely, the MC represented on the right-hand side of Fig. 7.10 is periodic of period $d = 2$, as it can be easily realized observing that any closed path has length $\ell = 6\,r + 4\,h$, from some non-negative integers r, h.

7.2.8 Asymptotic behavior of discrete-time MC

As seen, the time-dependent behavior of the state probability vector in any step n can be univocally determined from the initial state probability vector $\boldsymbol{p}(0)$ and the transition matrix $\boldsymbol{P}$ through (7.28), i.e., $\boldsymbol{p}(n) = \boldsymbol{p}(0)\boldsymbol{P}^n$. However, it is often of interest to investigate the *asymptotic distribution* of the process state, that is to say the distribution of X_n when n is arbitrarily large. The asymptotic state probability vector[4] is denoted by $\boldsymbol{\pi} = [\pi_0, \pi_1, \ldots, \pi_N]$ where π_j is the probability that process is in state j at some time arbitrarily far into the future, defined as

$$\pi_j = \lim_{n \to \infty} P_{i,j}(n)\,, \quad j \in \mathcal{S}\,, \tag{7.60}$$

when the limit exists and does not depend on i. We note that the asymptotic state probability, when it exists, is independent of the initial state probability vector $\boldsymbol{p}(0)$. Basically, π_j gives the probability that, sampling the process in a random instant arbitrarily far in the future, it is observed in state j.

Determining the asymptotic state probability vector $\boldsymbol{\pi}$ is one of the most important part of MC analysis. In fact, it captures the characteristics of the process when it reaches a "steady-state" behavior, i.e., after the effect of the initial conditions on the process evolution has faded away. Moreover, it is possible to prove that, whenever a MC possesses a proper[5] asymptotic state probability vector, the process is *ergodic*, so that time and statistical measures are almost surely equal (see (2.235)). In this case, the asymptotic probability π_j can be interpreted as the relative frequency of visits of state j by the process, so that the number of visits in n steps is

[4]See footnote 2 at page 427.

[5]A probability mass distribution is *proper* when it sums to one.

approximately equal to $n\pi_j$ as n grows to infinity. Hence, an empirical interpretation of π_j is as the fraction of time that the process spends in state j in the long run. This result, which will be motivated in the proof of Theorem 7.8, plays a major role in many models. For instance, if each visit to state j is associated to some kind of "cost" c_j, then the long run average cost per unit time (i.e., per step) associated with this MC is

$$\bar{C} = \sum_{j=0}^{N} \pi_j c_j \, .$$

Fortunately, the analysis of the asymptotic statistical behavior of MCs is often simpler than the process time-evolution analysis and can be performed by adopting two different approaches.

The first approach consists in the direct application of definition (7.60), which requires the analysis of the transition matrix $\boldsymbol{P}(n) = \boldsymbol{P}^n$ as n approaches infinity. However, this method is not always practical, since the computation of the transition probabilities in n steps often requires numerical methods.

The second approach consists in studying the *stationary* or invariant state probability vector $\tilde{\boldsymbol{p}}$ of the process, defined as

$$\tilde{\boldsymbol{p}} = \tilde{\boldsymbol{p}}\boldsymbol{P} \, .$$

It is possible to prove that, whenever the asymptotic state probability vector $\boldsymbol{\pi}$ exists, then it is equal to the stationary state probability vector $\tilde{\boldsymbol{p}}$, though the converse is not always true. These concepts are thoroughly discussed in this section.

7.2.8.1 Asymptotic state probability First of all, we note that states belonging to an open class have zero asymptotic state probability, since the MC will sooner or later leave forever that class for a closed class. Furthermore, a MC with more than one closed class does not admit a unique asymptotic state probability or, in other terms, it is not ergodic. In fact, the MC will remain *entrapped* in the states of the first closed class visited by the MC during its time-evolution, so that the asymptotic state probability will depend upon the initial state and the specific realization of the MC. Therefore, it makes sense to restrict our attention to irreducible MCs only, which have a single closed class with all recurrent states. Second, we observe that, in order for an asymptotic state probability to exist, the states cannot be periodic. To realize this, let consider a process with period $d > 1$. For any state j, $P_{j,j}(n)$ does not admit limit as n approaches infinity because it keeps oscillating between non-zero and zero values, depending on whether or not n is a multiple of d, so that the limit (7.60) does not exist. Thus, we conclude that only irreducible and aperiodic MCs admit unique asymptotic state probability. The following important theorem formalizes and completes this reasoning.

Theorem 7.8 *In an irreducible and aperiodic MC the asymptotic state probabilities π_j always exist and are independent of the probability vector $\mathbf{p}(0)$ of the initial state X_0. Furthermore, one of the two following cases applies:*

1. *either the process is transient or null recurrent, in which case $\lim_{n\to\infty} P_{i,j}(n) = 0$ for all i and j and there exists no proper asymptotic state probability;*

2. *or else the process is positive recurrent, in which case*

$$\pi_j = \lim_{n\to\infty} P_{i,j}(n) = \frac{1}{\tau_j} > 0 \quad i, j \in \mathcal{S} \, .$$

Proof The proof of the first point for an irreducible MC with all transient states follows directly from Proposition 7.5, according to which, for any pair of states i and j, we have

$$\sum_{n=0}^{\infty} P_{i,j}(n) < \infty$$

so that $\lim_{n\to\infty} P_{i,j}(n) = 0$. Let us now consider the case of an irreducible MC with recurrent states. For each state j, the mean recurrence time τ_j is given by (7.50). Because of the Markov property, anytime the process enters state j, the evolution from that instant onward is independent of the past history of the process and, in particular, of the previous visits to state j. Since the average number of steps between two consecutive visits to state j is τ_j, we conclude that the mean fraction of time μ_j that the process spends in state j is equal to

$$\mu_j = \frac{1}{\tau_j} . \tag{7.61}$$

We now introduce the rv $v_j(m)$ that accounts for the total number of visits to state j during time instants $0, 1, 2, \ldots, m-1$, for a limited number of steps m. Using the indicator function (2.163), we can write

$$v_j(m) = \sum_{n=0}^{m-1} \chi\{X_n = j\} . \tag{7.62}$$

The actual fraction of time that the MC spends in state j in m steps is, hence, equal to

$$\frac{v_j(m)}{m} , \tag{7.63}$$

whereas its *mean* is obtained by taking the conditional expectation of (7.63), given the initial state $X_0 = i$, i.e.,

$$\mu_j = \mathrm{E}\left[\frac{v_j(m)}{m}\middle| X_0 = i\right] . \tag{7.64}$$

Replacing (7.62) into (7.64) we get

$$\begin{aligned} \mu_j &= \mathrm{E}\left[\frac{1}{m}\sum_{n=0}^{m-1} \chi\{X_n = j\} \middle| X_0 = i\right] \\ &= \frac{1}{m}\sum_{n=0}^{m-1} \mathrm{E}\left[\chi\{X_n = j\} | X_0 = i\right] \\ &= \frac{1}{m}\sum_{n=0}^{m-1} \mathrm{P}\left[X_n = j | X_0 = i\right] \\ &= \frac{1}{m}\sum_{n=0}^{m-1} P_{i,j}(n) . \end{aligned} \tag{7.65}$$

Taking the limit for $m \to \infty$ we derive

$$\mu_j = \lim_{m\to\infty} \frac{1}{m}\sum_{n=0}^{m-1} P_{i,j}(n) . \tag{7.66}$$

Comparing (7.66) with (7.61), we conclude that

$$\lim_{m\to\infty} \frac{1}{m} \sum_{n=0}^{m-1} P_{i,j}(n) = \frac{1}{\tau_j} . \tag{7.67}$$

Now, we need to recall the following basic result. Suppose $a_0, a_1, a_2, \ldots$ is a sequence of real numbers. Then, it can be proved by elementary algebra that if the partial averages of the sequence converge to a in the form

$$\lim_{m\to\infty} \frac{1}{m} \sum_{n=0}^{m-1} a_n = a , \tag{7.68}$$

then also

$$\lim_{n\to\infty} a_n = a . \tag{7.69}$$

Since the sequence of real values $a_n = P_{i,j}(n)$ satisfies (7.68) with $a = 1/\tau_j$, as settled by (7.67), by applying (7.69) we conclude that

$$\lim_{n\to\infty} P_{i,j}(n) = \pi_j = \frac{1}{\tau_j} . \tag{7.70}$$

If state j is null recurrent its mean return time is $\tau_j = \infty$ and, hence, π_j is zero. Conversely, if state j is positive recurrent, i.e., $\tau_j < \infty$, then $\pi_j = 1/\tau_j > 0$, which concludes the proof. □

According to Theorem 7.8, an irreducible and aperiodic MC always admit a single asymptotic state probability vector $\boldsymbol{\pi}$. The elements of this vector can either be all zero (MC is transient or null-recurrent) or all positive (MC is positive recurrent), whereas it is not possible that some elements are zero and others are positive. As a consequence, to determine whether or not the MC is positive recurrent, it suffices to check if the asymptotic probability of *any* state of the MC is positive or zero.

Remark From (7.60), $\boldsymbol{\pi}$ can be easily computed by iteratively multiplying $\boldsymbol{P}^n$ by $\boldsymbol{P}$, until the series of matrix powers $\boldsymbol{P}, \boldsymbol{P}^2, \boldsymbol{P}^3, \ldots, \boldsymbol{P}^n$ converges (with reasonable precision) to the matrix with all rows equal to $\boldsymbol{\pi}$. Equivalently, as all rows of $\boldsymbol{P}^\infty$ are equal, from the Chapman-Kolmogorov equation (7.28), for any value of $\boldsymbol{p}(0)$ it is

$$\boldsymbol{\pi} = \lim_{n\to\infty} \boldsymbol{p}(n) = \lim_{n\to\infty} \boldsymbol{p}(0)\boldsymbol{P}^n = \boldsymbol{p}(0) \lim_{n\to\infty} \boldsymbol{P}^n \tag{7.71}$$

so that we have

$$\boxed{\boldsymbol{\pi} = \boldsymbol{p}(0)\boldsymbol{P}^\infty} \tag{7.72}$$

7.2.8.2 Stationary state probability

Definition 7.8 *A state probability vector* $\tilde{\boldsymbol{p}}$ *is said to be* stationary *if, starting from* $\tilde{\boldsymbol{p}}$*, the state probability vector does not change in any future step, i.e.,*

$$\boldsymbol{p}(n) = \tilde{\boldsymbol{p}}\, \boldsymbol{P}(n) = \tilde{\boldsymbol{p}} \quad \textit{for any} \quad n > 0 ,$$

or, equivalently (see (7.28)),

$$\boxed{\tilde{p} = \tilde{p}\, P\, .} \tag{7.73}$$

The stationary and asymptotic state probability vectors of an irreducible MCs are intimately related, as stated by the following theorem.

Theorem 7.9 *An irreducible, aperiodic, positive recurrent MC admits a unique stationary state probability vector $\tilde{\boldsymbol{p}}$ which equals the asymptotic state probability vector $\boldsymbol{\pi}$, i.e.,*

1. *$\boldsymbol{\pi}$ is a stationary state probability vector: $\boldsymbol{\pi} = \boldsymbol{\pi}\, \boldsymbol{P}$;*
2. *$\boldsymbol{\pi}$ is the only stationary state probability vector: if $\tilde{\boldsymbol{p}} = \tilde{\boldsymbol{p}}\, \boldsymbol{P}$ then $\tilde{\boldsymbol{p}} = \boldsymbol{\pi}$.*

Proof We first prove that $\boldsymbol{\pi}$ is a stationary state probability vector. Starting from the fundamental Chapman-Kolmogorov equations $\boldsymbol{p}(n+1) = \boldsymbol{p}(n)\boldsymbol{P}$, we can write

$$\boldsymbol{p}(n+1) - \boldsymbol{p}(n) = \boldsymbol{p}(n)\,(\boldsymbol{P} - \boldsymbol{I})\, . \tag{7.74}$$

Taking the limit for $n \to \infty$, from (7.71) we have $\boldsymbol{p}(n+1) \to \boldsymbol{\pi}$ and $\boldsymbol{p}(n) \to \boldsymbol{\pi}$, so that the right-hand side of (7.74) becomes

$$\lim_{n\to\infty} \boldsymbol{p}(n)\,(\boldsymbol{P} - \boldsymbol{I}) = \boldsymbol{\pi}\,(\boldsymbol{P} - \boldsymbol{I}) = \boldsymbol{0}\, ,$$

where $\boldsymbol{I}$ is the identity matrix and $\boldsymbol{0}$ is the all-zero row vector. The above equation hence gives

$$\boldsymbol{\pi} = \boldsymbol{\pi}\, \boldsymbol{P}\, ,$$

which proves the first point.

Now, suppose that $\tilde{\boldsymbol{p}}$ is a stationary state probability vector. Then, we have

$$\tilde{\boldsymbol{p}} = \tilde{\boldsymbol{p}}\, \boldsymbol{P}(k) = \tilde{\boldsymbol{p}}\, \boldsymbol{P}^k$$

for any k, where the last equality follows from the Chapman-Kolmogorov equation (7.19). Letting k approach infinity, we have

$$\tilde{\boldsymbol{p}} = \lim_{k\to\infty} \tilde{\boldsymbol{p}}\, \boldsymbol{P}^k = \tilde{\boldsymbol{p}}\, \boldsymbol{P}^{\infty} = \tilde{\boldsymbol{p}}\,(\boldsymbol{1}\, \boldsymbol{\pi})$$

where the last equality follows from (7.72). The normalization condition $\tilde{\boldsymbol{p}}\, \boldsymbol{1} = \sum_{j=0}^{N} \tilde{p}_j = 1$ finally yields

$$\tilde{\boldsymbol{p}} = (\tilde{\boldsymbol{p}}\, \boldsymbol{1})\, \boldsymbol{\pi} = \boldsymbol{\pi}$$

which proves the second point. □

According to Proposition 7.9, an alternative way to obtain the asymptotic probability vector $\boldsymbol{\pi}$ of an irreducible, aperiodic and positive recurrent MC consists in determining the stationary state probability vector by solving the system of equations

$$\boxed{\boldsymbol{\pi} = \boldsymbol{\pi}\, \boldsymbol{P}} \tag{7.75}$$

where the unknowns are the elements of the vector $\boldsymbol{\pi} = [\pi_0, \pi_1, \ldots, \pi_N]$. The above system can be rewritten in homogeneous form as

$$\boldsymbol{\pi}(\boldsymbol{I} - \boldsymbol{P}) = \boldsymbol{0}. \tag{7.76}$$

Hence, $\boldsymbol{\pi}$ can be obtained by solving the system of linear equations expressed by (7.76). However, such a system does not admit a unique solution, since the matrix $\boldsymbol{I} - \boldsymbol{P}$ is not full rank, as it can be immediately realized by observing that

$$(\boldsymbol{P} - \boldsymbol{I})\mathbf{1} = \boldsymbol{P}\mathbf{1} - \mathbf{1} = \mathbf{1} - \mathbf{1} = \mathbf{0}^{\mathrm{T}}$$

being $\boldsymbol{P}\mathbf{1} = \mathbf{1}$ for the stochastic property (7.11) of $\boldsymbol{P}$. Therefore, (7.76) is a system in $N+1$ unknowns and N linearly independent equations that, hence, admits infinite solutions in the form $c\boldsymbol{p}$ where c is a non-zero multiplicative constant. To determine the value of c for which $c\boldsymbol{p}$ is actually the asymptotic state probability vector of the MC we add another equation to the system, which corresponds to the *normalization condition*:

$$\boxed{\boldsymbol{\pi}\mathbf{1} = \sum_{j=0}^{N} \pi_j = 1.} \tag{7.77}$$

In conclusion, the asymptotic state probability vector $\boldsymbol{\pi}$, when it exists, can be found by solving the following system of equations

$$\boxed{\begin{cases} \boldsymbol{\pi}(\boldsymbol{I} - \boldsymbol{P}) & = \boldsymbol{0} \\ \boldsymbol{\pi}\mathbf{1} & = 1 \end{cases}} \Longrightarrow \begin{cases} \pi_0(1 - P_{0,0}) & = \sum_{i \in \mathcal{S}, i \neq 0} \pi_i P_{i,0} \\ \pi_1(1 - P_{1,1}) & = \sum_{i \in \mathcal{S}, i \neq 1} \pi_i P_{i,1} \\ & \vdots \\ \pi_j(1 - P_{j,j}) & = \sum_{i \in \mathcal{S}, i \neq j} \pi_i P_{i,j} \\ & \vdots \\ \pi_N(1 - P_{N,N}) & = \sum_{i \in \mathcal{S}, i \neq N} \pi_i P_{i,N} \\ \sum_{j=0}^{N} \pi_j & = 1 \end{cases} \tag{7.78}$$

where one of the first $N+1$ equations is actually redundant, being a linear combination of the others.[6]

The equations in the system (7.78) are called *probability flow balance equations*, or simply *balance equations*, for they lend themselves to a fascinating interpretation in terms of probability flow conservation law. To better explain this concept, let us consider the generic jth

[6]When the state space is infinite ($N = \infty$) it is possible that some unknowns remain after applying the classic substitution method to solve system (7.78). A possible strategy to overcome this impasse is resorting to the z-transforms method, for which we refer the reader to [2].

balance equation:

$$\boxed{\pi_j\left(1 - P_{j,j}\right) = \sum_{i=0, i\neq j}^{N} \pi_i P_{i,j}\,.} \tag{7.79}$$

Now, the asymptotic probability that the MC moves from state j into another state $i \neq j$, in one step, is given by

$$\begin{aligned}\lim_{n\to\infty} \mathrm{P}\left[X_{n+1} \neq j, X_n = j\right] &= \lim_{n\to\infty} \sum_{i=0, i\neq j}^{N} \mathrm{P}\left[X_{n+1} = i | X_n = j\right] \mathrm{P}\left[X_n = j\right] \\ &= \lim_{n\to\infty} \mathrm{P}\left[X_n = j\right] \sum_{i=0, i\neq j}^{N} P_{j,i} \\ &= \pi_j (1 - P_{j,j})\end{aligned}$$

which corresponds to the left-end side of (7.79). On the other hand, the asymptotic probability that the MC moves in one step into state j from a different state is given by

$$\lim_{n\to\infty} \mathrm{P}\left[X_{n+1} = j, X_n \neq j\right] = \lim_{n\to\infty} \sum_{i=0, i\neq j}^{N} \mathrm{P}\left[X_{n+1} = j | X_n = i\right] \mathrm{P}\left[X_n = i\right] = \sum_{i=0, i\neq j}^{N} \pi_i P_{i,j}\,,$$

which equals the right-hand side of (7.79). Therefore, (7.79) can be rewritten as

$$\lim_{n\to\infty} \mathrm{P}\left[X_{n+1} = j, X_n \neq j\right] = \lim_{n\to\infty} \mathrm{P}\left[X_{n+1} \neq j, X_n = j\right]$$

that represents an equilibrium condition (in a statistical sense) for the probability of state j: asymptotically (i.e., in steady-state), the probability that the MC enters a given state j from any other state must equal the probability that the MC leaves state j to any other state. Making a parallel with a fluid system, where each state of the MC is seen as a basin, whereas the probability of entering or leaving the state is associated to the flux in and out of the basin, then (7.79) actually requires the balance of the flux entering and leaving state j, thus the name *balance equation.*

Example 7.2 O Consider the MC with transition matrix

$$\boldsymbol{P} = \begin{bmatrix} 0 & 0.5 & 0.5 \\ 1 & 0 & 0 \\ 1 & 0 & 0 \end{bmatrix}.$$

Inspecting the structure of $\boldsymbol{P}$ or, analogously, the associated transition diagram, it is easy to realize that the process is irreducible, positive recurrent, and periodic with period $d = 2$. Therefore, it does not admit a unique asymptotic state probability vector. As a matter of fact, multiplying $\boldsymbol{P}$ by itself multiple times we obtain the following sequence of power matrices

$$\boldsymbol{P}^2 = \begin{bmatrix} 1 & 0 & 0 \\ 0 & 0.5 & 0.5 \\ 0 & 0.5 & 0.5 \end{bmatrix}, \boldsymbol{P}^3 = \begin{bmatrix} 0 & 0.5 & 0.5 \\ 1 & 0 & 0 \\ 1 & 0 & 0 \end{bmatrix}, \boldsymbol{P}^4 = \begin{bmatrix} 1 & 0 & 0 \\ 0 & 0.5 & 0.5 \\ 0 & 0.5 & 0.5 \end{bmatrix}, \boldsymbol{P}^5 = \begin{bmatrix} 0 & 0.5 & 0.5 \\ 1 & 0 & 0 \\ 1 & 0 & 0 \end{bmatrix}.$$

We immediately realize that the limit for $n \to \infty$ of $\boldsymbol{P}^n$ does not exist. Moreover, also the limit of $\boldsymbol{p}(n) = \boldsymbol{p}(0)\boldsymbol{P}(n)$ does not exist, except when $\boldsymbol{p}(0)$ is equal to the *stationary* state probability vector $\tilde{\boldsymbol{p}} = [0.5, 0.25, 0.25]$, for which we have

$$\tilde{\boldsymbol{p}}\,\boldsymbol{P}^n = \tilde{\boldsymbol{p}}\,, \quad n = 1, 2, 3, \ldots\,.$$

Therefore, an irreducible, positive recurrent and *periodic* MC does not admit a unique asymptotic probability vector, but it may admit a stationary probability vector. We also note that $\boldsymbol{P}^{2n}$ actually converges to the matrix

$$\lim_{n\to\infty} \boldsymbol{P}^{2n} = \begin{bmatrix} 1 & 0 & 0 \\ 0 & 0.5 & 0.5 \\ 0 & 0.5 & 0.5 \end{bmatrix}$$

where rows, however, are not all equal. Therefore, the state probability vector on even time instants, $\boldsymbol{p}(2n)$, actually converges toward the asymptotic probability vector given by $\lim_{n\to\infty} \boldsymbol{p}(0)\boldsymbol{P}(2n)$ that, however, depends on the initial state probability vector $\boldsymbol{p}(0)$.

Example 7.2 P Consider the MC with transition matrix

$$\boldsymbol{P} = \begin{bmatrix} 0 & 0.5 & 0.5 \\ 1 & 0 & 0 \\ 0.5 & 0 & 0.5 \end{bmatrix}.$$

The process is irreducible, positive recurrent, and aperiodic (as guaranteed by the self-transition into state 2). Therefore, according to Theorem 7.8, it does admit a unique asymptotic probability vector. To determine such a vector, we resort to the method of successive matrix powers, which yields

$$\boldsymbol{P}^2 = \begin{bmatrix} 0.75 & 0 & 0.25 \\ 0 & 0.5 & 0.5 \\ 0.25 & 0.25 & 0.5 \end{bmatrix}, \; \boldsymbol{P}^4 = \begin{bmatrix} 0.625 & 0.0625 & 0.3125 \\ 0.125 & 0.375 & 0.5 \\ 0.3125 & 0.25 & 0.4375 \end{bmatrix},$$

$$\boldsymbol{P}^8 = \begin{bmatrix} 0.4961 & 0.1406 & 0.3633 \\ 0.2812 & 0.2734 & 0.4453 \\ 0.3633 & 0.2227 & 0.4141 \end{bmatrix}, \; \boldsymbol{P}^{16} = \begin{bmatrix} 0.4176 & 0.1891 & 0.3933 \\ 0.3782 & 0.2135 & 0.4083 \\ 0.3933 & 0.2042 & 0.4026 \end{bmatrix},$$

$$\boldsymbol{P}^{32} = \begin{bmatrix} 0.4006 & 0.1996 & 0.3998 \\ 0.3993 & 0.2005 & 0.4003 \\ 0.3998 & 0.2001 & 0.4001 \end{bmatrix}, \; \boldsymbol{P}^{64} = \begin{bmatrix} 0.4 & 0.2 & 0.4 \\ 0.4 & 0.2 & 0.4 \\ 0.4 & 0.2 & 0.4 \end{bmatrix}.$$

All rows of the last matrix are equal, hence we conclude that $\boldsymbol{\pi} = [0.4, 0.2, 0.4] = \left[\frac{2}{5}, \frac{1}{5}, \frac{2}{5}\right]$.

Example 7.2 Q Let us consider again the two-state Markov chain of Example 7.2 G, also known as Gilbert model, whose state-transition diagram and transition matrix is reported in Fig. 7.2. Suppose we are now interested in determining the percentage of time in which the facility is busy (state 1) and the probability that the facility is found busy at time $K = 10000$, provided that the facility was initially idle (state 0). These quantities can be determined by resorting to the asymptotic analysis of the MC presented in this section. To determine the fraction of time in which the facility is busy we observe the process for a large number k of consecutive steps. Let r_k count the number of steps in which the facility is busy during the observation window of length k, i.e.,

$$r_k = \sum_{n=1}^{k} \chi\{X_n = 1\}\,.$$

An estimate of the fraction of time during which the facility is busy is given by

$$\frac{r_k}{k} = \frac{\sum_{n=1}^{k} \chi\{X_n = 1\}}{k} \tag{7.80}$$

which corresponds to the time average of the rp $\chi\{X_n = 1\}$, $n = 1, 2, \ldots$. Assuming $0 < pq < 1$, the MC is aperiodic and positive recurrent and, therefore, it is also ergodic.[7] Hence, for sufficiently large k, the estimate (7.80) can be approximated by the statistical expectation of $\chi\{X_n = 1\}$, that is

$$p_1(n) = \mathrm{E}\left[\chi\{X_n = 1\}\right] = \frac{p}{p+q} + (1-p-q)^n \left(p_0(1) - \frac{p}{p+q}\right) \tag{7.81}$$

where we have used the result (7.34) derived in Example 7.2 G. We note that, when $|1-p-q| < 1$, the right-most term in (7.81) goes to zero as n approaches infinity, so that asymptotically we have

$$\lim_{n\to\infty} p_1(n) = \frac{p}{p+q} \tag{7.82}$$

that corresponds to the fraction of time in which the facility is busy. Although in this simple case we have been able to determine the asymptotic probability directly from $\boldsymbol{p}(n)$, in general this is rather difficult. However, for large values of n, the probability $p_j(n)$ approaches the asymptotic probability π_j, which can be easily computed by solving the system of balance equations (7.78), which yields

$$\begin{cases} \pi_0 p = \pi_1 q \\ \pi_0 + \pi_1 = 1 \end{cases} \implies \pi_1 = \frac{\pi_0 p}{q} = \frac{(1-\pi_1)p}{q} \implies \pi_1 = \frac{p}{p+q},$$

as given by (7.82). Therefore, assuming that $K = 10000$ steps are sufficient for the system to reach the steady-state behavior, the probability π_1 can be interpreted as the chance that, observing the facility at time K, it is found in busy state. Otherwise, we should resort to evaluating $p_1(k)$ by (7.81).

Example 7.2 R Consider a manufacturing process that requires blocks of raw material to pass through three consecutive stages of refinement, numbered 0, 1 and 2. Each stage can be repeated for multiple work-cycles, each of duration T. After a cycle in stage j, the process continues to the next stage with probability p_j, whereas it remains in stage j for another cycle with probability $1 - p_j$, independently of the number of work-cycles already performed. When stage 2 is finally completed, the process may start anew from stage zero with a new block of raw material. We are interested in determining the fraction of time the refinement process remains in each of the stages. Furthermore, assuming that each work-cycle in stage j costs ε_j units of energy, we wish to determine the average energy cost of one entire refinement process.

Let X_n denote the stage of the refinement process at the nth work-cycle. It is easy to realize that X_n is the simple MC represented in Fig. 7.11.

The asymptotic state probability vector $\boldsymbol{\pi} = [\pi_0, \pi_1, \pi_2]$ of this process gives the fraction of time the refinement process remains in each stage, whereas the average energy cost of the entire refinement process corresponds to $\varepsilon_0\pi_0 + \varepsilon_1\pi_1 + \varepsilon_2\pi_2$. Therefore, it remains to determine the asymptotic probability vector $\boldsymbol{\pi}$. To this end, we solve the system of balance equations (7.78) that, in this specific case, can be written as

$$\begin{cases} \pi_0 p_0 = \pi_2 p_2 \\ \pi_2 p_2 = \pi_1 p_1 \\ \pi_0 + \pi_1 + \pi_2 = 1 \end{cases} \qquad \begin{cases} \pi_2 = \frac{p_0}{p_2}\pi_0 \\ \pi_1 = \frac{p_2}{p_1}\pi_2 \\ \pi_0 + \pi_1 + \pi_2 = 1 \end{cases} \qquad \begin{cases} \pi_2 = \frac{p_0}{p_2}\pi_0 \\ \pi_1 = \frac{p_0}{p_1}\pi_0 \\ \pi_0\left(1 + \frac{p_0}{p_2} + \frac{p_0}{p_1}\right) = 1 \end{cases}$$

[7] The case with $p = q = 0$ is of no practical interest, corresponding to a MC with two isolated states. When $p = q = 1$, instead, the process is periodic with period $d = 2$ and, hence, it does not admit a unique asymptotic state probability vector.

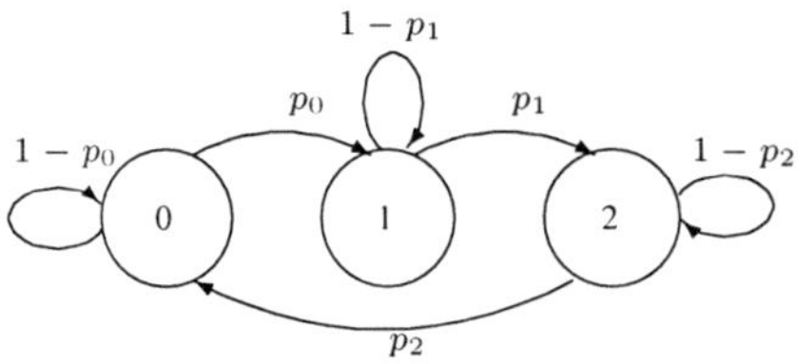

$$P = \begin{bmatrix} 1-p_0 & p_0 & 0 \\ 0 & 1-p_1 & p_1 \\ p_2 & 0 & 1-p_2 \end{bmatrix}$$

Figure 7.11 Transition diagram and transition matrix of the MC of Example 7.2 R.

whose solution is

$$\begin{cases} \pi_0 = \dfrac{p_2\, p_1}{p_1\, p_2 + p_0\, p_1 + p_0\, p_2}\,; \\ \pi_1 = \dfrac{p_0\, p_2}{p_1\, p_2 + p_0\, p_1 + p_0\, p_2}\,; \\ \pi_2 = \dfrac{p_0\, p_1}{p_1\, p_2 + p_0\, p_1 + p_0\, p_2}\,. \end{cases} \tag{7.83}$$

Example 7.2 S Consider again the MC of Example 7.2 H, represented in Fig. 7.4. The time-dependent behavior of the probability $p_j(n)$ for the states $j = 0$ and $j = 5$ when n increases was reported in the plot of Fig. 7.5, for two different settings of the initial state. We observed that the curves asymptotically tend to a limiting value that did not depend on the initial setting. We now have the tools for determining such asymptotic probabilities. Writing the system of balance equations (7.78) for this MC, we have

$$\begin{cases} \frac{1}{2}\pi_0 = \frac{1}{2}\pi_1 \\ \frac{1}{2}\pi_1 = \frac{1}{3}\pi_2 \\ \frac{2}{3}\pi_2 = \frac{1}{4}\pi_3 \\ \frac{3}{4}\pi_3 = \frac{1}{5}\pi_4 \\ \frac{4}{5}\pi_4 = \frac{1}{5}\pi_5 \\ \sum_{j=0}^{4}\pi_j = 1 \end{cases} \Longrightarrow \begin{cases} \pi_1 = \pi_0 \\ \pi_2 = \frac{3}{2}\pi_1 \\ \pi_3 = \frac{8}{3}\pi_2 \\ \pi_4 = \frac{15}{4}\pi_3 \\ \pi_5 = 4\pi_4 \\ \pi_0 + \pi_1 + \pi_2 + \pi_3 + \pi_4 + \pi_5 = 1 \end{cases} \Longrightarrow \begin{cases} \pi_1 = \pi_0 \\ \pi_2 = \frac{3}{2}\pi_0 \\ \pi_3 = 4\,\pi_0 \\ \pi_4 = 15\,\pi_0 \\ \pi_5 = 60\,\pi_0 \\ \pi_0\frac{165}{2} = 1 \end{cases}$$

from which we derive

$$\pi = \left[\frac{2}{165}, \frac{2}{165}, \frac{1}{55}, \frac{8}{165}, \frac{2}{11}, \frac{24}{33},\right]. \tag{7.84}$$

Remark It may be worth writing in clear letters that, even though the states probabilities become constant as n approaches infinity, it does not mean that the process stops moving after a given time instant. As a matter of fact, the process always varies, according to the transition matrix P. The only difference between short-term and long-term behavior of the process is that, in the short-term, the rate at which the process visits the different states is generally influenced by the initial conditions, whereas the statistics of the process evolution in the long-run does not depend any more on the initial state probabilities. As an example, we plot in Fig. 7.12 two realizations $X_n(\omega_1)$ (solid line) and $X_n(\omega_2)$ (dashed line) of the process considered in Examples 7.2 H and 7.2 S. $X_n(\omega_1)$ was obtained by setting the initial state in zero, whereas $X_n(\omega_2)$ started its evolution at state 5.

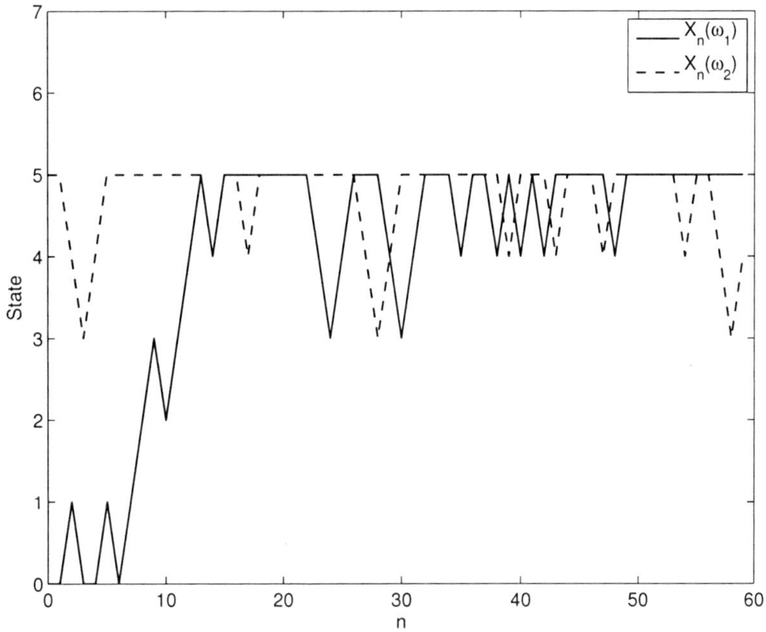

Figure 7.12 Two realizations of the process X_n of Example 7.2 S.

We observe that the two curves are initially distinct, since the time evolution is clearly influenced by the starting state, whereas after a few steps the curves overlap and intersect each other. From this point onward, it is no longer possible to infer the initial state of the process by observing its time-evolution, hence we say that the process has reached a steady-state behavior.

7.3 CONTINUOUS-TIME MARKOV CHAINS

Most of the theory developed for discrete-time MCs also applies, with just marginal adjustments, to continuous-time MCs. Therefore, in this section we will not put excessive emphasis on the aspects that are in common with the discrete-time case and that have been detailed in the previous section, whereas more attention will be paid to the aspects that characterize continuous-time MCs.

7.3.1 Definition of continuous-time MC

Definition 7.9 *A continuous-time rp* $\{X(t), t \geq 0\}$, *is a MC if for any integer* n, *for any ordered sequence of time instants* $0 < t_0 < t_1 < \ldots < t_n$, *and for any set* $\{\sigma_0, \sigma_1, \ldots, \sigma_n\}$

with $\sigma_i \in \mathcal{S}$, *where* $\mathcal{S} = \{0, 1, 2, \dots\}$ *is the state space, it holds*

$$\mathrm{P}\left[X(t_n) = \sigma_n | X(t_{n-1}) = \sigma_{n-1}, \dots, X(t_0) = \sigma_0\right] = \mathrm{P}\left[X(t_n) = \sigma_n | X(t_{n-1}) = \sigma_{n-1}\right] \tag{7.85}$$

We can recognize in (7.85) the *Markov property* for continuous time processes.

Example 7.3 A A blind drunk man, named Stunny, gets kicked out from a pub into a very long and narrow alley. Stunny is so drunk that he can barely control his step and he alternates periods in which he stands on the same position for a while with periods in which he slowly moves in a random direction along the alley. Let $X(t)$ denote the distance that separates Stunny from the pub entrance at time t, rounded to the meter. We assume that $X(t)$ can be modeled as a continuous-time rp, with countable state space $\mathcal{S} = \{0, 1, 2, \dots\}$. If Stunny's motion is driven only by its current distance from the pub entrance and it does not depend on the positions he occupied in the past nor on the time he has already spent in his current position, then $X(t)$ is a MC. In this case, the future evolution of the process from a given state does not depend on the path followed to reach that state, nor on the time Stunny has spent in his current state.

7.3.2 Transition probabilities of continuous-time MC

As for discrete-time MCs, the state space $\mathcal{S}$ will be hereinafter labelled by the nonnegative integers, i.e., $\mathcal{S} = \{0, 1, \dots, N\}$, where N can be infinite. The transition probability from state i to state j in a time interval of length u is denoted by

$$P_{i,j}(u) = \mathrm{P}\left[X(t_0 + u) = j | X(t_0) = i\right] . \tag{7.86}$$

In general, the transition probabilities may depend upon both the time interval u and the reference instant t_0, however we will limit our attention to *stationary* transition probabilities that are independent of t_0. In this case, the MC is termed *homogeneous*.

Example 7.3 B Let us further elaborate on Example 7.3 A. If Stunny's inebriation does not fade away in time, then the transition probabilities are stationary. In this case, the probability that, starting from a distance of $i = 3$ meters from the pub entrance, after a time interval of length u Stunny finds himself at a distance $j = 0$ is always equal to $P_{i,j}(u)$, independently of the time Stunny has already spent in the alley:

$$\mathrm{P}\left[X(t + u) = 0 | X(t) = 3\right] = P_{3,0}(u) , \quad t \geq 0 .$$

Conversely, if Stunny progressively sobers up, then the transition probabilities may change in time and the process would not be stationary anymore.

Following the footprints of Section 7.2.2, we define the *transition matrix* over a time interval u as:

$$\boldsymbol{P}(u) = \begin{bmatrix} P_{0,0}(u) & P_{0,1}(u) & \dots & P_{0,j}(u) & \dots & P_{0,N-1}(u) & P_{0,N}(u) \\ P_{1,0}(u) & P_{1,1}(u) & \dots & P_{1,j}(u) & \dots & P_{1,N-1}(u) & P_{1,N}(u) \\ \vdots & \vdots & \vdots & \vdots & \vdots & \vdots & \vdots \\ P_{i,0}(u) & P_{i,1}(u) & \dots & P_{i,j}(u) & \dots & P_{i,N-1}(u) & P_{i,N}(u) \\ \vdots & \vdots & \vdots & \vdots & \vdots & \vdots & \vdots \\ P_{N,0}(u) & P_{N,1}(u) & \dots & P_{N,j}(u) & \dots & P_{N,N-1}(u) & P_{N,N}(u) \end{bmatrix} \tag{7.87}$$

It is easy to realize that $\boldsymbol{P}(u)$ is a stochastic matrix, since it contains non negative elements and the elements of each row add up to one, i.e.,

$$\boldsymbol{P}(u)\mathbf{1} = \mathbf{1}\,. \tag{7.88}$$

In addition, we observe that letting u approach zero, the transition matrix approaches the identity matrix $\boldsymbol{I}$

$$\boldsymbol{P}(0) = \lim_{u\to 0^+} \boldsymbol{P}(u) = \boldsymbol{I}\,. \tag{7.89}$$

This result follows directly from the statistics of the sojourn times and the Chapman-Kolmogorov equations for continuous-time MC, which will be discussed in the next sections. However, the above relation lends itself to a simple intuitive interpretation, that is to say, over a zero time period the process remains in its current state with probability one.

7.3.3 Sojourn times of continuous-time MC

The Markov property is intimately related to the statistical distribution of the *sojourn time* associated to a given state j, defined as the time the process takes for a change of state. In fact, according to (7.85), the transition probability from a given state j into any other state in a finite time interval does not depend on the time the process has already spent in state j. In other terms, the sojourn time must satisfy the *memoryless property* and, as seen in (2.230), the only continuos distribution that satisfies this property is the exponential distribution.

Proposition 7.10 *The sojourn times s_j in a state $j \in \mathcal{S}$ of a continuous-time MC are statistically independent and exponentially distributed rvs with PDF*

$$p_{s_j}(a) = q_j \exp(-q_j\, a)\,; \quad a \geq 0\,, \tag{7.90}$$

where

$$q_j = \frac{1}{\mathrm{E}\left[s_j\right]} \tag{7.91}$$

is the parameter of the exponential PDF.

Proof Let $X(t_1, t_2) = j$ denote the event that the process *remains* in state j during the time interval $(t_1, t_2]$. The probability that the sojourn time s_j exceeds a can be expressed as

$$\mathrm{P}\left[s_j > a\right] = \mathrm{P}\left[X(0,a) = j | X(0) = j, X(0^-) \neq j\right] \tag{7.92}$$

where the condition $\{X(0) = j, X(0^-) \neq j\}$ indicates that the process entered state j exactly at time $t = 0$. Due to the Markov property, however, once given the state at time $t = 0$, the states of the process before this instant are irrelevant, so that (7.92) can be rewritten as

$$P[s_j > a] = P[X(0, a) = j | X(0) = j] . \quad (7.93)$$

Now, for any $a \geq 0$ and $b > 0$, (7.93) yields

$$P[s_j > a + b] = P[X(0, a + b) = j | X(0) = j] . \quad (7.94)$$

Partitioning the interval $(0, a + b]$ in the two consecutive and non-overlapping intervals $(0, a]$ and $(a, a + b]$, (7.94) can be written as

$$P[s_j > a + b] = P[X(a, a + b) = j, X(0, a) = j | X(0) = j]$$

and, applying properties of conditional probability, we get

$$P[s_j > a + b] = P[X(a, a + b) = j | X(0, a) = j, X(0) = j] \, P[X(0, a) = j | X(0) = j] . \quad (7.95)$$

Because of the Markov property, the condition $X(0) = j$ in the first conditional probability is irrelevant, so that (7.94) becomes

$$P[s_j > a + b] = P[X(a, a + b) = j | X(a) = j] \, P[X(0, a) = j | X(0) = j] . \quad (7.96)$$

Comparing (7.96) with (7.93), we realize that, being the transition probabilities stationary, the two conditional probabilities on the right-hand side of (7.96) can be replaced by $P[s_j > b]$ and $P[s_j > a]$, respectively, so that we obtain

$$P[s_j > a + b] = P[s_j > b] \, P[s_j > a] . \quad (7.97)$$

The above equation is equivalent to the memoryless property $P[s_j > a + b | s_j > b] = P[s_j > a]$, hence we conclude that the sojourn time s_j is exponentially distributed.

Finally, the independence of the sojourn times follows directly from the Markov property. In fact, once the process enters a new state, its evolution does not depend any more on the time it spent in the previous state. □

Example 7.3 C We can now make explicit an assumption that was kept silent in Example 7.3 A. In fact, in order for Stunny's random walk to be modeled as a MC it is mandatory that the periods in which Stunny remains in the same spot, which correspond to the sojourn times of the process $X(t)$ in different states, are independent rvs with exponential distributions. In any other case, the process $X(t)$ would not satisfy the Markov property. As a counter-example, assume that the pub's sign does not work properly and blinks at random intervals of 30, 60 or 120 seconds, with equal probability. Stunny moves a step towards or away from the pub's entrance every time the pub's sign blinks. Now, suppose that two students (of engineering), Andrei and Mark, observe Stunny's motion from a window open on the alley. After a while, both students have understood the characteristics of Stunny's motion. Suddenly, Andrei challenges Mark to bet that Stunny will still be in front of the pub entrance in 30 seconds since then.

Casting a glance out of the window, Mark sees that Stunny is currently in front of the pub entrance, but he does not know how long the man has been in that spot. Hence, he makes the following reasoning.

Let s_0 denote the sojourn time of Stunny in his current position. Mark knows that s_0 can either be equal to 30, 60 or 120 seconds. Then, let s_0' be the residual sojourn time, that is to say, the time before Stunny makes the next step. Mark calculates that $\mathrm{P}[s_0' < 30|s_0 = 30] = 1$, that is to say, if $s_0 = 30$ s Stunny will certainly move one step in the next 30 seconds. In the other two cases, instead, Stunny may or not move, depending on the time already spent in his current position. Without further information, Mark conjectures that, given s_0, the residual sojourn time is uniformly distributed in the interval $(0, s_0]$. Hence, he rapidly calculates that $\mathrm{P}[s_0' < 30|s_0 = 60] = 1/2$ and $\mathrm{P}[s_0' < 30|s_0 = 120] = 1/4$. Finally, assuming that s_0 takes the three values with equal probability, Mark concludes that Stunny will make a step away from its current position with probability $P = 1/3(1 + 1/2 + 1/4) = 7/12$, which is greater than one half. Hence, Mark accepted the bet, persuaded that his superior mathematical skills will increase his chance to win the bet. Unfortunately for him, Andrei is even smarter!

In fact, Andrei had kept note of Stunny's position from time to time, noting down the following observations $X(0) = 0$, $X(25) = 1$, $X(50) = 1$, $X(55) = 0$, $X(100) = 0$, $X(125) = 0$, where $t_0 = 125$ is the time of the bet. On the base of these observations, Andrei deduced that Stunny had been in the same spot for more than 70 seconds, but less then 75 seconds. Therefore, he concluded that the next blinking of the pub's sign would have occurred from 45 to 50 seconds later. Meanwhile, Stunny would have not moved from his current position. After this (extremely quick) reasoning, Andrei knows for sure that he will win the bet!

In this example, Andrei is advantaged because he can leverage on the past observations of Stunny's position to refine his guess on the future evolution of the process. This is possible only because the sojourn times in the different state are not exponentially distributed or, in other words, the process $X(t)$ is not Markovian. Conversely, if the process were Markovian, Andrei would not have any advantage over Mark from his past observations.

Starting at time 0 in state j, we now introduce the "residual sojourn time" in state j, denoted by s_j', as the time still remaining in state j when the process is observed in state j at a given instant t, so that we have

$$\mathrm{P}\left[s_j' > a\right] = \mathrm{P}\left[X(t, t+a) = j|X(t) = j\right] \tag{7.98}$$

where the notation $X(t, t+a) = j$ has been introduced in the proof of Proposition 7.10 to indicate that the process does not leave state j in the interval $(t, t+a]$. Note that, due to the memoryless property (2.230) of the exponential distribution of the sojourn time s_j, the residual sojourn time s_j' has the same exponential distribution of any sojourn times in state j, that is

$$p_{s_j}(a) = p_{s_j'}(a) = q_j e^{-q_j a}, \qquad a \geq 0. \tag{7.99}$$

In the following we will not distinguish between residual sojourn time or sojourn time, unless necessary.

Intuitively, the sojourn time in a given state j must be related to the tendency of the process to remain in that state, hence it should depend on the transition probability $P_{j,j}(t)$. This intuition is formalized by the following proposition.

Proposition 7.11 *In a continuous-time MC, the transition probability $P_{j,j}(h)$ over a short time interval of length h is equal to*

$$P_{j,j}(h) = \mathrm{P}\left[s_j > h\right] + o(h) = 1 - q_j\, h + o(h) \tag{7.100}$$

where $q_j = 1/\mathrm{E}[s_j]$ is the parameter of the exponential distribution of the sojourn time in state j and $o(h)$ is an infinitesimal of order higher than h, i.e.,

$$\lim_{h \to 0} \frac{o(h)}{h} = 0\,.$$

Proof A process that starts in state j at time 0 is observed in the same state j at time h if either it does not make any state-transition in the interval $(0, h]$, or it makes multiple transitions, the last of which brings it back to state j. By the Total Probability Theorem 2.12, we hence have

$$P_{j,j}(h) = \mathrm{P}[X(h) = j | X(0) = j] = \mathrm{P}[s_j > h] + \mathrm{P}[X(h) = j, s_j \le h | X(0) = j] \quad (7.101)$$

where $\mathrm{P}[s_j > h]$ is the probability that the process never leaves state j in the interval $(0, h]$, given that it is in j at time zero. Let us focus on the right-most term of (7.101), which accounts for the probability that the process makes multiple transitions before returning in j within a time interval h. For ease of notation, we will denote this probability by P_{mul}.

Applying the Total Probability Theorem with respect to s_j, we obtain

$$P_{\mathrm{mul}} = \int_0^h p_{s_j}(a)\, \mathrm{P}[X(h) = j | s_j = a, X(0) = j]\, da \qquad (7.102)$$

where $p_{s_j}(a)$ is the PDF of s_j given in (7.90). The term $\mathrm{P}[X(h) = j | s_j = a, X(0) = j]$ is the conditional probability of being in state j at time h given that the process leaves state j at time a. Denoting by i the state entered when the process leaves j, (7.102) can be written as

$$\begin{aligned} P_{\mathrm{mul}} &= \int_0^h p_{s_j}(a) \sum_{i \neq j} \mathrm{P}[X(h) = j, X(a) = i | s_j = a, X(0) = j]\, da \\ &= \int_0^h p_{s_j}(a) \sum_{i \neq j} \mathrm{P}[X(h) = j | X(a) = i]\, \mathrm{P}[X(a) = i | s_j = a, X(0) = j]\, da \\ &= \int_0^h p_{s_j}(a) \sum_{i \neq j} P_{i,j}(h - a) \tilde{P}_{j,i}\, da \end{aligned} \qquad (7.103)$$

where the second last row follows from the Markov property, whereas in the last row we have written $\tilde{P}_{j,i}$ in place of $\mathrm{P}[X(a) = i | s_j = a, X(0) = j]$, for ease of notation.[8] A necessary, but not sufficient, condition for the transition from i to j in a time interval $h - a$ to occur is that the sojourn time in state i is less than $h - a$. Therefore, we deduce that

$$P_{i,j}(h - a) \le \mathrm{P}[s_i \le h - a]\,. \qquad (7.104)$$

Using (7.104) in (7.103) we obtain

$$P_{\mathrm{mul}} \le \int_0^h p_{s_j}(a) \sum_{i \neq j} \mathrm{P}[s_i \le h - a]\, \tilde{P}_{j,i}\, da \qquad (7.105)$$

[8] We observe that $\tilde{P}_{j,i}$ gives the probability that, when it leaves state j, the process moves in state i. We will find again these conditional transition probabilities in a later section on embedded MCs.

and, recalling (7.90) we get

$$\begin{aligned} P_{\text{mul}} &\leq \int_0^h q_j e^{-q_j a} \sum_{i \neq j} \left(1 - e^{-q_i (h-a)}\right) \tilde{P}_{j,i} \, da \\ &= \int_0^h q_j e^{-q_j a} \, da \left(\sum_{i \neq j} \tilde{P}_{j,i} \right) - \sum_{i \neq j} \tilde{P}_{j,i} \, e^{-q_i h} q_j \int_0^h e^{-a(q_j - q_i)} \, da \, . \end{aligned} \tag{7.106}$$

Considering that, because of the normalization condition $\sum_{i \neq j} \tilde{P}_{j,i} = 1$, the bracketed sum in (7.106) equals to 1, and after some algebra we obtain

$$P_{\text{mul}} \leq 1 - e^{-q_j h} - \sum_{i \neq j} \tilde{P}_{j,i} \, e^{-q_i h} \frac{q_j}{q_j - q_i} + \sum_{i \neq j} \tilde{P}_{j,i} \frac{q_j}{q_j - q_i} e^{-q_j h} \, . \tag{7.107}$$

Now, using the Taylor series expansion $e^{-x} = 1 - x + o(x)$, the above inequality becomes

$$\begin{aligned} P_{\text{mul}} &\leq q_j \, h - \sum_{r \neq j} \tilde{P}_{j,r} \frac{q_j}{q_j - q_r} (q_j - q_r) \, h + o(h) \\ &= q_j \, h - q_j \, h \sum_{r \neq j} \tilde{P}_{j,r} + o(h) \, , \end{aligned} \tag{7.108}$$

and invoking again the normalization condition $\sum_{r \neq j} \tilde{P}_{j,r} = 1$ we finally obtain

$$P_{\text{mul}} = o(h) \, . \tag{7.109}$$

This result, which is interesting on its own right, basically states that the probability of multiple transitions in an short time h is infinitesimal of order higher than h. Then, the probability that a MC is still in state j at time h, given that it was in state j at time zero, is basically determined by the probability that the process does never leave that state in the short time interval h. Last, replacing (7.109) into (7.101) we have

$$P_{j,j}(h) = \mathrm{P}\left[s_j > h\right] + o(h)$$

and being

$$\mathrm{P}\left[s_j > h\right] = e^{-q_j h} = 1 - q_j h + o(h)$$

we get (7.100). □

Remark Proposition 7.11 basically says that, in a very short time, the process can make no transitions or, at most, a single transition. The probability that the MC makes two or more "almost simultaneous" transitions is negligible.

7.3.4 Chapman-Kolmogorov equations for continuous-time MC

As for discrete-time MCs, in the continuous time it is also possible to obtain the transition probabilities over a given time interval from the transmission probabilities over smaller intervals that partition the original one, as dictated by the following proposition.

Proposition 7.12 (Chapman-Kolmogorov equations) *For any pair of non-negative values u, v, the transition matrix in the time interval $u+v$ can be expressed as*

$$\boldsymbol{P}(u+v) = \boldsymbol{P}(u)\boldsymbol{P}(v) \tag{7.110}$$

that is to say,

$$P_{i,j}(u+v) = \sum_{s=0}^{N} P_{i,s}(u)P_{s,j}(v)\,, \quad i,j \in \mathcal{S}\,. \tag{7.111}$$

Proof The proof is omitted, being conceptually identical to that provided for discrete-time MCs. □

Equation (7.110) holds for any choice of the intervals u and v. In particular, by letting u approach zero, we get

$$\boldsymbol{P}(v) = \lim_{u\to 0} \boldsymbol{P}(u+v) = \left(\lim_{u\to 0} \boldsymbol{P}(u)\right)\boldsymbol{P}(v) = \boldsymbol{P}(0)\boldsymbol{P}(v)\,, \quad v \geq 0$$

which implicitly yields $\boldsymbol{P}(0) = \boldsymbol{I}$, as already observed.

7.3.5 The infinitesimal generator matrix $\boldsymbol{Q}$

Discussing discrete-time MC, we observed the primary role played by the one-step transition matrix $\boldsymbol{P}$ in characterizing the statistical evolution of the process. The counterpart of $\boldsymbol{P}$ for continuous time MC is played by the so-called *infinitesimal generator matrix*, or *intensity matrix*,

$$\boldsymbol{Q} = \begin{bmatrix} q_{0,0} & q_{0,1} & \cdots & q_{0,j} & \cdots & q_{0,N-1} & q_{0,N} \\ q_{1,0} & q_{1,1} & \cdots & q_{1,j} & \cdots & q_{1,N-1} & q_{1,N} \\ \vdots & \vdots & \ddots & \vdots & \ddots & \vdots & \vdots \\ q_{i,0} & q_{i,1} & \cdots & q_{i,j} & \cdots & q_{i,N-1} & q_{i,N} \\ \vdots & \vdots & \ddots & \vdots & \ddots & \vdots & \vdots \\ q_{N,0} & q_{N,1} & \cdots & q_{N,j} & \cdots & q_{N,N-1} & q_{N,N} \end{bmatrix}$$

which is defined as

$$\boxed{\boldsymbol{Q} = \lim_{h\to 0^+} \frac{\boldsymbol{P}(h) - \boldsymbol{I}}{h}\,.} \tag{7.112}$$

The elements in the main diagonal of $\boldsymbol{Q}$ are given by

$$q_{j,j} = \lim_{h\to 0^+} \frac{P_{j,j}(h) - 1}{h}\,, \tag{7.113}$$

provided that the limit exists and is finite. We observe that these terms are all negative. Using (7.100) into (7.113) and changing the sign of all the terms, we have

$$-q_{j,j} = \lim_{h\to 0^+} \frac{1 - P_{j,j}(h)}{h} = \lim_{h\to 0^+} \frac{q_j h + o(h)}{h} = q_j\,. \tag{7.114}$$

Hence, (7.100) can be rewritten as

$$P_{j,j}(h) = 1 + q_{j,j}\, h + o(h) \tag{7.115}$$

which links the elements in the main diagonal of the infinitesimal generator matrix $\boldsymbol{Q}$ with the self-transition probabilities.

The non-diagonal terms of $\boldsymbol{Q}$ are given by

$$q_{i,j} = \lim_{h\to 0^+} \frac{P_{i,j}(h)}{h}, \quad i \neq j, \tag{7.116}$$

provided that the limit exists and is finite. Rewriting (7.116) in terms of the transition probabilities in an infinitesimal time interval of length h, we obtain

$$P_{i,j}(h) = q_{i,j} h + o(h), \quad i, j \in \mathcal{S},\ j \neq i. \tag{7.117}$$

According to (7.117), for a very small time interval h the transition probability $P_{i,j}(h)$, for any i, j with $i \neq j$, is a linear function of h through the coefficient $q_{i,j}$ of the infinitesimal generator matrix, except for the correction terms $o(h)$ that becomes negligible as h approaches zero. For this reason, coefficients $q_{i,j}$ are said *transition intensities.*

Finally, we note that multiplying the matrix $\boldsymbol{Q}$ by the unit vector $\mathbf{1}$ we get

$$\boldsymbol{Q}\,\mathbf{1} = \lim_{h\to 0^+} \frac{\boldsymbol{P}(h)\,\mathbf{1} - \boldsymbol{I}\,\mathbf{1}}{h} = \lim_{h\to 0^+} \frac{\mathbf{1} - \mathbf{1}}{h} = \mathbf{0}^{\mathrm{T}} \tag{7.118}$$

where $\mathbf{0}$ is the all-zero row vector. In other words, the sum of the elements in each row of $\boldsymbol{Q}$ is zero, i.e.,

$$\sum_{j=0}^{N} q_{i,j} = 0, \tag{7.119}$$

from which we get

$$\sum_{j\neq i}^{N} q_{i,j} = -q_{i,i} = q_i. \tag{7.120}$$

It is possible to show that the non-diagonal transition intensities $q_{i,j}$ are *always* finite [5]. Consequently, when the number of states is finite, the diagonal elements $q_{i,i} = -q_i$ are also finite, whereas for MCs with infinite states, these elements may be infinite. A state j for which $q_j = \infty$ is said to be *instantaneous* because the sojourn time in such a state is zero with probability one. MCs with instantaneous states are of little interest to us, so that in the sequel we confine the discussion to MCs with only non-instantaneous states, which are termed *conservative.* We also observe from (7.90) that the mean sojourn time in state j becomes infinite when $q_j = 0$. In this case, state j is said to be an *absorbing state*, for the process never leaves such a state after the first visit.

7.3.6 Forward and Backward equations for continuous-time MC

In discrete-time MCs it is possible to determine the transition probabilities in any number of steps from the (one-step) transition matrix $\boldsymbol{P}$ through the Chapman-Kolmogorov equations, as expressed by (7.19). The following proposition formalizes a similar result for continuous-time MCs.

Proposition 7.13 *Given the infinitesimal generator matrix $\boldsymbol{Q}$, the transition matrix in the time interval t, $\boldsymbol{P}(t)$, is differentiable for all $t \geq 0$ and the derivative $\boldsymbol{P}'(t)$ is given by the so-called* forward Kolmogorov equation

$$\boldsymbol{P}'(t) = \boldsymbol{P}(t)\,\boldsymbol{Q} \tag{7.121}$$

or, equivalently, by the backward Kolmogorov equation

$$\boldsymbol{P}'(t) = \boldsymbol{Q}\,\boldsymbol{P}(t)\,. \tag{7.122}$$

Proof The proposition is clearly true for $t = 0$, since from (7.112) we have $\boldsymbol{Q} = \boldsymbol{P}'(0)$ and $\boldsymbol{P}(0) = \boldsymbol{I}$. For $t > 0$, the derivative of $\boldsymbol{P}(t)$ is

$$\boldsymbol{P}'(t) = \lim_{h\to 0} \frac{\boldsymbol{P}(t+h) - \boldsymbol{P}(t)}{h}\,. \tag{7.123}$$

Using the Chapman-Kolmogorov equation (7.110), we can write $\boldsymbol{P}(t+h) = \boldsymbol{P}(t)\boldsymbol{P}(h)$ that, replaced in (7.123), gives

$$\boldsymbol{P}'(t) = \lim_{h\to 0} \frac{\boldsymbol{P}(t)(\boldsymbol{P}(h) - \boldsymbol{I})}{h} = \boldsymbol{P}(t) \lim_{h\to 0} \frac{\boldsymbol{P}(h) - \boldsymbol{I}}{h} = \boldsymbol{P}(t)\,\boldsymbol{Q}\,,$$

which proves the forward equation. Similarly, by using $\boldsymbol{P}(t+h) = \boldsymbol{P}(h)\boldsymbol{P}(t)$ in (7.123) we get

$$\boldsymbol{P}'(t) = \lim_{h\to 0} \frac{(\boldsymbol{P}(h) - \boldsymbol{I})\boldsymbol{P}(t)}{h} = \boldsymbol{Q}\,\boldsymbol{P}(t)\,,$$

which proves the backward equation. □

The forward and backward equations are matrix differential equations in t. For finite-state processes ($N < \infty$), the transient solution of (7.121) and (7.122) takes the form

$$\boldsymbol{P}(t) = \exp(\boldsymbol{Q}\,t)\,, \quad t \geq 0 \tag{7.124}$$

where $\exp(\boldsymbol{Q}\,t)$ is the matrix exponential of $\boldsymbol{Q}t$, which can be expressed by using the Taylor series expansion of the exponential function in matrix form:

$$\exp(\boldsymbol{Q}\,t) = \boldsymbol{I} + \sum_{m=1}^{\infty} \frac{\boldsymbol{Q}^m\,t^m}{m!}\,. \tag{7.125}$$

Apart from the Taylor series expansion, the matrix exponential (7.124) can also be expressed as

$$\boxed{\boldsymbol{P}(t) = \exp(\boldsymbol{Q}\,t) = \lim_{m\to\infty} \left(\boldsymbol{I} + \frac{\boldsymbol{Q}\,t}{m}\right)^m}$$

which is the matrix equivalent of the well known limit $\lim_{m\to\infty}(1 + x/m)^m = e^x$.

Equation (7.124) proves the importance of the infinitesimal generator matrix $\boldsymbol{Q}$ in the characterization of the statistical properties of a continuous-time MC. In fact, through (7.124) it is possible to determine the transition probability matrix $\boldsymbol{P}(t)$ for any $t > 0$. Nonetheless, explicit solutions for $\boldsymbol{P}(t)$ in terms of the transition intensities $q_{i,j}$ are often very difficult to obtain, except for simple cases.

Example 7.3 D Consider a facility of some nature that can be either idle or busy. Let us denote by $X(t)$ the state of the facility at time t, so that $X(t) = 0$ if the facility is idle and $X(t) = 1$ if it is busy. Suppose that the idle and busy periods have random durations, with statistically independent exponential distribution with parameter λ and μ, respectively. Hence, the probability $P_{0,1}(h)$ of the system going from idle to busy in a very short time h is

$$P_{0,1}(h) = \mathrm{P}\left[s_0 \leq h\right] + o(h) = 1 - e^{-\lambda h} + o(h) = \lambda\, h + o(h)\,.$$

Similarly, the probability of going from busy to idle in a time h is

$$P_{1,0}(h) = \mathrm{P}\left[s_1 \leq h\right] + o(h) = 1 - e^{-\mu h} + o(h) = \mu\, h + o(h)\,.$$

Comparing these equations with (7.117), we realize that $X(t)$ can be described as a continuous-time MC with infinitesimal generator matrix

$$\boldsymbol{Q} = \begin{bmatrix} -\lambda & \lambda \\ \mu & -\mu \end{bmatrix}.$$

The forward equation (7.121), hence, turns out to be equal to

$$\begin{bmatrix} P'_{0,0}(t) & P'_{0,1}(t) \\ P'_{1,0}(t) & P'_{1,1}(t) \end{bmatrix} = \begin{bmatrix} P_{0,0}(t) & P_{0,1}(t) \\ P_{1,0}(t) & P_{1,1}(t) \end{bmatrix} \begin{bmatrix} -\lambda & \lambda \\ \mu & -\mu \end{bmatrix},$$

which gives rise to the following system of equations

$$\begin{cases} P'_{0,0}(t) = -\lambda\, P_{0,0}(t) + \mu\, P_{0,1}(t) \\ P'_{0,1}(t) = \lambda\, P_{0,0}(t) - \mu\, P_{0,1}(t) \\ P'_{1,0}(t) = -\lambda\, P_{1,0}(t) + \mu\, P_{1,1}(t) \\ P'_{1,1}(t) = \lambda\, P_{1,0}(t) - \mu\, P_{1,1}(t) \end{cases} \tag{7.126}$$

However, recalling that $P_{0,1}(t) = 1 - P_{0,0}(t)$ and $P_{1,0}(t) = 1 - P_{1,1}(t)$, the system (7.126) reduces to

$$\begin{cases} P'_{0,0}(t) & = -(\lambda + \mu) P_{0,0}(t) + \mu \\ P'_{1,1}(t) & = -(\lambda + \mu) P_{1,1}(t) + \lambda \end{cases}$$

Only the first equation needs to be solved since, by symmetry, the solution of $P_{1,1}(t)$ equals that of $P_{0,0}(t)$ after switching the roles of λ and μ. We have then to solve a first-order, non-homogeneous linear differential equation in the form

$$g'(t) = -\alpha g(t) + \beta\,, \quad t \geq 0\,. \tag{7.127}$$

Such a differential equation admits solution in the general form

$$g(t) = c(t) e^{-\alpha t} \tag{7.128}$$

where $c(t)$ needs to be chosen in order to satisfy the original differential equation. In our case, by replacing (7.128) in (7.127) we get

$$g'(t) = -\alpha c(t) e^{-\alpha t} + c'(t) e^{-\alpha t} = -\alpha g(t) + c'(t) e^{-\alpha t}$$

which requires

$$c'(t)e^{-\alpha t} = \beta \quad \Longrightarrow \quad c'(t) = \beta e^{\alpha t} \Longrightarrow \quad c(t) = \frac{\beta}{\alpha} e^{\alpha t} + \gamma \, .$$

where the constant γ can be determine from the initial condition $g(0) = c(0)$, which yields

$$\gamma = g(0) - \frac{\beta}{\alpha} \, .$$

Hence, the final solution of the differential equation is

$$g(t) = e^{-\alpha t} \left(\frac{\beta}{\alpha} e^{\alpha t} + g(0) - \frac{\beta}{\alpha} \right) = \frac{\beta}{\alpha} + \left(g(0) - \frac{\beta}{\alpha} \right) e^{-\alpha t} \, . \tag{7.129}$$

Making the substitutions $g(t) \to P_{0,0}(t)$, $\alpha \to (\lambda + \mu)$ and $\beta \to \mu$ and recalling the initial condition $P_{0,0}(0) = 1$ as stated by (7.89), we finally get the solution of our forward equations

$$P_{0,0}(t) = \frac{\mu}{\lambda + \mu} + \frac{\lambda}{\lambda + \mu} e^{-(\lambda + \mu)t} \, ,$$

$$P_{1,1}(t) = \frac{\lambda}{\lambda + \mu} + \frac{\mu}{\lambda + \mu} e^{-(\lambda + \mu)t} \, .$$

7.3.7 Embedded Markov chain

We have seen that a continuous-time MC $X(t)$ remains in a given state j_1 for a random time, with exponential distribution with parameter q_{j_1}. When a transition occurs, the process jumps to another state $j_2 \neq j_1$ where it stays for a random time with exponential distribution with parameter q_{j_2}. Sampling the continuous-time MC in the instants immediately after a change of state, we obtain a sequence of states $j_1, j_2, j_3, \ldots$ of a process that is *embedded* in the original process and that evolves in discrete steps (with random duration). Let $\tilde{X}_n$ denote the state of such a process at the nth sample. It is easy to realize that, when no state is absorbing, the sampled process $\tilde{X}_n$ can be seen as a discrete-time MC, which is called *embedded MC*.

Remark The embedded MC is defined over a discrete time domain that is obtained by sampling the original (continuous) time domain with nonuniform sampling period. As a matter of fact, each step of the embedded MC actually takes an exponentially-distributed random time, with parameter depending on the departing state.[9]

The transition probabilities matrix $\tilde{\boldsymbol{P}}$ of the embedded MC $\tilde{X}_n$ can be directly determined from the infinitesimal generator matrix $\boldsymbol{Q}$ of the original process $X(t)$. Let $\tilde{P}_{i,j}$ denote the one-step transition probability of $\tilde{X}_n$, that is to say, the probability that $X(t)$ directly enters j from i, given that a transition occurs. First of all, we observe that the embedded process is sampled only when the original process makes a transition to another state. Consequently, we

[9] In case the original process possesses some absorbing state, a transition out of any of such absorbing states would take infinite time. For this reason, absorbing states need to be treated separately.

have $\tilde{P}_{i,i} = 0$ for any i, that is to say, the embedded MC does not admit self-transitions.[10] For $j \neq i$, instead, we have

$$\tilde{P}_{i,j} = \lim_{h \to 0^+} \mathrm{P}\left[X(t+h) = j | X(t+h) \neq i, X(t) = i\right] \tag{7.130}$$

where the condition $\{X(t+h) \neq i, X(t) = i\}$, for $h \to 0$, denotes a transition out of state i occurring at time t. The right-hand side of (7.130) can be written as

$$\frac{\mathrm{P}\left[X(t+h) = j, X(t+h) \neq i, X(t) = i\right]}{\mathrm{P}\left[X(t+h) \neq i, X(t) = i\right]}$$

and, observing that the event $X(t+h) \neq i$ includes $X(t+h) = j$, the previous equation becomes

$$\begin{aligned}
\frac{\mathrm{P}\left[X(t+h) = j, X(t) = i\right]}{\mathrm{P}\left[X(t+h) \neq i, X(t) = i\right]} &= \frac{\mathrm{P}\left[X(t+h) = j | X(t) = i\right] \mathrm{P}\left[X(t) = i\right]}{\mathrm{P}\left[X(t+h) \neq i | X(t) = i\right] \mathrm{P}\left[X(t) = i\right]} \\
&= \frac{\mathrm{P}\left[X(t+h) = j | X(t) = i\right]}{\mathrm{P}\left[X(t+h) \neq i | X(t) = i\right]} \\
&= \frac{P_{i,j}(h)}{1 - P_{i,i}(h)} .
\end{aligned}$$

Letting h approach zero, we finally obtain

$$\tilde{P}_{i,j} = \lim_{h \to 0^+} \frac{P_{i,j}(h)}{1 - P_{i,i}(h)} = \lim_{h \to 0^+} \frac{P_{i,j}(h)/h}{(1 - P_{i,i}(h))/h} = \frac{q_{i,j}}{q_j} \tag{7.131}$$

where the last step follows from (7.116) and (7.113). Hence, the transition probabilities of the embedded MC are uniquely determined by the transition intensities of the continuous-time MC. As can be noted, the probability that the embedded MC makes a transition from i to j is proportional to $q_{i,j}$, and this reinforces the interpretation of $q_{i,j}$ as a measure of the "intensity" of passage from i to j. Alternatively, from (7.131) we can express the transition intensities $q_{i,j}$ in terms of the transition probabilities $\tilde{P}_{i,j}$ of the embedded MC as

$$q_{i,j} = q_i \tilde{P}_{i,j} . \tag{7.132}$$

Being $q_i = 1/\mathtt{m}_{s_i}$ we can say that q_i is a measure of the tendency of the process to leave state i. Moreover, relation (7.132) shows that the transition intensity $q_{i,j}$ from i to j equals the overall intensity q_i at which the process exits from state i times the conditional probability that the process moves into state j, given that a transition out of state i occurs.

In summary, the transition probability matrix $\boldsymbol{P}(t)$ and the infinitesimal generator matrix $\boldsymbol{Q}$ provide two equivalent descriptions of a continuous-time MC. By using $\boldsymbol{P}(t)$ it is possible to determine the state probability at any time t, whereas $\boldsymbol{Q}$ describes the process evolution as a sequence of states i in which the process remains for an exponentially-distributed sojourn time s_i with parameter q_i before jumping to the next state j with probability $\tilde{P}_{i,j} = q_{i,j}/q_i$. In fact,

[10] Absorbing states are an exception to this rule. In fact, for any absorbing state j it is necessary to set $\tilde{P}_{j,j} = 1$ in order to preserve the stochastic property of the transition matrix $\tilde{\boldsymbol{P}}$. Hence, the main diagonal of $\tilde{\boldsymbol{P}}$ will have all zero terms except in correspondence of the absorbing states. Continuous-time MCs with absorbing states are of marginal interest to the scope of this book and will not be further considered.

relations (7.131) and (7.132) make it possible to define a continuous-time MC in terms of the average sojourn time $\mathtt{m}_{s_j}$ in each state j and the transition matrix $\tilde{\boldsymbol{P}}$ of the embedded MC.

Example 7.3 E Mark, an employee of the human resource office of a big company, has the task of screening the job applications that arrive to the company. Mark processes one application at a time. As a first step, Mark gives a quick pass to the application to check whether the candidate profile's satisfies the basic requirements of the company. If the check fails, Mark spends some time to write a kind rejection letter to the candidate and, then, he starts the process anew with another application taken from the hip. Conversely, if the profile passes the first check, then Mark contacts the candidate for an interview. If the interview is not successful, Mark performs the same sad task of writing a rejection letter and, then, passes to another application. Instead, if the interview is successful, then Mark will complete the paperwork for hiring the person. Hence, the process starts again with another application. Let us number the different tasks of Mark's job as follows:

0) first screening of an application;

1) writing a rejection letter;

2) making an interview;

3) completing paperwork.

Suppose that the execution of each task takes a random time, with exponential distribution. The average time spent in each task, expressed in minutes, is

$$\mathtt{m}_{s_0} = 2\,, \quad \mathtt{m}_{s_1} = 3\,, \quad \mathtt{m}_{s_2} = 60\,, \quad \mathtt{m}_{s_3} = 1200\,.$$

Furthermore, suppose that only 10% of the application passes the first screening, and only 20% of the interviews are successful.

Let us denote by $X(t)$ the task that Mark is doing at time t. It is easy to verify that the process $X(t)$ is a continuous-time MC. To realize this, let us consider the discrete-time process $\tilde{X}_n$ obtained by sampling $X(t)$ immediately after the nth task change. From the description of the scenario, we can easily determine the state (i.e., task) transition probabilities of $\tilde{X}_n$, namely

$$\tilde{P}_{0,1} = 0.9\,, \tilde{P}_{0,2} = 0.1\,, \quad \tilde{P}_{1,0} = 1\,, \quad \tilde{P}_{2,0} = 0.8\,, \quad \tilde{P}_{2,1} = 0.2\,, \quad \tilde{P}_{3,0} = 1\,,$$

whereas all the other state transitions have zero probability. Since, given the current state of the process, the past history does not provide any additional information on the future evolution of the process, we conclude that $\tilde{X}_n$ is a discrete-time MC, with transition probability matrix

$$\tilde{\boldsymbol{P}} = \begin{bmatrix} 0 & 0.9 & 0.1 & 0 \\ 1 & 0 & 0 & 0 \\ 0.8 & 0 & 0 & 0.2 \\ 1 & 0 & 0 & 0 \end{bmatrix}.$$

Since the sojourn times in each state j are iid exponential rvs with mean $\mathtt{m}_{s_j}$, then $X(t)$ is a continuous-time MC, whereas $\tilde{X}_n$ is its embedded MC. Hence, recalling that $q_j = 1/\mathtt{m}_{s_j}$, the non-diagonal transition intensities $q_{i,j}$ can be determined through (7.132), thus giving

$$q_{0,1} = \frac{0.9}{2}\,, q_{0,2} = \frac{0.1}{2}\,, \quad q_{1,0} = \frac{1}{3}\,, \quad q_{2,0} = \frac{0.8}{60}\,, \quad q_{2,1} = \frac{0.2}{60}\,, \quad q_{3,0} = \frac{1}{1200}\,,$$

so that the intensity matrix $\boldsymbol{Q}$ is equal to

$$\boldsymbol{Q} = \begin{bmatrix} -\frac{1}{2} & \frac{0.9}{2} & \frac{0.1}{2} & 0 \\ \frac{1}{3} & -\frac{1}{3} & 0 & 0 \\ 0 & \frac{0.8}{60} & -\frac{1}{60} & \frac{0.2}{60} \\ \frac{1}{1200} & 0 & 0 & -\frac{1}{1200} \end{bmatrix} .$$

Remark The embedded MC $\tilde{X}_n$ associated to a continuous-time MC $X(t)$ shall not be confused with the discrete-time MC $X_n = X(nT)$ that can be obtained by sampling the original process with a constant period T. In fact, the duration of a "step" of the embedded MC equals the sojourn time of $X(t)$ in a given state and, therefore, it is random. Conversely, the sampled process $X_n = X(nT)$ is defined over a discrete-time domain, where the sampling period is constant and equal to T. When T is sufficiently small, the sampled process $X(nT)$ can be used as a discrete-time approximation of the original process $X(t)$.

7.3.8 Flow diagram of continuous-time MC

As seen in Sec. 7.2.5, the transition matrix $\boldsymbol{P}$ of a discrete-time MC can be graphically represented by a transition diagram. The same concept also applies to the infinitesimal generator matrix $\boldsymbol{Q}$, which can be represented by means of the so-called *flow diagram*, a directed graph having a node for each state $j \in \mathcal{S}$ and an arc from any state i to any state j for which $q_{i,j} > 0$. Note that, based on this definition, self-transitions are not considered, being $q_{j,j} \leq 0$ for any j. Each arc from i to j is given a weight corresponding to the transition intensity $q_{i,j}$.

An example of flow diagram and transition diagram of the associated embedded MC is given in Fig. 7.13.

Remark We observe that any path that can be traced on the flow diagram of $X(t)$ can also be drawn on the transition diagram of the associated embedded MC $\tilde{X}_n$ and *vice-versa*. In other words, the flow diagram of $X(t)$ and the transition diagram of the corresponding embedded process $\tilde{X}_n$ have exactly the same topology, the only difference being the weights given to the arcs that are the elements of $\boldsymbol{Q}$ in the first case and of $\tilde{\boldsymbol{P}}$ in the latter.

7.3.9 State probability of continuous-time MC

So far we have focused on the transition probability functions $P_{i,j}(t)$, which give the conditional probability that the MC is in state j at time t given that the process is in state i at time zero. We now turn our attention to the *state* probability, that is to say, the probability that the process is in a given state j at time t:

$$p_j(t) = \mathrm{P}\left[X(t) = j\right] . \tag{7.133}$$

The distribution of $X(t)$ will be often described by means of the *state probability vector* $\boldsymbol{p}(t) = [p_0(t), p_1(t), \ldots, p_N(t)]$, whose elements are non-negative and add up to one, as

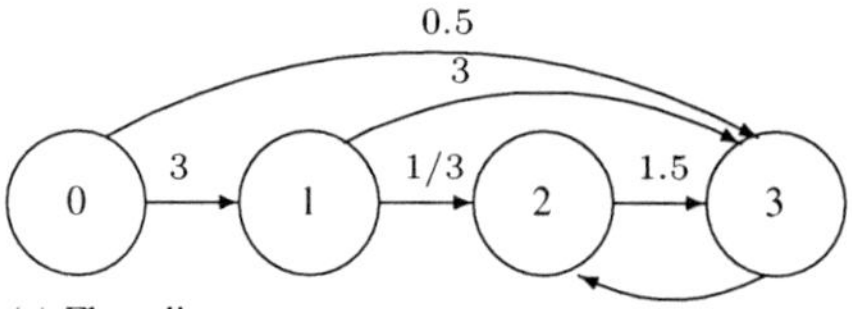

(a) Flow diagram

$$Q = \begin{bmatrix} -3.5 & 3 & 0 & 0.5 \\ 0 & -10/3 & 1/3 & 3 \\ 0 & 0 & -1.5 & 1.5 \\ 0 & 0 & 0.5 & -0.5 \end{bmatrix}$$

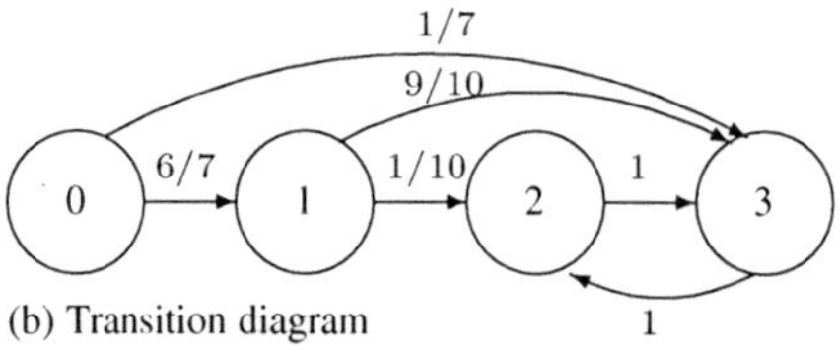

(b) Transition diagram

$$\tilde{P} = \begin{bmatrix} 0 & 6/7 & 0 & 1/7 \\ 0 & 0 & 1/10 & 9/10 \\ 0 & 0 & 0 & 1 \\ 0 & 0 & 1 & 0 \end{bmatrix}$$

Figure 7.13 Example of a) flow diagram for the continuous-time MC with associated infinitesimal generator matrix $\boldsymbol{Q}$ and b) corresponding transition diagram and transition matrix $\tilde{\boldsymbol{P}}$ of the embedded MC.

dictated by the normalization condition

$$\sum_{j=0}^{N} p_j(t) = \boldsymbol{p}(t)\,\mathbf{1} = 1\,.$$

For continuous-time MC, the state probability vector at any time t can be determined from the initial state probability vector $\boldsymbol{p}(0)$ and the transition matrix $\boldsymbol{P}(t)$ at time t, through the Chapman-Kolmogorov equation (7.111). In fact, applying the Total Probability Theorem 2.12 we can express $p_j(t)$ as

$$p_j(t) = \mathrm{P}\left[X(t) = j, \bigcup_{i \in \mathcal{S}} \{X(u) = i\}\right] = \mathrm{P}\left[\bigcup_{i \in \mathcal{S}} \{X(t) = j, X(u) = i\}\right] \tag{7.134}$$

for any $0 < u < t$. Since the events $\{X(u) = i\}$ for different values of i are disjoint, (7.134) can be expressed as

$$\begin{aligned} p_j(t) &= \sum_{i \in \mathcal{S}} \mathrm{P}\left[X(t) = j | X(u) = i\right] \mathrm{P}\left[X(u) = i\right] \\ &= \sum_{i \in \mathcal{S}} p_i(u) P_{i,j}(t-u) \end{aligned}$$

that in matrix form becomes

$$\boldsymbol{p}(t) = \boldsymbol{p}(u)\boldsymbol{P}(t-u) \quad 0 < u < t\,.$$

In particular, by setting $u = 0$ we get

$$\boldsymbol{p}(t) = \boldsymbol{p}(0)\boldsymbol{P}(t)\,, \tag{7.135}$$

which is the equivalent of (7.27) for continuous-time MC. Replacing $\boldsymbol{P}(t)$ with (7.124), we obtain

$$\boxed{\boldsymbol{p}(t) = \boldsymbol{p}(0)\exp(\boldsymbol{Q}\,t)\,,} \tag{7.136}$$

which is the time-continuous equivalent of (7.28). In principle, the time evolution of the state probability vector $\boldsymbol{p}(t)$ can be determined from (7.136), given the initial state probability vector $\boldsymbol{p}(0)$ and the infinitesimal generator matrix $\boldsymbol{Q}$. The same result can be expressed in differential form, as follows. The derivative of $\boldsymbol{p}(t)$ is given by

$$\boldsymbol{p}'(t) = \lim_{h\to 0}\frac{\boldsymbol{p}(t+h)-\boldsymbol{p}(t)}{h} = \lim_{h\to 0}\frac{\boldsymbol{p}(t)\left(\boldsymbol{P}(h)-\boldsymbol{I}\right)}{h} = \boldsymbol{p}(t)\lim_{h\to 0}\frac{\boldsymbol{P}(h)-\boldsymbol{I}}{h}\,.$$

According to definition (7.112), the last limit is equal to $\boldsymbol{Q}$, so that we obtain

$$\boxed{\boldsymbol{p}'(t) = \boldsymbol{p}(t)\boldsymbol{Q}} \tag{7.137}$$

that in scalar form can be written as

$$\boxed{p_j'(t) = \sum_{i=0,i\neq j}^{N} p_i(t)q_{i,j} - p_j(t)q_j\,;} \tag{7.138}$$

with $j \in \mathcal{S}$. In analogy to what seen for discrete-time MCs, equations (7.138) are called *balance equations*, since they are a generalization of the balance equations (7.79) derived for the discrete-time case. To explain this assert, we resort again to the fluid system model, which provides a intuitive interpretation of the differential equations (7.138). Let us then associate the continuous-time MC with a flow-dynamic system, where each state i is associated to a basin that contains $p_i(t)$ units of some liquid. Fluid flows from state $i =\neq j$ into j with an average rate of $p_i(t)q_{i,j}$ units of liquid per unit time, whereas it flows out of state j to other states with an overall average rate $p_j(t)q_j$. In this view, (7.138) simply states that the rate of variation of the liquid in basin j, i.e., the value of $p_j'(t)$, is equal to the difference of balance between the inbound flow $\sum_{i=0,i\neq j}^{N} p_i(t)q_{i,j}$ and the outbound flow $p_j(t)q_j$. Therefore, (7.138) is a generalization of the balance equation (7.79) we derived for discrete-time MC.

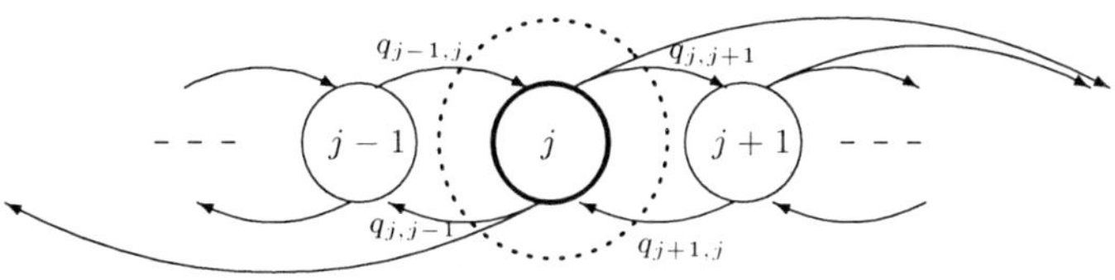

Figure 7.14 Graphical method to write the balance equations for a MC.

A simple alternative method to obtain the right hand side of (7.138) consists in drawing an imaginary boundary around the state j under consideration, as illustrated in Fig. 7.14 for a simple case, and then calculating the probability flows crossing that boundary. Each arc that crosses the boundary gives a contribution equal to its "weight" multiplied by the probability of the state that it emanates from. The sign of the contribution is positive for arcs that enter state j and negative for arcs that leave that state. Thus, we have a simple inspection technique to write the differential equations that define the time-dependent evolution of the state distribution of the MC.

Example 7.3 F Let us consider again Example 7.3 D, where a facility alternates between idle and busy states, remaining in each state a random time with exponential distribution with parameters λ and μ, respectively. We have seen that the forward equation in this case gives

$$\begin{aligned} P_{0,0}(t) &= \frac{1}{\lambda+\mu}\left(\mu + \lambda e^{-(\lambda+\mu)t}\right), & P_{0,1}(t) &= \frac{\lambda}{\lambda+\mu}\left(1 - e^{-(\lambda+\mu)t}\right), \\ P_{1,0}(t) &= \frac{\mu}{\lambda+\mu}\left(1 - e^{-(\lambda+\mu)t}\right), & P_{1,1}(t) &= \frac{1}{\lambda+\mu}\left(\lambda + \mu e^{-(\lambda+\mu)t}\right). \end{aligned} \tag{7.139}$$

Given the initial state probability vector $\boldsymbol{p}(0) = [p_0(0), p_1(0)]$, the probability of state zero at time t is

$$p_0(t) = p_0(0)P_{0,0}(t) + p_1(0)P_{1,0}(t) \tag{7.140}$$

and, replacing $p_1(0)$ with $1 - p_0(0)$, we get

$$p_0(t) = p_0(0)\left(P_{0,0}(t) - P_{1,0}(t)\right) + P_{1,0}(t) = p_0(0)e^{-(\lambda+\mu)t} + \frac{\mu}{\lambda+\mu}\left(1 - e^{-(\lambda+\mu)t}\right). \tag{7.141}$$

The time-dependent probability of state zero given in (7.141) is plotted in Fig. 7.15 together with its complementary probability $p_1(t)$. Square and circle marks are used to distinguish between state zero and one, respectively. The graph reports three sets of curves obtained by setting the initial state probability vector to $\boldsymbol{p}(0) = [1, 0]$ (solid lines), $\boldsymbol{p}(0) = [0.5, 0.5]$ (dotted lines) and $\boldsymbol{p}(0) = [\mu/(\lambda+\mu);\ \lambda/(\lambda+\mu)]$ (dashed lines). Observing the figure, we can immediately realize three simple but important facts.

1. Probabilities $p_0(t)$ and $p_1(t)$ obtained with equal initial setting are perfectly symmetric with respect to the horizontal line that cuts the graph in half. This is nothing else than the graphical representation of the fact that $p_0(t) = 1 - p_1(t)$.
2. For large values of t, the functions $p_0(t)$ and $p_1(t)$ tend to asymptotic values that do not depend on the initial state probability vector $\boldsymbol{p}(0)$.
3. If the initial state probability vector is equal to the asymptotic state probability vector, then the probability functions do not change in time. (Clearly, the state of the process keeps oscillating between zero and one over time, whereas the probability of observing the MC in either state in a time instant arbitrarily far into the future remains constant.)

Example 7.3 G Consider a MC $X(t)$ such that from any state i it is only possible to move into the successive state $j = i + 1$. Furthermore, suppose that the transition intensities are all equal to λ, so that the infinitesimal generator matrix is

$$\boldsymbol{Q} = \begin{bmatrix} -\lambda & \lambda & 0 & 0 & 0 & \cdots \\ 0 & -\lambda & \lambda & 0 & 0 & \cdots \\ 0 & 0 & -\lambda & \lambda & 0 & \cdots \\ \vdots & \vdots & \vdots & \vdots & \vdots & \ddots \end{bmatrix}.$$

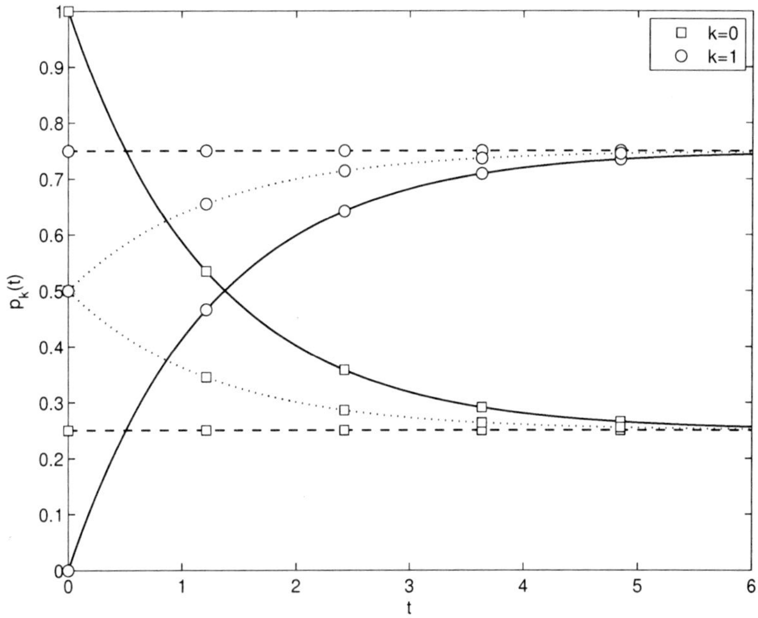

Figure 7.15 State probabilities $p_0(t)$ and $p_1(t)$ as a function of time for the two-state MC of Example 7.3 F with $\lambda = 0.6$ and $\mu = 0.2$.

The state flow diagram is shown below.

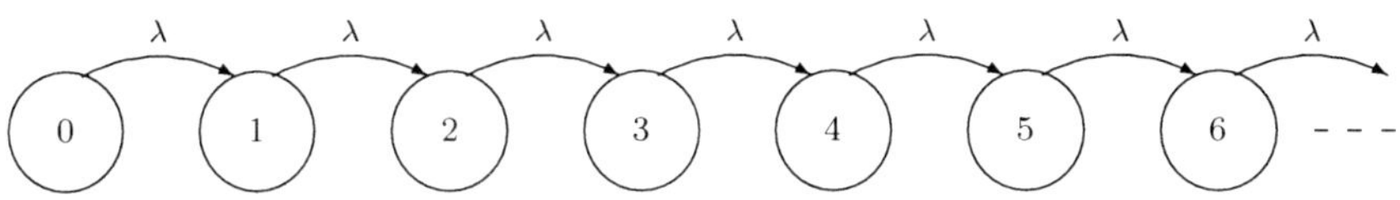

In this case, balance equations (7.138) yield

$$\begin{aligned} p_0'(t) &= -\lambda p_0(t)\,, \\ p_j'(t) &= \lambda p_{j-1}(t) - \lambda p_j\,, \quad j = 1, 2, \ldots\,. \end{aligned} \tag{7.142}$$

To solve this system of differential equations we introduce the auxiliary function $g_j(t)$ defined as

$$g_j(t) = e^{\lambda t} p_j(t)\,; \quad j = 0, 1, 2, \ldots \tag{7.143}$$

whose derivative is

$$g_j'(t) = \lambda e^{\lambda t} p_j(t) + e^{\lambda t} p_j'(t)\,.$$

Using (7.142) in the above equation, we get for $j = 0$

$$g_0'(t) = \lambda e^{\lambda t} p_0(t) - \lambda e^{\lambda t} p_0(t) \quad \Longrightarrow \quad g_0'(t) = 0 \tag{7.144}$$

and for $j \geq 1$

$$g_j'(t) = \lambda e^{\lambda t} p_j(t) + \lambda e^{\lambda t} p_{j-1}(t) - \lambda e^{\lambda t} p_j \quad \Longrightarrow \quad g_j'(t) = \lambda g_{j-1}(t). \tag{7.145}$$

Let us now suppose that $X(0) = 0$, so that the initial state probabilities are $p_0(0) = 1$ and $p_j(0) = 0$ for $j \geq 1$. According to (7.143), we have $g_0(0) = 1$ and $g_j(0) = 0$. Under these initial conditions, (7.144) gives $g_0(t) = 1$, whereas the differential equations (7.145) iteratively yield

$$g_1(t) = \lambda t, \quad g_2(t) = \frac{(\lambda t)^2}{2!}, \cdots, g_j(t) = \frac{(\lambda t)^j}{j!}.$$

Hence, from (7.143) we obtain

$$p_j(t) = e^{-\lambda t} \frac{(\lambda t)^j}{j!}; \quad j = 0, 1, 2, \ldots$$

that corresponds to the Poisson PMD with parameter λt. In practice, $X(t)$ is a Poisson process of intensity λ and $\boldsymbol{p}(t)$ is just the probability vector of the number of arrivals in the interval $[0, t)$, which is known to be Poisson distributed with parameter λt.

The same result can also be obtained in a different way. In fact, this MC has the particularity that from any state j the process can only move to the successive state $j + 1$. Therefore, given that $X(0) = 0$, we can have $X(t) = j$ only if the process has performed exactly j transitions in the time interval $[0, t]$. Let us denote by τ_k the exact instant in which the kth transition takes place. Thus, the probability that the process is in state j at time t can be expressed as

$$p_j(t) = \mathrm{P}\left[\tau_j \leq t, \tau_{j+1} > t\right] \tag{7.146}$$

where $\{\tau_j \leq t\}$ indicates the event that state j is reached before or at t, whereas $\{\tau_{j+1} > t\}$ indicates that the process leaves state j after t.

Since the time spent in a given state i corresponds to the sojourn time s_i, we have

$$\tau_j = \sum_{i=0}^{j-1} s_i.$$

Recalling that the rvs s_i are statistically independent, with exponential distribution with parameter $q_i = \lambda$, the time τ_j is a rv with Erlang distribution of index j and parameter λ, whose PDF is equal to

$$p_{\tau_j}(a) = \frac{\lambda(\lambda a)^{j-1}}{(j-1)!} e^{-\lambda a}; \quad a \geq 0.$$

Writing τ_{j+1} as $\tau_j + s_j$ and applying the Total Probability Theorem 2.12 on τ_j we finally get

$$\begin{aligned} p_j(t) &= \mathrm{P}\left[\tau_j \leq t, \tau_j + s_j > t\right] = \int_0^\infty p_{\tau_j}(a)\, \mathrm{P}\left[\tau_j \leq t, \tau_j + s_j > t | \tau_j = a\right] da \\ &= \int_0^t p_{\tau_j}(a)\, \mathrm{P}\left[s_j > t - a\right] da = \int_0^t \frac{\lambda(\lambda a)^{j-1}}{(j-1)!} e^{-\lambda a} e^{-\lambda(t-a)} da \\ &= e^{-\lambda t} \int_0^t \frac{\lambda(\lambda a)^{j-1}}{(j-1)!} da = e^{-\lambda t} \frac{(\lambda t)^j}{j!}. \end{aligned} \tag{7.147}$$

◦

7.3.10 Classification of continuous-time MC

Continuous-time MCs can be classified according to the same criteria already discussed for discrete-time MC, except for marginal adjustments. In particular, the classification of states in *transient* and *recurrent* that we have discussed for discrete-time MCs can be transferred via the embedded MC to a continuous process [7]. Conversely, the classification of states as periodic and aperiodic is meaningless in the continuous-time domain.

For the reader's convenience, we rapidly revise the state classification of continuous-time MCs. The reader interested in more in-depth study on this topic is referred to, e.g., [7].

Accessible and communicating states. State j is said to be *accessible* (or *reachable*) from state i if $P_{i,j}(t) > 0$ for some $t > 0$. This relation is denoted as $i \to j$. In other words, state j is accessible from i if the flow diagram contains a directed path from i to j.

States i and j are said to *communicate* when they are mutually accessible, i.e., $i \to j$ and $j \to i$. This relation is denoted as $i \leftrightarrow j$. With reference to the flow diagram, i and j communicate if there exists (at least) one closed path that includes states i and j.

Communication classes. The communication relation induces a partition of the states of the MC in equivalence *classes*, so that a class collects all and only the states that communicate with one another. Class $\mathcal{S}_c \subseteq \mathcal{S}$ is called *closed* if all the transitions that originate from $\mathcal{S}_c$ also terminate into $\mathcal{S}_c$, i.e.,

$$\forall i \in \mathcal{S}_c,\ \forall j \in \mathcal{S} - \mathcal{S}_c \Longrightarrow q_{i,j} = 0\,.$$

Conversely, class $\mathcal{S}_c$ is called *open* if there are transitions originating from $\mathcal{S}_c$ and ending into some states of $\mathcal{S} - \mathcal{S}_c$, that is

$$\exists\, i \in \mathcal{S}_c\,,\ \exists\, j \in \mathcal{S} - \mathcal{S}_c \quad \text{such that} \quad q_{i,j} > 0\,.$$

A MC is called *irreducible* if all its states belong to one closed class, that is, they all communicate. Otherwise, i.e., when the MC contains multiple (closed or open) communication classes, the MC is called *reducible*. The state flow diagram of an irreducible MC is strongly connected, i.e., for any pair of distinct nodes i, j there exists a path from i to j.

Recurrence. We denote by $H_{i,j}(t)$ the conditional probability that the first *return* into j occurs before t, given that $X(0) = i$. Note that $H_{i,j}(\infty)$ is the probability that the process starting in i will ever make a transition into j, whereas $H_{j,j}(\infty)$ is the probability that, starting from j, the process will ever return into j after having visited some other state. State j is said to be *recurrent* if $H_{i,j}(\infty) = 1$, and *transient* otherwise. For a recurrent state j, we can define the *mean recurrence time* τ_j as

$$\tau_j = \int_0^\infty (1 - H_{j,j}(t))dt\,. \tag{7.148}$$

If $\tau_j < \infty$, state j is said to be *positive recurrent*, whereas if $\tau_j = \infty$ the state is said to be *null recurrent*. It is possible to prove that the positive and null recurrence are class properties, that is, if state j is positive (null) recurrent, then any other state i in the same communication class is also positive (null) recurrent. Moreover, as for discrete-time MCs, it is possible to show that the states of a closed class with a finite number of states are all positive recurrent (see [7]).

Example 7.3 H With reference to the MC represented in Fig. 7.13, we can see that the only communicating states are 2 and 3, which then form a closed class. The other states belong to open classes and the MC is reducible.

Example 7.3 I The MC in Fig. 7.16 is irreducible, since all of its states communicate with one another. We can note that it is always possible to find a continuous path in the flow diagram from any state to any other state, i.e., the graph is strongly connected.

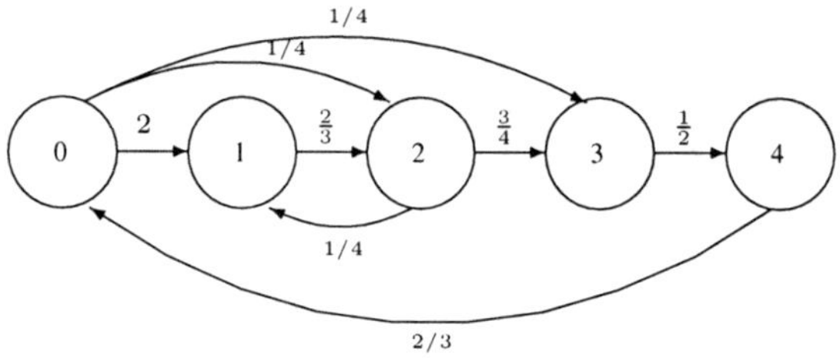

Figure 7.16 Flow diagram of the MC of Example 7.3 I.

7.3.11 Asymptotic behavior of continuous-time MC

The time-dependent behavior of the MC, i.e., the way in which the state probability vector distribution evolves over time from a given initial state probability vector $\boldsymbol{p}(0)$, is given by the matrix equation (7.136), $\boldsymbol{p}(t) = \boldsymbol{p}(0)\exp(\boldsymbol{Q}t)$. However, it is often difficult to find a closed form expression for $\boldsymbol{p}(t)$ that hence needs to be determined through numerical methods. Fortunately, in many cases the time-dependent analysis is of little interest, whereas it is more relevant to analyze the process when it has reached an asymptotic behavior, i.e., when the effect of the initial state state probability of the process is no longer discernible. Adopting the same notation introduced for discrete-time MC, the asymptotic or *steady-state* probability of state j is defined as

$$\pi_j = \lim_{t\to\infty} \mathrm{P}\left[X(t) = j\right] = \lim_{t\to\infty} p_j(t)\,, \quad j \in \mathcal{S}\,. \tag{7.149}$$

The conditions for the existence of the limits (7.149) are basically the same that have been discussed for the discrete-time MC, and are summarized by the following theorem, whose proof is similar to that of Theorem 7.8.

Theorem 7.14 *In an irreducible MC, the asymptotic probabilities π_j always exist and are independent of the initial state probability vector $\mathbf{p}(0)$ of $X(0)$. Furthermore, one of the two following cases applies:*

1. *either the process is transient or null recurrent, in which case $\pi_j = 0$ for all j;*

2. *or the process is positive recurrent, in which case*

$$\pi_j > 0 \quad \textit{for all } j.$$

It can also be shown that a Markov process that admits a proper asymptotic probability[11] is *ergodic*. In this case, the asymptotic probability π_j of state j is equal to the *fraction of time* that the process spends in state j or, from another point of view, to the probability that sampling the process in a random instant arbitrarily far in the future it is found in state j.

Example 7.3 J Consider again Example 7.3 C and suppose that Stunny actually remains in the same spot for a random time with exponential distribution that depends only on his current position. In other terms, the process $X(t)$ that describes the distance of the man from the pub entrance at the instant t is assumed to be a MC. Suppose that Andrei and Mark are giving a party in their flat and challenge their friends with a gamble on which distance will separate Stunny from the pub entrance an hour later. Most of the other students just make a random guess, but Andrei and Mark that had been observing Stunny for a long period of time T during which they had tracked the overall time T_j spent by Stunny at any distance j from the pub entrance. From this data, they estimate the asymptotic probability that Stunny is at distance j as $\pi_j \simeq \frac{T_j}{T}$. Knowing that in a hour the process $X(t)$ will reach its steady-state behavior, losing memory of its current state, they bet on the distance with the highest asymptotic probability, thus increasing their chance to win the bet!

The asymptotic state probability vector can be easily determined from the forward equation (7.137). By definition, in fact, the steady state probabilities do not change over time, so that the derivative of the state distribution vector $\boldsymbol{p}(t)$ has to approach the all-zero horizontal vector $\mathbf{0}$ as t grows to infinity, i.e., it must be

$$\lim_{t\to\infty} \boldsymbol{p}'(t) = \mathbf{0}\,.$$

Then, taking the limit for $t \to \infty$ of the forward equation (7.137) we get

$$\lim_{t\to\infty} \boldsymbol{p}'(t) = \lim_{t\to\infty} \boldsymbol{p}(t)\boldsymbol{Q} = \boldsymbol{p}(\infty)\,\boldsymbol{Q} = \mathbf{0}\,.$$

Replacing $\boldsymbol{p}(\infty)$ with the asymptotic state probability vector $\boldsymbol{\pi}$, we thus obtain

$$\boxed{\boldsymbol{\pi}\,\boldsymbol{Q} = \mathbf{0}} \tag{7.150}$$

which is the equivalent of (7.76) for continuous-time MCs. The generic jth row of (7.150) takes the form

$$\pi_j\, q_j = \sum_{i=0, i\neq j}^{N} \pi_i q_{i,j}\,, \tag{7.151}$$

[11] See footnote 5 at page 441.

which, from (7.138), is the *balance equation* at steady state. Recalling the parallel between a continuous-time MC and a flow-dynamic system proposed in Section 7.3.9, we can regard the left-hand side of (7.151) as the long-run average rate at which the "liquid" pours out from state j. The right-hand side, in turn, can be intended as the long-run average rate at which liquid flows into state j from any other state. Thus, relation (7.151) simply states that, in steady-state, the inbound and outbound (probability) flows for any state j are in perfect balance. We observe that, with reference to the inspection method described in Section 7.3.9 to write the balance equations (7.138), the balance condition (7.151) means that, in steady state, the net flow across any closed boundary must be zero.

The matrix form of the balance equation expressed by (7.150) corresponds to a system of $N+1$ linear and homogeneous equations in the form (7.151), where the unknowns are the $N+1$ asymptotic probabilities $\pi_0, \pi_1, \ldots, \pi_N$. However, one of such equation can be obtained as linear combination of the others, as it can be immediately realized recalling that $\boldsymbol{Q}$ is not full rank because the elements of each row add up to zero. Then, the system is solved by infinitely many vectors in the form $c\boldsymbol{p}$, where c is a non-zero real constant. To find the unique asymptotic state probability vector of the MC, it suffices to couple (7.150) with the *normalization condition*, which requires

$$\boxed{\boldsymbol{\pi}\mathbf{1} = \sum_{j=0}^{N} \pi_j = 1\,.} \tag{7.152}$$

In conclusion, the asymptotic state probability vector $\boldsymbol{\pi}$, in all the cases where it exists, can be found by solving the following system of equations

$$\boxed{\begin{cases} \boldsymbol{\pi}\boldsymbol{Q} &= \mathbf{0} \\ \boldsymbol{\pi}\mathbf{1} &= 1 \end{cases}} \implies \begin{cases} \pi_0\, q_0 &= \sum_{i \in \mathcal{S}, i \neq 0} \pi_i q_{i,0} \\ \pi_1\, q_1 &= \sum_{i \in \mathcal{S}, i \neq 1} \pi_i q_{i,1} \\ &\vdots \\ \pi_j\, q_j &= \sum_{i \in \mathcal{S}, i \neq j} \pi_i q_{i,j} \\ &\vdots \\ \pi_N\, q_N &= \sum_{i \in \mathcal{S}, i \neq N} \pi_i q_{i,N} \\ \sum_{j=0}^{N} \pi_j &= 1 \end{cases} \tag{7.153}$$

where one of the first $N+1$ equations is actually redundant, being a linear combination of the others.

Remark It might be worth noting that the transition matrix $\boldsymbol{P}(t)$ of an irreducible MC asymptotically converges toward a matrix with all rows equal to the asymptotic state probability vector π, i.e.,

$$\lim_{t\to\infty} \boldsymbol{P}(t) = \boldsymbol{P}(\infty) = \mathbf{1}\,\boldsymbol{\pi} = \begin{bmatrix} \pi_0 & \pi_1 & \pi_2 & \cdots & \pi_N \\ \pi_0 & \pi_1 & \pi_2 & \cdots & \pi_N \\ \vdots & \vdots & \vdots & \cdots & \vdots \\ \pi_0 & \pi_1 & \pi_2 & \cdots & \pi_N \end{bmatrix}$$

or, analogously,

$$\lim_{t\to\infty} P_{i,j}(t) = \pi_j \; ; \quad j = 0, 1, 2, \ldots .$$

This result follows directly from (7.135) and (7.149).

Example 7.3 K Let us consider again the two-state MC of Example 7.3 F, with infinitesimal generator matrix

$$\boldsymbol{Q} = \begin{bmatrix} -\lambda & \lambda \\ \mu & -\mu \end{bmatrix} .$$

In Example 7.3 F we found that the state probability vector at time t had elements

$$p_0(t) = p_0(0)e^{-(\lambda+\mu)t} + \frac{\mu}{\lambda+\mu}\left(1 - e^{-(\lambda+\mu)t}\right) ,$$
$$p_1(t) = 1 - p_0(t) .$$

By letting t approach infinity, we then obtain the asymptotic probabilities, which turn out to be equal to

$$\pi_0 = \frac{\mu}{\lambda+\mu}$$
$$\pi_1 = 1 - \pi_0 = \frac{\lambda}{\lambda+\mu} .$$

Note, that π does not depend on the probability vector $\boldsymbol{p}(0)$ of the initial state. The same result can be obtained by solving the flow balance system (7.153). In fact, we have

$$\begin{cases} \pi_0\,\lambda = \mu\,\pi_1 \\ \pi_0 + \pi_1 = 1 \end{cases} \implies \begin{cases} \pi_1 = \dfrac{\lambda}{\mu}\pi_0 \\ \pi_0 + \dfrac{\lambda}{\mu}\pi_0 = 1 \end{cases} \implies \begin{cases} \pi_0 = \dfrac{\mu}{\lambda+\mu} \\ \pi_1 = \dfrac{\lambda}{\lambda+\mu} \end{cases}$$

7.4 BIRTH-DEATH PROCESSES

7.4.1 Definition of BDP

A special position in the family of continuous-time MCs is occupied by the so-called *birth-death processes* that provide the building blocks of the elementary queueing theory which will be presented in Chapter 8. A birth-death process (BDP) is a continuous-time MC in which transitions occur only between adjacent states, i.e., from state $j > 0$ the next transition can only be to the "preceding" state $j - 1$ or to the "successive" state $j + 1$. It is also possible to define discrete-time BDPs, the only difference being the possibility of making self-transitions. In the following we consider continuous-time BDPs only, though an almost parallel treatment exists for the discrete-time case.

In the specific terminology used when dealing with a BDP, $X(t)$ denotes the number of members of a given *population* at the instant t. A transition from state j to state $j + 1$ will signify a *birth* of a new member of the population, whereas a transition from j to $j - 1$ will denote a *death* in the population. Although we use the terms population, birth, and death, the model in effect is not limited to biological populations: for example we could consider a call center where the population is the number of calls in service, or waiting to be served, a birth is the arrival of a new call, and a death is the end of a call.

In general, the population of a BDP is assumed to vary from zero to infinity. However, it is sometimes appropriate to limit the maximum number of members of the population to N. As we will see in detail in Chapter 8, BDPs with unlimited population are excellent models for a certain class of queueing systems, denoted by $\mathrm{M/M/m}$, whereas the processes with limited population are used to model another class of queueing systems, denoted by $\mathrm{M/M/m/K}$. For consistency with most literature on the subject, in the sequel we present the theory for the case with unlimited population, from which the case with limited population can be derived with only marginal adjustments.

The evolution of a BDP over a sufficiently small time interval h can be described through the probabilities of the following transition events:

- exactly one birth in $(t, t + h]$

$$P_{j,j+1}(h) = \mathrm{P}\left[X(t+h) = j+1 | X(t) = j\right] = \lambda_j\, h + o(h)\,, \quad j = 0, 1, 2, \ldots \tag{7.154}$$

- exactly one death in $(t, t + h]$

$$P_{j,j-1}(h) = \mathrm{P}\left[X(t+h) = j-1 | X(t) = j\right] = \mu_j\, h + o(h)\,, \quad j = 1, 2, \ldots \tag{7.155}$$

- exactly zero births and deaths in $(t, t + h)$

$$\begin{aligned} P_{0,0}(h) &= \mathrm{P}\left[X(t+h) = 0 | X(t) = 0\right] = 1 - \lambda_0\, h + o(h) \\ P_{j,j}(h) &= \mathrm{P}\left[X(t+h) = j | X(t) = j\right] = 1 - (\mu_j + \lambda_j)\, h + o(h)\,, \quad j = 1, 2, \ldots\,. \end{aligned} \tag{7.156}$$

where $o(h)$ is an infinitesimal of order higher than h that accounts for the probability of multiple birth and/or death events in the time interval h. Parameters λ_j and μ_j are termed *birth-rate* and *death-rate* for state j, respectively. Note that, from any state $j > 0$, the process can either move to the "successive" state $j+1$ or to the "preceding" state $j-1$, whereas from state $j = 0$ the process can only move forward, since deaths are not allowed, i.e., $\mu_0 = 0$.

By comparing the above equations with (7.115) and (7.117), obtained for a continuos-time MC, we can associate the birth-rates and death-rates with the transitions intensities $q_{i,j}$ as follows

$$\begin{aligned} \lambda_j &= q_{j,j+1}\,, \\ \mu_j &= q_{j,j-1}\,, \\ \lambda_j + \mu_j &= q_j = -q_{j,j}\,. \end{aligned} \tag{7.157}$$

Therefore, the infinitesimal generator matrix of a generic BDP takes the form of a square matrix of infinite size, with all-zero elements except for those in the main, upper and lower diagonals:

$$Q = \begin{bmatrix} -\lambda_0 & \lambda_0 & 0 & 0 & 0 & \cdots \\ \mu_1 & -(\lambda_1+\mu_1) & \lambda_1 & 0 & 0 & \cdots \\ 0 & \mu_2 & -(\lambda_2+\mu_2) & \lambda_2 & 0 & \cdots \\ 0 & 0 & \mu_3 & -(\lambda_3+\mu_3) & \lambda_3 & \cdots \\ \vdots & \vdots & \vdots & \vdots & \vdots & \ddots \end{bmatrix}. \tag{7.158}$$

Fig. 7.17 shows the corresponding flow diagram.

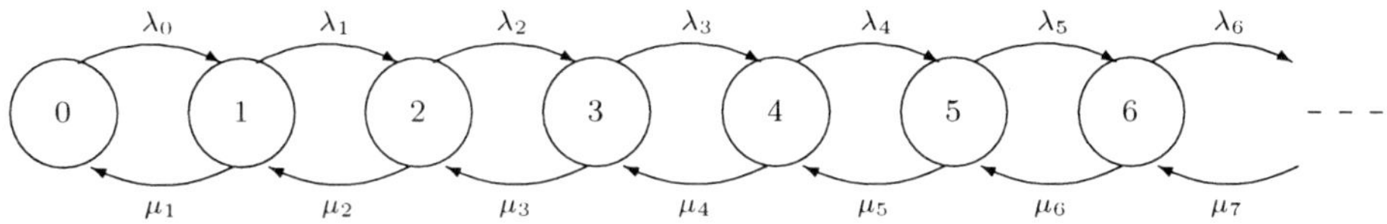

Figure 7.17 Flow diagram of a BDP.

As usual, the interest in studying a BDP is in the probability $p_j(t)$ that the population size is j at some time t. Furthermore, the analysis of the asymptotic behavior of the process, that is to say, of the asymptotic probabilities $\lim_{t\to\infty} p_j(t) = \pi_j$, is one of the most important aspects of BDP theory. Since BDPs are just a specific type of MC, all the results concerning BDPs can be directly obtained from the general theory of continuous-time MCs presented in the previous section, by simply considering the transition intensities $q_{i,j}$ given by (7.157). However, the derivation of these equations for BDPs turns out to be particularly simple, so that we prefer to repeat the calculation from scratch. In this way, we can revise and remark some basic principles that might be more difficult to capture in the general treatment of continuous-time MCs.

7.4.2 Time-dependent behavior of BDP

We begin by determining the Chapman-Kolmogorov relation for BDPs. Let us suppose that the population counts $j > 0$ members at time $t + h$, that is to say $X(t + h) = j$. This is possible if any of the three following events occurred:

1. the population size at time t was j and no births or deaths occurred in the interval $(t, t + h)$;
2. the population size at time t was $j - 1$ and a birth occurred in the interval $(t, t + h)$;
3. the population size at time t was $j + 1$ and a death occurred in the interval $(t, t + h)$.

These three events are mutually exclusive. Furthermore, any other event that yields $X(t+h) = j$ given that $X(t) = j$ involves multiple birth and/or death events in the interval $(t, t + h)$ that, according to the definition of BDP, has negligible probability $o(h)$ for small h. Applying the Total Probability Theorem 2.12 on $X(t)$, we thus have

$$p_j(t+h) = p_j(t)P_{j,j}(h) + p_{j-1}(t)P_{j-1,j}(h) + p_{j+1}(t)P_{j+1,j}(h) + o(h)\,, \quad j \geq 1\,. \tag{7.159}$$

Repeating the reasoning for $j = 0$, we get the boundary equation

$$p_0(t+h) = p_0(t)P_{0,0}(h) + p_1(t)P_{1,0}(h) + o(h)\,. \tag{7.160}$$

Replacing in the above equations the expression of the transition probabilities given by (7.154), (7.155), (7.156), we obtain for any $j > 0$

$$p_j(t+h) = p_j(t)(1-\lambda_j\, h-\mu_j\, h)+p_{j-1}(t)\lambda_{j-1}\, h+p_{j+1}(t)\mu_{j+1}\, h+o(h)\,, \quad j \geq 1\,, \tag{7.161}$$

and for $j = 0$

$$p_0(t+h) = p_0(t)(1-\lambda_0\, h) + p_1(t)\mu_1\, h + o(h)\,. \tag{7.162}$$

By subtracting $p_j(t)$ from both sides of each equation and dividing by h, we have

$$\frac{p_j(t+h)-p_j(t)}{h} = \begin{cases} -p_j(t)(\lambda_j+\mu_j) + p_{j-1}(t)\lambda_{j-1} + p_{j+1}(t)\mu_{j+1} + \dfrac{o(h)}{h} & ,\ j \geq 1 \\ -p_0(t)\lambda_0 + p_1(t)\mu_1 + \dfrac{o(h)}{h} & ,\ j = 0 \end{cases} \tag{7.163}$$

Taking the limit for $h \to 0$ we finally obtain the following *differential equations*

$$\boxed{p_j'(t) = \begin{cases} \lambda_{j-1}\, p_{j-1}(t) + \mu_{j+1}\, p_{j+1}(t) - (\lambda_j+\mu_j)\, p_j(t) & ,\ j \geq 1 \\ -\lambda_0\, p_0(t) + \mu_1\, p_1(t) & ,\ j = 0 \end{cases}} \tag{7.164}$$

which are the balance equations of the BDP.

These results can also be obtained by following the same simple method described in Section 7.3.9. The method refers the flow diagram associated to the MC, that in the case of BDPs is reported in Fig. 7.18. We have that the rate $p_j'(t)$ at which the probability of state j varies over time is equal to the difference between the inbound and outbound flows for state j. Since in a BDP state j can be "connected" to states $j-1$ and $j+1$ only, then the total *inbound probability flow* is equal to $\lambda_{j-1}p_{j-1}(t) + \mu_{j+1}p_{j+1}(t)$, whereas the *outbound probability flow* is given by $(\lambda_j + \mu_j)p_j(t)$. Then, the balance equation yields

$$\frac{dp_j(t)}{dt} = \lambda_{j-1}p_{j-1}(t) + \mu_{j+1}p_{j+1}(t) - (\lambda_j+\mu_j)p_j(t)\,, \tag{7.165}$$

which is equal to (7.164) for $j > 0$. The case for $j = 0$ is conceptually identical.

Unfortunately, the differential equations (7.164), in general, do not admit exact solutions except for some particularly simple cases. For instance, explicit solutions can be found for the two-state MC discussed in Example 7.3 F, which can be seen as an extremely simple BDP with population size limited to 1, or for the Poisson process of Example 7.3 G that provides an example of *pure-birth process* with homogeneous birth rate $\lambda_j = \lambda$ for any j. Exact solutions can also be found for a pure-birth process with general birth rates λ_j, which is discussed in Examples 7.4 A, 7.4 B and 7.4 C below. Example 7.4 D, instead, addresses the case of a birth-death process with homogenous birth-rate λ and death-rate μ, for which state probabilities are derived by means of numerical methods.

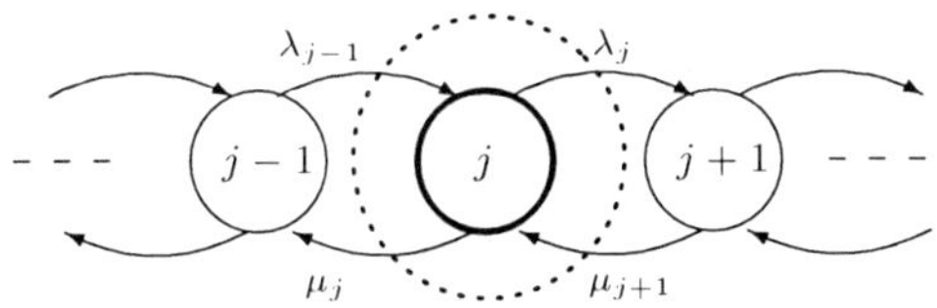

Figure 7.18 Graphical method to write the balance equations for BDPs.

Example 7.4 A Fig. 7.19 shows the state flow diagram of a *pure-birth* process with birth-rates $\lambda_0, \lambda_1, \lambda_2, \ldots$.

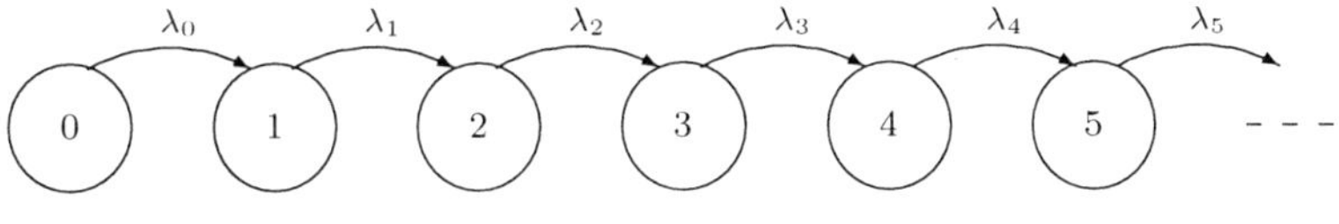

Figure 7.19 Pure-birth process with generic birth-rates.

The infinitesimal generator matrix is

$$\boldsymbol{Q} = \begin{bmatrix} -\lambda_0 & \lambda_0 & 0 & 0 & 0 & \cdots \\ 0 & -\lambda_1 & \lambda_1 & 0 & 0 & \cdots \\ 0 & 0 & -\lambda_2 & \lambda_2 & 0 & \cdots \\ \\ 0 & 0 & 0 & \ddots & \ddots & \ddots \end{bmatrix}.$$

It is clear that the state $X(t)$ grows indefinitely, since the size of the population can only increase. In this case, the balance equations (7.165) take the form

$$\begin{aligned} p_0'(t) &= -\lambda_0 p_0(t)\,, \\ p_j'(t) &= -\lambda_j p_j(t) + \lambda_{j-1} p_{j-1}(t)\,, \quad j = 1, 2, \ldots . \end{aligned} \tag{7.166}$$

The first equation in (7.166) gives

$$p_0(t) = p_0(0) e^{-\lambda_0 t}\,.$$

The solution to the second equation can be found by solving iteratively the following equation

$$p_j(t) = e^{-\lambda_j t} \left[\lambda_{j-1} \int_0^t p_{j-1}(a) e^{\lambda_j a} da + p_j(0) \right]\,, \quad j = 1, 2, \ldots$$

as it can be verified by substitution in (7.166).

Example 7.4 B In Example 7.3 G we derived state probabilities of a Poisson process starting from the PDF of the sojourn times. The same method can also be applied to the case of a pure-birth process with generic birth-rates. Let τ_k denote the exact instant in which the population reaches k members. Since the population begins with zero members and grows by exactly one member at each transition (birth), then we have

$$\tau_k = \sum_{i=0}^{k-1} s_i$$

where s_i is the sojourn time in state i, with exponential PDF

$$p_{s_i}(a) = \lambda_i\, e^{-\lambda_i\, a}\,, \qquad a \geq 0\,. \tag{7.167}$$

Starting from $X(0) = 0$, we then have

$$p_k(t) = \mathrm{P}\left[X(t) = k\right] = \mathrm{P}\left[\tau_k \leq t, \tau_{k+1} > t\right] \tag{7.168}$$

where the term on the right-hand side is the probability that the process reaches state k before or at instant t and leaves it after t. Writing τ_{k+1} as $\tau_k + s_k$ and applying the Total Probability Theorem 2.12 on τ_k we get

$$\begin{aligned} p_k(t) &= \int_0^{\infty} \mathrm{P}\left[\tau_k \leq t, \tau_k + s_k > t | \tau_k = a\right] p_{\tau_k}(a) da \\ &= \int_0^{t} \mathrm{P}\left[s_k > t - a\right] p_{\tau_k}(a) da\,. \end{aligned} \tag{7.169}$$

Recalling the exponential form of $p_{s_i}(a)$, see (7.167), (7.169) can be written as

$$p_k(t) = \frac{1}{\lambda_k} \int_0^t p_{s_k}(t-a) p_{\tau_k}(a) da = \frac{1}{\lambda_k} \left(p_{\tau_k} * p_{s_k}\right)(t) \tag{7.170}$$

where $(g * f)(a)$ denotes the convolution between the functions $g(a)$ and $f(a)$ as defined in (2.42). Since the sojourn times are statistically independent, then $p_{\tau_k}(a)$ can be expressed as the multi-fold convolution of the functions $p_{s_i}(a)$, with $i = 0, 1, \ldots, k-1$, and

$$p_k(t) = \frac{1}{\lambda_k} \left(p_{s_0} * p_{s_1} * \cdots * p_{s_{k-1}} * p_{s_k}\right)(t)\,, \qquad t \geq 0\,.$$

In the frequency domain, the above equation yields

$$\Psi_k(f) = \frac{1}{\lambda_k} \prod_{i=0}^{k} \Psi_{s_i}(f) \tag{7.171}$$

where $\Psi_k(f)$ is the Ftf of $p_k(t)$, whereas $\Psi_{s_i}(f)$ denotes the Ftf of $p_{s_i}(a)$ given by[12]

$$\Psi_{s_i}(f) = \frac{\lambda_i}{\lambda_i + j\, 2\pi\, f}\,, \tag{7.172}$$

where j is the imaginary factor. Replacing (7.172) in (7.171) we get

$$\Psi_k(f) = \frac{\prod_{i=0}^{k-1} \lambda_i}{\prod_{i=0}^{k} (\lambda_i + j\, 2\pi\, f)}\,. \tag{7.173}$$

[12]The Ftf $\Psi_x(f)$ of the PDF $p_x(a)$ of a rv x, when it exists, is called *characteristic function* and provides a complete statistical description of x.

Now suppose that the birth-rates λ_i, for $i = 0, 1, \ldots$, are all distinct. In this case, (7.173) can be expressed as

$$\Psi_k(f) = \sum_{i=0}^{k} \frac{A_i}{\lambda_i + j\,2\pi\, f} \tag{7.174}$$

where coefficients A_i can be determined by multiplying both sides of (7.173) by $(\lambda_i + j\,2\,\pi\, f)$ and, then, evaluating the equation in $f = j\lambda_i/(2\,\pi)$, which gives

$$A_i = \frac{\prod_{r=0}^{k-1} \lambda_r}{\prod_{r=0, r\neq i}^{k} (\lambda_r - \lambda_i)} \qquad i = 0, 1, \ldots, k\,. \tag{7.175}$$

Replacing (7.175) into (7.174) and taking the inverse Ftf, we finally get

$$p_k(t) = \sum_{i=0}^{k} A_i e^{-\lambda_i\, t} = \sum_{i=0}^{k} \frac{\prod_{r=0}^{k-1} \lambda_r}{\prod_{r=0, r\neq i}^{k} (\lambda_r - \lambda_i)} e^{-\lambda_i\, t}\,. \tag{7.176}$$

Equation (7.176) is valid under the assumption that the birth-rates $\lambda_0, \ldots, \lambda_k$ are all distinct. However, this restriction can be relaxed at the cost of introducing a more cumbersome notation. Let $\tilde{\lambda}_1, \tilde{\lambda}_2, \cdots, \tilde{\lambda}_{n_k}$ be the n_k *distinct* values taken by the birth-rates $\lambda_0, \ldots, \lambda_k$. Furthermore, let m_i be the multiplicity of $\tilde{\lambda}_i$, i.e., the number of birth-rates equal to $\tilde{\lambda}_i$. Clearly, we have $n_k \leq k + 1$ and $\sum_{i=1}^{n_k} m_i = k + 1$. In this case, (7.173) and, in turn, (7.174) become

$$\Psi_k(f) = \frac{\prod_{i=1}^{n_k} \tilde{\lambda}_i^{m_i}}{\lambda_k \prod_{i=1}^{n_k} \left(\tilde{\lambda}_i + j\,2\pi\, f\right)^{m_i}} = \sum_{i=1}^{n_k} \frac{\tilde{A}_i}{\left(\tilde{\lambda}_i + j\,2\pi\, f\right)^{m_i}} \tag{7.177}$$

where coefficients $\tilde{A}_i$ are given by

$$\tilde{A}_i = \frac{\prod_{r=1}^{n_k} \tilde{\lambda}_r^{m_r}}{\lambda_k \prod_{r=1, r\neq i}^{n_k} \left(\tilde{\lambda}_r - \tilde{\lambda}_i\right)^{m_r}}\,, \qquad i = 1, \ldots, n_k\,.$$

Each term on the right-hand side of (7.177) is the Ftf of an Erlang distribution, except for a multiplying constant. Hence, we finally get

$$p_k(t) = \sum_{i=1}^{n_k} \frac{\tilde{A}_i\, t^{m_i - 1}\, e^{-\tilde{\lambda}_i\, t}}{(m_i - 1)!}\,. \tag{7.178}$$

Example 7.4 C Consider the pure-birth process with birth-rates $\lambda_k = 1/(k+1)$ for $k = 0, 1, \ldots$, whose flow diagram is shown in Fig. 7.20

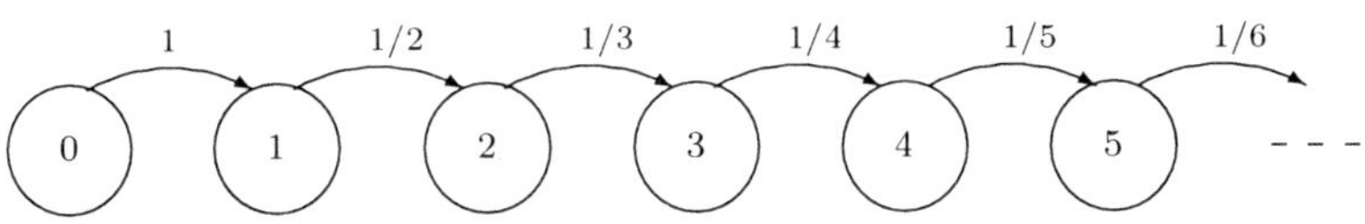

Figure 7.20 Pure-birth process with birth-rates $\lambda_k = 1/(k+1)$ of Example 7.4 C.

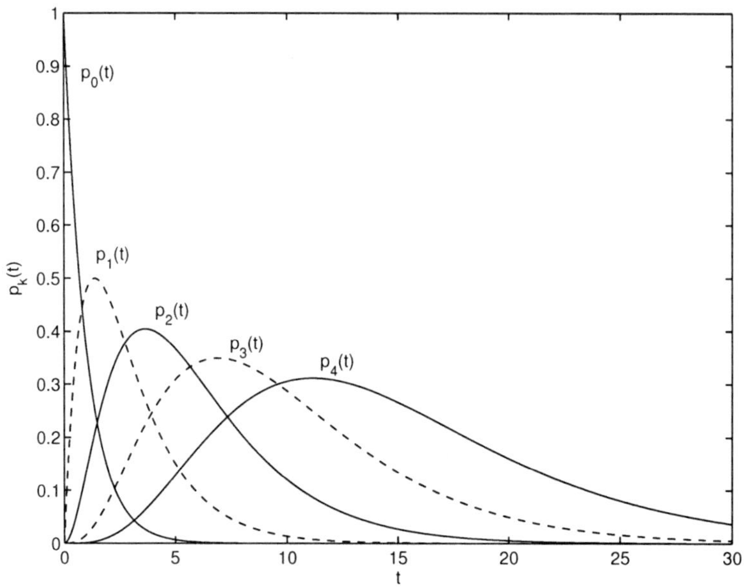

Figure 7.21 State probabilities $p_k(t)$ as a function of time for a pure-birth process with birth-rates $\lambda_k = 1/(k+1)$, $k = 0, 1, \ldots$.

Since the birth-rates are all distinct, the state probabilities can be obtained from (7.176) by using $1/(k+1)$ in place of λ_k. After some algebra, we get

$$p_k(t) = \sum_{\ell=0}^{k} \frac{(k+1)(\ell+1)^{k-1} e^{-\frac{t}{\ell+1}}}{(-1)^{k-\ell}(k-\ell)!(\ell-1)!} . \tag{7.179}$$

The state probabilities $p_k(t)$ for the first states are plotted in Fig. 7.21 as function of time t. We observe that all the state probabilities tend to zero as t approaches infinity, which is expected since all states are transient. Furthermore, we observe that, starting from a population with zero members, the size of the population tends to increase over time. For instance, the probability that the population size is still zero at time $t = 5$ seconds is very low, being $p_0(5) \simeq 0$, whereas at time $t = 10$ seconds, the size of the population will have likely grown to two or more members, as $p_0(10) \simeq 0, p_1(10) \simeq 0$ and $p_2(10) > 0$.

Example 7.4 D The flow diagram of Fig. 7.22 refers to the simple case of a BDP in which all birth rates are equal to λ and all death rates are equal to μ (except for state zero that, as usual, has zero death rate). As we will see in Chapter 8, this BDP with constant parameters is of primary importance since it models the behavior of the $\mathrm{M/M/1}$ queue, which is probably the simplest non-trivial model of a queueing system. Despite the rather elementary structure of this process, the time-dependent behavior of its state probabilities turns out to be rather complex to derive. Here we just report the final result, referring to [2] for its derivation. Assuming the process starts from state k_0, i.e., $p_{k_0}(0) = 1$ and $p_k(0) = 0$ for any

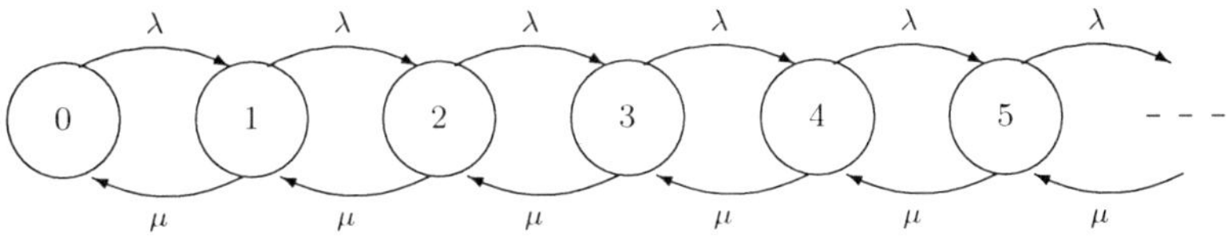

Figure 7.22 Birth-death process with homogeneous birth-rates λ and death-rates μ.

$k \neq k_0$, the probability of being in state k at the instant t is given by

$$p_k(t) = e^{-(\lambda+\mu)t} \Bigg[\rho^{\frac{k-k_0}{2}} I_{k-k_0}(2\mu\sqrt{\rho}\,t) + \rho^{\frac{k-k_0-1}{2}} I_{k+k_0+1}(2\mu\sqrt{\rho}\,t) + (1-\rho)\rho^k \sum_{j=k+k_0+2}^{\infty} \rho^{\frac{-j}{2}} I_j(2\mu\sqrt{\rho}\,t) \Bigg] \tag{7.180}$$

where

$$\rho = \frac{\lambda}{\mu}, \tag{7.181}$$

and $I_k(x)$ is the Bessel function of the first kind of order k, defined as

$$I_k(x) = \sum_{m=0}^{\infty} \frac{(x/2)^{k+2m}}{(k+m)!\,m!}, \quad k \geq -1\,. \tag{7.182}$$

In Fig. 7.23 we plot the probability of the states $k = 0, 1, 2$ and 3 as a function of time. Solid lines have been obtained by choosing the initial state $k_0 = 0$, whereas dashed lines have been obtained with an initial state $k_0 = 2$. In both cases, the birth and death rates are $\lambda = 0.2$ and $\mu = 1$, respectively. We observe that asymptotically all the probabilities tend to constant values, independently of the initial state. This is because the BDP considered in the example is an irreducible and positive recurrent MC (see Sec. 7.3.10) that, according to Theorem 7.14, does admit an asymptotic state probability vector, whose derivation is the topic of the rest of this section.

7.4.3 Asymptotic behavior of BDP

As we have learnt from the examples of the previous section, the solution of the system of differential equations (7.164) quickly becomes unmanageable when we consider non-trivial settings of the birth and death rates. Fortunately, the asymptotic behavior of a BDP can be studied in a rather simple and elegant way for a much larger set of cases. Adopting the same notation used for the asymptotic analysis of a general MC, the asymptotic probability of a state j, assuming it exists, is defined by (7.149). Moreover, as for continuous-time MCs, it is possible to prove that a BDP that admits proper asymptotic state probabilities[13] is ergodic.

[13]See footnote 5 at page 441.

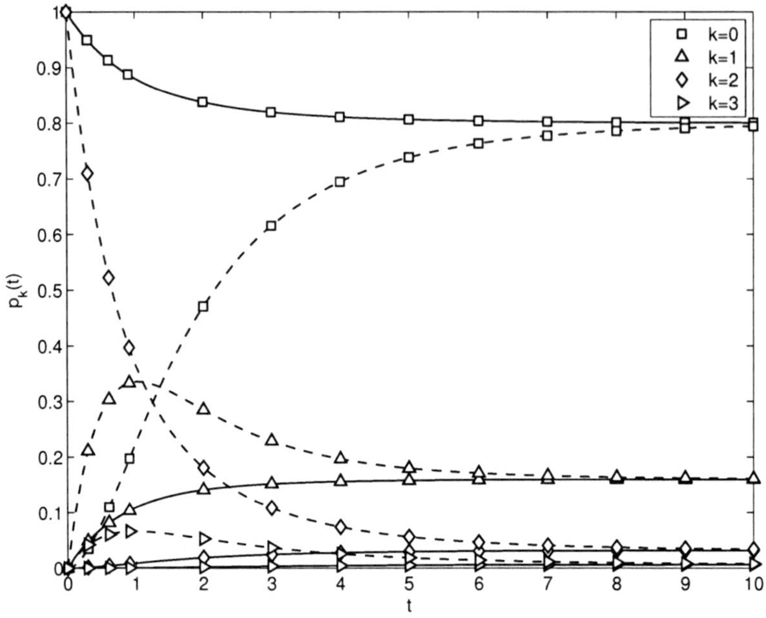

Figure 7.23 State probabilities $p_k(t)$ as a function of time for the pure-birth process of Example 7.4 D.

Accordingly, π_j can be interpreted as the probability that the population contains j members at some arbitrary time in the distant future, or as the percentage of time that the process spends in state j, in the long-run.

The existence of the asymptotic probabilities is subject to the same conditions discussed for continuous-time MCs and expressed in Theorem 7.14, which basically states that a BDP admits a unique asymptotic state probability vector $\boldsymbol{\pi}$, with $\pi_j > 0$ for $j = 0, 1, \ldots$, if the process is positive recurrent. The most natural approach to determine the asymptotic state probability vector would require to check first whether the process is positive recurrent and, in this case, to proceed with the computation of the probabilities (7.149). However, in the study of BDPs it turns out to be more convenient to adopt the opposite approach in which we *first* determine the asymptotic state probability vector $\boldsymbol{\pi}$ and, *next*, determine the conditions under which the vector $\boldsymbol{\pi}$ is actually a proper probability vector.

Derivation of the asymptotic state probabilities. We hence suppose that the limit in (7.149) exists and is positive. In this case, each state probability $p_j(t)$ tends to a constant value π_j as the time t approaches infinity, so that the limit of the time derivative $p'_j(t)$ as $t \to \infty$ is equal to zero, i.e.,

$$\begin{aligned} \lim_{t\to\infty} p_j(t) &= \pi_j\,, \\ \lim_{t\to\infty} p'_j(t) &= 0\,. \end{aligned} \tag{7.183}$$

Replacing these limits in the Chapman-Kolmogorov equations (7.164) we then get a set of *balance equations* for the BDP, which take the form

$$\begin{aligned} 0 &= \lambda_{j-1}\,\pi_{j-1} + \mu_{j+1}\,\pi_{j+1} - (\lambda_j + \mu_j)\,\pi_j\,, && j \geq 1\,, \\ 0 &= \mu_1\,\pi_1 - \lambda_0\,\pi_0\,, && j = 0\,. \end{aligned} \tag{7.184}$$

The asymptotic state probability vector $\boldsymbol{\pi}$ has to satisfy (7.184) for $j = 0, 1, 2, \ldots$ or, in other words, it has to solve the system of equations

$$\begin{cases} 0 &= \mu_1\,\pi_1 - \lambda_0\,\pi_0 \\ 0 &= \lambda_0\,\pi_0 + \mu_2\,\pi_2 - (\lambda_1 + \mu_1)\,\pi_1 \\ 0 &= \lambda_1\,\pi_1 + \mu_3\,\pi_3 - (\lambda_2 + \mu_2)\,\pi_2 \\ 0 &= \lambda_2\,\pi_2 + \mu_4\,\pi_4 - (\lambda_3 + \mu_3)\,\pi_3 \\ &\vdots \end{cases} \tag{7.185}$$

which can be rewritten in a more convenient form as

$$\begin{cases} 0 &= \mu_1\,\pi_1 - \lambda_0\,\pi_0 \\ \mu_1\,\pi_1 - \lambda_0\,\pi_0 &= \mu_2\,\pi_2 - \lambda_1\,\pi_1 \\ \mu_2\,\pi_2 - \lambda_1\,\pi_1 &= \mu_3\,\pi_3 - \lambda_2\,\pi_2 \\ \mu_3\,\pi_3 - \lambda_2\,\pi_2 &= \mu_4\,\pi_4 - \lambda_3\,\pi_3 \\ &\vdots \end{cases} \tag{7.186}$$

Observing (7.186) we note that the term on the right-hand side of each equation is equal to the term on the left-hand side of the following equation. Then, adding the first row to the second row and simplifying the terms that appear on opposite sides of the equation we get

$$\begin{cases} 0 &= \mu_1\,\pi_1 - \lambda_0\,\pi_0 \\ 0 &= \mu_2\,\pi_2 - \lambda_1\,\pi_1 \\ \mu_2\,\pi_2 - \lambda_1\,\pi_1 &= \mu_3\,\pi_3 - \lambda_2\,\pi_2 \\ \mu_3\,\pi_3 - \lambda_2\,\pi_2 &= \mu_4\,\pi_4 - \lambda_3\,\pi_3 \\ &\vdots \end{cases} \tag{7.187}$$

Repeating iteratively this operation on the following rows, we get

$$\begin{cases} 0 &= \mu_1\,\pi_1 - \lambda_0\,\pi_0 \\ 0 &= \mu_2\,\pi_2 - \lambda_1\,\pi_1 \\ 0 &= \mu_3\,\pi_3 - \lambda_2\,\pi_2 \\ 0 &= \mu_4\,\pi_4 - \lambda_3\,\pi_3 \\ &\vdots \end{cases} \implies \begin{cases} \mu_1\,\pi_1 &= \lambda_0\,\pi_0 \\ \mu_2\,\pi_2 &= \lambda_1\,\pi_1 \\ \mu_3\,\pi_3 &= \lambda_2\,\pi_2 \\ \mu_4\,\pi_4 &= \lambda_3\,\pi_3 \\ &\vdots \end{cases} \tag{7.188}$$

In it interesting to observe that this system of equations can be easily obtained by direct inspection of the flow diagram of the BDP reported in Fig. 7.24.

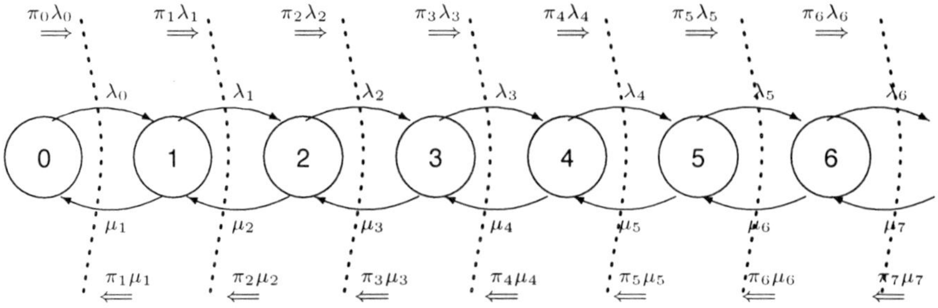

Figure 7.24 Flow diagram of a generic BDP.

In fact, we have already noted that, in the long-run, the net probability-flow across any closed boundary must be zero. In the case of a BDP, each state $j > 0$ can "exchange" probability flows with the "successive" and "preceding" states only. Therefore, the flow conservation principle can be expressed by imposing the flow balance through each vertical line separating adjacent states (see Fig. 7.24). Equating flows across this boundary, we get a set of equations in the form

$$\pi_j\lambda_j = \pi_{j+1}\mu_{j+1}\,, \qquad j = 0, 1, 2, \ldots \tag{7.189}$$

which correspond to the balance equations in the system (7.188). From (7.189) we get

$$\pi_{j+1} = \frac{\lambda_j}{\mu_{j+1}}\pi_j\,, \qquad j = 0, 1, 2, \ldots \tag{7.190}$$

that, replaced in (7.188), gives

$$\begin{cases} \pi_1 &= \frac{\lambda_0}{\mu_1}\,\pi_0 \\ \pi_2 &= \frac{\lambda_1}{\mu_2}\,\pi_1 \\ \pi_3 &= \frac{\lambda_2}{\mu_3}\,\pi_2 \\ \pi_4 &= \frac{\lambda_3}{\mu_4}\,\pi_3 \\ &\vdots \end{cases} \tag{7.191}$$

Replacing iteratively the unknown on the right-hand side of each equation with the expression on the right-hand side of the previous equation, we get

$$\begin{cases} \pi_1 &= \frac{\lambda_0}{\mu_1}\,\pi_0 \\ \pi_2 &= \frac{\lambda_0\lambda_1}{\mu_1\mu_2}\,\pi_0 \\ \pi_3 &= \frac{\lambda_0\lambda_1\lambda_2}{\mu_1\mu_2\mu_3}\,\pi_0 \\ \pi_4 &= \frac{\lambda_0\lambda_1\lambda_2\lambda_3}{\mu_1\mu_2\mu_3\mu_4}\,\pi_0 \\ &\vdots \end{cases} \tag{7.192}$$

from which the general solution of (7.184) is shown by induction to be

$$\pi_j = \frac{\lambda_0\lambda_1\cdots\lambda_{j-1}}{\mu_1\mu_2\cdots\mu_j}\pi_0 \qquad j = 1, 2, \ldots\,. \tag{7.193}$$

We have thus expressed the asymptotic state probabilities π_j in terms of a single unknown constant π_0:

$$\boxed{\pi_j = \pi_0 \prod_{k=0}^{j-1} \frac{\lambda_k}{\mu_{k+1}}, \quad j = 1, 2, \ldots .} \tag{7.194}$$

Once again, the unknown constant π_0 can be determined by imposing the *normalization condition*

$$\sum_{j=0}^{\infty} \pi_j = 1 . \tag{7.195}$$

Using (7.194) in (7.195) we get

$$\pi_0 + \sum_{j=1}^{\infty} \left(\pi_0 \prod_{k=0}^{j-1} \frac{\lambda_k}{\mu_{k+1}} \right) = 1$$

from which

$$\boxed{\pi_0 = \frac{1}{1 + \displaystyle\sum_{j=1}^{\infty} \prod_{k=0}^{j-1} \frac{\lambda_k}{\mu_{k+1}}} .} \tag{7.196}$$

As will be seen in Chapter 8, this simple product solution for π_j is a *principal* equation in elementary queueing theory, from which it is possible to derive the solution of most of the queueing models considered in this book.

Existence of the asymptotic state probabilities. It is now time to turn our attention to the *existence* of the steady-state or limiting probabilities. Theorem 7.14 states that the asymptotic state probabilities exist and they are unique if the process is irreducible. Furthermore, the asymptotic probabilities π_j are all positive (non-null) if the process is positive recurrent. These requirements clearly place a condition upon the birth and death rates of the process. Looking at the flow diagram of Fig. 7.24, it is easy to realize that the process is irreducible if and only if all the birth and death rates are non-zero, i.e.,

$$\lambda_j > 0 \quad \text{and} \quad \mu_{j+1} > 0 \quad \text{for}, \quad j = 0, 1, 2, \ldots . \tag{7.197}$$

To guarantee that the process is positive recurrent we can simply impose that the probabilities given by (7.194) are all positive: $\pi_j > 0$. Due to the product form of (7.194), this condition is verified when $\pi_0 > 0$. The particular role played by state zero in the existence of the asymptotic solution is related to the fact that state zero is a *reflecting barrier*: the process can move indefinitely toward states with higher index but it is lower limited by state zero. Condition $\pi_0 > 0$, therefore, guarantees that the process returns to visit state 0 from time to time in the long run and that it does not drift indefinitely toward higher states.

According to (7.196), the condition $\pi_0 > 0$ is satisfied when

$$1 + \sum_{j=1}^{\infty} \prod_{k=0}^{j-1} \frac{\lambda_k}{\mu_{k+1}} < \infty . \tag{7.198}$$

For example, a sufficient condition for (7.198) to hold is that for some k_0

$$\frac{\lambda_k}{\mu_k} < 1 \qquad \forall\ k \geq k_0 \,. \tag{7.199}$$

In fact, interpreting the process as the size of a given population, the condition (7.199) simply claims that the population size does not grow indefinitely only if, as the number of members k exceeds a certain threshold k_0, the rate of births becomes strictly lower than the rate of deaths. This condition is actually verified in many real-life systems. For instance, a population of rabbits in a given area can rapidly grow up to the point that the scarcity of food determines an increase in the death rate (or in the rate of departure of members from the community) that inverts the tendency. A similar situation arises in many other types of resource-limited systems, where the size of the "population" beyond a certain value determines a scarcity of resources that, in turn, results in a reduction of the birth rate.

Example 7.4 E Consider a BDP with a finite number N of states, so that the birth rate for state $N-1$ is zero. The corresponding flow diagram is shown in Fig. 7.25.

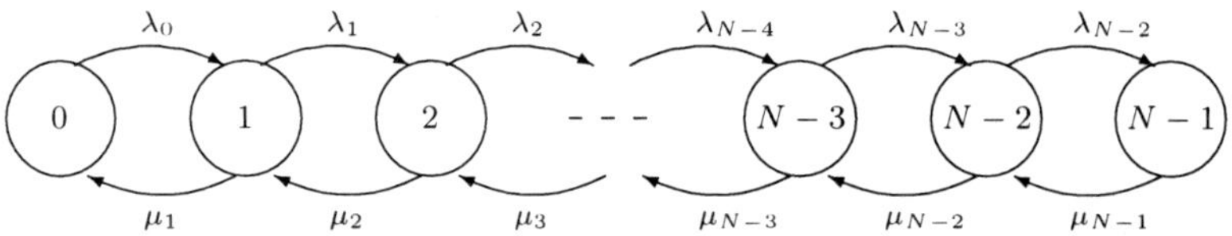

Figure 7.25 Flow diagram of the BDP of Example 7.4 E.

Suppose that birth-rates and death-rates are such that

$$\lambda_j = \mu_{j+1} \,, \quad j = 0, 1, 2, \ldots, N-2 \,. \tag{7.200}$$

The BDP is clearly irreducible and and, as long as N is finite, it is positive recurrent. Then, setting the balance equation for state zero we obtain

$$\pi_0 \lambda_0 = \pi_1 \mu_1 \implies \pi_0 = \pi_1 \,,$$

where we have simplified out λ_0 and μ_1 on the basis of (7.200). Repeating recursively the balance equation for states $j = 1, 2, \ldots, N-1$ we easily realize that the asymptotic state probability vector of the BDP is uniform, that is to say

$$\pi_j = \frac{1}{N} \,, \quad j = 0, 1, 2, \ldots, N-1 \,.$$

Therefore, observing the BDP in a random time arbitrarily far into the future, we can find it in any state with equal probability. Furthermore, the average size of the population in the long-run is equal to $\mathtt{m}_x = \frac{N-1}{2}$ and its variance is $\sigma_x^2 = \frac{N^2-1}{12}$. We observe that, letting N grow to infinity, each element of the steady-state probability vector tends progressively to zero, whereas both the mean and the variance of the population' size grow to infinity, with the variance growing faster than the mean. Therefore, as the number of states grows, the prediction on the actual value of the process state becomes more and more "uncertain". It can be shown that, with infinite states, the process becomes null recurrent, so that it does not admit any longer a proper asymptotic distribution. This case corresponds to the borderline case of a queueing system that is offered a workload exactly equal to its service capacity, as will be seen in Chapter 8.

Example 7.4 F Let us consider again the simple BDP with birth-rate λ and death-rate μ for each state, introduced in Example 7.4 D and represented in Fig. 7.22. We can now determine the asymptotic state probabilities of the process. By inspecting the flow diagram of Fig. 7.22 we can write the following balance equations

$$\begin{cases} \pi_1 &= \dfrac{\lambda}{\mu}\pi_0 \\ \pi_2 &= \dfrac{\lambda}{\mu}\pi_1 \\ &\vdots \end{cases} \tag{7.201}$$

from which we get the general product form

$$\pi_{j+1} = \left(\frac{\lambda}{\mu}\right)^j \pi_0 = \rho^j \pi_0 \qquad j = 0, 1, 2, \ldots \tag{7.202}$$

where $\rho = \lambda/\mu$. The value of π_0 can be obtained by imposing the normalization condition that yields

$$\pi_0 \sum_{j=0}^{\infty} \rho^j = 1 \quad \Longrightarrow \quad \pi_0 \frac{1}{1-\rho} = 1 \quad \Longrightarrow \quad \pi_0 = 1 - \rho\,. \tag{7.203}$$

Replacing (7.203) in (7.202) we get

$$\pi_j = (1-\rho)\rho^j\,, \qquad j = 0, 1, 2, \ldots\,. \tag{7.204}$$

Hence, the asymptotic PMD of the state process is geometric with parameter $1-\rho$ and statistical mean

$$\mathtt{m}_x = \frac{\rho}{1-\rho}\,. \tag{7.205}$$

Example 7.4 G Consider the BDP with flow diagram as in Fig. 7.26, characterized by constant birth rate λ and death rate $\mu_i = i\mu$ that increases proportionally to the size i of the population. Evaluating (7.196) for this specific case we obtain

$$\pi_0 = \frac{1}{1 + \displaystyle\sum_{j=1}^{\infty} \prod_{k=0}^{j-1} \frac{\lambda}{(k+1)\mu}} = \frac{1}{1 + \displaystyle\sum_{j=1}^{\infty} \frac{\lambda^j}{j!\mu^j}} = \frac{1}{\displaystyle\sum_{\ell=0}^{\infty} \frac{\lambda^\ell}{\ell!\mu^\ell}} = \frac{1}{e^{\frac{\lambda}{\mu}}} = e^{-\frac{\lambda}{\mu}}$$

whereas (7.194) yields

$$\pi_j = \pi_0 \prod_{k=0}^{j-1} \frac{\lambda}{(k+1)\mu} = \frac{\lambda^j}{j!\mu^j} e^{-\frac{\lambda}{\mu}}\,.$$

In conclusion, at steady-state the size of the population is distributed as a Poisson rv with parameter λ/μ. Note that this result holds for any choice of the parameters λ and μ, provided that they are positive. As

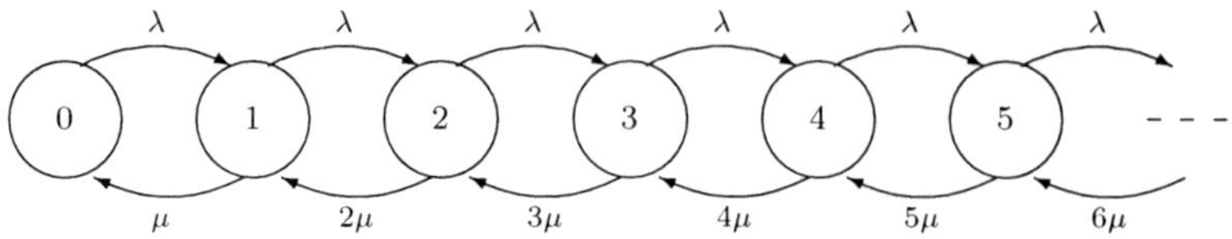

Figure 7.26 Flow diagram of a BDP with birth-rates λ and death-rates $\mu_j = j\mu$ for $j = 1, 2, \ldots$ ($\mathrm{M/M/\infty}$ queueing system).

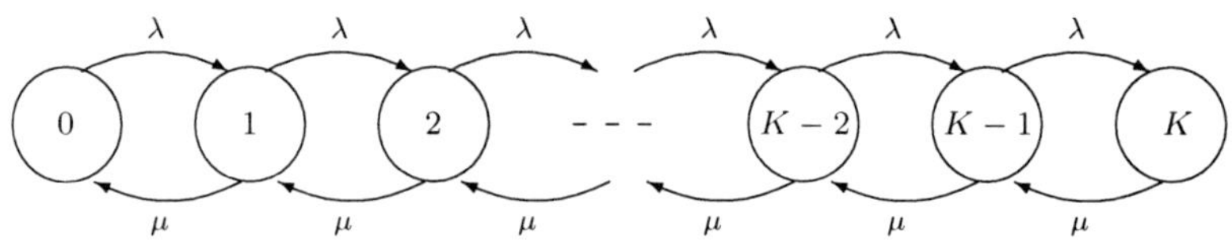

Figure 7.27 Flow diagram of a BDP with finite state space ($\mathrm{M/M/1/K}$ queuing system).

we will see in Chapter 8, this BDP models the state (i.e., number of customers) of an $\mathrm{M/M/\infty}$ queueing system.

Example 7.4 H The state diagram of Fig. 7.27 is an example of BDP with finite state space. This process models the evolution of a population whose size cannot exceed K members. The birth and death rates are all equal to λ and μ, respectively, except for state zero whose death rate is (as usual) zero, and state K, whose birth rate is also zero. Specializing (7.196) to this case we obtain

$$\pi_0 = \frac{1}{1 + \sum_{j=1}^{K} \prod_{k=0}^{j-1} \frac{\lambda}{\mu}} = \frac{1}{1 + \sum_{j=1}^{K} \frac{\lambda^j}{\mu^j}} = \frac{1}{\sum_{\ell=0}^{K} \left(\frac{\lambda}{\mu}\right)^{\ell}} = \frac{1 - \frac{\lambda}{\mu}}{1 - \left(\frac{\lambda}{\mu}\right)^{K+1}}$$

and

$$\pi_j = \pi_0 \prod_{k=0}^{j-1} \frac{\lambda}{\mu} = \frac{\left(1 - \frac{\lambda}{\mu}\right) \frac{\lambda^j}{\mu^j}}{1 - \left(\frac{\lambda}{\mu}\right)^{K+1}} \qquad j = 0, 1, 2, \ldots, K\,.$$

Therefore, the asymptotic state probability vector of the size of the population is, in this case, geometric truncated at K. This BDP models the $\mathrm{M/M/1/K}$ queueing system that will be dealt with in Chapter 8.

REFERENCES

1. A. A. Markov, *Extension of the limit theorems of probability theory to a sum of variables connected in a chain.* The Notes of the Imperial Academy of Sciences of St. Petersburg, VIII series, Physio-Mathematical College, vol. XXII, No. 9, Dec. 5, 1907.

2. L. Kleinrock, *Queueing systems.* vol I - Theory. John Wiley & Sons, New York, 1975.

3. P. V. Mieghem, *Performance analysis of communications networks and systems.* Cambridge University Press, United Kingdom, 2006.

4. H. M. Taylor, S. Karlin, *An introduction to stochastic modeling.* Academic Press, San Diego, California. 3rd ed., 1998.

5. S. Karlin, H. M. Taylor *A Second Course in Stochastic Processes.* Academic Press, San Diego, California. 1981.

6. E. Parzen, *Stochastic Processes.* Holden Day, Inc., 1962.

7. S. M. Ross, *Applied probability models with optimization applications.* Dover Publications, Inc., New York, 1969.

8. E. N. Gilbert, *Capacity of a burst-noise channel,* Bell Syst. Tech. J., vol. 39, pp. 1253-1265, Sept. 1960.

Problems

Poisson and exponential random variables

7.1 A shop is visited by x customers per day, where x is a Poisson rv with parameter λ. Each customer, independently of the others, purchases a random number g of items, with g having PMD $p_g(k) = \gamma(1-\gamma)^k$, $k = 0, 1, \ldots$. The cost c of an item can be modeled as an exponential rv with parameter μ. However, when a customer buys more than G items, he/she gets a d percent discount on the bill. Determine what follows.

a) The mean number of paying customers.

b) The mean number of sold items.

c) The mean outlay for any customer.

d) The probability that a customer spends more than S units of currency.

7.2 Consider a system where items arrive according to a Poisson process with parameter $\lambda = 2$ item/sec. Let $X(t)$ denote the number of items arrived in the time interval $[0, t)$ for any $t \geq 0$. Assuming that $X(0) = 0$, determine the following probabilities.

a) $\mathrm{P}[X(1) = 2]$;

b) $\mathrm{P}[X(1) = 2, X(3) = 6]$;

c) $\mathrm{P}[X(1) = 2 | X(3) = 6]$;

d) $\mathrm{P}[X(3) = 6 | X(1) = 2]$.

7.3 A system is subject to shocks that occur according to a Poisson process with parameter λ. Assume that the system stops working after an integer number k of consecutive shocks. Determine the PDF of the time T at which the system breaks down.

7.4 A system is subject to shocks that occur in time according to the arrivals of a Poisson process with parameter λ. The system survives to each shock with probability α, independently of how many shocks it has already undergone. Conversely, each shock may break down the system with probability $1-\alpha$. Find the probability that the system is still running at time t.

7.5 Let $X(t)$ be a Poisson process with parameter $\lambda = 3$ arrival/s. Find the probability of having 2 arrivals in the time interval $(0, 1]$, given that there are 5 arrivals in the time interval $(0, 3]$.

7.6 Consider a system where items arrive according to a Poisson process with parameter λ. Items are collected in a storage facility that can host up to Q objects. When the storage facility is full, i.e., exactly Q items have been collected, the system empties its storage facility in zero time and starts collecting items anew. Let $N(t)$ be the number of items in the system at time t, with $N(0) = 0$. Furthermore, let $T = \min\{t \geq 0 : N(t) = Q\}$ the time at which the storage facility fills up for the first time. Prove that

$$\mathrm{E}[T] = \frac{Q}{\lambda},$$

and

$$\mathrm{E}\left[\int_0^T N(t)dt\right] = \frac{1+2+\ldots+Q-1}{\lambda} = \frac{Q(Q-1)}{2\lambda}.$$

Discrete-time Markov Chain

7.7 A discrete-time process is a Markov chain if and only if

a) the state of the process at each time instant does not depend on the values assumed in any other instant;

b) the future states does not depend on the current state;

c) the future states does not depend on the past states;

d) the future states does not depend on the current state, given the past states;

e) the future states does not depend on the past states, given the current state.

7.8 Assuming X_n is a discrete-time MC, which of the following equations are correct and which are wrong?

a) $P[X_3 = 6|X_2 = 5, X_1 = 4, X_0 = 2] = P[X_3 = 6|X_2 = 5]$

b) $P[X_3 = 6|X_1 = 4, X_0 = 2] = P[X_3 = 6|X_0 = 2]$

c) $P[X_3 = 6|X_1 = 4, X_0 = 2] = P[X_3 = 6|X_1 = 4]$

d) $P[X_3 = 6|X_2 = 5, X_0 = 6] = P[X_3 = 6]$

e) $P[X_2 = 5|X_3 = 6, X_0 = 2] = P[X_3 = 6|X_2 = 5]$

f) $P[X_2 = 5|X_3 = 6, X_0 = 2] = P[X_3 = 6|X_2 = 5] / P[X_3 = 6|X_0 = 2] P[X_0 = 2]$

7.9 Prove that the two formulations of the Markov property provided by (7.1) and (7.2) are equivalent.

7.10 Say which of the following processes is as a discrete-time MC. Motivate your claim.

a) The process that counts the number of successes in a row of a Bernoulli process.

b) The process that counts the size of a population of cells in successive days, knowing that in a day each cell may generate another cell with probability p, independently of the other cells.

c) The process that describes the state of a manufacturing machine at the beginning of each hour, knowing that the machine can only be in one of three states, namely power off, stand-by, and active, and that after three hours of activity the machine needs to be powered off for another hour.

7.11 A stochastic matrix is

a) a matrix with only positive elements;

b) a matrix with elements no greater than one, such that the elements of each row sum to one;

c) a square matrix with positive or zero elements, such that the elements of each row sum to one;

d) a square matrix with only positive elements, such that the elements of each row sum to one;

e) a square matrix with only positive elements, such that the elements of each column sum to one;

f) a square matrix with elements in $[0, 1]$, such that the elements of each column sum to one.

7.12 Classify the MC with the following transition matrix

$$\boldsymbol{P} = \begin{bmatrix} 0 & 1 & 0 \\ 0 & 0 & 1 \\ 1 & 0 & 0 \end{bmatrix}$$

Find the asymptotic and/or stationary state probability vector, if it exists.

7.13 Consider a dice game in which, at each round, the player throws as many dice as the sum of the outcomes of the previous roll. Assume that the game begins throwing six dice.

a) Determine whether or not the process that counts the number of dice rolled at each round is Markovian.

b) Determine whether or not the process that counts the number of six at each round is Markovian.

Justify your claims.

7.14 Consider a dice with six facets. Let X_n denote the facet turned up at the nth throw. Determine whether the process $Y_n = X_n + X_{n+1}$ is a Markov chain or not.

7.15 Determine the asymptotic state probability vector of the discrete-time MC with transition matrix

$$\boldsymbol{P} = \begin{bmatrix} 0 & 1/2 & 1/2 \\ 1/2 & 0 & 1/2 \\ 1/2 & 1/2 & 0 \end{bmatrix}$$

7.16 Consider the discrete-time MC with transition matrix

$$\boldsymbol{P} = \begin{bmatrix} 0.1 & 0.7 & 0.2 \\ 0.4 & 0 & 0.6 \\ 0 & 0.5 & 0.5 \end{bmatrix}$$

Assuming the initial state probability vector is $\boldsymbol{p}(0) = [0.3,\ 0.4,\ 0.3]$, apply the Chapman-Kolmogorov equations to determine the state probability vector for $n = 1, 2, 3, 4$. Repeat the computation assuming $\boldsymbol{p}(0) = [1,\ 0,\ 0]$. Finally, compute the asymptotic probability vector, if it exists.

7.17 Draw the transition diagram for each of the transition matrices below.

$$(a)\quad \boldsymbol{P} = \begin{bmatrix} 1/2 & 1/2 & 0 & 0 \\ 1/2 & 1/2 & 0 & 0 \\ 1/3 & 1/3 & 1/3 & 0 \\ 1/4 & 1/4 & 1/4 & 1/4 \end{bmatrix} \qquad (b)\quad \boldsymbol{P} = \begin{bmatrix} 0 & 1/2 & 0 & 1/2 \\ 1/2 & 0 & 1/2 & 0 \\ 0 & 1/2 & 0 & 1/2 \\ 1 & 0 & 0 & 0 \end{bmatrix}$$

$$(c)\quad \boldsymbol{P} = \begin{bmatrix} 1/2 & 1/2 & 0 & 0 \\ 0 & 1/2 & 1/2 & 0 \\ 0 & 0 & 1/2 & 1/2 \\ 1 & 0 & 0 & 0 \end{bmatrix}$$

How do you classify the three MCs? Why? Compute the steady state probability probability vector $\boldsymbol{\pi}$, whenever possible.

7.18 Consider the MC with transition probability

$$\boldsymbol{P} = \begin{bmatrix} 0 & 1/2 & 1/2 \\ 1/3 & 1/3 & 1/3 \\ 1/2 & 0 & 1/2 \end{bmatrix}$$

Classify the MC and determine the asymptotic probability vector, if possible.

7.19 Consider the MC with transition probability

$$\boldsymbol{P} = \begin{bmatrix} 0 & 1 & 0 \\ 1/2 & 0 & 1/2 \\ 0 & 1 & 0 \end{bmatrix}$$

Classify the MC and determine the asymptotic state probability vector, if possible.

7.20 A transmission device receives packets to be forwarded over a certain link. Time is divided in time slots of T seconds each, and the device can transmit a single packet per time slot. Packets can be are buffered at the device output port. If the buffer is empty at the beginning of a time slot, the slot remains idle, otherwise the node transmits the head-of-the-queue packet. This type of multiplexing is called statistical time-division multiplexing and the tansmitting device is generally referred to as *statistical multiplexer* (SMUX).

The process X_n that gives the number of buffered packets at the beginning of the nth slot evolves as described by the following equation

$$X_{n+1} = X_n + v_n - \chi\{X_n > 0\} \tag{7.206}$$

where v_n denotes the number of packets generated during the nth time-slot, whereas $\chi\{X_n > 0\}$ is an indicator rv that equals one if the nth slot carries a packet transmission (the buffer was not empty at the beginning of the slot) and zero otherwise. Assuming that $X_0 = 0$ and that the rvs $\{v_n\}$, $n = 0, 1, \ldots$ are iid with PMD $p_v(k)$, $k = 0, 1, 2, \ldots$, also independent of the number of buffered packets, prove that X_n is a MC and find its transition matrix $\boldsymbol{P}$.

7.21 Consider a variation of the SMUX device of Problem 7.20 in which packets are generaed and get served in pairs, so that the process Y_m that counts the number of packet in the buffer at the beginning of each slot evolves according to the following equation

$$Y_{m+1} = Y_m + 2v_m - 2\chi\{Y_m > 0\}$$

where v_m is statistically independent of Y_m. Suppose that the initial state of the process can either be $Y_0 = 0$ or $Y_0 = 1$, with equal probability. Determine whether Y_m is a MC and, in this case, classify it and write its transition matrix.

7.22 In an assembly line of a factory, a robot is capable of fixing a bolt every T seconds, when working properly. However, from time to time the robot breaks down and needs repairing. Suppose that the probability that the robot breaks down after a bolt fixing is p, independently of the past history. Also suppose that each repair takes a random number of periods T with geometric distribution of mean $1/q$. Determine the following quantities:

a) the distribution of the number of bolts fixed in each run, i.e., in the interval between two consecutive repairs;

b) the percentage of time in which the robot is out-of order;

c) the probability that the robot is out-of-order after $k = 10000$ periods of length T.

7.23 For a generic function $g(n)$ defined for $n \geq 0$, we define its z-transform as

$$\mathcal{Z}\{g\}(z) = \sum_{n=0}^{\infty} g(n) z^n \tag{7.207}$$

where z is a complex variable. With reference to the bidirectional random walk of Example 7.2 M develop the following points.

a) Prove that $\mathcal{Z}\{P_{0,0}\} = \frac{1}{\sqrt{1-4pqz^2}}$.

b) Using the result of the previous point, derive the exact expression of $\sum_{n=0}^{\infty} P_{0,0}(n)$.

c) Prove that $\mathcal{Z}\{f_{0,0}\} = 1 - \sqrt{1 - 4pqz^2}$.

d) Using the result of the previous point, derive the exact expression of $f_{0,0}$ and of $f_{0,0}(n)$ for any positive integer n.

7.24 Some data link protocols for binary transmissions over serial lines, e.g., the *high-level data link control* (HDLC) protocol, the beginning and the end of a data frame are delimited by a predefined sequence of bits, called *frame sync sequence* (FSS). The bit sequence 01111110 containing six adjacent 1 bits is commonly used as FSS. To ensure that this pattern never appears in normal data, the transmitter adds a 0 bit after every five consecutive 1's. In this manner, the FSS can only appear at the beginning and end of the frame, but never in the data part. The extra bits, called *stuffed bits*, are removed by the receiver before passing the data to the upper layer. Note that, to avoid ambiguities at the receiver, the stuffed bit is added after every sequence of five 1's, even if the following data bit is 0, which could not be mistaken for a sync sequence. In this way, the receiver can unambiguously distinguish stuffed bits from normal bits. This technique is called *zero-bit insertion* or, more generally, *bit-stuffing*.

Now, consider a binary message of infinite length that has to be transmitted over a serial line using the zero-bit insertion technique. Assuming that each bit of the message takes the value 0 with probability q and 1 with probability $p = 1 - q$, independently of the other bits, we wish to determine the asymptotic percentage of stuffed bits in the message transmitted over the line.

7.25 Consider a wireless system where a random number of mobile users share a common transmission channel with capacity of $C = 1$ Mbit/s. We assume that the new requests for data transmission generated over the time by the mobile users form a Poisson process with parameter λ [request/s]. Each request consists in the transmission of a random number of data bits, with distribution approximately exponential and mean value L bits. (For simplicity, assume that system can transmit even fractions of bit.) The channel allocation is operated by a central controller, named base station (BS) on a periodic base, with period $T = 10$ ms. At the beginning of each period, the BS divides the channel capacity equally among all the *active* users, that is to say, users that have data to transmit. Therefore, denoting by X_n the number of active users at the beginning of the nth allocation period, each user is given a budget of C/X_n bits untill the following allocation period.

a) Prove that X_n is a MC.

b) Determine the transition matrix $\boldsymbol{P}$.

Continuous-time Markov Chain

7.26 Prove that the transition matrix $\boldsymbol{P}(u)$ of a continuous-time MC is continuous in the non-negative axis.

7.27 Which of the following properties apply to an *infinitesimal generator matrix* $\boldsymbol{Q}$?

a) Elements are non negative, except for those in the main diagonal that may assume any value.

b) Elements are non negative, except for those in the main diagonal, so that the elements of each row sum to zero.

c) Each element may be positive or negative, given that the elements of each row sum to zero.

d) Each element may be positive or negative, given that the elements of each column sum to zero.

e) All elements are lower or equal to one, so that the elements of each row sum to one.

f) All elements are lower or equal to one, so that the elements of each column sum to one.

g) All elements are positive.

7.28 Determine the asymptotic state probability vector of the MC with infinitesimal generator matrix

$$\boldsymbol{Q} = \begin{bmatrix} -1 & 0.5 & 0.5 \\ 0.1 & -0.2 & 0.1 \\ 0.4 & 0 & -0.4 \end{bmatrix}$$

7.29 Consider the process $X(t)$ that counts the number of people queueing for a taxi at the exit of a small railway station. Is it a continuous-time MC? Motivate your answer.

7.30 A server processes tasks of three different classes, namely C_0, C_1 e C_2, according to the following priority policy:

- if the server is idle, a new task is immediately executed, independently of its class;

- if the server is executing a task of class C_j, then it will refuse any new task of class C_i with $i \leq j$, whereas it will suspend the execution of C_j and immediately process the new task C_i if $i > j$. When the execution of a higher priority task C_i is completed, the server resumes the execution of the previous task C_j exactly from the point where it was interrupted.

Note that a task can be interrupted multiple times during its execution. Furthermore, the execution of a task of class C_0 may be interrupted by the arrival of a task of class C_1 that is immediately served and that, in turn, may be interrupted by the arrival of a task of class C_2. The execution of a task of class $j = 0, 1, 2$, when not interrupted, takes a time τ_j, which is a rv with exponential distribution with parameter μ_j, independent of the execution time of the other tasks. The tasks of the three different classes arrive to the server according to three independent Poisson processes with parameters λ_0, λ_1 and λ_2 for the class C_0, C_1 and C_2, respectively. Develop the following points.

a) Define a continuous-time MC that describes the state of the server.

b) Find the asymptotic state probability vector, as a function with parameters λ_j e μ_j.

c) Find the probability that the execution of a task of class C_0 gets interrupted by the arrival of a task of class C_1.

7.31 Consider a postal office with two counters denote by C_1 and C_2 with separates waiting queues. The service times are iid rvs with exponential distribution with parameter $\mu = 1/4$ [customer/min]. Customers arrive to the postal office according to a Poisson process with rate $\lambda = 1/5$ [customer/min] and, once entered the office, they join the shortest queue or, in case the queues have equal length, they choose a queue at random. Furthermore, the last customer of a queue will move to the other queue (in zero time) whenever this action will advance its position. Modeling the system state at time t with the pair $w(t) = (x_1(t), x_2(t))$, where $x_1(t)$ and $x_2(t)$ are the number of customers in front of the counters C_1 and C_2, respectively, develop the following points.

a) Prove that $w(t)$ is a (two-dimensional) MC.

b) Write the probability flow balance equations.

7.32 Consider a machinery that cycles along three different states, namely 1) working, 2) broken, 3) under repair. Suppose that, when the machine is working, hazards occur according to a Poisson process with parameter λ. Furthermore, assume that the machinery remains in broken state for an exponential random time with parameter μ and that the repair takes an independent random time with exponential distribution with parameter ζ. Compute the asymptotic probability of the working state.

Birth-death process

7.33 Mark which of the following matrices can be the infinitesimal generator matrix of a BDP.

$$(a)\ \boldsymbol{Q} = \begin{bmatrix} 1 & 0 & 0 \\ 0 & 0.8 & 0.2 \\ 0.1 & 0.4 & 0.5 \end{bmatrix} \quad (b)\ \boldsymbol{Q} = \begin{bmatrix} 1 & -1 & 0 \\ 0 & -0.2 & 0.2 \\ 0 & 4 & -4 \end{bmatrix} \quad (c)\ \boldsymbol{Q} = \begin{bmatrix} -1 & 1 & 0 \\ 0 & -0.2 & 0.2 \\ 0 & 4 & -4 \end{bmatrix}$$

$$(d)\ \boldsymbol{Q} = \begin{bmatrix} -2 & 2 & 0 \\ 0 & -2 & 2 \\ 1 & 3 & -4 \end{bmatrix} \quad (e)\ \boldsymbol{Q} = \begin{bmatrix} -1 & 1 & 0 \\ 0.8 & -2 & 1.2 \\ 0 & 3 & -3 \end{bmatrix} \quad (f)\ \boldsymbol{Q} = \begin{bmatrix} -4 & 3 & 0 \\ 0.8 & -2 & 1.2 \\ 0 & 2 & -3 \end{bmatrix}$$

7.34 Consider a *pure-death process* with constant death-rates, that is to say, a BDP having birth-rates identically equal to zero and death-rates $\mu_j = \mu > 0$ for $j = 1, 2, \ldots$. Suppose that initially the population is composed of m members, i.e., $p_m(0) = 1$ and $p_j(0) = 0$ for any $j \neq m$. Draw the state flow diagram and determine the time-dependent behavior of the state probabilities for such a process. What's the asymptotic state probability vector?

7.35 Determine the asymptotic state probability vector of the MC with infinitesimal generator matrix

$$Q = \begin{bmatrix} -1 & 1 & 0 \\ 0.1 & -0.2 & 0.1 \\ 0 & 4 & -4 \end{bmatrix}$$

7.36 Consider a BDP with state space $\mathcal{S} = \{0, 1\}$ and transition intensities

$$q_{0,1} = \lambda; \qquad q_{1,0} = \mu .$$

Determine what follows.

a) The transition probabilities $P_{i,j}(h)$ in a very short time interval of duration h.

b) The asymptotic state probability vector $\boldsymbol{\pi}$.

7.37 Consider a BDP with birth rates equal to

$$\lambda_k = \begin{cases} (K-k)\lambda, & \text{for } k < K \\ 0, & \text{for } k > K \end{cases}$$

and death rates

$$\mu_k = \begin{cases} k\mu, & \text{for } k \leq K \\ 0, & \text{for } k > K \end{cases}$$

where K is a positive integer that represents the maximum size of the population. Determine the asymptotic state probability vector.

7.38 Consider a BDP with state space $\mathcal{S} = \{0, 1, \ldots, 5\}$ and parameters

$$\lambda_i = i + 1\,, \quad i = 0, 1, \ldots, 4\,,$$
$$\mu_i = 6 - i\,, \quad i = 1, 2, \ldots, 5\,.$$

Find the asymptotic state distribution.

7.39 Consider a BDP with infinite state space, constant death rate μ (except for state zero, whose death rate is zero) and state-dependent birth rate given by

$$\lambda_j = \frac{\lambda}{j+1}\,, \quad j = 0, 1, 2, \ldots\,,$$

with $\lambda > 0$.

a) Comment on the existence of the asymptotic state distribution.

b) Find the asymptotic state distribution vector $\boldsymbol{\pi}$.

c) Find the asymptotic mean state $\lim_{t\to\infty} \mathrm{E}[X(t)]$.

7.40 Consider a BDP with parameters

$$\lambda_j = \begin{cases} \lambda\,, & 0 \leq j \leq K\,; \\ 2\lambda\,, & j > K\,; \end{cases}$$

$$\mu_j = \mu\,, \qquad j = 1, 2, \ldots$$

a) Find the asymptotic distribution vector $\boldsymbol{\pi}$, assuming it exists.

b) Find the relationship that must exist among the parameters of the problem in order for the asymptotic distribution to exist.

CHAPTER 8

QUEUEING THEORY

Andrea Zanella

This chapter is devoted to *queueing theory*, a branch of probability theory that permits to derive a number of metrics of great relevance for the design and analysis of telecommunication systems. A *queueing system* is an abstract model of a structure in which some entities, called *customers*, raise random requests for some type of *service* provided by entities called *servers*. The physical nature of service and customers is of no relevance for queueing theory, what matters is only the statistical description of the customers arrival and service processes. Queueing theory enables simple analysis of the system behavior in many different situations, capturing the way system parameters affect the performance statistics, such as waiting time for a customer to be served, fraction of time a server is busy, probability for a customer to be denied service, and so on. Therefore, queuing theory can be effectively applied to analyze the performance of a given queueing system in different contexts, or to dimension system parameters in order to fulfill some target performance requirements. The art of applied queueing theory is to construct a model of the queueing structure that is simple enough to be mathematically tractable, yet containing sufficient details so that its performance analysis provides useful insights into the behavior of the real structure itself. The birth of queueing theory is generally associated to the work of A. K. Erlang in the early 1900s, who derived several important formulas for teletraffic engineering that today bear his name. Since then, the range of applications of queueing theory has grown to include not only telecommunications and computer science, but also manufacturing, air traffic control, military logistics, and many other areas that involve service systems whose demands are random. The increasing popularity of this discipline has brought along an impressive growth in size and depth of the related literature, which today offers an extensive and

Principles of Communications Networks and Systems N. Benvenuto, M. Zorzi

complete coverage of the subject from a variety of different perspectives and flavors. Due to the importance of the topic, we could not refrain from dedicating at least one chapter to cover the very basic concepts of queueing theory. In spite of their relative simplicity, the topics presented in the sequel are of fundamental importance because they can be used for a first performance analysis of a telecommunication system.

8.1 OBJECTIVE OF QUEUEING THEORY

Waiting in line is a very common and (often) annoying experience that can occur several times a day. We may wait in queue for paying at the supermarket counter, depositing a check at the bank window, entering the most fashionable discos and pubs, or talking with a human operator when calling the customer-care number of our car insurance. Besides these situations, where the presence of waiting queues is explicit and tangible, there are several other occasions where the access to the required service is still managed through waiting queues, though their presence remains rather oblivious to the user. This happens, for example, any time we chat with a friend over Internet, or send an e-mail, or surf the web. In fact, all these services are provided by means of telecommunication systems that are shared by a huge number of customers: whenever a service facility, say a transmission channel, is required by more customers than it can simultaneously serve, the workload in excess is queued in a waiting buffer and served when the resource facility becomes available.

In summary, we come across a queueing system anytime a given service is claimed by multiple customers, a situation that is very likely to occur in many occasions. The analysis of queueing systems is the subject of a branch of mathematics, called *queueing theory*. Some of the aspects addressed by queueing theory regard the mean waiting time for a customer to be served, the mean number of customers waiting in queue, the fraction of time during which the server is idle, and so on. The results provided by queueing theory find application in a number of different fields. For instance, queueing theory can be used to regulate the duration of red and green traffic lights, to determine the number of checkout counters in a supermarket or the number of seats in the waiting room of a dentist. In the field of telecommunication networks, queueing theory can be used to dimension parts of the system, such as the transmission rate of a channel, the amount of memory to be provided in a router, the number of operators required in a call center, and so on.

The following sections will introduce the basic principles of queueing theory and some fundamental results that are of particular interest in the analysis of telecommunication systems. The reader interested in a more advanced and deeper presentation of the topic is referred, for instance, to the classic books by Feller [5] and Cohen [3], whereas the two textbooks by Kleinrock [2, 9] offer a treatment of the subject with a more practical tone.

8.2 SPECIFICATIONS OF A QUEUEING SYSTEM

Before plunging into the study of queueing theory, we need to introduce some basic concepts regarding the structure and operation of a queueing system. Unfortunately, the symbols and vocabulary used within the body of queueing theory literature are rather heterogeneous and there is no common agreement on either of them. Therefore, this section will also serve

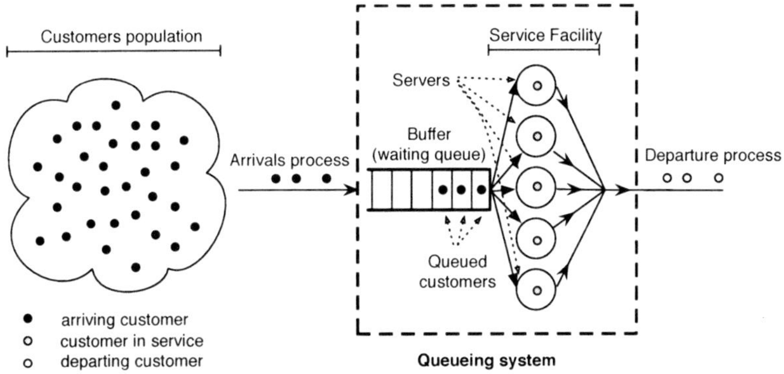

Figure 8.1 Model of a queueing system.

the purpose of defining the terminology and notation used in the sequel. For the reader's convenience, the main notation has also been collected in Tab. 8.1.

A generic queueing system (QS) is represented in Fig. 8.1, where a population of *customers* requires some type of service provided by a *service facility*, which consists of one or more entities called *servers*. Customers that arrive to the QS form the *arrival process* that, in general, is random. Once entered the QS, customers are immediately dispatched to available servers, if any. In case all the servers are busy serving customers, new arriving customers enters a *waiting queue* (or *buffer*), that is a storing area where customers are temporarily hosted waiting for service to become available. Each server can attend a single customer at a time and, when the service is completed, the customer leaves the system releasing the server for a new customer.

Here we limit our attention to systems with the so-called *work conserving* property, according to which a server can never remain idle when there are customers waiting for service. Using a figurative example, a work conserving system does not have "lazy" servers, which may decide to take some rest even when there are customers waiting for service. Experience suggests that not all the QSs that we encounter in our everyday life possess this appreciable property. In fact, in many cases, a server might be prevented from accepting customers for certain periods of time due to failures, periodic maintenance or other causes (such as chit-chatting with colleagues, taking a rest at the coffee machine, pretending to be busy in other tasks and so on). QSs that do not exhibit the work conserving property can be modeled by assuming that the service facility undergoes periods of *vacation* according to some statistic. The reader interested in QSs with service vacations is referred, for instance, to [16].

Before going in more depth in our analysis of QSs, we wish to highlight that queueing theory has a much larger scope than the analysis of real-life QSs, though this last aspect represents its most common and natural application. In fact, a QS is first of all a *mathematical model*, that is an abstract representation of specific aspects of a real system that makes it possible to investigate some features of the real system by means of rigorous mathematical and engineering methods. Accordingly, the roles of customer, server and waiting queue can be given to different elements of a real system, depending on the goal we wish to attain through the model. Examples 8.2 A–8.2 C shall help clarify this important aspect.

Table 8.1 Queueing system notation. The same symbols with the omission of the time references (n and t) are used when referring to asymptotic values.

Arrival process	C_n	nth customer
	t_n	arrival instant of C_n
	$\tau_n = t_n - t_{n-1}$	nth interarrival time
Departure process	d_n	departure instant of C_n
	$r_n = \tau_n - \tau_{n-1}$	nth interdeparture time
Service facility	m	number of servers in the service facility
	μ	service rate (for a single server)
	$\mathrm{m}\mu$	service capacity
	Q	waiting queue (storage) capacity
	$K = Q + \mathrm{m}$	system storage capacity
Occupancy measures	$z(t)$	number of customers in service (or number of servers in use) at time t
	$q(t)$	number of customers waiting in queue at time t
	$x(t) = z(t) + q(t)$	number of customers in the system at time t
Time measures	y_n	service time of C_n
	w_n	waiting time in queue of C_n
	$s_n = y_n + w_n$	system time of C_n
Traffic measures	λ	arrival rate
	η	throughput, or departure rate
	$\mathrm{m}\mu$	service capacity
Normalized measures	$\mathsf{G} = \frac{\lambda}{\mu}$	offered traffic
	$\mathsf{S} = \frac{\eta}{\mu}$	useful traffic
	$\rho = \frac{\lambda}{\mathrm{m}\,\mu}$	load factor, or utilization factor
Blocking systems	P_{blk}	call-blocking probability
	$\lambda_a = (1 - P_{\mathrm{blk}})\lambda$	customer admission rate
	$\rho_a = (1 - P_{\mathrm{blk}})\rho$	effective load factor

Example 8.2 A Having dinner in a restaurant can be an occasion to make a life experience of a QS. Consider a couple of lovebirds that wish to have a romantic dinner. They will probably be eager to seat at a dining table but, then, they will not mind to wait a little bit longer for food. Therefore, what they perceive as *waiting time* is the time before getting seated at a dining table: what follows, that is the time taken for placing the order to the waiter, consuming the dinner, getting the bill and leaving the table, is all part of the *service time*. This situation can be modeled by a multi-server QS, where each dining table plays the role of a server, the groups of people coming for dinner are the customers and the restaurant entrance, where people wait to be seated, provides the waiting queue. Resorting to queueing theory, the manager of the restaurant can determine, for example, the mean number of people packed in the restaurant entrance waiting to be seated and the mean amount of time they are required to wait for a table. Furthermore, he may also determine which actions need to be undertaken in order to modify these and other "performance measures." For instance, he may find the number of tables that shall be added in order to reduce by half the mean waiting time of the customers, or the number of waiters and cooks to hire for reducing the service time.

Example 8.2 B The call center of an on-line car insurance can be viewed as a multi-server QS with infinite waiting queue: the incoming calls are the customers, the human operators that pick up the calls play the role of servers (i.e., the service consists in clearing a customer call), whereas the automatic responder that kindly asks you to hold and wait for the first operator to become available provides the waiting queue.

Example 8.2 C A point–to–point data link can be considered as a single-server QS with finite queue. In this case, the data packets that have to be transmitted over the link are the customers, the link itself is the server (i.e., the transmission of data over the link is the service), whereas the waiting queue is the slice of memory (buffer) where the data packets that need to be forwarded through the link are temporarily held whenever the link is already engaged in transmitting previous packets. Note that the buffer size is always limited in practice. Therefore, if the packet generation rate exceeds the packet transmission link capacity for a long period of time, the buffer (waiting queue) will eventually fill up and the QS will start dropping packets, i.e., refusing newly arriving customers.

Example 8.2 D A private branch exchange (PBX) is a private telephone switchboard that offers a number of telephone services, such as short extensions for internal calls within a premise. Furthermore, the PBX is in charge of forwarding the outbound calls to the public switched telephone network (PSTN). Usually the number n of internal lines of a PBX is larger than the number m of lines toward the PSTN, so that the PBX performs an $\mathrm{n} \times \mathrm{m}$ multiplexing. Therefore, there is the possibility that a call toward an external number gets blocked because all the outbound lines are already engaged. This scenario can be modeled as a multi–server QS, in which the customers are represented by the outbound calls that arrive at the PBX, each of the m lines toward the PSTN plays the role of a single server, and there is no waiting queue. Note that when all the m lines toward the PSTN are engaged, any further outbound call that shall arrive to the PBX will be *blocked*.

In order to completely specify a QS it is necessary to define four features, namely: i) the arrival process of customers to the system, ii) the service process of each customer, iii) the queueing structure, and iv) the service discipline. Let us consider each of these aspects separately.

8.2.1 The arrival process

Customers belonging to a given population arrive to the QS according to a random arrival process. In this section we assume an infinite population of customers, so that the arrival process is not affected by the number of past arrivals. The customers that reach the QS are

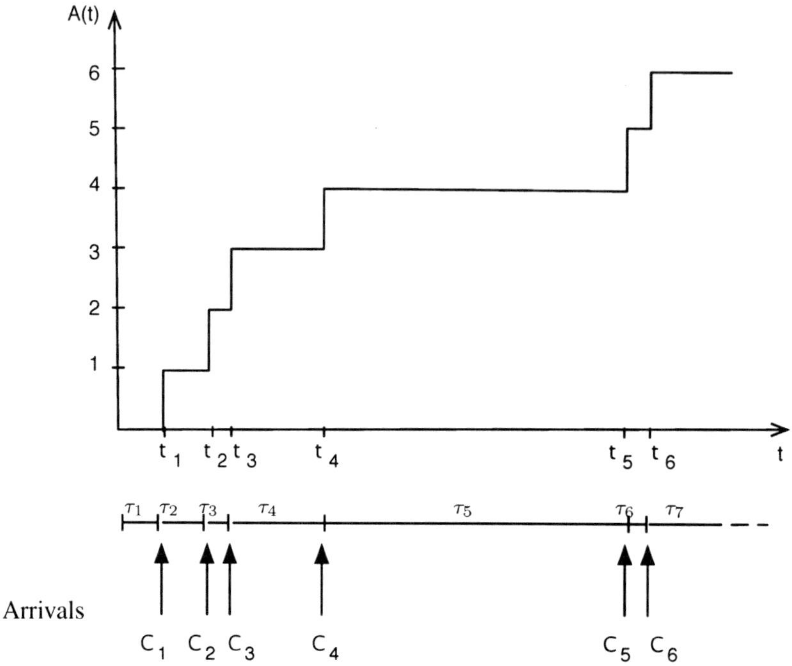

Figure 8.2 Example of customers arrival process $A(t)$.

denoted by $\{C_n\}$, where the integer $n \geq 1$ is the sequential order of arrival. The arrival time of C_n is denoted by t_n, whereas the time between the arrival of customers C_{n-1} and C_n is called *interarrival time* and denoted by $\tau_n = t_n - t_{n-1}$, where we conventionally set $t_0 = 0$. The arrival times are usually considered as iid rvs, with common CDF

$$P_\tau(a) = \mathrm{P}\left[\tau_n \leq a\right] , \quad a \geq 0 , \tag{8.1}$$

and PDF

$$p_\tau(a) = \frac{d\,P_\tau(a)}{da} , \quad a \geq 0 . \tag{8.2}$$

As illustrated in Fig. 8.2, the arrival process can be associated to a counting process $\{A(t), t \geq 0\}$ defined as

$$A(t) = \int_0^t \sum_{n=1}^{\infty} \delta(u - t_n)\, du \tag{8.3}$$

where $\delta(u)$ is the Dirac delta function (see Example 2.1 B). The process $A(t)$ counts the number of arrivals up to (and including) instant t. Note that, if interarrival times $\{\tau_n\}$ are iid and with finite mean $\mathtt{m}_\tau = \mathrm{E}\left[\tau\right]$ then, the counting process is homogeneous, and according to (2.231), the average number of customers that arrive at the service facility in a time unit, namely the *arrival rate*, can be expressed as the reciprocal of $\mathtt{m}_\tau$, that is[1]

$$\lambda = \frac{1}{\mathtt{m}_\tau}\,. \tag{8.4}$$

We here limit our attention to the classic case of *homogeneous Poisson process* for which, as seen in Theorem 2.22, the interarrival times $\{\tau_n\}$ are iid exponential rvs with parameter λ that, according to (8.4), is also equal to the process arrival rate (2.223). In this case, the interarrival times will have PDF

$$p_\tau(a) = \lambda e^{-\lambda a}\,, \qquad a \ge 0\,, \tag{8.5}$$

and CDF

$$P_\tau(a) = 1 - e^{-\lambda a}\,, \qquad a \ge 0\,. \tag{8.6}$$

8.2.2 The service process

The service facility consists of one or multiple (even infinite) *servers* that work in parallel and independently of one another, each attending a single customer at a time. In general, each server may supply a different service, though in this book we suppose that all servers provide statistically undistinguishable service. For a customer C_n, the *service time* y_n is the time C_n spends in the service facility. Service times are assumed to be iid rvs, also independent of the arrival process, with PDF $p_y(a)$, $a \ge 0$. Let $\mathtt{m}_y = \mathrm{E}\left[y_n\right]$ be the mean service time (assume finite) for each server. Then, according to (2.231), the *service rate* of a server, i.e., the average number of customers that can be served per time unit, is equal to

$$\mu = \frac{1}{\mathtt{m}_y}\,. \tag{8.7}$$

In particular, the service rate of a system with m servers working in parallel is equal to $\mathrm{m}\mu$ customers per time unit and is also called *service capacity* of the QS.

The most common service time distributions are the followings.

Deterministic – Service time is constant and equal to $1/\mu$, i.e.,

$$p_y(a) = \delta(a - 1/\mu)\,, \qquad a \ge 0\,, \tag{8.8}$$

and

$$P_y(a) = 1(a - 1/\mu)\,, \qquad a \ge 0\,. \tag{8.9}$$

Exponential – Service time is exponentially distributed with parameter μ, with PDF

$$p_y(a) = \mu e^{-\mu a}\,, \qquad a \ge 0\,, \tag{8.10}$$

and CDF

$$P_y(a) = 1 - e^{-\mu a}\,, \qquad a \ge 0\,. \tag{8.11}$$

Erlang-r – Service time has Erlang distribution (2.186) with index r and parameter $r\mu$, with PDF

$$p_y(a) = \frac{(r\mu)^r a^{r-1}}{(r-1)!} e^{-r\mu a}\,, \quad a \ge 0\,. \tag{8.12}$$

[1] Throughout this chapter we assume all counting processes are homogeneous so that (2.231) holds.

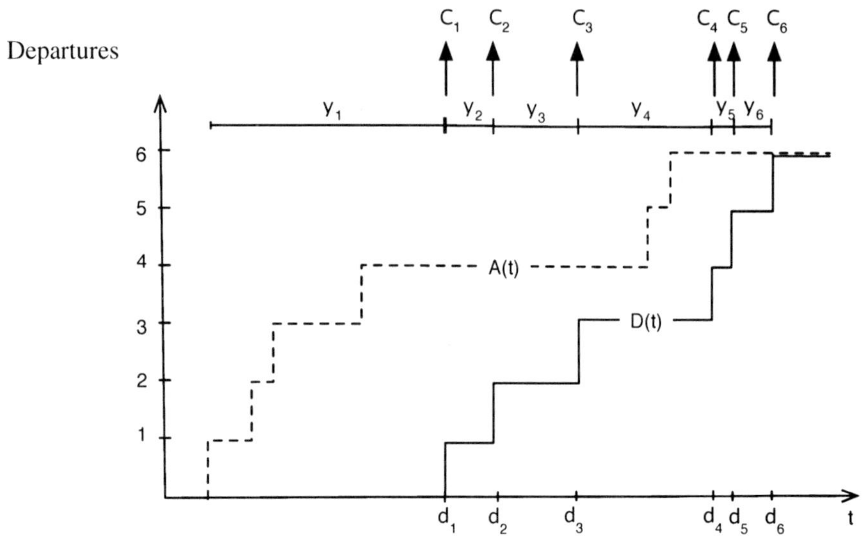

Figure 8.3 Example of customers departure process $D(t)$.

In this case, the service facility can be imagined as a cascade of r different stages of service that a customer must pass before a new customer is admitted into service, where each stage takes an exponentially distributed service time with parameter $r\mu$, independent of the other stages. Hence, the CDF of y can be expressed as

$$P_y(a) = 1 - \sum_{n=0}^{r-1} \frac{(r\mu a)^n}{n!} e^{-r\mu a}, \quad a \geq 0. \tag{8.13}$$

As illustrated in Fig. 8.3, the customers that leave the system after completion of the service form the *departure process* that can be associated to a counting process $\{D(t), t \geq 0\}$ defined as

$$D(t) = \int_0^t \sum_{n=1}^{\infty} \delta(u - d_n)\, du\,, \tag{8.14}$$

where d_n represents the departure instant of C_n. The *interdeparture time* r_n is defined as

$$r_n = d_n - d_{n-1}\,, \tag{8.15}$$

and its mean is denoted by $\mathtt{m}_r = \mathrm{E}[r_n]$ (assumed stationary and finite).

8.2.3 The queueing structure

The random nature of the arrival and service times may determine the accumulation of customers in the waiting queue of the QS. The maximum number of customers that can be buffered in the waiting queue is named *queue capacity* and denoted by Q, whereas the maximum number of customers that can be held in the system, either in queue or in service, is named *system storage capacity* and given by

$$K = \mathrm{m} + Q \,. \qquad (8.16)$$

A QS with limited storage capacity $K < \infty$ is called a *blocking system* because it refuses entry to customers that arrive when the QS is already full. This phenomenon was first observed in telephone networks, where it may happen that a call request gets blocked by a telephone switchboard because the lines are all engaged by other calls. For this reason, the probability P_{blk} for a customer to be refused admission to the system is often referred to as *call-blocking* probability.[2] In blocking systems, the average rate λ_a at which customers are *admitted* into the QS is lower than the arrival rate λ, since part of the arriving customers is blocked before entering the QS, as schematically represented in Fig. 8.4. A simple flow-conservation argument permits to conclude that, in the long run, λ_a is related to λ and P_{blk} through the following equation

$$\lambda_a = \lambda \left(1 - P_{\mathrm{blk}}\right) \,. \qquad (8.17)$$

In Fig. 8.4 λ_d represents the customer discard rate, given by

$$\lambda_d = \lambda - \lambda_a = \lambda P_{\mathrm{blk}} \,. \qquad (8.18)$$

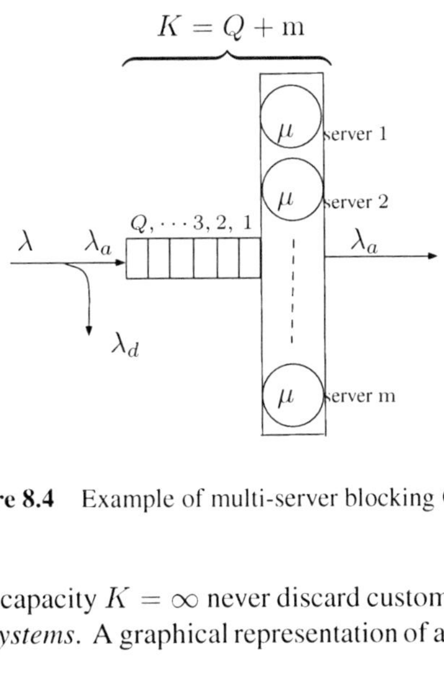

Figure 8.4 Example of multi-server blocking QS.

QSs with infinite storage capacity $K = \infty$ never discard customers and, for this reason, are referred to as *non-blocking systems*. A graphical representation of a single-server non-blocking QS is given in Fig. 8.5.

8.2.4 The service discipline

The order in which customers are taken from the waiting queue to be served is called *service discipline*, also referred to as *queuing discipline*. The most common service discipline is first-come first-served (FCFS), also known as first-in first-out (FIFO). According to this discipline,

[2]Other common terms for P_{blk} are outage-, dropping-, discard- or loss-probability, depending on the type of service provided by the system and on the cause of the service denial.

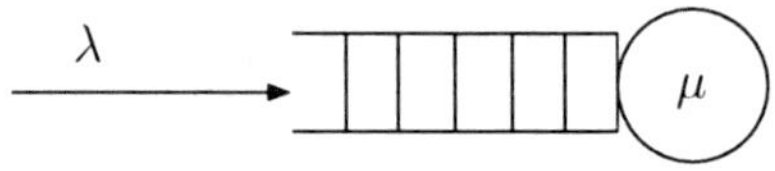

Figure 8.5 Example of single-server non-blocking QS.

customers are allowed into service in the same order as they arrive to the QS: newly arrived customers join the tail of the waiting queue, whereas the head-of-the-queue customer is served next. FCFS discipline is applied in many common situations such as, for instance, when people politely queue at the checkout counter of the university cafeteria or at the entrance of a movie theater. Another common discipline is last-come first-serve (LCFS), or last-in first-out (LIFO), where the newly arrived customers are placed at the head of the queue, pushing backward along the queue the customers that were already queued. LIFO is used, for example, in stack systems, where the last element inserted at the top of the stack is the first to be pulled out. Priority queueing is yet another discipline (or, better, a class of disciplines) in which service is differentiated according to the different priority of customers. In the remainder of this chapter we will treat the FCFS service discipline only.

8.2.5 Kendall notation

A very convenient way to specify a QS is provided by the *Kendall notation*, which consists in a six-part descriptor $A/B/C/K/N - S$, where

A: specifies the statistic of the interarrival time;

B: specifies the statistic of the service process;

C: is the number of servers;

K: is the system storage capacity;

N: is the size of the customer population;

S: denotes the service discipline (FCFS, LCFS, and so on).

The last three symbols are optional and, in case of omission, it is implicitly assumed $K = \infty$, $N = \infty$ and FCFS service discipline. Symbols A and B typically assume one of the following classic values:

$$\begin{array}{lll} \mathrm{M} & - & \textit{Memoryless} \text{ or } \textit{Markovian}, \text{ denotes the exponential distribution (see (2.161))}; \\ \mathrm{D} & - & \textit{Deterministic}, \text{ indicates a constant value}; \\ \mathrm{E_r} & - & \textit{Erlang-}r, \text{ denotes an Erlang distribution with index } r \text{ (see (2.186))}; \\ \mathrm{G} & - & \textit{Generic}, \text{ represents a generic distribution.} \end{array} \tag{8.19}$$

As a rule of thumb, QSs of type $\mathrm{M/M/-}$, named *Markovian QSs*, are much easier to study than QSs with other distributions. In particular, $\mathrm{M/M/1}$ yields a fairly simple analysis. On the other hand, results obtained for M/G/1 are more general in scope, since they apply to systems with any service time distribution. However, this comes at the cost of a more complex analysis. Lastly, the results available for G/G/1, which apply to any single-server system regardless of the interarrival and service time statistics, are just a few: even the mean waiting time is not exactly known, in general.

The following examples illustrate how to determine the suitable QS in different real life cases.

Example 8.2 E Referring to Example 8.2 A, assume that customers arrive to the restaurant according to a Poisson process and that the service times can be modeled as iid rvs with Erlang-r distribution. Hence, denoting by m the number of dining tables, the system can be studied by using $\mathrm{M/E_r/m}$.

Example 8.2 F We have already established that the call-center of Example 8.2 B can be described as a multi-server QS. Now, depending on the assumptions that we make and the simplifications that we are willing to accept in our abstraction, we can use different models for the system. For instance, if we assume that the calls arrive according to a Poisson process and that the time taken for an operator to clear a customer call can be approximated by a rv with exponential distribution, then the more suitable model for the system is the M/M/m, where m is the total number of operators. In particular, if there is a single operator we can consider the M/M/1. However, if we assume that customers are classified either as *normal* or *top-level* according to the revenue they bring to the company, and that top-level customers are served before normal customers, then the simple M/M/1 is no longer suitable to describe the system because it does not capture this aspect. We shall hence resort to the M/M/m with priority QS, whose rather advanced analysis is not treated in this book.

Example 8.2 G Consider the point–to–point data link of Example 8.2 C and suppose that packets are generated according to a Poisson process. Denoting by R the transmission rate of the link (in bit/s), the time taken to forward a packet of ℓ bits (i.e., the service time) is given by $y = \ell/R$. Assume that packet size can be modeled as a rv that takes values in a discrete set $\mathcal{L}$ with PMD $p_\ell(k)$, $k \in \mathcal{L}$. Then y takes values in the *discrete* set $\mathcal{A} = \{k/R : k \in \mathcal{L}\}$ with PMD $p_y(k/R) = p_\ell(k)$. In this case, the distribution of the packet size determines the distribution of the service time. Let us consider three typical cases.

a) If all the packets have equal size L, the service time is a constant equal to $y = L/R$. In this case, M/D/1 is a suitable system model.

b) If the packet size can take only a finite number r of values, i.e., $\mathcal{L} = \{L_1, \ldots, L_r\}$, the service time will take values in the finite set $\mathcal{A} = \{L_i/R : L_i \in \mathcal{L}\}$ and its distribution, in general, will not be of type (8.19). Lacking a more suitable model, we can only resort to the results obtained for M/G/1. Clearly, this is also the case for any other non-classic distribution of the service time.

c) If the packet size can be assumed to be geometrically distributed with mean $\mathtt{m}_\ell = 1/p$, then the service time distribution may be approximated by an exponential distribution with parameter $\mu = R/\mathtt{m}_\ell$, provided that μ is large enough (see Problem 8.3). The system can then be modeled as M/M/1.

Example 8.2 H Assume that the packets in Example 8.2 G are generated with a constant rate, as in case of video or audio streaming. The corresponding QS is, hence, characterized by a constant interarrival time. In this case, the system can be modeled as D/D/1, D/G/1 and D/M/1 for options a), b) and c), respectively.

Example 8.2 I The PBX of Example 8.2 D can be modeled as M/M/m/m, being m the number of outbound lines. However, if we assume that the PBX is capable of holding up to Q outgoing calls in a waiting state when all the outbound lines are busy, then a suitable model for the system is $\mathrm{M/M/m/K}$, with K given by (8.16).

8.3 PERFORMANCE CHARACTERIZATION OF A QS

The performance of a QS can be quantified in different ways, depending on the goal of the analysis. In general, the performance measures are rps and, as such, should be described by time-dependent statistical distributions. However, we are mainly interested in the *asymptotic* or *stationary* behavior of the system, that is to say, on the statistic of the performance measures after the transient effects due to the system initialization have vanished and the system has reached a stationary statistical behavior.[3] In these conditions, the statistical distribution of the quantities of interest is no longer dependent on time and, hence, is much easier to determine. Furthermore, as seen in Chapter 7 for MCs and BDPs, when the asymptotic distribution exists the process is *ergodic*, so that the statistical distribution is (almost surely) equal to the empirical distribution obtained from any realization of the process. When even the derivation of the asymptotic distribution is not straightforward, we will be satisfied with the evaluation of the first and second order moments of the performance measures, which still provide useful information about the system's behavior.

The performance measures of main interest can be divided in three categories, *occupancy*, *time* and *traffic* which are detailed in the following.

8.3.1 Occupancy measures

The occupancy measures refer to the *number of customers* that occupy different parts of the system at a given time instant. Typical occupancy measures are:

$q(t)$ – the number of customers in the waiting queue at time t;
$z(t)$ – the number of customer in service at time t;
$x(t)$ – the number of customers in the whole system at time t.

The three measures are obviously related by the following equation

$$x(t) = q(t) + z(t)\,. \tag{8.20}$$

Note that $q(t)$ cannot exceed the queue capacity Q, whereas $z(t)$ cannot exceed the total number of servers m, i.e.,

$$\begin{aligned} 0 \le q(t) \le Q\,, \\ 0 \le z(t) \le \mathrm{m}\,. \end{aligned} \tag{8.21}$$

Furthermore, in work conserving systems (see page 501), customers are queued only when all servers are busy, so that we can also write

$$q(t) = \begin{cases} 0 & , \quad x(t) \le \mathrm{m} \\ x(t) - \mathrm{m} & , \quad x(t) > \mathrm{m} \end{cases} \tag{8.22}$$

[3]The terms *steady-state* and *regime* behavior are also used to denote the asymptotic behavior of a QS.

and

$$z(t) = \begin{cases} x(t) & , \quad x(t) \leq \mathrm{m} \\ \mathrm{m} & , \quad x(t) > \mathrm{m}. \end{cases} \tag{8.23}$$

As we will see in detail later, $x(t)$ plays a primary role in the QS analysis and the derivation of the statistical distribution of $x(t)$ is of chief importance in Queueing Theory. In particular, the asymptotic behavior of $x(t)$ is related to the fundamental concept of *system stability*.

Definition 8.1 *A QS is said to be* stable *if $x(t)$ does admit proper*[4] *asymptotic PMD $p_x(k)$, $k = 0, 1, \ldots$, independent of the initial state $x(0) = k_0$. Conversely, if the asymptotic distribution of $x(t)$ is identically zero or depends on the initial state, the system is* unstable.

As we will see later, for Markovian queues the rp $x(t)$ is actually a BDP. In this case, Definition 8.1 simply says that an $\mathrm{M}/\mathrm{M}/-$ QS is stable when $x(t)$ is positive recurrent and, hence, ergodic (see Theorem. 7.14). Hence, in a stable Markovian QS, there is a positive probability that the system empties in the long-run (see Sec. 7.4.3), so that the number of customers in the QS does not grow indefinitely in time and the mean number of customers in the QS is asymptotically finite.

Non-blocking systems. We observe that, for non-blocking QSs, the stability condition simply states that the system can reach a stable behavior only if the customers arrive, on average, at a lower rate than they can be served, i.e.,

$$\boxed{\lambda < \mathrm{m}\mu\,.} \tag{8.24}$$

On the contrary case, i.e., if $\lambda > \mathrm{m}\mu$, the queue will grow indefinitely long for large t because customers arrive faster than they can be served. After some time, the server facility will work at service capacity $\mathrm{m}\mu$ without being able to keep the pace of customer arrivals, with the result that customers will indefinitely accumulate in the waiting queue at the average rate of $\lambda - \mathrm{m}\mu$. The QS is, hence, unstable. A graphical example of this situation is shown in Fig. 8.6, where we plot the number of customers $x(t) = A(t) - D(t)$ as a function of time in an $\mathrm{M}/\mathrm{M}/1$ QS with λ just 1% larger than μ. The straight dashed-line represents the expected number of customers in the QS as a function of time, given by $\mathrm{E}\left[x(t)\right] = (\lambda - \mu)t$.

Remark The case $\lambda = \mu$ is critical because it generally leads to an unpredictable system's behavior, with large variations of the number of customers in the QS over time. As we will see later when discussing $\mathrm{M}/\mathrm{M}/1$ queue, the situation with $\lambda = \mu$ corresponds to the case of a null recurrent MC, as discussed in Example 7.4 E.

Blocking systems. Blocking QSs are *always stable* because the number of customers is upper limited by the (finite) storage capacity K of the QS, so that $x(t)$ cannot grow indefinitely. For Markovian blocking QS, this result corresponds to the fact that $x(t)$ is an irreducible BDP with finite state space and, as seen in Sec. 7.3.10, it is always positive recurrent.

[4]See footnote 5 at page 441

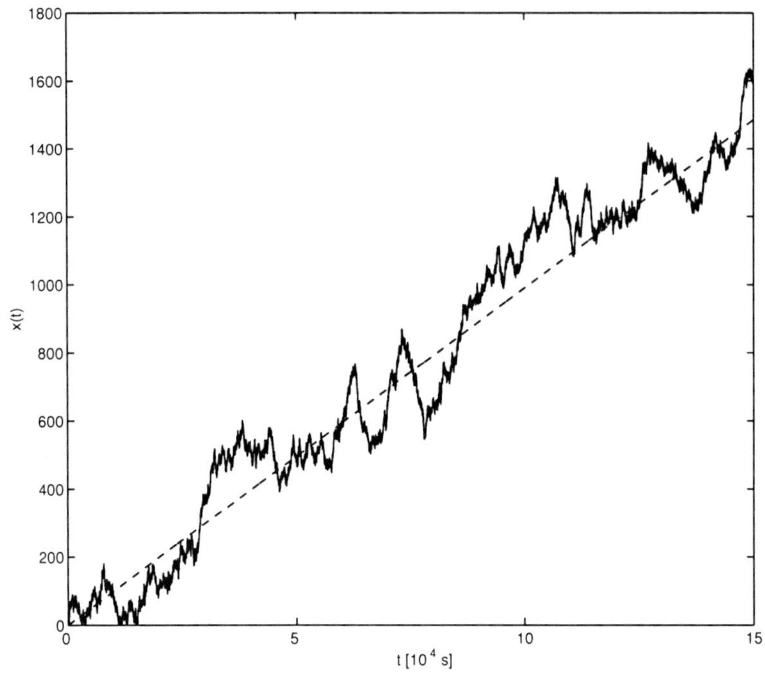

Figure 8.6 Number of customers versus time in an overloaded $\mathrm{M/M/1}$ QS with $\lambda = 1.01$ [customer/s] and $\mu = 1$ [customer/s].

8.3.2 Time measures

The time measures refer to the time spent by the generic customer C_n in the different parts of the system. The most important time measures are

w_n – the *waiting time*, or queueing time, is the time spent by C_n in the waiting buffer before entering service;

s_n – the *system time*, is the overall time C_n spends in the system, from the arrival to the departure instants, i.e.,

$$s_n = d_n - t_n \,. \tag{8.25}$$

These measures are related by

$$s_n = w_n + y_n \,, \tag{8.26}$$

where y_n is the service time of C_n (see Section 8.2.2).

For a single server QS, Fig. 8.7 illustrates how the occupancy and time measures vary according to the arrivals and departures of customers. The arrows at the bottom of the figure represent the arrivals of customers to the system, which are separated by the interarrival times τ_n. The solid line in the plot gives the number $x(t)$ of customers in the system at time t, whereas the dashed line shows the number $q(t)$ of queued customers. At each time instant t, the vertical distance between the two lines corresponds to the number $z(t)$ of customers in

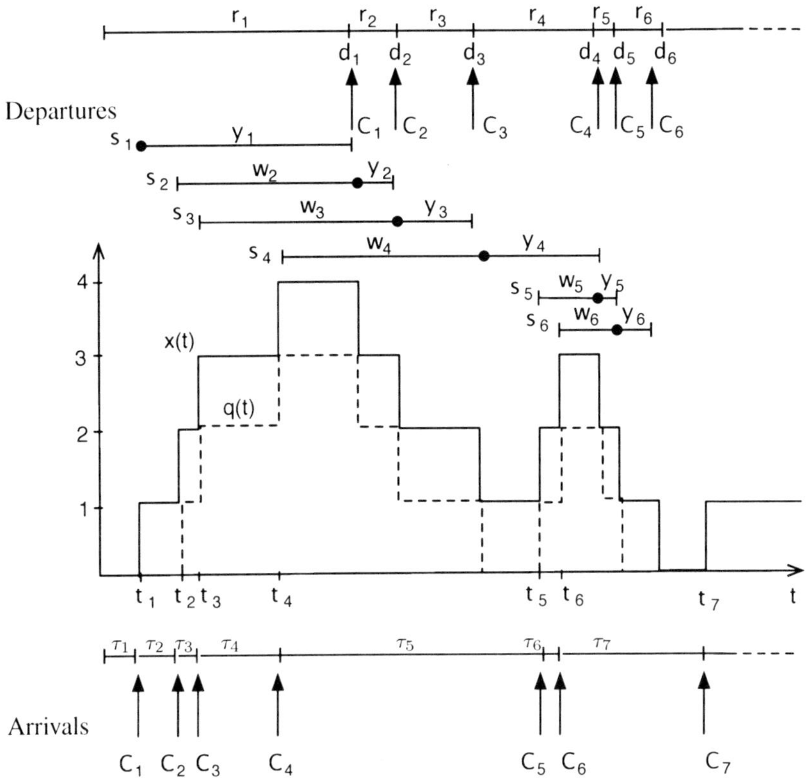

Figure 8.7 Example of customers arrival and departure in a queue system with m = 2 servers.

service at time t, which in case of a single-server QS can only be either one or zero. The number of customers in the system increases by one unit at each arrival and decreases by one unit at each departure. The system time s_n is reported as a horizontal interval atop the plot. The bullet indicates the instant at which the customer enters service and divide the system time s_n in its two components, namely the queueing time w_n and the service time y_n, according to (8.26).

8.3.3 Traffic measures

The last family of interest concerns the *workload* offered and carried out by the service facility and consists of the following five performance measures.

λ – The *arrival rate*, is the mean number of customer arrivals to the QS in the unit of time.

G – The *offered traffic*, is the mean number of arrivals to the QS in the mean service time. In other words, G is equal to the customer arrival rate λ normalized to the service rate μ, i.e.,

$$\boxed{\mathtt{G} = \lambda\, \mathtt{m}_y = \frac{\lambda}{\mu}\,.} \tag{8.27}$$

η – The *throughput*, or *departure rate*, is the mean number of customers that leave the QS per time unit. For all the QSs considered in this book, it holds

$$\eta = \frac{1}{\mathtt{m}_r}\,, \tag{8.28}$$

where $\mathtt{m}_r$ is the mean interdeparture time (8.15).

S – The *useful traffic*, is the mean number of departures from the QS in the mean service time. In other words, S is the throughput η normalized to the service rate μ, i.e.,

$$\mathtt{S} = \frac{\eta}{\mu} = \eta\, \mathtt{m}_y\,. \tag{8.29}$$

ρ – The QS *load factor* ρ, or *utilization factor* or *traffic intensity*. For a single-server system, ρ is defined as

$$\rho = \frac{\mathtt{m}_y}{\mathtt{m}_\tau} = \frac{\lambda}{\mu} = \lambda\, \mathtt{m}_y\,. \tag{8.30}$$

For multi-server QSs, under the work-conserving assumption of page 501, ρ is obtained by dividing (8.30) by the number of servers m, that is to say

$$\boxed{\rho = \frac{\lambda\, \mathtt{m}_y}{\mathrm{m}} = \frac{\lambda}{\mathrm{m}\,\mu}\,.} \tag{8.31}$$

Using (8.27), the load factor can also be written as

$$\rho = \frac{\mathtt{G}}{\mathrm{m}}\,. \tag{8.32}$$

As for the other performance measures, also the traffic measures are inter-related, as explained below.

Non-blocking QSs. From (8.31) we see that the stability condition (8.24) for a non-blocking QS can be written as

$$\rho < 1\,. \tag{8.33}$$

In this case the mean number of queued customers does not grows to infinity, and a simple flow-conservation argument leads us to conclude that the throughput must equal the arrival rate. Conversely, when $\rho > 1$, after a sufficiently long time the QS will hold enough customers to saturate the service facility, so that the throughput will equal the service capacity $\eta = \mathrm{m}\mu$, which explains why systems working in this condition are said to be *saturated*. However, it is clear that such a condition cannot be permanently held in practice, since working at service capacity comes at the cost of an infinitely long waiting time for customers to get served, a condition that, in general, is not acceptable by any kind of customer.

Wrapping up the discussion, for non-blocking QS we have

$$\eta = \begin{cases} \lambda & , \rho < 1 \\ \mathrm{m}\mu & , \rho \geq 1 \end{cases} \tag{8.34}$$

or, dividing both terms of (8.34) by μ and recalling (8.27),

$$\mathtt{S} = \begin{cases} \mathtt{G} & , \rho < 1 \\ \mathrm{m} & , \rho \geq 1. \end{cases} \tag{8.35}$$

Blocking QSs. The situation is significantly different for blocking QSs that are *always stable*, independently of the value of the load factor ρ. In this case, the throughput (8.28) will equal the average rate at which customers are *admitted* to the system, that is to say

$$\eta = \lambda_a = \lambda(1 - P_{\mathrm{blk}}), \tag{8.36}$$

or from (8.29)

$$\mathtt{S} = \lambda_a \mathtt{m}_y = \mathtt{G}(1 - P_{\mathrm{blk}}). \tag{8.37}$$

The load factor instead affects the call-blocking probability P_{blk} that, in fact, increases with λ in such a way that the average admission rate λ_a never exceeds the service capacity $\mathrm{m}\mu$.

Remark The call blocking probability is positive for *any* value of $\rho > 0$, not only for $\rho \geq 1$. In fact, it is possible that the QS temporarily fills its storing capacity because of random fluctuations of the arrival process or sporadic increase in the service times. In this case, the QS will not admit other customers until some buffering space becomes available. This event is more likely to occur for large values of ρ, though it may happen even when the system is, on average, lightly loaded.

Example 8.3 A The manager of the restaurant of Example 8.2 A wishes to have an idea of the number of tables he has to provide for avoiding long waiting queues of customers. He expects an arrival rate of $\lambda = 10$ couple/hour. Based on past experience, he can also anticipate that a couple remains seated at dining table for approximately one hour and a half. The offered traffic is hence equal to $\mathtt{G} = \lambda \mathtt{m}_y = 10 \times 1.5 = 15$ customers. Modeling the system as a non-blocking QS with m servers, the manager deduces that, for the system to be stable, he needs to provide a number of tables m such that the load factor ρ becomes strictly lower than one. According to (8.31), the load factor is equal to

$$\rho = \frac{15}{\mathrm{m}},$$

so that, to satisfy the stability condition (8.33) for non-blocking QSs, the manager has to provide at least $\mathrm{m} = 16$ dining tables.

8.4 LITTLE'S LAW

Little's law based on a simple flow conservation principle is a simple but fundamental result of queueing theory that applies to a broad set of QSs. In its general formulation, Little's law refers to a stream of customers that arrive to a generic "structure" according to some law and leave the structure after a certain time. No particular statistical assumptions are made on the interarrival and on the permanence times, except that both must have finite mean. Under these conditions, we have the following result:

Theorem 8.1 (Little's law) *The mean number of customers $\bar{X}$ within a flow-conserving structure is equal to the arrival rate of customers to the structure, $\bar{\Lambda}$, times the average time $\bar{S}$ spent by a customer in the structure:*

$$\boxed{\bar{X} = \bar{\Lambda}\,\bar{S}\,.} \tag{8.38}$$

Proof We will not provide a formal proof of this law, for which we refer to the original manuscript by Little [10]. Rather, we sketch a very intuitive reasoning. Let $A(t)$ and $D(t)$ denote the arrival and departure processes, respectively. The average arrival rate of customers to the structure up to time t is hence equal to

$$\Lambda(t) = \frac{A(t)}{t}\,, \tag{8.39}$$

while the number of customers in the structure at time t is $X(t) = A(t) - D(t)$ and the average time spent by a customer in the structure up to time t is equal to

$$S(t) = \frac{\int_0^t X(\tau)d\tau}{A(t)}\,. \tag{8.40}$$

Now, let $\bar{\Lambda}$, $\bar{X}$, and $\bar{S}$ denote the long-run average arrival rate, number of customers within the structure, and system time, respectively, as illustrated in Fig. 8.8. In formula, we have

$$\bar{\Lambda} = \lim_{t\to\infty} \Lambda(t) = \lim_{t\to\infty} \frac{A(t)}{t}\,, \tag{8.41}$$

$$\bar{X} = \lim_{t\to\infty} X(t) = \lim_{t\to\infty} \frac{\int_0^t X(\tau)d\tau}{t}\,, \tag{8.42}$$

$$\bar{S} = \lim_{t\to\infty} S(t) = \lim_{t\to\infty} \frac{\int_0^t X(\tau)d\tau}{A(t)}\,. \tag{8.43}$$

Dividing both numerator and denominator of (8.43) by t and using (8.41) and (8.42), we hence obtain

$$\bar{S} \simeq \frac{\bar{X}}{\bar{\Lambda}}\,, \tag{8.44}$$

which yields directly to (8.38).

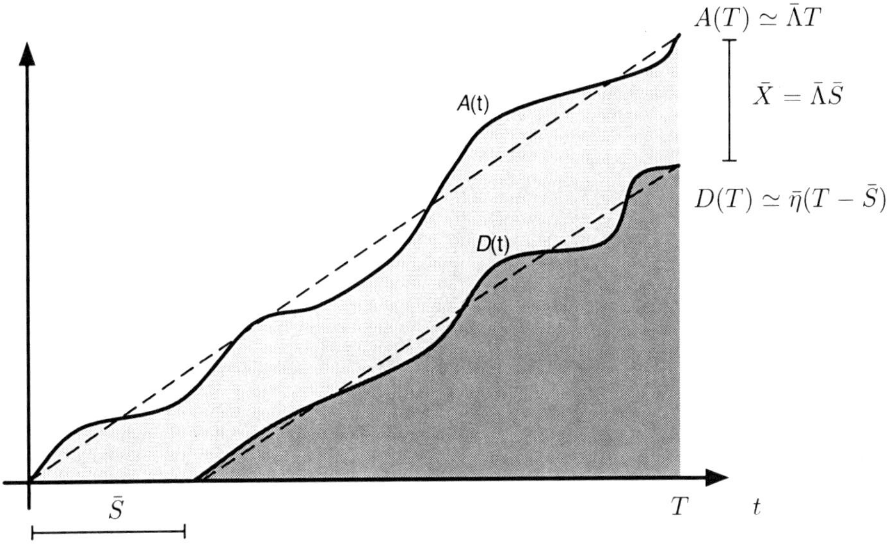

Figure 8.8 Illustration of Little's law.

An alternative proof is proposed in [4], where it is observed that, being $\bar{S} < \infty$ by assumption, then the average departure rate $\bar{\eta}$ must equal the average arrival rate $\bar{\Lambda}$. Now, the total number of customers that have departed from the system up to time T can be approximated to $D(T) \simeq \bar{\eta}(T - \bar{S}) = \bar{\Lambda}(T - \bar{S})$, where we have taken into account that each customer remains in the system, on average, a time $\bar{S}$. The average number of customers in the system at time T is hence given by

$$\bar{X} \simeq \frac{\int_0^T A(\tau)d\tau - \int_0^T D(\tau)d\tau}{T} \simeq \frac{A(T)T - D(T)T}{T}$$
$$= A(T) - D(T) = \bar{\Lambda}T - \bar{\eta}(T - \bar{S}) = \bar{\Lambda}T - \bar{\Lambda}(T - \bar{S}) = \bar{\Lambda}\,\bar{S}\,.$$

□

Although Little's law has been formulated in terms of time averages, the ergodicity of (most) QSs permits to replace empirical means with their statistical counterparts, thus giving the following fundamental expression of Little's result

$$\mathrm{E}\left[X(t)\right] = \mathrm{E}\left[\Lambda(t)\right]\mathrm{E}\left[S(t)\right]\,. \tag{8.45}$$

It is important to remark that Little's law can be applied to *any* closed system. In particular, applying Little's law to different parts of a QS we obtain the following important relations.

i) The "structure" is the entire QS (waiting queue and service facility):

$$\boxed{\mathtt{m}_x = \lambda\,\mathtt{m}_s\,.} \tag{8.46}$$

ii) The "structure" is the waiting queue only:

$$\boxed{\mathtt{m}_q = \lambda\, \mathtt{m}_w\,.} \tag{8.47}$$

iii) The "structure" is the service facility only:

$$\boxed{\mathtt{m}_z = \lambda\, \mathtt{m}_y\,.} \tag{8.48}$$

Further insight We observe that (8.48) together with (8.31) yields

$$\rho = \frac{\mathtt{m}_z}{\mathrm{m}}\,, \tag{8.49}$$

which justifies the interpretation of ρ as the fraction of time that each server of the QS is serving customers, i.e., the percentage of "utilization" of each server. In particular, when $\mathrm{m} = 1$, $\mathtt{m}_z$ is also equal to the asymptotic probability that the QS is not empty. In fact, we have

$$\begin{aligned}\mathtt{m}_z &= \lim_{t\to\infty} \mathrm{P}\left[z(t) = 1\right] = \lim_{t\to\infty} \mathrm{P}\left[x(t) > 0\right] \\ &= \mathrm{P}\left[\text{system is not empty}\right] = 1 - p_x(0)\,.\end{aligned} \tag{8.50}$$

Therefore, for any $\mathrm{G}/\mathrm{G}/1$ QS, from (8.49) we have

$$\rho = \mathrm{P}\left[\text{system is not empty}\right] = 1 - p_x(0)\,, \tag{8.51}$$

which offers still another interpretation of ρ as the fraction of time a single-server non-blocking system is not empty. Clearly, all these results hold under the assumption that the system is stable, so that, according to Definition 8.1, $x(t)$ admits proper asymptotic distribution $p_x(k)$.

Remark When applying Little's law to blocking systems ($\mathrm{G}/\mathrm{G}/\mathrm{m}/\mathrm{K}$), the arrival rate to be considered is the rate λ_a at which customers are admitted into the system, as given by (8.17).

Example 8.4 A A call center receives, on average, 20 calls per hour. The mean time for an operator to clear a call is 5 minutes. Each operator costs to the employer $c = 3/5$ units of some currency per minute. To make profit, each minute of call shall not cost, on average, more than $E = 1/3$ units of currency. We want to determine the minimum number of operators required for customers not to wait forever and the maximum number of operators for the company not to go bankrupt. To this end, we model the system as a $\mathrm{G}/\mathrm{G}/\mathrm{m}$, with $\lambda = 20/60$ customer/minute and $\mathtt{m}_y = 5$ minutes. Applying Little's law in the form of (8.48) we get $\mathtt{m}_z = \lambda \mathtt{m}_y = 5/3$. For the system to be stable, we require $\rho = \mathtt{m}_z/\mathrm{m} < 1$, so that we need at least $\mathrm{m} = 2$ operators. Now, assume that the call-center employs $\mathrm{m} \geq 2$ operators, which cost $\mathrm{m} \times c \times T$ units of currency over a time interval of T minutes. If T is sufficiently long to smooth out the fluctuations of the arrival and service processes, then the average number of calls processed in this period can be approximated to λT. Each call occupies an operator for an average time of $\mathtt{m}_y = 5$ minutes. The average cost per minute of call, denoted by e, is hence equal to

$$e = \lim_{T\to\infty} \frac{\mathrm{m}\, c\, T}{\lambda T \mathtt{m}_y} = c\frac{\mathrm{m}}{\lambda\, \mathtt{m}_y} = \frac{c}{\rho}\,.$$

Thus, to have profit it must be $e < E$, which yields

$$\mathrm{m} < \frac{\lambda \mathtt{m}_y E}{c} = \frac{5/3 \times 1/3}{3/5} = \frac{25}{27}\,.$$

Since the system stability imposes $\mathrm{m} \geq 2$, the call-center does not have any chance to remain in business in this situation! As a matter of fact, we can easily realize that, for the system to be stable and in business, the cost of each operator per unit time shall not exceed the threshold E. Imposing $e < E$, we have

$$\frac{c}{\rho} < E \Rightarrow c < E\rho < E\,,$$

where the last inequality follows from the stability condition (8.33).

Example 8.4 B The point–to–point data connection of Example 8.2 C is required to serve a mean traffic flow of $\lambda_b = 2.5$ Mbit/s. The packet size is fixed and equal to $L = 1250$ bytes. Adopting a conservative approach, we wish to dimension the minimum bit rate R_b of the transmission link for which the link is used no more than 60% of the time. To this end, we model the system as G/G/1 where the customers are the arriving packets and the service facility consists in the packet transmitter. Accordingly, the average arrival rate is $\lambda = \lambda_b/L = 2.5 \times 10^6/10^4 = 250$ pck/s and the mean service time is $\mathtt{m}_y = L/R_b$. Furthermore, applying (8.48), we have

$$\mathtt{m}_z = \lambda \mathtt{m}_y = \lambda \frac{L}{R_b}\,.$$

If the system is stable, the percentage of time the server is in use corresponds to the asymptotic probability that $z(t) > 0$ and, being $\mathrm{m} = 1$, it is equal to $\mathtt{m}_z$. Therefore, the condition $\mathtt{m}_z \leq 0.6$ yields

$$R_b \geq \frac{\lambda L}{0.6} = \frac{\lambda_b}{0.6} = 4.1667\ \text{Mbit/s}\,.$$

Note that we could reach the same result in a more direct way by simply requiring that 60% of the link bitrate is sufficient to carry the mean offered traffic, that is to say imposing $0.6\, R_b > \lambda_b$.

8.5 MARKOVIAN QUEUEING MODELS

We begin our analysis of QSs from the important class of Markovian queueing models M/M/–, characterized by interarrival and service times with exponential distributions.

We observe that the number $x(t)$ of customers in the QS increases by one at every arrival and decreases by one every time a customer completes its service and leaves the QS. If we see k customers in the QS at a given instant t_0, then the next arrival will occur after the so-called *residual interarrival time* τ', whereas the next departure will be observed when one of the customers under service completes its service. The remaining service time for a customer in service is called *residual service time* and denoted by y'. Due to the memoryless property of the exponential distribution (2.230), the residual times τ' and y' have the same distribution as any other interarrival time τ_n and service time y_n, respectively. Therefore, the time evolution of $x(t)$ from the instant t_0 onward only depends on the number k of customers observed in the system at time t_0, whereas the instants of arrival of such customers and the time they have already spent in the system are of no relevance for predicting the evolution of $x(t)$ for $t \geq t_0$. In other words, $x(t)$ exhibits the *Markov property* (7.85), which makes the process a continuous-time Markov chain (MC). In addition, it is possible to prove that $x(t)$ increases or decreases by a single unit at a time, so that it is actually a birth-death process (BDP) (see Sec. 7.4).

Remark Markovian QSs exhibit the following important and general properties:

- The state of Markovian QSs at time t can be summarized in a *single variable*, namely the number $x(t)$ of customers in the QS at time t.

- The process $x(t)$ exhibits the *Markov property* and may increase or decrease by a single unit at a time, which makes it a *Birth-death process*.

A formal proof of this statement will be provided for the simple case of $\mathrm{M/M/1}$ and, then, extended for similarity to the other QSs considered in this section, without proof.

Note that these properties are not longer true when either the interarrival times or the service times are not exponentially distributed, for in that case the statistics of the next arrival or departure event will depend upon the time elapsed since the last arrival or the time already spent by customers in service.

In the remainder of this section we will cover the basic Markovian QSs in increasing order of complexity and generality. We start from the simple $\mathrm{M/M/1}$, whose study provides the foundation upon which we will build the analysis of all the other Markovian QSs, and we conclude the section with the general $\mathrm{M/M/m/K}$, from which we can obtain all the other Markovian QSs as special cases. The analysis of each QS will proceed according to a fixed scheme: we first determine the birth and death rates for the BDP $x(t)$ that represents the state of the QS, then we determine the asymptotic PMD of $x(t)$ by inspection of the associated transition diagram, subsequently we discuss the stability of the QS by analyzing the conditions under which the asymptotic distribution exists, and finally we derive the asymptotic distributions of the occupancy and time measures.

For reader convenience we collect the main results at the end of the chapter in Tab. 8.2 and Tab. 8.3 for non-blocking and blocking QSs, respectively.

8.5.1 The $\mathrm{M/M/1}$ queueing system

The $\mathrm{M/M/1}$ model, illustrated in Fig. 8.9, refers to a single server QS with infinite storage capacity and FCFS service discipline. Customers arrive to the system according to a Poisson process with arrival rate λ, so that the interarrival times τ_n are statistically iid rvs with exponential distribution as in (8.5) and (8.6). The service times y_n are also statistically iid exponential rvs with service rate μ as in (8.10) and (8.11). Under these assumptions, we can easily prove that $x(t)$ is a BDP, as formally stated by the following proposition.

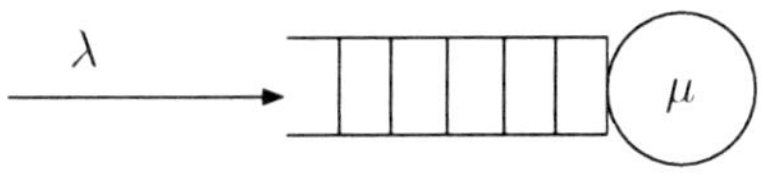

Figure 8.9 The M/M/1 queueing system.

Proposition 8.2 *The number of customers $x(t)$ in M/M/1 evolves as a BDP with birth rate λ and death rate μ for any state, except state zero whose death rate is zero.*

Proof As seen in Sec. 7.4, we need to prove that: i) $x(t)$ is a MC, and ii) in a very short time interval of length h, the only events that may occur with non negligible probability are either a single arrival or a single departure, whereas the probability of multiple arrivals or departures or even one arrival and one departure in h is $o(h)$.

Concerning the first point, we observe that, due to the memoryless property of the (residual) interarrival and service times, the knowledge of the past evolution of the process does not help refining the prevision of the future evolution from the current state. In other words, $x(t)$ exhibits the Markov property (7.85) and, hence, it is a MC.

Now, it remains to prove that the infinitesimal generator matrix $\boldsymbol{Q}$ has the form (7.158), with all zero elements, except for those in the main, upper and lower diagonals. Let $A(h)$ denote the number of arrivals in an interval of length h. Since the arrival process is Poisson, arrivals are spaced by interarrival times τ_n that are iid rvs with CDF (8.6). Therefore, we observe *one or more arrivals* in h if at least the first interarrival time τ_n is less than or equal to h. The probability of this event is thus equal to

$$\mathrm{P}\left[A(h) \geq 1\right] = \mathrm{P}\left[\tau_n \leq h\right] = 1 - e^{-\lambda h}. \tag{8.52}$$

Replacing the exponential function with its first order Taylor series expansion

$$e^{-\lambda h} = \sum_{k=0}^{\infty} \frac{(-\lambda h)^k}{k!} = 1 - \lambda h + o(h)\,, \tag{8.53}$$

we get

$$\mathrm{P}\left[A(h) \geq 1\right] = \lambda h + o(h)\,, \tag{8.54}$$

where $o(h)$ is an infinitesimal of higher order than h. Analogously, we have *two or more arrivals* in h when at least the first and the second arrivals occur within h, i.e., $\tau_n + \tau_{n+1} \leq h$. The probability of this event is

$$\begin{aligned} \mathrm{P}\left[A(h) \geq 2\right] &= \mathrm{P}\left[\tau_n + \tau_{n+1} \leq h\right] \\ &\leq \mathrm{P}\left[\tau_n \leq h, \tau_{n+1} \leq h\right] \\ &= \mathrm{P}\left[\tau_n \leq h\right] \mathrm{P}\left[\tau_{n+1} \leq h\right]\,, \end{aligned} \tag{8.55}$$

where the inequality in the second row comes from the fact that two non-negative numbers are always less than or equal to their sum, whereas the last row follows from the statistical independence of τ_n and τ_{n+1}. Now, using (8.6) in (8.55) and replacing the exponentials with their first order Taylor series expansion (8.53) we obtain

$$\mathrm{P}\left[A(h) \geq 2\right] \leq (\lambda h + o(h))(\lambda h + o(h)) = \lambda^2 h^2 + o(h) = o(h)\,. \tag{8.56}$$

We can hence conclude that the event of simultaneous arrivals can be neglected, whereas the probability of a single arrival in h is equal to

$$\mathrm{P}\left[A(h)=1\right]=\mathrm{P}\left[A(h)\geq 1\right]-\mathrm{P}\left[A(h)\geq 2\right]=\lambda h+o(h)\,,$$

and, consequently, the probability of no arrivals in h is

$$\mathrm{P}\left[A(h)=0\right]=1-\lambda h+o(h)\,.$$

A similar argument can be applied to prove that the probability of simultaneous departures is negligible. In fact, we observe that customers are served one at a time by the single server. If the server works at service capacity μ, that is, it is always busy serving customers without any idle period between two consecutive customers, then there will be a departure every service time y_n. Let $D_{\max}(h)$ be the number of departures in a time interval of length h when the server works at capacity.[5] The probability of *one or more departures* in h is given by

$$\mathrm{P}\left[D_{\max}(h)\geq 1\right]=\mathrm{P}\left[y_n<h\right]=1-e^{-\mu h}=\mu\,h+o(h)\,. \tag{8.57}$$

However, the probability of *two or more* departures in h is negligible, being

$$\begin{aligned}\mathrm{P}\left[D_{\max}(h)\geq 2\right]&=\mathrm{P}\left[y_n+y_{n+1}<h\right]<\mathrm{P}\left[y_n<h\right]\mathrm{P}\left[y_{n+1}<h\right]\\&=(\mu h+o(h))(\mu h+o(h))=o(h)\,.\end{aligned} \tag{8.58}$$

Therefore, in a short time interval h we can observe only individual departures, which occur with probability

$$\mathrm{P}\left[D_{\max}(h)=1\right]=\mathrm{P}\left[D_{\max}(h)\geq 1\right]-\mathrm{P}\left[D_{\max}(h)\geq 2\right]=\mu h+o(h)\,, \tag{8.59}$$

provided that the system is not empty at the beginning of the interval. Finally, we observe that the probability of arrivals and departures in the same interval h is still $o(h)$ because

$$\begin{aligned}&\mathrm{P}\left[A(h)\geq 1,D_{\max}(h)\geq 1\right]=\mathrm{P}\left[A(h)\geq 1\right]\mathrm{P}\left[D_{\max}(h)\geq 1\right]\\&\quad=(\lambda h+o(h))(\mu h+o(h))=\lambda\mu h^2+o(h)=o(h)\,.\end{aligned} \tag{8.60}$$

In conclusion, only individual arrivals or departures may occur in a short time interval with non-negligible probability. Therefore, $x(t)$ evolves as a BDP with birth probability in a short time interval of length h equal to

$$\mathrm{P}\left[x(t+h)=k+1|x(t)=k\right]=\mathrm{P}\left[A(h)=1\right]=\lambda h+o(h) \tag{8.61}$$

and death probability equal to

$$\mathrm{P}\left[x(t+h)=k|x(t)=k+1\right]=\mathrm{P}\left[D_{\max}(h)=1\right]=\mu h+o(h) \tag{8.62}$$

[5] It may be worth remarking that $D_{\max}(h)$ differs from the general departure process $D(h)$ since it assumes that the server is always busy serving customers. In fact, a server works at service capacity only when there are customers in the system. If the system is empty, the server will remain idle till the next arrival and, in this case, the gap between two consecutive departures is longer than the service time. Hence, the process $D_{\max}(t)$ is well defined only assuming the implicit condition that $x(t)>0$ for all t.

for $k = 0, 1, 2, \ldots$. Comparing (8.62) with (7.156), we conclude that

$$\boldsymbol{Q} = \begin{bmatrix} -\lambda & \lambda & 0 & 0 & 0 & \cdots \\ \mu & -(\lambda+\mu) & \lambda & 0 & 0 & \cdots \\ 0 & \mu & -(\lambda_2+\mu) & \lambda & 0 & \cdots \\ 0 & 0 & \mu & -(\lambda+\mu) & \lambda & \cdots \\ \vdots & \vdots & \vdots & \vdots & \vdots & \ddots \end{bmatrix}. \tag{8.63}$$

□

As seen in Section 7.4, the BDP $x(t)$ can be described by means of the flow diagram shown in Fig. 8.10.

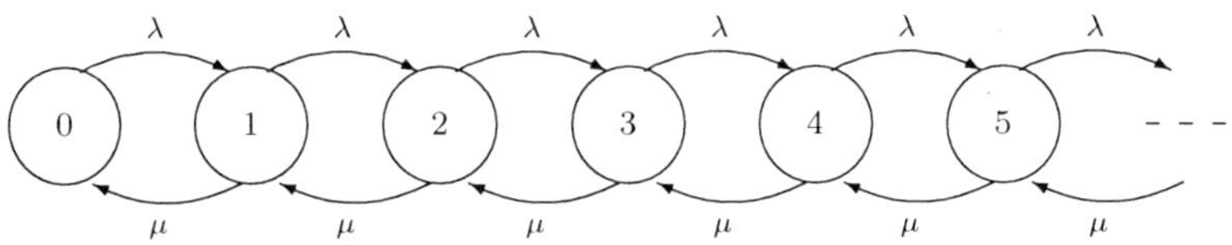

Figure 8.10 Flow diagram of M/M/1.

Inspecting Fig. 8.10, and recalling the flow-dynamic interpretation of the flow diagram proposed in Section 7.3.9, we can write the differential equations (7.164) that yield to the time-dependent form of the state probability vector. In the case of M/M/1, the differential equations (7.164) take the form

$$\begin{aligned} p_k'(t) &= \lambda p_{k-1}(t) + \mu p_{k+1}(t) - (\lambda+\mu) p_k(t) \\ p_0'(t) &= -\lambda p_0(t) + \mu p_1(t)\,. \end{aligned} \tag{8.64}$$

The time-dependent behavior of the state probability vector for this simple BDP has been thoroughly investigated in Example 7.4 D at page 482. For the scope of this book, however, the time-dependent behavior of the QS is of margin al interest, whereas the asymptotic analysis of the QS behavior is more relevant. As mentioned, in fact, in stable QSs all the quantities of interest asymptotically reach a *stationary* distribution that does not change in time. In this condition, the QS is said to work in *steady-state*. The asymptotic PMD of $x(t)$ is denoted as[6]

$$p_x(k) = \lim_{t\to\infty} \mathrm{P}\left[x(t) = k\right] \tag{8.65}$$

provided that the limit exists. Following the rationale presented in Section 7.4.3, we can express the equations (7.188) for the M/M/1 QS as

[6]Note that, in Chapter 7, we encountered and discussed the concept of asymptotic and stationary state probability vector for discrete-time and continuous time MC and for BDP. In that occasion, the asymptotic probability of the process being in state k was denoted by π_k. Here, however, we are interested in the asymptotic distribution of *all* the performance measures, not only $x(t)$. Hence, we prefer to adopt the notation $p_x(k)$ that explicitly indicates the variable x we are referring to.

$$\begin{cases} \mu\, p_x(1) &= \lambda p_x(0) \\ \mu\, p_x(2) &= \lambda\, p_x(1) \\ \mu\, p_x(3) &= \lambda\, p_x(2) \\ &\vdots \end{cases} \tag{8.66}$$

which corresponds to the balance of the probability-flows between any pair of consecutive state in the associated transition diagram, as illustrated in Fig. 8.11.

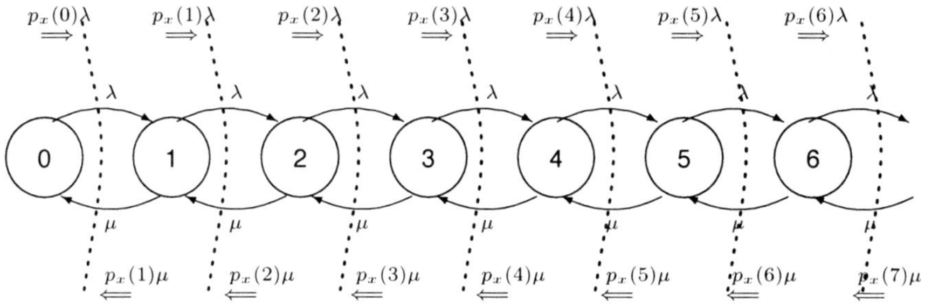

Figure 8.11 Probability-flow diagram of $\mathrm{M/M/1}$.

Rearranging the terms in each row of (8.66), we get

$$\begin{cases} p_x(1) &= \dfrac{\lambda}{\mu} p_x(0) \\ p_x(2) &= \dfrac{\lambda}{\mu}\, p_x(1) \\ p_x(3) &= \dfrac{\lambda}{\mu}\, p_x(2) \\ &\vdots \end{cases} \tag{8.67}$$

Finally, replacing iteratively the right-hand side of each row into the following row, we obtain a system where all the unknowns are expressed in terms of $p_x(0)$:

$$\begin{cases} p_x(1) &= \dfrac{\lambda}{\mu} p_x(0) \\ p_x(2) &= \left(\dfrac{\lambda}{\mu}\right)^2 p_x(0) \\ p_x(3) &= \left(\dfrac{\lambda}{\mu}\right)^3 p_x(0) \\ &\vdots \end{cases} \tag{8.68}$$

Observed that each equation in (8.68) is in the form

$$p_x(k) = \left(\frac{\lambda}{\mu}\right)^k p_x(0) = \rho^k\, p_x(0)\,, \qquad k = 0, 1, 2, \ldots\,. \tag{8.69}$$

The unknown constant $p_x(0)$ is determined by imposing the normalization condition, which yields

$$\sum_{k=0}^{\infty} p_x(k) = \sum_{k=0}^{\infty} \rho^k \, p_x(0) = p_x(0)\frac{1}{1-\rho} = 1\,, \tag{8.70}$$

where the last step requires $\rho < 1$, otherwise the geometric series diverges. We then find again the stability condition (8.33) under which $x(t)$ is a positive recurrent BDP and, as such, it does admit a proper asymptotic PMD. Hence, we have

$$p_x(0) = 1 - \rho\,. \tag{8.71}$$

Replacing (8.71) into (8.69), we finally obtain the asymptotic PMD of the number of customers in $\mathrm{M/M/1}$[7]

$$\boxed{p_x(k) = (1-\rho)\rho^k\,, \qquad k = 0, 1, 2, \ldots} \tag{8.72}$$

which turns out to be geometric with parameter $1 - \rho$, provided that $\rho < 1$.

Occupancy measures. The PMD (8.72) is geometric so that, recalling Example 2.6 G at page 85, the mean $\mathtt{m}_x$ and variance σ_x^2 of the number of customers in the system for $t \to \infty$ are given by

$$\boxed{\mathtt{m}_x = \lim_{t\to\infty} \mathrm{E}\left[x(t)\right] = \frac{\rho}{1-\rho} = \frac{\lambda}{\mu - \lambda}} \tag{8.73}$$

and

$$\sigma_x^2 = \lim_{t\to\infty} \mathrm{E}\left[(x(t) - \mathtt{m}_x)^2\right] = \frac{\rho}{(1-\rho)^2} = \frac{\lambda\mu}{(\mu-\lambda)^2}\,. \tag{8.74}$$

We observe that both $\mathtt{m}_x$ and σ_x^2 diverge as $\rho \to 1$. Hence, the value of $x(t)$ has a great amount of variability in the immediate neighborhood of $\rho = 1$, that is to say, the number of customers in the system may greatly vary over time when the load factor approaches the stability threshold $\rho = 1$. The relation (8.73) is graphically shown in Fig. 8.12, where the dashed curves, labelled $k_{0.05}$ and $k_{0.95}$, correspond to the 5% and 95% percentiles, respectively, and delimit the range of values that $x(t)$ can take with 90% of probability:

$$\mathrm{P}\left[x \le k_{0.05}\right] \le 0.05\,, \qquad \mathrm{P}\left[x \le k_{0.95}\right] \ge 0.95\,.$$

As emphasized by the vertical arrows that extend from the lower to the upper percentile, the range of values that x may assume with high probability rapidly enlarges as ρ gets close to 1.

The asymptotic PMD of the other occupancy measures, namely the number of customers in service $z = \lim_{t\to\infty} z(t)$ and in the waiting queue $q = \lim_{t\to\infty} q(t)$ can be easily deduced from $p_x(k)$. The statistic of z is trivial. Since there is a single server, the value of z can either be zero or one, depending on whether the server is idle or busy serving a customer. However, due to the work-conserving assumption (see page 501), the server is idle only when there are no customers in the system, an event that occurs with probability $p_x(0) = 1 - \rho$. Consequently,

[7] We observe that (8.72) may have been directly obtained by specializing (7.194) and (7.196) to $\mathrm{M/M/1}$, that is to say, using $\lambda_k = \lambda$ and $\mu_{k+1} = \mu$, for any $k \ge 0$.

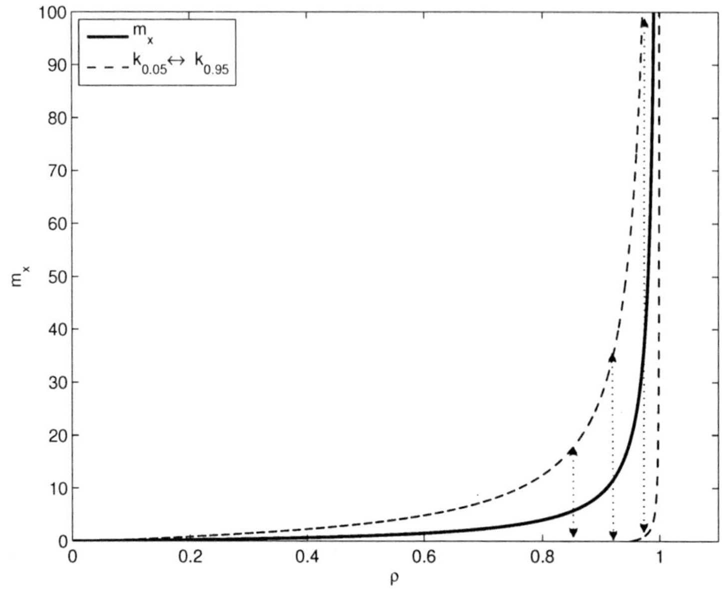

Figure 8.12 Mean number of customers as a function of ρ in $M/M/1$ (solid line) and 5% and 95% percentiles (dashed lines).

the chance that the server is busy is equal to the probability that there is at least one customer in the system and it is precisely equal to the load factor ρ. In summary, we have

$$\boxed{\begin{aligned} p_z(0) &= 1 - \rho \\ p_z(1) &= \rho \end{aligned}} \tag{8.75}$$

and the expected value is given by

$$\boxed{\mathtt{m}_z = p_z(1) = \rho\,.} \tag{8.76}$$

The derivation of the PMD of the number of customers q in the waiting queue is slightly more sophisticated. Now, q can be expressed in terms of x as

$$q = x - \chi\{x > 0\} = \begin{cases} 0 & , \ x = 0 \\ x - 1 & , \ x > 0 \end{cases} \tag{8.77}$$

where $\chi\{\cdot\}$ is the indicator function (2.163). The CDF of q is then given by

$$\mathrm{P}[q \leq k] = \mathrm{P}[x \leq k + 1] = \sum_{r=0}^{k+1} p_x(r) = \sum_{r=0}^{k+1} (1 - \rho)\rho^r = 1 - \rho^{k+2}\,, \qquad k = 0, 1, 2, \ldots \tag{8.78}$$

whereas the PMD turns out to be given by

$$p_q(k) = \begin{cases} 1-\rho^2 & , \quad k=0 \\ (1-\rho)\rho^{k+1} & , \quad k=1,2,\ldots \end{cases} \tag{8.79}$$

The mean number of queued customers $\mathtt{m}_q$ can be derived by applying the expectation operator to (8.77), which gives

$$\mathtt{m}_q = \mathrm{E}\left[x - \chi\{x>0\}\right] = \mathrm{E}[x] - \mathrm{E}\left[\chi\{x>0\}\right] = \mathtt{m}_x - \mathrm{P}[x>0] = \mathtt{m}_x - \rho \tag{8.80}$$

where the first step follows from the linearity of expectation, whereas in the second step, from (2.165) we have replaced $\mathrm{E}\left[\chi\{x>0\}\right]$ with $\mathrm{P}[x>0]$. Using (8.73) in (8.80) we finally have

$$\mathtt{m}_q = \frac{\rho^2}{1-\rho}. \tag{8.81}$$

Clearly, the mean number of queued customers also shows the same rapid increase of $\mathtt{m}_x$ as $\rho \to 1$.

Time measures. We now turn our attention to the PDF of the system time s and the waiting (or queueing) time w. Consider a customer C_n that arrives to the QS at time t_n. Let $x(t_n^-)$ denote the number of customers already in the QS in the instant immediately preceding the arrival of C_n. Furthermore, let $s|_k$ denote the system time of C_n given that $x(t_n^-) = k$. If $k = 0$, then C_n immediately enters service, otherwise it will join the tail of the queue. Therefore, we can write

$$s|_k = \begin{cases} y & , \quad k=0 \\ y_1' + y_2 + \ldots + y_k + y & , \quad k>0 \end{cases} \tag{8.82}$$

where y_1' is the residual service time of the customer being served at the arrival of C_n, while $y_2, y_3, \ldots, y_k$ are the service times of the $k-1$ customers already in the queue, and y is the service time of C_n. From the discussion at the beginning of Section 8.5, the residual service time y_1' is exponentially distributed, as the other service times $y_2, y_3, \ldots, y_k$, and y. The conditional system time $s|_k$ is hence equal to the sum of $k+1$ independent exponential rvs with parameter μ and, consequently, it has an Erlang distribution of index $k+1$ and parameter μ, whose PDF is given by (see (2.186))

$$p_{s|_k}(a) = \frac{\mu(\mu a)^k}{k!} e^{-\mu a} 1_0(a), \tag{8.83}$$

where $1_0(a)$ is the step-function defined in Example 2.1 A.[8] The unconditional PDF of s can then be obtained from (8.83) by applying the Total Probability Theorem 2.12

$$p_s(a) = \sum_{k=0}^{\infty} p_{s|_k}(a)\, \mathrm{P}\left[x(t_n^-) = k\right]. \tag{8.84}$$

[8]For the sake of compactness, in the following we often drop the $1_0(a)$ notation in the PDF and CDF expressions of time measures, which can only take non-negative values.

To evaluate (8.84) we need to know the asymptotic PMD of $x(t_n^-)$. In principle, the probability that an arriving customer finds other k customers in the system may differ from the asymptotic probability $p_x(k)$ of state k.[9]

Fortunately, it is possible to prove that when customer arrivals occur according to a Poisson process, the PMD of $x(t_n^-)$ is equal to the asymptotic PMD of $x(t)$. This property, known as *Poisson Arrivals See Time Averages* (PASTA), is formally stated in the following proposition, whose rather elaborated proof can be found, for instance, in [17].

Proposition 8.3 (PASTA) *In a QS where future arrival times and service times of previously arrived customers are independent (lack of anticipation assumption), the asymptotic distributions of the process state seen by arriving customers, $x(t_n^-)$, and by a random observer, $x(t)$, are equal if and only if the arrival process is Poisson. In this case, the asymptotic fraction of time that the process spends in state k is equal to the fraction of Poisson arrivals that find the process in state k, that is to say*

$$\lim_{n\to\infty} \mathrm{P}\left[x(t_n^-) = k\right] = \lim_{t\to\infty} \mathrm{P}\left[x(t) = k\right] = p_x(k)\,, \quad k = 0, 1, 2, \ldots. \tag{8.85}$$

Comforted by PASTA theorem, we can now use $p_x(k)$ given by (8.72) in place of $\mathrm{P}\left[x(t_n^-) = k\right]$ in (8.84), thus obtaining

$$\begin{aligned} p_s(a) &= \sum_{k=0}^{\infty} p_{s|_k}(a) p_x(k) \\ &= \sum_{k=0}^{\infty} \frac{\mu(\mu a)^k}{k!} e^{-\mu a} \rho^k (1-\rho) \\ &= \mu(1-\rho) e^{-\mu a} \sum_{k=0}^{\infty} \frac{(\mu \rho\, a)^k}{k!} \,. \end{aligned} \tag{8.86}$$

Recognizing the last summation as the Taylor series expansion of the exponential function $e^{\mu\rho\, a}$, we readily have

$$\boxed{\begin{aligned} p_s(a) &= \mu(1-\rho) e^{-\mu(1-\rho)a} \\ &= (\mu - \lambda) e^{-(\mu-\lambda)a} \,. \end{aligned}} \tag{8.87}$$

Hence, from (2.160), the asymptotic PDF of the system time turns out to be exponential with parameter $\mu(1-\rho) = \mu - \lambda$ and CDF

$$\mathrm{P}\left[s \le a\right] = 1 - e^{-\mu(1-\rho)a} \,. \tag{8.88}$$

The mean system time, therefore, is equal to

$$\boxed{\mathtt{m}_s = \frac{1}{\mu(1-\rho)} = \frac{1}{\mu - \lambda}\,,} \tag{8.89}$$

[9]For instance, consider D/D/1 with a constant interarrival time τ_c and a constant service time $y_c < \tau_c$. In this case, the server cyclically alternates busy periods of length y_c and idle periods of length $\tau_c - y_c$. Thus, a random observer will either see a single customer in the QS, with probability y_c/τ_c, or no customers, with probability $1 - y_c/\tau_c$. However, observing the QS at the arrival instant of a new customer, we would always find the system empty, that is $\mathrm{P}\left[x(t_n^-) = 0\right] = 1$. In this case, the PMD of the system occupancy $x(t_n^-)$ seen by an arriving customer differs from the asymptotic PMD of the number of customers $x(t)$ seen by a random observer.

as it could have been directly obtained from $\mathtt{m}_x$ through Little's equation (8.46). From (8.89), $\mathtt{m}_s$ presents the same behavior of $\mathtt{m}_x$ with respect to ρ, growing very quickly as $\rho \to 1$. This behavior, illustrated in Fig. 8.13 for different values of μ, is typical of most QSs and reflects the basic tradeoff between ρ and $\mathtt{m}_s$: higher ρ comes at the cost of longer $\mathtt{m}_s$. As a result, any attempt at using the service close to its full capacity ($\rho \to 1$) brings along a rapid increase in the system time $\mathtt{m}_s$.[10]

Remark The rapid worsening of some system performance measures near the critical value $\rho = 1$ can be ascribed to the *variability* in both the interarrival and service times: distributions of these rvs with a smaller variability will result in a reduction of the mean waiting time.

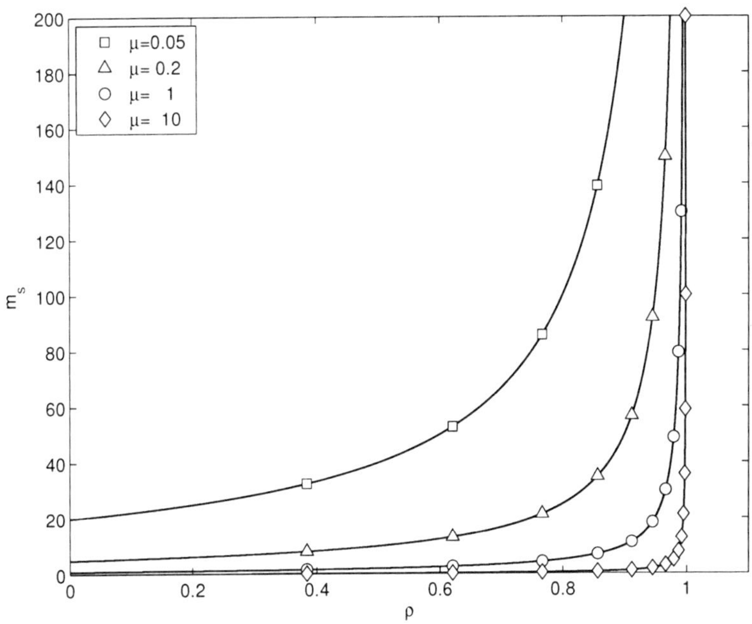

Figure 8.13 Mean system time $\mathtt{m}_s$ as a function of ρ in M/M/1, for different service rates μ.

The asymptotic PDF of the waiting time w can be obtained following basically the same rationale used for the system time. The waiting time $w|_k$, given that the number of customers in the system at the arrival time is $x(t_n^-) = k$, can be expressed as

$$w|_k = \begin{cases} 0 & , \quad k = 0 \\ y_1' + y_2 + \ldots + y_k & , \quad k > 0 \end{cases} \tag{8.90}$$

[10]The only exception to this fundamental rule is represented by the deterministic D/D/m, which remains stable even in the critical case of $\rho = 1$. In this QS, in fact, arrivals and departures are perfectly synchronized in such a way that for any customer arriving to the QS another customer leaves that QS, so that the service capacity is fully exploited and customers never wait in queue. If such a perfect synchronization between arrivals and departures is lost, the QS will inexorably drift towards instability.

where the terms are as in (8.82). Hence, when $k > 0$, $w|_k$ is the sum of k iid exponential rvs with parameter μ, which results in an Erlang rv of index k. The unconditional PDF of w can thus be expressed as

$$p_w(a) = \delta(a)p_x(0) + \sum_{k=1}^{\infty} \frac{\mu(\mu a)^{k-1}}{(k-1)!} e^{-\mu a} p_x(k) , \qquad a \geq 0 , \tag{8.91}$$

where the impulsive term accounts for the probability that the waiting time is zero because the system is empty at the arrival instant. Replacing (8.72) in (8.90) we get

$$\begin{aligned} p_w(a) &= (1-\rho)\delta(a) + \sum_{k=1}^{\infty} \frac{\mu(\mu a)^{k-1}}{(k-1)!} e^{-\mu a} \rho^k (1-\rho) \\ &= (1-\rho)\delta(a) + \rho\mu(1-\rho)e^{-\mu a} \sum_{r=0}^{\infty} \frac{(\mu\rho a)^r}{r!} . \end{aligned} \tag{8.92}$$

The last sum in (8.92) is the Taylor series expansion of $e^{\mu\rho a}$, thus yielding

$$\boxed{p_w(a) = (1-\rho)\delta(a) + \rho\mu(1-\rho)e^{-\mu(1-\rho)a} 1_0(a) .} \tag{8.93}$$

Further insight We have already noticed that the first term in (8.93) accounts for the probability that a customer enters directly in service because it finds the QS empty. The second term, instead, accounts for the PDF of w given that the service is busy at the customer's arrival time, an event which occurs with probability ρ. In this case, the time that a newly arrived customer waits in queue before being served is equal to the remaining system time s^* of the customer immediately ahead on the waiting queue. Because of the PASTA property, however, the statistic seen by an arriving customer is the same seen by a random observer, so that s^* is exponential distributed with parameter $\mu(1-\rho)$ as any other system time s. In conclusion, the waiting time w can also be expressed as

$$w = \chi\{x > 0\}\, s , \tag{8.94}$$

where x and s are independent rvs distributed as in (8.72) and (8.87), respectively.

The asymptotic CDF of w can be obtained by integrating (8.93) or directly from (8.94), giving

$$\begin{aligned} P[w \leq a] &= P[s\chi\{x > 0\} \leq a] \\ &= P[s\chi\{x > 0\} \leq a | x > 0]\, P[x > 0] + P[s\chi\{x > 0\} \leq a | x = 0]\, P[x = 0] \\ &= P[s \leq a]\,\rho + P[0 \leq a]\,(1-\rho) = \rho - \rho e^{-(\mu-\lambda)a} + 1 - \rho \\ &= 1 - \rho e^{-\mu(1-\rho)a} . \end{aligned} \tag{8.95}$$

The mean waiting time, finally, can be obtained from (8.94), yielding

$$\mathrm{m}_w = E[\chi\{x > 0\}\, s] = E[\chi\{x > 0\}]\, E[s] = \rho\, \mathrm{m}_s . \tag{8.96}$$

Replacing (8.89) into (8.96) we obtain another expression of the mean waiting time

$$\boxed{\mathtt{m}_w = \frac{\rho}{\mu(1-\rho)} = \frac{\lambda}{\mu(\mu-\lambda)}\,.} \tag{8.97}$$

The same result could have been reached by applying Little's law (8.47), or simply taking the expectation of (8.26), from which we have

$$\mathtt{m}_w = \mathrm{E}\left[s-y\right] = \mathrm{E}\left[s\right] - \mathrm{E}\left[y\right] = \mathtt{m}_s - \mathtt{m}_y = \frac{1}{\mu-\lambda} - \frac{1}{\mu} = \frac{\lambda}{\mu(\mu-\lambda)}\,.$$

Remark From (8.89) and (8.97) we note that, for a certain offered traffic $\mathtt{G}$, the mean system and waiting times of a customer decrease as the service capacity μ increases. Hence, a faster server yields better performance in terms of $\mathtt{m}_s$ and $\mathtt{m}_w$ than a slower one, even though the utilization factor ρ is the same. From another perspective, if we wish to maintain the same $\mathtt{m}_s$ against an increase of α percent of the customer arrival rate, $\lambda \to (1+\alpha)\lambda$, the service capacity shall undergo an increase of β percent, with

$$\beta = \frac{\lambda}{\mu}\alpha < \alpha\,. \tag{8.98}$$

The departure process from the $\mathrm{M/M/1}$ *queue.* We conclude our analysis of $\mathrm{M/M/1}$ by presenting a remarkable and very useful theorem due to Burke [2].

Theorem 8.4 *The departure process $D(t)$ of a stable $\mathrm{M/M/1}$, in stationary conditions, is a Poisson process with rate λ. Hence, from Theorem 2.22 the interdeparture times $r_n = d_n - d_{n-1}$ are iid exponential random variables with parameter λ.*

Proof For the complete proof of the Burke's theorem we refer the reader to the original script [2]. Here we will be satisfied by proving that the interdeparture times are exponential, whereas the other aspects of the proof will be only sketched.

It is possible to prove that, asymptotically, the probability of the number of customers in the QS observed by arriving customers is equal to the probability of the number of customers in the QS left behind by departing customers. In fact, any arriving customer that finds the QS in state k yields a transition from state k to $k+1$, whereas a departing customer that leaves k customers behind in the QS yields to a transition from $k+1$ to k. Asymptotically, the number of transitions from k to $k+1$ must equal the number of transitions from $k+1$ to k, on average, so that the fraction of customers finding the system in state k must equal the fraction of customers leaving the system in state k. Hence, the system occupancy statistic seen by arriving and departing customers is the same. A formal proof of this property can be found in [3, 2].

The PASTA theorem, in turn, states that the PMD of the number of customers in the system seen by arriving customers is equal to the asymptotic PMD of x. Hence, asymptotically, the PMD of the number of customers left in a $\mathrm{M/M/1}$ QS by a departing customer is equal to the PMD of the number of customers in the QS, that is to say

$$\lim_{n\to\infty} \mathrm{P}\left[x(d_n^+) = k\right] = p_x(k) = (1-\rho)\rho^k\,, \qquad k = 0, 1, 2, \ldots\,.$$

If the QS is not empty after the departure of C_n, then the next customer in queue, C_{n+1}, is immediately served and the next departure will occur after its service time y_{n+1}. Conversely, if C_n leaves behind an empty system, then the next departure will occur after the next customer arrives into the system and completes its service. The distribution of the interdeparture time r_n can then be expressed as

$$\begin{aligned} \mathrm{P}\left[r_n \le a\right] &= \mathrm{P}\left[r_n \le a | x > 0\right] \mathrm{P}\left[x > 0\right] + \mathrm{P}\left[r_n \le a | x = 0\right] \mathrm{P}\left[x = 0\right] \\ &= \mathrm{P}\left[y_n \le a\right] \rho + \mathrm{P}\left[y_n + \tau'_{n+1} \le a\right] (1 - \rho) \end{aligned} \tag{8.99}$$

where τ'_{n+1} is the residual interarrival time, which has an exponential distribution with parameter λ. The right-most probability in (8.99) can be obtained by using the Total Probability Theorem 2.12

$$\begin{aligned} \mathrm{P}\left[y_n + \tau'_{n+1} \le a\right] &= \int_0^a p_{y_n}(u) \,\mathrm{P}\left[\tau'_{n+1} \le a - u\right] du = \int_0^a \mu e^{-\mu u}\left(1 - e^{-\lambda(a-u)}\right) du \\ &= 1 - e^{-\mu a} - \mu e^{-\lambda a} \int_0^a e^{-(\mu-\lambda)u} du = 1 - e^{-\mu a} - \mu \frac{e^{-\lambda a} - e^{-\mu a}}{\mu - \lambda} . \end{aligned} \tag{8.100}$$

Replacing (8.100) in (8.99) we get

$$\begin{aligned} \mathrm{P}\left[r_n \le a\right] &= \rho(1 - e^{-\mu a}) + (1 - \rho)(1 - e^{-\mu a}) - (1 - \rho)\mu \frac{e^{-\lambda a} - e^{-\mu a}}{\mu - \lambda} \\ &= 1 - e^{-\mu a} - e^{-\lambda a} + e^{-\mu a} = 1 - e^{-\lambda a} \end{aligned}$$

which proves that r_n is exponential with parameter λ. The statistical independence of the interdeparture times follows from the Markov property of the departure process, which is a consequence of the memoryless nature of the arrival and service time distributions, as proven in [2]. □

Burke's theorem can be generalized to $\mathrm{M/M/m}$, which still exhibits a Poisson departure process of the same rate as the arrival process (provided that the stability condition (8.33) is satisfied). Burke also established that the departure process is statistically independent of any other process in the system and $\mathrm{M/M/m}$ is the only FCFS QS with this property. Therefore, Burke's theorem permits to decouple the input and output processes of each queue, drastically simplifying the analysis of chains and networks of $\mathrm{M/M/m}$ QSs. This topic is rather advanced and it will not be further investigated in this book. The interested reader is addressed, for instance, to [1].

8.5.2 The $\mathrm{M/M/m}$ queuing system

The $\mathrm{M/M/m}$ model generalizes $\mathrm{M/M/1}$ by assuming the presence of m servers that work in parallel, as illustrated in Fig. 8.14. The customers still arrive according to a Poisson process of rate λ and get served following a FCFS discipline. The servers work in parallel and independently of one another and the service times are statistically iid exponential rvs with service rate μ. An arriving customer is queued only when it finds all the servers busy, otherwise it immediately starts to be served by one of the available servers.

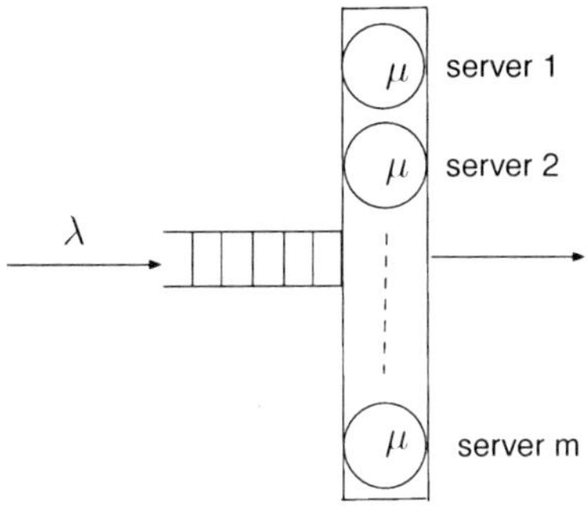

Figure 8.14 The M/M/m QS.

The number $x(t)$ of customers in the system evolves as a BDP with constant birth rate λ and death-rates that depend on the number of customers in service. To prove this claim we first observe that, with $x(t) = k > 0$ customers in the QS, the number of customers in service is $k_s = \min(k, \mathrm{m})$ whereas the remaining $k - k_s$ customers, if any, are in the waiting queue. Denoting by $y'_1, y'_2, \ldots, y'_{k_s}$ the residual service times of the k_s customers in service, the next departure occurs after a time equal to the least residual service time, $e_{k_s} = \min\left(y'_1, y'_2, \ldots, y'_{k_s}\right)$. The CDF of e_{k_s} is given by

$$\begin{aligned} \mathrm{P}\left[e_{k_s} \le a\right] &= 1 - \mathrm{P}\left[e_{k_s} > a\right] = 1 - \mathrm{P}\left[\min\left(y'_1, y'_2, \ldots, y'_{k_s}\right) > a\right] \\ &= 1 - \mathrm{P}\left[y'_1 > a, y'_2 > a, \ldots, y'_{k_s} > a\right] \end{aligned} \tag{8.101}$$

where the last step follows from the fact that the least of a set of values is larger than a threshold only if all the values exceed that threshold. Since the servers work independently one another, the residual service times are also statistically independent and, hence, (8.101) can be written as

$$\mathrm{P}\left[e_{k_s} \le a\right] = 1 - \prod_{\ell=1}^{k_s} \mathrm{P}\left[y'_\ell > a\right] . \tag{8.102}$$

Recalling that the residual service times are exponentially distributed with PDF as in (8.10), from (8.102) we get

$$\mathrm{P}\left[e_{k_s} \le a\right] = 1 - \left(e^{-\mu a}\right)^{k_s} = 1 - e^{-\mu k_s a}, \qquad a \ge 0, \tag{8.103}$$

which is an exponential distribution with parameter μk_s. Now, applying the same argument that has been used in (8.62) to derive the death probability in M/M/1, here we conclude that the probability of a departure in a time interval h, given $x(t) = k > 0$, is equal to

$$\mathrm{P}\left[x(t+h) = k-1 | x(t) = k\right] = \mathrm{P}\left[e_{k_s} \le h\right] + o(h) = k_s \mu h + o(h) \tag{8.104}$$

where the infinitesimal term $o(h)$ accounts for the probability of multiple (arrival or departure) events. The death rate μ_k from a given state k is hence equal to $k_s \mu$, so that we have

$$\mu_k = \begin{cases} k\,\mu &, \ k = 0, 1, \ldots, \mathrm{m} \\ \mathrm{m}\,\mu &, \ k > \mathrm{m}\,. \end{cases} \tag{8.105}$$

As can be seen, the death-rate is proportional to the number of *busy servers*. Consequently, the rate at which customers leave the system grows linearly with the number $x(t)$ of customers in the system as long as $x(t) \leq \mathrm{m}$, whereas when $x(t) > \mathrm{m}$ the servers are all busy and the departure rate remains limited to $\mathrm{m}\,\mu$. The flow-diagram associated to M/M/m is shown in Fig. 8.15.

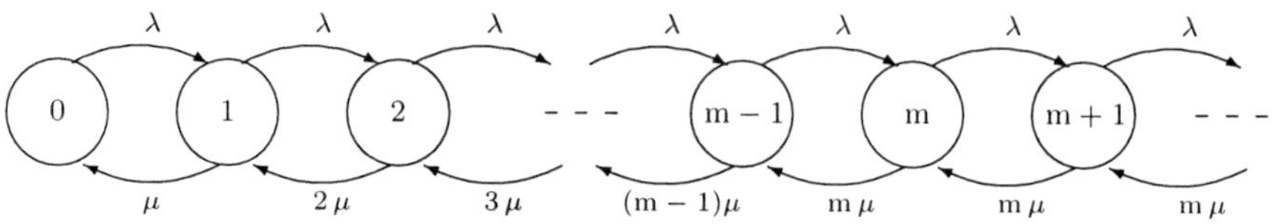

Figure 8.15 Flow-diagram of M/M/m.

The asymptotic PMD $p_x(k)$ of the number of customers in M/M/m can be obtained by solving the system of balance equations (7.188) between any pair of adjacent states. Inspecting the flow-diagram of Fig. 8.15, we easily obtain the following system of equations

$$\begin{cases} \lambda\, p_x(k-1) = k\mu\, p_x(k) & , \; 0 < k \leq \mathrm{m} \\ \lambda\, p_x(k-1) = \mathrm{m}\mu\, p_x(k) & , \; k > \mathrm{m} \end{cases} \tag{8.106}$$

which can be rewritten in the recursive form

$$\begin{cases} p_x(k) = \dfrac{\lambda}{k\mu}\, p_x(k-1) = \dfrac{\mathtt{G}}{k}\, p_x(k-1) & , \; 0 < k \leq \mathrm{m} \\ p_x(k) = \dfrac{\lambda}{\mathrm{m}\mu}\, p_x(k-1) = \dfrac{\mathtt{G}}{\mathrm{m}}\, p_x(k-1) & , \; k > \mathrm{m} \end{cases} \tag{8.107}$$

where, from (8.27), $\mathtt{G} = \frac{\lambda}{\mu}$. Unfolding the equations recursively back to $p_x(0)$, we obtain

$$\begin{cases} p_x(k) = \dfrac{\mathtt{G}}{k}\dfrac{\mathtt{G}}{k-1}\cdots\dfrac{\mathtt{G}}{1}\, p_x(0) = \dfrac{\mathtt{G}^k}{k!} p_x(0) & , \; 0 < k \leq \mathrm{m} \\ p_x(k) = \left(\dfrac{\mathtt{G}}{\mathrm{m}}\right)^{k-\mathrm{m}} p_x(\mathrm{m}) = \left(\dfrac{\mathtt{G}}{\mathrm{m}}\right)^{k-\mathrm{m}} \dfrac{\mathtt{G}^\mathrm{m}}{\mathrm{m}!} p_x(0) & , \; k > \mathrm{m}\,. \end{cases} \tag{8.108}$$

The value of $p_x(0)$ is given by the normalization condition

$$\begin{aligned} 1 &= \sum_{k=0}^{\infty} p_x(k) = \sum_{k=0}^{\mathrm{m}-1} p_x(k) + \sum_{k=\mathrm{m}}^{\infty} p_x(k) \\ &= p_x(0)\left(\sum_{k=0}^{\mathrm{m}-1} \frac{\mathtt{G}^k}{k!} + \frac{\mathtt{G}^\mathrm{m}}{\mathrm{m}!}\sum_{k=\mathrm{m}}^{\infty}\left(\frac{\mathtt{G}}{\mathrm{m}}\right)^{k-\mathrm{m}}\right) \\ &= p_x(0)\left(\sum_{k=0}^{\mathrm{m}-1} \frac{\mathtt{G}^k}{k!} + \frac{\mathtt{G}^\mathrm{m}}{\mathrm{m}!}\frac{\mathrm{m}}{\mathrm{m}-\mathtt{G}}\right) \end{aligned} \tag{8.109}$$

where the last step requires $\mathtt{G} < \mathrm{m}$, that is $\lambda < \mathrm{m}\mu$. From (8.109) we hence have

$$p_x(0) = \frac{1}{\displaystyle\sum_{r=0}^{\mathrm{m}-1} \frac{\mathtt{G}^r}{r!} + \frac{\mathtt{G}^{\mathrm{m}}}{\mathrm{m}!} \frac{\mathrm{m}}{\mathrm{m} - \mathtt{G}}}. \tag{8.110}$$

Replacing (8.110) into (8.108), we finally obtain the expression of the asymptotic PMD of the number of customers in M/M/m system,

$$p_x(k) = \begin{cases} \dfrac{\dfrac{\mathtt{G}^k}{k!}}{\left(\displaystyle\sum_{r=0}^{\mathrm{m}-1} \frac{\mathtt{G}^r}{r!} + \frac{\mathtt{G}^{\mathrm{m}}}{\mathrm{m}!} \frac{\mathrm{m}}{\mathrm{m} - \mathtt{G}}\right)}, & 0 \le k < \mathrm{m} \\[2ex] \dfrac{\dfrac{\mathtt{G}^{\mathrm{m}}}{\mathrm{m}!}\left(\dfrac{\mathtt{G}}{\mathrm{m}}\right)^{k-\mathrm{m}}}{\left(\displaystyle\sum_{r=0}^{\mathrm{m}-1} \frac{\mathtt{G}^r}{r!} + \frac{\mathtt{G}^{\mathrm{m}}}{\mathrm{m}!} \frac{\mathrm{m}}{\mathrm{m} - \mathtt{G}}\right)}, & k \ge \mathrm{m} \end{cases} \tag{8.111}$$

or, in a more compact form,

$$p_x(k) = \frac{\dfrac{\mathtt{G}^k}{k_s!\,\mathrm{m}^{k-k_s}}}{\displaystyle\sum_{r=0}^{\mathrm{m}-1} \frac{\mathtt{G}^r}{r!} + \frac{\mathtt{G}^{\mathrm{m}}}{\mathrm{m}!} \frac{\mathrm{m}}{\mathrm{m} - \mathtt{G}}}, \qquad k = 0, 1, \ldots . \tag{8.112}$$

where we recall that $k_s = \min(k, \mathrm{m})$.

It may be worth remarking that (8.112) requires $\mathtt{G} < \mathrm{m}$ that, according to (8.32), is equivalent to $\rho < 1$, which is the stability condition for M/M/m, as anticipated by (8.33) and (8.24). Hence, the system is stable only if the arrival rate is lower than the service capacity. We also observe that, as for M/M/1, $p_x(k)$ depends on λ and μ only through their ratio.

An important performance index in M/M/m is the so-called *queueing probability* or *call-waiting probability*, $C(\mathrm{m}, \mathtt{G})$, defined as the probability that an arriving customer is queued in the waiting buffer because all the servers are already occupied. The PASTA theorem guarantees that arriving customers see the steady-state system occupancy, so that the queueing probability is equal to

$$C(\mathrm{m}, \mathtt{G}) = \mathrm{P}\left[x \ge \mathrm{m}\right] = \sum_{k=\mathrm{m}}^{\infty} p_x(k) = p_x(\mathrm{m}) \sum_{r=0}^{\infty} \left(\frac{\mathtt{G}}{\mathrm{m}}\right)^r = \frac{p_x(\mathrm{m})\mathrm{m}}{\mathrm{m} - \mathtt{G}} \tag{8.113}$$

where in the forth step we have used the second equation of (8.108). Replacing $p_x(\mathrm{m})$ with the value given by (8.111) we thus have

$$C(\mathrm{m}, \mathtt{G}) = \frac{\dfrac{\mathtt{G}^{\mathrm{m}}}{\mathrm{m}!}}{\dfrac{\mathrm{m} - \mathtt{G}}{\mathrm{m}} \displaystyle\sum_{r=0}^{\mathrm{m}-1} \frac{\mathtt{G}^r}{r!} + \frac{\mathtt{G}^{\mathrm{m}}}{\mathrm{m}!}} = \frac{\dfrac{(\mathrm{m}\rho)^{\mathrm{m}}}{\mathrm{m}!}}{(1-\rho) \displaystyle\sum_{r=0}^{\mathrm{m}-1} \frac{(\mathrm{m}\rho)^r}{r!} + \frac{(\mathrm{m}\rho)^{\mathrm{m}}}{\mathrm{m}!}}, \tag{8.114}$$

which is known in the literature as *Erlang-C formula*. The formula can be used, for instance, to compute the probability that no trunk (i.e., server) is available for an arriving call (customer) in a system of m trunks, so that the call is forced to wait in queue. Similarly, the Erlang-C formula may be used to determine the probability that a call to a call-center with m operators is put on hold because all the operators are busy. The Erlang-C formula is plotted in Fig. 8.16 as a function of the offered traffic $\mathtt{G}$, for different values of m.

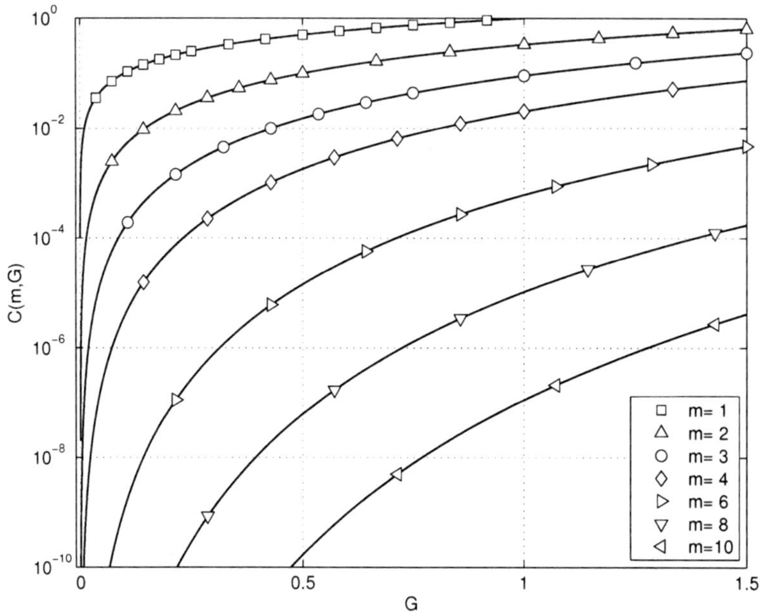

Figure 8.16 Queueing probability (Erlang-C formula) versus the offered traffic $\mathtt{G} = \mathrm{m}\rho = \lambda/\mu$ in $\mathrm{M/M/m}$ for different values of the number m of servers.

Occupancy measures. The mean number of customers in the system, $\mathtt{m}_x$, is equal to

$$\mathtt{m}_x = \sum_{k=1}^{\infty} k p_x(k) = \sum_{k=1}^{\mathrm{m}} k p_x(k) + \sum_{k=\mathrm{m}+1}^{\infty} k p_x(k) \tag{8.115}$$

where we have split the summation into two, reflecting the bimodal form of $p_x(k)$ as given by (8.111). For simplicity, we consider each part separately. Using (8.108), the first sum can be expressed as

$$\begin{aligned}\sum_{k=1}^{\mathrm{m}} k p_x(k) &= \sum_{k=1}^{\mathrm{m}} k \frac{\mathtt{G}^k}{k!} p_x(0) = \mathtt{G} \sum_{k=1}^{\mathrm{m}} \frac{\mathtt{G}^{k-1}}{(k-1)!} p_x(0) = \mathtt{G} \sum_{k=1}^{\mathrm{m}} p_x(k-1) \\ &= \mathtt{G}\,\mathrm{P}[x < \mathrm{m}] = \mathtt{G}(1 - \mathrm{P}[x \geq \mathrm{m}]) = \mathtt{G}(1 - C(\mathrm{m}, \mathtt{G}))\end{aligned} \tag{8.116}$$

where in the last step we have used the definition (8.114). For the second sum in (8.115), we have

$$\begin{aligned}\sum_{k=\mathrm{m}+1}^{\infty} kp_x(k) &= \sum_{r=1}^{\infty}(r+\mathrm{m})p_x(\mathrm{m}+r) = \sum_{r=1}^{\infty} rp_x(\mathrm{m}+r) + \mathrm{m}\sum_{r=1}^{\infty} p_x(\mathrm{m}+r) \\ &= p_x(\mathrm{m})\sum_{r=1}^{\infty} r\left(\frac{\mathtt{G}}{\mathrm{m}}\right)^r + \mathrm{m}\,\mathrm{P}\left[x>\mathrm{m}\right]\end{aligned} \tag{8.117}$$

where, in the last step, we have used the second equation of (8.108). Inspecting (8.117), we realize that the sum is the mean of a geometric rv with parameter $1-\frac{\mathtt{G}}{\mathrm{m}}$, divided by $1-\frac{\mathtt{G}}{\mathrm{m}}$, whereas the term $\mathrm{P}\left[x>\mathrm{m}\right]$ is equal to $C(\mathrm{m},\mathtt{G})$ subtracted of the term $\mathrm{P}\left[x=\mathrm{m}\right]$. Hence, (8.117) becomes

$$\sum_{k=\mathrm{m}+1}^{\infty} kp_x(k) = p_x(\mathrm{m})\frac{\frac{\mathtt{G}}{\mathrm{m}}}{\left(1-\frac{\mathtt{G}}{\mathrm{m}}\right)^2} + \mathrm{m}[C(\mathrm{m},\mathtt{G}) - p_x(\mathrm{m})]$$

that, using (8.113), yields

$$\sum_{k=\mathrm{m}+1}^{\infty} kp_x(k) = C(\mathrm{m},\mathtt{G})\left(\frac{\mathtt{G}}{\mathrm{m}-\mathtt{G}} + \mathtt{G}\right). \tag{8.118}$$

Now, using (8.116) and (8.118) into (8.115), we obtain

$$\boxed{\mathtt{m}_x = \mathtt{G} + C(\mathrm{m},\mathtt{G})\frac{\mathtt{G}}{\mathrm{m}-\mathtt{G}}.} \tag{8.119}$$

Note that $\mathtt{m}_x$ still tends to infinity as $\mathtt{G}\to\mathrm{m}$, that is $\rho\to 1$, as observed in $\mathrm{M/M/1}$.

The PMD of the number of customers in service, z, can be directly obtained from the PMD of x. In fact, as long as $x<\mathrm{m}$, all the customers in the system are actually under service and $z=x$, whereas $z=\mathrm{m}$ whenever $x\geq\mathrm{m}$. Therefore, the PMD of z is given by

$$\boxed{p_z(k) = \begin{cases}\mathrm{P}\left[x=k\right] = \dfrac{\mathtt{G}^k}{k!}p_x(0) & , \; 0\leq k<\mathrm{m} \\ \mathrm{P}\left[x\geq\mathrm{m}\right] = C(\mathrm{m},\mathtt{G}) & , \; k=\mathrm{m}.\end{cases}} \tag{8.120}$$

The mean number of busy servers is given by

$$\mathtt{m}_z = \sum_{k=1}^{\mathrm{m}} kp_z(k) = \sum_{k=1}^{\mathrm{m}-1} kp_x(k) + \mathrm{m}C(\mathrm{m},\mathtt{G}) = \sum_{k=1}^{\mathrm{m}} kp_x(k) - \mathrm{m}p_x(\mathrm{m}) + \mathrm{m}C(\mathrm{m},\mathtt{G}). \tag{8.121}$$

The sum on the right-hand side of (8.121) has been encountered during the derivation of $\mathtt{m}_x$. Hence, using (8.116) we get

$$\begin{aligned}\mathtt{m}_z &= \mathtt{G}(1-C(\mathrm{m},\mathtt{G})) - \mathrm{m}p_x(\mathrm{m}) + \mathrm{m}C(\mathrm{m},\mathtt{G}) \\ &= C(\mathrm{m},\mathtt{G})(\mathrm{m}-\mathtt{G}) + \mathtt{G} - \mathrm{m}p_x(\mathrm{m}).\end{aligned} \tag{8.122}$$

Finally, using the second equation of (8.108), we obtain

$$\boxed{\mathtt{m}_z = \mathtt{G} = \frac{\lambda}{\mu}\,.} \tag{8.123}$$

As usual, another way to derive (8.123) is through the Little's law in the form (8.48).

At this point, the PMD of the number of queued customers q can be derived right away. In fact, the queue is empty whenever x is less than or equal to m, so that

$$p_q(0) = \mathrm{P}\left[x \le \mathrm{m}\right] = 1 - C(\mathrm{m},\mathtt{G}) + p_x(\mathrm{m}) = 1 - \frac{\mathtt{G}}{\mathrm{m}}C(\mathrm{m},\mathtt{G})\,, \tag{8.124}$$

where, from (8.113), we replaced $p_x(\mathrm{m})$ with $C(\mathrm{m},\mathtt{G})(1-\mathtt{G}/\mathrm{m})$. Conversely, when x exceeds m, then the customers in excess are held in the queue so that we have

$$p_q(k) = p_x(\mathrm{m}+k) = \left(\frac{\mathtt{G}}{\mathrm{m}}\right)^k p_x(\mathrm{m}) = \left(\frac{\mathtt{G}}{\mathrm{m}}\right)^k\left(1-\frac{\mathtt{G}}{\mathrm{m}}\right)C(\mathrm{m},\mathtt{G})\,, \qquad k = 1,2,\ldots\,. \tag{8.125}$$

Summing up, we have

$$\boxed{p_q(k) = \begin{cases} 1 - \dfrac{\mathtt{G}}{\mathrm{m}}C(\mathrm{m},\mathtt{G}) & ,\ k = 0 \\[2ex] \dfrac{(\mathrm{m}-\mathtt{G})C(\mathrm{m},\mathtt{G})\mathtt{G}^k}{\mathrm{m}^{k+1}} & ,\ k = 1,2,\ldots \end{cases}} \tag{8.126}$$

The mean occupancy of the waiting queue can be directly obtained using (8.126). However, a simpler way consists in using (8.20) to express $\mathtt{m}_q$ as

$$\mathtt{m}_q = \mathtt{m}_x - \mathtt{m}_z \tag{8.127}$$

and, then, applying (8.119) and (8.123) to get

$$\boxed{\mathtt{m}_q = C(\mathrm{m},\mathtt{G})\frac{\mathtt{G}}{\mathrm{m}-\mathtt{G}}\,.} \tag{8.128}$$

Time measures. In multi-server systems, the most interesting time measure is the time w a customer is forced to wait in queue before a server becomes available. An arriving customer is immediately served when the number of customers in the system is less than m, so that at least one server is available. Otherwise, when all the servers are occupied, the arriving customer is queued. The PDF of w can be expressed by applying the Total Probability Theorem 2.12 with respect to the number k of customers found in the system at the arriving instant t_n,

$$p_w(a) = \sum_{k=0}^{\infty} p_w(a|x=k)p_x(k)\,, \tag{8.129}$$

where we have used x in place of $x(t_n^-)$ because of the PASTA theorem. As mentioned, the waiting time is certainly zero if there are idle servers at the arriving instant, i.e.,

$$p_w(a|x=k) = \delta(a)\,, \quad \text{for} \quad 0 \le k < \mathrm{m}. \tag{8.130}$$

If all the servers are occupied, the arriving customer is queued. Suppose that, at the arriving instant, there are $k \geq \mathrm{m}$ customers in the system, out of which m are in service, and $k - \mathrm{m} \geq 0$ are in the waiting queue. The newly arrived customer will have to wait for $k - \mathrm{m}$ departures in order to reach the head of the queue, plus an extra departure to get in service. During this time period all the servers are busy, so that the departure rate is equal to $\mathrm{m}\mu$, as can also be realized by inspecting the flow diagram of Fig. 8.15. Therefore, w can be expressed as the sum of $k - \mathrm{m} + 1$ interdeparture times, which are statstically iid exponential rvs with parameter $\mathrm{m}\mu$. The conditional PDF of w is, thus, an Erlang rv of index $k - \mathrm{m} + 1$ and parameter $\mathrm{m}\mu$,

$$p_w(a|x = k) = \frac{\mathrm{m}\mu(\mathrm{m}\mu a)^{k-\mathrm{m}}}{(k - \mathrm{m})!} e^{-\mathrm{m}\mu a}, \quad a \geq 0. \tag{8.131}$$

Replacing (8.131) and (8.130) into (8.129), we obtain

$$\begin{aligned} p_w(a) &= \sum_{k=0}^{\mathrm{m}-1} \delta(a) p_x(k) + \sum_{k=\mathrm{m}}^{\infty} p_x(k) \frac{\mathrm{m}\mu(\mathrm{m}\mu a)^{k-\mathrm{m}}}{(k-\mathrm{m})!} e^{-\mathrm{m}\mu a} \\ &= \delta(a) \sum_{k=0}^{\mathrm{m}-1} p_x(k) + p_x(\mathrm{m}) \sum_{k=\mathrm{m}}^{\infty} \left(\frac{\mathtt{G}}{\mathrm{m}}\right)^{k-\mathrm{m}} \frac{\mathrm{m}\mu(\mathrm{m}\mu a)^{k-\mathrm{m}}}{(k-\mathrm{m})!} e^{-\mathrm{m}\mu a} \\ &= \delta(a)\, \mathrm{P}\left[x < \mathrm{m}\right] + p_x(\mathrm{m}) \mathrm{m}\mu e^{-(\mathrm{m}\mu - \lambda)a}. \end{aligned} \tag{8.132}$$

Now, according to (8.113) we have $\mathrm{P}\left[x < \mathrm{m}\right] = 1 - C(\mathrm{m}, \mathtt{G})$ and $p_x(\mathrm{m}) = C(\mathrm{m}, \mathtt{G})(1 - \mathtt{G}/\mathrm{m}) = C(\mathrm{m}, \mathtt{G})(\mathrm{m}\mu - \lambda)/(\mathrm{m}\mu)$, so that (8.132) yields

$$p_w(a) = [1 - C(\mathrm{m}, \mathtt{G})]\delta(a) + C(\mathrm{m}, \mathtt{G})(\mathrm{m}\mu - \lambda) e^{-(\mathrm{m}\mu - \lambda)a}, \qquad a \geq 0. \tag{8.133}$$

As usual, the impulsive term accounts for the probability that an arriving customer is immediately served. The second term, instead, is the product of the queueing probability $C(\mathrm{m}, \mathtt{G})$ times the PDF of an exponential rv $\hat{w}$ with parameter $\mathrm{m}\mu - \lambda$: $\hat{w}$ is actually the waiting time of an arriving customer, given that it finds all the servers busy, so that $p_{\hat{w}}(a) = p_w(a|x \geq \mathrm{m})$. Therefore, the queueing time can be also expressed as

$$w = \chi\left\{x \geq \mathrm{m}\right\} \hat{w}. \tag{8.134}$$

Taking the expectation of (8.134) we get the mean waiting time

$$\mathtt{m}_w = C(\mathrm{m}, \mathtt{G})\, \mathtt{m}_{\hat{w}} = \frac{C(\mathrm{m}, \mathtt{G})}{\mathrm{m}\mu - \lambda} \tag{8.135}$$

where we have used

$$\mathtt{m}_{\hat{w}} = \frac{1}{\mathrm{m}\mu - \lambda} \tag{8.136}$$

which is the mean waiting time of an arriving customer, given that it gets queued. Recalling that $\mathtt{G} = \lambda/\mu$ we hence have

$$\mathtt{m}_w = \frac{C(\mathrm{m}, \mathtt{G})}{\mu(\mathrm{m} - \mathtt{G})}. \tag{8.137}$$

From (8.26), the system time s can be expressed as

$$s = w + y \tag{8.138}$$

where y is exponentially distributed with parameter μ. The PDF of s is given by the convolution of $p_y(a)$ and $p_w(a)$ that, after some easy algebra, turns out to be

$$\boxed{p_s(a) = \mu e^{-\mu a}\left(1 + C(\mathrm{m}, \mathtt{G})\frac{1 - (\mathrm{m} - \mathtt{G})e^{-\frac{\mathrm{m}-\mathtt{G}-1}{\mathrm{m}}a}}{\mathrm{m} - \mathtt{G} - 1}\right), \quad a \geq 0\,.} \tag{8.139}$$

Moreover, taking the expectation of all terms in (8.138), we immediately obtain

$$\boxed{\mathtt{m}_s = \frac{C(\mathrm{m}, \mathtt{G})}{\mu(\mathrm{m} - \mathtt{G})} + \frac{1}{\mu},} \tag{8.140}$$

where we have used (8.137) and the fact $\mathtt{m}_y = 1/\mu$.

Remark It is interesting to analyze the performance measures of $\mathrm{M/M/m}$ when varying the number of servers, m and the service rate μ in such a way that the service capacity $\mathrm{m}\mu$ remains constant. In Fig. 8.17(a) we report the queueing probability $C(\mathrm{m}, \mathtt{G})$ versus $\rho = \mathtt{G}/\mathrm{m} = \lambda/(\mathrm{m}\mu)$ (upper graph), whereas Fig. 8.17(b) shows the mean queueing time $\mathtt{m}_w$ (solid lines) and the mean system time $\mathtt{m}_s$ (dashed lines) versus ρ. All curves have been obtained by setting $\lambda = 1$, so that the service capacity $\mathrm{m}\mu$ basically corresponds to the reciprocal of ρ. The number of servers m has been varied as indicated in the figures legend. Inspecting the two figures we notice that, ρ being the same, an increase of m yields a reduction of both $C(\mathrm{m}, \mathtt{G})$ and $\mathtt{m}_w$, on the one hand, but an increase of $\mathtt{m}_s$, on the other hand. Loosely speaking, we can conclude that, the service capacity being the same, a QS with many slow servers offers better queue management (lower values of $C(\mathrm{m}, \mathtt{G})$ and $\mathtt{m}_w$) than a QS with few fast servers. However, the second QS is better in terms of mean system time.

We conclude this section noting that the performance measures of $\mathrm{M/M/1}$ can be obtained from those of $\mathrm{M/M/m}$ by setting $\mathrm{m} = 1$. In particular, the queueing probability becomes $C(1, \lambda/\mu) = \lambda/\mu$, which is the probability that the single server of $\mathrm{M/M/1}$ is busy.

Example 8.5 A Consider a PBX with m outgoing trunks. Assume that calls arrive to the PBX according to a Poisson process of rate $\lambda = 0.02$ [call/s] and that each call engages a trunk for an exponential time of mean $\mathtt{m}_y = 75$ s, so that the ratio $\lambda/\mu = 1.5$. Furthermore, suppose that calls that cannot be forwarded because all the m trunks are busy are punt on hold and served according to a FCFS policy. Such a system can be hence modeled as $\mathrm{M/M/m}$, with $\mathtt{G} = 1.5$. We immediately realize that, to guarantee the system stability (8.24), we need to provide *at least* $\mathrm{m} = 2$ trunks. Furthermore, if we wish to keep the queueing probability for an incoming call below a given threshold, say 10^{-4}, we see from Fig. 8.16 that we need to provide $\mathrm{m} = 9$ trunks.

Example 8.5 B The manager of a research lab needs additional computational resources in order to meet the demand for computing power of m new researchers. The lab manager can choose among three different options: 1) buy a supercomputer (SC) with a single waiting queue and a single CPU capable

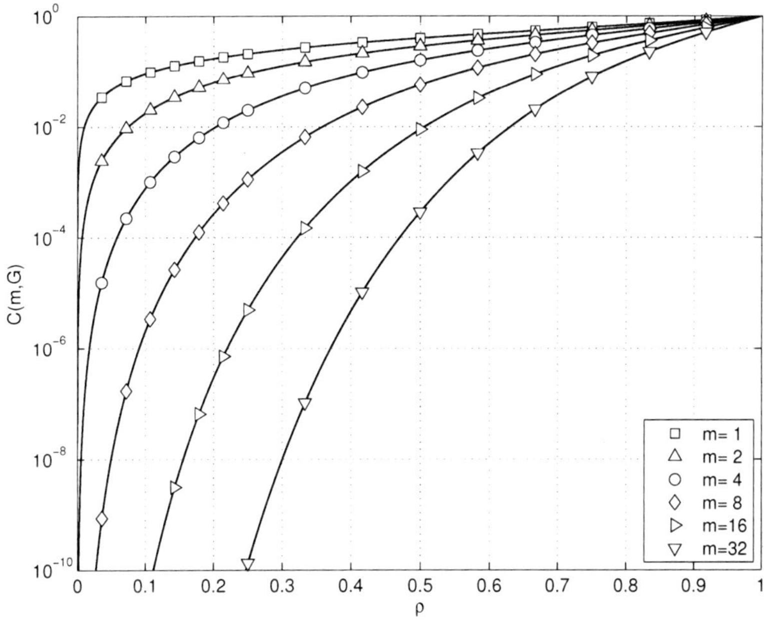

(a)

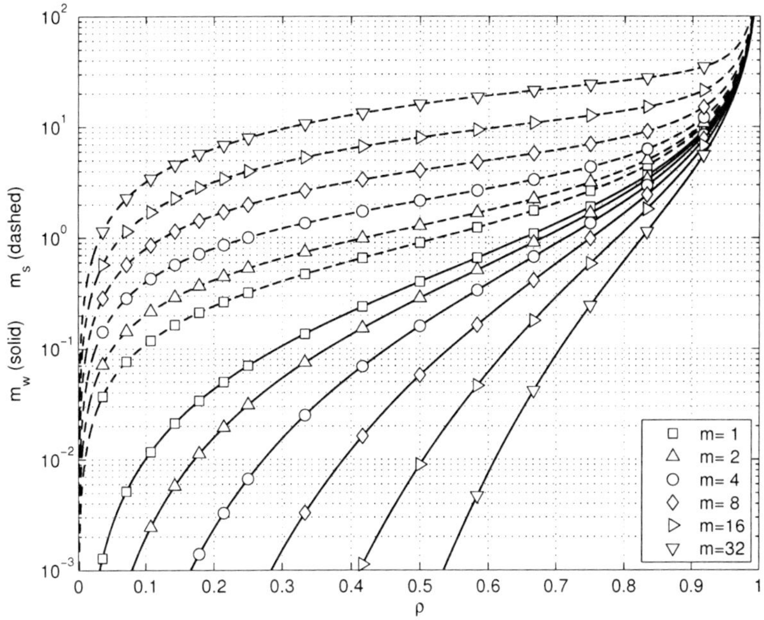

(b)

Figure 8.17 (a) queueing probability $C(\mathrm{m}, \mathtt{G})$ versus load factor ρ and (b) mean waiting time $\mathtt{m}_w$ (solid) and system time $\mathtt{m}_s$ (dashed) versus load factor $\rho = \mathtt{G}/\mathrm{m}$, in $\mathrm{M/M/m}$ for different values of the number m of servers.

of executing mR floating point operations per second (FLOPS); 2) buy a computer cluster (CC) with a single waiting queue and m CPUs that work in parallel, each capable of performing R FLOPS; 3) provide each researcher with its own personal computer (PC), with individual waiting queue and CPU capable of R FLOPS, without allowing other users to work on this machine.

Now, assume that each researcher generates computing tasks according to a Poisson process of rate λ [task/s] and that the execution of each task requires a random number ℓ of FLOPS with approximately exponential distribution of mean L, independently of the other tasks. Therefore, the execution of a task by a server with processing rate R FLOPS will require a random time with exponential distribution of mean $\mathtt{m}_y = 1/\mu = L/R$. Under these assumptions, the three options can be modeled respectively as: (SC) one $M/M/1$ with arrival rate $m\lambda$ and service rate $m\mu$; (CC) one $M/M/m$ with arrival rate $m\lambda$ and service rate μ for each server; (PC) a group of m $M/M/1$ QSs, each with arrival rate λ and service rate μ.

The lab manager wishes to compare the mean system time $\mathtt{m}_s$ of each solution, when varying the offered load λ. By using (8.89) and (8.140) we obtain the curves shown in Fig. 8.18, where we have assumed $m = 20$ and $\mu = 1$.

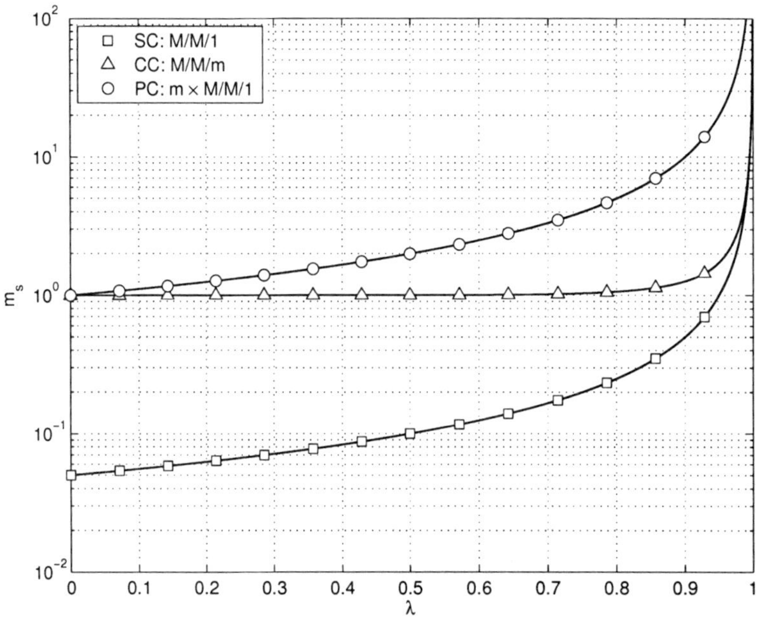

Figure 8.18 Mean system time versus arrival rate λ for the three queues of Example 8.5 B, with $m = 20$, $\mu = 1$.

As we can notice, the PC option is the worst solution. This is because, with m parallel (and non-interacting) $M/M/1$ QSs, it is always possible that some servers have queued tasks while others are idle. In the other two cases, instead, tasks are buffered in the same waiting queue and served as soon as some service facility becomes available. Therefore, tasks are never buffered if there is some computational resource available. The best performance in terms of $\mathtt{m}_s$ is provided by the SC option, using a super-server capable of processing tasks m times faster than in the other cases. As a matter of fact, for light loads or during periods of time with only a small number of tasks in the system, $\mathtt{m}_s$ is mainly due to the service time, which is m times shorter for SC than for CC. Clearly, this advantage progressively reduces as the traffic load increases and the systems gets close to saturation.

A similar analysis can be applied to other scenarios. For instance, considering the checkin counters at the airport, we can easily conclude that it is more convenient to have a single waiting queue with m checkin counters that work in parallel rather than a separate queue for each of the m checkin counters. Unfortunately, the best performing solution, which consists in having a single hostess capable of doing the paperwork m times faster than her colleagues, is not easy to realize in this case!

8.5.3 The $\mathrm{M}/\mathrm{M}/1/K$ queueing system

The $\mathrm{M}/\mathrm{M}/1/K$ model generalizes $\mathrm{M}/\mathrm{M}/1$ by limiting to K the number of customers that can be simultaneously held in the system. In other words, the waiting queue has a finite storage capacity of $K-1$ customers, so that the system can hold at most K customers, including the customer under service. Customers that arrive when the system is full are denied admission to the system and will be dropped immediately, without being served. This event is referred to as *blocking* or *dropping*.

It may be worth remarking that newly arriving customers will continue to be generated according to a Poisson process of rate λ, irrespective of the number of customers in the QS, but only those who find fewer than K customers already in the QS are actually *admitted* into the system, whereas the others are refused. Therefore, the *average* rate λ_a at which customers are admitted into the system is just a fraction of the arrival rate λ, as schematically illustrated in Fig. 8.19.

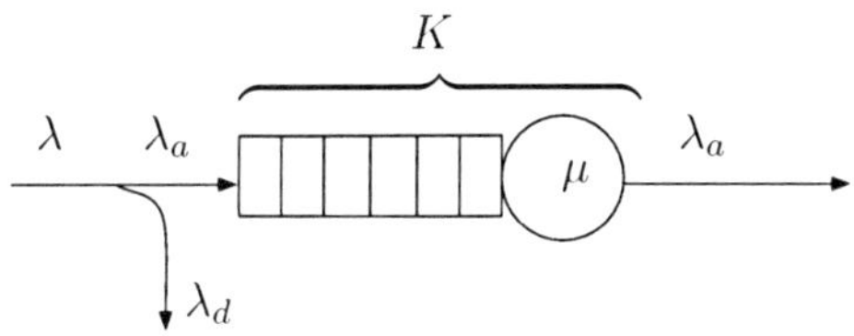

Figure 8.19 The $\mathrm{M}/\mathrm{M}/1/\mathrm{K}$ queueing system.

The asymptotic analysis of the $\mathrm{M}/\mathrm{M}/1/K$ behavior is very similar to that developed for $\mathrm{M}/\mathrm{M}/1$. In fact, we easily realize that $x(t)$, the number of customers in the system at time t, evolves as in $\mathrm{M}/\mathrm{M}/1$, with the only difference that it is upper limited by the (finite) system capacity K. Hence, $x(t)$ is a BDP with *finite* state space $\mathcal{S}=\{0,1,\ldots,K\}$ and birth-rate λ for any state, except state K that has birth rate equal to zero. Customers that are in the system get served one at a time, with service rate μ, so that the death-rate is equal to μ for any state, except state zero that has a zero death rate. The flow-diagram associated to this BDP is shown in Fig. 8.20.

Remark Reader shall not confuse the birth rate λ with the average admission rate λ_a. We remark that, as long as the QS is not full, the birth rate is λ, whereas it becomes zero when the system is full. Hence, the birth rate is state dependent. Conversely, the admission rate λ_a

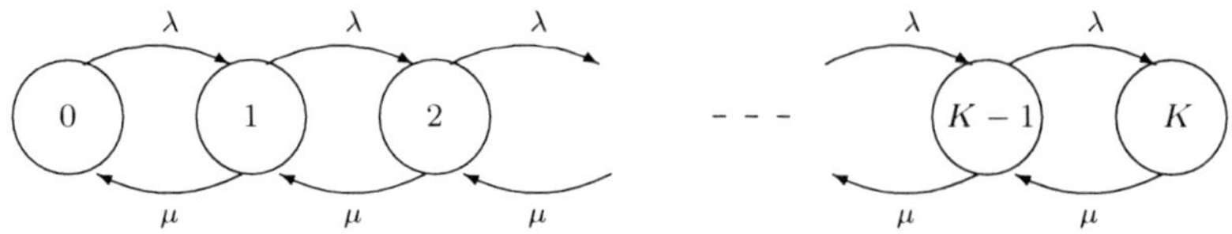

Figure 8.20 Flow-diagram of $\mathrm{M/M/1}/K$.

gives the mean number of customers admitted in the system in the long run and, hence, it is an asymptotic measure that does not depend on the current state of the QS.

Following the same rationale developed for $\mathrm{M/M/1}$, we can determine the asymptotic distribution of the number of customers in the system and of the other performance measures.

From the probability-flow balance for each pair of consecutive states in Fig. 8.20, we get a system of equations in the form

$$\mu\, p_x(k) = \lambda\, p_x(k-1)\,, \qquad k = 1, 2, \ldots, K\,; \tag{8.141}$$

which can be written as

$$\begin{aligned} p_x(k) &= \frac{\lambda}{\mu}\, p_x(k-1) \\ &= \rho\, p_x(k-1)\,, \qquad k = 1, 2, \ldots, K\,, \end{aligned} \tag{8.142}$$

with ρ as in (8.31). Unfolding (8.142) recursively back to $p_x(0)$, we get

$$p_x(k) = \rho^k\, p_x(0)\,, \qquad k = 1, 2, \ldots, K\,, \tag{8.143}$$

where $p_x(0)$ can be obtained through the normalization condition, which yields

$$p_x(0) = \frac{1-\rho}{1-\rho^{K+1}}\,. \tag{8.144}$$

Replacing (8.144) into (8.143) we finally obtain

$$\boxed{p_x(k) = \frac{(1-\rho)\rho^k}{1-\rho^{K+1}} = \frac{\left(1-\dfrac{\lambda}{\mu}\right)\left(\dfrac{\lambda}{\mu}\right)^k}{1-\left(\dfrac{\lambda}{\mu}\right)^{K+1}}\,, \qquad k = 0, 1, 2, \ldots, K\,.} \tag{8.145}$$

Note that the asymptotic PMD of the number of customers in $\mathrm{M/M/1}/K$ is geometric with parameter $1-\rho$ truncated at K. This result holds for any value of ρ, confirming that $\mathrm{M/M/1}/K$

is always stable since it admits a proper asymptotic PMD for any value of the load factor ρ. What is directly affected by ρ is, instead, the call-blocking probability P_{blk}, which is equal to the asymptotic probability that the system if fully occupied, i.e.,

$$\boxed{P_{\text{blk}} = p_x(K) = \frac{(1-\rho)\rho^K}{1-\rho^{K+1}}\,.} \tag{8.146}$$

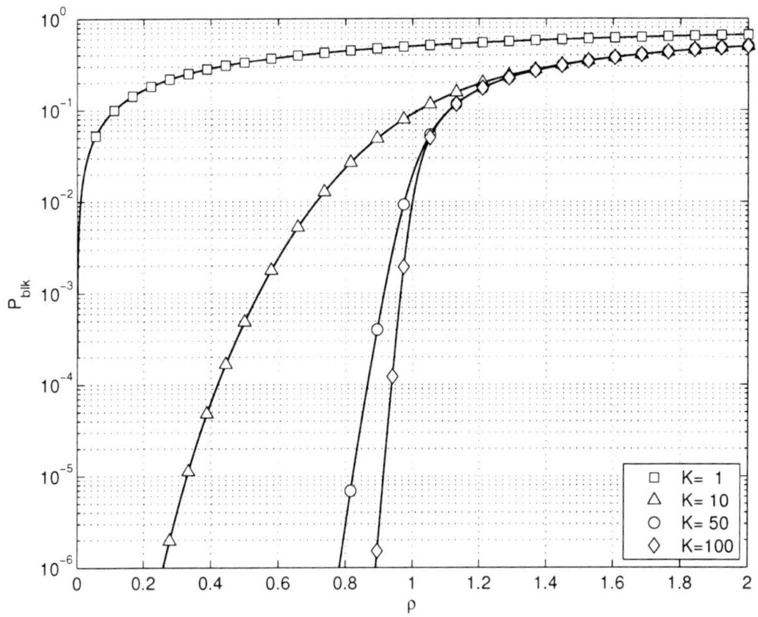

Figure 8.21 Call-blocking probability P_{blk} versus the load factor ρ in M/M/1/K, for different values of the storage capacity K.

The call-blocking probability as a function of ρ is reported in Fig. 8.21 for different values of K. As expected, P_{blk} rapidly increases when $\rho \to 1$, irrespective of the value of K. However, it shall be noted that $P_{\text{blk}} \neq 0$ also when ρ is well below one, since the random fluctuations of the arrival process may determine a temporary overload in the system, with the consequence that some customers are refused entry.

We can now formally prove the initial statement that $\mathrm{M/M/1/}K$ is a generalization of $\mathrm{M/M/1}$. As a mater of fact, letting K approach infinity, (8.145) returns (8.72), provided that $\rho < 1$. In some cases, it may be convenient to use $\mathrm{M/M/1}$ in place of the more appropriate $\mathrm{M/M/1/}K$ to model a blocking system. Suppose, for instance, that we wish to determine the storage capacity K^* for which the fraction of blocked customers in $\mathrm{M/M/1/}K$ drops below a given threshold P_{thr}, that is to say

$$K^* = \min\{K : P_{\text{blk}} \leq P_{thr}\}\,. \tag{8.147}$$

Using (8.146) in (8.147) we get

$$K^* = \min\left\{K : \frac{(1-\rho)\rho^K}{1-\rho^{K+1}} \leq P_{thr}\right\}. \tag{8.148}$$

Unfortunately, the inequality in (8.148) cannot be explicitly solved in K, so that K^* can only be determined by numerical methods. Conversely, using M/M/1, we may find a closed-form expression that approximates (8.148). To this end, we approximate P_{blk} given by (8.146) with the *tail probability* $\tilde{P}_{\text{blk}}$ that the number of customers in M/M/1 exceeds the threshold K, i.e., from (8.72)

$$\tilde{P}_{\text{blk}} = \text{P}\left[x > K\right] = 1 - \sum_{r=0}^{K} \rho^r(1-\rho) = \rho^{K+1}. \tag{8.149}$$

Using $\tilde{P}_{\text{blk}}$ in place of P_{blk} into (8.147), we get

$$\tilde{K}^* = \min\left\{K : \tilde{P}_{\text{blk}} \leq P_{thr}\right\} \implies \tilde{K}^* = \min\left\{K : \rho^{K+1} \leq P_{thr}\right\} \tag{8.150}$$

whose solution yields

$$\tilde{K}^* = \left\lceil \frac{\log(P_{thr})}{\log(\rho)} - 1 \right\rceil \tag{8.151}$$

where $\lceil x \rceil$ represents the smallest integer equal to or greater than x. Clearly, this solution makes sense only if $\rho < 1$, otherwise M/M/1 would be unstable. In Fig. 8.22 we compare the approximated value (8.151) with the exact solution (8.148) obtained through numerical methods. As it can be noted, the approximation is fairly good, in particular when the threshold P_{thr} is small.

Occupancy measures. Deriving the expression of the mean number of customers in the system, $\mathtt{m}_x$, is simple but rather cumbersome. From (8.146) we get

$$\mathtt{m}_x = \sum_{k=0}^{K} k p_x(k) = \frac{1-\rho}{1-\rho^{K+1}} \sum_{k=1}^{K} k\rho^k = \frac{\rho(1-\rho)}{1-\rho^{K+1}} \sum_{k=1}^{K} k\rho^{k-1}. \tag{8.152}$$

The argument in the right-most summation can be rewritten as the derivative of ρ^k with respect to ρ, so that the sum becomes

$$\sum_{k=1}^{K} k\rho^{k-1} = \sum_{k=1}^{K} \frac{d}{d\rho}\rho^k. \tag{8.153}$$

Given that the summation is limited, we are allowed to interchange sum and derivative, thus obtaining

$$\begin{aligned} \frac{d}{d\rho}\sum_{k=1}^{K}\rho^k &= \frac{d}{d\rho}\frac{1-\rho^{K+1}}{1-\rho} = \frac{-(K+1)\rho^K + (K+1)\rho^{K+1} + 1 - \rho^{K+1}}{(1-\rho)^2} \\ &= \frac{1-(K+1)\rho^K + K\rho^{K+1}}{(1-\rho)^2}, \end{aligned} \tag{8.154}$$

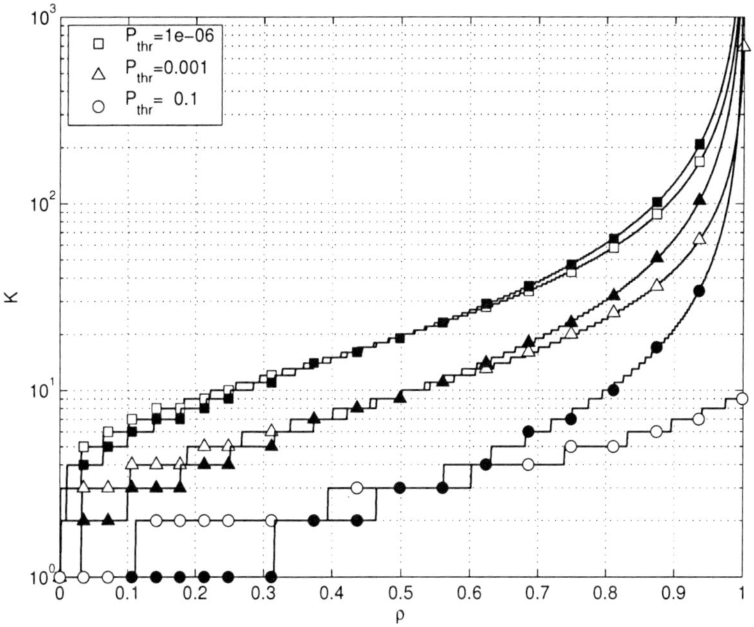

Figure 8.22 Minimum storage capacity K such that $P_{\rm blk} \leq P_{thr}$ in M/M/1/K for three values of P_{thr}: comparison between the exact solution K^* (white marks) and the approximated solution $\tilde{K}^*$ (black marks) when varying the load factor ρ.

and, replacing (8.154) in (8.152) we finally have

$$\boxed{\mathtt{m}_x = \frac{\rho}{1-\rho} - \frac{(K+1)\rho^{K+1}}{1-\rho^{K+1}}\,.} \tag{8.155}$$

The asymptotic PMD of the number of customers in the server, z and in queue, q, can be obtained by following the same scheme used in M/M/1. The value of z can either be zero or one, depending on whether the system is empty or not. Therefore, we have

$$\boxed{p_z(k) = \begin{cases} \dfrac{1-\rho}{1-\rho^{K+1}} & , \quad k=0 \\ \dfrac{\rho(1-\rho^K)}{1-\rho^{K+1}} & , \quad k=1\,. \end{cases}} \tag{8.156}$$

Hence, the expected value of z is equal to

$$\mathtt{m}_z = \frac{\rho(1-\rho^K)}{1-\rho^{K+1}} \tag{8.157}$$

that, recalling (8.146), can also be expressed as

$$\boxed{\mathtt{m}_z = \rho(1-P_{\rm blk})\,.} \tag{8.158}$$

Conversely to what found in $M/M/1$, in this case $\mathtt{m}_z$ is equal to a *fraction* of ρ, proportional to the acceptance probability $1 - P_{\text{blk}}$.

Using (8.158) in Little's law (8.48) we easily derive the average *admission* rate λ_a of customers into the system and, in turn, into the server, which is given by

$$\lambda_a = \frac{\mathtt{m}_z}{\mathtt{m}_y} = \rho(1 - P_{\text{blk}})\mu = \lambda\,(1 - P_{\text{blk}})\,, \tag{8.159}$$

which corresponds to (8.17).

The distribution of the number of customers q in the waiting queue can be obtained as done for $M/M/1$. In fact, q is always one unit less than the number of customers in the system x, if any, so that we can write

$$q = x - \chi\,\{x > 0\}\ . \tag{8.160}$$

Therefore, for $k = 1, 2, \ldots, K - 1$ we have

$$p_q(k) = p_x(k+1) = \frac{(1-\rho)\rho^{k+1}}{1-\rho^{K+1}}\,, \tag{8.161}$$

whereas, for $k = 0$ we have

$$p_q(0) = p_x(0) + p_x(1) = \frac{(1-\rho) + \rho(1-\rho)}{1-\rho^{K+1}} = \frac{1-\rho^2}{1-\rho^{K+1}}\,. \tag{8.162}$$

In summary, the PMD of q is equal to

$$\boxed{p_q(k) = \begin{cases} \dfrac{1-\rho^2}{1-\rho^{K+1}}\,, & k = 0 \\[2ex] \dfrac{(1-\rho)\rho^{k+1}}{1-\rho^{K+1}}\,, & k = 1, 2, \ldots, K-1 \end{cases}} \tag{8.163}$$

The expectation of q can be obtain from the PMD (8.163) or, more simply, by taking the expectation of both sides of (8.160), which gives

$$\mathtt{m}_q = \mathrm{E}\left[x - \chi\,\{x > 0\}\right] = \mathtt{m}_x - \mathrm{P}\left[x > 0\right] = \mathtt{m}_x - 1 + p_x(0) \tag{8.164}$$

that is

$$\boxed{\mathtt{m}_q = \frac{\rho^2(1 + (K-1)\rho^K - K\rho^{K-1})}{(1-\rho^{K+1})(1-\rho)}\,.} \tag{8.165}$$

Time measures. The analysis of the queueing and system times in $M/M/1/K$ closely resembles that developed for $M/M/1$, with the complication that we have now to focus only on the customers that are actually *admitted* into the system. Suppose hence that the customer C_n is admitted into the system and let x_a denote the number of customers that C_n finds in the system. The subscript a to x stresses the fact that we are considering the system occupancy only at the arrival instants of *admitted* customers. The system time of C_n can then be expressed as

$$s = y + \sum_{r=1}^{x_a} y_r\,, \tag{8.166}$$

where, as usual, y denotes the service time of C_n whereas $y_1, y_2, \ldots, y_{x_a}$ are the service times of the x_a customers that precede C_n in the service list, included the residual service time of the customer under service, if any. Conditioning on $x_a = k$, (8.166) yields the PDF (8.83). The unconditioned distribution is obtained by averaging (8.83) over the distribution $\mathrm{P}[x_a = k]$ of x_a. According to the PASTA theorem, the number of customers in the system observed by *any* arriving customer is distributed as in (8.145). However, a customer is accepted into the system only if it finds a system occupancy strictly less than K. Hence, conditioning on the fact that the customer is accepted, that is to say, that the number of customers found in the system is less than K, the probability of observing k customers is equal to

$$\mathrm{P}[x_a = k] = \mathrm{P}[x = k | x < K] = \frac{\mathrm{P}[x = k, x < K]}{\mathrm{P}[x < K]} = \frac{p_x(k)}{1 - P_{\mathrm{blk}}} = \frac{\rho^k(1-\rho)}{1-\rho^K} \tag{8.167}$$

for $k = 0, 1, \ldots, K-1$. From (8.84),

$$\begin{aligned} p_s(a) &= \sum_{k=0}^{K-1} p_{s|_k}(a)\,\mathrm{P}[x_a = k] = \sum_{k=0}^{K-1} \frac{\mu(\mu a)^k}{k!} e^{-\mu a} \frac{\rho^k(1-\rho)}{1-\rho^K} \\ &= \frac{\mu(1-\rho)}{1-\rho^K} e^{-\mu a} \sum_{k=0}^{K-1} \frac{(\mu\rho a)^k}{k!} = \frac{\mu-\lambda}{1-\rho^K} e^{-\mu a} \sum_{k=0}^{K-1} \frac{(\lambda a)^k}{k!} . \end{aligned} \tag{8.168}$$

Unfortunately, the summation cannot be expressed in a more compact form. However, observing that such a sum is the Taylor series expansion of $e^{\lambda a}$, truncated after the first K terms, we can rewrite (8.168) in the following form

$$\boxed{p_s(a) = \mu(1-\rho)e^{-\mu(1-\rho)a} \frac{\displaystyle\sum_{k=0}^{K-1} \frac{(\lambda a)^k}{k!} e^{-\lambda a}}{1-\rho^K} .} \tag{8.169}$$

If $\rho < 1$, as K approaches infinity the fraction on the right-hand side of (8.169) tends to one and (8.169) tends to (8.87).

The mean system time may be obtained from the PDF (8.169) or, more simply, using Little's law. Bearing in mind that the mean arrival rate in Little's law refers to the rate at which customers actually enter the system, from (8.46) with λ replaced by $\lambda_a = \lambda(1 - P_{\mathrm{blk}})$, we derive

$$\mathtt{m}_s = \frac{\mathtt{m}_x}{\lambda(1 - P_{\mathrm{blk}})} , \tag{8.170}$$

and, using (8.155),

$$\boxed{\mathtt{m}_s = \frac{1 - (K+1)\rho^K + K\rho^{K+1}}{\mu(1-\rho)(1-\rho^K)} .} \tag{8.171}$$

In Fig. 8.23 we report the mean system time $\mathtt{m}_s$ as a function of ρ, with $\mu = 1$ and different values of K. For the sake of comparison, we also show with a dashed line $\mathtt{m}_s$ in the limiting case of $K \to \infty$, which corresponds to the system time of $\mathrm{M/M/1}$ (see (8.89)). We observe that, as expected, $\mathtt{m}_s$ is always limited and tends to $\frac{K}{\mu}$, as $\rho \to \infty$.

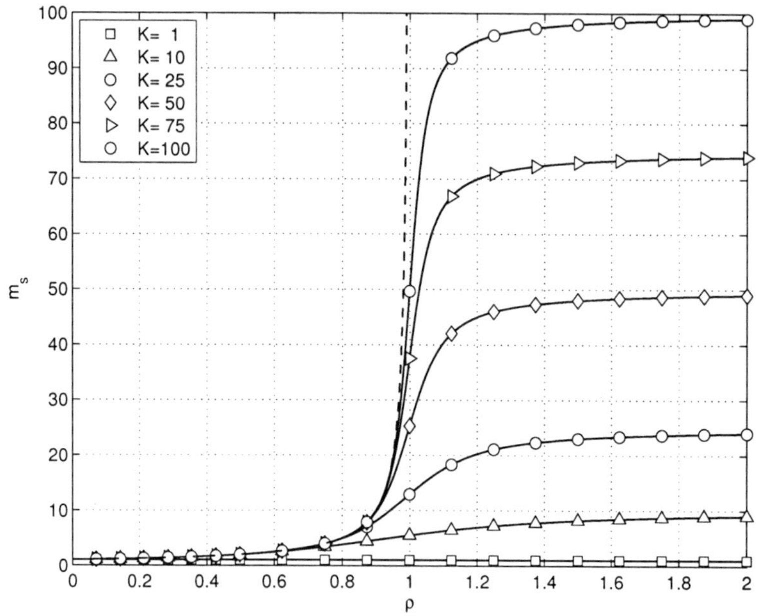

Figure 8.23 Mean system time $\mathtt{m}_s$ versus ρ in M/M/1/K, with $\mu = 1$ and different values of the storage capacity K.

In a very similar manner, the asymptotic PDF of the waiting time w can be expressed as

$$p_w(a) = \mathrm{P}\left[x_a = 0\right]\delta(a) + \sum_{r=1}^{K-1} \mathrm{P}\left[x_a = r\right] \frac{\mu(\mu a)^{r-1}}{(r-1)!} e^{-\mu a} . \tag{8.172}$$

Skipping the intermediate steps of the derivation, we go directly to the final expression

$$\boxed{p_w(a) = \frac{1}{1-\rho^K}\left((1-\rho)\delta(a) + \rho(\mu-\lambda)e^{-(\mu-\lambda)a} \sum_{h=0}^{K-2} \frac{(\lambda a)^h}{h!} e^{-\lambda a}\right), \quad a \geq 0 .} \tag{8.173}$$

Moreover, the mean waiting time $\mathtt{m}_w = \mathtt{m}_s - \mathtt{m}_y$ is equal to

$$\boxed{\mathtt{m}_w = \frac{(K-1)\rho^{K+1} - K\rho^k + \rho}{\mu(1-\rho)\left(1-\rho^K\right)} .} \tag{8.174}$$

Traffic measures. We conclude the analysis of $\mathrm{M/M/1/}K$ with the following remark.

Remark The stability of $\mathrm{M/M/1/}K$ for any value of the offered traffic $\mathtt{G}$ is payed back in terms of system throughput. In fact, according to (8.37), the useful traffic $\mathtt{S}$ of $\mathrm{M/M/1/}K$ is equal to

$$\mathtt{S} = \mathtt{G}\frac{1-\rho^K}{1-\rho^{K+1}} ,$$

which is strictly less than $\mathtt{G}$ for *any* value of $\mathtt{G}$. Conversely, M/M/1 requires $\mathtt{G} < 1$ to be stable, but when this condition is met, the useful traffic is exactly equal to $\mathtt{G}$ (see (8.35)).

8.5.4 The $\mathrm{M/M/m/m}$ queuing system

The $\mathrm{M/M/m/m}$ model refers to a multi-server QS with no facility to wait, i.e., no queue. In other words, the storage capacity K is equal to the number of servers in the system: if an arriving customer finds all the servers busy, it leaves the system without waiting for service, as schematically represented in Fig. 8.24. For this reason, $\mathrm{M/M/m/m}$ is sometimes referred to as *pure-loss QS*.

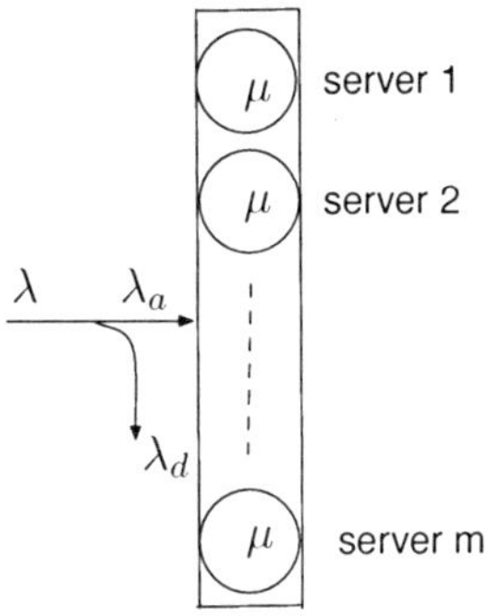

Figure 8.24 Pure-loss $\mathrm{M/M/m/m}$ queueing system.

This QS was originally used by A. K. Erlang to investigate the distribution of busy channels in telephony systems. A telephone exchange with m trunks can handle up to m incoming calls at once. If all the trunks are busy, the newly arriving calls are denied service. This may happen, for instance, in GSM cellular systems, where each GSM cell is capable of handling a limited number of simultaneous calls. Beyond this number, any further call request gets refused with a courtesy message like "We are sorry: the service is not available at the moment. Please, try again later." Therefore, customers that find some server available are immediately admitted into service, whereas those that find all the server stations occupied are dropped from the system. As usual, the service times are assumed to be independent exponential rvs with mean $1/\mu$. From the discussion of $\mathrm{M/M/m}$ and $\mathrm{M/M/1/K}$, we easily realize that the number $x(t)$ of customers in the system can be modeled as a BDP, with birth rates

$$\lambda_k = \begin{cases} \lambda & , \quad k = 0, 1, \ldots, \mathrm{m} - 1 \\ 0 & , \quad k = \mathrm{m} \end{cases} \tag{8.175}$$

and death-rates

$$\mu_k = k\mu \quad , \qquad k = 0, 1, \ldots, \mathrm{m} \, , \tag{8.176}$$

whose flow diagram is shown in Fig. 8.25.

The asymptotic behavior of $\mathrm{M/M/m/m}$ is basically a mixture of that of $\mathrm{M/M/m}$ and $\mathrm{M/M/1/}K$. In fact, the number of customers in the system evolves as in $\mathrm{M/M/m}$, as long as

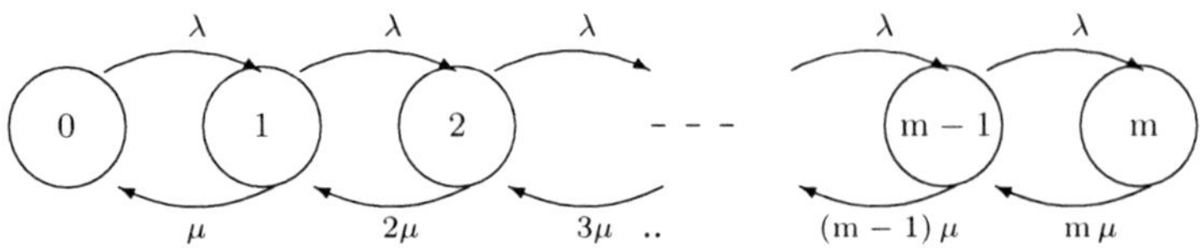

Figure 8.25 Flow diagram of $\mathrm{M/M/m/m}$.

the storage capacity $K = \mathrm{m}$ is not exhausted, whereas when all the servers are busy, the system drops newly arriving customers as in $\mathrm{M/M/1}/K$. Surprisingly, this mixed behavior does not yield to a more complex analysis. On the contrary, the asymptotic analysis of $\mathrm{M/M/m/m}$ is even simpler than that of $\mathrm{M/M/1}/K$ and $\mathrm{M/M/m}$.

The flow balance equations between adjacent states in the diagram of Fig. 8.25 have the form

$$\lambda p_x(k-1) = k\mu p_x(k) \quad , \qquad k = 1, \ldots, \mathrm{m}. \tag{8.177}$$

Unfolding each equation recursively back to $p_x(0)$ we obtain

$$p_x(k) = \left(\frac{\lambda}{\mu}\right)^k \frac{1}{k!} p_x(0) = \frac{\mathtt{G}^k}{k!} p_x(0) \,, \qquad k = 1, \ldots, \mathrm{m} \,, \tag{8.178}$$

where $\mathtt{G}$ is given by (8.27). The value of $p_x(0)$ is determined by the normalization condition,

$$1 = \sum_{k=0}^{\mathrm{m}} p_x(k) = p_x(0) \sum_{k=0}^{\mathrm{m}} \frac{\mathtt{G}^k}{k!} \,. \tag{8.179}$$

Therefore,

$$p_x(0) = \frac{1}{\displaystyle\sum_{r=0}^{\mathrm{m}} \frac{\mathtt{G}^r}{r!}} \,, \tag{8.180}$$

and from (8.178)

$$\boxed{p_x(k) = \frac{\mathtt{G}^k}{k! \displaystyle\sum_{r=0}^{\mathrm{m}} \frac{\mathtt{G}^r}{r!}} = \frac{(\mathrm{m}\rho)^k}{k! \displaystyle\sum_{r=0}^{\mathrm{m}} \frac{(\mathrm{m}\rho)^r}{r!}} \,, \qquad k = 0, 1, 2, \ldots, \mathrm{m} \,,} \tag{8.181}$$

where we used (8.32). Notice that (8.181) is a Poisson PMD with parameter $\mathtt{G} = \mathrm{m}\,\rho$, truncated at m. Hence, $x(t)$ does admit proper asymptotic PMD for *any values* of the offered traffic $\mathtt{G}$, that is to say, $\mathrm{M/M/m/m}$ is always stable (see Definition 8.1).

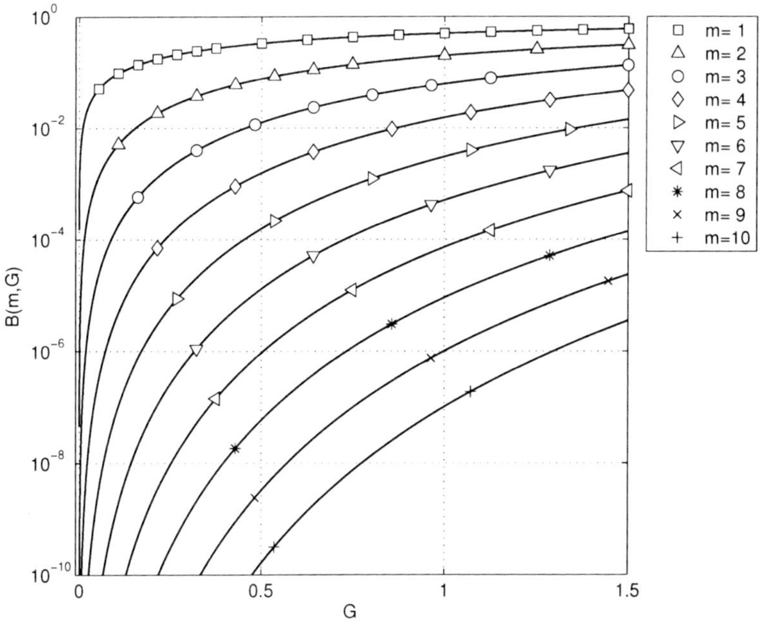

Figure 8.26 Call-blocking probability (Erlang-B formula) versus offered traffic $\mathtt{G} = \mathrm{m}\rho = \lambda/\mu$ in $\mathrm{M/M/m/m}$ for different values of the number m of servers.

The quantity of interest in $\mathrm{M/M/m/m}$ analysis is the call-blocking probability P_{blk}, which in this case is known as *Erlang-B formula* and indicated by $B(\mathrm{m}, \mathtt{G})$. Due to PASTA Theorem 8.3, $B(\mathrm{m}, \mathtt{G})$ corresponds to $p_x(k)$ for $k = \mathrm{m}$, hence from (8.181) we obtain

$$B(\mathrm{m}, \mathtt{G}) = \frac{\dfrac{\mathtt{G}^{\mathrm{m}}}{\mathrm{m}!}}{\displaystyle\sum_{r=0}^{\mathrm{m}} \frac{\mathtt{G}^r}{r!}} = \frac{\dfrac{(\mathrm{m}\rho)^{\mathrm{m}}}{\mathrm{m}!}}{\displaystyle\sum_{r=0}^{\mathrm{m}} \frac{(\mathrm{m}\rho)^r}{r!}} . \qquad (8.182)$$

Note that the Erlang-B formula (8.182) holds for any value of G.

Further insight The Erlang-B formula was first published by Erlang back in 1917 [4]. Successively, Pollaczek, Palm and others showed that (8.182) actually holds for any $\mathrm{M/G/m/m}$, i.e., for *any* service time distribution, provided that the input is Poisson with rate λ (see [14], page 170).

The call-blocking probability (8.182) is shown in Fig. 8.26 as a function of the offered traffic G, for different values of m. Fig. 8.27 compares the Erlang-B curves (dashed lines) with the Erlang-C curves (solid lines) given by (8.114). We notice that, G and m being the same, the call-blocking probability $B(\mathrm{m}, \mathtt{G})$ of $\mathrm{M/M/m/m}$ is lower than the queueing probability $C(\mathrm{m}, \mathtt{G})$ of $\mathrm{M/M/m}$. Therefore, under the same work load and the same service capacity, the

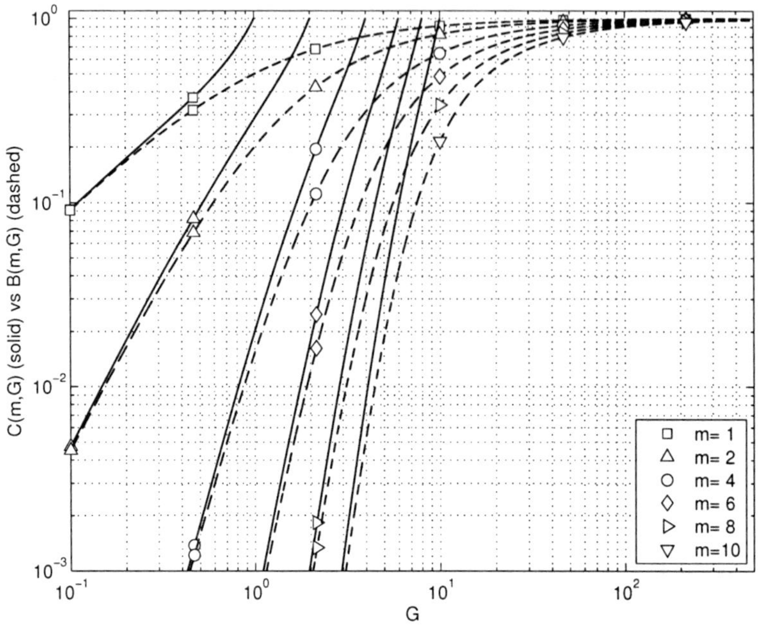

Figure 8.27 Call-blocking probability $B(\mathrm{m}, \mathtt{G})$ (dashed) and queueing probability $C(\mathrm{m}, \mathtt{G})$ (solid) versus offered traffic $\mathtt{G}$ for different values of the number m of servers.

probability that a customer finds a server available is higher for a blocking system than for a non-blocking one. Clearly, this performance gain is only apparent, since it is obtained at the cost of rejecting customers when the servers are all busy.

Occupancy measures. Since M/M/m/m does not have any queueing facility, the number of customers in queue is clearly zero and the number of servers in use z is equal to x. Therefore, the PMD of z is the same as given in (8.181). Moreover, we have

$$\mathtt{m}_x = \mathtt{m}_z = \sum_{k=1}^{\mathrm{m}} k p_x(k) = \mathtt{G} \sum_{k=0}^{\mathrm{m}-1} \frac{\mathtt{G}^k}{k!} p_x(0) \tag{8.183}$$

which yield

$$\mathtt{m}_x = \mathtt{m}_z = \mathtt{G}(1 - B(\mathrm{m}, \mathtt{G})) = \frac{\lambda}{\mu}\left(1 - \frac{1}{\sum_{k=0}^{\mathrm{m}} \frac{\lambda^k}{k!\mu^k}}\right). \tag{8.184}$$

Time measures. The system time of an accepted customer corresponds to its service time y, so that we have

$$p_s(a) = p_y(a) = \mu e^{-\mu a}, \quad a \geq 0, \tag{8.185}$$

and

$$\boxed{\mathtt{m}_s = \mathtt{m}_y = \frac{1}{\mu}\,.} \tag{8.186}$$

Note that, applying Little's law (8.46), we obtain the average rate λ_a at which customers are admitted into the system that is

$$\lambda_a = \frac{\mathtt{m}_x}{\mathtt{m}_s} = \lambda[1 - B(\mathrm{m}, \mathtt{G})] = \frac{\displaystyle\sum_{r=0}^{\mathrm{m}-1} \frac{\mathtt{G}^r}{r!}}{\displaystyle\sum_{r=0}^{\mathrm{m}} \frac{\mathtt{G}^r}{r!}}\,. \tag{8.187}$$

Traffic measures. As observed for M/M/1/K, the stability of M/M/m/m for any value of the offered traffic $\mathtt{G}$ is payed back in terms of throughput. In fact, according to (8.37), the useful traffic of M/M/m/m is given by

$$\mathtt{S} = \mathtt{G}\left(1 - \frac{\dfrac{\mathtt{G}^{\mathrm{m}}}{\mathrm{m}!}}{\displaystyle\sum_{r=0}^{\mathrm{m}} \frac{\mathtt{G}^r}{r!}}\right),$$

which is strictly less than $\mathtt{G}$.

8.5.5 The $\mathrm{M/M/m}/K$ queuing system

The $\mathrm{M/M/m}/K$ model, with $K \geq \mathrm{m}$, refers to a multi-server QS with limited buffering capacity $Q = K - \mathrm{m} \geq 0$ and is represented in Fig. 8.28.

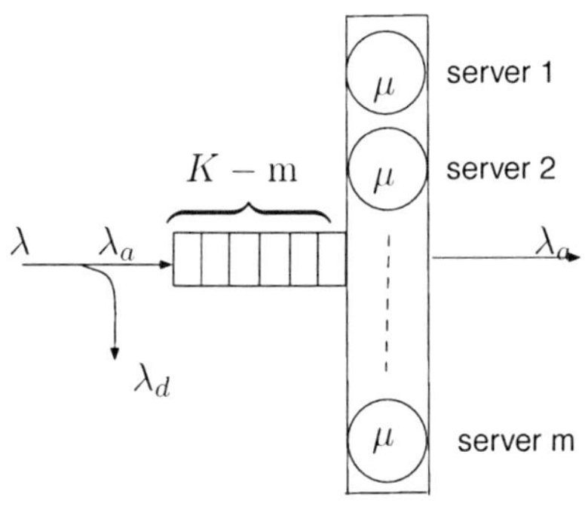

Figure 8.28 The M/M/m/K queueing system.

This QS generalizes all the other Markovian QSs seen so far that can be obtained from $\mathrm{M/M/m}K$ by appropriately setting the values of m and K. Similarly to M/M/m/m, it is possible to show that the number of customers in the system is a BDP with birth and death rates

$$\lambda_k = \begin{cases} \lambda & , \ k = 0, 1, \ldots, K-1 \\ 0 & , \ k = K \end{cases} \qquad \mu_k = \begin{cases} k\mu & , \ k = 1, 2, \ldots, \mathrm{m} \\ \mathrm{m}\mu & , \ k = \mathrm{m}+1, \ldots, K \end{cases} \tag{8.188}$$

and whose flow diagram is shown in Fig. 8.29.

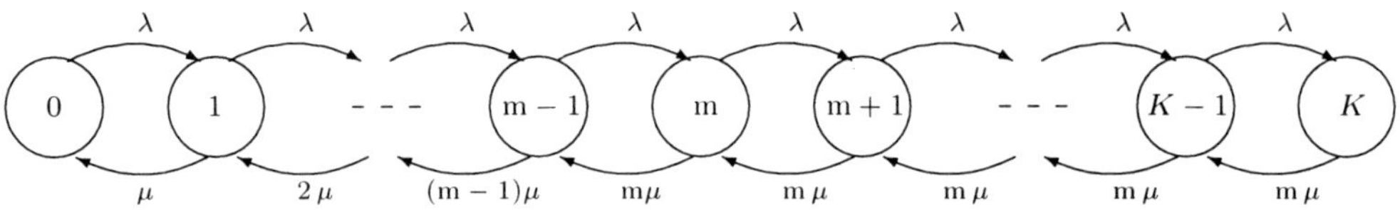

Figure 8.29 Flow-diagram of M/M/m/K.

By solving the system of flow balance equations we obtain the following asymptotic PMD

$$p_x(k) = \begin{cases} \dfrac{\mathtt{G}^k}{k!} p_x(0) & , \ k = 0, 1, \ldots, \mathrm{m}\,, \\ \dfrac{\mathtt{G}^{\mathrm{m}}}{\mathrm{m}!}\left(\dfrac{\mathtt{G}}{\mathrm{m}}\right)^{k-\mathrm{m}} p_x(0) & , \ k = \mathrm{m}+1, \ldots, K\,, \end{cases} \tag{8.189}$$

with

$$p_x(0) = \left[\sum_{r=0}^{\mathrm{m}} \frac{\mathtt{G}^r}{r!} + \frac{\mathtt{G}^{\mathrm{m}}}{\mathrm{m}!} \frac{\mathtt{G} - (\mathtt{G}/\mathrm{m})^{K-\mathrm{m}+1}}{\mathrm{m} - \mathtt{G}}\right]^{-1}. \tag{8.190}$$

From (8.190) we can immediately obtain the queueing and the call-blocking probabilities, denoted by P_{que} and P_{blk}, respectively. The *queueing probability* is given by

$$P_{\text{que}} = \mathrm{P}\left[\mathrm{m} \le x < K\right] = \sum_{k=\mathrm{m}}^{K-1} p_x(k)$$

that, using (8.190), yields

$$P_{\text{que}} = \frac{\left(1 - (\mathtt{G}/\mathrm{m})^{K-\mathrm{m}}\right)\dfrac{\mathtt{G}^{\mathrm{m}}}{\mathrm{m}!}}{\dfrac{\mathrm{m} - \mathtt{G}}{\mathrm{m}} \displaystyle\sum_{r=0}^{\mathrm{m}-1} \frac{\mathtt{G}^r}{r!} + \left(1 - (\mathtt{G}/\mathrm{m})^{K-\mathrm{m}+1}\right)\frac{\mathtt{G}^{\mathrm{m}}}{\mathrm{m}!}}. \tag{8.191}$$

Letting $K \to \infty$ in (8.191) we obtain the queueing probability of $\mathrm{M/M/m}$, as given by the Erlang-C formula (8.114). The *call-blocking* probability $P_{\rm blk} = p_x(K)$, instead, is equal to

$$P_{\rm blk} = \frac{\mathtt{G}^K}{\mathrm{m}^{K-\mathrm{m}}\mathrm{m}! \displaystyle\sum_{r=0}^{\mathrm{m}-1} \frac{\mathtt{G}^r}{r!} + \mathtt{G}^{\mathrm{m}} \frac{\mathrm{m}^{K+1-\mathrm{m}} - \mathtt{G}^{K+1-\mathrm{m}}}{\mathrm{m} - \mathtt{G}}} \tag{8.192}$$

that, using (8.32), can also be expressed as

$$P_{\rm blk} = \frac{\rho^K}{\displaystyle\frac{\mathrm{m}!}{\mathrm{m}^{\mathrm{m}}} \sum_{r=0}^{\mathrm{m}-1} \frac{(\mathrm{m}\rho)^r}{r!} + \frac{\rho^{\mathrm{m}} - \rho^{K+1}}{1-\rho}}. \tag{8.193}$$

Setting $K = \mathrm{m}$ in (8.192) we obtain the Erlang-B formula (8.182), whereas evaluating (8.193) for $\mathrm{m} = 1$ it returns the $P_{\rm blk}$ of $\mathrm{M/M/1}/K$, given by (8.146). Fig. 8.30 shows the call-blocking probability given by (8.192) as a function of the offered traffic $\mathtt{G}$, for different values of m. The solid curves refer to $\mathrm{M/M/m/m}$, with no buffer, whereas the dashed and dotted lines have been obtained for systems with buffer size Q equal to 1 and 25, respectively. It is clear that when $\mathtt{G}$ is light, $P_{\rm blk}$ decreases substantially with an increase in m. However, the value of m does not change significantly the situation when $\mathtt{G}$ is large. Interestingly, increasing Q by even a single unit yields a reduction of $P_{\rm blk}$, provided that the $\mathtt{G}$ is not very large.

Occupancy measures. The PMDs of the number of customers in the waiting queue and in service are not difficult to obtain, though the expressions are long and not much revealing. Therefore, we here limit our attention to the mean values of these quantities. More precisely, we derive only expressions of the mean number of customers in service, $\mathtt{m}_z$, and in the waiting queue, $\mathtt{m}_q$. The expression of $\mathtt{m}_x$, being simply equal to the sum of the other these quantities, is no reported.

Asymptotically, the mean number of busy servers is given by

$$\mathtt{m}_z = \sum_{r=1}^{\mathrm{m}} r p_x(r) + \mathrm{m} \sum_{r=\mathrm{m}+1}^{K} p_x(r) \tag{8.194}$$

that, applying (8.189), yields

$$\mathtt{m}_z = \frac{\mathtt{G}\left(\displaystyle\sum_{\ell=0}^{\mathrm{m}-1} \frac{\mathtt{G}^\ell}{\ell!} + \frac{\mathtt{G}^{\mathrm{m}}}{\mathrm{m}!} \frac{1-\rho^{K-\mathrm{m}}}{1-\rho}\right)}{\displaystyle\sum_{\ell=0}^{\mathrm{m}-1} \frac{\mathtt{G}^\ell}{\ell!} + \frac{\mathtt{G}^{\mathrm{m}}}{\mathrm{m}!} \frac{1-\rho^{K-\mathrm{m}+1}}{1-\rho}}. \tag{8.195}$$

Notice that, when $\rho < 1$, letting $K \to \infty$ in (8.195) the powers with exponent K tend to zero, so that numerator and denominator of the fraction simplify out and $\mathtt{m}_z \to \mathtt{G} = \frac{\lambda}{\mu}$, which is the result (8.123) found for $\mathrm{M/M/m}$.

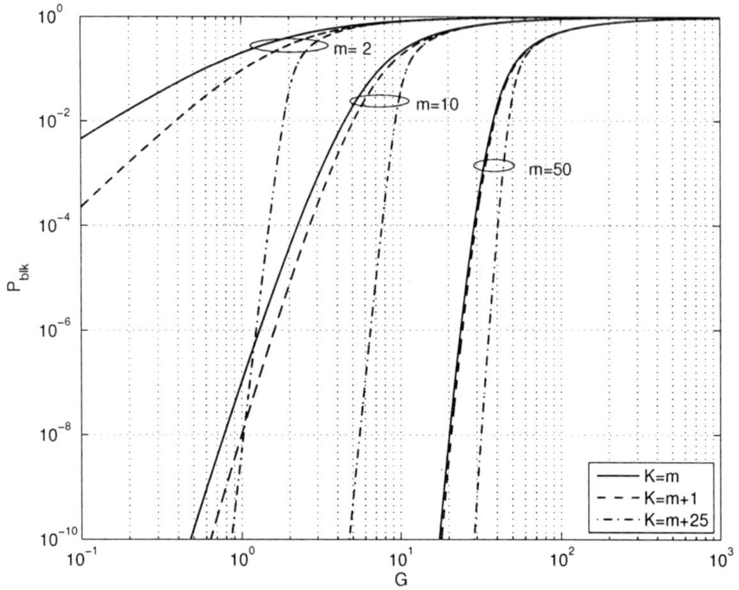

Figure 8.30 Call-blocking probability versus offered traffic $\mathtt{G} = \mathrm{m}\rho = \lambda/\mu$ in M/M/m/K, for different values of number m of servers and with $K = \mathrm{m}$ (solid), $K = \mathrm{m}+1$ (dashed), and $K = \mathrm{m}+25$ (dash-dotted).

The mean number of customers in the waiting queue is

$$\mathtt{m}_q = \sum_{r=1}^{K-\mathrm{m}} r p_x(\mathrm{m}+r) = \sum_{r=1}^{K-\mathrm{m}} r\rho^r \frac{\mathtt{G}^{\mathrm{m}}}{\mathrm{m}!} p_x(0) = \rho \frac{\mathtt{G}^{\mathrm{m}}}{\mathrm{m}!} p_x(0) \sum_{r=1}^{K-\mathrm{m}} r\rho^{r-1} . \tag{8.196}$$

Solving the finite sum on the right-hand side of (8.196) as in (8.154), we thus obtain

$$\boxed{\mathtt{m}_q = \frac{\rho \mathtt{G}^{\mathrm{m}} \left(1 - (K - \mathrm{m} + 1)\rho^K + K\rho^{K-\mathrm{m}+1}\right)}{\mathrm{m}!(1-\rho^2) \sum_{r=0}^{\mathrm{m}-1} \frac{\mathtt{G}^r}{r!} + (1+\rho)\mathtt{G}^{\mathrm{m}}\left(1 - \rho^{K-\mathrm{m}+1}\right)} .} \tag{8.197}$$

Time measures. The asymptotic probability distributions of the system and waiting time in queue in $\mathrm{M/M/m}K$ can be derived by following the same rationale presented in the previous section. We here report only the PDF of w, which is generally the most interesting measure when dealing with multi-server QSs.

First of all we recall that, in blocking systems, the analysis of the time measures has to be limited to the customers that are accepted into the QS, since it does not make sense to consider system time or waiting time in queue for customers that are blocked. Therefore, under the condition that an arriving customer is accepted into the system, the PMD of the number of customers x_a that it finds already in the system is given by

$$\mathrm{P}[x_a = r] = \mathrm{P}[x = r | x < K] = \frac{p_x(r)}{1 - P_{\mathrm{blk}}}, \qquad r = 0, 1, \ldots, K-1, \tag{8.198}$$

where $p_x(a)$ and P_{blk} are given in (8.189) and (8.192), respectively. Conditioning on $x_a = r$, the queueing time w is zero if $r < \mathrm{m}$, whereas it has an Erlang distribution of index $r + 1 - \mathrm{m}$ and parameter $\mathrm{m}\mu$ if $\mathrm{m} \le r \le K - 1$. The PDF of w is hence equal to

$$\begin{aligned} p_w(a) &= \delta(a)\,\mathrm{P}[x_a < m] + \sum_{\ell=0}^{K-\mathrm{m}-1} \frac{\mathrm{m}\mu(\mathrm{m}\mu a)^\ell}{\ell!} e^{-\mathrm{m}\mu a} p_{x_a}(\mathrm{m}+\ell) \\ &= \delta(a)\frac{1 - P_{\mathrm{que}}}{1 - P_{\mathrm{blk}}} + \sum_{\ell=0}^{K-\mathrm{m}-1} \frac{\mathrm{m}\mu(\mathrm{m}\mu a)^\ell}{\ell!} e^{-\mathrm{m}\mu a} \rho^\ell \frac{p_x(\mathrm{m})}{1 - P_{\mathrm{blk}}} \\ &= \delta(a)\frac{1 - P_{\mathrm{que}}}{1 - P_{\mathrm{blk}}} + \mathrm{m}\mu e^{-\mathrm{m}\mu a} \frac{\mathrm{G}^{\mathrm{m}}}{\mathrm{m}!} \frac{p_x(0)}{1 - P_{\mathrm{blk}}} \sum_{\ell=0}^{K-\mathrm{m}-1} \frac{(\lambda a)^\ell}{\ell!}. \end{aligned} \tag{8.199}$$

Replacing in (8.199) the expressions for P_{blk}, P_{que} and $p_x(0)$ given by (8.190), (8.191) and (8.192), respectively, we would obtain the final expression of $p_w(a)$, that, however, cannot be written in a much more compact form, so that we prefer to stop the derivation here.

The mean waiting time can be obtained by using Little's law, paying attention to consider the rate λ_a (8.17) at which customers are *admitted* into the system.

8.6 THE $\mathrm{M/G/1}$ QUEUING SYSTEM

In this section we extend our analysis to $\mathrm{M/G/1}$, which generalizes $\mathrm{M/M/1}$ by dropping the assumption of Markovian service times. Therefore, $\mathrm{M/G/1}$ applies to any single-server system with Poisson arrivals of rate λ, infinite buffer capacity, FCFS service discipline and iid service times with arbitrary distribution.

Since the service times are not necessarily memoryless, the residual service time of the customer in service at time t (if any) depends on the time it has already spent in service (*service age*). Therefore, the number of customers in the QS at time t, $x(t)$, is not longer sufficient to statistically characterize the evolution of the system from time t. In other words, the process $x(t)$ is not, in general, a continuous-time Markov chain.

Nonetheless, the asymptotic behavior of the system can be still analyzed in a rather simple manner, provided that we observe the system at particular instants, in such a way that the *sampled process* exhibits the Markov property. This method of analysis is known as the *embedded Markov chain* (see Sec. 7.3.7) and it is mainly due to Kendall [6] and Palm.

The most convenient choice of the sampling epochs turns out to be the set of departure instants $\{d_n\}$. Thus, let $X_n = x(d_n^+)$ denote the number of customers left in the system by the nth customer, C_n, when it leaves the queue after the completion of its service. Intuitively, looking at the system content immediately after the departure of a customer, the "residual" service time for the customer that is going to enter in service next is actually equal to its "whole" service time, that is to say, the service age is zero. Therefore, X_n is indeed sufficient to statistically characterize the number of customers in the system at the next sampling instant, i.e., X_{n+1}, and can hence be considered as the system state in these particular time instants.

Fortunately, the statistical characterization of the QS at the departure instants happens to be valid for *any* observation instant. In fact, asymptotically, the number X_n of customers in the system when a departure occurs is distributed as $x(\tau_m^-)$, i.e., the number of customers observed in the system by an arriving customer [3]. The PASTA Theorem 8.3, in turn, states that Poisson arrivals see the asymptotic distribution of the system state, i.e., that $x(\tau_m^-)$ is distributed as $x(t)$ for $t \to \infty$. Hence, we conclude that the distribution of X_n for $n \to \infty$ actually corresponds to the asymptotic distribution of $x(t)$, which is the distribution of the system content seen by a random observer.

We then proceed with the analysis of the process X_n, whose evolution in one step can be expressed as follows

$$\boxed{X_{n+1} = X_n - \chi\{X_n > 0\} + v_{n+1}\,,} \tag{8.200}$$

where v_{n+1} denotes the number of customer arrivals in the service time y_{n+1} of the $n+1$th customer C_{n+1}, whereas $\chi\{\cdot\}$ is the indicator function (2.163). Fig. 8.31 exemplifies (8.200) in two different cases, namely after the departure of C_n the system is empty, $X_n = 0$, or contains some customers, $X_n > 0$.

Recalling the properties of Poisson arrivals (see page 100), it is easy to realize that rvs $\{v_i\}$ are also iid, whose PMD $p_v(k)$ can be determined through the Total Probability Theorem 2.12. Moreover, upon conditioning on the service time, $y = a$, the number of arrivals v is distributed as a Poisson rv with parameter λa, so that, as seen in Example 2.6 F, we have

$$\mathrm{P}[v = k | y = a] = \frac{(\lambda a)^k}{k!} e^{-\lambda a}\,, \qquad k = 0, 1, \ldots. \tag{8.201}$$

whereas the conditional mean and statistical power are

$$\mathrm{E}[v|y = a] = \lambda a\,, \qquad \mathrm{E}\left[v^2|y = a\right] = \lambda a + \lambda^2 a^2\,, \tag{8.202}$$

respectively. Averaging (8.201) over the PDF of y we hence obtain

$$p_v(k) = \mathrm{P}[v = k] = \int_0^\infty \frac{(\lambda a)^k}{k!} e^{-\lambda a} p_y(a) da\,, \qquad k = 0, 1, \ldots. \tag{8.203}$$

Moreover, from (8.202) we get the mean

$$\mathtt{m}_v = \int_0^\infty \mathrm{E}[v|y = a]\, p_y(a) da = \int_0^\infty \lambda a p_y(a) da = \lambda \mathtt{m}_y \tag{8.204}$$

and statistical power

$$\mathtt{M}_v = \int_0^\infty \mathrm{E}\left[v^2|y = a\right] p_y(a) da = \int_0^\infty (\lambda a + \lambda^2 a^2) p_y(a) da = \lambda \mathtt{m}_y + \lambda^2 \mathtt{M}_y\,. \tag{8.205}$$

Clearly, the variance of v is given by

$$\sigma_v^2 = \mathtt{M}_v - \mathtt{m}_v^2 = \lambda \mathtt{m}_y + \lambda^2 \mathtt{M}_y - \lambda^2 {\mathtt{m}_y}^2 = \lambda \mathtt{m}_y + \lambda^2 \sigma_y^2\,. \tag{8.206}$$

Note that mean and variance of y are sufficient to determine mean and variance of v.

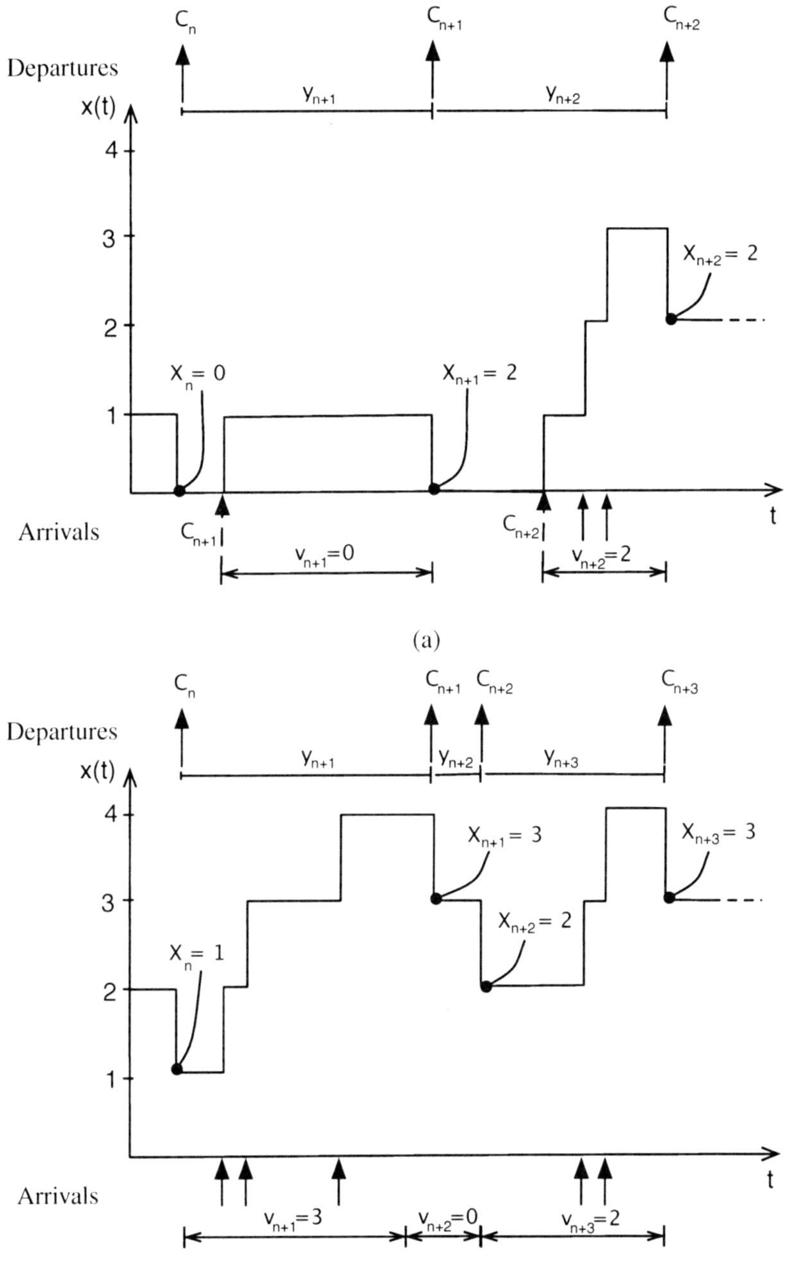

Figure 8.31 Example of state update of M/G/1 when (a) $X_n = 0$ and (b) $X_n > 0$.

Inspecting (8.200), it is easy to realize that, given the value of X_n, v_{n+1} becomes statistically independent of the past history of the process. Therefore, the state at the next observation instant, X_{n+1}, can be uniquely evaluated, in a statistical sense, from the current state X_n of the system, regardless of the past history of the process. In other words, the process X_n is an *embedded MC* (see Section 7.3.7). Then, the transition matrix $\boldsymbol{P}$ of the embedded MC can be directly obtained from (8.200). Starting from state $X_n = 0$, at the next instant the state will be $X_{n+1} = j \geq 0$ with probability $p_v(j)$, whereas from state $X_n = i > 0$, the next transition can either be into state $i - 1$, with probability $p_v(0)$, or into state $j \geq i$ with probability $p_v(j - i + 1)$. Using notation (7.5), we hence have

$$P_{i,j} = \begin{cases} p_v(j) & , \ i = 0, j \geq 0 \\ p_v(j - i + 1) & , \ i > 0, j \geq i - 1 \end{cases} \tag{8.207}$$

so that the transition matrix is of the form

$$\boldsymbol{P} = \begin{bmatrix} p_v(0) & p_v(1) & p_v(2) & p_v(3) & p_v(4) & p_v(5) & \cdots \\ p_v(0) & p_v(1) & p_v(2) & p_v(3) & p_v(4) & p_v(5) & \cdots \\ 0 & p_v(0) & p_v(1) & p_v(2) & p_v(3) & p_v(4) & \cdots \\ 0 & 0 & p_v(0) & p_v(1) & p_v(2) & p_v(3) & \cdots \\ 0 & 0 & 0 & p_v(0) & p_v(1) & p_v(2) & \cdots \\ \vdots & \vdots & \vdots & \vdots & \vdots & \vdots & \ddots \end{bmatrix} . \tag{8.208}$$

According to Theorem 7.9, a MC does admit a proper asymptotic distribution only if it is aperiodic, irreducible and positive recurrent. Recalling Definition 7.4 at page 433, we see that a sufficient condition for the MC with probability matrix (8.208) to be irreducible is that $p_v(1) > 0$ and $p_v(2) > 0$. In this case, in fact, there exists a transition from each state of the MC to the adjacent states, so that all states form a single communication class and the MC is irreducible. Furthermore, as observed at page 440, if $p_v(0) > 0$ then MC is also aperiodic, since it admits self-transitions. From (8.203), these conditions are always satisfied when $p_y(a)$ is not impulsive in zero, i.e., the event $y = 0$ has zero probability.

Hence, assuming that the MC is irreducible and aperiodic, according to Theorem 7.8 at page 442, it will admit a single asymptotic state probabilities vector $\boldsymbol{\pi}$ that will either have all zero elements, if the MC is transient or null-recurrent, or all positive elements, if the MC is positive recurrent. Therefore, the conditions that make the MC positive recurrent can be determined by requiring that the asymptotic probability of *any* of its state and, in particular, of state zero, is positive, that is to say, imposing $\pi_0 > 0$. Now, in single-server systems G/G/1, from (8.51) we can write

$$\pi_0 = \mathrm{P}[X = 0] = 1 - \rho . \tag{8.209}$$

Hence, the condition $\pi_0 > 0$ for the process to be positive recurrent is equivalent to $\rho < 1$, which is the stability condition for any non-blocking QS. Another interpretation of the stability condition can be obtained by using (8.30) in (8.209), which yields

$$\pi_0 = 1 - \lambda \mathtt{m}_y , \tag{8.210}$$

according to which M/G/1 can reach an asymptotic behavior only if $\lambda \mathtt{m}_y < 1$, i.e., the average number of arrivals in a generic service time is less than one. The case $\lambda \mathtt{m}_y = 1$ is borderline and leads to a recurrent *null* MC, where all states have zero asymptotic probability.

Now, provided that it exists, the asymptotic PMD of the number of customers in the system can be determined by using the methods discussed in Section 7.2.8. This approach is thoroughly investigated, for instance, in [3, 13]. In the sequel, instead, we content ourselves with the analysis of mean and variance of the system quantities that can be performed in a very simple manner by using (8.200).

Occupancy Measures. To determine the asymptotic mean of the number of customers in the system, $\mathtt{m}_x$, we square both sides of (8.200), obtaining

$$\begin{aligned} X_{n+1}^2 &= (X_n - \chi\{X_n > 0\} + v_{n+1})^2 \\ &= X_n^2 + \chi\{X_n > 0\}^2 + v_{n+1}^2 - 2X_n\chi\{X_n > 0\} - 2v_{n+1}\chi\{X_n > 0\} + 2X_n v_{n+1} \\ &= X_n^2 + \chi\{X_n > 0\} + v_{n+1}^2 - 2X_n - 2v_{n+1}\chi\{X_n > 0\} + 2X_n v_{n+1}\,, \end{aligned} \tag{8.211}$$

where the last step follows from

$$\chi\{X_n > 0\}^2 = \chi\{X_n > 0\}\,, \qquad \text{and} \qquad X_n\chi\{X_n > 0\} = X_n\,.$$

Taking the expectation of both sides of (8.211), we have

$$\mathrm{E}\left[X_{n+1}^2\right] = \mathrm{E}\left[X_n^2\right] + \mathrm{P}\left[X_n > 0\right] + \mathtt{M}_v - 2\,\mathrm{E}\left[X_n\right] - 2\mathtt{m}_v\,\mathrm{P}\left[X_n > 0\right] + 2\,\mathrm{E}\left[X_n\right]\mathtt{m}_v\,. \tag{8.212}$$

Then, letting n approach infinity and leveraging on the process stationarity (due to ergodicity), we get

$$\mathtt{M}_x = \mathtt{M}_x + (1-\pi_0) + \mathtt{M}_v - 2\mathtt{m}_x - 2\mathtt{m}_v(1-\pi_0) + 2\mathtt{m}_x\mathtt{m}_v\,, \tag{8.213}$$

from which we obtain

$$\mathtt{m}_x = \frac{(1-\pi_0)(1-2\mathtt{m}_v) + \mathtt{M}_v}{2(1-\mathtt{m}_v)} = \frac{(1-\pi_0)(1-2\mathtt{m}_v) + \mathtt{m}_v^2 + \sigma_v^2}{2(1-\mathtt{m}_v)}\,, \tag{8.214}$$

where we have used $\mathtt{M}_v = \mathtt{m}_v^2 + \sigma_v^2$. Now, according to (8.210) and (8.204), we have $1 - \pi_0 = \lambda\mathtt{m}_y = \mathtt{m}_v$ that, replaced in (8.214), gives

$$\mathtt{m}_x = \frac{\mathtt{m}_v}{2} + \frac{\sigma_v^2}{2(1-\mathtt{m}_v)}\,. \tag{8.215}$$

Finally, replacing $\mathtt{m}_v$ and σ_v^2 with the expressions provided in (8.204) and (8.206), respectively, we derive the so-called *Pollaczek-Khintchine (P-K) mean-value formula*

$$\boxed{\mathtt{m}_x = \frac{\lambda\mathtt{m}_y}{2} + \frac{\lambda\mathtt{m}_y + \lambda^2\sigma_y^2}{2(1-\lambda\mathtt{m}_y)}} \tag{8.216}$$

which clearly holds when $\lambda\mathtt{m}_y < 1$. Recalling (8.30), the P-K mean value formula can also be written as

$$\mathtt{m}_x = \frac{\rho}{2} + \frac{\rho + \lambda^2\sigma_y^2}{2(1-\rho)} = \rho + \frac{\lambda^2\mathtt{M}_y}{2(1-\rho)}\,. \tag{8.217}$$

Note that (8.216) provides an expression for the mean number of customers in the system at departure instants. However, as already discussed, this is also the mean number of customers in the system at any instant. We observe that $\mathtt{m}_x$ depends only upon the first two moments of the service time, namely $\mathtt{m}_y$ and σ_y^2. Furthermore, note that $\mathtt{m}_x$ grows linearly with σ_y^2. Therefore, under the same load factor ρ, the mean number of customers in a single-server QS increases with the "degree of randomness" of the service time. These considerations are reflected in the curves of Fig. 8.32 that show $\mathtt{m}_x$ as a function of the load factor ρ. The curves have been obtained by setting $\mathtt{m}_y = 1$ and varying λ, for three different choices of the service time distribution, namely deterministic ($\sigma_y^2 = 0$), exponential ($\sigma_y^2 = 1$) and Erlang-2 ($\sigma_y^2 = 2$). Note that, as expected, $\mathtt{m}_x$ rapidly increases as ρ approaches the critical stability-threshold $\rho = 1$. Furthermore, for the same ρ, $\mathtt{m}_x$ increases with σ_y^2. Hence, as mentioned at the beginning of the chapter, the randomness in some aspects of a QS generally translates into worse system performance.

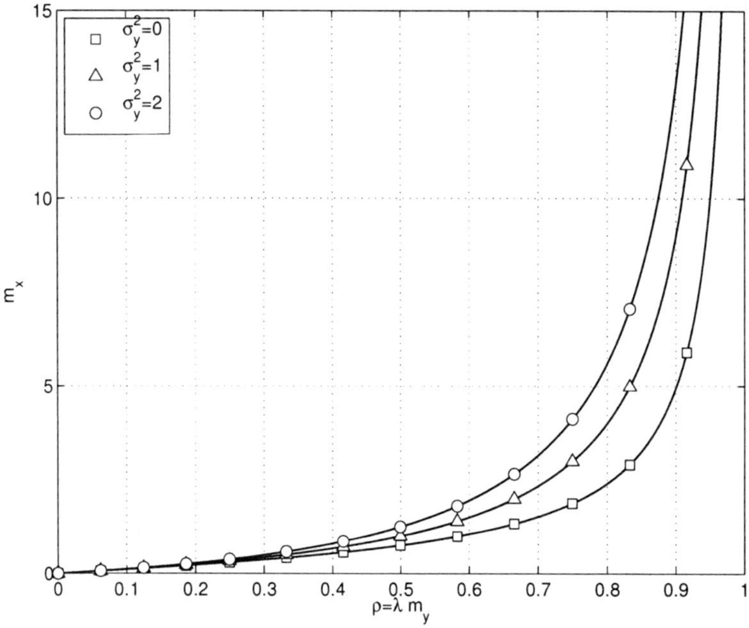

Figure 8.32 P-K mean value formula versus the load factor ρ in M/G/1, for $\mathtt{m}_y = 1$ and different values of σ_y^2.

The mean number of customers in service can be readily obtained, for instance, applying Little's law (8.48) that, provided the system, is stable, yields

$$\boxed{\mathtt{m}_z = \lambda \mathtt{m}_y = \rho\,.} \tag{8.218}$$

The last occupancy metric, $\mathtt{m}_q$, can be expressed as the difference between the mean number of customers in the system, $\mathtt{m}_x$, and those in service, $\mathtt{m}_z$, which gives

$$\boxed{\mathtt{m}_q = \frac{\lambda \mathtt{m}_y + \lambda^2 \sigma_y^2}{2(1-\lambda \mathtt{m}_y)} - \frac{\lambda \mathtt{m}_y}{2} = \frac{\lambda^2 \mathtt{M}_y}{2(1-\rho)}\,.} \tag{8.219}$$

Time measures. Asymptotically, the mean values of the system and queueing time can be directly obtained from the occupancy measures through Little's law. More specifically, applying (8.46) we have

$$\boxed{\mathtt{m}_s = \frac{\mathtt{m}_y}{2} + \frac{\mathtt{m}_y + \lambda\sigma_y^2}{2(1-\lambda\mathtt{m}_y)} = \mathtt{m}_y + \frac{\lambda\mathtt{M}_y}{2(1-\rho)}\,,} \tag{8.220}$$

whereas using (8.47) we get

$$\boxed{\mathtt{m}_w = \frac{\mathtt{m}_y + \lambda\sigma_y^2}{2(1-\lambda\mathtt{m}_y)} - \frac{\mathtt{m}_y}{2} = \frac{\lambda\mathtt{M}_y}{2(1-\rho)}\,.} \tag{8.221}$$

Example 8.6 A M/M/1 can be seen as a special case of M/G/1. Therefore, specifying the results obtained in the general case for a service time with PDF $p_y(a) = \mu e^{-\mu a}$, we shall find the same solutions derived in Section 8.5.1 for M/M/1.

We begin by deriving the distribution of the number of arrivals v in a generic service time y. Evaluating (8.203) for an exponential distribution, we get

$$p_v(k) = \int_0^\infty \frac{(\lambda a)^k}{k!} e^{-\lambda a} \mu e^{-\mu a}\, da = \lambda^k \mu \int_0^\infty \frac{a^k}{k!} e^{-(\lambda+\mu)a}\, da\,, \tag{8.222}$$

and multiplying and dividing of the right-hand side by $(\lambda+\mu)^{k+1}$ we get

$$p_v(k) = \frac{\lambda^k \mu}{(\lambda+\mu)^{k+1}} \int_0^\infty \frac{(\lambda+\mu)^{k+1} a^k}{k!} e^{-(\lambda+\mu)a}\, da\,. \tag{8.223}$$

Since the integrand function in (8.223) is the PDF of an Erlang distribution with parameter $k+1$, for the marginal rule its integral equals one so that (8.223) becomes

$$p_v(k) = \frac{\lambda^k}{(\lambda+\mu)^k} \frac{\mu}{\lambda+\mu} = \left(\frac{\lambda}{\lambda+\mu}\right)^k \left(1 - \frac{\lambda}{\lambda+\mu}\right) \tag{8.224}$$

which is a geometric PMD with parameter

$$q = 1 - \frac{\lambda}{\lambda+\mu} = \frac{\mu}{\lambda+\mu}\,. \tag{8.225}$$

Accordingly, mean and variance of v are equal to

$$\mathtt{m}_v = \frac{1-q}{q} = \frac{\lambda}{\mu} = \rho\,, \qquad \sigma_v^2 = \frac{1-q}{q^2} = \frac{\lambda(\lambda+\mu)}{\mu^2} = \rho^2 + \rho\,. \tag{8.226}$$

The transition matrix (8.208) has the form

$$\boldsymbol{P} = \begin{bmatrix} q & q(1-q) & q(1-q)^2 & q(1-q)^3 & q(1-q)^4 & q(1-q)^5 & \cdots \\ q & q(1-q) & q(1-q)^2 & q(1-q)^3 & q(1-q)^4 & q(1-q)^5 & \cdots \\ 0 & q & q(1-q) & q(1-q)^2 & q(1-q)^3 & q(1-q)^4 & \cdots \\ 0 & 0 & q & q(1-q) & q(1-q)^2 & q(1-q)^3 & \cdots \\ 0 & 0 & 0 & q & q(1-q) & q(1-q)^2 & \cdots \\ \vdots & \vdots & \vdots & \vdots & \vdots & \vdots & \ddots \end{bmatrix}\,. \tag{8.227}$$

From the probability-flow balance for each state, we obtain the following equations

$$\begin{cases} p_x(0)(1-q) &= p_x(1)q \\ p_x(1)(1-q(1-q)) &= p_x(0)q(1-q) + p_x(2)q \\ p_x(2)(1-q(1-q)) &= p_x(0)q(1-q)^2 + p_x(1)q(1-q)^2 + p_x(3)q \\ p_x(3)(1-q(1-q)) &= p_x(0)q(1-q)^3 + p_x(1)q(1-q)^3 + p_x(2)q(1-q)^2 + p_x(4)q \\ \vdots & \end{cases} \tag{8.228}$$

whose general form for $j \geq 1$ is

$$p_x(j)(1-q(1-q)) = p_x(0)q(1-q)^j + \sum_{i=1}^{j-1} p_x(i)q(1-q)^{j+1-i} + p_x(j+1)q\,. \tag{8.229}$$

The solution of this system of equations is given by

$$p_x(j) = \frac{1-q}{q}p_x(j-1) = \frac{(1-q)^j}{q^j}p_x(0)\,, \quad j \geq 1\,, \tag{8.230}$$

as can be easily verified by direct substitution or, more elegantly, by using an induction argument. To determine the unknown constant $p_x(0)$ we apply the normalization condition which gives

$$p_x(0) = 2 - \frac{1}{q} = 2 - \frac{\lambda+\mu}{\mu} = \frac{\mu-\lambda}{\mu} = 1 - \frac{\lambda}{\mu}\,. \tag{8.231}$$

The result holds true only if $q < 1/2$, that is to say $\lambda < \mu$, which corresponds to the stability condition of $\mathrm{M/M/1}$. Replacing (8.231) into (8.219) we finally have

$$p_x(j) = \left(\frac{\lambda}{\mu}\right)^j \left(1 - \frac{\lambda}{\mu}\right), \quad j \geq 0\,, \tag{8.232}$$

which is the geometric PMD with parameter $1-\rho$ that we found in Section 8.5.1.

We also verify that the P-K mean-value formula returns the mean of the geometric distribution (8.232). In fact, replacing (8.226) into (8.215) we obtain

$$\mathtt{m}_x = \frac{\lambda/\mu}{2} + \frac{\lambda/\mu + \lambda^2/\mu^2}{2(1-\lambda/\mu)} = \frac{\lambda}{2\mu} + \frac{\lambda(\mu+\lambda)}{2\mu(\mu-\lambda)} = \frac{\lambda(\mu-\lambda)+\lambda(\mu+\lambda)}{2\mu(\mu-\lambda)} = \frac{\lambda}{\mu-\lambda}\,. \tag{8.233}$$

8.7 THE $\mathrm{M/D/1}$ QUEUING SYSTEM

The $\mathrm{M/D/1}$ model is a particular case of $\mathrm{M/G/1}$, where the service times are constant, i.e., $y = T$. In this case, the statistics of the service time y become

$$p_y(a) = \delta(a-T)\,, \qquad \mathtt{m}_y = T\,, \qquad \sigma_y^2 = 0\,. \tag{8.234}$$

Since the arrival process is Poisson with rate λ, the number v of arrivals in a service time is a Poisson rv with parameter λT, whose PDF is

$$p_v(k) = \mathrm{P}\left[v = k | y = T\right] = \frac{(\lambda T)^k}{k!}e^{-\lambda T}\,, \qquad k = 0, 1, \ldots\,. \tag{8.235}$$

From Example 2.6 F, mean and variance of v are both equal to

$$\mathtt{m}_v = \sigma_v^2 = \lambda \mathtt{m}_y = \lambda T\,. \tag{8.236}$$

The transition matrix $\boldsymbol{P}$ turns out to be equal to

$$\boldsymbol{P} = \begin{bmatrix} e^{-\mu T} & \mu T e^{-\mu T} & \frac{(\mu T)^2}{2} e^{-\mu T} & \frac{(\mu T)^3}{3!} e^{-\mu T} & \frac{(\mu T)^4}{4!} e^{-\mu T} & \frac{(\mu T)^5}{5!} e^{-\mu T} & \cdots \\ e^{-\mu T} & \mu T e^{-\mu T} & \frac{(\mu T)^2}{2} e^{-\mu T} & \frac{(\mu T)^3}{3!} e^{-\mu T} & \frac{(\mu T)^4}{4!} e^{-\mu T} & \frac{(\mu T)^5}{5!} e^{-\mu T} & \cdots \\ 0 & e^{-\mu T} & \mu T e^{-\mu T} & \frac{(\mu T)^2}{2} e^{-\mu T} & \frac{(\mu T)^3}{3!} e^{-\mu T} & \frac{(\mu T)^4}{4!} e^{-\mu T} & \cdots \\ 0 & 0 & e^{-\mu T} & \mu T e^{-\mu T} & \frac{(\mu T)^2}{2} e^{-\mu T} & \frac{(\mu T)^3}{3!} e^{-\mu T} & \cdots \\ 0 & 0 & 0 & e^{-\mu T} & \mu T e^{-\mu T} & \frac{(\mu T)^2}{2} e^{-\mu T} & \cdots \\ \vdots & \vdots & \vdots & \vdots & \vdots & \vdots & \ddots \end{bmatrix}. \tag{8.237}$$

From (8.50), in non-blocking single-server systems we have $\mathtt{m}_z = 1 - p_x(0)$ and, from Little's law (8.48), $\mathtt{m}_z = \lambda \mathtt{m}_y$. Therefore, the asymptotic probability of state zero (empty system) is given by

$$p_x(0) = 1 - \mathtt{m}_z = 1 - \lambda T = 1 - \rho\,, \tag{8.238}$$

provided that $\rho < 1$. The asymptotic probability of the other states can be obtained by solving the system of flow balance equations. The derivation, however, is rather complex and cumbersome, so we report only the final result

$$\boxed{p_x(k) = (1-\rho)\left(e^{k\rho} + \sum_{r=1}^{k-1} (-1)^{k-r} e^{r\rho} \frac{(r\rho)^{k-r-1}}{(k-r)!}[k - r(1-\rho)]\right), \quad k = 0, 1 \ldots.} \tag{8.239}$$

The mean number of customers in the system is given by the P-K mean-value formula (8.216) that, in this case, returns

$$\mathtt{m}_x = \frac{\lambda T}{2} + \frac{\lambda T}{2(1-\lambda T)} = \frac{\lambda T(2 - \lambda T)}{2(1-\lambda T)} \tag{8.240}$$

or, using (8.31),

$$\boxed{\mathtt{m}_x = \frac{\rho(2-\rho)}{2(1-\rho)}\,.} \tag{8.241}$$

The mean number of customers in service is

$$\boxed{\mathtt{m}_z = \rho = \lambda T} \tag{8.242}$$

The mean number of queued customers $\mathtt{m}_q = \mathtt{m}_x - \mathtt{m}_z$ is finally equal to

$$\boxed{\mathtt{m}_q = \frac{\rho(2-\rho)}{2(1-\rho)} - \rho = \frac{\rho^2}{2(1-\rho)}\,.} \tag{8.243}$$

Finally, once again, applying Little's law (8.46) and (8.47) we can obtain the time measures by dividing the corresponding occupancy measures by the arrival rate λ, thus giving

$$\boxed{\mathtt{m}_s = \frac{T(2-\rho)}{2(1-\rho)},} \tag{8.244}$$

and

$$\boxed{\mathtt{m}_w = \frac{T\rho}{2(1-\rho)}.} \tag{8.245}$$

REFERENCES

1. G. Bolch, S. Greiner, H. de Meer, K. S. Trivedi, *Queueing Networks and Markov Chains: Modeling and Performance Evaluation with Computer Science Applications.* Wiley-Interscience; 2nd edition, April 14, 2006.
2. P. J. Burke, *The Output of a Queuing System*, Operations Research, Vol. 4, No. 6, pp. 699-704, December, 1956.
3. J. W. Cohen, *The Single Server Queue.* Wiley Interscience, New York, 1969.
4. A. K. Erlang, *Solution of some Problems in the Theory of Probabilities of Significance in Automatic Telephone Exchanges.* Elektrotkeknikeren, vol 13, 1917.
5. W. Feller, *An Introduction to Probability Theory and Its Applications - Volume I & II.* John Wiley, New York, 2nd edition, 1968.
6. D. G. Kendall, *Some problems in the Theory of Queues.* Journal of the Royal Statistical Society, Ser. B, 13, pp. 151–185, 1951.
7. D. G. Kendall, *Stochastic processes occurring in the Theory of Queues and their analysis by the method of the imbedded Markov Chain.* Annals of Mathematical Statistics, 24, pp. 338–354, 1953.
8. L. Kleinrock, *Queueing systems. Vol I - Theory.* John Wiley & Sons, New York, 1975.
9. L. Kleinrock, *Queueing systems. Vol II - Computer Applications.* John Wiley & Sons, New York, 1976.
10. J. D. C. Little, *A Proof of the Queueing Formula* $L = \lambda W$. Operations Research, 9, pp. 383-387, 1961.
11. A. A. Markov, *Extension of the limit theorems of probability theory to a sum of variables connected in a chain.* Reprinted in Appendix B of: R. Howard. *Dynamic Probabilistic Systems,* volume 1: Markov Chains. John Wiley and Sons, 1971.
12. P. V. Mieghem, *Performance analysis of communications networks and systems.* Cambridge University Press, United Kingdom, 2006.
13. A. Papoulis, *Probability, random variables and stochastic processes.* McGraw-Hill Kogakusha, Tokyo, 9th edition, 1965.
14. Ross, S. M., *Stochastic processes.* Wiley, New York, 1983.
15. H. M. Taylor, S. Karlin, *An introduction to stochastic modeling.* Academic Press, San Diego, California. 3rd edition, 1998.
16. N. Tian, Z. G. Zhang *Vacation Queueing Models: Theory and Applications.* Springer; 1st edition, August 23, 2006.
17. R. W. Wolff, *Poisson arrivals see time averages.* Operations Research n. 2, vol. 30, pp. 223-231, Mar.-Apr., 1982.

Table 8.2 Main results for non-blocking QSs.

Non-blocking QSs	performance measures	reference	page
M/M/1	stability: $\rho = \lambda/\mu < 1$		525
	$p_x(k) = \rho^k(1-\rho),\ k = 0, 1, \dots$	(8.72)	525
	$p_z(0) = 1 - \rho, \quad p_z(1) = \rho$	(8.75)	526
	$p_q(j) = \begin{cases} 1-\rho^2 & , j = 0 \\ \rho^{j+1}(1-\rho) & , j = 1, 2, \dots \end{cases}$	(8.79)	527
	$p_s(a) = (\mu - \lambda)e^{-(\mu-\lambda)a}1_0(a)$	(8.87)	528
	$p_w(a) = (1-\rho)\delta(a) + \rho\mu(1-\rho)e^{-\mu(1-\rho)a}1_0(a)$	(8.93)	530
	$\mathtt{m}_x = \frac{\rho}{1-\rho}$	(8.73)	525
	$\mathtt{m}_z = \rho$	(8.76)	526
	$\mathtt{m}_q = \frac{\rho^2}{1-\rho}$	(8.81)	527
	$\mathtt{m}_s = \frac{1}{\mu-\lambda}$	(8.89)	528
	$\mathtt{m}_w = \frac{\lambda}{\mu(\mu-\lambda)}$	(8.97)	531
M/M/m	stability: $\mathtt{G} = \lambda/\mu < \mathrm{m}$		534
	$C(\mathrm{m}, \mathtt{G}) = \left(\frac{\mathtt{G}^{\mathrm{m}}}{\mathrm{m}!}\right) \Big/ \left(\frac{\mathrm{m}-\mathtt{G}}{\mathrm{m}} \sum_{r=0}^{\mathrm{m}-1} \frac{\mathtt{G}^r}{r!} + \frac{\mathtt{G}^{\mathrm{m}}}{\mathrm{m}!}\right)$	(8.114)	535
	$p_x(k) = \frac{\frac{\mathtt{G}^k}{k_s!\,\mathrm{m}^{k-k_s}}}{\sum_{r=0}^{\mathrm{m}-1} \frac{\mathtt{G}^r}{r!} + \frac{\mathtt{G}^{\mathrm{m}}}{\mathrm{m}!}\frac{\mathrm{m}}{\mathrm{m}-\mathtt{G}}},\ k_s = \min(k, \mathrm{m}),\ k = 0, 1, \dots$	(8.112)	535
	$p_z(k) = \begin{cases} \frac{\mathtt{G}^k}{k!}p_x(0) & , 0 \le k < \mathrm{m} \\ C(\mathrm{m}, \mathtt{G}) & , k = \mathrm{m} \end{cases}$	(8.120)	537
	$p_q(k)\left(\frac{\mathtt{G}}{\mathrm{m}}\right)^k\left(1 - \frac{\mathtt{G}}{\mathrm{m}}\right)C(\mathrm{m}, \mathtt{G}), \qquad k = 1, 2, \dots$	(8.126)	538
	$p_s(a) = \mu e^{-\mu a}\left(1 + C(\mathrm{m}, \mathtt{G})\frac{1-(\mathrm{m}-\mathtt{G})e^{-\frac{\mathrm{m}-\mathtt{G}-1}{\mathrm{m}}a}}{\mathrm{m}-\mathtt{G}-1}\right)1_0(a)$	(8.139)	540
	$p_w(a) = [1 - C(\mathrm{m}, \mathtt{G})]\delta(a) + C(\mathrm{m}, \mathtt{G})(\mathrm{m}\mu - \lambda)e^{-(\mathrm{m}\mu-\lambda)a}1_0(a)$	(8.133)	539
	$\mathtt{m}_x = \mathtt{G} + C(\mathrm{m}, \mathtt{G})\frac{\mathtt{G}}{\mathrm{m}-\mathtt{G}}$	(8.119)	537
	$\mathtt{m}_z = C(\mathrm{m}, \mathtt{G})$	(8.123)	538
	$\mathtt{m}_q = C(\mathrm{m}, \mathtt{G})\frac{\mathtt{G}}{\mathrm{m}-\mathtt{G}}$	(8.128)	538
	$\mathtt{m}_s = \frac{C(\mathrm{m}, \mathtt{G})}{\mu(\mathrm{m}-\mathtt{G})} + \frac{1}{\mu},$	(8.140)	540
	$\mathtt{m}_w = \frac{C(\mathrm{m}, \mathtt{G})}{\mu(\mathrm{m}-\mathtt{G})}$	(8.137)	539
M/G/1	stability: $\rho = \lambda \mathtt{m}_y < 1$		562
	$\mathtt{m}_x = \frac{\lambda \mathtt{m}_y}{2} + \frac{\lambda \mathtt{m}_y + \lambda^2\sigma_y^2}{2(1-\lambda \mathtt{m}_y)}$	(8.216)	563
	$\mathtt{m}_z = \lambda \mathtt{m}_y$	(8.218)	564
	$\mathtt{m}_q = \frac{\lambda \mathtt{m}_y + \lambda^2\sigma_y^2}{2(1-\lambda \mathtt{m}_y)} - \frac{\lambda \mathtt{m}_y}{2} = \frac{\lambda^2 \mathtt{M}_y}{2(1-\rho)}$	(8.219)	564
	$\mathtt{m}_s = \frac{\mathtt{m}_y}{2} + \frac{\mathtt{m}_y + \lambda\sigma_y^2}{2(1-\lambda \mathtt{m}_y)}$	(8.220)	565
	$\mathtt{m}_w = \frac{\mathtt{m}_y + \lambda\sigma_y^2}{2(1-\lambda \mathtt{m}_y)} - \frac{\mathtt{m}_y}{2}$	(8.221)	565
M/D/1	stability: $\rho = \lambda T < 1$		567
	$p_x(k) = (1-\rho)\left(e^{k\rho} + \sum_{r=1}^{k-1}(-1)^{k-r}e^{r\rho}\frac{(r\rho)^{k-r-1}}{(k-r)!}[k - r(1-\rho)]\right)$	(8.239)	567
	$\mathtt{m}_x = \frac{\rho(2-\rho)}{2(1-\rho)}$	(8.241)	567
	$\mathtt{m}_z = \rho$	(8.242)	567
	$\mathtt{m}_q = \mathtt{m}_q = \frac{\rho(2-\rho)}{2(1-\rho)} - \rho = \frac{\rho^2}{2(1-\rho)}$	(8.241)	567
	$\mathtt{m}_s = \frac{T(2-\rho)}{2(1-\rho)}$	(8.244)	568
	$\mathtt{m}_w = \frac{T\rho}{2(1-\rho)}$	(8.245)	568

Table 8.3 Main results for blocking QSs.

Blocking QSs	performance measures	reference	page
M/M/1/K	$P_{\text{blk}} = \frac{(1-\rho)\rho^K}{1-\rho^{K+1}}$	(8.146)	545
	$p_x(k) = \frac{(1-\rho)\rho^k}{1-\rho^{K+1}}\ k = 0, 1, \ldots, K$	(8.145)	544
	$p_z(k) = p_z(k) = \begin{cases} \dfrac{1-\rho}{1-\rho^{K+1}} & , k = 0 \\ \dfrac{\rho(1-\rho^K)}{1-\rho^{K+1}} & , k = 1 \end{cases}$	(8.156)	547
	$p_q(k) = \begin{cases} \dfrac{1-\rho^2}{1-\rho^{K+1}} & , k = 0 \\ \dfrac{(1-\rho)\rho^{k+1}}{1-\rho^{K+1}} & , k = 1, 2, \ldots, K-1 \end{cases}$	(8.163)	548
	$p_s(a) = \mu(1-\rho)e^{-\mu(1-\rho)a} \frac{\sum_{k=0}^{K-1} \frac{(\lambda a)^k}{k!} e^{-\lambda a}}{1-\rho^K} 1_0(a)$	(8.169)	549
	$p_w(a) = \frac{(1-\rho)\delta(a)+\rho(\mu-\lambda)e^{-(\mu-\lambda)a}\sum_{h=0}^{K-2}\frac{(\lambda a)^h}{h!}e^{-\lambda a}}{1-\rho^K}$	(8.173)	550
	$\mathtt{m}_x = \frac{\rho}{1-\rho} - \frac{(K+1)\rho^{K+1}}{1-\rho^{K+1}}$	(8.155)	547
	$\mathtt{m}_z = \rho(1 - P_{\text{blk}})$	(8.158)	547
	$\mathtt{m}_q = \frac{\rho^2(1+(K-1)\rho^K - K\rho^{K-1})}{(1-\rho^{K+1})(1-\rho)}$	(8.165)	548
	$\mathtt{m}_s = \frac{1-(K+1)\rho^K + K\rho^{K+1}}{\mu(1-\rho)(1-\rho^K)}$	(8.171)	549
	$\mathtt{m}_w = \frac{(K-1)\rho^{K+1} - K\rho^k + \rho}{\mu(1-\rho)(1-\rho^K)}$	(8.174)	550
M/M/m/m	$B(\mathrm{m}, \mathtt{G}) = \frac{\frac{\mathtt{G}^{\mathrm{m}}}{\mathrm{m}!}}{\sum_{r=0}^{\mathrm{m}} \frac{\mathtt{G}^r}{r!}}$	(8.182)	553
	$p_x(k) = p_z(k) = \frac{\mathtt{G}^k}{k! \sum_{r=0}^{\mathrm{m}} \frac{\mathtt{G}^r}{r!}}, k = 0, 1, 2, \ldots, \mathrm{m},$	(8.181)	552
	$p_s(a) = p_y(a) = \mu e^{-\mu a} 1_0(a)$	(8.185)	554
	$\mathtt{m}_x = \mathtt{m}_z = \mathtt{G}(1 - B(\mathrm{m}, \mathtt{G}))$	(8.184)	554
	$\mathtt{m}_s = \mathtt{m}_y = \frac{1}{\mu}$	(8.186)	555
M/M/m/K	$P_{\text{que}} = \frac{(1-(\mathtt{G}/\mathrm{m})^{K-\mathrm{m}})\frac{\mathtt{G}^{\mathrm{m}}}{\mathrm{m}!}}{\frac{\mathrm{m}-\mathtt{G}}{\mathrm{m}}\sum_{r=0}^{\mathrm{m}-1}\frac{\mathtt{G}^r}{r!} + (1-(\mathtt{G}/\mathrm{m})^{K-\mathrm{m}+1})\frac{\mathtt{G}^{\mathrm{m}}}{\mathrm{m}!}}$	(8.191)	556
	$P_{\text{blk}} = \frac{\mathtt{G}^K}{\mathrm{m}^{K-\mathrm{m}}\mathrm{m}!\sum_{r=0}^{\mathrm{m}-1}\frac{\mathtt{G}^r}{r!} + \mathtt{G}^{\mathrm{m}}\frac{\mathrm{m}^{K+1-\mathrm{m}} - \mathtt{G}^{K+1-\mathrm{m}}}{\mathrm{m}-\mathtt{G}}}$	(8.192)	557
	$p_x(k) = \begin{cases} \left[\displaystyle\sum_{r=0}^{\mathrm{m}} \frac{\mathtt{G}^r}{r!} + \frac{\mathtt{G}^{\mathrm{m}}}{\mathrm{m}!}\frac{\mathtt{G} - (\mathtt{G}/\mathrm{m})^{K-\mathrm{m}+1}}{\mathrm{m}-\mathtt{G}}\right]^{-1} & , k = 0 \\ p_x(0)\mathtt{G}^k/k! & , k = 1, \ldots, \mathrm{m} \\ p_x(0)\mathtt{G}^k/(\mathrm{m}!\mathrm{m}^{k-\mathrm{m}}) & , k = \mathrm{m}+1, \ldots, K \end{cases}$	(8.189) (8.190)	556 556
	$p_w(a)\delta(a)\frac{1-P_{\text{que}}}{1-P_{\text{blk}}} + \mathrm{m}\mu e^{-\mathrm{m}\mu a}\frac{\mathtt{G}^{\mathrm{m}}}{\mathrm{m}!}\frac{p_x(0)}{1-P_{\text{blk}}}\sum_{\ell=0}^{K-\mathrm{m}-1}\frac{(\lambda a)^\ell}{\ell!}$	(8.173)	550
	$\mathtt{m}_z = \frac{\mathtt{G}\left(\sum_{\ell=0}^{\mathrm{m}-1}\frac{\mathtt{G}^\ell}{\ell!} + \frac{\mathtt{G}^{\mathrm{m}}}{\mathrm{m}!}\frac{1-\rho^{K-\mathrm{m}}}{1-\rho}\right)}{\sum_{\ell=0}^{\mathrm{m}-1}\frac{\mathtt{G}^\ell}{\ell!} + \frac{\mathtt{G}^{\mathrm{m}}}{\mathrm{m}!}\frac{1-\rho^{K-\mathrm{m}+1}}{1-\rho}}$	(8.195)	557
	$\mathtt{m}_q = \frac{\rho\mathtt{G}^{\mathrm{m}}\left(1-(K-\mathrm{m}+1)\rho^K + K\rho^{K-\mathrm{m}+1}\right)}{\mathrm{m}!(1-\rho^2)\sum_{r=0}^{\mathrm{m}-1}\frac{\mathtt{G}^r}{r!} + (1+\rho)\mathtt{G}^{\mathrm{m}}(1-\rho^{K-\mathrm{m}+1})}$	(8.197)	558
	$\mathtt{m}_s = \frac{1-(K+1)\rho^K + K\rho^{K+1}}{\mu(1-\rho)(1-\rho^K)}$	(8.171)	549
	$\mathtt{m}_w = \frac{(K-1)\rho^{K+1} - K\rho^k + \rho}{\mu(1-\rho)(1-\rho^K)}$	(8.174)	550

Problems

Preliminaries

8.1 A link with transmission rate $R_b = 1$ Mbit/s is used to forward packets having constant size $L = 1000$ byte. Find the time $\mathtt{m}_y$ taken to complete the transmission of a single packet.

8.2 A link with transmission rate $R_b = 1$ Mbit/s is used to forward packets having random size ℓ, which is assumed to be uniformly distributed in the discrete set $\mathcal{L} = \{1, 2, \ldots, 100\}$ kbyte.

a) Find the mean transmission time $\mathtt{m}_y$ of a packet along the line.

b) Find the probability that the transmission time of a packet is larger than 500 ms.

8.3 A link with transmission rate R_b is used to forward packets having random size ℓ, which is assumed to have geometric PMD

$$p_\ell(k) = p(1-p)^{k-1}, \quad k \in \mathcal{L} = \{1, 2, \ldots\} .$$

Prove that, if $m_\ell = 1/p$ is large enough, the distribution of the packet transmission times can be approximated to an exponential distribution with parameter $\mu = R_b/m_\ell$.

8.4 A dentist clinic opens up for business at $t = 0$. Customers arrive according to a Poisson rp with rate λ. Each medical treatment takes a random time y. Find the probability P and that the second arriving customer will not have to wait and also find the average value $\mathtt{m}_w$ of his waiting time for the two following cases:

a) $y = c$ =constant.

b) y is exponentially distributed with parameter $\mu = 1/c$.

Little's law

8.5 A hotel clerk takes, on average, $\mathtt{m}_y = 3$ minutes to attend to a single customer. Assuming that customers arrive at the clerk desk with average rate of $\lambda = 0.1$ customer/min, determine the fraction of time the clerk is idle.

8.6 Consider a transmission system where the packet transmission along the line takes a random time, with uniform distribution in the interval $[1, 1000]$ μs. Knowing, that the mean waiting time in queue is $\mathtt{m}_w = 500\,\mu s$ and that, on average, there are $\mathtt{m}_q = 0.2$ packets in the waiting buffer, determine what follows.

a) The mean number of packets in the system.

b) The fraction of time the line is engaged.

8.7 A router is busy processing data packets for 80% of the time. On average $\mathtt{m}_q = 3.2$ packets are waiting for service. Find the mean waiting time $\mathtt{m}_w$ of a packet given that the mean processing time is $\mathtt{m}_y = 1$ s.

Queueing models

8.8 Consider an $M/M/1$ QS with arrival rate $\lambda = 10$ customers per second and mean number of customers in the system $\mathtt{m}_x = 5$. Find the following quantities.

a) The mean service rate μ.

b) The mean system time $\mathtt{m}_s$.

c) The probability that a customer spends in the QS more than $2\mathtt{m}_y$.

8.9 Consider an M/M/1 QS with service rates $\mu = 10$ customers per second. In stationary regime, the server is busy for 90% of time. Determine what follows.

a) The arrival rate λ.

b) The mean number of customers in queue.

c) The probability that the number of queued customers is at least 20% greater than its mean.

d) The probability that a customer spends more than 2 seconds in the system.

8.10 Consider a structure consisting of two M/M/1 QSs working in parallel, with service rates μ_1 and μ_2, respectively. The overall arrival rate is Poisson with parameter λ. Each arriving customer, independently of the others, is dispatched to QS one with probability p_1 and to QS two with probability $p_2 = 1 - p_1$.

a) What is the stability condition for the whole structure?

b) Find p_1 such that the mean time a customer remains in the structure is minimized.

8.11 Customers arrive to a dentist's surgery according to a Poisson process with parameter λ. The dentist examines one customer at a time. Each examination takes a random time y with mean $\mathtt{m}_y = 5$ minutes and exponential PDF, independently of the number of customers in the (huge) waiting room and of the arrival process.

a) Find the arrival rate λ for which the mean waiting time of a customer in the waiting room is less than or equal to 20 minutes.

b) For such a λ, determine the mean number of customers in the waiting room.

c) Find the probability that there are no more than 2 customers in the waiting room, in stationary conditions.

d) Now, suppose that at time $t_0 = 6:30$ pm the ingress door is closed and no other customers are admitted into the dentist's surgery. Suppose that there are still $x(t_0) = 10$ customers in the dentist's surgery. What's the mean time required to the dentist to finish the work?

e) What becomes such a time in case $x(t_0)$ is random, with asymptotic PMD $p_x(k)$?

8.12 Consider an M/M/1 system with parameters λ and μ in which customers are impatient. Specifically, a customer that finds k other customers already in the system joins the queue with probability $\exp(-\alpha k/\mu)$ or leaves with complementary probability $1 - \exp(-\alpha k/\mu)$, where $\alpha \geq 0$.

a) Find the asymptotic probability $p_x(k)$ in terms of $p_x(0)$.

b) Give an expression for $p_x(0)$ in terms of the system parameters.

c) Find the stability condition assuming $\mu > 0$ and $\alpha > 0$.

d) Find the asymptotic distribution of x and its mean when $\alpha \to \infty$.

8.13 Let us model a router by an M/M/1 system with mean service time equal to $\mathtt{m}_y = 0.5$ s.

a) How many packets per second can be processed for a given mean system time of 2.5 s?

b) What is the increase in mean system time if the arrival rate increases by 10%?

8.14 Consider a system with an infinite population of users. Each user can be in two states, namely idle and active. In idle state, a user does not perform any action. In active state, instead, the user transmits one packet over a shared transmission channel. When the packet transmission is completed the user leaves the system. Users activate in a Poisson fashion, i.e., the time intervals between successive users activations are iid rvs with exponential distribution of parameter λ. The channel has a bit rate of R_b bit/s, which is equally divided among all the active users. Therefore, the transmission rate of any active user at time t will be equal to

$$r(t) = \frac{R_b}{x(t)},$$

where $x(t)$ denotes the number of active users at time t. Each packet has a random size with exponential distribution and mean L. Find what follows.

a) The stationary PMD and mean of $x(t)$ and the system stability conditions.

b) The mean system time of a customer.

c) Assuming that, after user U becomes active, the system stops accepting new users, i.e., no other users are allowed to activate. Find the mean time after which user U completes its transmission.

8.15 The information stand of a vendor in a trade fair is visited by customers that arrive according to a Poisson process of rate $\lambda = 0.2$ customer/min. Upon arrival, each customer, independently of the others and of the number of customers already in the stand, decides with probability p to talk to an employee to have additional information about the commercialized products, whereas with probability $1 - p$ she only takes a quick look at the information material made available in large exhibition area of the stand. The stand is attended by m employees and the time required by an employee to serve a customer is random, with exponential distribution of mean $\mathtt{m}_y = 10$ minutes. In case all the employees are busy, customers that wish additional information orderly wait in a single queue for the first available employee. Customers that are not interested to talk to an employee, instead, are not required to queue and can freely move in the exhibition area of the stand. In this case, a customer stays in the stand a random time, uniformly distributed in the interval $U = [2, 4]$ min. Determine what follows.

a) The maximum value of p for the waiting time in queue to not grow indefinitely in case the stand is attended by a single employee, i.e., $\mathrm{m} = 1$.

b) The percentage of time the employee is busy serving customers, for $p = 0.25$ and $\mathrm{m} = 1$.

c) The probability that a customer waits in queue more than 30 min, with $p = 0.25$ and $\mathrm{m} = 1$.

d) The mean time a customer that talks to an employee stays in the stand, with $p = 0.75$ and $\mathrm{m} = 3$.

e) The mean time any customer stays in the stand, with $p = 0.75$ and $\mathrm{m} = 3$.

8.16 Consider a call-center with $\mathrm{m} = 5$ operators that receives, on average, 20 calls per hour. Calls that cannot be immediately served are held in a waiting queue. Each operator costs to the employer $c = 1/5$ units of some currency per minute. To make profit, each minute of call shall not cost, on average, more than $E = 1/3$ units of currency. Determine the range of values of the mean service time $\mathtt{m}_y$ for which system is stable and the call-center remains on business.

8.17 Consider an $\mathrm{M/M/2}$ system with arrival rate λ and service rate μ, where $\lambda < 2\mu$.

a) Find the differential equation that govern the time-dependent probabilities $p_j(t)$.

b) Find the asymptotic probabilities π_j.

8.18 Consider a call center with two phone lines for service. Assume that calls arrive according to a Poisson process with rate λ. During some measurements it was observed that both the lines are busy 50% of the time, whereas the mean call holding time was 1 minute. Calculate the call queueing probability in the case that the mean call holding time increases from 1 minute to 2 minute.

8.19 Consider the entrance of a highway with $N = 5$ tollbooths. Tollbooths are of two types, namely A and B. Tollbooths of type A are enabled for automatic fast payment by radio transmission, whereas those of type B are reserved for cash payment. Accordingly, vehicles are divided in type A or B depending on whether they are enabled or not for fast payment. Suppose that the time taken by a vehicle to cross a tollbooth is a rv, with exponential distribution of mean $\mathtt{m}_{y_A} = 1$ seconds and $\mathtt{m}_{y_B} = 24$ seconds for tollbooth of type A and B, respectively. In the rush hour, the vehicles arrive to the highway entrance according to a Poisson process with parameter $\lambda = 0.1$ vehicle/s. On average, 20% of the vehicles are of type A. Vehicles choose at random among the tollbooths of their own type, irrespective of the queues length. Assuming that there is a single tollbooth of type A and four of type B, determine the steady-state values of the following metrics:

a) the mean number of vehicles of type A and type B waiting in queue to enter the highway;

b) the mean queueing time of any vehicle.

c) Finally, find the number of tollbooths of type A that minimizes the mean queueing time of any vehicle.

8.20 Determine the probability that an arriving customer is refused entry in a $\mathrm{M/M/1}/K$ system under the condition that

a) the previous customer was accepted in the system;

b) the previous customer was also rejected entry to the system.

8.21 Consider an $\mathrm{M/M/1}/K$ system with arrival rate $\lambda > 0$ and service rate $\mu > 0$, such that $\rho = \lambda/\mu < 1$. Determine the minimum value of the system storage capacity K to have at most p percent of lost customers, with $p \in \{10^{-1}, 10^{-2}, 10^{-3}\}$ and $\rho \in \{0.7, 0.9\}$.

8.22 Consider an $\mathrm{M/M/m/m}$ system with $\mathrm{m} \geq 1$, arrival rate λ and $\mathtt{G} = 5$.

a) Evaluate $B(\mathrm{m}, \mathtt{G})$ for $\mathrm{m} \in \{1, 2, 3, \ldots, 20\}$

b) How many servers do we need to have $B(\mathrm{m}, \mathtt{G}) \leq 10^{-6}$.

c) What's the value of λ for which we have $B(\mathrm{m}, \mathtt{G}) = 10^{-6}$ with $\mathrm{m} = 9$ and $\mu = 12$ [customer/s]?

d) How many customers do we have in the system in the previous case?

8.23 Consider a factory where an array of $N = 10$ machines work in parallel some raw materials to obtain some product. The processing of a block of raw material by any machinery takes an independent random time with exponential distribution with parameter $\mu = 1$. The raw materials arrive to the machine array according to a Poisson process with parameter λ and it is dispatched to the first available machinery. If all machines are busy, the arriving raw materials have to be dropped. We are interested on the stationary working regime of the system.

a) Suggest a valuable QS to model this scenario.

b) Determine the value of λ for which the mean number of servers in use is $N/2$.

c) Determine the value of λ for which the mean number of servers in use is $N \times 0.9$.

d) Determine the fraction of raw material dropped in the two previous cases.

e) Assume to add a buffer place to the system. What is now the arrival rate λ needed to have, on average, $N \times 0.9$ servers in use? What is the loss probability of raw material in this case?

8.24 **($\mathrm{M/M}/\infty$ model)** Consider a trade fair where visitors arrive according to a Poisson process of rate $\lambda = 240$ visitors per hour. The time a visitor spends in the trade fair can be modeled as an exponential rv with mean $\mathtt{m}_y$ hours, independent of the number of visitors. Let $x(t)$ be the number of customers in fair at time t.

a) Prove that $x(t)$ is a BDP and determines the birth and death rates.

b) Find the stationary PMD and mean of $x(t)$ and the system stability conditions.

c) Find the probability that the fair is visited by more than M customers in the first 30 minutes after its opening.

d) Find the stationary probability of having more than M customers in the fair at any given time.

8.25 Consider a system with Poisson arrival process with parameter λ and service times with a certain PDF $p_y(a)$, $a \geq 0$. Find the probability P that an arriving customer is immediately served and the mean queueing time $\mathtt{m}_w$ for the following distributions of the service time:

1. constant, with $y = C$;
2. uniform in $[0, 2C]$;
3. exponential with $\mathtt{m}_y = C$.

Which case is preferable?

8.26 An airplane takes exactly 6 minutes to land after it receives the clear-to-land signal from the airport's traffic control. Suppose that airplanes arrive according to a Poisson process, with rate $\lambda = 6$ arrivals per hour, and that only one airplane can land at a time.

a) How long can an airplane expect to circle before getting the clear-to-land signal?

b) How many airplane are expected to be waiting for such a signal, in stationary conditions?

8.27 Consider a single serve queue with exponential service time with parameter μ and Poisson arrivals with rate λ. New customers are sensitive to the length of the queue in the sense that a newly arriving customer that finds i other customers in the system will join the queue with probability $p_i = (i+1)^{-1}$, whereas with probability $1 - p_i$ it departs and does not return. Find the stationary probability distribution of this QS. (Suggestion: see also Problem 7.39)

CHAPTER 9

DATA LINK LAYER

Michele Rossi

In digital telecommunication systems the link layer is responsible for the transmission of data streams over point to point links. It performs framing of higher layer packets into link layer packet data units (PDUs) and can perform packet error recovery by exploiting ARQ mechanisms. In addition, the link layer also implements channel access procedures, which are often tightly coupled with the adopted ARQ scheme. Operations at this layer are essential for reliable and efficient transmissions over noisy and possibly shared links. We observe that the quality of service (QoS) at the application layer is highly dependent on the combination of techniques that we use at this layer as well as on the way we configure them. In particular, the following techniques play a fundamental role in the link layer: 1) the selected retransmission policy, 2) the feedback strategy that is used, at the receiver, to request the retransmissions of lost packets, 3) the channel access mechanism, 4) adaptive behavior in terms of data frame length (as a function of the error rate) as well as 5) advanced features such as combined ARQ and coding (for throughput enhancement). This chapter will give a solid introduction to points 1–4, will discuss some techniques related to point 5 and will cover the data link layer techniques used in Ethernet, IEEE 802.11 and Bluetooth.

Principles of Communications Networks and Systems N. Benvenuto, M. Zorzi

9.1 INTRODUCTION

The link layer is responsible for data transmission and recovery over point to point links. It performs framing of higher layer packets into usually smaller link layer packets, hereafter referred to as PDUs, which are transmitted over the channel according to a given channel access policy. In case of erroneous reception, these packets can be retransmitted through the exploitation of ARQ mechanisms. Of course, this entails some form of feedback through which the receiver notifies the transmitter about those packets that were corrupted by channel impairments or interference from other users. This feedback is achieved through dedicated acknowledgment messages (ACKs). The mechanisms implemented at the link layer are instrumental to achieving good communication performance, as they affect the delay and reliability at nearly all layers of the protocol stack, see Fig. 1.13. In fact, channel access policies dictate how users access the channel and, as such, are directly responsible for the level of congestion (multi-user interference) that is experienced over the link. A congested channel, in turn, causes packet collisions, which translate into a reduction of the throughput and an increase of the delay, as corrupted packets must be retransmitted for reliability.

The objective of this chapter is to thoroughly analyze the major link layer protocols and techniques used in communication systems. We provide a unified analysis of these mechanisms and systematically discuss (and contrast) their performance in different propagation environments. In doing so, we use accessible analytical methods and simple models for links, packet arrivals and retransmission backoff policies. More complex characterizations are not considered here and more involved protocol variants are also left aside. We however believe that the material covered in this chapter is still of utmost importance as it constitutes the foundation for the understanding of more complex systems.

The chapter starts with a simple analysis of time division and frequency division multiple access schemes. We then discuss in some depth polling protocols, random access protocols and carrier sense multiple access techniques. We subsequently compare these schemes in terms of throughput and delay. After this, we introduce the three major ARQ techniques: stop and wait ARQ (SW-ARQ), go back N ARQ (GBN-ARQ) and selective repeat ARQ (SR-ARQ), which are compared in terms of their throughput performance. We then discuss the importance of selecting the right packet size for link layer protocols and show, through a simple analytical model, how it can be calculated as a function of the link packet error rate. At the end of the chapter we discuss three important local area network (LAN) standards: Ethernet, IEEE 802.11 and Bluetooth, with a focus on their link layer mechanisms.

Assumptions Next, we give a list of basic assumptions that will be made in the analysis of all channel access and ARQ techniques that we address in this chapter.

BA1. User data (coming from higher layers) is fragmented and collected into packet data units (PDUs) of L bits (PDU length), where L_O bits (overhead) are allocated for the header and L_D bits (data) for the payload, i.e., $L = L_O + L_D$.

BA2. The channel transmission bit rate R_b bit/s is fixed. Thus, the packet transmission time is a constant,

$$t_{\mathrm{P}} = \frac{L}{R_b} \,. \tag{9.1}$$

We define the transmission time for the data portion of a packet as

$$t_{\mathrm{DATA}} = \frac{L_D}{R_b} . \qquad (9.2)$$

Moreover, as ACK messages are also transmitted at the same bit rate R_b, we define their transmission time as

$$t_{\mathrm{A}} = \tilde{t}_{\mathrm{A}} t_{\mathrm{P}} \qquad (9.3)$$

where $\tilde{t}_{\mathrm{A}} \in (0, 1]$.

BA3. Terminals can, at any given time, either be in transmitting or receiving mode but not both concurrently.[1] In addition, we neglect the switching time between transmitting and receiving states.

BA4. The system is fully reliable, i.e., in case a packet is erroneously received due to collisions or channel noise, we allow an infinite number of retransmission attempts for this packet. This assumption is made here for the sake of analytical simplicity. We observe, however, that due to practical constraints on the maximum tolerable packet delay, actual systems operate considering a finite number of retransmission attempts.

BA5. The channel propagation delay, τ_p, is assumed constant and equal for all users. We define $\tilde{\tau}_p$ as the channel propagation delay normalized to the packet transmission time,

$$\tilde{\tau}_p = \frac{\tau_p}{t_{\mathrm{P}}} . \qquad (9.4)$$

BA6. The processing time needed to check the correctness of a received packet and generate the corresponding acknowledgment packet is considered to be negligible.

BA7. The feedback channel only introduces delay, without causing packet losses, i.e., acknowledgments are received without errors and their transmission (receiver $\rightarrow$ transmitter) takes $t_{\mathrm{A}} + \tau_p$ seconds. This is of course a simplifying assumption that amounts to having a separate channel to handle acknowledgment traffic.

9.2 MEDIUM ACCESS CONTROL

In this section we present and analyze some important MAC mechanisms for multi user systems where there is a channel that can be simultaneously used by a number of users placed in close proximity of each other. The problem to be solved is to access this medium so that the transmissions from different users will minimally interfere with each other and still all users will be able to reliably communicate with their intended destinations. There are various ways in which this can be done. Our choice, in this chapter, was to subdivide existing access methods into three classes: 1) *deterministic*, 2) *demand based* and 3) *random*.

[1]This assumption will be relaxed for the analysis of CSMA/CD systems, where terminals can detect collisions during the transmission phase.

Deterministic Protocols Deterministic protocols are characterized by the fact that users know exactly when they can transmit as some schedule has been carefully assigned to them by a controller. In this case, the shared channel can be allotted to the different users through the assignment of channel resources, namely, time slots (time division multiple access (TDMA)), frequency bands (frequency division multiple access (FDMA)) or codes (code division multiple access (CDMA)). In all these cases, the assigned channel resources are orthogonal among users, which means that different users that transmit in different time slots, frequency bands or that use different codes for their transmission do not interfere with each other, at least in the presence of an ideal channel. The problem, in this case, is to properly assign the resources through a schedule, avoiding that multiple users access the same resource at the same time. Once this schedule is found, it is pre-distributed to the users and then used. From this, it is clear that deterministic protocols are centralized as a controller is needed to allocate and release channel resources. This can be done in a dynamic manner as a function of QoS requirements, traffic load and number of users. Also, note that, as long as the assigned resources are orthogonal, multiple users can safely access the channel *concurrently*. This is, for example, the case for non-overlapping frequency bands or different orthogonal codes. In TDMA systems, concurrent transmissions are of course not possible unless different codes are allotted to different users within the same slot, leading to hybrid CDMA/TDMA schemes.

Demand-Based Protocols The second class of protocols, referred to here as demand-based, still involves the coordinated assignment of channel resources. In this case the channel is used by a single user at a time and decisions about channel assignments, i.e., *which user will be using the channel in the next time slot/time frame*, are dynamically made in either a *centralized* or *distributed* manner. In the former case, a controller decides which users will be using the channel in the next time frame, optionally it can also decide the length of the time frame based on, e.g., user priorities and traffic loads. This decision is made through an exchange of messages between the user and the controller, where the user notifies the controller about the number of packets in its queue and its transmission priority. Hence, the controller decides which user should be served and for how long, according to some policy. In the distributed case, a special control packet (referred to as *token*) is transmitted by the user that releases the channel to the user that will be using the channel in the next time frame. For this scheme to work properly a transmission order must be known by all users, so all of them will know which is the next user in the transmission sequence.

Random Access Protocols The third class of protocols, random access, follows a completely different paradigm and users decide whether they should access the channel in a given time slot without any coordination. The advantage of this is that the controller is no longer needed as users schedule transmissions at random. This allows a truly distributed implementation, even in networks where the topology and the number of users cannot be known by all users, or where gathering this information entails a high communication cost. In this class of protocols we will analyze and discuss pure random access protocols, i.e., ALOHA and its slotted variant, and the carrier sense multiple access (CSMA) family of protocols, which is currently used for Ethernet and wireless local area networks (WLANs).

System Model for Channel Access Schemes In what follows we introduce a general system model for the analysis of the channel access schemes that we treat in this chapter.

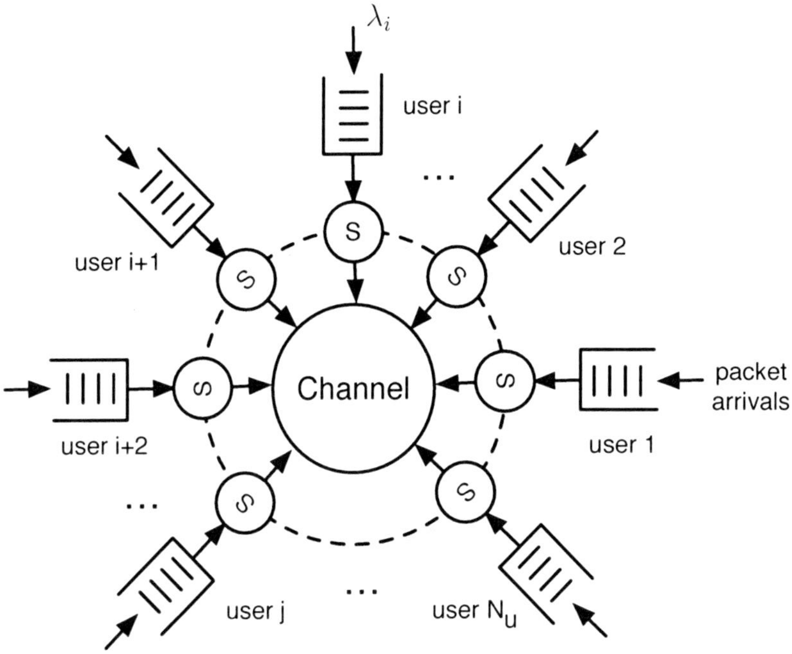

Figure 9.1 System model for the analysis of channel access schemes.

As depicted in Fig. 9.1 we consider a number of users that communicate through a shared channel. Depending on the access scheme under consideration this number will be finite or, for the sake of a tractable analysis, infinite. Each user $i \in \{1, 2, \ldots, N_u\}$ has a waiting queue that will be assumed of finite or infinite length, depending on the considered mathematical framework. Upon arrival, a packet waits in the corresponding queue until all packets ahead of it are transmitted. Attached to each queue there is a server (labeled as S in the figure), which implements the service policy. In Fig. 9.1 the servers have been connected through a dotted line to indicate that the service policies of different users can be either independent or coordinated. In the independent case, each user decides on his own when to access the channel (possibly based on his local estimation of the channel status), at which point he transmits the first packet in his queue. An example for this case is given by random access protocols. In the coordinated case the channel access activity is coordinated by a central controller that decides the transmission schedule for all users. Examples for this case are given by TDMA, FDMA and polling. In addition to BA1–BA7, for the analysis of channel access schemes we make the following assumptions.

MA1. We consider a shared channel, where all users can communicate with each other.

MA2. We assume that the packet arrival process for each user $i \in \{1, 2, \ldots, N_u\}$ is described by a Poisson process with rate $\lambda_i = \lambda$ packet/s (see Fig. 9.1 and Section 2.7.2). In addition, we further assume that arrivals at different users are statistically independent.

MA3. Whenever a given user i transmits a packet having as destination user j, it sends this packet over the shared communication channel. This packet will be concurrently received by all remaining users after a channel propagation delay τ_p.

MA4. If the transmissions of two or more packets partially overlap, these transmissions will interfere with each other. In this case we assume that these interfering packets are all erroneously received and we name such an event collision.

MA5. When the transmission of a packet does not overlap with any other transmissions we assume that this packet is successfully received by all remaining users after a channel propagation delay τ_p.

In what follows, we define the performance measures that will be characterized in this chapter. The notation that will be used throughout the chapter was introduced in Tab. 8.1.

Performance Measures

- The delay t_{delay} is defined as the time elapsed between the arrival of a given packet at the queue of a generic user $i \in \{1, 2, \ldots, N_u\}$ and the instant when this packet is entirely received at the intended destination $j \neq i$. In general, t_{delay} can be expressed as follows:

$$t_{\text{delay}} = s + \tau_p \tag{9.5}$$

  where:

  1. s is the *system time* defined as the time elapsed between the instant when the packet arrives at a user queue and the instant when the transmission of this packet is successfully completed.
  2. τ_p is the constant propagation delay after which the reception of this packet is complete at all users.

  The *average delay* $D = \mathrm{E}\left[t_{\text{delay}}\right]$ is given by:

$$\boxed{D = \mathtt{m}_s + \tau_p} \tag{9.6}$$

  where $\mathtt{m}_s = \mathrm{E}\left[s\right]$.

  In those cases where the packet transmission time is constant it is convenient to write s as

$$s = t_{\text{retx}}^{\text{tot}} + w + t_{\text{P}} \tag{9.7}$$

  where:

  1. $t_{\text{retx}}^{\text{tot}}$ is the total *retransmission delay*, i.e., the time taken for the retransmissions of the packet until its successful delivery. The delay due to the first transmission of the packet is not included in this term, while it includes the delay due to the last (successful) transmission of the packet. $t_{\text{retx}}^{\text{tot}}$ term will be zero for all schemes that do not retransmit corrupted or lost packets (see assumption BA4).

2. w is the queueing delay, given by the time spent by a given packet from its arrival at a user queue to the instant when the transmission of this packet begins.
3. t_P is the time taken by the first transmission of the packet.

The *average delay* in this case is given by:

$$D = \mathtt{m}_{\text{retx}} + t_P + \mathtt{m}_w + \tau_p \tag{9.8}$$

where $\mathtt{m}_w = \mathrm{E}[w]$ and $\mathtt{m}_{\text{retx}} = \mathrm{E}[t_{\text{retx}}^{\text{tot}}]$.

- The average packet arrival rate for the system is

$$\lambda_t = \sum_{i=1}^{N_u} \lambda_i = N_u \lambda \tag{9.9}$$

We remark that for those channel access schemes that do not perform retransmissions, the arrival rate λ_i is given by the rate at which new packets arrive at the link layer of user i. On the other hand, when retransmissions are used, λ_i is given by the sum of 1) the rate at which new packets arrive at the link layer and 2) the rate at which packets are retransmitted.

- The *normalized offered traffic* is defined as (see also (8.27) for $\mathtt{m}_y = t_P$)

$$\mathtt{G} = \lambda_t t_P = N_u \lambda t_P \tag{9.10}$$

where the second equality comes from (9.9). In other words $\mathtt{G}$ represents the number of packets transmitted over the channel in a packet transmission period t_P.

- The *system load factor* ρ is defined as

$$\rho = \frac{\lambda_t}{\mathrm{m}\mu} \tag{9.11}$$

where m is the number of servers and μ is the service rate (see (8.7)).

- The average throughput is referred to as η and is characterized in (8.28). The throughput is also referred to as *departure rate*, as it is the rate at which the system is able to successfully transmit packets. In this chapter we focus on the computation of the *normalized throughput* $\mathtt{S} \in [0, 1]$, defined as

$$\mathtt{S} = \eta t_P \tag{9.12}$$

Note that, for non-blocking and stable (i.e., when $\rho < 1$) single server ($\mathrm{m} = 1$) systems we have that $\eta = \lambda_t$ (see (8.34)) and

$$\mathtt{S} = \rho = \mathtt{G} \tag{9.13}$$

where this last equality follows from (8.32) and (8.35).

- Throughout the chapter the symbol $\sim$ will refer to normalized quantities with respect to $t_{\rm P}$.

9.2.1 Deterministic Access: TDMA and FDMA

Next, we discuss deterministic access protocols. These are also referred to as *conflict free* as they are designed to ensure that a user transmission does not interfere with any other ongoing transmission and is thus successful (in the absence of noise). This is accomplished through the allocation of the channel resource to the users in such a way there is no interference among the channel resources allotted to different users. A notable advantage of these protocols is that it is easy to ensure fairness among users and control the packet delay. The controller can in fact periodically monitor the QoS for all connections and modify the schedule according to a QoS goal. This is important for realtime applications. The next three subsections describe static allocation strategies, i.e., where channel resources are assigned *a priori* (i.e., at network design time) and kept unchanged during the operation of the system. In this case there is no overhead due to control messages, as the allocation never changes in time. However, this solution lacks flexibility and might present some performance reduction in terms of throughput. As an example, when a rigid allocation is considered some resources (e.g., time slots) might go unused even though some users have some data to transmit. This loss of performance will be further discussed and quantified in Section 9.2.8. More advanced TDMA schemes obviate this problem by using, e.g., transmission priorities that depend on the outstanding packets at the users. These schemes are not treated in this chapter.

9.2.2 Time Division Multiple Access

In TDMA the time axis is subdivided into time slots of the same duration. These are thus preassigned to the different users in the system. Each user can transmit freely in the assigned slot, as during this slot the entire channel resources are assigned to this user. The slot assignment follows a predetermined pattern which is repeated periodically. Specifically, if there are N_u users in the system, they are served according to a predetermined order in the TDMA slots $1, 2, \ldots N_u$: these N_u slots form a so called TDMA *cycle* or *frame*. After this, the process is repeated with a new TDMA frame and so on.

In the most basic TDMA schedule each user has exactly one slot per frame allocated to it. TDMA schemes can be inefficient when some users do not have data to send as their slots will go unused in that case. Notably, there exist more sophisticated solutions where multiple slots can be allotted to a given user. However, this can be implemented at the price of some extra communication overhead. In general, the TDMA paradigm relies on a central controller to distribute the slot assignments. Further details and performance evaluation for these schemes are given in [1].

Throughput To characterize the performance of a TDMA schedule consider a system with N_u users each with packet arrival rate λ. The packet transmission time $t_{\rm P}$ (see (9.1)) corresponds to the duration of a TDMA slot. The service at the different users is coordinated by a single centralized entity and the system can thus be seen as composed of multiple input queues (one per user) and a single server ($\rm m = 1$). Using (9.9) we have that $\lambda_{\rm t} = \lambda N_u$. Looking at the TDMA system as a single queue containing the packets of all users, we have that this queue is served at the rate of one packet per TDMA slot, i.e., $\mu = 1/t_P$. Hence, from (9.11) we obtain the system load factor as

$$\rho = \frac{\lambda_t}{\mu} = \lambda N_u t_P \, . \tag{9.14}$$

From (9.13), we obtain the normalized throughput as:

$$\boxed{S_{TDMA} = \rho = \lambda N_u t_P} \tag{9.15}$$

Delay The delay is computed using (9.7). Note that t_{retx}^{tot} is zero for the considered TDMA scheme due to assumption MA5 and the fact that at most one packet is (successfully) sent within each TDMA slot, i.e., packets never collide. The queueing delay w is given by $w = w_1 + w_2$ where w_1 accounts for the fact that in TDMA whenever a packet arrives at a given user it has to wait for a certain number of TDMA slots before the control returns to this user. From here, w_2 is the time taken for the transmission of all packets already stored in the queue. These two terms are obtained as follows:

- For w_1, we assume that all frames have the same duration and that new arrivals are uniformly distributed within a TDMA *frame*. The mean of a discrete rv uniformly distributed in $\{1, 2, \ldots, N_u\}$ is $N_u/2$. Hence, we have

$$\mathtt{m}_{w_1} = \frac{N_u L}{2R_b} = \frac{N_u t_P}{2} \, . \tag{9.16}$$

- For w_2, if we look at the waiting queue of a specific TDMA user, we see that this queue is served at the fixed rate of one packet per frame, i.e., it experiences a deterministic service time of $N_u t_P$ seconds. This reflects the fact that users must wait for their allotted slot to transmit. If we assume infinite storage capabilities, each queue can be seen as an M/D/1 system and, from (8.245) with $\mathtt{m}_y = N_u t_P$, we obtain

$$\mathtt{m}_{w_2} = \frac{\rho}{2(1-\rho)} N_u t_P \tag{9.17}$$

  where ρ is given in (9.14).

Thus, the mean queueing delay $\mathtt{m}_w$ is obtained combining (9.16) and (9.17) as:

$$\mathtt{m}_w = \mathtt{m}_{w_1} + \mathtt{m}_{w_2} = \frac{N_u t_P}{2} + \frac{\rho}{2(1-\rho)} N_u t_P \tag{9.18}$$

D_{TDMA} is thus calculated using (9.8) with $\mathtt{m}_{retx} = 0$:

$$\begin{aligned} D_{TDMA} &= \mathtt{m}_w + t_P + \tau_p \\ &= \frac{N_u t_P}{2} + \frac{\rho N_u t_P}{2(1-\rho)} + t_P + \tau_p \\ &= t_P \left(\frac{N_u}{2(1-\rho)} + 1 + \tilde{\tau}_p \right) \end{aligned} \tag{9.19}$$

where $\tilde{\tau}_p$ is the normalized propagation time as defined in (9.4). Using (9.15) the normalized average delay can be finally written as

$$\widetilde{D}_{\text{TDMA}} = \frac{N_u}{2(1 - S_{\text{TDMA}})} + 1 + \tilde{\tau}_p \tag{9.20}$$

9.2.3 Frequency Division Multiple Access

According to FDMA, the available frequency band is subdivided into a number of sub-bands and each sub-band is allotted to a single user. Hence, each user tunes its transmit and receive filters in order to successfully send and receive in its own sub-band, while filtering out what is transmitted elsewhere. An advantage of this scheme is its simplicity. Its main disadvantage is, as said above, that part of the channel resources is wasted when some users do not have data to transmit.

Throughput The system has a fixed rate R_b bit/s, which is equally subdivided among the N_u users in the system, i.e., each of them gets R_b/N_u bit/s. As the sub-bands are disjoint we have no interference among users, which can be seen as N_u independent queueing systems with their own separate input processes. For the N_u queues we assume that they have unlimited buffering space with the same arrival rate of λ packet/s. Each user transmits at the constant rate $\mu = R_b/(LN_u)$ packet/s on its sub-band. Applying (9.11) to the single waiting queue gives

$$\rho = \frac{\lambda}{\mu} = \frac{\lambda N_u L}{R_b} = \lambda N_u t_{\text{P}} \ .$$

We note that each queue is an M/D/1 system with utilization factor ρ. Thus, from (9.13) we have:

$$S_{\text{FDMA}} = \rho = \lambda N_u t_{\text{P}} \tag{9.21}$$

Note that S_{FDMA} in (9.21) equals S_{TDMA} in (9.15).

Delay The delay is calculated using (9.5). If we refer to x as the number of packets in a single queue system, including those packets waiting in the queue and in service, from the analysis of M/G/1 we have (see (8.217)):

$$\begin{aligned} \mathtt{m}_x &= \rho + \frac{\lambda^2 \mathtt{m}_y{}^2}{2(1-\rho)} = S_{\text{FDMA}} + \frac{\lambda^2 \mathtt{m}_y{}^2}{2(1 - S_{\text{FDMA}})} \\ &= S_{\text{FDMA}} + \frac{(\lambda N_u t_{\text{P}})^2}{2(1 - S_{\text{FDMA}})} \end{aligned} \tag{9.22}$$

where the second equation follows from (9.21) and the third from the fact that, for the single waiting queue the average service time is $\mathtt{m}_y = N_u t_{\text{P}}$. Using Little's result (8.47) we have $\mathtt{m}_x = \lambda \mathtt{m}_s$. Hence, from (9.6) we obtain the delay as

$$D_{\text{FDMA}} = \mathtt{m}_s + \tau_p = \frac{\mathtt{m}_x}{\lambda} + \tau_p \tag{9.23}$$

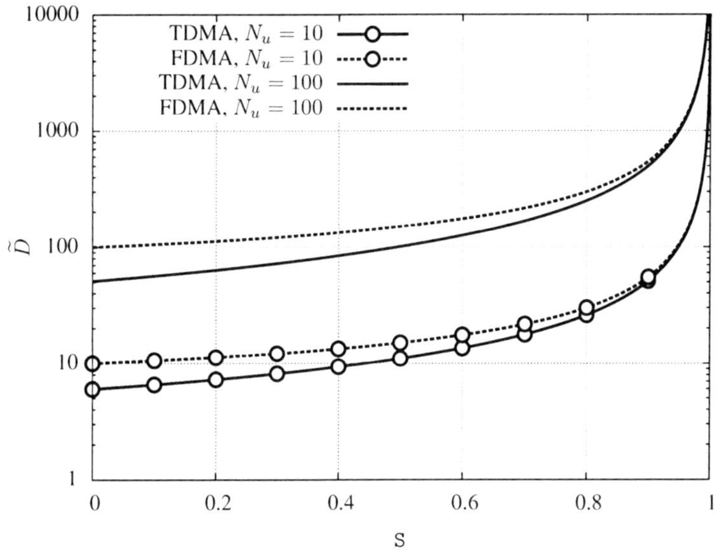

Figure 9.2 Normalized delay $\widetilde{D}$ versus the throughput S; $\tilde{\tau}_p = 0.01$ and $N_u \in \{10, 100\}$.

that using (9.22) gives:

$$\boxed{\widetilde{D}_{\text{FDMA}} = \frac{N_u}{2(1 - \text{S}_{\text{FDMA}})} + \frac{N_u}{2} + \tilde{\tau}_p} \qquad (9.24)$$

9.2.4 Comparison between TDMA and FDMA

Comparing the results in (9.19) and (9.24) we observe that

$$\boxed{\widetilde{D}_{\text{FDMA}} = \widetilde{D}_{\text{TDMA}} + \frac{N_u}{2} - 1} \qquad (9.25)$$

Hence, FDMA has longer delays than TDMA for $N_u \geq 2$, i.e., for all practical cases, and this result holds irrespective of the average packet arrival rate λ. In addition, the difference in terms of delay between the two schemes grows linearly with the number of users.

Example Results Fig. 9.2 shows the normalized delay $\widetilde{D}$ versus the throughput S for TDMA and FDMA considering $\tilde{\tau}_p = 0.01$ and $N_u \in \{10, 100\}$. As expected, an increasing number of users leads to longer delays for both access schemes.

9.2.4.1 Practical considerations From a practical standpoint, TDMA requires guard intervals between slots, which account for time synchronization errors. FDMA instead requires guard bands between adjacent frequency sub-bands, so as to mitigate the interference between adjacent channels. Both access schemes have their advantages and disadvantages.

TDMA The main advantages of TDMA are that it does not require frequency guard bands, there is no need for precise narrowband filters, due to the slot synchronism it is easy for mobile or base stations to initiate and execute handovers and guard times between time slots allow to reduce the impact of 1) clock instability and 2) differences in transmission, propagation and processing delays. On the downside, accurate clock synchronization is required for the entire network, which might be an issue when the network is large and spread over multiple transmission hops.

FDMA Low cost hardware technology can be used for filters and FDMA does not require network-wide clock synchronization. On the downside, it requires selective filtering to reduce the adjacent channel interference and within each sub-channel the transmission rate is fixed and usually small, which inhibits the flexibility in bit rate allocation. The subdivision of the frequency band into sub-bands can also be exploited for the transmission of a single data flow, according to the frequency division multiplexing (FDM) technique. If each of the sub-bands is sufficiently narrow, FDM has the double advantage of significantly reducing the signal distortion due to propagation effects[2] and requiring very efficient transmitting and receiving algorithms.[3] This technique is exploited by transmission systems based on orthogonal frequency division multiplexing (OFDM) (see Fig. 5.54), including IEEE 802.11a/g, which will be reviewed at the end of this chapter.

The most suitable access technique in general depends on application requirements as well as on hardware considerations. We shall note that TDMA and FDMA are often used together in modern cellular systems. As an example, OFDM with multiple access (OFDMA) is a multi-user version of OFDM where transmissions from different users alternate in the frequency domain, i.e., using different sub-carriers, and time, i.e., interleaving their transmissions within the same sub-carriers according to a time division scheme.

9.2.5 Demand-based Access: Polling and Token Ring

Polling systems In polling systems N_u users share the same channel and only one user can be served at any given time. Usually, these N_u users must alternate their transmissions according to some predefined transmission order and with non-zero switchover times. More formally, a polling system can be defined as "a system of multiple queues accessed in cyclic order by a single server" [2]. In particular, users are served in cyclic order until they have data to transmit and when the queue of a given user is empty, the control is immediately passed to the next user in the sequence. In the last century, polling models have been used to study a large number of systems. These cover various aspects of the industrial production such as the British cotton industry of the late 1950s and traffic signal control in 1960s. With the development of digital computers, in the 1970s polling models where studied to transfer data from terminals to a central server [3]. In the 1980s, this same model was exploited for the study of token passing schemes and in 1985 for resource arbitration and load sharing among multiprocessor computers [4]. For further historical remarks see the discussion in [2] and the references therein.

[2] In radio channels these are due to the so called frequency selective multi-path fading.

[3] These use a single receive filter, perform all the processing within the sub-bands in the digital domain and do no need channel equalizers.

A possible diagram of a polling system with N_u users is plotted in Fig. 9.1: here the users are served sequentially and, when a given user is served, the oldest packet in its waiting queue is sent over the channel. Polling systems are a broad class of MAC schemes that also comprises token ring protocols. Our choice for this chapter is to introduce token ring protocols below and then give a general but accessible analysis of polling and token ring. This analysis is sufficient to reveal the general behavior of polling and token ring schemes in terms of throughput and delay. More complex derivations go beyond the scope of this book and can be found in [2, 3, 5, 6].

Token ring The token ring technique is conceptually very similar to polling and can be described through an analogous analytical framework. From a practical point of view these systems are however very different. In fact, users on a token ring are (at least logically) organized in a ring topology and access to the channel is controlled by passing a permission token aroung the ring (this technique is also referred to as "token passing"). Hence, while in polling systems there exists a central unit (the server) that is in charge of switching the communication among users, in token ring networks the channel access is totally distributed. At initialization time, a designated user generates a free token which travels around the ring until a user ready to transmit changes it to busy and puts its packet(s) onto the ring. At the end of its transmission, the sending user passes the access permission to the next user by generating a new free token. The transmission of data around the ring is unidirectional and packets flow sequentially from user to user according to their order in the ring. Token ring, as we shall see below, has theoretical advantages over the CSMA with collision detection (CSMA/CD) of Ethernet.

A diagram of a generic token ring network is shown in Fig. 9.3: in Case a, user 1 has the token and transmits its packet(s) to user 2, which is the next user in the ring. In Case b, upon the completion of its transmission, user 1 passes the token to user 2. However, this user does not have own data to send and thus it immediately passes the token to user 3. In the last Case c, user 1 has the token and communicates with user 4. The intermediate users in the ring relay the packet(s) sent by user 1.

Review of important token ring networks Next, we briefly discuss the main types of networks exploiting the token passing technique:

1. **The attached resource computer network (ARCNET)** LAN protocol was the first widely available networking system for microcomputers; it was realized in 1976 by engineer John Murphy at Datapoint Corporation and became popular in the 1980s. The token-passing bus protocol was subsequently applied to enable communication among users for file-serving and distributed processing. ARCNET terminals were connected according to an "interconnected stars" cabling topology, that allowed user addition and removal without affecting the network connectivity. This made also easier to isolate and diagnose failures in a complex network. Various revisions of the technology were developed and in the early 1990s the transmission speed was pushed up to 100 Mbit/s. The system was however dismissed soon after due to the introduction into the market of the newer and less expensive 100 Mbit/s Ethernet technology.

2. **Token Bus [7]** networks implement the token ring protocol on a coaxial cable. A token is passed around the users and only the user having the token can transmit. In case a

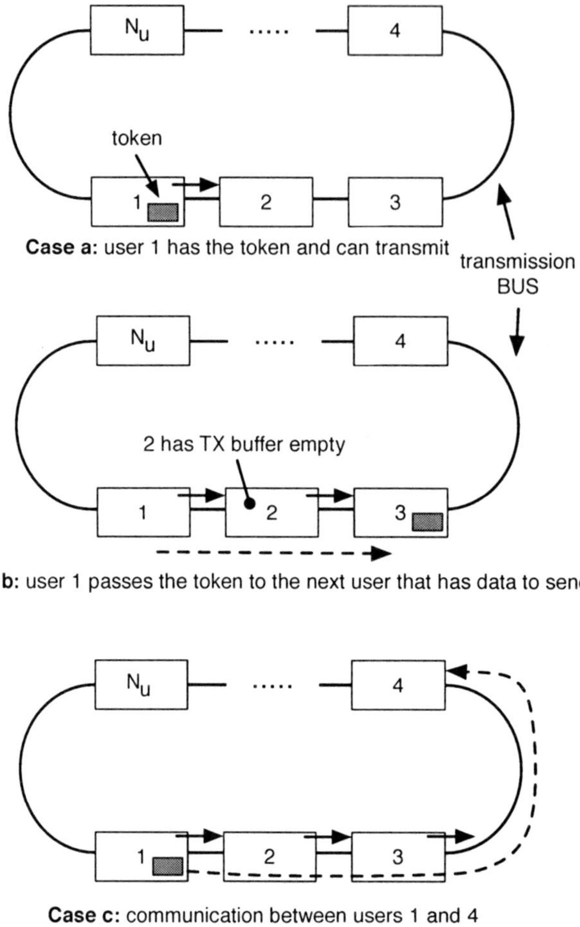

Figure 9.3 Diagram of a token ring system with N_u stations (users).

user does not have data to send, it passes the token to the next user on the virtual ring. Accordingly, each user must know the address of its neighbors in the ring. A special protocol is thus required to notify all users when users join or leave the ring, so that their identity and placement are always known. The token bus was standardized by the IEEE 802.4 working group [7].

3. **Token Ring / IEEE 802.5 [8].** In these networks users are connected in a ring topology and each user can directly hear transmissions only from its immediate neighbor. Permission to transmit is granted by a message (token) that circulates around the ring. The so called token ring network was developed by IBM in the 1970s. Subsequently, the Internet engineering task force (IETF) standardized the IEEE 802.5 specification, which is almost identical to IBM's original token ring and fully compatible to it. IBM's token ring network specifies a star, where all users are connected to a multistation access

unit (MSAU). Differently, IEEE 802.5 does not specify a topology, even though most IEEE 802.5 implementations are based on stars. Other minor differences exist, including transmission media type, routing information field size and supported transmission speeds. For further information on the IEEE 802.5 standard see [8].

4. **The fiber distributed data interface (FDDI) [9]** is a transmission system for LANs developed in the middle of the 1980s by the American national standards institute (ANSI). The transmission medium is an optical fiber (though copper cable is also supported), and can extend in range up to 200 kilometers and cover thousands of users. The FDDI protocol was derived from the IEEE 802.4 token bus timed token protocol. An FDDI network features two token rings, one of which for possible backup in case the primary ring fails. The primary ring offers bit rates up to 100 Mbit/s. However, when a network does not need the secondary ring for backup purposes, it can use it to carry data as well. This effectively extends the overall bit rate to 200 Mbit/s. FDDI has a larger maximum-frame size with respect to standard 100 Mbit/s Ethernet and this allows a higher throughput. The definition of FDDI is contained in four standards. For the physical layer we have the physical medium dependent (PMD) specification (which provides means for communicating over point-to-point links) and the physical layer protocol (PHY) specification (which handles synchronization between higher layer data and control symbols). The data link layer includes the MAC and the LLC specifications.

Performance measures Common objective when analyzing these systems is the derivation of:

1. The queueing delay w.
2. The system time s.
3. The *polling cycle time*, t_c, i.e., the time taken between two subsequent visits of the server to the same user's queue (in successive polling cycles).

In the next sections we present a general analysis for both polling and token ring systems. For a comprehensive and advanced treatment of these protocols the interested reader is referred to [3] and [5].

Classification of system models In the literature various policies were considered for polling and token ring systems. These can be classified as follows [5]:

1. **One message buffer systems.** According to this model at most one packet can be stored in the waiting queue of each user. Hence, new packets that find the buffer occupied are discarded, whereas a new packet can only arrive upon the departure of the packet currently stored in the queue.

2. **Infinite buffer systems.** Any number of packets can be stored at each user queue. In this case four types of service discipline have been studied:

   2.1 *Exhaustive service.* The server continues to serve a given user until its buffer is emptied. Packets arriving during any given serving period are also served within the same period before switching to a new user.

2.2 *Gated service.* The server only serves the packets that are waiting in the queue in the instant when it is polled. Hence, packets arriving during any given service period are queued and served during the subsequent polling cycle.

2.3 *Limited service.* According to this policy a user is served until 1) its queue is emptied or 2) a maximum number of packets is served for this user (and thus removed from its buffer), whichever occurs first.

2.4 *Decrementing service.* The decrementing (also referred to as *semiexhaustive*) polling scheme works as follows. When at a polling instant the buffer is found nonempty, the server continues to serve this buffer until the number of packets therein decreases to one less than what found at the polling instant.

Assumptions In addition to BA1–BA7, MA1–MA5, for the analysis in the following sections we make the following assumptions.

AP1. We consider a symmetric system, i.e., the statistics governing service and switchover times are equal for all users.

AP2. When the service for a given user i terminates, a switchover time, Δ_i, is needed to pass the control to the next user $i \bmod N_u + 1$ in the sequence. This time depends on the considered system and includes: 1) the time taken for the transmission of a packet that assigns the control to the next user in the sequence and 2) synchronization and processing delays. Moreover, we define the normalized switchover time for user i as:

$$\tilde{\Delta}_i = \frac{\Delta_i}{t_{\mathrm{P}}} . \tag{9.26}$$

AP3. The switchover time is equal for all users, i.e., $\Delta_i = \Delta$, $i = 1, 2, \ldots, N_u$.

AP4. We assume that service and switchover times are statistically independent and identically distributed across different polling events.

9.2.5.1 *Analysis of One Message Buffer Systems* The analysis that follows is based on that of [5, Chapter 2] and considers one message buffer systems with generic distributions for service and switchover times.

We consider a system of N_u users, numbered as $1, 2, \ldots, N_u$, which are served in cyclic order. We define as *polling cycle* the time between the instant user 1 is polled and the instant user N_u finishes its own service period (including the switchover time) and the server returns to user 1. We further define Δ_c^m, Q_c^m and B_c^m as the the total switchover time, the number of packets served and the total service time for a given polling cycle m, respectively. The *service time* is the time spent in transmitting packets during the polling cycle. Due to assumption AP4 we have that $\mathrm{E}\left[\Delta_c^m\right] = \mathrm{E}\left[\Delta_c\right]$, $\mathrm{E}\left[Q_c^m\right] = \mathrm{E}\left[Q_c\right]$ and $\mathrm{E}\left[B_c^m\right] = \mathrm{E}\left[B_c\right]$, $\forall\, m$. Moreover, due to assumption AP3 we have

$$\mathrm{E}\left[\Delta_c\right] = \sum_{i=1}^{N_u} \mathrm{E}\left[\Delta_i\right] = N_u \Delta . \tag{9.27}$$

Before proceeding with the analysis we need to introduce two definitions and recall an important result:

Definition 9.1 (regeneration point) *Let $X(t)$ be a generic rp and $t^\star$ be a sampling instant. $t^\star$ is a regeneration point for $X(t)$ if the statistics governing the future evolution of $X(t)$ only depends on the value of $X(t^\star)$, i.e., on the state of the rp at $t^\star$, and given the value of $X(t^\star)$ is conditionally independent of any past value of $X(t')$, with $t' < t^\star$.*

Definition 9.2 (stopping time) *Let X_n, $n \geq 1$, be a sequence of rvs $X_1, X_2, \ldots$. A stopping time s with respect to the sequence $\{X_n\}$ is a rv such that the event $\{s = n\}$, with $n \geq 1$, is statistically determined by (at most) the total information known up to time n, i.e., by $X_1, X_2, \ldots, X_n$. Thus, the event $\{s = n\}$ is statistically independent of $X_{n+1}, X_{n+2}, \ldots$.*

Theorem 9.1 (Wald's equation) *If s is a stopping time with respect to an i.i.d. sequence of rvs $\{X_n, n \geq 1\}$, with $\mathrm{E}[s] < +\infty$ and $\mathrm{E}[X] < +\infty$, then*

$$\mathrm{E}\left[\sum_{n=1}^{s} X_n\right] = \mathrm{E}[s]\,\mathrm{E}[X] \ . \tag{9.28}$$

Proof An accessible and rigorous proof can be found in [13, Chapter 3, p. 38]. □

Theorem 9.1 is a generalization of the fact that when the rv s is statistically independent of $\{X_n\}$ we have

$$\mathrm{E}\left[\sum_{n=1}^{s} X_n\right] = \mathrm{E}[s]\,\mathrm{E}[X] \ . \tag{9.29}$$

Calculation of delay and throughput performance Asymptotically, at instant t_n, let $X_{i,n}$ be the number of outstanding packets at the queue of user i (this number can either be one or zero, as queues can host at most one packet). We have that the instants at which polling cycles begin (i.e., user 1 is polled) *are not* regeneration points for $\{X_{i,n}\}$. However, the instants where 1) user 1 is polled and 2) all buffers are empty are regeneration points. This follows from the memoryless property of the arrival process, assumed to be Poisson distributed. In fact, when the system is sampled empty, the statistics of $X_{i,n}$ follows an exponential distribution that does not depend on $\{X_{i,n-1}, X_{i,n-2}, \ldots\}$. We refer to the lapse of time between two subsequent regeneration points as a *regenerative cycle*. Note that the statistics governing regenerative cycles does not change across cycles (as polling, service and arrival processes remain unaltered) and, due to their construction, different regenerative cycles are statistically independent.

Let M be the number of polling cycles in a regenerative cycle and let t_c^m be the duration of the mth polling cycle in this sequence, with $m = 1, 2, \ldots, M$. M is by definition a stopping time for $\{t_c^m\}$. Thus, from (9.28) the average duration of a regenerative cycle can be written as

$$\mathrm{E}\left[\sum_{m=1}^{M} t_c^m\right] = \mathrm{E}[t_c]\,\mathrm{E}[M] \tag{9.30}$$

Further, M is also a stopping time for $\{Q_c^m\}$, with Q_c^m the number of messages served in the mth polling cycle. Hence, the average number of packets served in a regenerative cycle is:

$$\mathrm{E}\left[\sum_{m=1}^{M} Q_c^m\right] = \mathrm{E}[Q_c]\,\mathrm{E}[M] \ . \tag{9.31}$$

Now, we define ϑ as the average number of packets in a given queue during a regenerative cycle, which coincides with the probability that whenever a user is polled its queue is non empty. Still, using (9.28),

$$\begin{aligned}\vartheta &= \frac{\mathrm{E}\left[\text{Number of packets served in a regenerative cycle}\right]}{\mathrm{E}\left[\text{Number of users polled in a regenerative cycle}\right]} \\ &= \frac{\mathrm{E}\left[\sum_{m=1}^{M} Q_c^m\right]}{N_u\,\mathrm{E}\left[M\right]} = \frac{\mathrm{E}\left[Q_c\right]}{N_u}\end{aligned} \tag{9.32}$$

where we recall that N_u is the number of users polled in a polling cycle, whereas $N_u\,\mathrm{E}\left[M\right]$ is the average number of users polled in a regenerative cycle. Referring to the total switchover time and the total service time in the polling cycle m as Δ_c^m and B_c^m, respectively, it follows that the mean duration of the mth polling cycle is $\mathrm{E}\left[t_c^m\right] = \mathrm{E}\left[\Delta_c^m + B_c^m\right]$. Thus, it holds

$$\mathrm{E}\left[\sum_{m=1}^{M} t_c^m\right] = \mathrm{E}\left[\sum_{m=1}^{M} (\Delta_c^m + B_c^m)\right] = \mathrm{E}\left[\sum_{m=1}^{M} \Delta_c^m\right] + \mathrm{E}\left[\sum_{m=1}^{M} B_c^m\right] . \tag{9.33}$$

Due to our assumption AP4 we have $\mathrm{E}\left[\Delta_c^m\right] = \mathrm{E}\left[\Delta_c\right]$, $m = 1, 2, \ldots$. Moreover, we observe that M is a stopping time for the random sequence $\{\Delta_c^m\}$ as the event $\{M = n\}$ is statistically independent of $\Delta_c^{n+1}, \Delta_c^{n+2}, \ldots$. Hence, we can apply (9.28) and write

$$\mathrm{E}\left[\sum_{m=1}^{M} \Delta_c^m\right] = \mathrm{E}\left[M\right]\mathrm{E}\left[\Delta_c\right] . \tag{9.34}$$

The total service time in a regenerative cycle is obtained as

$$\mathrm{E}\left[\sum_{m=1}^{M} B_c^m\right] = \mathrm{E}\left[M\right]\mathrm{E}\left[Q_c\right] t_{\mathrm{P}} \tag{9.35}$$

which amounts to multiplying the constant packet transmission time t_{P} by the mean number of packets transmitted during a polling cycle $\mathrm{E}\left[Q_c\right]$ and by the mean number of polling cycles in a regenerative cycle $\mathrm{E}\left[M\right]$.

Using (9.34) and (9.35) into (9.33) we obtain

$$\mathrm{E}\left[\sum_{m=1}^{M} t_c^m\right] = \mathrm{E}\left[M\right]\mathrm{E}\left[\Delta_c\right] + t_{\mathrm{P}}\,\mathrm{E}\left[M\right]\mathrm{E}\left[Q_c\right] \tag{9.36}$$

Hence, from (9.30) we obtain the mean polling cycle time as

$$\mathrm{E}\left[t_c\right] = \mathrm{E}\left[\Delta_c\right] + t_{\mathrm{P}}\,\mathrm{E}\left[Q_c\right] = N_u\Delta + t_{\mathrm{P}} N_u \vartheta \tag{9.37}$$

where the second equality follows from (9.27) and (9.32). An alternative expression of $\mathrm{E}\left[t_c\right]$ is found after introducing the throughput, see (9.45).

Throughput We now proceed with the calculation of the throughput η (see (8.28)), and relate it to the system time $\mathtt{m}_s$. At any given user, the queue alternates between the full state, of average duration $\mathtt{m}_s$, and the empty state of average duration equal to the average inter-arrival time of the Poisson packet arrival process, $1/\lambda$. Thus, the mean interdeparture time $\mathtt{m}_r$ (see (8.15)) is given by

$$\mathtt{m}_r = \mathtt{m}_s + 1/\lambda \,. \tag{9.38}$$

The throughput η for a given user in the system is calculated as $1/\mathtt{m}_r$ (i.e., the number of packets served per second for that user, see (8.28)). Hence, $\eta = N_u/(\mathtt{m}_s + 1/\lambda)$ is the overall throughput, expressed in packets served per second. Using (9.12), the normalized throughput can be written as

$$\mathrm{S_{POLL}} = \frac{N_u t_\mathrm{P}}{\mathtt{m}_s + 1/\lambda} \,. \tag{9.39}$$

An alternative expression for the throughput is given by the ratio between the mean number of packets served during a polling cycle $\mathrm{E}[Q_c]$ (see (9.31)) and its average duration $\mathrm{E}[t_\mathrm{c}]$ (see (9.30)), multiplied by t_P to obtain the normalized throughput

$$\boxed{\mathrm{S_{POLL}} = \frac{\mathrm{E}[Q_c]\, t_\mathrm{P}}{\mathrm{E}[t_\mathrm{c}]} = \frac{\mathrm{E}[Q_c]\, t_\mathrm{P}}{N_u\Delta + t_\mathrm{P}\,\mathrm{E}[Q_c]}} \tag{9.40}$$

where the second equality follows from (9.37). Note that in (9.40) $\mathrm{E}[Q_c]\, t_\mathrm{P}$ returns the average amount of time used for the transmission of packets in a polling cycle.

Delay Equalling (9.39) and (9.40), using (9.37), and solving for $\mathtt{m}_s$ gives

$$\mathtt{m}_s = N_u t_\mathrm{P} - \frac{1}{\lambda} + \frac{N_u^2\Delta}{\mathrm{E}[Q_c]} \,. \tag{9.41}$$

From (9.6) the average delay for the polling access scheme is found as:

$$D_\mathrm{POLL} = \mathtt{m}_s + \tau_p = N_u t_\mathrm{P} - \frac{1}{\lambda} + \frac{N_u^2\Delta}{\mathrm{E}[Q_c]} + \tau_p \,. \tag{9.42}$$

The normalized delay is

$$\boxed{\widetilde{D}_\mathrm{POLL} = N_u - \widetilde{\left(\frac{1}{\lambda}\right)} + \frac{N_u^2\tilde{\Delta}}{\mathrm{E}[Q_c]} + \tilde{\tau}_p} \tag{9.43}$$

In what follows we write $\widetilde{D}_\mathrm{POLL}$ as a function of $\mathrm{S_{POLL}}$. To this end, we rewrite the mean polling cycle time $\mathrm{E}[t_\mathrm{c}]$ as

$$\mathrm{E}[t_\mathrm{c}] = \mathrm{E}[\Delta_c] + t_\mathrm{P}\,\mathrm{E}[Q_c] = \mathrm{E}[\Delta_c] + \mathrm{E}[t_\mathrm{c}]\,\mathrm{S_{POLL}} \tag{9.44}$$

where the second equality follows from (9.40). Solving (9.44) for $\mathrm{E}[t_\mathrm{c}]$ and using (9.27) leads to

$$\mathrm{E}[t_\mathrm{c}] = \frac{N_u\Delta}{1 - \mathrm{S_{POLL}}} \,. \tag{9.45}$$

Note that when the offered traffic $\mathtt{G} = 0$ we have that $\mathtt{S}_{\mathrm{POLL}} = 0$ (for a stable system $\mathtt{G} = \mathtt{S}$, see (9.13)) and we get $\mathrm{E}[t_c] = N_u\Delta$. Now, expressing $\mathrm{E}[Q_c]$ in (9.43) according to (9.40) and using (9.45) for $\mathrm{E}[t_c]$ we get

$$\boxed{\widetilde{D}_{\mathrm{POLL}} = N_u - \widetilde{\left(\frac{1}{\lambda}\right)} + \frac{N_u(1-\mathtt{S}_{\mathrm{POLL}})}{\mathtt{S}_{\mathrm{POLL}}} + \tilde{\tau}_p} \tag{9.46}$$

Further performance measures Another interesting performance metric is the probability that a given packet arrives at a nonempty user queue and is therefore discarded, P_{blk}. Due to the Poisson nature of the arrival process, P_{blk} is given by the long run average fraction of time that a user buffer is nonempty, i.e.,

$$P_{\mathrm{blk}} = \frac{\mathtt{m}_s}{\mathtt{m}_r} = \frac{\mathtt{m}_s}{\mathtt{m}_s + 1/\lambda} \tag{9.47}$$

where the second equality follows from (9.38). A direct comparison of (9.39) and (9.47) allows us to rewrite $\mathtt{S}_{\mathrm{POLL}}$ as

$$\mathtt{S}_{\mathrm{POLL}} = \lambda N_u t_{\mathrm{P}}(1 - P_{\mathrm{blk}}) \tag{9.48}$$

showing that the throughput $\mathtt{S}_{\mathrm{POLL}}$ is a fraction $(1 - P_{\mathrm{blk}})$ of the total arrival rate λN_u (also note the similarity with (8.17)). If we apply Little's result (8.46) to the packets that are not discarded upon their arrival, using S_{POLL} from (9.39), we have that the number of packets x in the system at an arbitrary time has mean value

$$\mathtt{m}_x = \frac{\mathtt{S}_{\mathrm{POLL}}}{t_{\mathrm{P}}}\mathtt{m}_s = \frac{N_u}{\mathtt{m}_s + 1/\lambda}\mathtt{m}_s = N_u P_{\mathrm{blk}} \tag{9.49}$$

where the last equality follows from (9.47). (9.49) can be rewritten as

$$\mathtt{m}_x = N_u - \frac{N_u(1/\lambda)}{\mathtt{m}_s + 1/\lambda} \tag{9.50}$$

where the second term in the right-hand side is the average number of empty queues in the system. Note that all the above performance measures depend on $\mathrm{E}[Q_c]$, a quantity that is in general difficult to compute. Before presenting the quite involved analysis for the evaluation of $\mathrm{E}[Q_c]$ we illustrate some performance results which make use of expression (9.64) for $\mathrm{E}[Q_c]$.

Example results In Fig. 9.4 we show the normalized delay $\widetilde{D}_{\mathrm{POLL}}$ versus $\mathtt{S}_{\mathrm{POLL}}$ for the polling access scheme. The tradeoff curves are plotted considering the normalized traffic load $\mathtt{G}$ as the independent parameter, see (9.10). Moreover, performance measures were obtained by calculating $\mathrm{E}[Q_c]$ through (9.64) and using (9.40) and (9.43). The system parameters for Fig. 9.4 are $N_u = 100$, $\tilde{\tau}_p = 0.01$ and $\tilde{\Delta} \in \{0.01, 0.1, 1, 2\}$ (see (9.26)). First of all, we observe that for given Δ and t_{P} the maximum duration of a polling cycle (see (9.37)) is $N_u t_{\mathrm{P}} + N_u\Delta$, which occurs at high traffic loads, where all buffers are found nonempty by the server. In this case the system is said to be *saturated* and the corresponding normalized throughput, $\mathtt{S}_{\mathrm{POLL}}^{\mathrm{sat}}$, is given by the ratio between the total transmission time in a polling cycle, $N_u t_{\mathrm{P}}$, and the length of the polling cycle itself $N_u t_{\mathrm{P}} + N_u\Delta$:

$$\mathtt{S}_{\mathrm{POLL}}^{\mathrm{sat}} = \frac{N_u t_{\mathrm{P}}}{N_u t_{\mathrm{P}} + N_u\Delta} = \frac{1}{1+\tilde{\Delta}} \,. \tag{9.51}$$

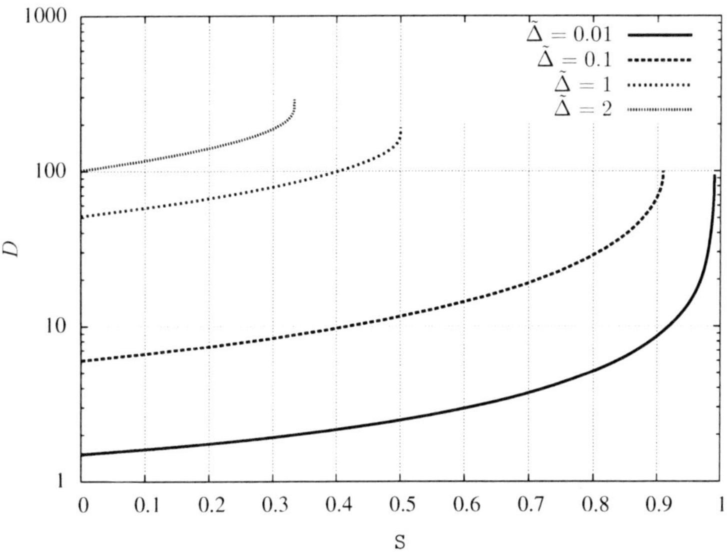

Figure 9.4 Normalized delay $\widetilde{D}_{\rm POLL}$ versus normalized throughput $\mathtt{S}_{\rm POLL}$ for the polling access scheme. $\tilde{\tau}_p = 0.01$, $N_u = 100$ and $\tilde{\Delta} \in \{0.01, 0.1, 1, 2\}$.

Thus, for, e.g., $\tilde{\Delta} = 1$ and $\tilde{\Delta} = 2$ the maximum throughput is $1/2$ and $1/3$, respectively, as confirmed by the results in Fig. 9.4. From this figure we additionally observe that for any given $\tilde{\Delta}$, for $\mathtt{G} \to 0$, there is a minimum delay of $N\Delta/2$. This is because when $\mathtt{G}$ is small the total service time for any given polling cycle approaches zero and, upon its arrival at a given user, a packet only has to wait for the server to poll the corresponding queue; this waiting period has mean value $N\Delta/2$. As expected, $\widetilde{D}_{\rm POLL}$ increases for increasing $\mathtt{G}$. Also, we note that $\widetilde{D}_{\rm POLL}$ is upper bounded by $(N-1) + N\tilde{\Delta}$, which occurs when the packet arrives at the end of the current service period and the system is saturated. In fact, in one message buffer systems packets have zero queueing delay and, upon their arrival, they only need to wait for the next polling instant in order to be served.

Derivation of $\mathrm{E}[Q_c]$ *for the case of constant parameters* The analysis that follows was originally derived in [10] (in the context of a machine repairman model) and then re-proposed in [5].

First, we obtain the PMD of the number of packets Q_c served in a polling cycle. We define the queue state of user i as u_i, where $u_i = 1$ if the queue at this user is full and $u_i = 0$ if the queue is empty. With $\tau_i(m)$ we indicate the polling instant for user i in the mth polling cycle. With $\tau_i^+(m)$ we instead indicate the instant where the service for user i ends. Note that the polling instant of user $i+1$ in the mth polling cycle is $\tau_{i+1}(m) = \tau_i^+(m) + \Delta$. Further, we refer to $u_i(m)$ as the queue state of user i at time $\tau_i(m)$ and with $\mathbf{U}_i(m)$ we denote the sequence

$$\mathbf{U}_i(m) = [u_1(m), u_2(m), \ldots, u_i(m), u_{i+1}(m-1), u_{i+2}(m-1), \ldots, u_{N_u}(m-1)] \ . \quad (9.52)$$

Moreover, we define $P_{\mathbf{U}_i(m)}(\mathbf{a})$, where $\mathbf{a} = [a_1, a_2, \ldots, a_{N_u}]$, $a_i \in \{0,1\}$, as the probability that the server observes the sequence $\mathbf{U}_i(m) = \mathbf{a}$ for the queue states of the N_u users at their polling instants up to and including $\tau_i(m)$. We observe that the transition of the system from $\mathbf{U}_{i-1}(m)$ to $\mathbf{U}_i(m)$ only depends on $\mathbf{U}_{i-1}(m)$ and on the packet arrivals at user i in the time interval $[\tau_i^+(m-1), \tau_i(m)]$. We note that

1. if $u_i(m) = 1$ this means that the server observed a packet in the queue of user i at time $\tau_i(m)$. Hence, this packet is served in the current polling period for user i, $[\tau_i(m), \tau_i^+(m)]$.
2. If $u_i(m) = 0$ the queue of user i is empty at time $\tau_i(m)$. In this case the service for this user is immediately concluded, i.e., $\tau_i(m)$ coincides with $\tau_i^+(m)$ and we have to wait the switchover period $[\tau_i^+(m), \tau_{i+1}(m)]$ before moving to the next user.
3. The process $\{\mathbf{U}_i(m), m \geq 1\}$ is a Markov chain having 2^{N_u} distinct states in $\{0,1\}^{N_u}$.

We now compute some useful probabilities for this Markov chain. First of all, note that $u_i(m) = 0$ if and only if there are no arrivals at user i in the interval $[\tau_i^+(m-1), \tau_i(m)]$. The duration of this interval is given by the sum of two contributions: the fist one is given by $N_u\Delta$, which is the total switchover time between $\tau_i^+(m-1)$ and $\tau_i(m)$, whereas the second contribution is given by $t_{\rm P}(\sum_{j=i+1}^{N_u} u_j(m-1) + \sum_{j=1}^{i-1} u_j(m))$, which is the total average transmission time for all the users served in $[\tau_i^+(m-1), \tau_i(m)]$. Note that the transmission time for each user is $t_{\rm P}$ if the corresponding queue was nonempty and zero otherwise. Given this, we obtain the probability that $u_i(m) = 0$ as

$$\mathrm{P}\left[u_i(m) = 0\right] = \exp\left[-\lambda N_u \Delta - \lambda t_{\rm P}\left(\sum_{j=i+1}^{N_u} u_j(m-1) + \sum_{j=1}^{i-1} u_j(m)\right)\right]. \tag{9.53}$$

Remark (on the buffer dynamics across cycles) Note that at time $\tau_i^+(m)$ the buffer of user i is always empty. In fact, if it was nonempty at time $\tau_i(m)$ it is served during $[\tau_i(m), \tau_i^+(m)]$, on the other hand, if it was empty at time $\tau_i(m)$ the service is immediately concluded and the server passes the control to the next user in the sequence and $\tau_i^+(m) = \tau_i(m)$. Note that this also implies that the value of $u_i(m-1)$ does not affect $u_i(m)$.

Now, we write two important flow balance equations that will be key in finding the asymptotic probabilities of $\{\mathbf{U}_i(m)\}$ (see also (7.188)). The first equation is

$$\begin{aligned} &P_{\mathbf{U}_i(m)}(a_1, \ldots, a_{i-1}, 0, a_{i+1}, \ldots, a_{N_u}) \\ &= \exp\left[-\lambda N_u \Delta - \lambda t_{\rm P}\left(\sum_{j=i+1}^{N_u} u_j(m-1) + \sum_{j=1}^{i-1} u_j(m)\right)\right] \\ &\cdot \sum_{a_i=0}^{1} P_{\mathbf{U}_{i-1}(m)}(a_1, \ldots, a_{i-1}, a_i, a_{i+1}, \ldots, a_{N_u})\,. \end{aligned} \tag{9.54}$$

where we calculate $P_{\mathbf{U}_i(m)}(a_1, \ldots, a_{i-1}, 0, a_{i+1}, \ldots, a_{N_u})$ as the product of two terms. The first term corresponds to the probability that $u_i(m) = 0$ and is given by (9.53). The second term accounts for the probabilities that the remaining queues j with $j \neq i$ are respectively in states a_j with $j = 1, 2, \ldots, i-1, i+1, \ldots, N_u$. Note that the state of queue $j \neq i$ in $\mathbf{U}_{i-1}(m)$ remains unchanged during the polling period of user i ($\mathbf{U}_{i-1}(m) \to \mathbf{U}_i(m)$), whereas a_i, which represents the value of $u_i(m-1)$, must be summed over all possible values because, as we have seen in the previous remark, $u_i(m-1)$ does not affect $u_i(m)$. The second equation evaluates the probability for a similar event with $u_i(m) = 1$,

$$\begin{aligned} &P_{\mathbf{U}_i(m)}(a_1, \ldots, a_{i-1}, 1, a_{i+1}, \ldots, a_{N_u}) \\ &= \left\{1 - \exp\left[-\lambda N_u \Delta - \lambda t_{\mathrm{P}}\left(\sum_{j=i+1}^{N_u} u_j(m-1) + \sum_{j=1}^{i-1} u_j(m)\right)\right]\right\} \\ &\quad \cdot \sum_{a_i=0}^{1} P_{\mathbf{U}_{i-1}(m)}(a_1, \ldots, a_{i-1}, a_i, a_{i+1}, \ldots, a_{N_u}) \, . \end{aligned} \tag{9.55}$$

Now, we consider the asymptotic probability

$$P_{\mathbf{U}_i}(\mathbf{a}) = \lim_{m \to +\infty} P_{\mathbf{U}_i(m)}(\mathbf{a}) \, . \tag{9.56}$$

Thus, by taking the limit for $m \to +\infty$ of (9.54) and (9.55) we obtain

$$\begin{aligned} P_{\mathbf{U}_i}(a_1, \ldots, a_{i-1}, 0, a_{i+1}, \ldots, a_{N_u}) &= \exp\left[-\lambda N_u \Delta - \lambda t_{\mathrm{P}}\left(\sum_{j=1, j\neq i}^{N_u} u_j\right)\right] \\ &\quad \cdot \sum_{a_i=0}^{1} P_{\mathbf{U}_{i-1}}(a_1, \ldots, a_{i-1}, a_i, a_{i+1}, \ldots, a_{N_u}) \end{aligned} \tag{9.57}$$

$$\begin{aligned} P_{\mathbf{U}_i}(a_1, \ldots, a_{i-1}, 1, a_{i+1}, \ldots, a_{N_u}) &= \left\{1 - \exp\left[-\lambda N_u \Delta - \lambda t_{\mathrm{P}}\left(\sum_{j=1, j\neq i}^{N_u} u_j\right)\right]\right\} \\ &\quad \cdot \sum_{a_i=0}^{1} P_{\mathbf{U}_{i-1}}(a_1, \ldots, a_{i-1}, a_i, a_{i+1}, \ldots, a_{N_u}). \end{aligned} \tag{9.58}$$

It can be verified, by direct substitution in (9.57) and (9.58), that the solution of the above equations is

$$P_{\mathbf{U}_i}(\mathbf{a}) = \begin{cases} K & , \quad a_j = 0, \ \forall\, j \\ K \displaystyle\prod_{j=0}^{\sum_{r=1}^{N_u} a_r - 1} \left\{\exp[\lambda(N_u \Delta + j t_{\mathrm{P}})] - 1\right\} & , \quad \sum_{r=1}^{N_u} a_r > 0 \end{cases} \tag{9.59}$$

where K is a constant such that

$$\sum_{(a_1, \ldots, a_{N_u}) \in \{0,1\}^{N_u}} P_{\mathbf{U}_i}(a_1, \ldots, a_{N_u}) = 1 \, . \tag{9.60}$$

We observe that $P_{\mathbf{U}_i}(\mathbf{a})$ in (9.59) does not depend on i and this is due to the symmetry of the system. Moreover, $P_{\mathbf{U}_i}(\mathbf{a})$ is expressed only as a function of $\sum_{r=1}^{N_u} a_r$, which is nothing but

Q_c, the number of packets served in a polling cycle. Considering that there are $\binom{N_u}{n}$ sequences $[a_1, \ldots, a_{N_u}]$ containing exactly n ones and $N_u - n$ zeros, it follows that the probability that $Q_c = n$ is obtained as

$$\mathrm{P}\left[Q_c = n\right] = \begin{cases} K & , \ n = 0 \\ K\binom{N_u}{n}\prod_{j=0}^{n-1}\left\{\exp[\lambda(N_u\Delta + jt_{\mathrm{P}})] - 1\right\} & , \ 1 \le n \le N_u \end{cases} \tag{9.61}$$

where K is such that (9.61) satisfies the normalization condition

$$\sum_{n=0}^{N_u} \mathrm{P}\left[Q_c = n\right] = 1\,. \tag{9.62}$$

Hence

$$K^{-1} = 1 + \sum_{n=1}^{N_u}\binom{N_u}{n}\prod_{j=0}^{n-1}\left\{\exp[\lambda(N_u\Delta + jt_{\mathrm{P}})] - 1\right\}. \tag{9.63}$$

From (9.61) and (9.63), the mean number of packets served in a polling cycle is

$$\begin{aligned} \mathrm{E}\left[Q_c\right] &= \sum_{n=1}^{N_u} n\,\mathrm{P}\left[Q_c = n\right] \\ &= \frac{\sum_{n=1}^{N_u} n\binom{N_u}{n}\prod_{j=0}^{n-1}\left\{\exp[\lambda(N_u\Delta + jt_{\mathrm{P}})] - 1\right\}}{1 + \sum_{n=1}^{N_u}\binom{N_u}{n}\prod_{j=0}^{n-1}\left\{\exp[\lambda(N_u\Delta + jt_{\mathrm{P}})] - 1\right\}} \\ &= \frac{N_u\sum_{n=0}^{N_u-1}\binom{N_u-1}{n}\prod_{j=0}^{n}\left\{\exp[\lambda(N_u\Delta + jt_{\mathrm{P}})] - 1\right\}}{1 + \sum_{n=1}^{N_u}\binom{N_u}{n}\prod_{j=0}^{n-1}\left\{\exp[\lambda(N_u\Delta + jt_{\mathrm{P}})] - 1\right\}} \end{aligned} \tag{9.64}$$

where the second equality follows from the identity $n\binom{N_u}{n} = N_u\binom{N_u-1}{n-1}$. From (9.32) we finally obtain the probability ϑ that a packet is found at a polling instant

$$\begin{aligned} \vartheta &= \frac{\mathrm{E}\left[Q_c\right]}{N_u} \\ &= \frac{\sum_{n=0}^{N_u-1}\binom{N_u-1}{n}\prod_{j=0}^{n}\left\{\exp[\lambda(N_u\Delta + jt_{\mathrm{P}})] - 1\right\}}{1 + \sum_{n=1}^{N_u}\binom{N_u}{n}\prod_{j=0}^{n-1}\left\{\exp[\lambda(N_u\Delta + jt_{\mathrm{P}})] - 1\right\}}\,. \end{aligned} \tag{9.65}$$

9.2.5.2 Analysis of Infinite Buffer Systems We consider a polling system with N_u users which are served in the order $1, 2, \ldots, N_u$, one user at a time according to an exhaustive service policy (see page 591). Note that the total service time for user i during a polling cycle depends on the number of packets waiting in the queue of user i when it is polled as, according to the exhaustive service discipline, they will all be served before passing the control to the next user $i \bmod N_u + 1$. In what follows we give a simple analysis of infinite buffer systems. For the sake of tractability we omit the proof of important results that will be instrumental to derive the delay performance. A rigorous and complete analysis of infinite buffer systems can be found in [5, Chapters 3 to 5].

For the analysis in this section we consider assumptions AP1–AP4. We recall that we consider the system stable ($\rho < 1$), i.e., the number of packets served on average equals the number of packet arrived during a polling cycle, i.e., (9.13) holds.

Polling cycle time The mean polling cycle time is still given by (9.45).

Throughput The normalized offered traffic is $\mathtt{G} = \lambda N_u t_{\mathrm{P}}$. Using (9.13) we obtain the normalized throughput $\mathtt{S}_{\mathrm{POLL}}$ as

$$\boxed{\mathtt{S}_{\mathrm{POLL}} = \lambda N_u t_{\mathrm{P}}} \tag{9.66}$$

Delay For the delay analysis we use the *pseudo-conservation law* due to Boxma and Groenendijk [11], that applies to exhaustive polling systems with infinite buffers, a single server and Poisson arrivals at all stations. This law applies to general systems where arrival rates and switchover times can differ for each user. Specializing it to our case (constant arrival rates and switchover times, equal for all user) leads to the following expression for the mean queueing time:

$$\mathtt{m}_w = \frac{N_u \lambda t_{\mathrm{P}}^2 + \Delta(N_u - \mathtt{G})}{2(1 - \mathtt{G})} \; . \tag{9.67}$$

Note that when the switchover time is zero, $\Delta = 0$, the mean queueing time becomes that of an M/G/1 queueing system with input rate $N_u\lambda$ and constant service rate $1/t_{\mathrm{P}}$ for which $\mathtt{M}_y = t_{\mathrm{P}}^2$, as given by (8.221). This demonstrates that, when $\Delta = 0$, the queues of the N_u users can be seen as a single equivalent queue with arrival rate $N_u\lambda$.

According to (9.8) and $\mathtt{m}_{\mathrm{retx}} = 0$,[4] the packet delay is given by

$$\boxed{\widetilde{D}_{\mathrm{POLL}} = \frac{\mathtt{S}_{\mathrm{POLL}} + \widetilde{\Delta}(N_u - \mathtt{S}_{\mathrm{POLL}})}{2(1 - \mathtt{S}_{\mathrm{POLL}})} + 1 + \tilde{\tau}_p} \tag{9.68}$$

where we used (9.66) and the fact that $\mathtt{G} = \mathtt{S}_{\mathrm{POLL}}$.

Example results In Fig. 9.5 we plot the normalized delay $\widetilde{D}_{\mathrm{POLL}}$ as a function of $\mathtt{S}_{\mathrm{POLL}}$ ($= \mathtt{G}$) for one message buffer (OMB) and infinite buffer (IB) systems, as given by (9.46) and (9.68), respectively. The system parameters for this plot are $N_u = 100$, $\tilde{\tau}_p = 0.01$ and $\tilde{\Delta} \in \{0.01, 0.1, 1\}$. At small $\mathtt{S}$ the two buffer models show similar performance. In fact, the two access policies are equivalent as long as long as no packets are dropped from the buffers in the OMB case. However, as $\mathtt{S}$ approaches one, the delay of infinite buffer systems increases indefinitely whereas that of one message buffer systems has a maximum. In addition, the maximum achievable throughput for OMB systems depends on the duration of the switchover

[4]Note that in the considered polling system collisions never occur as the channel is accessed by one user at a time; thus $\mathtt{m}_{\mathrm{retx}} = 0$.

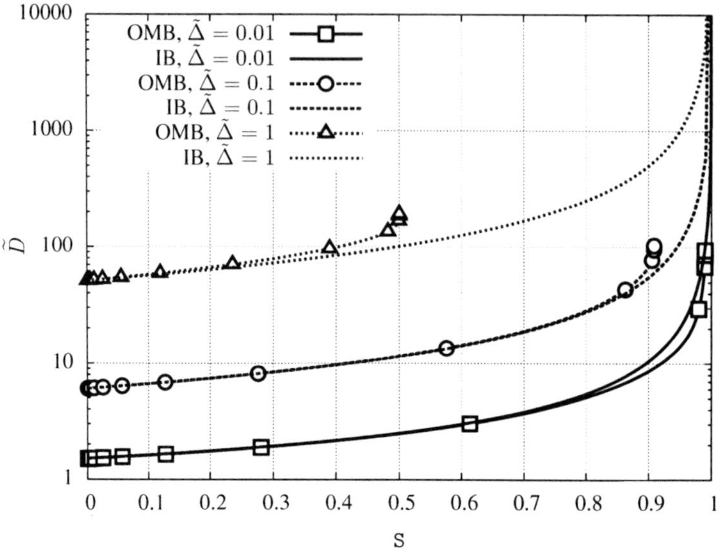

Figure 9.5 Normalized delay $\widetilde{D}_{\text{POLL}}$ versus normalized throughput $\mathtt{S}_{\text{POLL}}$ for one message buffer (OMB) and infinite buffer (IB) systems: $\tilde{\tau}_p = 0.01$, $N_u = 100$ and $\tilde{\Delta} \in \{0.01, 0.1, 1\}$.

time Δ, while for IB schemes it gets very close to one, independently of Δ. In fact, in IB systems, when the system is close to saturation (i.e., $\mathtt{S} \to 1$) the switchover delay in a polling cycle, $N_u\Delta$, become negligible with respect to the time spent in serving the user queues. Indeed, all queues are served until they are emptied and this takes an increasingly large amount of time when S (G) approaches one.

9.2.5.3 Stability conditions A few remarks on stability conditions are in order. We recall the notation of page 593 with $X_{i,n}$ being the number of packets at each queue i at instant t_n. The instants in which user i is polled and all queues are empty, i.e., $X_{i,n} = 0$, are regeneration points for the system. Let Δt_i be the time interval between two regeneration points for queue i. The polling system is stable if and only if $\lim_{a\to+\infty} \mathrm{P}[\Delta t_i > a] = 0$ and $\mathrm{E}[\Delta t_i] = \int_0^{+\infty} \mathrm{P}[\Delta t_i > a]\, da < +\infty$. Under these conditions, the mean buffer size at each user is finite and, in turn, the mean polling cycle time is also finite. The converse is not necessarily true. In fact, for limited service systems (see page 592) some buffers might grow indefinitely even though the polling cycle time is still finite. Necessary and sufficient conditions for the stability of the four service disciplines are [2]:

$$\begin{aligned} \text{exhaustive and gated:} \quad & \mathtt{G} < 1 \\ \text{limited:} \quad & \mathtt{G} + \lambda N_u \Delta < 1 \\ \text{decrementing:} \quad & \mathtt{G} + \lambda(1 - \lambda t_{\text{P}}) N_u \Delta < 1 \end{aligned} \tag{9.69}$$

where λ and Δ are respectively the packet arrival rate and the switchover time for each user, whereas G is the normalized offered traffic (see (9.10)). Note that the switchover time Δ does not play any role in the stability condition for the exhaustive and gated service policies. This is why, in these cases, when the system is close to saturation the time spent by the server in passing the control to the next station becomes negligible with respect to the time spent serving the user queues. We also note that the above concept of stability does not apply to one message buffer systems. In fact, in these systems the buffer contains at most one packet and, in turn, the polling cycle time is always finite with maximum value $N_u t_P + N_u \Delta$. In these systems there is however a positive probability of discarding packets, as we have seen in (9.47).

9.2.6 Random Access Protocols: ALOHA and Slotted ALOHA

Random access techniques arise from packet radio networks, where the transmission channel is shared among users. The peculiarities of these access techniques are:

1. The users autonomously decide the transmission instants for their packets.
2. Once a user accesses the channel, it transmits using the entire available rate.
3. If multiple users transmit concurrently so that their transmissions overlap in time, all or some of the messages at the intended receivers might be undecodable. In this case, we say that *collisions* affected the transmission and retransmission of the collided packets is needed for full reliability: an ARQ technique is usually coupled with the MAC to this purpose.

We note that in some cases overlapping packets do not cause a collision. This might be due to different reasons. First, any particular packet might be received with a larger power with respect to the others; so the receiver successfully synchronizes on the strongest signal and treats the rest as noise. Second, there might be very little time overlap among the transmitted packets. Third, some packets might be recovered due to the use of powerful error correction codes. In addition to packet collisions, errors in received packets can be also due to channel noise.

The advantages of random access techniques are that: 1) users do not need a controller for resource assignment that makes it possible to have fully decentralized solutions and 2) the channel is used aggressively as users try to access it as soon as they have new data to send; this is especially advantageous at low traffic loads (remind the drawback of TDMA that, for a large number of users, wastes channel resources waiting for the assigned slot). The downside of random access techniques is mainly related to the collision problem, whose impact on throughput and delay performance becomes larger with increasing traffic load.

Assumptions for the analysis of random access protocols For random access schemes we make the following additional assumptions, with respect to BA1–BA7 in Section 9.1.

RA1. The channel is accessed by an infinite number of users: those with a packet to transmit will contend for the channel, the remaining users will stay idle.

RA2. Packet arrivals are modeled through a Poisson process of parameter λ packet/s as follows. Whenever a new packet arrives, it is inserted in the queue of one of the users that have no data traffic to transmit (i.e., one of the idle users). This simplifies the analysis as the queueing delay is always zero.

RA2. All users transmit their data over the channel independently, i.e., without the need for a centralized control to separate their packets in time.

RA4. The only cause of packet errors is due to packet collisions which generate multiuser interference, whereas the channel is assumed to be noiseless. In other words, packet errors are never due to impairments induced by the transmission channel.

RA5. Whenever the transmission of multiple packets overlap, we assume that the resulting interference is destructive, i.e., all overlapping packets are lost with probability one. This is a somewhat pessimistic assumption for the reasons that we discussed above.

RA6. In case the channel is sensed before transmission attempts, the time taken to sense the channel is assumed to be negligible.

RA7. The compound packet arrival process, given by the superposition of new packet arrivals and retransmissions, is assumed to be a Poisson process. This assumption is not rigorously verified for the protocols that we discuss below and it is considered for the sake of analytical tractability. However, as demonstrated by the simulation results in [14], this assumption leads to very accurate approximations of throughput and delay performance.

9.2.6.1 ALOHA ALOHA is a very simple access protocol where each user sends data whenever it has a packet to send. If the packet successfully reaches the destination, hence there are no collisions, the transmission is completed. On the other hand, in case of incorrect reception, the packet is sent again after a random delay called backoff time. This protocol was originally developed at the University of Hawaii by N. Abramson [15] for use with satellite communication systems in the Pacific.

In a wireless broadcast system (where one source user sends information to all remaining users) or a half-duplex two-way link (where users communicate alternatively in time, never concurrently), ALOHA works very well. However, as networks become more complex, for example in a LAN involving multiple sources and destinations or wireless systems where multiple transmitters share the same physical medium, the collision problem becomes more and more important. The result is a degradation of throughput, delay and energy efficiency performance. In fact, when two or more packets collide, the corresponding data is likely to be lost, which means that the channel resources used in this communication are wasted (throughput reduction) and the collided packets must be retransmitted, leading to longer delays and a higher energy consumption.

Definition 9.3 (backlogged users and rescheduled transmissions) *We define a user as* backlogged *if its previously transmitted packet collided and this user is now in the process of retransmitting the collided packet. We define a packet as* rescheduled *if it was previously transmitted and collided during all its previous transmission attempts; this packet due to assumption BA4 will be scheduled again for transmission according to the protocol rules.*

Definition 9.4 (backoff time) *We define the* backoff time *for a given transmitted packet as the time elapsed between the instant where the transmitter is notified by the receiver that this packet is collided and the subsequent retransmission of the packet.*

Throughput For simplicity, we assume an infinite number of users (RA1 above). To have a tractable analysis, we assume that the backoff time for collided packets is described by the rv τ_{rtx} with exponential PDF

$$p_{\tau_{\mathrm{rtx}}}(a) = \beta e^{-\beta a} \ , \ a \geq 0 \tag{9.70}$$

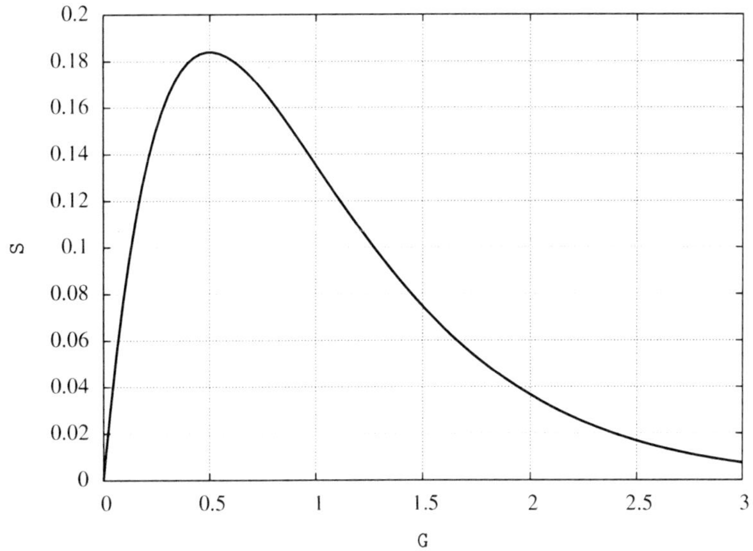

Figure 9.6 Normalized throughput versus the normalized offered traffic for pure ALOHA.

where $\beta > 0$ is a system parameter. Moreover, we assume that the backoff times of different transmitted packets are independent and identically distributed. Hence, the overall packet rate of the system is given by the Poisson process of rate λ (governing packets arrivals) plus the packets due to retransmissions (backlogged users). From RA7, at a given time instant, if there are n backlogged users we have that the packet transmission process, including new transmissions as well as retransmissions, can be approximated by a further Poisson process of rate $\lambda_t(n) = \lambda + n\beta$ as n are the users with a packet to retransmit at rate β. For the following analyses, we assume that n varies slowly with time. This is of course an approximation as n decreases by one when a collided packet is correctly transmitted and increases whenever a packet collisions occurs. However, for small β (long backoff times before attempting a retransmission) this assumption leads to precise results (see [6, p. 288]). Now, if a packet is sent by a given user at time t, any other transmission starting between $t - t_P$ and $t + t_P$ will lead to a collision. In other words, we have a successful transmission if and only if there are no further transmissions for an interval of $2t_P$ seconds. This interval is referred to as "vulnerable period" for the transmitted packet, as only during this lapse of time new arrivals cause a collision. Thus, the success probability corresponds to the probability of no further arrivals for the Poisson process of rate $\lambda_t(n)$ within an interval of lenght $2t_P$ and is given by[5]

$$P_{succ} = e^{-\lambda_t(n)2t_P} \ . \tag{9.71}$$

From (9.10) and RA2, the normalized offered traffic $\mathtt{G}(n)$ is given by $\mathtt{G}(n) = \lambda_t(n)t_P$. As $\mathtt{G}(n)$ is the fraction of time the channel is busy with the transmission of data packets but only

[5] We observe that this analysis is approximated as n is considered constant for an interval of $2t_P$ seconds. When β is small, i.e., users wait a long time before attempting a retransmission, the effect of this assumption is also small [6].

a fraction of them (P_{succ}) is successfully received, the normalized throughput of the ALOHA system is thus found as $\mathtt{S}_{\text{ALOHA}} = \mathtt{G}(n)P_{\text{succ}}$. Thus, using (9.71) we can write

$$\boxed{\mathtt{S}_{\text{ALOHA}} = \mathtt{G}(n)e^{-2\mathtt{G}(n)}} \tag{9.72}$$

In Fig. 9.6 we show $\mathtt{S}_{\text{ALOHA}}$ as a function of $\mathtt{G}(n)$. The throughput presents a maximum of $1/(2e) \simeq 0.18$ which is obtained for $\mathtt{G}(n) = 0.5$, as can be verified by equalling the first order derivative of (9.72) to zero and solving for $\mathtt{G}(n)$, i.e.,

$$\frac{d\mathtt{S}_{\text{ALOHA}}}{d\mathtt{G}(n)} = e^{-2\mathtt{G}(n)}(1 - 2\mathtt{G}(n)) = 0 \Rightarrow \mathtt{G}(n) = \frac{1}{2} \,. \tag{9.73}$$

From Fig. 9.6 we can see that an increasing $\mathtt{G}(n)$ leads to a better occupation of the channel and thus the throughput also increases. However, beyond $\mathtt{G}(n) = 0.5$ packet collisions dominate the throughput performance.

We observe that this simple analysis, though useful to describe the relationship between offered traffic $\mathtt{G}$ and throughput $\mathtt{S}$, does not describe the dynamics of the system. In fact, as n changes, $\mathtt{G}(n)$ changes too and this leads to a feedback effect by which the number of backlogged users will also change.

Delay The average delay is computed using (9.8) according to the following steps:

1. *Mean queueing delay* $\mathtt{m}_w$. Due to RA2, $\mathtt{m}_w = 0$, as a new packet is assigned to an idle user.

2. *Decomposition of* $\mathtt{m}_{\text{retx}}$. Let n_{retx} be the average number of retransmissions per packet and t_{retx} be the total time taken by each retransmission. According to our backoff time model, t_{retx} is independent and identically distributed for each packet. Hence, we have that

$$\mathtt{m}_{\text{retx}} = \mathtt{m}_{n_{\text{retx}}} \mathtt{m}_{t_{\text{retx}}} \,. \tag{9.74}$$

3. *Computation of* $\mathtt{m}_{n_{\text{retx}}}$. RA7 plays a fundamental role. In fact, it assures that the probability of success P_{succ} for any given packet is the same regardless of whether the packet is new or it has collided a number of times. From (9.71) and (9.72) we can write $P_{\text{succ}} = \mathtt{S}_{\text{ALOHA}}/\mathtt{G}(n)$ or

$$\frac{\mathtt{G}(n)}{\mathtt{S}_{\text{ALOHA}}} = \frac{1}{P_{\text{succ}}} \tag{9.75}$$

which can be seen as the average number of transmissions needed to successfully deliver a packet. Hence, the average number of retransmissions is

$$\mathtt{m}_{n_{\text{retx}}} = \frac{\mathtt{G}(n)}{\mathtt{S}_{\text{ALOHA}}} - 1 \tag{9.76}$$

as any packet is by definition successfully delivered at its last transmission.

4. *Computation of* $\mathtt{m}_{t_{\text{retx}}}$. The transmission of each packet takes t_{P} seconds and τ_p further seconds are needed for its reception at the receiver. In case a packet collides due to

concurrent transmissions we assume that the transmitter is notified through an acknowledgment (ACK) that is received error free (see BA7 in Section 9.1).[6] The transmission of the acknowledgment (ACK) takes t_A seconds and τ_p further seconds are needed for its reception at the transmitter. Upon the reception of the ACK, the transmitter waits a random backoff time τ_{rtx} which is picked from the exponential distribution of parameter β, see (9.70).[7] This leads to an average backoff period of $E[\tau_{rtx}] = 1/\beta$, after which the transmitter resends the collided packet. Summing these contributions we obtain $\mathtt{m}_{t_{retx}}$ as

$$\mathtt{m}_{t_{retx}} = t_P + 2\tau_p + t_A + \frac{1}{\beta}. \tag{9.77}$$

Thus, $\mathtt{m}_{retx}$ is given by

$$\mathtt{m}_{retx} = \mathtt{m}_{n_{retx}}\mathtt{m}_{t_{retx}} = \left(\frac{\mathtt{G}(n)}{\mathtt{S}_{ALOHA}} - 1\right)\left(t_P + 2\tau_p + t_A + \frac{1}{\beta}\right). \tag{9.78}$$

From the calculations in the previous steps and (9.8) we obtain the mean delay as

$$\begin{aligned} D_{ALOHA} &= \mathtt{m}_{n_{retx}}\mathtt{m}_{t_{retx}} + t_P + \tau_p \\ &= \left(\frac{\mathtt{G}(n)}{\mathtt{S}_{ALOHA}} - 1\right)\left(t_P + 2\tau_p + t_A + \frac{1}{\beta}\right) + t_P + \tau_p \\ &= \left(e^{2\mathtt{G}(n)} - 1\right)\left(t_P + 2\tau_p + t_A + \frac{1}{\beta}\right) + t_P + \tau_p \end{aligned} \tag{9.79}$$

where the second equality follows from (9.72). The normalized delay is

$$\boxed{\widetilde{D}_{ALOHA} = \left(e^{2\mathtt{G}(n)} - 1\right)\left(1 + 2\tilde{\tau}_p + \tilde{t}_A + \widetilde{\left(\frac{1}{\beta}\right)}\right) + 1 + \tilde{\tau}_p} \tag{9.80}$$

Example Results Fig. 9.7 shows the tradeoff between normalized delay $\widetilde{D}_{ALOHA}$ and throughput $\mathtt{S}_{ALOHA}$ for the ALOHA protocol. All curves in this figure are plotted using the offered traffic $\mathtt{G}$ as the independent parameter. For this figure, and for all the remaining figures in this chapter, we considered a link layer transmission bit rate $R_b = 11$ Mbit/s, whereas the size of PDUs and ACKs is 1500 and 40 bytes, respectively. These settings give $t_P = 1.04$ ms and $t_A = 27.74$ μs. The remaining parameters were $\tilde{\tau}_p \in \{0.01, 0.1, 1\}$ and $\beta \in \{1/(100t_P), 1/(10t_P), 1/t_P\}$. From Figs. 9.7(a) and 9.7(b) we see how these parameters impact the delay, leading to a rather expected behavior. We observe that there exists an initial (stable) region within which both delay and throughput increase for increasing $\mathtt{G}$. As $\mathtt{G}$ continues to increase (unstable region) the throughput starts decreasing while the delay is still increasing and becomes unbounded as the throughput approaches zero.

[6] Another possibility, often preferred in practice, is that the receiver notifies the transmitter sending an ACK only when the packet is successfully received, whereas no feedback is sent in case of collision. The analysis of this policy is entirely similar to the one discussed in this chapter.

[7] Backing off after the reception of the ACK is necessary to desynchronize different transmitters of the collided packets, so as to avoid further collisions during their subsequent retransmission attempts.

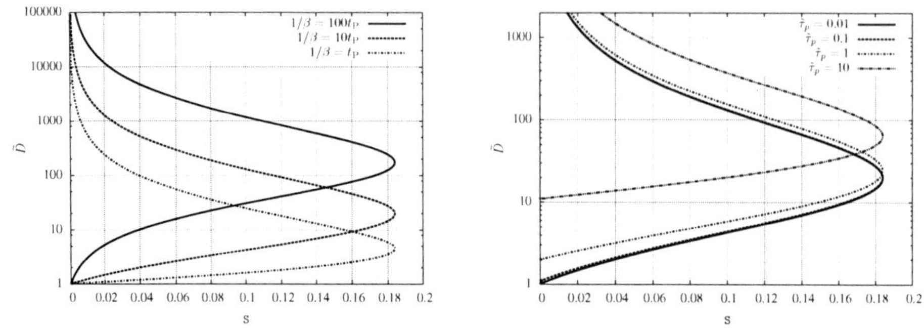

(a) $\tilde{\tau}_p = 0.01$ and varying β.

(b) $\beta = 1/(10t_{\mathrm{P}})$ and varying $\tilde{\tau}_p$.

Figure 9.7 Normalized delay $\widetilde{D}_{\mathrm{ALOHA}}$ versus normalized throughput $\mathrm{S}_{\mathrm{ALOHA}}$ for two setups: a) $\tilde{\tau}_p = 0.01$ and varying β and b) $\beta = 1/(10t_{\mathrm{P}})$ and varying $\tilde{\tau}_p$.

9.2.6.2 Slotted ALOHA The poor performance of ALOHA with a large number of contending users is due to the fact that there is no coordination among them, as they access the channel whenever they have data to transmit. This is good at low offered traffic but creates many collisions when the offered traffic increases beyond a certain threshold. This problem can be mitigated by: 1) slotting the time with a slot time equals to the packet transmission time and 2) synchronizing the users so that they perform their packet transmissions at the beginning of these slots. In particular, a packet arriving in the middle of a time slot must wait until the end of this slot for its actual transmission over the channel. This is exactly what is done in slotted ALOHA. Hence, the vulnerable period that for ALOHA was $2t_{\mathrm{P}}$ for slotted ALOHA is reduced to t_{P} (a single slot). Putting it differently, we have a successful transmission if and only if a single packet is scheduled for transmission within a time slot.

Throughput The throughput is computed as the fraction of successful slots, i.e., of slots where a single packet is scheduled for transmission. As in ALOHA the input rate λ is increased to $\lambda_{\mathrm{t}}(n) = \lambda + n\beta$, where n is the number of backlogged users, and the transmission process (including new packets as well as retransmissions) is again a Poisson process with rate $\lambda_{\mathrm{t}}(n)$. The probability of success is thus given by

$$P_{\mathrm{succ}} = e^{-\lambda_{\mathrm{t}}(n)t_{\mathrm{P}}} = e^{-\mathtt{G}(n)} \; . \tag{9.81}$$

where the normalized offered traffic is still equal to $\mathtt{G}(n) = \lambda_{\mathrm{t}}(n)t_{\mathrm{P}}$. The normalized throughput of slotted ALOHA is computed as $\mathtt{G}(n)P_{\mathrm{succ}}$ or

$$\boxed{\mathrm{S}_{\mathrm{SALOHA}} = \mathtt{G}(n)e^{-\mathtt{G}(n)}} \tag{9.82}$$

Analogously to what done for the ALOHA system, taking the first order derivative of $\mathrm{S}_{\mathrm{SALOHA}}$ with respect to $\mathtt{G}(n)$ and equalling it to zero we have

$$\frac{d\mathrm{S}_{\mathrm{SALOHA}}}{d\mathtt{G}(n)} = e^{-2\mathtt{G}(n)}(1 - \mathtt{G}(n)) = 0 \Rightarrow \mathtt{G}(n) = 1 \; . \tag{9.83}$$

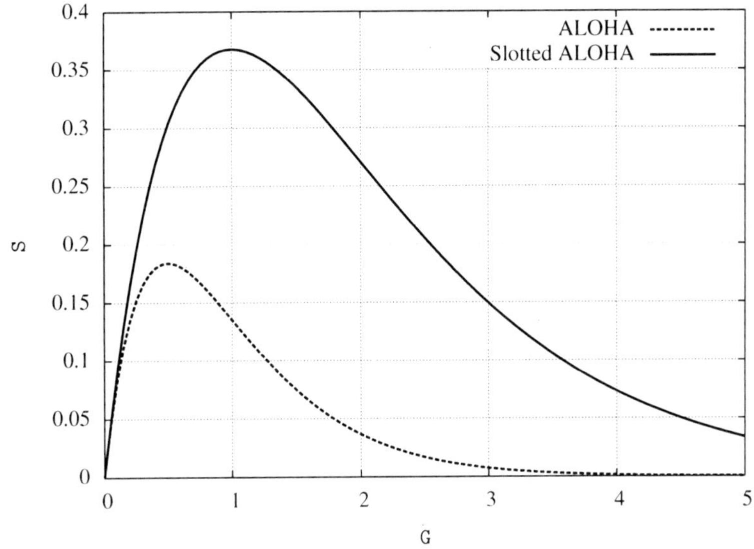

Figure 9.8 Normalized throughput versus normalized offered traffic for ALOHA and slotted ALOHA.

The throughput in this case is maximized when $\mathtt{G}(n) = 1$, which means that slotted ALOHA is able to deal with higher offered traffic. In addition, the maximum normalized throughput is $\mathtt{S}_{\text{SALOHA}} = 1/e \simeq 0.37$ which is twice that achievable with pure ALOHA. This is shown in Fig. 9.8, where the normalized throughput of pure ALOHA is also plotted for comparison.

Delay For the delay analysis we use (9.8) following the four step procedure illustrated in page 606. In what follows we specialize this procedure to Slotted ALOHA only computing the quantities that differ with respect to those obtained for ALOHA.

1. *Computation of* $\mathtt{m}_w$. For $\mathtt{m}_w$ we note that a packet arriving in the middle of a time slot must wait until the end of this slot for its actual transmission over the channel. Thus, considering uniform arrivals we have that

$$\mathtt{m}_w = \frac{t_{\text{P}}}{2} \ . \tag{9.84}$$

2. *Computation of* $\mathtt{m}_{n_{\text{retx}}}$. As for ALOHA we have

$$\frac{\mathtt{G}(n)}{\mathtt{S}_{\text{SALOHA}}} = \frac{1}{P_{\text{succ}}} \tag{9.85}$$

which represents the average number of transmissions per packet. Hence, $\mathtt{m}_{n_{\text{retx}}}$ is

$$\mathtt{m}_{n_{\text{retx}}} = \frac{\mathtt{G}(n)}{\mathtt{S}_{\text{SALOHA}}} - 1 \ . \tag{9.86}$$

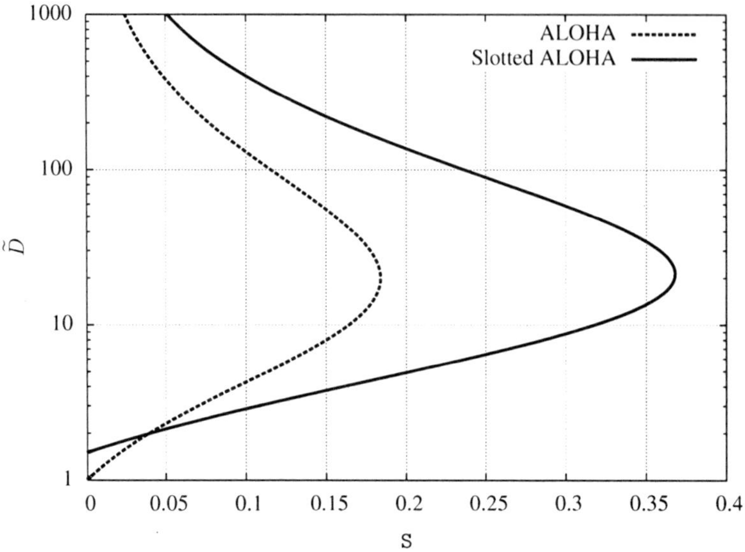

Figure 9.9 Normalized delay $\widetilde{D}$ versus normalized throughput S: comparison between ALOHA and slotted ALOHA with $\tilde{\tau}_p = 0.01$ and $1/\beta = 10t_P$.

3. *Computation of* $\mathtt{m}_{t_{\text{retx}}}$. With respect to (9.77) we have an additional delay term $\mathtt{m}_w = t_P/2$, which is due to the fact that packets scheduled for retransmission need to wait for the end of the slot where their retransmission timer expires. Thus we can write

$$\begin{aligned}\mathtt{m}_{t_{\text{retx}}} &= t_P + \mathtt{m}_w + 2\tau_p + t_A + \frac{1}{\beta} \\ &= t_P + \frac{t_P}{2} + 2\tau_p + t_A + \frac{1}{\beta}\,.\end{aligned} \tag{9.87}$$

From (9.8), (9.84), (9.86) and (9.87) the mean delay is found as

$$\begin{aligned}D_{\text{SALOHA}} &= \mathtt{m}_{n_{\text{retx}}}\mathtt{m}_{t_{\text{retx}}} + t_P + \mathtt{m}_w + \tau_p \\ &= \left(\frac{\mathtt{G}(n)}{\mathtt{S}_{\text{ALOHA}}} - 1\right)\left(t_P + \frac{t_P}{2} + 2\tau_p + t_A + \frac{1}{\beta}\right) + t_P + \frac{t_P}{2} + \tau_p \\ &= \left(e^{\mathtt{G}(n)} - 1\right)\left(\frac{3}{2}t_P + 2\tau_p + t_A + \frac{1}{\beta}\right) + \frac{3}{2}t_P + \tau_p\end{aligned} \tag{9.88}$$

where the third equality follows from (9.82). The normalized delay is

$$\boxed{\widetilde{D}_{\text{SALOHA}} = \left(e^{\mathtt{G}(n)} - 1\right)\left(\frac{3}{2} + 2\tilde{\tau}_p + \tilde{t}_A + \widetilde{\left(\frac{1}{\beta}\right)}\right) + \frac{3}{2} + \tilde{\tau}_p} \tag{9.89}$$

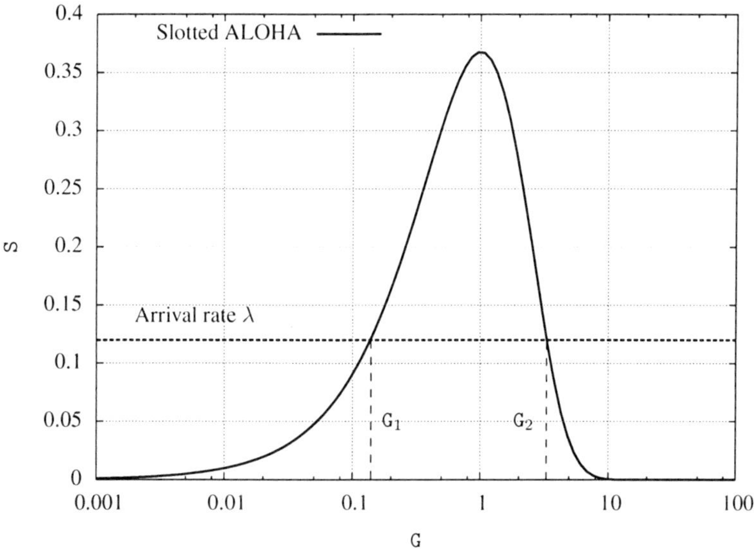

Figure 9.10 Normalized throughput versus normalized offered traffic in logarithmic scale for slotted ALOHA.

Example Results Fig. 9.9 shows the normalized delay versus normalized throughput for ALOHA and slotted ALOHA. For this graph we considered the link layer transmission parameters of Fig. 9.7, while $\tilde{\tau}_p = 0.01$ and $1/\beta = 10t_{\rm P}$. Again, from Figs. 9.8 and 9.9 we observe a first stable region, for which both delay and throughput are increasing functions of $\mathtt{G}$, followed by an unstable region, where the throughput decreases for increasing $\mathtt{G}$. We also note that when the offered traffic is very small ($\mathtt{G} \ll 1$) pure ALOHA outperforms slotted ALOHA in terms of delay. In fact, slotted ALOHA forces the transmission of all packets according to slot boundaries, and this leads to an additional delay term.

9.2.6.3 Stability of ALOHA and Slotted ALOHA Now, we qualitatively discuss the stability problem of ALOHA and slotted ALOHA. In what follows, we refer to the slotted ALOHA protocol, bearing in mind that the following discussion also holds for ALOHA. Consider Fig. 9.10 where we plot (9.82) in logarithmic scale to emphasize what happens at small offered traffic. Assume that the user packet arrival rate λ is smaller than the maximum throughput e^{-1}. In this case, we have that the horizontal line of the user packet arrival rate λ crosses the throughput $\mathtt{S}$ is two points. Consider the point in the leftmost part of the graph and call $\mathtt{G}_1$ the offered traffic at which $\mathtt{S}$ crosses λ. If $\mathtt{G}(n) = \mathtt{G}_1$ the system is at equilibrium as the throughput equals the arrival rate λ. However, small fluctuations of $\mathtt{G}(n)$ in practice can always occur. In particular, if $\mathtt{G}(n)$ grows a little above $\mathtt{G}_1$, we have that $\mathtt{S}$ grows a little beyond λ and, in this case, packets leave the system faster than they arrive. In turn, this will cause a reduction of $\mathtt{G}(n)$ and the system will go back to the operating point $(\mathtt{G}, \mathtt{S}) = (\mathtt{G}_1, \lambda)$. When, however, a large variation in the offered traffic occurs, so that $\mathtt{G}(n)$ increases beyond $\mathtt{G}_2$ then we have that $\mathtt{S}$ becomes smaller than λ. In this case packets leave the system at a rate smaller than the arrival rate and this will contribute to a further increase of $\mathtt{G}(n)$. The system in this case becomes congested and the throughput rapidly approaches zero.

In ALOHA we observe that the instantaneous $\mathtt{G}(n)$ goes beyond $\mathtt{G}_2$ with probability one as time tends to infinity, as even large variations in the offered traffic occur with nonzero probability. This is the reason why we say that these access protocols are unstable. It is therefore necessary, for proper operation, to stabilize these schemes. From the above discussion it should be clear that it is impossible to stabilize the system if we assume that the transmission policy is independent of the system state. Some strategy should instead be implemented to limit the maximum number of retransmission attempts, i.e., the instantaneous $\mathtt{G}(n)$. A possible and simple method to do so is now discussed. Assume that it is possible to coordinate the backlogged users and consider the following threshold policy: at the beginning of any given slot i, say, at most σ backlogged users may retransmit their packet. In detail, when the number of backlogged users is smaller than σ, all of them will *retransmit with probability* ν in the next slot. When this number is instead larger than σ, exactly σ users will be picked from the whole population of backlogged users and each of them will retransmit in the next slot with probability ν. All other backlogged users refrain from transmitting during the next slot. For a proof of the stabilizing properties of this threshold policy we refer the reader to [1].

We note that even though the above threshold based rule is able to stabilize the system, it is unclear how it could be implemented efficiently. In fact, it is impractical that all users keep track of those that are backlogged as this would require a large transmission overhead and, besides this, is not an easy task to implement. Fortunately, there exist stabilizing policies that do not require to know the exact number of backlogged users. Basically, these policies update the transmission probabilities according to what happened in the previous slot as $\nu_{k+1} = f(\nu_k, \text{feedback at slot } k)$, for a proper choice of the function $f(\cdot)$. These policies have maximum stable throughput e^{-1}, see [16]. See also [6, Section 4.2, p. 283] for a further discussion on stabilizing techniques.

In any event, even using stabilizing policies with slotted ALOHA the maximum normalized throughput is only e^{-1}, which is essentially due to the fact that users disregard the actual system state when they decide to transmit.

9.2.7 Carrier Sense Multiple Access

The poor performance of ALOHA access protocols is due to the fact that users transmit their packets irrespective of the behavior of the other users, i.e., without checking whether their transmission is likely to collide with packets that are concurrently transmitted. CSMA protocols put a remedy to this by adopting a "listen before talk" approach, where the channel is sensed before transmitting. If any given user detects channel activity then it defers its transmission to a later time. If, instead, the channel is sensed idle by this user, it goes ahead and transmits its packet. As we show shortly, this is particularly effective for networks where the channel propagation delay τ_p is small compared to the packet transmission time t_{P}.

We observe, however, that collisions are still possible. As an example, consider the case where two users sense the channel "nearly" at the same time: they will find the channel idle and thus they will both transmit their packets and cause a collision. The word "nearly" that we used in the previous sentence deserves an explanation. It is in fact unlikely that two users will sense the channel exactly at the same time. There is, however, a vulnerability period of τ_p seconds (the channel propagation delay) during which collisions are still possible. In fact, if the first user transmits at a given time instant t_1, its message will be received at time $t_1 + \tau_p$ and, any other user that senses the channel between t_1 and $t_1 + \tau_p$ will find it idle. This second user, upon sensing the channel idle, transmits its own packet that will however collide with the one being transmitted. This simple example illustrates the dependence of the performance of CSMA on τ_p. This dependence will be quantified by the analysis of the next section.

CSMA strategies are all based on the above rationale. The main differences among the many solutions proposed in the literature lie in how users react upon sensing the channel busy. Most of these solutions were first proposed and analyzed by Tobagi and Kleinrock [14, 18, 19]. In what follows we discuss and, where possible, analyze the three CSMA protocols presented in [14] as well as CSMA/CD [20], that is used in Ethernet systems [12], and CSMA with collision avoidance (CSMA/CA), that is used in IEEE 802.11 wireless LANs [21].

9.2.7.1 Nonpersistent CSMA The first CSMA scheme that we discuss is called nonpersistent CSMA (NP-CSMA) and is illustrated in Fig. 9.11. According to NP-CSMA, whenever a packet is ready for transmission (block labeled "new packet" in the figure) the transmitter performs channel sensing. If the channel is found busy the packet transmission is deferred to some later time according to a given retransmission delay or backoff time τ_{rtx} with PDF $p_{\tau_{\mathrm{rtx}}}(x)$ as in (9.70). When this time expires, the transmitter will sense again the channel and make another decision. When the channel is eventually sensed idle the packet is transmitted. At this point, two events can occur: 1) the packet is successfully delivered to the destination which, in turn, sends an acknowledgment (ACK) thus completing the transmission procedure for this packet, or 2) the packet collides. In this case, the transmitter will perform again the backoff procedure to retransmit the packet. The analysis that follows is based on [14].

Assumption RA8. As for the analysis of ALOHA access protocols we define a n augmented packet arrival rate $\lambda_{\mathrm{t}}(n) = \lambda + n\beta$, where n is the number of backlogged users, which is assumed slowly varying. Thus, from now on we will omit the n in the normalized offered traffic $\mathtt{G}(n) = \lambda_{\mathrm{t}}(n)t_{\mathrm{P}}$ and in the user packet arrival rate $\lambda_{\mathrm{t}}(n)$, using $\mathtt{G}$ and λ_{t} instead.

Remark $\mathtt{G}$ denotes the arrival rate of new and rescheduled packets at the transmitter. However, observe that all arrivals do not necessarily result in actual transmissions, in fact, a packet that finds the channel busy is rescheduled without being sent. Hence, $\mathtt{G}$ is the offered traffic and only a fraction of it represents the packets that are actually transmitted over the channel.

Throughput As illustrated in Fig. 9.12(a) time is subdivided into *cycles*, each composed of a *busy period* (BP), where transmission activity is present in the channel, followed by an *idle period* (IP), of length, respectively, B and I.

The statistics of I is derived from the Poisson assumption for the packet arrival process (RA2). In detail, the probability that there are no scheduled transmissions in a time period of a seconds corresponds to the probability that there are zero arrivals for the Poisson process of rate λ_{t} (that also accounts for pending retransmissions) in this time period. Hence, from (8.6) we have

$$P_I(a) = \mathrm{P}\left[I \leq a\right] = 1 - e^{-\lambda_{\mathrm{t}} a}\ ,\ a \geq 0 \tag{9.90}$$

from which the PDF is computed as (see also (8.5))

$$p_I(a) = \lambda_{\mathrm{t}} e^{-\lambda_{\mathrm{t}} a}\ ,\ a \geq 0\,. \tag{9.91}$$

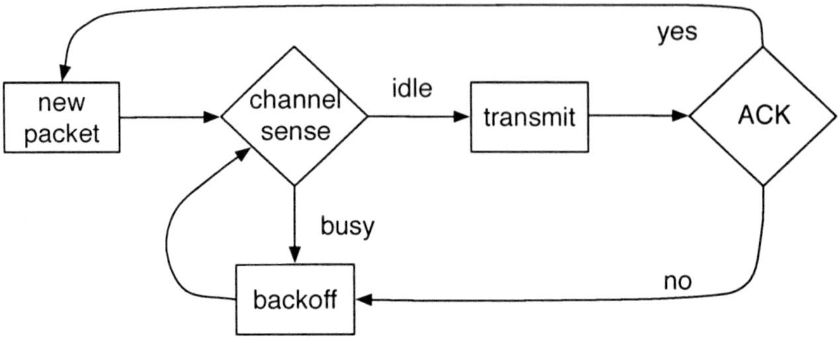

Figure 9.11 Illustration of NP-CSMA and 1P-CSMA protocols.

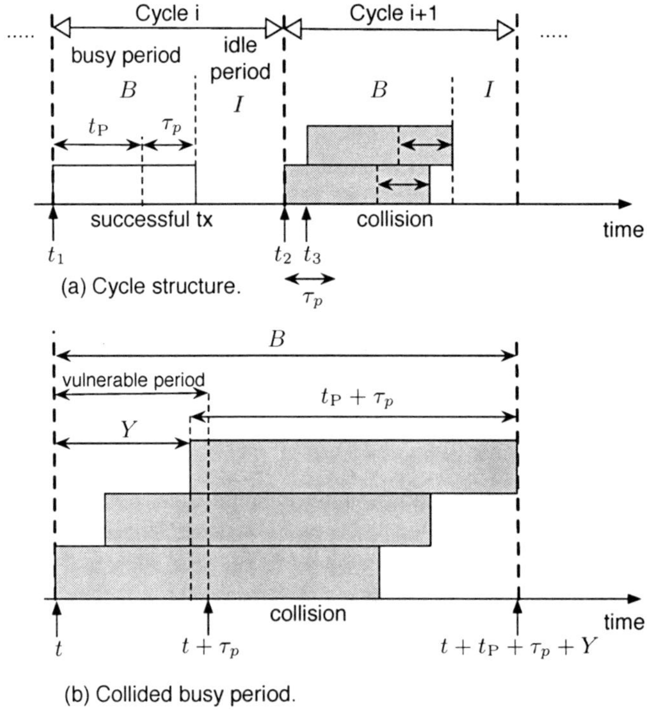

Figure 9.12 Cycle structure for the analysis of NP-CSMA.

Thus, I is exponentially distributed with mean[8]

$$\mathtt{m}_I = \mathrm{E}\left[I\right] = \frac{1}{\lambda_{\mathrm{t}}} \,. \tag{9.92}$$

[8]This is the mean inter-arrival time PDF of a Poisson process.

A BP is classified as successful (BP-S) or collided (BP-C), of length denoted, respectively as B_s and B_c, with

$$\mathrm{BP} = \begin{cases} \text{BP-S} & , \text{ if a single packet is transmitted within } B = B_\mathrm{s} \text{ seconds} \\ \text{BP-C} & , \text{ if multiple packets are transmitted within } B = B_\mathrm{c} \text{ seconds} \end{cases} . \tag{9.93}$$

The (useful) part of a BP dedicated to the successful transmission of one packet is referred to as useful sub period (USP) and its length is denoted by U. Hence, it is

$$U = \begin{cases} 0 & , \text{ if BP = BP-C} \\ t_\mathrm{P} & , \text{ if BP = BP-S} \end{cases} . \tag{9.94}$$

In Fig. 9.12(a), cycle i contains a BP-S of length $B_\mathrm{s} = t_\mathrm{P} + \tau_p$ with $U = t_\mathrm{P}$. In the figure, upon sensing the channel free, a first user schedules its transmission at time t_1. The transmission of its packet takes t_P seconds and, once this transmission is complete, τ_p additional seconds are needed for the packet to reach the intended destination. Cycle $i+1$ of Fig. 9.12(a) contains a BP-C of length $B_\mathrm{c} = (t_3 + t_\mathrm{P} + \tau_p) - t_2$ seconds with $U = 0$. For cycle $i+1$, at time t_2 a second user starts transmitting a new packet and a further third user schedules its transmission in the interval $[t_2, t_2 + \tau_p]$. Due to the finite propagation delay τ_p, this third user detects the channel idle in this interval and thus starts its own transmission. This transmission will however collide with the packet being transmitted by the second user. From this example, we see that a BP-C occurs when more than one user transmits at the beginning of the BP within the so called vulnerable period (VP) of length τ_p, see also Fig. 9.12(b).

If t is the transmission time of the first packet in a BP (see Fig. 9.12(b)), the probability that the BP is successful (i.e., BP = BP-S) is given by the probability that there are no arrivals within the VP $[t, t+\tau_p]$, or

$$\begin{aligned} P_\mathrm{s} &= \mathrm{P}\left[\text{BP = BP-S}\right] \\ &= \mathrm{P}\left[\text{No arrivals in } \tau_p \text{ seconds}\right] \\ &= e^{-\lambda_\mathrm{t}\tau_p} = e^{-\mathtt{G}\tilde{\tau}_p} \end{aligned} \tag{9.95}$$

Analogously, we define

$$\begin{aligned} P_\mathrm{c} &= \mathrm{P}\left[\text{BP = BP-C}\right] \\ &= \mathrm{P}\left[\text{At least one arrival in } \tau_p \text{ seconds}\right] \\ &= 1 - P_\mathrm{s} = 1 - e^{-\mathtt{G}\tilde{\tau}_p} . \end{aligned} \tag{9.96}$$

where we used (9.4) and (9.10). From (9.94), (9.95) and (9.96) the mean of U is obtained as

$$\mathtt{m}_U = \mathrm{E}\left[U\right] = t_\mathrm{P} P_\mathrm{s} + 0 P_\mathrm{c} = t_\mathrm{P} P_\mathrm{s} . \tag{9.97}$$

Now, we evaluate $\mathtt{m}_B = \mathrm{E}\left[B\right]$. First we define Y as the time elapsed between the transmission instants of the first and the last packets in a BP-C. In Fig. 9.12(a) it is $Y = t_3 - t_2$. Note that $Y \leq \tau_p$ as after τ_p seconds from the transmission of the first packet of a BP further transmission attempts are interdicted by the channel sense mechanism. The CDF of Y, i.e., the probability that Y is smaller than or equal to y, is

$$\begin{aligned} P_Y(y) &= \mathrm{P}\left[Y \le y\right] \\ &= \mathrm{P}\left[\text{No packet arrivals in } [t+y, t+\tau_p]\right] \\ &= \mathrm{P}\left[\text{No packet arrivals during } \tau_p - y \text{ seconds}\right] \\ &= e^{-\lambda_t(\tau_p - y)} \quad , \ 0 \le y \le \tau_p \ . \end{aligned} \tag{9.98}$$

From (9.98), the PDF of Y is

$$p_Y(y) = \frac{dP_Y(y)}{dy} = e^{-\lambda_t \tau_p}\delta(y) + \lambda_t e^{-\lambda_t(\tau_p - y)} \ , \ 0 \le y \le \tau_p \tag{9.99}$$

where $\delta(y)$ is defined in Example 2.1 B. Moreover,

$$\begin{aligned} \mathtt{m}_Y = \mathrm{E}\left[Y\right] &= \int_0^{\tau_p} y p_Y(y) dy \\ &= \tau_p - \frac{1 - e^{-\lambda_t \tau_p}}{\lambda_t} \\ &= t_P\left(\tilde{\tau}_p - \frac{1 - e^{-\tilde{\tau}_p \mathtt{G}}}{\mathtt{G}}\right) . \end{aligned} \tag{9.100}$$

From the definition of Y it is

$$B_c = Y + t_P + \tau_p \tag{9.101}$$

and

$$B_s = t_P + \tau_p \ . \tag{9.102}$$

Hence, it results

$$\begin{aligned} \mathtt{m}_B = \mathrm{E}\left[B\right] &= (t_P + \tau_p)P_s + (\mathrm{E}\left[Y | \mathrm{BP} = \mathrm{BP\text{-}C}\right] + t_P + \tau_p)P_c \\ &= \mathtt{m}_Y + t_P + \tau_p \\ &= t_P\left(1 + 2\tilde{\tau}_p - \frac{1 - e^{-\tilde{\tau}_p \mathtt{G}}}{\mathtt{G}}\right) \end{aligned} \tag{9.103}$$

where we have used (9.100). From (9.12), the throughput is the fraction of time the channel is occupied with successful transmissions. Here the throughput is obtained as the ratio between the mean useful time $\mathtt{m}_U$ for a generic cycle and its average duration given by the sum of its mean busy and idle periods, $\mathtt{m}_B + \mathtt{m}_I$.[9] The final expression for the normalized throughput is obtained using this rationale together with (9.92), (9.97) and (9.103)

$$\boxed{\begin{aligned} \mathtt{S}_{NP-CSMA} &= \frac{\mathtt{m}_U}{\mathtt{m}_U + \mathtt{m}_I} \\ &= \frac{\lambda_t t_P e^{-\lambda_t \tau_p}}{\lambda_t(t_P + 2\tau_p) + e^{-\lambda_t \tau_p}} \\ &= \frac{\mathtt{G}e^{-\tilde{\tau}_p \mathtt{G}}}{\mathtt{G}(1 + 2\tilde{\tau}_p) + e^{-\tilde{\tau}_p \mathtt{G}}} \end{aligned}} \tag{9.104}$$

[9] At the beginning of each cycle the process statistics restarts and is independent of the past transmission history; hence the time when a new cycle begins is called regeneration (or renewal) instant. This allows us to find the normalized throughput as in (9.104), in terms of ratio of means. This rationale is justified and made rigorous by the renewal theory. For an introductory treatise on this subject the interested reader is referred to [13].

Note that when $\tilde{\tau}_p \to 0$ the normalized throughput becomes

$$\tilde{\tau}_p \to 0 \Rightarrow \mathtt{S}_{NP-CSMA} = \frac{\mathtt{G}}{1+\mathtt{G}} \,. \tag{9.105}$$

Thus for $\tilde{\tau}_p \to 0$ (the best possible case for CSMA) and $\mathtt{G} \to +\infty$ then $\mathtt{S}_{NP-CSMA} \to 1$, i.e., the channel is fully utilized. This can be seen in Fig. 9.13, where we plot S as a function of G for different values of the normalized propagation delay $\tilde{\tau}_p$. The throughput of slotted ALOHA is also shown for comparison. Clearly, when $\tilde{\tau}_p$ is small NP-CSMA significantly outperforms slotted ALOHA. This shows that in terrestrial wireless networks such as, e.g., wireless local area network (WLAN), where $\tau_p \ll t_{\mathrm{P}}$, the use of NP-CSMA with its "listen before talk" approach is indeed effective. However, when $\tilde{\tau}_p$ approaches one, i.e., the channel propagation delay becomes comparable to or larger than the packet transmission time, the throughput of slotted ALOHA is higher than that of NP-CSMA. In this case, sensing the channel is useless and should be avoided as it does not provide meaningful information about ongoing packet transmissions. An example of the latter situation is given by satellite channels, which are characterized by large end to end delays.[10]

We finally observe that the curves in Fig. 9.13 for $\tilde{\tau}_p \neq 0$ are similar in shape to those obtained for ALOHA. Thus, NP-CSMA suffers from the same instability problems of ALOHA as G increases.

Delay For the delay analysis we use (9.8) following the four step procedure illustrated in page 606. In what follows we specialize this procedure to CSMA computing the quantities that differ with respect to those obtained for ALOHA.

1. *Computation of* $\mathtt{m}_w$. For the calculation of $\mathtt{m}_w$ we must introduce a new quantity called channel busy probability.

Definition 9.5 (channel busy probability) *We define P_{busy} as the probability that, upon its arrival, any given packet (either new or rescheduled) finds the channel busy.*

P_{busy}, is obtained following the approach of [14, Appendix B] through a renewal argument similar to the one used to evaluate (9.104). From Fig. 9.12(b), we recall that the first τ_p seconds of any given cycle are always sensed idle by the channel sense mechanism. Accordingly, in each cycle there are on average $\mathtt{m}_B - \tau_p$ seconds during which the channel is sensed busy. Due to the PASTA property of Poisson processes, see (8.85), P_{busy} is found as the ratio between $\mathtt{m}_B - \tau_p$ and the average duration of a cycle $\mathtt{m}_B + \mathtt{m}_I$, i.e.,

$$\begin{aligned} P_{\mathrm{busy}} &= \frac{\mathtt{m}_B - \tau_p}{\mathtt{m}_B + \mathtt{m}_I} = \frac{t_{\mathrm{P}} + \tau_p - \frac{1-e^{-\lambda_t \tau_p}}{\lambda_t}}{t_{\mathrm{P}} + 2\tau_p - \frac{1-e^{-\lambda_t \tau_p}}{\lambda_t} + \frac{1}{\lambda_t}} \\ &= \frac{\mathtt{G}(1+\tilde{\tau}_p) - (1-e^{-\tilde{\tau}_p \mathtt{G}})}{\mathtt{G}(1+2\tilde{\tau}_p) + e^{-\tilde{\tau}_p \mathtt{G}}} \,. \end{aligned} \tag{9.106}$$

[10] In fact, most communications satellites are located in the geostationary orbit at an altitude of approximately 35786 km above the equator and radio waves travel at about the speed of light, 300000 km/s. Hence, if a transmitter is located on the equator and communicates with a satellite directly above it then the propagation delay is $\tau_p \simeq 120$ ms, which is much larger than the packet transmission time of typical wireless communication systems where the transmission of a packet of 1000 bits at 1 Mbit/s takes about $t_{\mathrm{P}} \simeq 1$ ms.

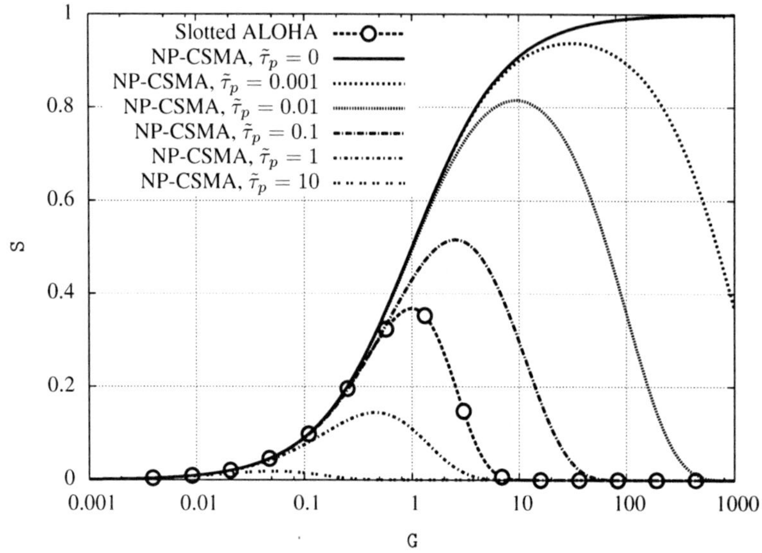

Figure 9.13 Normalized throughput versus normalized offered traffic for NP-CSMA. The normalized throughput of slotted ALOHA is also plotted for comparison.

According to NP-CSMA, we transmit whenever the channel is found idle and we instead defer the transmission to a later time when it is found busy. As in ALOHA, random backoff times τ_{rtx} are picked from an exponential distribution with parameter β, which leads to an average backoff period of $\mathrm{E}[\tau_{\text{rtx}}] = 1/\beta$ seconds. Finding the channel idle may require multiple channel sense operations. In particular, the channel is found idle at the $i+1$th channel sense with probability $P^i_{\text{busy}}(1 - P_{\text{busy}})$. Each of the previous i backoffs yields an average delay of $\mathrm{E}[\tau_{\text{rtx}}] = 1/\beta$ seconds. Thus, the corresponding total average sensing period is $i\,\mathrm{E}[\tau_{\text{rtx}}]$ seconds, while the average period until the channel is sensed idle is

$$\begin{aligned}\mathtt{m}_w &= \sum_{i=0}^{\infty}\left[iP^i_{\text{busy}}(1 - P_{\text{busy}})\,\mathrm{E}[\tau_{\text{rtx}}]\right]\\ &= \frac{P_{\text{busy}}}{\beta(1 - P_{\text{busy}})}\\ &= \frac{\mathtt{G}(1+\tilde{\tau}_p) - (1 - e^{-\tilde{\tau}_p\mathtt{G}})}{\beta(1+\mathtt{G}\tilde{\tau}_p)}\,.\end{aligned} \tag{9.107}$$

2. *Computation of* $\mathtt{m}_{t_{\text{retx}}}$. For $\mathtt{m}_{t_{\text{retx}}}$ we use (9.87) with $\mathtt{m}_w$ given by (9.107).

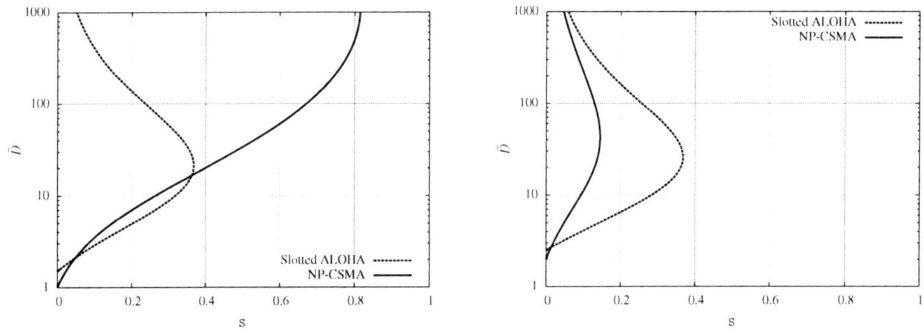

(a) $\tilde{\tau}_p = 0.01$ (b) $\tilde{\tau}_p = 1$

Figure 9.14 Normalized delay versus normalized throughput: comparison between NP-CSMA and slotted ALOHA for $1/\beta = 10t_{\mathrm{P}}$.

Applying (9.8) together with the previous results we obtain the mean delay for NP-CSMA

$$D_{\mathrm{NP-CSMA}} = \mathtt{m}_{n_{\mathrm{retx}}}\mathtt{m}_{t_{\mathrm{retx}}} + t_{\mathrm{P}} + \mathtt{m}_w + \tau_p$$
$$= \left(\frac{\mathtt{G}}{\mathrm{S}_{\mathrm{NP-CSMA}}} - 1\right)\left(t_{\mathrm{P}} + \mathtt{m}_w + 2\tau_p + t_{\mathrm{A}} + \frac{1}{\beta}\right) + t_{\mathrm{P}} + \mathtt{m}_w + \tau_p \,. \tag{9.108}$$

The normalized delay is given by

$$\boxed{\widetilde{D}_{\mathrm{NP-CSMA}} = \left(\frac{\mathtt{G}}{\mathrm{S}_{\mathrm{NP-CSMA}}} - 1\right)\left(1 + \tilde{\mathtt{m}}_w + 2\tilde{\tau}_p + \tilde{t}_{\mathrm{A}} + \widetilde{\left(\frac{1}{\beta}\right)}\right) + 1 + \tilde{\mathtt{m}}_w + \tilde{\tau}_p} \tag{9.109}$$

where $\tilde{\mathtt{m}}_w = \mathtt{m}_w/t_{\mathrm{P}}$ and $\mathtt{m}_w$ is given by (9.107).

Example Results The tradeoff between normalized delay and normalized throughput is shown in Fig. 9.14, where we compare NP-CSMA with slotted ALOHA for two values of $\tilde{\tau}_p$. The curves in this plot have been obtained by varying the offered traffic $\mathtt{G}$ in (9.104) and (9.109) as the independent variable. The channel transmission parameters are the same used for Fig. 9.7. From Fig. 9.14(a) we see that NP-CSMA performs substantially better than slotted ALOHA in terms of both throughput and delay when $\mathtt{G}$ is high. When $\mathtt{G} \ll 1$ slotted ALOHA is once again affected by the additional delay due to slotted transmission. For intermediate values of $\mathtt{G}$ slotted ALOHA performs slightly better in terms of delay than NP-CSMA as this is affected by a random backoff time whenever the channel is sensed busy. Overall, the backoff strategy is effective when $\mathtt{G}$ is high, whereas at low $\mathtt{G}$ the more aggressive retransmission policy of slotted ALOHA, where packets are transmitted regardless of the channel status, is advantageous. Fig. 9.14(b) confirms what we have seen in Fig. 9.13 for the throughput, i.e., that NP-CSMA is inefficient when the propagation delay is very large, e.g., when $\tau_p = t_{\mathrm{P}}$ ($\tilde{\tau}_p = 1$) as in the figure.

9.2.7.2 One Persistent CSMA In one persistent CSMA (1P-CSMA) upon sensing the channel busy, the channel sensing operation is immediately repeated, Fig. 9.11 applies here as well just bypassing the backoff block. This leads to an aggressive transmission behavior, according to which packet transmissions occur as soon as the channel is sensed idle. The rationale behind this is to reduce, as much as possible, the periods of inactivity of the channel in order to hopefully increase the throughput performance.

Throughput 1P-CSMA was first proposed and analyzed by Tobagi and Kleinrock in [14]. In the following, we refer to the simpler analysis of [20]. With reference to Fig. 9.15 the channel occupation is subdivided into cycles where each cycle is composed of a busy and an idle period. Differently from NP-CSMA, busy periods are further subdivided into a number of busy sub-periods (BSP) which can be either of type 1 (BSP-1) or of type 2 (BSP-2):

1. BSP-1 can either be successful (BSP-1-S) or collided (BSP-1-C) . It is successful when a single packet is transmitted in the BSP. Instead, it is collided when a single packet is transmitted at the beginning of the BSP and further interfering packets (one or more) are sent within the vulnerable period of this BSP, but after the transmission of the first packet. As an example, consider the leftmost BSP-1 of Fig. 9.15: this period is successful as no other packets are sent within the period. The rightmost BSP of Fig. 9.15 is also of type 1 but it is collided as 1) a single packet is sent at the beginning of the BSP and 2) two further (colliding) packets are sent in its vulnerable period, i.e., within τ_p seconds form the first packet transmission.

2. BSP-2 is always collided. In type 2 BSPs multiple colliding packets are sent at the beginning of the BSP, whereas additional packets may or may be not sent within its vulnerable period.

In addition, we use the rv Y to describe the time elapsed between the beginning of the transmission of the first and that of the last packet in a BSP: two examples of Y are given in Fig. 9.15.

Remark As an illustrative example of the protocol behavior, consider the leftmost BSP-1 in Fig. 9.15 and let t be the instant at which this period starts. The same reasoning holds for any other BSP. According to 1P-CSMA, any packet that arrives in the interval $[t_1, t_2] = [t + \tau_p, t + t_{\rm P} + \tau_p]$ (i.e., after the VP of BSP-1) will sense the channel busy and thus it must defer its transmission to when the channel is sensed idle again, i.e., at time $t + t_{\rm P} + \tau_p$. The number of packets that are accumulated during a BSP is equal to the number of arrivals in $t_2 - t_1 = t_{\rm P}$ seconds. In Fig. 9.15 exactly two packets arrive in $[t_1, t_2]$; these find the channel busy and are transmitted as soon as the channel is idle again, i.e., at time $t + t_{\rm P} + \tau_p$, where they collide. The rightmost BSP-1 in Fig. 9.15 is collided as three packets arrive within the VP of this BSP. [11]

For the characterization of the transmission process we use a discrete time Markov chain having three states: 0, 1 and 2. The transition diagram is given in Fig. 9.16. The CSMA system is in state 0 during idle periods, in state 1 during BSPs of type 1 and in state 2 during BSPs of type 2. Transitions among states occur at the start of a new period and depend on the number of packets arrived during the previous period, as we now explain. In state 0 no users have a

[11] This model is exact under assumption BA5, i.e., for a symmetric system where all users are perfectly synchronized.

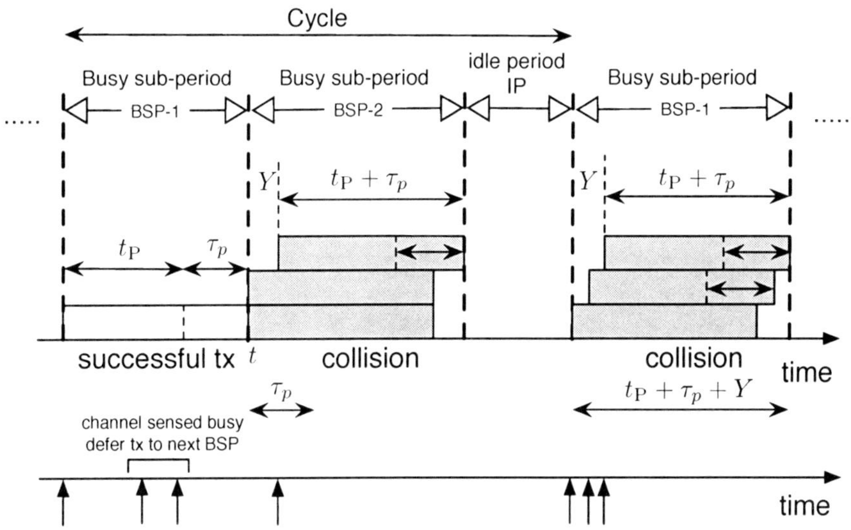

Figure 9.15 Cycle structure for the analysis of 1P-CSMA (vertical arrows indicate packet arrivals).

packet to send. Whenever a new packet arrives at a certain user, a first packet is sent over the channel, thus moving the system to state 1. Note that the transition probability $P_{02} = 0$ as, due to the Poisson assumption for the packet arrival process (RA2), the probability that two packets arrive exactly at the same time instant is zero. After an idle period there is a busy period and this period must be of type 1, thus $P_{00} = 0$ and $P_{01} = 1$. Moreover, from state 1 the system may remain in state 1 when a single packet arrives during the BSP, may move to state 2 in case multiple packets arrive during the BSP or go back to state 0 in case no packets arrive during the BSP. Similarly, from state 2 the state changes to state 1, remains in state 2 or goes to state 0 if one, multiple or zero packets arrive during the BSP, respectively.

Definition 9.6 (system state) *We characterize the state of the Markov chain as $X(t)$ and we use the notation $X(t) = i$ to indicate that the process is in state i, with $i \in \{0, 1, 2\}$, at time t. To simplify the notation in what follows we omit the time index t in $X(t)$.*

Definition 9.7 (asymptotic state characterization of X) *We define π_i, with $i \in \{0, 1, 2\}$, the asymptotic probability that $X = i$. From (7.61) we recall that these probabilities can be seen as the mean fraction of time that the system spends in any given state i as time goes to infinity. In the following analysis we refer to T_i as the duration of a BSP of type i.*

From the above discussion, in any given BSP a successful transmission is only possible if the system is in state 1 and no packets arrive during the first τ_p seconds of this sub-period. Due to assumption RA2 the probability of this event is $\exp\{-\lambda_t \tau_p\}$. The renewal argument used in (9.104) can be extended [13] to compute the throughput of 1P-CSMA under the Markov chain model

$$\boxed{\mathrm{S}_{\mathrm{1P-CSMA}} = \frac{t_P \pi_1 \exp\{-\lambda_t \tau_p\}}{\sum_{i=0}^{2} \pi_i \, \mathrm{E}[T_i]}} \tag{9.110}$$

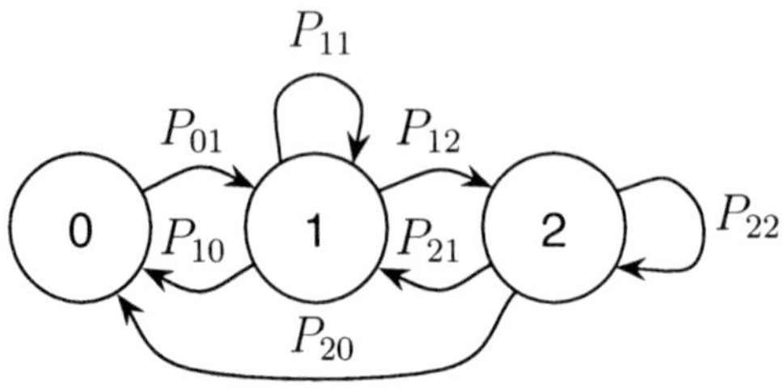

Figure 9.16 Transition diagram of the Markov chain for 1P-CSMA.

In fact, in (9.110) the numerator is the mean time to successfully transmit a packet, with $\pi_1 \exp\{-\lambda_t \tau_p\}$ being the probability that the system is in a successful BSP and t_P the packet transmission time. The denominator is instead the mean duration of a BSP. Next, we calculate the mean time spent in each state, $\mathrm{E}[T_i]$, and the asymptotic probabilities $\{\pi_i\}$.

Computation of $\mathrm{E}[T_i]$ We note that the idle periods of 1P-CSMA are exponentially distributed with mean $1/\lambda_t$, from (9.92). Thus for T_0 we have:

$$\mathtt{m}_{T_0} = \mathrm{E}[T_0] = \frac{1}{\lambda_t} \,. \tag{9.111}$$

For T_1 and T_2 we refer again to Fig. 9.15. All users that become ready (i.e., have a packet to transmit) during a given BSP will transmit at exactly the same instant at the beginning of the subsequent sub-period. Moreover, due to assumptions RA1 and RA2 of Section 9.2.6 the number of concurrent transmissions at the beginning of a BSP has no influence on the random variable Y. Again, the Poisson arrival process is memoryless, which implies that the arrival time of any packet within the vulnerable period of any given BSP is independent of how many packets were transmitted at the beginning of this sub-period. For this reason we have that the average duration of BSPs of type 1 and 2 is identical. Hence, it follows that

$$P_{1j} = P_{2j} \,, \; j \in \{0, 1, 2\} \tag{9.112}$$

and

$$\mathtt{m}_{T_1} = \mathtt{m}_{T_2} \tag{9.113}$$

where $\mathtt{m}_{T_1} = \mathrm{E}[T_1]$ and $\mathtt{m}_{T_2} = \mathrm{E}[T_2]$. Analogously to (9.101) and (9.102), we can write the duration of BSP-1 as

$$T_1 = t_P + \tau_p \tag{9.114}$$

and the duration of BSP-2 as

$$T_2 = Y + t_P + \tau_p \,. \tag{9.115}$$

$\mathrm{E}[Y]$ is computed exactly as in (9.100) and $\mathtt{m}_{T_1}$, $\mathtt{m}_{T_2}$ are obtained as $\mathtt{m}_B$ of (9.103). Thus we write

$$\mathtt{m}_{T_1} = \mathtt{m}_{T_2} = t_P \left(1 + 2\tilde{\tau}_p - \frac{1 - e^{-\tilde{\tau}_p \mathtt{G}}}{\mathtt{G}} \right) . \tag{9.116}$$

Computation of the asymptotic probabilities π_i From the transition diagram of Fig. 9.16 we can write the following flow balance equations (see (7.78))

$$\begin{aligned}\pi_0 &= \pi_0 P_{00} + \pi_1 P_{10} + \pi_2 P_{20} = \pi_1 P_{10} + \pi_2 P_{20} \\ \pi_2 &= \pi_0 P_{02} + \pi_1 P_{12} + \pi_2 P_{22} = \pi_1 P_{12} + \pi_2 P_{22} \\ \pi_0 &+ \pi_1 + \pi_2 = 1\end{aligned} \tag{9.117}$$

Solving this system of equations and using (9.112) we obtain the asymptotic state probabilities as:

$$\pi_0 = \frac{P_{10}}{1 + P_{10}}, \quad \pi_1 = \frac{P_{10} + P_{11}}{1 + P_{10}}, \quad \pi_2 = \frac{1 - P_{10} - P_{11}}{1 + P_{10}}. \tag{9.118}$$

For the calculation of P_{10} we must distinguish between successful and collided BSPs of type 1. P_{10} is given by the sum of two terms, $P_{10}^a + P_{10}^b$, where P_{10}^a is the probability that the BSP of type 1 is successful, i.e., that there are no arrivals in the VP of τ_p seconds, and that there are no arrivals in the following $t_{\rm P}$ seconds. Thus we can write:

$$\begin{aligned}P_{10}^a &= {\rm P}\left[\text{no arrivals in } t_{\rm P} \text{ seconds and successful BSP} | X = 1\right] \\ &= e^{-\lambda_{\rm t} t_{\rm P}} e^{-\lambda_{\rm t} \tau_p}.\end{aligned}$$

The calculation of P_{10}^b is slightly more involved and corresponds to the probability that the BSP of type 1 is collided, i.e., that at least one packet is transmitted within the vulnerable interval and that no further packets arrive in the following $t_{\rm P}$ seconds. To compute P_{10}^b assume that a BSP of type 1 starts at time t. Conditioning on $Y = y$, with $y \le \tau_p$, the transmission time of the last collided packet within the VP, the following BSP will be an idle period (i.e., of type 0) only if no packets arrive between time $t_1 = t + \tau_p$ and time $t_2 = t + t_{\rm P} + \tau_p + y$ (the end of the BSP).[12] This event occurs with probability $\exp\{-\lambda_{\rm t}(t_2 - t_1)\} = \exp\{-\lambda_{\rm t}(t_{\rm P} + y)\}$. Unconditioning we have:

$$\begin{aligned}P_{10}^b &= {\rm P}\left[\text{no arrivals in } t_{\rm P} \text{ seconds and collided BSP} | X = 1\right] \\ &= \int_0^{\tau_p} e^{-\lambda_{\rm t}(t_{\rm P}+y)} p_Y(y) dy.\end{aligned}$$

where the PDF of Y is given by (9.99). Finally, P_{10} is found as:

$$\begin{aligned}P_{10} &= {\rm P}\left[\text{no arrivals in } t_{\rm P} \text{ seconds and successful BSP} | X = 1\right] \\ &\quad + {\rm P}\left[\text{no arrivals in } t_{\rm P} \text{ seconds and collided BSP} | X = 1\right] \\ &= e^{-\lambda_{\rm t}(t_{\rm P}+\tau_p)} + \int_0^{\tau_p} e^{-\lambda_{\rm t}(t_{\rm P}+y)} \lambda_{\rm t} e^{-\lambda_{\rm t}(\tau_p - y)} dy \\ &= (1 + \lambda_{\rm t}\tau_p) e^{-\lambda_{\rm t}(t_{\rm P}+\tau_p)} \\ &= (1 + \tilde{\tau}_p \mathtt{G}) e^{-(1+\tilde{\tau}_p)\mathtt{G}}.\end{aligned} \tag{9.119}$$

For the transition probability from state 1 to itself we follow a similar rationale and write $P_{11} = P_{11}^a + P_{11}^b$. P_{11}^a is the probability that there are no arrivals within the vulnerable period

[12] Note that, having conditioned on $Y = y$ no packets can arrive in $[t + y, t + \tau_p]$ as y is by definition the arrival time for the last packet within the first τ_p seconds of the BSP.

and that there is exactly one arrival during the following $t_{\rm P}$ seconds, i.e.,

$$P_{11}^a = {\rm P}\left[\text{one arrival in } t_{\rm P} \text{ seconds and successful BSP}|X=1\right]$$
$$= \lambda_{\rm t} t_{\rm P} e^{-\lambda_{\rm t} t_{\rm P}} e^{-\lambda_{\rm t}\tau_p} .$$

For P_{11}^b at least one colliding packet arrives within the vulnerable period and we must have a single packet arrival between time $t_1 = t + \tau_p$ and time $t_2 = t + t_{\rm P} + \tau_p + y$, where t is the instant where the BSP begins. The probability of this event is $\lambda_{\rm t}(t_2 - t_1)\exp\{-\lambda_{\rm t}(t_2 - t_1)\} = \lambda_{\rm t}(t_{\rm P} + y)\exp\{-\lambda_{\rm t}(t_{\rm P} + y)\}$. In this case we have:

$$P_{11}^b = {\rm P}\left[\text{one arrival in } t_{\rm P} \text{ seconds and collided BSP}|X=1\right]$$
$$= \int_0^{\tau_p} \lambda_{\rm t}(t_{\rm P} + y) e^{-\lambda_{\rm t}(t_{\rm P}+y)} p_Y(y)dy .$$

Thus, P_{11} is found as

$$\begin{aligned} P_{11} &= {\rm P}\left[\text{one arrival in } t_{\rm P} \text{ seconds and successful BSP}|X=1\right] \\ &\quad + {\rm P}\left[\text{one arrival in } t_{\rm P} \text{ seconds and collided BSP}|X=1\right] \\ &= \lambda_{\rm t} t_{\rm P} e^{-\lambda_{\rm t} t_{\rm P}} e^{-\lambda_{\rm t}\tau_p} + \int_0^{\tau_p} \lambda_{\rm t}^2 (t_{\rm P} + y) e^{-\lambda_{\rm t}(t_{\rm P}+y)} e^{-\lambda_{\rm t}(\tau_p - y)} dy \\ &= \lambda_{\rm t} e^{-\lambda_{\rm t}(t_{\rm P}+\tau_p)} \left[t_{\rm P} + \lambda_{\rm t}\tau_p \left(t_{\rm P} + \frac{\tau_p}{2}\right)\right] \\ &= {\tt G} e^{-(1+\tilde{\tau}_p){\tt G}} \left[1 + \tilde{\tau}_p {\tt G}\left(1 + \frac{\tilde{\tau}_p}{2}\right)\right] . \end{aligned} \tag{9.120}$$

Last, P_{12} is obtained as $P_{12} = 1 - P_{11} - P_{10}$.

The normalized throughput is obtained combining (9.111), (9.116), (9.112), (9.118), (9.119), (9.120) into (9.110)

$$S_{\rm 1P-CSMA} = \frac{\lambda_{\rm t} t_{\rm P} e^{-\lambda_{\rm t}(t_{\rm P}+2\tau_p)}[1 + \lambda_{\rm t} t_{\rm P} + \lambda_{\rm t}\tau_p(1 + \lambda_{\rm t} t_{\rm P} + \lambda_{\rm t}\tau_p/2)]}{\lambda_{\rm t}(t_{\rm P} + 2\tau_p) - (1 - e^{-\lambda_{\rm t}\tau_p}) + (1 + \lambda_{\rm t}\tau_p)e^{-\lambda_{\rm t}(t_{\rm P}+\tau_p)}} . \tag{9.121}$$

That from (9.4) and (9.10) can be rewritten as:

$$\boxed{S_{\rm 1P-CSMA} = \frac{{\tt G} e^{-(1+2\tilde{\tau}_p){\tt G}}[1 + {\tt G} + \tilde{\tau}_p {\tt G}(1 + {\tt G} + \tilde{\tau}_p {\tt G}/2)]}{{\tt G}(1 + 2\tilde{\tau}_p) - (1 - e^{-\tilde{\tau}_p {\tt G}}) + (1 + \tilde{\tau}_p {\tt G})e^{-(1+\tilde{\tau}_p){\tt G}}}} \tag{9.122}$$

Note that when $\tilde{\tau}_p \to 0$ the normalized throughput becomes:

$$\tilde{\tau}_p \to 0 \Rightarrow S_{\rm 1P-CSMA} = \frac{{\tt G}(1 + {\tt G})}{1 + {\tt G}e^{\tt G}} . \tag{9.123}$$

Thus for $\tilde{\tau}_p \to 0$ (the best possible case for CSMA) and ${\tt G} \to +\infty$ then $S_{1P-CSMA} \to 0$. A comparison of the throughput of 1P-CSMA against that of NP-CSMA is given in Fig. 9.17. We recall that 1P-CSMA was proposed in an attempt to improve the performance of NP-CSMA by reducing idle periods. Indeed, the performance of 1P-CSMA is much lower than expected and worse than that of NP-CSMA. The aggressive transmission policy of 1P-CSMA is effective only for small offered traffic loads, where it slightly outperforms nonpersistent schemes.

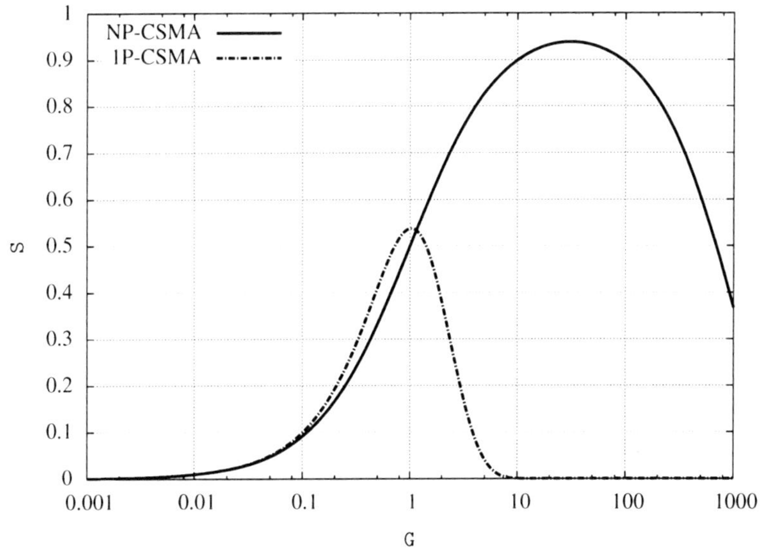

Figure 9.17 Normalized throughput versus normalized offered traffic: comparison between 1P-CSMA and NP-CSMA for $\tilde{\tau}_p = 0.001$.

Calculation of P_{busy} for 1P-CSMA Let us now consider the transmission of a given packet. As illustrated in Fig. 9.15, a cycle is composed of a single idle period and a number k of BSPs. In the time spanned by these BSPs the channel is sensed busy during $k\mathtt{m}_{T_1} - k\tau_p$ seconds, as there is a vulnerable period of τ_p seconds for each BSP during which the channel is sensed idle. Consider, for now, the average number of BSPs in a cycle $\mathtt{m}_k$ as a known quantity, it is

$$\mathtt{m}_B = \mathtt{m}_k \mathtt{m}_{T_1} \ . \tag{9.124}$$

Accordingly we have (see also Definition 9.5)

$$P_{\text{busy}} = \frac{\mathtt{m}_k(\mathtt{m}_{T_1} - \tau_p)}{\mathtt{m}_B + \mathtt{m}_I} = \frac{\mathtt{m}_B - \mathtt{m}_k \tau_p}{\mathtt{m}_B + \mathtt{m}_I} \ . \tag{9.125}$$

In fact, at the denominator $\mathtt{m}_I + \mathtt{m}_B$ is the cycle length, while at the numerator we have the total average length of the BSPs in a cycle, $\mathtt{m}_B$, subtracted of the average duration of all VPs, $\mathtt{m}_k \tau_p$. In what follows we calculate $\mathtt{m}_k$ and $\mathtt{m}_B$ ($\mathtt{m}_I$ is defined in (9.92)).

Computation of $\mathtt{m}_k$ Consider the first BSP of a given cycle. The probability that the next BSP will not be idle is $P_{11} + P_{12} = 1 - P_{10}$. Thus, the probability that in a cycle there are j busy BSPs followed by an idle sub-period is $(1 - P_{10})^{j-1} P_{10}$. This means that the number of BSPs in a cycle is geometrically distributed with mean

$$\mathtt{m}_k = \sum_{j=1}^{+\infty} j(1 - P_{10})^{j-1} P_{10} = \frac{1}{P_{10}} \ . \tag{9.126}$$

Computation of $\mathtt{m}_B$ using (9.124) and (9.126)

$$\mathtt{m}_B = \frac{\mathtt{m}_{T_1}}{P_{10}} \,. \tag{9.127}$$

Note that there is no need to distinguish between busy sub-periods of type 1 and 2 in this equation as for 1P-CSMA these BSPs have the same average duration, see (9.113).

Using (9.126) and (9.127) into (9.125) we finally obtain

$$\begin{aligned} P_{\text{busy}} &= \frac{\mathtt{m}_{T_1} - \tau_p}{\mathtt{m}_{T_1} - \frac{P_{10} t_{\text{P}}}{\mathtt{G}}} \\ &= \frac{\mathtt{G} + \tilde{\tau}_p \mathtt{G} - (1 - e^{-\tilde{\tau}_p \mathtt{G}})}{\mathtt{G} + 2\tilde{\tau}_p \mathtt{G} - (1 - e^{-\tilde{\tau}_p \mathtt{G}}) + (1 + \tilde{\tau}_p \mathtt{G}) e^{-(1+\tilde{\tau}_p)\mathtt{G}}} \,. \end{aligned} \tag{9.128}$$

Delay For the delay analysis we use (9.8) following the four step procedure illustrated in page 606. In what follows we specialize this procedure to 1P-CSMA only computing the quantities that differ with respect to those obtained for ALOHA.

1. *Computation of* $\mathtt{m}_w$*:* Each packet that finds the channel busy goes through a queueing delay w. In fact, whenever a packet arrives in the middle of a BSP and finds the channel busy, it must wait until the beginning of the next BSP for being transmitted. Under the condition that a given packet found the channel busy (either in state 1 or 2) the average waiting time until the channel is detected idle, i.e., the queueing delay, can be calculated as [14]

$$\text{E}\left[w|\text{busy}\right] = \frac{\text{E}\left[(t_{\text{P}} + Y)^2\right]}{2\,\text{E}\left[t_{\text{P}} + Y\right]} = \frac{t_{\text{P}}^2 + 2t_{\text{P}}\,\text{E}\left[Y\right] + \text{E}\left[Y^2\right]}{2(t_{\text{P}} + \text{E}\left[Y\right])} \,. \tag{9.129}$$

In turn, $\text{E}\left[Y^2\right]$ is obtained from the PDF of Y (9.99) as

$$\begin{aligned} \text{E}\left[Y^2\right] &= \lambda_{\text{t}} e^{-\lambda_{\text{t}}\tau_p} \int_0^{\tau_p} y^2 e^{\lambda_{\text{t}} y} dy \\ &= \lambda_{\text{t}} e^{-\lambda_{\text{t}}\tau_p} \left(\frac{e^{\lambda_{\text{t}}\tau_p}(2 - 2\lambda_{\text{t}}\tau_p + \lambda_{\text{t}}^2\tau_p^2) - 2}{\lambda_{\text{t}}^3} \right) \\ &= \tau_p^2 - \frac{2}{\lambda_{\text{t}}}\,\text{E}\left[Y\right] \end{aligned} \tag{9.130}$$

where $\text{E}\left[Y\right]$ is given by (9.100). Hence, together with (9.129) we have

$$\text{E}\left[w|\text{busy}\right] = \frac{t_{\text{P}}^2 + \tau_p^2 + 2(t_{\text{P}} - \lambda_{\text{t}}^{-1})\,\text{E}\left[Y\right]}{2(t_{\text{P}} + \text{E}\left[Y\right])} \tag{9.131}$$

and the average queueing delay is found unconditioning as

$$\mathtt{m}_w = \text{E}\left[w\right] = \text{E}\left[w|\text{busy}\right] P_{\text{busy}} \tag{9.132}$$

which is computed using (9.100), (9.128) and (9.131).

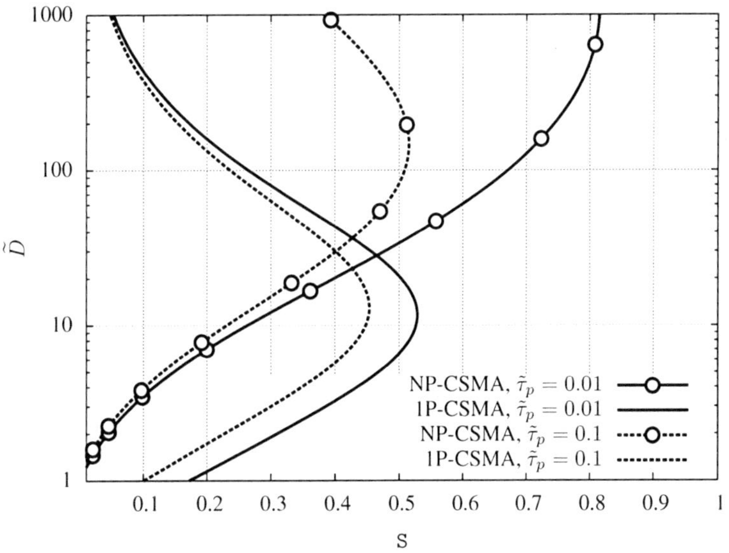

Figure 9.18 Normalized delay versus normalized throughput: comparison between 1P-CSMA and NP-CSMA for $1/\beta = 10t_\mathrm{P}$.

2. *Computation of* $\mathtt{m}_{t_\mathrm{retx}}$. For $\mathtt{m}_{t_\mathrm{retx}}$ we use (9.87) with $\mathtt{m}_w$ given by (9.132).

Applying (9.8) together with the previous results we obtain the mean normalized delay for 1P-CSMA as

$$\widetilde{D}_{\mathrm{1P-CSMA}} = \left(\frac{\mathtt{G}}{\mathtt{S}_{\mathrm{1P-CSMA}}} - 1\right)\left(1 + \tilde{\mathtt{m}}_w + 2\tilde{\tau}_p + \tilde{t}_\mathrm{A} + \widetilde{\left(\frac{1}{\beta}\right)}\right) + 1 + \tilde{\mathtt{m}}_w + \tilde{\tau}_p \tag{9.133}$$

where $\tilde{\mathtt{m}}_w = \mathtt{m}_w / t_\mathrm{P}$.

Example Results In Fig. 9.18 we show the normalized delay versus the normalized throughput for 1P-CSMA and NP-CSMA. All curves are plotted by varying the normalized offered traffic $\mathtt{G}$. The transmission parameters are those of Fig. 9.7, while $1/\beta = 10t_\mathrm{P}$ and $\tilde{\tau}_p$ is varied in the set $\{0.01, 0.1\}$. As can be seen from this plot, 1P-CSMA outperforms NP-CSMA at low $\mathtt{G}$ (where, due to its aggressive transmission behavior, 1P-CSMA has substantially shorter delays), whereas for higher values of $\mathtt{G}$ NP-CSMA gives better performance; this also results from the fact that, with respect to 1P-CSMA, NP-CSMA enters the unstable region for higher values of $\mathtt{G}$.

9.2.7.3 p persistent CSMA

p persistent CSMA (pP-CSMA) tries to reach a compromise between reduction of collision probability, as in NP-CSMA, and the reduction of idle times, as in 1P-CSMA. In this scheme the time axis is finely slotted, with a slot time equal to τ_p. We may consider, for simplicity, that all users are synchronized so that they begin their transmissions at the beginning of a slot. An illustration of the pP-CSMA protocol is shown in Fig. 9.19.

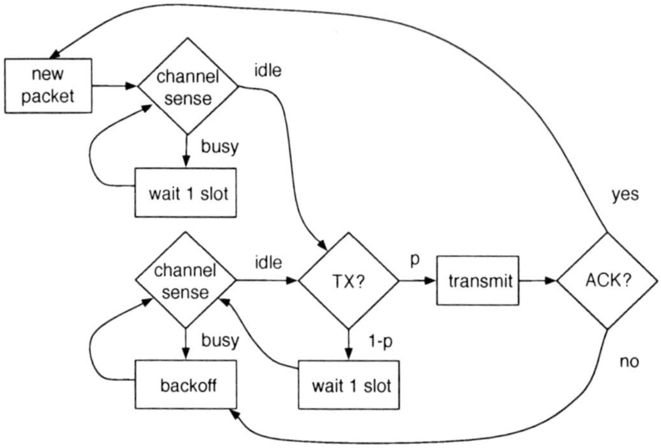

Figure 9.19 Illustration of the pP-CSMA protocol.

When a user is ready to send, it senses the channel. If the channel is sensed busy the user waits until the channel is found idle. When the channel is idle, the user transmits with probability p. With probability $1 - p$ it instead defers its transmission until the next slot. If the next slot is also idle, the user either transmits or defers again, with the same probabilities p and $1 - p$, respectively. This process is iterated until either the packet has been transmitted or another user has begun its own transmission. In the latter case, the channel is sensed busy and the user acts as if there had been a collision, i.e., it waits a random backoff time, according to a given distribution $p_{\tau_{\mathrm{rtx}}}(a)$, and starts over again.

This scheme is not analyzed here as its performance evaluation involves mathematical tools that go beyond the target of this book, see [14] for further details.

9.2.7.4 CSMA with Collision Detection The ALOHA family of protocols suffer from the interference due to concurrently transmitted packets. ALOHA suffers the most as no precautions are taken to counteract the collision problem. CSMA protocols reduce the packet collision probability through a "listen before talk" approach, i.e., by sensing the channel before initiating a new packet transmission. At first sight, their throughput performance seems to be the best possible as it only depends on the propagation delay τ_p, which is not under the user's control, at least for a given channel. However, a further increase in the throughput is possible through the observation that, whenever a collision occurs, the throughput is reduced by the length of the corresponding busy sub-period. Thus, in order to improve the throughput we must reduce the length of "collided" sub-periods. This is exactly the rationale behind the strategy of CSMA/CD.

In CSMA/CD a user, upon detecting a collision, first broadcasts[13] a *jamming tone* for t_j seconds after which it stops transmitting. The jamming tone is sent to alert all other users about the detected collision and has the effect of shortening the packet transmission phase as soon as a collision is detected. For the analysis of CSMA/CD we consider the diagram of Fig. 9.20, where the transmission cycle of the 1 persistent version of CSMA/CD is depicted.

[13]The broadcast is performed on the same channel where the data is transmitted.

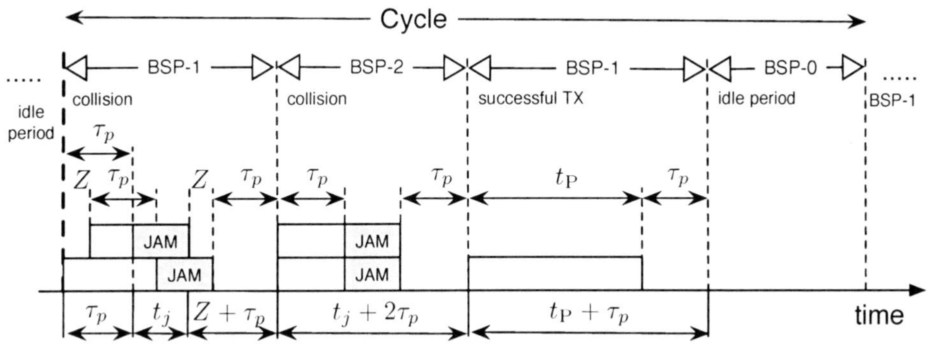

Figure 9.20 Cycle structure for the analysis of one persistent CSMA/CD (1P-CSMA/CD).

Consider the leftmost busy BSP first. At the beginning of this BSP a first packet is sent by user 1. However, a second user 2 also transmits its own packet at time $Z < \tau_p$, i.e., within the vulnerable period of this BSP, and this causes a collision. As in CSMA/CD each user continuously monitors collisions, in our example, user 2 detects an ongoing collision at time τ_p, when it receives the signal transmitted by user 1. At this time, user 2 starts transmitting a jamming tone, which lasts t_j seconds. User 1 also starts the transmission of a further jamming tone at time $Z + \tau_p$, i.e., after τ_p seconds from the beginning of the transmission of user 2. If we relax assumption BA5 of page 579, admitting that there are asymmetries in the underlying topology, all remaining users become aware of the ongoing collision only at time $Z + 2\tau_p + t_j$. In fact, in this case we need to wait the longest delay to be sure that all users are reached by at least one jamming tone. In our example a further user 3 might in fact be connected to user 1 but not to user 2 and this third user will complete the reception of the jamming tone sent by user 1 only at time $Z + 2\tau_p + t_j$.

Before analyzing the performance of the protocol, we note that CSMA/CD is often too complex to be implemented in most systems. In wired networks based on Ethernet [12] CSMA/CD is successfully used, as collision detection is easy to implement for this type of technology; it is in fact sufficient to check voltage levels on the transmission cable and a collision is detected whenever they differ from what was sent. In wireless systems, however, this technique is hard to use. First, many radio devices can either transmit or receive at any given time. This makes the channel half duplex and, in turn, interference detection is not possible. In addition, near field effects make it difficult to detect other interfering transmissions while transmitting. For the following analysis we extend assumption BA3 of page 579 admitting that users can detect collisions while transmitting their packets.

Throughput analysis We evaluate the throughput of the 1 persistent version of CSMA/CD. A transmission cycle is composed of a number of BSPs, during which packets either collide or are successfully transmitted, followed by an idle period, during which no packets are transmitted (see Fig. 9.20 for an example). The transmission process is still characterized by the three state Markov chain of Fig. 9.16, where the states still represents the number of packets transmitted at the beginning of the BSP. Busy sub-periods of type 0, 1 and 2 are associated with states 0, 1 and 2, respectively, see page 620. As in Section 9.2.7.2, with π_i and T_i, we denote, respectively, the asymptotic state probability and the time spent in state i, $i \in \{0, 1, 2\}$. The normalized throughput is found using (9.110). The quantities $\mathrm{E}[T_i]$ and π_i are calculated in the following.

Computation of $\mathrm{E}[T_i]$ From (9.92) T_0 is still exponentially distributed. Thus, we have $\mathrm{E}[T_0] = 1/\lambda_t$. To evaluate $\mathrm{E}[T_1]$ we must distinguish between successful and collided BSP of type 1. Given that the system is in state 1 (i.e., a single packet is transmitted at the beginning of the BSP) we have a successful BSP of type 1 with probability $\exp\{-\lambda_t \tau_p\}$, i.e., when no packets are transmitted during the vulnerable period of the BSP. For the collided BSP of type 1 with time t as its starting time, we define $t + Z$ as the time at which the first colliding packet is sent, where we recall that Z is a rv denoting the arrival time for the first colliding packet in the BSP. Hence, $\exp\{-\lambda_t z\}$ is the probability that no colliding packets are transmitted in $[t, t+z]$, i.e., that the first colliding packet is transmitted at some time $t' > z$. Hence, the PDF of Z is $p_Z(z) = \lambda_t \exp\{-\lambda_t z\}$. Now, we have that

$$T_1 = \begin{cases} t_P + \tau_p & , \text{ successful BSP} \\ 2\tau_p + t_j + Z & , \text{ collided BSP} \end{cases} . \tag{9.134}$$

$\mathrm{E}[T_1]$ is thus obtained as

$$\begin{aligned} \mathrm{E}[T_1] &= \mathrm{E}[T_1|\text{successful BSP}]\, \mathrm{P}[\text{successful BSP}] \\ &\quad + \mathrm{E}[T_1|\text{collided BSP}]\, \mathrm{P}[\text{collided BSP}] \\ &= (t_P + \tau_p)e^{-\lambda_t \tau_p} + \int_0^{\tau_p} (2\tau_p + t_j + z)\lambda_t e^{-\lambda_t z} dz \\ &= (t_P + \tau_p)e^{-\lambda_t \tau_p} + (2\tau_p + t_j)(1 - e^{-\tau_p \lambda_t}) + \frac{1 - e^{-\lambda_t \tau_p}}{\lambda_t} - \tau_p e^{-\lambda_t \tau_p} \\ &= t_P \left[e^{-\tilde{\tau}_p \mathsf{G}} + \left(2\tilde{\tau}_p + \tilde{t}_j + \frac{1}{\mathsf{G}} \right)(1 - e^{-\tilde{\tau}_p \mathsf{G}}) \right] \end{aligned} \tag{9.135}$$

with $\tilde{t}_j = t_j / t_P$. Moreover, from Fig. 9.20 it can be seen that $T_2 = t_j + 2\tau_p$ is constant, hence

$$\mathrm{E}[T_2] = t_j + 2\tau_p = t_P(\tilde{t}_j + 2\tilde{\tau}_p) . \tag{9.136}$$

Computation of the asymptotic state probabilities π_i The asymptotic state probabilities are still given by (9.117). In this case, however, here (9.112) does not hold. Hence, the solution is thus slightly more complicated and is given by

$$\begin{aligned} \pi_1 &= (P_{20} + P_{21})/\Psi \\ \pi_2 &= (1 - P_{10} - P_{11})/\Psi \\ \pi_0 &= (1 - \pi_1 - \pi_2) = ((1 - P_{11})P_{20} + P_{10}P_{21})/\Psi \end{aligned} \tag{9.137}$$

where the normalizing constant Ψ is

$$\Psi = (1 - P_{10} - P_{11})(1 + P_{20}) + (1 + P_{10})(P_{20} + P_{21}) . \tag{9.138}$$

For the transition probabilities we still have that $P_{02} = P_{00} = 0$. Thus $P_{01} = 1$. The calculation of P_{10} and P_{11} follows by distinguishing between two cases:

1) successful BSP. In case 1 the system moves to state 0 in the next BSP if no packets arrive during the entire transmission of the packet sent in the current BSP (i.e., during $t_{\rm P}$ seconds).

2) collided BSP. In case 2 (see, e.g., the leftmost BSP in Fig. 9.20), in order to move to state 0 there should not be packet arrivals between the beginning of the transmission of the first packet sent in the BSP and the end of the jamming tone transmitted by the user that sent this packet (i.e., during $\tau_p + t_j + Z$ seconds).

Thus, we define

$$T_1' = \begin{cases} T_1'(1) = t_{\rm P} & , \text{ successful BSP} \\ T_1'(2) = \tau_p + t_j + Z & , \text{ collided BSP} \end{cases} . \tag{9.139}$$

P_{10} is thus found as

$$\begin{aligned} P_{10} &= \mathrm{P}\left[\text{no arrivals in } T_1'(1) \text{ seconds and successful BSP} | X = 1\right] \\ &\quad + \mathrm{P}\left[\text{no arrivals in } T_1'(2) \text{ seconds and collided BSP} | X = 1\right] \\ &= e^{-\lambda_{\rm t} t_{\rm P}} e^{-\lambda_{\rm t} \tau_p} + \int_0^{\tau_p} e^{-\lambda_{\rm t}(\tau_p + t_j + z)} \lambda_{\rm t} e^{-\lambda_{\rm t} z} dz \\ &= e^{-(1+\tilde{\tau}_p)\mathtt{G}} + \frac{e^{-(\tilde{\tau}_p + \tilde{t}_j)\mathtt{G}}(1 - e^{-2\tilde{\tau}_p \mathtt{G}})}{2} . \end{aligned} \tag{9.140}$$

Following the same reasoning, for P_{11} we move to state 1 in the next BSP if a single packet arrives in $T_1'(1)$ seconds and $T_1'(2)$ seconds for cases 1 and 2, respectively. Hence, P_{11} is obtained as

$$\begin{aligned} P_{11} &= \mathrm{P}\left[\text{one arrival in } T_1'(1) \text{ seconds and successful BSP} | X = 1\right] \\ &\quad + \mathrm{P}\left[\text{one arrival in } T_1'(2) \text{ seconds and collided BSP} | X = 1\right] \\ &= \lambda_{\rm t} t_{\rm P} e^{-\lambda_{\rm t} t_{\rm P}} e^{-\lambda_{\rm t} \tau_p} + \int_0^{\tau_p} \lambda_{\rm t}(\tau_p + t_j + z) e^{-\lambda_{\rm t}(t_j + \tau_p + z)} \lambda_{\rm t} e^{-\lambda_{\rm t} z} dz \\ &= \mathtt{G} e^{-(1+\tilde{\tau}_p)\mathtt{G}} + \frac{1}{4} e^{-(\tilde{\tau}_p + \tilde{t}_j)\mathtt{G}} \left[(1 - e^{-2\tilde{\tau}_p \mathtt{G}})(1 + 2(\tilde{\tau}_p + \tilde{t}_j)\mathtt{G}) - 2\tilde{\tau}_p \mathtt{G} e^{-2\tilde{\tau}_p \mathtt{G}}\right] . \end{aligned} \tag{9.141}$$

P_{12} is obtained as $P_{12} = 1 - P_{10} - P_{11}$.

To evaluate P_{20} and P_{21} we follow a similar rationale as for state 1. Specifically, when in state 2, the system transitions to state 0 if there are zero arrivals in $\tau_p + t_j$ seconds, whereas it transitions to state 1 if there is a single arrival in the same period. This leads to

$$\begin{aligned} P_{20} &= \mathrm{P}\left[\text{zero arrivals in } \tau_p + t_j \text{ seconds} | X = 2\right] \\ &= e^{-\lambda_{\rm t}(\tau_p + t_j)} \\ &= e^{-(\tilde{\tau}_p + \tilde{t}_j)\mathtt{G}} \end{aligned} \tag{9.142}$$

and

$$\begin{aligned} P_{21} &= \mathrm{P}\left[\text{one arrival in } \tau_p + t_j \text{ seconds} | X = 2\right] \\ &= \lambda_{\rm t}(\tau_p + t_j) e^{-\lambda_{\rm t}(\tau_p + t_j)} \\ &= (\tilde{\tau}_p + \tilde{t}_j)\mathtt{G} e^{-(\tilde{\tau}_p + \tilde{t}_j)\mathtt{G}} . \end{aligned} \tag{9.143}$$

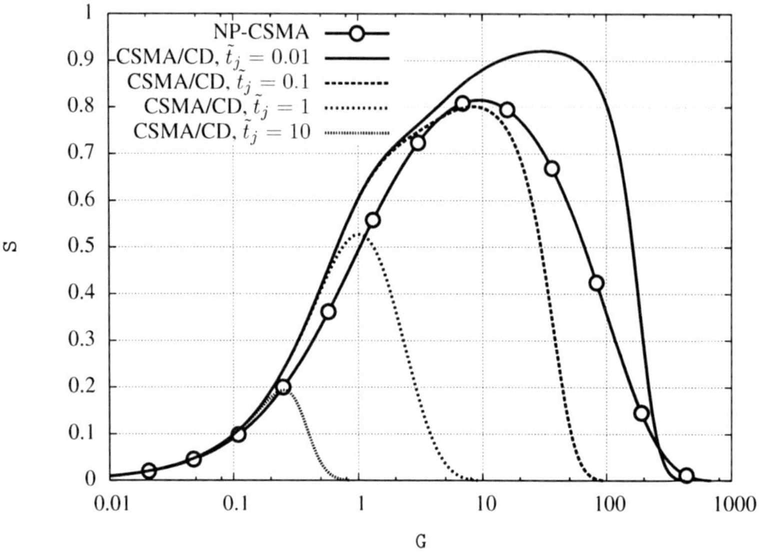

Figure 9.21 Normalized throughput versus normalized offered traffic: comparison between NP-CSMA and 1P-CSMA/CD for $\tilde{\tau}_p = 0.01$.

while P_{22} is found as $P_{22} = 1 - P_{20} - P_{21}$.

The normalized throughput is finally obtained putting together (9.110), (9.111), (9.135), (9.136) and (9.137)

$$\mathtt{S}_{1P-CSMA/CD} = \frac{(P_{20} + P_{21})e^{-\tilde{\tau}_p \mathtt{G}}}{K} \tag{9.144}$$

where

$$\begin{aligned} K = {} & \mathtt{G}^{-1}((1 - P_{11})P_{20} + P_{10}P_{21}) \\ & + \left[e^{-\tilde{\tau}_p \mathtt{G}} + (2\tilde{\tau}_p + \tilde{t}_j + \mathtt{G}^{-1})(1 - e^{-\tilde{\tau}_p \mathtt{G}}) \right] (P_{20} + P_{21}) \\ & + (2\tilde{\tau}_p + \tilde{t}_j)(1 - P_{10} - P_{11}) \,. \end{aligned} \tag{9.145}$$

Example Results In Fig. 9.21 we compare $\mathtt{S}_{1P-CSMA/CD}$ with $\mathtt{S}_{NP-CSMA}$ for a normalized propagation delay of $\tilde{\tau}_p = 0.01$.[14] Various curves are plotted for CSMA/CD by varying the normalized jamming time $\tilde{t}_j$. When $\tilde{t}_j$ is sufficiently small (e.g., $\tilde{t}_j = 0.01$) we have that 1P-CSMA/CD substantially outperforms NP-CSMA. However, as $\tilde{t}_j$ approaches one the collision detection mechanism becomes inefficient and NP-CSMA performs better than 1P-CSMA/CD.

Delay For the delay analysis we use (9.8) following the four step procedure illustrated in page 606 only computing the quantities that differ from those obtained in the previous analysis.

[14] An analysis of nonpersistent CSMA/CD (NP-CSMA/CD) is very involved and is not given here. We however note that the general observations that we made for 1P-CSMA/CD are also valid for NP-CSMA/CD.

In 1P-CSMA/CD each packet that arrives in the middle of a BSP and finds the channel busy incurs a queueing delay w. In fact, the packet in question must wait until the beginning of the next BSP for its actual transmission over the channel.

Definition 9.8 $P[\text{busy}|X=i]$ *indicates the probability that any given packet finds the channel* busy *assuming the system is in state* i *(with* $i \in \{0,1,2\}$*).*

Assumption RA9. For the following analysis we make the additional assumption that, whenever a packet (new or rescheduled) arrives in a BSP (of either type 1 or 2), its time of arrival is uniformly distributed within this sub-period. It can be seen that, while the uniform distribution of packet arrivals holds due to assumption RA2 for transmission cycles it does not hold for arrivals during BSPs. This assumption is made here to provide a simpler analysis that will nevertheless track sufficiently well the actual delay performance of the scheme. For more precise but more involved analyses the reader is referred to [22, 23].

In case the packet arrives during a BSP of type 0, it will find the channel idle and will be immediately transmitted: in this case the queueing delay is $w=0$ and $P[\text{busy}|X=0]=0$. Using RA9, in case the packet arrives during a BSP of type 1 it finds the channel busy with probability

$$P[\text{busy}|X=1] = \frac{E[T_1]-\tau_p}{E[T_1]} = \frac{\left[e^{-\tilde{\tau}_p \mathsf{G}} + \left(2\tilde{\tau}_p + \tilde{t}_j + \frac{1}{\mathsf{G}}\right)(1-e^{-\tilde{\tau}_p \mathsf{G}})\right] - \tilde{\tau}_p}{\left[e^{-\tilde{\tau}_p \mathsf{G}} + \left(2\tilde{\tau}_p + \tilde{t}_j + \frac{1}{\mathsf{G}}\right)(1-e^{-\tilde{\tau}_p \mathsf{G}})\right]} \tag{9.146}$$

using (9.135). Analogously, in case the packet arrives during a BSP of type 2 we have

$$P[\text{busy}|X=2] = \frac{E[T_2]-\tau_p}{E[T_2]} = \frac{\tilde{t}_j + \tilde{\tau}_p}{\tilde{t}_j + 2\tilde{\tau}_p} \tag{9.147}$$

using (9.136). Note that differently from 1P-CSMA, here $P[\text{busy}|X=1] \neq P[\text{busy}|X=2]$ as $E[T_1] \neq E[T_2]$. Now, we find the queueing delay w under the condition that a packet arrived during a BSP of type 1 and found the channel busy. We need to distinguish between two cases, namely, 1) successful BSP and 2) collided BSP. In case 1 the packet must wait a random time between 0 and $T_1'(1)$, depending on the arrival time of this packet within the BSP. In case 2) the packet must instead wait a random time between 0 and $T_1'(2)$. From (9.134) and (9.139) we see that:

$$T_1' = T_1 - \tau_p\,. \tag{9.148}$$

According to the arguments in [14] we have that

$$E[w|\text{ busy}, X=1] = \frac{E[(T_1')^2]}{2\,E[T_1']}\,. \tag{9.149}$$

Hence from (9.148) we have

$$E[T_1'] = E[T_1] - \tau_p = t_P\left[e^{-\tilde{\tau}_p \mathsf{G}} + \left(2\tilde{\tau}_p + \tilde{t}_j + \frac{1}{\mathsf{G}}\right)(1-e^{-\tilde{\tau}_p \mathsf{G}}) - \tilde{\tau}_p\right], \tag{9.150}$$

whereas $\mathrm{E}\left[(T_1')^2\right]$ is computed as

$$\begin{aligned}
\mathrm{E}\left[(T_1')^2\right] &= \mathrm{E}\left[(T_1')^2|\text{successful BSP}\right]\mathrm{P}\left[\text{successful BSP}\right] \\
&\quad + \mathrm{E}\left[(T_1')^2|\text{collided BSP}\right]\mathrm{P}\left[\text{collided BSP}\right] \\
&= (T_1'(1))^2 e^{-\lambda_t \tau_p} + \int_0^{\tau_p} (T_1'(2))^2 \lambda_t e^{-\lambda_t z} dz \\
&= t_P^2 e^{-\lambda_t \tau_p} + \int_0^{\tau_p} (\tau_p + t_j + z)^2 \lambda_t e^{-\lambda_t z} dz \\
&= t_P^2 e^{-\lambda_t \tau_p} + (1 - e^{-\lambda_t \tau_p})\left[\frac{2}{\lambda_t^2} + \frac{2(\tau_p + t_j)}{\lambda_t} + (\tau_p + t_j)^2\right] \\
&\quad - e^{-\lambda_t \tau_p}\left(3\tau_p^2 + 2\tau_p t_j + \frac{2\tau_p}{\lambda_t}\right) \\
&= t_P^2\left\{e^{-\tilde{\tau}_p \mathtt{G}} + (1 - e^{-\tilde{\tau}_p \mathtt{G}})\left[\frac{2}{\mathtt{G}^2} + \frac{2(\tilde{\tau}_p + \tilde{t}_j)}{\mathtt{G}} + (\tilde{\tau}_p + \tilde{t}_j)^2\right]\right. \\
&\quad \left. - e^{-\tilde{\tau}_p \mathtt{G}}\left(3\tilde{\tau}_p^2 + 2\tilde{\tau}_p \tilde{t}_j + \frac{2\tilde{\tau}_p}{\mathtt{G}}\right)\right\} .
\end{aligned} \tag{9.151}$$

The queueing delay when a packet arrives in the middle of a BSP of type 2 and finds the channel busy is found through a similar approach. In this case the packet has to wait a random time between 0 and $T_2' = t_j + \tau_p$. The average queueing delay delay is given by

$$\mathrm{E}\left[w|\text{ busy}, X = 2\right] = \frac{\mathrm{E}\left[(T_2')^2\right]}{2\,\mathrm{E}\left[T_2'\right]} = \frac{t_j + \tau_p}{2} = t_P \frac{(\tilde{t}_j + \tilde{\tau}_p)}{2} . \tag{9.152}$$

Hence, the average queueing delay is found as

$$\mathtt{m}_w = \sum_{i=1}^{2} \mathrm{E}\left[w|\text{ busy}, X = i\right] \mathrm{P}\left[\text{busy}|i\right] \pi_i . \tag{9.153}$$

Now, the time taken by the transmission of a collided packet during a BSP of type 1 is given by $T_1'(2) = t_j + \tau_p + Z$ (see leftmost BSP of Fig. 9.20). As Z for a collided BSP is always smaller than or equal to τ_p, we have that $T_1'(2) \leq t_j + 2\tau_p$. The transmission time for a packet sent in a BSP of type 2 is instead constant and equal to $t_j + \tau_p$. Accordingly, the average time taken by a collided transmission can be upper bounded as follows

$$\mathrm{E}\left[t_P|\text{collided BSP}\right] \leq \pi_1(t_j + 2\tau_p) + \pi_2(t_j + \tau_p) . \tag{9.154}$$

From the analysis in (9.74)–(9.78) together with (9.153) and (9.154) we write

$$\mathtt{m}_{\text{retx}} \leq \left(\frac{\mathtt{G}}{\mathtt{S}_{\text{CSMA/CD}}} - 1\right)\left(\pi_1(t_j + 2\tau_p) + \pi_2(t_j + \tau_p) + 2\tau_p + t_A + \frac{1}{\beta} + \mathtt{m}_w\right) \tag{9.155}$$

where $\mathtt{m}_w$ is defined in (9.153).

Thus, we obtain an upper bound (which is tight when $\tau_p \ll t_j$) for the delay

$$D_{\text{1P-CSMA/CD}} \leq \left(\frac{\mathtt{G}}{\mathtt{S}_{\text{1P-CSMA/CD}}} - 1\right)\Big(\pi_1(t_j + 2\tau_p) + \pi_2(t_j + \tau_p) + 2\tau_p + t_{\text{A}} + 1/\beta + \mathtt{m}_w\Big) + \mathtt{m}_w + t_{\text{P}} + \tau_p \ . \tag{9.156}$$

The normalized delay is

$$\boxed{\widetilde{D}_{\text{1P-CSMA/CD}} \leq \left(\frac{\mathtt{G}}{\mathtt{S}_{\text{1P-CSMA/CD}}} - 1\right)\left(\pi_1(\tilde{t}_j + 2\tilde{\tau}_p) + \pi_2(\tilde{t}_j + \tilde{\tau}_p) + 2\tilde{\tau}_p + \tilde{t}_{\text{A}} + \widetilde{\left(\frac{1}{\beta}\right)} + \tilde{\mathtt{m}}_w\right) + \tilde{\mathtt{m}}_w + 1 + \tilde{\tau}_p} \tag{9.157}$$

where $\tilde{\mathtt{m}}_w = \mathtt{m}_w/\tau_p$.

Example Results Fig. 9.22 shows the normalized delay, $\widetilde{D}_{\text{CSMA/CD}}$ (approximated using (9.157)) versus normalized throughput for 1P-CSMA/CD. All curves are plotted by varying the offered traffic $\mathtt{G}$. The transmission parameters are those of Fig. 9.7, β is set such that $1/\beta = 10t_{\text{P}}$, the normalized propagation delay is set to $\tilde{\tau}_p = 0.01$ and $\tilde{t}_j$ is varied in the set $\{0.01, 0.1, 1, 10\}$. As can be seen from this plot, CSMA/CD is very efficient in terms of delay and throughput when $\tilde{t}_j$ is small, e.g., $\tilde{t}_j = 0.01$, whereas NP-CSMA performs better for longer durations of the jamming tone.

Toward realistic backoff models In what follows we extend the above analysis to the case of a more realistic backoff model (which is used in many practical implementations of CSMA and CSMA/CD, including Ethernet) where the backoff time is doubled at each retransmission attempt (*exponential backoff*) unless a maximum backoff time $r_{\max}\tau_{\max}$ is reached. In more detail:

- For the first retransmission attempt of a rescheduled packet the backoff time is picked from a uniform distribution in $[0, \tau_{\max}]$.

- If this first retransmission fails, the backoff time for the second retransmission is picked from a uniform distribution in $[0, 2\tau_{\max}]$.

- In case $j-1$ subsequent retransmissions for the same packet are unsuccessful, the backoff time τ_j for the jth retransmission is chosen uniformly distributed in the interval $[0, \min(2^{j-1}, r_{\max})\tau_{\max}]$.

Assuming the average number of retransmissions, $\mathtt{G}/\mathtt{S} - 1$, as an integer we have that the average time spent by a packet in the backoff state $\mathtt{m}_{\text{backoff}} = \text{E}\left[\tau_{\text{backoff}}\right]$ can be approximated

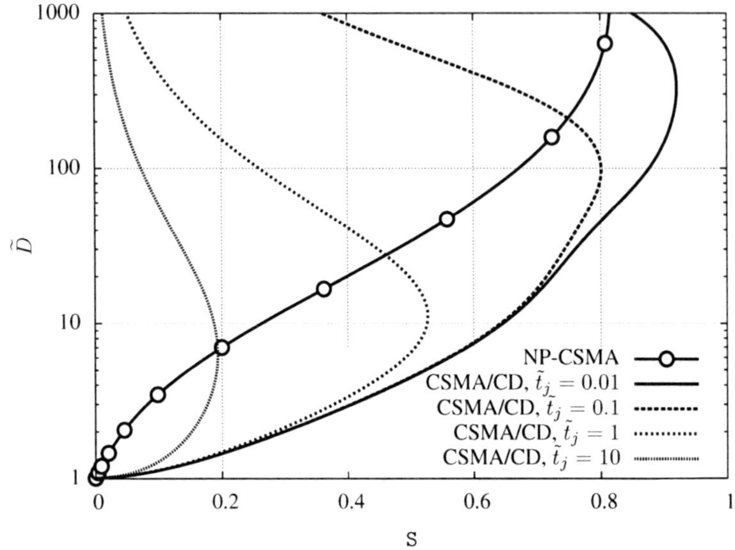

Figure 9.22 Normalized delay versus normalized throughput for 1P-CSMA/CD with $\tilde{\tau}_p = 0.01$, $1/\beta = 10t_P$ and $\tilde{t}_j \in \{0.01, 0.1, 1, 10\}$. $\widetilde{D}$ versus S for NP-CSMA with the same values of $\tilde{\tau}_p$ and β is also plotted for comparison.

as follows

$$\mathtt{m}_{\text{backoff}} \simeq \sum_{j=1}^{\frac{\mathtt{G}}{\mathtt{S}}-1} \mathrm{E}[\tau_j] = \sum_{j=1}^{\frac{\mathtt{G}}{\mathtt{S}}-1} 2^{j-1}\frac{\tau_{\max}}{2}$$
$$= \begin{cases} \dfrac{\tau_{\max}}{2}\left(2^{(\frac{\mathtt{G}}{\mathtt{S}}-1)}-1\right) & , \ \frac{\mathtt{G}}{\mathtt{S}}-1 \le r_{\max} \\ \dfrac{\tau_{\max}}{2}\left(2^{(\frac{\mathtt{G}}{\mathtt{S}}-1)}-1\right)+\left(\dfrac{\mathtt{G}}{\mathtt{S}}-1-r_{\max}\right)\dfrac{r_{\max}\tau_{\max}}{2} & , \ \frac{\mathtt{G}}{\mathtt{S}}-1 > r_{\max} \end{cases} \tag{9.158}$$

A more accurate calculation is as follows. Let $P_{\text{succ}}(j)$ be the probability that exactly j transmissions of the same packet are needed for its correct delivery. From (9.75), $\mathtt{S}/\mathtt{G}$ is the probability that a given packet transmission is successful. Thus, we have

$$\begin{aligned} \mathtt{m}_{\text{backoff}} &= \sum_{j=2}^{+\infty} P_{\text{succ}}(j)\min(2^{j-2}, r_{\max})\frac{\tau_{\max}}{2} \\ &= \sum_{j=2}^{+\infty}\left(1-\frac{\mathtt{S}}{\mathtt{G}}\right)^{j-1}\frac{\mathtt{S}}{\mathtt{G}}\min(2^{j-2}, r_{\max})\frac{\tau_{\max}}{2}\ . \end{aligned} \tag{9.159}$$

Note that in the above equation j starts from 2 as the backoff time for the first transmission of the packet is zero. Also, we use $j-2$ in place of $j-1$ in the calculation of the backoff time as j now represents the number of transmissions.

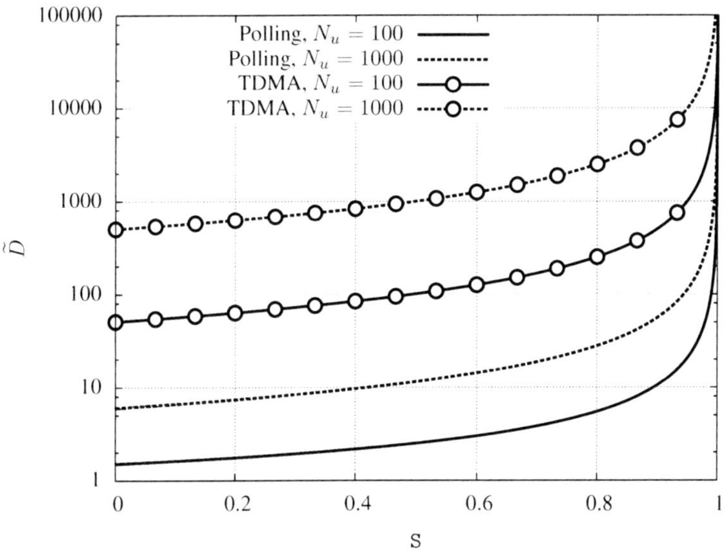

Figure 9.23 Comparison of polling and TDMA access schemes: normalized delay $\widetilde{D}$ versus normalized throughput $\mathtt{S}$ for $\tilde{\tau}_p = 0.01$, $N_u \in \{100, 1000\}$ and $\tilde{\Delta} = 0.01$ (polling).

The delay formula can be finally rewritten by replacing the term $(\mathtt{G}/\mathtt{S}_{\text{CSMA/CD}} - 1)(1/\beta)$ in (9.156) with $\text{E}\left[\tau_{\text{backoff}}\right]$, using either (9.158) or (9.159):

$$\widetilde{D}_{\text{1P-CSMA/CD}} \leq \left(\frac{\mathtt{G}}{\mathtt{S}_{\text{1P-CSMA/CD}}} - 1\right)\left(\pi_1(\tilde{t}_j + 2\tilde{\tau}_p) + \pi_2(\tilde{t}_j + \tilde{\tau}_p) + 2\tilde{\tau}_p + \tilde{t}_{\text{A}} + \tilde{\mathtt{m}}_w\right) + \tilde{\mathtt{m}}_{\text{backoff}} + \tilde{\mathtt{m}}_w + 1 + \tilde{\tau}_p \tag{9.160}$$

where $\tilde{\mathtt{m}}_{\text{backoff}} = \mathtt{m}_{\text{backoff}}/t_{\text{P}}$. Approximations (9.158) and (9.159) for the average backoff time can also be used for the analyses that were presented in the previous sections.

9.2.8 Performance Comparison of Channel Access Schemes

In the following we compare the performance of the channel access schemes that were analyzed in the previous sections. The transmission parameters R_b, t_{P} and t_{A} are the same as for Fig. 9.7. Moreover, the delay of the polling scheme is obtained using (9.68), i.e., for an infinite buffer.

TDMA vs polling In Fig. 9.23 we show the tradeoff between normalized delay $\widetilde{D}$ and normalized throughput $\mathtt{S}$ for TDMA (see (9.19)) and polling (see (9.68)). For both schemes we considered a normalized propagation delay of $\tilde{\tau}_p = 0.01$ and $N_u \in \{100, 1000\}$. For polling we considered a constant switchover time of $\tilde{\Delta} = 0.01$. As expected, the delay of both access systems increases for increasing N_u. Polling however performs better because in TDMA transmission slots are always allotted to the users, even when they have no data to transmit. In polling, instead, when users have no packets to transmit the channel is immediately

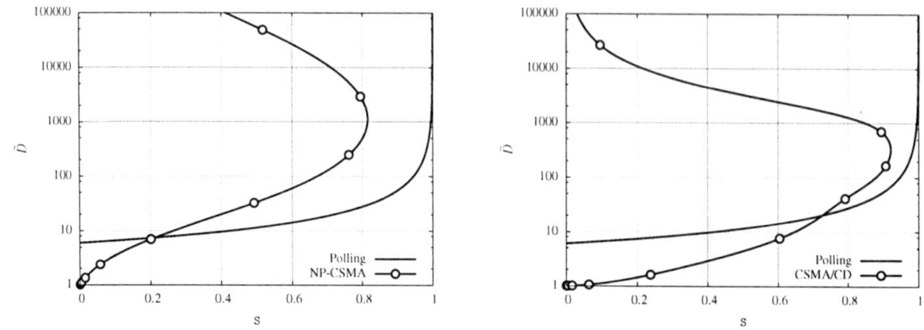

(a) Polling *vs* NP-CSMA (b) Polling *vs* CSMA/CD

Figure 9.24 Normalized delay $\widetilde{D}$ versus normalized throughput S for NP-CSMA, CSMA/CD and polling. The system parameters are $\tilde{\tau}_p = 0.01$, $1/\beta = 10t_{\mathrm{P}}$ (NP-CSMA and CSMA/CD), $\tilde{t}_j = 0.01$ (CSMA/CD), $N_u = 1000$ (polling) and $\tilde{\Delta} = 0.01$ (polling).

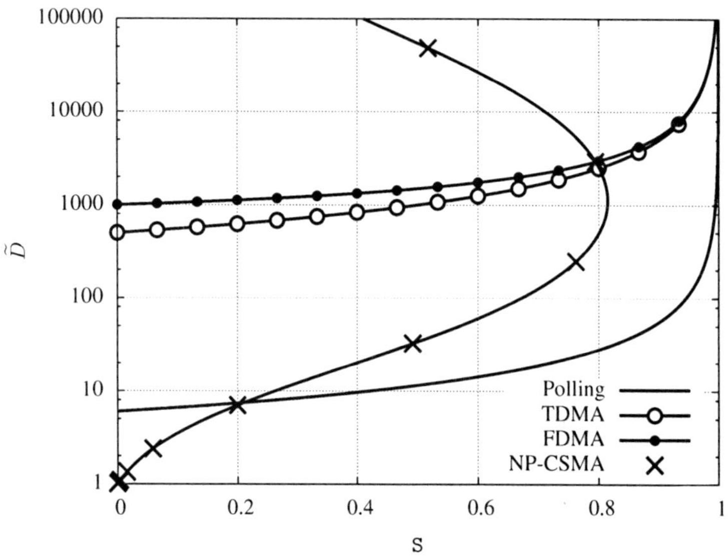

Figure 9.25 Comparison of channel access schemes: normalized delay $\widetilde{D}$ versus normalized throughput S for $\tilde{\tau}_p = 0.01$, $1/\beta = 10t_{\mathrm{P}}$ (NP-CSMA), $N_u = 1000$ (FDMA, TDMA and polling) and $\tilde{\Delta} = 0.01$ (polling).

allocated to the next user in the polling sequence. This only takes a single switchover time Δ, which in general is much smaller than the duration of a TDMA slot, t_{P}.

Polling vs CSMA Fig. 9.24 shows the tradeoff between normalized delay $\widetilde{D}$ and normalized throughput S for polling (see (9.68)) and CSMA (see (9.109) and (9.157)). For both schemes we considered a normalized propagation delay of $\tilde{\tau}_p = 0.01$. For polling we considered $N_u = 1000$ and $\tilde{\Delta} = 0.01$, whereas for CSMA schemes we set the average retransmission delay to $1/\beta = 10t_P$ (for both NP-CSMA and CSMA/CD) and the normalized jamming period to $\tilde{t}_j = 0.01$ (CSMA/CD). In general, CSMA schemes perform better at low normalized offered traffic G, where the channel is sensed idle with high probability and CSMA users, after a short channel sense period, perform their transmissions. With polling, instead, users have to wait for their transmission turn even though the preceding users in the polling sequence do not have packets to transmit. At large G the situation is however reversed. In fact, in this case CSMA is generally affected by collision events, which force the users to back off and retransmit the collided packets. Collisions instead never occur for polling, as transmissions are disjoint in time. For increasing G the advantage of scheduling in polling is clear as it allows to keep the delay low and to fully exploit the available channel resources, thus achieving a throughput very close to one. Moreover, the $\widetilde{D}$ – S curve of CSMA lies to the right with respect to the curve of NP-CSMA and this is due to the effectiveness of the CSMA collision detection mechanism.

Note that the price to pay for the higher performance of polling at high traffic is the presence of a controller (the server) or, in the case of a token ring access scheme, of a distributed token mechanism. These, however, cannot be utilized in all networks. As an example, in distributed (and possibly mobile) wireless networks, also referred to as *ad hoc* networks, users do not have any knowledge of the entire network topology, of the number of users in the system and so forth. The presence of a controller in this case is impossible as users often are not within the same transmission range. The utilization of a distributed token ring mechanism is also impractical as it would require a large communication overhead for the discovery of users, and the communication/maintainance of a shared transmission sequence. In token protocols, in fact, all users need to know the next user in the transmission sequence, which must be uniquely determined. This overhead nullifies the advantages of controlled access schemes and makes them impractical for distributed wireless networks.

Remark (CSMA/CD *vs* token ring) As confirmed by the results in Fig. 9.24(b), an important advantage of polling (e.g., token ring protocols) over CSMA/CD (e.g., Ethernet) is that it guarantees equitable access to the communication channel by every user in the network. In fact, when the number of packets that a user is allowed to transmit is bounded, even though some time is taken for the acquisition of the token, this will always be received by any user within a maximum delay; thus making the channel access *deterministic*. This feature is important for realtime communication systems, where the communication among peers is time constrained. For this reason, networks such as ARCNET and FDDI (see Section 9.2.5) were utilized in the past for real-time application such as in process control. Nowadays, however, the availability of switched gigabit Ethernet and QoS capabilities in Ethernet switches, the limited cost of this equipment and its ubiquity have largely eliminated the token ring technology from the market. These concepts will be addressed in greater detail at pages 660 and 658.

Comparison of polling, TDMA, FDMA and CSMA This last comparison is plotted in Fig. 9.25, where we show the normalized delay $\widetilde{D}$ versus the normalized throughput S for polling, TDMA, FDMA and CSMA. Transmission parameters are the same used for Figs. 9.23 and 9.24. As motivated in Section 9.2.4, TDMA outperforms FDMA for all values of S. As per our discussion above, polling performs better than both TDMA and FDMA as channel resources are never allocated for those users that have no packets to transmit. CSMA performs better than the other access schemes at low traffic. In fact, at low G CSMA users transmit immediately after having sensed the channel, which usually takes a small amount of time (much less than t_P). The unsynchronized access policy of CSMA however fails when G is high. In this case packet collisions prevail and the system becomes unstable, i.e., the delay diverges while the throughput tends to zero, as discussed in Section 9.2.6.3. Instability instead does not occur for deterministic and controlled access schemes as they adopt collision-free transmission policies. In fact, their throughput tends to one for increasing G as the channel never goes underutilized due to idle times or collisions.

9.3 AUTOMATIC RETRANSMISSION REQUEST

The main functionality of the link layer is that of providing means for accessing the channel and limiting, as much as possible, collisions among the packets transmitted over the link by different users. Beside this, the link layer is also responsible for the correct delivery of the transmitted data to the intended destination, i.e., for the data reliability over point-to-point links and this is achieved using ARQ techniques. The fundamental idea behind ARQ is to use acknowledgments sent from the receiver to the sender to inform the latter about the status of the PDUs transmitted over the forward channel (either correctly received or erroneous). The basic diagram for ARQ schemes is shown in Fig. 9.26: at the sender side the data packets coming from higher layers are fragmented and reassembled into link layer packets of usually smaller size. These are numbered (a unique sequence number is assigned to each packet) and passed to the transmission queue (TX queue in the figure). Packet transmissions can either occur on a slot basis or asynchronously, depending on the MAC scheme used underneath (e.g., CSMA or TDMA). The channel will introduce errors that might be due to multi-user interference, attenuation of the signal as well as a combination of the two; the transmitted data will be affected by these errors at the receiver side. Usually, cyclic redundancy check (CRC) fields are appended to each link layer packet so that its integrity can be verified at the receiver: this block is labeled "PDU error check" in Fig. 9.26. If a given packet is found in error, the receiver sends back to the sender a Ênegative acknowledgment (NAK), a positive acknowledgment (ACK) is instead sent in case the packet is correctly received. If the amount of interference introduced by the channel is so high that the packet header is undecodable, the corresponding packet is completely lost and we have a *missing PDU* event. The receiver can thus go through the received data and detect missing packets by observing holes in the received packet sequence (this technique is referred to in the literature as gap detection (GD) and the transmission of negative acknowledgments (NAKs) can also be elicited by the detection of missing PDUs. We observe that ACKs and NAKs are usually smaller in size than link layer PDUs and they are often protected by error correcting codes. All of this makes their error probability negligible with respect to that experienced by the PDUs transmitted over the forward link and this motivates our assumption BA7 in Section 9.1.

Upon the transmission of a new packet from the transmission queue ("TX queue" in the figure), the sender moves it onto the retransmission queue ("RETX queue" in the figure). Packets residing in the RETX queue usually have higher transmission priority than new PDUs. Whenever a NAK is received for a given packet, that packet is picked from the RETX queue and is transmitted again increasing the retransmission count for this packet and resetting the associated timeout timer, if any. When an ACK is eventually received for this PDU, the corresponding memory in the RETX queue if freed and the packet is marked as correctly delivered. The delay experienced by PDUs over the forward channel, referred to as forward channel transmission delay[15], is given by the sum of the packet transmission delay $t_{\rm P}$, the propagation delay τ_p and the processing delays at both sender and receiver. In the analysis that follows we neglect these last two delay terms, see assumption BA6 of Section 9.1.

Analogous facts hold true for the feedback delay. At the receiving user, correctly received PDUs are usually stored in a resequencing buffer. PDUs remain in this buffer until missing PDUs are correctly received and are passed to higher layers in order and without holes in their sequence numbers; this feature is referred to as *in-order delivery*. This process introduces a further delay term that is referred to as *resequencing delay*. As an alternative to this, the receiver could also operate according to an *out-of-order delivery* policy. In this case PDUs are passed to higher levels as soon as they are received, irrespective of possible errors, holes in the sequence or out-of-order events. In this case, an error indication is inserted in each packet to allow a correct error management at the application layer. In-order delivery is recommended when flow control algorithms are used at the transport layer (e.g., TCP flow control, used for file transfer), whereas out-of-order delivery is recommended for realtime applications (e.g., audio/video streaming). The latter usually have strict delay requirements and out-of-order delivery allows to promptly pass PDUs to the video/audio player thus effectively minimizing latency. This will help in preventing the playout buffer, i.e., the buffer used at the application for the buffering of incoming video/audio streaming flows, from getting empty. Further, the data contained in missing PDUs can be interpolated by the application at playout time.

A maximum number of retransmissions per packet might also be accounted for at the sender side to further limit the delay experienced by link layer packets. To sum up, the overall delay experienced by link layer PDUs over the forward channel is given by the sum of the *forward channel transmission delay* and the *resequencing delay*. The delay experienced by acknowledgment messages is instead given by the feedback channel propagation delay (τ_p) summed to the transmission delay of acknowledgment packets ($t_{\rm A}$) and their processing times at both sender and receiver.

Further optimizations of ARQ techniques (not all treated in this chapter) include: 1) the utilization of FEC jointly with retransmission schemes, 2) the use of cumulative feedback messages that acknowledge multiple PDUs (rather than using a feedback message per packet) and 3) the adaptation of the packet size as a function of the channel error probability.

The exact way in which retransmissions are managed depends on the error retransmission policy implemented at the sender. In the next sections we discuss and analyze three different retransmission policies: Stop and Wait (SW-ARQ), Go Back N (GBN-ARQ) and Selective Repeat (SR-ARQ). For the analyses in this chapter we make the following further assumptions:

LLA1. PDUs transmitted over the forward channel are corrupted with probability p. Moreover, errors are independent and identically distributed (i.i.d.) among PDUs, i.e., packets are in error with the same probability p independently of all previous error events. ACKs and NAKs are assumed to be error free.

LLA2. For the sake of analytical tractability, we assume that when a PDU is correctly transmitted a new one is always present in the transmission queue (see Fig. 9.26) which exactly describes a continuous packet source.

[15]Note that this delay contribution does not include the time that a packet spends in the transmission queue from its arrival to its first transmission over the channel.

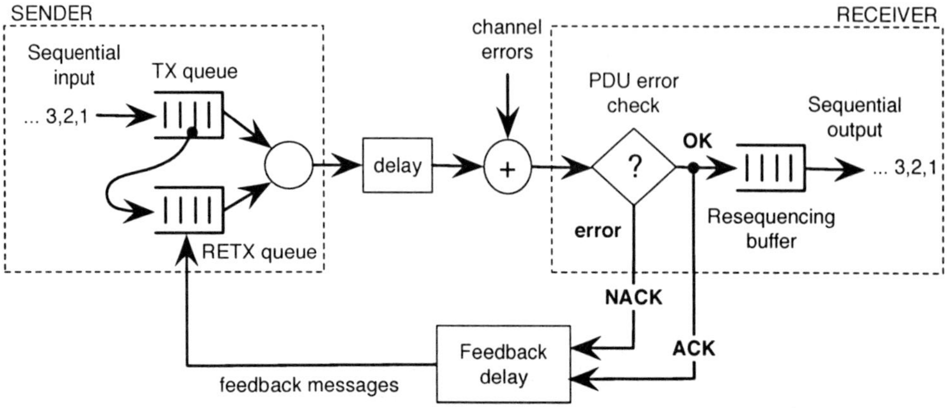

Figure 9.26 Diagram for ARQ schemes.

An observation about packet errors is in order. Packets may be corrupted and in this case at the receiver these packets are received but the corresponding CRC check fails (indicating that some of the bits in these packets were corrupted by the channel). Alternatively, packets may be lost and this occurs under strong channel interference. In the former case, ARQ algorithms at the receiver notify the sender through negative acknowledgments (NAKs). In the latter, instead, timeout timers at the transmitter are the main technique to recover from losses (as packets are not received). While we will refer to these two different types of errors in the description of the protocols, we will not distinguish between them in the related mathematical analyses, by assuming instead that losses are always due to packet corruption. This is done to keep the analysis simple and does not affect the generality of our results and observations.

Before delving into the analysis of the protocols we introduce the concepts of window size, sequence numbers, round trip time (RTT) and pipe capacity.

Definition 9.9 (sequence numbers) *In ARQ protocols, PDUs are numbered sequentially according to their arrival order at the link layer and sequence numbers, SN, are sent along with the corresponding PDUs, by including them into their packet headers. Sequence numbers are used to request retransmissions in case of lost or corrupted packets at the receiver and to reorder the received packets for those protocols where out-of-order reception can occur.*

Definition 9.10 (window size) *The ARQ window size W is defined as the maximum number of unacknowledged packets that a sender can transmit over its communication link. These are also referred to as outstanding packets.*

In general, the sender keeps track of the two variables SN_{first} and SN_{last}, where SN_{first} is the smaller sequence number among all outstanding packets and SN_{last} is the sequence number of the last packet transmitted. A sequence number, SN_{next}, is also maintained at the sender and indicates the identifier of the next packet to be transmitted. A logical representation of the sender transmission process is given in Fig. 9.27. This diagram refers to the case where transmitted packets are selectively acknowledged by the receiver. PDUs with $SN < SN_{\text{first}}$ have been already sent and acknowledged, thus they will no longer be sent by the ARQ

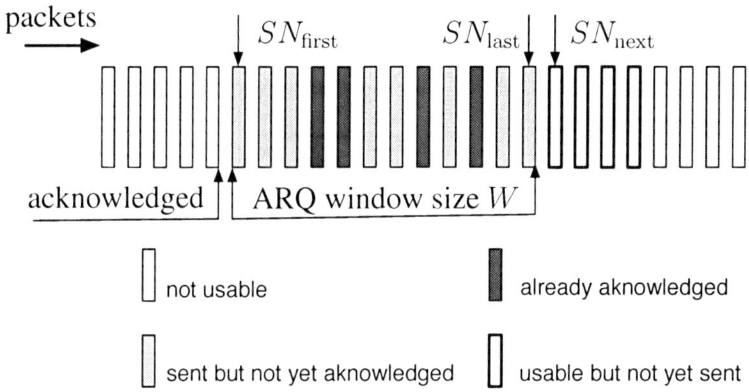

Figure 9.27 ARQ window: sender view of the transmission process.

protocol. The ARQ window is used to govern the packet transmission process, by limiting to W the number of unacknowledged packets that can be transmitted over the channel at any one time. Thus, the sender is allowed to transmit up to and including W data packets starting from $SN_{\rm first}$. Four of them (black filled in the figure) are acknowledged and the reception of the corresponding ACKs allows the transmission of four further data packets (labeled usable but not yet sent in the figure).

Definition 9.11 (round trip time) *Let A be a transmitting node that is communicating with a receiving node B using an ARQ strategy. Consider a generic packet sent by node A. Upon receiving this packet node B sends back to node A a status message (ACK or NAK, according to the packet status at the receiver). We define* round trip time, $t_{\rm RTT}$, *as the time elapsed between the beginning of the transmission of the packet and the instant when the corresponding status message is completely received at node A.*

Definition 9.12 (pipe capacity) *For a given communication between a sender and a receiver we define the* pipe capacity *of the channel as the product of the transmission rate* R_b *and the round trip time* $t_{\rm RTT}$. *Its value represents the number of bits that can be transmitted back-to-back before the ACK for the first bit sent is received at the sender. In other words, the pipe capacity corresponds to the amount of data (measured in bits) that can be sent over a certain network circuit at any given time without being acknowledged.*

Performance Measures. In this chapter, we adopt a more convenient definition for the normalized throughput and introduce a new performance metric called transmission efficiency.

- The *normalized throughput* of ARQ systems, $S_{\rm ARQ}$, is defined as the ratio of information bits successfully accepted by the receiver per unit time to the total number of bits that could be transmitted per unit time (considering a continuous transmission at the full rate R_b). Focusing on the correct reception of a tagged PDU, we can equivalently compute the throughput as the number of bits contained in this PDU divided by the number of bits that have been transmitted over the forward channel for the correct reception of this PDU. Also, dividing both terms by the transmission rate R_b we have that $S_{\rm ARQ}$ can be obtained as the ratio of $t_{\rm P}$ and $\mathtt{m}_{t_{\rm T}} = \mathrm{E}\left[t_{\rm T}\right]$, where $t_{\rm T}$ is the amount of time during which the forward channel has been occupied "exclusively" for the successful delivery of this PDU. That is, any other PDU transmitted during this period of time is discarded at the receiver. Formally,

$$\boxed{S_{\rm ARQ} = \frac{t_{\rm P}}{\mathtt{m}_{t_{\rm T}}}} \tag{9.161}$$

- The *transmission efficiency*, $\eta_{\rm ARQ}$, is given by the product of the normalized throughput and the fraction $t_{\rm DATA}/t_{\rm P}$, i.e.,

$$\boxed{\eta_{\rm ARQ} = S_{\rm ARQ}\frac{t_{\rm DATA}}{t_{\rm P}}} \tag{9.162}$$

  $\eta_{\rm ARQ}$ is a rescaled version of $S_{\rm ARQ}$ which additionally accounts for the inefficiency due to the presence of packet headers. In fact, packet headers do not contribute to the useful throughput, i.e., to the transfer of actual data between the sender and the receiver.

The analyses in this chapter are based on [24].

9.3.1 Stop and Wait ARQ

SW-ARQ is the simplest ARQ retransmission policy. It consists of sending a packet and waiting for the corresponding ACK/NAK or the expiration of a timeout timer. Specifically, if the packet is correctly received, the receiver generates an ACK and transmits it back to the sender. According to assumption LLA2, upon the reception of this ACK at the sender side a new packet is generated and transmitted over the channel. If the packet is erroneously received, the receiver generates a NAK message and sends it back to the transmitter. Upon the reception of this NAK, the sender retransmits the lost packet and the procedure is reiterated. If the feedback from the receiver (either ACK or NAK) does not arrive before the expiration of a timeout timer of $t_{\rm o}$ seconds, the sender retransmits the PDU. As we shall see below, this scheme is inefficient for links with large pipe capacities. The basic diagram for SW-ARQ is shown in Fig. 9.28. The total time to transmit a packet and receive the corresponding ACK (RTT) for an error free channel is:

$$t_{\rm RTT} = t_{\rm P} + t_{\rm A} + 2t_{\rm proc} + 2\tau_p \tag{9.163}$$

where τ_p is the propagation delay (the same delay is experienced over forward and feedback channels) and $t_{\rm proc}$ is the processing delay (assumed identical at transmitter and receiver). From now on, according to BA6, we will consider $t_{\rm proc} = 0$.

Remark (setting the retransmission timeout of SW-ARQ) Note that a proper setting of the timeout timer requires that

$$t_{\rm o} \geq t_{\rm RTT} \tag{9.164}$$

otherwise the timeout for any given packet would expire before the reception of the corresponding ACK and this leads to incorrect protocol operation. In the subsequent analysis we

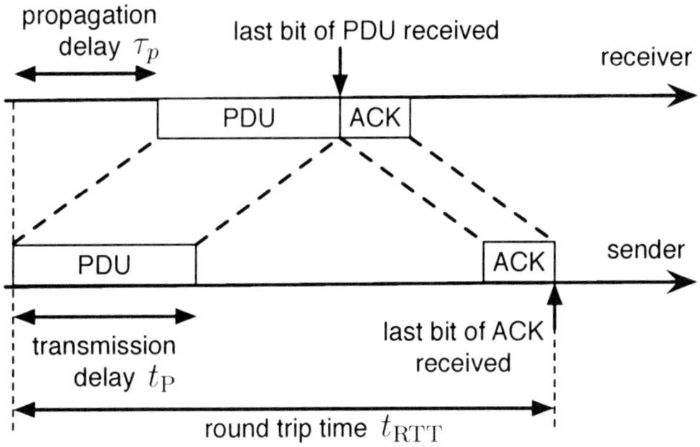

Figure 9.28 Timing diagram for SW-ARQ.

assume $t_o = t_{\rm RTT}$. Note in this case timeouts and NAKs have the same effect as they will elicit a retransmission at the same exact time, i.e., exactly $t_{\rm RTT}$ seconds after the transmission of an erroneous packet. This simplifies the analysis as we do not need to distinguish between the two error recovery mechanisms. For the sake of analytical tractability, this assumption will be also made for the following analyses in this chapter.

Throughput The total retransmission time after i subsequent retransmissions of a given packet is $t_i = i t_{\rm RTT}$. In this case, each packet entirely occupies the channel from the instant of its first transmission to the reception of the corresponding ACK. Thus, $t_{\rm T}$ is obtained as

$$t_{\rm T} = t_{\rm RTT} + t_i = t_{\rm RTT}(1+i)\ ,\ i \geq 0\ . \tag{9.165}$$

The probability that i retransmissions are needed for the correct delivery of a given packet is $p^i(1-p)$. Hence, $\mathtt{m}_{t_{\rm T}}$ is

$$\begin{aligned} \mathtt{m}_{t_{\rm T}} &= t_{\rm RTT} + \sum_{i=1}^{+\infty} t_i p^i (1-p) = t_{\rm RTT} + (1-p) t_{\rm RTT} \sum_{i=1}^{+\infty} i p^i \\ &= t_{\rm RTT}\left[1 + \frac{p}{1-p}\right] = \frac{t_{\rm P} + t_{\rm A} + 2\tau_p}{1-p} \end{aligned} \tag{9.166}$$

where the first term ($t_{\rm RTT}$) of the first line accounts for the last transmission of the PDU, whereas the second term accounts for possible retransmissions. The normalized throughput of SW-ARQ as a function of the packet error probability p is thus obtained from (9.161) as

$$\boxed{\mathrm{S}_{\rm SW-ARQ}(p) = \frac{t_{\rm P}}{\mathtt{m}_{t_{\rm T}}} = \frac{t_{\rm P}(1-p)}{t_{\rm P} + t_{\rm A} + 2\tau_p}} \tag{9.167}$$

which can be rewritten as

$$\boxed{\mathsf{S}_{\mathrm{SW-ARQ}}(p) = \frac{1-p}{1+\tilde{t}_{\mathrm{A}} + 2\tilde{\tau}_p}} \tag{9.168}$$

From (9.162) the transmission efficiency η can be obtained as

$$\eta_{\mathrm{SW-ARQ}} = \frac{t_{\mathrm{DATA}}}{\mathrm{m}_{t_{\mathrm{T}}}} = \frac{t_{\mathrm{P}}}{\mathrm{m}_{t_{\mathrm{T}}}} \frac{t_{\mathrm{DATA}}}{t_{\mathrm{P}}} = \mathsf{S}_{\mathrm{SW-ARQ}}(p) \frac{t_{\mathrm{DATA}}}{t_{\mathrm{P}}} \,. \tag{9.169}$$

Remark (discussion of (9.169)) For error free channels ($p = 0$) using (9.167) we have that

$$\mathsf{S}_{\mathrm{SW-ARQ}}(0) = \frac{t_{\mathrm{P}}}{t_{\mathrm{P}} + t_{\mathrm{A}} + 2\tau_p} = \frac{t_{\mathrm{P}}}{t_{\mathrm{RTT}}} \,. \tag{9.170}$$

The above equation (9.170) shows that the throughput of SW-ARQ is inversely proportional to the round trip time and this is true even for error free channels. This makes SW-ARQ inefficient over links with large pipe capacity. This will be quantitatively characterized below in our comparison of the presented ARQ schemes, see Fig. 9.32.

The effect of channel errors is that of reducing both S and η by a factor $1 - p$.

9.3.2 Go Back N ARQ

GBN-ARQ is slightly more involved and was proposed to increase the throughput for links with larger pipe capacities. The go back number $N \geq 1$ corresponds to the window of the GBN protocol. Hence, for any packet with $SN = i$, packet $i + N$ cannot be sent before packet i has been acknowledged. For an error free channel, the total time to transmit a packet and receive the corresponding ACK (RTT) is still given by t_{RTT} in (9.163).

Differently from SW-ARQ, in GBN-ARQ several PDUs can be sent without waiting for the next packet to be requested and this increases the transmission efficiency of the protocol, as idle times are reduced. The receiver accepts packets only in the correct order and sends request numbers RN back to the transmitter using feedback messages (ACKs or NAKs). In detail, if the last packet is correctly received in order (i.e., without leaving holes in the received packet sequence) and has $SN = i$, the receiver sends an ACK with $RN = i + 1$ to request the transmission of packet with $SN = i + 1$. ACKs are cumulative, i.e., an ACK carrying the sequence number $RN = i+1$ acknowledges all PDUs having $SN < i+1$. On the other hand, when the receiver gets an incorrect packet with $SN = i$ or when it gets a packet with $SN > i$ and the last packet received in-order has sequence number $SN = i - 1$, it sends back to the sender a NAK with $RN = i$. Once received at the sender, this NAK will elicit the in-order retransmission of all PDUs starting from $SN = i$.

We note that the feedback channel may also be unreliable and in this case ACKs or NAKs might be lost or corrupted and, as a consequence, the sender is not notified about the errors at the receiver. Also, lost feedback messages are not retransmitted and, in turn, this uncertainty could last for ever. To cope with this, a timeout timer is maintained at the sender for each outstanding packet.

The protocol rules for GBN-ARQ at the sender are as follows:

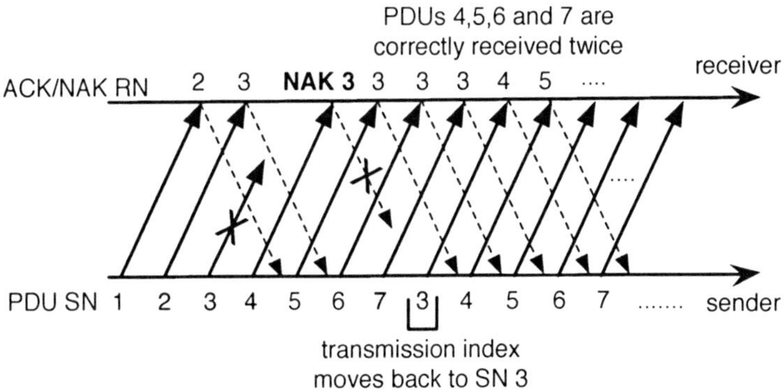

Figure 9.29 Retransmission policy example for GBN-ARQ with $N = 5$.

1. If $SN_{\text{next}} < SN_{\text{first}} + N$ and if a new packet is available for transmission, get this packet from the input buffer, assign the sequence number SN_{next} to this packet, transmit it and initialize a new timeout timer for this packet. Update SN_{next} as $SN_{\text{next}} \leftarrow SN_{\text{next}} + 1$.

2. If an error free ACK is received containing $RN_{\text{next}} > SN_{\text{first}}$, set $SN_{\text{first}} = RN_{\text{next}}$. This effectively moves upwards the lower end of the ARQ window.[16] In addition, at the sender delete the timeout timers associated with all PDUs with $SN < RN_{\text{next}}$, as these PDUs were all acknowledged by this ACK.

3. If an error free NAK is received containing the request number RN_{next}, set $SN_{\text{first}} = RN_{\text{next}}$ and $SN_{\text{next}} = RN_{\text{next}}$. Moreover, the transmission process is moved back to SN_{next}, i.e., the next transmitted PDUs will be $SN_{\text{next}}, SN_{\text{next}} + 1, SN_{\text{next}} + 2, \ldots$.

4. When a timeout for an outstanding packet expires, i.e., there are N outstanding PDUs (that is $SN_{\text{next}} - SN_{\text{first}} = N$) and none of them is acknowledged before the expiration of the timeout timer associated with the first outstanding packet (with sequence number SN_{first}), the sender sets $SN_{\text{next}} = SN_{\text{first}}$ and the transmission process is moved back to SN_{next}.

In Fig. 9.29 we show an example of a GBN-ARQ policy with $N = 5$. In this example the packet with $SN = 3$ is lost. The receiver sends an ACK with $RN_{\text{next}} = 3$ upon receiving PDU with $SN = 2$ and sends a NAK with $RN_{\text{next}} = 3$ when it receives the first out-of-order PDU (with $SN = 4$). However, the NAK is lost as well and the sender retransmit the lost packet thanks to the timeout mechanism. In detail, after the transmission of PDUs with sequence numbers $4, 5, 6$ and 7, including the lost PDU, there are exactly 5 outstanding packets not acknowledged. Hence, after the transmission of packet with $SN = 7$, the timeout timer for the first outstanding packet (having $SN = 3$) expires and the transmission process is moved back to this PDU, i.e., the transmission process restarts from packet with $SN = 3$, followed by PDUs with sequence numbers $4, 5, 6, 7, \ldots$. In addition, some packets (PDUs with sequence numbers $4, 5, 6$ and 7) are transmitted twice even though they were not in error. From this

[16]From points 1 and 2 we see why go back N error recovery schemes are often called ARQ sliding window protocols. In fact, the ARQ window effectively slides upwards as the process evolves.

example we see that the main drawback of GBN-ARQ is that whenever a given packet is detected in error, the receiver also rejects the following $N-1$ packets even though they may be correctly received. This results in their retransmission and in a severe degradation of the throughput, especially when t_{RTT} is large.

Remark (setting the retransmission timeout of GBN-ARQ) For a correct setting of the timeout timer t_o we must have that $t_o \geq t_{\mathrm{RTT}}$ as otherwise only part of each round trip time will be filled with packet transmissions and this will leads to an under utilization of the channel resources. t_o is by definition related to the window N as $t_o = Nt_{\mathrm{P}}$.[17] Hence, we have that:

$$t_o \geq t_{\mathrm{RTT}} \Rightarrow N \geq 1 + \frac{t_{\mathrm{A}} + 2\tau_p}{t_{\mathrm{P}}} \tag{9.171}$$

which can be rewritten as

$$N \geq 1 + \tilde{t}_{\mathrm{A}} + 2\tilde{\tau}_p \ . \tag{9.172}$$

with $\tilde{t}_{\mathrm{A}}$ and $\tilde{\tau}_p$ defined as in (9.3) and (9.4), respectively. For the subsequent analysis we assume $t_o = t_{\mathrm{RTT}}$ or equivalently that $N = 1 + \tilde{t}_{\mathrm{A}} + 2\tilde{\tau}_p$.

Throughput The time taken by the retransmission of any given packet is $t_o = Nt_{\mathrm{P}}$. Thus, the time taken by $i \geq 1$ retransmissions of the same packet is $t_i = it_o$, while the last transmission of the packet (that is by definition successful) only takes t_{P} seconds. Note that in GBN-ARQ any other packets that are transmitted between the first and the last transmission of this packet do not contribute to the throughput as these will all be retransmitted after its successful transmission, irrespective of whether they are successfully transmitted or not. Thus, t_{T} is obtained as

$$t_{\mathrm{T}} = t_{\mathrm{P}} + t_i = t_{\mathrm{P}}(1 + iN) \ , \ i \geq 0 \ . \tag{9.173}$$

As for SW-ARQ, the probability that any given packet requires exactly i retransmissions for its correct delivery is $p^i(1-p)$. Hence, we can write

$$\begin{aligned} \mathtt{m}_{t_{\mathrm{T}}} &= t_{\mathrm{P}} + \sum_{i=1}^{+\infty} t_i p^i (1-p) = t_{\mathrm{P}} + (1-p) t_o \sum_{i=1}^{+\infty} i p^i \\ &= t_{\mathrm{P}} + t_o \frac{p}{1-p} = \frac{(N-1)p t_{\mathrm{P}} + t_{\mathrm{P}}}{1-p} \ . \end{aligned} \tag{9.174}$$

The normalized throughput of GBN-ARQ as a function of the packet error probability p is thus obtained as

$$\boxed{\mathtt{S}_{\mathrm{GBN-ARQ}}(p) = \frac{t_{\mathrm{P}}}{\mathtt{m}_{t_{\mathrm{T}}}} = \frac{1-p}{(N-1)p+1}} \tag{9.175}$$

If $t_{\mathrm{DATA}} = D/R_b$ is the transmission time for the packet load, the transmission efficiency $\eta_{\mathrm{GBN-ARQ}}$ is calculated as

$$\eta_{\mathrm{GBN-ARQ}} = \mathtt{S}_{\mathrm{GBN-ARQ}}(p) \frac{t_{\mathrm{DATA}}}{t_{\mathrm{P}}} \ . \tag{9.176}$$

[17] We assume that the timeout for a given packet expires when this packet and $N-1$ further PDUs are outstanding and the transmission of these N PDUs takes Nt_{P} seconds.

Remark (discussion of (9.175)) Note that for error free channels ($p = 0$) using (9.175) we have that

$$S_{\text{GBN-ARQ}}(0) = 1 \tag{9.177}$$

which shows that, in the absence of channel errors, GBN-ARQ is efficient even for links with large pipe capacities. However, in the presence of channel errors the protocol shows inefficiencies as some packets are needlessly retransmitted. This causes a degradation of the throughput that can be avoided using the more advanced selective repeat retransmission policy that we discuss in the next section.

Besides the GBN-ARQ protocol that we treated here, go back N covers an entire class of algorithms, the reader is referred to [6] for a deeper discussion on the matter.

9.3.3 Selective Repeat ARQ

A last and more advanced ARQ algorithm is SR-ARQ. As in GBN-ARQ, packets are numbered sequentially and sequence numbers are included in the packet headers. Also in this case, several PDUs can be sent without waiting for the corresponding acknowledgments and this increases the transmission efficiency of the protocol (through a reduction of idle times). In addition, in SR-ARQ correctly received PDUs are never transmitted multiple times to the destination; this leads to a better throughput compared to GBN-ARQ, as new packets could be transmitted instead of resending correctly received PDUs. The number of unacknowledged (outstanding) PDUs that can be sent by the protocol is dictated by the ARQ window size W. Upon the reception of a new PDU, the receiver sends a feedback message, that is either an ACK or a NAK, for correct and erroneous packets, respectively. Unlike GBN-ARQ, the receiving entity continues to accept and acknowledge (through ACK and NAK) PDUs sent after an initial error. The transmitter uses the following variables and data structures: 1) SN_{next}, the sequence number of the next packet to be transmitted, 2) SN_{last}, the highest sequence number transmitted so far, 3) W the ARQ window size, 4) a transmission buffer, containing PDUs with new data coming from higher layers, 5) a retransmission buffer, containing already transmitted PDUs, 6) N_{out} the number of outstanding PDUs that also corresponds to the number of PDUs in the retransmission buffer, 7) a timeout timer associated with each outstanding PDU, which is started when the packet is transmitted and expires after t_o seconds.

The transmission process is as follows:

1. Upon the reception of a new feedback message do the following:

   1.1. If this message is an ACK for a given packet with $SN = i$, then remove this packet from the retransmission buffer, delete the associated timeout timer and update $SN_{\text{next}} \leftarrow SN_{\text{last}} + 1$. After this, 1.1.1) if the transmission buffer is not empty pick a new packet with $SN = SN_{\text{next}}$ from this buffer, update $SN_{\text{last}} \leftarrow SN_{\text{last}} + 1$, transmit this PDU, move it into the retransmission buffer and initialize a new timeout timer for this PDU. N_{out} remains unvaried as one packet was acknowledged and a new one was transmitted. 1.1.2) if the transmission buffer is empty, update N_{out} as $N_{\text{out}} \leftarrow N_{\text{out}} - 1$.

   2.1. If this message is a NAK for a given packet with $SN = i$, then set $SN_{\text{next}} = i$, pick packet with $SN = SN_{\text{next}}$ from the retransmission buffer, reset the associated

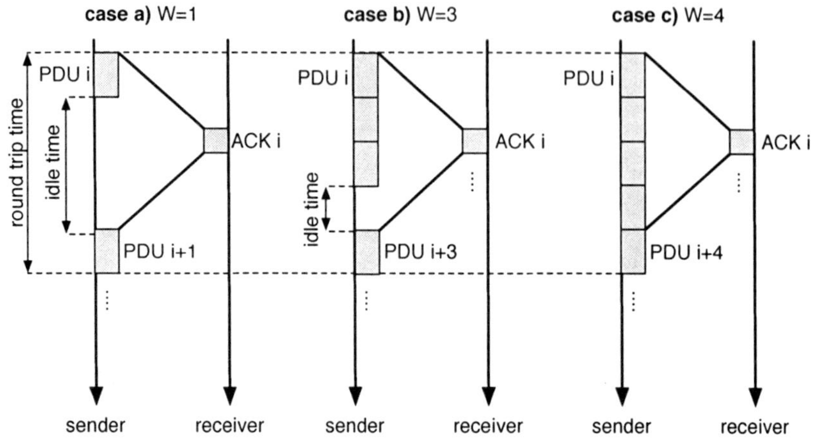

Figure 9.30 Pipelining in ARQ protocols: effect of the ARQ window size on the throughput performance. Only in case c) the window size suffices to fully utilize the channel capacity.

timeout timer and retransmit this PDU. N_{out} remains unvaried as the packet in question is still unacknowledged. SN_{last} is also unvaried in this case.

2. If no feedback message is received and the sender is ready to transmit a new packet do the following:

   2.1. If the transmission buffer is not empty and $N_{\text{out}} < W$ then set $SN_{\text{next}} \leftarrow SN_{\text{last}} + 1$, pick a new packet with $SN = SN_{\text{next}}$ from the transmission buffer, transmit it, move it into the retransmission buffer, initialize a new timeout timer for this PDU, set $SN_{\text{last}} \leftarrow SN_{\text{last}} + 1$ and $N_{\text{out}} \leftarrow N_{\text{out}} + 1$.

   2.2. If the transmission buffer is empty or $N_{\text{out}} \geq W$ do nothing.

3. If the timeout timer for a given outstanding packet with $SN = i$ expires, set $SN_{\text{next}} = i$, pick the packet with $SN = SN_{\text{next}}$ from the retransmission buffer, transmit it and reset the associated timeout timer. N_{out} and SN_{last} remain unvaried.

Remark (sliding window) With a SW-ARQ protocol, the sender waits for an acknowledgment message after the transmission of each packet. Hence, from what we have seen above, we can conclude that SW-ARQ has a sliding window of size one, i.e., at most one outstanding packet is permitted at any given time. This might be however much less than the channel's capacity. In fact, for maximum channel utilization the amount of link layer data in transit (in bits) at any given time should be equal to the pipe capacity $R_b t_{\text{RTT}}$. The key feature of sliding window protocols is that they exploit pipelined communication to better utilize the channel capacity. In GBN-ARQ the sender can send a maximum of N packets without acknowledgment, where N is the window size of the protocol. If $NL \geq R_b t_{\text{RTT}}$ (or equivalently $Nt_{\text{P}} \geq t_{\text{RTT}}$) GBN-ARQ fully utilizes the available channel resources and the channel pipe is filled at any time instant. Moreover, in GBN-ARQ the packets in the same window always form a continuous sequence of PDUs (without holes in their sequence numbers). Analogously, in SR-ARQ

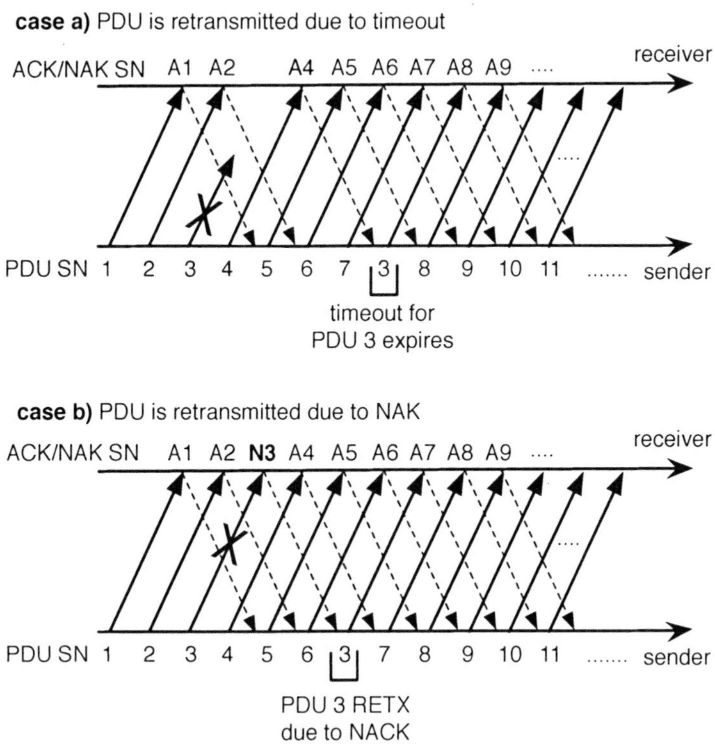

Figure 9.31 Retransmission policy examples for SR-ARQ with $N = 5$.

the channel pipe is always filled if $Wt_P \geq t_{RTT}$. However, in this case the PDUs contained in a window do not necessarily form a continuous sequence. This is because in SR-ARQ PDUs are selectively acknowledged and retransmitted without necessarily having to retransmit all packets in the current window when a loss occurs. Since all PDUs currently within the sender's window may be lost or corrupted during their transmission, the sender of both SR and GBN-ARQ schemes must keep all these PDUs in its memory for possible retransmission (the retransmission queue is used to this end). Thus, the minimum size for the retransmission buffer corresponds to the length of the window for both protocols. In Fig. 9.30 we show three possible cases corresponding to: a) SW-ARQ, b) SR-ARQ with $Wt_P < t_{RTT}$ and c) $Wt_P = t_{RTT}$; only in the last case c) is the channel fully utilized. As shown in this figure the SR-ARQ of case b) is inefficient as the channel can only be partially filled with packet transmissions. This leads to idles times and ultimately to a reduction of the throughput.[18]

In Fig. 9.31 we continue the error recovery example of Fig. 9.29 showing how lost or corrupted PDUs are retransmitted by SR-ARQ. In the first case a) (subfigure on the top) packet 3 is lost, the ARQ sender continues to transmit new PDUs until the number of outstanding packets

[18]We note that the throughput calculation in this chapter only holds when $Wt_P \geq t_{RTT}$.

equal the ARQ window size (of 5 PDUs). At this point, the timeout timer for the lost packet fires and this packet is retransmitted. Differently from GBN-ARQ, after the retransmission of packet 3 the transmission process continues from packet 8, as according to SR-ARQ correctly received packets are never retransmitted. In case b) (subfigure on the bottom) packet 3 is corrupted and this is detected at the receiver, where a negative acknowledgment is sent for this packet. This NAK is received before the expiration of the timeout timer associated with the corrupted packet and elicits its retransmission. We observe that, differently from GBN-ARQ, PDUs $4, 5, 6$ and 7 are transmitted once, which increases the channel efficiency as the channel is now used for new packet transmissions.

Remark (receiver logic) In SW and GBN-ARQ the receiver does not need a buffer. In fact, in the former protocol the receiver handles one packet at a time, whereas in the latter cumulative ACKs are sent so that any incoming packet is kept and passed to higher layers only if its SN corresponds to the expected SN in the transmission sequence (considering in-order delivery). Incoming PDUs are discarded otherwise. In SR-ARQ a buffer at the receiver is instead required. In fact, for improved throughput performance, SR-ARQ allows the reception of out-of-order PDUs and these have to be locally stored at the link layer until the missing packets in the sequence are correctly received. The actual size of the receiver buffer depends on the statistics the the resequencing delay at the receiver, see [25, 26].

Throughput For the calculation of the normalized throughput of SR-ARQ we proceed as follows. The time taken by the retransmission of any given packet is given by t_P. Differently from SW-ARQ, idle times for the the reception of feedback messages are not to be accounted for. Differently from GBN-ARQ, each retransmission only affects the packet in error and not the remaining PDUs in the same window. Thus, $i \geq 1$ retransmissions of the same packet occupy the channel for $t_i = it_\mathrm{P}$ seconds. As above, the probability that any given packet requires exactly i retransmissions for its correct delivery is $p^i(1-p)$, while the last transmission of the packet (that is by definition successful) only takes t_P seconds. In SR-ARQ any given packet occupies the channel only during its first transmission and possible retransmissions, while any other packet that is successfully sent in between contributes to the throughput. Thus, the channel occupation time t_T can be expressed in terms of the number of retransmissions i as

$$t_\mathrm{T} = t_\mathrm{P} + t_i = t_\mathrm{P}(1+i) \; , \; i \geq 0 \; . \tag{9.178}$$

$\mathtt{m}_{t_\mathrm{T}}$ is obtained as

$$\begin{aligned} \mathtt{m}_{t_\mathrm{T}} &= t_\mathrm{P} + \sum_{i=1}^{+\infty} t_i p^i (1-p) = t_\mathrm{P} + (1-p) t_\mathrm{P} \sum_{i=1}^{+\infty} i p^i \\ &= \frac{t_\mathrm{P}}{1-p} \; . \end{aligned} \tag{9.179}$$

The normalized throughput of SR-ARQ as a function of the packet error probability p is thus obtained as:

$$\boxed{\mathrm{S}_{\mathrm{SR-ARQ}}(p) = \frac{t_\mathrm{P}}{\mathtt{m}_{t_\mathrm{T}}} = 1 - p} \tag{9.180}$$

If $t_{\text{DATA}} = D/R_b$ is the transmission time for the packet load, the transmission efficiency $\eta_{\text{SR-ARQ}}$ is calculated as

$$\eta_{\text{SR-ARQ}} = \text{S}_{\text{SR-ARQ}}(p)\frac{t_{\text{DATA}}}{t_{\text{P}}} . \tag{9.181}$$

Remark (discussion of (9.180)) For error-free channels, i.e., $p = 0$, we have that

$$\text{S}_{\text{SR-ARQ}}(0) = 1 \tag{9.182}$$

which shows that SR-ARQ is efficient over error free links with large pipe capacity. Note also that for an ideal feedback channel with zero delay we have that $t_{\text{A}} + 2\tau_p = 0$, i.e., ACKs are immediately received upon the completion of the transmission of the last bit of the corresponding PDUs. In this case, (9.171) returns $N = 1$ and GBN-ARQ, SW-ARQ and SR-ARQ behave in the same exact manner (note also that (9.175) with $N = 1$ equals (9.180)). On the other hand, when $N > 1$ and $p > 0$ SR-ARQ has a higher throughout. In fact,

$$\text{S}_{\text{SR-ARQ}}(p) > \text{S}_{\text{GBN-ARQ}}(p) \Leftrightarrow 1 < (N-1)p + 1 \Leftrightarrow (N-1)p > 0 . \tag{9.183}$$

The higher throughput of SR-ARQ is due to the fact that correctly received PDUs are never retransmitted and this leads to a more efficient utilization of the channel.

9.3.4 Performance Comparison of ARQ Schemes

In this section we compare the transmission efficiency, η, of the above ARQ protocols. As an example scenario, we consider a UMTS transmission link with a rate of $R_b = 128$ kbit/s and interleaving over four transmission frames (of 10 ms each), which gives a one way propagation delay of about $\tau_p = 40$ ms (this delay also includes processing time, which is not made explicit for simplicity). packet headers and payloads take 32 and 432 bits, respectively, whereas the size of acknowledgment messages is 30 bits, assuming that only packet headers are sent over the backward channel. We calculated the window size of GBN-ARQ as $N = \lceil t_{\text{RTT}}/t_{\text{P}} \rceil = 24$, which corresponds to the number of PDUs that are outstanding at any given time. Assuming that bit errors are i.i.d., the packet error probability p is related to the bit error probability P_{bit} through the following equation

$$\boxed{p = 1 - (1 - P_{\text{bit}})^L} \tag{9.184}$$

which is the probability that at least one of the bits in the packet is corrupted (packet error probability).

In Fig. 9.32 we plot the transmission efficiency of SW, GBN and SR-ARQ as a function of P_{bit}. As expected, SR-ARQ has the highest throughput and SW-ARQ the smallest. This is due to the rather large pipe capacity that is of about 10965 bits. In this case, SW-ARQ is unable to fully utilize this channel capacity. In fact, transmitting one packet per round trip time leads to utilizing the channel as in case a) of Fig. 9.30. GBN-ARQ performs close to SR-ARQ.

In Fig. 9.33 we changed the one way propagation delay to $\tau_p = 1\ \mu$s and kept all remaining parameters unchanged. This delay is typical of indoor wireless communications where the

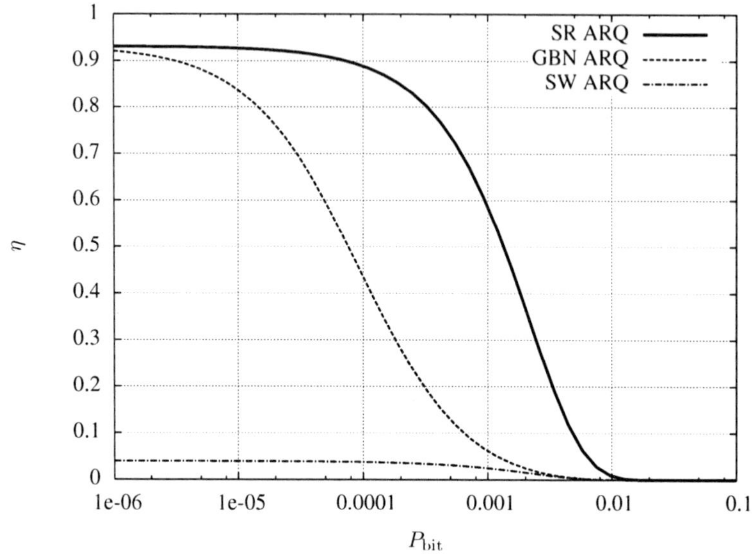

Figure 9.32 Transmission efficiency η for SW-ARQ, GBN-ARQ and SR-ARQ. $\tau_p = 40$ ms.

distance between transmitter and receiver is expected to be smaller than 100 meters. In this case the performance gap among protocols is substantially reduced. In particular, we observe that there exists a threshold value for P_{bit} (of about 10^{-4}) beyond which SW-ARQ outperforms GBN-ARQ. This is due to the fact that GBN becomes inefficient for larger error rates as every time an error occurs it retransmits the lost packet together with the following $N - 1$ packets in the corresponding window. This introduces some inefficiency which in the present results is reflected by the fact that a simpler SW-ARQ strategy performs better. For error rates smaller than the above P_{bit} threshold, GBN-ARQ performs better than SW-ARQ as the latter inherently introduces idle times between packet transmissions. However, we observe that the throughput difference for this vale of τ_p is not that substantial. In this case even the simpler SW strategy might be preferable due to its reduced complexity in terms of memory requirements.[19]

9.3.5 Optimal PDU Size for ARQ

Next, as an example, we consider the efficiency η of SR-ARQ and we show how it is possible to compute an optimal packet size as a function of the bit error rate P_{bit}, using as optimality criterion the maximization of η. Noting that $(L - L_O)/L = L_D/(L_D + L_O) = t_{\text{DATA}}/t_{\text{P}}$, we can write the efficiency of SR-ARQ as

$$\eta_{\text{SR-ARQ}} = \frac{t_{\text{DATA}}}{t_{\text{P}}}(1-p) = \frac{L - L_O}{L}(1 - P_{\text{bit}})^L \tag{9.185}$$

[19]As we shall see below, 802.11 wireless systems use SW-ARQ and, from our discussion, we see that this is a good choice due to the typically small propagation delay in Wireless Local Area Networks (WLANs), i.e., the type of networks 802.11 radios were designed for.

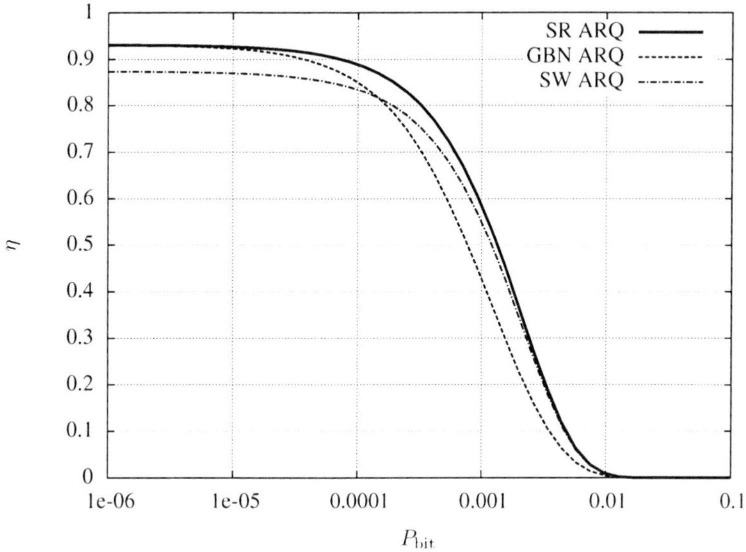

Figure 9.33 Transmission efficiency η for SW-ARQ, GBN-ARQ and SR-ARQ. $\tau_p = 1\ \mu$s.

where we used (9.184).

Now, equalling the first order derivative of (9.185) to zero and solving for L we obtain the optimal packet length, for given L_O and P_{bit}:

$$\begin{aligned} \frac{d\eta_{\text{SR-ARQ}}}{dL} = 0 \Rightarrow & \frac{(1-P_{\text{bit}})^L[L_O + L(L-L_O)\log(1-P_{\text{bit}})]}{L^2} = 0 \\ \Rightarrow L = & \frac{L_O \pm \sqrt{L_O^2 - \frac{4L_O}{\log(1-P_{\text{bit}})}}}{2} . \end{aligned} \tag{9.186}$$

Note that $(1 - P_{\text{bit}}) < 1$ and, in turn, the term $\log(1 - P_{\text{bit}})$ is smaller than zero, which implies that $L_O^2 - 4L_O/\log(1 - P_{\text{bit}})$ is strictly larger than L_O^2. Given this, the solution with the minus sign would lead to a negative packet length, which is infeasible. As a result, the optimal packet length L_{opt} is given by the solution with the plus sign in (9.186):

$$\boxed{L_{\text{opt}} = \frac{L_O + \sqrt{L_O^2 - \frac{4L_O}{\log(1-P_{\text{bit}})}}}{2}} \tag{9.187}$$

The optimal payload length is thus given by $L_{\text{opt}} - L_O$.

Example Results In Fig. 9.34 we show the transmission efficiency η as a function of the bit error rate P_{bit} for a first case where the packet payload length was chosen according to (9.187) (optimal payload size) and two further cases where the packet payload was set to 50 and 1500 bits. As expected, the optimal selection of the packet size leads to the maximum throughput at all error rates (for given L_O). Also, at small P_{bit} larger payloads lead to better

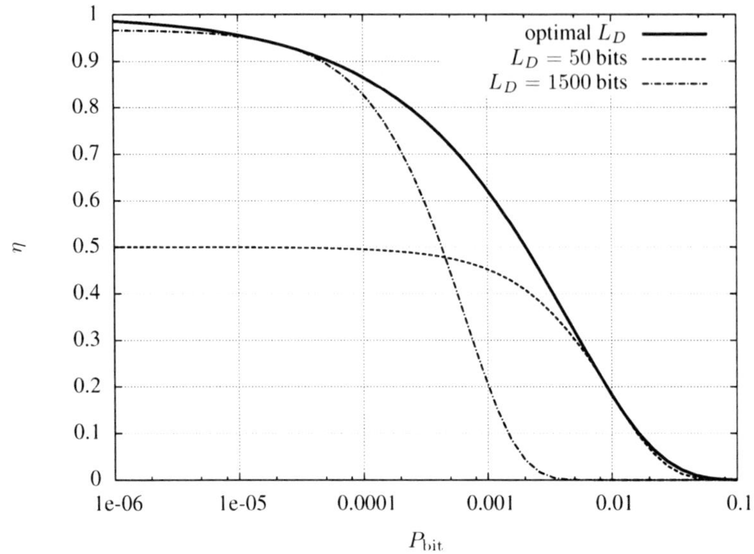

Figure 9.34 Transmission efficiency η for SR-ARQ: comparison between optimal and arbitrary selection of the PDU size.

throughput performance, whereas when P_{bit} increases it is convenient to use shorter packets. This reflects the tradeoff between overhead (given by the packet header field) and packet error probability. Specifically, when P_{bit} is small the packet error rate is small as well and the effect that dominates the throughput performance is that of decreasing the packet overhead. This amounts to reducing the fraction of control bits for each packet and thus, for given L_O, to using large payloads. When P_{bit} is high, it is instead more important to limit the packet error rate, as every erroneous packet (even in the case where a single bit is corrupted) will be retransmitted, thus wasting channel resources. The optimal tradeoff is captured by (9.187).

The above tradeoff is the main reason why cellular systems such as, e.g., UMTS [27] tend to use smaller packet lengths at the link layer (often smaller than 500 bits), whereas standard Ethernet [12] adopts packet sizes of up to 1518 bytes (excluding the preamble and the start-of-frame delimiter which take 8 more bytes). The former system has in fact a much higher bit error probability than the second (where the packet error rate due to channel impairments is negligible). Thus, the above reasonings apply. Algorithms for the selection of the optimal packet size in wireless channels through the exploitation of feedback messages from the receiver are treated in further detail in [28].

9.4 EXAMPLES OF LAN STANDARDS

In this section we discuss three popular LAN standards, namely, Ethernet [12], IEEE 802.11 [21] and Bluetooth [31]. In addition, we discuss advantages and drawbacks of Wireless Local Area Networks (WLANs) with respect to the wired Ethernet technology. Frequency regulations for the actual deployment and utilization of WLANs are also presented.

9.4.1 Ethernet

Ethernet is the most popular technology in use today for computer communications in LANs. It defines a number of wiring and signaling standards for the physical layer of the OSI model. In addition, it specifies a data link layer with the role of connecting different network elements through a common addressing format. The success of the Ethernet technology is due to its good balance among transmission rates, cost and ease of installation. The most widespread wired LAN technology consists of the combination of the twisted pair versions of Ethernet for connecting end systems to the network, with fiber optic Ethernet for the network backbone. This network setup has been used since 1980, largely replacing competing LAN standards such as token ring, FDDI, and ARCNET (see Section 9.2.5).

Ethernet was developed at Xerox PARC in 1973—1975. The first patent application of Ethernet, named *"Multipoint Data Communication System with Collision Detection"* was filed in 1975. The inventors in such an application were Robert M. Metcalfe, David R. Boggs, Charles P. Thacker and Butler W. Lampson. In the subsequent year, an initial version of the system was deployed at Xerox PARC and Metcalfe and Boggs described it in the seminal paper [12].

The institute for electrical and electronic engineers (IEEE) defines the Ethernet standard as IEEE Standard 802.3.

Physical Layer One of the most important parts of designing and installing a network is deciding the cabling medium and wiring topology to use. In what follows, we discuss several important media types:

M1. *10BASE-5* consists of a rigid and thick cable with a diameter of 0.375 inches and an impedance of 50 ohms. The "10" refer to the transmission speed of up to 10 Mbit/s, "BASE" is short for *baseband* signaling, and the "5" stands for the maximum segment length of 500 meters. This cabling technology was used to create the network backbone supporting as many as 100 users in a bus topology that could extend up to 500 meters. New network devices could be connected to the cable by drilling into the media with a device known as "vampire tap". Network devices must be installed only at precise 2.5 meters intervals. The 10BASE-5 technology was largely replaced by more cost effective technologies and is now obsolete.

M2. 10 *Mbit/s Ethernet*: in this category we have 10BASE-2 cabling, which offers the advantages of the 10BASE-5 bus topology with reduced cost and easier installation. 10BASE-2 cable is thinner and more flexible than 10BASE-5 and supports BNC connectors. On the other hand, it can only handles 30 users per network segment, which has a maximum length of 200 meters and network devices must be at least 1.5 meters apart. Another type of cabling is 10BASE-T (where the "T" stays for twisted pair cable), also referred to as unshielded twisted pair (UTP). 10BASE-T is a cable consisting of four wires (two twisted pairs) which is very similar to telephone cable in terms of appearance and shape of its end connectors (8P8C modular connectors often called RJ45). A 10BASE-T Ethernet network uses a star topology, with each user being connected directly to an Ethernet hub. The major limitations for this type of cable is the maximum cable length of 100 meters (due to attenuation phenomena), and that each user must have its own connection to the hub. UTP cables are thinner and more flexible than the coaxial cables used for the 10BASE-2 or 10BASE-5 standards. Moreover, they are graded into a number of categories, according to the supported transmission speed. 10BASE-2 cables require UTP of category 3 which can carry data at rates up to 10 Mbit/s.

M4. *Fast Ethernet*: the most common cabling in this class is 100BASE-TX. It uses category 5 UTP cabling to support transmission rates up to 100 Mbit/s. 100BASE-TX is based on star topologies, where all users are connected to a hub.

M4. *Gigabit Ethernet*: Ethernet over fiber is commonly used in structured cabling or enterprise data center applications. Fibers are however rarely used for the connection of end user systems due to cost and convenience reasons. Their advantages lie in performance, electrical isolation and distance (up to tens of kilometers). 10-gigabit Ethernet is becoming more popular in enterprise and carrier networks. Further developments are starting on 40 and 100 Gbit/s. 1000BASE-T cabling provides transmission rates up to 1 Gbit/s on UTP cables of type 5e. 100BASE-FX ("FX" stays for "fiber") and 1000BASE-LX ("LX" stands for "long" wavelength laser) respectively provide data rates of 100 Mbit/s and 1 Gbit/s. The 10-gigabit Ethernet family of standards defines media types for single-mode fiber (long haul), multi-mode fibre (up to 300 m), copper twisted pair (up to 100 m) and copper backplane (up to 1 m).

For a deeper discussion on hubs and star topologies see the paragraph "Interconnecting Ethernet networks" below.

In Ethernet, data packets transmitted on the communication channel are called *frames*. Frames start with a preamble of 7 bytes, followed by the *start frame delimiter* (1 byte) and the MAC destination and source addresses (6 bytes each). After this the frames contain the Ethernet type field (2 bytes), the payload field (between 46 and 1500 bytes), a CRC field (4 bytes). This adds up to packet sizes of 72–1526 bytes, depending on the size of the payload. After the transmission of a frame, the transmission medium should remain idle for a so called *inter-frame gap*, which corresponds to the time taken to transmit 12 bytes at the nominal transmission speed of the link. This respectively amounts to 9600 ns, 960 ns and 96 ns for Ethernet working at 10, 100 and 1000 Mbit/s.

Link Layer Ethernet was designed to allow the communication of users residing on a common bus through a CSMA/CD access scheme as follows:

1. If a data packet is ready for transmission then perform channel sensing.
2. If the channel is sensed idle transmit the data packet, else wait until the channel becomes idle, wait a further inter-frame gap period and transmit the packet.
3. If the packet is successfully transmitted then reset its retransmission counters and get ready for the transmission of a new packet. If instead a collision is detected (collision detection feature) then do the following:

   3a. Transmit a jamming signal for a further jamming interval so that all users connected to the transmission medium will also detect the collision.

   3b. Increment the retransmission counter and discard the packet if the maximum number of retransmissions is reached. Otherwise, wait a random backoff time, calculated according to the number of previous retransmissions, and retransmit the packet.

3c. Go to step 1.

Note that the success of the Ethernet technology is also due to the simplicity of its channel access technique. In fact, CSMA/CD is a random access scheme and, as such, it does not require a central entity to coordinate the transmissions of different users. This allows the system to be scalable. For example, adding a new user to the system only requires to connect the user to the shared medium and this user can communicate right away, without the need for any setup procedure. A polling protocol would instead require the registration of the user at the polling server, while a token ring would require updating the transmission sequence for all users.

Interconnecting Ethernet networks Ethernet networks are organized according to bus, star topologies or their combination. A bus topology consists of users connected together by a single cable ("the bus"). Each user is connected to the bus and directly communicates with all other users on the bus. As observed above, the main advantage of bus networks is that it is easy to add new users as it is sufficient to connect them to the cable. However, the main drawback is that any damage in the cable will cause all users to lose their connection to the network.

Ethernet was originally used with bus topologies, having CSMA/CD to coordinate the channel access of the users on the bus. However, while a simple passive wire was enough for small networks, it was not usable for larger systems, where a single damage in any portion of the cable could make the entire Ethernet segment unusable. In addition, in case of failure it was extremely difficult to isolate the problem.

Engineers soon realized that star topologies are a better option as a failure occurring in any of their connected sub-networks only affects this sub-network and not the whole system. Instead, the entire system is only affected when the failure occurs at the star point. For these reasons, reorganizing a large bus network by means of star topologies makes it easier to locate failures and makes the network more robust. In Ethernet the star points are called *Ethernet hubs*. Hubs send every packet to all their ports, thus the maximum achievable throughput is limited by that of the single link and all links must operate at the same speed.

In Ethernet long buses can be obtained through the use of *Ethernet repeaters*. These, simply regenerate the signal by forwarding it to the next network segment. Hence, a large network could be build mixing buses, repeaters and hubs (CSMA/CD is still used as the channel access technique). However, this big network will be a single collision domain, as a collision will propagate through the whole network thus reaching all users. Note that this limits the maximum achievable throughput. As an example, consider a network consisting of a hub (or a repeater) that connects two buses. Users A and B are in one bus, whereas the remaining two users C and D are on the second bus. Also, assume that A wants to communicate with B and C wants to communicate with D. In this case A and C must share the same communication channel as if they were on the same bus, as their respective signals are propagated to both buses and, in turn, only one of them can successfully transmit at any one time. One might however separate the two buses so that the signal in the first bus is not propagated to the second and vice versa. This is exactly what is done by *Ethernet bridges*.

Differently from Ethernet repeaters, which operate on electric signals, Ethernet bridges operate on protocol frames. Moreover, they only pass from one network segment to another valid Ethernet packets. Hence, collisions and packet errors are isolated. Moreover, bridges usually learn where the devices are by reading the MAC addresses of the packets transmitted on the connected network segments and do not forward packets to a network segment if this segment does not contain the destination node. Using Ethernet bridges in our above example would allow users A and C to transmit concurrently, thus doubling the throughput.

Ethernet switches are similar to Ethernet hubs in that they are used to connect multiple sub-networks into a star topology where the switch is the star point. However, as soon as the switch learns the MAC addresses associated with each of its ports it only forwards the packets to the necessary network segments. This communication paradigm is referred to as *switched Ethernet.*

9.4.2 Wireless Local Area Networks

The concept of WLANs dates back to the early 1970s. In the past decades WLANs received a lot of attention from academia and industries, the latter being interested in actual deployment for coverage extension of existing cellular networks or for the provisioning of packet based Internet access in public areas such as hotels, airports, etc. This type of usage scenario for WLAN is defined as *hot spot* coverage. WLANs are infrastructure-based wireless networks, as access points have to be deployed ahead of time and each of them is responsible for the users within its transmission range (or within its *coverage domain*). The coverage of multiple access points might overlap in space; in this case they usually transmit on channels operating over different frequency bands so as to minimize interference. Typical coverage ranges span between 10 and 100 meters.

The wireless bandwidth is a scarce resource that has to be shared among different systems ranging from broadcasting TV services, cellular networks, military applications, etc. Due to its importance, the usage of frequency bands is regulated by National and International institutions. In the United States, the body in charge of this is the federal communication commission (FCC). In Europe, the body in charge of regulating the access to the wireless spectrum is the european telecommunications standards institute (ETSI). These institutions have designated several frequency bands where a license is only needed to build the devices operating over them and not for their usage on these frequencies. These bands are known as industrial scientific and medical (ISM). The most common ISM bands for WLANs are: 2.4 GHz–2.483 GHz, 5.15 GHz–5.35 GHz, 5.725 GHz–5.825 GHz (Unites States), 5.47 GHz–5.725 GHz (European Union) and 902 MHz–928 MHz. WLANs have benefits and disadvantages with respect to wired Ethernet infrastructures. The major benefits of WLANs are:

B1. *Convenience*: the wireless nature of the communication allows users to exploit packet based Internet services at home as well as in public places such as airports, coffee shops, restaurant, hotels, etc, without the need for a wired infrastructure.

B2. *Cost*: the cost of setting up a WLAN is much smaller than that required to set up an equivalent wired network, i.e., one that serves the same number of users and in the same locations. In some cases, setting up a wired network is not convenient due to high installation cost, or even impossible, e.g., in historical buildings.

B3. *Expandability*: additional users are automatically served. In wired network serving an increasing number of users requires the deployment of further cables. Also, new access points can be easily installed to increase coverage or support a larger number of users within the same geographical area.

B4. *Nomadic use*: users on the move can connect to the network irrespective of their position within the WLAN.

B5. *Temporary installations*: wireless networks can be conveniently installed and removed.

The major disadvantages of WLANs are:

D1. *Security*: due to the inherently shared nature of the wireless medium, the transmission between any pair of WLAN users can be overheard by any other user within their communication range. A malicious user might thus attempt to decode the flow being transmitted and obtain sensitive information. To counteract this, several encryption methods were developed. These however: 1) slow down user association routines, that are required to set up links between users and access points, 2) require the transmission of a higher number of packets (higher interference and energy cost) and 3) consume energy at both transmitter and receiver to run encryption routines.

D2. *Range*: each access point has a coverage of about 10–100 meters, depending on the type of environment, i.e., on the presence of obstacles, other WLANs or transmission sources that create interference, etc. When the area to be served is large many access points are to be deployed even in those cases where the number of users in this area is small.

D3. *Reliability*: the wireless channel is prone to errors due to: 1) the degradations of the received signal quality that is caused by complex propagation effects and 2) the multi-user interference caused by users that concurrently transmit within the same geographical area. All of this leads to unreliable communication links. Even though the link layer counteracts channel errors through the usage of judicious channel access and ARQ schemes, in the presence of a large number of users and, especially when the distance from the access point is close to its coverage range, the signal quality is often poor and insufficient for a proper support of, e.g., audio/video streaming flows.

D4. *Transmission rates*: the transmission rate is dynamically adjusted by the sender so as to counteract the degraded signal quality when, e.g., a user gets farther away from its serving access point. A reduction of the transmission rate is also used in response to packet collisions. In any event, even at nominal maximum bit rates (typically ranging in 1–100 Mbit/s) the transmission rates of WLANs are considerably smaller than those offered by modern Ethernet networks (100 Mbit/s up to several Gbit/s).

9.4.3 IEEE 802.11

The IEEE 802.11 specification consists of a number of standards describing the physical and channel access techniques of the 802.11 radio technology. Using 802.11 it is possible to create high-speed wireless LANs, provided that the users are within communication range of the access points. In practice, 802.11 radios can be used to wirelessly connect laptop computers, desktop computers, mobile phones, personal digital assistants (PDAs) and any other device located within typical distances of 10–50 meters indoors or within several hundred metres outdoors. The name Wi-Fi (short for "Wireless Fidelity"), which is often used to refer to the 802.11 technology, is the name of the certification given by the Wi-Fi Alliance, the group that ensures compatibility between hardware devices that use the 802.11 standard. Wi-Fi providers are starting to cover areas that have a high concentration of users like train stations, airports, and hotels, with 802.11 networks. These access areas are referred to as "hot spots".

The available bit rates depend on the version of 802.11. For instance, 802.11b can provide rates up to 11 Mbit/s, whereas the maximum data rate of 802.11a is 54 Mbit/s. Note that the transmission rate also depends on the propagation environment, i.e., on the distance between transmitter and receiver as well as on the presence of obstacles and other transmitters in the same frequency band. Advantages and drawbacks of the 802.11 technology are the same we discussed above for WLANs. More on radio characteristics and link layer mechanisms of 802.11 is illustrated in what follows.

Physical Layer Since the first specification, released in 1997, several versions of 802.11 have been released. They mainly differ in the techniques used at the physical layer to transmit data. In what follows we discuss the following IEEE standards: 802.11, 802.11b, 802.11a, 802.11g and 802.11n. For a detailed treatment of the specific modulation techniques in these standards the reader is referred to a specialized book such as [29].

1. *IEEE 802.11*: the 802.11 working group in 1990 started to define a specification for the physical and the layers in the ISM band. The first IEE 802.11 specification was released in 1997. In this release of the specification, three different options for the physical layer were proposed: 1) infrared, 2) frequency hopping spread spectrum (FHSS) and 3) direct sequence spread spectrum (DSSS). Out of these three, DSSS was the one used and implemented in commercial products. Two different types of modulations of the signal were used: 1) differential binary phase shift keying (DBPSK) for the transmission rate of 1 Mbit/s and 2) differential quadrature phase shift keying (DQPSK) for the transmission rate of 2 Mbit/s. Differential modulation was used as it does not require carrier reconstruction as the symbol information is carried into the difference among received phases. Due to this, the receiver is significantly simpler to implement than ordinary phase shift keying. On the downside, however, it produces more erroneous demodulations. 11 different channels were defined in the 2.4 GHz ISM band for a total bandwidth occupation of 80 MHz. The transmission over a given channel occupies 22 MHz and a separation of at least 25 MHz is required so that two different channels do not interfere. Given this, only three channels, channel 1, 6 and 11 can be used in the same area without interfering. Note that operating on the ISM band was essential for the widespread use of the IEEE 802.11 technology. In fact, in this band a license is needed to build a radio device but it is not necessary for its use.

2. *IEEE 802.11b*: the aim of this working group was to increase the maximum bit rates of IEEE 802.11. The two bit rates of 5.5 and 11 Mbit/s were added by exploiting a so called pseudo random Barker sequence to spread the information over the 22 MHz bandwidth associated with a given transmission channel and obtain higher bit rates than IEEE 802.11. The number of channels, the bandwidth occupation per channel and the channel access subsystem remained unvaried with respect to legacy 802.11 (i.e., 1997 release). The IEEE 802.11b physical layer has a rate shift mechanism to fall back to 1 and 2 Mbit/s and therefore has backward compatibility with legacy IEEE802.11. The IEEE 802.11b standard was released in 1999.

3. *IEEE 802.11a*: the 802.11a work group released its specification in 2001. The main goals of this working group were to port 802.11 to the 5 GHz ISM band and further increase its physical transmission rates. The MAC layer was maintained equal to that of IEEE 802.11b. The maximum transmission rate of IEEE 802.11a is 54 Mbit/s with fall-back rates of 48, 36, 24, 18, 12, 9 and 6 Mbit/s. There are 300 MHz of

available bandwidth in the 5 GHz ISM band. Hence, 801.11a can support up to 8 non-overlapping channels, that can operate in the same geographical area without interfering with each other. This is instrumental to build a cellular-like structure with multiple partially overlapping 802.11 cells. With 8 non-overlapping channels (compared to the three available for 802.11) building such structures is facilitated. The increase in the transmission rates was possible through the usage of a technology called OFDM. The basic premise of OFDM is to split the total bandwidth B allocated to a transmission channel into N_{OFDM} sub-bands so that each sub-band takes B/N_{OFDM} Hz. Each sub-band is modulated independently of any other sub-band and its modulated flow is transmitted in parallel to that of the remaining $N_{\mathrm{OFDM}} - 1$ sub-bands. This increases the resilience of the transmitted data to the propagation phenomena that arise in indoor and mobile environments. In 802.11a a 20 MHz channel is subdivided into 64 sub-bands, of 312.5 kHz each. Only 48 sub-bands are used to transmit data, whereas 12 of them are zeroed to reduce adjacent channel interference and 4 are used to transmit pilot symbols for channel estimation, which are needed for a correct reconstruction of the transmitted signals at the receiver. Multiple data streams are separately transmitted on each sub-band according to a specific type of modulation (i.e., BPSK, QPSK, 16QAM or 64QAM, see [29]). Convolutional coding can also be added to the data flow, supported rates are: $1/3$, $2/3$ and $3/4$. Even though for a given distance the power attenuation is at least four times larger in the 5 GHz frequency band than in the 2.4 GHz range, 802.11a has been experimentally shown to have much better performance than IEEE 802.11b for distances up to 70 meters, i.e., its throughput is at least doubled with respect to that of IEEE 802.11b.

4. *IEEE 802.11g*: IEEE 802.11g was released in 2003. The main objective of working group g was to define an extension of IEEE 802.11b in the 2.4 GHz ISM band. The 802.11g radio technology enables the transmission of data at a maximum speed of 54 Mbit/s and, for interoperability with the 802.11b release, it also retains the transmission rates of IEEE 802.11b. IEEE 802.11g still exploits OFDM as the transmission technique and, as 802.11b, can exploit a maximum of three non-overlapping channels in the 2.4 GHz ISM band.

5. *IEEE 802.11n*: 802.11n standardization activities initiated in 2004 and the IEEE 802.11n standard was finally published in October 2009. Nevertheless, before the publication date major manufacturers released "pre-n" products based on a draft release of the standard. The aim of these vendors was that of pushing ahead with this new technology in an attempt to get early advantages in terms of bit rates. IEEE 802.11n builds on previous 802.11a/g standards and, as such, it also uses the OFDM transmission technology. In 802.11n, however, new features such as multiple-input multiple-output (MIMO) operation at the physical layer and frame aggregation at the MAC layer are added. 802.11n radios are expected to provide physical data rates up to 600 Mbit/s. MIMO relies on multipath signals and uses multiple antennas at both transmitter and receiver. The multipath propagation phenomenon is due to the reflection and diffraction of radio waves. In practice, reflected signals arrive at the receiver some time after the line of sight signal has been received. Hence, multiple copies are received for each transmitted symbol. 802.11b/a/g radios used spread spectrum and OFDM techniques to mitigate this interference. Instead, MIMO exploits the inherent diversity of the multipath signal to

Table 9.1 Comparison of 802.11 radio standards.

Version	Released	Freq. [GHz]	Max. rate [Mbit/s]	Mod.
802.11	1997	2.4	2	infrared, freq. hopping DSSS
802.11b	1999	2.4	11	DSSS
802.11a	2001	5	54	OFDM
802.11g	2003	2.4	54	OFDM
802.11n	2009	2.4 and 5	600	OFDM

increase a receiver's ability to recover the message information from the signal. Another feature of MIMO is the so called spatial division multiplexing (SDM), according to which multiple independent data streams are spatially multiplexed and simultaneously transmitted within one spectral channel of bandwidth. This can significantly improve the throughput. Each spatial stream requires a distinct antenna at both the transmitter and the receiver. Channel bonding is a further technique exploited by 802.11n at the physical layer. According to channel bonding two separate non-overlapping channels are used to transmit data, thus increasing the transmission rate. The 40 MHz mode uses two adjacent 20 MHz bands so that the physical transmission rate is doubled with respect to that of a single 20 MHz channel. We finally observe that the MIMO technology requires additional circuitry, including an analog-to-digital converter for each antenna and this translates into higher implementation costs.

A comparison of 802.11 radio standards is reported in Table Tab. 9.1; the maximum transmission rate of IEEE 802.11n is obtained using four antennas, where each spatial stream has a channel width of 40 MHz (doubled with respect to previous releases of IEEE 802.11).

Communication Modes the standard defines two communication modes: 1) *infrastructure-based* and 2) *infrastructureless*.

1) According to the infrastructure-based communication mode, users connect to a fixed base access point, forming a star communication topology. The infrastructure-based mode is commonly used to set up the so called Wi-Fi hot spots and the associated topology is referred to as basic service set (BSS). A drawback of infrastructure-based networks is the cost of buying and installing the access points. Its main advantage is that each access point has full knowledge of the status of the users within its transmission range and this can be used to optimize the access to the channel.

2) According to the infrastructureless communication mode, users communicate among each other according to an ad hoc paradigm, i.e., the network is fully distributed, a user can communicate with any other user within its communication range and all users can be exploited as routers in order to reach any other user in the network. Also, if any user has access to the wired Internet, it can be exploited to provide Internet access to

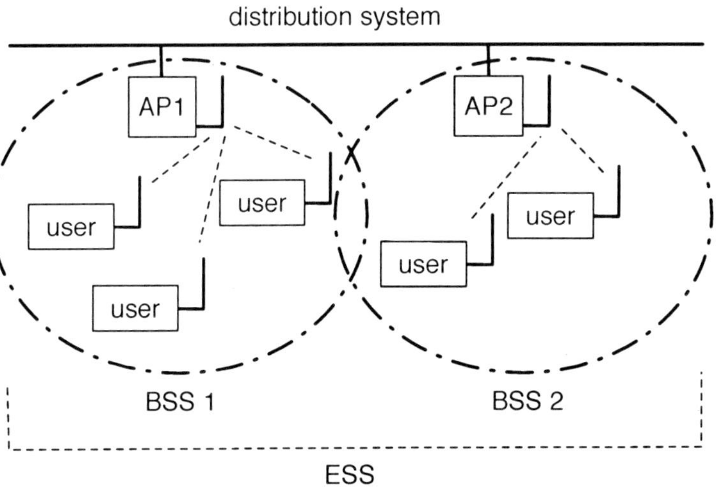

Figure 9.35 Example of ESS.

the whole ad hoc network. There are various performance issues associated with the ad hoc mode, many of which are still not completely solved. It is in fact challenging to route packets across a distributed wireless network, as the topology is not known a priori and users can be mobile, thus leading to a dynamic environment. Note that, in several situations such as inter-vehicular communication and disaster recovery, a network should be quickly set up and should be able to provide a minimum level of connectivity even without any infrastructure. The way to achieve this is to configure the network in ad hoc mode. According to the terminology in the standard, a network operating in the ad hoc mode is said to form an independent basic service set (IBSS).

In addition, as seen in Fig. 9.35 a WLAN based on the IEEE 802.11 wireless technology can be formed by a number of cells. Each cell is controlled by an access point (called AP in Fig. 9.35) that provides connectivity to all the users within its transmission range. Each cell forms a BSS. Usually, access points are directly connected to the wired Internet and are also connected with each other through a backbone. The backbone (called distribution system in Fig. 9.35) is typically Ethernet based but can also be a dedicated wireless link. The whole WLAN, including the access points, the users and the backbone, is seen as a single network and according to the standard is called extended service set (ESS).

Link Layer In the 1997 802.11 standard two access methods were defined: 1) point coordination function (PCF) and 2) distributed coordination function (DCF). The first uses a poll and response protocol to eliminate the possibility of contention for the medium. PCF is suitable for communications with a central unit (the access point) that coordinates the transmission of the users in its coverage area through a dedicated polling technique. PCF was however implemented in very few hardware devices and for this reason will not be covered any further in what follows. DCF is based on CSMA with collision avoidance (CA). In DCF, users send a new medium access control service data unit (MSDU) upon sensing the channel and detecting

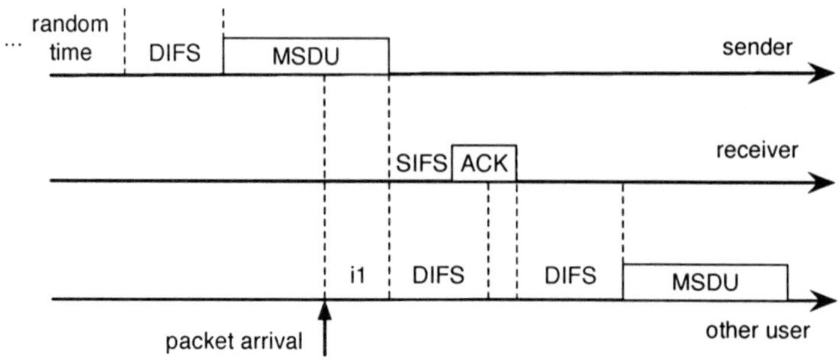

Figure 9.36 Basic access mode of IEEE 802.11 (a zero propagation delay is assumed).

that there is no other transmission in progress. If two users detect the channel as idle at the same time and both have data to send, a collision might occur. The 802.11 MAC defines a collision avoidance scheme to reduce the probability of collision. CA uses random backoff considering the time slotted. Each station has a backoff timer, expressed in units of slots, in the interval $[0, BW - 1]$, where BW is the backoff window. BW at the first transmission attempt of any given packet is set to BW_{min}. Whenever a user has a packet to transmit it senses the channel. If the channel is found idle it keep sensing it for an additional distributed inter frame space (DIFS), which is, e.g., 34 μs for 802.11a. After having sensed the channel idle for this additional DIFS, this user starts down-counting its backoff timer which is decremented as long as the channel is sensed idle and stopped when the channel is sensed busy. After being stopped, the backoff timer is reactivated as soon as the channel is sensed idle again for a time period longer than a DIFS. Only when the backoff timer fires can the user transmit its MSDU. If this MSDU collides, the backoff timer is doubled, i.e., the new BW is set to $\min(2BW, BW_{\text{max}})$ where BW_{max} is the maximum window size, and the above process (backoff timer initialization, channel sensing and down-counting) is repeated. This does not completely prevent packets from colliding. However, it reduces the collision probability as users exploit backoff timers to uniformly distribute their packet transmissions. Also, doubling the backoff after each unsuccessful transmission effectively accounts for the density of transmitting users. In fact, an increasing density corresponds to a higher collision rate. If a collision event occurs, the corresponding backoff timer is doubled so as to postpone its retransmission and this effectively decreases the instantaneous density of transmitting users. Thus, we conclude that backing off is a way of load control so as to counteract packet collisions.

Note that a radio device is usually equipped with a single antenna and in this case collisions cannot be detected while transmitting as the antenna is used in half-duplex mode, i.e., it can either transmit or receive at any one time. 802.11 was initially designed for single antenna systems and its MAC reflects this design principle. Hence, transmitted MSDUs must be explicitly acknowledged. The *basic access mode* of 802.11 adopts the two way handshake scheme shown in Fig. 9.36. The sender waits a random backoff time, according to the value of its backoff timer, then it waits a further DIFS and transmits its MSDU. This MSDU is received after a propagation delay (assumed to be zero in Fig. 9.36) and, at the receiver, a further short inter frame space (SIFS) is waited for before sending the corresponding ACK. Note that the SIFS is shorter than the DIFS so as to give priority to the transmission of ACKs, i.e., to the transmission from the receiving users. To see this, consider the user on the bottom of Fig. 9.36.

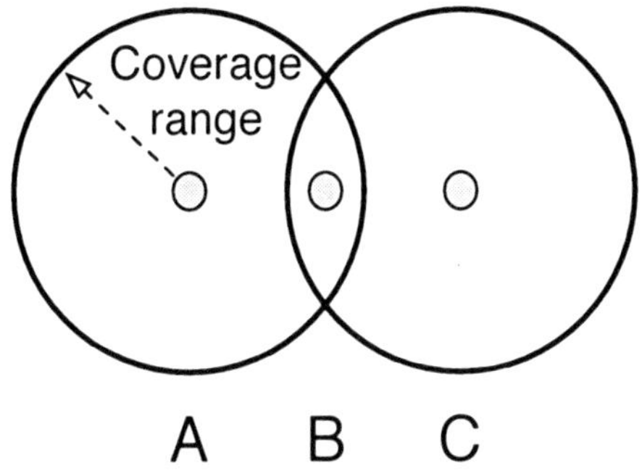

Figure 9.37 Illustration of the hidden terminal problem.

This user gets a packet arrival during the transmission of a data packet. It however defers the transmission of its new MSDU as it senses the channel busy (interval i1 in the figure). Note that this user will transmit its MSDU only after the transmission of the ACK from the receiver as the SIFS is smaller than the DIFS. Moreover, upon the completion of the ACK transmission, the channel must be sensed idle for a further DIFS before this user can transmit its MSDU.

Among the many problems that arise in wireless networks, the 802.11 MAC is particularly affected by the so called *hidden terminal problem*. We illustrate this problem through the simple network of Fig. 9.37. There are three users labeled A, B and C. B is within the transmission range of both A and C but A and C cannot hear each other. Let us assume that A is transmitting to B. If C also has a MSDU to transmit to B, according to the DCF scheme, it senses the channel and it finds it idle as it cannot hear A's transmission. As a consequence, user C starts transmitting its own packet as well. This packet however collides, at user B, with the MSDU transmitted by user A. The hidden terminal problem can be alleviated (but not completely solved) by the following virtual carrier sensing (VCS) scheme (often referred to as *floor acquisition*), that is also part of the 802.11 standard. VCS uses two further control packets: request to send (RTS) and clear to send (CTS), that are usually shorter than MSDUs. A user that wants to send a packet senses the channel as explained above and sends an RTS. Upon receiving this RTS the destination replies, after a SIFS, with a CTS to indicate that it is ready to receive the MSDU. Both RTS and CTS contain the total duration of the MSDU transmission, including the time for the transmission of the data packet and the reception of the following ACK. This information is used by all surrounding users to infer that a transmission is ongoing and defer their transmissions until its end. This is implemented through a timer called network allocation vector (NAV). In detail, as long as the NAV of any given user is greater than zero, this user must refrain from accessing the wireless channel. The use of RTS and CTS packets makes it possible to become aware of transmissions from hidden users, while also learning for how long the channel will be used by them.

A diagram of the four way handshake with virtual carrier sensing of IEEE 802.11 is plotted in Fig. 9.38: a DIFS is waited for before transmitting the RTS, whereas CTS, MSDU and ACK are all separated by short inter frame spaces. Other users set their NAVs whenever they overhear an RTS, a CTS or a MSDU and defer from transmitting until the end of the current channel contention.

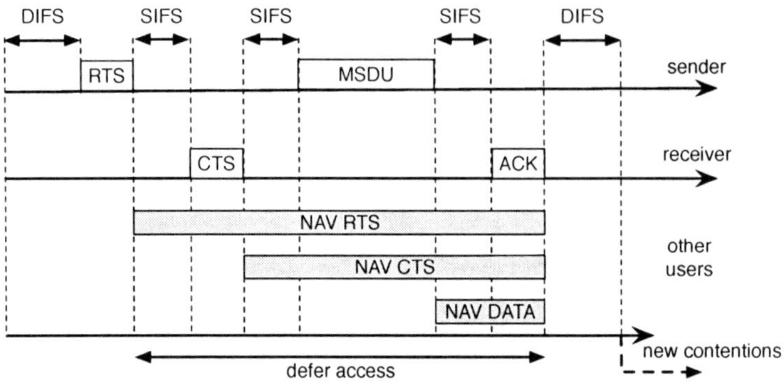

Figure 9.38 Four way handshake with virtual carrier sensing of IEEE 802.11 (a zero propagation delay is assumed). A DIFS is waited before starting the transmission of the RTS, whereas SIFSs separate the transmissions of CTS, MSDU and ACK. All other station sets their NAVs and defer from transmitting during the channel access.

As an illustrative example, consider again Fig. 9.37 with the three users now adopting the VCS mechanism. In this case, before starting its transmission to user B, user A sends an RTS having as destination user B. B responds to this RTS with a CTS that is also overheard by user C. Hence, C in this case is aware of the ongoing transmission between users A and B and of the expected duration of this communication. Moreover, C sets its NAV and accesses the channel only when this timer fires, i.e., at the end of the transmission from user A. In this specific example VCS can effectively solve the hidden terminal problem.

However, there are practical situations where VCS fails. Consider for instance the case where user C starts listening to the channel after the transmission of the CTS sent by user B. In this case, C would not be informed about the ongoing transmission and might interfere with A's transmission. This might occur in case of mobility, transmission ranges that vary with time or users that periodically put their radio to sleep for energy conservation. In practical settings, the VCS mechanism is far less effective than expected and in some cases only degrades the performance. For an experimental verification of this see, e.g., Chapter 3 of [30].

9.4.4 Bluetooth

Bluetooth is a technology for wireless communication that was designed with the main intention of replacing connection cables among cell phones, laptops and other devices such as headsets, personal computers, laptops, printers, global positioning system (GPS) receivers, digital cameras, video game consoles, etc. As such, it was designed for short range communication (power class dependent, allowing communication between 1 and 100 meters) and with the goal of being energy efficient, so that it could be used in handheld and portable devices. The specification of the Bluetooth transmission technology was the result of a joint effort of a group of companies. These companies in May 1998 conceived an organization called Bluetooth *Special Interest Group* (SIG), whose aim was the design, the definition and the promotion of a short range and power efficient wireless technology. The most influential companies (also called promoter members) within SIG were Ericsson Technology Licensing,

Intel Corporation, Microsoft (since 1999), Motorola (since 1999), Nokia Corporation, IBM and Toshiba Corporation.

As a curiosity, the name Bluetooth is taken from the tenth century Danish King Harald Blatand. This king was in fact instrumental in unifying warring factions of what is now Norway, Sweden and Denmark. This is in line with the main design goal of the Bluetooth technology that was designed to allow collaboration among different industrial environments such as computing, mobile phone and automotive markets.

Working groups The SIG had several working groups:

G1. *Air interface*: this group dealt with the specification of the Bluetooth radio, including carrier generation, digital modulation, coding, power strength control, etc.

G2. *Software*: this group dealt with the specification of a protocol stack suited to low power communication and optimized for realtime audio/video communications.

G3. *Interoperability*: this group defined user profiles that illustrated how the Bluetooth technology could be used for communication with other devices. In order to use Bluetooth, a device must be in fact compatible with certain Bluetooth profiles. These define the possible applications and uses of the technology. A few utilization examples are: 1) Data access point, 2) Ultimate headset, i.e., wireless connection of the headset with computers or mobile phones, 3) Cordless computer, i.e., wireless connection of mouse, keyboard and other external devices to home computers or laptops, 4) Business card exchange, 5) Ad hoc networking, i.e., distributed and autonomous communication among Bluetooth devices.

G4. *Compliance*: this work group dealt with the definition of the tests necessary to assess Bluetooth radio compliance. These are needed for a massive production, often from different companies, of the technology.

G5. *Legal*: this group dealt with legal affairs related to intellectual property matters.

G6. *Marketing*: this last group had the role of promoting the product and creating the need for its widespread use.

Design Considerations The main design considerations that were accounted for in the development of the Bluetooth radio were:

DC1. *Transmission rates*: provide relatively high transmission rates (nominally 1 Mbit/s).

DC2. *Energy efficiency*: realize an energy efficient communication stack, e.g., battery energy should be conserved when there is no data to transmit.

DC3. *Communication infrastructure*: the Bluetooth standard should work even without an existing communication infrastructure, i.e., Bluetooth devices should be able to recognize each other, coordinate among themselves in order to build a network and ultimately communicate with each other.

DC4. *Size*: the radio device should be small so that it can be integrated into mobile computers and cellphones.

DC5. *Utilization cost*: Bluetooth was designed to work in the license free ISM frequency band around 2.4 GHz. This means that a license is needed for building a Bluetooth device but not for its use.

DC6. *Monetary cost*: the Bluetooth radio chip should be inexpensive.

In what follows we describe the main characteristics of the Bluetooth technology.

Physical Layer The specification defines three power classes.

1) Class 1 covers the shorter personal area within a room, with a communication range of about 1 meter and transmit power of 1 mW (0 dBm).

2) Class 2 and a higher power level that can cover a medium range area, such as within a home, this has a communication range of about 10 meters and a transmit power of 2.5 mW (4 dBm).

3) Class 3 supports communications within about 100 meters and has a transmit power of 100 mW (20 dBm).

Bluetooth exploits the frequency hopping spread spectrum technology. According to FHSS, the time axis is subdivided into transmission slots of the same duration and the available bandwidth B is split into a number N_b of sub-bands. At each transmission slot, both transmitter and receivers synchronize their radios on the same band (that is a small portion of the entire bandwidth, i.e., B/N_b) and pick the band to use in the next transmission slot according to a predetermined and common pseudo-random sequence. This pseudo-random sequence is also referred to as *hopping sequence*. Note that this technique entails a tight synchronization between transmitter and receivers as, for a correct communication, they must always be working on the same frequency band. In Bluetooth the overall bandwidth B is equal to 83.5 MHz and is split into 79 sub-bands. The carrier frequency for band k is 2.402 GHz plus k MHz, with $k = 0, 1, \ldots, 78$. Guard intervals of 2.25 MHz are accounted for at each end of B so that Bluetooth transmissions will minimally interfere with other systems operating in the ISM band. Transmitter and the intended receivers hop from sub-band i to sub-band j with $i \neq j$ at 1600 hops/s, that corresponds to a transmission slot duration of 625 μs. FHSS mitigates channel impairments, i.e., the distortions in time and frequency that, in a mobile environment, are due to propagation phenomena such as reflections and diffraction of the signal. Moreover, FHSS makes the Bluetooth system robust with respect to interference among Bluetooth users and with respect to other technologies that operate in the same frequency band, e.g., IEEE802.11.

Communication Modes In the point-to-point communication mode there is a single Bluetooth device called the *master* that communicates with another Bluetooth device, called the *slave*. The master sets the hopping sequence, that is uniquely identified by its own clock and address, while the slave synchronizes its clock and hopping sequence with those of the master.

In the point-to-multipoint communication mode a single master communicates with up to 7 slaves. Again, the master dictates the hopping sequence, which is unique and followed by all slaves. A point-to-multipoint configuration including a master and multiple slaves is referred to as *piconet*. Piconet formation is performed according to a two steps procedure. First, the master must become aware of its neighboring users; this phase is called *device discovery*. Second, a negotiation is performed, during a so called *link establishment* phase, to set up communication links among the master and the new slaves. Communication among slaves is only possible through the master. In fact, the piconet has a star topology with the master in the center.

A last possibility is that of connecting multiple piconets together so as to form a distributed wireless network. This network setup is referred to as *scatternet*. In this case, e.g., a slave can be used as a gateway between different piconets. Note that, as piconets will have different time offsets and hopping sequences, a gateway user needs to maintain time offset and hopping sequence for each piconet it is connected to. This implies a higher energy consumption for gateway users. There are three ways to interconnect piconets so as to form a scatternet:

1. *Master-to-master*: in case two masters (of two different piconets) are neighbors, one of them becomes a slave of the other piconet and acts as a gateway.

2. *Gateway slaves*: in this case a user is neighbor of two masters and can be used to connect the corresponding piconets.

3. *Intermediate gateways*: the slaves of two masters are neighbors. In this case one of them becomes a master of a new piconet that includes the other slave as a gateway.

Device Discovery Device discovery is needed to set up a communication link between a master and a new slave. In device discovery between two users, one must be in *inquiry* mode (the discovering device) and the other must be in the *inquiry scan* mode. The inquirer transmits inquiry packets that ask users in the inquiry scan mode to identify themselves and provide information about their clock. This information will be used to set up the link. Inquiry packets are transmitted using a pre-defined inquiry hopping sequence with carrier frequencies changing every 312.5 μs, i.e., its hopping rate is twice that used for communication within the piconet. The device in inquiry scan mode adopts the same hopping sequence using a very slow rate, i.e., one frequency every 1.28 s. This makes it possible for the two devices to encounter in the same frequency band in a relatively short time. When a user in inquiry scan receives an inquiry packet it sends back to the inquirer a frequency hopping synchronization packet (FHS) packet that includes information about its own system clock. Upon the reception of this packet, the inquirer obtains clock information for the user that just replied and invites it to join its piconet. During this invitation (referred to as *paging*) the master sends another FHS packet so as to communicate its own hopping sequence to the user that is being invited to join the piconet. Only when this last packet is received can a user become a slave and start using the master's hopping sequence.

Link Layer As said above, the time axis is subdivided into transmission slots of 625 μs each. Slot boundaries are dictated by the master, whereas slaves synchronize their receiving circuitry on these boundaries accounting for the offset between their own clock and that of the master. The transmission of packets occurs according to a packet based TDD as follows. In the basic transmission mode, that consists of single slot packets, at most one packet is transmitted in each transmission slot. The master starts its own packet transmissions on even numbered transmission slots, whereas slaves can only transmit on odd numbered transmission slots. Moreover, a slave can transmit in a given slot only if it was polled by the master in the previous transmission slot. This leads to a TDD polling scheme according to which masters and slaves alternate their transmission activity every 625 μs.

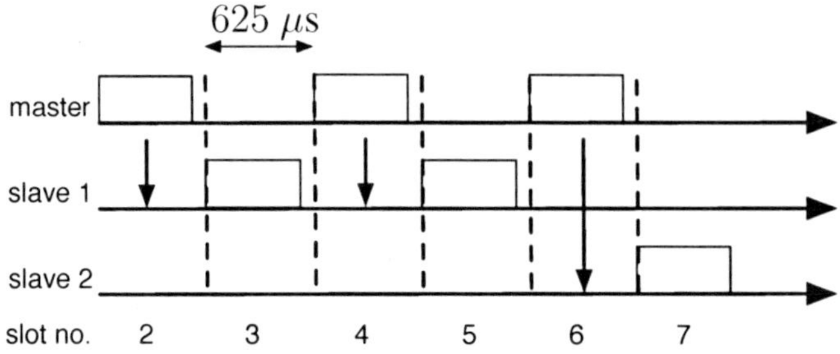

Figure 9.39 Example of packet based TDD for the Bluetooth wireless transmission system.

An example of this is given in Fig. 9.39. This figure is related to a piconet with a master and two slaves. The master sends packets in transmission slots 2, 4 and 6, whereas slaves transmit in odd numbered slots. Slave 1 is polled by the master in slots 2 and 4 and thus transmits in the subsequent slots 3 and 5, respectively. Slave 1 is polled in slot 6 and can transmit in the next slot 7. A different carrier frequency is used in each transmission slot, according to the hopping sequence dictated by the master. Both slaves synchronize their receiver with this sequence.

The Bluetooth standard defines several control and data packet formats. Considering single slot packets with minimum overhead (i.e., no FEC field is included in the payload) we have that each packet can carry 27 bytes of data. Thus, if a packet of this type is transmitted every two slots we reach a transmission rate of 27 bytes every two slots (1250 μs), which translates into a unidirectional transmission rate of 172.8 kbit/s for either uplink or downlink transmissions. Higher bit rates can be obtained using multi-slot packets as we now explain. In fact, as shown in Fig. 9.40, packets can also take 3 or 5 consecutive transmission slots; in these two cases the maximum amount of data carried by the payload amounts to 183 (3 slot packets) and 339 bytes (5 slot packets), which respectively lead to transmission rates of 585.6 kbit/s (one packet every four slots) and 723.2 kbit/s (one packet every six slots).[20] During the transmission of multi-slot packets the carrier frequency is kept unchanged. However, in the slot immediately following the transmission of a multi-slot packet the carrier frequency is the same as in the case of single packet transmissions. In other words, even though the carrier frequency remains constant, the hopping sequence generator is updated at each transmission slot as if single packets were transmitted.

The Bluetooth standard also defines synchronous and asynchronous links. In a synchronous connection oriented (SCO) link the master assigns transmission slots periodically to its slaves. In this case, one or multiple slots (in case of multi-slot packets) are reserved at fixed intervals. SCO links are useful for, e.g., video/audio streaming where the data flow arrives at regular intervals as dictated by the frame/sample rate. Alternatively, links can be set up according to the asynchronous connection-less (ACL) link model. In this last case the master polls a given slave whenever certain events occurs, i.e., the master has data to send to this slave. This link model is good for the transmission of asynchronous data traffic. A simple ARQ scheme is used for ACL links. In detail, whenever a packet transmitted by the master (resp. slave)

[20]The bit rate achievable for packets of five slots, i.e., 723.2 kbit/s, is the maximum unidirectional bit rate achievable for version 1.x of the Bluetooth system. Bluetooth 2.0 can reach higher transmission rates of up to 3 Mbit/s.

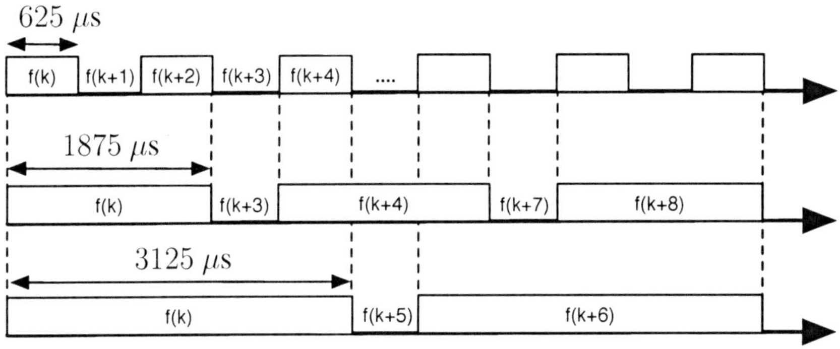

Figure 9.40 Bluetooth multi-slot packets. $f(k)$ is the carrier frequency determined by the hopping sequence generator for transmission slot k.

is corrupted, the slave (master) in the next available transmission slot sends a packet with a NAK bit set. This notifies the master (slave) about the error and elicits the retransmission of the corrupted packet. SCO and ACL links can be used concurrently, i.e., the master can allocate at the same time SCO and ACL links for the slaves in its piconet, according to their communication requirements.

Piconet Scheduling The slaves in a piconet can be served according to different policies. According to pure round robin (PRR) each slave is polled exactly once per cycle, where a cycle is the time taken to serve all slaves. The advantage of this scheme is its bounded maximum delay. However, as we have noticed for polling access schemes, for PRR channel resources are wasted in case some slaves do not have data to transmit (as the slots assigned to these users will go unused). A second technique, called limited round robin (LRR) is equal to PRR except for the fact that during a cycle multiple transmission slots can be allocated for each slave and there is a maximum number of slots per slave per cycle. LRR increases the throughput as more slots are allocated to those users that have more data to send. However, all slaves are polled for at least one slot in every cycle and this still wastes channel resources when they do not have data to transmit. The exhaustive round robin (ERR) policy polls slaves until they have data to send. This further increases the throughput but under certain cirumstances may lead to unfairness problems. More advanced policies include the adaptive round robin (ARR) mechanism. With ARR, unloaded slaves are skipped so as to avoid inefficiencies. This policy needs some priority and traffic control scheme so that previously unloaded users can be put back into the serving list. The drawback of this scheme is that unloaded users suffer from some delay before they can be served again when new traffic arrives.

Power Modes and Energy Conservation Energy conservation is facilitated by the following five power saving modes:

M1 *Active*: a slave in this state is always on and listening to all transmission slots. This state ensures maximum responsiveness but a slave in this state has a high energy consumption.

M2 *Sniff*: a slave becomes active periodically. In particular, the master agrees to poll the slave at regular intervals. The slave needs to check for incoming polling messages from the master only at the end of these intervals and can put its radio to sleep otherwise. Note that the polling sequence and transmission slot boundaries must be always tracked even during sleeping periods. The advantages of this power mode is that it allows a reduction of the energy consumption of the slaves, though they are less responsive in case the master needs to poll them outside the planned transmission slots.

M3 *Hold*: a slave can stop listening to the channel for a specified amount of time, that needs to be agreed beforehand with the master. Again, this mode allows to save energy. In addition, a slave can establish a connection with other piconets while holding on.

M4 *Park*: slaves in active, sniff and hold modes are considered *piconet members* by the master, i.e., these slaves are actively participating in the piconet communication. Instead, slaves in parked mode are no longer considered piconet members. They, however, still need to maintain the synchronism with the master and they do so by periodically listening to the master. This power mode is more energy efficient than those discussed above but a parked user must become a piconet member before it can start a new communication with the master. This implies a negotiation phase that takes additional time and impacts responsiveness. Each piconet can have at most one master, 7 members and 255 parked users.

M5 *Standby*: a user in this state no longer needs to maintain the synchronism with the master. A negotiation is required in order to become a piconet member and only then can a user start a new data transmission with the master.

The Bluetooth system has additional energy saving features. For example, to further reduce the energy expenditure and the interference among piconets every slave can evaluate the received signal strength of each packet it receives from the master and ask the master to adjust its transmission power so as to keep it at the minimum level required for an error free communication. Moreover, power control is fine grained as different transmission levels can be kept at the master for every slave in its piconet.

REFERENCES

1. R. Rom and M. Sidi, *Multiple Access Protocols: Performance and Analysis*. Springer-Verlag, New York, 1989.

2. H. Takagi, *Queuing Analysis of Polling Models,* ACM Computing Surveys, Vol. 20, No. 1, pp. 5–28, March 1988.

3. W. Bux, *Local-area subnetworks: A performance comparison,* IEEE Transactions on Communications, Vol. COM-29, No. 10, pp. 1465–1473, October 1981.

4. Y. T. Wang and R. J. T. Morris, *Load sharing in distributed systems,* IEEE Transactions on Computers, Vol. C-34, No. 3, pp. 204–217, March 1985.

5. H. Takagi, *Analysis of Polling Systems,* The MIT Press, 1986.

6. D. Bertsekas and R. Gallager, *Data Networks,* Prentice Hall, Second Edition, 1992.

7. J. Postel and J. Reynolds, *RFC 1042: A Standard for the Transmission of IP Datagrams over IEEE 802 Networks,* IETF Request For Comment, February 1988.

8. *IEEE 802.5 Token Ring Standard Part 5: Token Ring Access Method and Physical Layer Specification,* IEEE standards board, 1998.

9. F. Ross, *FDDI–A Tutorial,* IEEE Communications Magazine, Vol. 24, No. 5, pp. 10–17, May 1986.

10. C. Mack, T. Murphy and N. L. Webb, *The Efficiency of N Machines Uni-Directionally Patrolled by One Operative when Walking Time is Constant and Repair Times are Variable,* Journal of the Royal Statistical Society, Ser. B, Vol. 19, No. 1, pp. 173–178, 1957.

11. O. J. Boxma and W. P. Groenendijk, *Pseudo-Conservation Laws in Cyclic-Service Systems,* Journal of Applied Probability, Vol. 24, pp. 949–964, 1987.

12. R. M. Metcalfe and D. R. Boggs, *Ethernet: distributed packet switching for local computer networks,* Communications of the ACM, Vol. 19, No. 7, July 1976.

13. S. M. Ross, *Applied Probability Models with Optimization Applications*, Dover Publications, Inc., New York, 1992.

14. L. Kleinrock and F.A. Tobagi, *Packet Switching in Radio Channels: Part I – Carrier Sense Multiple-Access Modes and Their Throughput Delay Characteristics,* IEEE Transactions on Communications, Vol. COM-23, No. 12, pp. 1400–1416, December 1975.

15. N. Abramson, *The ALOHA system – another alternative for computer communications,* Proceedings of Fall Joint Computer Conference, AFIPS Conference, November 1970.

16. B. Hajek and T. Van Loon, *Decentralized Dynamic Control of a Multiaccess Broadcast Channel,* IEEE Transactions on Automatic Control, Vol. 27, No. 3, pp. 559–569, June 1982.

17. F. A. Tobagi and L. Kleinrock, *Packet Switching in Radio Channels: Part III – Polling and (Dynamic) Split-Channel Reservation Multiple Access,* IEEE Transactions on Communications, Vol. COM-24, No. 8, pp. 832–845, October 1976.

18. F. A. Tobagi and L. Kleinrock, *Packet Switching in Radio Channels: Part II – The Hidden Terminal Problem in Carrier Sense Multiple-Access and the Busy Tone Solution,* IEEE Transactions on Communications, Vol. COM-23, No. 12, pp. 1417–1433, December 1975.

19. F. A. Tobagi and L. Kleinrock, *Packet Switching in Radio Channels: Part IV – Stability Considerations and Dynamic Control in Carrier Sense Multiple-Access,* IEEE Transactions on Communications, Vol. COM-25, No. 10, pp. 1103–1119, October 1977.

20. K. Sohraby, M. L. Molle and A.N. Venetsanopoulos, *Comments on Throughput Analysis for Persistent CSMA Systems,* IEEE Transactions on Communications, Vol. COM-35, No. 2, pp. 240–243, February 1987.

21. M. S. Gast, *802.11 Wireless Networks: The Definitive Guide,* O'Reilly Networking, Second Edition, 2002.

22. T. Apostolopoulos and E. Protonotarios, *Queueing Analysis of Buffered CSMA/CD Protocols,* IEEE Transactions on Communications, Vol. COM-34, No. 9, pp. 898–905, September 1986.

23. T. Takine, Y. Takahashi and T. Hasegawa, *An Approximate Analysis of a Buffered CSMA/CD,* IEEE Transactions on Communications, Vol. 36, No. 8, pp. 932–941, August 1988.

24. S. Lin, D. J. Costello and M. J. Miller, *Automatic-repeat-request error-control schemes,* IEEE Communications Magazine, Vol. 12, No. 12, pp. 5–17, December 1984.

25. M. Rossi and M. Zorzi, *Analysis and Heuristics for the Characterization of Selective Repeat ARQ Delay Statistics over Wireless Channels,* IEEE Transactions on Vehicular Technology, Vol. 52, No. 5, pp. 1365–1377, September 2003.

26. M. Rossi, L. Badia and M. Zorzi, *SR ARQ Delay Statistics on N-State Markov Channels with Non-instantaneous feedback,* IEEE Transactions on Wireless Communications, Vol. 5, No. 6, pp. 1526–1536, June 2006.

27. H. Holma and A. Toskala, *WCDMA for UMTS: Radio Access For Third Generation Mobile Communications,* John Wiley & Sons, Third Edition, 2004.

28. E. Modiano, *An adaptive algorithm for optimizing the packet size used in wireless ARQ protocols,* Springer Wireless Networks, Vol. 5, No. 4, pp. 279–286, November 2004.

29. A. Goldsmith, *Wireless Communications,* Cambridge University Press, 2005.

30. Stefano Basagni Ed., *Mobile Ad Hoc Networking,* John Wiley & Sons, Ltd, 2004.

31. B. A. Miller and C. Bisdikian, *Bluetooth Revealed: The Insider's Guide to an Open Specification for Global Wireless Communications,* Prentice Hall, 2000.

Problems

Medium Access Control

9.1 Consider a number of nodes that communicate to a central server through a shared link exploiting an FDMA channel scheme. The transmission parameters are $\tilde{\tau}_p = 1\ \mu s$, packets are $L = 1000$ bytes long and are transmitted at the constant bit rate $R_b = 10$ Mbit/s. The packet arrival process at each node is Poisson with parameter λ.

a) Compute the maximum arrival rate that can be supported by this access scheme so that the average delay will be smaller than 100 ms when there are 10 nodes in the system.

b) Now, consider $\lambda = 10$ packet/s and compute the maximum number of nodes that can be supported by this system so that the total average delay for their delivery to the server is smaller than 100 ms.

9.2 Consider a wireless transmission system with N_u users, where each user transmits data packets of length $L = 1000$ bits at the constant bit rate $R_b = 1$ Mbit/s.

a) If $N_u = 10$, say which system is preferable in terms of average delay performance between TDMA and FDMA.

b) For the best system in terms of delay performance, calculate the maximum number of users that can be supported so as to keep the average delay below 100 ms; consider the above parameters, $\tilde{\tau}_p = 0.01$ and $\lambda = 10$ packet/s.

9.3 The access point (AP) of a wireless system uses TDMA to transmit data packets to the N_u users within coverage. The TDMA frame has N_u access slots, whereas packet arrivals at the AP for each user are Poisson distributed with parameter λ packet/s. The packet transmission time, t_P, is constant and equal to the TDMA slot duration. Consider the following two system configurations:

A incoming packets at the AP are queued in a single queue in the order in which they arrive. At each TDMA slot the head of line (HOL) packet is picked from the queue and transmitted over the channel, whereas no transmission occurs during this slot if the queue is found empty.

B N_u queues are maintained at the AP, one for each user in the system. Each time a packet directed to user i arrives at the AP, it is inserted in the corresponding queue. During slot i, the HOL packet is picked from queue i and transmitted over the channel, in the next slot the HOL packet of queue $(i + 1)$ `mod` N_u is transmitted and so on, i.e., queues are cyclically served, one per TDMA slot. No transmission occurs when a queue is found empty.

a) Say which of the two configurations above performs better in terms of normalized throughput.

b) Say which one provides better performance in terms of average queueing delay.

9.4 Consider the following access scheme that combines TDMA and a specialized version of slotted ALOHA. There are 30 users, separated into two groups of $n_1 = 10$ (group 1) and $n_2 = 20$ users (group 2). Even access slots (i.e., slot numbers $0, 2, 4, \ldots$) are reserved for users of group 1. Odd access slots (i.e., $1, 3, 5, \ldots$) are reserved for users of group 2. Contention within each group is resolved by the following specialized slotted ALOHA protocol: whenever a user has a packet to transmit it waits for the first available access slot, i.e., an even time slot if the user belongs to group 1 and an odd time slot if the user belongs to group 2, and with probability p_i (i is 1 or 2 according to the corresponding group label) transmits the packet, otherwise it defers its transmission to a later slot. Users always have a packet to send whenever there is a transmission opportunity (i.e., a slot is selected for transmission). Furthermore, when more than one packet is transmitted in a slot, a collision occurs and all packets transmitted in that slot are lost. A successful transmission event occurs when a single packet is transmitted in the slot. We define p_i as *optimal* if it maximizes the probability of a successful transmission in the slots alloted to group $i \in \{1, 2\}$.

Determine the normalized throughput (in packet/slot) of the system, assuming that every user has always something to send and that all users use the corresponding optimal access probability $p_i^\star$, where $i \in \{1, 2\}$ is their group label.

9.5 A wireless sensor network is composed of 100 memory constrained sensor nodes whose buffer at the MAC layer can host at most one data packet. Hence, at any given node, a new packet is discarded whenever upon its arrival it finds the buffer occupied. These nodes sense a physical phenomenon and generate packets according to a Poisson process with rate $\lambda = 0.2$ packet/s. Moreover, they communicate with a common data collection node through a shared wireless channel using a polling access scheme with switchover time $\Delta = 0.02$ s. The remaining channel parameters are: packet size $L = 80$ bytes and constant transmission rate $R_b = 255$ kbit/s.

a) Compute the mean polling cycle time for this system.

b) Compute the throughput of the system expressed in packet/s.

c) Compute the probability that a packet is discarded at any given node upon finding the buffer occupied.

9.6 Consider a cellular network where a special frequency band is reserved for mobile nodes to request new connections. Access to this special frequency band follows the ALOHA protocol. A connection request packet is $L = 125$ bytes long while the channel as a bit rate of $R_b = 100$ Kbit/s. What is the maximum number of connection requests per unit of time, $\lambda^\star$, that can be supported by this system?

9.7 Consider a slotted ALOHA system. Assume that the aggregate transmission attempts by the large number of users are modelled as a Poisson process with average rate of λ packet/s. Each slot has a constant duration equal to the packet transmission time t_{P}.

a) What is the probability that a transmission is successful on its first attempt?

b) What is the probability that it takes exactly k attempts for a transmission to be successful?

c) What is the average number of transmission attempts needed for a packet to be successful?

d) Suppose the receivers are improved so that, in case of parallel transmissions in the same slot, a maximum number of two concurrent transmissions can be received correctly. Collision takes place only if three or more packets are sent during the same slot. Express the normalized throughput as a function of the normalized offered traffic.

9.8 Consider a nonpersistent CSMA scheme.

a) Prove that the system throughput as a function of the normalized offered traffic $\mathtt{G}$ is a decreasing function of the normalized channel propagation delay.

b) Compute $\mathtt{S}$ for $\tilde{\tau}_p \in \{0.01, 2\}$ and $\mathtt{G} = 10$. Compare these results to the normalized throughput of ALOHA and find the value of $\tilde{\tau}_p$, referred to as $\tilde{\tau}_p^*$ for which the throughput of ALOHA overcomes that of nonpersistent CSMA.

9.9 Consider an FDMA system with K orthogonal sub-channels. Within each sub-channel the transmission follows a slotted ALOHA scheme. For the input process consider the infinite population case with offered normalized traffic $\mathtt{G}$.

a) Find the normalized throughput of this system when the K sub-channels are selected uniformly at random by any input packet.

b) Compare the normalized throughput $\mathtt{S}(K, \mathtt{G})$ with respect to the throughput of a single channel ALOHA system, $\mathtt{S}(1, \mathtt{G})$, find the throughput gain in terms of the ratio $\eta(K, \mathtt{G}) = \mathtt{S}(K, \mathtt{G})/\mathtt{S}(1, \mathtt{G})$ and find the minimum number of sub-channels K to use so that $\eta > 1$, for any $\mathtt{G}$. Finally, determine the maximum achievable gain, η, as a function of $\mathtt{G}$.

9.10 Consider $N_u = 10$ wireless nodes, all within the transmission range of one another, which all broadcast data on a shared channel in the following manner: for any given node a single broadcast packet of length $L = 1000$ bits is sent at any one time, packets arrive at an given node according to a Poisson process with rate λ packet/s and whenever a packet is ready for transmission, it is sent over the channel at a rate of $R_b = 1$ Mbit/s, irrespective of the channel status. Any time two or more packets overlap in time, they result in a collision, while packets sent in disjoint time intervals are successfully received by all nodes.

a) Find the value of λ that maximizes the normalized throughput of this system, i.e., the rate at which the N_u users successfully send packets over the shared channel.

b) Now, consider $\tilde{\tau}_p = 0.01$ and solve the above problem considering nonpersistent CSMA as the channel access policy. Compare the results with respect to those obtained with the previous access method.

Automatic Retransmission reQuest

9.11 Suppose that we have two computers interconnected by a 30 km long optical fiber pair (i.e., full-duplex) that transfers data at the rate of $R_b = 100$ Mbit/s (the same data rate is achieved over the forward and feedback channels). Due to the long fiber length, the bit error rate P_{bit} is quite high, i.e., $P_{\text{bit}} = 10^{-4}$. In addition, bit errors are i.i.d., data packets are $L = 4500$ bits long (including a CRC field of $L_{\text{CRC}} = 32$-bits and a header field of $L_{\text{HDR}} = 160$ bits) and the size of acknowledgments is 496 bits (including CRC and header fields). GBN-ARQ is used for error control at the link layer. Assume that light within this fiber propogates at a rate of $3 \cdot 10^5$ km/sec and that the processing time at the nodes is zero.

a) What is the probability that a packet and the corresponding acknowledgment are received correctly (referred to as "packet success rate" p)?

b) What window size should be used for GBN-ARQ?

c) What is the realized transmission efficiency $\eta_{\text{GBN-ARQ}}$ (i.e., the data bits that are successfully sent over the channel per second)?

9.12 A mobile phone communicates to a base station (BS) of a cellular system through a dedicated wireless link having the following characteristics:

- The data channel has a constant bit rate of $P_{\text{bit}} = 5 \cdot 10^{-4}$.
- Each link layer PDU has a size of $L = 366$ bits, including data, $L_D = 350$ bits, and header, $L_O = 16$ bits, fields.
- The round trip time at the data link layer, i.e., the time taken between the beginning of the transmission of a PDU and the complete reception of the related ACK message, is $RTT = 200$ ms. t_A and propagation delay τ_p are negligible.
- Link layer PDUs are transmitted at the constant rate $R_b = 256$ kbit/s.

a) Calculate the throughput expressed in terms of useful data bits transmitted per second for SW-ARQ and SR-ARQ. Why is SW-ARQ so inefficient?

b) Now, consider the following hybrid ARQ (HARQ) scheme. Each PDU is split into $K = 6$ data blocks of the same size and prior to its transmission is sent to an encoder which returns an encoded PDU of N data blocks. Encoded PDUs are thus sent over the channel using SR-ARQ. At the receiver, up to $N - K$ erroneous blocks of data can be corrected, in any order, for any PDU. Thus, if a PDU is received with a number of erroneous blocks smaller than or equal to $N - K$, the corresponding original K-block PDU is recovered with no need for further retransmissions. Otherwise, the PDU is deemed as *failed* and a negative acknowledgment is sent to the transmitter, requesting the retransmission of this PDU (i.e., of the corresponding N blocks of data). Calculate the throughput (useful data bits per second) of this HARQ scheme for $N = 7$ and $N = 8$ and compare the results with those obtained for SR-ARQ.

c) Consider SR-ARQ and the SR-HARQ of the previous point. Let $P_f(\text{ARQ})$ and $P_f(\text{HARQ})$ be the error probabilities for a plain link layer PDU (ARQ) and an encoded PDU (HARQ), respectively. Define $P_s(\text{ARQ}) = 1 - P_f(\text{ARQ})$ and $P_s(\text{HARQ}) = 1 - P_f(\text{HARQ})$. Referring to $\eta_{\text{SR-ARQ}}$ and $\eta_{\text{SR-HARQ}}$ as the throughput of ARQ and HARQ, respectively, say when the following inequality $\eta_{\text{SR-HARQ}} > \eta_{\text{SR-ARQ}}$ holds as a function of K, N, $P_s(\text{ARQ})$ and $P_s(\text{HARQ})$.

9.13 Consider a SR-ARQ data link scheme operating between a sender and a receiver. At the sender each higher layer packet, referred to as link layer SDU, is split into n link layer PDUs of the same size. Each time a PDU is transmitted, it is received in error with probability p, assumed constant, and is instead correctly received with probability $1 - p$. Each PDU is sent over the channel using a SR-ARQ considering a maximum of L transmission attempts, i.e., whenever a PDU is transmitted L times and it is still erroneous, it is discarded at the transmitter and tagged as *failed* at the receiver. We refer to *processed* as a PDU that is either *successfully* transmitted during one of its transmission attempts or *failed*, as this PDU in both cases will not be further considered for transmission. At the receiver, any given SDU is passed to the higher layers whenever all its PDUs are marked as *processed*.

a) Calculate the SDU error probability at the receiver as a function of p, n and L.

b) Let RTT be the round trip time at the link layer, i.e., the elapsed from the instant when a PDU is transmitted over the channel to the reception of the corresponding ACK/NACK message at the sender. Define T_{SDU} as the time taken for the transmission of a SDU, i.e., from the transmission instant of the first PDU to the one where all its PDUs are marked as *processed* at the receiver. Calculate $\text{E}[T_{\text{SDU}}]$ and $\text{P}[T_{\text{SDU}} = jRTT], j \geq 1$. For this calculation assume that $nt_{\text{P}} \ll RTT$. This means that, at the first transmission attempt, all PDUs composing a SDU are transmitted at the beginning of a round trip time. Moreover, if $k \leq n$ of them are erroneously received, they will all be transmitted at the beginning of the next round. This process is iterated until all PDUs are marked as *processed*.

9.14 A Selective Repeat ARQ operates as the link layer protocol of a satellite transmission channel. The satellite channel alternates between three states: state 1, where the bit error probability is $P^1_{\text{bit}} = 10^{-6}$, state 2, with $P^2_{\text{bit}} = 10^{-5}$ and state 3 with $P^3_{\text{bit}} = 10^{-3}$. The steady state probabilities that the channel is in each of the states are $\pi_1 = 0.5$, $\pi_2 = 0.2$ and $\pi_3 = 0.3$ for states 1, 2 and 3, respectively. The transmission rate at the link layer is constant and equal to $R_b = 1$ Mbit/s, whereas the size of the PDU headers is 100 bits. Also, assume that the channel state never changes during the transmission of link layer PDUs.

a) Find the average transmission efficiency, η, expressed in bit/s, of SR-ARQ when PDUs have a constant size of 500 bytes, including data and header fields.

b) Calculate the average transmission efficiency, η, expressed in bit/s, of a modified SR-ARQ scheme where the PDU size is optimally adapted as a function of the channel state.

9.15 A system with two nested stop-and-wait ARQ schemes is given. The inner scheme utilizes a hybrid ARQ protocol that allows up to 2 re-transmissions. Consider the inner ARQ scheme first. The probability that the first transmission of a PDU is successful is $p_1 = 0.2$, the probability that the first re-transmission of this PDU is successful given that the first transmission attempt failed is $p_2 = 0.6$, and the probability that the second retransmission is successful given that the first transmission and the first retransmission of the PDU were both unsuccessful is $p_3 = 0.9$.

The outer scheme is a SW-ARQ where each of its packets comprises $n = 10$ PDUs of the inner ARQ. Hence, this outer ARQ operates on blocks of n inner PDUs and retransmits the entire block whenever it contains one or more erroneous PDUs, until this block is successfully sent. In detail, (1) the outer ARQ sends a block of PDUs to the inner ARQ, (2) upon receiving this block the inner ARQ start transmitting each of the PDUs therein, according to its protocol, (3) the inner ARQ notifies the outer ARQ whenever a PDU is sent successfully, (4) if the inner ARQ detects an error after the second retransmission of a PDU, it marks this PDU as *failed* and notifies the outer ARQ about this event. (5) When the outer ARQ receives a notification for each of the PDUs in its current block, it decides to (5a) retransmit the entire block in case at least one of the PDUs in it is *failed* or (5b) pass a new block of n PDUs to the inner ARQ.

Assume, that the channel is always filled with transmitted PDUs, that all errors are detected and that the feedback channel is error free.

a) Calculate the normalized throughput for the inner ARQ

$$\mathsf{S}_{\text{inner}} = \frac{\text{E [correctly received PDUs]}}{\text{E [transmitted PDUs]}} .$$

b) Calculate the residual PDU error rate $P_e(1)$ after passing through the inner ARQ.

c) Calculate the probability $P_e(n)$ that the outer scheme requires the retransmission of a block of packets.

d) Calculate the average number of transmission attempts per outer block for the outer ARQ.

e) Calculate the normalized throughput of the combined ARQ scheme, $\mathsf{S}_{\text{combined}}$.

9.16 Consider a sender A that communicates over a wireless channel to a receiver B using SR-ARQ. The round trip time, intended as the time from the beginning of the transmission of a PDU at terminal A to the complete reception of the corresponding ACK at terminal B is $RTT = 500$ ms. The transmission bit rate is $R_b = 1$ Mbit/s. The size of PDU headers and payloads is respectively $L_O = 100$ bits and $L_D = 4900$ bits. The window size of the SR-ARQ protocol is $W = 50$ PDUs and the channel has a bit error probability of $P_{\text{bit}} = 10^{-4}$.

a) Compute the normalized throughput S of the given SR-ARQ.

b) Compute a new window size W' so that the channel pipe is entirely filled with transmitted PDUs and compute the normalized throughput S' for this case.

9.17 Consider the following communication scenario: a sender A transmits its data to a receiver B over a communication channel composed of two point-to-point links. The first link connects A to a router C, whereas the second link connects C to B. The channel parameters for the first link are: transmission bit rate $R_b^1 = 1$ Mbit/s, error rate $P_{\text{bit}}^1 = 0.0001$. The channel parameters for the second link are: transmission bit rate $R_b^2 = 500$ kbit/s, error rate $P_{\text{bit}}^2 = 0.0005$. The propagation delay is constant and equal for the two links, $\tau_p = 50\ \mu$s. The size of acknowledgment messages is $A = 500$ bits.

a) Assuming that two independent SW-ARQ protocols are used to assure an error free communication over the two links, compute the maximum achievable normalized throughput S_i, $i = 1, 2$, over each of the two links (A $\rightarrow$ C and C $\rightarrow$ B).

b) Compute the maximum transmission bit-rate, $B_{\max}$, from A $\rightarrow$ B that can be supported by the two ARQ protocols working in sequence.

9.18 Consider a GBN-ARQ scheme. PDUs arrive at the transmission buffer of the sender according to a Poisson process with rate λ packet/s. The transmission of a PDU takes one time unit and the corresponding acknowledgment is received $N - 1$ time units upon the completion of the PDU transmission. PDUs can be retransmitted due to two reasons:

1. A given PDU i can be rejected a the receiver due to errors in the forward channel. In this case the transmitted will send PDUs $i + 1, i + 2, \ldots, i + N - 1$ (if these are available) and, after that, it will go back retransmitting PDU $i, i + 1, i + 2, \ldots$. We assume that the PDU error rate over the forward channel is p_{fwd} and the error process is iid.

2. The acknowledgment message for any given PDU, transmitted over the feedback channel, can be erroneous as well. The acknowledgment error rate over the feedback channel is indicated as p_{fdb} and the error process over this channel is iid.

a) Calculate the average service time for a given PDU entering the ARQ system. (*Hint: the transmitter's queue can be modeled using the M/G/1 theory.*)

9.19 Consider a GBN-ARQ operating between two terminals A and B. Let β be the expected number of PDUs transmitted from A to B between the completion of the transmission (from A) of a given PDU and the instant where the feedback message for this PDU is entirely received at A. Let p be the probability that a PDU is erroneously received at B, assuming an iid PDU error process. Assume that a correct feedback is received for each PDU (ideal feedback channel) and that, whenever a negative acknowledgment is received for PDU i, terminal A goes back transmitting PDUs $i, i+1, i+2, \ldots$.

- **a)** Let γ be the expected number of PDUs transmitted from A to B for each PDU successfully received at B. Show that $\gamma = 1 + p(\beta + \gamma)$.
- **b)** Defining the normalized throughput as $\mathsf{S} = 1/\gamma$, find S as a function of β and p.

9.20 Consider an access point (AP) that transmits data traffic in downlink to three terminals A, B and C. The AP transmits a unicast data flow to each of the terminals using TDMA, where the TDMA frame is $n = 1000$ slots and the traffic directed to terminal A is sent during the first n_{A} slots of the frame, the traffic directed to terminal B during the following n_{B} slots and the traffic for terminal C uses the last n_{C} slots of the TDMA frame. The AP implements a fully reliable SR-ARQ scheme within each group of slots. In this way, we have three dedicated (no cross-interference) and fully reliable downlink channels, one for each terminal. Assume that the bit error probabilities for the three connections are $P_{\mathrm{bit}}^{\mathrm{A}} = 0.002$, $P_{\mathrm{bit}}^{\mathrm{B}} = 0.0005$ and $P_{\mathrm{bit}}^{\mathrm{C}} = 0.00001$ for terminal A, B and C, respectively. PDUs are $L = 100$-bit long and the transmission of each PDU takes an entire TDMA slot.

- **a)** Compute n_{A}, n_{B} and n_{C} so that the three terminals have the same effective throughput (i.e., the throughput measured after the SR-ARQ). In case this *fair schedule* entails the allocation of fractional number of slots, obtain a feasible allocation (where n_{A}, n_{B} and n_{C} are nonnegative integers) assigning an extra slot to the user with the highest bit error probability.

CHAPTER 10

NETWORK LAYERS

Leonardo Badia

This chapter describes the network operations taking place above the point-to-point data exchange, which realize remote communications between distant network elements. This requires introducing a further level of cooperation, where multiple users are able not only to coordinate their access to the medium, but also to reciprocally exchange services. To this end, they have to harmonize their operations, and in particular to use the same protocols. At the network layer they should be able to cooperate to establish an end-to-end communication made of interconnected links. This is the purpose of routing, which will be discussed first. To perform routing, the nodes need to be able to recognize each other and establish a remote communication. In the Internet, the node identification and the network management systems, are based on the well known IP. IP probably represents the most important part of the Internet Protocol suite, which however has another important counterpart in its transport layer protocols known as the TCP and the user datagram protocol (UDP). Finally, the application layer will be reviewed, mainly for what concerns its interaction with lower layers, which mostly happens in relation to the domain name resolution, and well-known applications such as the world wide web and email. Thus, this chapter is organized in five sections. First, we review basic notions, terminologies, and representations of communication networks which will be used in the rest of the chapter. This will find application in the second section, where we describe routing protocols, i.e., we translate theory into practical procedures to realize routing. Then, we discuss IP, along with its addressing issues, in the third section. The transport layer is instead the subject of the fourth section, where we investigate UDP, TCP, and flow control mechanisms. Finally, the last section focuses on the application layer, in particular the Internet domains and their naming system

Principles of Communications Networks and Systems N. Benvenuto, M. Zorzi

through the domain name server (DNS), as well as applications and protocols used everyday, such as the world wide web (WWW), the hyper-text transfer protocol (HTTP), email, and the simple mail transfer protocol (SMTP).

10.1 INTRODUCTION

At the physical layer, the main hurdle against communication is the nature of the channel, which suffers from noise, interference, inconstant behavior, and various other limitations. The data link layer addresses these channel impairments and tries to coordinate the access of multiple users. In this manner, the first two layers of the stack try to hide all the physical imperfections of communication, leaving a theoretically error-free channel to use for network communication. This is the scenario on which the upper layers operate to supply services among remote network elements.

Naturally, this abstraction is not perfect in real cases; some problems of lower layers can remain unsolved and affect the upper layers. Nevertheless, even if we assumed that the physical and data link layers were perfect, we would still have many open challenges at the network layer and above. Indeed, we should also change our perspective from solving inherent problems of single-link data communication, as previously done, to constructing end-to-end connections. This implies that different network parts now are requested to cooperate. While at the MAC sublayer the other users were competitors, at the network layer they become helpers who are necessary to aid the communication.

Before starting to describe the first of those upper layers, i.e., the network layer, we need some definitions and fundamental concepts. Thus, this introductory section will contain technical aspects about network terminology and characterization. In more detail, we give at first a review of the interconnecting devices, according to the layering rationale, which is useful to understand how the network elements are linked. Secondly, we formalize a general model of network structures. The extent of such a description is much broader than communication networks, and actually dates back prior to the pervasive diffusion of communication devices. Indeed, other kinds of networks, such as those of vehicle traffic or transportation systems, are older than those based on electronic communicating devices. For this reason, general network models, as well as theoretical results about them, were introduced when the Internet did not exist, and the description of data networks is strongly indebted to these theoretical results.

10.1.1 Switching and connecting

There exist a wide variety of interconnecting devices which enable data to be sent from one physical communication link (a cable, or a wireless link, or anything alike) to another. At the same time, there is also a number of names with which these devices are called. Even though the distinction among these names is sometimes subtle, it is important because it refers to the layering structure. Thus, the name of the interconnecting device implicitly describes at which layer it operates, which in turn determines exactly the information used in actuating the interconnection. We clarify this concept by looking directly at the examples, proceeding layer by layer.

Layer 1: Repeater, Hub An interconnection at layer 1 is realized by simply propagating the electrical signal further. This may require some signal amplification; a device realizing this is called a *repeater*. The name clarifies that it does nothing but sending the signal forward, enabling its propagation to a longer distance. If the signal is digital, the repeater may also eliminate distortion effects introduced by the channel. Such an operation is called *regeneration*. A regenerative repeater tries to clean the signal from noise before amplifying it (which would result in amplifying the noise as well).

Sometimes, multiple repeaters are integrated in a single device called *hub*; the difference between repeater and hub is in the number of interconnected lines, which is two for the former and higher for the latter. Actually, a hub does not even need to amplify the signal, the emphasis in this terminology is just in the ability of connecting multiple inputs.

Repeaters or hubs perform the simplest kind of interconnection. These devices are blind to the information content of what they are transmitting (i.e., repeating). They simply take the electrical signal and send it forward, just improving its physical quality. Upper layer devices use instead information with higher level of abstraction.

Layer 2: Bridge, Switch The devices operating on the data link layer are called *bridges*; similar to before, a device with the same functionality as a bridge, only with the additional requirement of multiple-input capability, is often called with a different name, i.e., *switch*. When layer 2 is considered, the requirement of connecting multiple lines is very common. Thus, the terms "bridge" and "switch" are often used as synonyms.

The interconnection realized by a switch involves multiple access, which is no longer regulated by physical (i.e., electrical) characteristics of the signal, but rather by the MAC sublayer. A switch does not only interpret the bits of a packet, but also its content. In practice, the switch extracts the MAC address of the destination, which is contained in the packet, and uses this piece of information to determine how to connect the lines.

However, bridges and switches cannot realize all kinds of interconnections. Especially, they do not have the capability of interconnecting all kinds of medium access protocols since the MAC addressing would be different. For example, a network operating on a hybrid wired/wireless physical layer, i.e., where some communication lines are cables and others are radio links, is clearly impossible to realize via repeaters, which do not interconnect different physical technologies. Switches can realize an interconnection only between certain kinds of different MAC sublayers, e.g., between Ethernet and Fast Ethernet. However, to enable an entirely general interface of different medium access mechanisms, we must transfer to the network layer.

Layer 3: Router The purpose of the network layer is to allow communication between different kinds of technologies in a single network construct. This is achieved by means of a universal reference system, for example realized through IP addressing, which will be discussed in Section 10.3. The interconnection devices at layer 3 are called *routers*, a term that will be referred to often in the rest of this chapter. In fact, as the name suggests, routers are in charge of directing the messages from source to destination, an operation called *routing*. For this reason, routers will be main characters during the next section, since they are the places where routing decisions are made. Actually, also hubs can sometimes operate some simple routing procedures, but these are very rough techniques which do not require interactions at network layer and especially operate with a specific MAC. Routers are instead capable of performing much more complex procedures, in an entirely general context. Note also that they always connect several lines at once, thus there is no need for a different name for single-input and multiple-input connecting devices.

Layer 4 and above: Gateway Devices realizing connections at layers higher than 3 are often generically denoted as *gateways*. They includes transport gateways, which operate at layer 4 and are able to interconnect parts of the network using a different transport protocol, as well as application gateways, which are able to understand the content of the exchanged data and translate it into a different format. In general, the term is used to denote all interconnecting devices which have higher capability than simply managing routing of packets; for example, certain gateways allow to better reach the connection end points, or to filter the communication to improve security; these functions are sometimes called *proxy* and *firewall*. Other gateways are able to perform network address translation (NAT), which will be discussed in Subsection 10.3.2.

10.1.2 Networks and network topology

The interconnections between communication devices create what is called a network. This may be a very complicate and intricate structure, or even a simple one. Actually, even two devices connected to each other may already form a network. However, networks can in turn be interconnected with each other, so as to create larger networks, which should actually be called networks of networks (and networks of networks of networks...). The Internet is a prime example of that; more than a network, it is a nested structure of networks of networks.

To take a general and abstract view of networks, it is common to refer to the topology concepts introduced in Chapter 1. Thus, we will speak in the following of nodes and links (or edges, or arcs) rather than some specific terms related to the layer we operate at. In particular, although we use the approach Euler adopted for a "bridge" problem, most of the time our nodes will not be bridges (in the sense of communication devices), but rather routers, as we focus on layer 3, i.e., the network layer. [1]

The network topology is described by how the edges interconnect the nodes. Several topological structures are possible, and some of them have a proper name, as discussed in Chapter 1. According to the physical nature of the links, it may be appropriate to use either directed or undirected edges, in which cases we will speak of directed or undirected graph, respectively. In the former case, the order of the pair representing the edge matters, so that (A, B) and (B, A) are different pairs. It is not even guaranteed that they both are included in the set $\mathcal{E}$. Undirected edges can instead be used in both directions.

Actually, an even more precise taxonomy would distinguish also *bidirectional* links, where links are directed, but they always exist in both directions. This can be regarded, in light of what discussed in Chapter 1, as a consequence of the duplex characteristics of the medium. In other words, directed links allow communication only in one direction; undirected links instead can either be regarded as half-duplex or full-duplex channels. In the latter case, they are actually equivalent to a pair of directed links in directions opposed to each other.

Note that the assumption of having edges in both directions is in certain cases quite natural. For instance, if a communication link between A and B is achieved by deploying a fixed cable, this link does not distinguish between directions, and therefore can identically serve as an edge

[1] In any event, in Euler's problem, the bridges (those over the river) where the *edges*, not the nodes.

either from A to B or from B to A. However, there are many remarkable cases in which this is not true. One possibility is that the connection means is anisotropic. Moreover, strong asymmetry between certain transmitter-receiver pairs is actually common in radio communication, for example in radio and TV broadcasting. Moreover, even when the transmission is not meant to be unidirectional, radio transmitters have different power levels. Small devices are usually severely limited in their transmission range. Thus, a proper modeling should consider the possibility of using links only in one direction. Nevertheless, hereafter we will blur the distinction between directed and undirected graphs. Although the results presented in the following mostly refer to undirected graphs, it is almost always straightforward, though tedious, to extend them to directed graphs.

Similarly, a possible extension is to include multiple edges between nodes. The definition of graph reported above considers only a single link between any two nodes.[2] Multiple connections actually occur in practice, since in more recent communication networks we may have many ways to connect nodes at the physical layer, or we use multiple interfaces. Especially, this is the case for optical fiber or wireless networks which use multiple channels on an FDMA basis. However, it is not difficult to extend the description made above to this case; we avoid to use multiple edges only for the sake of simplicity, but nothing forbids to include elements in $\mathcal{E}$ which refer to the same source-destination pair.[3]

We outline other terms related to the topological view of the network which will be used in the rest of this chapter. If $\mathcal{Q} \subseteq \mathcal{N}$, i.e., $\mathcal{Q}$ contains a subset of nodes, and $\mathcal{F} \subseteq \mathcal{E} \cap \mathcal{Q}^2$, then $\mathcal{F}$ is a subset of the edges in $\mathcal{E}$ which connect only nodes belonging to subset $\mathcal{Q}$. In this case, the graph $\mathcal{H} = (\mathcal{Q}, \mathcal{F})$, i.e., identified by the sets of nodes $\mathcal{Q}$ and edges $\mathcal{F}$ is said to be a *subgraph* of $\mathcal{G}$.

When discussing network algorithms, it is customary to refer to two nodes A and B as *neighbors* if an edge exists which interconnects them. Actually, this means that the edge is bidirectional or there are both edges from A to B and from B to A. In case only the directed link from A to B exists, but not vice versa, one may say that B is an *out-neighbor* of A, which is in turn an *in-neighbor* of B. In the following, we will refer mostly to the case of bidirectional links, but the whole rationale can be extended by replacing “neighbor” with “in-neighbor” or “out-neighbor” where appropriate, according to the context. Also, a node C which is not directly a neighbor of A, but is neighbor of a neighbor B of node A, can be called *two-hop* neighbor. This can be generalized to including k-hop neighbors for any positive integer k; in particular, the plain neighbors can be dubbed as *one-hop* neighbors.

Other important terms which refer to the network graph representation are *walk*, *path*, and *cycle*. A walk on a graph $\mathcal{G}$ is an ordered sequence $\mathcal{W} = (i_0, i_1, \ldots, i_L)$ of nodes belonging to $\mathcal{N}$ which satisfies the property that every pair of subsequent elements is connected through an edge, i.e., for every k with $1 \leq k \leq L$, it holds that $(i_{k-1}, i_k) \in \mathcal{E}$. An alternative definition, which can be shown to be entirely equivalent, is to regard the walk as a sequence of edges, rather than nodes, with the condition that the tail of the ith edge is the head of the $(i-1)$th one. Sometimes the number of involved hops, which is therefore L, is said to be the length of the walk measured in hops.

The nodes of the sequence are said to be *visited* by the walk. We can also say that the nodes are *traversed* during the walk; similarly to before, we can use this term in reference to links,

[2]Notably, the Königsberg's bridge problem involved instead multiple bridges between districts, so Euler's formulation of the problem considers this extension of the topological representation.

[3]Only, notations become more complicated; for example (A,B) is no longer a unique identifier for the edge.

rather than nodes; thus, link (i, j) is traversed if both nodes i and j are subsequently visited. A path $\mathcal{P}$ is a walk described by a sequence where all nodes are different, i.e., the nodes are visited without repetition, or formally

$$\mathcal{P} = (i_0, i_1, \ldots, i_L) \text{ where: } h \neq k \Leftrightarrow i_h \neq i_k\,. \tag{10.1}$$

A cycle (or loop) is instead a walk where the first and the last node coincide; as we excluded edges looping on the same node, a cycle must be at least of length 2. A path can never be a cycle, and vice versa, since the nodes in a path are all different, whereas in a cycle the first and last nodes are the same. According to the definitions above, if we select a subset of subsequent nodes from a walk, we obtain a (shorter) walk. Similarly, if we take a subset of subsequent nodes from a path, the result is again a path, since nodes cannot be repeated. It follows that path cannot, beyond *being* cycles, also *contain* cycles, meaning that any sequence of nodes contained in a path, being itself a path, is therefore not a cycle. This property is often referred as "paths are cycle-free."

A graph $\mathcal{G} = (\mathcal{N}, \mathcal{E})$ is said to be *connected* if, for any $i, j \in \mathcal{N}$ there is a path $\mathcal{P} = (i = i_0, i_1, \ldots, i_L = j)$ connecting them. A connected graph without any cycle is said to be a *tree*. Subgraphs of a tree $\mathcal{G} = (\mathcal{N}, \mathcal{E})$ which are connected are called *subtrees*. Subtrees are necessarily trees (they do not contain cycles, as $\mathcal{G}$ does not). If $\mathcal{G} = (\mathcal{N}, \mathcal{E})$ is a tree, the following facts are easy to prove:

- The number of edges $|\mathcal{E}|$ is equal to the number of nodes $|\mathcal{N}|$ minus 1, i.e., $|\mathcal{E}| = |\mathcal{N}| - 1$
- The only subtree of $\mathcal{G}$ such that the set of nodes is $\mathcal{N}$ is $\mathcal{G}$ itself. Indeed, any subgraph $\mathcal{G}'$ of the form $(\mathcal{N}, \mathcal{E}')$, where $\mathcal{E}' \subseteq \mathcal{E}$, is a tree only if $\mathcal{E}' = \mathcal{E}$, since removing an edge from $\mathcal{E}$ makes the graph no longer connected
- If $\mathcal{H} = (\mathcal{Q}, \mathcal{F})$ is a subtree of $\mathcal{G} = (\mathcal{N}, \mathcal{E})$ different from $\mathcal{G}$ itself, i.e., if the set of nodes $\mathcal{Q}$ is a proper subset of $\mathcal{N}$, then its set of edges $\mathcal{F}$ must coincide with $\mathcal{E} \cap \mathcal{Q}^2$, i.e., $\mathcal{F}$ contains all the links of $\mathcal{E}$ between nodes of $\mathcal{Q}$.

Examples of graphs, with a visual representation of nodes, edges, neighbor nodes, walks, paths, cycles, and trees are shown in Fig. 10.1.

If a graph $\mathcal{G} = (\mathcal{N}, \mathcal{E})$ has a subgraph $\mathcal{H} = (\mathcal{N}, \mathcal{F})$ which is a tree and whose set of nodes $\mathcal{N}$ coincides with that of $\mathcal{G}$, then $\mathcal{H}$ is said to be a *spanning tree* of $\mathcal{G}$. Sometimes, it is said that $\mathcal{H}$ is a spanning tree over the set of nodes $\mathcal{N}$, since spanning trees are often introduced with the purpose of "covering" (i.e., interconnecting without any cycle) a set of nodes; however, this can be done only when there is no ambiguity on the set of edges which are used to perform such covering. In other words, not only must the set of nodes $\mathcal{N}$ be specified, which is the same for $\mathcal{G}$ and $\mathcal{H}$, but also it must be clear that the set of edges $\mathcal{F}$ used by $\mathcal{H}$ is a subset of $\mathcal{E}$. Implicitly, we will extend this terminology also to subtrees. That is, if $\mathcal{G} = (\mathcal{N}, \mathcal{E})$ is a graph, $\mathcal{H} = (\mathcal{N}, \mathcal{F})$ is a spanning tree of $\mathcal{G}$ and $\mathcal{H}_1 = (\mathcal{N}_1, \mathcal{F}_1)$, where $\mathcal{N}_1 \subseteq \mathcal{N}$ and $\mathcal{F}_1 \subseteq \mathcal{F}$, is a subtree of $\mathcal{H}$, we will say that $\mathcal{H}_1$ is a spanning tree over the set of nodes $\mathcal{N}_1$. This is just a shortcut to say that $\mathcal{H}_1$ is a spanning tree of the subgraph $\mathcal{G}_1 = (\mathcal{N}_1, \mathcal{E}_1)$ of the original graph $\mathcal{G}$ where $\mathcal{F}_1 \subseteq \mathcal{E}_1 = \mathcal{E} \cap \mathcal{N}_1^2$.

Another useful fact can be therefore stated as follows.

Theorem 10.1 *Assume $\mathcal{G} = (\mathcal{N}, \mathcal{E})$ is a graph, and $\mathcal{H} = (\mathcal{N}, \mathcal{F})$ is a spanning tree over the set of nodes $\mathcal{N}$ (meaning that $\mathcal{F} \subseteq \mathcal{E}$). Assume also that $\mathcal{N}$ contains a subset $\mathcal{N}_1$, and*

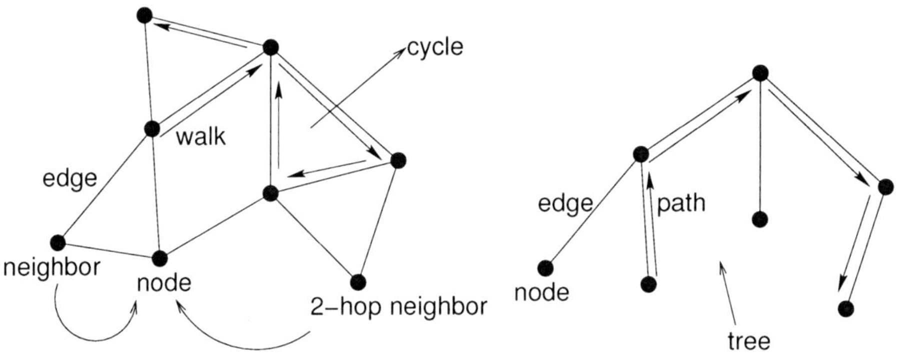

Figure 10.1 Examples of topological concepts.

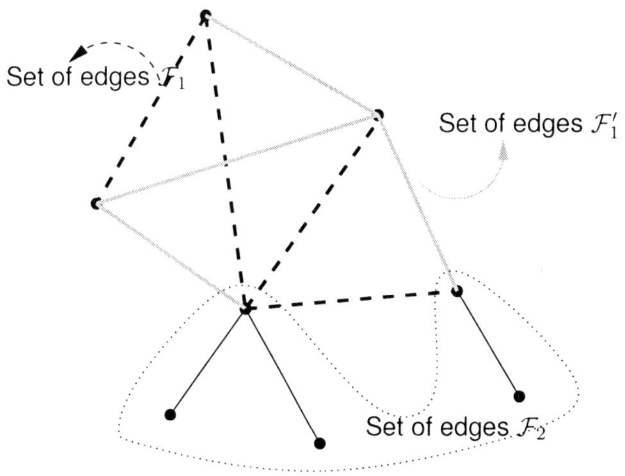

Figure 10.2 An example of application of Theorem 10.1. The spanning tree $\mathcal{H}$ over the set of nodes in this network has a subtree $\mathcal{H}_1 = (\mathcal{N}_1, \mathcal{F}_1)$ and therefore its edges are the superposition of the edges in $\mathcal{F}_1$ (dashed black lines) and $\mathcal{F}_2$ (solid black lines); if $\mathcal{H}_1' = (\mathcal{N}_1, \mathcal{F}_1')$ is another spanning tree over $\mathcal{N}_1$ (solid grey lines), the graph $\mathcal{H}' = (\mathcal{N}, \mathcal{F}_1' \cup \mathcal{F}_2)$ is also a spanning tree over $\mathcal{N}$.

$\mathcal{H}_1 = (\mathcal{N}_1, \mathcal{F}_1)$ is a spanning tree over $\mathcal{N}_1$ (meaning again that $\mathcal{F}_1 \subseteq \mathcal{E}$). Denote with $\mathcal{F}_2 = \mathcal{F} \setminus \mathcal{F}_1$. If we consider another spanning tree over $\mathcal{N}_1$, i.e., a tree $\mathcal{H}_1' = (\mathcal{N}_1, \mathcal{F}_1')$ such that $\mathcal{F}_1' \subseteq \mathcal{E} \cap \mathcal{N}_1^2$, we have that $\mathcal{H}' = (\mathcal{N}, \mathcal{F}_1' \cup \mathcal{F}_2)$ is also a spanning tree over $\mathcal{N}$ (an example is shown in Fig. 10.2).

Proof Both graphs $\mathcal{H}_1$ and $\mathcal{H}_1'$ are trees and hence they do not have cycles. Adding the nodes of $\mathcal{N}_2$ and edges contained in $\mathcal{F}_2$ to the graph $\mathcal{H}_1'$, thus obtaining $\mathcal{H}'$, cannot create any cycle. This is easy to see by considering an auxiliary representation of the graphs where all nodes of $\mathcal{N}_1$ are concentrated in a single node. This representation does not distinguish between $\mathcal{H}_1$

and $\mathcal{H}'_1$, since they become just one single node. Since extending $\mathcal{H}_1$ with the nodes of $\mathcal{N}_2$ and edges contained in $\mathcal{F}_2$ did not create any cycle by hypothesis, as $\mathcal{H}$ does not have cycles, so must happen for $\mathcal{H}'_1$. Therefore, $\mathcal{H}'$ has no cycles. Since it is also a connected graph, it is a spanning tree over the nodes of $\mathcal{N}$. □

10.2 ROUTING

The term "routing" describes the procedures to select paths (also called routes) in a network. A common example of routing in everyday life involves the selection of roads when driving towards a destination. Also information networks require routing decisions to guide the messages through the communication devices to the final destination.

The routing problem exists even outside the field of telecommunication networks; it was identified and studied by mathematicians in general contexts, which has given it a rigorous and theoretically solid basis. However, the strong impulse that the task of finding efficient routes has received in the last decades is motivated by the growth of telecommunication networks.

The routing problem will be regarded at many levels. First, we will consider the theory of *routing algorithms*. The term "algorithm" means that a set of rules is closely followed to solve a problem; this is a very broad and general meaning. Indeed, in the literature one may find both highly theoretical strategies and concrete real-world solutions, all going under the name of "routing algorithms." However, we will make a more sparing use of the term, just referring it with a limited and more abstract scope. Namely, we will mention "routing algorithms" only as sets of mathematical rules which apply to abstract representations of the networks. In this sense, only approaches taken from the mathematical background of routing theory will be referred to as routing algorithms. At first, we will consider this level of abstraction, where routing is performed on graph representation of the network topology.

As a further step towards bringing this background in the real world of communication and computer networks, we need to insert other technical specifications, which impose additional constraints and require to account for implementation aspects. The result is a set of more down-to-earth procedures, which we will name *routing strategies*. When this name will be used, it means that routing is still approached from a general perspective, but the description has a technical flavor related to details which are not present in abstract topologies.

Finally, when we present specific real-world implementations which apply to definite cases, we will more appropriately name them as *routing protocols*. For example, we will speak of *computational complexity* of a routing *algorithm*, in order to emphasize this mathematical aspect, whereas we will discuss practical advantages and disadvantages of routing *strategies*, and finally we will review real-world implementation of routing *protocols*.

Routing semantics: unicast, broadcast, multicast A classification which is important to be aware of involves routing semantics. In communication networks, the basic form of information exchange involves a single source-destination pair. This situation is referred to as a *unicast* communication. It is the case of a typical phone call, or a plain letter sent by mail. In a unicast communication, the route to find is therefore between the source and the destination.

However, it is also possible that the same message is meant to reach more than one destination. The most general case is achieved when we do not discriminate between the receivers, i.e., the message potentially addresses every node in the network. Such a situation is referred to as *broadcast*, which is the philosophy behind a tv show, or a newspaper article.

Sometimes, different cases with multiple users are introduced. In particular, *multicast* correspond to the case where the message is meant for a group of destinations, which however do not necessarily include all nodes. In this situation, the number of destinations is greater than one, as was in broadcast, but one requires to identify who is meant to be a destination and who is not; this latter part resembles unicast, which can be thought of as a special case of multicast where only one user is identified as a legitimate destination. A multicast communication is realized, for example, in a video-conference system or a pay-per-view tv show, where multiple users, but not all users, receive the communication. In practical cases, multicast can be often seen as an extension of broadcast. For example, valid destinations can be identified as a subset of the network nodes by a proper discrimination mechanism, i.e., a simple criterion which tells the ordinary nodes apart from the elite members of the legitimate destinations group (in the examples above, those who are authorized participants or have paid a subscription fee). Other cases also can be considered, where the number of destinations is potentially more than one, but smaller than the number of all nodes. They need special care to be analyzed, but for what concerns routing, there is not much difference with the general broadcase case. For this reason, in the following we will distinguish just unicast, meant as one-to-one communication, and broadcast, meant as one-to-many. As will be shown next, these two main cases pose a semantic difference on the concrete goal of the routing algorithm.

10.2.1 Routing objectives

From a high-level perspective, the routing algorithm should simply select a source-destination path according to some goodness criterion. Viable choices include, e.g., the overall reliability of the route, the absence of congestion on some traversed node, the low latency to go across the path. Instead of describing the positive aspects of a route, it is common to take the opposite viewpoint and consider its *costs*; such a characterization holds for the criteria mentioned above if we then consider failure probability, congestion, latency, as possible costs. Actually, costs can even be related to external non-technical aspects, for example monetary costs. The theoretical objective of routing is therefore a cost minimization. This implies that, for any given problem, we can re-scale the cost specifications by an arbitrary positive constant, and still have the same problem. However, defining how exactly state this cost minimization problem, and how to achieve it, is not simple, as will be argued below.

Minimum cost routes One problem of finding a "good" route (i.e., with low cost) is that the performance metrics that determine costs may be dynamic characteristics of the network and, even worse, the choices made by the routing algorithm may change them. For example, assume a source S and a destination D can be connected through two possible paths, say, $\mathcal{P}_1$ and $\mathcal{P}_2$.

This situation is reported in Fig. 10.3 for a practical case of interest: [4] a residential area S is connected to downtown D through two possible routes, a highway and a country road through several sub-urban areas. In such a case, the routing algorithm corresponds to the decision process made by the inhabitants of the residential area who work downtown to reach their workplace in the morning (and also to go home in the afternoon, when they do not work

[4] In the graphical representation, $\mathcal{P}_1$ and $\mathcal{P}_2$ may look like they both consist of a single hop, in spite of our previous assumption about the uniqueness of the edge between any pair of nodes. However, this is just made for the sake of presentation and does not affect the correctness of the reasoning.

overtime): which route is it better to take? Clearly, the same holds for analogous cases, such as having data that can be sent over two possible paths to their destination in the network.

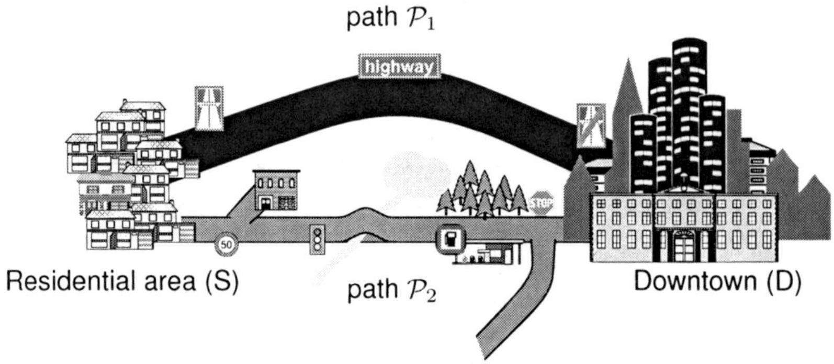

Figure 10.3 A routing problem in real life. Which path between $\mathcal{P}_1$ and $\mathcal{P}_2$ is better to go downtown?

One can choose $\mathcal{P}_1$ if it is safe to assume this alternative is less congested, or in general has a lower cost, than the other route $\mathcal{P}_2$. Only, if this strategy is blindly applied, i.e., all inhabitants go downtown through route $\mathcal{P}_1$, or all the data are sent along this path, the network will end up having a traffic jam over $\mathcal{P}_1$, while $\mathcal{P}_2$ will be congestion-free! To solve this problem, we should enable to update the selected route when a better one is available. This check is performed, for example, between periodic time intervals. In principle, this solves the problem but still can have some ping-pong (bouncing) effect between the updates. Considering again Fig. 10.3, this happens when switching most of the traffic to $\mathcal{P}_2$ causes $\mathcal{P}_1$ to become then better, so the problem occurs again, only with reversed roles between $\mathcal{P}_1$ and $\mathcal{P}_2$, at the next path evaluation. In order to avoid ping-pong effects, a hysteresis mechanism may be used, meaning that the choice of route $\mathcal{P}_1$ is changed in favor of a better route $\mathcal{P}_2$ only if the latter provides a significant advantage over $\mathcal{P}_1$, and viceversa.

A practical implementation of routing which is aware of network dynamics, although in a simplified manner, is as follows. We first assume that route costs are determined a priori. The path from S to D is instantaneously selected as the one with minimal cost. Then, after a given time interval, network dynamics are taken into account and costs are periodically updated, so that routes can be changed when appropriate. This means that a change occurs if not only a better route exists, but also its cost is lower than that of the old route minus a proper margin. For the whole time interval between updates, costs are regarded as constant values, and possible influences of the routing algorithm choices on the costs will be taken into account only at the next cost update. Indeed, this does not entirely guarantee that the solution will be stable. Actually, defining and verifying the stability of the routing strategy is a very complex issue, which goes beyond the scope of this chapter. However, such a mechanism is close to what can be implemented in practice and solves most of the instability problems (ping-pong effect) that may occur in communication networks.

The approach outlined above implies that a given algorithm is performed several times, i.e., every time a source initiates a transmission towards a given destination. Thus, there are in reality several instances of the algorithm, which behave as multiple concurrent agents.

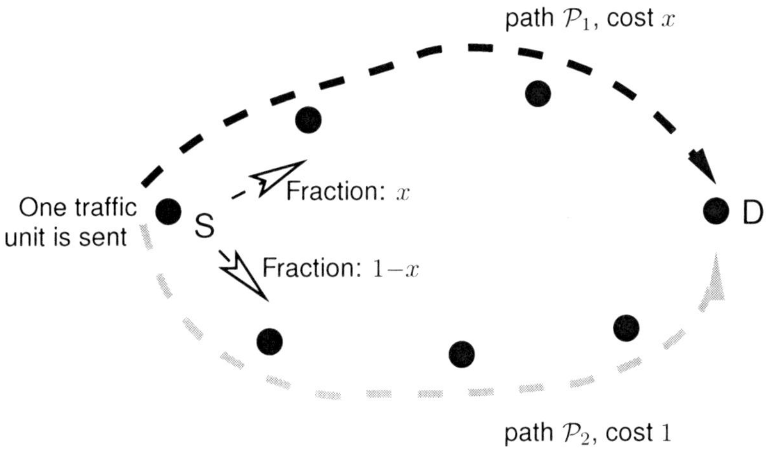

Figure 10.4 A routing problem where the cost of $\mathcal{P}_1$ is dependent on the traffic, while the cost of $\mathcal{P}_2$ is not. One unit of traffic is sent from S to D. (The paths traverse multiple nodes only for the sake of simplicity in the representation).

This approach still does not entirely solve the problem of path costs which are dependent on network-wide parameters. Consider the problem reported in Fig. 10.4. The scenario is the same reported in Fig. 10.3, but we take an abstract representation (although, similarly to Euler's representation of Königsberg bridges, it is topologically equivalent). We relate the costs of paths to their latency, which in turn is assumed to depend on the traffic congestion on the path.

Recall that S represents the residential area and D is the downtown, which is the destination of the commuters. Assume one unit of traffic needs to be sent from S to D. This represents the amount of workers going downtown every morning. This traffic can be split over two available paths $\mathcal{P}_1$ and $\mathcal{P}_2$; denote with x, with $0 \leq x \leq 1$, the traffic amount sent along $\mathcal{P}_1$, so that the amount sent along $\mathcal{P}_2$ is $1 - x$. The characteristics of the paths are as follows. Since $\mathcal{P}_1$ represents the highway, we can assume it has a cost which is directly proportional to the amount of traffic that is routed through it. Thus, we can simply write the cost as x (assuming for simplicity the proportionality constant as equal to 1). In fact, when the traffic is very low, a highway has a very fast traversing time, so the latency is also low. However, during rush hours congestion may cause the traffic to be delayed for longer times. Path $\mathcal{P}_2$ is instead a sub-urban route with a fixed latency cost, equal to 1. This is due to its speed limits considerably lower than those on the highway; in such a case, the latency does not depend on the network load but it is high under any traffic condition.

One immediate consideration is that, since the input traffic is taken equal to 1, the cost of $\mathcal{P}_2$ is always greater than or equal to that of $\mathcal{P}_1$ for any value of x. This implies that the "best" route is always $\mathcal{P}_1$, therefore it does not seem sensible for the routing algorithm to select $\mathcal{P}_2$. If this holds, $x = 1$, in other words, all the commuters take the highway. Thus, the total latency cost experienced by their community is 1. Strangely enough, this is not the best possible choice from a network-wide perspective. If the flow of workers were split between the two paths in identical amounts, i.e., $x = 1/2$, path $\mathcal{P}_1$ would become less congested and its cost would decrease. Even though the path cost did not improve for those workers selecting

$\mathcal{P}_2$, the overall cost would be $3/4$. This represents a significant improvement with respect to the previous case where all workers selected the apparently better route, which turned out to have a higher total cost. In other words, the combination of optimally chosen routes costs 33% more (1 instead of $3/4$).

This better way of selecting the routes can be seen as the outcome of a random decision. In practice, each worker flips a coin to decide whether to take the highway or the sub-urban route. Or it might be that some "self-sacrificing" commuters choose the slower route intentionally. In either case, it is counterintuitive that this way of selecting a path achieves a better result than doing it in a supposedly optimal manner. Actually, it is simple to verify that the choice of $x = 1/2$ is in reality the globally optimal choice. In fact, the overall cost is $x^2 - x + 1$, which is maximized for $x = 1/2$ according to the first order conditions. The reasons for this strange outcome are also grounded in how this "minimal" cost is evaluated. The main problem here is that when several agents try to independently determine an optimal solution to their own needs, the combined results can not be optimal at all in a global sense. This fact occurs often in practical circumstances of real life, and is mathematically formalized by a science known as game theory [1]. Even though a thorough discussion about this subject is out of the scope of the present book, this result, discovered already in 1920 by Pigou [2], serves as a caveat that the minimality of a route cost is not easy to determine, and that selecting routes through independent executions of a routing algorithm may not be always good. On the other hand, it can be proven [3] that Pigou's example, where the combination of independently chosen routes costs 33% more than the global optimum, represents the worst possible case. Thus, the inherent inefficiency of this implementation is, at most, a cost which is 33% higher, and in most cases is actually lower than that.

For simplicity reasons, the concept of optimality is often meant considering a single instance of source and destination, i.e., the optimal route from S to D is the one with minimum cost, even though, according to the discussion above, this is in reality not optimal if regarded from a global perspective. In the following, we will adopt this terminology as well, while being aware that this may not be optimal at all. However, while global optimality would be desirable, it is difficult to achieve since it requires a network-wide knowledge and a centralized supervision which are hardly realizable in practice. Thus, everything related to the concept of an optimal route is to be read hereafter in a local sense. Methodologies for a real optimization, intended as a search for a global optimum, can be derived in a more sophisticated manner through a mathematical theory known as optimal multicommodity flow, which is outside the scope of the present chapter. Further readings on this topic can be found in [4].

Costs of links and spanning trees A further common assumption about route costs is that, beyond being known a priori, they can be split at the link level, and the overall path cost is obtained as an aggregate of the costs of traversed links. Formally, the cost $\ell(e)$ is defined for every link $e \in \mathcal{E}$ as a function $\ell : \mathcal{E} \mapsto \mathbb{R}$. With a slight abuse of notation, if $e = (i, j)$ we will write its cost as $\ell(i, j)$ rather than the more cumbersome notation $\ell((i, j))$. Since we avoid multiple links between the same pair of nodes, this notation is legitimate.

Link cost is usually referred to as "distance" or "length" of an edge, as in problems of routing over roads, physical distance is one of the most common and reasonable choices. Such a custom is also implicit in our choice of the symbol "ℓ" (as in "length"). However, we will use the term "cost" whenever possible, since it is more general. For example, even in road networks, a driver may prefer to take a route without fees rather than a shorter highway with a toll; in this example, the cost does not coincide with the physical distance, but it is the monetary cost instead. The link costs can also be called *weights*, and a graph with a cost function associated with its links is frequently called a *weighted graph*.

The link costs induce the cost of a path $\mathcal{P} = \{i_0, \ldots, i_L\}$, which is denoted as $c(\mathcal{P})$. The path cost $c(\mathcal{P})$ only depends on the costs of the links involved, i.e., $\ell(i_{k-1}, i_k)$ with $k \in \{1, \ldots, L\}$.

Very frequently the costs are assumed to be *additive*, meaning that the aggregate cost of the path is their sum, i.e.,

$$c(\mathcal{P}) = \sum_{k=1}^{L} \ell(i_{k-1}, i_k) \tag{10.2}$$

For example, this assumption is correct for the physical distance in road networks. It is also appropriate if the cost metric ought to describe, as in the network of Fig. 10.4, the total latency of the path. Thus, the overall route cost is simply the sum of the delay costs of all links. Similar conditions can hold for other metrics, although sometimes an adjustment is needed. For example, if link reliability (or better, the unreliability) is defined through a numerical value representing the failure probability, the overall aggregate for the path is obtained through a product, not a sum. That is, if $p_{\text{fail}}^{(e)}$ is the probability that link e can fail, the reliability of the link is $1 - p_{\text{fail}}^{(e)}$. Now, if we assume the failure events to be independent, which is likely to be a reasonable assumption, the overall reliability $R_{\mathcal{P}}$ of a path $\mathcal{P} = \{i_0, \ldots, i_L\}$ can be obtained as

$$R_{\mathcal{P}} = \prod_{k=1}^{L} \left(1 - p_{\text{fail}}^{(i_{k-1}, i_k)}\right). \tag{10.3}$$

However, this can again be translated into a sum by taking the logarithms. Thus, a proper choice of the cost metric in this case is to take the cost of link (i, j) as

$$\ell(i, j) = -\sum_{k=1}^{L} \log(1 - p_{\text{fail}}^{(i_{k-1}, i_k)}). \tag{10.4}$$

In the following we will tacitly assume that costs are attributed to the links, and the overall cost of a path can be evaluated through a sum. Such a view simplifies the analysis of routing issues. For example, if congestion, or failure, or any other undesirable event happens over a link, only the cost of that link needs to be updated in order to reflect this, while the costs of surrounding links can be kept unchanged. At the same time, if the link costs are additive, a cost increase for a single link will correctly reflect on an increased cost for the whole path.

The assumption of additive cost has an important consequence:[5] we can prove an important result that will be used in the following of this section, called *optimality principle*. It states that every part of an optimal path is also an optimal path itself. The following theorem formalizes this fact.

Theorem 10.2 (Optimality principle) *If the optimal path from node A to node B traverses node C, the part of this path from node C onwards is also the optimal path from node C to node B.*

[5]Actually, this requirement can be relaxed and replaced by some weaker condition involving monotonicity of the costs and metric spaces. In practical cases, however, this generally means to require that costs are additive.

Proof The optimal path $\mathcal{P}$ from A to B is split by C in two parts, which we call $\mathcal{P}_1$ and $\mathcal{P}_2$. The theorem states that the optimal path from C to B is $\mathcal{P}_2$, which can be easily proven by contradiction. If the optimal path from C to B were $\mathcal{P}'_2 \neq \mathcal{P}_2$, we would have that $cost(\mathcal{P}'_2) < cost(\mathcal{P}_2)$. However, we could then reach B from A by using path $\mathcal{P}_1$ to reach C and then $\mathcal{P}'_2$ to reach B. As costs are additive, the overall cost of this joint path would be

$$cost(\mathcal{P}_1) + cost(\mathcal{P}'_2) < cost(\mathcal{P}_1) + cost(\mathcal{P}_2) = cost(\mathcal{P})$$

which contradicts the hypothesis that $\mathcal{P}$ is the optimal path from A to B. □

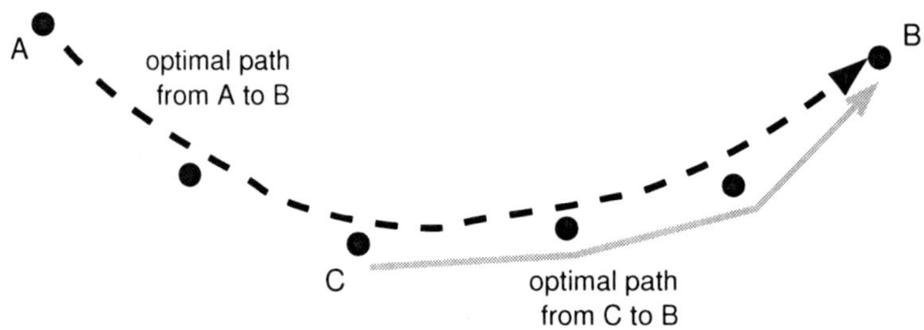

Figure 10.5 Application of the optimality principle. C is on the best route from A to B, thus the portion of this route from C to B is also the best path from C to B.

A practical instance of the optimality principle is reported in Fig. 10.5. This enables the following description of the outcome of the routing procedures. If two routes to (or from) the same node share a common part, there is no need to keep track of both paths in detail. For example, if the optimal path from A to B passes through node C, and so does the optimal path from A to B′, we can apply the optimality principle to adopt the following simplified representation. Instead of explicitly reporting all nodes belonging to the optimal paths from A to B and from A to B′, we just mention that both paths comprise the optimal path from A to C; then, the paths split towards B or B′, respectively.

As a generalization of this process, we can think of representing all optimal paths starting from the same node A to all other nodes in the network with a tree structure. This is commonly called a *shortest path spanning tree* rooted at node A, which contains all optimal paths from node A to all other nodes (or vice versa). In this case, we use the term "shortest" as commonly done by the related literature, i.e., in the sense of minimum cost. It may be better to call it "optimal," since, as already discussed, the cost function may actually differ from the physical distance; yet, also this terminology is slightly inappropriate, because of the differences between global and local optimality. In the end, there is no perfect term, so we will stick to "shortest," even though we implicitly assume that it does not necessarily refer to the physical distance.

Note also that it is appropriate to use the shortest path spanning tree only when unicast paths are sought. In order to represent the best way to achieve a broadcast connection whose source is A, i.e., a communication from node A to all other nodes *simultaneously*, a different structure should be used. This is the *minimum weight spanning tree*, which is a collection of paths having the lowest aggregate cost together. We stress that the difference between the shortest path and the minimum weight spanning trees is subtle, and the terminology does not

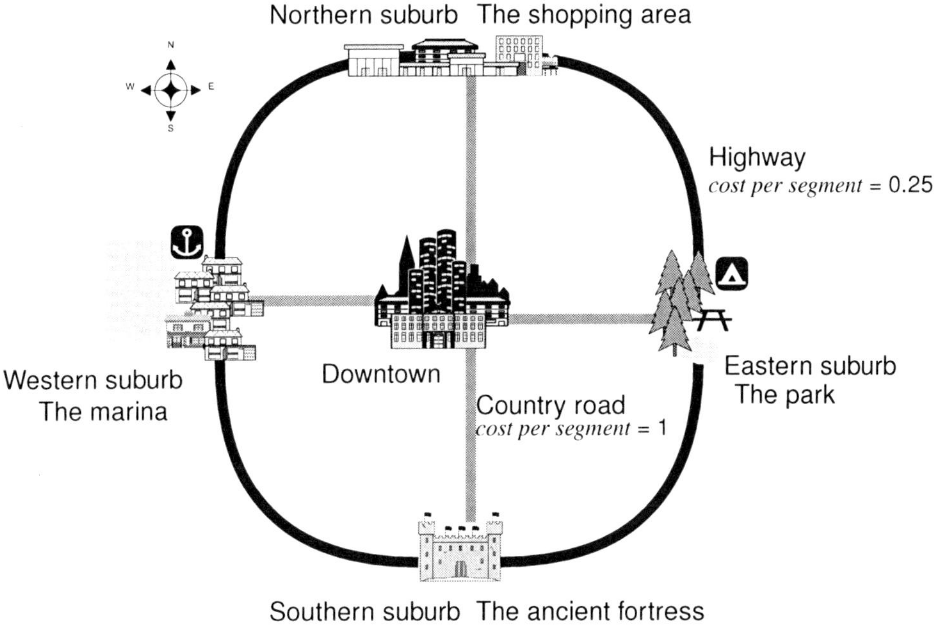

Figure 10.6 A city with four districts. The inhabitants and the tourist need to solve a unicast and a broadcast routing problem, respectively.

help. The attribute of "minimum weight" refers in this case to the whole tree, not to the path (in which case it would mean the same as "shortest path").

To better understand this difference, consider the following example, reported in Fig. 10.6. A city is surrounded by four different suburbs, located at the four cardinal points, i.e., north, south, east, and west. Each suburb has some inhabitants, which need to commute every day to downtown. To help the commuters, the municipality want to establish a new fast transportation system (this may be, for example, a set of bus lines or fast surface trains). This new transportation system must connect the city with all the suburbs. For simplicity, we assume that all the suburbs have the same number of inhabitants, and the same importance, meaning that none of them is necessarily privileged in being connected to the city, not even if the mayor leaves in one of them. However, it is clearly not admissible that one suburb is left out.

There are different ways to connect the city and the suburbs among each other. A straight route may be realized from any suburb to downtown, following a direct connection. Alternatively, the suburbs may be circularly interconnected, thus realizing any of four segments, e.g., between northern and eastern suburbs, between eastern and southern, and so on; only subsequent suburbs can be interconnected in this way, though (e.g., no segment between northern and southern suburbs). These two kinds of links have different costs, which for simplicity are all chosen as constants: each direct link costs 2, and each circular segment cost 1. A possible explanation is that, similarly to Fig. 10.4, the direct link is a country road, and the circular segments are parts of a highway system. Or, if links are railroads, the higher cost of direct links is because they need to intersect the roads entering town. For simplicity, this cost evaluation is the same if evaluated from the perspectives of the municipality and the commuters. This means that the link weights indicated above represent both the cost of deploying the infrastructure (which is paid by the municipality) and the price of the tickets over that particular interconnection for the passengers (which is paid by the commuters). Actually, this direct proportionality may even be sensible.

The goal of the municipality is to find the *best*, which is to be read as the *cheapest*, way to build this transportation system up. When discussing the possible options, two solution emerges. The association of the commuters would like to build the system only over the direct paths. Such a solution has overall cost equal to 8 and allows to reach downtown from every suburb with a ticket price of 2, the lowest possible, and the same for all suburbs. The financial advisor of the municipality believes instead that the best solution is to realize the system by connecting only one suburb with a direct link (it does not matter which one, actually), and further interconnect this suburb with the others through the circular segments. There are several ways to do this (in fact, 16 such solutions exist), all equivalent in terms of overall cost, which is 5. It should be noted, however, that in these solutions some commuters pay a higher cost than 2 to reach downtown.

How is it possible that different solutions are proposed, in spite of the network, and also the costs, being the same? The reason is that the routing semantics is different in the two perspectives. Although it is required that all suburbs are connected with downtown, this can be done in two basic ways: either connecting optimally each suburb individually taken, and then superimposing all the solutions, or realizing a globally optimal connection of the city and its suburbs.

The former approach is that taken by the commuters' association, which is like solving a unicast communication. In fact, the inhabitants of a suburb only want to reach downtown in an optimal manner, and essentially do not care if the path is not optimal for the inhabitants of another suburb. The financial advisor instead wants to minimize the global cost, and therefore tries to solve a broadcast communication problem. Thus, the shortest path spanning tree contains a representation of the routes desired by the commuters, i.e., it comprises only the direct paths, without using the circular segments. A minimum weight spanning tree can be instead any of the solutions found by the financial advisor, i.e., one direct link and three circular segments.

10.2.2 Routing algorithms

In this section we discuss the main theoretical algorithms for routing, which are the basis for the implementation of routing protocols in real networks.

First, we present two important algorithms to derive a shortest path spanning tree, proposed around the 1950s. The first one is the Bellman–Ford algorithm, named after Richard E. Bellman and Lestor R. Ford jr.,[6] the second is the Dijkstra algorithm, proposed by Edsger W. Dijkstra [7].

Both algorithms have the goal to find the "best" routing path in a network represented as a graph $\mathcal{G} = (\mathcal{N}, \mathcal{E})$. We assume that every arc (i, j) in $\mathcal{E}$ has a weight, denoted as $\ell(i, j)$,

[6]The algorithm was proposed first by Bellman [5] and subsequently by Ford and D. R. Fulkerson [6]. Only Ford is oddly mentioned in the literature since another related algorithm is called the Ford-Fulkerson algorithm.

representing the routing metric, i.e., the communication cost between nodes i and j. For completeness, if i and j are not connected through an arc of $\mathcal{E}$, we set $\ell(i, j) = \infty$. As already mentioned, although cost ℓ is often referred to as "distance," it does not necessarily refer to a physical or geometrical concept. In this sense, the Bellman–Ford and Dijkstra algorithms are often said to derive the shortest path (meant as the one with lowest cost) between any pair of nodes.

To understand what these algorithms do, it is not restrictive to give a specific name, e.g., to call it "A," to the destination node of the desired route. Thus, the problem solved by these algorithms can be stated as: "find the cheapest way to reach A." Since the result of the algorithm is a spanning tree connecting A with all nodes of the network, actually both algorithms find at once the best routes to node A from *all* other nodes.

Note that, as previously mentioned, we will restrict our attention to undirected graph. If this is the case, these algorithms also find all the shortest paths *from* a given node A to every other node. Otherwise, i.e., if link direction matters, a similar reasoning still holds; however, the two tasks (finding the best routes from all nodes to A versus the ones from A to all other nodes) have in general different outcomes and require to execute two different algorithm instances, where paths towards A, or starting from A, respectively, are considered.

We will also present algorithms for minimum weight spanning trees. As argued in the previous section, these are different from the optimal spanning tree found by the Bellman–Ford and Dijkstra algorithms, though the related algorithms share some similarities.

Finally, we remark that the algorithms we will discuss in the following make sense only in the case the starting graph representing the network is connected, otherwise the problem of interconnecting some nodes may be not solvable. In the following, we will always assume implicitly that this condition is verified, i.e., any graph $\mathcal{G} = (\mathcal{N}, \mathcal{E})$ we start from is connected.

The Bellman–Ford algorithm Before describing the algorithm in detail, some preliminaries are required. The Bellman–Ford algorithm derives its solution in an iterative fashion, by constructing sequence of nodes which are temporary best walks. We say "walks" since we do not necessarily forbid to visit the same node more than once. Thus, the algorithm seems at first to find walks, not paths. Recall that a path is always a walk (and therefore a shortest path is also an optimal walk), but not vice versa; this depends on whether the walk contains cycles or not (only in this latter case the walk is also a path).

Actually, we can distinguish three possible cases: i) any cycle that can be found in the network has overall weight (i.e., the sum of all link weights) which is less than zero; that is, all cycles in the network have *positive* weight; ii) the network contains at least one cycle with zero weight, but none with negative weight; iii) the network contains cycles with negative weight.

Although the physical meaning of case iii) may be questionable, it is theoretically possible, and is the most problematic to deal with for the algorithm. In this case, it is possible to enter a negative-weight cycle and loop through it indefinitely many times. The sum weight of any walk can be brought in this way as close as wanted to $-\infty$. In its original version, the Bellman–Ford algorithm therefore fails in case iii). This fact can be regarded from different perspectives. On the one hand, this is due to the algorithm being iterative; any temporary solution can be improved through cycling over a negative loop one more time. Thus, there is no final iteration. On the other hand, the algorithm will start finding walks which contain cycles, so they are of the "walk-but-not-path" kind.

However, apart from this case which is unlikely to happen in practical networks, the algorithm works fine in cases i) and ii). In fact, in case i) the algorithm only will find walks which are also path, and the reason is evident by contradiction. To prove it, assume that one of the best walks found, say from node B, contained a cycle. It is not restrictive to say then that the best walk from B reaches at first node C through a walk $\mathcal{W}_1$, enters a cycle $\mathcal{C}_1$ which returns to C, and finally reaches A through a walk $\mathcal{W}_2$. However, $\mathcal{C}_1$ must have overall weight larger than zero, so if we simply remove it, i.e., we reach A following $\mathcal{W}_1$ and then $\mathcal{W}_2$, we obtain a legitimate walk with lower cost from B to A. This contradicts the assumption that following $\mathcal{W}_1$, $\mathcal{C}_1$, and $\mathcal{W}_2$ is the best walk.

A similar reasoning applies to case ii). Only, this time we cannot say that the cost of $\mathcal{C}_1$ is positive cost, as it may be zero. However, in case ii) it is still guaranteed that any best walk is either a path itself, or there exist a path with identical cost, thus making it a best walk as well. The best walk which is also a shorter path is exactly obtained by removing all possible cycles with zero weight, as per the discussion above. Also, note that removing cycles with zero weight does not decrease the cost of the walk but makes it shorter in the sense of having fewer hops to traverse. This is important since, as will be evident in the following, the iterations in the Bellman–Ford algorithm proceeds by increasing the number of used hops, so that solutions with lower number of hops are found first. This implies that, even if cycles with zero weight are present in the network, the solutions found do not loop through them.

To sum up, in case iii) the algorithm cannot be applied; as a simple way to rule it out, we just require that the Bellman–Ford requires that the network does not contain any cycle with negative weight. In cases i) and ii) instead, we have that the walks found by the algorithm are also paths. This is why we will speak of *paths* found by the algorithm.

We now proceed with the real description of the algorithm. As mentioned, it works in an iterative fashion and sets a constraint on the length of the routes to node A, which is relaxed at every step; in other words, at iteration h it tries to connect each node to A along the best possible path using h hops or less. Then, h is increased by 1 and the procedure is repeated. In this way, the algorithm determines at first all optimal one-hop walks, then all optimal two-hop walks, and so on. As said above, these walks are also paths. Thus, if the shortest path from a given node to node A has length equal to K hops, it will be found at the Kth iteration.

However, the fact that the best path is found at the Kth iteration does not itself guarantee that the algorithm recognizes it as optimal. This is only discovered when the algorithm terminates; this means that the iteration does not produce any update, i.e., the optimal walk found using $h + 1$ hops does coincide with that using h hops. We need to prove that this termination condition works and the Bellman–Ford algorithm eventually terminates after at most $|\mathcal{N}|$ iterations. We will prove also these statements in a while.

Before proceeding, we describe the algorithm notation and the structures employed by it to find the optimal walks, and to determine that they are the shortest paths to A. The Bellman–Ford algorithm uses auxiliary variables to memorize the temporary best walk found from node i to node A and its cost. During the algorithm execution, these variables can be updated if needed. After the algorithm terminates, if it does, they are no longer temporary variables and become the actual best walk and the actual cost of the best walk.

The temporary best walk found from node i to node A is denoted as $\mathcal{P}_h(i)$, which is an ordered list with maximum length equal to h, and its cost is represented by $C_h(i)$. For simplicity, we omit the last node in $\mathcal{P}_h(i)$, since it will be always "A" for any i. For example, $\mathcal{P}_3(\text{B}) = (\text{B}, \text{C}, \text{D})$ and $C_3(\text{B}) = 25$ mean that the optimal walk from B to A passes through

C and D (thus involving 3 hops) and has cost 25. Finally, it is worth noting that the first node in $\mathcal{P}_h(i)$ is always i itself.

At its 0th iteration, i.e., at the initialization phase, the Bellman–Ford algorithm sets $\mathcal{P}_0(i) = \emptyset$ for all nodes, $C_0(\mathrm{A}) = 0$ and $C_0(i) = \infty$ for every $i \neq \mathrm{A}$. As an arc cost equal to ∞ meant that no arc existed, an infinite total walk cost $C_h(i)$ means that no walk from node i has been found up to the hth iteration.

The iterative cost update rule is as follows:

$$C_{h+1}(i) = \min_{j \in \mathcal{N}} \left(\ell(i,j) + C_h(j)\right) \tag{10.5}$$

and, if $C_{h+1}(i) < C_h(i)$, the walk is updated to

$$\mathcal{P}_{h+1}(i) = \left(j_{\text{opt}}, \mathcal{P}_h(j_{\text{opt}})\right) \tag{10.6}$$

where j_{opt} is the value of j found in (10.5), i.e.,

$$j_{\text{opt}} = \arg\min_{j \in \mathcal{N}} \Big(\ell(i,j) + C_h(j)\Big) \tag{10.7}$$

The termination condition stated above corresponds to terminating the iteration at step $h+1$ if

$$\forall i \in \mathcal{N}: \quad C_h(i) = C_{h-1}(i) \tag{10.8}$$

What we need to prove is that the algorithm works in finding the shortest paths and terminates in a finite number of steps, under the conditions mentioned above. We formalize this in the following theorem.

Theorem 10.3 *At each iteration h, the Bellman–Ford algorithm generates an optimal walk reaching node A using h hops. If all cycles in the network have non-negative costs, then all optimal walks found by the algorithms are also shortest paths and the algorithm terminates after at most $|\mathcal{N}|$ iterations, using the termination condition (10.8). In such a case, the last paths found before the algorithm terminates are indeed the ones with minimum cost, i.e., the shortest paths to A.*

Proof First of all, assuming that all cycles have non-negative cost means that we rule out the aforementioned case iii), where the algorithm fails. That is, we are restricting our scope to the cases i) and ii). The theorem, which we still have to prove, basically means that in these cases the algorithm indeed works. However, in the preliminaries of this discussion we already proved part of the theorem statement, i.e., that the algorithm always find walks which are also paths.

Thus, we just need to prove that these walks are optimal. More precisely, we will prove that the algorithm finds at the hth iteration the best possible walk using at most h hops. This can be proven by induction. It is trivial to verify that the first iteration of the algorithm updates the costs producing $C_1(i) = \ell(i, \mathrm{A})$, which is finite only if the arc (i, A) exists in set $\mathcal{E}$. In this case, the path $\mathcal{P}_1(i)$ is equal to (i), which actually represents (i, A) (recall that we omit the last node from this representation) and is, obviously, the optimal walk using at most one hop.

To prove the inductive step, assume that after h iterations we have found all the optimal paths using at most h hops. When the next iteration $h+1$ is considered, the optimal paths using at most $h+1$ hops are sought. Certain nodes do not find a better path to A (i.e., one with a lower cost) than that found using at most h hops, if they are allowed to use one more hop (i.e., $h+1$ hops). If *all* nodes are in this condition, according to (10.8), the algorithm ends. Thus, as long as the termination condition is correct, which will be proven shortly, the optimality of the paths is confirmed.

Otherwise, we can focus only on those nodes which improve their routes to A by using one more hop. Select one of those, and call it B. This node B can find a better walk to A using $h+1$ hops rather than using at most h hops. We need to prove that $\mathcal{P}_{h+1}(\mathrm{B})$ is actually such a better walk, and it is actually the best possible choice with $h+1$ hops at disposal. To prove it, observe that the best path starting from B and using $h+1$ hops must have a first visited node, say node C. The rest of the path, from node C to node A is h hop long. For the optimality principle, it is also the optimal path from C to A using at most h hops, and by the inductive hypothesis, it corresponds to $\mathcal{P}_h(\mathrm{C})$.

Thus, the optimal path using at most $h+1$ hops for node B is found by selecting C in the best possible manner, which is what (10.6) does, i.e., it selects the best neighbor such that the cost of reaching it and then proceeding along *the best path of the neighbor* is minimized. As a side consequence, this also proves that if none of the routes is updated at iteration $h+1$, all nodes found their best path using at most h hops. This validates the effectiveness of condition (10.8), which first of all justifies the reasoning made above, and also implies that once iteration $h+1$ has no effect, there is no need to go further with iteration $h+2$, where no update can be similarly found.

To conclude the proof, we just need to prove that the maximum number of iterations is $|\mathcal{N}|$. Recall that cycles are forbidden, as already discussed. Thus, the conclusion follows by the simple observation that any path cannot contain more than $|\mathcal{N}|$ hops without repeating at least one visited node, thus containing a cycle. To sum up: if no negative-cost cycles exist, the iterative mechanism of the Bellman–Ford algorithm (10.5)–(10.6) is valid, so is the termination rule (10.8), and finally the algorithm is guaranteed to terminate after at most $|\mathcal{N}|$ iterations. □

Finally, we can evaluate the computational complexity of the Bellman–Ford algorithm with a worst-case approach, as follows. The algorithm requires at most $|\mathcal{N}|$ iterations; at each iteration, the temporary path found is updated for $O(|\mathcal{N}|)$ nodes, and the update requires to evaluate a minimum among $O(|\mathcal{N}|)$ alternatives. Therefore, the worst-case complexity of the Bellman–Ford algorithm is $O(|\mathcal{N}|^3)$. However, this computation considers the (hopefully unlikely) case that the algorithm takes the highest possible number of iterations. In practical cases, the number of hops d on the "longest shortest path" (or, in better words, the optimal path with the highest number of hops) is considerably less than $|\mathcal{N}|$. The Bellman–Ford algorithm finds such an optimal path at iteration d, and terminates immediately after. Moreover, the cost update performed at each iteration can be shown to be $O(|\mathcal{E}|)$. Even though $|\mathcal{E}| \leq |\mathcal{N}|^2$, in many networks the former value is considerably smaller than the latter if many pairs of nodes are not connected through an edge. Thus, the resulting complexity should be more appropriately written as $O(d|\mathcal{E}|)$, where $d|\mathcal{E}|$ is significantly lower than $|\mathcal{N}|^3$ in several cases of interest.

The Dijkstra algorithm The Dijkstra algorithm is similar to the Bellman–Ford algorithm, even though it has lower worst-case computational complexity. It works only if the link costs are taken as non-negative, which is a common and often reasonable assumption for data networks. However, should it not hold, it would be necessary to revert to the Bellman–Ford algorithm.

Note that if the link costs are non-negative, all cycles automatically have non-negative cost (it is actually a more stringent condition). Thus, we do not need to rule out the case where the network has cycles with negative weight, as done for the Bellman–Ford algorithm, and it always verified that optimal walks are also optimal paths.

The Dijkstra algorithm uses again an iterative approach, evaluating path costs and updating them. The Bellman–Ford algorithm blindly iterated all temporary costs, and recognized there was no need for further updates only after an iteration which left all costs unchanged. The Dijkstra algorithm adopts a slightly different approach: it determines the shortest paths to A in increasing order of cost.

In other words, it seeks at first the node which can be connected to A by spending the lowest overall cost of the path. Indeed, there may be several nodes with the same minimal cost; if this is the case, the algorithm selects the one which uses fewer hops on the path, assuming that the number of hops is not already used as the routing metric. If there are still multiple choices, ties are broken in any manner, even randomly (the algorithm is unaffected by this). Thus, we can say that the algorithm starts by finding *the* node, which we call B_1, which has the lowest cost to reach A. Then, it is observed that the corresponding path between B_1 and A cannot be improved in any way, as will be proven next: it is, similarly to the previous theorem, a consequence of the optimality principle.

Thus, the idea is to label the node as "solved," since the algorithm has already found the best way to connect it to A (which is labeled as "solved" from the beginning). Differently from the Bellman–Ford algorithm, the Dijkstra algorithm immediately acknowledges that the path from B_1 to A is optimal. This concludes the first iteration.

At the next iteration, the algorithm searches for node B_2 which is the one, excluding B_1, which can reach A along the path with lowest cost. This node is also labeled as "solved." In general, at iteration h the algorithm seeks among the unlabeled nodes which is the one with lowest cost to reach A, and labels it as "solved." Thus, the algorithm labels all nodes in exactly $|\mathcal{N}|$ iterations.

How can the algorithm find the node to label at each iteration? At the first iteration, this is not difficult. The node B_1 with overall lowest cost to reach A is simply the one which minimizes the single-hop cost, see in the following theorem.

Theorem 10.4 *The node B_1 is connected to A on an optimal path consisting of a single hop, and is the node satisfying*

$$\ell(B_1, A) = \min_{j \in \mathcal{N}} \ell(j, A)$$

Proof The theorem is a trivial consequence of the optimality principle. If node B_1 had an optimal path to node A consisting of two or more hops, there would be a last node, which we call C, before reaching A along this path. Node C is different from B_1; yet, due to the optimality principle, the single link connecting C to A is an optimal path and has a cost lower than or equal to the cost of the optimal path from B_1 to A, since all link weights are non-negative. So, C should have been selected instead of B_1, as it has either lower cost to reach A, or the same cost but uses fewer hops, which contradicts the assumption of B_1 being the first choice of the algorithm. Hence, the optimal path with minimum cost among all optimal paths is a single-hop path. The rest trivially follows. In fact, once we proved it is a single-hop path, it must consist of the single hop with lowest link cost. □

The only missing ingredient of the idea behind the algorithm is that at each iteration h we can always find the shortest path which has the hth lowest cost among all shortest paths to A from the other nodes. To prove this, we first formalize the notation used by the Dijkstra algorithm and then we state the related theorem.

As done by the Bellman–Ford algorithm, we define, for every node i and every iteration h, a temporary path $\mathcal{P}_h(i)$; again, in the paths description we omit the last node of the path, since it is always A. At the same time, we also define a temporary cost value $C_h(i)$; these are the temporary optimal path reaching A from i at the h iteration, and its cost, respectively. Also, for each iteration h we define a set $\mathcal{S}_h$ containing all nodes which have been solved up to iteration h.

The initialization of the algorithm corresponds to impose $\mathcal{S}_0 = \{\mathrm{A}\}$, and, for every $i \neq \mathrm{A}$, $\mathcal{P}_1(i) = (i)$ and $C_1(i) = \ell(i, \mathrm{A})$. The iteration rule is as follows. At each step $h \geq 1$:

1. Find, among the nodes not yet solved, the next node B_h to label as solved, i.e.,

$$\mathrm{B}_h = \arg \min_{i \notin \mathcal{S}_{h-1}} C_h(i) \tag{10.9}$$

   and add it to the set of solved nodes, i.e., $\mathcal{S}_h = \mathcal{S}_{h-1} \cup \{\mathrm{B}_h\}$. If $\mathcal{S}_h = \mathcal{N}$, terminate the algorithm. Else, continue.

2. Update the optimal paths and their costs for the other nodes outside $\mathcal{S}_h$, as follows. For all $i \notin \mathcal{S}_h$:

$$C_{h+1}(i) = \min\left(C_h(i), \ell(i, \mathrm{B}_h) + C_h(\mathrm{B}_h)\right) \tag{10.10}$$

   and, if $C_{h+1}(i) < C_h(i)$, the path is updated to

$$\mathcal{P}_{h+1}(i) = \left(\mathrm{B}_h, \mathcal{P}_h(\mathrm{B}_h)\right) \tag{10.11}$$

By construction, the Dijkstra algorithm adds a new node to the set $\mathcal{S}_h$ with respect to set $\mathcal{S}_{h-1}$ at each iteration. Thus, it terminates after exactly $|\mathcal{N}| - 1$ iterations. Also, the procedures for cost and path updates are reminiscent of those used for the Bellman–Ford algorithm. We only need to prove that, for each node, the algorithm finds the shortest path to reach A. To this end, we introduce the following preliminary lemma.

Theorem 10.5 *If $i < j \leq |\mathcal{N}| - 1$, we have $C_h(B_i) \leq C_h(B_j)$ for any $h \geq i$.*

Proof The statement is equivalent to saying that at any iteration, the optimal path cost found by the algorithm for any solved node is lower than that found for all yet unsolved nodes (when these nodes are also solved, their costs are no longer updated, so the statement still holds). Since $i < j$, the thesis is surely true at iteration i; since node B_i becomes solved, whereas B_j still remains unsolved, $C_i(\mathrm{B}_i)$ cannot be greater than $C_i(\mathrm{B}_j)$.

For any further iteration $h > i$, the cost of B_i does not increase, that is, $C_h(\mathrm{B}_i) = C_i(\mathrm{B}_i)$. We can prove that, instead, $i < j$ implies $C_h(\mathrm{B}_j) \geq C_i(\mathrm{B}_i)$ by an inductive process starting from i. At the ith step the thesis was shown above to be true. Now, we assume it to be true for iteration $h \geq i$ and deduce it for iteration $h + 1$, which can be done by contradiction. Namely, assume that for all $j > i$ $C_h(\mathrm{B}_j) \geq C_i(\mathrm{B}_i)$ (this is the inductive hypothesis), but at the same time there exists $k > i$ such that $C_{h+1}(\mathrm{B}_k) < C_i(\mathrm{B}_i)$. Now, check the update rule (10.10) for node B_k. As postulated, the left-hand term is less than $C_i(\mathrm{B}_i)$; however, it is also equal to the minimum between two terms which are both larger than $C_i(\mathrm{B}_i)$, due to the inductive hypothesis. This is therefore a contradiction, which concludes the proof. □

We can now use this result to prove the general theorem.

Theorem 10.6 *The Dijkstra algorithm finds the best path from each node $i \in \mathcal{N}$ to node A.*

Proof We start proving that the Dijkstra algorithm finds at each iteration h the best path from each node $i \in \mathcal{N}$ to node A traversing only intermediate nodes belonging to set $\mathcal{S}_h$. There are two cases according to the "solved" status of i at iteration h.

Case (a): $i \in \mathcal{S}_h$, i.e., it is solved. This also means that i corresponds to some B_k with $k \leq h$. The statement to prove means that the algorithm gives the best path from B_k to A internally to set $\mathcal{S}_h$. Similarly to the theorem proved for the Bellman–Ford algorithm, we can proceed by using the optimality principle. The only difference with that theorem is that now we have to prove the additional statement that once a node is labeled as solved, there is no need to further update its path and cost; in other words, the optimal path from B_k to A does not traverse node B_j with $j > k$. This can be proven by contradiction as a consequence of the previous lemma. In fact, the algorithm can include node B_j in the path $\mathcal{P}_h(\mathrm{B}_k)$ only if $h \geq j$. But in this case, we fall under the previous lemma, i.e., $C_h(\mathrm{B}_k) \leq C_h(\mathrm{B}_j)$, whereas the update can take place only if $C_h(\mathrm{B}_k) > C_h(\mathrm{B}_j)$, which is a contradiction.

Case (b): $i \notin \mathcal{S}_h$, i.e., it is not solved. We can prove the statement by induction. At iteration $h = 0$ the thesis is verified, as the paths which only visit the nodes contained in $\mathcal{S}_0 = \{A\}$ are limited to the single-hop connections to A, which are those set by the initialization of the algorithm. Assuming that the statement holds for all iterations up to h, we prove it also holds for $h + 1$ as follows. The optimal path from i to A through $\mathcal{S}_{h+1}$ must have a first node C. If node C belongs to $\mathcal{S}_h$, then the optimal path has been already found at iteration h by the inductive hypothesis. Otherwise, which is the case when the optimal path changes between iteration h and $h + 1$, it must hold that $\mathrm{C} = \mathrm{B}_{h+1}$; thus, the optimal path from i to A through $\mathcal{S}_{h+1}$ can be split into two parts, i.e., the single hop from i to B_{h+1} and the path from B_{h+1} to A, which is optimal due to the optimality principle, and only traverses nodes belonging to $\mathcal{S}_h$, due to the inductive hypothesis. Indeed, this is what the Dijkstra algorithm does, i.e., at iteration $h + 1$, either it keeps the temporary path unchanged, or it updates them as passing through B_{h+1} and then proceeding optimally.

Now, we can wrap up these two cases and see that they imply the general thesis. In fact, after $|\mathcal{N}| - 1$ iterations all nodes of the network belong to $\mathcal{S}_{|\mathcal{N}|-1}$, i.e., $\mathcal{S}_{|\mathcal{N}|-1} = \mathcal{N}$. Thus, the algorithm eventually finds the shortest paths to A from all nodes. □

The computational complexity of the Dijkstra algorithm can be evaluated in a worst-case approach as follows. The iterations are exactly $|\mathcal{N}| - 1$. At each of them, the computation performed corresponds to a minimum which is evaluated among $O(|\mathcal{N}|)$ nodes (more precisely, the minimum at iteration h is evaluated among $|\mathcal{N}| - h$ nodes). Thus, the resulting worst-case complexity is $O(|\mathcal{N}|^2)$. At first glance, this complexity result seems to put Dijkstra algorithm in a more favorable light than Bellman–Ford, whose worst-case complexity is $O(|\mathcal{N}|^3)$. However, as discussed in the previous paragraph, the practical complexity of Bellman–Ford is often lower than this, and in many cases the two algorithms perform comparably.

Algorithms for minimum weight spanning trees There are also algorithms to build minimum weight spanning trees. Their approaches have some similarity with the algorithms to build shortest path spanning trees, even though the problem is different. In fact, a simple basic idea which is exploited by these algorithms is analogous to the optimality principle, and can be formulated as follows: a subtree of a minimum weight spanning tree is also a minimum weight spanning tree. More formally, we can state:

Theorem 10.7 *Assume that $\mathcal{G} = (\mathcal{N}, \mathcal{E})$ is a graph with some assigned weights to the edges, and $\mathcal{H} = (\mathcal{Q}, \mathcal{F})$ is a subgraph of $\mathcal{G}$ which is a minimum weight spanning tree over the set of nodes $\mathcal{Q}$. If $\mathcal{H}_1 = (\mathcal{Q}_1, \mathcal{F}_1)$ is a subtree of $\mathcal{H}$, this is a minimum weight spanning tree over the set of nodes $\mathcal{Q}_1$.*

Proof The fact that $\mathcal{H}_1$ is a subtree of $\mathcal{H}$ means that the edges in $\mathcal{F}_1$ are all contained in $\mathcal{F}$. In other words, the nodes in set $\mathcal{Q}_1$ still make a connected graph without cycles if linked by the edges in $\mathcal{F}_1$; the thesis claims this is also the one with lowest possible weight among the connected subgraphs of $\mathcal{G} = (\mathcal{N}, \mathcal{E})$ without cycles. Thus, we need to show that there is no better way (i.e., with lower overall weight) to connect the nodes in $\mathcal{Q}_1$ than using edges taken from $\mathcal{F}$ (which can be done in a unique way, described by set $\mathcal{F}_1$). This statement can be simply proven by contradiction, using the result of Theorem 10.1. Assume that one such better solution exists, and call it $\mathcal{H}'_1 = (\mathcal{Q}_1, \mathcal{F}'_1)$. The set of nodes in the graph $\mathcal{H}'_1$ is the same as in graph $\mathcal{H}_1$, but the set of edges is different. In particular, $\mathcal{F}'_1$ is not a subset of $\mathcal{F}$, and the overall weight of the edges in $\mathcal{F}'_1$ is lower than that of the edges in $\mathcal{F}$. If we define $\mathcal{F}_2 = \mathcal{F} \setminus \mathcal{F}_1$, the contradiction follows easily from Theorem 10.1, which states that $\mathcal{H}' = (\mathcal{Q}, \mathcal{F}'_1 \cup \mathcal{F}_2)$ is a spanning tree over the set of nodes $\mathcal{Q}$ and from the observation that its overall weight is strictly lower than that of $\mathcal{H}$, contradicting the hypothesis that $\mathcal{H}$ is a minimum weight spanning tree over $\mathcal{Q}$. □

This principle can be seen at work in many algorithms, one of which is called *Prim algorithm*. This algorithm is named after Robert C. Prim [8], even though it was first discovered by Czech scientist Vojtěch Jarník in 1930 [9]. However, Jarník published this result in Czech, so it was not well known to the English-speaking research community. This algorithm was also independently rediscovered by E. Dijkstra, and indeed there are similarities with this algorithm and the already discussed Dijkstra algorithm. For these reasons, the Prim algorithm is sometimes also called Prim–Jarník, or Prim–Dijkstra, or with any combination of these three names. Similarly to the Dijkstra algorithm, the Prim algorithm requires that the weights expressed by ℓ are non-negative.

To describe how the algorithm works, we adopt the same notation already employed before: given a graph $\mathcal{G} = (\mathcal{N}, \mathcal{E})$ with some associated weight ℓ, we want to find a minimum weight spanning tree $\mathcal{H} = (\mathcal{N}, \mathcal{F})$. The basic idea of the Prim algorithm is to exploit Theorem 10.7 in a fashion similar to the Dijkstra algorithm, i.e., creating a temporary tree containing nodes already *solved*, whose number is increased by one at each step with the addition of exactly one edge. For this reason, the algorithm is bound to terminate in $|\mathcal{N}| - 1$ steps.

We need a node to start with, akin to the Dijkstra algorithm, where such a node was called A. Actually, in the Dijkstra algorithm such a node had the physical meaning of the destination point of all unicast communications, or alternatively it could be seen as the source of a unicast communication towards all other nodes. Now, node A can still be regarded as the source of a broadcast communication. However, since the broadcast communication means that all the nodes are interconnected, the choice of the first node A is actually arbitrary.

The setup of the algorithm is analogous to the Dijkstra algorithm: this time, the set of solved nodes is a tree, which at step h is denoted as $\mathcal{H}_h = (\mathcal{Q}_h, \mathcal{F}_h)$, and is initialized to $\mathcal{H}_0 = (\{\mathrm{A}\}, \emptyset)$. Also, we define, for every node i, a cost variable $C_h(i)$ and a *parent node candidate* on the tree $U_h(i)$. These variables are initialized (for all $i \neq \mathrm{A}$, since for A they are irrelevant) as $U_1(i) = A$ and $C_1(i) = \ell(i, \mathrm{A})$. The iteration rule is as follows. At each step $h \geq 1$:

1. Find, among the nodes not yet solved, the next node B_h to label as solved, i.e.,

$$\mathrm{B}_h = \arg \min_{i \notin \mathcal{Q}_{h-1}} C_h(i) \tag{10.12}$$

   and add it to the subtree of solved nodes, i.e., $\mathcal{Q}_h = \mathcal{Q}_{h-1} \cup \{\mathrm{B}_h\}$ and $\mathcal{F}_h = \mathcal{F}_{h-1} \cup (U_{h-1}(\mathrm{B}_h), \mathrm{B}_h)$. If $\mathcal{Q}_h = \mathcal{N}$, terminate the algorithm. Else, continue.

2. Update the optimal paths and their costs for the other nodes outside $\mathcal{Q}_h$, as follows. For all $i \notin \mathcal{Q}_h$:

$$C_{h+1}(i) = \min\left(C_h(i), \ell(i, \mathrm{B}_h)\right) \tag{10.13}$$

   and, if $C_{h+1}(i) < C_h(i)$,

$$U_{h+1}(i) = \mathrm{B}_h \tag{10.14}$$

The evolution is very reminiscent of Dijkstra algorithm, the only differences being that we do not need to keep track of the whole path, but only of the other end of the link, i.e., the parent node on the tree (to this end, we use the variable $U_h(i)$), and a slightly different cost update rule; in fact, compare (10.13) with (10.10), and see that the former does not include a cost term for the path, but rather only the link cost is taken into account, due to the different objective that the Prim algorithm has.

The complexity of the Prim algorithm depends on the implementation. A straightforward version which proceeds analogously to the Dijkstra algorithm has a complexity of $O(|\mathcal{N}|^2)$. However, this value can be decreased with a proper data sorting and some artifices, so that finding the node to add at any iteration does not require to scan the whole set of unsolved nodes.

The explanation of the Prim algorithm is as follows. The initial tree $\mathcal{H}_0$ is clearly a minimum weight spanning tree for the set of nodes consisting of only A. Similarly, any other node B_h can be seen as covered by a minimum weight spanning tree whose set of nodes is the singleton $\{\mathrm{B}_h\}$ and whose set of edges is the empty set, so the tree cost is 0. Each time a new node is added to the minimum weight spanning tree covering set $\mathcal{Q}_{h-1}$, Theorem 10.7 can be invoked to verify that the result is also a minimum weight spanning tree covering the set of nodes $\mathcal{Q}_{h-1} \cup \{\mathrm{B}_h\}$, since it joins two minimum weight spanning trees with a minimum weight link.

Another algorithm based on the same principle is the *Kruskal algorithm*, proposed by Joseph Kruskal in 1956 [10]. This algorithm works in a similar fashion to the Prim algorithm, but instead of considering a single tree, it uses a "forest" $\mathcal{Y}$, i.e., a set of trees. The algorithm starts with $\mathcal{Y}$ containing $|\mathcal{N}|$ minimum weight spanning trees, one for each node of the network, and these trees only cover one node. In other words, $\mathcal{Y} = \{(\{i\}, \emptyset) : i \in \mathcal{N}\}$. The algorithm works iteratively with the objective of joining two trees to form a single new one, so that it can terminate when the forest contains a single tree. To describe the merging rule, take the auxiliary set $\mathcal{Z}$ which contains edges, and which is initialized as equal to $\mathcal{E}$.

Thus, the algorithm iterations proceed as follows. At each step h:

1. Find, in set $\mathcal{Z}$, the minimum weight edge $e_{\min}$, i.e.,

$$e_{\min} = \arg\min_{e \in \mathcal{Z}} \ell(e) \,. \tag{10.15}$$

2. Let $\mathcal{Z} = \mathcal{Z} \setminus \{e_{\min}\}$, i.e., remove $e_{\min}$ from $\mathcal{Z}$.

3. There are two possible cases: i) the two vertices i and j such that $e_{\min} = (i, j)$ belong to the same tree in $\mathcal{Y}$; then, return to step 1. ii) they belong to two different trees, denoted with $\mathcal{T}_1 = (\mathcal{Q}_1, \mathcal{F}_1)$ and $\mathcal{T}_2 = (\mathcal{Q}_2, \mathcal{F}_2)$; in other words, $e_{\min}$ connects one element of $\mathcal{Q}_1$ with one element of $\mathcal{Q}_2$; in this case, go to step 4.

4. The forest $\mathcal{Y}$ is equal to $\mathcal{Y}_0 \cup \{\mathcal{T}_1, \mathcal{T}_2\}$, where $\mathcal{Y}_0$ is another set of trees (not necessarily non-empty). Define $\mathcal{T}$ as the tree resulting by merging $\mathcal{T}_1$ and $\mathcal{T}_2$ with edge $e_{\min}$, i.e., $\mathcal{T} = (\mathcal{Q}_1 \cup \mathcal{Q}_2, \mathcal{F}_1 \cup \mathcal{F}_2 \cup \{e_{\min}\})$. Hence, update $\mathcal{Y}$ to become $\mathcal{Y}_0 \cup \mathcal{T}$. If now $\mathcal{Y}$ contains a single tree, the algorithm is terminated; otherwise, return to step 1.

A straightforward implementation of the Kruskal algorithm has the same complexity of the sorting algorithm applied to the set of the edges. In fact, there is no need to find $e_{\min}$ at each iteration, instead the set of the edges can be sorted preliminarily. As known from sorting algorithm theory, there are efficient procedures to sort $\mathcal{E}$ in $O(|\mathcal{E}| \log |\mathcal{E}|)$. Note that $|\mathcal{E}|$ is upperbounded by $|\mathcal{N}|^2$, though it may be significantly smaller, as already discussed for the Bellman–Ford algorithm. There are also some technical improvements, not discussed here for the sake of brevity, which can decrease the complexity in certain special cases.

10.2.3 Technical aspects of routing implementation

We now leave the mathematical theory of routing algorithms to deal with more concrete implementations of routing. When abstract results are brought into practice, it often happens that some additional elements come into play. First of all, communication networks have different characteristics. Thus, the implementation details may vary and result in different practical routing strategies. More in general, it may happen that a theoretically good solution becomes inefficient or, depending on the scenario, even inapplicable when brought to the real world; at the same time, another approach, possibly one supposed to be worse, may be preferred instead for practical reasons. In this subsection we will review some examples of these aspects.

Circuit-switched vs. packet-switched routing In Chapter 1, we introduced two very important concepts: the layer subdivision of the communication process and the switching methodology (where we distinguish between circuit-switched and packet-switched networks).

The layered structure enables the creation of paths over a network (seen as a graph) without knowing the nature of the links, in particular one does not need to know if they use circuits or packets. However, practical reasons suggest to separate these two cases. For this reason, for example, the routing strategies used for telephony networks, which are circuit-switched, are different from those employed over the packet-switched Internet. This happens because the two kinds of network switching have different time-scales for what concerns the update of the routing decision. Thus, even though they have the same routing objective, i.e., to select paths with low (possibly minimal) cost, they try to achieve it differently.

Circuit-switching networks should keep the same path for the whole communication. It is actually possible that a circuit is changed and the traffic is re-routed, but this should be regarded as an exceptional and costly event, which is performed when there is no other choice (e.g., one line breaks, so the circuit no longer exists). Packet-switched networks could instead determine the best route virtually on a per-packet basis.

Actually, we should distinguish between packet-switching based on datagrams and on virtual circuits. As defined in Chapter 1, datagrams are inherently unreliable. They can arrive out of order, or not arrive at all. Thus, if the communication packets are datagrams, nothing forbids to send them over different paths. Datagram networks are more flexible, but also more inherently prone to instability. As discussed in Subsection 10.2.1, updating the routes at every small variation of the link costs can lead to undesirable oscillations.

Virtual circuit networks, instead, mimic circuit-switched networks, so they try to behave as said above in reference to circuit-switching. However, they can update the established virtual circuits when the link costs change significantly. They do not need to react immediately, though, as the datagram networks do. In this way, routes are more stable, but can become inefficient without being noticed, and in general there may be a waste of network resources. Thus, for virtual circuit networks, the performance depends in the end on the frequency of cost updates, which is a tunable parameter; it should be neither too rapid, to avoid instability, nor too slow, to prevent resources from being wasted.

Since the discussion above does not distinguish between real and virtual circuits, one may theoretically apply these considerations to circuit-switched networks as well. However, in practice circuit-switched do not use the same rationale as virtual-circuit-based packet-switched networks, because they have been since long based on *hierarchical routing*, which will be discussed in the following.

Indeed, the manager of circuit-switched networks (in particular, telephony networks) may want to slightly deviate from the general objective of route optimality, as intended in the previous parts. In these networks, it is also important to achieve fast connection establishment; as we will see, hierarchical routing offers this advantage, at the price of a partially reduced goodness of the route (solely regarded from the routing metric standpoint).

Nevertheless, as discussed in the introduction, the recent research approaches mostly involve packet-switched networks, since circuit-switched networks are converging towards solutions which were initially created for packet-switched networks. As a consequence, the whole discussion of this section, which mostly involve packet-switched networks, still may apply to some extent to circuit-switched networks as well.

Choice of the routing metric While in the routing algorithm the existence of link costs can be simply postulated, in a real-world environment one may wonder which metric to select.

Two main performance metrics can be generally used to evaluate the performance of a communication system: *throughput* and *delay*. The former quantifies the amount of traffic (in bits per second) which reaches the destination in a time unit, the latter is the time (in seconds) spent in this process.

These metrics are not specifically related to routing. Similar (and often called with the same name) quantities are present in other layers of the protocol stack. For example, we have already mentioned throughput and delay in the previous chapters at the data link layer, and we will see them again in the transport layer. However, some facts are worth mentioning here, since they are specific to the network layer, or in general they come into play when we move from a single-link transmission to a data communication over a whole network.

First of all, differently from the data link layer, in a network context several links are traversed, with possibly different impacts on the resulting performance. Both delay and throughput are therefore to be regarded as *end-to-end* metrics. For the former metric, which in a network scenario is also often called *latency*, considering an end-to-end perspective means that different packets may experience considerably different delays (unless a circuit-switching or a virtual circuit approach is employed). Thus, the multiple delay terms need to be averaged over all packets. Note that, for certain kinds of traffic, other statistical characteristics, such as the standard deviation of the delay, may be more important than the average delay.

For what concerns the throughput, things are even more complex. In general, communication links may not be perfectly reliable, so that some data are lost in the transmission process. From an end-to-end perspective, we define throughput more properly as the amount of traffic, which can be delivered from source to destination without being lost somewhere in between. In other words, the throughput is the difference between offered load and unserved load. This is reminiscent of something already seen at the data link layer: also there, not all offered load contributed to throughput, and some unserved load was present. However, layer 2 also includes error control mechanisms, therefore one can assume that reliability is guaranteed. This still means that not all offered load can be accommodated, but nothing of the unserved load is lost. This last point does not hold, in general, for the *end-to-end* throughput. Networks consisting of several different links are more difficult to control and during the routing process one cannot guarantee that some paths become highly unreliable because of some nodes becoming disconnected from the network (e.g., turned off), or some links being interrupted (e.g., a physical break-up of cables). Thus, we must admit that some data can be lost, and the network layer does not have any way to deal with it.[7]

In particular, reliability of the links plays an important role in wireless networks. In wireline networks, the existence of the edges is defined by the presence of a deployed cable, and the losses over that cable can usually be assumed to be very small. In wireless networks, the definition of edges is more vague (in theory, every pair of nodes can communicate if enough power is spent), and at the same time not all edges have the same reliability.

For these reasons, throughput and delay in an end-to-end perspective are different from the same terms at the data link layer, although they are clearly related. And, similarly to what happened at lower layers, they are in a trade-off relationship between each other. At the network layer, one can try to maximize the global throughput or minimize the overall average delay. These objectives are actually contrasting, as in Chapter 7 it was seen that high throughput is achieved at the price of long queueing delays. However, the idea here is to identify a suitable trade-off point so that, in the end, packets are sent through the network to their final destination in the fastest possible way. If desired, more packets can be sent, but this will also increase the delay they experience. In any case, it seems reasonable to assume that fast lines are preferred in the path choice, as for example in real life selection of driving routes and in any other example discussed in Subsection 10.2.1.

If the delay is taken as the cost of the edges, each node at either end can periodically estimate the delay over an edge by means of special "probing" packets. One packet of this kind is often called ECHO. As the name suggests, the receiver of an ECHO packet must immediately reply to it with an answer packet, so that the transmitter of the probing packet, upon reception of

[7]However, as we will see in Section 10.4, the transport layer can solve this problem instead.

the answer, can measure the round trip time (rtt); the delay of the line is estimated as the rtt divided by 2.

Such procedures are used in physics, for instance to measure the speed of the sound. If a sound-reflecting wall is placed D meter apart from a source of a loud and short noise, e.g., a firing gun, we are able to estimate the round trip time of the sound as the delay between the instants in which we pull the gun trigger and we hear the echo.[8] Thus, the speed of sound is $2D$ divided by the rtt. Theoretically speaking, as proposed by Galileo [11], this technique could be used to measure the speed of light as well. At his time, the resulting method was unfortunately impractical since the light is too fast; modern technology may give some help, as discussed in [12].

However, while the air is isotropic for sound and light, at least with good approximation, data links are not necessarily perfectly bidirectional. Thus, the delay estimate obtained in this way for data links may be coarse. Actually, there are also some technicalities which are worth mentioning. A delay estimate may or may not take the node load into account. If only the latency over the line is required, the delay is computed from the instant the ECHO leaves the transmitter. Alternatively, the transmitter can also take into account its queueing delay, by starting the computation of the delay when the probing packet is generated; in this way the link cost estimate somehow reflects the load, as node congestion causes higher values of the routing metric derived in this manner.

In general, there is no unique answer to the problem of choosing whether and how to evaluate the link delay. Especially, the rtt of the probing packet may be affected by the processing time at the receiver. Due to these reasons, a simpler evaluation is often used. In many practical evaluations, the link cost is simply put equal to a constant term (without loss of generality, we can think of it as being equal to 1). When such a metric is adopted, it is said that the routing objective becomes to minimize the *hop count* of the path, since actually the cost of a path is simply equal to the number of hops involved. This is in most cases a sensible choice, since the physical propagation delay over the line represented by the edge is generally small compared to the processing time at the node, which is instead the dominant term in the delay experienced by a packet; so, as a rule of thumb, the smaller the number of traversed nodes, the better the route (i.e., the lower its delay cost).

The choice of hop count as the routing metric gives rise to another interpretation of the "shortest" (in the sense of optimal) path. Now the meaning of "short" no longer relates to the physical distance, but instead treats all hops as if they were of the same identical length. This approach is therefore an approximation, but is found to work reasonably well in practical cases. The cost of a path is also immediate to evaluate, as it can be directly inferred by tracking the traversed hops. A slightly more refined variant can be used to improve the realism, by considering, instead of the same cost for all edges, a finite set of values so that some edges are considered more expensive.

Finally, it is worth observing that, due to the considerations made above, the "hop count" approach is not a good choice for wireless networks, where reliability is an issue. Thus, the cost of the edges cannot be taken as a constant term, and more sophisticated techniques are required. Approaches used in practice are still based on the rationale of the exchange of probing packets; for a detailed review, the interested reader can consult [13, 14].

[8] Here "echo" is meant as the sound bouncing back, not an ECHO packet. However, ECHO packet are indeed named as their use recalls this procedure.

Centralized vs. distributed routing When a routing algorithm is selected for a given network, another important decision to make is whether it is operating in a distributed or a centralized manner.

A centralized implementation of the routing algorithm means that the routes are assigned by some central entity. This may require an exchange of messages so that this entity collects information about the whole network topology and is able to decide which routes to assign. Then, the central entity also needs to inform the nodes of the decision made. This can be done explicitly, for example by letting this central entity guide the routing operation of all nodes, even though such a procedure implies a significant control overhead.

A distributed implementation has the advantage of avoiding such exchanges, but it also needs to be carefully planned so that the nodes can operate in a coordinated manner, which is not simple. In this sense, certain algorithms are better suitable for a distributed implementation. In particular, the Bellman–Ford algorithm can be used to realize a distributed asynchronous routing strategy.

It is simple to extend this algorithm, as is described in Subsection 10.2.2, to a distributed synchronous version, so that all nodes can compute their "distance" (i.e., the cost of the path) towards a reference node A; in reality, the algorithm can be easily extended, by simply changing the destination node, so as to obtain the distance to every other node in the network. Thus, assuming the availability of a unique time reference, which would define some instants for the route updates every T_u seconds, i.e., at times $T_u, 2T_u, \ldots, hT_u, \ldots$, every node can infer an estimated distance to node A at these instants. Thus, denote with $D_i^{\rightarrow A}[h]$ the distance to A estimated by node i after update h. Every node i exchanges its own evaluation among its neighbors, which belong to set $\mathcal{J}_i = \{j \in \mathcal{N} : (i,j) \in \mathcal{E}\}$ and also performs an estimate of the link costs $\ell(i,j)$ for all (i,j) where $j \in \mathcal{J}_i$.

At the next update $h+1$, the distance estimates are updated to

$$D_i^{\rightarrow A}[h+1] = \min_{j \in \mathcal{J}_i} \left(\ell(i,j) + D_j^{\rightarrow A}[h] \right) \tag{10.16}$$

$$D_A^{\rightarrow A}[h+1] = 0 \,. \tag{10.17}$$

This procedure converges to reach the real distances no matter how the initialization at update 0 is set. However, since a universal time clock is actually very difficult to achieve, an asynchronous implementation of the above procedure can be considered. This means that, instead of sending a distance estimate to all neighbors at regular interval, the nodes can do it freely. Especially, it is appropriate that they inform their neighbors when a significant change of the distances occurs. It can be proven that even this very simple implementation converges to the correct distances and enables all nodes to find their best path to node A. This makes the distributed implementation of the Bellman–Ford algorithm very popular, as will be discussed next.

Local representation of the paths Regardless of the centralized or distributed implementation, the ultimate agents of the routing operations are the node themselves. Since a packet can traverse several nodes, each node must, upon reception of a packet, verify whether it is the last node of the path, i.e., the intended destination, or not; in this case, the packet is further processed, otherwise it must be sent along the intended path. The optimality principle guarantees that if the final destination is a given node D, the intermediate node can simply forward the packet along the best route to D.

For this reason, even though the nodes may be actually able to evaluate the whole best path, they could simply take advantage of this principle and keep memory only of the "next node" on the shortest path. Thus, the nodes do not store the whole shortest path, but they just know which is the best direction for any destination. This representation of the paths from a local perspective is different from a global overview of the route. This is actually true in many cases of the everyday life, for example when driving a car. The driver may know the whole route to its destination or can simply follow indicator signs pointing to the final goal at every crossroad; both approaches allow to reach the destination.

We can therefore proceed as follows. In a local view of the network every node only knows its "neighbors." Using only this knowledge, the routing decision can be represented through a *routing table*. Assume that the neighbors of node i are included in the set $\mathcal{J}_i$. The routing table is an array of pairs (k, j), where the first element $k \in \mathcal{N}$ can be any arbitrary node of the network, and $j \in \mathcal{J}_i$ is the next node along the best path to reach k. This is analogous to having an indicator sign at every cross road: the routing table tells that the traffic to k should be sent to a next hop j, and then, once it gets there, another routing table should be consulted to decide what to do next.

The collection of potential destination nodes in the routing table should be exhaustive, at least in theory. However, it is customary to have a list of known destinations paired with their next node, and the last entry of the routing table is just a "default" case, which contains the next node where to forward all of the remaining traffic. This often works well in practice, as the nodes may have a limited number of neighbors, thus most of the routing decisions can be represented by the default case.

In certain networks where the storage space is limited, it may be desirable to avoid that the forwarded packets spend time in queueing buffers. When multiple packets arrive at node i, instead of using always the same next hop j, which could lead to congestion on link (i, j) it may be better to send them elsewhere. A possible scheme, which is called *deflection routing*, uses the routing table only for the first received packet, which is therefore routed over the real best path, and simultaneously received packets are forwarded over another outgoing link selected at random (excluding the link they came from).

The whole idea of routing tables is appropriate to implement paths which are represented through a spanning tree, as the "next hop" to reach a given destination is unique and can be determined by looking at the spanning tree. An alternative representation is, once a route between S and D has been discovered, to include the path which is desirable to follow in the packet header. In this manner, every intermediate node just checks the header and forwards the packet to the node which is the next one on the indicated path. This approach, known as *source routing*, makes the routing operation more similar to virtual circuit routing. In fact, when routing tables are used, there is no guarantee that two packets sent to the same destination will follow the same path, as the routing table may get updated in the meantime. Instead, source routing overrides the routes which the intermediate nodes may think of choosing.

Even though source routing avoids the need for table lookup, it requires that all the intermediate nodes must agree on blindly following the instructions. This also fails if some nodes cannot use the required route (e.g., because one of the links is no longer available). Moreover, the header may significantly grow if the path is very long. Finally, the derivation of the path to follow requires a network-wide awareness which may be difficult to achieve without the exchange of several control packets, which cause additional overhead.

In general, a further common problem for both spanning tree and source routing is that they are able to select good paths only if the available information correctly reflects the real status of the network. In this sense, we can think of distinguishing between adaptive (or dynamic) routing, where this kind of information is updated with a certain frequency, and non-adaptive (or static) routing, where paths are selected in advance and regardless of the real network operating condition. Theoretically, static routing is less efficient, because it make routing decisions once and after that does not keep searching "good" routes. However, a detailed description of the network conditions, including the network topology (i.e., which nodes are connected through an edge) and the costs of the links, may be difficult to update and can require the exchange of several control packets which increase the messaging overhead, potentially causing congestion. Thus, in certain cases even a conceptually simple static routing strategy may be useful.

10.2.4 Routing strategies

In this subsection we will review some traditional routing strategies, where the algorithmic foundations described before are brought to a new level, i.e., including practical aspects and therefore deriving applied frameworks. Note that some of the strategies we introduce in the following can be used as the basis to derive more complicated strategies. That is, simple routing strategies $\mathtt{r}_1, \mathtt{r}_2, \ldots, \mathtt{r}_\mathtt{N}$, can be combined to form other strategies (which are, in reality, meta-strategies) where several cases are distinguished and strategy $\mathtt{r}_1$ is used in the first case, $\mathtt{r}_2$ is used in the second, and so on.

Flooding. The flooding strategy mimics a water inundation, since it blindly disseminates packets throughout the whole network. When a packet is to be sent to a destination, the source sends it to *all* of its neighbors. Each node receiving the packet does the same, i.e., it forwards it to every other host it is connected to, except the ones it received the packet from. In this way, the message will be eventually delivered to all reachable network parts.

Since flooding does not need to know the costs of the links and does not care how frequently the network conditions are measured, it can be classified as a static routing strategy. Flooding offers two immediate advantages, namely, its simplicity, which enables its implementation even in network nodes with limited routing capabilities, and its reliability, as if a path to the destination exists, the algorithm is bound to find it.

However, this high reliability of guaranteeing packet delivery also comes with delivering the very same packet to every other node, and in multiple copies, which is inefficient. Even worse, there is no stop to the procedure which keeps transmitting packets indefinitely.

For these reasons, some variations to the basic flooding scheme must be implemented in order to obtain a more viable strategy. For example, the packet can include a list, where every forwarder puts its identifier. In this way, a node can detect whether it already forwarded the packet in the past; eventually, this will end the otherwise unlimited flooding process. Another possibility is to include a counter (sometimes called Time-To-Live, or TTL) in the flooded packet. Such a value is initially set to the expected maximum path length to the final destination and is decremented by one at every hop. When the TTL counter of a packet reaches 0, that packet is no longer forwarded. This solution is simpler to implement than the list of all previously visited nodes, but requires a rough estimate of the expected path length. In the case this estimate is not available, a conservative (worst-case) dimensioning can be used. Note that the introduction of a TTL counter is more in general useful for any kind of routing to avoid obsolete packets going indefinitely around the network. This idea is also exploited by IP networks and will be discussed in Section 10.3.

Even with these modifications, flooding is often impractical, but can have application as a part of a more complex technique which, under given conditions, can perform a flooding-like dissemination of packets. For this reason, it is important to understand the rationale behind the flooding procedure.

In particular, simple protocols exist, which are based on flooding but improve it to the point it becomes possible to use it. These protocols may be desirable in specific situations, for example when the network consists of a very high number of nodes, which in turn have low complexity and reliability. These include scenarios with harsh environmental conditions, such as a military battlefield or remote sensing in wild natural surroundings, where the nodes can be destroyed or simply fail. At the same time, if the nodes are unable to perform complex routing tasks, flooding-based routing may become appealing.

Examples of such variants of flooding include the so-called *gossiping*, where the message is not forwarded to everybody, but only to a smaller subset of the neighbor set. This can be, for example, based on a given forwarding probability less than 1, in which case the term *probabilistic flooding* is adopted. Another similar variant is called *selective flooding* and works in case the nodes are able to roughly estimate the direction in which the information is coming. In this situation, it is useless to propagate the message backwards, so the node chooses to forward the message only to those neighbors which are supposed to provide a positive advancement to the final destination.

Distance Vector Distance Vector (DV) is a routing strategy, in which each router maintains a routing table with the direction and the distance to any node in a network. Every router needs to update its table by exchanging messages with neighboring nodes. This procedure serves to realize a distributed version of the Bellman–Ford algorithm, as previously described.

The basic requirement for DV routing is that each router knows its distance to each of its neighbors. In the simple case hop count is used as the routing metric, all neighbors have distance one, and no further measurement is required. The DV strategy then proceeds by spreading information about distances from 1-hop neighbors to 2-hop neighbors, and further to k-hop neighbors. This is achieved by having each node periodically propagate its routing table.

For example, denote the cost of the link between neighbor routers i and j as $\ell(i, j)$. If router j (but not i) is also neighbor of node k, j knows also $\ell(j, k)$. When j sends its routing table to its neighbors, i receives the information about $\ell(j, k)$ and can estimate its own distance to k as $\ell(i, j) + \ell(j, k)$. Thus, i evaluates this sum and puts it into its routing table as the distance to k, i.e., $\ell(i, k)$. A similar reasoning, with i and k reverted, can be done by node k. It is also clear that this information can easily propagate, since in the first transmission, two-hop neighbors are informed of their mutual distance. At the next transmission of the routing tables, three-hop neighbors become also aware of their distance, and so on.

However, this information spreading, based on propagating what received by neighboring nodes, reminds the spreading of a gossip rumor, or the children "Telephone game." The routers keep spreading their routing tables, which they evaluated based on the routing tables of their neighbors, without asking or checking whether the information received is correct. Thus, in case some link weight metrics change, DV routing can encounter serious problems.

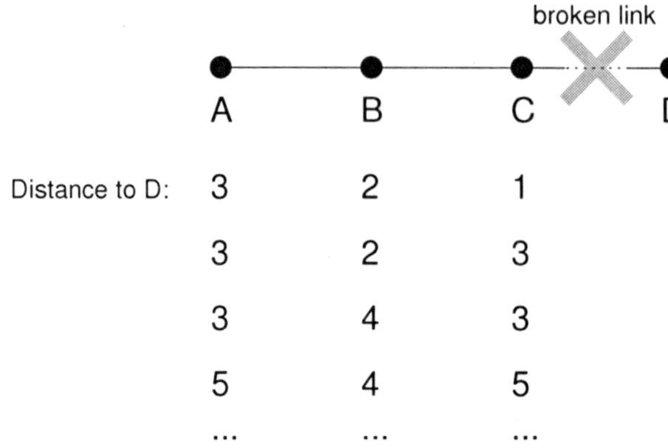

Figure 10.7 The count-to-infinity problem.

In particular, a problem known as *count-to-infinity* may arise. This problem is well-known to occur in naïve implementations of DV routing, and a typical example showing it can be done in a simple linear network topology A–B–C–D, using hop count as the routing metric. This last choice is not restrictive, but simplifies the reasoning. Assume the routing table propagation has been stabilized, so that every node in the line has a correct record of its distance to any other node. In particular, the distance to D is equal to 3 for A, 2 for B, 1 for C. Now, assume that node D becomes unreachable, as it is impossible to communicate through the former link between C and D. This situation is represented in Fig. 10.7.

Node C sees that D no longer sends routing table updates, but at the same time keeps receiving the routing table of B, where $\ell(\mathrm{B}, \mathrm{D})$ is listed as equal to 2. Little does C know that the route identified by B is the one passing through C itself, which has just become infeasible as C and D no longer communicate. For what C knows, it just assumes that B is on another path to reach D in two hops. Thus, C blindly updates $\ell(\mathrm{C}, \mathrm{D})$ to 3 hops, as the sum of $\ell(\mathrm{B}, \mathrm{D})$, which has been confirmed to be 2, and $\ell(\mathrm{B}, \mathrm{C})$ which is still equal to 1 as B and C are neighbors. Now, the distances to D are 3, 2, and 3, for A, B, and C, respectively. When B receives a routing update from C, it sees that $\ell(\mathrm{C}, \mathrm{D})$ has been set to 3. Again, B is unaware that this change is due to the previous routing table exchange, where B itself told C that its distance towards D is 2. In this case, B simply thinks that C knows a route which reaches D in three hops, so B updates $\ell(\mathrm{B}, \mathrm{D})$ to 4, which, as before, is the value sent by C as $\ell(\mathrm{C}, \mathrm{D})$ plus one, as C is a neighbor. At the next exchange, C will update $\ell(\mathrm{C}, \mathrm{D})$ to 5. At the same time, only at this time A will also change $\ell(\mathrm{A}, \mathrm{D})$; in fact, now B claims its distance to D to equal 4, while during all previous exchanges, B always claimed $\ell(\mathrm{B}, \mathrm{D})$ to be 2. As a result, we can see that the DV approach already requires some exchanges before the news of change reaches node A, and even more if A needs to propagate the news to somebody else. Even worse, the news is still incorrect, as in reality all distances to D should be set to infinity, as D is no longer reachable. Thus, it takes a very long time (virtually infinite) to reach the correct distance evaluation.

Link State As will be discussed in Subsection 10.2.5, a routing protocol based on the Distance Vector strategy was implemented in the ARPANET, the progenitor of the Internet. However, because of the count-to-infinity problem it was replaced in favor of another strategy called *Link State*, which is the basis for many routing protocols. The Link State strategy is based on the Dijkstra algorithm, whereas the Distance Vector was based on a distributed implementation of the Bellman–Ford algorithm. The advantage of using the Link State strategy does not arise here, however. Recall that even though the worst-case computational complexity is different, the two algorithms perform actually similarly in practice.

When the Link State strategy is executed, each node tries to learn periodically the network conditions. To this end, first of all it needs to discover its neighbors. Then, it measures the cost to reach its neighbors, which may be a delay estimate if the link weights represent latencies, or is simply a constant if the hop count metric is used. The name of the strategy indeed refers to the fact that a sufficiently recent evaluation of the surrounding links and their cost is required.

The information about the cost of the surrounding links is then exchanged by all nodes. In this way, the nodes can acquire knowledge about the whole network, so as to infer a network-wide map which is required to run the Dijkstra algorithm and discover the shortest path to reach all possible destinations.

In more detail, the Link State strategy evolves as follows. First, the information about the existence of neighbors, and the cost to reach them, is collected by all nodes. This can be done through a proper exchange of ECHO packets. As discussed in Subsection 10.2.3 this may or may not include a delay evaluation, according to whether or not such a term is included in the routing metric. Strictly speaking, there is no fixed rule as to when to perform the exchange of ECHO packets. They can be exchanged periodically, or can be triggered by some specific events, such as connection of new nodes or failure of some lines. The correctness of the choice depends on how likely these events are, and how variable the link costs are.

Once this information collection phase is completed, a second phase starts, where a global network exchange is performed so that the evaluations related to neighbor nodes, known only locally until this point, are disseminated to the whole network. This is performed through proper packets called *Link State* packets, where each node puts the previously collected information about the cost estimate to reach everyone of its neighbors.

Link State packets are disseminated via a basic structure which is nothing but the flooding strategy previously discussed. Indeed, every received packet is forwarded on all lines except the one it came from. However, this plain strategy would be rather inefficient, so some improvements are often implemented as follows.

First, sequence numbers are used to keep the flood in check. Flooding guarantees that the packet reaches the whole network, but it easily happens that duplicates arrive, even of older packets; sequence numbers guarantee that only the newer information is used. However, it can still happen that obsolete packets survive in the network, and can be erroneously mistaken for new ones (either because of a corrupt sequence number, or a wrap-up in the sequence number field). To avoid these problems, an additional *Age* field is also introduced. As discussed in the flooding strategy, this enables a time-out so that packets do not go through the network indefinitely. Finally, an additional improvement can be obtained if the nodes, instead of forwarding Link State packets immediately, as per the basic flooding strategy, buffer them for a while, so that they can send only the most recent of those received during the buffering interval.

Once the Link State information has reached all the nodes of the network, every node can compute its optimal path to any destination through the Dijkstra algorithm. Since this flood of Link State packets causes all nodes to acquire a global view of the network topology, the count-to-infinity problem is solved. However, other problems may still be present when the global network topological view acquired in such a manner is not coherent. Contradictory topological descriptions may arise when some nodes fail to recognize a link with their neighbors, or, conversely, they claim to have a link which no longer exists. When these problems arise, it may take long for the Dijkstra algorithm to converge to a solution. Finally, another drawback of the Link State strategy is that it may involve the exchange of cumbersome amounts of information, especially if the network is very large. Though improved, the basic dissemination mechanism is flooding, which severely increases the overhead.

Hierarchical routing Hierarchical routing is a composite strategy which can be useful when the network size increases. As mentioned in the previous subsection, it is also often utilized in circuit-switched networks.

When the network size is large, it may actually be convenient to regard it as the union of multiple *subnetworks*.[9] Such subnetworks can be connected through different devices, as explained in section 9.1. If a subnetwork has the size of a LAN, simple interconnection at the MAC sublayer can be done through bridges, or more sophisticated routing capabilities can be added if routers are employed. For wide area networks, it is recommended to utilize gateways, which are able to interconnect also higher layers.

Conceptually, a large network resulting from the interconnection of multiple subnetworks can still be seen as a graph where the interfaces (routers, bridges, gateways) are additional nodes. Performing one of the already mentioned routing strategy on this graph, results in a *nonhierarchical* routing.

Instead, a hierarchical view of the network can be introduced if the following rationale is applied. Two levels of routing are considered. In the higher level, each subnetwork is seen as a single node, and only the connections among different subnetworks are regarded as edges. This means that any pair of nodes/subnetworks is connected through an edge if and only if at least one edge exists, connecting two different nodes each belonging to either subnetwork. A similar approach has been previously used during the proof of Theorem 10.1. The resulting graph is called the *backbone* of the hierarchy.

At the lower level, subnetworks are considered as separate entities. The hierarchical routing strategy corresponds to performing a given routing strategy on the subnetworks to handle local traffic only, i.e., where both source and destination belong to the same subnetwork. Indeed, different subnetworks can adopt different routing strategies, according to their characteristics, e.g., size or connectivity. Traffic directed outside a subnetwork is instead redirected to one of the interfaces connecting the other subnetworks. When this is reached, a routing strategy (again, not necessarily the same used in the subnetwork) is used at the higher level, to determine the path to follow over the backbone, i.e., through the subnetworks, so as to reach the final destination. Note that, even though conceptually the forwarding of a packet through a backbone node is just one operation, in reality, this may require a series of local forwarding on the same subnetwork.

Hierarchical routing processes are common in many applications. For example, as already mentioned they are typical of telephone networks. Another case where hierarchical routing is adopted is the postal system. To deliver a letter, the post service has separate procedures for local (urban) and extra-urban destinations. In the former case, we can think of the letter as being sent through the city along a shortest path. Otherwise, the postal code is considered, in

[9]This term ought not be confused with "subnets" which are similar, though more specific, subdivision of a network which will be discussed in Section 10.3.

order to identify the destination city. Thus, first the letter is sent towards the destination city, and only there the street address is used to identify the final destination. Such examples also hint that the addressing system should reflect the hierarchy, as done by telephone numbers or postal addresses, where the local identifier of the destination is preceded by a prefix identifying the destination on the backbone network. This observation will become relevant to understand certain features of the IP addressing system in Subsection 10.3.2.

In these examples, the construction of the backbone is quite natural, as it depends on real aspects of telephone lines or cities. However, subnetworks may even be constructed intentionally in networks without any pre-determined structure. In particular, it may be recognized that, in a given graph, aggregates of nodes sharing a common characteristic (usually, this regards physical proximity) can be identified. These aggregates are called *clusters* and they may be found by automated procedures known as *clustering algorithms*. A survey on these techniques can be found in [15]. Again, we use the term "algorithm" as this is a quite theoretical topic which founds only one of its many application in communication networks. After having applied such algorithms, the subnetworks and the network backbone are identified by the clusters and the interconnections among them, respectively.

The advantage of using hierarchical routing is as follows. When a routing device needs to forward a packet according to a routing table, it does not need to have a separate entry for any node of the network. Actually, for local traffic this is still true since the local routing strategy proceeds as in the nonhierarchical case by finding the path to follow to reach any other destination within the subnetwork. However, when the traffic is sent outside the local subnetwork, it is sufficient to decide where to forward it according to the identifier of the destination subnetwork. Thus, the routing table considers a single entry for any possible destination subnetwork, which implies that the size of the routing table is considerably reduced.

At the same time, the forwarding procedure can be simplified if the addressing system takes into account the routing hierarchy. For example, one sensible possibility is that the initial part of the address is a prefix identifying the subnetwork, and the rest describes where actually the destination is located in the subnetwork. In this case, instead of looking at the whole address, the routers not belonging to the destination subnetwork can simply consider the initial part of the address. Then, they forward the message towards the final destination; in fact, to perform this task, they do not need to have knowledge of the precise position of the destination point.

However, these advantages come at a price, which is precisely that the path found by a hierarchical routing strategy is generally not the shortest (i.e., the minimum weight one). This is for example visible in Fig. 10.8; if the routing metric is simply the hop count, the shortest path from S to D is the solid black one, but since S and D belong to different subnetworks, the hierarchical routing strategy indicates to follow another path. More in general, the shortest path on the backbone (plus, possibly, the additional hops to reach the subnetwork boundaries) may be different from the optimal path, i.e., it may have a higher cost.

Finally, it is also possible to extend this rationale to networks with more than one level. There is no conceptual difference in nesting many subnetworks and also in performing a multiple-level hierarchy, and the advantages, but also the disadvantages, of such a procedure are repeated.

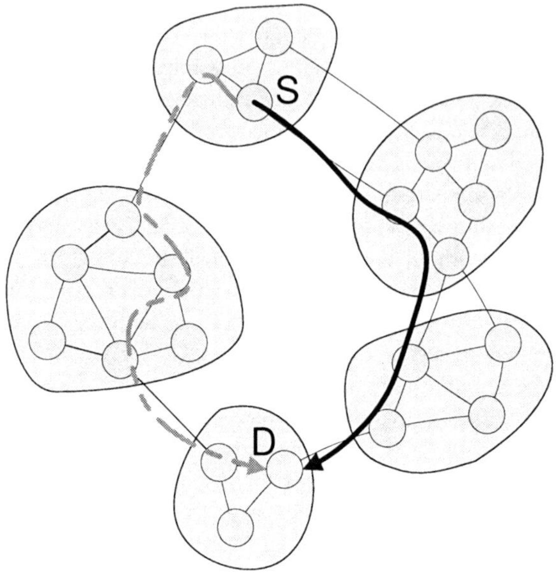

Figure 10.8 An example of hierarchical routing inefficiency. The path in black has overall lower cost, but the path in grey is selected instead, since the number of hops over the backbone is lower.

10.2.5 Routing protocols

This subsection mentions some examples of routing protocols which are (or were) used by real networks, to understand how the concepts mentioned before find practical application.

Routing protocols for the Internet Several routing protocols exist for wireline networks, in particular the Internet. In most cases, they can be seen as practical implementations of the routing strategies described previously. In particular, three protocols are worth mentioning: the Routing Information Protocol (RIP), the Open Shortest Path First (OPSF), and the Border Gateway Protocol (BGP).

Before discussing them, we anticipate some details of the Internet structure, a topic which will be discussed in more depth in Section 10.3. What is important to know for now is that the Internet can be seen as consisting of several network entities called *Autonomous Systems* (ASes). This determines a sort of implicit hierarchy, since the internal management of every AS is, as the name suggests, independent of the rest of the network.

Therefore, it is appropriate to distinguish between routing protocols used *within* a single autonomous system, and protocols for routing data between ASes. Since the interconnection among ASes often involve higher layers and therefore are realized by means of gateways, routing protocols of the former kind are called *interior gateway protocols*, whereas the latter type contains *exterior gateway protocols*.

Originally, the Internet used RIP as the interior gateway protocol. This protocol is based on distance vector routing, and employs hop count as the routing metric. Such a protocol was not designed having in mind large networks as those that can be found now as part of the Internet. For example, it used a maximum distance of 15 hops. Also, the distance vector routing strategy, on which RIP is based, shows significant weaknesses in large networks, as discussed above, in particular the count-to-infinity problem.

For these reasons, it was officially replaced by OSPF, which is based on link state routing. OSPF is also able to support different routing metrics. Finally, it is better suited for the inherent hierarchy of the Internet; as will be discussed in the following, the Internet addressing system moved towards supporting networks of variable size, and OSPF is able to handle this. Thus, this protocol is probably the most common one used as interior gateway protocol in the Internet. It is not the only one, though, since in the end ASes can be autonomously managed. For example, another alternative protocol, which is also based on link state routing, is the Intermediate System to Intermediate System (IS-IS) routing protocol. Differently from OSPF, IS-IS is not an official Internet standard, but is widely used in certain large service provider networks.

The recommended exterior gateway protocol, i.e., the one to use for routing data between ASes, is instead the Border Gateway Protocol (BGP). Actually, BGP can be used also as an internal gateway protocol, in which case it is dubbed as Interior Border Gateway Protocol (IBGP). BGP is often considered to be based on the Distance Vector strategy, though in reality it sometimes slightly deviate from standard DV routing. The point is that BGP does not use routing metrics as its only driving criterion. Instead, it makes routing decisions also based on more general network policies (or politics). For example, it may be undesirable to route through a certain neighboring AS, even though it is along the most efficient route. This may be due to legal reasons or other external specifications, such as not trusting the neighboring AS if it is located in a foreign country, or belongs to a competitor. To handle these cases, the BGP standard defines several routing criteria beyond the simple efficiency metrics, to find the most suitable path.

Since BGP is not purely based on Distance Vector, it avoids many problems plaguing DV routing. In particular, due to the policy requirements, BGP always checks the route in its integrity before forwarding packets, thereby avoiding the count-to-infinity problem, since a BGP router is always able to recognize that some line went down. However, this does not mean that BGP is free from instability problems, especially since it operates on a large structure, and in continuous growth, such as the Internet. Especially, one serious problem which is still present in the Internet is represented by the growth of routing tables, which causes the number of entries in BGP routers to be rapidly increasing. This problem is related to the overpopulation of the Internet, and analogous to other problems discussed in the next section is also connected to the addressing system issues.

Routing protocols for wireless networks In certain kinds of wireless networks, nodes are highly mobile. In particular, the main scenario of reference is that of Mobile Ad-hoc networks, a kind of non-infrastructured network where nodes have total freedom of movement.

In such a context, classic routing protocols encounter several problems; thus, different protocols are proposed to deal with these issues. In particular, some characteristic aspects of wireless routing are as follows. First of all, power consumption becomes an issue: since mobile nodes are battery-powered, the most important goal is to avoid using demanding links in terms of energy expenditure. Thus, the choice of hop count as the routing metric (i.e., the link weight) may not be entirely appropriate. In particular, the energy consumed to send a packet through a link is an increasing function of the distance (the link attenuation is roughly proportional to a power of the distance). As a practical consequence, using a long link is often worse than using two shorter links with an intermediate relay.

Moreover, node mobility implies a concrete risk of highly variable paths. This means that routing tables need frequent updates; in between, the routing information may be incorrect. There are some common ways to deal with this problem, i.e., different techniques to improve the routing table accuracy. They are usually subdivided into *proactive*, *reactive*, and *geographic* routing techniques.

The approach of proactive routing is simply to update the routing tables with sufficiently high frequency (because of this, they are also called table-driven protocols). As the name suggests, a procedure to discover a path to any possible destination is initiated in advance, before knowing whether this information will be required at all. Thus, a procedure called *route discovery* is involved, which consists of exchanging routing information among nodes. The advantage of a proactive approach is a prompt response when the routing information is needed. Conversely, it may cause high overhead, especially if route discoveries are very frequent. In particular, it is often the case that the route discovery procedure actually discovers many unnecessary routes, i.e., towards nodes which do not receive any packet (however, this cannot be predicted in advance).

The opposite philosophy is that of reactive routing, which is, "search a path only when needed." In other words, the route discovery procedure is never triggered in advance, but only when the path to the destination is unknown. In reality, reactive routing protocols do not discover a route for any packet, but exploit some artifices. First of all, when many packets are sent towards the same destination, there is obviously no need to initiate multiple route discovery procedures. Thus, what actually happens is that, once a path is found, it is kept in memory for a given time. When the expiration time is reached, the path is considered obsolete and a new one is sought through another route discovery procedure. Moreover, the nodes are hopefully smart enough to exploit the optimality principle seen in Subsection 10.2.1; thus, when node C participates in, or simply eavesdrops, a route discovery procedure between nodes S and D, and infers to be an intermediate node towards D, it can automatically learn how to reach D without initiating another route discovery.

Proactive and reactive routing use route discovery procedures with different timings. Geographic (also called position-based) routing, instead, does not use route discovery at all, but replaces it with an additional requirement, i.e., that nodes must be aware of their own geographic position, for example, thanks to a satellite positioning system. Additionally, when sending a packet a source node must also be aware of the position of the destination node. If this is possible, instead of keeping routes in memory, routing can be enabled by simply asking all nodes to send the packet to a neighbor which is closer (in the geographical sense) to the destination. There are several alternatives to determine which neighbor to select (the closest to the source, the closest to the destination, or the best node according to some other criteria), and also techniques to avoid dead-ends, i.e., where no neighbors closer to the destination exist. The price to pay to perform geographic routing is however the additional equipment required to determine the position. Satellite positioning devices are becoming cheaper, but their cost is still non-negligible; moreover, they do not work indoors. Other localization techniques are available, mostly based on cooperative exchange of received signal strength among neighbor nodes, but their precision is still coarse. Their further development is an open field of research.

It is worth mentioning that these three approaches are not necessarily mutually exclusive. Examples of *hybrid routing* protocols exist, where proactive, reactive, and/or geographic approaches are followed at the same time. Especially, since the conditions of mobile networks may be extremely dynamic, the rationale can be changed between reactive and proactive routing, according to how often the routes change and how often they are required.

We now mention some concrete examples of routing protocols for mobile wireless networks [16]. Examples of proactive protocols are the Distributed Source Distance Vector (DSDV) protocol and the Optimized Link State Routing (OLSR) protocol. They realize distributed implementations of the Distance Vector strategy, and the Link State strategy, respectively. Famous reactive protocols include instead the Ad-hoc On-demand Distance Vector (AODV) and the Dynamic Source Routing (DSR) protocols. From their names it is evident the former implements the DV routing strategy, whereas the latter does not use routing tables, but rather source routing. These protocols send routing discoveries only when required, and include rules to eavesdrop route discovery procedures performed by other nodes.

An example of hybrid routing with both a proactive and a reactive routing component is represented by the Zone Routing Protocol (ZRP). ZRP tries to combine proactive and reactive approaches by defining a "zone" around each node, which consists of its k closest neighbors, where k is a tunable parameter. The underlying concept is similar to the one described for hierarchical routing, with the only difference that within a zone a proactive routing approach is used, whereas inter-zone routing is reactive. Finally, there are several protocols based on geographic routing. Sometimes they are also using ideas taken from the proactive and reactive approaches. A review of geographic routing, and how it performs in wireless networks, can be found in [17].

10.3 THE INTERNET AND IP

Besides its technological aspects, the Internet is one of the most important mass phenomena of the last centuries. Started as ARPANET, an interconnecting structure among military and academic institutions in the United States, it has now become a pervasive aspect of everyday life for many citizens of developed and developing countries.

Structure of the Internet Even though it is commonly considered as such, it would be inappropriate to think of the Internet simply as a "network." The structure of the Internet is quite heterogeneous, since it comprises several systems with different properties, policies, and ownerships. Thus, it is better to consider it as a collection of multiple interconnected networks, without any pre-assigned structure, as basically every network can attach itself to the Internet as long as it complies with some basic requirements. The real structure of the Internet is therefore quite mutable and what could be its exact characterization from the modeling point of view is still subject of studies [18]. A point on which all models agree is that the structure of the Internet is nested, and several self-similar levels can be identified; for example, continental aggregations of networks are clearly present, which can be subdivided into regional intermediate networks, in turn composed of local networks. However, this (entirely arbitrary) hierarchy is not defined a priori, and other similar descriptions are possible.

Thus, the Internet can be generically represented as a network of subnetworks, called Autonomous Systems (ASes), which are connected to each other. The internal management of every AS is left to some local administrator, which is responsible for what happens within the

AS. However, all ASes connected to the Internet must adopt some common interfacing rules, which come into play at layer 3 (also involving, to some extent, the upper layers, in particular layer 4).

The management of the interconnection of the ASes is regulated by a network layer protocol, which is so fundamental for the Internet that it is simply called the Internet Protocol (IP). Actually, IP was born as a detachment to the network layer of the then-called TCP-IP, now referred to as Internet Protocol Suite, where TCP stands for Transmission Control Protocol. TCP is a protocol for a reliable transport layer, and will be discussed in the next section, together with its counterpart for unreliable transport, called User Datagram Protocol (UDP). However, IP is the part of this suite with the fundamental task of coordinating all the ASes with the same language, and in this respect it turned out to be an exceptional success. This happened in spite of some shortcomings which will be discussed in the following, and still represent an open challenge for the scientific community.

10.3.1 The Internet Protocol

The Internet adopts a connectionless, best-effort approach. It is therefore suitable for packet-switched communication. Connection-oriented communications are actually possible thanks to TCP, therefore they will be discussed in the next section. When discussing IP instead, we always assume that packets are unreliable and can be delivered out-of-order.

There are multiple versions of the Internet Protocol. In the rest of this chapter we will refer most of our attentions to the first important version of IP, which goes under the name of IP version 4 (IPv4 for short), and is still the most popular one and in widespread use. Other versions were expected to replace IPv4, in particular the more recent IPv6. However, the transition towards IPv6, seen by many as unavoidable, is still ongoing, and there are doubts about when it will end (if ever). For the moment, IPv4 and IPv6 co-exist, with IPv4 being predominant. Also, most of the properties we will discuss for IPv4 hold for IPv6 as well, just with some technical difference which does not substantially change the explanation. Therefore, in what follows, unless specified differently, "IP" implicitly stands for "IPv4."

Because of packet-switching and best-effort rationales, the basic units of communications through IP are packets called *datagrams*. The terms "packet" and "datagram" are synonyms over the Internet, but the latter better emphasizes the inherent unreliability of the communication. Indeed, IP does not care about in-order delivery, nor does it prevent packets between the same source-destination pair from following different routes.

However, IP is able to identify source and destination of a communication in a (theoretically) unique manner among the whole Internet, and allows them to recognize that the data they are exchanging has been fragmented into packets. To see how these properties are implemented, we need to look at the IP packet structure.

IP datagram structure A detailed map of the IP datagram is reported in Tab. 10.1.

The first field, *Version*, is always set to 4 for IPv4. The second 4-bit field is the *Internet Header Length* (IHL) which describes the length of the header, expressed in 32-bit words. It can also be seen as the point where the payload starts. The header size is variable due to the presence of optional fields (i.e., since the field *Option* has variable length). The minimum value is 5, the maximum is 15. The subsequent field, called *Differentiated Service*, evolved throughout the years from its original definitions. Its role ought to be to differentiate several

Table 10.1 The packet structure of IPv4.

	4 bits	4 bits	8 bits	3 bits	13 bits
0–31	Version	Header length	Differentiated service	Total length	
32–63	Identification			Flags	Fragment offset
64–95	Time to live		Protocol	Header checksum	
96–127	Source address				
128–159	Destination address				
160 +	Options (variable length)				
	Data				

kinds of traffic with different characteristics and priority, but it is supported only by a limited number of routers and commonly ignored by the others.

The 16-bit *Total Length* field defines the entire datagram (header plus data) size, expressed in bytes. The minimum is 20 and the maximum is 65535. Actually, in practice the maximum length is often set to 1500 bytes if the data link layer uses an Ethernet MAC. Longer blocks of data are fragmented, and the fragmentation is managed with some flags discussed in the following. The field *Identification* is used to identify fragments of the same block, which have the same value of this field. The *Flags* control the fragmentation as follows. The first bit is always zero, as it is still reserved (unused). The other two flags are called Don't Fragment (DF) and More Fragments (MF); the former tells that fragmentation cannot be used (i.e., fragmented data must be discarded) and the latter is set to 1 for all fragments but the last one. The *Fragment Offset* specifies the offset of a particular fragment relative to the beginning of the original datagram, expressed in 8-byte multiples.

The *Time-To-Live* (TTL) 8-bit field is used to limit the lifetime of datagrams, to avoid that obsolete packets still propagate throughout the network. Theoretically speaking, it counts time in seconds, but every fraction is rounded up. Since the typical latencies in communication networks are lower than one second, it just counts hops, decreasing by 1 at every transmission. The *Protocol* field clarifies the nature of the field *Data*. The number reported here is associated with the specific use for which the data are intended. Remember that, according to the encapsulation principle, the whole packet produced by a given layer (data+control information) are framed in the data field of the packet for lower index layer. The Protocol field is therefore interpreted by looking at a specific list of recognized protocols; such a list, maintained by the Internet assigned numbers authority (IANA), is contained as a lookup table in every IP router. According to this list, if the Data field is to be read, e.g., as TCP content, the field Protocol is set to 6; if it is UDP, the Protocol is set to 17. Actually, not only transport packets may be encapsulated here; for example, if the Data contains routing information to be read by the OSPF protocol, the Protocol field is equal to 89.

The *Header Checksum* serves for error detection in the header; mismatched packets are discarded. The data are not checked, since this task is left to the encapsulated protocol; as we will see later, both UDP and TCP have indeed checksum fields. This field needs to be recomputed at every hop since at least the Time-to-Live field will have a different value.

The 32-bit *Source address* and *Destination address* are probably the most important fields and will be discussed in more detail in the next subsection. After that, some additional *Options* can be specified, followed by the *Data*.

10.3.2 IP addressing system

One extremely important feature of the Internet Protocol is to provide a universal addressing system for the Internet. IPv4 addresses are commonly written in a notation called *dotted decimal*, where the four octets of bits are written as a separate value in decimal notation (thus, between 0 and 255) and separated by dots. For example, an IP address can be written as 192.0.2.45. Since IPv4 uses 32-bit addresses, the entire address space consists of 2^{32} (approximately 4.3 billions) addresses. However, as detailed in Tab. 10.2, some of them are reserved for special purposes, which slightly reduces the whole available space. For example, range 192.0.2.0–192.0.2.255 is to be used in academic textbooks (as done few lines above), and not available for real use, just like "555" telephone numbers. A more important range of reserved address is that of private networks, which will be mentioned later when discussing Network Address Translation (NAT). Finally, range 127.0.0.0—127.255.255.255 is used for the localhost loopback, which is discussed below.

Table 10.2 Special addresses

0.0.0.0 — 0.255.255.255	Zero addresses (unused)
10.0.0.0 — 10.255.255.255	Reserved for private networks (see NAT)
127.0.0.0 — 127.255.255.255	Localhost loopback
169.254.0.0 — 169.254.255.255	Reserved for other special uses
172.16.0.0 — 172.31.255.255	Reserved for private networks (see NAT)
192.0.2.0 — 192.0.2.255	Reserved for theoretical examples
192.18.0.0 — 192.19.255.255	Reserved for other special uses
192.88.99.0 — 192.88.99.255	Reserved for other special uses
192.168.0.0 — 192.168.255.255	Reserved for private networks (see NAT)

An IP address, at least ideally, is uniquely associated with a machine, or *host*, connected to the Internet. However, this kind of correspondence is neither injective nor surjective. In fact, an IP address is an identifier of a network interface, so a host connected on two networks must have two interfaces and therefore two IP addresses. Yet, this is true only for routers which are on the edge of the networks (that is, the border gateways mentioned at the end of the previous section), which interconnects two or more different networks and therefore must be present on all of them, whereas most of the hosts on the Internet have a single network interface and therefore a single IP address. However, another case of (partial) double addressing is introduced by address 127.0.0.1, also called *localhost*. This is reserved as an alias used to identify "this same host," meaning that every host sees itself as having two IP addresses, the real one, and localhost. Indeed, all IP addresses starting with 127 are unused, and all communications directed to them are redirected to localhost. This may be a useful shortcut to identify this very same computer without specifying its IP address, e.g., because the same instruction is to be repeated on several different machines, with different actual addresses. Finally, the correspondence between addresses and host is unbalanced also since there are many IP addresses which are not used. This may be surprising, since it is commonly claimed that there is an overall shortage of IP addresses. Actually, these two statements are not in contradiction, as will be explained in the following.

Classful addressing Communication through IP implies that the datagram contains the addresses of source and destination hosts. However, as previously discussed, hosts are not connected to the Internet as stand-alone entities; rather, they belong to a specific network. Hosts belonging to the same network should have similar addresses; this helps finding them over the Internet. Especially, a routing table cannot contain an entry for every host, or it would be too large. It is definitely better to have a single entry for a whole network, so as to enable hierarchical routing as described in Subsection 10.2.4. For implementation reasons, the initial part of the IP address should ideally label the network, and the final part is instead the identifier of the host within the network it belongs to. In other words, the network identifier ought to be a prefix of the whole address, so that a router adopting a hierarchical strategy can simply read this part only, which saves some processing time.[10] With this split between network and host identifier, some special values are reserved for specific uses. For example, the IP address with the host identifier entirely made of bits set to 0 is considered equivalent to the network identifier and is always assigned to the border gateway of the network. It cannot be therefore assigned to any host. The address where the host identifier is instead set to all 1-bits is also reserved and used for *broadcast* messages within the local network. We have already seen the need for non-selective broadcast forwarding of packets, for example in the flooding routing strategies. Other applications will be mentioned in the following.

Also note that, since the inherent hierarchy of the Internet is actually multi-level, it would be useful to have nested prefixes. Indeed, specific ranges of the addressing space can be put in relationship with certain geographic areas, but this is just coincidental. Most of these reasonings are no longer valid in recent times; while it was possible some time ago to recognize the country of a specific IP address, the ever increasing problems in the availability of IP addresses, of which we will speak in the following, have made this a mere speculation.

When reading IP addresses (e.g., the destination address of a given packet), it may be useful to know in advance some characteristics of the network where they belong. For instance, it may be desirable to know whether two numerically close IP addresses are associated with hosts on the same network, or their numerically similar value is just incidental. If the address contains the network identifier as a prefix, this simply means to check whether the two addresses have the same prefix. However, this would be possible only if the prefix size were known in advance. Instead, in general the subdivision into network and host identifiers is not uniform in the Internet; while all IP addresses have the same 32-bit format, these can be split into network and host identifier in several manners.

[10]In reality, certain devices read the bit in reverse order than human beings, i.e., they read the least-significant bit first. Clearly, in such a case this advantage is lost.

In the initial times of the Internet, this point was addressed by means of classful addressing. This is a special assignment of IP addresses to hosts, implying that the entire address space is divided into *classes*. A class is defined as a range of contiguous IP addresses, which characterizes the size of the network identifier for those addresses belonging to it. In other words, according to the requirements discussed above, the IP addresses contain a network identifier and a host descriptor which identifies the host within the network. How the IP address is split into these two parts is determined according to its class. In fact, since the splitting rule is the first thing to determine, and if a prefix rule is adopted, the class identifier should be a pre-prefix (i.e., a prefix nested in the first part of the larger prefix which is the network identifier).

It was therefore decided that different class types of the IP address are determined by the four most significant bits. Apart from special reserved combinations, five classes where defined, commonly identified with capital Latin letters, i.e., Class A, B, C, D, and E. Only classes A, B, and C were used for IP addressing as discussed above; class D was a special addressing range for multicast application, while class E addresses were reserved for further uses, still to be defined. Each class is different in which part of the address identifies the network and which one is the descriptor of the host. The whole specification is reported in Tab. 10.3. Note that this subdivision has to be further compared with Tab. 10.2, where one learns, for example, that three different class A networks (network 0, network 10, and network 127) are not actually available.

Table 10.3 Classful addressing scheme

Class	**Begins with**	**Address range**	**Network id / Host id**
Class A	0	0.0.0.0 — 127.255.255.255	8 bits / 24 bits
Class B	10	128.0.0.0 — 191.255.255.255	16 bits / 16 bits
Class C	110	192.0.0.0 — 223.255.255.255	24 bits / 8 bits
Class D	1110	224.0.0.0 — 239.255.255.255	for multicast applications
Class E	1111	240.0.0.0 — 255.255.255.255	for future uses

Class	**No. of networks**	**Max No. of their hosts**
Class A	$2^7 = 128$	$2^{24} = 16777216$
Class B	$2^{14} = 16384$	$2^{16} = 65536$
Class C	$2^{21} = 2097152$	$2^8 = 256$

It is common to see the four bits which identify the class as integrated in the network identifier, which results in referring to the class as a characteristic of the network itself. That is, networks whose addressing range falls within class A, B, or C are denoted as "class A networks," "class B networks," or "class C networks," respectively. The network class is also loosely associated with the size. In reality, by taking the first four bits of any IP address, and therefore determining its class, we only have an upper bound on the size of the network; the number of hosts on a network cannot be larger than the number of addresses, since the IP address is a unique identifier. However, there is no guarantee that all IP addresses within the range of the network are assigned to a concrete host.

Unfortunately, the classful addressing does not provide an efficient allocation of addresses if compared with the typical needs of organizations who require IP addressing for their local network hosts. Consider for example a university network. It is reasonable to assume that every computer (or printer, or similar device) in the laboratories and offices, not to count routers interconnecting buildings and laptops belonging to visitors, requires to be connected to the Internet and therefore needs an IP address. A class C space, with its 254 addresses, would be insufficient for all these hosts; however, a class B would provide 65534 addresses, which are likely to be excessive. Actually, there are a lot of small organizations which need around 1000 IP addresses; while a class C network is too small for them, a class B will overprovide them leaving more than 60 thousands addresses unused. Moreover, the number of class B networks is relatively limited, amounting to 16384 only. This is definitely lower than the number of universities or small organizations which may think of asking for a class B network because a class C network is too small. Similarly, even a class C network, which is the smallest choice, may be too large for a tiny LAN with a dozen of hosts (actually, there are even class B networks where the number of used IP address is not far from a few dozen). This explains why, even though in principle the IPv4 addressing space supports billions of hosts and millions of networks, only few of the former are actually used and few of the latter indeed fit the needs.

The problems of IPv4 addressing are the result of a bad design choice, since classes were for simplicity defined by splitting the four bytes of an IPv4 address into a network identifier and a host identifier whose sizes are multiple of one byte. This may speed computations but also gives inappropriate numbers for the network size (either too small or too large). More in general, the rapid growth of the Internet was underestimated, since the addressing space of IPv4 would have been very large had the Internet stayed, as the original ARPANET, a simple system of interconnecting military and academic research institutions mostly throughout Northern America, instead of becoming the world-wide success it is now. Also, for these same historical reasons, many of the networks who received class A or B numbers are located in the United States, thus the geographic distribution of large networks was highly uneven. Hence, the problem of scarce network addresses is serious especially in developing countries with high number of inhabitants who are potential Internet subscribers.

Subnets and classless addressing Another problem arises when an organization is already connected to the Internet and has requested, for example, a class B network address. Assume now that another part of the same organization intends to connect to the Internet. Typically, this is the case when another department of a university or a detached branch of an industrial organization requires a network address for their own hosts. Apart from the general difficulty of finding another network address, assigning it would be wasteful and difficult to manage. After all, there is already a class B network address registered to the same organization, even though the hosts are physically located somewhere else. And, as we discussed before, it is likely that a small fraction of the 65534 host addresses enabled by the class B network are actually in use.

Thus, the solution adopted is to split the entire network addressing space into smaller pieces, so as to reflect the subdivision of the organization into departments or branches. The resulting fractions of network are called *subnets*, thereby producing a system with yet another hierarchical level. If the main network is class B, the last 16 bits are the host identifier; we can split, again adopting a prefix, these bits into a subnet identifier and the identifier of the host over the subnet. Similar computations could be done for class A networks and even for class

C networks (provided that one wants to split such narrow address space into smaller pieces). Assume, just for the sake of exposition, that the original network is a class B one, and its 16 bits used for the host numbers are split so that the first 6 identify the subnet and the remaining 10 denote the host within a given subnet.

Now, we have three levels of hierarchy; incidentally, this can also be exploited in the routing table to better organize it. Routers which are external to the network simply address all the traffic towards it. However, when a packet reaches the router acting as the gateway for the main class B network, it needs to be directed towards the specific host internally within the main network. It would be cumbersome to have an entry in the routing table for all hosts of a class B network, which are potentially up to 65534, while with the numbers above there is slightly more than a thousand nodes to address within a subnet, and 64 possible subnet identifiers. This is fairly easier to manage via another hierarchical level in the routing procedure. Thus, the subnet subdivision enables a better size for routing table management, in spite of the original bad choices of the classful addressing.

The counterpart of this new level of hierarchy is that subnets needs to be somehow specified in the configuration of those routers who are internal to the main network, otherwise there is no way to know about their existence, since the whole address is just read as network identifier and host number, without any subnet specification. The idea here is to define a *subnet mask*, which tells the router which part of the address is to be read as subnet identifier. It is called a "mask" because it is simply a sequence of zeros and ones; since the network and subnet identifier are prefixes in the whole IP address, the mask is always a series of ones, followed by the proper number of zeros to pad the result to 32 bits. However, subnet masks are normally reported in the dotted decimal notation. Thus, in the numerical example above, the subnet mask, which consists of 22 bits equal to 1 followed by 10 zeros, would be written as 255.255.252.0. Another alternative style, which is shorter and cleaner, but less used, is to write the mask as $/Z$, where Z is the number of ones, in this example $/22$.

Since the subnet concept breaks the rigid structure of the classes in a somehow successful manner, it was thought that a simple solution, even though still insufficient, to the problem of IP address shortage could be to allow the subnet conventions to be adopted by any network in the Internet. This mechanism, called Classless InterDomain Routing (CIDR), suggests that all networks could be regarded as subnets in the sense explained above, and therefore IP addresses could be assigned to networks in chunks of variable size, without respecting the definitions of class A, B, and C. Actually, this is not entirely anarchic, since some properties are still required. Especially, even variable size networks should have a unique network identifier, which must be a prefix of any IP address belonging to the network. Therefore, the implementation is very much akin to those of the subnets described above. If a network needs a thousand hosts, it can receive an addressing range of 10 bits, and therefore an identifier X which is a binary string with length equal to 22 bits. A mask of $/22$ bits must be used for this network; this means that its addressing range (written with binary digits) starts at X–0000000000 (X followed by 10 zeros) and ends at X–1111111111 (X followed by 10 ones); actually, the first and last address cannot be used for any host. In dotted decimal notation, the addressing range may look like $x.y.z.0$ to $x.y.(z+3).255$, where z is not arbitrary but is necessarily a multiple of 4. Clearly, it is still possible to recognize old class A, B, and C networks since they do not change their addressing range and use a mask of 8, 16, and 24 bits.

IPv6 and NAT A temporary solution like CIDR avoided a collapse of the whole addressing system but is evidently imperfect. After all, it still leaves too many holes in the addressing ranges and does not avoid that big chunks of IPv4 addresses are unnecessarily assigned to big networks. All network researchers therefore agree that the IPv4 system, as is, requires more decisive adjustments. At the moment, two solutions exist, and the research community is still split between supporters of either solution.

The first solution is entirely radical: to replace the entire IPv4 system with a better one, a new version of the IP protocol called IP version 6 (IPv6 for short). IPv6 uses 16 bytes instead of 4, as IPv4 does; this leads some people to believe that the version number is equal to the number of bytes in the address length (with "6" being a short version of "16"). In reality, the label "version 5" was already assigned to an experimental flow-oriented streaming protocol called the Internet Stream Protocol, thus the next available number was 6. The larger addressing space of IPv6 is definitely larger than that of IPv4, since 128 bits can denote 2^{128} addresses, i.e., about 34×10^{37}, i.e., fifty thousands of septillions for any human being on Earth,[11] although most people probably do not even think of wanting one (still, IPv4 cannot give an IP address to everybody).

However, in spite of the 10 years elapsed since its standardization, the adoption of IPv6 is only in its infancy. Many hardcore IPv6 supporters blame for this the deployment of another solution, the Network Address Translation (NAT), which is ugly compared to the neat IPv6, but is now widely utilized, since it is easier and more practical to implement. Especially, since the problem of IP address shortage is very acute in developing countries, where the Internet is in rapid expansion, NAT has proven itself to be very popular there.

The basic idea of NAT is to create a separate (and virtual) address space which is used only within the local, also sometimes called "masqueraded," network. To this end, NAT requires an edge router acting as a translator between the virtual address space of the masqueraded network and the real address space of the Internet. For this reason, the router must also have at least one "real", i.e., public IP address; however, it is the only host of the local network with such a requirement.

The translation performed by this router according to the NAT technique consists of modifying the header of IP packets passing across. For outgoing packets, the modification is simply a replacement of the source address, which belongs to the virtual address space of the masqueraded network, with one of the public IP addresses of the router. Then, the end destination will see this IP as the source of the packet. Additionally, checksum fields need to be updated too.

The reverse translation is more difficult, as in this case the router must recognize that a packet containing its own destination address is in reality an incoming packet intended to reach a host within the local network. Thus, the edge router must replace its IP address in the destination field with the virtual IP address of the real intended destination. To this end, it must store beforehand basic data about each active connection, in order to be able to distinguish packets intended for different destinations. When an outgoing packet is processed, the router keeps memory of the destination address and other data, in particular the TCP/UDP port, a concept which will be defined and discussed in the next section. For the moment, the only thing needed to know about it is that ports are identifiers assigned by the transport layer to uniquely identify an end-to-end connection. Thus, the pair (address, port) is memorized by the router, which uses it to discern the end destination of reply packets.

[11]The world population is currently estimated around 6.7 billions. A septillion is 10^{24}.

In the simplest case, which is possible when the router has multiple public IP addresses, the router can easily associate each of them with a different connection. Thus, the router replaces the source IP address field of every outgoing packet with a different address chosen among its own public addresses, so that the same source in the local network is associated with the same public address. However, this limits the total number of ongoing connections. Especially, in the case the router has only one public IP address, which is referred to as "overloaded NAT" this solution would be of no use. However, since a connection is uniquely identified by a (address, port) pair, and observing that the number of ports is considerably large, an enhancement of the NAT can be envisaged for the overloaded case, by performing a translation of the source IP address as well as of the source port, which is sometimes called Port-based NAT or Network Address Port Translation (NAPT).

This procedure, as described, only allows a local node L in the masqueraded network to receive traffic from a host H if some packets have been previously sent *from* L to H, since in this way the router is able to establish the translation rules for packets coming from H to be redirected to L. To solve this problem, most routers performing NAT allow also the possibility of a permanent configuration of the NAT, often referred to as *static NAT*. Such a configuration, performed manually, enables traffic generated from an external host H to reach a specific host L in the masqueraded network directly.

Even though any set could be used as the virtual address space of the local network, there would be a problem if a communication request were directed to an external host having an IP address belonging to this set. Luckily enough, there are certain sets of IP address, called Private Network Addresses, which can be used to this end, as they are reserved and cannot be used as public IP addresses. Typically, NAT is then performed by using a Private Network Address, which is of the form 10.x.x.x, or 172.16.x.x through 172.31.x.x, or 192.168.x.x, as the virtual address space of the local network. In that space, also the edge router has one private virtual address, as well as being connected to the Internet.

The success of NAT is the main reason why the Internet is still expanding in spite of the limitations of IPv4-based addressing. However, the use of NAT has some consequences, from both theoretical and practical perspectives. In abstract, using NAT to masquerade a private network and enable its access to the Internet through a single IP address breaks the end-to-end semantics of IP. Moreover, if the port information is used, which belongs to the transport layer, also the separation between layers is broken. Hence, although useful, the modifications of the packet header performed by NAT violate the limits of the network layer. As a result, a destination receiving a packet which has been translated through NAT does not really know where the source is located, it only knows that it belongs to a local network; the route followed by the data can be tracked back only to the border router of this local network.

Moreover, as to exploit NAT it is necessary to pass through the edge router, local private networks can exert a form of control over the locally generated traffic. This can even be regarded as a benefit, for example in terms of increased security. Unauthorized local stations can be prohibited from generating traffic towards external hosts of the Internet. At the same time, changing the (address, port) pair may realize *de facto* a sort of hidden traffic control (e.g., blocking certain sites, or certain services). These functionalities can be performed also by other means and are in general not always desirable.

Finally, NAT only postpones, but does not definitely solve, the problem of IP address shortage. For IPv6 enthusiasts, the only final solution is IPv6; from their perspective, not only does NAT not represent a real solution, but also it delays the replacement of IPv4 with IPv6, thus causing in the end even bigger problems. On the other hand, it is even possible that other solutions, alongside the directions of NAT, will be envisioned in the future, which will allow to overcome the address shortage. It is also worth noting that further changes in the Internet structure are possible, so that the addressing problem can become more and more important, or even disappear. To sum up, this is a still open problem, with no clear and definite solution, which is likely to keep researchers busy at least for the next few years.

10.3.3 Control protocols

Other protocols are strictly related to IP and its addressing system. Strictly speaking, they are not entirely network layer protocols, rather they interface the network layer with other parts of the protocol stack. However, given their close relationship with the IP addressing system, it looks appropriate to discuss them here. They are the Address Resolution Protocol (ARP), which finds the MAC address associated to a given IP address and its counterparts which serve to assign IP addresses to hosts, in particular RARP and DHCP.

Address Resolution Protocol (ARP) Every terminal connected to the Internet is provided with an IP address. However, the layered structure of the network implies that the data link layer does not understand IP addresses. To actuate the communication, it is required to translate the IP address into the MAC address. Again, this passage through two different and supposedly unique addresses may seem redundant. However, it is required due to the presence of multiple layers. It also has an advantage, as the MAC address (which can be, e.g., an Ethernet or IEEE 802.11 address) enables the communication through this specific MAC only, while the IP address is general. This double address allows realizing local networks based on one specific MAC protocol, whose addresses are the only ones which the hosts understand, and at the same time these networks can be interconnected through IP even though they may be speaking different languages at the MAC layer.

Nevertheless, this solution poses an important question, i.e., how to map an IP address into the corresponding MAC address, and vice versa. Indeed, the IP packet only contains the IP address of the destination, which needs to be translated into a MAC address. One possible solution is to store a list of IP address/MAC address pairs. However, this may be very memory-consuming, as several such pairs should be memorized.

A better approach is realized via the so-called Address Resolution Protocol (ARP). To understand it, consider the following figurative example which deals with a real life situation. Alice has been invited to a costume party, where she hopes to encounter her friend Bob. However, she realizes that the costumes hide the features of the people, so she is not able to recognize Bob in the crowd. However, a simple solution is to shout "Bob!" and see if somebody turns. In this case, this person must be Bob, thus Alice can memorize his costume and knows whom to speak with.

This is what ARP basically does; clearly, translate in the proper context the costumes, the friend's name, and the calling in a loud voice. The ARP scheme can indeed be applied to many cases of double addressing between Layer 3 and 2. However, we will refer to it as a translator of IP addresses, since almost always Layer 3 addressing involves IP (IPv4 for the most). The idea behind the ARP is to raise a widespread request to contact a host with a certain IP address, and wait for it to reply. In practice, the source sends a broadcast request asking the entire local network who is assigned with a given IP address. The stations receiving this broadcast request simply ignore it if they have a different IP address; only the one matching the request answers, telling its MAC address. As a side note, one may observe that Alice's strategy works only if there is just one Bob at the party, and all the costumes are different. Similarly, ARP works only if MAC and IP addresses are unique; this is why uniqueness is such an important property of addressing systems.

Now, see the ARP operation in an example taken from real networking. Assume a LAN running Ethernet as the MAC protocol, which can be for example a computer lab in a university building. Each station in the lab is associated with a 48-bit Ethernet address. Assume now one of the machines in the lab, acting as the source and therefore denoted as S, wants to send a packet to another of the machines in the lab, called D, whose IP and MAC addresses are IP_D and MAC_D, respectively (replace these values with any string in the proper format, e.g., IP_D can be 192.0.2.4 and similarly for the MAC address).

The source S is in some way able to derive the IP address of D. This retrieval is actually performed by the DNS, which will be discussed later in this chapter; for the moment, just assume that IP_D known by S. However, the MAC address of D, MAC_D, is instead still unknown by the source. To find it, S simply sends a *broadcast* packet onto the local network. Recall that this is done by setting the destination address as D's network identifier followed by all remaining bits equal to 1; this is the same as IP_D with all unmasked bits set to 1. In other words, if D's network is a class C network as defined in Subsection 10.3.2, and therefore its mask is $/24$, then the network identifier, and therefore the broadcast address is simply $192.0.2.255$. The broadcast packet just asks the machine whose IP address is IP_D to show up. Upon reception of that packet, all terminals but the intended destination D discard the packet. Station D, instead, replies with a packet containing its MAC address, which is this time sent directly to S.

Such a structure is simple to implement, as there is no need to relate IP and MAC addresses somehow. The system manager simply assigns IP address (and a subnet mask, to derive the broadcast address) and the burden of relating the two addresses is left on the ARP. Also, some improvements are possible. Since it is likely that station D will be contacted soon in the future after the first packet exchange, node S will better keep the newly discovered MAC address of D stored for a while in a cache. This eliminates the need to send a broadcast packet every time. At the same time, the broadcast packet sent by S also includes the MAC address of S, so that D does not need to send an ARP request in turn. Indeed, D can immediately cache the MAC address of S already upon reception of the ARP packet.

In a similar manner, an ARP request known as Gratuitous ARP can be sent by a host when it boots. Such a request is not meant to solicit a reply, but merely serves to update the caches of other hosts in the local network with the MAC address of the newly booted host.

Finally, another extension of the ARP is required when source S and destination D do not belong to the same local network. As in the costume party example before, shouting (i.e., broadcasting a packet) can only be done locally. The border gateway of the local network where S belongs would not forward a broadcast ARP packet sent by S. In this case, however, the source is aware that the destination belongs to another local network, as this information is contained within the IP addresses. Assume that, as in the previous example, D belongs to a computer lab of one university, whereas S belongs to the network of another university, and its IP address is IP_S. Then, S checks its routing table and sends the packet to the router handling the remote traffic directed to the D's network, as it only knows the IP address of D which is IP_D. This router then repeats the procedure, possibly with new ARP requests; that is, it forwards the packet to the next hop, found on the routing table, and if the IP address to send the packet to is not within the ARP cache, it sends first an ARP request. Hop by hop, this procedure eventually finds the destination D and enables all intermediate nodes to learn and cache the MAC address required for the next hop communication.

RARP and DHCP The logical inverse of the ARP is a protocol obtaining IP addresses from known MAC addresses. There is a protocol, called indeed Inverse ARP, which does this thing in the same manner as ARP maps IP addresses into MAC addresses, but is used only in circuit-switched networks and mentioned here for completeness.

A different protocol, called Reverse ARP (RARP), also translates MAC addresses to IP addresses, but is used by a requesting station to obtain its own IP address. The motivation of this protocol resides in the need for an IP assignment when a new station is booted. Actually, there is no strict need to have such an IP assignment, as one could decide to assign a given IP permanently to a given station, so it is not necessary to assign an IP address to it each time it boots. However, there are several reasons not to do so. For example, it can be desirable to have the same startup sequence for many similar machines, as for example the computers of a lab; instead, giving each machine a personalized IP address requires to program all of them individually. Moreover, this solution is clearly impractical if there are several machines which are not connected to the local network most of the time, and/or they connect to different networks; this especially happens in local wireless networks where laptops or similar devices have a strongly intermittent connectivity.

Thus, when a station S connects to its local network, will typically ask for an IP address assignment. The RARP simply dictates S to broadcast its MAC address and demand to receive an IP address. In other words, similarly to the ARP procedure, S declares "my MAC address is MAC_S, is there anybody who knows my IP address?" However, such a procedure requires a RARP server which answers the request, and there must be one of them on every local network. In fact, again as broadcast messages can not be forwarded outside the local network, the RARP request must be solved locally.

For this reason, RARP was replaced by another bootstrap procedure called BOOTP. The artifice used by BOOTP to overcome the broadcast problem is to use UDP messages (actually, application-layer messages carried through UDP), which are forwarded over routers. However, this situation was still unsatisfactory, since BOOTP requires to configure the pairs of IP and MAC addresses by hand. For this reason, BOOTP has been obsoleted by the Dynamic Host Configuration Protocol (DHCP).

The main advantage of DHCP is to enable dynamic and automatic assignment of IP addresses. Manual and static assignments are possible, though. Moreover, the DHCP server can be placed on a different subnet; however, this still requires a DHCP relay agent on every local networks. Initially, the network administrator gives a list with a pool of IP addresses to the DHCP server. At the instant a new station connects to the local network, it asks the server to receive an IP address. The new station just needs to know the IP address of the DHCP server to perform the request.

Since the address pool is limited, the DHCP server may reclaim IP addresses; otherwise, a host receiving an IP assignment and then leaving the network will be associated with its IP address forever. Thus, a concept called *leasing* is used, i.e., the DHCP server grants a given address for a limited time interval, at the end of which the address can be renewed or revoked, so as to enable dynamic re-use of IP addresses. Still, certain DHCP servers, depending on the manufacturer, also enable to preserve the IP address, so that the same host is always assigned with the same address, if available, even in subsequent bootstraps.

10.4 THE TRANSPORT LAYER

As specified by its name, the Transport Layer is responsible for delivering data over the network, from the generating source to the specific process on the host computer requesting them. To this end, the transport layer separates data into packets, and sometimes introduces additional features, such as error control and flow control.

Network capacity We already mentioned in Subsection 10.2.3 several metrics evaluating the network layer performance, including throughput and delay (also called latency). As said, there is a trade-off between them. This means that layer 3 chooses an operating point of the network (for example, by establishing a given routing algorithm), and as a result a given pair of values is assigned to throughput and delay. Several pairs are indeed possible; however, either metric can be improved only if the other becomes worse.

Note that also packet loss probability is another important issue for routing. However, at layer 4 perfect reliability may be re-introduced through retransmissions by certain transport protocols such as TCP. Or, if the transport layer is not entirely reliable, we may simply set a target packet loss probability, and evaluate throughput and delay at this reliability level. Thus, hereafter we will leave reliability out of the discussion.

In general, the transport layer hinders multiple access and routing decisions; this means that it sets some network general conditions, among which the operating points of layer 3 and 2 are selected. In particular, although throughput and delay can be both improved, their relative performance limit can get a theoretical characterization as an inherent metric of the network itself, called network *capacity*.

There are several ways to look at this concept of "capacity." A simple formal definition of network capacity is: the amount of data that can be reliably delivered throughout the network in a time unit. More in general, the concept of network capacity is also connected to the aforementioned trade-off, and accounts for the fact that requiring a low delay in delivering the information also implies to reduce the throughput significantly. For the transport layer, this means that the traffic which can not be delivered in time should not even be admitted into the network. However, we can also have higher throughput if the traffic can tolerate higher delays.

A related way to quantify network capacity is formalized through the so-called *bandwidth-delay product* (BDP). To understand the terminology, note that, although this may be questionable, throughput is sometimes dubbed "bandwidth." Thus, the BDP simply corresponds to the product of throughput and average end-to-end latency; such a parameter is treated as an inherent characteristic of the network. It is measured in bits (i.e., bits per second times seconds) and corresponds to the amount of data which is in transit through the network, i.e., already transmitted but not yet received. It can be very high for networks whose links have high transmission rates but also long delays, such as those including satellite links; such networks are often dubbed "Long Fat Networks" (LFN), which is a joke acronym which approximately reads as "elephant."

The bandwidth-delay product is to be seen as a theoretical limit of the network capacity; the transport layer can set a lower delay, a lower throughput, or both, by regulating the offered load. In this sense, high bandwidth-delay product can be a problem for the transport layer.

If the transport layer needs to guarantee the reliability of the transported data, there will be some data which are passing through the network without being yet correctly received at the end destination. As long as the source is not informed of their correct reception, e.g., by an acknowledgment from the end destination, it must store them in a buffer, in case they need to be sent again. Thus, this buffer needs to be very large if the bandwidth-delay product is high. Moreover, the flow control operations of the transport layer may be difficult if the bandwidth-delay product is very large, as the regulation of the offered load has many degrees of freedom. These problems will be clear when discussing the transport layer protocol known as TCP.

Transport layer ports There are several names to dub the packets created by the Transport Layer, depending on the protocol in use; e.g., they may be named *segments* for TCP, or datagrams for UDP (this is the same name as that of IP packets; it is done on purpose, for reasons discussed below). There are other generic names for transport layer packets, the most common of which seems to be *unit*, also used as part of an acronym: transport protocol data unit (TPDU). In the following we will use this term when referring to a generic transport layer packet. Another terminology issue involves the end points of a communication at the transport layer, which are often referred to as *ports*. This name evokes the idea of an entry point for the connection at a specific host, but in reality it is a virtual construct to enable a logic classification of different communications going on between the hosts. Actually, we are deliberately mixing some different descriptions here, since this feature should be provided, according to the original ISO/OSI model, by the Session Layer (Layer 5). However, since we are skipping Layers 5 and 6 altogether (as discussed in Chapter 1), we will include this aspect in Layer 4, as is done by the TCP/IP model. A port is identified by a port number and the network address of the host.

Consider the situation over the Internet, with which we will deal almost exclusively for the rest of the section. While layer 3 only uses IP, at layer 4, two transport protocols can be used: TCP and UDP. The set of these three protocols is often referred to as Internet Protocol Suite, or TCP/IP suite. UDP is not mentioned in the name "TCP/IP" only because it is basically just a straightforward extension of IP to layer 4, with some additional header to respect the encapsulation paradigm; nevertheless, it is still important for certain purposes (e.g., realtime applications and multicast). Thus, UDP directly inherits from IP a best-effort packet based approach. Differently from UDP, TCP supports instead virtual circuits and aims at realizing reliable end-to-end communication, through ARQ and congestion control mechanisms that we will discuss in the following.

In this context, both TCP and UDP address the connection end point through a pair (IP address, port number). From a transport layer perspective, the combination of port number and the IP address is called a *socket*. In reality, the use of the IP address in the transport layer is breaking the layer structure. In theory, the layers should be entirely independent of each other, and the TCP/UDP ports, which represent the addressing system at layer 4, should not rely on IP addressing (which, as we mentioned in the previous section, has also its own problems). However, the point is that the Internet Protocol suite (or TCP/IP) was designed as a whole piece, and it was therefore thought to adopt this approach which, though breaking the layer structure, has some useful advantages. In particular, the IP address can be thought of as a prefix of the transport layer address, thus in this sense there is no need to repeat it again within the TCP or UDP header, as we will discuss later. However, this approach is different from what happens between layers 2 and 3, as for example Ethernet MAC addresses are entirely different from IP addresses.

The definition of the number codes for the ports is specified by the already mentioned IANA. A port identifier is 16 bits long, so there are 65536 available port numbers. However, some Internet services, which correspond to certain application layer protocols, are so important that they require a port number only for themselves. It was therefore decided to reserve low port numbers for these protocols, i.e., the range from 0 to 1023, which are called "well-known ports." About one third of this range is actually assigned, though not all of the protocols with such a reserved port number are widely used. A list of "better-known" ports, i.e., frequently used well-known ports, is reported in Tab. 10.4. Here, together with acronyms explained elsewhere in this chapter, there are also some important applications which are out of the scope of this book, e.g., the File Transfer Protocol (FTP) and the Secure Shell (SSH). Also, note that certain application layer protocols can be supported by either TCP or UDP, or both.

Higher port numbers should be used for user-defined applications, i.e., in theory they are for free use by all remaining applications. Actually, in certain cases IANA also standardized some of them. Moreover, even when there is no official reservation for a given number, the range is so wide that most of the times there is a *de facto* practice for using certain ports for specific application layer protocols.

Table 10.4 Some of the most commonly used well-known ports.

No.	Application	Supported by	Used for
20	FTP data	TCP	File transfer
21	FTP control	TCP	File transfer
22	SSH	TCP/UDP	Remote secure login
23	Telnet	TCP	Remote unencrypted login
25	SMTP	TCP	E-mail exchange
53	DNS	TCP/UDP	IP address translation
67	BOOTP and DHCP server	UDP	Assign IP address at booting
68	BOOTP and DHCP client	UDP	Assign IP address at booting
80	HTTP	TCP/UDP	WWW Interface
179	BGP	TCP	Exterior gateway routing

10.4.1 User Datagram Protocol

Since UDP is simpler, we will review it first. The D in the acronym "UDP", standing for "Datagram," admits that UDP packets are inherently unreliable; it is also the same name used for IP packets, implicitly admitting that UDP adds little to the underlying IP. Indeed, UDP does not provide virtual circuits, nor reliable communication; this does not mean that it is not possible to implement these functions somewhere else. Actually, the idea of UDP is that these features are delegated to the application layer.

The structure of the UDP packet is reported in Tab. 10.5. The header consists of just four fields, and two of them are not even mandatory. As discussed, the UDP header includes the source and destination port numbers, whereas the rest of the addressing is taken from the IP

Table 10.5 The packet structure of UDP.

	16 bits	16 bits
0–31	Source port	Destination port
32–63	UDP length	UDP checksum
64 +	Data	

header. However, the source port is required only for bidirectional communications, in case a quick reply is needed; thus, in general it is optional, and can even be set to zero.

The other two fields are the length of the datagram, including the UDP header, and a checksum value, which is optional. This bland form of error control is the only one the UDP does; certain real time applications do not even employ it, since they require the application layer to introduce additional reliability features. However, the checksum may be a useful feature and it is mandatory when UDP is used over IPv6.

Since most applications require a flow-controlled reliable stream of data, UDP is not suitable for them. However, even an unreliable service may have its applications, especially since it is faster and with less overhead than TCP. In other words, the network capacity achieved with UDP has higher throughput and/or lower latency. This is why UDP is used by real-time multimedia application; for example, UDP supports the real time protocol (RTP). Moreover, it can also provide multicast applications, whereas TCP cannot. It is worth noting that these kinds of applications, such as Internet telephony, real-time video, online games, can tolerate some packet loss; their top requirements are low latency and high throughput, and when a packet is lost, they have no time to wait for its retransmission, anyway. Finally, UDP is important for DNS, a fundamental application which will be discussed in the next section. Although DNS also can be supported through TCP, in most implementations the DNS service is provided based on UDP.

10.4.2 Transmission Control Protocol

Even though UDP has its application, TCP is the real protagonist of the transport layer. It is used for many application protocols, including those for web browsing, reliable file transfer, and email exchange. It was originally conceived in [19], together with IP, to form the original core of the Internet Protocol Suite, which is for this reason also named TCP/IP.

Especially, TCP includes the main feature of providing connection-oriented services over a connectionless IP network. To implement this feature, a key property of the TCP protocol is that every single byte exchanged through a TCP connection must be labeled with a 32-bit sequence number. Actually, this value is another limitation which has been surpassed by technological advancements. A sequence number consisting of 32 bits can unambiguously label 4 gigabytes, which was a huge amount of data in early Internet times, and therefore could take days, even weeks to be transmitted. Thus, the risk of having two identical sequence numbers on the network was extremely limited, as it would have occurred only if packets survived in the Internet for days. With fast transmission lines, which have speeds of the order of Gb/s, the whole sequence number space is cycled in few tens of seconds, and therefore ambiguities are possible; to avoid them, some tricks are required. This is similar to other solutions which were proposed in recent years to patch some performance issue of TCP.

Table 10.6 The packet structure of TCP.

	4 bits	4 bits	8 bits	16 bits
0–31	Source port			Destination port
32–63	Sequence number			
64–95	Acknowledgement number			
96–127	Data offset	Reserved	Flags	Window size
128–159	Checksum			Urgent pointer
160 +	Options (variable length)			
	Data			

Now, we are ready to consider the structure of the TCP packet, also often called TCP *segment*, which is reported in Tab. 10.6. Source and destination ports are reported in two 16-bit fields. Then, the sequence number follows, which, as discussed, is a 32-bit value. Since TCP aims at guaranteeing a reliable transport layer, acknowledgements are required for the transmitted packets; packets which are transmitted and are not acknowledged must be retransmitted.[12] As there are acknowledgements, there is also an acknowledgement number, which is reported in the following field. It would seem sensible that the acknowledgement number reported the number of the last correctly received packet, i.e., it said "I confirm reception of packet X." This is generally how acknowledgements are seen; however, in reality the acknowledgement number of TCP denotes the value of the next expected sequence number in the opposite direction, i.e., it actually says "I am now waiting to receive packet $X + 1$.". The reason to fill the field in this manner is as follows. Assume packet X is correctly received and the acknowledgement reported it by saying "I received packet X." If such a message is lost, there is no way for the transmitter to know whether packet X is received. Once the following acknowledgment were received (which then said "I confirm reception of packet $X + 1$."), the best guess of the transmitter is then that packet X is lost, which is not. Similar reasonings can also be induced by out-of-order segments. These problems are avoided if the acknowledgement says instead: "The next packet I am still missing is packet $X + 1$." If this message is lost, the next one will

[12] Actually, this is true only in principle; there are several features, such as the "delayed ACK" that will be mentioned in the following, which slightly change the practical implementation of this idea.

be "The next packet I am still missing is packet $X+2$," which confirms that packet X has been received. It is easy to check that also out-of-order segments do not represent a problem either. In this manner, when a packet is lost, it will be continuously requested by ACK messages, i.e., the receiver will continuously send "I am waiting for packet X." These duplicate ACKs (also called dupACKs) can however happen also due to some glitches in the delays experienced by different packets.

These fields are then followed by a Data Offset value, which specifies the length of the TCP header expressed in 32-bit words, i.e., at which point in the segment the data actually start. The minimum value is 5 and the maximum is 15. The 12 subsequent bits are control flags; the first 4 are still currently reserved for future use and must be set to 0. The remaining 8 bits specify certain control instructions; only the first two of them have been added over the time, and do not involve major variations, so that the TCP flag structure is basically still akin to its original design. From the most to the least significant, the flags are the CWR, ECE, URG, ACK, PSH, RST, SYN, FIN bits. The most important of them are the ACK bit, which tell the receiver that the "Acknowledgement" field is significant, and the SYN and FIN bits, which will be discussed next for the connection establishment and release.

The 16-bit "Window size" field specifies the flow control mechanism of TCP and will be discussed later. The Checksum field implements a simple error-checking of the header and data. Since TCP implements an ARQ mechanism, retransmission of lost data is therefore requested. The Urgent pointer denotes that certain data is urgent and may violate the basic flow control. Finally, further options and the data are introduced.

TCP connection management Since TCP provides connection-oriented transport, it requires proper control messages for connection establishment, control and release. Moreover, it also implicitly assumes that the connection is full-duplex, that is, both connection endpoints need to exchange data and support for bidirectional communication must be provided. Even though this is generally correct, it is also often verified that connections are strongly asymmetric, since the usual situation is that an application client contacts a server and requests data; after some negotiation phase, the data flow from the server to the client (downlink) is often much higher than that from the client to the server (uplink).

Nevertheless, assuming bidirectionality gives the advantage that when some data is sent, the reply in the opposite direction can be transmitted in a fast manner. If uplink and downlink connections were established separately, one would end up waiting for the other to be ready, and much more time would be wasted. Finally, as will be discussed in the next paragraph, the full-duplex connection establishment enables to perform flow control, which is another key feature of TCP.

This is the reason why acknowledgements are not transmitted as stand-alone packets, but are rather piggy-backed on the data sent in the opposite direction. Clearly, this implies that when the data flow in the opposite direction is extremely limited, as in the server-client example mentioned above, it becomes sometimes necessary to force an acknowledgement without any data. This is just one example why data are an optional component of TCP segments.

Coherently with this discussion, connections are established in TCP through a *three-way handshake* procedure. The normal sequence is that the endpoint requesting connection (the client) sends a TCP segment without data and the SYN and ACK bits set to 1 and 0, respectively. If no problem occurs, the server answers with both SYN and ACK set to 1, and the client finally acknowledges this with another segment where SYN is 0 and ACK is 1. There are other operations to deal with the cases where some segment gets lost or the server refuses the connection; solutions to handle these situations exist and can be managed by setting the proper flags in other TCP segments. However, we will not discuss them in further detail. The flag FIN is instead used for the connection release. Even though it is full-duplex, the connection is released as though it were made of two simplex connections. When a segment with FIN set to 1 is received, it is read as a notification that the sender has no more data to send. After its acknowledgement, this side of the connection is shut down, but nothing forbids the other to keep sending data (even though it should not expect any answer to them). The other direction of the connection is closed only when another segment with FIN flag set to 1 is sent. Thus, this can be sent sequentially, i.e., as another additional segment, or simultaneously to the acknowledgement of the first FIN. There is no difference, only in the latter case one useless transmission is saved.

TCP Flow Control TCP performs flow control through a window-based approach. This means that a transmitter is not allowed to send data at will, but it must be authorized in advance by the intended receiver of this data, and at the same time the receiver can also limit the amount of data it wants to receive. The basic reason to do so are the limited capabilities available at the receiver's side. Since the receiver has a finite buffer, and additionally it may have other limitations of computational nature, in TCP connections the receiver is required to advertise in advance its availability to receive data, and the transmitter is requested to comply with the advertised value. In other words, the transmitter can send data only through a *window* of finite size. Since sending more data than what the receiver can accept would result in losing part of them, the TCP transmission window should not exceed what is called the *receiver advertised window*. This latter value is notified by the receiver in the segment exchange, and is likely to be decreased by the transmission of further data and increased by their delivery at the receiver's application layer when they are ready to be processed. Actually, the receiver starts advertising its window size immediately in its first packet, i.e., the one where SYN and ACK are set to 1, to inform the transmitter of its available capability.

To avoid inefficiencies, the transmitter and the receiver should not cause changes of the advertised window too often, since these variations in the window size should be notified, therefore increasing the channel occupancy. The transmitter is therefore encouraged to aggregate data as much as possible. In particular, in certain applications, instead of sending every single byte as it is generated, it should send the first one and then buffer the others, and transmit them together. Such a procedure is called the *Nagle algorithm*. Similarly, a receiver with saturated buffer, which therefore has advertised a window size equal to 0, should wait a little before advertising a positive value, since there is a concrete risk that the buffer immediately becomes saturated again (a situation called *silly window syndrome*).

TCP Congestion Control However, it is worth noting that the injection of data in the network should not depend only on the receiver state. The network between the two connection endpoints is a dynamic system, which does not comprise just the transmitter and the receiver. It is well possible that, even though the receiver buffer is not full, data are lost somewhere during their transit through the network. In particular, these losses may be substantial, meaning that it is not a matter of a single unlucky packet not being delivered, but rather almost all the data sent beyond a certain amount are lost. There might be a given (maybe small) quantity of data going through, but the rest does not reach the destination. When this happens, the performance of the transport protocol drops tremendously. Such a network behavior, which may happen for several reasons, goes under the name of *congestion*. The concept of network congestion is just experimental, not etiologic; that is, in the end, one does not need to know why congestion arises, only to avoid its effects. And, since the advertised window just describes the receiver's status, losses can occur due to congestion even though the amount of data sent is below the advertised window.

The way to reduce this loss of data is simply to send fewer packets, since they are feeding the congestion. The more the data sent, the more difficult to solve the congestion. This is the reason why TCP considers, besides the flow control procedure based on a sliding window mechanism described above, also a *congestion control* strategy. The congestion control is then nothing but a mechanism to regulate the rate of packets entering the network, keeping the data flow under that size that would trigger congestion. In practice, this means that the management of the window size in TCP does not only account for the advertised window, but also considers a second window, called *congestion window*. The transmission window used by TCP is the minimum between these two windows.

While the advertised window (as its name tells) is regulated based on information directly sent by the receiver, the congestion window is to be deduced from external data. As there is nobody explicitly telling the transmitter how congested the network is, it has to figure it out somehow. The TCP sender therefore bases this decision on acknowledgements (or lack thereof) received for the data sent, and timers in the data exchange. The general idea is that, as long as data goes through normally, the congestion window is increased, which means that the advertised window tends to become the dominant constraint to decide how much data to send. When packets instead receive acknowledgements after a suspiciously long time, or are not acknowledged at all, the sender infers that a congestion situation is close, and reduces the congestion windows as a consequence. However, these operations are not identical; intuitively, when there is no sign of congestion, increases of the congestion window should be large when the window is small, and more gradual when it is already large. Conversely, decreases of the congestion window should be significant to get a prompt response against congestion.

This translates into an implementation of the congestion control mechanism of TCP which consists of different inter-related procedures to regulate the congestion window. The first one is called *slow start*, although this name is recognized by everybody as misleading. In fact, it regulates indeed the initial phase of the congestion window, but it usually causes it to grow very rapidly, i.e., in an exponential manner. However, previously proposed algorithms were even faster, so this actual one is "slow" compared to them. According to the slow start algorithm, when the connection is initiated, the congestion window CW is set to a value equal to 1 maximum segment size (hereafter denoted with MSS), or 2 MSS, depending on the implementation. The value of MSS might be, e.g., equal to 1024 bytes. During the slow start phase, CW increases by 1 MSS for each received ACK. In practice, CW doubles each round trip time if all packets are acknowledged, therefore causing exponential increase of CW. This behavior continues until CW becomes of the same size of the advertised window or a segment is lost. If this latter event takes place, half of the current CW is saved as a "slow start threshold" ϑ_{SS}, and slow start begins again from the initial value of CW (1 or 2 MSS). Again, CW increases exponentially, but when it reaches ϑ_{SS} it enters another algorithm, called *congestion avoidance*.

The congestion avoidance phase is entered when it is assumed that CW is close to its "right" size. In fact, during the slow start phase, a loss event happened when CW was doubled from an old (safe) value X to $2X$. The most suitable value, it is presumed, should be somewhere between X and $2X$. Thus, instead of doubling the window again, it is better to increase it gradually to better approach this value. Thus, during the congestion avoidance mode, each ACK increases CW by $1/CW$ MSS. If the whole set of segments transmitted in an rtt is acknowledged, the resulting increase of CW is therefore almost linear (i.e., 1 MSS per rtt). Again, if a loss occurs, ϑ_{SS} is set to $CW/2$ and CW is brought back to 1 MSS and another slow start phase begins.

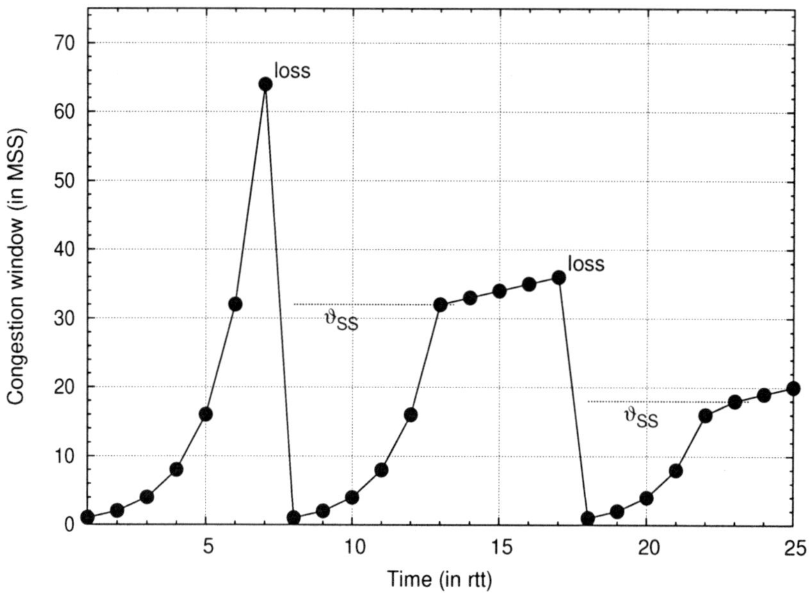

Figure 10.9 An example of evolution of the TCP congestion window.

An example of evolution is presented in Fig. 10.9. Here, the time axis is measured in rtt intervals. During the first 7 transmission rounds, the congestion window management follows the slow start algorithm, so that CW is doubled at every rtt. Then, a loss occurs (when $CW = 64$ MSS), so the slow start threshold is set to 32 MSS and CW is reset to 1 MSS. Again, slow start is executed, but when CW reaches 32 MSS, which is the size of ϑ_{SS}, the slow start phase ends and the congestion avoidance procedure is run instead, causing a linear increase of CW. This continues until the 17th round, when another loss occurs. Since $CW = 38$ MSS, ϑ_{SS} is now set to 19 MSS and another slow start phase begins.

It is important to note that this way of proceeding assumes that lost segments are due to network congestion. This is an acceptable assumption for many networks which are sufficiently reliable. However, when the quality of data link layer transmissions is poor, losses may also occur due to bad reception. As already said, this situation typically happens in wireless networks, therefore specific solutions to implement TCP over wireless represent an important research challenge.

Remarkably, the congestion control mechanism requires a practical criterion to determine whether segments are lost. The idea is that the acknowledgement for a given segment ought to be received after a rtt. Therefore, the transmitter employs a retransmission timeout (RTO), which roughly speaking is an estimate of the average rtt, to detect when the segment can be considered as lost. However, there are some subtleties in the definition of RTO. First, a smoothed average of the rtt is computed. Second, a margin is added based on the mean deviation (which is a cheap approximation of the standard deviation).

Other features which may be implemented in the TCP congestion control algorithm are the so-called *fast retransmit* and *fast recovery*. In the fast recovery procedure, if packets are lost in the congestion avoidance phase, CW is reduced to ϑ_{SS}, rather than the smaller initial value. Fast retransmit is instead a variation of the retransmission procedure for missing packets. According to the discussion above, packets must be retransmitted when they are not acknowledged before the RTO timer expires. Fast retransmit dictates to retransmit also packets for which three dupACKs are received, i.e., which are requested in four (one regular plus three dupACKs) acknowledgement messages.

Theoretical analysis of the transmission control protocol Various models have been proposed to analytically characterize the throughput of TCP. References [20, 21] model the behavior of TCP for channels affected by i.i.d. packet errors, capturing the effect of fast retransmit and neglecting the transmission dynamics induced by the timeout mechanism. While these models reveal the general relationship between the maximum window size $W_{\max}$, the round trip time RTT and the TCP packet loss probability p, they only provide accurate results for moderate error rates. In addition to this, the analysis in [22] also models the TCP timeout dynamics, leading to a closed-form expression of the TCP send rate R_S, which is very close to the TCP throughput in most cases, under the assumptions of i.i.d. packet errors and fixed round trip time. This model is the most widely used, due to its simplicity and its accuracy for i.i.d. channels, and gives the following expression for R_S

$$R_S = \min\left\{\frac{W_{\max}}{RTT}, \left(RTT\sqrt{\frac{2bp}{3}} + T_O \min\left\{1, 3\sqrt{\frac{3bp}{8}}\right\} p(1+32p^2)\right)^{-1}\right\} \tag{10.18}$$

where $W_{\max}$ is the maximum TCP window size expressed in number of TCP segments, RTT is the round trip time between the TCP sender and the TCP receiver expressed in seconds. The parameter b is used at the TCP receiver; in most implementation it is equal to 2. It has the following meaning: when packets are received correctly and in sequence (i.e., in the same order as they are transmitted), an ACK is sent every b TCP packets; this feature is referred to as "delayed ACK" and is implemented to limit the traffic transmitted over the return channel; p is the TCP packet error probability (the packet error process is assumed i.i.d.) and T_O is the average duration of timeout intervals, expressed in seconds. From (10.18) the expected number of TCP segments that are successfully transferred over the channel per second is returned as a function of $W_{\max}$, RTT, b, T_O and p. Note that $W_{\max}/RTT$ is the maximum send rate, which is achieved when $p = 0$. In this case, the only parameters affecting the rate at which TCP injects new packets into the network (e.g., its send rate) are the maximum window size and the round trip time. Indeed, a maximum number of $W_{\max}$ TCP segments can be sent during each round trip time when the transmission channel is error-free.

10.5 THE APPLICATION LAYER

The application layer involves all protocols that actuate remote process communications in their application form, although the original OSI model actually separates some functionalities in the Session and Presentation Layers. Now that we have reviewed all the underlying structure, we are ready to discuss some examples of Application Layer elements, which represent some familiar applications to the everyday computer user. They are the Domain Name Server (DNS), the World Wide Web (WWW), and the email exchange through the Simple Mail Transfer Protocol (SMTP).

10.5.1 Domain Name Server

At the application layer, DNS is a system used to assign identifiers for hosts connected to the Internet. The main purpose of DNS is to translate numerical identifiers into meaningful alphanumerical names. The latter are easier to remember for human users, and are more commonly adopted to denote the host in the everyday use. Also the reverse translation is provided, i.e., from a name to a numerical address. In other words, the DNS serves as the "phone book" for the Internet, as it translates the alphanumerical *hostname* of a computer into its IP address. An example of name provided by DNS is *telecom.dei.unipd.it*, which is the translation of the IP address 147.162.2.28.

Main properties of DNS A required characteristic of the DNS system is resilience to the underlying structure of the Internet. To this end, certain robustness criteria are implemented. Moreover, DNS implements a hierarchical structure (on which we will return in a while) both in the exterior aspects of the names and, more importantly, in their management. This enables a distributed implementation, which is the key property to guarantee robustness.

First of all, DNS must provide names, i.e., strings, which are themselves satisfying some criteria of transparency and robustness to changes in the underlying connectivity of the Internet. Since DNS provides an end interface with human users, any routing information should be hidden and replaced by other data which are easier to remember and interpret for human beings. In particular, IP addresses (to some extent) contain some references to the network class, identify the hosts, and are needed to determine a route, but these pieces of information are meaningless for the end users (at least, for most of them). Conversely, DNS gives a more figurative representation of Internet hosts, as it enables the description of other, more "evident," characteristics.

The first property which is immediately noticeable is the last part of the name, which usually describes either the geographical location (by country) or the general purpose of the host. The former case holds for names ending with a country identifier such as *.it*, *.de*, *.fr*; the latter is the case for names whose final part is, e.g., *.com*, *.edu*, *.org*. In certain countries, e.g., United Kingdom (*.uk*) or Japan (*.jp*) the final part of the address contains both data, since the final part describes the country and a "penultimate" part mimics the general purpose labels, as in *.co.uk* or *.ac.jp*. In general, this information about the geographical position is not always verified and therefore may be inaccurate; however, it is still useful for human beings.

Other data can be inserted in the host name in several ways, even though they are not always standard and sometimes require more than an educated guess. In the example above, a user who is familiar with certain conventions of naming can also learn that the host belongs to a university; as a tacit convention, most of the Italian university, but not all of them, have a string "uni" in their name. A more expert user may infer that "unipd" actually means "the University of Padua" since "PD" is the province code of Padua. With considerably more inside knowledge, a human user could also infer that "DEI" is the department acronym and the host somehow relates to telecommunications.

Certain aspects, such as the presence of dots, are not just conventional or accidental, but depend on the hierarchical structure of DNS, which involves more technical details, as will be clearer below.

The DNS hierarchy. If the domain name system were implemented with a single central name register, it would surely be so huge that updating and using it would be rather impractical, if not impossible. Therefore, the solution adopted is to realize a *distributed system*: after all, even phone books are available on a local (either city, province, or region) level.

DNS defines and utilizes several *name servers*; each name server is in charge of assigning a name to each domain under its authority. Within this structure, operations are performed *hierarchically*, as name servers have multiple levels of depth. Authoritative name servers for a specific domain can also identify a subset of it as a sub-domain (i.e., domain of a lower level in the hierarchy), each with a specific authoritative name server. This solution is also particularly useful to increase the robustness of DNS, as the failure of a name server does not affect other name assignments at a higher or separate level in the hierarchy.

To realize the hierarchy, the domain name space is organized as a tree, and the names have a well-known structure, where two or more alphanumerical identifiers (called *labels*) are separated by dots. For example, *telecom.dei.unipd.it* contains 4 labels. Each label describes a level of the tree; this hierarchy is to be read from right to left, i.e., in the example, the highest level is described by *it* and the lowest by *telecom*. The hierarchical structure of the labels mirrors that of the actual tree, whose root is the parent node of top-level domains, whose label can be, for example *it*, *de*, *fr* (describing a geographical area) or *com*, *edu*, *org* (referring to the purpose of the host).

Each further label to the left corresponds with a sub-domain of the domain above. For example, *unipd.it* is a sub-domain of *it*, *dei.unipd.it* is a sub-domain of unipd.it, and so on. It is worth noting that domains which are also sub-domains (i.e., with two or more labels) must be host themselves, with an assigned IP address; this is not true for first-level domains, i.e., there is no host in the Internet named *it* or *com*. The maximum number of levels which can be handled by a DNS is 127, and each label is limited to 63 bytes, although these boundaries are very loose in practice and actual names are usually much shorter. Certain domain can also impose more stringent limitations and specific conventions for the choice of labels. Also, in current implementation not all ASCII characters are permitted, just lower and upper case letters (although it is often recommended to use lower case only to avoid ambiguities), numerical digits, and a set of hyphen characters, such as a dash $-$, the underscore $\_$, and the tilde $\sim$. This prevents extended international alphabets (such as letters with grave or acute accents, or non-Latin letters). Internationalized domain name systems have been developed to work around this issue, and are about to be adopted.

Conventional labeling may also be used in practice, as previously discussed. Another common convention, for example, is to assign the label *www* to web sites (a concept which the reader should be familiar with; in any event, web sites will be discussed together with the WWW in Subsection 10.5.2). In case multiple web sites are present, *www* usually denotes a general purpose one, whereas specific web sites have their own label. This is also the case for the example mentioned above, as *www.dei.unipd.it* exists and contains the general web site of the DEI department, whereas *telecom.dei.unipd.it* is the specific web site of the telecom engineering group.

This structure is very flexible, as the naming responsibility for low level domains is in the end delegated to small servers. In the example mentioned above, the domain *dei.unipd.it* is relatively limited compared to its top-level domain. Indeed, it serves just the DEI department, so it is easy to assign a descriptive name to its sub-domains based on their specific local characteristics. In the example, easy to memorize labels have been assigned to research groups for their own website.

The one described above is just a basic DNS scheme. However, some extensions are possible and encouraged. For example, the correspondence between names and IP addresses is not always one-to-one. The same IP address can be associated with multiple names, for example to host many websites on the same machine. At the same time, many IP addresses can share the same name, to ensure fault tolerance. This is especially useful for important servers, including domain name servers themselves. In particular, the top-level DNS domain is managed by many very powerful servers, to avoid that, in case of failure, the entire naming system becomes unavailable.

Another functionality enabled by DNS is blacklisting certain hosts, e.g., sending malicious emails or spam. The blacklist can be stored and distributed in order to block malicious IP addresses.

10.5.2 Email exchange and the World Wide Web

Electronic mail (or email for short) and the World Wide Web (WWW) represent the applications most widely used by the everyday Internet user, and for the average user they are "the Internet." However, they have specific characteristics which will be now analyzed. In particular, one may not fail to notice that they rely, with some differences, on the domain names established through DNS. A simpleton may therefore say that WWW is "Internet with *http* in front" and the email is "Internet with @ in between." There is of course more than that, but these half-joke definitions actually contain part of the truth.

Email Since the very first implementation of computer networks, electronic mail, i.e., the automated exchange of text messages, was immediately envisaged as a very useful application. Even though it ultimately took some time to have the widespread diffusion of email messages, already in the early version of the Internet, the ARPANET, several protocols were developed to enable exchanges of textual files. The most common protocol used to this end, the Simple Mail Transfer Protocol (SMTP) was standardized during the 1970s, and, with some modifications and extensions, is still in use today.

The idea of SMTP is extremely simple: to enable exchange of text-based messages, where some strings can be read as commands, and the rest is the content of the message. From the technical point of view, the SMTP mostly provides an interpreter for text strings which are exchanged through a transport layer realized with TCP. A well-known port number equal to 25 is assigned to SMTP. The typical interaction is between an SMTP client and an SMTP server; some commands are defined to open the session, exchange operating parameters, specify and possibly verify recipients (also enabling multiple recipients, carbon-copies, and so on), transmit the message and close the session.

All the data exchanged can be regarded as pure text. However, part of it, such as the control messages and the strings in the header, is akin to the writings on the envelope in real mail messages: it is intended for the postman, not for the reader. But the matter is identical, both the address on the envelope and the content of the letter are plain text. An important example of this concept can be seen in the "email address" and the use of "@". This symbol, which reads "at" but has also several different pronunciation or nicknames throughout the world, is used in the email addresses of sender and destination. Email addresses are virtual locations from and to which email messages can be sent, and are denoted in the format *account@domain*. For example, if the account (i.e., the personal email address) of the user John Doe in the domain *dei.unipd.it* is *jdoe*, the resulting email address is *jdoe@dei.unipd.it*. This idea of using the symbol "@" to denote a user "located" at some host is not actually exclusive to email, but it is there that it is most commonly used. The SMTP requires an email address identifier for the sender and the receiver(s) of the message, and additionally may include e.g., carbon copy, blind carbon copy, and reply-to addresses.

SMTP is able to submit the message even when the destination host is not reachable. In this case, to let the destination host receive the message when the connection endpoint becomes alive, SMTP often interacts closely with other protocols, such as the Post Office Protocol version 3 (POP3) and the Internet Message Access Protocol (IMAP). These protocols are designed to manage the "virtual mail box," i.e., to enable the access to messages received during offline periods. The difference is that a POP3 server is used to download all email messages into some directory in the local host, whereas IMAP keeps them into memory in a proper server. Such a choice is usually preferred by users who often change the computer from which they connect.

The email content is theoretically based on ASCII text. However, the data in the message is pure payload, and can therefore be read as the receiving user prefers. This is similar to the envelope content in the real mail system; it does not need to be written in the same language of the address on the envelope, it may even not be text at all. A widely used solution to standardize different kinds of content rather than just plain text is represented by the Multipurpose Internet Mail Extensions (MIME), which specifies additional codes for audio, video, pictures and similar non-text contents. Finally, many email interpreters enable the possibility of introducing hypertextual commands in the email content, based on the Hyper-Text Transfer Protocol (HTTP) which will be discussed in the following, when we will deal with the World Wide Web.

The World Wide Web Also shortened to "Web," or "WWW", it is a gigantic system of interconnected hypertext documents which can be accessed through the Internet. It was invented at the end of the 1980s by Tim Berners-Lee at CERN, and its initial purpose was to exchange documents on particle physics. The key feature of the WWW is to act as a repository of documents. Moreover, although these documents are stored in textual format, they are not meant as plain text files, but rather hypertext. This implies two main differences: they can have different appearance from plain text, thanks to the introduction of formatting commands, and also do not require sequential reading, since they may contain links to other documents, in a non-sequential manner typical of hypertext.

The hypertext documents of the WWW, which are called *web pages* and are collected in repositories called *web sites*, have therefore *anchors* to each other, and may display additional embellishments such as character typesetting and simple text formatting. Web pages are retrieved through the Internet by exploiting a naming system based on Uniform Resource Locators (URLs). The URL system, in turn, heavily leverages the DNS mechanism. Indeed, in the everyday informal speech, most people say "Internet address" to mean "URL", not "IP address." This is because, as said, for the human users the string of bits of the IP address is hidden behind the DNS. Actually, only a part (though a large one) of the URL is a domain name. A typical URL, such as *http://telecom.dei.unipd.it/index.html* consists of: i) the name of the protocol, in this case HTTP; ii) the DNS name of the host which contains the page in memory; iii) the actual name of the page, which refers to a file stored in the destination host. Regarding this last part, note that the page name may contain a hierarchical folder structure on the destination host. For example, *http://telecom.dei.unipd.it/courses/index.html* refers to a web page called *index.html* in the folder *courses* (therefore distinct from the page *index.html* in the root folder).

The extension "html" (sometimes shortened to "htm" in operating systems that do not support extensions longer that three characters) denotes that the page is written with a language called hyper-text markup language (HTML). As the name suggests, this is a widespread instrument to encode hypertext, which is commonly used to represent web pages and define their appearance. This is often paired with the HTTP protocol, which is actually the only protocol we will discuss related to the web. Basically, the ultimate purpose of HTTP is to enable the visual representation on screen of the web pages written using HTML markup.

The well-known port 80 is used for HTTP, which is supported by both TCP and UDP. A more detailed description of HTTP is outside the scope of the present chapter. The main idea behind the protocol structure is however similar to what discussed for SMTP. The transferred data consist of ASCII characters, which can be read directly by human beings, even though the average user is supposed to leave their proper interpretation to a dedicated computer program called *browser*. The key feature of a browser is simply that it must be capable of displaying properly the web pages (as they are meant to be seen on the computer screen by their *web designer*). For this reason, there are several browser programs available, which are more or less equivalent even though most of them offer additional features to simplify or improve the activity of reading web pages (which goes under the name of "web browsing" and has become very widespread both for work reasons or as a hobby). For example, most browsers usually allow retrieving commonly displayed content more easily; this can be done both automatically, i.e., by caching some data which are requested often from certain sites, or upon a request of the user (a typical related operation is to "bookmark" some URLs which are frequently accessed).

We will not discuss HTML in depth either. From an abstract point of view it may be sufficient to regard it just as a markup convention. The content of the web page encoded with HTML consists of a mixture of two kinds of data: actual text and escape sequences, delimited by special characters, which the browser is able to interpret as commands (e.g., for changing the color or the font of a character, to indent a line, or to include a picture in one of some standard formats). Actually, this describes the main content stored in the part of the file called *body*. On top of this structure, some general control information is also added in the initial part of the page, which is called *head*, to display the web page in the proper manner and enable its management through HTTP. The basic instruments of HTML are indeed limited, while recent web pages are actually offering more complicated features and web masters are always trying to include additional ones, especially for what concerns graphics, sound, video, and interactivity with the users. Thus, it is worth mentioning that, as the complexity of the content provided over the WWW increases, dedicated *helper programs* or *plug-in* are developed to offer additional services in parallel, or embedded into, the browser programs. For a given web page, the browser recognizes the application to use, and invokes it, according again to the MIME format.

REFERENCES

1. G. Owen, *Game Theory*, Academic Press, 1995.
2. T. Roughgarden, *Selfish Routing and the Price of Anarchy*, MIT Press, 2005.
3. T. Roughgarden, E. Tardòs, "How bad is selfish routing?," *Journal of the ACM*, Vol. 49, No. 2, pp. 236–259, May 2002.
4. D. Bertsekas, R. Gallager, "Data Networks," Prentice Hall, Upper Saddle River, NJ, 2nd edition, chapters 5.4–6, 1992.
5. R. E. Bellman, *On a routing problem*, Quart. Appl. Math., Vol. 16, pp. 87–90, 1958.
6. L. R. Ford, Jr. and D. R. Fulkerson, *Flows in networks*, Princeton Univ. Press, Princeton, NJ, 1962.
7. E. W. Dijkstra, *A note on two problems in connection with graphs*, Numer. Math. Vol. 1, pp. 269–271, 1959.
8. R. C. Prim, *Shortest connection networks and some generalizations*, Bell Syst. Tech. J., Vol. 36, pp. 1389–1401, 1957.
9. V. Jarník, *O jistém problému minimalním*, Práce Moravské Přírodovědecké Společnosti, Vol. 6, pp. 57–63, 1930. (in Czech)
10. J. Kruskal, *On the shortest spanning subtree of a graph and the traveling salesman problem*, Proceedings of the American Mathematical Society, vol. 7, no. 1, pp. 48–50, 1956.
11. G. Galilei, *Dialogues concerning two new sciences*, 1638.
12. J. Lepak, M. Crescimanno, *Speed of light measurement using ping*, American Physical Society, 2002.
13. D. De Couto, D. Aguayo, J. Bicket, and R. Morris, *A high-throughput path metric for multi-hop wireless routing*, Wireless Networks, Vol. 11, No. 4, pp. 419–434, 2005.
14. R. P. Draves, J. Padhye, and B. D. Zill, *Routing in multi-radio, multi-hop wireless mesh networks*, Proceedings ACM Mobicom, pp. 114-128, 2004.
15. R. Xu, D. Wunsch II, *Survey of Clustering Algorithms*, IEEE Transactions on Neural Networks, Vol. 16, No. 3, pp. 645–678, 2005.
16. M. Abolhasan, T. Wysocki, E. Dutkiewicz, *A reviewer of routing protocols for mobile ad hoc networks*, Ad Hoc Networks, Elsevier, Vol. 2, No. 1, pp. 1–22, 2004.
17. M. Mauve, A. Widmer, H. Hartenstein, *A survey on position-based routing in mobile ad hoc networks*, IEEE Network, Vol. 15, No. 6, 30–39, 2001.
18. R. Pastor-Satorras, A. Vespignani, *Evolution and Structure of the Internet: A Statistical Physics Approach*, Cambridge University Press, Cambridge, United Kingdom, 2004.
19. V. Cerf, R. Kahn, *A Protocol for Packet Network Intercommunication*, IEEE Transactions on Communications, Vol. 22, No. 5, pp. 637–648, 1974.

20. "The performance of TCP/IP for networks with high bandwidth-delay products and random loss," IEEE/ACM Transactions on Networking, Vol. 5, No. 3, pp. 336–350, June 1997.

21. M. Mathis, T. Ott, J. Semke, and J. Mahdavi, "The macroscopic behavior of the TCP congestion avoidance algorithm," ACM SIGCOMM Computer Communication Review, Vol. 27, No. 3, pp. 67–82, July 1997.

22. J. Padhye, V. Firoiu, D. F. Towsley and J. F. Kurose, "Modeling TCP Reno performance: a simple model and its empirical validation," IEEE/ACM Transactions on Networking, Vol. 8, No. 2, pp. 133–145, April 2000.

INDEX

absorbing state, 459
ADC
  analog-to-digital converter, 131, 334
additive white Gaussian noise, 211
ad hoc networks, 639
A-law, 151
ALOHA
  delay, 606
  offered traffic, 605
  performance of, 607
  success probability, 605
  throughput, 604, 606
  vulnerable transmission period, 605
alphabet, 79
analog repeaters, 343
analog transmission
  output signal-to-noise ratio, 341
anti-aliasing filter, 62
antipodal signals, 273
aperiodic state, 440
arc, 7
arrival rate, 505, 513
arrival times, 98
a.s. constant, 80
asymptotic distribution, 441
autocorrelation function, 98
automatic retransmission request, 640
  Go Back N, 646
    retransmission timeout, 648
    sender protocol rules, 646
    throughput, 648
  Selective Repeat, 649
    sender protocol rules, 649
    throughput, 652
  Stop and Wait
    retransmission timeout, 644
    throughput, 645
  gap detection, 640
  in-order delivery, 641
  negative acknowledgment, 640
  normalized throughput, 643
  optimal packet length, 655
  out-of-order delivery, 641
  outstanding packets, 642
  performance evaluation, 653
  positive acknowledgment, 640
  resequencing delay, 641
  sequence numbers, 642
  stop and wait ARQ, 644
  throughput performance, 655
  transmission efficiency, 644
  window size, 642
autonomous system, 723
average cross power, 42

average length, 169
average power, 41
backbone, 718
backward Kolmogorov equation, 460
balance equation, 447, 474
balance equations, 446, 467, 485
band, 53
  of a filter, 54
bandwidth, 54
Baud, 288
bidirectional links, 686
bidirectional random walk, 438
binary channel, 293
binary modulation
  antipodal signals, 273
  basis, 269
  bit error probability, 271
  constellation, 269
  correlation coefficient, 268
  distance, 270
  orthogonal signals, 274
  probability of correct reception, 271
  probability of error, 271
  single filter receivers, 277
birth-death processes, 475
bit error probability, 251
bit mapper, 287
  inverse, 289
bit rate, 288
block code, 363
blocking system, 507
Boltzmann constant, 198
BPSK, 274, 315
bridge, 685
buffer, 501
bus, 8
call-blocking, 507
call-waiting probability, 535
capacity, 392
  of a numerical channel, 392
channel
  frequency selective, 329
  M-ary, 294
  binary symmetric, 294
  binary, 293
  memoryless, 296
  numerical, 294
Chapman-Kolmogorov equations, 425
  continuous-time MC, 457
class, 727
closed communication class, 432
cluster, 719
code division multiple access, 325
code
  equivalent, 372
  linear, 371
  rate, 364
  systematic, 364
codeword, 363
companding, 149
complex Gaussian rv, 95
  circularly symmetric —, 95
complex-valued rp, 97
compression
  μ-law, 151
  A-law, 151
compression-expanding, 149
conditional PDF, 88
connected graph, 688
conservative, 459
constant
  almost surely (a.s.) —, 80
constellation, 75
constraint length, 385
continuous time Markov chain, 451
continuous-time Markov chain
  recurrence, 471
convolution, 47
convolutional codes, 384
correlation
  coefficient, 268
  matrix, 93
  receiver, 293
  of two rvs, 92
coset, 375
counting process, 99
covariance
  matrix, 93
  of two rvs, 92
cross energy, 39
cross power spectral density, 111
CSMA
  nonpersistent CSMA, 613
CSMA/CD
  delay, 632, 635
  jamming tone, 628
  Markov chain, 629
  performance of, 635
  throughput, 629, 632
cycle, 688
cyclic
  code, 377
  convolution, 377
  shift, 377
cyclostationary, 105
DAC
  digital-to-analog converter, 132, 134, 334
data
  compression, 168
  link layer, 577
datagram, 724, 737
decision regions, 254
  optimum regions, 258, 260
decision
  probability of correct decision, 254
  probability of error, 255

decodability, 169
decoder, 134
decoding, 134
deflection routing, 713
delta
Dirac function, 29
demand based protocols
infinite buffer systems, 600
delay formula, 601
delay, 601
performance of, 601
polling cycle time, 601
throughput formula, 601
throughput, 601
one message buffer systems, 592
delay, 595
flow balance equations, 598
packet loss probability, 596
performance of, 596
polling cycle time, 594
throughput formula, 595
throughput, 595
stability, 602
switchover time, 592
departure rate, 514
dictionary, 166
digital
communication system, 287
modulation, 251
transmission, 251
analog repeaters, 343
output signal-to-noise ratio, 338
Dirac delta function, 29
direct sequence, 332
directed graph, 686
dispersion, 230
ECHO, 710
edge, 7
efficiency
of a finite rv, 165
EHF, 232
electrical power
density, 193
available, 196
average, 193
instantaneous, 193
units of measure, 194
Elias code, 178
email address, 749
embedded Markov chain, 559, 462, 562
encapsulation, 16
encoder, 134
encoding, 134
energy, 39
between two signals, 39
spectral density, 40
equality in the mean square sense, 92
ergodicity, 106, 510
Erlang-B formula, 553
Erlang-C formula, 536
events, 79
expectation, 82
fundamental theorem, 82
exponential rv, 84
FDMA
delay, 586
throughput, 586
filter, 46
interpolator —, 51
firewall, 686
flow diagram, 465
forward Kolmogorov equation, 460
Fourier
coefficients, 34
series, 33
transform, 31
frequency hopping, 323, 333
frequency response, 48
narrowband radio link, 234
optical fiber, 230
transmission line, 218
FSK, 271, 275, 319
full band, 53
full duplex, 686
fundamental frequency, 33
gamma distribution, 153
gateway, 686
Gaussian
rps, 98
rv, 81
Q function, 121
generating
matrix, 372
polynomial, 378
geographic routing, 722
Gilbert model, 429
Gram-Schmidt orthonormalization, 70
graph, 7
Gray encoding, 303
half duplex, 686
Heaviside
conditions, 59
step function, 28
HF, 232
hitting probability, 434
holder, 134
homogeneous, 421, 452
hop count, 711
hop, 8
hub, 685
Huffman coding, 174
hybrid routing, 722
hypothesis testing, 253
impulse, 29
response, 47
transmission line, 219

indicator
  function, 257
  rv, 84
infinitesimal generator matrix, 458
information, 157
information rate, 166, 392
inner product, 67
instantaneous, 459
instantaneous power, 41
intensity matrix, 458
interarrival time, 98, 504
interdeparture time, 506
Internet, 723
interpolation, 50
intersymbol interference, 292
inverse raised cosine function, 29
inverse square root raised cosine function, 29
irreducible MC., 433
irrelevance principle, 116
joint entropy, 162
jointly WSS, 111
Kendall notation, 508
Kraft inequality, 171
law of rare events, 91
LAN standards, 656
  802.11
    basic access, 666
    basic service set, 664
    communication modes, 664
    distributed inter frame space, 666
    extended service set, 665
    first specification, 662
    floor acquisition, 667
    hidden terminal, 667
    independent basic service set, 664
    link layer, 665
    network allocation vectors, 667
    physical layer, 662
    short inter frame space, 666
    virtual carrier sense, 667
  802.11a, 662
  802.11b, 662
  802.11g, 663
  802.11n, 663
  Bluetooth, 668
    active mode, 674
    design considerations, 669
    device discovery, 671
    frequency hopping, 670
    hold mode, 674
    link layer, 671
    multi-slot packets, 672
    park mode, 674
    physical layer, 670
    piconet scheduling, 673
    piconets, 670
    power modes, 673
    scatternets, 671
    sniff mode, 674
    special interest group, 668
    standby mode, 674
    transmission rates, 672
    working groups, 669
  Ethernet, 657
    1000BASE-LX, 658
    1000BASE-T, 658
    100BASE-FX, 658
    100BASE-TX, 658
    10BASE-2, 657
    10BASE-5, 657
    10BASE-T, 657
    bridges, 659
    bus topology, 659
    CSMA/CD, 658
    Fast Ethernet, 658
    Gigabit Ethernet, 658
    interconnecting network elements, 659
    jamming signal, 658
    link layer, 658
    packet frame, 658
    repeaters, 659
    star topology, 659
    switches, 660
  IEEE 802.11, 661
  WLAN, 660
    benefits of, 660
    disadvantages, 661
    european telecommunications standard institute, 660
    federal communication commission, 660
    industrial scientific and medical bands, 660
  transmission rates, 661
Laplacian distribution, 153
linear space, 66
  Euclidean space, 66
  Gram-Schmidt orthonormalization, 70
  Hermitian space, 67
  continuous time signal space, 69
  dimension, 68
  inner product, 67, 76
  norm, 67
  orthogonality, 68
  orthonormal basis, 68–69
  projection, 78
  properties, 66
  span, 67
linearity, 46
link, 8
link budget, 215
link state packets, 717
Little's law, 516
load, 192
load factor, 514
low density parity check, 393
marginal rules, 89
matrix

transition–, 453
maximum a posteriori, 261
maximum likelihood, 262
M-ary modulation, 279
bi-orthogonal signals, 285
lower bound, 281
orthogonal signals, 283
upper bound, 279
mean, 82
of a rp, 97
mean recurrence time, 437, 471
Markov chain
definition, 418, 451
discrete time, 418
ergodic, 441, 473
homogeneous–, 421, 452
Markov
property, 452
Markovian QSs, 508
medium access control, 579
FDMA, 586
TDMA, 584
collision, 582
collisions, 579
delay, 582
demand based protocols, 580
ARCNET protocol, 589
FDDI, 591
decrementing service, 592
exhaustive service, 591
gated service, 591
infinite buffer systems, 591
limited service, 592
one message buffer systems, 591
performance measures, 591
polling cycle time, 591
polling systems, 588
queueing delay, 591
system time, 591
token bus, 589
token ring, 589–590
deterministic protocols, 580, 584
performance of, 587
practical considerations, 587
feedback channel, 579
offered traffic, 583
packet arrival process, 581
packet transmission time, 583
performance measures, 582
performance of, 637
processing time, 579
propagation delay, 582
queueing delay, 582
random access protocols, 580, 603
backoff models, 635
collisions, 603
stability, 611
retransmission delay, 582
shared channel, 581
system arrival rate, 583
system load factor, 583
system model, 580
system time, 582
throughput, 583
waiting queue, 581
memoryless
binary symmetric channel, 335
source, 167
MF, 232
mid-riser, 137
minimum distance, 279
minimum distance criterion, 263
modulation, 251
differential PSK, 318
M-ary, 252
amplitude and phase modulation, 304
amplitude shift keying, 298
binary, 252
digital, 251
direct sequence, 332
frequency hopping, 333
frequency-shift keying, 319
ortogonal frequency division multiplexing, 329
phase-shift keying, 314
pulse amplitude, 298
quadrature amplitude, 304
spread spectrum, 329, 332
time hopping, 334
multiplexing
optical frequency division, 228
wavelength division, 228
mutual information, 391
μ law, 151
M/D/1, 566
M/G/1, 559
M/M/1, 520
M/M/1/K, 543
M/M/m, 532
M/M/m/K, 555
M/M/m/m, 551
M/M/m
stability condition, 535
Nagle algorithm, 742
NAT, 686
neighbor, 687
network, 686
network layers, 683
node, 7
noise figure, 204, 206
noise temperature, 199, 203
nominal information, 165
non-aliasing condition, 61
non-blocking systems, 507
non-uniform quantizer
signal-to-quantization noise ratio, 151
nonpersistent CSMA

channel busy probability, 617
delay formula, 619
delay, 617
performance of, 619
throughput formula, 616
throughput, 613
normal
rv, 81
normalization condition, 80, 446, 474, 487
normalized rv, 81
normalized complementary Gaussian distribution, 81
null recurrent, 437
numerical channel, 294
Nyquist criterion, 64, 292
occupancy, 510
offered traffic, 513
one message buffer systems
polling cycle, 592
one persistent CSMA
balance equations, 623
channel busy probability, 625
delay formula, 627
delay, 626
Markov chain, 620
system state, 621
throughput formula, 621, 624
throughput, 620
open communication class, 432
optical fiber
glass fiber, 228
graded-index fiber, 228
laser diode, 228
light emitting diode, 228
plastic fiber, 228
step-index fiber, 228
orthogonal signals, 274
orthogonal
rp with its conjugate, 112
rps, 111
ortogonal frequency division multiplexing, 329
PAM, 298
PASTA property, 528
parity check matrix, 373
parity symbols, 361
Parseval theorem, 39
path, 687
PBX, 503
PCM
2-PAM output signal-to-noise ratio, 342
output SNR, 338
PDF, 80
perfect channel, 394
periodic state, 440
pipe capacity, 643
Planck constant, 198
point processes, 98
Poisson process, 100
Pollaczek-Khintchine (P-K) mean-value formula, 563
port, 737
positive recurrent, 437
power gain, 201
power spectral density, 107
of thermal noise, 198
power
average, 41
cross, 42
instantaneous, 41
practical band, 57
practical bandwidth, 57
prefix code, 170
private branch exchange, 503
proactive routing, 722
probability
a priori, 252
flow balance equations, 446
mass distribution, 80
maximum a posteriori criterion, 261
maximum likelihood criterion, 262
minimum distance criterion, 263
of correct decision, 254
of error, 255
of transition, 293–294
stationary transition –, 421
transition –, 421
proxy, 686
PSD, 107
PSK, 73, 274, 314
pure-birth process, 478
pure-death process, 497
pure-loss QS, 551
QAM, 304
Q function, 121
QPSK, 73, 316
quantization error, 138
granular, 141
overload region, 141
saturation region, 141
quantization noise, 141
granular, 141
quantizer characteristic, 136
quantizer dynamic range, 138
quantizer
loading factor, 143
mid-riser characteristic, 138
mid-tread characteristic, 138
non-uniform, 144
performance, 142
quantization step size, 137
quantizer characteristic, 136
saturation value, 138
signal-to-quantization error ratio, 142
uniform, 137
quantum, 36
queue capacity, 506
queueing system, 501
queueing theory, 499

arrival process, 503
arrival rate, 505, 513
Kendall notation, 508
Little's law, 516
M/D/1, 566
M/G/1, 559
M/M/1, 520
M/M/1/K, 543
M/M/m, 532
M/M/m/K, 555
M/M/m/m, 551
Markovian queueing models, 519
offered traffic, 513
performance characterization, 510
service process, 505
throughout, 514
useful traffic, 514
queuing discipline, 507
queueing system stability, 511
radio link
directional antenna, 233
free space path loss, 234
frequency ranges, 232
Friis transmission equation, 234
horn antenna, 234
isotropic antenna, 233
narrowband channel model, 233
parabolic antenna, 234
raised cosine function, 29
random access protocols
ALOHA, 604
backlogged users, 604
backoff time, 604
carrier sense multiple access, 612
CSMA, 612
CSMA/CD, 628
one persistent CSMA, 619
p persistent CSMA, 627
slotted ALOHA, 608
random processes
orthogonal with its conjugate, 112
point, 98
random vector (rve), 86
rare events
law of, 91, 101
reactive routing, 722
realization, 79, 97
receiver
minimum distance implementation, 264
rect function, 28
rectangle function, 28
recurrent, 434
redundancy, 165
reflecting barrier, 487
regenerative repeaters, 343
repeater sections, 209
repeater, 684
response, 45
frequency —, 48
impulse —, 47
round trip time, 643
route discovery, 722
router, 685
routing algorithms, 690
routing strategies, 690
routing table, 713
rp, 96
rv, 79
sampling frequency, 50
sampling period, 49
sampling rate, 50
sampling theorem, 60
sampling, 49
saturated, 514
Schwarz inequality, 39
segment, 737, 740
sequence, 36
servers, 501
service capacity, 505
service discipline, 507
service rate, 505
Shannon-Fano code, 174
Shannon-Whittaker sampling theorem, 60
SHF, 232
sifting property, 30
sign function, 28
signal
basis, 79
constellation, 75
continuous time —, 27
discrete time —, 36
distance, 76
energy, 39, 76
inner product, 76
graphical representation of a —, 28
projection, 78
real-valued —, 28
vector representation, 75
signal-to-noise ratio, 211
electrical, 212
statistical, 211
sinc function, 29
slotted ALOHA
delay formula, 610
delay, 609
performance of, 611
success probability, 608
throughput formula, 608
throughput, 608
vulnerable transmission period, 608
socket, 737
sojourn time, 423, 453
source coding, 167
source routing, 713
source, 192
spanning tree, 688

specific attenuation, 219
spectral density
  energy, 40
  power —, 107
spectral efficiency, 326
spectral lines, 109
speech coding, 153
  coding by modelling, 156
  frequency domain coding, 156
  waveform coding, 156
speech
  unvoiced, 153
  voiced, 153
spread spectrum, 329, 332
square root raised cosine function, 29
standard deviation, 83
standard room temperature, 198
state probability vector, 427, 465
state space, 418
state transition probabilities in k steps, 421
state
  aperiodic, 440
  periodic, 440
stationary, 421, 510
statistical moment, 84
statistical multiplexer, 495
statistical power, 83
statistical power of a rp, 98
step function, 28
stochastic matrix, 423
strictly concave, 160
strictly convex, 160
subgraph, 687
subnet, 729
subnet mask, 730
subnetwork, 718
subtree, 688
sufficient statistics, 253
switch, 685
symbol error probability, 365
symbol rate, 288
syndrome, 375
  decoding, 375
system, 45
  linear, 46
  stability, 511
  storage capacity, 506
  time-invariant, 46
systematic code, 364
TCP window, 742
TDMA
  TDMA frame, 584
  TDMA schedule, 584
  delay, 585
  throughput, 584
theorem of irrelevance, 253, 267
thermal noise, 197
  noise temperature, 199
  power spectral density, 198
throughput, 514
time, 510
time hopping, 334
time invariance, 46
topology, 5, 686
traffic intensity, 514
traffic, 510
transformation, 45
transient, 434
transition
  band, 61
  intensities, 459
  matrix in k steps, 421
  matrix, 422, 453
  probabilities, 422
  probability, 293–294
transmission lines
  attenuation constant, 218
  attenuation, 219
  characteristic frequency, 219
  characteristic impedance, 218
  crosstalk, 221
  far-end crosstalk, 221
  gauge, 218
  near-end crosstalk, 221
  phase constant, 218
  propagation delay, 219
  specific attenuation, 219
transmission medium
  coaxial cable, 217
  optical fiber, 228
  radio link, 231
  shielded twisted-pair cable, 217
  twisted-pair cable, 217
  unshielded twisted-pair cable, 217
transmission system model, 211
transmission system
  AWGN model, 211
  narrowband model, 214
tree, 688
trellis diagram, 385
triang function, 29
triangle function, 29
turbo codes, 393
two-port linear network, 200
  cascade of, 206
  Friis equation, 208
  load matching, 202
  matched, 202
  noise figure, 204
  noise temperature, 203
  passive, 206
  power gain, 201
  signal-to-noise ratio, 211
two-terminal device, 192
  absence of distortion, 196
  active, 192

available power density, 196
load matching condition, 195
UHF, 232
uncertainty principle, 56
uncorrelated
rps, 111
undirected graph, 686
uniform quantizer
signal-to-quantization error ratio, 143
unit, 737
universal mobile telecommunications system, 325
useful traffic, 514
useless channel, 394
utilization factor, 514
vacation, 501
variance, 83
vertex, 7
VHF, 232
video coding, 156
Viterbi algorithm, 388
VLF, 232
vocoder, 156
Voronoi
cells, 263
regions, 263
waiting queue, 501
waiting time, 503
walk, 687
Walsh codes, 323
WGN, 111
white noise, 111
Woodward sine cardinal function, 29
word error probability, 365
work conserving, 501
workload, 513
WSS, 105

Lightning Source UK Ltd.
Milton Keynes UK
04 October 2010

160758UK00001B/3/P

9 780470 971918